FLUID MECHANICS
FOR ENGINEERS
IN SI UNITS

DAVID A. CHIN
University of Miami

330 Hudson Street, NY, NY 10013

Vice President and Editorial Director, ECS: Marcia J. Horton
Executive Editor: Holly Stark
Field Marketing Manager: Demetrius Hall
Senior Product Marketing Manager: Bram van Kempen
Marketing Assistant: Jon Bryant
Editorial Assistant: Amanda Brands
Acquisitions Editor, Global Edition: Sourabh Maheshwari
Senior Managing Editor: Scott Disanno
Production Project Manager: Greg Dulles
Assistant Project Editor, Global Edition: Vikash Tiwari
Senior Manufacturing Controller, Global Edition: Kay Holman
Program Manager: Erin Ault
Media Production Manager, Global Edition: Vikram Kumar
Director of Operations: Nick Sklitsis
Operations Specialist: Maura Zaldivar-Garcia
Cover Designer: Lumina Datamatics
Cover Photo: © bankerwin/Shutterstock.com
Manager, Rights and Permissions: Rachel Youdelman
Senior Project Manager, Rights and Permissions: Timothy Nicholls
Composition: GEX Publishing Services
Typeface: 10.5pt Times LT Pro

Many of the designations by manufacturers and seller to distinguish their products are claimed as trademarks. Where those designations appear in this book, and the publisher was aware of a trademark claim, the designations have been printed in initial caps or all caps.

The author and publisher of this book have used their best efforts in preparing this book. These efforts include the development, research, and testing of theories and programs to determine their effectiveness. The author and publisher make no warranty of any kind, expressed or implied, with regard to these programs or the documentation contained in this book. The author and publisher shall not be liable in any event for incidental or consequential damages with, or arising out of, the furnishing, performance, or use of these programs.

MATLAB is a registered trademark of The MathWorks, Inc., 3 Apple Hill Drive, Natick, MA 01760-2098.

Pearson Education Limited
KAO Two
KAO Park
Harlow
CM17 9NA
United Kingdom

and Associated Companies throughout the world

Visit us on the World Wide Web at:
www.pearsonglobaleditions.com

© Pearson Education Limited 2018

The rights of David A. Chin to be identified as the author of this work has been asserted by them in accordance with the Copyright, Designs and Patents Act 1988.

Authorized adaptation from the United States edition, entitled Fluid Mechanics for Engineers, First Edition, ISBN 978-0-13-380312-9, by David A. Chin, published by Pearson Education © 2017.

All rights reserved. No part of this publication may be reproduced, stored in a retrieval system, or transmitted in any form or by any means, electronic, mechanical, photocopying, recording or otherwise, without either the prior written permission of the publisher or a license permitting restricted copying in the United Kingdom issued by the Copyright Licensing Agency Ltd, Saffron House, 6-10 Kirby Street, London EC1N 8TS.

All trademarks used herein are the property of their respective owners. The use of any trademark in this text does not vest in the author or publisher any trademark ownership rights in such trademarks, nor does the use of such trademarks imply any affiliation with or endorsement of this book by such owners.

British Library Cataloguing-in-Publication Data

A catalogue record for this book is available from the British Library

ARP impression 98

Typeset by GEX Publishing Services
Printed and bound in Great Britain by Ashford Colour Press Ltd.

ISBN 10: 1-292-16104-3
ISBN 13: 978-1-292-16104-4

To Stephanie and Andrew.

"Wherever there is a human being, there is an opportunity for a kindness."

Seneca

Contents

Preface 11

Chapter 1 Properties of Fluids 17

1.1 Introduction 17
1.1.1 Nomenclature 19
1.1.2 Dimensions and Units 20
1.1.3 Basic Concepts of Fluid Flow 26
1.2 Density 27
1.3 Compressibility 32
1.4 Ideal Gases 36
1.4.1 Equation of State 36
1.4.2 Mixtures of Ideal Gases 37
1.4.3 Thermodynamic Properties 39
1.4.4 Speed of Sound in an Ideal Gas 44
1.5 Standard Atmosphere 44
1.6 Viscosity 46
1.6.1 Newtonian Fluids 46
1.6.2 Non-Newtonian Fluids 53
1.7 Surface Tension 55
1.8 Vapor Pressure 61
1.8.1 Evaporation, Transpiration, and Relative Humidity 63
1.8.2 Cavitation and Boiling 64
1.9 Thermodynamic Properties of Liquids 67
1.9.1 Specific Heat 67
1.9.2 Latent Heat 68
1.10 Summary of Properties of Water and Air 69
Key Equations in Properties of Fluids 70
Problems 72

Chapter 2 Fluid Statics 87

2.1 Introduction 87
2.2 Pressure Distribution in Static Fluids 88
2.2.1 Characteristics of Pressure 88
2.2.2 Spatial Variation in Pressure 89
2.2.3 Practical Applications 92
2.3 Pressure Measurements 101
2.3.1 Barometer 101
2.3.2 Bourdon Gauge 103
2.3.3 Pressure Transducer 104
2.3.4 Manometer 105
2.4 Forces on Plane Surfaces 110
2.5 Forces on Curved Surfaces 120

2.6 Buoyancy 127
2.6.1 Fully Submerged Bodies 127
2.6.2 Partially Submerged Bodies 132
2.6.3 Buoyancy Effects Within Fluids 138
2.7 Rigid-Body Motion of Fluids 139
2.7.1 Liquid with Constant Acceleration 141
2.7.2 Liquid in a Rotating Container 145
Key Equations in Fluid Statics 148
Problems 150

Chapter 3 Kinematics and Streamline Dynamics 177

3.1 Introduction 177
3.2 Kinematics 178
3.2.1 Tracking the Movement of Fluid Particles 181
3.2.2 The Material Derivative 188
3.2.3 Flow Rates 190
3.3 Dynamics of Flow along a Streamline 192
3.4 Applications of the Bernoulli Equation 202
3.4.1 Flow through Orifices 203
3.4.2 Flow Measurement 209
3.4.3 Trajectory of a Liquid Jet 214
3.4.4 Compressibility Effects 216
3.4.5 Viscous Effects 218
3.4.6 Branching Conduits 220
3.5 Curved Flows and Vortices 222
3.5.1 Forced Vortices 223
3.5.2 Free Vortices 226
Key Equations in Kinematics and Streamline Dynamics 229
Problems 232

Chapter 4 Finite Control Volume Analysis 256

4.1 Introduction 256
4.2 Reynolds Transport Theorem 257
4.3 Conservation of Mass 259
4.3.1 Closed Conduits 263
4.3.2 Free Discharges from Reservoirs 265
4.3.3 Moving Control Volumes 267
4.4 Conservation of Linear Momentum 268
4.4.1 General Momentum Equations 269
4.4.2 Forces on Pressure Conduits 273
4.4.3 Forces on Deflectors and Blades 281
4.4.4 Forces on Moving Control Volumes 282
4.4.5 Wind Turbines 288
4.4.6 Reaction of a Jet 293
4.4.7 Jet Engines and Rockets 296
4.5 Angular Momentum Principle 298
4.6 Conservation of Energy 307
4.6.1 The First Law of Thermodynamics 308

4.6.2 Steady-State Energy Equation 309
4.6.3 Unsteady-State Energy Equation 320
Key Equations in Finite Control Volume Analysis 323
Problems 327

Chapter 5 Differential Analysis 357

5.1 Introduction 357

5.2 Kinematics 358
5.2.1 Translation 358
5.2.2 Rotation 360
5.2.3 Angular Deformation 363
5.2.4 Linear Deformation 363

5.3 Conservation of Mass 365
5.3.1 Continuity Equation 365
5.3.2 The Stream Function 372

5.4 Conservation of Momentum 375
5.4.1 General Equation 376
5.4.2 Navier–Stokes Equation 379
5.4.3 Nondimensional Navier–Stokes Equation 381

5.5 Solutions of the Navier–Stokes Equation 385
5.5.1 Steady Laminar Flow Between Stationary Parallel Plates 385
5.5.2 Steady Laminar Flow Between Moving Parallel Plates 388
5.5.3 Steady Laminar Flow Adjacent to Moving Vertical Plate 391
5.5.4 Steady Laminar Flow Through a Circular Tube 394
5.5.5 Steady Laminar Flow Through an Annulus 396
5.5.6 Steady Laminar Flow Between Rotating Cylinders 399

5.6 Inviscid Flow 402
5.6.1 Bernoulli Equation for Steady Inviscid Flow 404
5.6.2 Bernoulli Equation for Steady Irrotational Inviscid Flow 407
5.6.3 Velocity Potential 409
5.6.4 Two-Dimensional Potential Flows 411

5.7 Fundamental and Composite Potential Flows 415
5.7.1 Principle of Superposition 415
5.7.2 Uniform Flow 417
5.7.3 Line Source/Sink Flow 418
5.7.4 Line Vortex Flow 421
5.7.5 Spiral Flow Toward a Sink 424
5.7.6 Doublet Flow 426
5.7.7 Flow Around a Half-Body 428
5.7.8 Rankine Oval 433
5.7.9 Flow Around a Circular Cylinder 437

5.8 Turbulent Flow 441
5.8.1 Occurrence of Turbulence 443
5.8.2 Turbulent Shear Stress 443
5.8.3 Mean Steady Turbulent Flow 445

5.9 Conservation of Energy 446
Key Equations in Differential Analysis of Fluid Flows 449
Problems 455

Chapter 6 Dimensional Analysis and Similitude 477

6.1 Introduction 477
6.2 Dimensions in Equations 477
6.3 Dimensional Analysis 481
6.3.1 Conventional Method of Repeating Variables 483
6.3.2 Alternative Method of Repeating Variables 486
6.3.3 Method of Inspection 487
6.4 Dimensionless Groups as Force Ratios 488
6.5 Dimensionless Groups in Other Applications 493
6.6 Modeling and Similitude 494
Key Equations for Dimensional Analysis and Similitude 506
Problems 507

Chapter 7 Flow in Closed Conduits 525

7.1 Introduction 525
7.2 Steady Incompressible Flow 526
7.3 Friction Effects in Laminar Flow 532
7.4 Friction Effects in Turbulent Flow 536
7.5 Practical Applications 544
7.5.1 Estimation of Pressure Changes 544
7.5.2 Estimation of Flow Rate for a Given Head Loss 546
7.5.3 Estimation of Diameter for a Given Flow Rate and Head Loss 547
7.5.4 Head Losses in Noncircular Conduits 548
7.5.5 Empirical Friction Loss Formulas 549
7.5.6 Local Head Losses 552
7.5.7 Pipelines with Pumps or Turbines 559
7.6 Water Hammer 560
7.7 Pipe Networks 565
7.7.1 Nodal Method 566
7.7.2 Loop Method 568
7.8 Building Water Supply Systems 573
7.8.1 Specification of Design Flows 574
7.8.2 Specification of Minimum Pressures 574
7.8.3 Determination of Pipe Diameters 576
Key Equations for Flow in Closed Conduits 583
Problems 587

Chapter 8 Turbomachines 608

8.1 Introduction 608
8.2 Mechanics of Turbomachines 609
8.3 Hydraulic Pumps and Pumped Systems 614
8.3.1 Flow Through Centrifugal Pumps 616
8.3.2 Efficiency 621
8.3.3 Dimensional Analysis 622
8.3.4 Specific Speed 626
8.3.5 Performance Curves 630
8.3.6 System Characteristics 632

8.3.7 Limits on Pump Location 635
8.3.8 Multiple Pump Systems 640
8.3.9 Variable-Speed Pumps 642

8.4 Fans 644
8.4.1 Performance Characteristics of Fans 644
8.4.2 Affinity Laws of Fans 645
8.4.3 Specific Speed 646

8.5 Hydraulic Turbines and Hydropower 648
8.5.1 Impulse Turbines 648
8.5.2 Reaction Turbines 654
8.5.3 Practical Considerations 658

Key Equations for Turbomachines 664
Problems 668

Chapter 9 Flow in Open Channels 693

9.1 Introduction 693

9.2 Basic Principles 694
9.2.1 Steady-State Continuity Equation 694
9.2.2 Steady-State Momentum Equation 694
9.2.3 Steady-State Energy Equation 711

9.3 Water Surface Profiles 724
9.3.1 Profile Equation 724
9.3.2 Classification of Water Surface Profiles 725
9.3.3 Hydraulic Jump 731
9.3.4 Computation of Water Surface Profiles 737

Key Equations in Open-Channel Flow 746
Problems 749

Chapter 10 Drag and Lift 759

10.1 Introduction 759

10.2 Fundamentals 760
10.2.1 Friction and Pressure Drag 762
10.2.2 Drag and Lift Coefficients 762
10.2.3 Flow over Flat Surfaces 765
10.2.4 Flow over Curved Surfaces 767

10.3 Estimation of Drag Coefficients 770
10.3.1 Drag on Flat Surfaces 770
10.3.2 Drag on Spheres and Cylinders 774
10.3.3 Drag on Vehicles 781
10.3.4 Drag on Ships 784
10.3.5 Drag on Two-Dimensional Bodies 785
10.3.6 Drag on Three-Dimensional Bodies 786
10.3.7 Drag on Composite Bodies 786
10.3.8 Drag on Miscellaneous Bodies 789
10.3.9 Added Mass 790

10.4 Estimation of Lift Coefficients 791
10.4.1 Lift on Airfoils 791
10.4.2 Lift on Airplanes 794
10.4.3 Lift on Hydrofoils 799

10.4.4 Lift on a Spinning Sphere in Uniform Flow 800
Key Equations for Drag and Lift 803
Problems 806

Chapter 11 Boundary-Layer Flow 827

11.1 Introduction 827
11.2 Laminar Boundary Layers 829
11.2.1 Blasius Solution for Plane Surfaces 829
11.2.2 Blasius Equations for Curved Surfaces 834
11.3 Turbulent Boundary Layers 836
11.3.1 Analytic Formulation 836
11.3.2 Turbulent Boundary Layer on a Flat Surface 837
11.3.3 Boundary-Layer Thickness and Shear Stress 844
11.4 Applications 845
11.4.1 Displacement Thickness 845
11.4.2 Momentum Thickness 849
11.4.3 Momentum Integral Equation 850
11.4.4 General Formulations for Self-Similar Velocity Profiles 854
11.5 Mixing-Length Theory of Turbulent Boundary Layers 856
11.5.1 Smooth Flow 856
11.5.2 Rough Flow 857
11.5.3 Velocity-Defect Law 858
11.5.4 One-Seventh Power Law Distribution 859
11.6 Boundary Layers in Closed Conduits 859
11.6.1 Smooth Flow in Pipes 860
11.6.2 Rough Flow in Pipes 861
11.6.3 Notable Contributors to Understanding Flow in Pipes 862
Key Equations for Boundary-Layer Flow 863
Problems 867

Chapter 12 Compressible Flow 884

12.1 Introduction 884
12.2 Principles of Thermodynamics 885
12.3 The Speed of Sound 891
12.4 Thermodynamic Reference Conditions 898
12.4.1 Isentropic Stagnation Condition 898
12.4.2 Isentropic Critical Condition 903
12.5 Basic Equations of One-Dimensional Compressible Flow 905
12.6 Steady One-Dimensional Isentropic Flow 907
12.6.1 Effect of Area Variation 907
12.6.2 Choked Condition 908
12.6.3 Flow in Nozzles and Diffusers 910
12.7 Normal Shocks 923
12.8 Steady One-Dimensional Non-Isentropic Flow 935
12.8.1 Adiabatic Flow with Friction 936
12.8.2 Isothermal Flow with Friction 949
12.8.3 Diabatic Frictionless Flow 951
12.8.4 Application of Fanno and Rayleigh Relations to Normal Shocks 957

12.9 Oblique Shocks, Bow Shocks, and Expansion Waves 962
12.9.1 Oblique Shocks 962
12.9.2 Bow Shocks and Detached Shocks 970
12.9.3 Isentropic Expansion Waves 972
Key Equations in Compressible Flow 977
Problems 984

Appendix A Units and Conversion Factors 999
A.1 Units 999
A.2 Conversion Factors 1000

Appendix B Fluid Properties 1003
B.1 Water 1003
B.2 Air 1004
B.3 The Standard Atmosphere 1005
B.4 Common Liquids 1006
B.5 Common Gases 1007
B.6 Nitrogen 1008

Appendix C Properties of Areas and Volumes 1009
C.1 Areas 1009
C.2 Properties of Circles and Spheres 1011
C.2.1 Circles 1011
C.2.2 Spheres 1012
C.3 Volumes 1012

Appendix D Pipe Specifications 1013
D.1 PVC Pipe 1013
D.2 Ductile Iron Pipe 1014
D.3 Concrete Pipe 1014
D.4 Physical Properties of Common Pipe Materials 1014

Bibliography 1015

Index 1026

Preface

Beginning with my formative years as a graduate student at Caltech and Georgia Tech, I have applied fluid mechanics in the context of many engineering disciplines. Also, having taken all of the graduate-level fluid mechanics courses in mechanical engineering, aerospace engineering, civil engineering, and geophysics, and having taught fluid mechanics for more than 30 years, I felt well qualified and motivated to author a fluid mechanics textbook for engineering students. The unique features of this textbook are that it: (1) focuses on the basic principles of fluid mechanics that engineering students are likely to apply in their subsequent required undergraduate coursework, (2) presents the material in a rigorous fashion, and (3) provides many quantitative examples and illustrations of fluid mechanics applications. Students in all engineering disciplines where fluid mechanics is a core course should find this textbook stimulating and useful. In some chapters, the nature of the material necessitates a bias towards practical applications in certain engineering disciplines, and the disciplinary area of the author also contributes to the selection and presentation of practical examples throughout the text. In this latter respect, practical examples related to civil engineering applications are particularly prevalent. To help students learn the material, interactive instruction, tutoring, and practice questions on selected topics are provided via Pearson Mastering Engineering™.

The content of a first course in fluid mechanics. This is a textbook for a first course in fluid mechanics taken by engineering students. The prerequisites for a course using this textbook are courses in calculus through differential equations, and a course in engineering statics. Additional preparatory coursework in rigid-body dynamics and thermodynamics are useful, but not essential. The content of a first course in fluid mechanics for engineers depends on the the curricula of the students taking the course and the interests of the instructor. For most first courses in fluid mechanics, the following topics are deemed essential: properties of fluids (Chapter 1), fluid statics (Chapter 2), kinematics and streamline dynamics (Chapter 3), finite-control-volume analysis (Chapter 4), dimensional analysis and similitude (Chapter 6), and flow in closed conduits (Chapter 7). Additional topics that are sometimes covered include: differential analysis (Chapter 5), turbomachines (Chapter 8), flow in open channels (Chapter 9), drag and lift (Chapter 10), boundary-layer flow (Chapter 11), and compressible flow (Chapter 12). The topics covered in this textbook are sequenced such that the essential topics are covered first, followed by the elective topics. The only exception to this rule is that the chapter on differential analysis (Chapter 5) is placed within the sequence of essential material, after the chapter on control-volume analysis (Chapter 4). This is done for pedagogical reasons since, if differential analysis is to be covered, this topic should be covered immediately after control-volume analysis. If an instructor chooses to omit differential analysis and move directly from control-volume analysis to any of the other essential or elective topics, then the book is designed such that there will be no loss of continuity and students will not suffer from not having covered differential analysis. However, coverage of boundary-layer flow is facilitated by first covering differential analysis. Some of the considerations to be taken into account in selecting elective topics to be covered in a first course in fluid mechanics are given below.

Turbomachines. Coverage of turbomachines is sometimes considered as a mandatory component of a first course in fluid mechanics, and this is particularly true in civil, environmental, and mechanical engineering curricula. Pumps are an integral component of many closed-conduit systems, and turbines are widely used to extract energy from flowing fluids such as water and wind. The essentials of (turbo-)pumps and turbines are covered. Useful topics that are related to pumps include identifying the type of pump that would be most efficient for any given application, and using performance curves to determine the operating point of a pump in a pipeline system. Important topics covered that are related to turbines include identifying the type of hydraulic turbine that would be most efficient in extracting hydropower for any given site condition, and estimating the energy that could be extracted based on given turbine specifications.

Open-channel flow. Open-channel flow is an essential subject area in civil and environmental engineering curricula. However, in these curricula, the subject of open-channel flow is not always covered in a first course in fluid mechanics, being frequently covered in a subsequent course on water-resources engineering. Students in mechanical engineering and related academic programs are less likely to be exposed to open-channel flow in subsequent coursework, and so introductory coverage of this material in a first course in fluid mechanics might be desirable. A feature of this textbook is that it covers the fundamentals of open-channel flow with sufficient rigor and depth that civil and environmental engineering students taking a follow-on course in water resources engineering would have sufficient preparation that they need not be re-taught the fundamentals of open-channel flow. Students in other disciplines, particulary in mechanical engineering, would be have sufficient background to solve a variety of open-channel flow problems from first principles.

Boundary-layer flow, drag, and lift. An understanding of boundary-layer flow is a prerequisite for covering the essential topics of drag and lift. However, there are many aspects of boundary-layer flow that are not directly relevant to understanding drag and lift, and detailed coverage of boundary-layer flow in advance of drag and lift could divert attention from the practical applications of drag and lift. Consequently, the essential elements of boundary-layer flow are presented in an abbreviated form in the chapter on drag and lift (Chapter 10), with much more detailed coverage of boundary-layer flow presented in the subsequent dedicated chapter (Chapter 11). This arrangement of topics facilitates choosing to cover drag and lift, but not to cover boundary-layer flows in detail in a first course in fluid mechanics. Using such an approach, Chapter 10 would be covered, Chapter 11 would be an elective chapter, and there is no discontinuity in the presentation of the material.

Compressible flow. The treatment of compressible flow in this textbook takes a step into the modern era by ceasing reliance on compressible-flow curves and compressible-flow tables, sometimes called gas tables, which have been a staple of the treatment of compressible flow in other elementary fluid mechanics texts. The rule that one should not read a number from a graph or read a number from a table when one knows the analytic equation from which the graph or table is derived is followed in this text. The practice of reading compressible-flow variables from graphs and tables is an approximate approach originated in an earlier era when the solution of implicit equations were problematic. With modern engineering calculation software, such as Excel and MATLAB®, solution of implicit equations are more easily and accurately done numerically on a personal computer.

Philosophy. A first course in fluid mechanics must necessarily emphasize the fundamentals of the field. These fundamentals include fluid properties, fluid statics, basic concepts of fluid flow, and the forms of the governing equations that are useful in solving practical problems. To assist students in solving practical problems, the most useful relationships are highlighted (shaded in blue) in the text, and the key equations in each chapter are listed at the end of the chapter. In engineering curricula, fluid mechanics is regarded as an engineering science that lays the foundation for more applied courses. Consequently, fundamentals of fluid mechanics that are not likely to be applied in subsequent courses taken by undergraduate engineering students are not normally covered in a first course in fluid mechanics. This philosophy has been adopted in designing the content of this textbook. For graduate students requiring more specialized knowledge of fluid mechanics, such as conformal mapping applications in ideal flow, geophysical fluid dynamics, turbulence theory, and advanced computational methods in fluid dynamics, a second course in fluid mechanics would be required. Notwithstanding the needs of graduate students specializing in areas closely related to pure fluid mechanics, this textbook provides the fundamentals of fluid mechanics with sufficient rigor that advanced courses in fluid mechanics need only build on the content of this book and need not reteach this material.

David A. Chin, Ph.D., P.E.
Professor of Civil and Environmental Engineering
University of Miami

Resources for Instructors and Students

- Pearson Mastering Engineering. This online tutorial homework program, www.masteringengineering.com, is available with *Fluid Mechanics for Engineers in SI Units*. It provides instructors customizable, easy-to-assign, and automatically graded homework and assessments, plus a powerful gradebook for tracking student and class performance. Tutorial homework problems emulate the instructor's office-hour environment. These in-depth tutorial homework problems are designed to coach students with feedback specific to their errors and optional hints that break problems down into simpler steps. This digital solution comes with Pearson eText, a complete online version of the book.

- Instructor's Solutions Manual. This supplement is available to adopters of this textbook in PDF format.

- Presentation Resource. All figures and tables from the textbook are available in PowerPoint format.

- Video Solutions. Provided with Pearson Mastering Engineering, Video Solutions offer step-by-step solution walkthroughs of representative homework problems from sections of the textbook. Make efficient use of class time and office hours by showing students the complete and concise problem solving approaches that they can access anytime and view at their own pace. The videos are designed to be a flexible resource to be used however each instructor and student prefers.

your work...

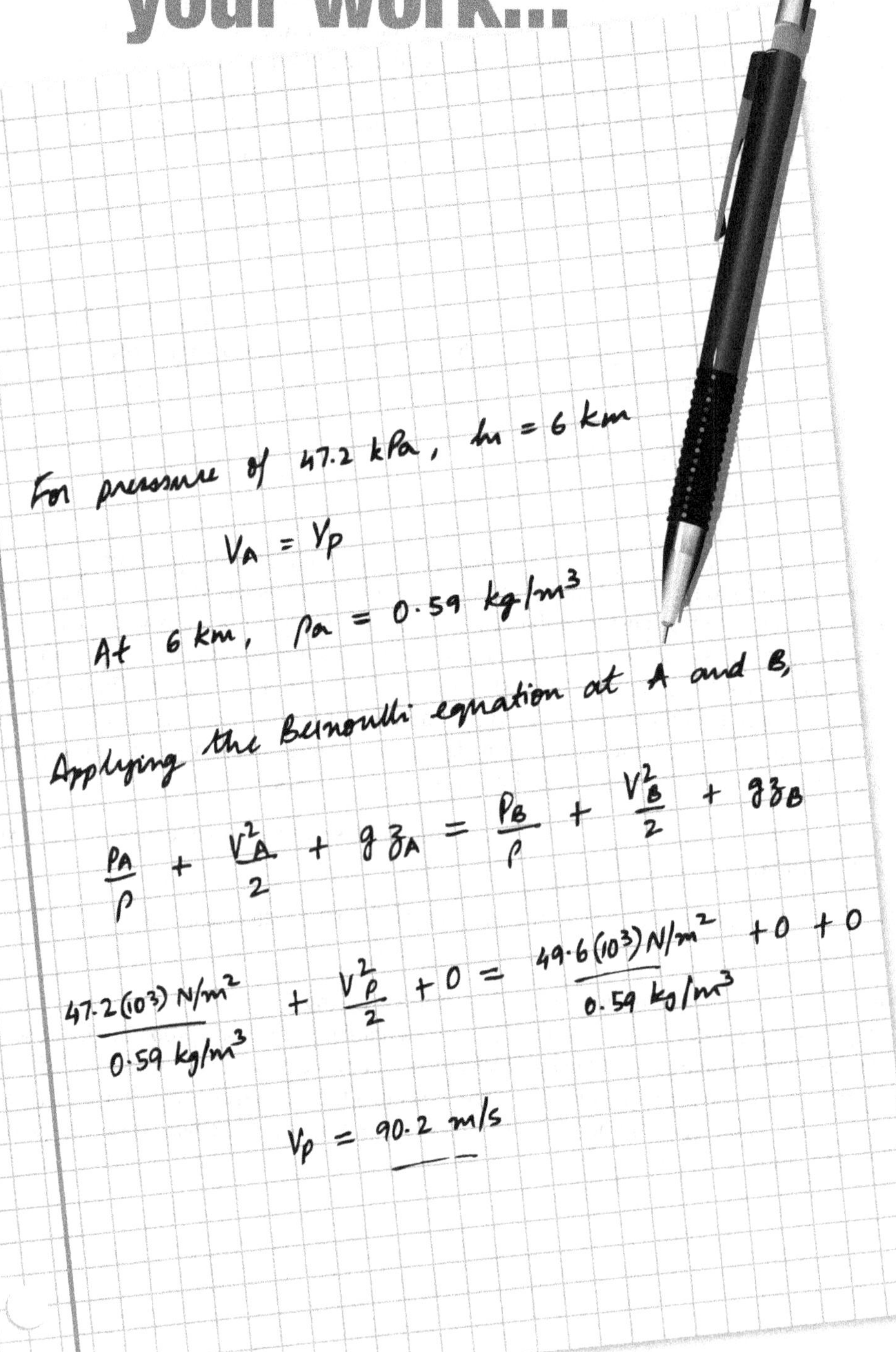
For pressure of 47.2 kPa, h = 6 km
VA = Vp
At 6 km, ρa = 0.59 kg/m3
Applying the Bernoulli equation at A and B,
pA/ρ + VA2/2 + g zA = pB/ρ + VB2/2 + g zB
47.2(103) N/m2 / 0.59 kg/m3 + Vp2/2 + 0 = 49.6(103) N/m2 / 0.59 kg/m3 + 0 + 0
Vp = 90.2 m/s

your answer specific feedback

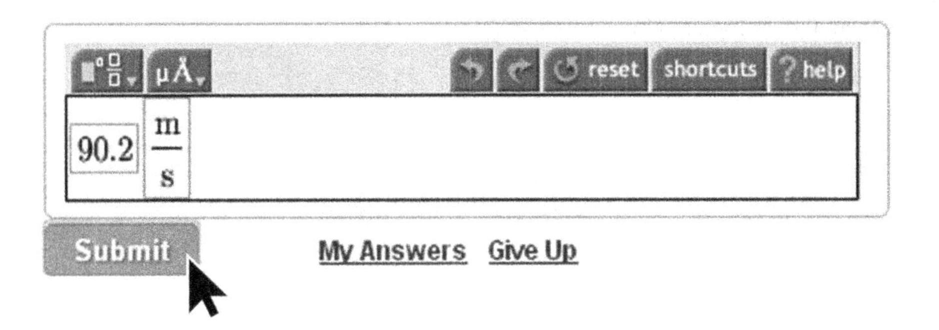

Incorrect; Try Again

It appears you have used the wrong value of density from the table. Check the density corresponding to the altitude.

www.MasteringEngineering.com

Acknowledgements

Pearson would like to thank and acknowledge the following for their contributions to the Global Edition.

Contributors

Rakesh Kumar Dhingra, Sharda University
Basant Singh Sikarwar, Amity University

Reviewers

Kanchan Chatterjee, Dr. B. C. Roy Engineering College
Rakesh Kumar Dhingra, Sharda University
Vibha Maru
Vipin Sharma, Delhi Technological University

Chapter 1

Properties of Fluids

LEARNING OBJECTIVES

After reading this chapter and solving a representative sample of end-of-chapter problems, you will be able to:

- Identify the characteristics of a fluid and describe the fundamental differences between solids, liquids, and gases.
- Understand dimensional homogeneity, fundamental dimensions, and systems of units.
- Understand the constitutive relationships and fluid properties relevant to engineering applications.
- Identify and readily quantify the key properties of water and air.

1.1 Introduction

Fluid mechanics is the study of the behavior of liquids and gases. The study of fluids at rest is called *fluid statics*, and the study of fluids in motion is called *fluid dynamics*. Applications of fluid mechanics are found in a variety of engineering disciplines. Aerospace engineering applications include the design of aircraft, aerospace vehicles, rockets, missiles, and propulsion systems. Biomedical applications include the study of blood flow and breathing. Civil engineering applications include the design of conveyance structures, dams, water-supply systems, oil and gas pipelines, wastewater processing systems, irrigation systems, and the determination of wind loads on buildings. Mechanical engineering applications include the design of plumbing systems, heating ventilation and air conditioning systems, lubrication systems, process-control systems, pumps, fans, turbines, and engines. Naval architecture applications include the design of ships and submarines. Aside from engineering applications of fluid mechanics, the earth sciences of hydrology, meteorology, and oceanography are based largely on the principles of fluid mechanics. A wide variety of fluid mechanics applications are apparent across many disciplines. However, the fundamentals of fluid mechanics that form the bases of these applications are relatively few, and the intent of this book is to cover these fundamentals.

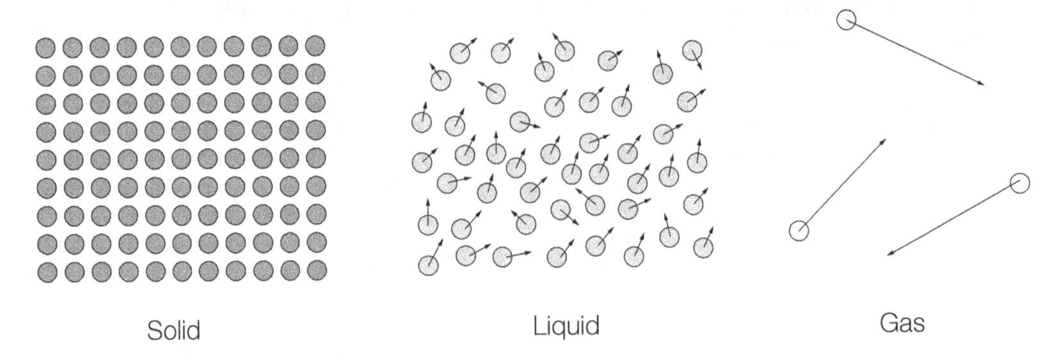

Figure 1.1: Molecular-scale views of solid, liquid, and gas

States of matter. The three states of matter commonly encountered in engineering are *solid*, *liquid*, and *gas*. Liquids and gases are both classified as fluids, and microscopic (molecular-scale) views of a solid, liquid, and gas are illustrated in Figure 1.1. Individual molecules (or atoms) in a solid are held together by relatively strong forces, and the molecules can only vibrate around an average position without any net movement. In contrast, the molecules in a liquid move relatively slowly past one another, and gas molecules move freely and at high speeds. In terms of the arrangement of molecules within the different phases of matter, in solids the molecules are closely packed in a regular pattern, in liquids the molecules are close together but do not have a fixed position relative to each other, and in gases the molecules are relatively far apart and move about independently of each other. Liquids and solids are sometimes referred to as *condensed-phase* matter because of the close spacing of their molecules.

Mechanical behavior of fluids. From a behavioral viewpoint, fluids are differentiated from solids by how they respond to applied stresses. Consider the volume of a substance acted upon by surface stresses as shown in Figure 1.2. The surface stresses can be expressed in terms of components that are normal and tangential to the surface of the specified volume; these components are called the *normal stress* and *shear stress* components, respectively. The normal stress causes the substance within the volume to compress (or expand) by a certain fixed amount, regardless of whether the substance is a fluid or a solid. However, a fluid will respond to an applied shear stress differently from a solid. A fluid will deform continuously under the action of an applied shear stress, whereas a solid will deform only

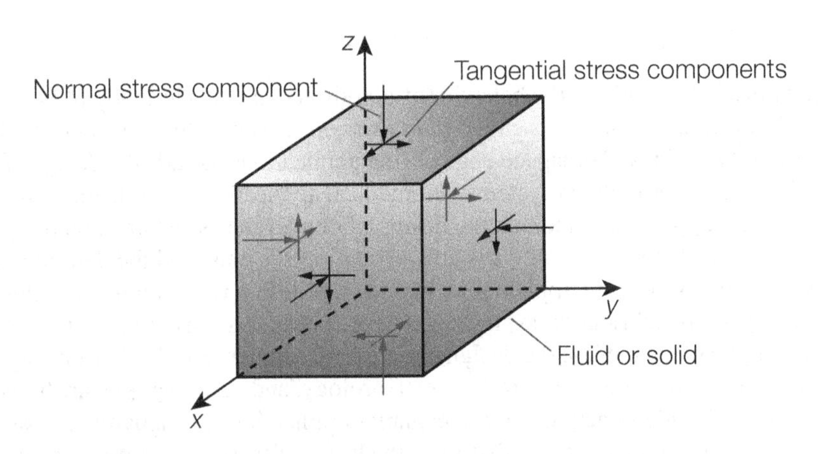

Figure 1.2: Surface stress components on a substance

by a finite fixed amount under the action of an applied shear stress. Continuous deformation under an applied shear stress is the property that differentiates a fluid from a solid. In fact, continuous deformation under the action of a shear stress is the defining behavior of a fluid.

Hybrid materials. Some materials are unusual in that they behave like a solid under some conditions and like a fluid under other conditions. Typically, these materials are solid-like when applied shear stresses are small and fluid-like when applied shear stresses are high. Examples include slurries, asphalt, and tar. The study of these types of hybrid materials is called *rheology*, which is often considered to be a field separate from fluid mechanics. In some cases, liquids and gases coexist, such as water containing air bubbles and water-steam mixtures. The flows of these mixtures are commonly called *multiphase flows*, and the study of these flows is a specialized area of fluid mechanics.

Physical differences between liquids, gases, and vapors. Liquids and gases are both fluids but with primary physical differences—a gas will expand to completely fill the volume of any closed container in which it is placed, whereas a liquid will retain a relatively constant volume within any container in which it is placed. This difference in behavior is caused by the relatively strong cohesive forces between molecules in a liquid, which tend to hold them together, compared with the weak forces between molecules in a gas, which allow them to move relatively independently of each other. Liquids will generally form a free surface in a gravitational field if unconfined from above. A *vapor* is a gas whose temperature and pressure are such that it is very near the liquid phase. Thus, steam is considered to be a vapor because its state is normally not far from that of water, whereas air is considered to be a gas because the states of its gaseous components are normally very far from the liquid phase.

Continuum approximation. Fluids as well as solids are made up of discrete molecules, and yet it is commonplace to disregard the discrete molecular nature of a fluid and view it as a continuum. The *continuum idealization* allows us to treat fluid properties as varying continually in space with no discontinuities. This idealization is valid as long as the size of the fluid volume is large compared to the space between molecules in the fluid. Under normal temperatures and pressures, the spacing of molecules in a fluid is on the order of 10^{-6} mm for gases and 10^{-7} mm for liquids. Hence, the continuum model is applicable as long as the characteristic length scale of the fluid volume is much larger than the characteristic spacing between molecules. The continuum approximation is sometimes considered applicable for volumes as small as 10^{-9} mm^3. It is interesting to note that the spacing between molecules in a liquid is not much different from the spacing between molecules in a solid. However, the molecules in liquids are less restrained in their ability to move relative to each other. In the case of gases, the characteristic spacing between molecules is sometimes measured by the *mean free path* of the molecules, which is the average distance traveled by a molecule between collisions. Under standard atmospheric conditions, the mean free path of molecules in air is on the order of 6.4×10^{-5} mm. At very high vacuums or at very high elevations, the mean free path may become large; for example, it is about 10 cm for atmospheric air at an elevation of 100 km and about 50 m at an elevation of 160 km. Under these circumstances, *rarefied gas flow theory* should be used and the impact of individual molecules should be considered.

1.1.1 Nomenclature

Fluid mechanics can be divided into three branches: statics, kinematics, and dynamics. *Fluid statics* is the study of the mechanics of fluids at rest, *kinematics* is the study of the geometry of fluid motion, and *fluid dynamics* is the study of the relationship between fluid motion and the forces acting on the fluid. Fluid dynamics is further divided into several specialty areas.

The study of fluid dynamics when the fluid is incompressible and frictionless is called *hydrodynamics*. Fluids that are incompressible and frictionless are called *ideal fluids*. In contrast to ideal fluids, real fluids have some degree of compressibility and internal friction. The study of liquid flows in pipes and open channels is sometimes called *hydraulics*, a term that some civil engineers associate with the description of flow based on empirical relationships rather than the fundamental physical laws on which fluid mechanics is based. *Gas dynamics* deals with the flow of fluids that undergo significant density changes, such as the flow of gases through nozzles at high speeds, and *aerodynamics* deals with the flow of gases (especially air) over bodies such as aircraft, rockets, and automobiles at high or low speeds.

Computational fluid mechanics. In many cases, the governing equations of fluid mechanics cannot be solved analytically, and numerical methods are used to determine the flow conditions at selected locations in the flow domain. The application of numerical methods to solve the governing equations of fluid mechanics is called *computational fluid mechanics*. Such applications are endemic to the field of aerospace engineering, although these techniques are also used for advanced applications in other engineering disciplines.

1.1.2 Dimensions and Units

Dimensions are physical measures by which variables are expressed, and examples of dimensions are mass, length, and time. *Units* are names assigned to dimensions, and examples of units are the kilogram (a unit of mass) and the meter (a unit of length). The seven fundamental dimensions in nature and their base units in the *Système International d'Unités (SI system)* are listed in Table 1.1. Additional units that are sometimes taken as fundamental are the unit of a plane angle (radian, rad), and the unit of a solid angle (steradian, sr). However, these units are properly classified as derived units in the SI system. The SI system of units is an *absolute system of units*, because it does not involve a fundamental dimension of force, which is a gravity effect.

Gravitational units. A *gravitational system of units* uses force as a fundamental dimension. The dimensions of force, mass, length, and time are related by Newton's law, which states that

$$F = ma \tag{1.1}$$

where a force F causes a mass m to accelerate at a rate a. In a gravitational system, F and m are not independent dimensions and the relationship between F and m is fixed by specifying the numerical value of a, which is commonly taken as unity in defining fundamental dimensions. A gravitational system in common use in the United States is the *U.S. Customary*

Table 1.1: Fundamental Dimensions and Units

Dimension	SI Unit	Symbol	USCS Unit	Symbol
Mass	kilogram	kg	–	–
Force	–	–	pound	lb
Length	meter[1]	m	foot	ft
Time	second	s	second	sec
Temperature	kelvin	K	rankine	°R
Electric current	ampere	A	ampere	A
Luminous intensity	candela	cd	candela	cd
Amount of substance	mole	mol	mole	mol

[1]The official spelling is "metre." In the United States, "meter" is used.

System (USCS) in which the fundamental dimension of force has a unit of pound (lb). The fundamental dimensions of the USCS are listed in Table 1.1 along with those of the SI system.

Dimensions in fluid mechanics applications. In fluid mechanics applications, the SI fundamental dimensions that are generally used include mass [M], length [L], time [T], temperature [Θ], and amount of substance [mol]. In the USCS system, force [F] replaces mass [M] as a fundamental dimension. Fundamental dimensions are sometimes referred to as *primary dimensions*, with dimensions derived from combinations of primary dimensions being referred to as *secondary dimensions*. In this text, square brackets are used to illustrate the dimensions of a given variable. For example, the statement "v is the velocity [LT^{-1}]" means that the velocity denoted by v has dimensions of length divided by time.

Dimensional homogeneity. All equations derived from fundamental physical laws must be *dimensionally homogeneous*. If an equation is dimensionally homogeneous, then all terms in a summation must have the same dimensions, which also means that terms on both sides of an equal sign must have the same dimensions.

EXAMPLE 1.1

Application of Newton's second law to the settling of a spherical particle in a stagnant fluid yields the theoretical relationship

$$mg - \frac{\pi}{8} C_{\rm D} \rho V^2 D^2 = m \frac{dV}{dt}$$

where m is the mass of the particle [M], g is the acceleration due to gravity [LT^{-2}], $C_{\rm D}$ is a (dimensionless) drag coefficient [-], ρ is the density of the fluid [ML^{-3}], V is the settling velocity [LT^{-1}], and t is time [T]. Determine whether the given equation is dimensionally homogeneous.

SOLUTION

Expressing the variables in the given equation in terms of their dimensions yields

$$mg - \frac{\pi}{8} C_{\rm D} \rho V^2 D^2 = m \frac{dV}{dt} \quad \rightarrow \quad [\mathrm{M}] \left[\frac{\mathrm{L}}{\mathrm{T}^2}\right] - [\text{-}] \left[\frac{\mathrm{M}}{\mathrm{L}^3}\right] \left[\frac{\mathrm{L}}{\mathrm{T}}\right]^2 [\mathrm{L}]^2 = [\mathrm{M}] \frac{\left[\frac{\mathrm{L}}{\mathrm{T}}\right]}{[\mathrm{T}]}$$

$$\rightarrow \quad \frac{\mathrm{ML}}{\mathrm{T}^2} + \frac{\mathrm{ML}}{\mathrm{T}^2} = \frac{\mathrm{ML}}{\mathrm{T}^2}$$

Because each term in the given equation has the same dimensions, the equation is dimensionally homogeneous.

The requirement of dimensional homogeneity is particularly useful in checking the derivation of equations obtained by algebraic manipulation of other dimensionally homogeneous equations. This is because any equation derived from a set of dimensionally homogeneous equations must itself be dimensionally homogeneous.

Table 1.2: SI Derived Units

Unit Name	Quantity	Symbol	In Terms of Base Units
degree Celsius	temperature	°C	K
hectare	area	ha	10^4 m^2
hertz	frequency	Hz	s^{-1}
joule	energy, work, quantity of heat	J	N·m
liter	volume	L	10^{-3} m^3
watt	power	W	J/s
newton	force	N	kg·m/s^2
pascal	pressure, stress	Pa	N/m^2

SI Units. Some key conventions in the SI systems that are relevant to fluid mechanics applications are given below.

- In addition to the base SI units, a wide variety of units derived from the base SI units are also used. A few commonly used derived units are listed in Table 1.2.
- When units are named after people, such as the newton (N), joule (J), and pascal (Pa), they are capitalized when abbreviated but not capitalized when spelled out. The abbreviation capital L for liter is a special case, used to avoid confusion with one (1).
- In accordance with Newton's law (Equation 1.1), 1 N is defined as the force required to accelerate a mass of 1 kg at 1 m/s^2; hence,

$$1\ \text{N} = 1\ \text{kg} \times 1\ \text{m/s}^2$$

- A nonstandard derived unit of force that is commonly used in Europe is the *kilogram force* (kgf), where 1 kgf is the gravitational force on a 1 kg mass, where the gravitational acceleration is equal to the standard value of 9.80665 m/s^2; therefore, 1 kgf = 9.80665 N. It is not uncommon in Europe to see tire pressures quoted in the nonstandard unit of kgf/cm^2. It is also common in Europe to express weights in *kilos*, where 1 kilo = 1 kgf.
- The unit of absolute temperature is the kelvin,[1] which is abbreviated K without the degree symbol. In engineering practice, the degree Celsius (°C) is widely used in lieu of the kelvin and the relationship between these temperature scales is given by

$$T_\text{K} = T_\text{C} + 273.15 \tag{1.2}$$

where T_K and T_C are the temperatures in kelvins and degrees Celsius, respectively. Note that 1 K = 1°C, and an ideal gas theoretically has zero energy when the temperature is equal to 0 K. The reference quantity 273.15 K in Equation 1.2 is exactly 0.01 K below the triple point of water.

- The units of second, minute, hour, day, and year are correctly abbreviated as s, min, h, d, and y, respectively.

[1]Named in honor of the Irish and British physicist and engineer William Thomson (also known as The Lord Kelvin) (1824–1907).

Table 1.3: Prefixes to SI Units

Factor	Prefix	Symbol	Factor	Prefix	Symbol
10^{12}	tera	T	10^{-3}	milli	m
10^{9}	giga	G	10^{-6}	micro	μ
10^{6}	mega	M	10^{-9}	nano	n
10^{3}	kilo	k	10^{-12}	pico	p

- In using prefixes with SI units, multiples of 10^3 are preferred in engineering usage, with other multiples avoided if possible. Standard prefixes and their associated symbols are given in Table 1.3. Note that the prefix "centi," as in *centimeter*, is not a preferred prefix because it does not involve a multiple of 10^3. Unit prefixes are typically utilized when the magnitude of a quantity is more than 1000 or less than 0.1. For example, 2100 Pa can be expressed as 2.1 kPa and 0.005 m as 5 mm.

USCS Units. USCS units are sometimes called *English units*, *Imperial units*, or *British Gravitational units*. Some key conventions in the USCS system that are relevant to fluid mechanics applications are given below.

- The USCS system is a gravitational system of units in which the unit of length is the foot (ft), the unit of force is the pound (lb), the unit of time is the second (s), and the unit of temperature is the degree Rankine[2] (°R). In engineering practice, the degree Fahrenheit (°F) is widely used in lieu of the degree Rankine and the relationship between these temperature scales is given by

$$T_{\mathrm{R}} = T_{\mathrm{F}} + 459.67 \tag{1.3}$$

  where T_{R} and T_{F} are the temperatures in degrees Rankine and degrees Fahrenheit, respectively. Note that $1°\mathrm{R} = 1°\mathrm{F}$, and an ideal gas theoretically has zero energy when the temperature is equal to $0°\mathrm{R}$.

- Other fundamental units that are not usually encountered in fluid mechanics applications are the same for the USCS and SI systems, specifically the units of electric current (ampere, A), luminous intensity (candela, cd), and amount of substance (mole, mol).

- In the USCS system, the unit of mass is the slug, which is a derived unit from the fundamental unit of force, which is the pound. The slug is defined as the mass that accelerates at 1 ft/sec^2 when acted upon by a force of 1 pound; hence,

$$1 \text{ slug} = \frac{1 \text{ lb}}{1 \text{ ft/sec}^2}$$

- The abbreviation for pound is sometimes equivalently expressed as "lbf" rather than "lb" to emphasize that the pound is a unit of force (1 lb = 1 lbf). The pound force (lbf) in the USCS system is a comparable quantity to the newton (N) in the SI system, where 1 lbf $\approx$ 4.448 N.

- The USCS units of second, minute, hour, day, and year are correctly abbreviated as sec, min, hr, day, and yr, respectively. However, it is not uncommon to use the SI abbreviations (s, min, h, d, and y, respectively) when otherwise using USCS units.

[2]Named in honor of the Scottish physicist and engineer William John Macquorn Rankine (1820–1872).

English Engineering units. The English Engineering system of units is almost identical to the USCS, with the main differences being that both force and mass are taken as fundamental dimensions in the English Engineering system and the pound mass (lbm) is used as the unit of mass. The relationship between the pound mass (lbm) and the slug is

$$1 \text{ lbm} = 1 \text{ slug} \times 32.174 \text{ ft/sec}^2$$

This relationship is derived from the basic definition that a gravity force of 1 lbf will accelerate a mass of 1 lbm at a rate of 32.174 ft/sec^2, which is the acceleration due to gravity. A mass of 1.0000 slug is equivalent to 32.174 lbm.

Conversion between units. It is generally recommended that engineers have a sense of the conversion factors from one system of units to another, especially for the most commonly encountered dimensions and units. Such conversion factors can be found in Appendix A.2. Some fields of engineering commonly use mixed units, where some quantities are traditionally expressed in USCS units and other quantities are traditionally expressed in SI units. A case in point is in applications related to the analysis and design of air-handling units, where airflow rates are commonly expressed in CFM (= ft^3/min) and power requirements are expressed in kW. When mixed units are encountered in a problem, it is generally recommended to convert all variables to a single system of units before beginning to solve the problem. This text uses SI units.

Conventions and constants. In cases where large numbers are given, it is common practice not to use commas, because in some countries, a comma is interpreted as a decimal point. A recommended practice is to leave a space where the comma would be; for example, use 25 000 instead of 25,000. Acceleration due to gravity, g, is used in the analysis of many fluid flows, and by international agreement, standard gravity, g, at sea level is 9.80665 m/s^2. Actual variation in g on Earth's surface is relatively small and is usually neglected. To illustrate the variability, g is approximately equal to 9.77 m/s^2 on the top of Mount Everest and is approximately 9.83 m/s^2 at the deepest point in Earth's oceans; hence, the deviation is less than 0.4% from standard gravity. It is sometimes convenient to represent the units of g as N/kg rather than m/s^2, particularly in dimensional analysis applications. In analyzing fluid behavior, reference is commonly made to *standard temperature and pressure*. By convention, standard temperature is 15°C and standard pressure is 101.3 kPa. These standard conditions roughly represent average atmospheric conditions at sea level at 40° latitude.

Physical appreciation of magnitudes. In engineering applications, it is important to have a physical appreciation of the magnitudes of quantities, at least to make an assessment of whether calculated results and designs are physically realistic. With this in mind, the following approximate relationships between SI units and USCS units might be helpful.

- FORCE: A force of 1 N is roughly equal to $\frac{1}{4}$ lb, which is approximately the weight of a small apple. A weight of 1 lb is roughly equal to 4 N. In many cases, force units of kilonewtons are more appropriate.

- PRESSURE: A pressure of 1 Pa is roughly equal to 10^{-4} lb/in^2. The pressure unit of pascal (Pa) is too small for most pressures encountered in engineering applications. Units of kilopascal or megapascal are usually more appropriate, where 1 kPa $\approx$ 0.1 lb/in^2 and 1 MPa $\approx$ 100 lb/in^2. The pressure unit of "atmosphere" (atm) is a convenient unit in many applications, because 1 atm is equal to standard atmospheric pressure at sea level (= 101.3 kPa). Pressure units of atm, bar, and kgf/cm^2 are approximately equal because 1 atm ≈ 1.01 bar ≈ 1.03 kgf/cm^2.

- VOLUME: A volume of 1 m^3 is roughly equal to 35 ft^3. The unit of m^3 is quite large for some applications. Units of liters (L), where 1 L = 0.001 m^3, are frequently used when dealing with smaller volumes. A volume of 1 L is roughly equal to $\frac{1}{4}$ gal, and 1 L is approximately equal to 1 quart.

- VOLUME FLOW RATE: The conventional SI unit of (volume) flow rate is m^3/s, the conventional USCS unit of flow rate is ft^3/s (cfs), and 1 m^3/s ≈ 35 cfs. These conventional units are used to represent fairly large flows such as those found in rivers and streams. Smaller flow rates such as airflow rates in building ventilation systems are typically expressed in ft^3/min (CFM) or m^3/min, and liquid flow rates in pipelines are commonly expressed in L/min, L/s, or gallons per minute (gpm). Note that 1 m^3/s = 60 000 L/min = 1000 L/s ≈ 2120 CFM ≈ 15 850 gpm.

Consideration of significant digits. The number of significant digits in a number reflects the accuracy of the number. Because the last significant digit in a number is regarded as uncertain (±1), numbers with one significant digit can have a maximum error of 100%, numbers with two significant digits can have a maximum error of 10%, numbers with three significant digits can have a maximum error of 1%, and so on. In engineering applications, measured numbers seldom have accuracies greater than 0.1%, and such numbers are represented by no more than four significant digits. In performing calculations, one cannot arrive at a result that is more accurate (in terms of percentage error) than the numbers used in calculations to arrive at that number; hence, the final result of calculations cannot have more significant digits than the numbers used in calculating that result. The following three rules are useful: (1) For multiplication and division, the number of significant digits in the calculated result is equal to the number of significant digits in the least accurate number used in the calculation; (2) for addition and subtraction, the number of significant decimal places in the result equal that of the least number of significant decimal places in the added/subtracted numbers; and (3) where multiple operations are involved, extra (nonsignificant) digits in the intermediate calculations are retained and the final result is rounded to the appropriate number of significant digits based on the accuracy of the numbers used in the calculations. Because the solution of most problems in fluid mechanics involves multiple operations, retaining nonsignificant digits in intermediate quantities and rounding the final result to the appropriate number of significant digits is the most common practice.

EXAMPLE 1.2

(a) In the analysis of building ventilation, airflows are commonly expressed in units of ft^3/min or CFM. If the fresh airflow into a particular building space is 800 CFM, what is the airflow rate in m^3/s? (b) A relationship between the energy per unit weight, h_p, added by a pump and the flow rate, Q, through the pump is given by

$$h_p = 22.3 + 7.54 \times 10^5 Q^2 \tag{1.4}$$

where h_p is in N·m/N or m and Q is in m^3/s. What is the equivalent relationship if h_p is expressed in ft and Q in gallons per minute (gpm)?

SOLUTION

Commonly used conversion factors are found in Appendix A.2. The basic conversion factors to be used here are 1 ft = 0.3048 m, 1 min = 60 s, and 1 (US) gal = 3.785 L = 3.785 × 10^{-3} m^3.

(a) Using the basic conversion factors,

$$800 \text{ ft}^3/\text{min} = 800 \frac{\text{ft}^3}{\text{min}} \times 0.3048^3 \frac{\text{m}^3}{\text{ft}^3} \times \frac{1}{60} \frac{\text{min}}{\text{s}} = 0.3775 \text{ m}^3/\text{s} \approx \mathbf{0.378 \text{ m}^3/\text{s}}$$

In general, a converted quantity should have approximately the same accuracy as that of the original quantity. Therefore, the number of significant digits in the converted quantity should be the same in the converted and original quantities. Intermediate conversion calculations and conversion factors must be at least as accurate as the quantity being converted.

(b) Using the basic conversion factors,

$$1.00 \text{ m}^3/\text{s} = 1.00 \frac{\text{m}^3}{\text{s}} \times \frac{1}{3.785 \times 10^{-3}} \frac{\text{gal}}{\text{m}^3} \times 60 \frac{\text{s}}{\text{min}} = 1.585 \times 10^4 \text{ gpm}$$

This gives the conversion factor for m^3/s to gpm. Applying this conversion factor along with 1 ft = 0.3048 m to the given empirical equation, where h_p is in ft and Q is in gpm, gives the following equivalent equation in the modified units:

$$(0.3048\, h_\text{p}) = 22.3 + 7.54 \times 10^5 \left(\frac{Q}{1.585 \times 10^4}\right)^2 \quad \rightarrow \quad \mathbf{h_\text{p} = 73.2 + 9.85 \times 10^{-3} Q^2} \tag{1.5}$$

Therefore, Equations 1.4 and 1.5 are equivalent, with the exception that h_p and Q in Equation 1.4 are in m and m^3/s, respectively, whereas in Equation 1.5, h_p and Q are in ft and gpm, respectively. The coefficients in both equations have the same number of significant digits and therefore yield results of comparable accuracy.

1.1.3 Basic Concepts of Fluid Flow

Fluid flows are influenced by a variety of forces, with the dominant forces usually including pressure forces, gravity forces, and drag forces caused by fluid motion relative to solid boundaries. Whenever a moving fluid is in contact with a solid surface, the velocity of the fluid in contact with the solid surface must necessarily be equal to the velocity at which the solid surface is moving. This is called the *no-slip condition*, and the region within the fluid close to a solid surface where the velocity of the fluid is affected by the no-slip condition is called the *boundary layer*. Viscosity is the fluid property that causes the formation of a boundary layer. Although all fluid flows bounded by solid surfaces have boundary layers, in some cases, there are (outer) regions of the flow field where the viscosity of the fluid exerts a negligible influence on the fluid motion. Such flows are called *inviscid flows*, where the word "inviscid" means "without viscosity." In addition to the no-slip condition at a solid boundary, there also exists a *no-temperature-jump condition*. This requires the temperature of the fluid in contact with a solid boundary to be equal to the temperature of the boundary itself.

Classification of fluid flows. Classification of fluid flows is the conventional approach to simplifying the analysis fluid flows. Flows within various classifications are typically characterized by the unimportance of some forces, which leads to simplifications in the governing equations. A variety of flow classifications are of importance in engineering analyses. A fluid flow can be classified as a *viscous flow* when the viscosity of the fluid exerts a significant influence on the flow and as *inviscid flow* when the viscosity has a negligible influence on the flow. A fluid flow can be classified as *laminar flow* when random perturbations in velocity do not occur and as *turbulent flow* when random perturbations in the velocity field do occur. A fluid flow can be classified as an *internal flow* when the flow is confined within solid boundaries and as an *external flow* when the flow is unconfined around a solid object. Examples of internal flows are flows in pipes and ducts, and examples of external flows are flows around buildings and the wings of airplanes (i.e., airfoils). Internal flows are dominated by the influence of viscosity throughout the flow field, whereas in external flows, the viscous effects are limited to boundary layers near solid surfaces and to wake regions downstream of bodies. A fluid flow is classified as *incompressible flow* when the density of the fluid remains approximately constant throughout the flow field. The fluid flow is classified as *compressible flow* when the density of the fluid within the flow field varies significantly in response to pressure variations. Liquid flows and gas flows at speeds much less than the speed of sound are typically taken as incompressible flows. The classification of fluid flows, implementation of simplifications, and derivation of consequent relationships that are useful in analysis and design is the tact followed in this text on applied fluid mechanics.

Role of fluid properties. The behavior of a fluid depends on its properties, and in a fluid mechanical sense, fluids only differ from each other to the extent their properties are different. The physical properties of fluids that are important in most engineering applications include density, viscosity, compressibility, surface tension, saturation vapor pressure, and latent heat of vaporization. These properties as well as others are commonly referred to as *thermodynamic properties*, because they are used in quantifying the heat content of fluids and the conversion of energy between different forms. Fluids can be either liquids or gases, and fluid properties are sometimes given under conditions referred to as *standard temperature and pressure* (STP). For air, STP is generally taken as 15°C and 101.3 kPa. The definitions of commonly used fluid properties, along with their utilization in various engineering applications, are presented in the following sections.

1.2 Density

The *density* (or *mass density*) of a fluid, ρ, is defined as the mass of fluid per unit volume [ML^{-3}]; hence,

$$\text{density, } \rho = \frac{\text{mass of substance}}{\text{volume of substance}} \tag{1.6}$$

The densities of most gases are directly proportional to pressure and inversely proportional to temperature, whereas the densities of most liquids are relatively insensitive to pressure but depend on temperature. In comparison to gases, liquids are commonly regarded as incompressible. A 1% change in density of water at 101.3 kPa requires a change in pressure of about 21.28 MPa. In contrast, a 1% change in the density of air at 101.3 kPa requires a change in pressure of only 1.01 kPa. Liquids are about three orders of magnitude more dense than gases, with mercury being one of the denser liquids (ρ = 13 580 kg/m^3) and hydrogen being the least dense gas (ρ = 0.0838 kg/m^3).

Density of water and other liquids. At temperatures in the range of 0–100°C and at a standard atmospheric pressure of 101.3 kPa, water exists in the liquid state. The densities of pure water at temperatures between 0°C and 100°C are given in Appendix B.1, and the densities of several other commonly encountered liquids are given in Appendix

B.4. For most liquids, the density decreases monotonically as the temperature increases. However, pure water has the unusual property of the density increasing with temperatures between 0°C and 4°C and then decreasing with temperature higher than 4°C; hence, water has its maximum density at 4°C. This unique property of water explains why when temperatures drop to near freezing (i.e., near 0°C) over a lake or another water body, the colder, less dense water "floats" to the top, causing ice to form from the top down rather than the bottom up as would occur if water had the monotonic density properties of most other liquids, which would lead to the denser, colder water being on the bottom. The density of water as a function of temperature in the ranges of 0–100°C and 0–10°C is illustrated in Figures 1.3(a) and 1.3(b), respectively. It is apparent from these figures that the peak in the density of water is not noticeable on the scale of density variations over the temperature range of 0–100°C, but is readily apparent on the scale of density variations over the temperature range of 0–10°C. An approximate analytic expression (within ±0.2%) for the density of water, ρ_w, at a temperature, T, in the range of 0–100°C is

$$\rho_\text{w} = 1000 - 0.0178\,|T - 4|^{1.7} \tag{1.7}$$

where ρ_w is in kg/m^3 and T is in °C. The addition of salt to water increases the density of the water, suppresses the temperature at which the maximum density occurs, and suppresses the freezing point of the water. The effect of salt on suppressing the freezing point of water is utilized when "road salt" is applied to prevent the formation of ice on roads. The effect of salt on increasing the density of water explains why seawater intrudes below fresh water in coastal areas. Seawater is a mixture of pure water and various salts, and the salt content of seawater is commonly measured by the salinity, S, which is defined by

$$\text{salinity, } S = \frac{\text{weight of dissolved salt}}{\text{weight of mixture}} \tag{1.8}$$

The average salinity of seawater is typically taken as 0.035, which is commonly expressed as 35‰, the symbol ‰ meaning "parts per thousand." At this salinity, the average density of seawater is 1030 kg/m^3. The effect of salt content on density is vividly illustrated in Figure 1.4, which shows water with different salt concentrations that have been carefully poured in layers on top of each other. The different layers are identified using a different dye color in each layer. Of course, the salt concentration is lowest in the top layer and highest in the bottom layer.

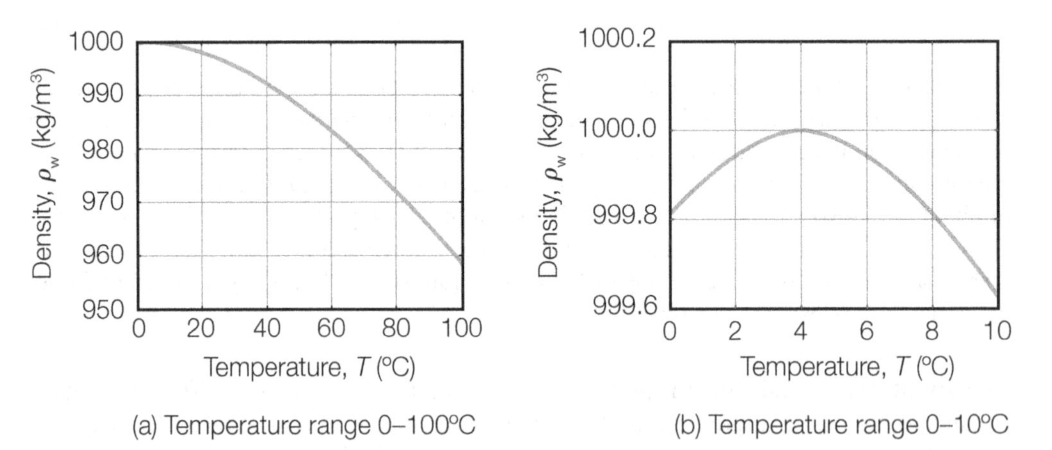

Figure 1.3: Density of water as a function of temperature

Figure 1.4: Layers of water with different salt content
Source: ginton/Fotolia.

Density of air and other gases. Standardized properties of air are commonly used in engineering analyses. By volume, standard (dry) air contains approximately 78.09% nitrogen, 20.95% oxygen, 0.93% argon, 0.039% carbon dioxide, and small amounts of other gases. Atmospheric air also contains a variable amount of water vapor, with an average volumetric content of around 1%. At the standard atmospheric pressure of 101.3 kPa, the density of standard air at a temperature of 15°C is 1.225 kg/m^3. The density of standard air as a function of temperature at a pressure of 101.3 kPa is given in Appendix B.2. Standard atmospheric pressure (101.3 kPa) is typically used to approximate conditions at sea level. The density of air at elevations above sea level typically decrease with increasing altitude, which is the net result of decreasing pressure and decreasing temperature with altitude. Although the variation of air density with altitude varies with location, the *standard atmosphere* is used in engineering analyses to approximate the variation of air density with altitude. The variation of air density with altitude in a standard atmosphere is given in Appendix B.3. Gases at states far removed from their liquid states are commonly approximated as ideal gases, and the variation of density with temperature and pressure is commonly approximated by the ideal gas law, which is discussed in more detail in Section 1.4. For ready reference, the densities of several gases that are commonly encountered in engineering applications are given in Appendix B.5, where these densities correspond to standard atmospheric pressure and a temperature of 20°C.

Specific weight. The *specific weight* (or *weight density*) of a fluid, γ, is defined as the weight per unit volume and is related to the density by

$$\text{specific weight, } \gamma = \rho g = \frac{\text{weight of substance}}{\text{volume of substance}} \tag{1.9}$$

where g is the acceleration due to gravity, which can be taken as its standard value of 9.80665 m/s^2. As a reference point, at a temperature of 20°C and a pressure of 101.3 kPa, the specific weight of water is 9790 N/m^3 and the specific weight of air is 11.8 N/m^3. Referring to γ as the weight density and ρ as the mass density can be helpful in characterizing the relationship between these two closely related fluid properties.

Specific gravity. The *specific gravity*, SG, of a liquid is defined as the ratio of the density of the liquid to the density of pure water at some specified temperature, usually 4°C. The *specific gravity* is also called the *relative density*, with the latter term being more widely used in the United Kingdom and the former term more widely used in the United States. The definition of the specific gravity (= relative density) is given by

$$\text{specific gravity, SG} = \frac{\rho}{\rho_{\text{H}_2\text{O}@4^\circ\text{C}}} = \frac{\text{density of liquid}}{\text{density of water at }4^\circ\text{C}} \tag{1.10}$$

where the density of pure water at 4°C is equal to 1000 kg/m^3. In some specialized applications, a reference temperature of 15.56°C is used instead of 4°C, a practice that is particularly common in the petroleum industry. However, because the density of water is 1000 kg/m^3 at 4°C and 999.04 kg/m^3 at 15.56°C, the adjusted reference temperature changes the specific gravity by less than 0.1%. The specific gravity of a gas is the ratio of its density to that of either hydrogen or air at some specified temperature and pressure; there is no general agreement on these standards, and the specific gravity of a gas is a seldom-used quantity. Note that the specific gravity of a substance is a dimensionless quantity. The specific gravities of several substances used in engineering applications are listed in Table 1.4 in decreasing order of magnitude. Not shown in Table 1.4 are the specific gravities of various crude oils, which vary depending on the source. Crude oils in the western United States typically have specific gravities in the range of 0.87–0.92, those in the eastern United States have specific gravities around 0.82, and Mexican crude oil has specific gravities around 0.97. Distillates of oil such as gasolines, kerosenes, and fuel oils have specific gravities in the range of 0.67–0.98.

Specific volume. The *specific volume*, v, is the volume occupied by a unit mass of fluid. It is commonly applied to gases and expressed in m^3/kg. The specific volume is related to the density by

$$\text{specific volume, } v = \frac{1}{\rho} = \frac{\text{volume of substance}}{\text{mass of substance}} \tag{1.11}$$

The specific volume is not a commonly used property in fluid mechanics; it is more commonly used in the field of thermodynamics.

Table 1.4: Typical Specific Gravities of Selected Engineering Materials

Substance*	Specific Gravity (SG)	Substance*	Specific Gravity (SG)
Gold	19.3	Seawater	1.025
Uranium	18.7	**Water**	**0.998**
Mercury	13.6	SAE 10-W motor oil	0.92
Lead	11.4	Dense oak wood	0.93
Copper	8.91	Ice (at 0°C)	0.916
Steel	7.83	Benzene	0.879
Cast iron	7.08	Crude oil	0.87
Aluminum	2.64	Ethyl alcohol	0.790
Concrete (cured)	2.4	Gasoline	0.72
Blood (at 37°C)	1.06	Balsa wood	0.17

*Liquids are at 20°C unless otherwise stated.

EXAMPLE 1.3

A closed cylindrical storage tank with a diameter of 3 m and and a height of 2.1 m is intended to store gasoline. Leakage of rainwater into the tank has resulted in a 1.0-m layer of water on the bottom of the tank, a 0.5-m layer of gasoline in the middle, and a 0.6-m layer of air at atmospheric pressure shown in Figure 1.5. The storage tank weighs 0.5 kN, and all fluids within the tank are at 20°C. (a) Estimate the weight of the tank and its contents when it is full of gasoline. (b) Estimate the weight of the tank and its contents under the condition shown in Figure 1.5.

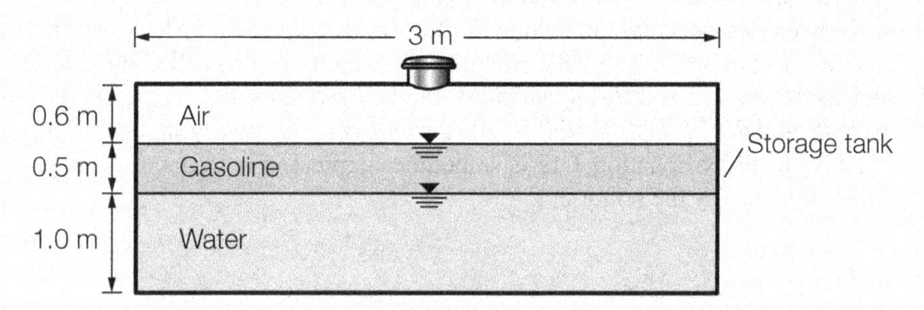

Figure 1.5: Storage tank with water and gasoline

SOLUTION

From the given data: $D = 3$ m, $h = 2.1$ m, $h_{\text{w}} = 1.0$ m, $h_{\text{g}} = 0.5$ m, $h_{\text{a}} = 0.6$ m, and $W_{\text{tank}} = 0.5$ kN. The following fluid properties (at 20°C) are obtained from Appendix B: $\rho_{\text{w}} = 998.2$ kg/m^3, $\rho_{\text{g}} = 680$ kg/m^3, and $\rho_{\text{a}} = 1.204$ kg/m^3. The subscripts "w," "g," and "a" refer to water, gasoline, and air, respectively. Taking $g = 9.807$ m/s^2, the specific weights of the fluids are given by

$$\gamma_{\text{w}} = \rho_{\text{w}} g = 9.789 \text{ kN/m}^3, \qquad \gamma_{\text{g}} = \rho_{\text{g}} g = 6.669 \text{ kN/m}^3, \qquad \gamma_{\text{a}} = \rho_{\text{a}} g = 0.01181 \text{ kN/m}^3$$

The cross-sectional area, A, of the tank, the total volume, V, of the tank, and the volumes occupied by water, gasoline, and air are derived from the given data as follows:

$$A = \pi D^2/4 = 7.069 \text{ m}^2, \qquad V = Ah = (7.069)(2.1) = 14.84 \text{ m}^3$$

$$V_{\text{w}} = Ah_{\text{w}} = (7.069)(1.0) = 7.069 \text{ m}^3, \qquad V_{\text{g}} = Ah_{\text{g}} = (7.069)(0.5) = 3.535 \text{ m}^3$$

$$V_{\text{a}} = Ah_{\text{a}} = (7.069)(0.6) = 4.241 \text{ m}^3$$

(a) The weight, W, of the tank plus its contents when the tank is full of gasoline is given by

$$W = W_{\text{tank}} + \gamma_{\text{g}} V = 0.5 + (6.669)(14.84) = \mathbf{99.5\ kN}$$

(b) The weight, W, of the tank plus its contents under the condition shown in Figure 1.5 is given by

$$\begin{aligned} W &= W_{\text{tank}} + \gamma_{\text{w}} V_{\text{w}} + \gamma_{\text{g}} V_{\text{g}} + \gamma_{\text{a}} V_{\text{a}} \\ &= 0.5 + (9.789)(7.069) + (6.669)(3.535) \text{ kN} + (0.01181)(4.241) = \mathbf{93.3\ kN} \end{aligned}$$

1.3 Compressibility

The *compressibility* of fluids are typically characterized by the *bulk modulus of elasticity*, E_v [FL^{-2}], defined by the relation

$$\text{bulk modulus of elasticity, } E_v = -\left.\frac{dp}{dV/V}\right|_{T_0} \approx -\left.\frac{\Delta p}{\Delta V/V}\right|_{T_0} \tag{1.12}$$

where Δp is the pressure increment [FL^{-2}] that causes a volume $V[L^3]$ of a substance to change by $\Delta V[L^3]$ and T_0 is the constant temperature maintained during the imposition of the pressure increment. A negative sign appears in Equation 1.12 because an increase in pressure causes a decrease in volume. In the limit as $\Delta p \to 0$, Δp becomes the differential dp and ΔV becomes dV. The bulk modulus of elasticity is also called by several other names, such as the *volume modulus of elasticity*, the *bulk modulus*, the *coefficient of compressibility*, the *bulk modulus of compressibility*, and the *bulk compressibility modulus*. The bulk modulus, E_v, as defined by Equation 1.12 is sometimes expressed in terms of density, ρ, and density differential, $d\rho$, by the relation

$$E_v = \left.\frac{dp}{d\rho/\rho}\right|_{T_0} \tag{1.13}$$

The bulk modulus of a fluid is a similar property to the modulus of elasticity of a solid, where the bulk modulus relates to the strain of a volume and the modulus of elasticity (without the word "bulk") relates to the strain of a length. Large values of the bulk modulus of elasticity are usually associated with a fluid being incompressible, although true incompressibility would correspond to an infinitely large bulk modulus. In most engineering problems, the bulk modulus at or near atmospheric pressure is the one of interest. Standard atmospheric pressure at sea level is 101.3 kPa. The bulk modulus of liquids is a function of temperature and pressure, and values of E_v for water at conditions close to atmospheric pressure are shown in Appendix B.1, where it is apparent that water has a minimum compressibility at around 50°C. A typical value for the bulk modulus of cold water is 2200 MPa, and increasing the pressure of water by 6900 kPa compresses it to only 0.3% of its original volume; hence, the incompressibility assumption seems justified. The compressibility of liquids covers a wide range; for example, mercury has a compressibility 8% of that of water, and nitric acid is six times more compressible than water. The compressibility characteristics of common liquids and gases are shown in Appendices B.4 and B.5, respectively. Generally, gases are very compressible, with the bulk modulus of a typical gas at room temperature being around 0.1 MPa, compared with a bulk density of 2200 MPa for water.

Compressibility of an ideal gas. The ideal gas law can be used to derive the compressibility of gases that can be approximated as being ideal. Using the ideal gas equation and Equation 1.13 as the definition of E_v gives

$$p = \rho RT \quad \to \quad \left.\frac{dp}{d\rho}\right|_{T_0} = RT_0 = \left.\frac{p}{\rho}\right|_{T_0} \quad \to \quad E_v = \left.\frac{dp}{d\rho/\rho}\right|_{T_0} = p \tag{1.14}$$

Hence, the bulk modulus of elasticity of an ideal gas at any given pressure is numerically equal to the pressure. This result is in agreement with the previously stated result that the bulk modulus of a typical gas at room temperature is around 0.1 MPa, given that standard atmospheric pressure is 101.3 kPa.

Relationship of bulk modulus to speed of sound. Sound is defined as a pressure wave of infinitesimal magnitude. The speed of sound, c, can be expressed in terms of the ratio of the change in pressure to the change in density, $\mathrm{d}p/\mathrm{d}\rho$, at a constant entropy, s. For liquids (but not gases), small changes in pressure usually yield small changes in temperature, and the following approximation is applicable

$$c = \sqrt{\left.\frac{\mathrm{d}p}{\mathrm{d}\rho}\right|_s} \quad \rightarrow \quad c \approx \sqrt{\left.\frac{\mathrm{d}p}{\mathrm{d}\rho}\right|_{T_0}} = \sqrt{\frac{E_\mathrm{v}}{\rho}} \quad \text{(for liquids)} \tag{1.15}$$

where c, E_v, and ρ for the given liquid are all evaluated at the temperature T_0. For water at 15°C, the speed of sound is approximately 1450 m/s.

Classification of fluids and flows by compressibility. Although truly incompressible fluids do not exist in nature, fluids are classified as being *incompressible* when the change in density with pressure is so small as to be negligible. This is usually the case with liquids; however, gases are also considered to be effectively incompressible under flow conditions where the pressure variation is small compared with the absolute pressure. Such flow conditions are generally called *incompressible flows*. The flow of air in a ventilating system is an example of a case of incompressible flow where the air can be treated as incompressible because the pressure variation is so small that the change in density is negligible. In contrast to the incompressible flow of air through a ventilating system, in cases where air (or some other gas) is flowing at very high velocity through a long conduit or there is significant heat exchange between the gas and its surroundings, the drop in pressure might be so great that the change in density cannot be ignored; this is a case of *compressible flow*.

Compressibility effects in water. For most practical purposes, water can be taken as incompressible. However, in some cases, the pressure variation experienced by the water is so high that the compressibility of water must be taken into consideration. A case in point is the effect of rapid valve closure in a conduit containing flowing water, which results in an effect called *water hammer*. The term "water hammer" is derived from the hammering sound caused by the generated pressure wave within the conduit.

EXAMPLE 1.4

Water in municipal water distribution systems typically experiences pressures in the range of 140–700 kPa, yet the density is usually taken as a constant in engineering calculations. Estimate the maximum percentage error that is expected by assuming a constant density.

SOLUTION

The compressibility of water, E_v, can be taken as 2150 MPa, and according to Equation 1.13,

$$E_\mathrm{v} \approx \frac{\Delta p}{\Delta\rho/\rho}$$

which can be rearranged to yield

$$\frac{\Delta\rho}{\rho} \approx \frac{\Delta p}{E_\mathrm{v}} = \frac{700 - 140}{2.15 \times 10^6} = 0.000260$$

The percentage change in density corresponding to a pressure change of 560 kPa (= 700 kPa – 140 kPa) is 0.026%. Therefore, the maximum percentage error in assuming a constant density is **0.026%**.

Compressibility effects in air and other gases. Compressibility has a significant influence on the motion of objects moving through a stationary gas such as air when the velocity of the object is on the same order of magnitude as the speed of sound in the gas. The parameter used to measure the relative importance of compressibility is the Mach number, Ma,[3] defined by

$$\text{Ma} = \frac{V}{c} \tag{1.16}$$

where V is the velocity of a moving object in a stationary fluid or the velocity of a moving fluid and c is the speed of sound within the fluid at the given temperature. The conventional rule in applying Ma to measure the importance of compressibility is that the compressibility of the fluid is important when Ma ≥ 0.3. The speed of sound in air is on the order of 340 m/s, depending on the temperature, so Ma $= 0.3$ corresponds to approximately 100 m/s. Typically, for airflows at velocities less than 100 m/s, pressure changes are small and changes in density due to compressibility effects are generally less than 3%. As a consequence, air is usually assumed to be incompressible in systems where the flow velocities are less than 100 m/s. As a reference point, the fastest race lap ever recorded at the Indianapolis 500 is 105 m/s, with higher speeds attained on the straightaways.

Density changes due to combined pressure and temperature changes. The fractional change in the density of a fluid per unit change in pressure at any given temperature, T_0, is given by the *isothermal compressibility*, α, defined as

$$\alpha = \frac{1}{E_v} = \left.\frac{\mathrm{d}\rho/\rho}{\mathrm{d}p}\right|_{T_0} \tag{1.17}$$

The fractional change in the density per unit change in temperature at any given pressure, p_0, is given by the *coefficient of volume expansion*, β, which is sometimes called the *coefficient of thermal expansion*, and is defined as

$$\beta = -\left.\frac{\mathrm{d}\rho/\rho}{\mathrm{d}T}\right|_{p_0} \tag{1.18}$$

where the negative sign is used to ensure that β is positive, because an increase in temperature typically (although not always) results in a decrease in density. It is also useful to remember that application of Equation 1.18 to an ideal gas yields $\beta = 1/T$. Combining the fluid properties α and β, as defined by Equations 1.17 and 1.18, gives the following approximation to the fractional change in density resulting from a pressure change, Δp, and a temperature change, ΔT:

$$\frac{\Delta\rho}{\rho} \approx \alpha\Delta p - \beta\Delta T \tag{1.19}$$

Density variations in fluids are typically more sensitive to temperature than to pressure.

[3]Named in honor of the Austrian physicist Ernst Mach (1838–1916).

EXAMPLE 1.5

The deepest known point in the world's oceans is located in the Mariana Trench in the Pacific Ocean, with an estimated depth of around 10.9 km. The temperature and pressure at the bottom of the Mariana Trench is estimated to be 3°C and 108.5 MPa, respectively. If temperature and pressure at the surface of the ocean above the Mariana Trench is 20°C and 101 kPa, respectively, and salinity changes in the water column are neglected, estimate the percentage change in the seawater density between the ocean surface and the bottom of the trench.

SOLUTION

From the given data: $T_1 = 20$°C, $p_1 = 101$ kPa, $T_2 = 3$°C, and $p_2 = 1.085 \times 10^5$ kPa. The isothermal compressibility, α, and the coefficient of volume expansion, β, of seawater can be estimated from the properties of water given in Appendix B.1 as follows:

Temperature (°C)	E_v (10^6 kPa)	$\alpha = 1/E_\text{v}$ (10^{-6} kPa)	β (10^{-3} K^{-1})
3	2.04	0.489	0.068
20	2.18	0.459	0.207
Average	2.16	0.474	0.138

Taking the average values of α and β between the top and bottom of the water column and substituting these values and the given values of the other parameters in Equation 1.19 yields

$$\frac{\Delta\rho}{\rho} \approx \alpha\Delta p - \beta\Delta T = (0.474 \times 10^{-6})(108500 - 101) - (0.138 \times 10^{-3})(3 - 20) \quad \rightarrow$$

$$\frac{\Delta\rho}{\rho} \approx 0.0490 = 4.9\%$$

Therefore, assuming that there are no changes in salinity and that α and β for seawater can be approximated by α and β for fresh water, the density at the bottom of the Mariana Trench is estimated to be approximately **4.9%** higher than the density at the overlying ocean surface.

Density variations not related to compressibility. It is possible for a fluid to be incompressible and yet still have a spatially variable density. Such situations exist in liquids where the density is a function of temperature and/or a function of the concentration of a dissolved constituent. For example, density variations in the groundwater underlying coastal areas can be caused by salinity intrusion from the ocean, and density variations created by discharging fresh water into saltwater environments can cause density variations that have a significant influence on the flow. In both of the aforementioned cases, the density variations are not related to compressibility.

1.4 Ideal Gases

When the state of a gas is far removed from the liquid phase, the behavior of the gas closely approximates the behavior of an *ideal gas*. Ideal gases are sometimes called *perfect gases*, and such gases are characterized by their molecules being sufficiently far apart that intermolecular forces are negligible. An ideal gas obeys the relationship

$$pV = nR_u T \tag{1.20}$$

where p is the absolute pressure [FL^{-2}], V is the volume occupied by the gas [L^3], n is the number of moles of the gas [mol], R_u is the *universal gas constant* [$\text{E}^4\ \text{mol}^{-1}\Theta^{-1}$], and T is the absolute temperature [Θ]. The conventional value of the universal gas constant is $R_u = 8.31424621$ J/mol·K, which is actually derived for hydrogen. For real gases, R_u is not strictly constant, however, for monatomic and diatomic gases the variation from the conventional value is very small. Equation 1.20 is commonly called the *ideal gas law*, *perfect gas law*, or *equation of state for an ideal gas*.

Avogadro's law. *Avogadro's law* states that all gases under the same conditions of temperature and pressure contain the same number of molecules per unit volume. Quantitatively, the ideal gas law, as given by Equation 1.20, also requires that any (ideal) gas under the same conditions of temperature and pressure contains the same number of molecules per unit volume, which is in exact agreement with Avogadro's law. Avogadro's number is defined as the number of molecules in a mole of a gas, which is generally taken as 6.023×10^{23} molecules/mole.

Occurrence of ideal gases. For most gases encountered in engineering applications, it is a reasonable approximation to assume that the gas behaves like an ideal gas. For example, at temperatures greater than -50°C, air can be quite closely approximated as an ideal gas, provided the pressure is not extremely high. The approximation of gases as ideal gases is used throughout this text.

Vapors. As the pressure is increased and/or the temperature is decreased, a gas becomes a vapor, and as gases approach the liquid phase, the property relations become much more complicated than Equation 1.20. Thus, the density and other properties must be obtained from vapor tables and charts. Such tables and charts exist for steam, ammonia, sulfur dioxide, Freon,[5] and other vapors common in engineering use. Dense gases such as water vapor in steam power plants and refrigerant vapor in refrigerators should not be treated as ideal gases because they usually exist at a state near the liquid state.

1.4.1 Equation of State

An *equation of state* expresses the relationship between a property and the variables that define the state of a substance. The equation of state commonly used to characterize the behavior of ideal gases is derived from the ideal gas law and relates the density of the gas to the temperature and pressure. Defining the gas density, ρ, by

$$\rho = \frac{nM}{V} \tag{1.21}$$

where M is the molar mass of the gas (formerly called the *molecular weight*) [M mol^{-1}], Equations 1.20 and 1.21 combine to give the following expression for the gas density:

$$\rho = \frac{pM}{R_u T} \tag{1.22}$$

[4]E denotes a dimension of energy [=FL].

[5]Freon is a trade name of a product (refrigerant) developed by DuPont, Inc.

Hence, the density of any (ideal) gas at any temperature and pressure can be computed if the molar mass is known. In fluid mechanics applications of the ideal gas law, it is common practice to combine the universal gas constant, R_u, and the molar mass, M, into a single quantity, R, called the *gas constant*, such that R is defined by

$$R = \frac{R_u}{M} \tag{1.23}$$

Therefore the gas constant, R, is unique to each gas and is usually listed among the properties of that gas, whereas the universal gas constant is approximately the same for all gases. Combining Equations 1.21–1.23 gives the following commonly used expression for the density of an ideal gas:

$$\rho = \frac{p}{RT} \tag{1.24}$$

This equation is sometimes referred to as the *equation of state for ideal gases*, which also refers to the equivalent expression given by Equation 1.20. It is apparent from Equation 1.22 that under fixed conditions of temperature (T) and pressure (p), the density of a gas is proportional to its molar mass (M). Both hydrogen (M = 2 g/mol) and helium (M = 4 g/mol) are less dense than air (which consists predominantly of nitrogen) at the same temperature and pressure. Seemingly, it would be better to use hydrogen than helium in balloons to achieve better buoyancy in air. However, hydrogen is extremely flammable and helium is inert, hence our preference for helium in balloons! When helium is used to elevate hot air balloons, the density of the helium is reduced by heating the helium in the balloon, usually with an open flame.

1.4.2 Mixtures of Ideal Gases

When a gas consists of a mixture of different gases (e.g., air containing mixture of nitrogen, oxygen, and other gases), the behavior of each component gas can be analyzed by assuming that the other gases are not present. This approach is justified based on *Dalton's law of partial pressures*, which states that each gas exerts its own pressure, called the *partial pressure*, as if the other gases were not present. Therefore, for any given volume of a gas mixture, the pressure exerted by the mixture is the sum of the (partial) pressures exerted by each of the gas components. In some cases, a mixture of gases is treated as a single gas (i.e., a pure substance) with an equivalent molar mass. Although this approach is not exactly valid, because the relative molar concentrations change with temperature and pressure, for a common gas mixture such as air, the mixture ratio remains approximately constant at a pressure of 101.3 kPa and temperatures in the range of −113°C to 1900°C, with the equivalent molar mass of standard air generally taken as 28.96 g/mol. The gas constant for air, R_{air}, is derived from the universal gas constant, R_u, and the composite molar mass of air, M_{air} (= 28.96 g/mol), such that

$$R_{air} = \frac{R_u}{M_{air}} = \frac{8.3142 \text{ J/mol·K}}{28.96 \text{ g/mol}} = 0.2871 \text{ J/g·K} = 287.1 \text{ J/kg·K}$$

Mixtures of gases are sometimes expressed in terms of the fraction of volume occupied by each of the constituent gases; for example, air is sometimes said to consist of 78% nitrogen, 21% oxygen, and 1% other gases. The volume fractions in a mixture can be derived directly from the mole fractions, as illustrated in Figure 1.6 for the case of a three-gas mixture. It is apparent from Figure 1.6 that if the mole fraction of component i of a gas mixture is n_i/n_T, where n_i is the number of moles of component i and n_T is the total number of moles (i.e., $n_T = \sum_{i=1}^{N} n_i$), then the volume fraction V_i/V is given by

$$\frac{V_i}{V} = \frac{n_i}{n_T} \tag{1.25}$$

where V_i is the volume of component i, V is the total volume, and all volumes are calculated at the same temperature and pressure.

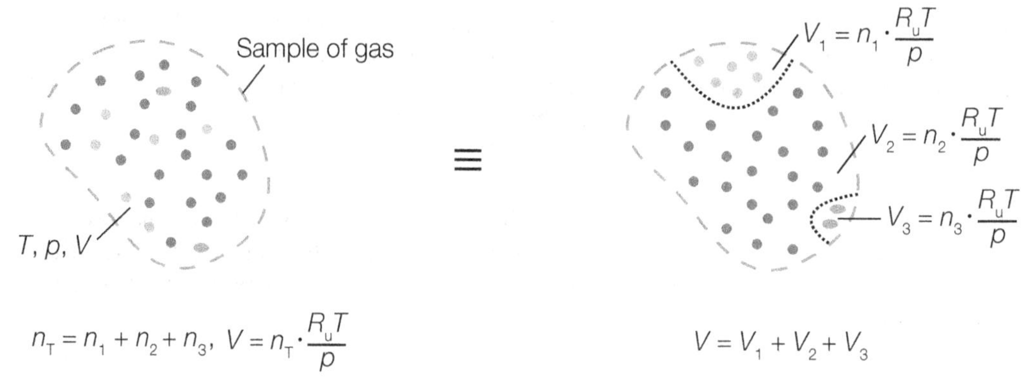

Figure 1.6: Mixture of gases

EXAMPLE 1.6

Typical ambient air at 20°C and 101.3 kPa (atmospheric pressure) contains 21% oxygen by volume. The molar mass of the ambient air is 28.96 g, and the molar mass of oxygen is 32 g. In a particular case, 20 N of ambient air is pumped into a tank with a capacity of 21.2 L. (a) What volume of ambient air is compressed? (b) What is the gauge pressure of the air in the tank? Assume that the temperature of the compressed air in the tank is 20°C. (c) What is the partial pressure and specific weight of the ambient oxygen? (d) What weight of oxygen is put in the tank? (e) What is the partial pressure and specific weight of the oxygen in the tank?

SOLUTION

(a) From the given ambient air data, T = 20°C and p = 101.3 kPa. The specific weight of air, γ, can be calculated using the ideal gas law expressed in the form

$$\gamma = \frac{pmg}{RT} \tag{1.26}$$

For the ambient air, $p = 101.3 \times 10^3$ Pa, m = 28.96 g/mol = 0.02896 kg/mol, g = 9.807 m/s^2, R = 8.314 J/mol·K, and T = 273 + 20 = 293 K. Substituting these data into Equation 1.26 gives the specific weight of air as

$$\gamma = \frac{(101.3 \times 10^3)(0.02896)(9.807)}{(8.314)(293)} = 11.81 \text{ N/m}^3$$

Because the weight of ambient air compressed (pumped into the tank) is 20 N, the volume, V, of ambient air pumped into the tank is given by

$$V = \frac{20}{\gamma} = \frac{20}{11.81} = \mathbf{1.693\ m^3}$$

(b) The process of compressing a fixed mass of gas at a constant temperature is described by the process equation

$$p_1 V_1 = p_2 V_2 \tag{1.27}$$

In this case, p_1 = 101.3 kPa, V_1 = 1.693 m^3, and V_2 = 21.2 L = 0.0212 m^3. Substituting these data into Equation 1.27 gives the pressure of air in the tank, p_2, as

$$p_2 = \frac{p_1 V_1}{V_2} = \frac{(101.3)(1.693)}{0.0212} = 8090 \text{ kPa}$$

Hence, the gauge pressure of the air in the tank is 8090 kPa – 101.3 kPa = **7989 kPa.**

(c) If p_{O_2} is the partial pressure of oxygen (O_2) in any volume, V, of ambient air at a pressure p, then the ideal gas law gives the following relationship:

$$\frac{pV}{n_{air}} = \frac{p_{O_2}V}{n_{O_2}} = RT \quad \rightarrow \quad \frac{p_{O_2}}{p} = \frac{n_{O_2}}{n_{air}} \tag{1.28}$$

where n_{air} and n_{O_2} are the number of moles of air and the number of moles of O_2, respectively, in a volume V of ambient air. Substituting $p = 101.3$ kPa and $n_{O_2}/n_{air} = 0.21$ into Equation 1.28 gives the partial pressure of O_2 in the ambient air as

$$p_{O_2} = (101.3)(0.21) = \mathbf{21.3\ kPa}$$

The specific weight of O_2 in the ambient air, γ_{O_2}, is given by the ideal gas law (Equation 1.26), and taking $p_{O_2} = 21.3 \times 10^3$ Pa, $m = 32$ g/mol $= 0.032$ kg/mol, $g = 9.807$ m/s^2, $R = 8.314$ J/mol · K, and $T = 273 + 20 = 293$ K yields

$$\gamma_{O_2} = \frac{(21.3 \times 10^3)(0.032)(9.807)}{(8.314)(293)} = \mathbf{2.74\ N/m^3}$$

(d) The weight of O_2 in the tank is given by

$$\begin{aligned}\text{weight of } O_2 \text{ in the tank} &= \text{specific weight of } O_2 \text{ in air} \times \text{volume of air} \\ &= 2.74 \times 1.693 = \mathbf{4.639\ N}\end{aligned}$$

(e) The partial pressure of O_2 in the tank can be calculated using the relation

$$p_1V_1 = p_2V_2 \tag{1.29}$$

where p_1 and V_1 are the partial pressure of O_2 and volume of ambient air, respectively, and p_2 and V_2 are the partial pressure of O_2 and volume of the tank, respectively. Taking $p_1 = 21.3$ kPa, $V_1 = 1.693$ m^3, and $V_2 = 0.0212$ m^3, Equation 1.29 gives

$$p_2 = \frac{(21.3)(1.693)}{0.0212} = \mathbf{1700\ kPa}$$

The specific weight of O_2 in the tank is given by

$$\text{specific weight of } O_2 \text{ in tank} = \frac{\text{weight of } O_2 \text{ in tank}}{\text{volume of tank}} = \frac{4.639\ \text{N}}{0.0212\ \text{m}^3} = \mathbf{219\ N/m^3}$$

1.4.3 Thermodynamic Properties

Ideal gases must obey the *first law of thermodynamics*, which states that the change in internal energy within a system containing a fixed mass of fluid is equal to the heat added to the system minus the work done by the system. A system containing a fixed mass of fluid is commonly called a *closed system*. The first law of thermodynamics is commonly expressed in the differential form

$$du = dQ - p\,dV \tag{1.30}$$

where du is the change in internal energy [E], dQ is the heat added [E], p is the pressure within the system [FL^{-2}], and dV is the change in volume of the system [L^3]. For ideal gases, the internal energy, u, is a measure of molecular activity that depends only on the temperature of the gas. If an ideal gas is expanded or compressed at a constant temperature (i.e., $\mathrm{d}u = 0$), then the corresponding heat transfer, Q_{12}, into the system as the gas changes from State 1 to State 2 is derived by combining Equation 1.30 with the ideal gas law (Equation 1.20), which yields

$$Q_{12} = \int_{V_1}^{V_2} p\,\mathrm{d}V = \int_{V_1}^{V_2} \frac{nR_\mathrm{u}T}{V}\,\mathrm{d}V = nR_\mathrm{u}T \int_{V_1}^{V_2} \frac{1}{V}\,\mathrm{d}V = nR_\mathrm{u}T \ln\frac{V_2}{V_1} \tag{1.31}$$

Therefore, if an ideal gas contained in a closed system undergoes an isothermal (and reversible) change from a volume V_1 to a volume V_2, the heat transfer to the system is given by

$$Q_{12} = nR_\mathrm{u}T \ln\frac{V_2}{V_1} \quad \text{(isothermal, reversible process)} \tag{1.32}$$

EXAMPLE 1.7

A closed system contains 1.1 m^3 of air at a temperature of 15°C and a pressure of 150 kPa. If the system is expanded isothermally and reversibly to a volume of 2.5 m^3, what heat must be added to the system to maintain isothermal conditions?

SOLUTION

From the given data: $V_1 = 1.1$ m^3, $T_1 = 15$°C $= 288$ K, $p_1 = 150$ kPa, and $V_2 = 2.5$ m^3. Applying the ideal gas law (Equation 1.20) gives

$$p_1V_1 = nR_\mathrm{u}T_1 \quad \rightarrow \quad nR_\mathrm{u}T_1 = (150 \times 10^3)(1.1) = 1.65 \times 10^5 \text{ J}$$

Because the process is isothermal and reversible, $nR_\mathrm{u}T$ remains constant, and Equation 1.32 gives

$$Q_{12} = nR_\mathrm{u}T \ln\frac{V_2}{V_1} = (1.65 \times 10^5)\ln\frac{2.5}{1.1} = 1.35 \times 10^5 \text{ J} = 135 \text{ kJ}$$

Therefore, to maintain isothermal conditions during the expansion process, **135 kJ** of heat must be added to the air within the system.

Polytropic process equation. Under more general conditions, when a defined mass of gas contained within a closed system is expanded or compressed, the relationship between density and pressure in the final state depends on the conditions in the initial state, as well as the conditions imposed as the gas moves from one state to the other. A fundamental equation derived from the first law of thermodynamics that is used to describe the behavior of ideal gases in various processes is given by

$$pv^n = p_1v_1^n = \text{constant} \quad \text{or} \quad \frac{p}{p_1} = \left(\frac{\rho}{\rho_1}\right)^n = \text{constant} \tag{1.33}$$

where p is the absolute pressure [FL^{-2}], v is the specific volume ($= 1/\rho$) [L^{3}M^{-1}], ρ is the mass density [ML^{-3}], and n is a dimensionless constant that may have any nonnegative value from zero to infinity, depending on the process to which the gas is subjected. Any thermodynamic process that obeys Equation 1.33 is called a *polytropic process*, and the exponent

n in Equation 1.33 is called the *polytropic index*. It is unfortunate that the thermodynamic literature universally uses the symbol n to represent both the amount of substance (in moles) and the polytropic index, and this convention is followed here. The polytropic process equation given by Equation 1.33 is particularly useful in describing the change in gas properties during expansion and compression processes that involve heat transfer.

Types of processes. If the change process for a defined mass of gas is at a constant pressure, the process is called *isobaric*. If the change process for a defined mass of gas is at a constant temperature, the process is called *isothermal*. If there is no heat transfer to or from the gas, the process is called *adiabatic*. A frictionless and reversible adiabatic process is called an *isentropic process*. The relationship between internal energy (i.e., u in Equation 1.30) and temperature change is generally parameterized by the *specific heat*, where the specific heat of a substance is defined as the amount of thermal energy per unit mass required to raise the temperature of the substance by 1 degree. The value of the polytropic index, n, used in Equation 1.33 is related to the specific heat as follows:

$$n = \begin{cases} 0 & \text{isobaric process} \\ 1 & \text{isothermal process} \\ k & \text{frictionless and reversible adiabatic process (= isentropic process)} \\ < k & \text{expansion with friction} \\ > k & \text{compression with friction} \end{cases} \tag{1.34}$$

where k is the *specific heat ratio* defined as the ratio of the specific heat at constant pressure to the specific heat at constant volume such that

$$\text{specific heat ratio, } k = \frac{c_p}{c_v} \tag{1.35}$$

where c_p is the specific heat at constant pressure $[\mathrm{EM}^{-1}\Theta^{-1}]$ and c_v is the specific heat at constant volume $[\mathrm{EM}^{-1}\Theta^{-1}]$. The specific heat ratio, k, is sometimes called the *adiabatic exponent*, and generally, $k > 1$ (because $c_p > c_v$). For expansion with friction, n is less than k, and for compression with friction n is greater than k. It is apparent that the process equation given by Equation 1.33 with the exponent n given by Equation 1.34 describes many different types of processes. In fact, this is the reason Equation 1.33 is called the polytropic process equation.

Specific heats of ideal gases. For gases $c_v < c_p$, because if a gas is heated at a constant volume, less energy is required to raise the temperature than if the gas is heated at constant pressure. Values of c_p and c_v are only a function of temperature, and c_p and c_v for several common gases can be found in Appendix B.5. The gas constant, R, is related to the specific heats by the relation

$$R = c_p - c_v \tag{1.36}$$

For air and diatomic gases at usual temperatures, typical approximations are $k = 1.40$, $c_p = 1005$ J/kg·K, and $c_v = 718$ J/kg·K. A gas for which c_v is a constant is said to be *calorifically perfect* (Massey and Ward-Smith, 2012), and ideal gases are assumed to be calorifically perfect. For all real gases, c_p and c_v increase gradually with temperature and k decreases gradually with temperature.

EXAMPLE 1.8

A tank with a movable cover contains 1.2 kg of air as shown in Figure 1.7. The pressure exerted by the tank cover on the air is approximately 200 kPa. The initial temperature of the air in the tank is 30°C, and heat is added to the air in the tank to raise its temperature to 300°C. (a) Estimate the amount of heat added. (b) What is the percentage change in volume corresponding to this temperature change?

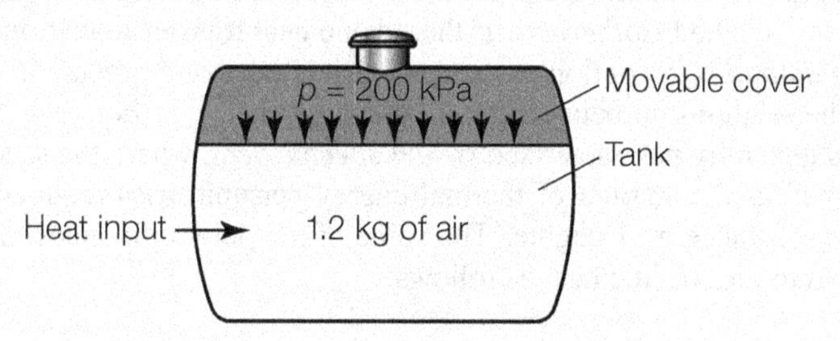

Figure 1.7: Heat addition to air in a tank

SOLUTION

From the given data: $m = 1.2$ kg, $p = 200$ kPa, $T_1 = 30°\text{C} = 303$ K, and $T_2 = 300°\text{C} = 573$ K. The pressure of the air in the tank remains constant.

(a) Assume that the specific heat at a constant pressure of 200 kPa is approximately equal to the specific heat at a constant pressure of 101.3 kPa, which is standard atmospheric pressure. The specific heat, c_p, of air at 30°C is 1007 J/kg·K, and the specific heat at 300°C is 1044 J/kg·K (from Appendix B.2); so the average specific heat, $\bar{c}_p$, between 30°C and 300°C can be estimated as

$$\bar{c}_p = \frac{1007 + 1044}{2} = 1026 \text{ J/kg·K}$$

The heat, Q_{12}, added to raise the temperature of the air in the tank from 30°C to 300°C can be estimated as

$$Q_{12} = m\bar{c}_p(T_2 - T_1) = (1.2)(1026)(573 - 303) = 3.32 \times 10^5 \text{ J} = \mathbf{332\ kJ}$$

(b) The percentage change in volume can be derived from the ideal gas law. Let V_1 and V_2 be the volumes corresponding to T_1 and T_2, respectively; then

$$m = \frac{pV_1}{RT_1} = \frac{pV_2}{RT_2} \quad \rightarrow \quad \frac{V_1}{T_1} = \frac{V_2}{T_2} \quad \rightarrow \quad \frac{V_2}{V_1} = \frac{T_2}{T_1} = \frac{573}{303} = 1.89$$

$$\frac{V_2 - V_1}{V_1} \times 100 = \left[\frac{V_2}{V_1} - 1\right] \times 100 = [1.89 - 1] \times 100 = 89\%$$

Therefore, the percentage change in volume is **89%**.

Derived process equations. Equations 1.22 and 1.33 can be combined to yield other useful relations; for example,

$$\frac{T_1}{T_2} = \left(\frac{v_1}{v_2}\right)^{n-1} = \left(\frac{\rho_2}{\rho_1}\right)^{n-1} = \left(\frac{p_2}{p_1}\right)^{\frac{n-1}{n}} \tag{1.37}$$

are particularly useful relationships in various applications. Differentiating Equation 1.33 yields

$$npv^{n-1}\mathrm{d}v + v^n \mathrm{d}p = 0 \tag{1.38}$$

and combining Equations 1.12 and 1.38 yields the following expression for the bulk modulus, E_v, of a perfect gas:

$$E_\mathrm{v} = np \tag{1.39}$$

Therefore, in an isothermal process, $E_\mathrm{v} = p$ and in an isentropic process, $E_\mathrm{v} = kp$. In both cases, the bulk modulus increases as the pressure increases. If compression of the gas occurs at a constant temperature, then the process is isothermal ($n = 1$) and E_v is called the *isothermal bulk modulus*. If compression of the gas occurs without the addition of heat and there is no friction, then the process is adiabatic ($n = k$) and E_v is called the *adiabatic bulk modulus*.

EXAMPLE 1.9

A 1 m^3 volume of carbon dioxide (CO_2) at 15°C and 101.3 kPa is compressed isentropically to a pressure of 500 kPa. After this compression, the volume of CO_2 is maintained constant and the CO_2 is allowed to cool down to a temperature of 15°C. (a) What is the temperature of the CO_2 after the isentropic compression? (b) What is the pressure in the compressed volume after it cools down to 15°C?

SOLUTION

From the given data: $V_1 = 1$ m^3, $T_1 = T_3 = 15°\text{C} = 288$ K, $p_1 = 101.3$ kPa, and $p_2 = 500$ kPa. The subscripts denote the states such that 1 = initial state, 2 = state after isentropic compression, and 3 = state after cooling. The thermodynamic properties of CO_2 are $c_p = 858$ J/kg·K, $c_v = 670$ J/kg·K, $k = c_p/c_v = 1.28$, and $R = c_p - c_v = 188$ J/kg·K (from Appendix B.5).

(a) The initial density of CO_2 (in State 1), ρ_1, is given by the ideal gas law as

$$\rho_1 = \frac{p_1}{RT_1} = \frac{101.3 \times 10^3}{(188)(288)} = 1.871 \text{ kg/m}^3$$

The relationship between variables before and after isentropic compression is given by Equation 1.37 with $n = k = 1.28$, which yields

$$\frac{T_1}{T_2} = \left(\frac{p_2}{p_1}\right)^{\frac{n-1}{n}} \quad \rightarrow \quad \frac{288}{T_2} = \left(\frac{500}{101.3}\right)^{\frac{1.28-1}{1.28}} \quad \rightarrow \quad T_2 = 203 \text{ K} = \mathbf{-70°C}$$

$$\rho_2 = \frac{p_2}{RT_2} = \frac{500 \times 10^3}{(188)(203)} = 13.10 \text{ kg/m}^3$$

(b) After cooling, $\rho_3 = \rho_2 = 13.10$ kg/m^3, and the pressure, p_3, is given by the ideal gas law, such that

$$p_3 = \rho_3 R T_3 = (13.10)(188)(288) = 7.09 \times 10^5 \text{ Pa} = \mathbf{709\ kPa}$$

1.4.4 Speed of Sound in an Ideal Gas

Using the relationship between density and pressure in ideal gases, as given by Equation 1.24, and applying Equations 1.15 and 1.39 gives the speed of sound, c, in an ideal gas as

$$c = \sqrt{RTk} \tag{1.40}$$

where R is the gas constant, T is the absolute temperature, and k is the specific heat ratio. Equation 1.40 is commonly used to calculate the speed of sound in air.

EXAMPLE 1.10

Air at 20°C enters a conduit from a large open area and attains a speed of 150 m/s within the conduit. Determine the speed of sound in the flowing air and assess whether compressibility effects should be taken into account in modeling the flow of the air into the conduit.

SOLUTION

From the given data: $T = 20°\text{C} = 293$ K and $V = 150$ m/s. For standard air, $k = 1.40$ and $R = 287$ J/kg·K. The speed of sound in the air can be calculated using Equation 1.40, which gives

$$c = \sqrt{kRT} = \sqrt{(1.40)(287)(293)} = \mathbf{343\ m/s}$$

The importance of compressibility on the fluid flow is measured using the Mach number, Ma, defined by Equation 1.16, which gives

$$\text{Ma} = \frac{V}{c} = \frac{150}{343} = 0.437$$

Because Ma > 0.3 **compressibility is important** and must be considered to model the flow of air into the conduit accurately.

1.5 Standard Atmosphere

The U.S. *standard atmosphere* is used in various applications, such as in the design of aircraft, missiles, and spacecraft. The standard atmosphere in the United States extends to at least 86 km above the surface of Earth for most atmospheric properties and as far as 1000 km for some properties. The U.S. standard atmosphere approximates mean conditions at a latitude of 45° North in July. Variations in temperature and pressure in the U.S. standard atmosphere are shown in Figure 1.8 and are tabulated in Appendix B.3. Widely utilized properties of air at sea level in the standard atmosphere are given in Table 1.5. Care should be taken in applying standard sea-level atmospheric conditions in engineering analyses, because actual average atmospheric conditions can vary significantly with location. For

Table 1.5: Air Properties at U.S. Standard Atmosphere at Sea Level

Property	Value
Temperature	288.15 K (= 15°C)
Pressure	101.33 kPa
Density	1.225 kg/m^3

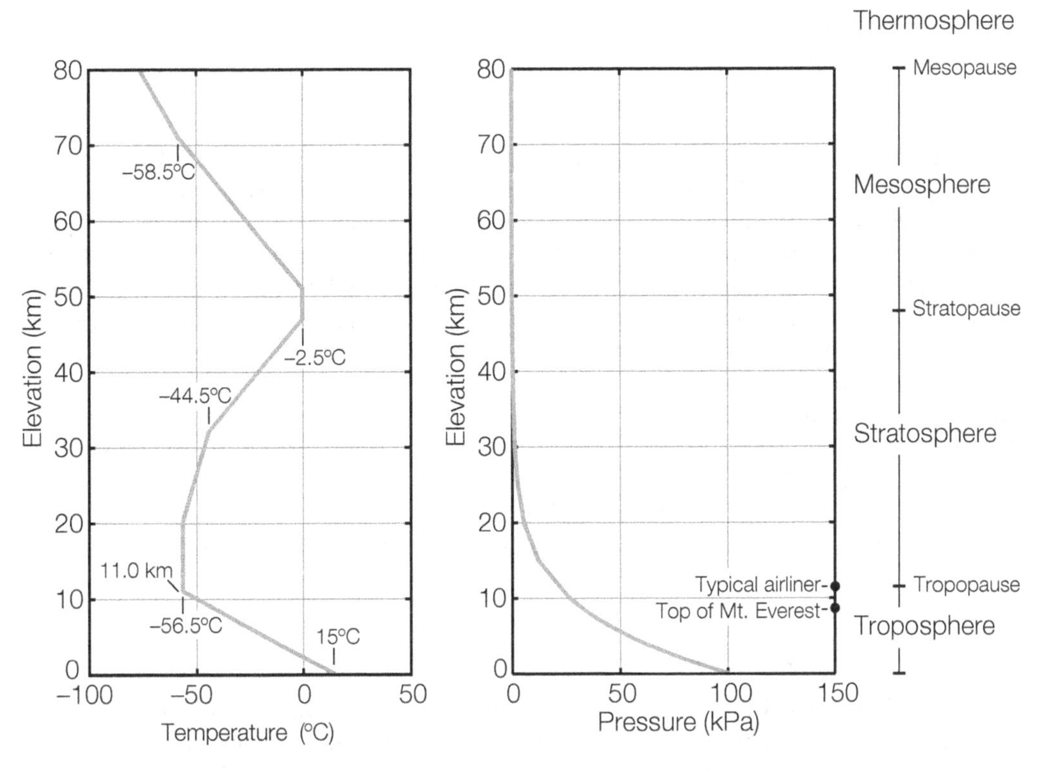

Figure 1.8: U.S. Standard Atmosphere

example, the average temperature in the atmosphere at sea level in northern Maine is much different from that in southern Florida. At higher elevations, differences in average atmospheric condition as a function of location are reduced compared with those at sea level, but differences still remain.

Layers in the atmosphere. The *troposphere* is the lowest layer in Earth's atmosphere and contains approximately 80% of the atmosphere's mass and 99% of its water vapor and aerosols. Within the troposphere, the temperature decreases with increasing height at a *lapse rate* of approximately −6.5°C/km; the average height of the troposphere is approximately 11 km in the middle latitudes. Commercial aircraft typically cruise near the top of the troposphere, where the temperature, pressure, and air density are −56.5°C, 22.6 kPa, and 0.37 kg/m^3, respectively. These air properties can be compared with corresponding standard values at sea level in the U.S., which are 15°C, 101.3 kPa, and 1.23 kg/m^3. The atmospheric layer above the troposphere is the *stratosphere*, within which the temperature increases with increasing height, in contrast to what occurs in the troposphere. The stratosphere varies in vertical extent but is typically within the range of 11–51 km above Earth's surface at middle latitudes. Within the stratosphere, atmospheric pressure becomes nearly zero. The *mesosphere* is the layer above the stratosphere in which the temperature again decreases with increasing height. The upper boundary of the mesosphere, called the *mesopause*, can be the coldest naturally-occurring place on Earth with temperatures below −143°C. The mesosphere is typically within the range of 51–71 km above Earth's surface at middle latitudes. The *thermosphere* or *ionosphere* is the layer directly above the mesosphere within which ultraviolet radiation causes ionization. The thermosphere typically begins at

71 km above Earth. In the *exosphere*, beginning at 500–1000 km above the Earth's surface, the atmosphere turns into space.

Related phenomena. Standard atmospheric pressure changes from 101.325 kPa at sea level to 89.88 kPa and 54.05 kPa at altitudes of 1000 m and 2000 m, respectively. The typical atmospheric pressure in Denver at elevation 1610 m is 83.4 kPa. The decline of atmospheric pressure with elevation can affect many aspects of daily life. For example, cooking takes longer at high altitudes because water boils at a lower temperature at lower atmospheric pressures. Nose bleeding is a common experience at high altitudes because the difference between the blood pressure and the atmospheric pressure is larger in this case, and the delicate walls of veins in the nose are often unable to withstand this extra stress. For a given temperature, the density of air is lower at high altitudes; thus, a given volume contains less air and less oxygen. Consequently, humans tend to tire more easily and to experience breathing problems at high altitudes. Those acclimatized to such situations tend to develop more efficient lungs. The lower air densities associated with lower air pressures at high altitudes also yield less lift on airfoils at a given speed; hence, airplanes require longer runways for takeoff.

1.6 Viscosity

A fluid is differentiated from a solid by its ability to resist an applied shear stress. Whereas a solid can resist an applied shear stress by deforming a certain amount, a fluid deforms continuously under the influence of shear stress, no matter how small the shear stress. In solids, stress is proportional to strain, but in fluids, stress is proportional to strain rate. In qualitative terms, the "viscosity" of a fluid refers to the fluid's resistance to flow, and fluids with higher viscosities do not flow as readily as fluids with lower viscosities under the same applied (shear) stress. In quantitative terms, the viscosity of a fluid is formally measured by its *dynamic viscosity*, which is the proportionality constant between the shear stress and the strain rate of a fluid element.

1.6.1 Newtonian Fluids

Consider the rectangular fluid element illustrated in Figure 1.9, where the shape of the element at time t is ABCD and the shape of the element at time $t + \Delta t$ is A′B′C′D′. The element shown in Figure 1.9 is acted upon by shear stresses, τ [FL^{-2}], on the upper and lower surfaces, and the angle $\Delta\theta$ [rad] is the shear strain in the time interval Δt [T] resulting from the

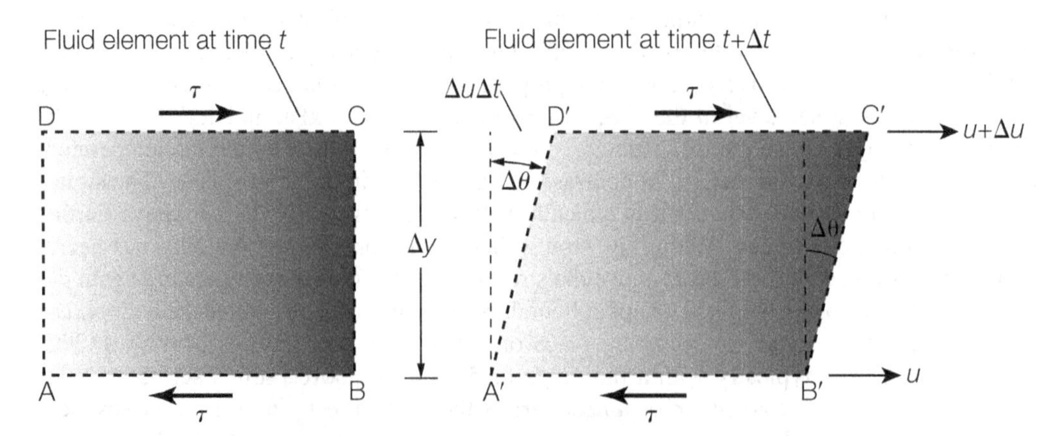

Figure 1.9: Effect of shear stress on fluid element

applied shear stress. If the velocity of the bottom surface of the fluid element is u [LT^{-1}] and the velocity of the top surface is $u + \Delta u$ [LT^{-1}], then for small values of $\Delta\theta$,

$$\Delta\theta \approx \tan\Delta\theta = \frac{(\Delta u)(\Delta t)}{\Delta y} \tag{1.41}$$

The *dynamic viscosity*, μ [$ML^{-1}T^{-1}$], is defined as the proportionality constant between the shear stress, τ, and the strain rate, $\Delta\theta/\Delta t$, such that

$$\tau = \mu\frac{\Delta\theta}{\Delta t} \tag{1.42}$$

The dynamic viscosity, μ, is also called the *coefficient of viscosity*, *absolute viscosity*, or simply the *viscosity*. Combining Equations 1.41 and 1.42 gives

$$\tau = \mu\frac{\Delta u}{\Delta y} \tag{1.43}$$

The term $\Delta u/\Delta y$, which approximates the strain rate, is the *velocity gradient* in the y-direction, because it is equal to the change in velocity, Δu, for a given change, Δy, in distance along the y-direction. Taking the limit as $\Delta y \to 0$ yields

$$\tau = \mu\frac{\mathrm{d}u}{\mathrm{d}y} \tag{1.44}$$

This linear proportionality between the shear stress, τ, and the strain rate, $\mathrm{d}u/\mathrm{d}y$, only exists for a class of fluids called *Newtonian fluids*. So Equation 1.44 is sometimes called *Newton's equation of viscosity* or *Newton's law of viscosity*, because it was first suggested by Sir Isaac Newton. It should be noted that utilization of the word "law" to describe Equation 1.44 is generally inappropriate because is not a natural law, but rather an empirical constitutive equation. Examples of Newtonian fluids include water, air, oil, gasoline, alcohol, kerosene, benzene, and glycerine. In Newtonian fluids, the absolute viscosity does not change with the rate of deformation.

Variables affecting the magnitude of the viscosity. For liquids, the viscosity is practically independent of pressure, and the effect of any small variation in pressure on viscosity is usually disregarded except at extremely high pressures. In the case of gases, absolute viscosity is relatively insensitive to moderate pressure changes. For example, increasing the pressure of air from 101.3 kPa to 5.07 MPa at 20°C increases the absolute viscosity by about 10%. For both liquids and gases, the absolute viscosity is sensitive to changes in temperature. As the temperature increases, the viscosities of all liquids decrease, whereas the viscosities of all gases increase. This is because the force of cohesion, which diminishes with temperature, predominates in liquids, whereas in gases, the predominant factor affecting viscosity is the interchange of molecules between layers of different velocities. Molecular interchange between layers causes a shear and produces a friction force. At higher temperatures, molecular activity increases, causing the viscosity of gases to increase with temperature.

Viscosity of water and other fluids. The viscosity of water varies nonlinearly between 1.78 mPa·s at 0°C and 0.282 mPa·s at 100°C, with a viscosity of 1.00 mPa·s at 20°C. The (dynamic) viscosity of water as a function of temperature is given in Appendix B.1. A commonly used viscosity unit is the *poise*[6] [P], which is equivalent to 0.1 Pa·s, and a centipoise [cP] is one-hundredth of a poise. This is a convenient unit because the viscosity of water at 20°C is 1.002 cP. In current practice, use of the cP unit is discouraged in favor of the Pa·s unit. The viscosities of several fluids at standard atmospheric pressure (101 kPa) and 20°C are listed in Table 1.6. It is commonplace to refer to small-viscosity lubricating oils as being "light" and large-viscosity lubricating oils as being "heavy." In general, liquids are more viscous than gases, although the viscosity of gases is a seldom used quantity.

[6]Named in honor of the French anatomist Jean Louis Poiseuille (1799–1869).

Table 1.6: Viscosities at Standard Atmospheric Pressure and 20°C (unless otherwise stated)

Fluid	Dynamic Viscosity, μ (Pa·s)	Fluid	Dynamic Viscosity, μ (Pa·s)
Glycerin:		**Water:**	
−20°C	134.0	**0°C**	**0.0018**
0°C	10.5	**20°C**	**0.0010**
20°C	1.52	**100°C** (liquid)	**0.00028**
40°C	0.31	**100°C** (vapor)	**0.000012**
Engine oil:		Blood, 37°C	0.00040
SAE 10W	0.10	Gasoline	0.00029
SAE 10W-30	0.17	Ammonia	0.00015
SAE 30	0.44	Air	0.000018
SAE 50	0.86	Hydrogen, 0°C	0.0000088
Mercury	0.0015		
Ethyl alcohol	0.0012		

Application to flow between flat plates (lubrication). A common application of Newton's law of viscosity is to calculate the shear force exerted by a fluid on a solid surface when the fluid is moving relative to the surface. An example of such a case is illustrated in Figure 1.10(a), which depicts a fluid contained between two flat plates a distance h apart, where the top surface is moving at a constant velocity V and the bottom surface is stationary. This case could also be representative of cases in which both plates are moving, with the top plate moving at a constant velocity V relative to the bottom plate. The forces acting on an element of fluid between the plates are illustrated in Figure 1.10(b), where the fluid element has dimensions $\Delta x \times \Delta y \times \Delta z$, the upstream and downstream pressures on the fluid element are p_1 and p_2, respectively, the shear forces on the bottom and top surfaces are τ_1 and τ_2, respectively, and the flow direction is normal to the (vertical) direction in which gravity acts.

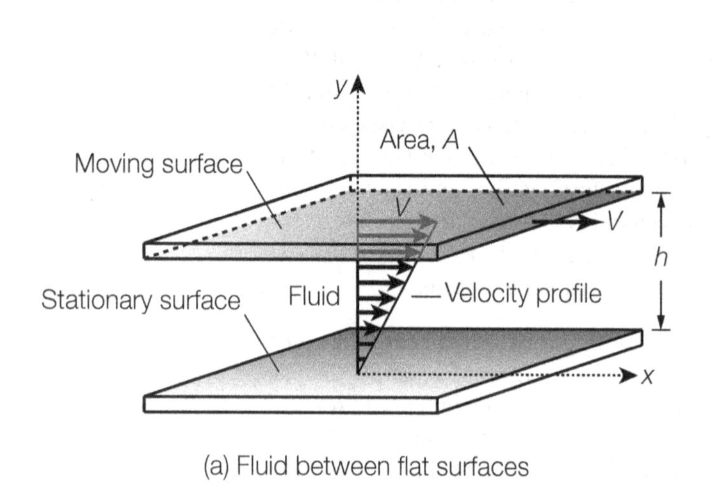

(a) Fluid between flat surfaces

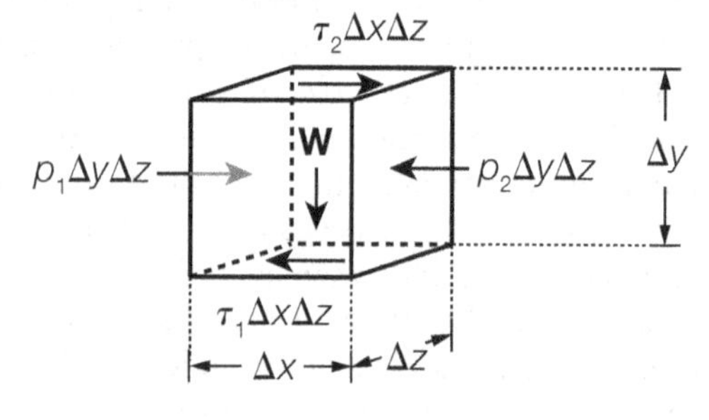

(b) Forces on fluid element

$$|\boldsymbol{F}| = \tau_0 A = \left(\mu \frac{du}{dy}\right)\bigg|_{y=h} A = \left(\mu \frac{V}{h}\right) A$$

F

(c) Force on top surface

Figure 1.10: Fluid between flat plates

Because the fluid element is not accelerating, the sum of the forces in the flow direction must be equal to zero, which requires that

$$(p_1 - p_2)\Delta y \Delta z + (\tau_2 - \tau_1)\Delta x \Delta z = 0 \quad (1.45)$$

If there is no pressure gradient in the flow direction, then $p_1 = p_2$ and Equation 1.45 simplifies to $\tau_1 = \tau_2$, which means that the shear stress remains constant in the Δy-direction and hence, the velocity gradient must also remain constant in accordance with Newton's law of viscosity. This can be expressed analytically as

$$\tau = \text{constant} = \mu \frac{du}{dy} \rightarrow \frac{du}{dy} = \text{constant} \quad (1.46)$$

which means that the velocity profile must be linear, as illustrated in Figure 1.10(a). Therefore, the shear stress, τ_0, exerted by the fluid on the top plate in Figure 1.10(a) is given by

$$\tau_0 = \mu \left. \frac{du}{dy} \right|_{y=h} = \mu \frac{V}{h} \quad (1.47)$$

Hence the force, F, exerted on the top plate is illustrated in Figure 1.10(c) and is given by

$$F = \tau_0 A = \mu \frac{V}{h} A \quad (1.48)$$

There are many practical applications in which Equation 1.48 can be used, such as in cases where a fluid (liquid) lubricant is contained between moving parts of a machine and between rotating concentric cylinders. In applying the formulation given by Equation 1.48, it is important to remember that this formulation strictly applies only to cases in which there is no pressure gradient in the flow direction, the flow direction is normal to the gravitational force, and conditions are at a steady state. In the variant cases, other formulations must be used.

EXAMPLE 1.11

SAE 30 oil at 20°C is used as a lubricant to slide a large flat plate over a stationary flat surface. A 0.5-mm thickness of oil is contained between the plate and the stationary surface, and the plate is moved at a velocity of 0.7 m/s. Estimate the shear stress on the plate and on the stationary surface. In what directions do these shear stresses act?

SOLUTION

From the given data: $h = 0.5$ mm and $V = 0.7$ m/s. For SAE 30 oil at 20°C, $\mu = 440$ mPa·s (from Appendix B.4). The magnitude of the shear stress, τ_0, on the plate and the stationary surface is given by Equation 1.47 as

$$\tau_0 = \mu \frac{V}{h} = (440 \times 10^{-3}) \frac{0.7}{0.5 \times 10^{-3}} = 616 \text{ N/m}^2 = \mathbf{616\ Pa}$$

Although the shear stresses on the moving plate and the stationary surface are equal in magnitude, they do not act in the same direction. The shear stress on the (top) moving plate acts in the **opposite direction** of the plate movement. The shear stress on the (bottom) stationary surface acts in the **same direction** as the plate movement.

Application to flow in open channels (erosion). The shear stress on the boundary of an open channel containing flowing water determines the stability of deposited sediment and native geologic material on the boundary of the channel. For example, in the design of drainage channels, civil engineers must ensure that the channel lining material is adequate to resist the boundary shear stress when the channel is carrying the design flow. In the design of sanitary sewers, the concern is the opposite; in this case, engineers must ensure that deposited sediments are scoured when the sewer is carrying the design flow.

EXAMPLE 1.12

In open channels of large width, the flow velocity at any given point within the channel is commonly assumed to be unaffected by the distance from the sides of the channel, depending only on the distance from the bottom of the channel. The velocity distribution, $u(y)$, in an open channel of large width is commonly estimated using the relation

$$u(y) = 5.75\sqrt{gy_0S_0}\log\left(\frac{30y}{k_s}\right)$$

where y is the distance from the bottom of the channel, y_0 is the depth of flow, S_0 is the slope of the channel, and k_s is the height of the roughness elements on the bottom of the channel. (a) Estimate the shear stress 1 cm above the bottom of a channel, where the slope of the channel is 0.5%, the depth of flow is 2.5 m, and the characteristic roughness height is 1 cm. (b) If the average shear stress on the bed material is equal to the theoretical stress 1 cm above the bottom of the channel and the critical shear stress required to move the bed material in the channel is 1 N/m^2, determine whether the channel is in a stable condition. Assume that the temperature of the water is 20°C.

SOLUTION

Taking $y_0 = 2.5$ m, $S_0 = 0.005$, and $k_s = 0.01$ m, the velocity distribution is given by

$$\begin{aligned} u(y) &= 5.75\sqrt{gy_0S_0}\log\left(\frac{30y}{k_s}\right) = 5.75\sqrt{(9.81)(2.5)(0.005)}\log\left(\frac{30y}{0.01}\right)\ \text{m/s} \\ &= 2.01(3.48 + \log y)\ \text{m/s} = 6.99 + 0.873\ln y\ \text{m/s} \end{aligned}$$

(a) At 20°C, $\mu = 0.00100$ Pa·s $= 0.00100$ N·s/m^2, and the shear stress, τ_0, 1 cm from the bottom of the channel is given by

$$\begin{aligned} \tau_0 &= \mu \left.\frac{du}{dy}\right|_{y=0.01\text{m}} = \mu \left.\frac{d}{dy}(6.99 + 0.873\ln y)\right|_{y=0.01\text{m}}\ \text{N/m}^2 \\ &= 0.00100\left[\frac{0.873}{y}\right]_{y=0.01\text{m}}\ \text{N/m}^2 = 0.0873\ \text{N/m}^2 \end{aligned}$$

Therefore, the average shear stress 1 cm above the bottom of the channel is **0.0873 Pa**.

(b) Because the shear stress required to move the bed material is 1 Pa and the actual shear stress caused by the flowing water is 0.0873 Pa, the channel is **stable**.

Kinematic viscosity. A fluid property that is closely related to the (dynamic) viscosity, μ [$ML^{-1}T^{-1}$], is the *kinematic viscosity*, ν [L^2T^{-1}], which is defined by the relation

$$\nu = \frac{\mu}{\rho} \tag{1.49}$$

where ρ is the density of the fluid [ML^{-3}]. The kinematic viscosity is called "kinematic" because it is quantified in terms of the kinematic dimensions of length (L) and time (T) and does not involve the dynamic dimension of mass (M). Recall that *kinematics* is defined as the study of the geometry of motion. It is sometimes more convenient to work with ν as the fluid property rather than μ and ρ separately, particularly in cases where μ and ρ occur only in the combination μ/ρ. The values of μ and ρ for water as a function of temperature are given in Appendix B.1 and can be combined to yield the kinematic viscosity, ν, as a function of temperature. For liquids, values of ν can usually be taken as being independent of pressure, except at extremely high pressures. In contrast to liquids, the kinematic viscosity of gases varies strongly with pressure because of changes in density. A commonly used kinematic viscosity unit is the *stoke*[7] [St]. The stoke is equivalent to 10^{-4} m^2/s (= 1 cm^2/s), and a centistoke is one-hundredth of a stoke. The viscosity of water at 20°C is approximately 1 centistoke (1 cSt). In current practice, use of the cSt unit is discouraged in favor of using either m^2/s or mm^2/s.

Measurement of viscosity. Measurements of viscosity are commonly made with *viscometers*, which relate controlled measurements of viscous flows to the theoretical flow equation applicable to the particular configuration of the viscometer. There are many different types of viscometers, such as the *rotating drum viscometer*, the *capillary-tube viscometer*, and the *falling-ball viscometer*. In most engineering applications, the viscosity of the working fluid is available in the open literature and measurement of viscosity is not necessary.

Empirical equations for viscosity of liquids. In the case of liquids, the dynamic viscosity, μ, generally decreases with increasing temperature and the relationship between viscosity and temperature can usually be approximated by *Andrade's equation*, which is given by

$$\mu = a \cdot 10^{\frac{b}{T-c}} \quad \text{or} \quad \mu = a \cdot \exp\left(\frac{b}{T-c}\right) \tag{1.50}$$

where a, b, and c are constants and T is the temperature [K]. For water, using the first expression in Equation 1.50, $a = 2.414 \times 10^{-5}$ Pa·s, b = 247.8 K, and c = 140 K yields values of μ with errors less than 2.5% within the temperature range of 0–370°C (Touloukian et al., 1975). An alternative empirical expression that is claimed to have better performance than Equation 1.50 is given by (White, 2011)

$$\ln\frac{\mu}{\mu_0} = a + b\left(\frac{T_0}{T}\right) + c\left(\frac{T_0}{T}\right)^2 \tag{1.51}$$

where T is the absolute temperature [K], the subscript 0 indicates the reference state, and a, b, and c are constants. For water, taking T_0 = 273.16 K, $\mu_0 = 1.792 \times 10^{-3}$ Pa·s, $a = -1.94$, $b = -4.80$, and $c = 6.74$ yields values of μ with errors less than $\pm 1\%$ within the normal temperature range 0–100°C of liquid water.

[7]Named in honor of the English physicist Sir George Stokes (1819–1903).

Empirical equations for viscosity of gases. In the case of gases, the dynamic viscosity, μ, generally increases with increasing temperature and the relationship between viscosity and temperature can usually be approximated by the *Sutherland equation*, which is given by

$$\mu = \frac{dT^{\frac{3}{2}}}{T + e} \tag{1.52}$$

where d and e are constants. Both Equations 1.50 and 1.52 can be used to interpolate viscosities between two temperatures at which the viscosities are known, where the bounding conditions are used to determine the empirical constants. For air under standard atmospheric conditions, $d = 1.458 \times 10^{-6}$ kg/(m·s·K$^{\frac{1}{2}}$) and $e = 110.4$ K. If Equation 1.52 is applied at a reference condition where $\mu = \mu_0$ and $T = T_0$, then Equation 1.52 can be expressed in the form

$$\frac{\mu}{\mu_0} = \left(\frac{T}{T_0}\right)^{\frac{3}{2}} \frac{T_0 + e}{T + e} \tag{1.53}$$

Typically, T_0 is taken as 273 K and μ_0 is the corresponding dynamic viscosity. As an alternative to the Sutherland equation (Equation 1.52 or Equation 1.53), a power law relationship is sometimes assumed, where

$$\frac{\mu}{\mu_0} = \left(\frac{T}{T_0}\right)^n \tag{1.54}$$

where the exponent, n, is commonly taken as 0.7.

EXAMPLE 1.13

The measured viscosity of air at 0°C and 400°C is 17.29 μPa·s and 32.61 μPa·s, respectively. (a) Use the Sutherland equation with conventional parameter values to determine the percentage error in the estimated viscosity at 400°C. (b) Repeat part (a) using the power-law relationship instead of the Sutherland equation.

SOLUTION

From the given data: $T_0 = 0°\text{C} = 273$ K, $T_1 = 400°\text{C} = 673$ K, $\mu_0 = 17.29$ μPa·s, and $\mu_1 = 32.61$ μPa·s.

(a) For the Sutherland equation, Equation 1.53, the conventional parameter value is $e = 110.4$ K. Application of Sutherland equation gives

$$\frac{\mu}{\mu_0} = \left(\frac{T}{T_0}\right)^{\frac{3}{2}} \frac{T_0 + e}{T + e} \quad \rightarrow \quad \frac{\mu}{17.29} = \left(\frac{673}{273}\right)^{\frac{3}{2}} \frac{273 + 110.4}{673 + 110.4} \quad \rightarrow \quad \mu = 32.75\ \mu\text{Pa·s}$$

The percentage error in this estimate is (32.75 – 32.61)/32.61 × 100 = **0.43%**.

(b) For the power-law relationship, Equation 1.54, the conventional parameter value is $n = 0.7$. Application of the power-law relationship gives

$$\frac{\mu}{\mu_0} = \left(\frac{T}{T_0}\right)^n \quad \rightarrow \quad \frac{\mu}{17.29} = \left(\frac{673}{273}\right)^{0.7} = 32.52\ \mu\text{Pa·s}$$

The percentage error in this estimate is (32.52 – 32.61)/32.61 × 100 = **−0.29%**.

1.6.2 Non-Newtonian Fluids

Fluids in which the shear stress is not linearly proportional to the strain rate are called *non-Newtonian fluids*, because they do not obey Newton's law of viscosity. The relationship between shear stress and strain rate in non-Newtonian fluids typically takes one of the following forms:

$$\tau = K\left(\frac{du}{dy}\right)^n = \underbrace{\left[K\left|\frac{du}{dy}\right|^{n-1}\right]}_{\text{apparent viscosity}}\frac{du}{dy} \quad \text{or} \quad \tau = \mu_{\text{ap}}\frac{du}{dy} \tag{1.55}$$

where K is called the *consistency index* and n is called the *power-law index* or the *flow behavior index* (Pritchard, 2011). Equation 1.55 reduces to the Newtonian-fluid relationship when $n = 1$ and $K = \mu$. Examples of non-Newtonian fluids include slurries, colloidal suspensions, polymer solutions, blood (in smaller arteries), and liquid plastics. The relationship between shear stress, τ, and strain rate, du/dy, in non-Newtonian fluids is sometimes expressed in terms of Newton's law of viscosity, Equation 1.44, but with a variable viscosity called the *apparent viscosity*, μ_{ap}. The behavior of various types of non-Newtonian fluids are compared with the behavior of Newtonian fluids in Figure 1.11. Fluids in which the apparent viscosity increases with increasing strain rate (i.e., $n > 1$) are called *dilatant fluids* or *shear-thickening fluids*, and such fluids include two-phase mixtures such as water-sand mixtures, drilling fluids, mud, quicksand, and many other mixtures of solid and liquid particles. Fluids in which the apparent viscosity decreases with increasing strain rate (i.e., $n < 1$) are called *pseudoplastic fluids* or *shear-thinning fluids*, and such fluids include some paints (e.g., latex), liquid cement, colloidal suspensions, paper pulp in water, polymer solutions, molten plastics, blood plasma, fluids with suspended particles, adhesives, inks, syrups, milk, and molasses. Materials that can resist finite shear stresses (such as a solid) but deform continuously when the shear stress exceeds a critical value are called *Bingham plastics*[8]. The relationship between the shear stress, τ, and the strain rate, du/dy, for a Bingham plastic can be described by

$$\tau = \tau_{\text{yield}} + \mu_{\text{b}}\frac{du}{dy} \tag{1.56}$$

where τ_{yield} is the threshold *yield stress* required for the initiation of continuous strain and μ_{b} is the dynamic viscosity of the Bingham plastic material. Materials that behave like Bingham plastics include tar, asphalt, drilling mud, clay suspensions, slurries, some greases, water suspensions of fly ash, and sewage sludge. The behavior of the Bingham plastic shown in

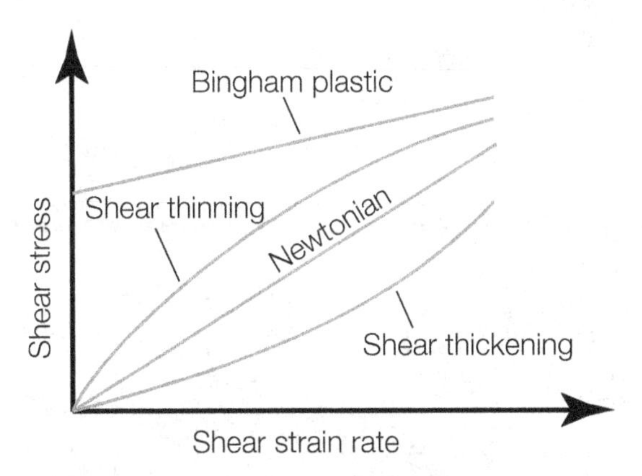

Figure 1.11: Shear stress versus strain rate in fluids

[8]Named in honor of the American chemist Eugene C. Bingham (1878–1945).

Figure 1.11 exhibits Newtonian behavior beyond the critical shear stress. However, some Bingham plastics exhibit non-Newtonian behavior beyond the critical shear stress. Fluids in which the apparent viscosity decreases with time under a constant applied shear stress are called *thixotropic fluids*, and fluids in which the apparent viscosity increases with time under a constant applied shear stress are called *rheopectic fluids*; many paints are thixotropic fluids. A fluid that partially or fully returns to its original shape after an applied stress is released is called *viscoelastic*; bitumen, nylon, flour dough, and many biological fluids are viscoelastic. Most engineering applications involve Newtonian fluids, and the study of fluid mechanics is generally taken as the study of the behavior of Newtonian fluids; the study of the behavior of non-Newtonian fluids is generally included in the field of *rheology*.

Example of non-Newtonian fluid. Molten rock is an example of a non-Newtonian fluid. Molten basalt shown in Figure 1.12 is common in lava flows from volcanoes. Basalt typically has less than 52% by weight of silica (SiO_2), and because of its low silica content, basalt has a low viscosity. As a consequence of this low viscosity, basaltic lava can flow quickly and easily move more than 20 km from a vent. The low viscosity typically allows volcanic gases to escape without generating enormous eruption columns. However, basaltic lava fountains and fissure eruptions still form explosive fountains hundreds of meters tall.

Inviscid fluids. An *inviscid fluid* is defined as a fluid in which the viscosity is equal to zero. Although inviscid fluids do not exist in reality, the motion of many fluids can be approximated by assuming that the fluid is inviscid. However, this approximation can only be made away from solid boundaries, since at solid boundaries (such as the walls of flow conduits), particles of fluid adhere to these surfaces and their velocities are zero—this is called the *no-slip condition*, which occurs in all viscous fluids and would not occur if the fluid were inviscid.

Figure 1.12: Viscous molten rock
Source: Hawaiian Volcano Observatory, USGS.

1.7 Surface Tension

Liquids have cohesion and adhesion, both of which are forms of molecular attraction. *Cohesion* is the intermolecular attraction of liquid molecules for each other, and *adhesion* is the attraction of molecules of a certain liquid for molecules of another substance, where that other substance is usually a solid. Cohesion between molecules of a liquid enables the liquid to resist tensile stress, whereas adhesion between molecules of a liquid and a solid enables the liquid to adhere to a solid. The cohesive forces between (liquid-phase) molecules on the surface of a liquid form a "cohesive skin" or "cohesive membrane" that covers the surface of the liquid. Molecules on the surface of a liquid have attractive forces to other molecules in the liquid phase that are much greater than attractive forces to the gas-phase molecules above the liquid surface. In liquid droplets, these intermolecular forces pull the surface molecules inward to minimize the surface area of the liquid. The intermolecular forces responsible for surface tension are illustrated in Figure 1.13, from which it is apparent that intermolecular attraction between the molecules of a liquid is the primary cause of liquid-liquid surface tension, and the intermolecular attraction between molecules of a liquid and a solid is the primary cause of liquid-solid surface tension. In general, specification of surface tension between solids and liquids or between liquids and other liquids also requires stating the substances involved. Commonly observed phenomena that are attributed to surface-tension effects include water dripping from an open faucet as near-spherical droplets, mercury droplets forming nearly spherical balls that can be rolled like a solid ball on a smooth surface, small (light) insects walking on water, razor blades and small steel needles supported on the surface of a water body, and the ability to carefully fill a cup with water until the water surface is about 3 mm above the rim of the cup. Engineering applications in which surface-tension effects are important include processes involving bubble formation and growth, as well as liquid films on solid surfaces.

Definition of surface tension. The *surface tension*, σ, is commonly defined as the intermolecular attraction per unit length along any line in the liquid surface and can be expressed as

$$\sigma = \frac{F}{L} \tag{1.57}$$

where F is the surface tension force over a distance L in the surface of the fluid. Fundamentally, the surface tension is more appropriately defined as the amount of energy per unit interface area required to overcome the molecular attraction within the interface. This concept is illustrated in Figure 1.14 for an element of a surface interface with dimensions $L \times dx$.

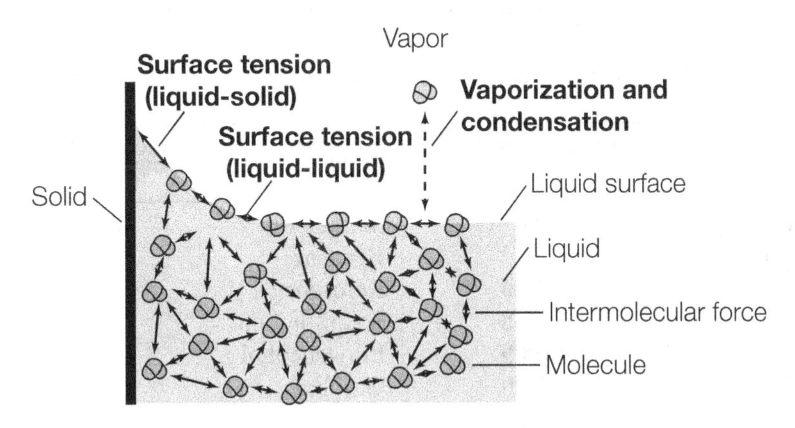

Figure 1.13: Intermolecular forces and surface tension

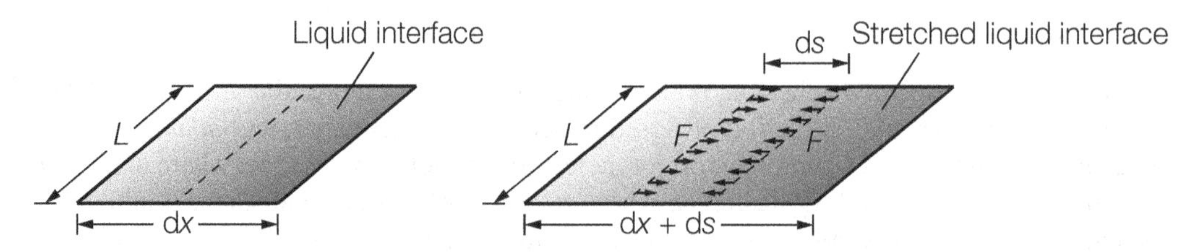

Figure 1.14: Liquid interface and surface tension

If the element of the surface interface is expanded by an amount ds and the molecular attractive force over a length (perpendicular to ds) is equal to F, then the energy required to expand the surface area is equal to $F\,\mathrm{d}s$. Because the surface area of the expanded element is $L\,\mathrm{d}s$, the energy per unit surface area required to expand the interface is given by

$$\text{energy per unit interface area} = \frac{F\,\cancel{\mathrm{d}s}}{L\,\cancel{\mathrm{d}s}} = \frac{F}{L} = \sigma \tag{1.58}$$

This result confirms that the definition of the surface tension given by Equation 1.57 can be interpreted as the energy per unit interface area required to overcome the molecular attraction. In standard SI units, the surface tension is commonly expressed in N/m, which could alternatively be expressed as J/m^2 to make the meaning of surface tension more explicit. The surface tension of a liquid-gas interface generally depends on the temperature.

Surface tension of liquids in contact with air. The surface tension of a liquid is defined only at a liquid-liquid or liquid-gas interface, and the contacting liquid or gas must generally be specified along with the magnitude of the surface tension. The surface tensions of several fluids in contact with air at standard atmospheric pressure (101 kPa) and 20°C are listed in Table 1.7 in decreasing order of magnitude. The surface tension of pure water in contact with air varies from 75.6 mN/m at 0°C to 58.9 mN/m at 100°C, with a surface tension of 72.8 mN/m at 20°C. The surface tension of water in contact with air as a function of temperature between 0°C and 100°C is given in Appendix B.1. The surface tension of most organic compounds falls within the range of 25–40 mN/m, and the surface tension of liquid metals are typically in the range of 300–600 mN/m. Surface tension generally decreases with temperature, with the decrease being nearly linear with absolute temperature. Surface tension at the *critical temperature* is equal to zero, where the critical temperature is defined as the temperature at which the liquid and its vapor can coexist. The tension that exists between two immiscible liquids at their interface is commonly called the *interfacial tension.*

Table 1.7: Surface Tension of Various Fluids in Contact with Air*

Fluid	Surface Tension, σ (mN/m)	Fluid	Surface Tension, σ (mN/m)
Mercury	484	SAE 30 oil	35
Water:		Benzene	28.9
0°C	**75.6**	Kerosene	26.8
20°C	**72.8**	Soap solution	25
100°C	**58.9**	Ethyl alcohol	23
Glycerin	63.0	Gasoline	22
Blood, 37°C	58		

*At standard atmospheric pressure and 20°C unless otherwise stated.

Effects of impurities. The surface tension of liquids can be changed considerably by impurities. Certain chemicals, called *surfactants*, can be added to a liquid to decrease its surface tension. For example, soaps and detergents lower the surface tension of water and enable it to penetrate the small openings between fibers for more effective washing. As another example, adding dishwashing liquid to water that is supporting a small needle by surface tension will cause the needle to lose its surface support and sink. Dissolved salts such as sodium chloride (NaCl) raise the surface tension of water.

Liquid droplets and bubbles in air and other gases. Liquid droplets in air are encountered in many engineering applications, with liquid bubbles in air encountered less frequently. The pressure inside droplets and bubbles is generally greater than the pressure outside droplets and bubbles because of surface tension. Consider the free body of a half-droplet of radius R as shown in Figure 1.15(a). If the pressure inside the droplet is p_1 and the pressure outside the droplet is p_2 and the weight of the fluid in the droplet is neglected, then for equilibrium, the sum of the forces is equal to zero such that

$$\text{droplet:} \quad p_1(\pi R^2) - p_2(\pi R^2) - 2\pi R \cdot \sigma = 0 \quad \rightarrow \quad p_1 - p_2 = \frac{2\sigma}{R} \tag{1.59}$$

which indicates that the pressure inside the droplet (p_1) is greater than the pressure outside the droplet (p_2) by an amount $2\sigma/R$. In the case of a bubble, there is a thin layer of liquid between the inside and outside of the bubble, thus creating two surfaces in which surface tension acts (i.e., the inside surface and the outside surface). Consider the free body of a half-bubble of radius R as shown in Figure 1.15(b), where equilibrium requires that

$$\text{bubble:} \quad p_1(\pi R^2) - p_2(\pi R^2) - 2 \times 2\pi R \cdot \sigma = 0 \quad \rightarrow \quad p_1 - p_2 = \frac{4\sigma}{R} \tag{1.60}$$

where p_1 and p_2 are the pressures inside and outside the bubble, respectively. Equation 1.60 shows that the pressure inside the bubble is greater than the pressure outside the bubble by an amount equal to $4\sigma/R$. Also, comparing Equations 1.60 and 1.59, it is apparent that the pressure inside a bubble is two times higher than the pressure inside a droplet of the same size.

Gas bubbles in liquids. Gas bubbles in liquids are encountered in many applications. An example of such an application is in the oxygenation of liquids where air or oxygen is "bubbled" into a reservoir containing a liquid. The injection of gas bubbles is also used to induce

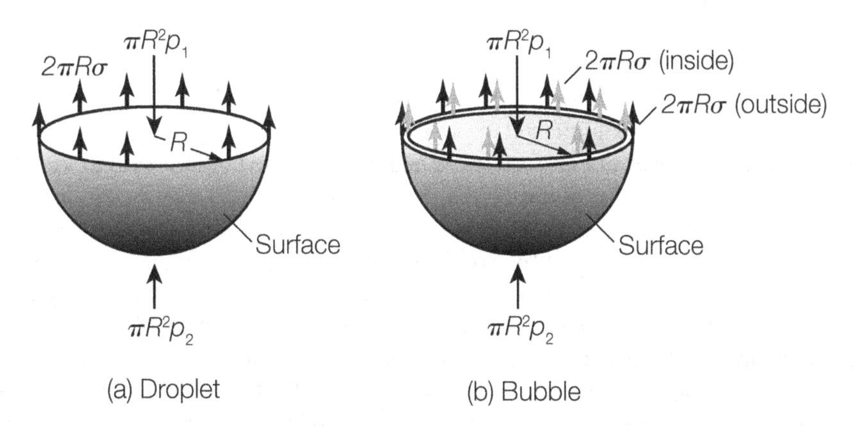

Figure 1.15: Pressure differences in droplets and bubbles

circulation in otherwise stagnant bodies of water such as lakes and reservoirs. The free body of a half gas bubble is the same as the free body of a half-droplet as shown in Figure 1.15(a), where p_1 is the pressure in the liquid outside the bubble, p_2 is the pressure of the gas inside the bubble, σ is the surface tension at the gas-liquid interface, and R is the radius of the gas bubble. The pressure difference Δp between the inside and outside of the bubble is therefore given by Equation 1.59, which can be expressed as

$$\Delta p = \frac{2\sigma}{R} \tag{1.61}$$

where the pressure inside the bubble is Δp higher than outside the bubble. The pressure difference, Δp, must be taken into account when sizing the bubble injection mechanism, because a gas pressure of $p_0 + \Delta p$ is required to generate the bubble, where p_0 is the pressure in the liquid at the location bubbles are to be generated.

EXAMPLE 1.14

A tank of wastewater is oxygenated by bubbling pure oxygen (O_2) from the bottom of the tank. The (hydrostatic) pressure of the water at the bottom of the tank is 130 kPa, the temperature of the water is 25°C, and the average diameter of the O_2 bubbles is approximately 2 mm. Atmospheric pressure is 101.3 kPa. At what pressure relative to atmospheric pressure is O_2 being injected into the tank? You may assume that the surface tension at the O_2-water interface is the same as at the air-water interface.

SOLUTION

From the given data: $p_w = 130$ kPa, $p_{atm} = 101.3$ kPa, $T = 25$°C, $D = 2$ mm, and $R = D/2 = 1$ mm. For the water-air interface at 25°C, $\sigma = 72.0$ mN/m. The difference in pressure, Δp, between the O_2 inside the bubble and the surrounding water is given by Equation 1.61, where

$$\Delta p = \frac{2\sigma}{R} = \frac{2(72.0 \times 10^{-3})}{1 \times 10^{-3}} = 144 \text{ Pa} = 0.144 \text{ kPa}$$

Because atmospheric pressure is 101.3 kPa and the pressure outside the bubble is 130 kPa, the pressure inside the bubble relative to atmospheric pressure, p_b, is given by

$$p_b = (p_w + \Delta p) - p_{atm} = (130 + 0.144) - 101.3 = 28.8 \text{ kPa}$$

Therefore, O_2 must be injected in the bottom of the tank at a pressure of approximately **28.8 kPa** above atmospheric pressure.

Liquid droplets on solid surfaces. In cases where a liquid droplet contacts a solid surface, the angle formed between the liquid and the solid surface is called the *contact angle* or *wetting angle*, θ, which is illustrated in Figure 1.16. When $\theta < 90°$, the liquid is said to *wet* the solid; otherwise, when $\theta > 90°$, the liquid is said to be *nonwetting*. The wetting angle, θ, is a property of the contact fluids (e.g., water and air) and the contact solid (e.g., glass) and is usually sensitive to the condition of the solid surface, such as its roughness, cleanliness, and temperature. Examples of solid surfaces that are wetted by water in air are soap and glass, whereas wax is nonwetting by water in air. The effect of the condition of a solid surface on wetting can be observed when a car is waxed. Water usually wets the car before it is waxed, but forms "beads" on the car after it is waxed. Surfaces wetted by water are generally referred to as *hydrophilic surfaces*, and surfaces that are not wetted by water are *hydrophobic surfaces*. The wetting angle for mercury-air-glass is 130°; hence, mercury does not wet glass.

Capillarity. *Capillarity* results from the interaction of fluids with small-diameter tubes and is due to both cohesion and adhesion. When cohesion is of less importance than adhesion, a liquid wets a solid surface it touches and rises at the point of contact; if cohesion predominates, the liquid surface will depress at the point of contact. The intermolecular (solid-liquid) force between a liquid and a solid surface is illustrated in Figure 1.13. The rise or fall of a liquid in a small-diameter tube inserted into a liquid is called the *capillary effect*, and such narrow tubes or confined flow channels are sometimes called *capillaries*. The capillary effect is partially responsible for the rise of water to the top of tall trees; it influences the movement of liquids in unsaturated soils; and it is responsible for the rise of oils in the wicks of lanterns, where the rise of a fluid from a liquid surface into a woven material is called *wicking*. Water wicking into a paper towel and ink wicking into a felt-tip pen are additional examples. The curved liquid surface that develops in a tube is called the *meniscus*. Capillary rise in a tube is illustrated in Figure 1.17(a), where the pressure in the air above the meniscus is equal to atmospheric pressure, p_{atm}. At the level of the liquid surface in the reservoir (into which the capillary tube is inserted), the pressure is equal to atmospheric pressure, p_{atm}, and the pressure within the tube at the same elevation as the reservoir-liquid surface is also equal to atmospheric pressure, p_{atm}. The fact that the pressures at the same elevation inside and outside the tube are equal can be derived from pressure distribution in static fluids, which is covered in detail in the next chapter. Considering the column of liquid in the tube as shown in Figure 1.17(b) (and setting the sum of the forces on the column of liquid equal to zero) requires that

$$\cancel{p_{\text{atm}}\pi r^2} + 2\pi r\sigma\cos\theta - \cancel{p_{\text{atm}}\pi r^2} - \pi r^2 h\gamma = 0 \qquad (1.62)$$

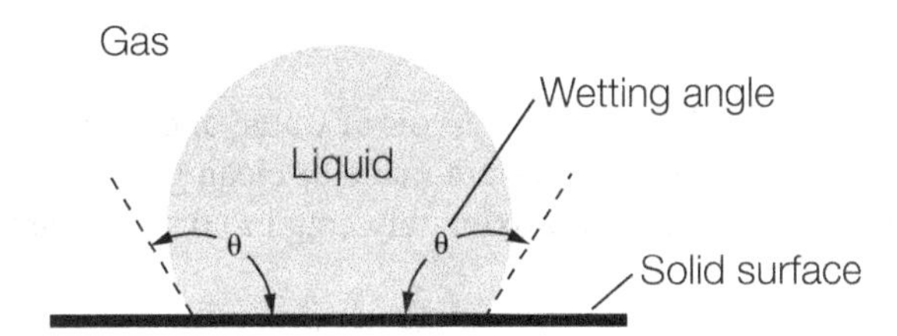

Figure 1.16: Wetting angle

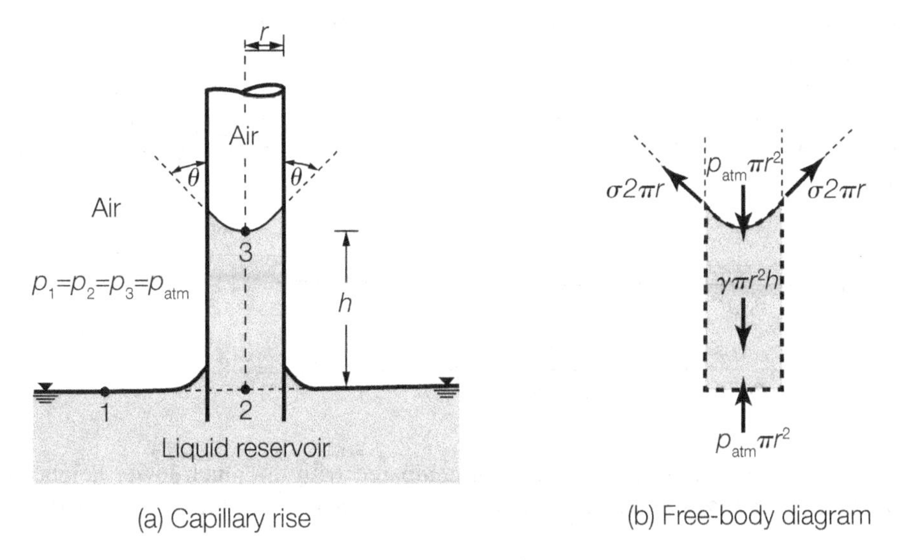

(a) Capillary rise

(b) Free-body diagram

Figure 1.17: Capillarity

where forces in the upward direction are taken as positive, θ is the wetting angle [degrees], γ is the specific weight of the liquid [FL^{-3}], r is the radius of the tube [L], and the surface of the liquid in which the capillary tube is immersed is located at a distance h below the meniscus. Simplifying and rearranging Equation 1.62 yields

$$h = \frac{2\sigma\cos\theta}{\gamma r} \tag{1.63}$$

Equation 1.63 can be used to compute either the capillary rise or depression in a glass tube. Liquids that rise in a glass capillary tube (such as water) have contact angles less than 90° (i.e., $\theta < 90°$), and liquids that are depressed in a glass capillary tube (such as mercury) have contact angles greater that 90° (i.e., $\theta > 90°$). Note that the meniscus lifts a small volume of liquid near the tube walls, in addition to the volume $\pi r^2 h$ used in deriving Equation 1.63. For larger tube diameters, with smaller capillary rise heights, this small additional volume can become a large fraction of $\pi r^2 h$; so Equation 1.63 overestimates the amount of capillary rise or depression, particularly for large-diameter tubes. In practice, the shape of the meniscus ceases to be spherical for tube diameters greater than around 3 mm (Massey and Ward-Smith, 2012), and the capillary effect of water is usually negligible in tubes whose diameter is greater than 10 mm. When pressure measurements are made using capillary tubes in manometers and barometers, it is important to use sufficiently large tubes to minimize the capillary effect.

EXAMPLE 1.15

A capillary tube of diameter 2 mm is placed in a reservoir of water, and a similar capillary tube is placed in a reservoir of methanol. If both tubes are made of clean glass and the water-glass and methanol-glass contact angles are both approximately equal to 0°, compare the heights of rise in both capillary tubes.

SOLUTION

From the given data: D = 2 mm = 0.002 m, and $\theta_\text{w} = \theta_\text{m} = 0°$. The radius of the capillary tube is $r = D/2$ = 0.001 m. The properties of water (H_2O) and methanol (CH_3OH) can be derived from Appendix B.4 as σ_w = 0.073 N/m, ρ_w = 998 kg/m^3, σ_m = 0.0225 N/m, and ρ_m = 791 kg/m^3, where the subscripts "w" and "m" refer to water and methanol, respectively. Using these data, the heights of rise of water and methanol, h_w and h_m, are derived from Equation 1.63 as follows:

$$h_\text{w} = \frac{2\sigma_\text{w}\cos\theta_\text{w}}{\rho_\text{w} g r} = \frac{2(0.073)\cos 0°}{(998)(9.807)(0.001)} = 0.0149\ \text{m} = 14.9\ \text{mm}$$

$$h_\text{m} = \frac{2\sigma_\text{m}\cos\theta_\text{m}}{\rho_\text{m} g r} = \frac{2(0.0225)\cos 0°}{(791)(9.807)(0.001)} = 0.0058\ \text{m} = 5.8\ \text{mm}$$

Therefore, water will rise to a height of 14.9 mm compared with the much lower height of 5.8 mm for methanol. The lower surface tension of methanol is primarily responsible for this difference.

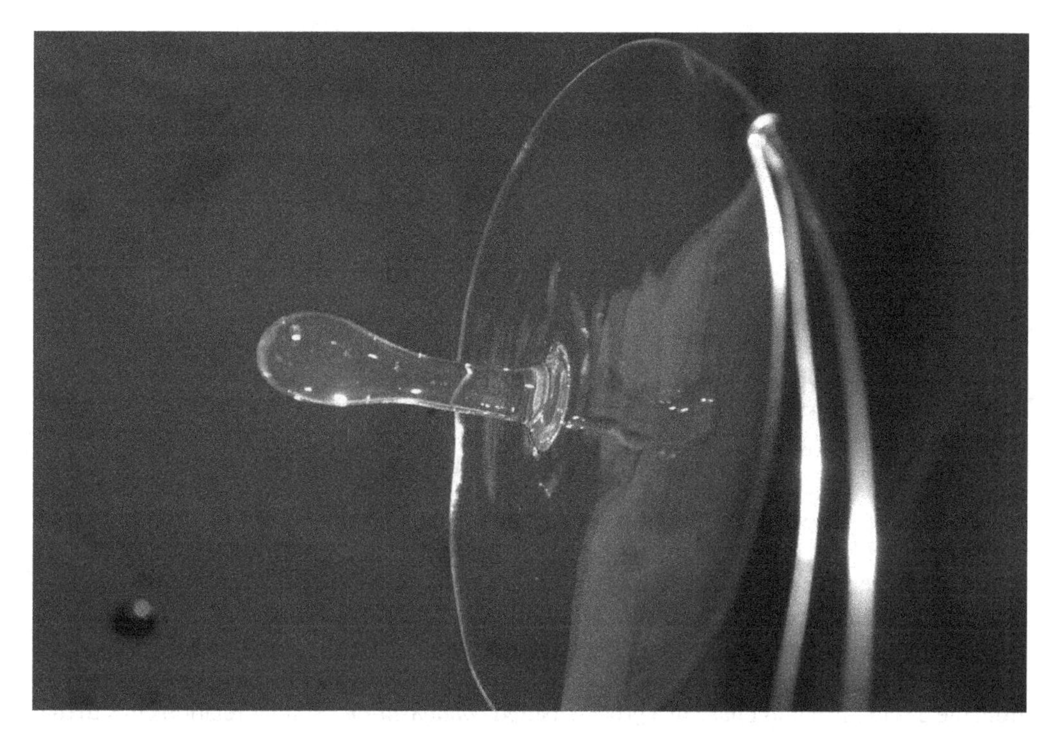

Figure 1.18: Surface tension in space
Source: Astronaut Donald R. Pettit/NASA.

Surface tension in microgravity. Surface tension dominates the physics of fluids in microgravity environments, such as on the International Space Station (ISS). A demonstration of a surface-tension effect on the ISS is shown in Figure 1.18, where a water film created within a metal loop (by cohesion of the water to the metal) is displaced outward. The displaced water is held together by surface tension. Interestingly, the fluid in the film was caused to rotate by heating part of the film with a flashlight, which caused an imbalance in surface tension (because surface tension is a function of temperature), which caused the water film to rotate within the loop in an attempt to restore the balance. This movement caused by a surface-tension imbalance is called *Marangoni convection*. Such effects are rarely observed on Earth due to the more dominant effect of gravity.

1.8 Vapor Pressure

On the surface of a liquid at the liquid-gas interface, there is a continuous interchange of molecules between the liquid and gaseous phases of the liquid substance; this process is illustrated in Figure 1.13. The molecules of the liquid substance in the gaseous phase exert a partial pressure in the gaseous environment above the liquid. The *saturation vapor pressure* or *saturated vapor pressure* of a fluid is defined as the partial pressure of the gaseous phase of the fluid that is in contact with the liquid phase of the fluid when there is no net exchange of mass between the two phases; that is, equilibrium conditions exist. The magnitude of the saturation vapor pressure increases with molecular activity and is therefore a function of temperature.

Relationship between vapor pressure and temperature. A typical relationship between the saturation vapor pressure and temperature of a liquid is illustrated graphically in Figure 1.19, where this particular relationship is that of water. The solid line in Figure 1.19 gives the saturation vapor pressure as a function of temperature, where the saturation vapor pressure

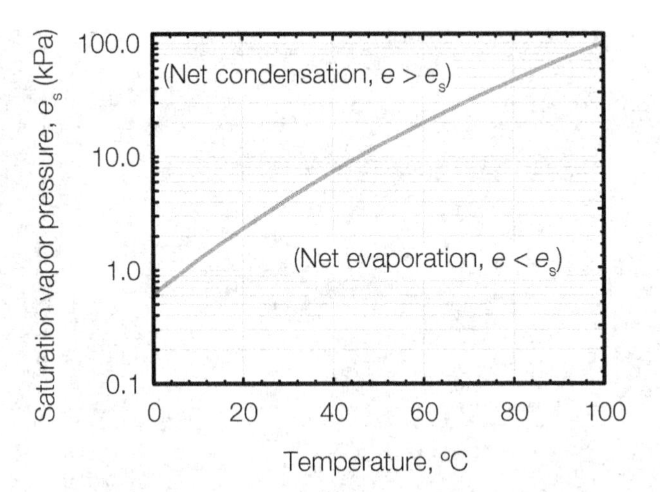

Figure 1.19: Saturation vapor pressure versus temperature for water

is commonly represented as e_s. At any given temperature, if the actual vapor pressure, e, is less than the saturation vapor pressure, e_s, then there is a net movement of molecules from the liquid to the gas phase in a process called *evaporation*. If the actual vapor pressure, e, is greater than the saturation vapor pressure, e_s, then there is a net movement of molecules from the gas to the liquid phase in a process called *condensation*. Similar relationships to that shown in Figure 1.19 for water also exist for other substances. The actual vapor pressure, e, of a substance is seldom exactly equal to its saturation vapor pressure; therefore, equilibrium between the liquid and vapor phases of a substance seldom exists. The saturation vapor pressure is a function of temperature, with the saturation vapor pressure generally increasing with increasing temperature. The increase in vapor pressure with temperature is a consequence of the velocity of the liquid molecules increasing with temperature, thereby increasing their ability to move from the liquid to the gaseous phase.

Relationship between saturation vapor pressure and volatility. Substances with large saturation vapor pressures tend to evaporate rapidly when open to the atmosphere, because the rate of evaporation is proportional to the difference between the saturation vapor pressure and the actual vapor pressure, and the actual vapor pressures of most substances tend to remain low in well-ventilated environments. Consequently, the saturation vapor pressure is used as a measure of the *volatility* of a substance. The saturation vapor pressure of several substances are contrasted in Table 1.8, where the saturation vapor pressures are listed in increasing order of magnitude. Listed substances with vapor pressures higher than that of water (i.e., $>$ 2 kPa) seldom exist as liquids under usual atmospheric conditions.

Table 1.8: Saturation Vapor Pressure, e_s, of Various Fluids*

Fluid	e_s (kPa)	Fluid	e_s (kPa)
Ethylene glycol	0.500	Acetaldehyde	98.7
Xenon difluoride	0.600**	Butane	220
Water	**2.3**	Formaldehyde	435.7
Propanol	2.4	Propane	1013**
Methyl isobutyl ketone	2.66**	Carbonyl sulfide	1255
Freon 113	39.9	Carbon dioxide	5700**

*At 20°C unless otherwise noted. **At 25°C.

Saturation vapor pressure of water. The saturation vapor pressure of pure water varies between 0.611 kPa at 0°C and 101.3 kPa at 100°C, with a value of 2.34 kPa at 20°C. The saturation vapor pressure of water as a function of temperature is tabulated in Appendix B.1. In actuality, the vapor pressure of water in the atmosphere normally contributes no more than about 2 kPa to the total atmospheric pressure of about 101 kPa, with the remainder of the atmospheric pressure contributed by other gases, primarily nitrogen (N_2) and oxygen (O_2). A cycle of mass transfer between the liquid and gaseous phase of water occurs almost daily on Earth. During the cooler nighttime hours, the saturation vapor pressure of water is reduced below the actual vapor pressure, causing liquid precipitation called *dew*, while during the warmer daytime hours, the saturation vapor pressure typically rises above the actual vapor pressure, causing evaporation.

1.8.1 Evaporation, Transpiration, and Relative Humidity

The process of evaporation of water from the surface of Earth and from the stomatae of plants is a key component of the hydrologic cycle. The rate of evaporation is an important quantity in many agricultural operations, where the combination of plant transpiration and soil-water evaporation rates determine irrigation requirements. Plant transpiration and soil-water evaporation are usually lumped into a single process called *evapotranspiration*, and the rate of evapotranspiration is directly proportional to the difference between the saturation vapor pressure and the actual vapor pressure of water in the atmosphere above the plants. In hydrologic applications, the saturation vapor pressure is commonly denoted by e_s and the actual vapor pressure is commonly denoted by e. A property that is related to both e_s and e is the *relative humidity*, RH, which is defined by the relation

$$\text{RH} = \frac{e}{e_s} \times 100 \tag{1.64}$$

where RH is a percent. Clearly, evaporation can only occur when the relative humidity is less than 100%.

EXAMPLE 1.16

The air temperature in an agricultural area during a typical summer day is shown in the following table. If the relative humidity at noon is 70%, estimate the time of day the relative humidity becomes 100%.

Time	Midnight	3 a.m.	6 a.m.	9 a.m.	Noon	3 p.m.	6 p.m.	9 p.m.
Temperature (°C)	20	18.3	20	23.3	32.2	33.9	26.7	22.7

SOLUTION

The saturation vapor pressure of water as a function of temperature is given in Appendix B.1. At noon, the temperature is 32.2°C and the saturation vapor pressure (from Appendix B.1) is 4.93 kPa. Because the relative humidity is 70%, the actual vapor pressure, e, is given by

$$e = (0.7)(4.93) = 3.45 \text{ kPa}$$

Assuming that the actual vapor pressure of water in the atmosphere remains constant, at other times of the day, the relative humidity, RH, can be taken as

$$\text{RH} = \frac{3.45}{e_s} \times 100$$

where e_s is the saturation vapor pressure at the ambient air temperature. Reading the saturation vapor pressures as a function of temperature from Appendix B.1 yields the following variation in relative humidity, beginning at noon.

Time	Noon	3 p.m.	6 p.m.	9 p.m.
Temperature (°C)	32.2	33.9	26.7	22.7
e_s (kPa)	4.93	5.47	3.53	2.79
RH (%)	70	63.1	96.3	100

The relative humidity at 9 p.m. is calculated to be more than 100%, which will result in condensation to maintain the relative humidity at 100%. Therefore, the relative humidity becomes equal to 100% at some time between 6 p.m. and 9 p.m. Linearly interpolating the saturation vapor pressure between 6 p.m. and 9 p.m. indicates that at approximately 6:20 p.m., the saturation vapor pressure is 3.45 kPa and the humidity equals 100%.

1.8.2 Cavitation and Boiling

Whenever the pressure within a liquid is less than or equal to the saturation vapor pressure of the liquid, vapor cavities spontaneously form within the liquid; this process is called *cavitation*. These vapor cavities, which are also called *vapor bubbles* or *cavitation bubbles*, are in equilibrium with the surrounding liquid when the vapor pressure inside the cavity is equal to the saturation vapor pressure of the liquid.

Boiling. Boiling is the spontaneous formation of vapor cavities within a liquid, and a liquid boils when the saturation vapor pressure of the liquid is equal to the pressure within the liquid. Two ways of attaining a boiling condition are (1) by raising the temperature of the liquid at a constant pressure and (2) by decreasing the pressure in the liquid at a constant temperature. The first process is what is commonly observed when a liquid is heated to its boiling point, and the second process occurs when flowing fluids are accelerated and local pressures are reduced below the vapor pressure of the liquid.

Boiling caused by raising the temperature. Boiling of water in a pot placed on a stove burner is an everyday example of the cavitation phenomenon caused by raising the temperature of a liquid. The pressure within the water (in the pot) remains approximately constant and equal to the local atmospheric pressure, but the saturation vapor pressure of the water increases as the temperature of the water is increased. If the atmospheric pressure is approximately 101 kPa, the saturation vapor pressure becomes equal to the pressure within the water (101 kPa) when the temperature of the water reaches 100°C, at which point vapor cavities are spontaneously formed and water boils at 100°C. Attempts to raise the water temperature above 100°C will result in much larger cavities that are unsustainable in the pot. Keep in mind that as the water temperature is being raised, the rate of evaporation of the water is also increasing, because the rate of evaporation is proportional to the difference between the saturation vapor pressure and the actual vapor pressure. The actual vapor pressure typically remains constant at around 2 kPa. At locations where atmospheric pressure is less than 101 kPa, such as at higher elevations, the temperature required for the saturation vapor pressure of water to equal atmospheric pressure is less than 100°C; thus, water boils at a lower temperature than 100°C. For example, in Denver, Colorado, where the approximate elevation is 1524 m and atmospheric pressure is around 84.3 kPa, water boils at 94°C. At the summit of Mount Everest, with an approximate elevation of 8800 m and atmospheric pressure of 30 kPa, water boils at 69°C. To offset the effects of reduced atmospheric pressure and to cook food faster, a pressure cooker (i.e., a pot with a sealed top and a pressure relief valve) can be used to increase the temperature at which water boils, by increasing the pressure in the airspace above the water. For a pressure cooker with a typical airspace pressure of 300 kPa, water boils at 134°C.

EXAMPLE 1.17

Atmospheric pressure in a mountain region is equal to 82 kPa. Estimate the boiling point of water. Compare your result with the boiling point of water at sea level, where the atmospheric pressure can be assumed to be 101 kPa.

SOLUTION

At a saturation vapor pressure of 82 kPa, interpolation of the tabulated relationship between saturation vapor pressure and temperature in Appendix B.1 gives the corresponding water temperature as 93.8°C. Therefore, assuming that the pressure within the water body is approximately equal to the atmospheric pressure, the water will boil at **93.8°C**. At sea level, the pressure within a water body is approximately equal to 101 kPa and the corresponding boiling point is **100°C**.

Cavitation. Boiling caused by the reduction of pressure in a flowing fluid is more commonly referred to as cavitation. In this context, cavitation is caused by reducing the pressure within a liquid while maintaining an approximately constant temperature. This mechanism of cavitation frequently occurs when liquids flow through low-pressure regions in closed conduits, and cavitation also commonly occurs in hydroelectric turbines. Other mechanisms of cavitation also exist. *Fixed cavitation* occurs when a liquid flows past a solid body, creating a zone of low pressure behind the body in which the pressure is less than the saturation vapor pressure, thereby causing a vapor cavity to form behind the body. *Vortex cavitation* occurs in the high-velocity, low-pressure core of a vortex; this type of cavitation is commonly observed in *tip vortices* generated by boat propellers. Cavitation in tip vortices decreases the forces exerted on rudders and decreases the thrust and performance of the propeller. *Vibratory cavitation* is generated by pressure waves in liquids. Pressure waves are pressure fluctuations with high pressure followed by low pressure; cavitation can occur during the low pressure part of the cycle. Typically, cavitation is a transient phenomenon where vapor cavities are generated as a fluid passes through a low pressure zone, and the cavities collapse when the liquid moves into zones of higher pressure.

EXAMPLE 1.18

Water at 20°C is pumped through a closed conduit and experiences a range of pressures. What is the minimum allowable pressure to prevent cavitation within the conduit?

SOLUTION

The saturation vapor pressure of water at 20°C is 2.337 kPa to prevent cavitation (Appendix B.1). Hence, the (absolute) water pressure in the conduit should be maintained above **2.337 kPa** to prevent cavitation.

Negative effects of cavitation. Severe damage to conduits and hydromachinery can occur when vapor-filled cavities are formed in regions of lower pressure and are subsequently transported to regions of higher pressure, with the vapor cavities collapsing in regions of higher pressure. This type of cavitation is called *traveling cavitation*, and the rotors of pumps and turbines are particularly susceptible to damage by traveling cavitation. The collapsing of vapor-filled cavities is a process that is physically similar to squashing a balloon until it collapses. The release of vapor from collapsing cavities can generate vapor-jet velocities as high as 110 m/s and cause pressures as high as 50 MPa (Suslick, 1989), although pressures in the range of 800–1400 MPa have also been reported in the vicinity of collapsing vapor cavities (Potter et al., 2013; Elger et al., 2013; Massey and Ward-Smith, 2012). The damage caused by traveling cavitation in hydraulic machinery and hydraulic structures is called *pitting*, and an example of pitting on a propeller is shown in Figure 1.20. The upstream side of a propeller, called the *suction side*, experiences lower pressures than the downstream side of the propeller, called the *pressure side*. Cavitation is more likely to occur on the outer edge of the suction side of a propeller, because the pressure is lower on this side of the propeller, the speed of the surface of the propeller increases with distance from the hub (recall $v = r\omega$), and the pressure in the water adjacent to a propeller surface decreases with increasing speed of the surface. In the case of propellers, the damage due to pitting causes the blades to become less efficient and hence reduce thrust, a condition commonly referred to as *thrust breakdown*. In most cases where pitting due to cavitation occurs in hydromachines, the damage is repaired when it occurs. If left unrepaired, pitting in parts of a hydromachine can lead to drops in efficiency and, ultimately, major damage to the rest of the machine. In hydromachinery such as pumps and turbines, cavitation can be detected by increased vibration levels and large increases in noise, particularly in the moderately high frequency range of 15–100 kHz. Aside from damage to solid surfaces caused by collapsing vapor-filled cavities, these cavities created by the cavitation process can alter the flow characteristics substantially and can cause machinery that is designed to operate in a liquid continuum to operate inefficiently, because this continuum is now filled with vapor cavities.

Figure 1.20: Pitting in propellers
Source: U.S. National Park Service.

Positive effects of cavitation. Cavitation can also have beneficial effects, as in *ultrasonic cleaning* where vibratory cavitation generates vapor bubbles that have a cleaning effect when they collapse. In a military application, *supercavitating torpedoes* have sufficient speed and shape to generate a large vapor bubble that surrounds the torpedo, thereby significantly reducing the contact area with water, which allows the torpedo to move at a much higher speed than if it were completely surrounded by liquid water. Cavitation is also used in *shock wave lithotripsy*, which is a medical procedure for destroying kidney stones.

Cavitation number. The potential for cavitation in flowing liquid is commonly measured by the *cavitation number*, Ca, defined as

$$\mathrm{Ca} = \frac{p - p_\mathrm{v}}{\frac{1}{2}\rho V^2} \tag{1.65}$$

where p is the pressure within the liquid, p_v is the saturation vapor pressure of the liquid at its actual temperature, ρ is the density of the liquid, and V is the speed of the liquid.

1.9 Thermodynamic Properties of Liquids

Two important thermodynamic properties of liquids are specific heat and latent heat. Specific heat relates the change in internal energy in a liquid to the increase in temperature of the liquid, whereas latent heat relates the change in thermal energy to the change in phase of a liquid.

1.9.1 Specific Heat

The specific heat of a liquid is defined as the change in internal energy of the liquid per unit mass of the liquid per unit change in temperature. This relationship is commonly expressed in the form

$$\Delta u = mc\Delta T \tag{1.66}$$

where Δu is the change in internal (thermal) energy, m is the mass of the liquid, c is the specific heat, and ΔT is the change in temperature. For gases, two specific heats are defined, c_p and c_v, where c_p is the specific heat at constant pressure and c_v is the specific heat at constant volume. However, in contrast to gases where typically $c_p \neq c_v$, in liquids, $c_p \approx c_v$ and there is little variation of specific heat with temperature. The specific heat of liquids is commonly represented simply as c. Water is usually taken to have a specific heat of 4.18 kJ/kg·K.

EXAMPLE 1.19

A tank contains 100 L of water at 20°C. Estimate the heat that must be added to the water to raise its temperature to 25°C.

SOLUTION

From the given data: V = 100 L = 0.1 m^3, T_1 = 20°C, and T_2 = 25°C. At 20°C, the density of water is 998.2 kg/m^3. Assume that the specific heat of the water can be taken as c = 4.18 kJ/kg·K. Because the mass of water in the tank is equal to ρV, the heat, H, that must be added to raise the temperature is given by

$$H = \rho V c(T_2 - T_1) = (998.2)(0.1)(4.18)(25 - 20) = 2086 \text{ kJ} \approx \mathbf{2.09\ MJ}$$

1.9.2 Latent Heat

Latent heat refers to the energy per unit mass that is absorbed or released by a fluid upon a change in phase of the fluid at a constant temperature and pressure. The *latent heat of vaporization*, L_v, is the amount of heat required to convert a unit mass of a fluid from the liquid to the vapor phase at a given temperature [EM^{-1}]. This quantity is also called the *heat of vaporization* or the *enthalpy of vaporization*, with the latter term more commonly used in the field of thermodynamics. For water, the latent heat of vaporization ranges from 2.499 MJ/kg at 0°C to 2.256 MJ/kg at 100°C, with a value of 2.452 MJ/kg at 20°C. Values of the latent heat of vaporization as a function of temperature are given in Appendix B.1. The latent heat of vaporization is widely used in applications relating to evaporation and plant transpiration, where the net solar and long-wave energy fluxes at the surface of Earth, in J/(m^2·d), result in a mass flux of liquid terrestrial water to water vapor in the atmosphere. The latent heat released into the atmosphere when water condenses is the primary source of the power of thunderstorms and hurricanes. The *latent heat of fusion*, L_f, is the amount of heat required to convert a unit mass of a solid to a unit mass of liquid at the melting point [EM^{-1}]. The latent heat of fusion is also called the *enthalpy of fusion*. The latent heat of fusion of water at 0°C is 334 J/kg. For any substance, the latent heat of fusion is generally less than the latent heat of vaporization.

EXAMPLE 1.20

The net solar and long-wave energy flux measured during July in Miami, Florida, averages 19 MJ/(m^2·d). Estimate the resulting evaporation from an open water body.

SOLUTION

From the given data: E = 19 MJ/(m^2·d). For water at 20°C, L_v = 2.452 MJ/kg and ρ = 998 kg/m^3. The energy required to evaporate 1 m^3 of water at 20°C is equal to ρL_v, where

$$\rho L_v = (998)(2.452) = 2450 \text{ MJ/m}^3$$

The amount of evaporation resulting from the energy flux, E, is therefore given by $E/\rho L_v$, where

$$\frac{E}{\rho L_v} = \frac{19}{2450} \text{ m/d} = 7.76 \text{ mm/d}$$

The total evaporation in July from an open water body in Miami can therefore be estimated as 7.76 mm/d × 31 d = **241 mm.**

1.10 Summary of Properties of Water and Air

Many engineering applications require utilization of the properties of water and air. In cases where temperature changes are not a significant factor, a temperature of 20°C is commonly assumed. The corresponding properties of water and air are given in Table 1.9.

Table 1.9: Summary of Physical Properties of Water and Air*

Property	Symbol	Units	Water	Air
Density	ρ	kg/m^3	998	1.225
Specific weight	γ	kN/m^3	9.79	0.0120
(Dynamic) viscosity	μ	N·s/m^2	1.00×10^{-3}	0.0182×10^{-3}
(Kinematic) viscosity	ν	m^2/s	1.00×10^{-6}	–
Bulk modulus	E_v	kPa	2.15×10^6	–
Surface tension	σ	N/m	72.8×10^{-3}	N/A
Saturation vapor pressure	e_s	kPa	2.34	N/A
Heat of vaporization	L_v	MJ/kg	2.453	–
Heat of fusion (at 0°C)	L_f	MJ/kg	0.334	–
Specific heat at constant pressure	c_p	J/(kg·K)	4180	1000
Specific heat at constant volume	c_v	J/(kg·K)	4180	715
Ratio of specific heats	k	–	1.0	1.4
Gas constant	R	J/(kg·K)	N/A	287.1

*At 20°C and 101.3 kPa where applicable.

Key Equations in Properties of Fluids

The following list of equations is particularly useful in solving problems related to the physical properties of fluids. If one is able to recognize these equations and recall their appropriate use, then the learning objectives of this chapter have been met to a significant degree.

DENSITY

Definition: $$\rho = \frac{\text{mass of substance}}{\text{volume of substance}}$$

Water: $$\rho_{\text{w}} = 1000 - 0.0178\,|T - 4|^{1.7}$$

Salinity: $$S = \frac{\text{weight of dissolved salt}}{\text{weight of mixture}}$$

Specific weight: $$\gamma = \rho g = \frac{\text{weight of substance}}{\text{volume of substance}}$$

Specific gravity: $$\text{SG} = \frac{\rho}{\rho_{\text{H}_2\text{O@4°C}}} = \frac{\text{density of liquid}}{\text{density of water at 4°C}}$$

Specific volume: $$v = \frac{1}{\rho} = \frac{\text{volume of substance}}{\text{mass of substance}}$$

COMPRESSIBILITY

Bulk modulus of elasticity: $$E_{\text{v}} = -\frac{\text{d}p}{\text{d}V/V} \approx -\frac{\Delta p}{\Delta V/V}; \qquad E_{\text{v}} = \frac{\text{d}p}{\text{d}\rho/\rho}$$

Speed of sound: $$c = \sqrt{\frac{\text{d}p}{\text{d}\rho}} = \sqrt{\frac{E_{\text{v}}}{\rho}}$$

DENSITY CHANGES

With temperature and pressure: $$\frac{\Delta\rho}{\rho} \approx \alpha\Delta p - \beta\Delta T; \qquad \alpha = \frac{1}{E_{\text{v}}} = \frac{\text{d}\rho/\rho}{\text{d}p}; \quad \beta = -\frac{\text{d}\rho/\rho}{\text{d}T}$$

PROPERTIES OF IDEAL GASES

Ideal gas law: $$pV = nR_{\text{u}}T; \qquad \rho = \frac{p}{RT}$$

General process equations: $$pv^n = p_1v_1^n = \text{constant} \quad \text{or} \quad \frac{p}{p_1} = \left(\frac{\rho}{\rho_1}\right)^n = \text{constant}$$

$$\frac{T_1}{T_2} = \left(\frac{v_1}{v_2}\right)^{n-1} = \left(\frac{\rho_2}{\rho_1}\right)^{n-1} = \left(\frac{p_2}{p_1}\right)^{\frac{n-1}{n}}$$

Isothermal, reversible process: $$Q_{12} = nR_{\text{u}}T\ln\frac{V_2}{V_1}$$

Specific heat ratio: $k = \frac{c_p}{c_v}$

Gas constant: $R = c_p - c_v$

Bulk modulus: $E_\text{v} = np$

Speed of sound: $c = \sqrt{RTk}$

VISCOSITY

Newton's equation of viscosity: $\tau = \mu \frac{\mathrm{d}u}{\mathrm{d}y}$

Kinematic viscosity: $\nu = \frac{\mu}{\rho}$

Andrade's equation (liquids): $\mu = a \cdot 10^{\frac{b}{T-c}}$ or $\mu = a \cdot \exp\left(\frac{b}{T-c}\right)$

Empirical equation (liquids): $\ln \frac{\mu}{\mu_0} = a + b\left(\frac{T_0}{T}\right) + c\left(\frac{T_0}{T}\right)^2$

Sutherland equation (gases): $\mu = \frac{dT^{\frac{3}{2}}}{T+e}$ or $\frac{\mu}{\mu_0} = \left(\frac{T}{T_0}\right)^{\frac{3}{2}} \frac{T_0+e}{T+e}$

Empirical equation (gases): $\frac{\mu}{\mu_0} = \left(\frac{T}{T_0}\right)^n$

Non-Newtonian fluids: $\tau = K\left(\frac{\mathrm{d}u}{\mathrm{d}y}\right)^n = \left[K\left|\frac{\mathrm{d}u}{\mathrm{d}y}\right|^{n-1}\right]\frac{\mathrm{d}u}{\mathrm{d}y}$ or $\tau = \mu_\text{ap}\frac{\mathrm{d}u}{\mathrm{d}y}$

SURFACE TENSION

Definition: $\sigma = \frac{F}{L}$

Droplet: $p_1 - p_2 = \frac{2\sigma}{R}$

Bubble: $p_1 - p_2 = \frac{4\sigma}{R}$

Capillary rise: $h = \frac{2\sigma \cos\theta}{\gamma r}$

VAPOR PRESSURE

Relative humidity: $\text{RH} = \frac{e}{e_\text{s}} \times 100$

Cavitation number: $\text{Ca} = \frac{p - p_\text{v}}{\frac{1}{2}\rho V^2}$

Applications of the above equations are provided in the text, and additional problems to practice using these equations can be found in the following section.

PROBLEMS

Section 1.1: Introduction

1.1. The lift force, F_L [F], exerted on an object with a plan area A [L^2] by a fluid with an approach velocity V [LT^{-1}] and density ρ [ML^{-3}] is usually derived using the relation

$$F_L = C_L \tfrac{1}{2}\rho V^2 A$$

where C_L is an empirical constant called the *lift coefficient*. (a) What are the units of C_L if standard SI units are used for F_L, ρ, V, and A? (b) What adjustment factor would be applied to C_L if standard USCS units were used for F_L, ρ, V, and A?

1.2. A force balance in a particular fluid flow is combined with Newton's second law to yield the equation

$$\rho\frac{d^2z}{dt^2} + a\frac{dz}{dt} + bz = c$$

where ρ, z, and t are dimensional variables with the following dimensions: ρ [ML^{-3}], z [L], and t [T]. (a) Determine the dimensions of the system parameters a, b, and c. (b) If standard SI units are to be used in the given equation and values of ρ, z, and t are provided in g/cm^3, mm, and h, respectively, what conversion factors must be applied to these variables before they are used in the equation?

1.3. An equation that is commonly used to describe the volume flow rate of water in an open channel is given by

$$Q = \frac{1}{n}\frac{A^{\frac{5}{3}}}{P^{\frac{2}{3}}}S_0^{\frac{1}{2}}$$

where Q is the volume flow rate in the channel [L^3T^{-1}], n is a constant that characterizes the roughness of the channel surface [dimensionless], A is the flow area [L^2], P is the perimeter of the flow area that is in contact with the channel boundary [L], and S_0 is the slope of the channel [dimensionless]. This equation is usually applied using SI units, where Q is in m^3/s, A is in m^2, and P is in m. (a) Is the given equation dimensionally homogeneous? (b) If the equation is not dimensionally homogeneous, what conversion factor must be inserted after the equal sign for the equation to work with Q in ft^3/s, A in ft^2, and P in ft?

1.4. Give the dimensions and typical SI units of the following quantities commonly used in engineering: energy, force, heat, moment, momentum, power, pressure, strain, stress, and work.

1.5. Use prefixes to express the following quantities with magnitudes in the range of 0.01–1000: (a) 6.27×10^7 N, (b) 7.28×10^5 Pa, (c) 4.76×10^{-4} m^2, and (d) 8.56×10^5 m.

1.6. Convert the following quantities to SI units: 15 gallons per minute, 99 miles per hour, 20 feet per second, 150 cubic feet per minute, 1540 gallons, 28 acres, and 600 horsepower.

1.7. Derive the following conversion factors: (a) horsepower to watts and (b) pounds per square inch to pascals.

Section 1.2: Density

1.8. What is the largest percentage error in the density of liquid water that can be made if the density is always assumed to be equal to 990 kg/m^3?

1.9. If 4 L of a liquid with density 1020 kg/m^3 is mixed with 6 L of a liquid with density 940 kg/m^3, what is the density of the mixture?

1.10. (a) What is the specific weight of water at 0°C, 20°C, and 100°C? (b) What is the specific gravity of water at 0°C, 20°C, and 100°C?

1.11. A 450-L storage tank is completely filled with water at 25°C. (a) If the top of the storage tank is left open to the atmosphere and the water in the tank is heated to 80°C, what volume of water will spill out of the tank? (b) If the water is cooled back down to 25°C, by what percentage will the weight of water in the tank be reduced from its original weight? Neglect the expansion of the tank when the water is heated.

1.12. If the specific weight of a substance is 15 kN/m^3, what is its density and specific gravity?

1.13. If the specific gravity of a substance is 2.5, what is its density and specific weight?

1.14. The concentration of a dissolved or suspended substance in a liquid is commonly expressed as the ratio of the mass of the substance to the mass of the mixture of the liquid and the substance. The concentration unit of parts per million (ppm) is an example of such a ratio. Determine the relationship between the mass ratio, the specific gravity of the pure liquid, and the specific gravity of the mixture.

1.15. If the density of a substance is 810 kg/m^3, what is its specific gravity and its specific weight?

1.16. A storage reservoir contains 250 kg of a liquid that has a specific gravity of 2. What is the volume of the storage reservoir?

1.17. An empty container weighs 10 N, and when filled with kerosene at 20°C, it weighs 50 N. Estimate the volume of kerosene required to fill the container. What is the mass of kerosene required to fill the container?

Section 1.3: Compressibility

1.18. Show that the relationship between the bulk modulus and the fractional change in density, as given by Equation 1.13, can be derived from the definition of the bulk modulus in terms of the volumetric strain, as given by Equation 1.12.

1.19. What magnitude of pressure change would be necessary to change the density of water by 1% at 20°C?

1.20. Water at 20°C is pumped into a spherical tank of diameter 3 m, and pumping is stopped when the water just fills the tank. The tank manufacturer claims that the tank can withstand an internal gauge pressure of up to 9 MPa and that the deformation of the tank is negligible for pressures up to 9 MPa. What additional mass of water could be pumped into the tank and still maintain the integrity of the tank?

1.21. As the pressure in a liquid reservoir is increased from 150 kPa to 28000 kPa, the volume of liquid in the reservoir decreases from 1.500 m^3 to 1.450 m^3. Estimate the bulk modulus of the liquid.

1.22. The pressure on 10 m^3 of benzene at 20°C is increased by 10 MPa. Estimate the expected change in the volume of benzene.

1.23. Water at 20°C is poured into a pot and brought to a boiling point at 100°C at a constant atmospheric pressure of 101.3 kPa. (a) Use the coefficient of volume expansion to estimate the percentage change in the density of the water. (b) Estimate the percentage change in the depth of water in the pot.

1.24. Ethylene glycol, which is commonly used as a coolant in car radiators, has a coefficient of volume expansion of around 5.7×10^{-4} K^{-1}. Estimate the percentage change in density when the temperature of ethylene glycol is raised from 10°C to 90°C. Assume that the pressure remains constant.

1.25. A plastic container is completely filled with gasoline at 15°C. The container specifications indicate that it can endure a 1% increase in volume before rupturing. Within the temperatures likely to be experienced by the gasoline, the average coefficient of volume expansion of gasoline is 9.5×10^{-4} K^{-1}. Estimate the maximum temperature rise that can be endured by the gasoline without causing a rupture in the storage container.

1.26. Compare the speed of sound in mercury with the speed of sound in air. Assume that both liquids are at 20°C.

1.27. Measurements indicate that the acoustic velocity in a dense liquid is 1800 m/s. If the specific gravity of the liquid is 1.9, what is the bulk modulus of the liquid?

Section 1.4: Ideal Gases

1.28. The limiting condition for the continuum approximation to be applicable in an ideal gas is sometimes defined as the condition at which the gas contains less than 10^{10} molecules per mm^3. (a) For an ideal gas in which the temperature is 15°C, what will be the pressure at the limiting condition? Take into consideration that Avogadro's number is 6.023×10^{23} molecules/mole. (b) What term is used to describe a gas in which the continuum approximation is not valid?

1.29. Compare the density of helium to the density of air at a pressure of 101 kPa and a temperature of 25°C. What are the specific volumes of these two gases at the given temperature and pressure?

1.30. Air is commonly assumed to be an ideal gas; in addition, the properties of air are frequently obtained from tabulated values. Compare the densities of air at temperatures in the range of −40 to 1000°C at standard atmospheric pressure as given in Appendix B.2 with the densities estimated using the ideal gas law. Based on your results, assess the accuracy of estimating the air density at atmospheric pressure using the ideal gas law.

1.31. If pure oxygen is compressed such that its density and pressure are 5 kg/m^3 and 450 kPa, respectively, estimate the temperature of the oxygen.

1.32. A steel tank is filled with helium at a temperature of 17°C and a pressure of 550 kPa. If the tank has a volume of 3 m^3, what is the mass, density, and weight of helium in the tank?

1.33. A steel tank is to be sized to contain 12 kg of pure oxygen at a temperature of 27°C and a pressure of 15 MPa. It is desired that the tank have a cylindrical shape, with a length that is three times its diameter. What are the required dimensions of the tank?

1.34. The mass of air in a tank is estimated as 12 kg, and the temperature and pressure of the air in the tank are measured as 67°C and 210 kPa, respectively. Estimate the volume of air in the tank.

1.35. The mass of compressed air in a 210-L steel tank is determined by weighing the tank plus its contents and then subtracting the known weight of the empty tank. The mass of compressed air in the tank is found to be equal to 3.2 kg. If the tank is located in a room where the temperature is maintained at 25°C, estimate the pressure of the air in the tank.

1.36. A 0.1-m^3 tank is filled with air at a temperature and pressure of 20°C and 400 kPa, respectively. What is the weight of air in the tank?

1.37. A large classroom is 10 m wide and 12 m long, and the distance from the floor to the ceiling is 4 m. The air in the room is at standard atmospheric pressure, and the temperature in the room is 20°C. (a) Estimate the mass of air (in kg) and the weight of air (in kN) in the room. (b) If the room is cooled down to 10°C, what is the percentage change in the mass of air in the room?

1.38. A steel tank contains air at a temperature and pressure of 20°C and 600 kPa, respectively. If the temperature of the air in the tank increases to 30°C, what is the change in pressure? What would be the change in pressure if the gas in the tank were helium?

1.39. A car tire has low air pressure at 130 kPa, and the pressure needs to be raised to 210 kPa. If the volume of the tire is 15 L and the temperature of the air in the tire is 30°C, estimate the mass of air that needs to be added. Assume that the air temperature and the volume of the tire remain constant.

1.40. A car tire has a volume of 20 L and has a recommended (gauge) inflation pressure of 210 kPa at a temperature of 25°C. (a) If driving on the highway on a hot day causes the temperature of the air in the tire to increase to 65°C, what will be the (gauge) air pressure in the tire? (b) If under the condition in part (a) air is let out of the tire to restore the tire pressure to 210 kPa, what will be the (gauge) air pressure in the tire when the air cools down to 25°C? Assume that the tire volume remains constant and that atmospheric pressure is 101.3 kPa. (Note: Tire pressures are stated as gauge pressures, and the absolute pressure is equal to the gauge pressure plus the atmospheric pressure.)

1.41. The National Football League (NFL) requires that footballs used in games be inflated to a pressure in the range of 86.2–93.1 kPa. In a 2015 playoff game between the New England Patriots and the Indianapolis Colts, the footballs used in the game were inflated in a room with an air temperature of 20°C; at game time, the temperature on the field was recorded as 10°C. Upon checking the footballs prepared by the New England Patriots, the referees found that the pressures in the footballs were only around 72.4 kPa. Atmospheric pressure can be assumed to have been approximately 101.4 kPa. (a) What is the minimum pressure that would be expected in the footballs if only the temperature change is taken into account? (b) What on-field temperature would be required for the pressure in the footballs to be 72.4 kPa? (c) Propose a theory of how the pressures in the footballs ended up being 72.4 kPa. Assume that the volume of a football does not change significantly for pressures in the range of 72.4–93.1 kPa.

1.42. Before embarking on a 40-minute drive on the highway, a motorist adjusts his tire pressure to 207 kPa. At the end of his trip, the motorist measures his tire pressure as 241 kPa. Assume that the volume of the tire remains constant during the trip. (a) Estimate the percentage increase in the temperature of the air in the tire. (b) If the initial temperature of the air in the tire is assumed to be equal to the ambient air temperature of 25°C, what is the estimated temperature of the air in the tire at the end of the trip.

1.43. (a) A spherical balloon with a diameter of 8 m is filled with helium at 22°C and 210 kPa. Determine the number of moles and the mass of helium in the balloon. (b) When the air temperature in an automobile tire is 27°C, the pressure gauge reads 215 kPa. If the volume of the tire is 0.030 m^3, determine the pressure rise in the tire when the air temperature in the tire rises to 53°C. (Note: Volume of a sphere is $\pi D^3/6$; relative molecular mass of He is 4.003; universal gas constant, R_u, is 8312 J/kmol·K.)

1.44. A helium balloon is to be used to lift a lab rat that weighs 1.5 N. The density of the air is 1.17 kg/m^3, atmospheric pressure is 100 kPa, the air temperature is 25°C, and the balloon weighs 0.5 N. What mass of helium (in kilograms) must be put in the balloon to lift the rat?

1.45. Consider the case of an air bubble released from the bottom of a 12-m-deep lake as shown in Figure 1.21. The bubble is filled with air, it has an initial diameter of 6 mm, no air is lost or gained in the bubble as it rises, and the air in the bubble and the surrounding lake water maintain a constant temperature of 20°C. Atmospheric pressure, p_atm, on the surface of the lake is 101.3 kPa, and the pressure, p, at any depth, z, below the surface of the lake is given by $p = p_\mathrm{atm} + \gamma z$, where γ is the specific weight of the water in the lake. Estimate the diameter of the bubble when it surfaces.

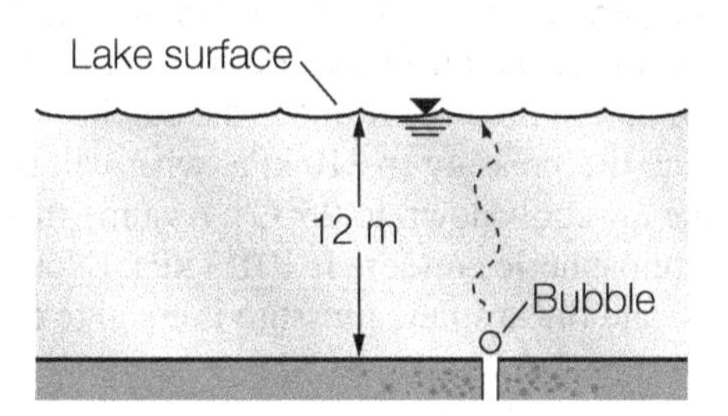

Figure 1.21: Bubble rising in a lake

1.46. A 1.1-m^3 volume of air at a pressure of 101 kPa is compressed to a volume of 0.45 m^3. (a) What is the pressure in the compressed volume if the compression process is isentropic? (b) What is the pressure if the compression process is isothermal?

1.47. The atmospheric pressure in dry air at 20°C is 101.3 kPa, and a sample of the air indicates that it is 20% oxygen (O_2) and 80% nitrogen (N_2) by volume. (a) Estimate the partial pressure of O_2 and N_2 in the air. (b) Estimate the density of the air.

1.48. A 2.0-m^3 volume of pure oxygen at a temperature and pressure of 20°C and 100 kPa, respectively, is expanded to a volume of 4.0 m^3. (a) What is the pressure in the expanded volume if the expansion process is isentropic? What are the initial and final densities of the gas? (b) What is the pressure in the expanded volume if the expansion process is isothermal? What are the initial and final densities of the gas? (c) For the expansion process described in part (b), what amount of heat must be added to the gas?

1.49. Air flows into the nozzle shown in Figure 1.22, where the inflow air has a temperature and pressure of 27°C and 101 kPa, respectively, and the outflow air has a temperature of −73°C. Flow within the nozzle can be assumed to be adiabatic and frictionless. Estimate the pressure and density of the air exiting the nozzle.

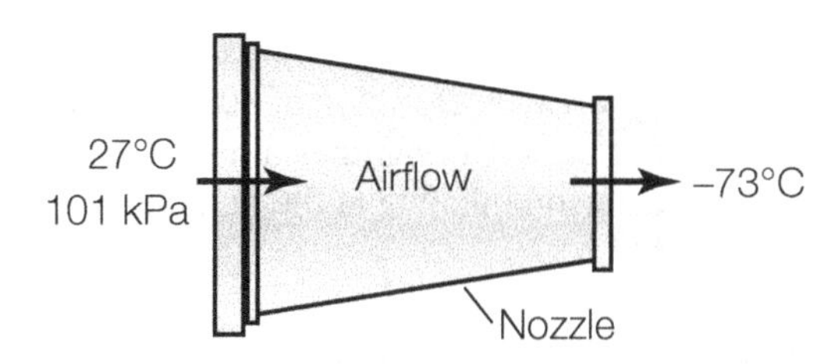

Figure 1.22: Flow of air through a nozzle

1.50. A cylinder contains 0.3 m^3 of air at 20°C and 120 kPa pressure. With a piston mechanism, the air in the cylinder is compressed isentropically to a pressure of 700 kPa. What is the temperature of the air after compression?

1.51. A siren emits sound waves to alert people in the surrounding community of the occurrence of a tornado. If the temperature of the air is 22°C, approximately how long does it take the sound to travel 1.1 km?

1.52. Compare the speed of sound in pure hydrogen with the speed of sound in air. Assume that both gases are at 20°C.

1.53. Remote measurements of the properties of a mystery gas find that the molecular weight of the gas is 35 g/mol and the specific heat of the gas at constant pressure is 1025 J/kg·K. Estimate the speed of sound in this gas at a temperature of 22°C. Assume that the behavior of the gas can be approximated by that of an ideal gas.

Section 1.5: Standard Atmosphere

1.54. Use the standard atmosphere to estimate the temperature and pressure at the top of Mount Everest, which is approximately 8840 m above sea level. Compare your estimated values with values reported in the open literature.

1.55. (a) Use the air density profile in the standard atmosphere as given in Appendix B.3 to estimate the weight of air above 1 m^2 of Earth's surface at sea level. (b) Use this result to estimate the pressure exerted on the surface of Earth by the standard atmosphere and compare this result with the conventional sea-level atmospheric pressure of 101.3 kPa. Explain the extent to which these pressures are in agreement. (c) If the total mass of air above 1 m^2 was placed in a rectangular box with a base area of

$1\ m^2$ and the air in the box was at standard sea-level temperature and pressure, what would be the height of the box?

1.56. The lowest pressures at sea level are generally associated with hurricanes and tornados, and the lowest recorded pressure generated by any of these natural phenomena is reported to be 87.06 kPa. What elevation within the standard atmosphere would have this same pressure?

1.57. The summit of Mount Rainier (in Washington) experiences an average high temperature of -7.2°C and an average low temperature of -15°C. Compare this range of temperatures with the temperature expected in the standard atmosphere. What is the expected atmospheric pressure on the summit of Mount Rainier?

1.58. Compare the mass of oxygen per unit volume of air at sea level with the corresponding value at a mountain resort at an elevation of 3000 m. If a person takes in the same volume of air with each breath and breathes at the same rate, what percentage reduction in oxygen is the person inhaling at the mountain resort? Assume a standard atmosphere.

1.59. A Boeing 777 aircraft has windows of dimension 270 mm $\times$ 380 mm, and the cabin pressure during flight is typically controlled at around 100 kPa. If the aircraft is cruising at an altitude of 11 km, estimate the net force on each window.

1.60. The Boeing 787 Dreamliner has a design cruising speed of 913 km/h at an altitude of 10 700 m. Assuming a standard atmosphere, what is the Mach number at which the aircraft flies under cruising conditions? Should the compressibility of air be taken into account in modeling the flight of this aircraft? Explain.

1.61. A research aircraft with sensitive external instrumentation has a design cruising speed of 885 km/h and must not fly with a Mach number greater than 0.85 to avoid compressibility effects compromising the functioning of the instrumentation. Estimate the maximum altitude at which this aircraft should cruise.

Section 1.6: Viscosity

1.62. Tabulate the kinematic viscosity of water, ν, as a function of temperature.

1.63. A fluid with a specific gravity of 0.92 has a kinematic viscosity of $5 \times 10^{-4}\ m^2/s$. What is the dynamic viscosity of the fluid?

1.64. Methane at a temperature of 25°C and pressure of 110 kPa has a dynamic viscosity of 12 μPa·s. Estimate the kinematic viscosity at this same temperature and pressure.

1.65. Benzene at a temperature of 20°C flows at a high speed over a smooth flat plate and exerts a shear stress of 0.5 Pa on the plate. Estimate the velocity gradient at the surface of the plate and the velocity 2 mm away from the surface.

1.66. A fluid with a viscosity of 0.300 Pa·s flows between parallel plates as illustrated in Figure 1.23. The top and bottom plates each have an area of 1.5 m^2, and the distance between the plates is 200 mm. The distribution of velocity within the fluid is given by

$$u = 0.8(1 - 100y^2)$$

where u is the velocity in m/s and y is the distance from the centerline in meters. (a) Determine the shear stress on the top and bottom plates. (b) Determine the shear forces exerted by the fluid on the top and bottom plates.

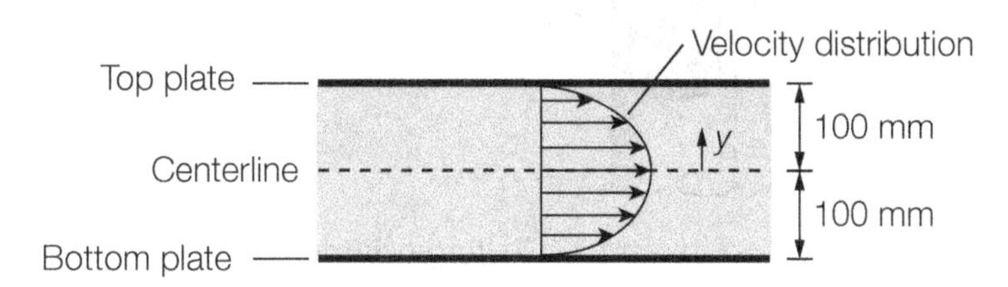

Figure 1.23: Fluid flow between parallel plates

1.67. A lubricant is contained between two concentric cylinders over a length of 1.3 m. The inner cylinder has a diameter of 60 mm, and the spacing between the cylinders is 0.6 mm. If the lubricant has a dynamic viscosity of 0.82 Pa·s, what force is required to pull the inner cylinder at a velocity of 1.7 m/s along its axial direction? Assume that the outer cylinder remains stationary and that the velocity distribution between the cylinders is linear.

1.68. A fluid is constrained between two 75-cm-long concentric cylinders, where the diameter of the inner cylinder is 15 cm and the diameter of the outer cylinder is 15.24 cm. The inner cylinder rotates at 200 rpm, and the viscosity of the fluid is 0.023 Pa·s. (a) Determine the force that is exerted on the outer cylinder. (b) Determine the torque and power needed to rotate the inner cylinder.

1.69. A viscometer is constructed with two 30-cm-long concentric cylinders, one 20.0 cm in diameter and the other 20.2 cm in diameter. A torque of 0.13 N·m is required to rotate the inner cylinder at 400 rpm. Calculate the viscosity of the fluid.

1.70. SAE 30 oil at 20°C is contained within the 2-mm space between a rotating cylinder and a fixed cylindrical reservoir as shown in Figure 1.24. If the 0.5-m-diameter rotating cylinder is required to turn at a rate of 3 rpm, what torque is required to rotate the cylinder?

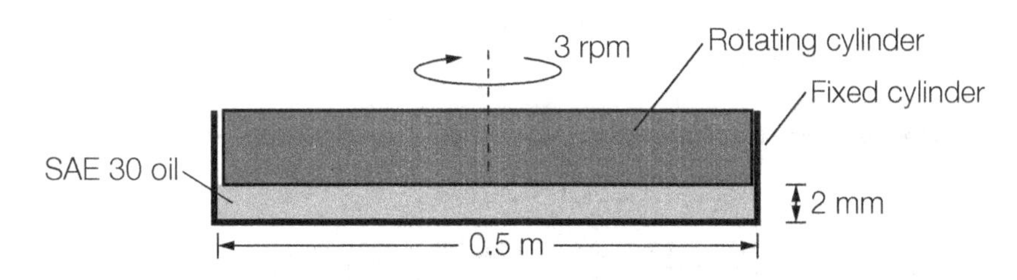

Figure 1.24: Rotating cylinder

1.71. A cylinder with a mass of 0.8 kg, a diameter of 50 mm, and a length of 10 cm is to be dropped down a shaft of diameter 53 mm as shown in Figure 1.25. The annular region between cylinders is filled with SAE 30 oil having a viscosity of 0.29 Pa·s. The inner cylinder is dropped with an initial velocity of zero and ultimately achieves a constant speed called the *terminal speed.* (a) Determine the terminal speed. (b) Determine the time required to attain the terminal speed.

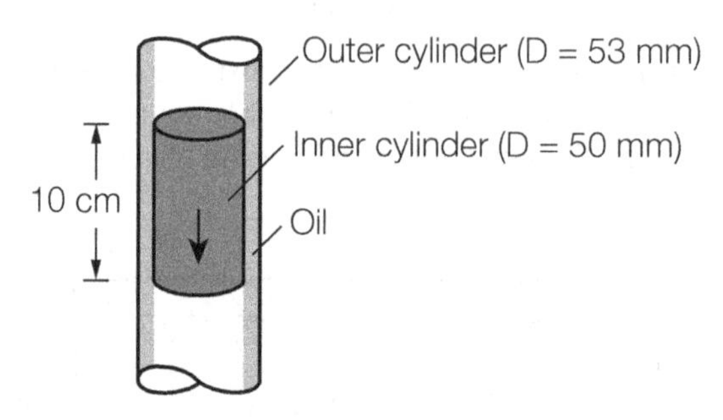

Figure 1.25: Falling cylinder

1.72. A 0.4-mm-diameter cable is to be pulled through a 1.2-mm-diameter glycerin-filled cavity as shown in Figure 1.26. The glycerin to be used has a dynamic viscosity of 1.5 Pa·s, and the cable can resist a 84-N tensile force before it begins to fail. If the cable is to be pulled at a velocity of 1.3 m/s, what is the maximum length of the cavity that can be used to avoid tensile failure in the cable? Assume that the cable can be kept at the center of the cavity.

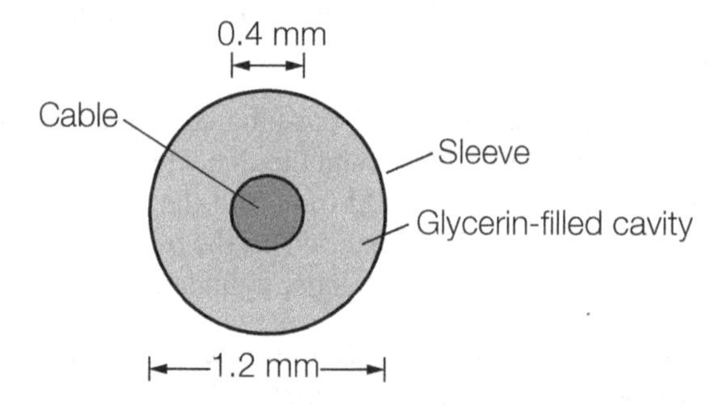

Figure 1.26: Pulling a cable through a glycerin-filled cavity

1.73. A flat plate with an area of 1.7 m^2 is moved at a velocity of 1.5 m/s over a stationary plate as shown in Figure 1.27. Two lubricants are contained in the 0.7-mm space between the plates. The bottom 0.4 mm contains a lubricant with a dynamic viscosity of 0.2 Pa·s, and the top 0.3 mm contains a lubricant with a dynamic viscosity of 0.3 Pa·s. (a) What is the velocity at the interface between the two lubricants? (b) What force is required to move the top plate?

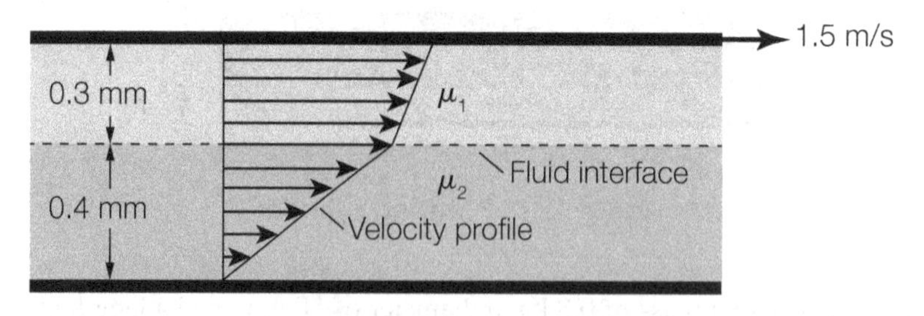

Figure 1.27: Two lubricants between flat plates

1.74. A thin plate is pulled at a speed of 9 m/s between two fixed stationary plates as shown in Figure 1.28. Lubrication for the moving plate is provided by SAE 50 oil at 20°C. The moving plate is 1 m long and 1.28 m wide, and the spacings between the top and bottom plates are 40 mm and 25 mm, respectively. Estimate the force required to move the plate.

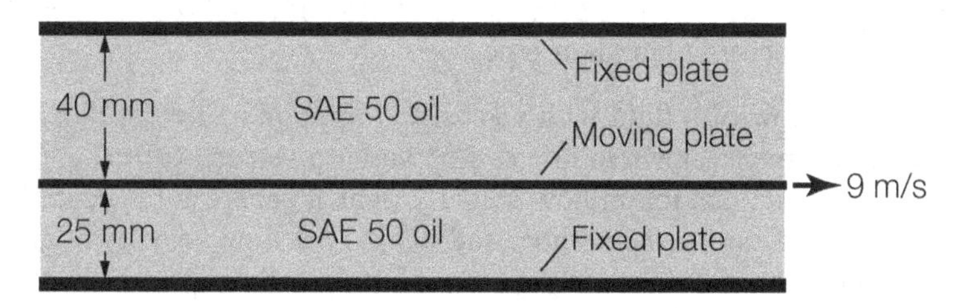

Figure 1.28: Moving plate between two fixed plates

1.75. Water at 20°C flows down the inclined channel shown in Figure 1.29, where the velocity distribution is given by

$$u = 1.2y(1 - y)$$

where u is the velocity in m/s and y is the distance from the bottom of the channel in meters. What is the magnitude of the shear stress exerted by the flowing water on the bottom of the channel?

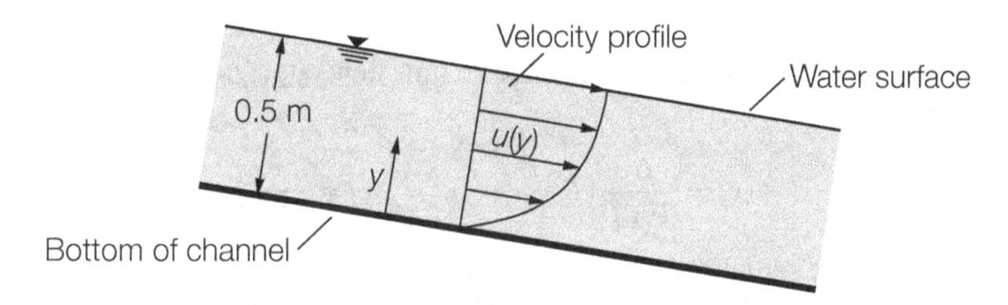

Figure 1.29: Water flowing down an open channel

1.76. A block of weight W slides down an inclined plane that is lubricated by a thin film of oil as shown in Figure 1.30. The film contact area is A, and its thickness is h. Assuming a linear velocity distribution in the film, derive an expression for the terminal (zero-acceleration) speed, V, of the block. Find the terminal speed of the block if the block mass is 6 kg, A = 35 cm^2, θ = 15°, and the film is 1-mm-thick SAE 30 oil at 20°C.

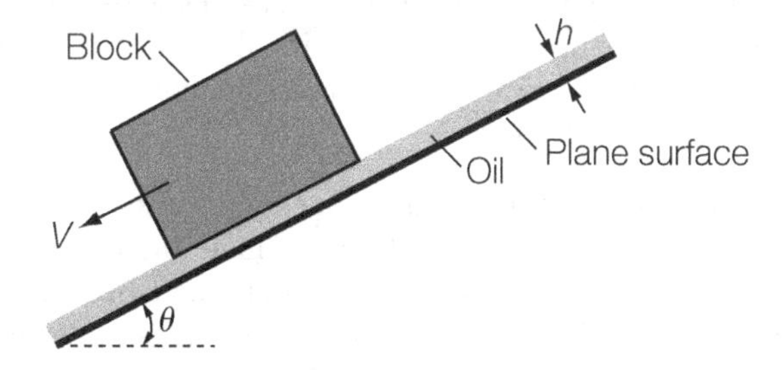

Figure 1.30: Weight on a lubricated plane

1.77. If the velocity distribution in a pipe of radius R is given by

$$V(r) = V_0 \left(1 - \frac{r^2}{R^2}\right)$$

where V_0 is the velocity at the pipe centerline and r is the distance from the center of the pipe, find (a) the shear stress as a function of r and (b) the shear force on the pipe boundary per unit length of pipe.

1.78. Consider the flow of a fluid with viscosity μ through a circular pipe. The velocity profile in the pipe is shown in Figure 1.31 and is given as $u(r) = u_{\text{max}}(1 - r^n/R^n)$, where u_{max} is the maximum flow velocity (which occurs at the centerline), r is the radial distance from the centerline, and $u(r)$ is the flow velocity at any position r. Develop an expression for the drag force exerted on the pipe wall by the fluid in the flow direction per unit length of pipe.

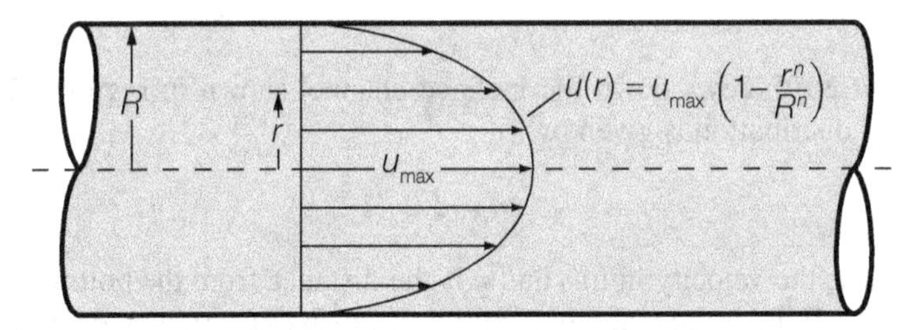

Figure 1.31: Velocity distribution

1.79. The velocity distribution, $u(r)$, and the volume flow rate, Q, of a fluid through a pipe are given by

$$u(r) = \frac{\Delta p}{16\mu L}(D^2 - 4r^2), \qquad Q = \frac{\pi}{128\Delta p}\mu L D^4$$

where r is the radial distance from the center of the pipe, Δp is the pressure drop over a distance L along the pipe, μ is the dynamic viscosity of the fluid, and D is the diameter of the pipe. Consider a pipe of diameter 1 cm that is designed to transport SAE 10 oil. If the volume flow rate is maintained at a fixed value, what would be the percentage difference in pressure drop per unit length of pipe if SAE 30 oil was used instead of SAE 10 oil? What would be the percentage change in shear stress per unit length?

1.80. (a) Use Andrade's equation to estimate the viscosity of water at standard atmospheric pressure for temperatures in the range of 0–100°C and compare these estimates with the tabulated values given in Appendix B.1. What is the maximum percentage difference between the estimated and tabulated viscosities within this temperature range? (b) Repeat part (a) using the alternative empirical expression given by Equation 1.51. What equation would you recommend for estimating the viscosity of water in the temperature range of 0–100°C?

1.81. The dynamic viscosity of pure nitrogen is estimated to be 16.40 μPa·s at 0°C and 20.94 μPa·s at 100°C. Use these data to estimate the dynamic viscosity at 50°C using (a) linear interpolation, (b) the Sutherland equation, and (c) the power-law relationship. (d) From the tabulated value of the dynamic viscosity of nitrogen at 50°C given in Appendix B.6, identify which of the estimation methods used in parts a–c yields the most accurate estimate of the dynamic viscosity at 50°C.

Section 1.7: Surface Tension

1.82. Would you expect that the surface tension of water in contact with air is the same as the surface tension of water in contact with pure oxygen? Explain.

1.83. A steel pin is supported on a water surface as shown in Figure 1.32. Find the relationship between the deflection angle, θ, and the weight of the pin. If the maximum deflection angle that can be sustained is 10°, what is the maximum pin size (i.e., diameter, length) that can be supported by the water?

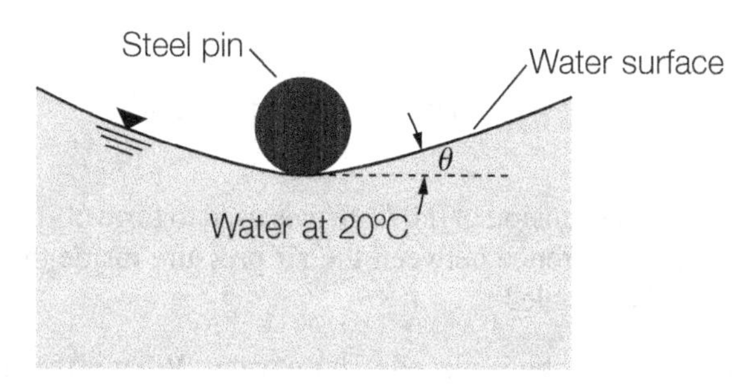

Figure 1.32: Steel pin on water surface

1.84. Solid spheres of varying sizes can be supported to varying extents on the surface of a liquid, depending on the weight and size of the sphere and the surface tension of the liquid. Consider the case where the liquid is water at 20°C. (a) What diameter lead sphere can be supported on the surface of the water? (b) What diameter concrete sphere can be supported?

1.85. Raindrops generally vary in size from 0.5–4 mm. Determine the corresponding range of increased pressure of water within a raindrop compared with the (atmospheric) pressure outside the raindrop. Assume that the temperature of the water in a raindrop is 20°C.

1.86. Using a spray nozzle, small droplets of glycerin are formed in the atmosphere. The droplets have a diameter of 0.6 mm, and the temperature of glycerin is 20°C. If atmospheric pressure is taken as 101.3 kPa, what is the absolute pressure within the droplets?

1.87. The free-body diagram of a half-droplet of radius R is shown in Figure 1.33. In this case, p_1 is the pressure at the center of the droplet, p_2 is the pressure outside the droplet, σ is the surface tension at the liquid-air interface, and $\mathbf{W}$ is the weight of the liquid in the half-droplet. (a) Show that the pressure difference $p_1 - p_2$ is given by

$$p_1 - p_2 = \frac{2\sigma}{R} - \frac{2}{3}\gamma R \tag{1.67}$$

where γ is the specific weight of the liquid. Contrast this result with the conventional pressure-difference equation for liquid droplets in air. (b) A manufacturing process utilizes 1.5-mm-diameter droplets of SAE 30 oil at 20°C in air. Contrast the pressure difference between the air and the center of the droplet calculated using the conventional equation with that calculated using Equation 1.67. Assess whether neglecting the weight of the liquid is justified in this case.

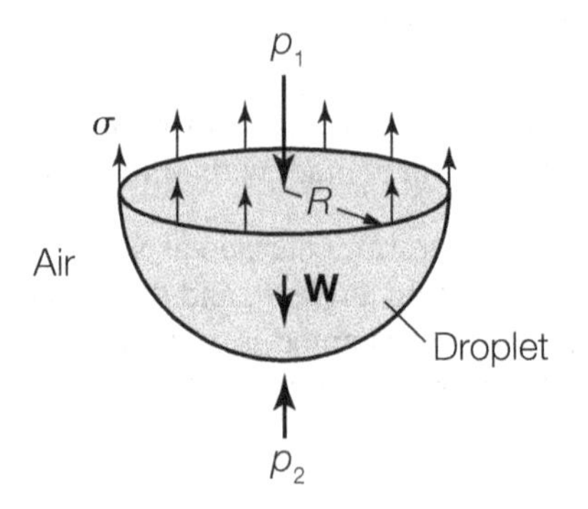

Figure 1.33: Half-droplet free body

1.88. A liquid with a surface tension of 0.072 N/m is used to form 60-mm-diameter bubbles in air. What is the difference between the air pressure inside the bubble and the air pressure outside the bubble?

1.89. A clean glass capillary tube is placed in pure water. What is the smallest diameter of the capillary tube that can be used such that the capillary rise will be no more than 6 mm?

1.90. A 1.5-mm-diameter capillary tube is inserted in a liquid, and it is observed that the liquid rises 15 mm in the tube and has a contact angle of 15° with the surface of the glass tube. If a hydrometer indicates that the liquid has a specific gravity of 0.8, what is the surface tension of the liquid? Would you expect that this same surface tension would be found if the experiment was done using a tube material other than glass? Explain.

1.91. A 1-mm-diameter glass capillary tube is inserted in a beaker of mercury at 20°C. Previous experimenters report that the contact angle between mercury and the glass material is 127°. What is the expected depth of depression of mercury in the capillary tube?

1.92. Determine the maximum diameter of a glass capillary tube that can be used to cause a capillary rise of benzene that exceeds four tube diameters. Assume a temperature of 25°C and assume that the contact angle of benzene on glass is approximately 15°.

1.93. Derive an expression for the capillary rise height h for a liquid of surface tension σ and contact angle θ between two vertical parallel plates a distance W apart as shown in Figure 1.34. What will h be for water at 20°C if $W = 0.5$ mm?

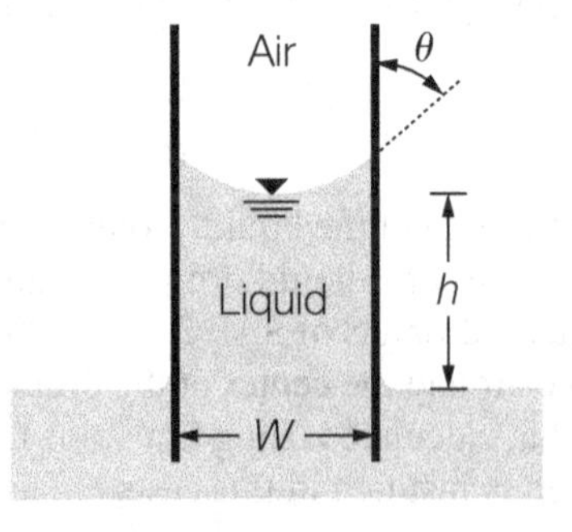

Figure 1.34: Capillary rise between plates

1.94. The pressure, p, in a liquid just below the meniscus in a capillary tube with a rise height, h, can be estimated by the hydrostatic pressure relation

$$p = p_0 - \gamma h \tag{1.68}$$

where p_0 is the pressure at the liquid surface in the source reservoir and γ is the specific weight of the liquid. Consider a capillary tube that is inserted in a reservoir of water at 20°C, where the water surface in the reservoir is exposed to an atmospheric pressure of 101.3 kPa. The surface tension of the water is 73 mN/m, and the contact angle of the water with the capillary-tube material is 5°. Given that the smallest workable diameter of the capillary tube is the diameter that would cause vaporization (i.e., cavitation) of the water at the meniscus of the liquid in the capillary tube, what is the limiting diameter of the capillary tube?

1.95. The difference in pressure across the meniscus of a liquid in a capillary tube can be estimated by the relation

$$p_0 - p = \frac{2\sigma}{R} \tag{1.69}$$

where p_0 is the pressure above the meniscus (usually atmospheric pressure), p is the pressure below the meniscus, σ is the surface tension of the liquid, and R is the radius of curvature of the meniscus. (a) Combine Equations 1.68 and 1.69 to determine the relationship between the curvature of the meniscus and the rise height in a capillary tube. (b) If water at 10°C has a rise height of 75 mm in a capillary tube, what is the estimated radius of curvature of the meniscus?

Section 1.8: Vapor Pressure

1.96. A storage tank is partially filled with gasoline at 20°C. If the air above the gasoline is pumped out using a vacuum pump, what is the minimum absolute pressure that can be attained in the open space above the gasoline?

1.97. (a) If air at 25°C is 25% oxygen and 75% nitrogen and atmospheric pressure is 101.3 kPa, estimate the density of air at 25°C. (b) If the air described in part (a) is compressed into a 1.5-m^3 tank with an absolute pressure of 210 kPa and temperature of 20°C, what is the weight of the air in the tank? (c) If the relative humidity of air is measured as 85% at 25°C, estimate the vapor pressure of water and the temperature at which water condensation will begin to occur.

1.98. The outside temperature and humidity in the morning are given as 25.6°C and 75%, respectively. If the moisture content of the air remains the same throughout the day, estimate the humidity at noon when the temperature is 32.2°C.

1.99. The temperature and humidity of outside air are 25°C and 80%, respectively, and the interior of a building is cooled such that condensation forms on the glass windows. Estimate the temperature inside the building.

1.100. What would be the boiling point of water in a high mountain area where the atmospheric pressure is 90 kPa?

1.101. Water at 50°C is poured into an open steel tank where the air above the water surface is at an atmospheric pressure of 101.3 kPa. The tank is sealed, and the air in the tank is evacuated with a vacuum pump. A pressure gauge installed in the tank measures the air pressure in the tank relative to atmospheric pressure. What is the gauge reading when the water in the tank begins to boil?

1.102. The pressure of water adjacent to any given location on a rotating propeller is inversely proportional to the speed of the propeller at that location. (a) Explain why cavitation adjacent to a propeller is more likely to occur near the tips of the propeller rather than near the hub of the propeller. (b) A study of a particular propeller indicates that the pressure near the tip of the propeller under operational conditions will be around 5 kPa when the temperature of the water is 20°C. Is cavitation likely to occur? Explain.

1.103. A group of nomads in a mountainous terrain do not want to settle at any location where the boiling point of water is less than 92°C. What is the highest elevation at which they should consider settling?

1.104. What minimum allowable pressure will prevent cavitation of water in a pipeline system that transmits water at 35°C? How is this minimum pressure requirement different from that in a pipeline containing gasoline at 20°C?

1.105. A pressure gauge on the suction side of a large cooling tower water pump shows an absolute pressure of 7 kPa when the pump is operating under normal conditions. What is the maximum allowable temperature of the water entering the pump that will prevent the occurrence of cavitation?

1.106. An underground storage tank is initially completely filled with gasoline at 20°C and sealed. If gasoline is withdrawn from the sealed tank (to fill up a car) and a space is created in the tank above the surface of the gasoline, what is the pressure in the space? What types of molecules are contained in the space?

1.107. A peanut farmer in south Georgia has drilled a well and found water 12 m below the ground. To use the water for irrigation, the farmer extends a pipe down the well and connects the pipe to the intake of a pump located on the ground surface. The temperature of the water in the well is 25°C. It is known from engineering analysis that the gauge pressure, p, at a height z above the water surface is given by

$$p = -\gamma\left(1 + 0.24\frac{Q^2}{gD^5}\right)z$$

where γ is the specific weight of the water, Q is the volume flow rate, and D is the diameter of the pipe. If atmospheric pressure is 101 kPa, Q is 50 L/min, and D is 50 mm, determine the height z above the water surface where the water will begin to vaporize (i.e. cavitate). Based on your result, will the farmer's irrigation system work? Explain.

1.108. Model tests on a supercavitating torpedo in seawater at 20°C indicate that the minimum pressure, p_{min}, on the surface of the torpedo is related to the speed, V, of the torpedo by the relation

$$p_{\text{min}} = 120 - 0.402V^2$$

where p_{min} is in kPa and V is in m/s. What is the minimum torpedo speed required for cavitation to begin on the surface of the torpedo?

Section 1.9: Thermodynamic Properties of Liquids

1.109. Estimate the evaporation rate of water that results from a net solar radiation of 20 MJ/(m^2·d). Assume that the temperature of the water body is 20°C.

1.110. If water in a lake evaporates at a rate of 4.06 mm/d without any change in temperature, estimate the net incident radiation. The lake temperature is 15°C.

1.111. Thunderstorms occur when water in the atmosphere condenses in a cloud to form raindrops. Estimate the rate at which energy is supplied to a thunderstorm when water condenses at the rate of 10 kg/s. The air temperature in the cloud is 5°C.

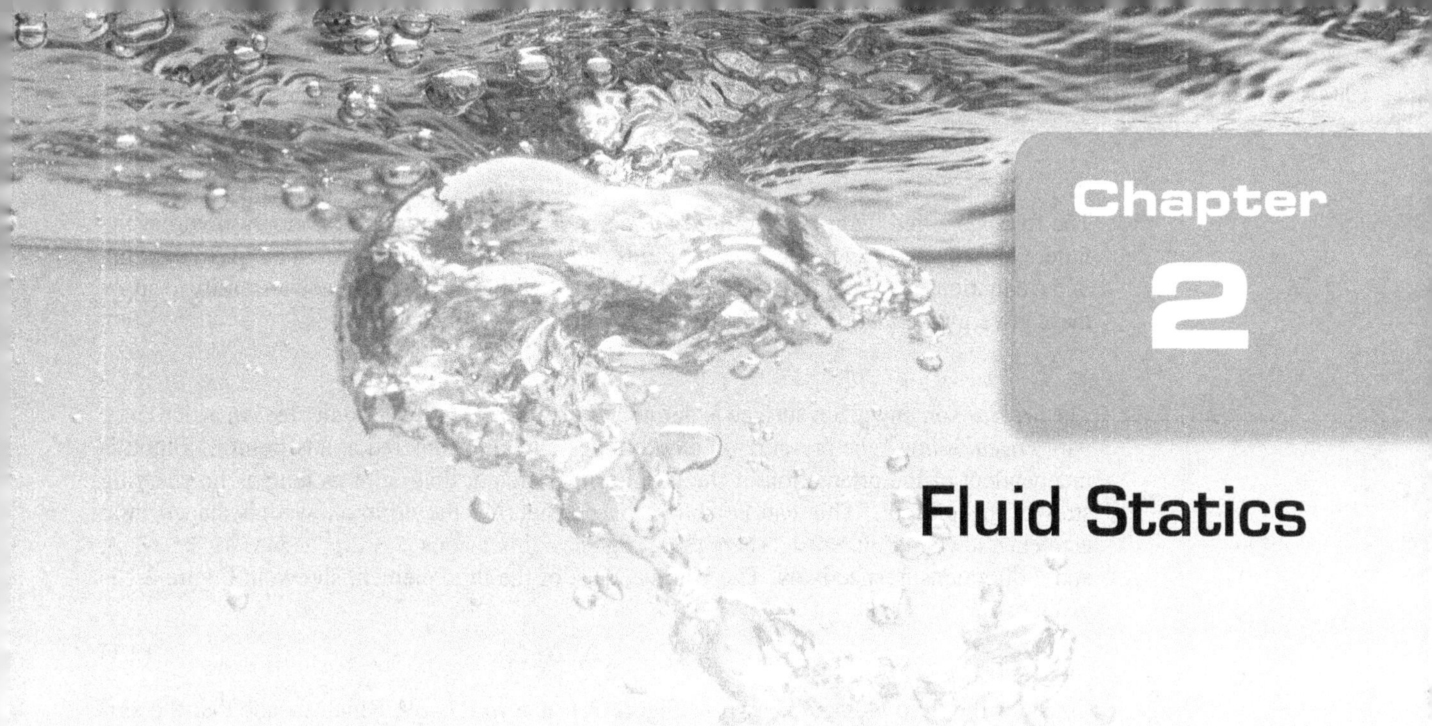

Chapter 2

Fluid Statics

LEARNING OBJECTIVES

After reading this chapter and solving a representative sample of end-of-chapter problems, you will be able to:

- Recognize and apply the key equations governing the spatial distribution of pressure in static fluids.
- Understand the mechanics of hydraulic machinery that utilize the principles of hydrostatics.
- Understand the difference between gauge pressure and absolute pressure.
- Understand the basic principles of pressure measurement and manometry.
- Calculate the hydrostatic forces on plane and curved surfaces.
- Analyze and evaluate the equilibrium forces on structures with hydrostatic loads.
- Analyze and evaluate buoyancy forces and assess the stability of submerged and floating objects.
- Understand the meaning, occurrence, and governing equations of rigid-body fluid motions.

2.1 Introduction

Fluid statics is the study of forces exerted by fluids at rest. In engineering applications, such forces exerted by static fluids are of concern because they can overturn dams, burst pressure vessels, and break lock gates on canals. Fluid statics is sometimes referred to as *hydrostatics* when the fluid is a liquid and as *aerostatics* when the fluid is a gas. However, the term "hydrostatics" is also commonly used to mean the study of fluids at rest, regardless of whether the fluid is a liquid or gas. The study of fluid statics is sometimes expanded to include the study of fluid behavior in a moving container, where the entire fluid mass is moving along with the container and there is no relative motion between fluid particles within the fluid mass. Such fluid motions, called *rigid-body motions*, are covered in this chapter.

2.2 Pressure Distribution in Static Fluids

The forces exerted by any static fluid are derived from the pressure distribution within the fluid. Consequently, analyses of static fluids and their effects are based on an understanding of the equations governing the pressure distribution within static fluids, and an ability to apply these governing equations to determine the forces exerted by static fluids.

2.2.1 Characteristics of Pressure

The *pressure* on any given surface is defined as the normal force per unit area. In accordance with *Pascal's law*,[9] the pressure on any infinitesimal area centered at any point in a fluid is independent of the orientation of the infinitesimal area at that point as long as no shearing stresses are present. This can be shown by considering the arbitrary wedge-shaped fluid element shown in Figure 2.1, where the wedge has dimensions Δx, Δy, and Δz in the x-, y-, and z-directions, respectively. The volume, ΔV, of the fluid element shown in Figure 2.1 is given by

$$\Delta V = \frac{1}{2}\Delta x \Delta y \Delta z \tag{2.1}$$

Consider the fluid element shown in Figure 2.1 as a *free body*, which means that the surrounding fluid is replaced by its effect, which in this case are pressures exerted on the faces of the free body. Furthermore, because the fluid is not moving, the sum of the forces must be equal to zero in each of the coordinate directions. Equilibrium in the y-direction requires that

$$p_y \Delta x \Delta z - p \Delta s \Delta x \sin\theta = 0 \tag{2.2}$$

where p_y and p are the pressure on the face in the xz plane and the pressure on the hypotenuse face, respectively, and Δs is the length of the hypotenuse face. Because $\sin\theta = \Delta z/\Delta s$, Equation 2.2 becomes

$$p_y \Delta x \Delta z - p \Delta s \Delta x \left(\frac{\Delta z}{\Delta s}\right) = 0 \quad \rightarrow \quad p_y - p = 0 \quad \rightarrow \quad p_y = p \tag{2.3}$$

This result shows that the pressure on the face in the xz plane is equal to the pressure on the inclined face. It is important to remember that this formulation implicitly assumes that the

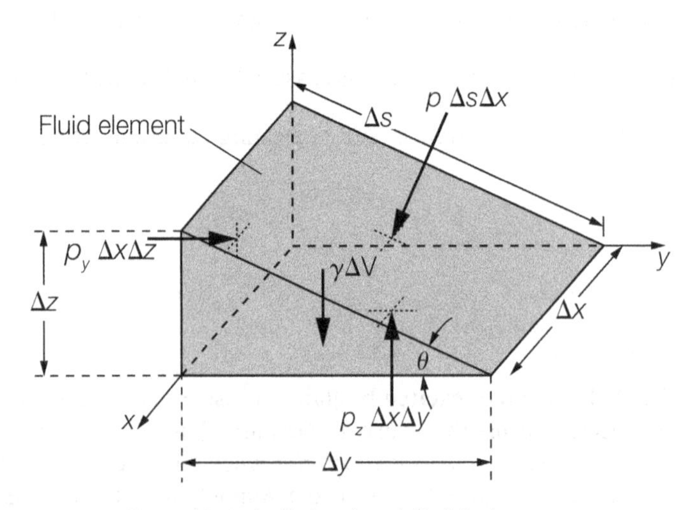

Figure 2.1: Infinitesimal fluid element

[9]Named in honor of the French mathematician and philosopher Blaise Pascal (1623–1662).

volume of fluid being considered is infinitely small, such that the pressure can be assumed to have a constant value over each face. For equilibrium in the z-direction,

$$p_z \Delta x \Delta y - \gamma \left(\frac{1}{2} \Delta x \Delta y \Delta z \right) - p \Delta s \Delta x \cos\theta = 0 \tag{2.4}$$

where p_z is the pressure on the face in the xy plane and γ is the specific weight of the fluid. Because $\cos\theta = \Delta y / \Delta s$, Equation 2.4 becomes

$$p_z \Delta x \Delta y - \gamma \left(\frac{1}{2} \Delta x \Delta y \Delta z \right) - p \Delta s \Delta x \left(\frac{\Delta y}{\Delta s} \right) = 0 \tag{2.5}$$

Simplifying Equation 2.5 and taking the limit as the fluid volume becomes infinitely small gives

$$\lim_{\Delta z \to 0} \left[p_z - \frac{1}{2} \gamma \Delta z - p = 0 \right] \quad \to \quad p_z - p = 0 \quad \to \quad p_z = p \tag{2.6}$$

Equations 2.3 and 2.6 collectively show that the pressure on an infinitesimally small fluid element in a static fluid is independent of the orientation of the face on which the pressure acts, because $p_y = p_z = p$. This analysis could be similarly repeated by considering a wedge with its vertical face in the yz plane (instead of the xz plane), which would yield $p_x = p$, where p_x is the pressure on a wedge face in the yz plane. Combining these results gives $p_x = p_y = p_z = p$. This shows that the pressure, p, at any point in a static fluid can be taken as a scalar quantity.

2.2.2 Spatial Variation in Pressure

Consider the control volume within a static fluid as illustrated in Figure 2.2(a), where the dimensions of the control volume are $\Delta x \times \Delta y \times \Delta z$. If the pressure at the center of the control volume is p and the specific weight of the fluid is γ, then the forces acting in the yz plane are illustrated in Figure 2.2(b), and a force balance in the y-direction yields

$$\left(p - \frac{\partial p}{\partial y} \frac{\Delta y}{2} \right) \Delta x \Delta z - \left(p + \frac{\partial p}{\partial y} \frac{\Delta y}{2} \right) \Delta x \Delta z = 0 \quad \to \quad \frac{\partial p}{\partial y} = 0 \tag{2.7}$$

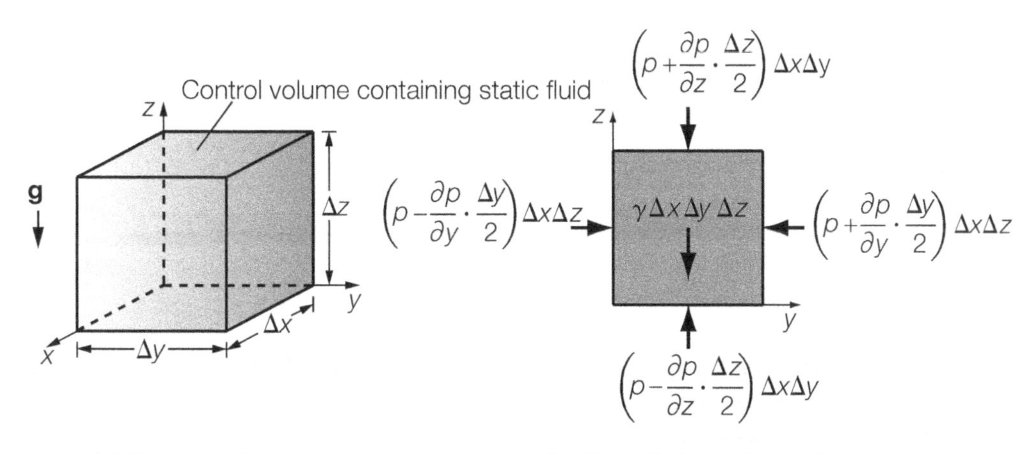

Figure 2.2: Force balance in a static fluid

indicating that there is no variation in pressure in the (horizontal) y-direction. The force balance in the x-direction is similar to the force balance in the y-direction and yields

$$\left(p - \frac{\partial p}{\partial x}\frac{\Delta x}{2}\right)\Delta y \Delta z - \left(p + \frac{\partial p}{\partial x}\frac{\Delta x}{2}\right)\Delta y \Delta z = 0 \quad \rightarrow \quad \frac{\partial p}{\partial x} = 0 \tag{2.8}$$

indicating that there is no variation in pressure in the (horizontal) x-direction. Collectively, Equations 2.7 and 2.8 indicate that *there is no pressure variation in the horizontal (xy) plane in a static fluid*. A force balance in the (vertical) z-direction yields

$$\left(p - \frac{\partial p}{\partial z}\frac{\Delta z}{2}\right)\Delta x \Delta y - \left(p + \frac{\partial p}{\partial z}\frac{\Delta z}{2}\right)\Delta x \Delta y - \gamma \Delta x \Delta y \Delta z = 0 \quad \rightarrow \quad \frac{\partial p}{\partial z} = -\gamma \tag{2.9}$$

where the negative sign on the left-hand side of Equation 2.9 indicates that the pressure, p, decreases with increasing elevation, z. According to Equations 2.7 and 2.8, the pressure, p, is independent of x and y and therefore depends only on the vertical coordinate z. Consequently, the partial derivative in Equation 2.9 can be replaced by the total derivative and written as

$$\frac{\mathrm{d}p}{\mathrm{d}z} = -\gamma \tag{2.10}$$

This equation describes the pressure variation within all static fluids. The pressure difference, Δp, between points at elevations z_1 and z_2 can be determined by integration of Equation 2.10 as

$$\Delta p = -\int_{z_1}^{z_2} \gamma \, \mathrm{d}z \tag{2.11}$$

For compressible fluids such as gases, γ is normally expressed as a function of p or z. For incompressible fluids, which includes most liquids, the fluid density is constant, and Equation 2.10 can be integrated to yield

$$p = -\gamma z + \text{constant} \tag{2.12}$$

This equation is commonly expressed in the following forms:

$$p + \gamma z = \text{constant} \quad \text{or} \quad \frac{p}{\gamma} + z = \text{constant} \tag{2.13}$$

The quantity $p/\gamma + z$ is called the *piezometric head*, and Equation 2.13 states that the piezometric head is constant throughout any static incompressible fluid. A pressure distribution that is described by Equation 2.13 is called a *hydrostatic pressure distribution*. Most fluids have some degree of compressibility, so Equation 2.13 is not exactly correct. However, over relatively small changes in elevation, z, such as on the order of several meters, the corresponding pressure changes usually do not cause significant changes in γ.

EXAMPLE 2.1

Consider the open bottle filled with water at 20°C as shown in Figure 2.3. The pressure on the water surface is equal to the atmospheric pressure of 101 kPa, and the water surface is 150 mm above the bottom of the bottle. Determine the pressures at Points A, B, and C, where Points A and B are at the midheight of the bottle and Point C is on the bottom.

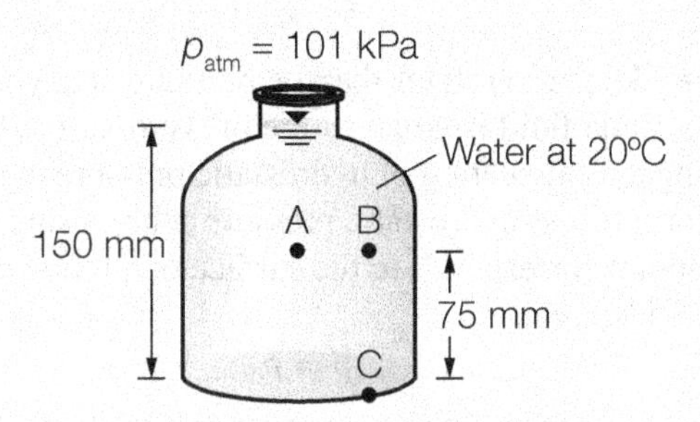

Figure 2.3: Water in a bottle

SOLUTION

From the given data: $p_{\text{atm}} = p_{\text{surf}} = 101$ kPa, $z_{\text{surf}} = 150$ mm = 0.150 m, $z_{\text{A}} = z_{\text{B}} = 75$ mm = 0.075 m, and $z_{\text{C}} = 0$ m. For water at 20°C, $\rho = 998.2$ kg/m^3 (from Appendix B.1). Taking g = 9.807 m/s^2 gives the specific weight, γ, as

$$\gamma = \rho g = (998.2)(9.807) = 9789 \text{ N/m}^3 = 9.789 \text{ kN/m}^3$$

Because water is incompressible, the pressure distribution is hydrostatic and is described by Equation 2.13. The constant in Equation 2.13 can be determined by applying Equation 2.13 on the water surface, which yields

$$p + \gamma z = p_{\text{surf}} + \gamma z_{\text{surf}} = 101 + (9.789)(0.150) = 102.5 \text{ kPa}$$

Therefore, in accordance with Equation 2.13, $p + \gamma z$ must be equal to 102.5 kPa at all locations within the water bottle. Applying this condition to locations A, B, and C yields the following results:

$$p_{\text{A}} + \gamma z_{\text{A}} = 102.5 \quad \rightarrow \quad p_{\text{A}} + (9.789)(0.075) = 102.5 \quad \rightarrow \quad p_{\text{A}} = 101.7 \text{ kPa}$$

$$p_{\text{B}} + \gamma z_{\text{B}} = 102.5 \quad \rightarrow \quad p_{\text{B}} + (9.789)(0.075) = 102.5 \quad \rightarrow \quad p_{\text{B}} = 101.7 \text{ kPa}$$

$$p_{\text{C}} + \gamma z_{\text{C}} = 102.5 \quad \rightarrow \quad p_{\text{C}} + (9.789)(0) = 102.5 \quad \rightarrow \quad p_{\text{C}} = 102.5 \text{ kPa}$$

As expected, the calculated pressures at A and B are the same because they are in the same horizontal plane; both pressures are equal to **101.7 kPa**. The pressure at C is equal to **102.5 kPa**. Note: In cases involving liquids that are exposed to the atmosphere, it is usually more convenient to work with pressures relative to atmospheric pressure (called *gauge pressures*) rather than with absolute pressures. Also, the specific weight of water at 20°C was calculated as 9.789 kN/m^3; this is worth remembering because it comes up often in engineering calculations.

2.2.3 Practical Applications

Consider the case illustrated in Figure 2.4, where the point P is located a distance h below a liquid surface and a distance z above a fixed datum. If the pressure at the liquid surface above P is equal to p_0 and the pressure at P is equal to p_P, then applying Equation 2.13 at both the liquid surface and P yields

$$p_0 + \gamma(z + h) = p_\mathrm{P} + \gamma z \tag{2.14}$$

which simplifies to

$$p_\mathrm{P} = p_0 + \gamma h \tag{2.15}$$

This commonly used equation states that the pressure at any point located a distance h below another point in a static fluid is equal to the pressure at the higher point plus γh. For most problems involving the calculation of hydrostatic pressure in incompressible fluids, Equation 2.15 is the best form of the hydrostatic pressure relationship for direct application. In cases where the higher point corresponds to the surface of a fluid exposed to atmospheric pressure, p_atm, Equation 2.15 becomes

$$p = p_\mathrm{atm} + \gamma h \tag{2.16}$$

It has previously been shown (in Section 2.2.2) that within any static fluid the pressure remains constant in any given horizontal plane. Application of this result to the case illustrated in Figure 2.4 requires that the pressures at Points P, Q, and R be the same, in which case

$$p_\mathrm{P} = p_0 + \gamma h = p_1 + \gamma h_1 + p_2 + \gamma h_2 \tag{2.17}$$

where h_1 and h_2 are the depths Points Q and R below the liquid surfaces where the pressures are p_1 and p_2, respectively, and γ is the density of the fluid that is continuous between the chambers shown in Figure 2.4. It should be noted that the pressures p_0, p_1, and p_2 can be due to either pressurized gas or the presence of an immiscible liquid above the interface.

Absolute pressure versus gauge pressure. Pressure distributions described by Equation 2.13 or, equivalently, Equation 2.15 are called *hydrostatic pressure distributions.* Pressure measured relative to that in a perfect vacuum (i.e., absolute zero pressure) is called *absolute pressure.* Pressure measured relative to the local atmospheric pressure is called *gauge pressure*, because practically all pressure gauges register zero when open to the atmosphere. Absolute pressure is generally positive, whereas gauge pressure is positive if the absolute pressure exceeds the local atmospheric pressure and is negative when the absolute pressure is less than the local atmospheric pressure. A negative gauge pressure is sometimes expressed as a positive *suction pressure* or a positive *vacuum pressure.* Atmospheric pressure is sometimes referred to as the *barometric pressure*,and standard atmospheric pressure at sea level

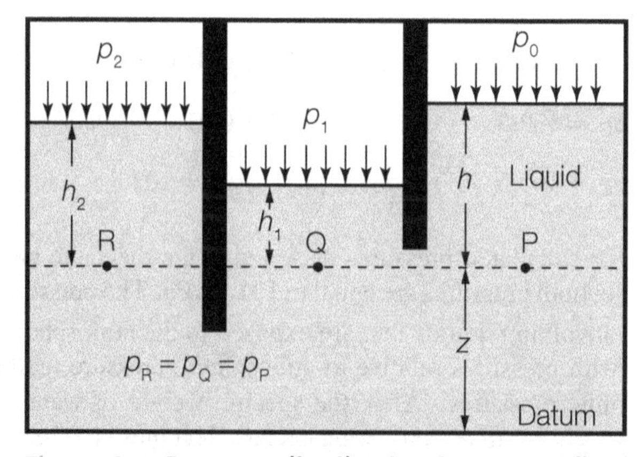

Figure 2.4: Pressure distribution in a static fluid

is 101.325 kPa. The relationship between absolute pressure, p_{abs}, gauge pressure, p_{gauge}, and local atmospheric pressure, p_{atm}, is given by

$$p_{gauge} = p_{abs} - p_{atm} \tag{2.18}$$

In thermodynamic analyses, it is essential to use absolute pressure because most thermal properties are functions of the absolute pressure of the fluid. For example, absolute pressure must be used in the ideal gas equation (Equation 1.20) and in most problems involving gases and vapors. Gauge pressures are used mostly in problems involving liquids. An exception is when the absolute pressure in a liquid approaches or equals the saturation vapor pressure, in which case the pressure in the liquid is usually expressed as an absolute pressure.

EXAMPLE 2.2

Consider the two tanks shown in Figure 2.5. Both tanks are pressurized, with one tank partially filled with gasoline at 20°C and the other tank filled with air at 20°C. The measured gauge pressures at A and B are 80 kPa and 130 kPa, respectively, when the atmospheric pressure is 100 kPa. (a) What is the absolute pressure at A and B? (b) Estimate the gauge pressures at C and D. (c) If atmospheric pressure changes to 80 kPa, what will be the gauge pressures at A, B, C, and D?

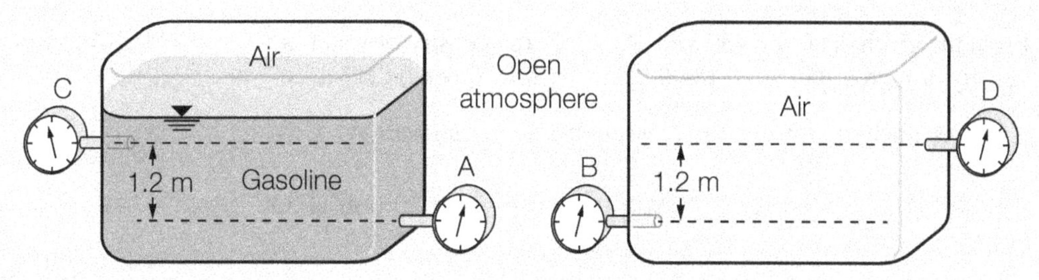

Figure 2.5: Pressurized tanks

SOLUTION

From the given data: $p_{Ag} = 80$ kPa, $p_{Bg} = 130$ kPa, $p_{atm} = 100$ kPa, and $\Delta z = 1.2$ m. For gasoline at 20°C, $\gamma = 6.67$ kN/m^3 (from Appendix B.4).

(a) The absolute pressures at A and B, p_{Aa} and p_{Ba}, respectively, are calculated as follows:

$$p_{Aa} = p_{Ag} + p_{atm} = 80 + 100 = \mathbf{180\ kPa},$$

$$p_{Ba} = p_{Bg} + p_{atm} = 130 + 100 = \mathbf{230\ kPa}$$

(b) The gauge pressure at C, p_{Cg}, can be estimated from the gauge pressure at A, p_{Ag}, using the hydrostatic relation

$$p_{Cg} = p_{Ag} - \gamma\Delta z = 80 - (6.67)(1.2) = \mathbf{72\ kPa}$$

Because pressure variations in gases are negligible over elevation differences on the order of meters, the gauge pressure at D is approximately equal to the gauge pressure at B, which is **130 kPa**.

(c) Gauge pressures are relative to atmospheric pressure. If the atmospheric pressure declines by 100 kPa−80 kPa = 20 kPa, then because the absolute pressures within the tanks remain constant, the gauge pressures will all increase by 20 kPa. Hence, the new gauge pressures at A, B, C, and D are as follows:

$$p_{Ag} = \mathbf{100\ kPa}, \quad p_{Bg} = \mathbf{150\ kPa}, \quad p_{Cg} = \mathbf{92\ kPa}, \quad p_{Dg} = \mathbf{150\ kPa}$$

Pressure head. Pressure is sometimes expressed as the height of a static column of fluid that would cause the given pressure. For a pressure p, the corresponding height of the fluid column, h, is given by

$$\text{pressure head, } h = \frac{p}{\gamma} \tag{2.19}$$

where γ is the specific weight of the fluid. The height h is called the *pressure head.* The reference fluid used in calculating the pressure head varies with the application. In applications involving the flow of water, the reference fluid is usually water at 20°C. It is sometimes convenient to express pressures occurring in one fluid in terms of the height of another fluid. For example, standard atmospheric pressure is sometimes expressed as "760 mm of mercury."

EXAMPLE 2.3

Pressure in a pipe at a certain location in a water distribution system is 480 kPa. (a) What is the pressure head in units of m H_2O? (b) What is the pressure head in units of m Hg? (c) If a hose was attached to the pipe and extended vertically, how high would water rise in the hose?

SOLUTION

From the given data: p = 480 kPa. At 20°C, the density of water is $\gamma_\text{w} = 9.789$ kN/m^3 and the density of mercury is $\gamma_\text{m} = 132.9$ kN/m^3 (from Appendix B.4).

(a) As a height of water, the pressure head, h_w, is given by

$$h_\text{w} = \frac{p}{\gamma_\text{w}} = \frac{480}{9.789} = \mathbf{49.0\ m\ H_2O}$$

(b) As a height of mercury, the pressure head, h_m, is given by

$$h_\text{m} = \frac{p}{\gamma_\text{m}} = \frac{480}{132.9} = \mathbf{3.6\ m\ Hg}$$

(c) Because the liquid in the pipeline is water, in the attached hose, water would rise to a height equal to the pressure head in m H_2O, which is **49.0 m.**

Surface-tension effect on pressure head. The pressure head, h, given by Equation 2.19 is typically assumed to be equal to the height of rise of a liquid in an open tube attached to a conduit or reservoir containing a liquid at gauge pressure p. This assumption neglects the effect of surface tension on the rise height in the connected tube, which is usually an appropriate assumption. Taking the additional rise height due to surface tension into account, the rise height, h', of a liquid in an open tube of diameter D connected to a conduit or reservoir in which the liquid is at gauge pressure p is given by

$$h' = \frac{p}{\gamma} + \frac{4\sigma\cos\theta}{\gamma D} \tag{2.20}$$

where σ is the surface tension of the liquid and θ is the contact angle between the liquid and the tube material.

EXAMPLE 2.4

A 2-mm-diameter clean glass tube is attached to a storage tank containing water at 15°C, and the water is observed to rise to a height of 82 mm in the attached tube. (a) What is the contribution of surface tension to the rise height? (b) What is the pressure head in the storage tank? (c) What is the minimum diameter tube that can be used to limit the surface tension effect to less than 0.1% of the rise height in the tube?

SOLUTION

From the given data: $D = 2$ mm and $h' = 82$ mm. For water at 15°C, $\rho = 999.1$ kg/m^3, $\gamma = \rho g = 9798$ N/m^3, $\sigma = 73.5$ mN/m, and for water in contact with clean glass, $\theta \approx 0°$.

(a) The contribution, Δh, of surface tension to the rise height in the tube is given by

$$\Delta h = \frac{4\sigma \cos\theta}{\gamma D} = \frac{4(0.0735)\cos 0°}{(9798)(0.002)} = 1.5 \times 10^{-2}\ \text{m} \quad \rightarrow \quad \Delta h = \mathbf{15\ mm}$$

(b) Because the observed rise height is h', the pressure head, h, is obtained as follows:

$$h' = h + \Delta h \quad \rightarrow \quad h = h' - \Delta h = 82\ \text{mm} - 15\ \text{mm} \quad \rightarrow \quad h = \mathbf{67\ mm}$$

(c) When the surface tension contributes 0.1% to the rise height,

$$\frac{\Delta h}{67 + \Delta h} = 0.001 \quad \rightarrow \quad \Delta h = 6.707 \times 10^{-2}\ \text{mm}$$

Let D_0 be the tube diameter at which $\Delta h = 6.707 \times 10^{-2}$ mm; then

$$\frac{4\sigma\cos\theta}{\gamma D_0} = 6.707 \times 10^{-5}\ \text{m} \quad \rightarrow \quad \frac{4(0.0735)\cos 0°}{(9798)D_0} = 6.707 \times 10^{-5}\ \text{m}$$

$$\rightarrow \quad D_0 = 0.447\ \text{m} = \mathbf{447\ mm}$$

Hydraulic machinery. The hydrostatic pressure distribution forms the basis of many practical devices, such as hydraulic brakes, lifts, jacks, presses, and other hydraulic machinery. The basic principle on which hydraulic machinery is based is illustrated in Figure 2.6, where a force F_1 is exerted on the piston at Point 1, which is connected via a *hydraulic fluid* to a piston at Point 2, which is at the same level as Point 1. If the cross-sectional area of the piston at Point 1 is A_1 and the cross-sectional area of the piston at Point 2 is A_2, then for a hydrostatic pressure distribution in the hydraulic fluid (which requires that pressures at Points 1 and 2 be equal),

$$\frac{F_1}{A_1} = \frac{F_2}{A_2} \longrightarrow F_2 = F_1\frac{A_2}{A_1} \tag{2.21}$$

where F_2 is the force that is exerted on the piston at Point 2. Because $A_2 \gg A_1$ in Equation 2.21, it is clear that $F_2 \gg F_1$. This result demonstrates that it is possible to apply a small force (F_1) to exert a large force (F_2). This occurs because, even though the exerted and applied pressures are the same, the applied pressure corresponds to a small force over a small

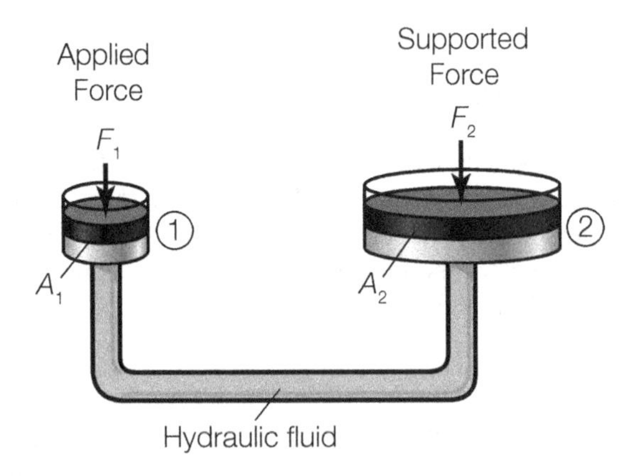

Figure 2.6: Hydraulic system

area, and the exerted pressure corresponds to a large force over a large area. In hydraulic machinery, the force F_2 is commonly called the *output force* and the applied force, F_1, is commonly called the *input force*. The ratio of the output force to the input force is called the *mechanical advantage*, such that

$$\text{mechanical advantage} = \frac{\text{output force}}{\text{input force}} \tag{2.22}$$

Forces can be applied manually or by using some other means such as compressed air. An example of a hydraulic system that uses compressed air to apply a force is the *hydraulic lift*, which uses a mechanism similar to that shown in Figure 2.6. Using a hydraulic lift, a relatively heavy car can be lifted by exerting a relatively small force on a small piston that is connected by hydraulic fluid to a larger piston that supports the car. Hydraulic brakes work by the same principle, with the force applied by a foot in automobiles and by an electromechanical device in airplanes. Power steering in cars is another hydraulic system in everyday use. Equation 2.21 can also be used when Points 1 and 2 are at different elevations and hydrostatic pressure variations are neglected. Neglecting hydrostatic pressure variations in hydraulic machinery is usually justifiable because the applied pressures in these systems are usually very high and much greater than the hydrostatic pressure variations. For example, applied pressures in automobile hydraulic brakes are typically on the order of 10 MPa, pressures in aircraft hydraulic brakes are typically on the order of 40 MPa, and pressures in hydraulic jacks are typically on the order of 70 MPa.

EXAMPLE 2.5

Consider the hydraulic jack shown in Figure 2.7 where an applied force F is being used to lift a weight of 10 kN. The lift piston has a diameter of 80 mm, and the plunger piston has a diameter of 28 mm. The pivot of the jack is located 50 mm from the plunger shaft, and the applied force is 400 mm from the plunger shaft. The hydraulic fluid has a density of 920 kg/m^3. Estimate the applied force, F, required to lift a weight of 10 kN.

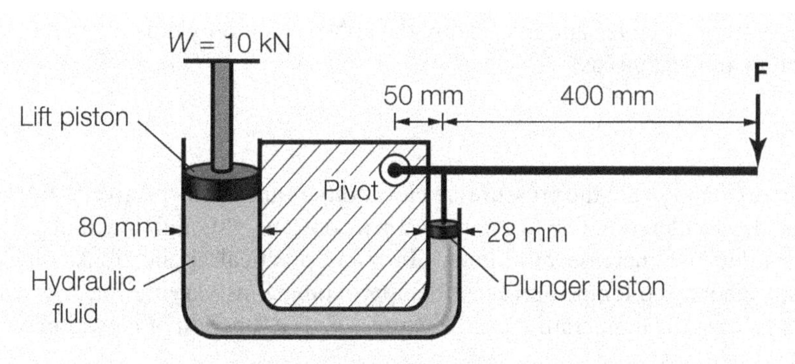

Figure 2.7: Hydraulic jack

SOLUTION

From the given data: $W = 10\,\text{kN}$, $D_1 = 80\,\text{mm}$, $D_2 = 28\,\text{mm}$, $L_1 = 50\,\text{mm}$, and $L_2 = 400\,\text{mm}$. Assume that the variation of hydrostatic pressure within the hydraulic fluid is negligible compared with the applied pressures. Using the given data,

$$A_1 = \frac{\pi D_1^2}{4} = \frac{\pi(0.080)^2}{4} = 5.027 \times 10^{-3}\,\text{m}^2, \quad A_2 = \frac{\pi D_2^2}{4} = \frac{\pi(0.028)^2}{4} = 6.158 \times 10^{-4}\,\text{m}^2$$

$$p_1 = \frac{W}{A_1} = \frac{10}{5.027 \times 10^{-3}} = 1.989 \times 10^3\,\text{kPa}$$

Because the hydrostatic pressure variation is neglected, the pressure under the lift piston, p_1, is equal to the pressure under the plunger piston, p_2, and the force, F_2, exerted by the plunger piston is calculated as follows:

$$p_2 = p_1 = 1.989 \times 10^3\,\text{kPa} \;\rightarrow\; F_2 = p_2 A_2 = (1.989 \times 10^3)(6.158 \times 10^{-4}) \;\rightarrow\; F_2 = 1.225\,\text{kN}$$

Taking moments about the pivot gives

$$F_2 \cdot L_1 = F \cdot (L_1 + L_2) \quad \rightarrow \quad (1.225)(50) = F(50 + 400) \quad \rightarrow \quad F = 0.1361\,\text{kN}$$

Therefore, an applied force of approximately **0.136 kN** is required to lift a weight of 10 kN. This represents a mechanical advantage of approximately 10/0.136 = 74:1.

Hydrostatic pressure variation in gases. Because the specific weights of gases are relatively small when compared with those of liquids, the pressure gradient in the vertical direction within a gas, as given by Equation 2.10, is generally small. Consequently, the pressure in gases is commonly taken as being approximately constant over vertical length scales on the order of tens of meters or less. In cases where vertical length scales are large, the density of a gas as a function of temperature and pressure can be estimated using the ideal gas law, which requires that

$$\rho = \frac{p}{RT} \tag{2.23}$$

where p is the pressure [FL^{-2}], R is the gas constant [$\text{EM}^{-1}\Theta^{-1}$], and T is the absolute temperature [Θ]. Combining Equations 2.10 and 2.23 yields

$$\frac{\mathrm{d}p}{\mathrm{d}z} = -\frac{gp}{RT}$$

Separating variables and integrating the above equation yields the vertical pressure distribution in an ideal gas as

$$\ln\frac{p_2}{p_1} = -\frac{g}{R}\int_{z_1}^{z_2}\frac{\mathrm{d}z}{T} \tag{2.24}$$

where p_1 and p_2 are the pressures at elevations z_1 and z_2, respectively. Equation 2.24 assumes that the acceleration due to gravity, g, is a constant, although it actually decreases by 0.03% per kilometer increase in altitude. In many practical applications, such as within Earth's atmosphere, the temperature can be approximated as varying linearly with elevation, z, in which case the temperature can be expressed as a function of elevation by

$$T = T_0 - bz \tag{2.25}$$

where T_0 is the temperature at $z = 0$ and b is the rate at which the temperature decreases with elevation, commonly known as the *lapse rate*. For the U.S. standard atmosphere, within the troposphere, which extends from sea level (z = 0 m) to an elevation of 11 km above sea level (z = 11 km), the parameters in Equation 2.25 are T_0 = 288.15 K (15°C) and b = 0.00650 K/m (= 6.5°C/km). For elevations in the range of 11–20 km, the temperature is constant at −56.5°C, and for elevations above 20 km, the temperature begins to rise again. Combining Equations 2.24 and 2.25 yields

$$p = p_0\left(1 - \frac{bz}{T_0}\right)^{\frac{g}{Rb}} = p_0\left(\frac{T}{T_0}\right)^{\frac{g}{Rb}} \tag{2.26}$$

where p_0 is the pressure at $z = 0$ and p is the pressure at elevation z. For air under standard conditions, the exponent in Equation 2.26 is given by

$$\text{for standard atmosphere:}\quad \frac{g}{Rb} = 5.26 \tag{2.27}$$

If Equation 2.26 is applied under standard atmospheric conditions, then it can be shown that atmospheric pressure changes by 0.1% over an elevation difference of 10 m and changes by 2% over an elevation difference of 1 km. These results provide sufficient justification for neglecting changes in atmospheric pressure over relatively small elevation differences. Combining Equations 2.26 and 2.23 the air density, ρ, can be expressed as

$$\rho = \rho_0\left(1 - \frac{bz}{T_0}\right)^{\frac{g}{Rb}-1} \tag{2.28}$$

where ρ_0 is the air density at $z = 0$, which is usually taken as the standard value of 1.2255 kg/m^3. It is generally important to remember that atmospheric conditions can be highly variable on a daily basis, with standard atmospheric conditions representing long-term averages.

EXAMPLE 2.6

Consider a region of Earth where the troposphere has a sea-level temperature and pressure of 15°C and 101.3 kPa, respectively, and a lapse rate of 7.40°C/km. At all elevations within the troposphere, the air consists of 21% oxygen by volume. Measurements by NASA indicate that an adult person at rest breathes in air at a rate of around 7.5 L/min and needs approximately 0.84 kg/d of oxygen to survive. Estimate the maximum elevation at which a person at rest could survive without needing a supplemental supply of oxygen.

SOLUTION

From the given data: T_0 = 15°C = 288.15 K, p_0 = 101.3 kPa, and b = 7.40 K/km. For oxygen, R_{oxy} = 259.8 J/kg·K, and for air, R_{air} = 287.1 J/kg·K (from Appendix B.5). For the given lapse rate,

$$\frac{g}{R_{air}b} = \frac{9.807}{(287.1)(7.40 \times 10^{-3})} = 4.62$$

Taking z as the elevation in kilometers, the following conditions exist at an elevation z:

$$T = T_0 - bz = 288.15 - 7.40z \text{ K}$$

$$p = p_0\left(\frac{T}{T_0}\right)^{\frac{g}{R_{air}b}} = 101.3\left(\frac{288.15 - 7.40z}{288.15}\right)^{4.62} \text{ kPa} = 1.013 \times 10^5(1 - 0.0257z)^{4.62} \text{ Pa}$$

$$\rho_{O_2} = \frac{p}{R_{oxy}T} = \frac{101.3 \times 10^5(1 - 0.0257z)^{4.62}}{259.8(288.15 - 7.40z)} \text{ kg/m}^3 = \frac{0.390(1 - 0.0257z)^{4.62}}{288.15 - 7.40z} \text{ kg/L}$$

For a person at rest at an elevation z:

$$\text{volume of air inhaled per day} = 7.5\frac{\text{L}}{\text{min}} \times 60\frac{\text{min}}{\text{h}} \times 24\frac{\text{h}}{\text{day}} = 10\,800 \text{ L}$$

$$\text{volume of O}_2 \text{ inhaled per day} = 0.21(10\,800) = 2268 \text{ L}$$

$$\text{mass of O}_2 \text{ inhaled per day} = \rho_{O_2}(2268) = \frac{0.390(1 - 0.0257z)^{4.62}}{288.15 - 7.40z}(2268)$$

$$= \frac{884(1 - 0.0257z)^{4.62}}{288.15 - 7.40z} \text{ kg}$$

Because 0.84 kg/day is required for survival, if z_c is the critical elevation for survival, then

$$0.84 = \frac{884(1 - 0.0257z_c)^{4.62}}{288.15 - 7.40z_c}$$

which yields z_c = 11.7 km. Hence, the maximum elevation for sufficient oxygen is **11.7 km**. Interestingly, commercial airliners typically fly at an elevation that is lower than this critical elevation.

Constant temperature condition. In the special case where the temperature is constant and equal to T_0, $b = 0$ in Equation 2.25 and the exponent in Equation 2.26 becomes undefined. In this case, the pressure distribution can be derived directly from Equation 2.24 by taking $T = T_0$, which yields

$$p_2 = p_1 \exp\left[-\frac{g(z_2 - z_1)}{RT_0}\right] \tag{2.29}$$

Constant temperature conditions are characteristic of the lower stratosphere for elevations in the range of 11.0–20.1 km above Earth's surface, where the temperature is approximately constant at −56.5°C. Constant temperature conditions can also occur as a transient state in the lower troposphere (i.e., the atmospheric layer between the surface of Earth and the stratosphere). This takes place when the ground cools and causes temperatures to remain constant, or even increase, with elevation above ground level. A temporary state in which the temperature increases with elevation above the ground is called an *inversion*.

EXAMPLE 2.7

A transient condition occurs at a particular location where the air temperature remains constant at 25°C up to an elevation of 2500 m above the ground. In contrast, under average conditions, the air temperature has a lapse rate of 7.00°C/km. (a) Compare the percentage decrease in air pressure across the transient isothermal layer with the percentage decrease under average conditions. (b) Repeat part (a) considering density instead of pressure.

SOLUTION

From the given data: $T_1 = 25°\text{C} = 298.2$ K, $\Delta z = 2500$ m, and $b = 7.00$°C/km. For air, $R = 287.1$ J/kg·K. With the given lapse rate, the temperature, T_2, at an elevation 2500 m above ground and the constant g/Rb are given by

$$T_2 = T_1 - b\Delta z = 298.2 - (7.00\times10^{-3})(2500) = 281.9 \text{ K}, \quad \frac{g}{Rb} = \frac{9.807}{(287.1)(7.00\times10^{-3})} = 4.88$$

(a) Under isothermal conditions, the percentage change in the pressure over a 2500 m change in elevation is derived from Equation 2.29 as follows:

$$\text{pressure change} = \frac{p_2 - p_1}{p_1} \times 100 = \left\{\exp\left[-\frac{g\Delta z}{RT_1}\right] - 1\right\} \times 100$$

$$= \left\{\exp\left[-\frac{(9.807)(2500)}{(287.1)(298.2)}\right] - 1\right\} \times 100 = \mathbf{-24.9\%}$$

Under average conditions, the percentage change in the pressure over a 2500 m change in elevation is derived from Equation 2.26 as follows:

$$\text{pressure change} = \frac{p_2 - p_1}{p_1} \times 100 = \left[\left(\frac{T_2}{T_1}\right)^{\frac{g}{Rb}} - 1\right] \times 100$$

$$= \left[\left(\frac{281.9}{298.1}\right)^{4.88} - 1\right] \times 100 = \mathbf{-25.6\%}$$

Therefore, based on these results, the pressure change is slightly less under isothermal conditions:

(b) Under isothermal conditions, the percentage change in density is the same as the percentage change in pressure as shown by the following calculation:

$$\text{density change} = \frac{\rho_2 - \rho_1}{\rho_1} \times 100 = \frac{\dfrac{p_2}{RT_1} - \dfrac{p_1}{RT_1}}{\dfrac{p_1}{RT_1}} \times 100 = \frac{p_2 - p_1}{p_1} \times 100 = \mathbf{-24.9\%}$$

Under average conditions, the percentage change in density is derived from Equation 2.28 as follows:

$$\text{density change} = \frac{\rho_2 - \rho_1}{\rho_1} \times 100 = \left[\left(\frac{T_2}{T_1}\right)^{\frac{g}{Rb}-1}\right] \times 100$$

$$= \left[\left(\frac{281.9}{298.1}\right)^{4.88-1}\right] \times 100 = \mathbf{-20.9\%}$$

Therefore, the reduction in density over the 2500-m layer is greater under isothermal conditions than under average conditions.

Practical Considerations

It is apparent that temperature, pressure, and air density all decrease with increasing altitude above sea level, with consequent practical effects such as a decrease in the boiling point of water and a decrease in the mass of oxygen per unit volume. The elevations of the ten highest cities in the world with populations greater than 100 000 are given in Table 2.1. Native residents of these cities tend to have higher lung capacities that are adapted to the lower oxygen concentrations in the air. It is interesting to note that eight of the top-ten highest cities are in Bolivia or Peru, which are neighboring countries.

Table 2.1: Large Cities at High Elevation

City	Country	Elevation (m)	Population	Estimation Year
El Alto	Bolivia	4150	1 184 392	2010
Potosí	Bolivia	4090	170 000	2007
Shigatse	China	3836	100 000	2011
Juliaca	Peru	3825	225 146	2007
Puno	Peru	3819	120 229	2007
Oruro	Bolivia	3706	250 700	2011
Lhasa	China	3658	373 000	2009
La Paz	Bolivia	3640	845 480	2010
Cusco	Peru	3399	358 052	2011
Huancayo	Peru	3052	425 000	2012

2.3 Pressure Measurements

There are a variety of devices for measuring pressure, with mechanical and electrical pressure gauges[10] being most common. Manual pressure gauges based on measuring fluid levels are used primarily in research applications. In principle, pressure cannot be measured directly and all pressure measurement instruments measure a difference in pressure, usually the difference between the actual pressure and atmospheric pressure.

2.3.1 Barometer

A *barometer* is an instrument used to measure the absolute pressure of the atmosphere; consequently, the atmospheric pressure is sometimes referred to as the *barometric pressure*. Two types of barometers are illustrated in Figure 2.8. In the *barometer* shown in Figure 2.8(a), the open end of a tube is immersed in a liquid reservoir that is open to the atmosphere. This configuration can be attained by initially filling the tube with a liquid while the tube is kept submerged in a reservoir containing the liquid and then turning the tube upside down while keeping the open end in the liquid reservoir. If the tube is long enough such that a liquid surface occurs within the tube, the atmospheric pressure, $p_{\rm atm}$, is related to the height of the liquid in the tube by

$$p_{\rm atm} = \gamma y + p_{\rm vapor} \tag{2.30}$$

where γ is the specific weight of the liquid in the barometer, y is the rise height, and $p_{\rm vapor}$ is the saturation vapor pressure of the liquid. The near vacuum above the liquid in the tube

[10]The spellings *gauge* and *gage* are used interchangeably in American practice, and both are acceptable.

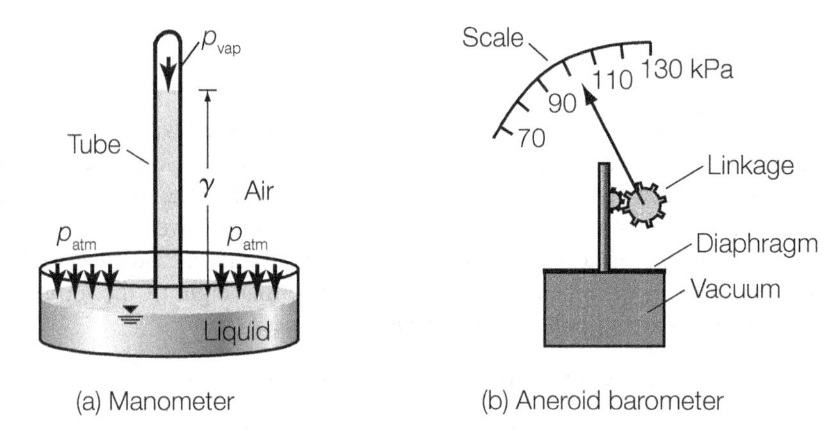

Figure 2.8: Types of barometers

is known as the *Torricellian vacuum*.[11] The liquid used in barometers of this type is usually mercury, because its density is large enough to enable a reasonably short tube to be used and because its saturation vapor pressure is negligibly small at ordinary temperatures. For mercury at 20°C, γ = 133.3 kN/m^3 and p_{vapor} = 0.16 Pa, which is much less than typical values of p_{atm}, which are on the order of 10^5 Pa; hence, p_{vapor} is typically neglected in Equation 2.30. For accurate measurements with a mercury barometer, corrections are necessary to account for the variation of the density of mercury with temperature, the thermal expansion of the (usually brass) scale, and surface-tension effects. Great care should be exercised when handling mercury because it is a very toxic substance. An *aneroid*[12] *barometer* does not use a liquid; it is illustrated in Figure 2.8(b). This type of barometer measures the difference in pressure between the atmosphere and an evacuated cylinder using a sensitive elastic diaphragm and linkage system.

Standard atmospheric pressure. Standard sea-level atmospheric pressure can be expressed in a variety of equivalent ways. For example, standard atmospheric pressure at sea level can be equivalently expressed as 101.325 kPa, 760 mm of mercury (Hg), and 10.34 m of water. Notice that if water was used as the barometer fluid, the barometer would have to be more than 10.34 m high, compared with a mercury barometer that requires a minimum height of only 0.76 m; this is primarily why mercury is the preferred manometer fluid.

Units. A commonly used pressure unit is the *standard atmosphere*, which is defined as the pressure produced by a column of mercury 760 mm in height at 0°C under standard gravitational acceleration (9.807 m/s^2). The unit "millimeters of mercury," denoted by mm Hg, is also called the *torr*. Atmospheric pressure is sometimes expressed in millibars (mb), where 1 bar = 100 kPa (1 mb = 0.1 kPa); hence, standard atmospheric pressure is 1013.25 mb. The pressure in the eye of a hurricane is commonly expressed in mb, with lower pressures associated with higher wind speeds. As examples, the Category 5 hurricanes Katrina and Andrew, which devastated New Orleans and South Florida, respectively, had minimum eye pressures of approximately 920 mb.

[11] Named in honor of the Italian physicist and mathematician Evangelista Torricelli (1608–1647).
[12] The word "aneroid" means "not using a liquid."

EXAMPLE 2.8

The average annual temperature and pressure at Blue Mountain Peak in Jamaica are reported as 5°C and 580 mm Hg, respectively. What is the average annual pressure in kilopascals and millibars?

SOLUTION

From the given data: p = 580 mm Hg. Because 760 mm Hg = 101.325 kPa = 1013.25 mb, the given pressure is converted to other units as follows:

$$p = 580 \times \frac{101.325}{760} = \mathbf{77.33\ kPa}, \qquad p = 580 \times \frac{1013.25}{760} = \mathbf{773.3\ mb}$$

Note that the conversion of pressures is independent of temperature, which reflects the fact that the height of mercury when used to represent atmospheric pressure is standardized to 0°C.

2.3.2 Bourdon Gauge

The most widely used mechanical device for measuring pressure is the *Bourdon-tube pressure gauge*,[13] illustrated in Figure 2.9. The Bourdon-tube gauge is directly connected to the fluid, and the fluid fills a metal tube having an elliptical cross section. The elliptical tube is mechanically attached to a pointer that registers zero (gauge) pressure when the tube is empty. When the fluid fills the tube and the pressure is increased, the elliptical cross section moves toward becoming circular, the tube straightens, and the attached pointer moves to indicate the pressure. The pressure scale on a Bourdon-tube pressure gauge is generally calibrated by the manufacturer, and the gauge is reliable as long as it is not subjected to excessive pressure

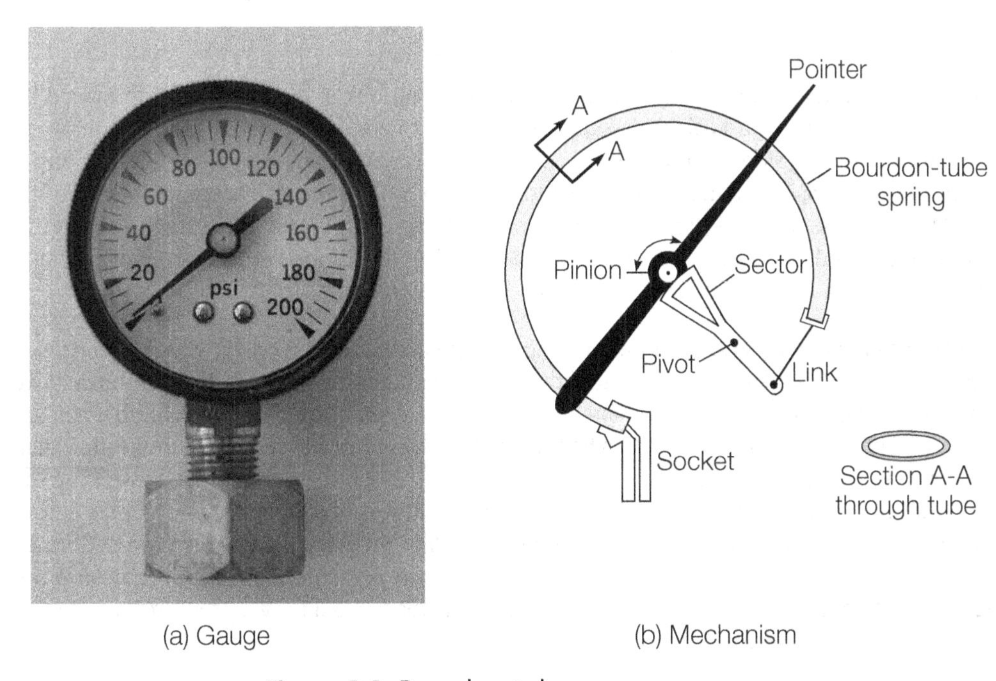

Figure 2.9: Bourdon-tube pressure gauge

[13]Named in honor of the French engineer Eugène Bourdon (1808–1884).

pulses. Using metal tubes of varying stiffness, the range of pressures that can be measured by Bourdon gauges is varied. Bourdon gauges can also be calibrated to measure negative gauge pressures, and such gauges are called *vacuum gauges*. Gauges calibrated to measure both positive and negative pressures are called *compound gauges*. If a pressure higher than the intended maximum is applied to a Bourdon gauge, the tube may be strained beyond its elastic limit and the gauge must be recalibrated.

Practical Considerations

The pressure indicated by a Bourdon gauge is the pressure at the center of the gauge, and an appropriate correction must be applied to the measured pressure in cases where the pressure is required at some other location. For example, if a Bourdon gauge is connected to a pipe to measure the pressure of a fluid contained in the pipe, then the pressure difference between the center of the pipe and the center of the Bourdon gauge must be taken into account. Hence, if a Bourdon gauge is attached to a pipe such that the center of the gauge is Δz above the center of the pipe, then the pressure at the center of the pipe, p, is related to the pressure measured by the gauge, p_{gauge}, using the relation

$$p = p_{\text{gauge}} + \gamma \Delta z \qquad (2.31)$$

where γ is the specific weight of the fluid in the pipe. To avoid the necessity for the pressure correction $\gamma \Delta z$, the Bourdon gauge is sometimes mounted such that the center of the gauge is at the same elevation as the location at which the pressure is to be measured. For example, if the pressure at the center of a pipe is being measured, the gauge can be mounted on the side of the pipe rather than on the top of the pipe. Such considerations are relevant only to measuring the pressures in liquids; when the Bourdon gauge is used to measure gas pressures, corrections for differences in elevation are usually negligible.

2.3.3 Pressure Transducer

Pressure transducers are devices that measure the variations in electrical signals caused by pressure variations. In a typical pressure transducer, one side of a small diaphragm is exposed to the fluid system and the flexure of the diaphragm caused by pressure variations is measured by a sensing element connected to the other side of the diaphragm. The electrical signals from pressure transducers can be recorded using computer data acquisition systems, and the measured pressures can be displayed digitally. There are several types of pressure transducers, with a variety of designs for the sensing element.

Strain-gauge transducers. A common type of sensing element is one in which a resistance wire strain gauge is attached to the diaphragm. As the diaphragm flexes, the resistance wire changes length, and the resulting change in resistance is manifested as a measurable change in voltage across the resistance wire. The transducer is calibrated to read pressures rather than voltages. *Gauge pressure transducers* use atmospheric pressure as a reference by venting the back side of the pressure-sensing diaphragm to the atmosphere, giving zero signal output at atmospheric pressure regardless of altitude. *Absolute pressure transducers* are calibrated to have zero signal output at full vacuum. *Differential pressure transducers* measure the pressure difference between two locations directly instead of using two pressure transducers and taking their difference.

Piezoelectric transducers. *Piezoelectric transducers*, also called *solid-state pressure transducers*, work on the principle that an electric potential is generated in a crystalline substance when it is subjected to mechanical pressure.This phenomenon, first discovered by brothers

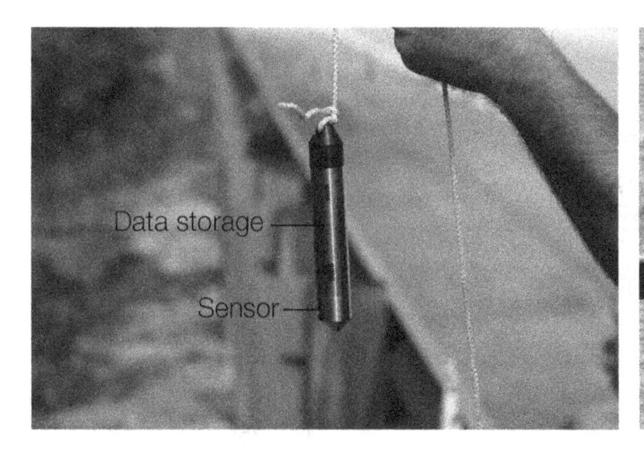

(a) Transducer

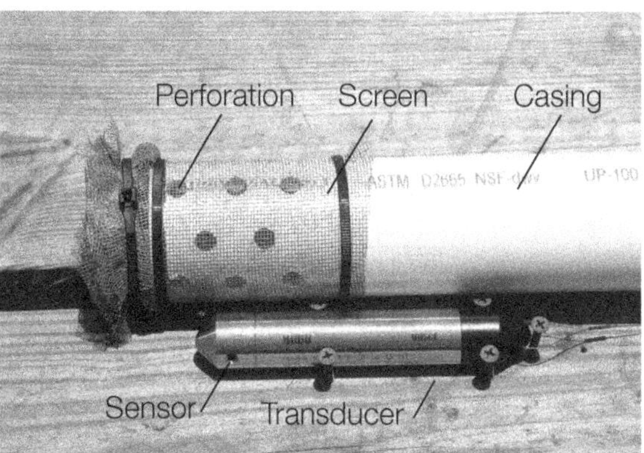

(b) Transducer and casing

Figure 2.10: Gauge pressure transducer

Pierre and Jacques Curie in 1880, is called the *piezoelectric effect*. Piezoelectric pressure transducers, typically made of quartz crystals, have a much faster frequency response compared with diaphragm units. Piezoelectric transducers are most suitable for high-pressure applications, but they are generally not as sensitive as diaphragm-type transducers, especially at low pressures.

Transducer used in hydrologic applications. A gauge pressure transducer of the type commonly used in hydrologic applications to measure the depth of water above the transducer is shown in Figure 2.10. The key components of the transducer shown in Figure 2.10(a) are the sensor that measures the pressure and the data-storage housing in which the recorded data is stored for subsequent downloading. Figure 2.10(b) shows the transducer beside the well casing in which it is inserted for field application. Key components of the well casing are the perforations to allow the surrounding water to enter the casing, and a screen to keep out debris. The sensor is typically placed below the lowest expected water stage, and is calibrated before installation by submerging the sensor to known depths and recording the corresponding voltage signals. The data logger uses this depth-voltage relationship to convert voltage readings to water depths.

2.3.4 Manometer

Manometers are instruments that use liquids in vertical or inclined tubes to measure pressures or pressure differences. Three commonly utilized manometers are the piezometer, the U-tube manometer, and the inclined-tube manometer.

2.3.4.1 Piezometer

A *piezometer* is simply a vertical tube that is open to the fluid at the bottom. The pressure at the bottom of the tube, p_0, is measured by the height of rise of the fluid in the tube, h_0, using the hydrostatic pressure relationship

$$p_0 = \gamma h_0 \tag{2.32}$$

where γ is the specific weight of the fluid. Examples of piezometer applications are shown in Figure 2.11. Piezometers are commonly used in groundwater hydrology to measure pressures in groundwater as shown in Figure 2.11(a), where the piezometer normally consists of a vertical pipe with a small open interval over which the pressure is to be measured. Typical piezometer diameters in this case are on the order of 50 mm. It is interesting to note that in the confined aquifer shown in Figure 2.11(a), the height of water in the piezometer can rise above the ground surface; so any ground-level well that penetrates the confined aquifer will

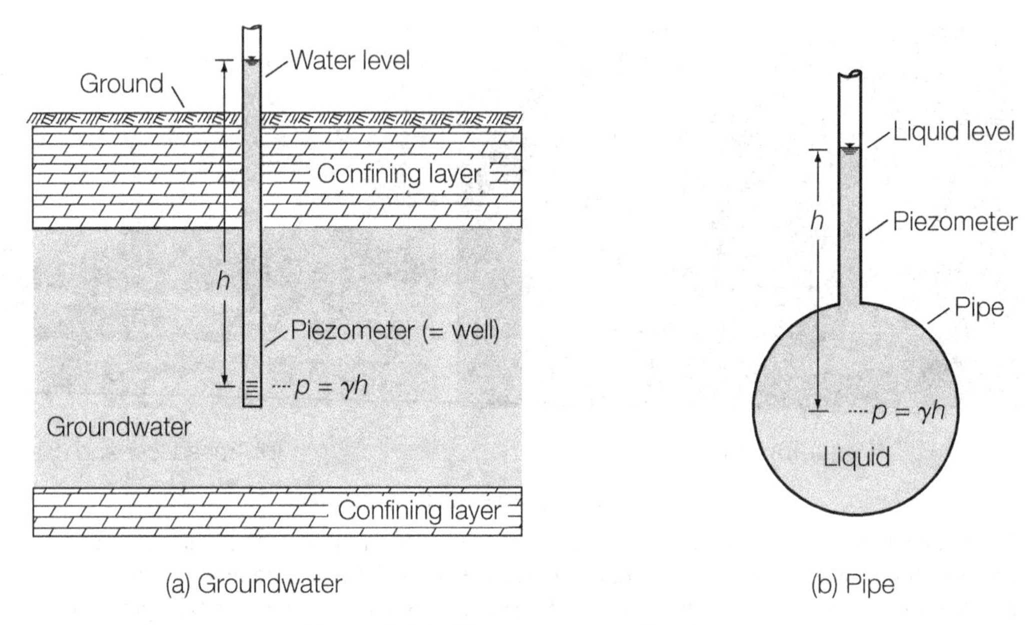

Figure 2.11: Piezometer applications

result in flowing water through the ground-level opening of the well. Such wells are called *artesian wells*. Piezometers can also be used to measure pressures of confined liquids, such as in the pipe shown in Figure 2.11(b). In this case, the piezometer is attached to the top of the pipe and the pressure, p, at the center of the pipe is equal to γh, where γ is the specific weight of the fluid in the pipe and h is the height of the liquid in the piezometer above the center of the pipe. Piezometer applications in closed conduits are typically limited to measuring relatively small pressures because large rise heights in these piezometers are not practical to measure. In measuring the pressure in flowing fluids, care should be taken that the piezometer opening to the pipe is normal to the flow direction and does not obstruct the flowing fluid and that the opening is relatively small, being typically less than 3 mm and preferably about 1 mm in diameter. Sometimes piezometer connections are placed circumferentially around a pipe in the form of a *piezometer ring* to average out flow irregularities. For general piezometer applications in static fluids, the diameter of the piezometer should be greater than 15 mm to limit capillary error (Douglas et al., 2001).

Limitations. Limitations on using piezometers are that they can only be used for measuring pressures that are greater than atmospheric pressure (otherwise, air would be sucked into the system); the magnitudes of the measured pressures must be such that the liquid heights in the piezometer are practical; and the fluid in which the pressure is being measured must be a liquid, not a gas.

2.3.4.2 U-tube and inclined-tube manometers

U-tube and inclined-tube manometers differ from piezometers in that the attached tubes are not vertical and generally contain a liquid other than that in the connected fluid system. Examples of U-tube and inclined-tube manometers are shown in Figure 2.12, where the U-tube manometer is being used to measure the pressure at A and the inclined-tube manometer is being used to measure the difference in pressure between A and B. "U-tube" manometers are so named because their tubes have a "U" shape, and "inclined-tube" manometers are so named because they have a straight, inclined tube as part of their configuration. For the U-tube manometer shown in Figure 2.12, the pressure at A, denoted by p_A, can be determined

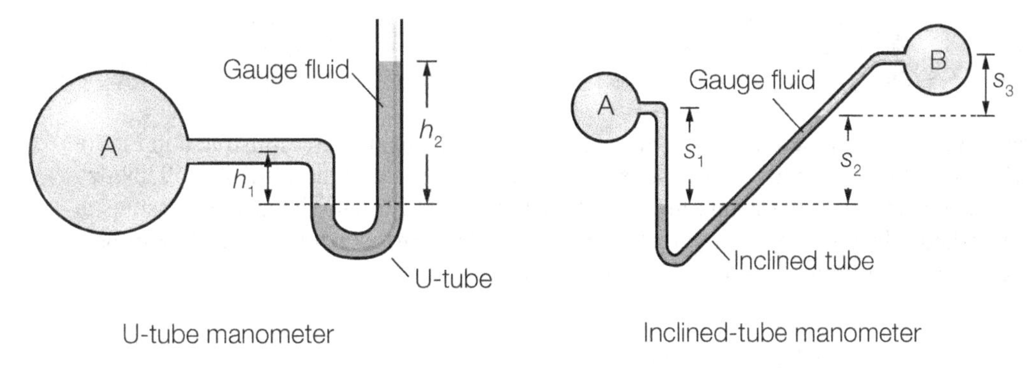

Figure 2.12: Types of manometers

from the measurements h_1 and h_2 using the fact that the pressure is distributed hydrostatically in the manometer, which requires that

$$p_A = p_{atm} + \gamma_2 h_2 - \gamma_1 h_1 \tag{2.33}$$

where p_{atm} is the atmospheric pressure at the open end of the manometer and γ_1 and γ_2 are the specific weights of the liquids in the manometer having heights h_1 and h_2, respectively. Equation 2.33 is derived by tracking the changes in hydrostatic pressure within the manometer as one moves between adjacent liquid interfaces. In a similar manner, for the inclined-tube manometer, the pressure difference between B and A, denoted by $p_B - p_A$, can be determined from the measurements s_1, s_2, and s_3 using the relations

$$p_A + \gamma_1 s_1 - \gamma_2 s_2 - \gamma_3 s_3 = p_B \quad \rightarrow \quad p_B - p_A = \gamma_1 s_1 - \gamma_2 s_2 - \gamma_3 s_3 \tag{2.34}$$

where γ_1, γ_2, and γ_3 are the specific weights of the liquids in the manometer having heights s_1, s_2, and s_3, respectively. Equations 2.33 and 2.34 demonstrate the general approach for relating pressures at the beginning and end of a manometer to liquid heights within the manometer; this approach works for any manometer configuration. The general approach starts with the pressure at one end of the manometer; then the hydrostatic pressure increments within the manometer are added or subtracted until the other end is reached; the resulting expression is then equated to the pressure at the other end of the manometer. It is useful to remember that the pressure increments are necessarily positive going downward and negative going upward.

General approach to relating fluid elevation to pressure. The relationship between elevations of the fluid interfaces in a manometer and the pressures at both ends of the manometer is generally determined using the hydrostatic pressure distribution. The pressure at one end of a manometer, p_0, is related to the pressure at the other end of the manometer, p_1, by the hydrostatic relationship

$$p_0 + \sum_{i=1}^{N} \gamma_i \Delta z_i = p_1 \tag{2.35}$$

where N is the number of layers of fluid within the manometer, γ_i is the specific weight of the ith layer, and Δz_i is the difference in elevation between the top and bottom of the ith layer. In cases where the pressure at one end of the manometer is known (e.g., p_0 is known), the pressure at the other end of the manometer, p_1, can be calculated directly from the manometer fluid elevations and specific weights, Δz_i and γ_i, using Equation 2.35. In cases where the pressures at both ends of the manometer are unknown, the difference in pressure, $p_0 - p_1$, can be calculated directly from the manometer fluid elevations and specific weights using Equation 2.35. Application of Equation 2.35 is illustrated by the following example.

EXAMPLE 2.9

The water pressure in a pipeline is to be measured by a U-tube manometer illustrated in Figure 2.13, where Point D is open to the atmosphere. The specific weight of the water is 9.79 kN/m^3, and the specific weight of the gauge fluid is 29.6 kN/m^3. (a) Estimate the gauge pressures at Points A, B, and C by applying the hydrostatic pressure relation to each fluid layer. (b) Estimate the gauge pressure at Point A using Equation 2.35 directly.

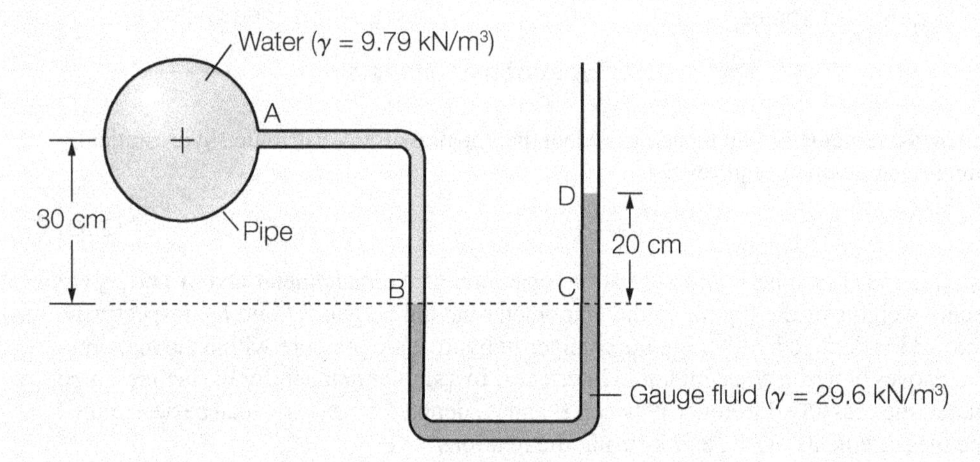

Figure 2.13: U-tube manometer

SOLUTION

From the given data: $\gamma_{\text{AB}} = 9.79$ kN/m^3, $\gamma_{\text{BC}} = \gamma_{\text{CD}} = 29.6$ kN/m^3, $p_{\text{D}} = p_{\text{atm}}$, $\Delta z_{\text{AB}} = 0.30$ m, $\Delta z_{\text{BC}} = 0$ m, and $\Delta z_{\text{CD}} = 0.20$ m. The pressure distribution in each fluid layer is hydrostatic in accordance with Equation 2.15.

(a) The following relationships are derived using the hydrostatic pressure distribution in each fluid layer:

$$p_{\text{C}} = p_{\text{D}} + \gamma_{\text{CD}}\Delta z_{\text{CD}} = p_{\text{atm}} + 29.6(0.20) \quad \rightarrow \quad p_{\text{C}} - p_{\text{atm}} = \mathbf{5.92\ kPa}$$

$$p_{\text{B}} = p_{\text{C}} + \gamma_{\text{BC}}\Delta z_{\text{BC}} = (p_{\text{atm}} + 5.92) + 29.6(0) \quad \rightarrow \quad p_{\text{B}} - p_{\text{atm}} = \mathbf{5.92\ kPa}$$

$$p_{\text{A}} = p_{\text{B}} - \gamma_{\text{AB}}\Delta z_{\text{AB}} = (p_{\text{atm}} + 5.92) - 9.79(0.30) \quad \rightarrow \quad p_{\text{A}} - p_{\text{atm}} = \mathbf{2.98\ kPa}$$

(b) Using Equation 2.35 directly, the gauge pressure at Point A, $p_{\text{A}} - p_{\text{atm}}$, is given by

$$p_{\text{A}} + \gamma_{\text{AB}}\Delta z_{\text{AB}} + \gamma_{\text{BC}}\Delta z_{\text{BC}} - \gamma_{\text{CD}}\Delta z_{\text{CD}} = p_{\text{atm}}$$

$$p_{\text{A}} + 9.79(0.30) + 29.6(0) - 29.6(0.20) = p_{\text{atm}} \quad \rightarrow \quad p_{\text{A}} - p_{\text{atm}} = \mathbf{2.98\ kPa}$$

Differential manometers. Manometers that measure differences in pressure are called *differential manometers.* When large pressure differences are being measured, a heavy manometer fluid such as mercury is desirable, whereas for small pressure differences, a light fluid such as oil (or even air) is preferable.

EXAMPLE 2.10

A differential manometer, which is used to measure the difference in pressure between two water transmission lines, is illustrated in Figure 2.14. If the specific weight of the water is 9.79 kN/m^3 and the specific weight of the gauge fluid is 29.6 kN/m^3, what is the difference in pressure between the two pipelines?

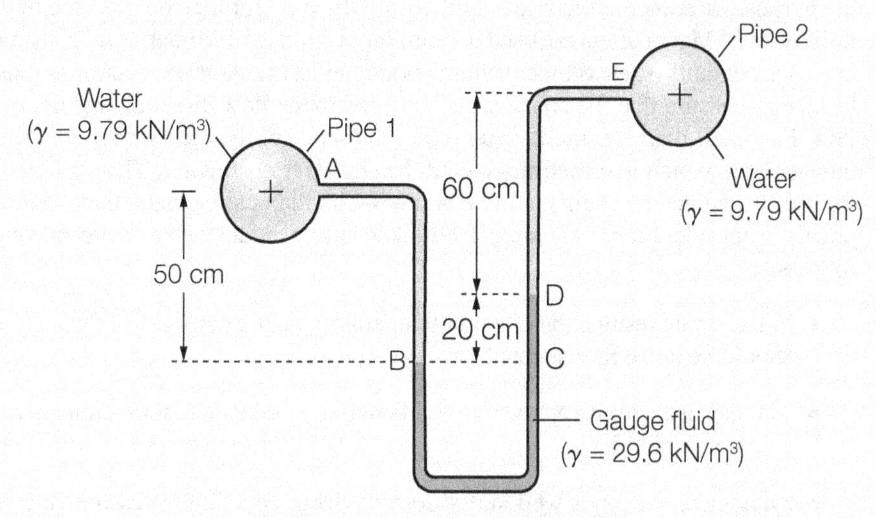

Figure 2.14: Differential manometer

SOLUTION

From the given data: $\gamma_{AB} = \gamma_{DE} = 9.79$ kN/m^3, $\gamma_{BC} = \gamma_{CD} = 29.6$ kN/m^3, $\Delta z_{AB} = 0.50$ m, $\Delta z_{CD} = 0.20$ m, and $\Delta z_{DE} = 0.60$ m. Applying the hydrostatic pressure distribution in each fluid layer (i.e., applying Equation 2.35) and noting that the pressure at Points B and C are the same, because they are at the same elevation, gives

$$p_A + \gamma_{AB}\Delta z_{AB} - \gamma_{CD}\Delta z_{CD} - \gamma_{DE}\Delta z_{DE} = p_E$$

$$p_A + (9.79)(0.50) - (29.6)(0.20) - (9.79)(0.60) = p_E \quad \rightarrow \quad p_A - p_E = 6.90 \text{ kPa}$$

Therefore, the difference in pressure between the pipe connections at Points A and E is **6.90 kPa**, with the higher pressure at the connection located at A.

Practical Considerations

U-tube manometers can be used to measure moderately high pressures and can be used in cases of negative gauge pressure. When large pressures are to be measured, several U-tube manometers can be connected in series. The usual motivation for using an inclined-tube manometer is to obtain a more accurate measurement of the location of the fluid interface contained within the inclined tube. Hence, inclined-tube manometers are more commonly used when small pressure changes are involved. For a given vertical change in the interface elevation, Δz, the interface will move a greater distance, Δs, along the inclined tube such that $\Delta s = \Delta z / \sin\theta$, where θ is the inclination of the tube. To avoid increased uncertainty in locating the exact position of the interface, it is generally recommended that $\theta \geq 5°$. Both U-tube and inclined-tube manometers are often made of glass or plastic, and the most common manometer liquid is mercury, although other liquids such as water, alcohol, carbon tetrachloride, or oil are sometimes used. The fluid in the manometer is commonly called the *gauge fluid*, and a gauge fluid must be immiscible with other fluids in contact with it. When multiple gauge fluids are used, the fluids must be layered such that the densest fluid is on the bottom and the fluids decrease in density from the bottom up. If this ordering is not followed, the gauge fluids will naturally adjust themselves to this order. High-density gauge fluids serve the purpose of reducing the rise height within the tube for a given pressure in the connected fluid system. Manometers are used in a similar fashion to piezometers to measure the pressure in closed conduits (such as pipes) with flowing fluids; the same connection precautions should be taken to ensure that the connection is normal to the flow direction and not obstructing the flow. For small-diameter manometer tubes, the shape of the meniscus might need to be taken into account, which is sometimes called the *capillary correction*. The capillary correction is frequently avoided by using manometer tubes with diameters greater than 15 mm. Additional practical considerations that should be taken into account when using manometers are as follows:

- For accurate results, the effect of temperature on the densities of the manometer liquids should be taken into account.

- The density within each gauge fluid should be constant; for example, water must not contain air bubbles.

- Manometers are not well suited to measuring very high pressures that are changing rapidly with time. Mechanical or electronic pressure measurement devices might be preferable in those cases.

For measuring very small pressure differences, *micro-manometers* are sometimes used. In these types of manometers, the location of the meniscus is typically measured using a small telescope with cross hairs. Such readings usually take time, and micro-manometers are appropriate mostly for measuring pressures that are at a steady state.

2.4 Forces on Plane Surfaces

Hydrostatic forces on plane (i.e., flat) surfaces are important factors in the design of storage tanks, ships, dams, and other similar structures that are in contact with fluids. Such cases typically involve static liquids in which the pressure varies significantly with depth. Consider the general case of a plane (flat) surface within a static liquid, as illustrated in Figure 2.15. The pressure, p, on the area $\mathrm{d}A$ located at a vertical distance h below the liquid surface is given by

$$p = p_0 + \gamma h \tag{2.36}$$

where p_0 is the pressure at the liquid surface. It is convenient to utilize the y-coordinate in the surface plane as illustrated in Figure 2.15(a). If the y-axis intersects the liquid surface at an angle θ, then the resultant hydrostatic pressure force, F, acting normal to the plane surface area, A, is given by

$$F = \int_A p\,\mathrm{d}A = \int_A (p_0+\gamma h)\,\mathrm{d}A = \int_A p_0\,\mathrm{d}A + \int_A \gamma y \sin\theta\,\mathrm{d}A = p_0 A + \gamma \sin\theta \int_A y\,\mathrm{d}A \tag{2.37}$$

which can be written as

$$F = (p_0 + \gamma \bar{y} \sin\theta)A \tag{2.38}$$

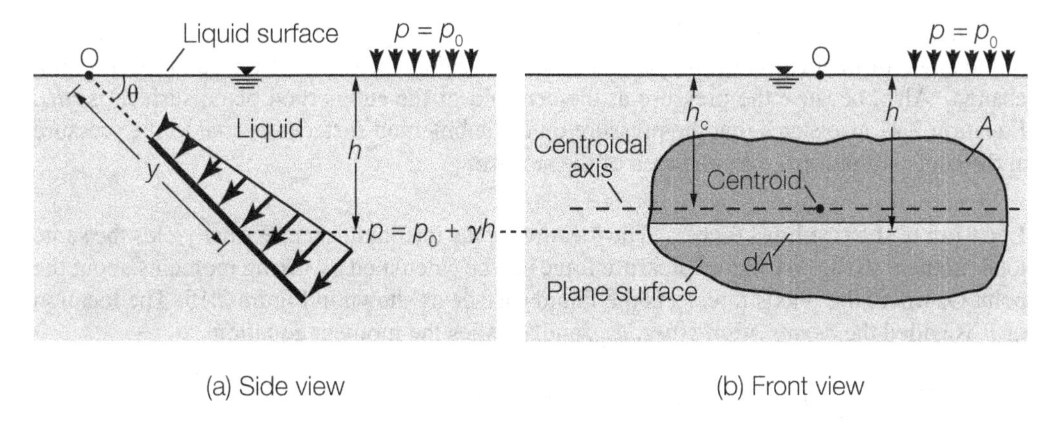

Figure 2.15: Force on a plane surface

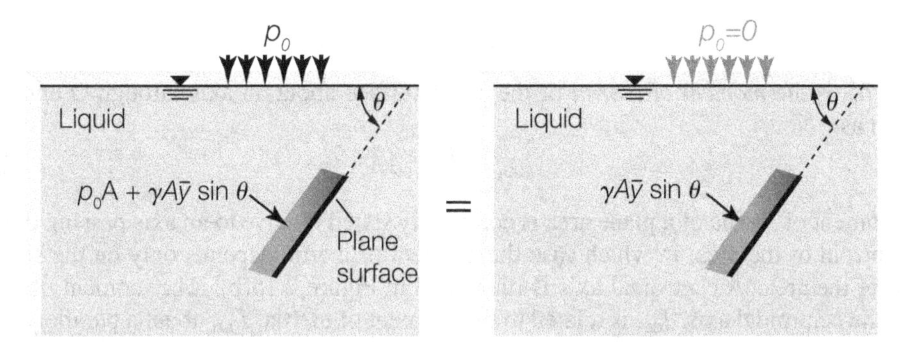

Figure 2.16: Hydrostatic force on a plane surface due to liquid only

where $\bar{y}$ is the distance to the centroid of the plane surface measured along the y-axis and given by

$$\bar{y} = \frac{1}{A}\int_A y\,\mathrm{d}A \tag{2.39}$$

In most cases, the hydrostatic force of interest is the force exerted by the liquid that is in contact with the plane surface. If the liquid was not present (e.g., if the liquid was drained to a level below the plane surface), then the force on the surface would be equal to p_0A. Hence, the force exerted by the liquid on the plane surface is equal to the total force given by Equation 2.38 minus p_0A. Alternatively, the net hydrostatic force exerted on the plane surface by the liquid can be obtained by simply taking $p_0 = 0$. A scenario in which it is not appropriate to take $p_0 = 0$ occurs when the surface of the liquid is actually the interface with another liquid. In such cases, p_0 is equal to the pressure at the liquid interface. Using Equation 2.38, the hydrostatic force on a plane surface can be expressed as

$$F = \begin{cases} (p_0 + \gamma\bar{y}\sin\theta)A, & p_0 \neq 0 \\ \gamma A\bar{y}\sin\theta, & p_0 = 0 \end{cases} \tag{2.40}$$

Equation 2.40 can be expressed in an alternative form by noting that $\bar{y}\sin\theta$ is equal to the vertical depth of the centroid, h_c; hence, Equation 2.40 can be expressed as

$$F = \begin{cases} (p_0 + \gamma h_c)A, & p_0 \neq 0 \\ \gamma A h_c, & p_0 = 0 \end{cases} \tag{2.41}$$

It is clear from Equation 2.41 that the force on a submerged plane surface is independent of the angle of inclination of the plane surface as long as the depth of the centroid does not change. Also, because the pressure at the centroid of the submerged plane surface is γh_c, Equation 2.41 also states that the pressure on the submerged surface is equal to the pressure at the centroid multiplied by the area of the surface.

Location of the resultant force. The location of the resultant force, F, that yields the same total moment as the distributed pressure force can be calculated by taking moments about the point O, where the y-axis intersects the liquid surface as shown in Figure 2.15. The location of F is called the *center of pressure*, $y_{\rm cp}$, and satisfies the moment equation

$$F y_{\rm cp} = \int_A yp \, {\rm d}A = \int_A y(p_0 + \gamma y \sin\theta) \, {\rm d}A = p_0 \int_A y \, {\rm d}A + \gamma \sin\theta \int_A y^2 \, {\rm d}A$$

which can be compactly expressed as

$$F y_{\rm cp} = p_0 A \bar{y} + \gamma \sin\theta I_{00} \tag{2.42}$$

where I_{00} is the *moment of inertia* of the plane surface about an axis through O and I_{00} is defined as

$$I_{00} = \int_A y^2 \, {\rm d}A \tag{2.43}$$

The moment of inertia of a plane area is commonly stated relative to an axis passing through the centroid of the area, in which case the moment of inertia depends only on the size and shape of the area. A centroidal axis is illustrated in Figure 2.15(b). The moment of inertia about the centroidal axis, $I_{\rm cc}$, is related to the moment of inertia, I_{00}, about a parallel axis by the relation

$$I_{00} = I_{\rm cc} + Ad^2 \tag{2.44}$$

where d is the distance between the centroidal axis and the parallel axis. The relation given by Equation 2.44 is commonly called the *parallel axis theorem* or the *parallel-axis-transfer theorem*. In the present case, $d = \bar{y}$, and Equation 2.44 can be written as

$$I_{00} = I_{\rm cc} + A\bar{y}^2 \tag{2.45}$$

The moments of inertia of several plane areas about their centroidal axes are given in Appendix C. Combining Equations 2.42 and 2.45 yields

$$F y_{\rm cp} = p_0 A \bar{y} + \gamma \sin\theta (I_{\rm cc} + A\bar{y}^2) = A\bar{y}(p_0 + \gamma \sin\theta \bar{y}) + \gamma \sin\theta I_{\rm cc} \tag{2.46}$$

Substituting Equation 2.38 for the hydrostatic force, F, gives

$$y_{\rm cp} = \bar{y} + \frac{\gamma \sin\theta I_{\rm cc}}{(p_0 + \gamma \bar{y} \sin\theta)A} \tag{2.47}$$

In the common case where the hydrostatic force of interest is due only to the liquid above the surface, $p_0 = 0$. Hence, the location of the center of pressure of a hydrostatic force can be presented generally as

$$y_{\rm cp} = \begin{cases} \bar{y} + \dfrac{\gamma \sin\theta I_{\rm cc}}{(p_0 + \gamma \bar{y} \sin\theta)A}, & p_0 \neq 0 \\[2ex] \bar{y} + \dfrac{I_{\rm cc}}{A\bar{y}}, & p_0 = 0 \end{cases} \tag{2.48}$$

This equation shows that the center of pressure, $y_{\rm cp}$, is below the centroid, $\bar{y}$, of the plane area, because the quantity added to $\bar{y}$ in Equation 2.48 is generally positive.

EXAMPLE 2.11

The rectangular gate shown in Figure 2.17(a) has dimensions 3 m × 2 m and is pin-connected at B. If the surface on which the gate rests at A is frictionless, what is the reaction at A?

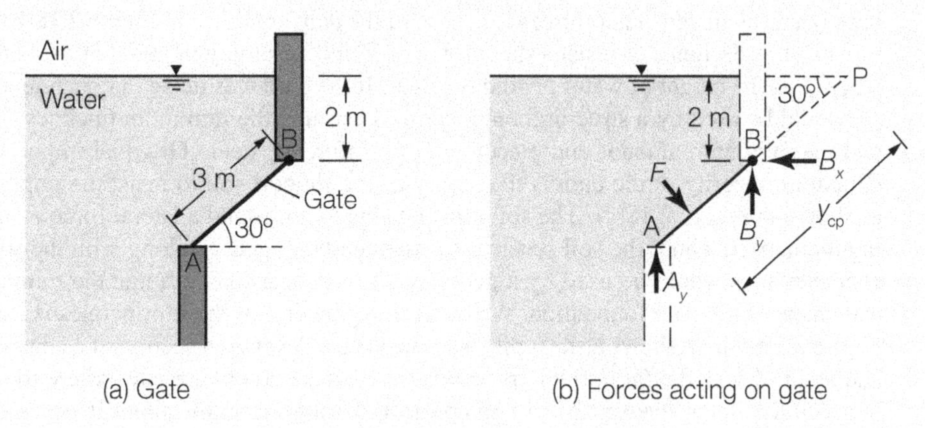

Figure 2.17: Rectangular gate

SOLUTION

A free-body diagram of the gate is shown in Figure 2.17(b), where B_x and B_y are the components of the support force at B (the pin is assumed to be frictionless; hence, there is no fixed moment at B), A_y is the normal reaction of the support surface at A (the surface at A is frictionless; hence, the support force is normal to the surface), and F is the hydrostatic force on the gate exerted by the water. The hydrostatic force on the gate is the same as if the water surface extended to the point P. Because the gate is rectangular, the distance, $\bar{y}$, of the centroid of the gate from P is given by

$$\bar{y} = \text{PB} + \frac{3}{2}\text{ m} = \frac{2}{\sin 30^\circ} + \frac{3}{2} = 5.5\text{ m}$$

and the geometric properties of the gate are

$$A = (2)(3) = 6\text{ m}^2,\ I_{xx} = \frac{bd^3}{12} = \frac{(2)(3)^3}{12} = 4.5\text{ m}^4,\ y_{\text{cp}} = \bar{y} + \frac{I_{xx}}{A\bar{y}} = 5.5 + \frac{4.5}{(6)(5.5)} = 5.64\text{ m}$$

Taking the specific weight of water as $\gamma = 9.79$ kN/m^3, the hydrostatic force, F, on the gate is given by

$$F = \gamma A\bar{y}\sin\theta = (9.79)(6)(5.5)\sin 30^\circ = 162\text{ kN}$$

The reaction force at A, represented by A_y, can be determined by taking moments about B, in which case

$$F[y_{\text{cp}} - \text{PB}] = A_y(3\cos 30^\circ) \quad\rightarrow\quad 162\left[5.64 - \frac{2}{\sin 30^\circ}\right] = A_y(3\cos 30^\circ) \quad\rightarrow\quad A_y = 102\text{ kN}$$

The gate therefore exerts a force of **102 kN** on the support at A.

Practical Considerations

The forces exerted on a structure determine the support forces and moments that must be provided to keep the structure standing. To illustrate this concept, consider the dam shown in Figure 2.18 where the resultant hydrostatic force, F, and its distance below the surface, $y_{\rm cp}$, have been calculated. Therefore, at the base of the dam as shown in Figure 2.18(b), the structure must be designed to resist a shear force of F and a bending moment of $M = F(h - y_{\rm cp})$, where h is the height of water behind the dam. If the dam was made of concrete, then F and M would be used by a structural engineer to determine the minimum thickness of the dam and the minimum amount and placement of reinforcing steel. Of equal importance to the structural integrity of the dam is the ability of the support soil to resist the imposed forces, as shown in Figure 2.18(c). The soil must be able to withstand a lateral force F and a bending moment M about the soil surface. Consequently, F and M along with the soil strength characteristics would be used by a geotechnical engineer to determine the minimum depth and shape of the dam foundation. A practical application of these principles is illustrated in Figure 2.19, which shows a sheet pile wall between a dewatered section of a construction site and an un-dewatered (at normal groundwater level) section of the site, where the difference between the water levels is Δh. In this project, dewatering of the site was necessary to construct the foundation of a building. To provide sufficient resistance to the hydrostatic force and corresponding overturning moment, structural and geotechnical engineers had to ensure that the sheet pile wall was sufficiently strong and was driven to sufficient depth to resist the hydrostatic load.

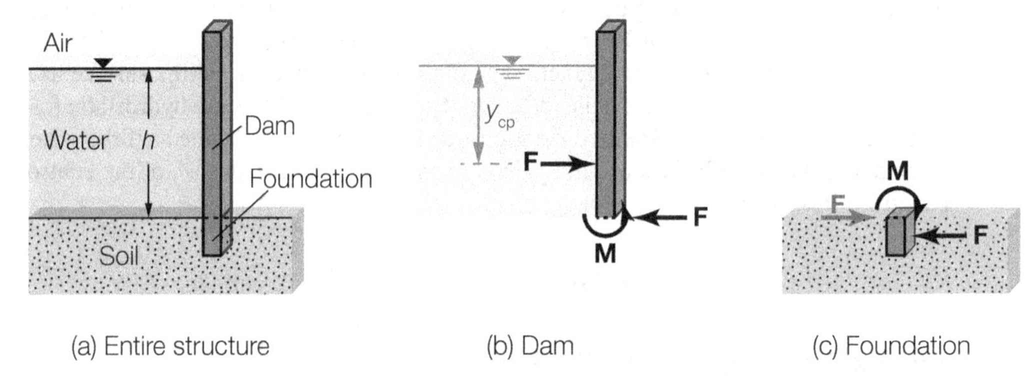

Figure 2.18: Support forces on a dam

Figure 2.19: Sheet-pile wall separating an un-dewatered and dewatered section of a construction site

EXAMPLE 2.12

A 10-m-wide salinity control gate shown in Figure 2.20 is used to control the subsurface movement of salt water into inland areas. (a) Determine the magnitude and location of the net hydrostatic force on the gate. (b) If the gate is mounted on rollers with a coefficient of friction equal to 0.2 and the gate weighs 10 kN, calculate the force required to lift the gate. The density of seawater at 20°C can be taken as 1025 kg/m^3.

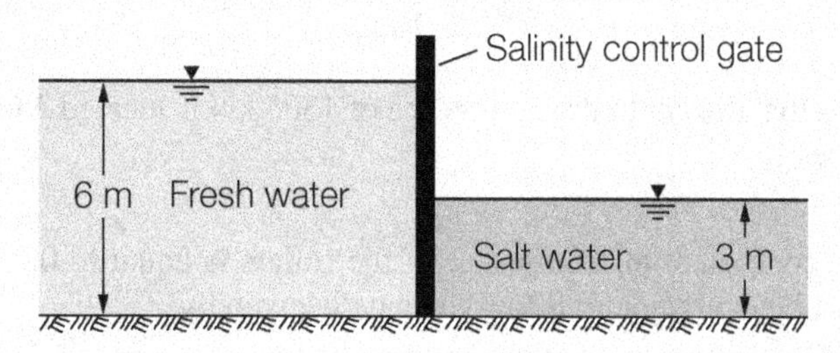

Figure 2.20: Salinity control gate

SOLUTION

From the given data: $\gamma_f = 9.79$ kN/m^3, $A_f = (6)(10) = 60$ m^2, $\bar{h}_f = 0.5(6) = 3$ m, $\gamma_s = (1025)(9.807) = 10\,050$ N/m^3 $= 10.1$ kN/m^3, $A_s = (3)(10) = 30$ m^2, and $\bar{h}_s = 0.5(3) = 1.5$ m.

(a) On the freshwater side of the gate, the total hydrostatic force, F_f, is given by Equation 2.41 as

$$F_f = \gamma_f A_f \bar{h}_f = (9.79)(60)(3) = 1760 \text{ kN}$$

On the saltwater side, the total hydrostatic force, F_s, is given by

$$F_s = \gamma_s A_s \bar{h}_s = (10.1)(30)(1.5) = 455 \text{ kN}$$

The net hydrostatic force, F, on the gate is given by

$$F = F_f - F_s = 1760 - 455 = \mathbf{1305\ kN}$$

The location, y_{cp}, of the resultant hydrostatic force on each side of the gate is calculated using Equation 2.48, where

$$y_{cp} = \bar{y} + \frac{I_{cc}}{A\bar{y}}$$

On the freshwater side of the gate, $\bar{y} = 3$ m, $I_{cc} = (10)(6)^3/12 = 180$ m^4, and $A = 60$ m^2; therefore, the center of pressure, y_{cp}, is given by

$$y_{cp}\big|_{fresh} = 3 + \frac{180}{(60)(3)} = 4 \text{ m}$$

On the saltwater side of the gate, $\bar{y} = 1.5$ m, $I_{cc} = (10)(3)^3/12 = 22.5$ m^4, and $A = 30$ m^2, and the center of pressure, y_{cp}, is given by

$$y_{cp}\big|_{salt} = 1.5 + \frac{22.5}{(30)(1.5)} = 2 \text{ m}$$

The location of the net hydrostatic force, y_0, can be determined by taking moments about the surface of the freshwater, which gives

$$Fy_0 = F_f\, y_{cp}|_{fresh} - F_s(3 + y_{cp}|_{salt}) = (1760)(4) - (455)(3+2) = 4765 \text{ kN·m}$$

Because $F = F_f - F_s = 1760 - 455 = 1305$ kN,

$$y_0 = \frac{4765}{1305} = 3.65 \text{ m}$$

which means that the net hydrostatic force of 1305 kN is located **3.65 m** below the freshwater surface.

(b) Because the coefficient of friction, μ, of the rollers is equal to 0.2, the frictional force, F_μ, that must be overcome in lifting the gate is given by

$$F_\mu = \mu F = 0.2(1305) = 261 \text{ kN}$$

The total force to lift the gate is equal to the weight of the gate (10 kN) plus the frictional force, F_μ. Therefore, the force required to lift the gate is 10 kN + 261 kN = **271 kN**.

Lateral location of center of pressure: Basic formulation. The center of pressure on a plane surface that is submerged in a liquid is located at the point (x_{cp},y_{cp}) relative to coordinate axes that are centered on the liquid surface and contained within the plane of the surface. The location of the center of pressure relative to axes centered at Point O on the liquid surface is illustrated in Figure 2.21, where the location of the centroid ($\bar{x}$,$\bar{y}$) of the plane surface is also shown. The magnitude, F, of the resultant force acting at the center of pressure was derived previously and is given by Equation 2.40; the y coordinate of the center of pressure, y_{cp}, is given by Equation 2.48. In cases where the plane surface is symmetrical about the y-axis, $x_{cp} = 0$, and in cases where the plane surface is not symmetrical about the y-axis,

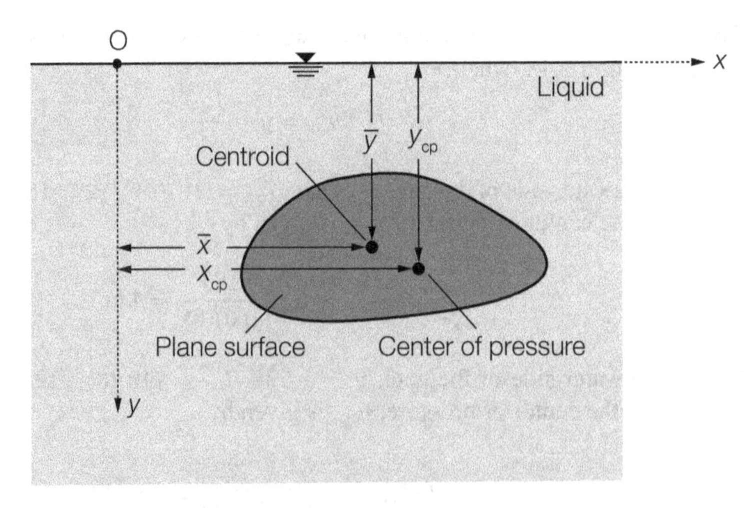

Figure 2.21: Location of the center of pressure on a plane surface

$x_{\mathrm{cp}} \neq 0$. The lateral location of the center of pressure on a submerged plane surface, x_{cp}, can be determined by using the moment equation

$$x_{\mathrm{cp}} F = \int_A x p \, \mathrm{d}A \tag{2.49}$$

where F is the total hydrostatic force on the plane surface, A is the area of the plane surface, and p is the hydrostatic pressure on an element of area $\mathrm{d}A$. Substituting the expression for the hydrostatic pressure distribution on a plane area $\mathrm{d}A$ into Equation 2.49 and taking $p_0 = 0$ yields

$$x_{\mathrm{cp}} F = \int_A x(\gamma y \sin\theta) \, \mathrm{d}A$$

which can be expressed as

$$x_{\mathrm{cp}}(\gamma A \bar{y} \sin\theta) = \gamma \sin\theta \int_A xy \, \mathrm{d}A = \gamma \sin\theta I_{xy} \tag{2.50}$$

where I_{xy} is the *product moment of inertia* of the area A with respect to the x- and y-axes. In accordance with the parallel axis theorem, the product moment of inertia about the centroidal axes of any area A, denoted by I_{xyc}, is related to the product moment of inertia, I_{xy}, about any other set of axes by the relation

$$I_{xy} = I_{xyc} + A\bar{x}\bar{y} \tag{2.51}$$

where I_{xyc} is the product moment of inertia relative to the centroidal axes and $\bar{x}$ and $\bar{y}$ are the coordinates of the centroid relative to the actual x- and y-axes being used. Equation 2.51 assumes that the centroidal x- and y-axes are parallel to the x- and y-axes used to define the plane area. Combining Equations 2.51 and 2.50 yields the x-coordinate of the center of pressure as

$$x_{\mathrm{cp}} = \bar{x} + \frac{I_{xyc}}{A\bar{y}} \tag{2.52}$$

It is noteworthy that if the area, A, is symmetrical with respect to an axis passing through the centroid, the resultant force must lie along the line $x = \bar{x}$, because I_{xyc} is equal to zero in this case. Under this condition, the center of pressure is directly below the centroid on the y-axis.

Lateral location of center of pressure: Alternative formulation. As an alternative to the formulation given by Equation 2.52, the lateral location of the center of pressure, x_{cp}, of an area A can sometimes be more conveniently determined using the following formulation of the moment equation

$$x_{\mathrm{cp}} F = \int_A x_{\mathrm{p}} p \, \mathrm{d}A = \int_{y_1}^{y_2} x_{\mathrm{p}}(\gamma y \sin\theta) \, \mathrm{d}A \tag{2.53}$$

where x_{p} is the x-coordinate of the centroid of the elemental area $\mathrm{d}A$. Equation 2.53 can be applied to elemental strips that are at the same depth (y) from the liquid surface, in which case x_{p} is the distance to the centroid of each elemental strip at a distance y below the surface. For example, if the width, b, of the area element is expressed as a function of y, then $\mathrm{d}A = b(y)\mathrm{d}y$ and Equation 2.53 can be expressed as

$$x_{\mathrm{cp}} F = \int_{y_1}^{y_2} x_{\mathrm{p}}(\gamma y \sin\theta) b(y) \, \mathrm{d}y \tag{2.54}$$

where it is important to remember that $b(y)$ represents "b as a function of y," not b multiplied by y.

EXAMPLE 2.13

The gate shown in Figure 2.22 is to be opened using a motor. What torque must be supplied by the motor to open the gate?

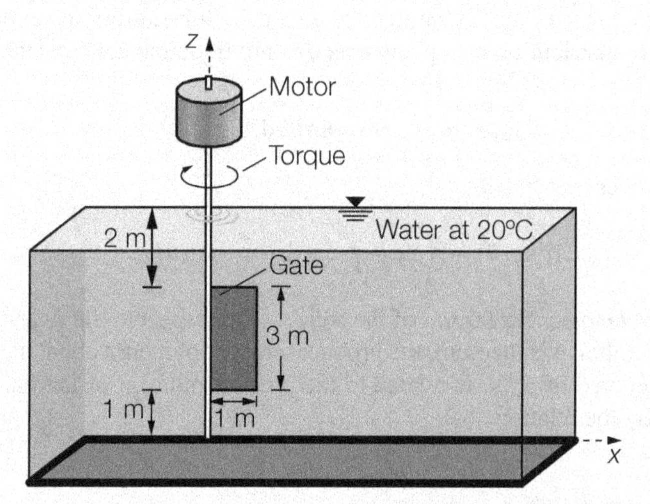

Figure 2.22: Gate to be opened by a motor

SOLUTION

Use the x- and z-axes shown in Figure 2.22; at 20°C, the specific weight of water $\gamma = 9.79\ \text{kN/m}^3$. From the given dimensions of the rectangular gate, $b = 1$ m and $d = 3$ m; hence,

$$A = bd = (1)(3) = 3\ \text{m}^2, \qquad \bar{x} = \frac{b}{2} = \frac{1}{2}\ \text{m} = 0.5\ \text{m}$$

$$\bar{y} = 2\ \text{m} + \frac{d}{2} = 2\ \text{m} + \frac{3}{2}\ \text{m} = 3.5\ \text{m}, \qquad I_{xyc} = 0\ \text{(property of a rectangle)}$$

$$x_{\text{cp}} = \bar{x} + \frac{I_{xyc}}{A\bar{y}} = 0.5\ \text{m} + \frac{0}{(3)(3.5)} = 0.5\ \text{m}, \qquad F = \gamma A\bar{y} = (9.79)(3)(3.5) = 103\ \text{kN}$$

$$T = Fx_{\text{cp}} = (103)(0.5) = 51.5\ \text{kN·m}$$

Therefore, the torque required to open the gate is **51.5 kN·m**. A motor capable of producing this torque should be selected.

Layers of different liquids. The formulations presented here can be applied to cases in which, instead of having one homogeneous liquid, there are multiple layers of different immiscible liquids. In such cases, the force on the plane surface and its location are calculated separately for each homogeneous layer using the derived expressions for homogeneous liquids. The total force is then the sum of the individual forces, and the (effective) location of the total force is at the point where it yields the same moment as the sum of the moments exerted by the forces in the individual layers.

EXAMPLE 2.14

A storage tank contains 3 layers of liquids as shown in Figure 2.23. A 0.5-m-thick layer of SAE 30 oil overlays a 0.6-m layer of water, which overlays a 0.65-m layer of glycerin. All liquids are at 20°C. The wall of the tank is 2-m long. Estimate the magnitude and location of the resultant force on the wall of the tank.

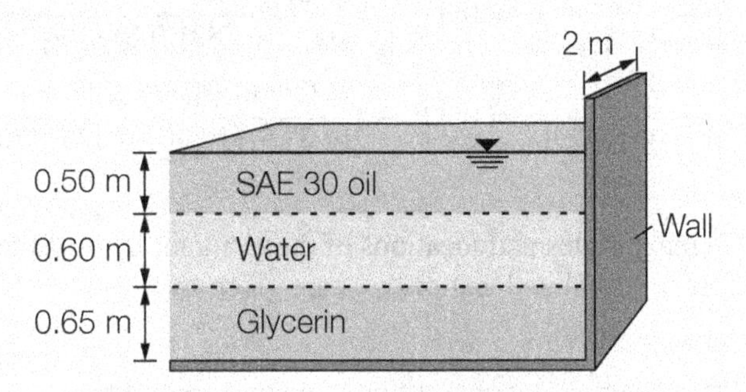

Figure 2.23: Three liquid layers in a storage tank

SOLUTION

From the given data: $\Delta z_1 = 0.50$ m, $\Delta z_2 = 0.60$ m, $\Delta z_3 = 0.65$ m, and $L = 2$ m. From the liquid properties given in Appendix B.4, $\rho_1 = 918$ kg/m^3, $\rho_2 = 998.2$ kg/m^3, and $\rho_3 = 1260$ kg/m^3. The corresponding specific weights are $\gamma_1 = 9.003$ kN/m^3, $\gamma_2 = 9.789$ kN/m^3, and $\gamma_3 = 12.36$ kN/m^3. The following preliminary calculations are useful:

$$A_1 = L\Delta z_1 = (2)(0.50) = 1.0 \text{ m}^2, \qquad A_2 = L\Delta z_2 = (2)(0.60) = 1.2 \text{ m}^2$$

$$A_3 = L\Delta z_3 = (2)(0.65) = 1.3 \text{ m}^2, \qquad \bar{y}_1 = \frac{\Delta z_1}{2} = \frac{0.5}{2} = 0.25 \text{ m}$$

$$\bar{y}_2 = \frac{\Delta z_2}{2} = \frac{0.60}{2} = 0.30 \text{ m}, \qquad \bar{y}_3 = \frac{\Delta z_3}{2} = \frac{0.65}{2} = 0.325 \text{ m}$$

$$I_{c1} = \frac{L(\Delta z_1)^3}{12} = \frac{(2)(0.50)^3}{12} = 2.083 \times 10^{-2} \text{ m}^4$$

$$I_{c2} = \frac{L(\Delta z_2)^3}{12} = \frac{(2)(0.60)^3}{12} = 3.600 \times 10^{-2} \text{ m}^4$$

$$I_{c3} = \frac{L(\Delta z_3)^3}{12} = \frac{(2)(0.65)^3}{12} = 4.577 \times 10^{-2} \text{ m}^4$$

$$p_{02} = \gamma_1\Delta z_1 = (9.003)(0.50) = 4.502 \text{ kPa}$$

$$p_{03} = p_{02} + \gamma_2\Delta z_2 = 4.502 + (9.789)(0.60) = 10.38 \text{ kPa}$$

Using the given and calculated parameters, the resultant hydrostatic force on the wall exerted by each liquid layer is calculated from Equation 2.40 as follows:

$$F_1 = \gamma_1 A_1 \bar{y}_1 = (9.003)(1.0)(0.25) = 2.251 \text{ kN}$$

$$F_2 = [p_{02} + \gamma_2\bar{y}_2]A_2 = [4.502 + (9.789)(0.30)](1.2) = 8.926 \text{ kN}$$

$$F_3 = [p_{03} + \gamma_3\bar{y}_3]A_3 = [10.38 + (12.36)(0.325)](1.3) = 18.72 \text{ kN}$$

The location of each resultant force relative to the top of the liquid layer is calculated from Equation 2.48 as follows:

$$y_{cp1} = \bar{y}_1 + \frac{I_{c1}}{A_1\bar{y}_1} = 0.25 + \frac{2.083 \times 10^{-2}}{(1.0)(0.25)} = 0.3333 \text{ m}$$

$$y_{cp2} = \bar{y}_2 + \frac{\gamma_2 I_{c2}}{[p_{02} + \gamma_2\bar{y}_2]A_2} = 0.3 + \frac{(9.789)(3.600 \times 10^{-2})}{[4.502 + (9.789)(0.3)](1.2)} = 0.3395 \text{ m}$$

$$y_{cp3} = \bar{y}_3 + \frac{\gamma_3 I_{c3}}{[p_{03} + \gamma_3\bar{y}_3]A_3} = 0.325 + \frac{(12.36)(4.577 \times 10^{-2})}{[10.38 + (12.36)(0.325)](1.3)} = 0.3552 \text{ m}$$

Using the calculated magnitudes and locations of the resultant hydrostatic forces in each layer, the total resultant force, F, and its location, y_{cp}, are given by

$$F = F_1 + F_2 + F_3 = 2.251 + 8.926 + 18.72 = 29.90 \approx 29.9 \text{ kN}$$

$$y_{cp} = \frac{F_1 y_{cp1} + F_2(\Delta z_1 + y_{cp2}) + F_3(\Delta z_1 + \Delta z_2 + y_{cp3})}{F}$$

$$= \frac{(2.251)(0.3333) + (8.926)(0.50 + 0.3395) + (18.72)(0.50 + 0.60 + 0.3552)}{29.90} = 1.17 \text{ m}$$

Therefore, the resultant force on the wall is approximately **29.9 kN** and is located approximately **1.17 m** below the surface of the liquid in the tank.

2.5 Forces on Curved Surfaces

In many engineering applications, the surface of interest is nonplanar (i.e., curved); such surfaces include tanks, dams, and pipes. An example of a dam is shown in Figure 2.24, where the upstream face of the dam is curved such that the hydrostatic forces on the dam are supported by the walls of the canyon in which the dam is located.

Horizontal component of the hydrostatic force. Consider the general case of a curved surface in a static liquid, as illustrated in Figure 2.25. If the liquid between the curved surface and the liquid surface is taken as a free body, then the (horizontal) hydrostatic pressures above the plane AA on both sides of the free body are equal and opposite. Therefore, the hydrostatic force, F_{AB}, on the vertical projection of the surface (AB) must be equal to the horizontal component of the reaction force of the curved surface. Denoting the reaction force of the curved surface by R, with components R_x and R_y, balancing forces in the x-direction yields

$$R_x = F_{AB} = (p_0 + \gamma\bar{y}_v)A_v \tag{2.55}$$

where p_0 is the pressure on the liquid surface, A_v is the projected area of the curved surface onto the vertical plane, and $\bar{y}_v$ is the vertical distance from the liquid surface to the centroid of A_v. In most cases, only the hydrostatic force due to the liquid above the curved surface is of interest, in which case $p_0 = 0$ is used in Equation 2.55. This condition is illustrated in Figure 2.26. Cases in which $p_0 \neq 0$ occur when the surface of the liquid is actually the interface with

Figure 2.24: Dam structure
Source: Bureau of Reclamation, U.S. Department of the Interior.

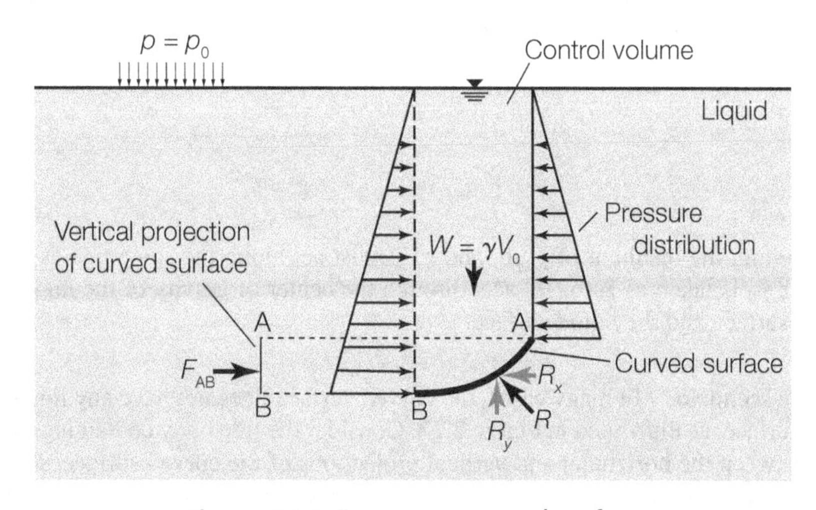

Figure 2.25: Force on a curved surface

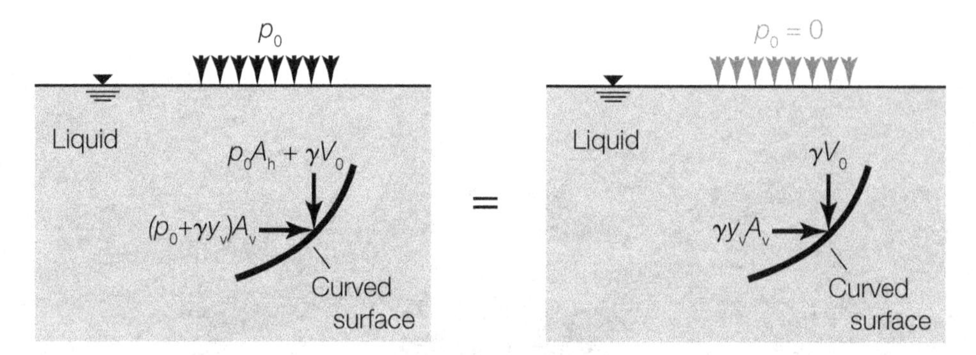

Figure 2.26: Net hydrostatic force on a curved surface

another liquid. In such cases, p_0 is equal to the pressure at the liquid interface. The general expression for the horizontal component of the hydrostatic force on a curved surface is given by

$$R_x = \begin{cases} (p_0 + \gamma\bar{y}_v)A_v, & p_0 \neq 0 \\ \gamma A_v \bar{y}_v, & p_0 = 0 \end{cases} \tag{2.56}$$

For equilibrium, the force on the vertical projection of the curved surface, F_{AB}, and R_x must act in the same straight line (i.e., they are collinear). Therefore, R_x must act through the center of pressure of the projection of the curved surface in the vertical plane.

Vertical component of the hydrostatic force. The vertical component of the force on a curved surface can be obtained by balancing forces in the vertical direction, which yields

$$R_y = p_0 A_h + W = p_0 A_h + \gamma V_0 \tag{2.57}$$

where A_h is the projected area of the curved surface onto the horizontal plane, W is the weight of the liquid between the curved surface and the liquid surface, and V_0 is the corresponding volume of liquid. In most applications, the pressure on the liquid surface is equal to atmospheric pressure and the hydrostatic force due only to the liquid is being sought; in these cases, p_0 can be taken as effectively equal to zero. The general expression for the vertical component of the hydrostatic force on a curved surface is therefore given by

$$R_y = \begin{cases} p_0 A_h + \gamma V_0, & p_0 \neq 0 \\ \gamma V_0, & p_0 = 0 \end{cases} \tag{2.58}$$

For equilibrium, the resultant of $p_0 A_h$ and γV_0 must act along the same line as R_y, and in cases where p_0 is taken as zero, R_y acts through the center of gravity of the liquid between the curved surface and the liquid surface.

Alternative scenario. In many cases, the curved surface does not have any liquid directly above the surface, as illustrated in Figure 2.27. Consider the free body containing the volume of liquid between the horizontal and vertical projections of the curved surface. Equilibrium in the x-direction requires that the reaction force on the curved surface, R_x, be equal to the hydrostatic force on the vertical projection of the curved surface, F_v; therefore,

$$R_x = F_v = (p_0 + \gamma\bar{y}_v)A_v \tag{2.59}$$

where p_0 is the pressure on the liquid surface, A_v is the area of the vertical projection of the curved surface, and $\bar{y}_v$ is the distance of the centroid of A_v from the liquid surface. Equation 2.59 is the same as Equation 2.56, so Equation 2.56 is the general equation for calculating the horizontal force on a curved surface. Equilibrium in the vertical (y-) direction for the case shown in Figure 2.27 requires that

$$F_h - W - R_y = 0 \tag{2.60}$$

where F_h is the force exerted on the horizontal projection of the curved surface and is given by

$$F_h = (p_0 + \gamma h)A_h \tag{2.61}$$

where h is the depth of the horizontal projection from the liquid surface and A_h is the area of the horizontal projection. The weight of the liquid in the free body, W, is given by

$$W = \gamma V \tag{2.62}$$

where V is the volume of the free body. Combining Equations 2.60–2.62 yields

$$R_y = F_h - W = p_0 A_h + \gamma h A_h - \gamma V = p_0 A_h + \gamma(hA_h - V) \tag{2.63}$$

The term in parentheses is equal to the volume, V_0, between the curved surface and the extended liquid surface as indicated in Figure 2.27. Therefore, Equation 2.63 can be written as

$$R_y = p_0 A_h + \gamma V_0 \tag{2.64}$$

where V_0 is the volume between the curved surface and the extended liquid surface. Equation 2.64 is the same as Equation 2.58, provided that V_0 is taken as the volume between the curved surface and the extended liquid surface. Therefore, Equation 2.58 is the general equation for calculating the vertical force on a curved surface.

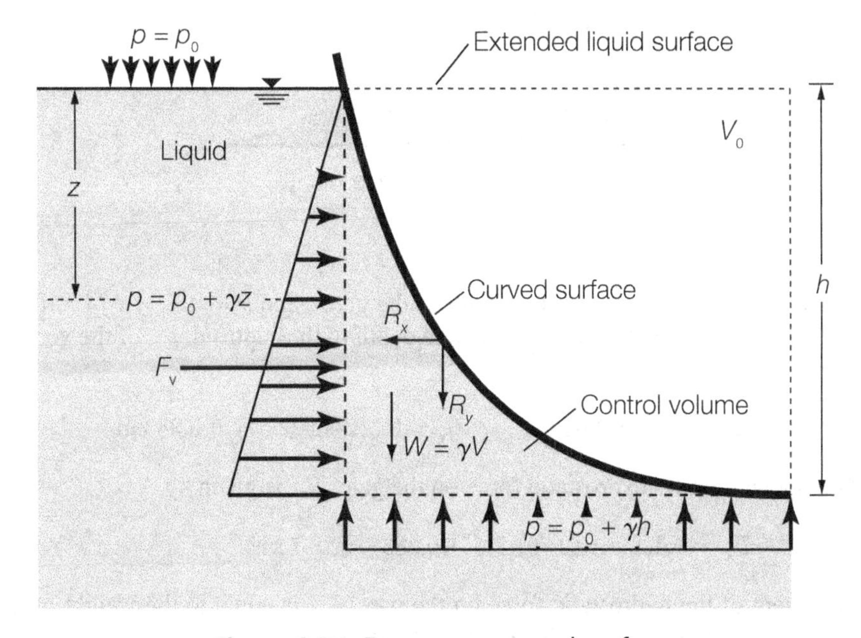

Figure 2.27: Force on a curved surface

Resultant hydrostatic force. The resultant hydrostatic force, **R**, on a curved surface can be represented by its magnitude, R, and orientation, θ, which can both be determined from the vertical and horizontal components of the hydrostatic force on a curved surface, R_x and R_y, respectively, using the relations

$$R = \sqrt{R_x^2 + R_y^2}, \quad \text{and} \quad \theta = \tan^{-1}\left(\frac{R_y}{R_x}\right) \tag{2.65}$$

The line of action of **R** can be determined by equating the moment of **R** about any point to the sum of the moments of all the forces that are used to calculate **R**. These forces generally include p_0A_h, γV_0, $\gamma A_\text{v}y_\text{v}$, R_x, and R_y, where each of these forces has been previously defined. When the curved surface is formed by a circular arc, the resultant hydrostatic force, **R**, always passes through the center of the circle from which the arc is formed. This is so because all pressure forces on the surface of the arc are normal to the surface and must therefore necessarily pass through the center point.

EXAMPLE 2.15

Water at 20°C is restrained behind a spillway using the radial (Tainter) gate illustrated in Figure 2.28. If the gate is 10-m wide and the water level is at the midpoint of the gate, calculate the hydrostatic force on the gate.

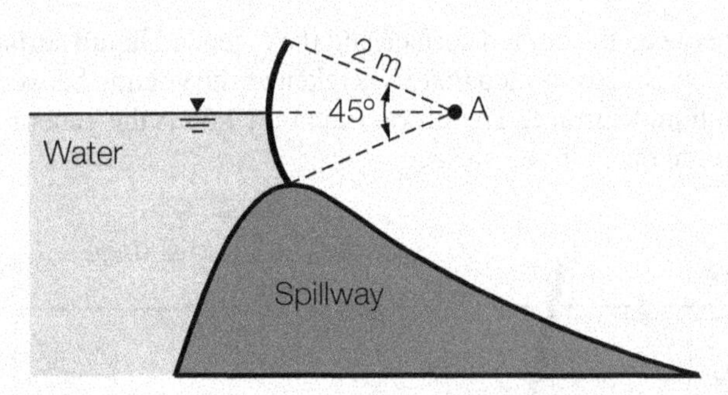

Figure 2.28: Tainter gate

SOLUTION

From the given data: r = 2 m, w = 10 m (= width of the gate), and θ = 45° = $\pi/4$ radians. For water at 20°C, γ = 9.79 kN/m^3. The area, A_v, and depth to the centroid, $\bar{y}_\text{v}$, of the vertical projection of the gate are calculated as

$$A_\text{v} = (2\sin 22.5°)(10) = 7.65\ \text{m}^2, \qquad \bar{y}_\text{v} = 0.5(2\sin 22.5°) = 0.383\ \text{m}$$

The horizontal component of the hydrostatic force on the gate, F_x, is given by

$$F_x = \gamma A_\text{v}\bar{y}_\text{v} = (9.79)(7.65)(0.383) = 28.7\ \text{kN}$$

The vertical component of the hydrostatic force on the gate, F_y, is equal to the weight of the water that would occupy the volume, V_0, illustrated in Figure 2.29(a).

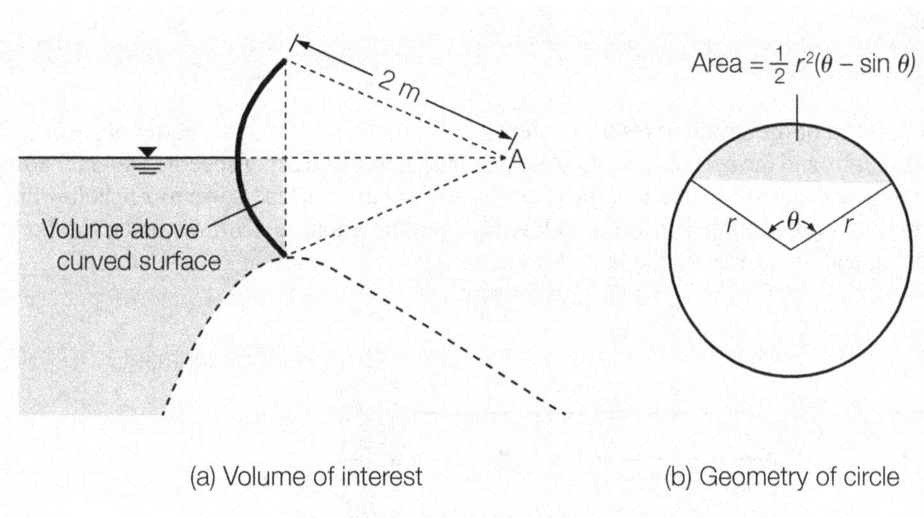

Figure 2.29: Geometric properties of a circle

Based on the geometry of a circle, illustrated in Figure 2.29(b),

$$V_0 = \left[\frac{1}{4}r^2(\theta - \sin\theta)\right] w = \left[\frac{1}{4}(2^2)\left(\frac{\pi}{4} - \sin\frac{\pi}{4}\right)\right](10) = 0.783 \text{ m}^3$$

The vertical component of the hydrostatic force on the gate is given by

$$F_y = \gamma V_0 = (9.79)(0.783) = 7.66 \text{ kN}$$

The total hydrostatic force, F, is then given by

$$F = \sqrt{F_x^2 + F_y^2} = \sqrt{(28.7)^2 + (7.66)^2} = 29.7 \text{ kN}$$

The Tainter gate must therefore support a hydrostatic force of **29.7 kN**.

Forces on three-dimensional surfaces. The previous analyses consider only curved surfaces that can be described in two dimensions, in which case the horizontal and vertical forces occur in the same plane. For irregular surfaces that cannot be described in two-dimensional space, the x-component of the force on the curved surface is in a different plane than the y-component of the force on the curved surface, in which case these forces cannot be vectorially summed to find a single resultant force. In other words, the force components must be maintained as separate entities.

Layers of different liquids. The formulations presented here can also be applied to cases in which there are layers of different immiscible liquids. In such cases, the force on the curved surface and its location are calculated separately for each homogeneous layer using the derived expressions for homogeneous liquids. The total force is the sum of the individual forces, and the location of the resultant force yields the same moment as the sum of the moments exerted by the forces in the individual layers.

EXAMPLE 2.16

The top 1 m of the stratified reservoir shown in Figure 2.30 contains a liquid of specific weight 9.50 kN/m^3, and the bottom 2 m of the reservoir contains a liquid of specific weight 9.80 kN/m^3. The 3-m-wide lateral segment of the reservoir has a planar surface down to 2 m below the liquid surface and the bottom 1 m has a radius of curvature equal to 1 m. Estimate the horizontal, vertical, and total force on the lateral segment.

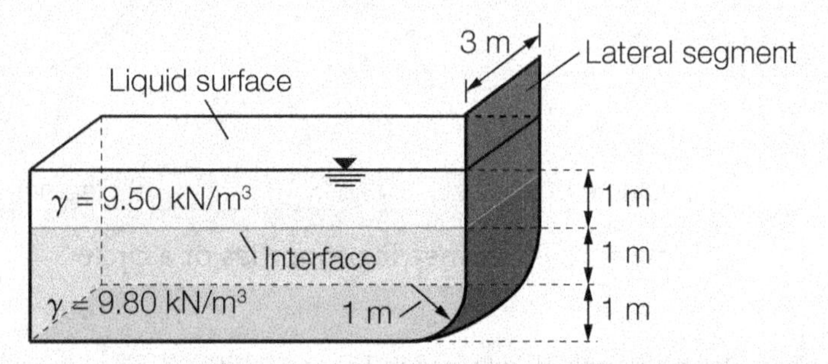

Figure 2.30: Layers of liquids in a reservoir

SOLUTION

Denote the liquids in the upper and lower layers with subscripts U and L, respectively. Then

$$\text{horizontal force due to upper layer, } F_{\text{hU}} = \gamma_{\text{U}}\bar{y}_{\text{Uv}}A_{\text{Uv}}$$

$$\text{horizontal force due to lower layer, } F_{\text{hL}} = (\gamma_{\text{U}}h_{\text{U}} + \gamma_{\text{L}}\bar{y}_{\text{Lv}})A_{\text{Lv}}$$

$$\text{total horizontal force, } F_{\text{h}} = F_{\text{hU}} + F_{\text{hL}}$$

$$\text{total vertical force, } F_{\text{v}} = \gamma_{\text{U}}V_{\text{U}} + \gamma_{\text{L}}V_{\text{L}}$$

$$\text{resultant force, } F = \sqrt{F_{\text{h}}^2 + F_{\text{v}}^2}$$

From the given data: $\gamma_{\text{U}} = 9.50$ kN/m^3, $A_{\text{Uv}} = 1$ m $\times$ 3 m $= 3$ m^2, $\bar{y}_{\text{Uv}} = 0.5$ m, $h_{\text{U}} = 1$ m, $\gamma_{\text{L}} = 9.80$ kN/m^3, $A_{\text{Lv}} = 2$ m $\times$ 3 m $= 6$ m^2, $\bar{y}_{\text{Lv}} = 1.0$ m, $V_{\text{U}} = 1$ m $\times$ 1 m $\times$ 3 m $= 3$ m^3, and $V_{\text{L}} = (1 \text{ m} \times 1 \text{ m} + \pi/4 \times 1^2) \times 3 \text{ m} = 5.356$ m^3. Using these data, the forces are given by

$$F_{\text{hU}} = (9.50)(0.5)(3) = 14.3 \text{ kN}$$

$$F_{\text{hL}} = [(9.50)(1) + (9.80)(1)](6) = 115.8 \text{ kN}$$

$$F_{\text{h}} = 14.3 \text{ kN} + 115.8 \text{ kN} = \mathbf{130.1 \text{ kN}}$$

$$F_{\text{v}} = (9.50)(3) + (9.80)(5.356) = \mathbf{81.0 \text{ kN}}$$

$$F = \sqrt{130.1^2 + 81.0^2} = \mathbf{153 \text{ kN}}$$

In the context of structural design, the magnitudes of these forces would subsequently be used in the design of the structural support members.

2.6 Buoyancy

The *buoyant force* is defined as the net upward force exerted on a body immersed in a fluid. Quantitatively, the buoyant force is equal to the net hydrostatic force exerted on the surface of the body by the surrounding fluid. The equations used to calculate the buoyant force are derived from the equations used to calculate forces on curved surfaces.

2.6.1 Fully Submerged Bodies

Consider the fully submerged body shown in Figure 2.31, where the volume of fluid above the upper surface of the body, above AB, is equal to V_0 and the volume of the body is V. The hydrostatic force on the upper surface of the submerged body is equal to the weight of the fluid above the upper surface, $F_{\rm U}$; hence,

$$F_{\rm U} = \gamma_{\rm f} V_0 \tag{2.66}$$

where $\gamma_{\rm f}$ is the specific weight of the fluid. The hydrostatic force on the lower surface of the submerged body is equal to the weight of the fluid above the lower surface, $F_{\rm L}$, where

$$F_{\rm L} = \gamma_{\rm f}(V_0 + V) \tag{2.67}$$

The net hydrostatic force, F, on the body is therefore given by

$$F = F_{\rm L} - F_{\rm U} = \gamma_{\rm f}(V_0 + V) - \gamma_{\rm f} V_0 = \gamma_{\rm f} V \tag{2.68}$$

This result indicates that the buoyant force on a submerged body is equal to the weight of the fluid displaced by the body. Denoting the buoyant force by $F_{\rm b}$ gives

$$F_{\rm b} = \gamma_{\rm f} V \tag{2.69}$$

This relationship is commonly known as *Archimedes' principle* or the *Principle of Archimedes*.[14] The buoyant force, $F_{\rm b}$, acts vertically upward and passes through the centroid of the volume of fluid displaced by the submerged body, where the centroid of the displaced volume is sometimes referred to as the *center of buoyancy*. Note that the center of buoyancy

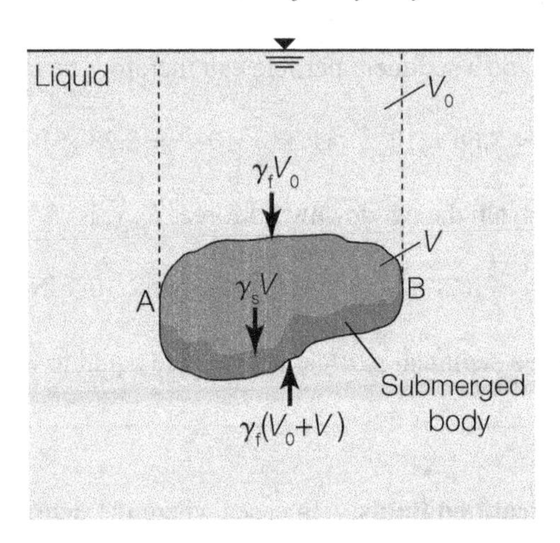

Figure 2.31: Forces on a submerged body

[14]Named in honor of the Greek philosopher, mathematician, and engineer Archimedes (287–212 BC).

need not coincide with the center of gravity of the submerged body. The buoyant force is countered by the weight, W, of the body that acts vertically downward, where

$$W = \gamma_s V \tag{2.70}$$

and γ_s is the specific weight of the submerged body. The net downward force, F_{net}, acting on a submerged body is therefore given by

$$F_{net} = (\gamma_s - \gamma_f)V \tag{2.71}$$

Clearly, a body will sink when its density is greater than that of the surrounding fluid ($\gamma_s > \gamma_f$) and will float when its density is less than that of the fluid ($\gamma_s < \gamma_f$). In the case of floating bodies, the buoyant force is equal to the weight of the volume of fluid displaced.

EXAMPLE 2.17

A spherical sediment particle on the bottom of a river has a diameter of 3 mm and a specific gravity of 2.65. If the coefficient of friction is 0.37, estimate the shear force on the bottom of the channel that is required to move the particle.

SOLUTION

From the given data: D = 3 mm = 0.003 m, SG = 2.65, and $\mu_f = 0.37$. For water at 20°C, $\gamma_f = 9790$ N/m^3. The buoyant force, F_b, on the spherical particle is given by

$$F_b = \gamma_f V = \gamma_f \left[\frac{1}{6}\pi D^3\right] = (9790)\left[\frac{1}{6}\pi(0.003)^3\right] = 1.38 \times 10^{-4} \text{ N}$$

The weight, W, of the sediment particle is given by

$$W = \gamma_s V = \text{SG} \cdot \gamma_f V = 2.65 F_b = 2.65(1.38 \times 10^{-4}) = 3.66 \times 10^{-4} \text{ N}$$

The net (downward) force, F_{net}, on a sediment particle can therefore be calculated as

$$F_{net} = W - F_b = 3.66 \times 10^{-4} - 1.38 \times 10^{-4} = 2.28 \times 10^{-4} \text{ N}$$

The friction force, F_f, resulting from the net downward force, F_{net}, is

$$F_f = \mu_f F_{net} = (0.37)(2.28 \times 10^{-4}) = 8.44 \times 10^{-5} \text{ N}$$

The shear force needed to move a sediment particle is therefore equal to $\mathbf{8.44 \times 10^{-5}}$ **N.**

Buoyancy of solids in stratified fluids. In cases where the density of a fluid surrounding a body is not constant but varies with depth, such as in a stratified fluid, the magnitude of the buoyant force on the body is still equal to the weight of the displaced fluid. However, the line of action of the buoyant force does not pass through the centroid of the displaced volume; it passes through the center of gravity of the displaced volume.

Buoyancy of fluids in other fluids. Buoyancy effects are not limited to solid bodies, because the buoyancy effect also occurs when a body of fluid is surrounded by a fluid of different density. A practical example of the buoyancy effect of one fluid upon another is when treated domestic wastewater (fresh water) is discharged into the ocean via an ocean outfall located on the ocean floor. Because fresh water is less dense than seawater, the fresh water discharge rises as it mixes with the ocean water. Another example is the rise of heated air when discharged into cooler air, because warm air is less dense than the surrounding cooler air. For ideal gases under the same conditions of temperature and pressure, the density of a gas is proportional to its molecular weight. Therefore, pure gases with molecular weights less than 28.96 g/mol will rise in air (e.g., helium, 4 g/mol; methane, 16 g/mol), whereas pure gases with greater molecular weights (e.g., oxygen, 32 g/mol; chlorine, 71 g/mol) will sink in air.

Stability of submerged bodies. A submerged body is *stable* if any displacement from its equilibrium position induces a *righting moment* that causes the body to return to its initial (equilibrium) position. This is illustrated in Figure 2.32 for the case of a spherical submerged body with its weight centered in the lower portion of the body. The buoyant force effectively acts at the centroid of the displaced fluid volume, which is called the *center of buoyancy*, and is denoted by C in Figure 2.32; the center of buoyancy coincides with the centroid of the submerged body. The weight of the submerged body effectively acts at the center of gravity of the body, which is denoted by G in Figure 2.32. It is apparent from Figure 2.32(b) that any displacement of the submerged body from its equilibrium position will induce a righting moment that acts to return the body to its initial (equilibrium) position. It is further apparent that any submerged body in which the center of buoyancy is above the center of gravity will be stable. A practical application of this principle is in submarines, where a stable design for a submarine calls for the engines and crew cabins to be located in the lower half of the submarine to shift the weight to the bottom as much as possible. Displacement of a submerged or floating body from its normal equilibrium position is commonly referred to as *tilt* or *list*, and bodies can be characterized as "tilting" or "listing" when such deviations occur.

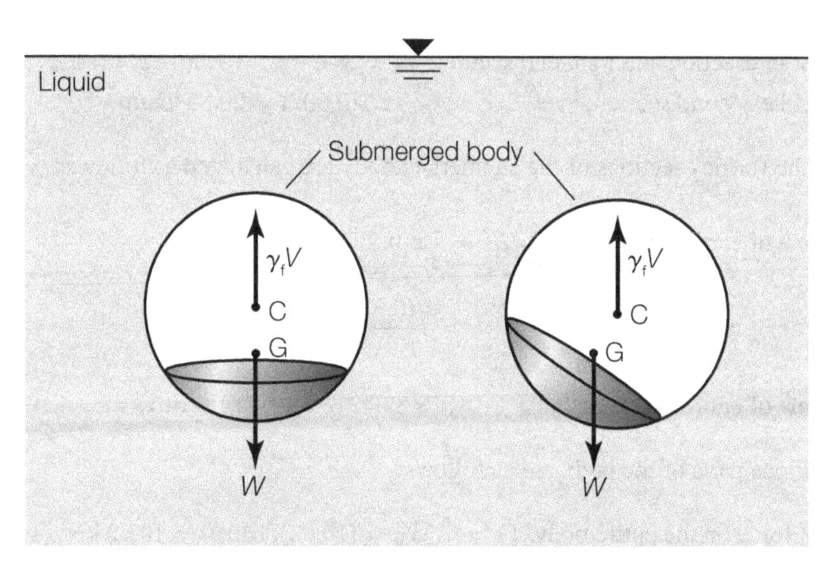

(a) Stable position (b) Displaced position

Figure 2.32: Stability of a submerged body

EXAMPLE 2.18

A body submerged in seawater has a cylindrical midsection and hemispherical end sections as shown in Figure 2.33. The midsection and one of the hemispherical end sections have an effective specific gravity of 1.3, and the other hemispherical end section has an effective specific gravity of 2.1. Because this body is unstable in the horizontal position, the body is kept horizontal by a support structure located at a distance d from the centroid (Point G) of the body. Determine the required support force and the distance d.

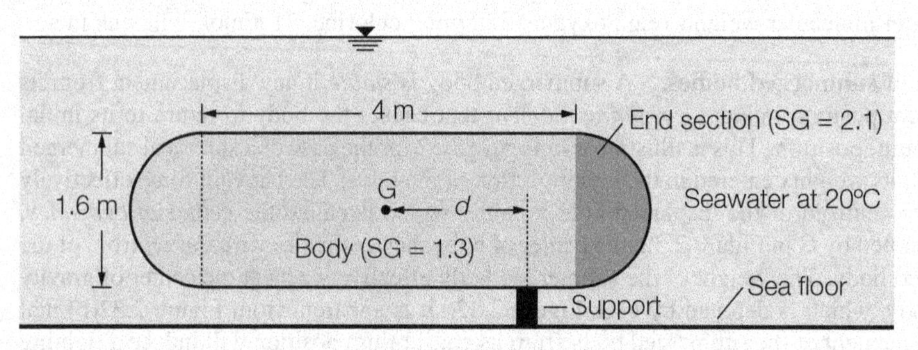

Figure 2.33: Unstable body with support

SOLUTION

From the given data: D = 1.6 m, $R = D/2 = 0.8$ m, L = 4 m, $\text{SG}_1 = 1.3$, and $\text{SG}_2 = 2.1$. For seawater at 20°C, γ_{sw} = 10.03 kN/m^3 (from Appendix B.4). For fresh water at 4°C, γ_{w} = 9.807 kN/m^3. The volume of a hemisphere of radius R is $2\pi R^3/3$, and the centroid of a hemisphere is located $3R/8$ from its base. The specific weights of the two parts of the body are calculated as follows:

specific weight of midsection and light end section, $\gamma_{\text{m}} = \text{SG}_1 \cdot \gamma_{\text{w}} = 1.3(9.807) = 12.75\ \text{kN/m}^3$
specific weight of heavy end section, $\gamma_{\text{e}} = \text{SG}_2 \cdot \gamma_{\text{w}} = 2.1(9.807) = 20.59\ \text{kN/m}^3$

The volumes of the various sections of the submerged body are calculated as follows:

$$\text{volume of end section, } V_{\text{e}} = \frac{2}{3}\pi R^3 = \frac{2}{3}\pi(0.8)^3 = 1.072\ \text{m}^3$$

$$\text{volume of midsection, } V_{\text{m}} = \frac{\pi D^2}{4}L = \frac{\pi(0.8)^2}{4}(4) = 8.042\ \text{m}^3$$

$$\text{volume of entire body, } V_{\text{b}} = V_{\text{m}} + 2V_{\text{e}} = 8.042 + 2(1.072) = 10.19\ \text{m}^3$$

The forces on various parts of the body are as follows:

$$\text{buoyant force on the entire body, } F_{\text{b}} = \gamma_{\text{sw}}V_{\text{b}} = (10.03)(10.19) = 102.2\ \text{kN}$$

$$\text{weight of light end section, } W_{\text{h1}} = \gamma_{\text{m}}V_{\text{e}} = (12.75)(1.072) = 13.67\ \text{kN}$$

$$\text{weight of heavy end section, } W_{\text{h2}} = \gamma_{\text{e}}V_{\text{e}} = (20.59)(1.072) = 22.07\ \text{kN}$$

$$\text{weight of midsection, } W_{\text{m}} = \gamma_{\text{m}}V_{\text{m}} = (12.75)(8.042) = 102.5\ \text{kN}$$

If the support force is F, then equilibrium of forces in the vertical direction requires that

$$F = W_{\mathrm{m}} + W_{\mathrm{h1}} + W_{\mathrm{h2}} - F_{\mathrm{b}} = 102.5 + 13.67 + 22.07 - 102.2 = 36.07\ \mathrm{kN}$$

Considering the submerged body as a free body and taking moments about the centroid and accounting for the fact that the weight of the midsection and the buoyant force on the entire body both act through the centroid of the body, the moment equation gives

$$W_{\mathrm{h1}} \cdot \left[\frac{L}{2} + \frac{3}{8}R\right] + F \cdot d = W_{\mathrm{h2}} \cdot \left[\frac{L}{2} + \frac{3}{8}R\right]$$

$$(13.67) \cdot \left[\frac{4}{2} + \frac{3}{8}(0.8)\right] + (36.07) \cdot d = (22.07) \cdot \left[\frac{4}{2} + \frac{3}{8}(0.8)\right] \quad \rightarrow \quad d = 0.536\ \mathrm{m}$$

Therefore, a support force of approximately **36.1 kN** located at a distance of **0.536 m** from the centroid of the submerged body keeps the body stable in the horizontal position.

Buoyancy effects on humans. The density of an average human being is around 1000 kg/m^3, the density of fresh water at 20°C is 998 kg/m^3, and the density of seawater at 20°C is around 1025 kg/m^3. Hence, the average person will sink in fresh water but float in seawater—a nice experiment when looking for something to do at the beach! Also, if an average person has a weight of 800 N, then assuming a person's density is 1000 kg/m^3, a person's average volume is 0.0849 m^3. Because the specific weight of air at 20°C is 1.22 kg/m^3, the buoyancy force on an average human being on the surface of Earth is around 0.98 N. Hence, in a vacuum, an average person would weigh about 0.98 N more than in Earth's atmosphere. Weightlessness that astronauts experience in space is similar (although not identical) to the weightlessness that an astronaut would experience underwater wearing a space suit of sufficient volume such that the weight of the astronaut plus the space suit is equal to the weight of the volume of water displaced. In fact, NASA astronauts train for extravehicular activities (also known as "space walks") in just such an environment at the Neutral Buoyancy Laboratory (NBL) located at the Johnson Space Center in Houston, Texas. An image of an astronaut training in the zero-buoyancy pool at the NBL is shown in Figure 2.34. For specific tasks, the training pool contains a mock-up to simulate the astronaut's activities in space; the NBL pool is 61.5 m long, 31 m wide, and 12 m deep.

Figure 2.34: Astronaut at National Buoyancy Laboratory
Source: Science Source.

2.6.2 Partially Submerged Bodies

If the weight of a body is less than the weight of the liquid displaced by the entire body, then the body will rise to the surface of the liquid and float. Under equilibrium conditions, the weight of the liquid displaced by the submerged portion of a floating body will be equal to the weight of the body itself. For a floating body, the buoyant force, F_B, is given by

$$F_B = \gamma_f V_f \tag{2.72}$$

where γ_f is the specific weight of the fluid and V_f is the volume of fluid displaced by the floating body, which is also equal to the volume of the body that is submerged. The equilibrium condition can be expressed as

$$W = F_B \quad \rightarrow \quad W = \gamma_f V_f \tag{2.73}$$

where W is the weight of the floating body. Under equilibrium conditions, the buoyant force and the weight of the floating body must be collinear. If they are not collinear, there is a net moment acting on the floating body, which then would not be in equilibrium.

Estimation of submerged volume. Equation 2.73 can be expressed in the form

$$\bar{\rho}_b g V_b = \rho_f g V_f \tag{2.74}$$

where $\bar{\rho}_b$ is the average density of the floating body, V_b is the total volume of the floating body, g is gravity, and ρ_f is the density of the fluid in which the body is submerged. Rearranging Equation 2.74 gives

$$\frac{V_f}{V_b} = \frac{\bar{\rho}_b}{\rho_f} \tag{2.75}$$

Equation 2.75 states that the volume fraction of the floating body that is submerged is equal to the ratio of the average density of the floating body to the density of the fluid. Hence, less fluid is displaced as the average density of the body is reduced (e.g., when ships expel ballast water). It is also clear from Equation 2.75 that a body will float with a portion above the surface when $\bar{\rho}_b/\rho_f < 1$, sink when $\bar{\rho}_b/\rho_f > 1$, and be neutrally buoyant when $\bar{\rho}_b/\rho_f = 1$. This result is used to practical effect by submarines, which take water into their ballast tanks (to increase their average density) to facilitate submerging and diving, expel water from their ballast tanks (to decrease their average density) to facilitate ascending, and keep the right amount of water in the ballast tanks to facilitate efficient cruising at or below the water surface.

EXAMPLE 2.19

A barge is 10 m long, 5 m wide and 3.5 m deep, and it weighs 1500 kN. If the barge sails in fresh water, how much of the barge will be below the water surface?

SOLUTION

From the given data: ℓ = 10 m, w = 5 m, h = 3.5 m, and W = 1500 kN. For water, γ_f = 9.79 kN/m^3. Let V be the volume of water displaced by the barge; hence, for equilibrium,

$$W = \gamma_f V \quad \rightarrow \quad V = \frac{W}{\gamma_f} = \frac{1500}{9.79} = 153 \text{ m}^3$$

If x is the depth of the barge below the waterline, then

$$10 \times 5 \times x = 153 \text{ m}^3 \quad \rightarrow \quad x = 3.06 \text{ m}$$

Thus, **3.06 m** of the barge is below the water surface and $3.5 - 3.06 = 0.44$ m is above the water surface.

Hydrometer. A *hydrometer* is an instrument that is commonly used to measure the specific gravity of a liquid. Typical hydrometers are small glass tubes that are less than 30 cm long, with large- and small-diameter cylindrical sections that are less than 3 cm and 1 cm in diameter, respectively, as shown schematically in Figure 2.35. The small-diameter cylindrical section of the hydrometer is called the *stem*. Hydrometers generally contain a ballast (i.e., added weight) to ensure that the instrument is submerged in pure water to a sufficient degree that the small-diameter tube protrudes at the surface. The small-diameter tube contains a scale that can be used to measure the degree of submergence. Some hydrometers also have a built-in thermometer so that the temperature at which measurements are taken can be recorded; such hydrometers are called *thermohydrometers*. If a hydrometer is placed in pure water, then equilibrium requires that

$$W = \gamma_\mathrm{w} V_0 \tag{2.76}$$

where W is the total weight of the hydrometer, γ_w is the specific weight of water, and V_0 is the volume of the hydrometer submerged when the hydrometer is placed in pure water. If the hydrometer is placed in a fluid with specific weight γ_f and the hydrometer sinks by an amount Δh, from its equilibrium position in water, then

$$W = \gamma_\mathrm{f}(V_0 + A_0 \Delta h) \tag{2.77}$$

where A_0 is the cross-sectional area of the stem. Combining Equations 2.76 and 2.77 and noting that $\gamma = \rho g$ gives

$$\mathrm{SG}_\mathrm{f} = \frac{\rho_\mathrm{f}}{\rho_\mathrm{w}} = \frac{V_0}{V_0 + A_0 \Delta h} \tag{2.78}$$

where SG_f is the specific gravity of the fluid and ρ_f and ρ_w are the densities of the fluid and pure water, respectively. Because V_0 and A_0 are physical properties of the hydrometer, Equation 2.78 gives the relationship between the relative submergence of the hydrometer, Δh, and the specific gravity, SG_f, of the fluid. Hence, based on Equation 2.78, the stem of the hydrometer can be calibrated to directly measure the specific gravity of the fluid. Equation 2.78 is valid whether the fluid is less dense than water ($\mathrm{SG}_\mathrm{f} < 1,\ \Delta h > 0$) or the fluid is denser than water ($\mathrm{SG}_\mathrm{f} > 1,\ \Delta h < 0$). Hydrometers can also be used to measure the relative amounts of different liquids in a mixture, because the density of a mixture is related to the relative amounts of mixed fluids and their pure densities. Applications of hydrometers in this regard include measuring the amount of antifreeze in car radiators and measuring the charge in a car battery, because the density of the fluid in the battery changes as sulfuric acid (H_2SO_4) is consumed or produced.

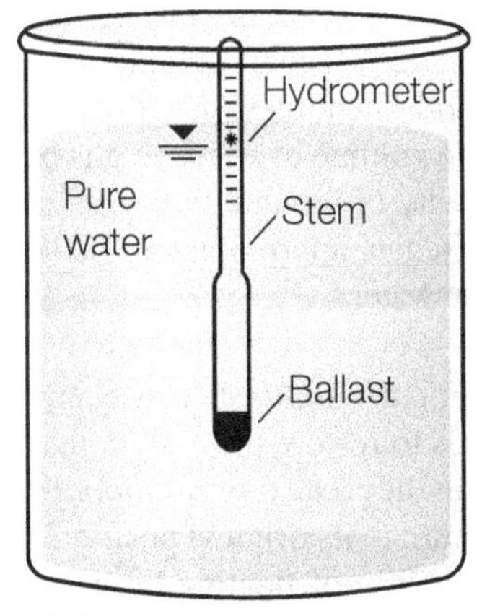

(a) Calibration in pure water

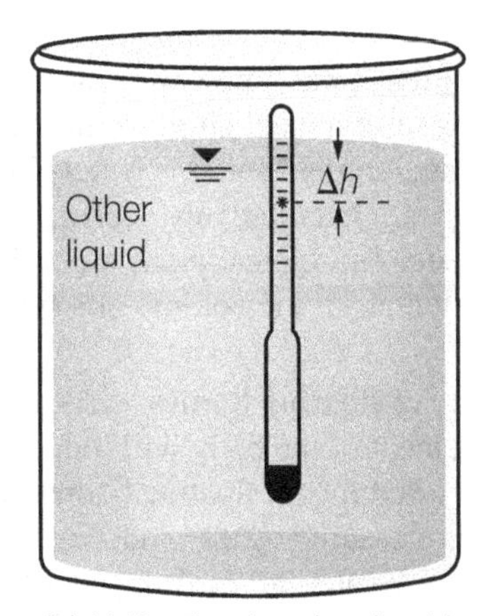

(b) Utilization in other liquid

Figure 2.35: Hydrometer

EXAMPLE 2.20

An antifreeze manufacturer knows that the right blend is obtained when the final product has a specific gravity of 0.92, and the company uses a hydrometer to check the specific gravity of the final product. The standard company hydrometer has a stem diameter of 8 mm, a stem length of 10 cm, and a weight of 0.300 N. The pure-water mark on the hydrometer is 8 cm below the top of the stem and is aligned with the liquid surface when the hydrometer is placed in pure water at 4°C. How far above the pure-water mark should the mark be that is aligned with the surface of the final product antifreeze?

SOLUTION

From the given data: $\text{SG}_\text{f} = 0.92$, $D = 8$ mm, $L = 10$ cm, and $W = 0.300$ N. For water at 4°C, $\gamma_\text{w} = 9807\ \text{N/m}^3$. Using these data, the stem cross-sectional area, A_0, and the volume of the hydrometer below the pure-water mark, V_0, are given by

$$A_0 = \frac{\pi D^2}{4} = \frac{\pi(8)^2}{4} = 50.27\ \text{mm}^2$$

$$V_0 = \frac{W}{\gamma_\text{w}} = \frac{0.300}{9807} = 3.059 \times 10^{-5}\ \text{m}^3 = 30\,590\ \text{mm}^3$$

The displacement of the hydrometer from the pure-water line, Δh, is determined from Equation 2.78 as follows:

$$\text{SG}_\text{f} = \frac{V_0}{V_0 + A_0\Delta h} \quad \rightarrow \quad 0.92 = \frac{30\,590}{30\,590 + 50.27\Delta h} \quad \rightarrow \quad \Delta h = 52.9\ \text{mm}$$

Therefore, in a fluid of specific gravity 0.92, the fluid surface should be aligned with a mark on the hydrometer stem that is approximately **5.3 cm** above the pure-water mark.

Buoyancy-related natural phenomena. A classic floating body encountered by many civil engineering students in the United States is the concrete canoe. These concrete canoes are shaped such that the weight of the canoe and the people in the canoe can be supported by the volume (of lake water) that is displaced by the canoe. Interestingly, many laypeople are taken aback by the mention of a concrete canoe (because they think of concrete as a dense material that sinks); however, these individuals will not give a second thought about going on a cruise in a steel ship! A floating iceberg is a natural phenomenon with which many people are familiar, and it is widely known that most of the volume of an iceberg is below the water level. Quantitatively, icebergs typically have a density of around 900 kg/m^3 and float in seawater that typically has a density of 1025 kg/m^3. Hence, approximately 900/1025, or seven-eighths, of the volume of a typical iceberg is below water.

Stability of floating bodies. All floating bodies are *vertically stable*, because any vertical displacement of the body will induce a net force that tends to restore the body to its original position. Stability of floating bodies is typically assessed in the context of *rotational stability*, which is measured by the tendency of a rotated body to return to its original position. Usually the word "rotational" is dropped when discussing the stability of floating bodies, and this practice will be followed here. If a righting moment develops when a floating body lists, the body will be stable regardless of whether the center of buoyancy is above or below the center of gravity. Conversely, if an overturning moment develops when a floating body lists, the body is unstable. Examples of stable and unstable floating bodies are shown in Figure 2.36. If the center of buoyancy, B, is directly above the center of gravity, G, the body is always stable. However, the location of B below G in floating bodies does not guarantee instability as it does for fully submerged bodies. This is because the position of the center of buoyancy B can move relative to a floating body as it tilts, due to its shape, whereas for a fully submerged body, the position of B is fixed relative to the body. This is illustrated in Figure 2.36(a), which is stable even though B is below G. This situation is representative of boats and ships, although the cross sections of ships are more streamlined than shown in Figure 2.36(a). As the ship tilts to the right in a similar manner to Figure 2.36(a), the center of buoyancy, B, moves to the right further than the line of action of the body weight, W; so the buoyancy provides the righting moment. A measure of stability for floating bodies is the *metacentric height*, GM, which is the distance between the center of gravity, G, and the *metacenter*, M, where the metacenter is defined as the intersection of the lines of action of the buoyant force through the body before and after rotation. The angle of rotation of a floating body is commonly called the *rolling angle* or the *angle of heel*, and the location of the metacenter generally depends on this angle. However, the metacenter may be considered a fixed point for most hull shapes for small rolling angles up to about 20°. A floating body is stable if M is above G (and thus GM is positive), and unstable if M is below G (and thus GM is negative). In the latter case, the weight and the buoyant force acting on the tilted body generate an overturning moment instead of a restoring moment, causing the body to capsize; the floating body shown in Figure 2.36(b) has a negative metacentric height. The magnitude of the metacentric height GM is a measure of the stability: the larger the magnitude of GM, the more stable the floating body. An approximate relation for estimating the metacentric height is

$$\mathrm{GM} = \frac{I_{00}}{V_{\mathrm{sub}}} - \mathrm{GB} \tag{2.79}$$

where I_{00} is the centroidal moment of inertia of the shape of the area created by the intersection of the floating body and the water surface (i.e., the *waterline area*), V_{sub} is the volume of the submerged portion of the body, and GB is the distance from the center of gravity, G, of the floating body to the center of buoyancy, B. The distance GB is taken as positive when B is below G. When B is above G, GB is negative, and Equation 2.79 indicates that GM is always positive; hence, the body is always stable when B is above G. Note that I_{00}/V_{sub} is

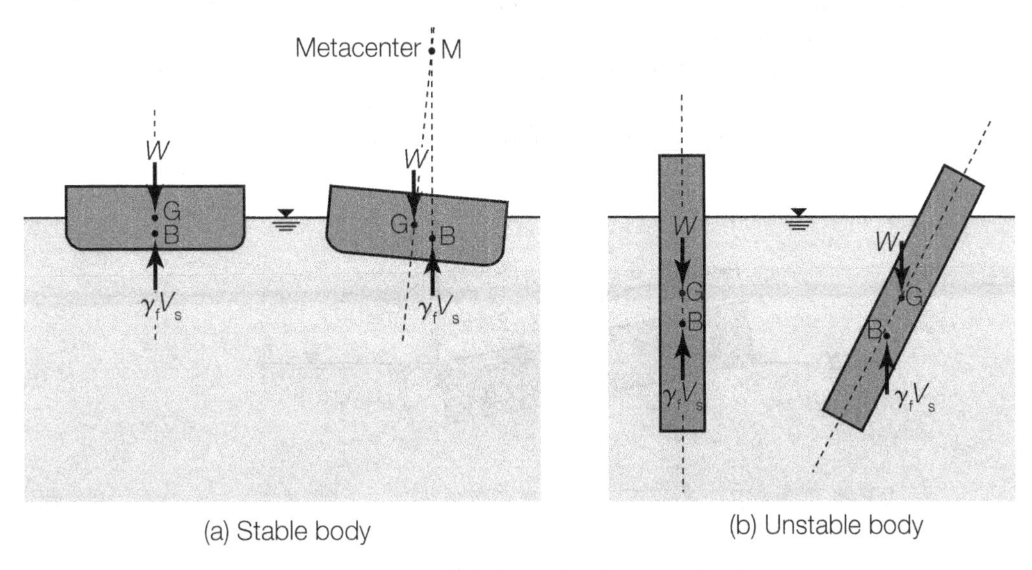

(a) Stable body

(b) Unstable body

Figure 2.36: Stability of a floating body

always positive. For small rolling angles, θ, the righting or overturning moment, M_θ, can be approximated using the relation

$$M_\theta \approx \gamma_f V_{sub} \cdot GM \cdot \theta \tag{2.80}$$

where γ_f is the specific weight of the fluid in which the body is floating, and θ is the rolling angle in radians. For accurate estimates of M_θ, more elaborate analyses are generally required.

Practical Considerations

It is recommended that small vessels have a minimum metacentric height of 0.46 m and that large ships have a minimum metacentric height of 1.07 m (Avallone and Baumeister, 1996). However, too large of a metacentric height creates uncomfortable rocking motions that cause sea sickness. In warships and racing yachts, stability is more important than comfort, and such vessels have large metacentric heights. If the liquid in the hull of a ship is not constrained, the center of gravity of the floating body will move toward the center of buoyancy when the ship rolls, thus decreasing the righting couple and the stability. For this reason, floating vessels usually store liquid ballast or fuel oil in tanks or bulkhead compartments. Similarly, overloaded ferry boats must be careful that all passengers do not gather on one side of the upper deck. To avoid creating unstable conditions, container ships must be careful not to stack containers so high on the deck that the center of gravity of the ship is raised to an unstable level. Placing cargo further from the centerline decreases the vessel's moment of inertia with little sacrifice of stability.

EXAMPLE 2.21

A rectangular block of uniform material is displaced in water at 20°C as shown in Figure 2.37. The block has a length of 800 mm, a width of 300 mm, and a height of 50 mm. A uniform vertical load of 20 N/m is applied at P. (a) What is the weight of the block? (b) Estimate the metacentric height of the floating block before the load at P is applied. (c) If the load at P is suddenly removed, what is the righting moment before the block starts to move? Compare the magnitude of the righting moment with the approximate value derived using Equation 2.80.

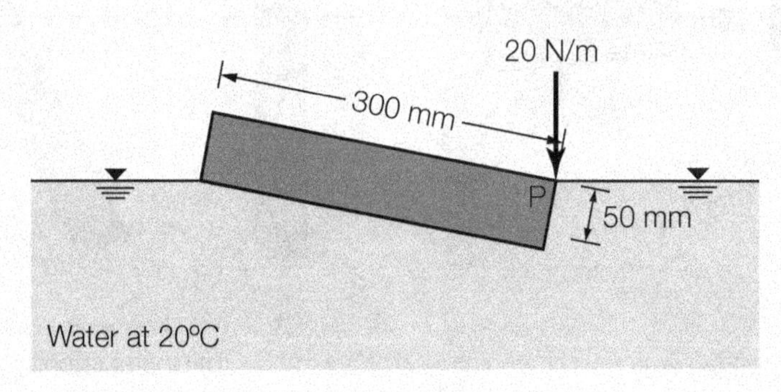

Figure 2.37: Floating block

SOLUTION

From the given data: $L = 800$ mm, $b = 300$ mm, $d = 50$ mm, and $w = 20$ N/m. The portion of the block below water and the locations of the forces acting on the block are shown in Figure 2.38.

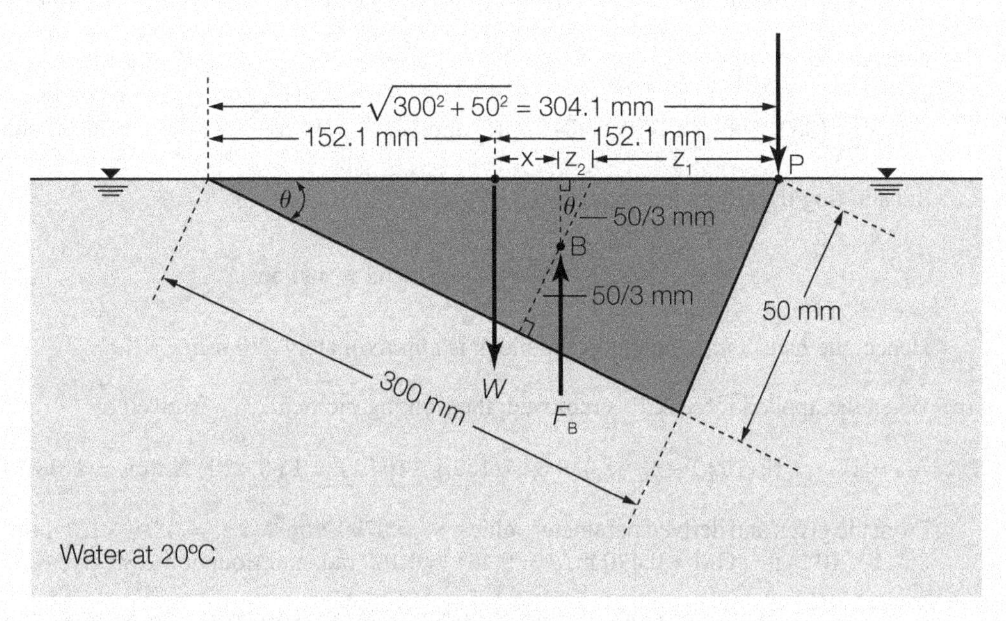

Figure 2.38: Floating block

The force P is given by $P = 20$ N/m $\times$ 0.8 m $= 16$ N.

(a) The weight of the block can be calculated by requiring that the sum of the moments about the center of buoyancy (Point B) be equal to zero. Point B is located at the centroid of the triangle shown in Figure 2.38. The following geometric relations are apparent:

$$\theta = \tan^{-1}\left(\frac{50}{300}\right) = 9.46^\circ, \qquad z_1 = \frac{100}{\cos 9.46^\circ} = 101.4 \text{ mm}$$

$$z_2 = \frac{50}{3}\sin 9.46^\circ = 2.74 \text{ mm}, \qquad z_1 + z_2 = 101.4 \text{ mm} + 2.74 \text{ mm} = 104.1 \text{ mm}$$

Taking moments about B gives

$$W(152.1 - z_1 - z_2) = P(z_1 + z_2) \quad \rightarrow \quad W(152.1 - 104.1) = 16(104.1) \quad \rightarrow \quad W = \mathbf{34.7\,N}$$

(b) The metacentric height GB is given by Equation 2.79 as

$$\text{GM} = \frac{I_{00}}{V_\text{sub}} - \text{GB} \tag{2.81}$$

and the moment of inertia of the block about the water surface, I_{00}, is given by

$$I_{00} = \frac{bd^3}{12} = \frac{(800)(300)^3}{12} = 1.800 \times 10^9 \text{ mm}^4$$

Calculation of V_{sub} requires knowing the depth of submergence, h, under equilibrium conditions. Taking the specific weight of water as $\gamma_w = 9790$ N/m^3, when the block is floating in its stable position,

$$\gamma_w(0.3)(0.8)h = W \quad \rightarrow \quad (9790)(0.3)(0.8)h = 34.7 \quad \rightarrow \quad h = 14.77 \text{ mm}$$

hence,

$$V_{sub} = (300)(800)(14.77) = 3.544 \times 10^6 \text{ mm}^3, \qquad \text{GB} = \frac{50}{2} - \frac{14.77}{2} = 17.62 \text{ mm}$$

Substituting these data into Equation 2.81 yields

$$\text{GM} = \frac{1.800 \times 10^9}{3.544 \times 10^6} - 17.62 = 490 \text{ mm}$$

Hence, the metacentric height of the block is approximately **490 mm.**

(c) When the applied force at P is removed, the righting moment, M_θ, is given by

$$M_\theta = Wx = W(152.1 - z_1 - z_2) = 34.7(152.1 - 104.1) = 1.66 \times 10^3 \text{ N·mm} = \mathbf{1.66\ N{\cdot}m}$$

From the given and derived parameter values, $\gamma_f = 9790$ N/m^3, $V_{sub} = 3.544 \times 10^6$ mm^3 = 3.544×10^{-3} m^3, GM = 0.490 m, $\theta = 9.46° = 0.165$ rad, and Equation 2.80 gives

$$M_\theta \approx \gamma_f V_{sub} \cdot \text{GM} \cdot \theta = (9790)(3.544 \times 10^{-3})(0.490)(0.165) = \mathbf{2.81\ N{\cdot}m}$$

Therefore, in this case, the approximation given by Equation 2.80 yields only a rough estimate of the restoring moment.

Stable positions. Many floating bodies have more than one stable position. For example, a ship typically has more than one stable floating position. It is stable when floating right-side up; it also is stable when floating upside down.

2.6.3 Buoyancy Effects Within Fluids

Buoyancy effects within fluids occur whenever the density within a fluid varies. Such density variations can occur because of differential heating of the fluid or as a consequence of the discharge of a fluid of one density into a fluid of another density. These differences in density cause relative movements of the fluid due to buoyancy, and such movements are commonly called *convection currents.*

Convection process. To illustrate the convection process, consider the case where a fluid is heated at a given location. Because the density of a fluid typically decreases with increasing temperature, the heated fluid becomes less dense than the surrounding fluid and the buoyancy effect causes the heated volume to rise. The rising fluid volume continues to rise as long as its temperature exceeds that of the surrounding fluid, and the heated/rising fluid volume is replaced by cooler fluid as it moves upward; this is the process of *free convection.* Fluid motion generated by a lighted cigarette (and made visible by the accompanying smoke) is an example of free convection.

Atmospheric convection. The convection process occurs on a large scale within Earth's atmosphere where the lower part of the atmosphere is continually being mixed by convection currents caused by differential heating of Earth's surface. The decrease in pressure with altitude above the ground has a significant effect on convection within the atmosphere. An air volume heated at the surface of Earth rises due to buoyancy. This air volume encounters a reduced-pressure environment as it rises, which causes the volume to expand and the air within the volume to cool. Because air is a poor conductor of heat, the expansion is approximately adiabatic, and assuming that the heated air is dry, such that no heat is generated by condensing water, the standard pressure variation within the atmosphere causes the temperature within the heated volume to decrease at a rate of approximately 0.01 K/m. This dry adiabatic lapse rate is greater than the typical lapse rate in the surrounding atmosphere of 0.0065 K/m; hence, the rising volume cools more rapidly than the surrounding air. At the point where the temperature of the rising air volume equals that of the surrounding air, the volume ceases to rise and the atmosphere becomes stable again. In cases where the atmospheric lapse rate is greater than the dry adiabatic lapse rate, the heated air volume remains warmer and less dense than its surroundings, which is an unstable situation that frequently results in thunderstorms.

2.7 Rigid-Body Motion of Fluids

Rigid-body motion occurs when all parts of a body move together, such as would occur if the body were composed of solid material. Rigid-body motion is sometimes referred to as *solid-body motion*. Rigid-body motions sometimes occur in fluids, particularly in cases where the fluid is in a container that is moving with a constant velocity, moving with a constant acceleration, or rotating. The key characteristic of rigid-body fluid motion is that there is no relative motion between fluid particles, and such fluids are said to be in *relative equilibrium*. Examples of rigid-body motion of fluids include the motion of fluids contained and being transported in tanker trucks, and the motion of fluids contained in rotating cylinders.

Governing equations. Consider an infinitesimal volume of fluid within a control volume having dimensions $\Delta x \times \Delta y \times \Delta z$ with a pressure, p, at the center of the volume as shown in Figure 2.2. This is the same infinitesimal control volume that was used in determining the pressure distribution within a static fluid by equating the sum of forces in each coordinate direction to zero. In the analysis of rigid-body motion, the sum of the forces on the infinitesimal fluid element is not equal to zero, but is equal to the elemental mass multiplied by the acceleration the fluid mass is undergoing. In referring to Figure 2.2, the net forces in the x-, y-, and z-directions, denoted by δF_x, δF_y, and δF_z, respectively, are given by

$$\delta F_x = \left(p - \frac{\partial p}{\partial x}\frac{\Delta x}{2}\right)\Delta y\Delta z - \left(p + \frac{\partial p}{\partial x}\frac{\Delta x}{2}\right)\Delta y\Delta z = -\frac{\partial p}{\partial x}\Delta x\Delta y\Delta z \tag{2.82}$$

$$\delta F_y = \left(p - \frac{\partial p}{\partial y}\frac{\Delta y}{2}\right)\Delta x\Delta z - \left(p + \frac{\partial p}{\partial y}\frac{\Delta y}{2}\right)\Delta x\Delta z = -\frac{\partial p}{\partial y}\Delta x\Delta y\Delta z \tag{2.83}$$

$$\delta F_z = \left(p - \frac{\partial p}{\partial z}\frac{\Delta z}{2}\right)\Delta x\Delta y - \left(p + \frac{\partial p}{\partial z}\frac{\Delta z}{2}\right)\Delta x\Delta y - \rho g\Delta x\Delta y\Delta z = -\left(\frac{\partial p}{\partial z} + \rho g\right)\Delta x\Delta y\Delta z \tag{2.84}$$

According to Newton's law, the net force on a fluid element is equal to the product of the mass and the acceleration of the fluid element; hence,

$$\underbrace{\delta\mathbf{F}}_{\text{net force}} = \underbrace{(\rho\Delta x\Delta y\Delta z)}_{\text{mass}} \times \underbrace{\mathbf{a}}_{\text{acceleration}} \tag{2.85}$$

where $\delta\mathbf{F}$ is a vector with components δF_x, δF_y, and δF_z and $\mathbf{a}$ is the acceleration vector. Combining Equations 2.82–2.85 and dividing by $\Delta x\Delta y\Delta z$ gives the following governing equation for the pressure distribution in fluids that are moving as rigid bodies:

$$\nabla p + \rho g\mathbf{k} = -\rho\mathbf{a} \tag{2.86}$$

where $\mathbf{k}$ is the unit vector in the z-direction and ∇p is the gradient of p in Cartesian coordinates defined by

$$\nabla p = \frac{\partial p}{\partial x}\mathbf{i} + \frac{\partial p}{\partial y}\mathbf{j} + \frac{\partial p}{\partial z}\mathbf{k} \tag{2.87}$$

Equation 2.86 is commonly expressed in the form

$$\nabla p = \rho(\mathrm{g} - \mathbf{a}) \tag{2.88}$$

where g is the gravity vector, which is equal to $-g\mathbf{k}$. Equation 2.88 can be regarded as the general equation describing the pressure variation within a fluid that is static relative to its container, where the container is moving with an acceleration $\mathbf{a}$. It is known from vector calculus that the gradient of a scalar function is equal to a vector that is aligned in a direction normal to the contours of the scalar function. Consequently, Equation 2.88 shows that contours of constant pressure are normal to the vector g – $\mathbf{a}$. Therefore, in cases where $\mathbf{a}$ and g are collinear, the fluid surface must be normal to the direction of g, which means that the fluid surface must be horizontal. In cases where the container is accelerating in a nonvertical direction, the fluid surface will necessarily be inclined and normal to the g – $\mathbf{a}$ vector.

Cases of radial symmetry. In cases where there is radial symmetry, it is usually more convenient to apply Equation 2.88 using cylindrical coordinates, in which case

$$\nabla p = \frac{\partial p}{\partial r}\mathbf{e}_r + \frac{1}{r}\frac{\partial p}{\partial \theta}\mathbf{e}_\theta + \frac{\partial p}{\partial z}\mathbf{e}_z \tag{2.89}$$

where $\mathbf{e}_r$, $\mathbf{e}_\theta$, and $\mathbf{e}_z$ are unit vectors in the r-, θ-, and z-directions, respectively.

Case of a stationary container. In the case where the fluid container is stationary, $\mathbf{a} = 0$, and Equation 2.88 gives

$$\nabla p = -\rho g\mathbf{k} \Rightarrow \frac{\partial p}{\partial x} = 0, \quad \frac{\partial p}{\partial y} = 0, \quad \frac{\partial p}{\partial z} = -\gamma \tag{2.90}$$

which are the same equations derived previously for a static fluid.

Case of free fall. A liquid in free fall without any drag accelerates downward with $a_x = 0$ and $a_z = -g$. Under this condition, the components of Equation 2.88 are

$$\frac{\partial p}{\partial x} = 0, \quad \frac{\partial p}{\partial y} = 0, \quad \frac{\partial p}{\partial z} = -\rho(g - g) = 0 \tag{2.91}$$

Thus, the pressure variation in all three coordinate directions is equal to zero. Therefore, according to this formulation, a fluid droplet in free fall (without drag) in the atmosphere has a constant pressure within the droplet equal to atmospheric pressure. The pressure within the droplet would need to be slightly higher than atmospheric pressure to account for surface tension holding the droplet together.

EXAMPLE 2.22

A tank is filled with water at 20°C to a depth of 2 m. The tank is placed on an elevator that undergoes a variety of motions. Determine the gauge pressure at the bottom of the tank for the following elevator motions: (a) stationary, (b) accelerating upward at 4 m/s^2, (c) accelerating downward at 4 m/s^2, and (d) free-falling.

SOLUTION

From the given data: $h = 2$ m. For water at 20°C, $\rho = 998.2$ kg/m^3. Because the elevator acceleration, a, has the same (vertical) line of action as gravity, Equation 2.88 can be expressed as

$$\frac{\mathrm{d}p}{\mathrm{d}z} = \rho(-g-a) \quad \rightarrow \quad \Delta p = -\rho(g+a)\Delta z \tag{2.92}$$

where Δp is the pressure at the bottom of the tank minus the pressure at the water surface, which is equal to the gauge pressure at the bottom of the tank, $g = 9.807$ m/s^2, a is the elevator acceleration, with a being positive when the acceleration is in the upward direction, and Δz is the elevation of the bottom of the tank minus the elevation of the water surface. Substituting known quantities into Equation 2.92 gives

$$\Delta p = -(998.2)(9.807+a)(-2)\ \text{Pa} \quad \rightarrow \quad \Delta p = 1.996 \times 10^3(9.807+a)\ \text{Pa}$$
$$\rightarrow \quad \Delta p = 1.996(9.807+a)\ \text{kPa}$$

This equation can be used to directly estimate the gauge pressure, Δp, at the bottom of the tank for various accelerations of the elevator. The results are given in the following table.

Case	Condition	a (m/s^2)	Δp (kPa)
(a)	stationary	0	**19.6**
(b)	upward acceleration	4	**27.6**
(c)	downward acceleration	−4	**11.6**
(d)	free fall	−9.807	**0.0**

Therefore, the gauge pressure at the bottom of the tank varies in the range of 0–27.6 kPa, depending on the motion of the elevator.

2.7.1 Liquid with Constant Acceleration

Consider the case of constant acceleration in the xz plane as shown in Figure 2.39, which could correspond to the motion of a liquid contained in a tanker truck that is moving with a constant acceleration. Under this condition, the components of Equation 2.88 are

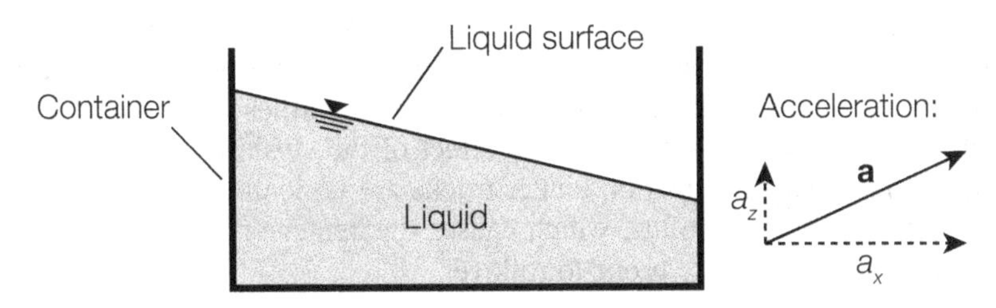

Figure 2.39: Liquid with constant acceleration

$$\frac{\partial p}{\partial x} = -\rho a_x, \quad \frac{\partial p}{\partial y} = 0, \quad \frac{\partial p}{\partial z} = -\rho(g + a_z) \tag{2.93}$$

where a_x and a_z are the components of the acceleration of the liquid container. Based on Equation 2.93, the change in pressure, $\mathrm{d}p$, between two points that are $\mathrm{d}x$ and $\mathrm{d}z$ apart is given by

$$\mathrm{d}p = -\rho a_x \, \mathrm{d}x - \rho(g + a_z) \, \mathrm{d}z \tag{2.94}$$

which integrates to

$$\Delta p = -\rho a_x \Delta x - \rho(g + a_z)\Delta z \tag{2.95}$$

where the prefix Δ indicates the (finite) change in the quantity following the prefix. On the liquid surface, the pressure is constant; therefore, between any two points on the surface $\Delta p = 0$, and Equation 2.95 gives

$$\frac{\Delta z}{\Delta x} = -\frac{a_x}{g + a_z} \tag{2.96}$$

indicating that the liquid surface is flat with a slope of $-a_x/(g + a_z)$. A liquid completely filling a closed tank does not have a free surface. However, the planes of constant pressure would still be inclined to the x-direction at an angle of $\tan^{-1}[-a_x/(g + a_z)]$.

Pressure distribution. It can be inferred from the above analysis that the slope of any constant-pressure contour within the liquid is planar with the same slope as the surface of the liquid. It can also be deduced from Equation 2.95 (and is given explicitly by Equation 2.93) that the vertical pressure gradient within the liquid is given by

$$\frac{\partial p}{\partial z} = -\rho(g + a_z) \tag{2.97}$$

which indicates that the pressure distribution is not hydrostatic in the vertical direction, which would require the vertical pressure gradient to be equal to $-\rho g$. It can further be shown from Equation 2.95 that the subsurface pressure gradient in the direction normal to the liquid surface is given by

$$\frac{\mathrm{d}p}{\mathrm{d}s} = -\rho\sqrt{a_x^2 + (g + a_z)^2} \tag{2.98}$$

where s is the coordinate measured normal to the liquid surface. It is apparent that the pressure distribution is not hydrostatic in the s-direction. Interestingly, the pressure could be assumed to be hydrostatic in the s-direction provided that the acceleration due to gravity was taken as $\sqrt{a_x^2 + (g + a_z)^2}$. Nevertheless, it is generally recommended to work with vertical (z) and horizontal (x) axes rather than the s-axis.

Practical Considerations

Tanker trucks transporting liquids are typically harder to control than trucks transporting solids, primarily due to the sloshing of the liquid in both backward-and-forward and side-to-side motions. The tanks on some tanker trucks contain baffles to minimize the additional forces on the truck caused by sloshing. The effect of the sloshing motion is further compounded by the fact that the tanks on tanker trucks are typically mounted higher than the "beds" of trucks transporting solids, which causes the center of gravity of the truck to be higher and makes the truck more prone to rollover.

EXAMPLE 2.23

The gasoline tanker truck shown in Figure 2.40 has an elliptical tank that is 2.50 m wide, 3.70 m high, and 12.0 m long. The gasoline carried by the tanker typically has a specific gravity of approximately 0.72. For the case in which the tank is half full, estimate the overturning moment about the base of the tank that is caused by the shifting of gasoline when the truck rounds a corner at 6.71 m/s and the corner has a radius of curvature equal to 45.7 m.

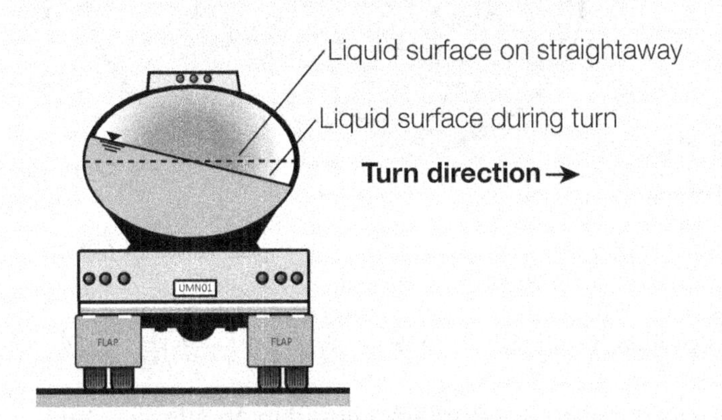

Figure 2.40: Liquid in tanker truck

SOLUTION

Assign the following variable names to the given data: W = 2.50 m, H = 3.70 m, L = 12.0 m, SG = 0.72, v = 6.71 m/s, and R = 45.7 m. Taking the density of water as 998 kg/m^3 (at 20°C), the density of gasoline, ρ_{gas}, is given by

$$\rho_{\text{gas}} = \text{SG}\,\rho_{\text{water}} = 0.72(998) = 719\ \text{kg/m}^3$$

The lateral acceleration, a_x, is the centripetal acceleration as the truck turns and is calculated as

$$a_x = \frac{v^2}{R} = \frac{6.71^2}{45.7} = 0.985\ \text{m/s}^2$$

Calculation of the overturning moment caused by a shifting liquid in an elliptical tank requires the determination of the center of gravity of the displaced liquid. This calculation is facilitated by relationships derived by Romero et al. (2007) and shown in Figure 2.41.

The variables a_t and b_t are the lengths of the principal axes in the y- and x-directions, respectively; r is the axis ratio defined as $r = a_\text{t}/b_\text{t}$; G_0 is the location of the center of gravity of the original (horizontal-surface) liquid at coordinates $(0, y_0)$; G_1 is the location of the center of gravity of the displaced (inclined-surface) liquid at coordinates (x_c, y_c); O is the location of the bottom of the tank at coordinates $(0, -a_\text{t})$; and F_x F_y are the components of the forces on the displaced liquid, calculated by

$$F_x = ma_x, \quad F_y = mg$$

where m is the mass of the liquid in the tank; a_x is the x component of the liquid acceleration; and M_O is the overturning moment at the base of the tank caused by the liquid.

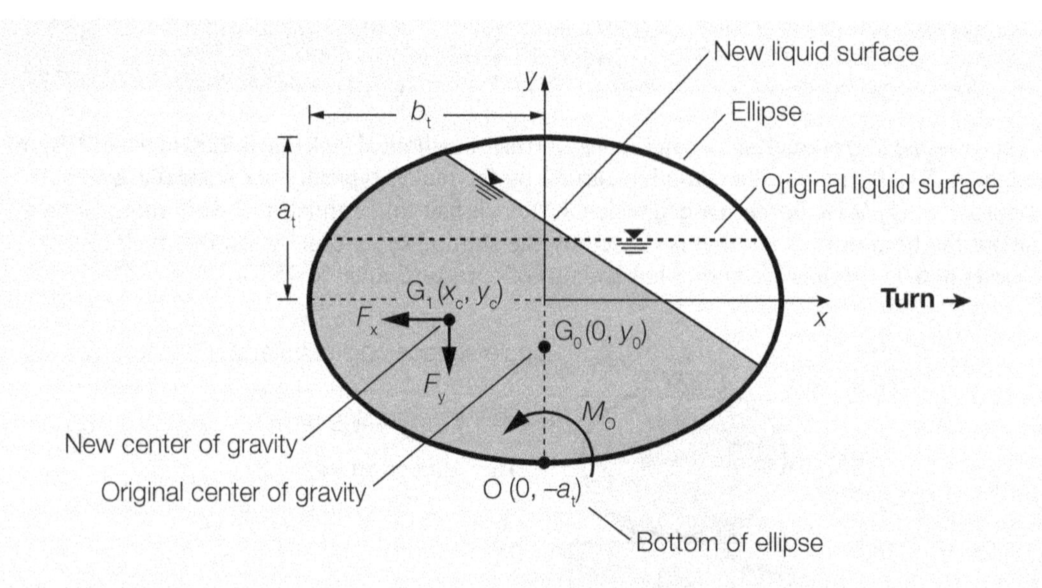

$$x_c = \frac{y_0}{r}\cos\left[\tan^{-1}\left(-\frac{rg}{a_x}\right)\right]; \quad y_c = -y_0 \sin\left[\tan^{-1}\left(-\frac{rg}{a_x}\right)\right]; \quad M_O = F_x(y_c + a_t) - F_y(x_c)$$

Figure 2.41: **Displaced liquid in an elliptical tank**

From the given data: $a_t = H/2 = 3.70/2 = 1.85$ m, $b_t = W/2 = 2.50/2 = 1.25$ m, and $r = a_t/b_t = 1.85/1.25 = 1.48$. Because the tank is initially half full, the submerged area is a half-ellipse, and using the geometric properties of a half-ellipse (Appendix C) gives the y-coordinate of the centroid, y_0, the volume of the liquid, V, and the mass of the liquid, m as

$$y_0 = -\frac{2H}{3\pi} = -\frac{2(3.70)}{3\pi} = -0.785 \text{ m}$$

$$V = \tfrac{1}{8}\pi HWL = \tfrac{1}{8}\pi(3.70)(3.50)(12.0) = 61.0 \text{ m}^3$$

$$m = \rho_{\text{gas}}V = (719)(61.0) = 43\,859 \text{ kg}$$

Substituting these values into the relationships given in Figure 2.41 yields

$$x_c = \frac{y_0}{r}\cos\left[\tan^{-1}\left(-\frac{rg}{a_x}\right)\right] = \frac{-0.785}{1.48}\cos\left[\tan^{-1}\left(-\frac{(1.48)(9.81)}{0.985}\right)\right] = -0.0358 \text{ m}$$

$$y_c = -y_0 \sin\left[\tan^{-1}\left(-\frac{rg}{a_x}\right)\right] = -(-0.785)\sin\left[\tan^{-1}\left(-\frac{(1.48)(9.81)}{0.985}\right)\right] = -0.783 \text{ m}$$

$$F_x = ma_x = (43\,859)(0.985) = 43\,200 \text{ N} = 43.2 \text{ kN}$$

$$F_y = mg = (43\,859)(9.81) = 430\,300 \text{ N} = 430.3 \text{ kN}$$

$$M_O = F_x(a_t + y_c) - F_y(x_c) = 43.2(1.85 - 0.783) - 430.3(-0.0358) = 61.5 \text{ kN·m}$$

Therefore, the overturning moment induced by the liquid in the tank as the truck rounds the corner is **61.5 kN·m**, which can be contrasted with an overturning moment of zero when the truck is moving straight ahead.

2.7.2 Liquid in a Rotating Container

When a liquid is placed in a container that is rotating at a constant rate, the viscosity of the liquid causes the liquid to rotate and eventually reach a state where there is no relative motion between the container and the liquid. Under this condition, both the liquid and the container behave as a single rigid body, and this type of motion is called *rigid-body rotation*. To analyze the motion of a liquid in a rotating container, it is more effective to apply Equation 2.88 using cylindrical coordinates. If the rate of rotation is ω, then the acceleration, **a**, of a liquid element at a distance r from the center of rotation is given by

$$\mathbf{a} = -r\omega^2\mathbf{e}_r \tag{2.99}$$

where $\mathbf{e}_r$ is the unit vector in the r-direction. Substituting Equation 2.99 into Equation 2.88 with the pressure gradient expressed in cylindrical coordinates (Equation 2.89) gives

$$\frac{\partial p}{\partial r} = \rho r\omega^2, \quad \frac{\partial p}{\partial \theta} = 0, \quad \frac{\partial p}{\partial z} = -\rho g \tag{2.100}$$

Based on Equation 2.100, the change in pressure, $\mathrm{d}p$, between two points that are $\mathrm{d}r$ and $\mathrm{d}z$ apart is given by

$$\mathrm{d}p = \rho r\omega^2\,\mathrm{d}r - \rho g\,\mathrm{d}z \tag{2.101}$$

To determine the equation of a constant-pressure contour, take $\mathrm{d}p = 0$ and denote the z-coordinate of a constant-pressure contour by z_p; hence, Equation 2.101 gives

$$\frac{\mathrm{d}z_p}{\mathrm{d}r} = \frac{r\omega^2}{g} \tag{2.102}$$

which integrates to yield

$$z_p = \frac{\omega^2}{2g}r^2 + z_0 \tag{2.103}$$

where z_0 is the elevation of the constant-pressure contour at $r = 0$. Because the liquid surface is a constant-pressure contour, Equation 2.103 gives the equation of the liquid surface, where z_0 is the elevation of the liquid surface at the center of the rotating container. It is apparent from Equation 2.103 that the shape of the liquid surface is parabolic.

Pressure distribution. Equation 2.101 can be integrated directly to give the pressure distribution as

$$p_2 - p_1 = \frac{\rho\omega^2}{2}(r_2^2 - r_1^2) - \rho g(z_2 - z_1) \tag{2.104}$$

which gives the pressure p_2 at any location (r_2,z_2) in terms of a given pressure p_1 at location (r_1,z_1). It is further apparent from Equation 2.104 that at any value of r, the pressure is distributed hydrostatically in the vertical (z) direction.

Free surface in a cylindrical container. If the initial depth of liquid in a cylindrical container is h_i, then the volume of liquid, V, in the cylinder is given by

$$V = \pi R^2 h_\mathrm{i} \tag{2.105}$$

where R is the radius of the cylinder. The volume of liquid in the cylinder can also be calculated from Equation 2.103, which gives

$$V = \int_0^R 2\pi r z_p \, dr = \int_0^R 2\pi r \left(\frac{\omega^2}{2g} r^2 + z_0 \right) dr = \pi R^2 \left(\frac{\omega^2 R^2}{4g} - z_0 \right) \tag{2.106}$$

Combining Equations 2.105 and 2.106 by eliminating V gives

$$z_0 = h_\mathrm{i} - \frac{\omega^2 R^2}{4g} \tag{2.107}$$

Substituting Equation 2.107 into 2.103 gives the shape of the free surface in terms of h_i as

$$z_p = h_\mathrm{i} - \frac{\omega^2}{4g}(R^2 - 2r^2) \tag{2.108}$$

This relationship is particularly useful for determining the rotational speed (ω) at which a liquid will spill from its container. Also, because the volume of a paraboloid is equal to one-half the area of its base area times its height, Equation 2.108 can be used to show that h_i is halfway between the high and low points of the free surface. Hence, with rigid-body rotation, the center of the liquid drops by $\omega^2 R^2/(4g)$ and the outer edge of the liquid rises by this same amount.

EXAMPLE 2.24

Water at 20°C is contained within a 5-cm-diameter cylinder that is 10 cm high as shown in Figure 2.42. The cylinder is rotated about its central axis at a rate of 200 rpm. (a) What is the maximum initial depth of water in the container such that no spillage occurs? (b) For the initial depth calculated in part (a), what is the difference in pressure between the center bottom of the cylinder and the side bottom of the cylinder? (c) What is the minimum initial depth of liquid such that the liquid surface will not intersect the bottom before it spills? (d) For the initial depth calculated in part (c), what is the difference in pressure between the side bottom of the cylinder and the water surface?

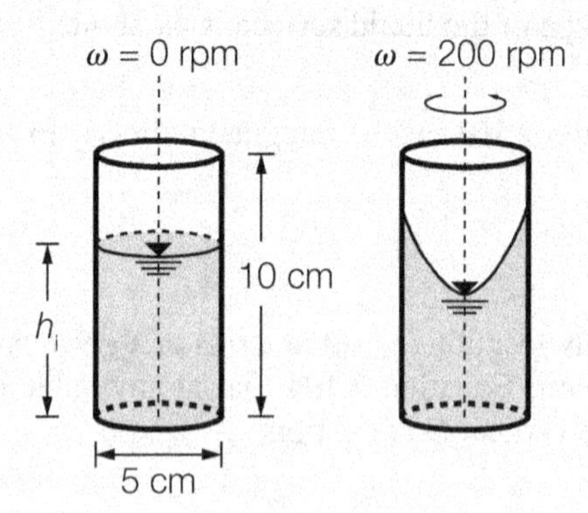

Figure 2.42: Liquid in a rotating cylinder

SOLUTION

From the given data: R = 2.5 cm = 0.025 m, H = 10 cm = 0.10 m, and ω = 200 rpm = 20.94 rad/s. For water at 20°C, ρ = 998.2 kg/m^3.

(a) When the cylinder is rotating, the elevation of the liquid surface is given by Equation 2.108. When spillage occurs, $z_p = H = 0.10$ m at $r = R = 0.025$ m. Rearranging Equation 2.108 gives

$$h_i = z_p + \frac{\omega^2}{4g}(R^2 - 2r^2) = z_p + \frac{\omega^2}{4g}(R^2 - 2R^2)$$
$$= z_p - \frac{\omega^2}{4g}(R^2) = 0.10 - \frac{20.94^2}{4(9.81)}(0.025^2) = 0.093 \text{ m}$$

Hence, an initial liquid depth of 9.3 cm in the 10-cm-deep cylinder will result in incipient spillage at 200 rpm. The maximum depth to avoid spillage is therefore equal to **9.3 cm**.

(b) The variation of pressure within a rotating cylinder is given by Equation 2.104. In this case, p_2 is the pressure at the side bottom of the cylinder, p_1 is the pressure at the center bottom of the cylinder, $r_2 = 0.025$ m, $r_1 = 0$, $z_1 = z_2$, and Equation 2.104 gives

$$p_2 - p_1 = \frac{\rho\omega^2}{2}(r_2^2 - r_1^2) - \rho g(z_2 - z_1) = \frac{(998.2)(20.94)^2}{2}(0.025^2 - 0^2) - 0 = \mathbf{137\ Pa}$$

(c) Because the initial water depth, h_i, in a rotating cylinder is always halfway between the high and low points of the free surface, when the cylinder is initially half full, the liquid surface will intersect the bottom at the same time it spills. Hence, the minimum depth of liquid such that it will spill before the liquid surface intersects the bottom is 10 cm/2 = **5 cm**.

(d) The variation of pressure is given by Equation 2.104. In this case, p_2 is the pressure at the side bottom of the cylinder, p_1 is the pressure at water surface, $r_2 = r_1 = 0.025$ m, $z_2 = 0$ m, $z_1 = 0.10$ m, and Equation 2.104 gives

$$p_2 - p_1 = \frac{\rho\omega^2}{2}(r_2^2 - r_1^2) - \rho g(z_2 - z_1) = 0 - (998.2)(9.807)(0 - 0.10) = \mathbf{979\ Pa}$$

Practical Considerations

A common application of a liquid in a rotating cylinder is a *centrifugal separator*, in which heavier particles contained within the liquid move outward and lighter particles are displaced inward. A similar principle is used in a *cyclone separator*, which removes suspended particles in air flows.

Key Equations in Fluid Statics

The following list of equations is useful in solving problems related to fluid statics. If one is able to recognize these equations and recall their appropriate use, then the learning objectives of this chapter have been met to a significant degree. Derivations of these equations, definitions of the variables, and detailed examples of usage can be found in the main text.

PRESSURE DISTRIBUTION IN STATIC FLUIDS

General equation:
$$\frac{dp}{dz} = -\gamma$$

Incompressible fluid:
$$p + \gamma z = \text{constant}, \qquad p_P = p_0 + \gamma h$$

Gauge pressure:
$$p_{\text{gauge}} = p_{\text{abs}} - p_{\text{atm}}$$

Pressure head:
$$h = \frac{p}{\gamma}$$

Surface-tension effect on pressure head:
$$h' = \frac{p}{\gamma} + \frac{4\sigma\cos\theta}{\gamma D}$$

In ideal gas or atmosphere:
$$\ln\frac{p_2}{p_1} = -\frac{g}{R}\int_{z_1}^{z_2}\frac{dz}{T}$$

$$p = p_0\left(1 - \frac{bz}{T_0}\right)^{\frac{g}{Rb}} = p_0\left(\frac{T}{T_0}\right)^{\frac{g}{Rb}}$$

$$\rho = \rho_0\left(1 - \frac{bz}{T_0}\right)^{\frac{g}{Rb}-1}$$

$$p_2 = p_1\exp\left[-\frac{g(z_2 - z_1)}{RT_0}\right] \qquad (\text{for } b = 0)$$

PRESSURE MEASUREMENTS

Barometer:
$$p_{\text{atm}} = \gamma y + p_{\text{vapor}}$$

Bourdon gauge:
$$p = p_{\text{gauge}} + \gamma\Delta z$$

Manometers:
$$p_0 + \sum_{i=1}^{N}\gamma_i\Delta z_i = p_1$$

FORCES ON PLANE SURFACES

Force:
$$F = \begin{cases} (p_0 + \gamma h_c)A, & p_0 \neq 0 \\ \gamma A h_c, & p_0 = 0 \end{cases}$$

Center of pressure (y):

$$y_{cp} = \begin{cases} \bar{y} + \dfrac{\gamma \sin\theta I_{cc}}{(p_0 + \gamma\bar{y}\sin\theta)A}, & p_0 \neq 0 \\ \bar{y} + \dfrac{I_{cc}}{A\bar{y}}, & p_0 = 0 \end{cases}$$

FORCES ON CURVED SURFACES

Force (x):

$$R_x = \begin{cases} (p_0 + \gamma\bar{y}_v)A_v, & p_0 \neq 0 \\ \gamma A_v \bar{y}_v, & p_0 = 0 \end{cases}$$

Force (z):

$$R_z = \begin{cases} p_0 A_h + \gamma V_0, & p_0 \neq 0 \\ \gamma V_0, & p_0 = 0 \end{cases}$$

BUOYANCY

Fully submerged body, buoyant force: $F_b = \gamma_f V$

Fully submerged body, net force: $F_{net} = (\gamma_s - \gamma_f)V$

Partially submerged body: $W = \gamma_f V_f$

Fraction of volume submerged: $\dfrac{V_f}{V_b} = \dfrac{\bar{\rho}_b}{\rho_f}$

Hydrometer: $SG_f = \dfrac{\rho_f}{\rho_w} = \dfrac{V_0}{V_0 + A_0 \Delta h}$

Metacentric height: $GM = \dfrac{I_{00}}{V_{sub}} - GB$

RIGID-BODY MOTION OF FLUIDS

General equation: $\nabla p = \rho(\mathbf{g} - \mathbf{a})$

Constant acceleration: $\Delta p = -\rho a_x \Delta x - \rho(g + a_z)\Delta z$

Surface of fluid: $\dfrac{\Delta z}{\Delta x} = -\dfrac{a_x}{g + a_z}$

Fluid in rotating cylinder:

$$z_p = \frac{\omega^2}{2g} r^2 + z_0$$

$$p_2 - p_1 = \frac{\rho\omega^2}{2}(r_2^2 - r_1^2) - \rho g(z_2 - z_1)$$

Cylindrical container: $z_p = h_i - \dfrac{\omega^2}{4g}(R^2 - 2r^2)$

Applications of the above equations are provided in the text, and additional problems to practice using these equations can be found in the following section.

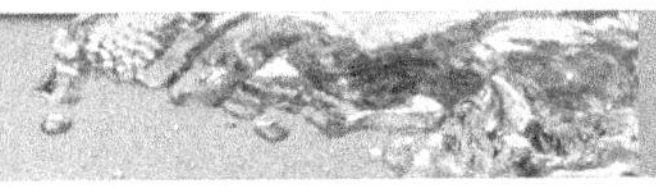

PROBLEMS

Section 2.2: Pressure Distribution in Static Fluids

2.1. A glycerin storage tank has been contaminated with crude oil as shown in Figure 2.43. The tank is open to the atmosphere, where the atmospheric pressure is 101 kPa. Within the tank, a 0.7-m-thick layer of crude oil floats on the glycerin that is 2.25 m deep. Both liquids are at 20°C. Determine the pressure on the bottom of the tank, both as an absolute pressure and as a gauge pressure.

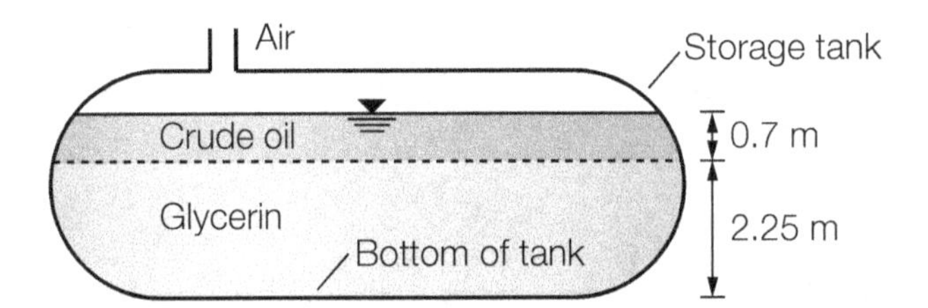

Figure 2.43: Contaminated storage tank

2.2. (a) At what depth below the surface of a water body will the (gauge) pressure be equal to 200 kPa? (b) If a 1.55-m-tall person orients himself vertically underwater in a pool, what pressure difference does he feel between his head and his toes? Assume water at 20°C.

2.3. A liquid is stratified such that the specific gravity of the fluid at the surface is 0.98 and the specific gravity at a depth of 12 m is equal to 1.07. Assuming that the specific gravity varies linearly between the liquid surface and a depth of 12 m, determine the pressure at a depth of 12 m. State whether this is a gauge pressure or an absolute pressure.

2.4. (a) A large tank contains water at 20°C, and the absolute pressure 12 m below the surface of the water is measured as 200 kPa. Estimate the atmospheric pressure above the tank. (b) If the water in the tank was replaced by a liquid with a specific gravity of 0.85, what absolute pressure and what gauge pressure would be measured 6 m below the surface of the liquid?

2.5. A pipeline leads from a reservoir to a closed valve as illustrated in Figure 2.44. Calculate the gauge and absolute pressures at the valve. Assume water at 20°C and standard atmospheric pressure.

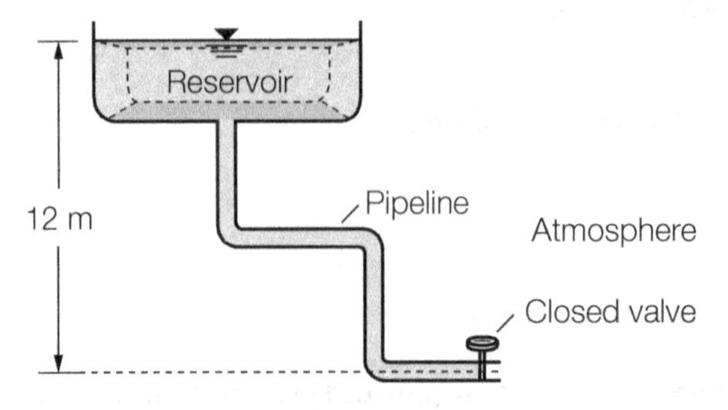

Figure 2.44: Pipeline system

2.6. The pressure in the airspace above an oil (SG = 0.80) surface in a tank is 14 kPa. Find the pressure 1.5 m below the surface of the oil.

2.7. Three tanks contain water at 20°C and air as shown in Figure 2.45. The three tanks have open connections to each other, and the air (gauge) pressure in tank B is 6 kPa. What are the air pressures in tanks A and C?

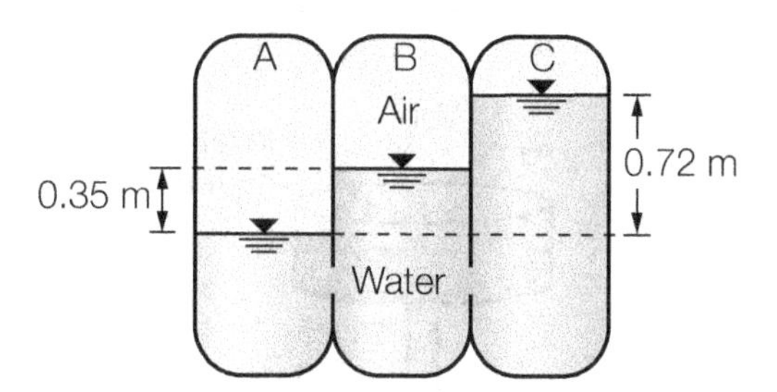

Figure 2.45: Connected tanks

2.8. A 6-mm-diameter air bubble is released at a location 25 m below the surface of a lake, and the temperature of the lake is approximately uniform at 20°C. Estimate the diameter of the bubble as it approaches the water surface. Assume standard sea-level atmospheric conditions above the surface of the lake.

2.9. A bubble is released from an ocean vent located 20 m below the ocean surface. Atmospheric pressure above the ocean surface is equal to 101.3 kPa, and the ocean temperature down to a depth of 20 m is approximately constant at 20°C. Estimate the ratio of the density of the air in the bubble at a depth of 20 m to the density of the air in the bubble just as it reaches the ocean surface.

2.10. A container that is open to the atmosphere stores two immiscible liquids. The liquid on top has a thickness of 7 m and specific weight of 9 kN/m^3, and the liquid on the bottom has a thickness of 2.3 m. If a transducer measures a (gauge) pressure of 92 kPa on the bottom of the tank, estimate the specific gravity of the liquid on the bottom. Must the liquid on the bottom necessarily be denser than the liquid on the top? Explain.

2.11. Water pressure at a pipeline junction measures 450 kPa. What is the corresponding pressure head (expressed as a height of water)?

2.12. What is the pressure head (of water) corresponding to a pressure of 810 kPa? What depth of mercury at 20°C will be required to produce a pressure of 810 kPa?

2.13. Find the pressure head in millimeters of mercury (Hg) equivalent to 80 mm of water plus 60 mm of a fluid whose specific gravity is 2.90. The specific weight of mercury can be taken as 133 kN/m^3. Assume a temperature of 20°C.

2.14. Find the pressure head corresponding to an atmospheric pressure of 101.3 kPa. Give your answer in terms of millimeters of mercury.

2.15. A 8-mm-diameter glass tube is attached to a pressurized storage tank containing water at 25°C. The water in the tube is observed to rise to a height of 85 mm above the attachment point. (a) What correction to the rise height must be made to account for surface tension? (b) What is the pressure head in the reservoir where the tube is attached?

2.16. It is common practice to use elevated reservoirs to maintain the pressures in municipal water supply systems. Such a system is illustrated in Figure 2.46. If the pressure in the pipeline is to be maintained in the range of 350–500 kPa and the reservoir is not to be less than half full, estimate the height of the midpoint of the reservoir above the pipeline and the minimum required space between the midpoint and the top of the reservoir. The pressure in the air above the tank is maintained at atmospheric pressure.

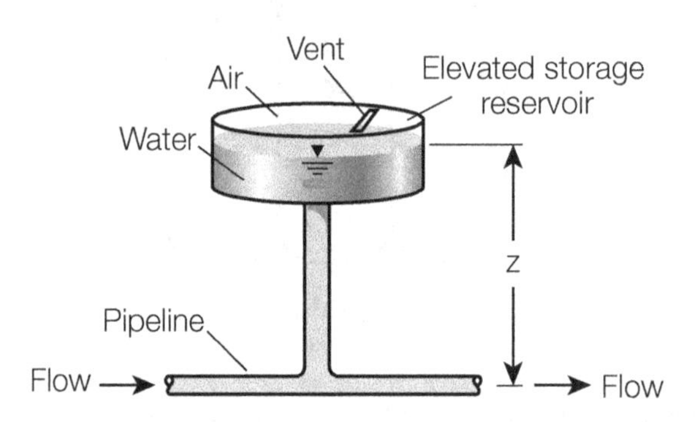

Figure 2.46: Storage reservoir

2.17. Blood pressure in humans is normally expressed as a ratio x/y, where x is the maximum arterial pressure in mm Hg, called the *systolic pressure*, and y is the minimum arterial pressure in mm Hg, called the *diastolic pressure*. A typical blood pressure is 120/70. Blood pressure readings are normally taken as the same level as the heart, and blood at 37°C has a density of around 1060 kg/m^3. (a) Consider a (tall) person whose head is 0.48 m above her heart and whose toes are 1.46 m below her heart. Assuming static conditions, compare the blood pressure in her head to the blood pressure in her toes. (b) If a tube were connected to the artery in which the blood pressure was being measured, what would be the maximum height that blood would rise in the tube?

2.18. Fluid is to be infused intravenously into the arm of a patient by placing the fluid in a bottle and suspending the bottle at a sufficient height above the patient's arm that the fluid can be fed by gravity into the body. Under typical conditions, an arm-level (gauge) pressure of 150 mm Hg is required to provide a sufficient flow rate of fluid. If the intravenous fluid has a density of 1025 kg/m^3, how far above arm level must the fluid level in the bottle be held?

2.19. The head of an adult male giraffe is typically 6 m above the ground. The normal blood pressure of a giraffe (at heart level) is typically stated as 280/180, where 280 the maximum arterial pressure in mm Hg and 180 is the minimum arterial pressure in mm Hg. The density of a giraffe's blood is approximately 1060 kg/m^3. (a) What is the change in the blood pressure in the giraffe's head, in millimeters of mercury, as it moves from grazing on a tall tree to drinking from a pond at ground level? (b) Assuming that there is a static distribution of blood pressure in the giraffe's body and that its heart is at approximately the mid-elevation of its body, estimate the maximum blood pressure in its head.

2.20. The hydraulic system shown in Figure 2.47 uses compressed air at a pressure of 310 kPa to lift a load on a platform. The area of the piston connected to the chamber of compressed air is 8 cm^2, the weight of the piston is 55 N, the area of the platform is 600 cm^2, the weight of the platform is 820 N, the bottom of the platform is located 1.2 m above the bottom of the piston, and the density of the hydraulic fluid is

900 kg/m^3. (a) What force is exerted by the compressed air on the piston? (b) What weight mounted on the platform can be lifted by the compressed air? (c) If the piston is displaced by 12 cm, what is the displacement of the platform?

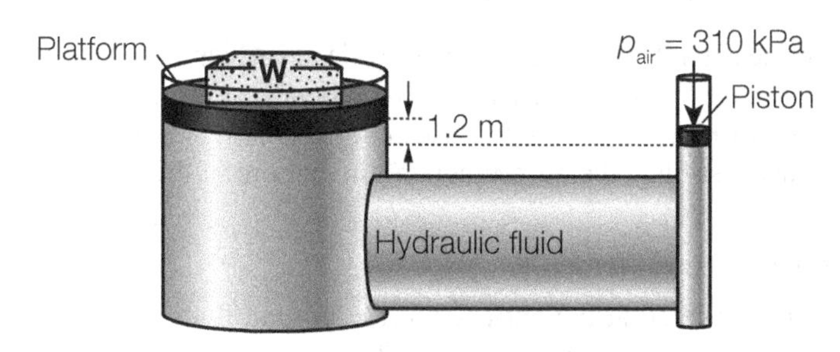

Figure 2.47: Hydraulic lift

2.21. Two piston diameters are being considered for use in a hydraulic system: a 25-mm-diameter piston and a 100-mm-diameter piston. If an applied force of 500 N to the 25-mm piston is found to be satisfactory, what force on the 100-mm piston at the same location would be required so as not to compromise the performance of the hydraulic system?

2.22. The summit of Mount Rainier (in Washington) is at an elevation of 4342 m and is reported to experience an average annual temperature of -11°C and a typical atmospheric pressure of 58 kPa. Compare these measured conditions with the temperature and pressure expected using the standard atmosphere given by Equations 2.25 and 2.26.

2.23. Within the standard atmosphere, the stratosphere exists between elevations 11–20 km and is characterized by a constant temperature of -56.5°C. The standard-atmosphere pressure at the bottom of the stratosphere (at elevation 11 km) is equal to 22.63 kPa, and the acceleration due to gravity, g, within the stratosphere varies within the narrow range of 9.776–9.761 m/s^2. Assuming that air within the stratosphere behaves like an ideal gas, use the given data to estimate the theoretical pressure at the top of the stratosphere. Compare your result with the standard-atmosphere pressure at 20 km (given in Appendix B.3).

2.24. Assume standard conditions of temperature and pressure at sea level, a lapse rate of 6.5°C/km, and a hydrostatic distribution of pressure in the atmosphere. Using these assumptions, calculate the theoretical temperature and pressure at 1-km elevation intervals between elevations 0 km and 11 km. Compare your result with the U.S. standard atmosphere given in Appendix B.3. What is the maximum temperature difference in °C and the maximum pressure difference in kPa?

2.25. Atmospheric pressure at ground level is due to the weight of the overlying atmosphere. The pressure and temperature at ground level are 101.3 kPa and 20°C, respectively, and the lapse rate is estimated as 6.3°C/km. If the top of the atmosphere is estimated as the elevation at which the pressure is 1 Pa, estimate the thickness of the atmosphere.

2.26. Blue Mountain Peak is the highest point in Jamaica with an elevation of 2256 m. The average temperature at Blue Mountain Peak is 5°C, and average climatic conditions at sea level in Jamaica are a temperature of 27°C and an atmospheric pressure of 101 kPa. (a) Estimate the atmospheric pressure at Blue Mountain Peak. (b) At what temperature will water boil at Blue Mountain Peak?

2.27. Show that the parameter g/Rb that is used to calculate the pressure distribution in the troposphere can be taken as 5.26 for the standard atmosphere. Use the standard atmosphere to estimate the temperature at which water boils in the capital city of La Paz (Bolivia), which is 3640 m above sea level.

2.28. A gold mine extends to a level 4 km below land surface. The land surface is approximately at sea level, and a vertical air shaft connects the atmosphere at the surface to the lower elevations of the mine. Assuming standard atmospheric conditions at the land surface, estimate the air pressure at the bottom of the mine. State any assumptions.

Section 2.3: Pressure Measurements

2.29. A barometer located at the entrance to the ground floor of the Burj Khalifa building in Dubai estimates a pressure of 100.8 kPa on an average day in August, when the temperature is 37°C. The height of the Burj Khalifa is reported to be 829.8 m. Estimate the barometric pressure at the top of the building.

2.30. A light airplane uses an aneroid barometer to measure changes in altitude. At one elevation, the pressure is 96 kPa, and at a higher elevation, the pressure is 85.3 kPa. Estimate the difference in altitude corresponding to these measurements. Assume that the airplane is operating in a standard atmosphere.

2.31. Consider the case of a manometer attached to a tank containing kerosene at 20°C as shown in Figure 2.48. Atmospheric pressure is 101.3 kPa. For the given conditions, what would be the pressure reading on the Bourdon gauge mounted at Point P?

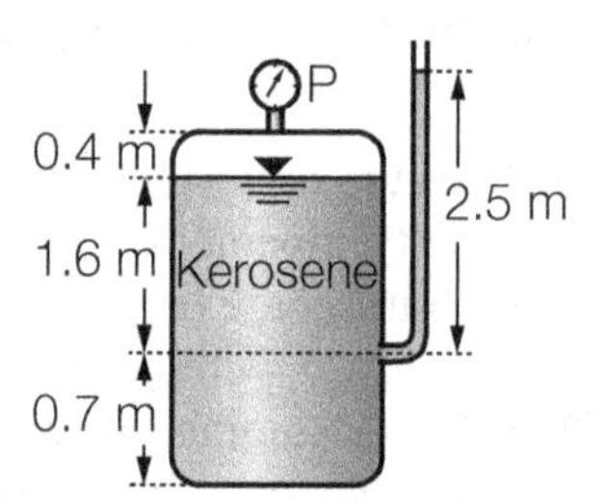

Figure 2.48: Manometer connected to a gasoline tank

2.32. A manometer is attached to a tank containing two immiscible liquids as shown in Figure 2.49. The top (light) liquid has a specific gravity of 0.8 and a thickness of 0.3 m, and the bottom (dense) liquid has a specific gravity of 2.8 and a thickness of 0.3 m. The manometer is connected to the bottom fluid at an elevation 0.1 m below the fluid interface. (a) Determine the elevation, Δz, of the liquid in the manometer above the interface. (b) What is the gauge pressure on the bottom of the tank?

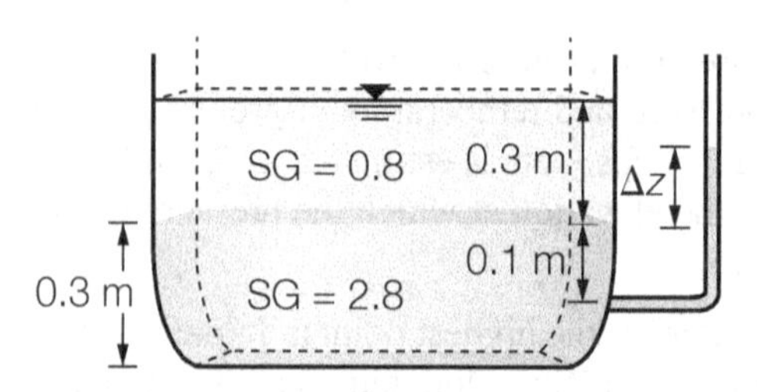

Figure 2.49: Manometer connected to a tank containing two fluids

2.33. The U-tube manometer shown in Figure 2.50 is used to measure the pressure in the pipeline at A. The water is at 30°C, and the specific weight of the gauge fluid is 45 kN/m^3. For the given fluid heights, determine the pressure in the pipeline.

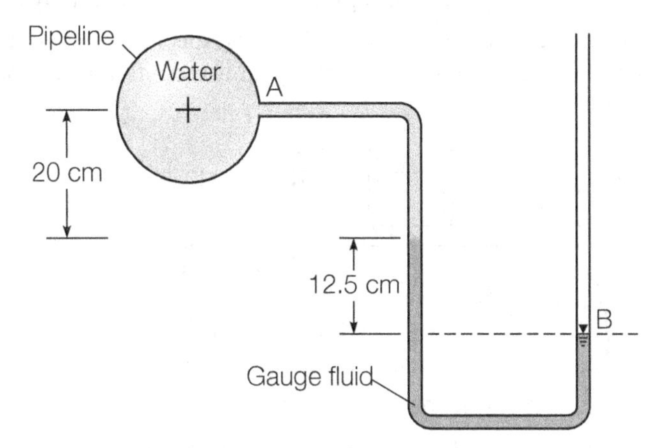

Figure 2.50: U-tube manometer

2.34. The U-tube manometer shown in Figure 2.51 is used to measure the pressure exerted by the compressed air in a storage reservoir containing SAE 10 oil. For the measurements shown, what is the gauge pressure exerted by the compressed air? Assume that all liquids are at a temperature of 20°C.

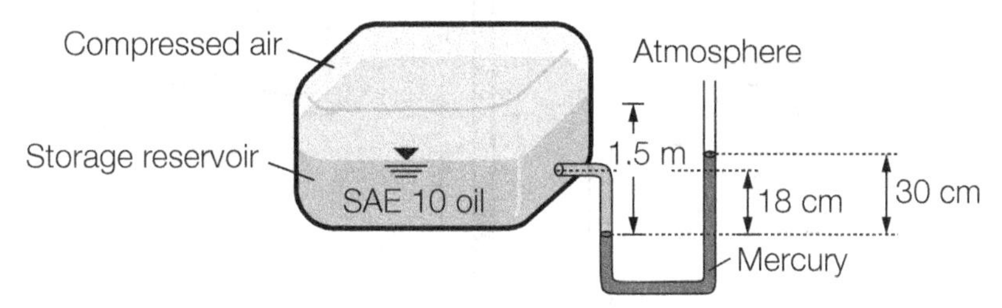

Figure 2.51: Measurement of pressure of compressed air

2.35. Consider the differential manometer shown in Figure 2.52. Express the pressure difference between Points A and B in terms of $\gamma_{\mathrm{w}}, \gamma_{\mathrm{f}}, h_1, h_2$, and h_3.

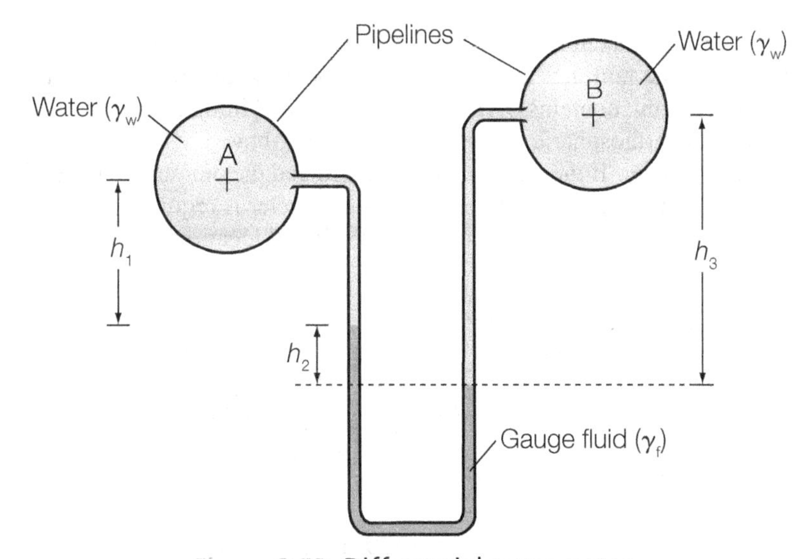

Figure 2.52: Differential manometer

2.36. The differential manometer shown in Figure 2.53 is used to measure the pressure drop across an orifice plate in a pipe. Because the pressure drop across an orifice plate is proportional to the flow rate through the opening in the plate, such an arrangement is commonly used to measure the volume flow rate in pipes. In the present case, the fluid flowing in the pipe is water at 30°C and the specific weight of the gauge fluid is 20.5 kN/m^3. For the given fluid heights, what is the difference in pressure between Points 1 and 2?

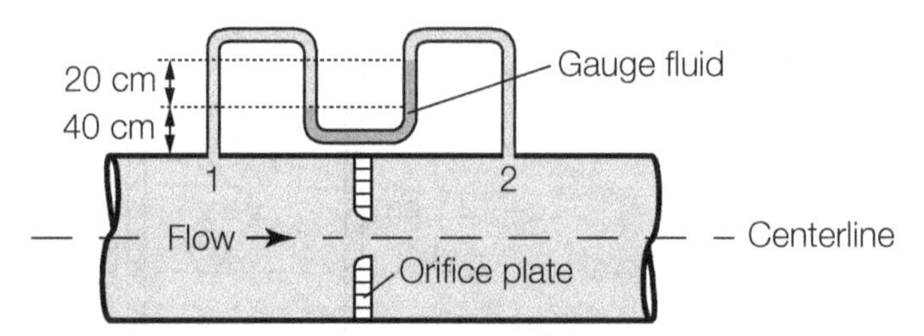

Figure 2.53: Measurement of pressure across an orifice plate

2.37. Determine the pressure difference between the water pipe and the oil pipe shown in Figure 2.54.

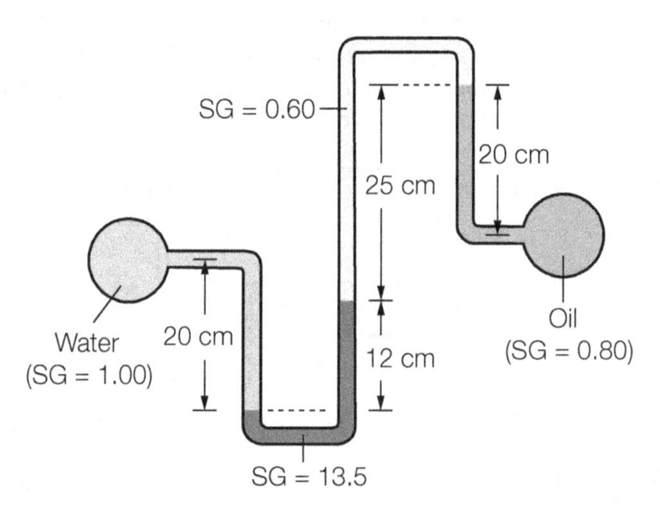

Figure 2.54: Manometer

2.38. An inclined manometer is attached to a cylindrical storage tank as shown in Figure 2.55, where the tank has a diameter of 1 m and the manometer has a diameter of 10 mm. The tank contains SAE 30 oil at 20°C, and initially the air above the oil in the tank is at atmospheric pressure, as is the air above the liquid surface in the inclined manometer. If the storage tank is sealed and the pressure in the tank increases by 200 Pa, what angle, θ, of the inclined manometer is required such that the liquid in the manometer moves 200 mm from its original location?

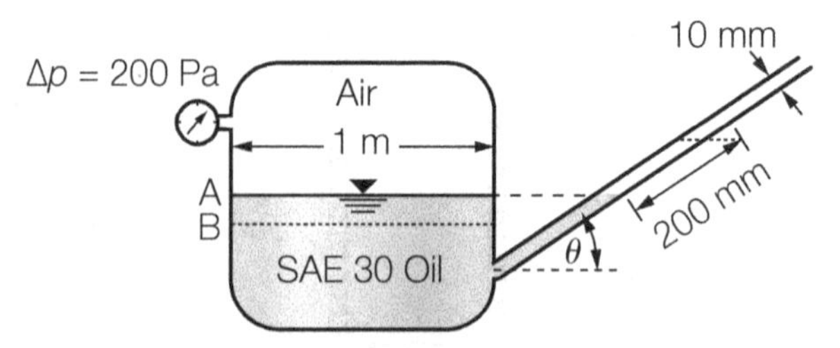

Figure 2.55: Inclined manometer

2.39. The pressure of water flowing through a pipe is measured by the inclined differential manometer shown in Figure 2.56. For the values given, calculate the pressure in the pipe.

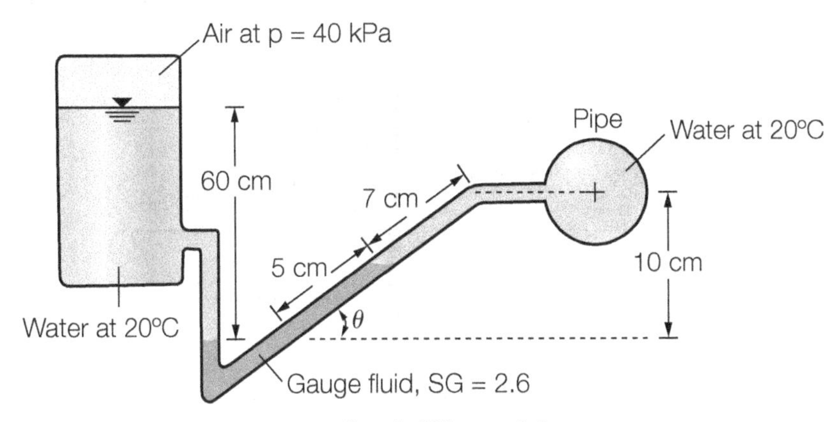

Figure 2.56: Inclined differential manometer

Section 2.4: Forces on Plane Surfaces

2.40. A 3.2 m × 4.1 m rectangular gate is mounted vertically in the side of a water storage reservoir such that the 3.2-m side of the gate is parallel to the water surface. The gate weighs 20 kN and is mounted in vertical guides such that the gate can be opened by pulling upward with cables attached to the top of the gate. The coefficient of friction between the guides and the gate is 0.05. When the water level is 2 m above the top of the gate, what force is required to lift the gate?

2.41. A section of a wall is to be constructed by pouring liquid concrete into a form in the shape of a wall as shown in Figure 2.57. The wall section is 4 m high, 3.5 m long, and 0.3 m thick. The specific gravity of liquid concrete can be estimated as 2.5, and before the concrete dries, the pressure distribution in the liquid concrete can be assumed to be hydrostatic. The support beams on each side of the formwork should be located such that they are aligned with the resultant hydrostatic force on the formwork. How far above the bottom of the wall should the support beams be located? What should be the lateral location of the support beams?

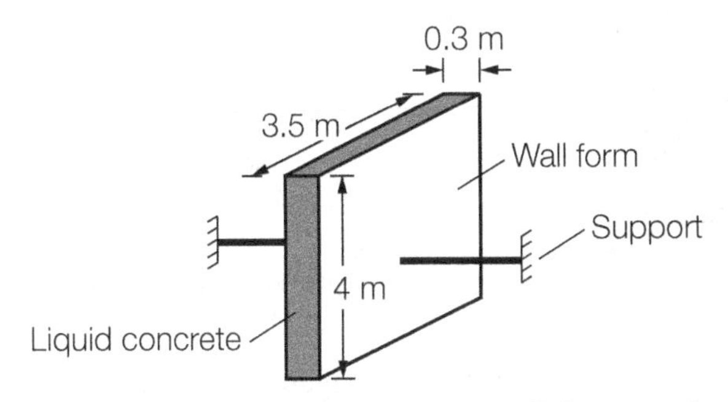

Figure 2.57: Liquid concrete in wall formwork

2.42. A 2 m × 3 m rectangular gate is located on the sloping side of a water reservoir such that the 2-m side of the gate is parallel to the water surface. The side of the reservoir (and the gate) slopes at an angle of 60° to the horizontal, and the top of the gate is 2.5 m vertically below the water surface. Estimate the resultant hydrostatic force on the gate and the effective location of this resultant force, as measured vertically downward from the water surface.

2.43. The 2.2-m-diameter pipe illustrated in Figure 2.58 drains water from a reservoir, where the crown of the pipe is 3 m below the surface of the reservoir. If the pipe entrance is covered with a 2.2-m-diameter gate hinged at A, determine the magnitude and location of the net hydrostatic force on the gate. What torque would need to be applied at A to open the gate?

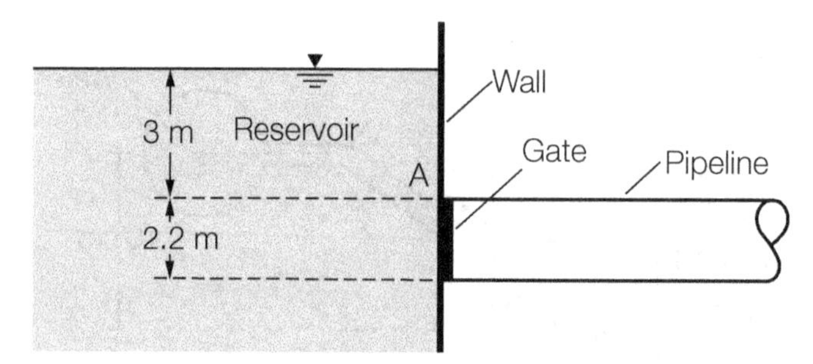

Figure 2.58: Submerged gate

2.44. A vertical rectangular gate of height h and width w is installed such that the top of the gate is a distance d below the water surface in a reservoir. Calculate the hydrostatic force on the gate and show that the location of the center of pressure is given by

$$y_{\rm cp} = d + \frac{L(3d + 2h)}{6d + 3h}$$

2.45. A reservoir is to be built to contain a slurry as shown in Figure 2.59. It is proposed that the retaining wall of the reservoir be 3 m high and 1 m thick and be made of concrete with a density of 2800 kg/m^3. A review of site conditions indicates that the coefficient of friction between the concrete wall and the underlying soil is equal to 0.35. It is expected that the slurry will have a density of 1500 kg/m^3. (a) Estimate the height, h, of slurry that will cause the wall to fail by sliding over the soil, (i.e., shear failure). (b) Estimate the height, h, of slurry that will cause the wall to topple over by rotating about the point P. Which is the more likely failure mode?

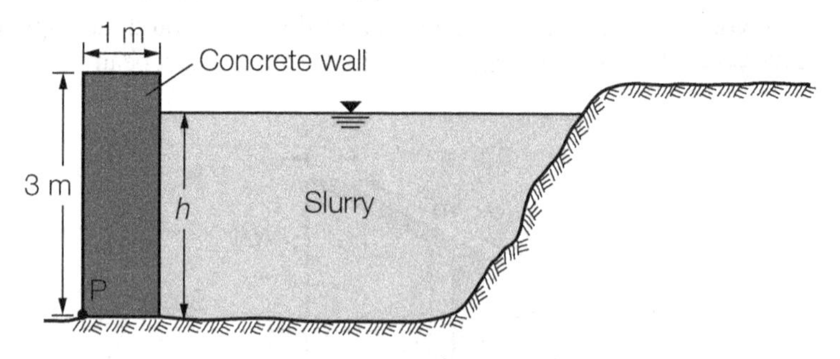

Figure 2.59: Slurry wall

2.46. The proposed concrete dam shown in Figure 2.60 has a bottom width of 25 m, it has a top width of 5 m, and the water level behind the dam is to be 4 m below the top of the dam. The specific gravity of the concrete used in the dam is 2.4. The water depth downstream of the dam is 3 m. Water seepage under the dam causes an uplift pressure distribution that varies linearly from the upstream hydrostatic pressure to the downstream hydrostatic pressure. Estimate the height, h, of water behind the dam that will cause the dam to overturn. Assume water at 20°C.

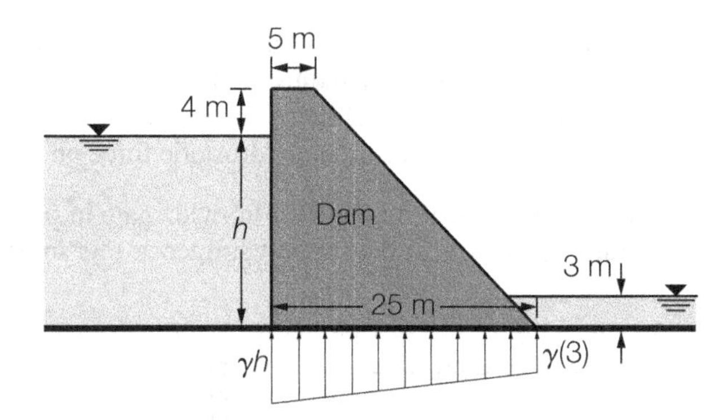

Figure 2.60: Concrete dam with uplift pressure

2.47. A vertical circular gate of diameter D is installed such that the top of the gate is a distance d below the water surface in a reservoir. Calculate the hydrostatic force on the gate and show that the location of the center of pressure is given by

$$y_{\text{cp}} = D + \frac{D(8d + 5D)}{16d + 8D}$$

2.48. A 2.5-m-diameter gate is connected to a water reservoir as shown in Figure 2.61. The gate is oriented at 35° to the horizontal, the centroid of the gate is 1.5 m vertically below the water surface, and the weight of the gate is 500 kN. The gate is pin-connected to the reservoir at the top of the gate (at P) and is opened by applying a vertical force at the bottom of the gate (at Q). (a) What is the magnitude of the resultant hydrostatic force and its location relative to the top of the gate? (b) What is the magnitude of the force required to open the gate?

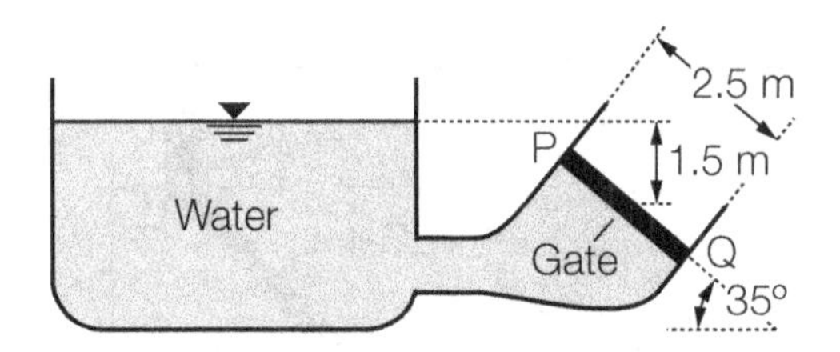

Figure 2.61: Gate connected to a reservoir

2.49. A semicircular steel gate is installed in the side of a water storage reservoir that has a side slope of 50° as shown in Figure 2.62. The gate has a radius of 3 m, the top of the gate is supported by a horizontal shaft attached to the reservoir, and the bottom of the gate is supported at the point P with a force F_P. When the depth of water in the tank is 15 m, what support force is necessary to keep the gate from opening?

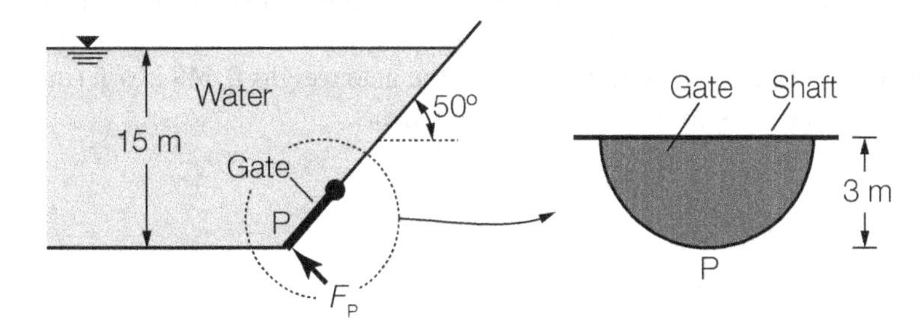

Figure 2.62: Gate in the side of a reservoir

2.50. A circular hatch is located in a sloping wall of a water storage reservoir, where the wall slopes at 35° to the horizontal, the radius of the hatch is 420 mm, and the center of the hatch is 3 m below the water surface (measured along the sloping wall). Find the magnitude and location of the resultant hydrostatic force on the hatch.

2.51. Determine the force at P needed to hold the 4-m-wide gate in position as shown in Figure 2.63. You may assume that the support surface at P is smooth (i.e., frictionless).

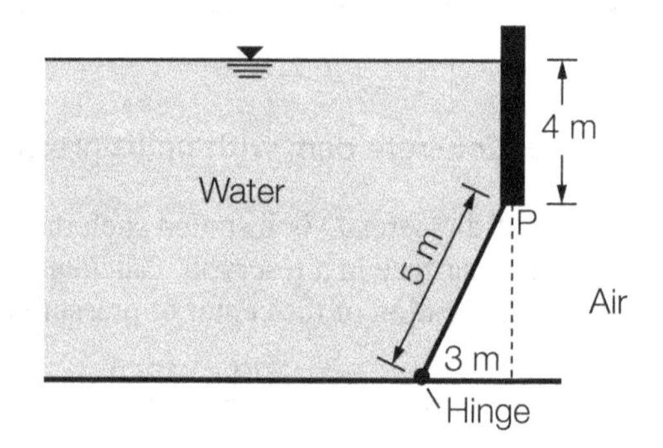

Figure 2.63: Force on a gate

2.52. The 500-kg, 5-m-wide rectangular gate shown in Figure 2.64 is hinged at B and makes an angle of 45° with the floor at A. (a) If the gate is opened by applying a normal force at the center of the gate, determine the force required to open the gate. (b) If you could choose any location to apply the force to open the gate, what location would you choose and what force would need to be applied?

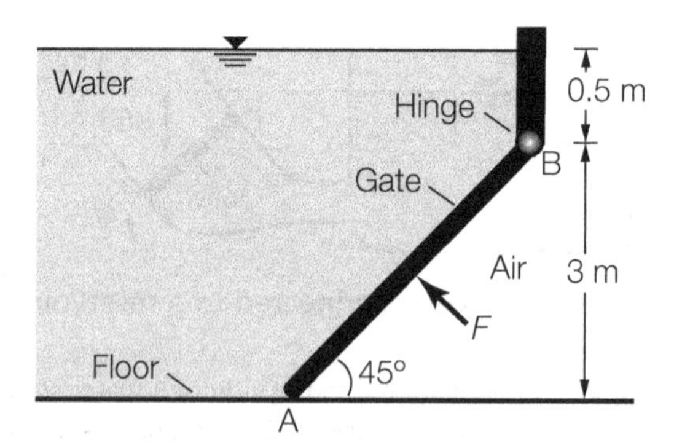

Figure 2.64: Gate

2.53. The rectangular gate shown in Figure 2.65 is intended to control the flow of seawater into a freshwater reservoir. If the fresh water has a density of 998 kg/m^3, the seawater has a density of 1025 kg/m^3, and the gate weighs 0.448 kN/m (into the page), find the depth of seawater when the gate opens.

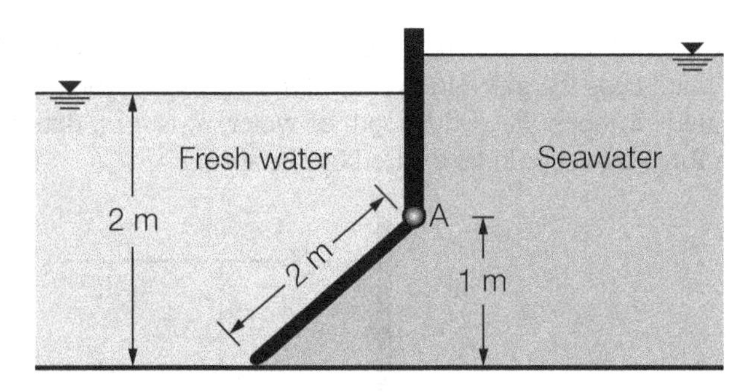

Figure 2.65: Salinity control gate

2.54. The water in a reservoir is contained by an elliptical gate illustrated in Figure 2.66, where D is the diameter of the pipe leading to the gate and D and $D/\sin\theta$ are the lengths of the minor and major principal axes of the elliptical gate, respectively. If $D = 1.2$ m, $\theta = 30°$, and the water surface in the reservoir is 9 m above the centerline of the gate, determine the resultant hydrostatic force on the gate and the location of the center of pressure. What moment at Point P would be required to keep the gate closed? Neglect the weight of the gate.

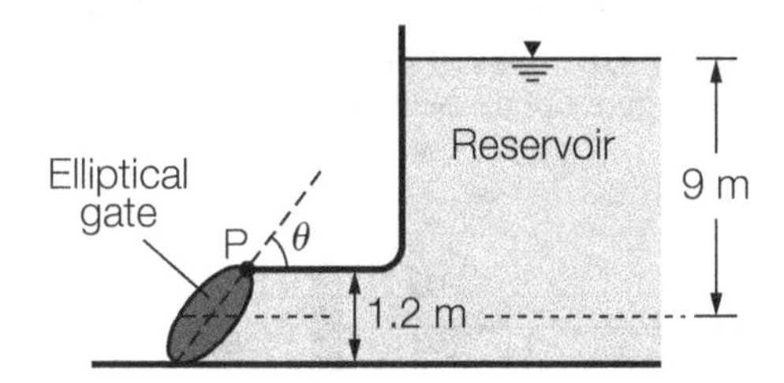

Figure 2.66: Gate in a reservoir

2.55. A square flap gate is shown in Figure 2.67, where the side dimensions of the gate (b) are 1.5 m, the gate weighs 8 kN, the gate is hinged 1 m above its center (a = 0.25 m), and the face of the gate is sloped 4° from the vertical. To what depth will water rise behind the gate before it opens?

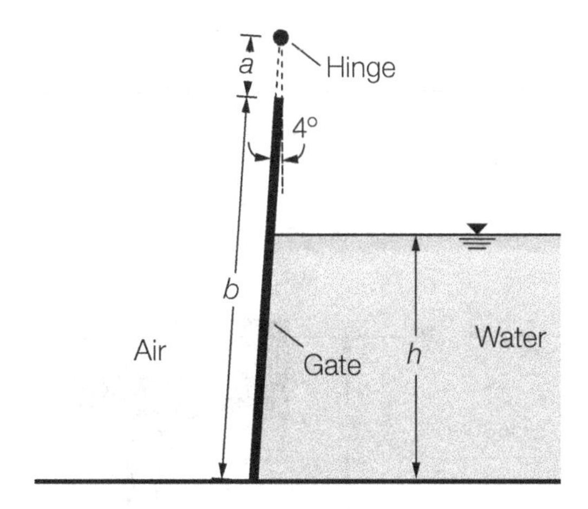

Figure 2.67: Flap gate

2.56. The 2-m-wide gate shown in Figure 2.68 is pinned at P and is connected to a restraining spring that keeps the gate closed until the hydrostatic force is sufficient to open it. If the gate is to open when the depth of water, y, is 4 m, determine the required restraining force that should be exerted by the spring.

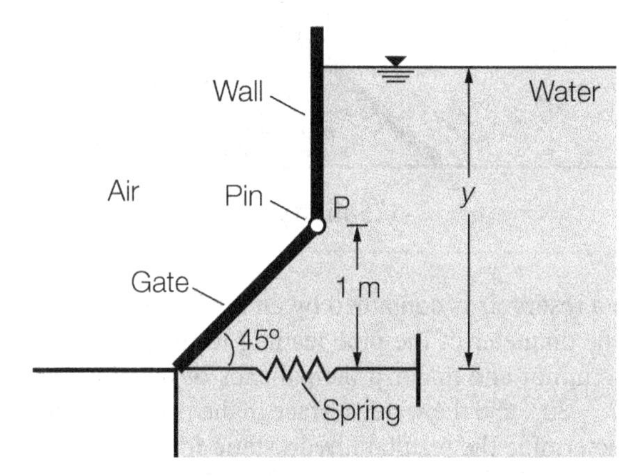

Figure 2.68: Gate with restraining spring

2.57. Gate AB in Figure 2.69 is 1.52 m wide and opens to let fresh water out when the ocean tide is dropping. The hinge at A is 0.61 m above the freshwater level. At what ocean level h will the gate open? Neglect the weight of the gate.

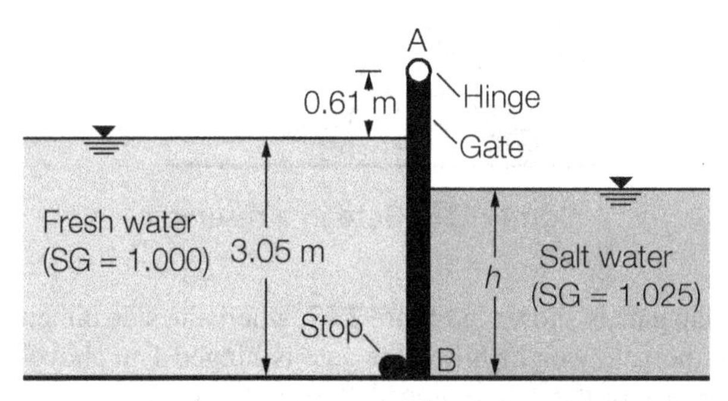

Figure 2.69: Tide gate

2.58. A hinged gate is located under an elevated dock in a freshwater reservoir as shown in Figure 2.70. The hinge is located 0.5 m from the bottom of the reservoir, and the top of the gate is 0.7 m above the hinge. The width of the gate is 3 m. (a) What depth of water, h, will cause the gate to open? Assume water at 20°C. (b) Show that the depth of liquid required to open the gate is independent of the density of the liquid. Note that the gate consists of a single 1.2 m × 3 m rectangular piece and that the dock offers no resistance to the rotation of the gate.

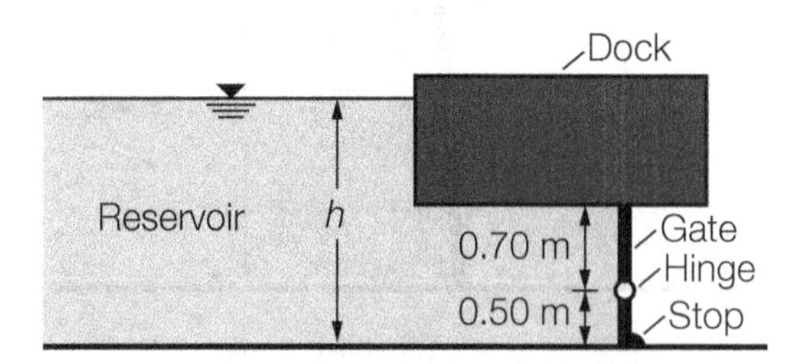

Figure 2.70: Hinged gate under a dock

2.59. The two sides of a V-shaped water trough are hinged together at the bottom where they meet, as shown in Figure 2.71, making an angle of 45° with the ground from both sides. Each side is 0.75 m long, and the two sides are held together by cables placed every 6 m along the length of the trough. Calculate the tension in each cable when the trough is filled to the level of the cable.

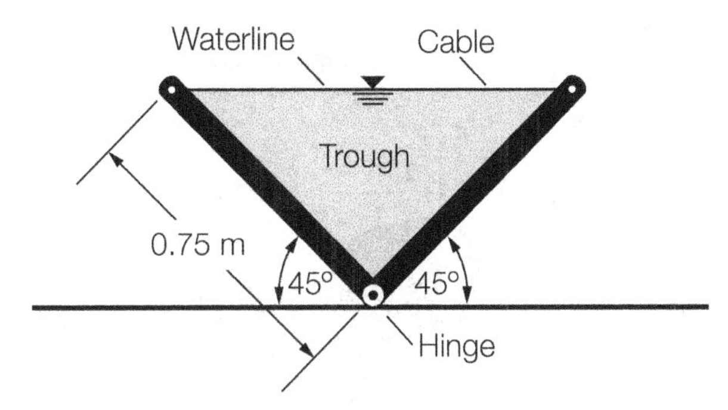

Figure 2.71: V-shaped water trough

2.60. Consider the pressurized storage tank shown in Figure 2.72, where water at 20°C is contained by a rectangular gate that is 2 m long by 3 m wide and is open to the atmosphere on one side. The side of the tank in contact with water makes an angle of 30° with the horizontal, and above the gate support is 2 m of inclined surface and 1 m of vertical surface below the water surface. The air above the water surface is at a gauge pressure of 300 kPa. What is the magnitude of the net hydrostatic force on the gate? At what depth below the water surface does the resultant hydrostatic force act?

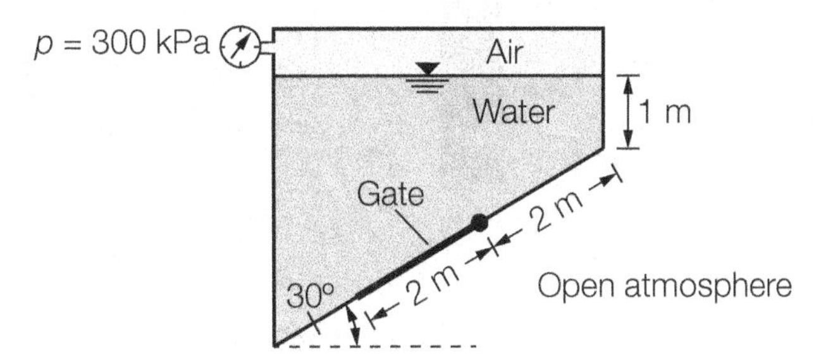

Figure 2.72: Pressurized storage tank

2.61. Find the magnitude and location of the resultant hydrostatic force on the gate shown in Figure 2.73. Also find the magnitude of the moment about axis XX that would be required to open the gate. The pressure above the water surface is equal to atmospheric pressure, as is the pressure on the other side of the gate.

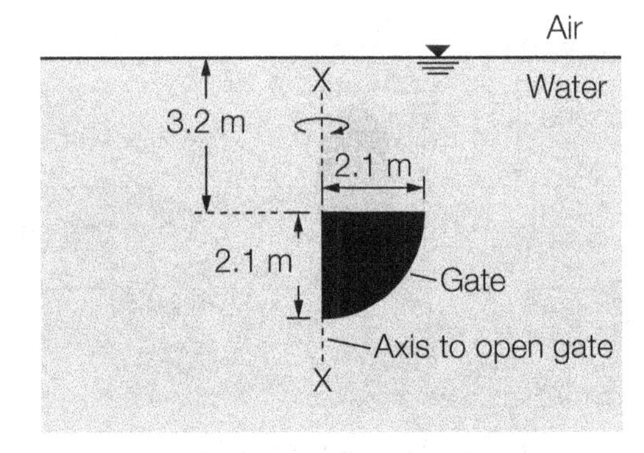

Figure 2.73: Submerged gate

2.62. Consider the semi-elliptical gate shown in Figure 2.74. Determine the magnitude and location of the resultant hydrostatic force on the gate and find the magnitude of the moment about axis XX that would be required to open the gate. The pressure above the water surface is equal to atmospheric pressure, as is the pressure on the other side of the gate.

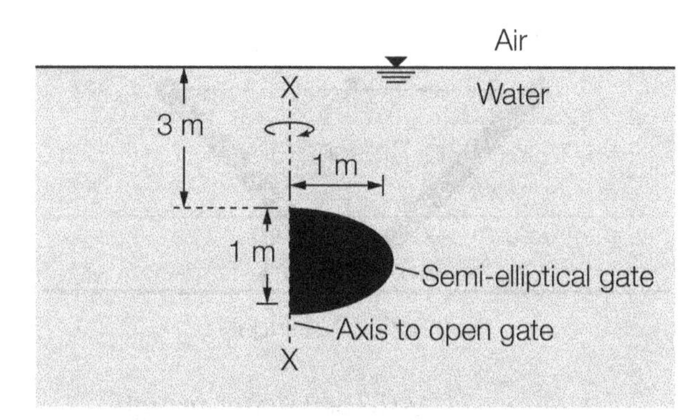

Figure 2.74: Submerged semi-elliptical gate

2.63. Consider the submerged gate in a stratified liquid as shown in Figure 2.75. Find the magnitude and location of the resultant hydrostatic force on the gate.

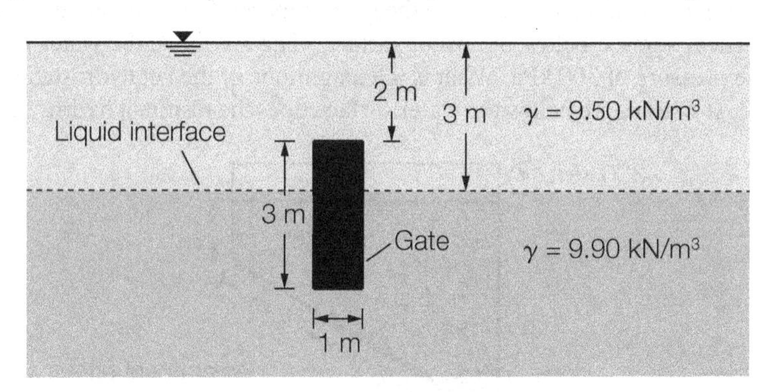

Figure 2.75: Submerged gate

2.64. Determine the magnitude and location of the resultant hydrostatic force on the elliptical gate shown in Figure 2.76. The center of the ellipse coincides with the interface between the two fluids.

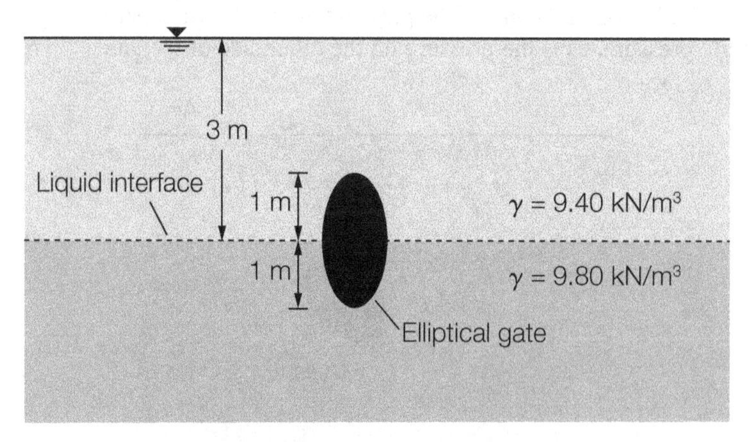

Figure 2.76: Submerged elliptical gate

Section 2.5: Forces on Curved Surfaces

2.65. Pressurized water fills the tank in Figure 2.77. Compute the net hydrostatic force on the conical surface ABC.

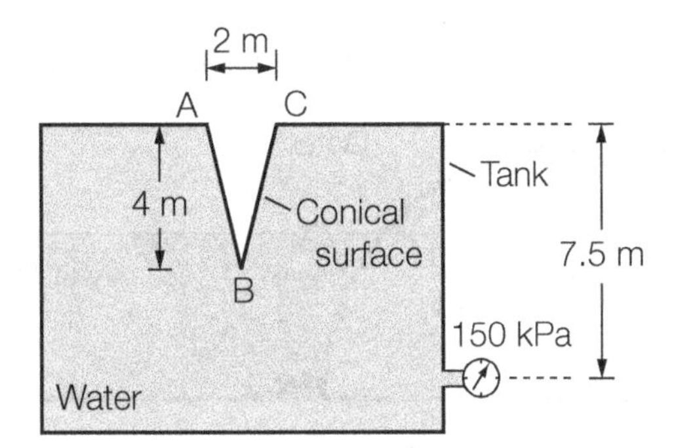

Figure 2.77: Force on a conical surface

2.66. The step dam shown in Figure 2.78 is to be constructed on an earth foundation that can support a maximum shear force of 2500 kN. The length of the dam is 10 m, and the step height and depth are to be equal. What maximum step height can be used without exceeding the support capacity of the foundation? Under this limiting condition, what is the vertical hydrostatic force on the dam? Assume water at 20°C.

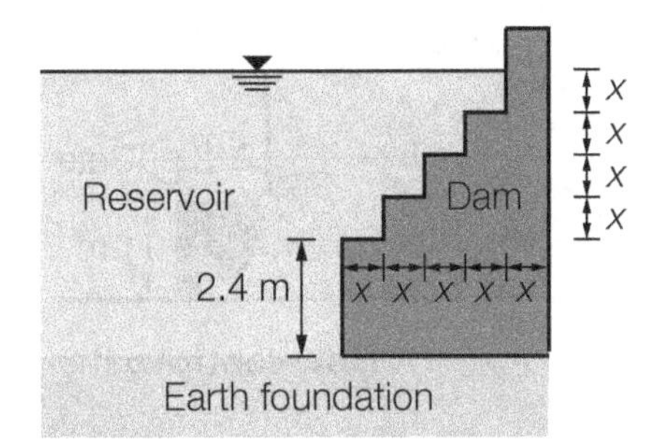

Figure 2.78: Step dam

2.67. A 5-m-long seawall separates a freshwater body from a saltwater body as shown in Figure 2.79. The wall is a total of 4 m high, with the top half of the wall being semicircular. Under design conditions, the surface of the freshwater body is at the top of the wall and the surface of the saltwater body is at the midheight of the wall. What is the net hydrostatic force on the wall under the design condition? Assume that both the fresh water and salt water are at 20°C.

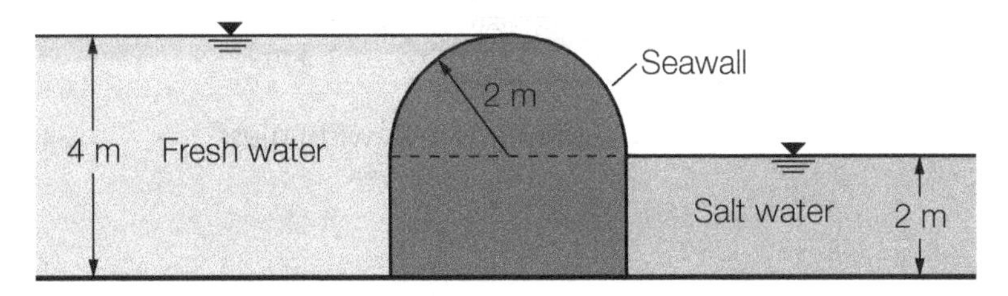

Figure 2.79: Seawall at the design condition

2.68. A gate in the shape of a quarter of a circle divides a freshwater body from a salt-water body as shown in Figure 2.80. The gate has a radius of 3.5 m and a length (perpendicular to the page) of 4.8 m. (a) Determine the height, h, of salt water that will cause the horizontal hydrostatic forces on both sides of the gate to be equal. (b) Determine the height of salt water that will cause the vertical hydrostatic forces on the gate to be equal. Assume that both the fresh water and salt water are at 20°C.

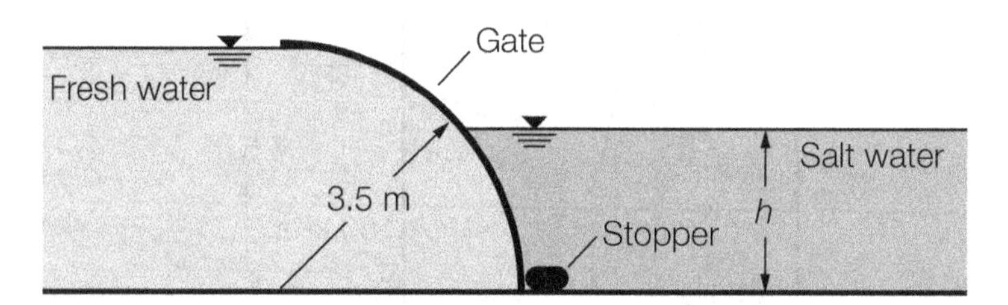

Figure 2.80: Gate separating fresh water from salt water

2.69. The flow control gate shown in Figure 2.81 has the shape of a quadrant of a circle of radius 1 m and a weight of 40 kN per meter of length perpendicular to the page. The gate is pin-connected at P. Determine the height, h, of water in the reservoir behind the gate that will cause the gate to open. Assume water at 20°C.

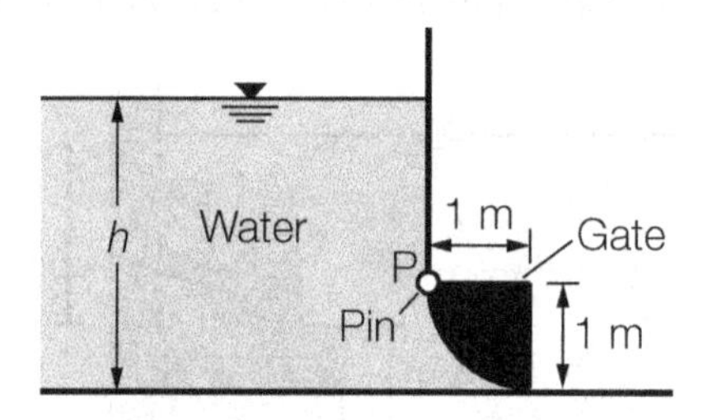

Figure 2.81: Flow control gate

2.70. An 8-m-deep by 20-m-long aquarium is to be designed with a glass viewing area at the bottom. If the viewing section is shaped like the quadrant of a circle (Figure 2.82), calculate the hydrostatic force on the viewing glass.

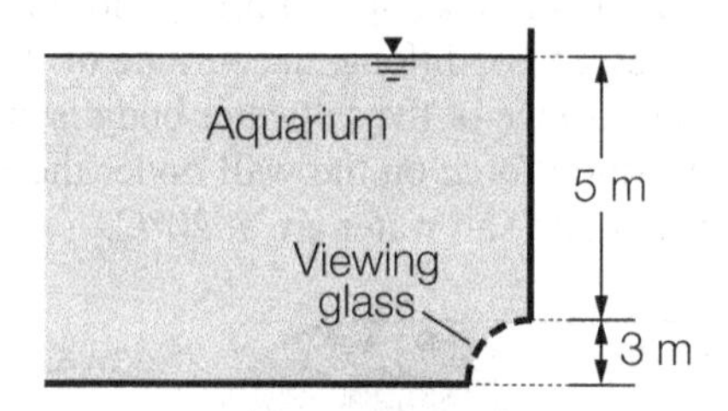

Figure 2.82: Submerged viewing glass

2.71. The gate shown in Figure 2.83 is used to contain water in an aquarium. The gate extends from A to C, consists of an arc of a circle between A and B (the viewing window), and is planar between B and C. The gate is 2 m wide, is pinned at A, and rests on a frictionless surface at C. For the water-surface elevation shown in Figure 2.83, determine the hydrostatic force on the gate.

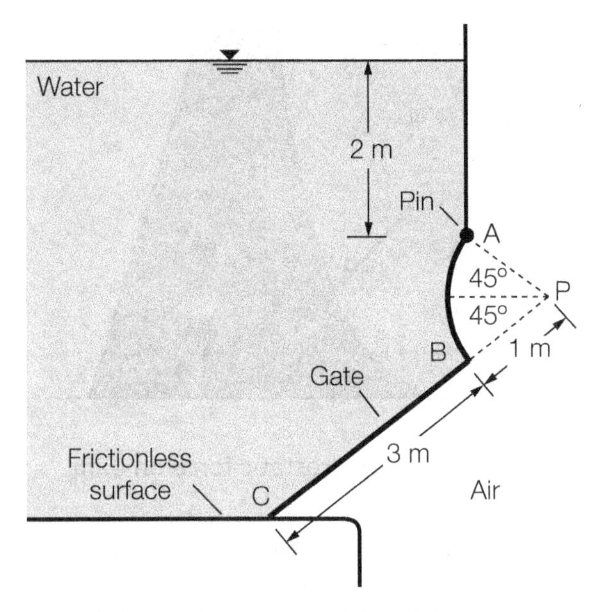

Figure 2.83: Aquarium gate

2.72. Water is contained in a reservoir by the Tainter gate illustrated in Figure 2.84. The gate has the shape of a quadrant of a circle, it is 5 m wide, it weighs 10 kN, and its center of gravity is at Point G. Determine the net hydrostatic force on the gate and the magnitude and direction of the moment that must be applied at P to open the gate.

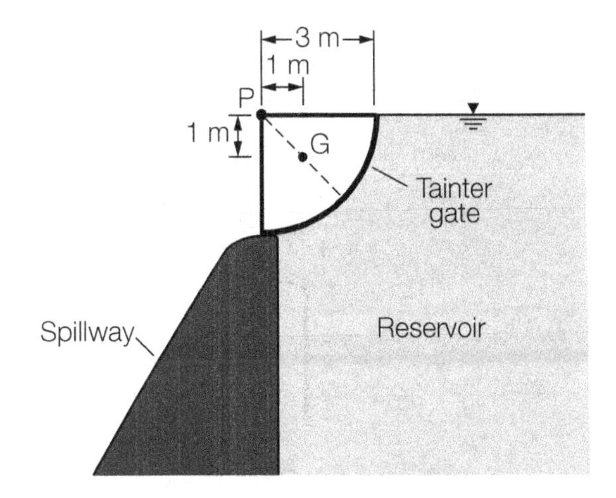

Figure 2.84: Radial gate

2.73. The face of the dam shown in Figure 2.85 is curved according to the relation $y = x^2/2.4$, where y and x are in meters. The height of the water surface above the horizontal plane through A is 15.25 m. Calculate the resultant force due to the water acting on a unit breadth of the dam and determine the position of the point B at which the line of action of this force cuts the horizontal plane through A.

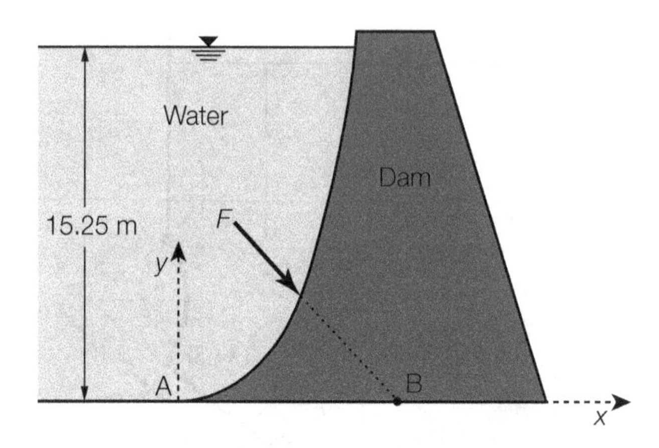

Figure 2.85: Hydrostatic force on dam

2.74. An aquarium proposes using a 1-m-diameter hemispherical viewing glass in its piranha exhibit as shown in Figure 2.86. The center of the viewing glass is proposed to be 5 m below the water surface. The viewing glass is to be fixed to the wall using rivets, each capable of supporting 100 N of normal force and 5 N of shear force. What is the minimum number of rivets required? If a flat circular viewing glass is used instead, what is the minimum number of rivets required? What is the ratio of the hydrostatic force on the top half of the viewing glass to the hydrostatic force on the bottom half of the viewing glass? Comment on how you would distribute the rivets around the viewing glass.

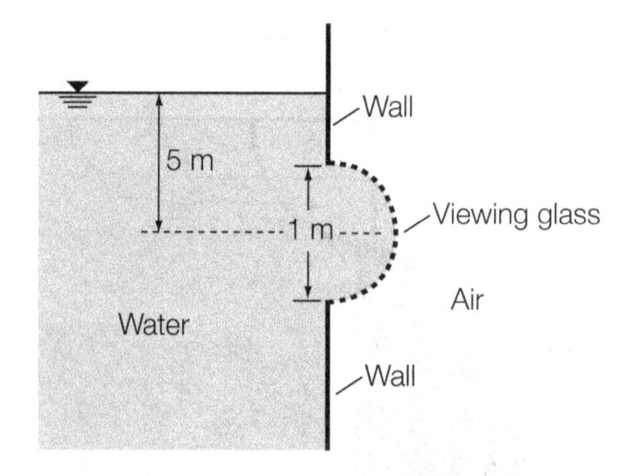

Figure 2.86: Aquarium

2.75. A water trough of a semicircular cross section of radius 0.5 m consists of two symmetric parts hinged together at the bottom as shown in Figure 2.87. The two parts are held together by cables placed every 3 m along the length of the trough. Calculate the tension in each cable when the trough is filled to the rim.

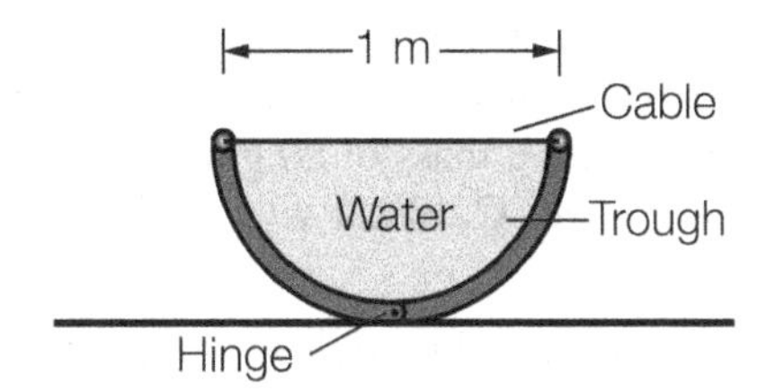

Figure 2.87: Semicircular channel

2.76. Gate AB in Figure 2.88 is a three-eighths circle, is 3 m wide into the page, is hinged at B, and is resting against a smooth wall at A. The specific weight of the seawater being retained by the gate is 10.05 kN/m^3. Compute the reaction forces at Points A and B.

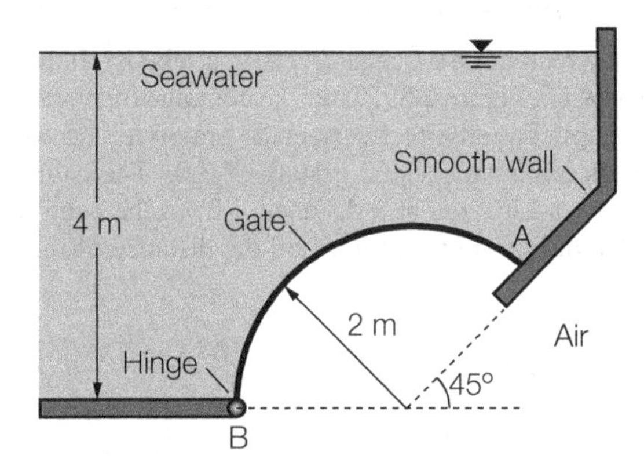

Figure 2.88: Submerged gate

2.77. A natural gas pipeline has a diameter of 410 mm and a wall thickness of 4.2 mm. If the pressure of the gas in the pipeline is equal to 800 kPa, estimate the average circumferential stress in the pipe wall.

Section 2.6: Buoyancy

2.78. The weight of a solid object in air is measured as 40 N, and the weight of this same object in water at 20°C is measured as 25 N. Estimate the specific weight and volume of the object.

2.79. Archimedes' principle was developed as a by-product of a royal project undertaken by Archimedes to determine whether the crown of King Hiero II of Syracuse, Sicily, was made of pure gold. Knowing the specific gravity of gold and being able to weigh the king's crown in both air and water, Archimedes, in effect, developed a relationship between the specific gravity of an object and the ratio of the weight of an object in water to the weight of that same object in air. (a) Determine this relationship. (b) Apply this relationship to determine the specific gravity of the object described in Problem 2.78.

2.80. A hot air balloon has a diameter of 15 m, and the weight of the balloon plus attached load is 2 kN. To what temperature must the air in the balloon be heated to achieve liftoff? Assume standard sea-level conditions in the surrounding air.

2.81. A 3-m-diameter balloon with a hard (nonexpandable) shell is filled with helium and released into the atmosphere. The balloon plus the helium has a mass of 8 kg. At what elevation will the balloon stabilize in a standard atmosphere?

2.82. A hot air balloon is to carry a weight of 1.5 kN over an area that has a typical atmospheric pressure of 101 kPa and a typical summer temperature of 20°C. The balloon is made of a material that has a mass of 80 g/m^2, and the air in the balloon is to be heated to a temperature of 80°C. What will be the diameter of the balloon under stable conditions?

2.83. The drag force, F_D, on a spherical particle settling with velocity, v, in a fluid can be approximated by

$$F_\mathrm{D} = 3\pi\mu v D$$

where μ is the viscosity of the fluid and D is the particle diameter. If a 2-mm-diameter particle with a specific gravity of 2.65 is stirred up from the bottom of a river, estimate the sedimentation velocity, v. (*Hint:* A particle settles with a constant velocity when the sum of the forces on the particle is equal to zero.)

2.84. A cabin used by recreational scuba divers in a lake is 10 m long by 15 m wide by 4 m high as shown in Figure 2.89. The lake contains fresh water at 20°C. The lighter section of the cabin has an effective specific gravity of 1.5, and a heavier section of the cabin has an effective specific gravity of 3.0. The cabin is kept horizontal by a linear support wall located at a distance x from the centroidal axis of the cabin. Determine the required support force and the distance of the support wall from the centroidal axis.

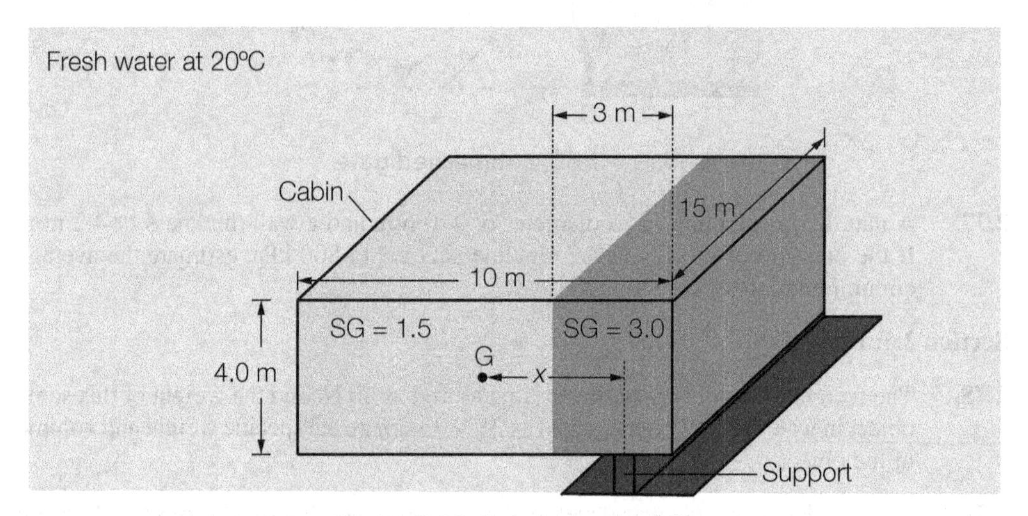

Figure 2.89: Submerged cabin

2.85. A tank contains a layer of SAE 30 oil floating on a layer of water, where both liquids are at 20°C. A cube of dimensions 0.15 m × 0.15 m × 0.15 m is placed in the tank, and it is observed that the cube locates itself such that its top and bottom faces are parallel to the interface between fluids and 15% of the height of the cube is located in the oil layer. Estimate the density of the cube.

2.86. The volume of seawater displaced by ships is usually controlled by the intake and discharge of ballast water, which increases and decreases the weight of the ship, respectively. Show that the percentage change in displaced seawater is equal to the percentage change in the weight of the ship.

2.87. The hollow steel sphere shown in Figure 2.90 is to be used to house instruments for oceanic research and is to be sized such that 90% of its volume is below the surface of the ocean. The wall thickness is to be 25 mm, the weight of the instrumentation to be contained in the sphere is 500 N, and the density of the steel to be used in manufacturing the sphere is 8000 kg/m^3. What should be the diameter of the sphere?

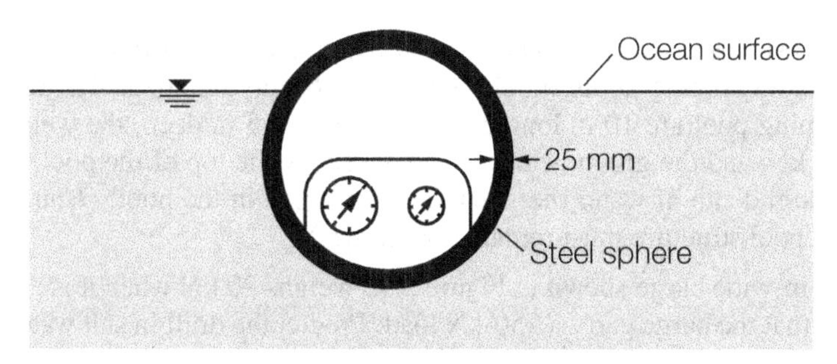

Figure 2.90: Instrumentation sphere

2.88. The cylindrical object shown in Figure 2.91 is to be sized such that it floats in a 1-m-deep pool of water with a minimum clearance of 100 mm from the bottom of the pool. The specific gravity of the object is estimated as 0.85. What is the maximum allowable diameter of the object?

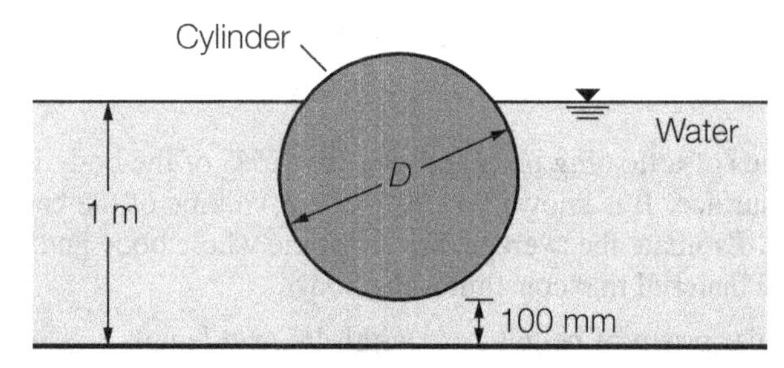

Figure 2.91: Cylindrical object floating in water

2.89. The standard canoe model from a manufacturer weighs 810 N, and an ultralight (and ultrastrong) model of the same size and shape weighs 220 N. Both models are capable of carrying the same load. For any given load, determine the additional volume of water displaced by the heavier model.

2.90. If a body of specific gravity SG_1 is placed in a liquid of specific gravity SG_2, what fraction of the total volume of the body will be above the surface of the liquid? If an iceberg has a specific gravity of 0.95 and is floating in seawater with a specific gravity of 1.25, what fraction of the iceberg is above water?

2.91. A new tactic in naval warfare is being proposed in which ships are sunk by a high-density release of bubbles from below the ship to reduce the effective density of the water surrounding the ship. This tactic is illustrated in Figure 2.92. If a ship has a specific gravity of 0.8 and bubbles with a diameter of 10 mm are released, what is the number of bubbles per m^3 required to sink the ship? Does your answer depend on the shape of the ship? Explain.

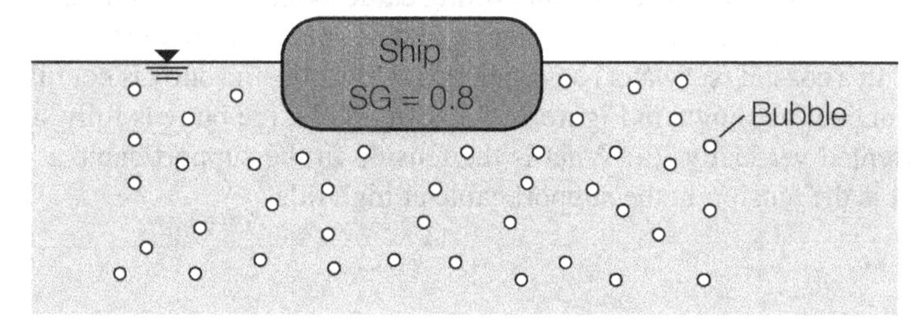

Figure 2.92: Bubbles beneath a ship

2.92. It is conventional wisdom that in areas where the groundwater is close to the land surface, swimming pools should not be completely emptied. The dimensions of a swimming pool are 10 m long by 5 m wide by 2.5 m deep, the weight of the pool is 500 kN, and the groundwater is 1.25 m below the top of the pool. Determine the minimum depth of water that must be maintained in the pool. (*Hint*: The net force on the pool structure must remain downward.)

2.93. The 3-m-wide barge shown in Figure 2.93 weighs 20 kN when it is empty. It is proposed that the barge carry a 250-kN load. Predict the draft in salt water (SG = 1.03).

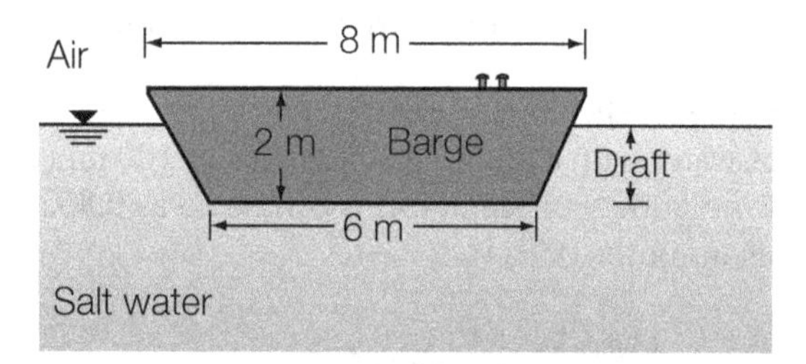

Figure 2.93: Buoyancy

2.94. Observations of a floating body indicate that 75% of the body is submerged below the water surface. It is known that 90% of the volume of the body consists of open (air) space. Estimate the average density of the whole body and the average density of the solid material that constitutes the body.

2.95. A rectangular prism of height 2 m, width W, and length L is placed in a stratified liquid as shown in Figure 2.94. The top layer of the liquid has a specific gravity of 1.2 and a thickness of 1.2 m, and the bottom layer has a specific gravity of 1.6. (a) Determine the minimum specific gravity of the body for which the body will fully penetrate the top layer. (b) Determine the depth of penetration, if any, into the bottom layer when the body has a specific gravity equal to 1.0.

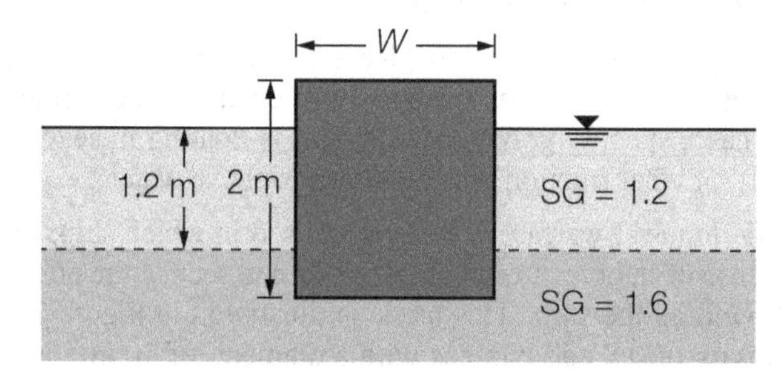

Figure 2.94: Floating body in a stratified fluid

2.96. Consider a buoy attached to a supporting cable as shown in Figure 2.95. The buoy is 3 m long, it has a diameter of 200 mm, and its specific gravity relative to seawater is 0.60. In a coastal (seawater) environment at low tide, the buoy is not fully submerged and inclines as shown in Figure 2.95. At high tide, the buoy is fully submerged and is oriented vertically. (a) What is the tension in the support cable at low tide? (b) What is the tension in the support cable at high tide?

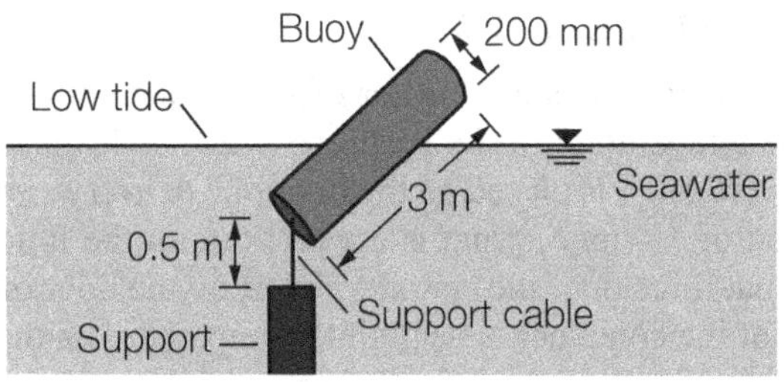

Figure 2.95: Buoy at low tide

2.97. A hydrometer with a stem diameter of 12 mm weighs 0.246 N. An engineer places the hydrometer in pure water at 20°C and marks the place on the stem corresponding to the water surface. When the engineer places the hydrometer in a test liquid, the mark is 2 cm above the surface of the liquid. Estimate the specific gravity of the test liquid.

2.98. A hydrometer with a stem diameter of 9 mm is placed in distilled water, and the volume of the hydrometer below the water surface is estimated to be 20 cm^3. If the hydrometer is placed in a liquid with a specific gravity of 1.2, how far above the liquid surface will the distilled water mark be located?

2.99. A cylinder of diameter 0.70 m and height 0.60 m is placed in a liquid as shown in Figure 2.96. The cylinder has a specific gravity of 0.65, and the liquid has a specific gravity of 0.90. Determine whether the cylinder is stable at its initial orientation.

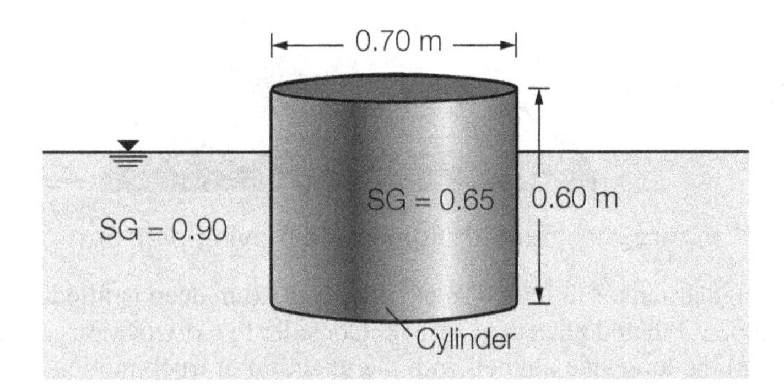

Figure 2.96: Cylinder in a liquid

2.100. The canoe shown in Figure 2.97 has a length of 4 m, and with one rower in the canoe, the dimensions of the submerged portion of the canoe are as shown as well. Estimate the maximum height of the center of gravity of the canoe plus the rower (relative to the bottom of the canoe) for the canoe to be stable.

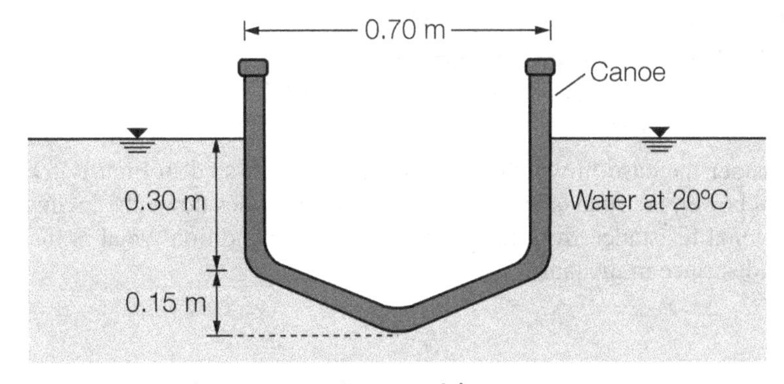

Figure 2.97: Canoe with one person

Section 2.7: Rigid-Body Motion of Fluids

2.101. A cylindrical container has a diameter of 0.4 m and contains kerosene to a depth of 0.6 m. The temperature of the kerosene is 20°C. If the container, with its long axis oriented vertically, is placed on the floor of a delivery elevator that ascends with an acceleration of 1.5 m/s^2, what is the pressure in the fluid on the bottom of the container? What force does the container exert on the floor of the elevator? Assume that the mass of the container is negligible compared with that of the kerosene.

2.102. A water truck is mounted with a cylindrical tank that has a diameter of 2 m and a length of 10 m. The long axis of the tank is oriented with the direction of truck motion. If the tank is filled with water at 20°C and the truck accelerates at a rate of 2 m/s^2, estimate the difference in magnitude between the resultant hydrostatic force on the front and back of the tank.

2.103. A 4-m-diameter, 2-m-deep cylindrical tank containing a caustic liquid with a density of 1040 kg/m^3 is placed on a moving platform as shown in Figure 2.98. The tank is filled to a depth of 1.2 m. (a) What is the maximum allowable acceleration of the moving platform that will prevent the liquid from spilling out of the tank? (b) At this limiting condition, what is the (gauge) pressure at the bottom front of the tank and what is the gauge pressure at the bottom back of the tank?

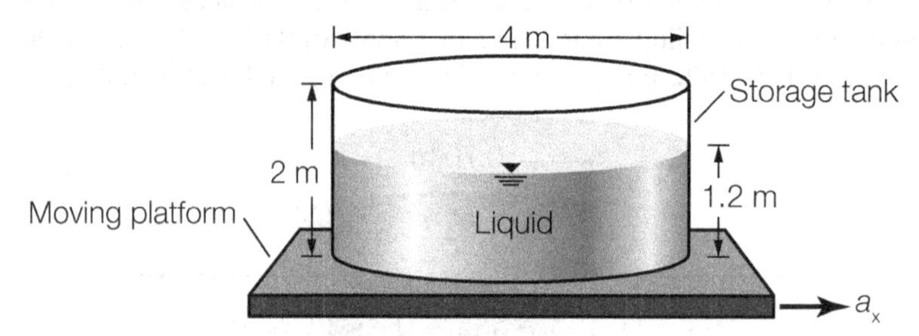

Figure 2.98: Tank of liquid on a moving platform

2.104. A rectangular tank 3 m long, 0.8 m wide, and 1.6 m deep is filled with kerosene to a depth of 1.2 m and placed on a truck. Consider two orientations of the tank on the truck: (a) the long side aligned with the direction of truck motion and (b) the short side aligned with the direction of truck motion. Which orientation would allow the greatest truck acceleration without spillage? What is that acceleration?

2.105. Sensors in a decelerating tanker truck indicate that the liquid in the tank has a 10° slope. At what rate is the truck decelerating?

2.106. If a tanker truck accelerates from rest to a highway speed of 90 km/h in 10 seconds, what slope of (unconfined) liquid would be expected to occur in the tank? Assume that the truck accelerates at a constant rate.

2.107. The tanker truck shown in Figure 2.99 accelerates at a rate of 5 m/s^2 down a 25° incline. Determine the slope of the liquid surface in the tank. Is the slope of the liquid surface greater than or less than the slope of the incline?

2.108. Consider the case in which the tanker truck illustrated in Figure 2.99 completely loses power, the road is iced over, and the truck slides down the incline without any frictional resistance from the road. Under this condition, what is the slope of the liquid surface in the tank?

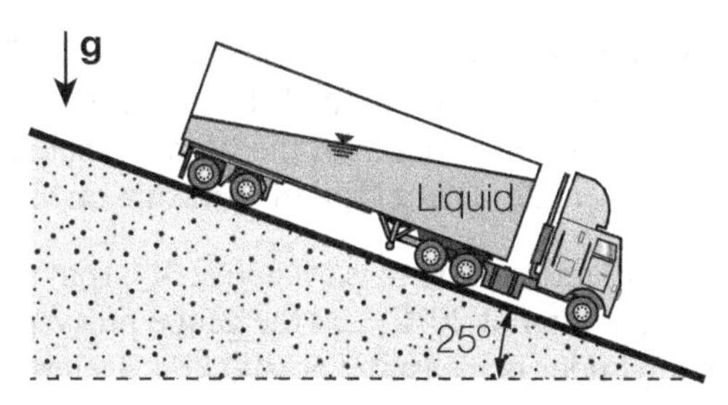

Figure 2.99: Tanker truck accelerating down an incline

2.109. Water at 20°C is contained in a U-tube as shown in Figure 2.100. Atmospheric pressure is 101 kPa. (a) What rate of acceleration toward the right would cause the liquid level in the left-hand arm of the U-tube to be 40 mm higher than the liquid level in the right-hand arm of the U-tube? (b) What rate of rotation around the z-axis would cause cavitation to occur at the center of the U-tube (i.e., at Point P)?

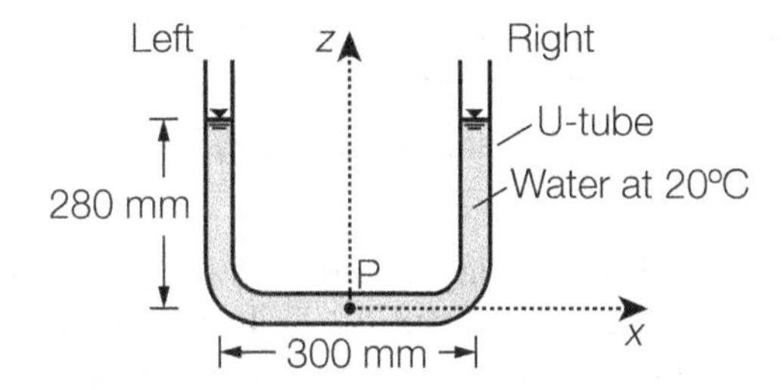

Figure 2.100: U-tube containing water

2.110. If the U-tube shown in Figure 2.100 is rotated about a vertical axis that is 50 mm to the right of Point P and parallel to the z-axis, what rate of rotation would cause the water levels in the arms of the U-tube to differ by 40 mm?

2.111. The U-tube shown in Figure 2.101 is in the shape of a semicircle between A and B, where the semicircle has a radius of 0.2 m and is centered at P. The U-tube is filled with a liquid such that the levels are the same at A and B, and then the tube is sealed at A. The liquid remains open to the atmosphere at B. The U-tube is then rotated at 450 rpm about a vertical axis that is 0.1 m from P. (a) Identify the location within the U-tube where the pressure is a minimum. (b) Plot the gauge pressure in the U-tube as a function of the radial distance from the axis of rotation. (c) If the liquid is water at 20°C and atmospheric pressure is 101.3 kPa, determine the minimum absolute pressure in the U-tube and assess whether cavitation will occur under the given conditions.

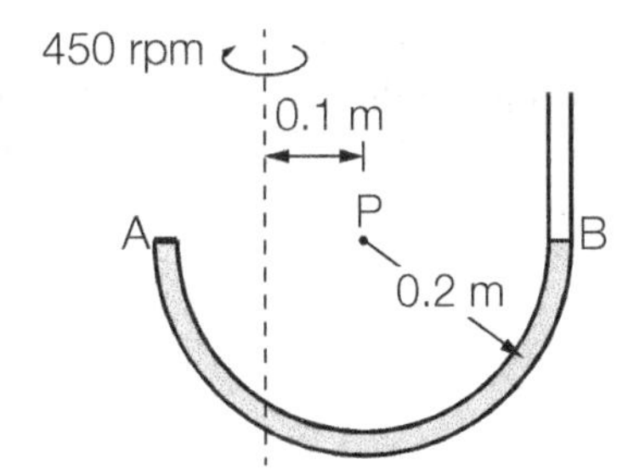

Figure 2.101: Circular U-tube

2.112. A barrel is filled with SAE 10 oil at 20°C and then sealed. The barrel has a diameter of 0.6 m, has a height of 2 m, and is rotated about its central axis at a rate of 40 rad/s. At any given height above the bottom of the barrel, what is the difference between the pressure at the center of the barrel and the pressure on the perimeter of the barrel?

2.113. A sealed cylinder of diameter 0.50 m is rotated at 400 rpm as shown in Figure 2.102. The cylinder contains water at 20°C, and the gauge pressure on the periphery of the top surface of the cylinder is measured as 200 kPa. Determine the force on the top of the cylinder.

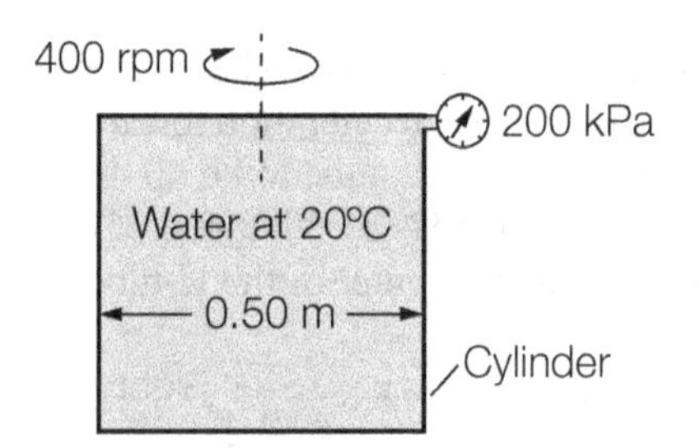

Figure 2.102: Rotating cylinder

2.114. A cylindrical reservoir containing a liquid can be used to determine the rotational speed of a platform as shown in Figure 2.103. In this particular case, a 4-cm-diameter cylindrical reservoir is placed at the center of a rotating platform and the range of the liquid surface in the reservoir is observed to be 1.5 cm. Determine the rate at which the platform is rotating in revolutions per minute.

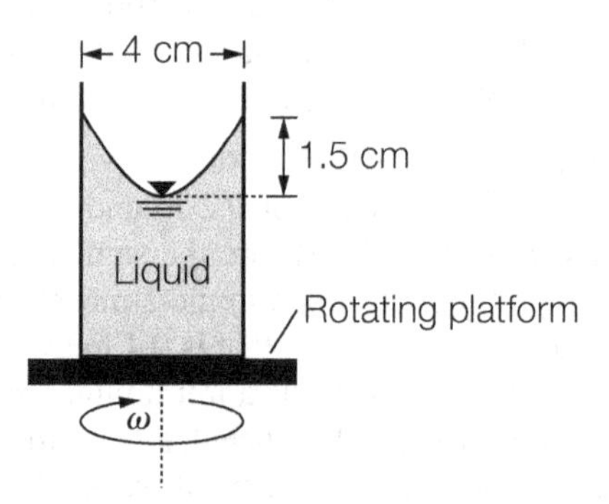

Figure 2.103: Liquid reservoir on a rotating platform

2.115. A 1.6-m-diameter cylinder is filled with a liquid to a depth of 1.1 m and rotated about its center axis. (a) Assuming that the cylinder is tall enough for the liquid not to spill, at what rotational speed will the liquid surface intersect the bottom of the cylinder? (b) If the cylinder is rotated at 60 rpm, what is the minimum height of the cylinder that prevents spillage?

Chapter 3

Kinematics and Streamline Dynamics

LEARNING OBJECTIVES

After reading this chapter and solving a representative sample of end-of-chapter problems, you will be able to:

- Understand the kinematic variables used to quantify the translation of fluids.
- Calculate and plot pathlines, streamlines, and streaklines.
- Understand the meaning of the material derivative and its application.
- Calculate flow rates and average velocities.
- Understand the derivation of the Bernoulli equation and the restrictions to its use.
- Identify the preferred forms of the Bernoulli equation for liquids and gases.
- Apply the Bernoulli equation to flow measurement devices and jet trajectories.
- Understand the impact of compressibility and viscous effects on application of the Bernoulli equation.
- Apply the Bernoulli equation to branched conduits.
- Understand the dynamics of curved flows and vortices.

3.1 Introduction

A fluid can be considered as a continuum of fluid "particles," where the motion of each fluid particle is described by Newton's second law. From this perspective, the law of motion that governs a fluid particle is the same law of motion that governs a solid particle. In the case of fluid particles, spatial variations in pressure, frictional forces between adjacent fluid particles, and gravity forces are the primary forces causing motion; the motion of solid particles is frequently dominated by frictional (drag) forces caused by the relative motion of the solid body to a surrounding fluid. The description of fluid motion relies on an understanding of the kinematics of motion. In cases where frictional forces between adjacent fluid particles are negligible, the equation of (fluid) motion along a particle pathway, called a *streamline*, is greatly simplified, and the analytic description of this type of motion is useful in many engineering applications. The fundamental translational kinematic properties of fluids, along with the dynamics of motion along streamlines, are covered in this chapter.

Kinematic properties of fluid flow. The field of *kinematics* is the study of the geometry of motion, and the fundamental kinematic properties describe the translation, rotation, linear deformation, and angular deformation of a given mass, which can be either a fluid or a solid. The kinematic description of fluid motion consists of specifying how the fundamental kinematic properties of a fluid element vary in space and time. The dimensions of kinematic properties generally involve only length and time. Derived kinematic properties that are commonly used to describe fluid flow include the average velocity, volume flow rate, and mass flow rate; all of these properties relate to the translation of fluid particles. Fluid flows are commonly classified as being either *laminar* or *turbulent*. The kinematic difference between these types of flows is that turbulent flows exhibit random fluctuations in velocity at fixed locations within the velocity field, whereas laminar flows do not exhibit such random velocity fluctuations. The kinematics of turbulent flow fields are frequently described using the time-averaged velocities within the flow field.

Streamline dynamics. The field of *dynamics* is the study of the relationship between the applied forces and the resulting motion caused by the applied forces. The basic equation describing the dynamics of fluid flow is Newton's second law, which states that the sum of the forces acting on any fluid element is equal to the mass of the element multiplied by its acceleration. Under steady-state conditions, the path followed by individual fluid elements is called a *streamline*, and studying the dynamics of fluid elements along streamlines can greatly simplify the analysis of fluid flow. The study of fluid flow along streamlines, although restricted mostly to steady-state flows and flows in which viscous and compressibility effects are negligible, finds wide application in engineering. The dynamics of motion along streamlines in nonviscous fluids is described by the *Bernoulli equation*, which is an integrated form of Newton's second law that is particularly well suited for the analysis of fluid flows. This chapter covers the fundamental derivation of the Bernoulli equation, the restrictions associated with using the Bernoulli equation, and several applications of the Bernoulli equation to cases of engineering interest.

3.2 Kinematics

The term *kinematics* refers to the study of the geometry of motion; hence, the term *fluid kinematics* refers to the study of geometry of fluid motions. Fundamental kinematic motions can be separated into translation, rotation, linear strain, and shear strain. Translational motion is of primary interest in finite-control-volume formulations, whereas rotation, linear strain, and shear strain are used more in differential formulations. Translational kinematic properties of fluid motion are covered in this chapter, and rotation, linear strain, and shear strain along with their applications are described in Chapter 5, which deals with the differential analysis of fluid motion. The kinematic variables that describe the translation of a fluid element are position, velocity, and acceleration.

Position vector. Position is typically defined by the *position vector* **r**, which is related to the Cartesian coordinates (x, y, z) by

$$\mathbf{r} = x\mathbf{i} + y\mathbf{j} + z\mathbf{k} \tag{3.1}$$

where **i**, **j**, and **k** are unit vectors in the positive x-, y-, and z-directions, respectively.

Velocity. The velocity, **v**, of a fluid particle whose trajectory is described by the position vector $\mathbf{r}(t)$ is given by

$$\mathbf{v} = \frac{d\mathbf{r}}{dt} \tag{3.2}$$

and the velocity vector can be expressed in component form as

$$\mathbf{v} = v_x\mathbf{i} + v_y\mathbf{j} + v_z\mathbf{k} \tag{3.3}$$

where v_x, v_y, and v_z are the components of the velocity vector $\mathbf{v}$ in the x-, y-, and z- coordinate directions, respectively. Equations 3.1–3.3 can be combined to yield the following relationship between the velocity components and the Cartesian coordinates:

$$v_x = \frac{\mathrm{d}x}{\mathrm{d}t}, \quad v_y = \frac{\mathrm{d}y}{\mathrm{d}t}, \quad v_z = \frac{\mathrm{d}z}{\mathrm{d}t} \tag{3.4}$$

Because the velocity is a vector, it has both magnitude and direction. The magnitude of the velocity, V, is given by

$$V = |\mathbf{v}| = \sqrt{v_x^2 + v_y^2 + v_z^2} \tag{3.5}$$

Technically, the velocity of a fluid particle is a vector ($\mathbf{v}$) and the speed of the fluid particle is a scalar (V). However, in many applications, especially in cases where the direction of flow is apparent, such as rectilinear flow in a prismatic conduit, the speed of the flow is also called the *velocity*. In cases where the velocity is expressed as a function of space and time, such as in the form $\mathbf{v}(\mathbf{r}, t)$, $\mathbf{v}(\mathbf{r}, t)$ is called the *velocity field*. A location in a velocity field where the velocity is equal to zero is called a *stagnation point*, and a velocity field may have multiple stagnation points.

Acceleration. The acceleration of a fluid particle, **a**, is related to the velocity, **v**, and position vector, **r**, of the particle by

$$\mathbf{a} = \frac{\mathrm{d}\mathbf{v}}{\mathrm{d}t} = \frac{\mathrm{d}^2\mathbf{r}}{\mathrm{d}t^2} \tag{3.6}$$

Because the velocity, $\mathbf{v}$, is a function of x, y, z, and t, Equation 3.6 can be expressed in expanded form using the chain rule of partial differentiation, which yields

$$\mathbf{a} = \frac{\mathrm{d}\mathbf{v}}{\mathrm{d}t} = \frac{\partial \mathbf{v}}{\partial t} + \frac{\partial \mathbf{v}}{\partial x}\frac{\mathrm{d}x}{\mathrm{d}t} + \frac{\partial \mathbf{v}}{\partial y}\frac{\mathrm{d}y}{\mathrm{d}t} + \frac{\partial \mathbf{v}}{\partial z}\frac{\mathrm{d}z}{\mathrm{d}t} \tag{3.7}$$

Combining Equations 3.7 and 3.4 yields the following scalar relations between the acceleration and velocity in a flowing fluid:

$$a_x = \frac{\partial v_x}{\partial t} + v_x\frac{\partial v_x}{\partial x} + v_y\frac{\partial v_x}{\partial y} + v_z\frac{\partial v_x}{\partial z} \tag{3.8}$$

$$a_y = \frac{\partial v_y}{\partial t} + v_x\frac{\partial v_y}{\partial x} + v_y\frac{\partial v_y}{\partial y} + v_z\frac{\partial v_y}{\partial z} \tag{3.9}$$

$$a_z = \underbrace{\frac{\partial v_z}{\partial t}}_{\text{local acceleration}} + \underbrace{v_x\frac{\partial v_z}{\partial x} + v_y\frac{\partial v_z}{\partial y} + v_z\frac{\partial v_z}{\partial z}}_{\text{advective acceleration}} \tag{3.10}$$

where a_x, a_y, and a_z are the components of the acceleration vector, **a**, in the x-, y-, and z-coordinate directions, respectively. For each component of the acceleration given in Equations 3.8–3.10, the first term on the right-hand side accounts for the acceleration caused by local changes in velocity with time and is called the *local acceleration*; the last three terms on the right-hand side account for the acceleration caused by spatial variations in the flow velocity $\mathbf{v}$. These last three terms are collectively called the *advective acceleration* and are sometimes called the *convective acceleration*. Equations 3.8–3.10 can be expressed in vector notation as

$$\mathbf{a} = \frac{\partial \mathbf{v}}{\partial t} + (\mathbf{v} \cdot \nabla)\mathbf{v} \tag{3.11}$$

where ∇ is the gradient operator defined as

$$\nabla(\cdot) = \frac{\partial(\cdot)}{\partial x}\mathbf{i} + \frac{\partial(\cdot)}{\partial y}\mathbf{j} + \frac{\partial(\cdot)}{\partial z}\mathbf{k}$$

Vector notation is commonly used to represent the equations of fluid motion, primarily because it offers a compact way of representing the three-component scalar equations that are usually necessary to describe fluid motion.

EXAMPLE 3.1

Consider a case in which the velocity distribution of air within a large control volume is given by

$$\mathbf{v} = 2(2y^2 - 1)\mathbf{i} + (2x - 5)\mathbf{j} + 9z\mathbf{k} \text{ m/s}$$

where x, y, and z are Cartesian coordinates in meters. (a) Determine the magnitude and direction of the velocity at the origin, (b) calculate the acceleration field and determine the magnitude and direction of the acceleration at the origin, (c) find the location of the stagnation point within the velocity field, and (d) find the location where the acceleration is equal to zero.

SOLUTION

From the given velocity field, the velocity components are given by

$$v_x = 2(2y^2 - 1), \qquad v_y = 2x - 5, \qquad v_z = 9z$$

(a) The velocity vector at the origin is obtained by taking $(x, y, z) = (0, 0, 0)$, which yields

$$\mathbf{v} = 2[2y^2 - 1]\mathbf{i} + [2x - 5]\mathbf{j} + 9z\mathbf{k} = 2[2(0)^2 - 1]\mathbf{i} + [2(0) - 5]\mathbf{j} + 9(0)\mathbf{k} = -2\mathbf{i} - 5\mathbf{j}$$

Hence, the velocity vector at the origin is in the xy plane, because the z component of the velocity is equal to zero. The magnitude and direction of the velocity at the origin (in the xy plane) are given by

$$|\mathbf{v}| = \sqrt{(-2)^2 + (-5)^2} = 5.39 \text{ m/s}, \qquad \theta_{xy} = \tan^{-1}\left(\frac{-5}{-2}\right) = 248^\circ$$

Therefore, the velocity has a magnitude of **5.39 m/s** and makes an angle of **248°** with the positive x-axis; the angle is measured counterclockwise in the xy plane.

(b) The components of the acceleration field are determined using Equations 3.8–3.10, which yield

$$\begin{aligned}
a_x &= 0 + 2(2y^2 - 1)(0) + (2x - 5)(8y) + (9z)(0) = 8y(2x - 5) \\
a_y &= 0 + 2(2y^2 - 1)(2) + (2x - 5)(0) + (9z)(0) = 4(2y^2 - 1) \\
a_z &= 0 + 2(2y^2 - 1)(0) + (2x - 5)(0) + (9z)(9) = 81z
\end{aligned}$$

Therefore, the acceleration field is given by

$$\mathbf{a} = \mathbf{8y(2x - 5)i + 4(2y^2 - 1)j + 81zk \text{ m/s}^2}$$

The acceleration at the origin is determined from the equation of the acceleration field by setting $x = y = z = 0$, which yields $\mathbf{a} = -4\mathbf{j}$. Therefore, the magnitude of the acceleration is **4 m/s²** and the direction is **along the negative y-axis.**

(c) The coordinates of the stagnation point (x_s, y_s, z_s) must necessarily satisfy the relations

$$\begin{aligned} v_x &= 0 \quad \rightarrow \quad 2(2y_s^2 - 1) = 0 \quad \rightarrow \quad y_s = 0.707\text{ m} \\ v_y &= 0 \quad \rightarrow \quad 2x_s - 5 = 0 \quad \rightarrow \quad x_s = 2.5\text{ m} \\ v_z &= 0 \quad \rightarrow \quad 9z_s = 0 \quad \rightarrow \quad z_s = 0\text{ m} \end{aligned}$$

Therefore, the coordinates of the stagnation point are **(2.5 m, 0.707 m, 0 m)**.

(d) The location where the acceleration, **a**, is equal to zero (x_0, y_0, z_0) must necessarily satisfy the relations

$$\begin{aligned} a_x &= 0 \quad \rightarrow \quad 8y_0(2x_0 - 5) = 0 \quad \rightarrow \quad x_0 = 2.5\text{ m} \\ a_y &= 0 \quad \rightarrow \quad 4(2y_0^2 - 1) = 0 \quad \rightarrow \quad y_0 = 0.707\text{ m} \\ a_z &= 0 \quad \rightarrow \quad 81z_0 = 0 \quad \rightarrow \quad z_0 = 0\text{ m} \end{aligned}$$

Therefore, the coordinates of the location where the acceleration is equal to zero are **(2.5 m, 0.707 m, 0 m)**.

In this particular case, both the velocity and acceleration are equal to zero at the same location in the flow field. Under general circumstances, the locations of zero velocity and zero acceleration need not coincide.

3.2.1 Tracking the Movement of Fluid Particles

Particle tracking is used to aid in both flow visualization and analysis by tracking the movement of fluid elements (also known as "fluid particles") within a flow field. Using particle-tracking data, pathlines, streamlines, and streaklines are produced and used to better understand the flow field being studied. An example of a particle-tracking application is shown in Figure 3.1 where the airflow around a Brazuca-style soccer ball (used in the 2014 World Cup tournament) is being tested in a wind tunnel at NASA's Ames Research Center at Moffett Air Field in California. Airflow around the ball is visualized when smoke particles are released in the air and the flow field is illuminated with laser light. Investigators use the smoke patterns

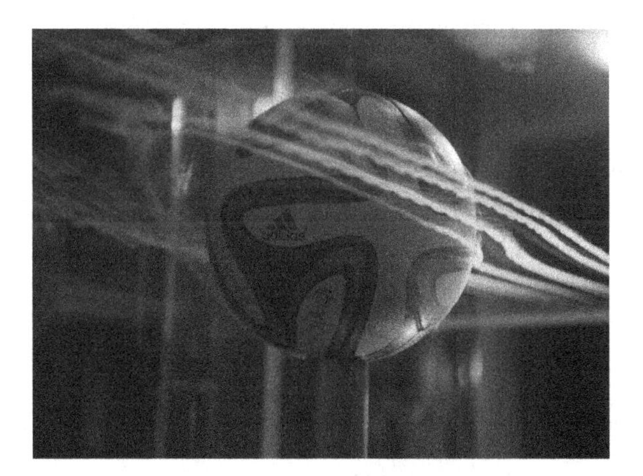

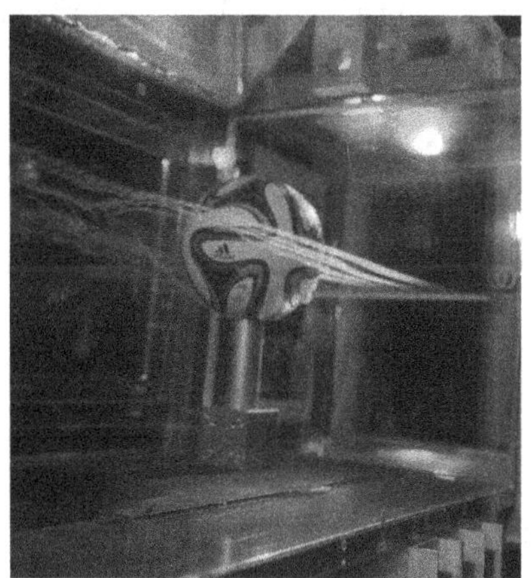

Figure 3.1: Particle tracking in airflow around a soccer ball
Source: Ames Research Center/NASA.

to identify the speed of the ball at which the smoke pattern suddenly changes, which occurs when the *knuckling effect* is greatest. For the Brazuca, this occurred at an airspeed of around 48 km/h. In practical terms, a Brazuca ball kicked at around 48 km/h follows its most irregular (knuckling) flight path.

Pathlines. A *pathline* is the trace made by a single particle over a period of time, and it shows the direction of the velocity of the particle at successive instants of time. Pathlines can be calculated numerically from a known velocity field using the relation

$$\mathbf{r} = \mathbf{r}_{\text{start}} + \int_{t_{\text{start}}}^{t} \mathbf{v}\,\mathrm{d}t \tag{3.12}$$

where $\mathbf{r}$ is the location of a point on the pathline at time t, $\mathbf{r}_{\text{start}}$ is the location of the same at time t_{start}, and $\mathbf{v}$ is the velocity field. At any instant of time on a pathline,

$$\frac{\mathrm{d}\mathbf{r}}{\mathrm{d}t} = \mathbf{v} \quad \rightarrow \quad \frac{\mathrm{d}x}{\mathrm{d}t} = v_x, \quad \frac{\mathrm{d}y}{\mathrm{d}t} = v_y, \quad \frac{\mathrm{d}z}{\mathrm{d}t} = v_z \tag{3.13}$$

This equation is particularly useful in determining an analytic expression for a pathline within a given velocity field. The trace followed by particles of injected smoke from a single upstream port in Figure 3.1 is an example of a pathline.

EXAMPLE 3.2

Consider a two-dimensional velocity field in which the velocity vector, $\mathbf{v}$, is given by

$$\mathbf{v} = 2x^2\mathbf{i} + 3y\mathbf{j} \text{ m/s}$$

where x and y are Cartesian coordinates in meters and $x, y \geq 0$. Determine the equation of the pathline followed by a fluid particle released at the point (1 m, 1 m). For what duration of time does the released particle remain within the domain of $0 \leq x \leq 10$ m, $0 \leq y \leq 10$ m. Plot the pathline within this domain.

SOLUTION

From the given velocity field, $v_x = 2x^2$ and $v_y = 3y$. Applying Equation 3.13 using v_x gives

$$\frac{\mathrm{d}x}{\mathrm{d}t} = v_x \quad \rightarrow \quad \frac{\mathrm{d}x}{\mathrm{d}t} = 2x^2 \quad \rightarrow \quad \frac{\mathrm{d}x}{x^2} = 2\,\mathrm{d}t \quad \rightarrow \quad \int \frac{\mathrm{d}x}{x^2} = 2\int \mathrm{d}t \quad \rightarrow \quad -\frac{1}{x} = 2t + C_1$$

Applying the initial condition that $x = 1$ when $t = 0$ yields the value of the constant C_1 as follows:

$$-\frac{1}{1} = 2(0) + C_1 \quad \rightarrow \quad C_1 = -1$$

Therefore, the variation of the x-coordinate with time for a point on the pathline is given by

$$-\frac{1}{x} = 2t + (-1) \quad \rightarrow \quad x = \frac{1}{1 - 2t}$$

Repeating the above analysis for the y component of flow gives

$$\frac{\mathrm{d}y}{\mathrm{d}t} = v_y \quad \rightarrow \quad \frac{\mathrm{d}y}{\mathrm{d}t} = 3y \quad \rightarrow \quad \frac{\mathrm{d}y}{y} = 3\,\mathrm{d}t \quad \rightarrow \quad \int \frac{\mathrm{d}y}{y} = 3\int \mathrm{d}t \quad \rightarrow \quad \ln y = 3t + C_2$$

Applying the initial condition that $y = 1$ when $t = 0$ yields the value of the constant C_2 as follows:

$$\ln(1) = 3(0) + C_2 \quad \rightarrow \quad C_2 = 0$$

The variation of the y-coordinate with time for a point on the pathline is therefore given by

$$\ln y = 3t + 0 \quad \rightarrow \quad y = e^{3t}$$

In summary, the above results show that the equation of the pathline can be expressed in parametric form as $\boldsymbol{x = 1/(1-2t)}$ and $\boldsymbol{y = e^{3t}}$. A particle released at (1 m, 1 m) remains in the prescribed domain as long as

$$\frac{1}{1-2t} \leq 10 \text{ m} \quad \text{and} \quad e^{3t} \leq 10 \text{ m} \quad \rightarrow \quad t \leq 0.45 \text{ s} \quad \text{and} \quad t \leq 0.767 \text{ s}$$

Therefore, both conditions are satisfied as long as $\boldsymbol{t \leq 0.45}$ **s**; hence, this is the time that a released particle will remain in the prescribed domain. The pathline is plotted in Figure 3.2 for values of t in the range $0 \leq t \leq 0.45$ s.

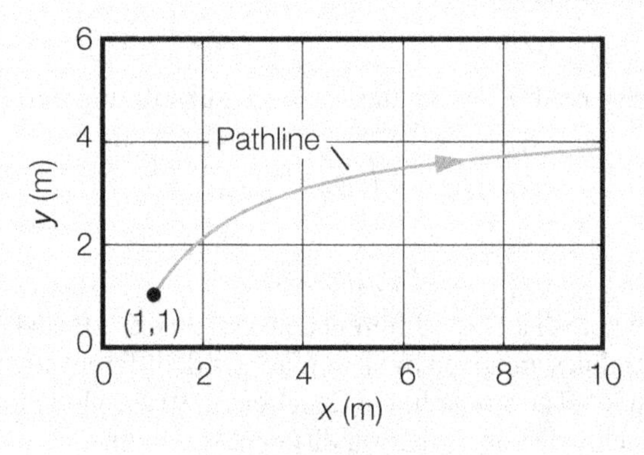

Figure 3.2: Pathline

Streamlines. *Streamlines* show the mean direction of movement of a number of fluid particles at the same instant of time; in other words, a streamline is a curve that is everywhere tangent to the instantaneous local velocity vector. The arc length along a streamline, **dr**, and the velocity vector, **v**, can be expressed as

$$\mathbf{dr} = \mathrm{d}x\,\mathbf{i} + \mathrm{d}y\,\mathbf{j} + \mathrm{d}z\,\mathbf{k} \tag{3.14}$$

$$\mathbf{v} = v_x\mathbf{i} + v_y\mathbf{j} + v_z\mathbf{k} \tag{3.15}$$

where $(\mathrm{d}x, \mathrm{d}y, \mathrm{d}z)$ and (v_x, v_y, v_z) are components of **dr** and **v**, respectively. Because the streamline must be parallel to the local velocity vector (by definition of the streamline), the components of **dr** must be proportional to those of **v**; hence, the equation of the streamline is given by

$$\frac{\mathrm{d}x}{v_x} = \frac{\mathrm{d}y}{v_y} = \frac{\mathrm{d}z}{v_z} \tag{3.16}$$

In the case of two-dimensional flow, Equation 3.16 can be expressed in the form

$$\left(\frac{\mathrm{d}y}{\mathrm{d}x}\right)_{\text{streamline}} = \frac{v_y}{v_x} \tag{3.17}$$

If the velocity field is known in terms of x and y, then Equation 3.17 can be integrated to give the equations of the streamlines. Pathlines and streamlines are identical under steady-flow conditions when there are no fluctuating velocity components (i.e., when the flow is laminar). The tracks of injected smoke shown in Figure 3.1 are streamlines. In the case of turbulent flow, pathlines and streamlines are not coincident—the pathlines are very irregular, and the streamlines are everywhere tangent to the local mean temporal velocity.

EXAMPLE 3.3

A steady-state two-dimensional velocity field is described in Quadrant I of the Cartesian plane by

$$\mathbf{v} = 3x\mathbf{i} + 2y\mathbf{j}\ \text{m/s} \tag{3.18}$$

where x and y are the coordinates in meters, and $x, y \geq 0$. Determine the equations of the streamlines in the flow domain. Plot the streamlines to indicate the directions of flow along each of the plotted streamlines.

SOLUTION

From the given velocity field: $v_x = 3x$ and $v_y = 2y$. Substituting into Equation 3.16 gives

$$\frac{dy}{v_y} = \frac{dx}{v_x} \quad \rightarrow \quad \frac{dy}{2y} = \frac{dx}{3x} \quad \rightarrow \quad \int \frac{dy}{2y} = \int \frac{dx}{3x} + C' \quad \rightarrow \quad \frac{1}{2}\ln y = \frac{1}{3}\ln x + C'$$

which simplifies to

$$y = Cx^{\frac{2}{3}} \tag{3.19}$$

where C is a constant; note that C is a different constant than C'. Based on the derived result, the equations of the streamlines are given by $\boldsymbol{y = Cx^{\frac{2}{3}}}$, with different streamlines having different values of the constant, C. The streamlines (Equation 3.19) are plotted in Figure 3.3 for various values of C, where each value of C yields a different streamline.

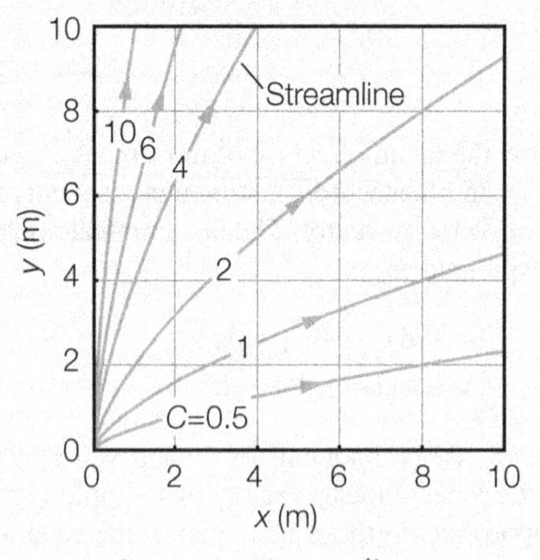

Figure 3.3: Streamlines

The direction of flow along each streamline is indicated by arrows, where the flow directions are determined with the aid of the given velocity field (Equation 3.18), which shows that both components of the velocity are generally positive when $x, y \geq 0$. In the flow pattern shown in Figure 3.3, the fluid is emanating from the origin and then bending in the direction of the positive x-axis.

A concept that is closely related to streamlines is that of a *streamtube*, which is defined as an imaginary conduit bounded by streamlines. Fluid flow does not occur across streamlines (because by definition, the velocity vector is tangent to a streamline), so any fluid bounded by a defined set of streamlines remains within those streamlines as if the streamlines formed a solid tube. Hence, the name "streamtube."

Streaklines. A *streakline* is the trace made by particles that originate at a particular injection point and can be derived from the velocity field using the relation

$$\mathbf{r} = \mathbf{r}_0 + \int_{t_0}^{t} \mathbf{v}\, dt \tag{3.20}$$

where $\mathbf{r}$ is the position of a point on the streakline at time t, $\mathbf{r}_0$ is the location of the injection point, t_0 is the time the tracer is injected, and $\mathbf{v}$ is the velocity field. Streaklines are used more in laboratory studies than in practical applications. Individual smoke lines shown in Figure 3.1 are examples of streaklines.

EXAMPLE 3.4

An unsteady-state two-dimensional velocity field is described in Quadrant I of the Cartesian plane by

$$\mathbf{v} = (u_0 + a \sin \omega t)\mathbf{i} + (v_0 + b \cos \omega t)\mathbf{j} \text{ m/s}$$

where t is the time is seconds, a and b are the amplitudes of the temporal velocity fluctuations, $\omega = 2\pi/T$ is the frequency of the velocity fluctuations, T is the period of the fluctuations, and u_0 and v_0 are drift velocities. Consider the particular case where a = 1 m/s, b = 1.5 m/s, T = 10 s, and $u_0 = v_0 = 1$ m/s. Estimate the streakline within the domain $0 \leq x \leq 10$ m and $0 \leq y \leq 10$ m for particles released continuously from (x, y) = (0 m, 0 m) starting at $t = 0$ s. Use 1-second time increments in estimating the streakline.

SOLUTION

Using the given parameters of the velocity field, the components of the velocity are

$$v_x = u_0 + a \sin \omega t = 1 + (1) \sin \left(\frac{2\pi}{10}\right) t \quad \rightarrow \quad v_x = 1 + \sin(0.6283t) \text{ m/s}$$

$$v_y = v_0 + b \cos \omega t = 1 + (1.5) \cos \left(\frac{2\pi}{10}\right) t \quad \rightarrow \quad v_y = 1 + 1.5 \cos(0.6283t) \text{ m/s}$$

For a particle at any given location in the flow field, the movement of the particle is described by

$$\frac{dx}{dt} = 1 + \sin(0.6283t) \quad \rightarrow \quad \int_{x_i}^{x_{i+1}} dx = \int_{t_i}^{t_{i+1}} [1 + \sin(0.6283t)]\, dt$$

$$\frac{dy}{dt} = 1 + 1.5 \cos(0.6283t) \quad \rightarrow \quad \int_{y_i}^{y_{i+1}} dy = \int_{t_i}^{t_{i+1}} [1 + 1.5 \cos(0.6283t)]\, dt$$

Performing the integrations leads to the following useful expressions for the incremental movement of a fluid particle in the x- and y-coordinate directions during the time interval, $\Delta t = t_{i+1} - t_i$:

$$\Delta x = \Delta t - 1.592[\cos(0.6283t_{i+1}) - \cos(0.6283t_i)]$$

$$\Delta y = \Delta t + 2.387[\sin(0.6283t_{i+1}) - \sin(0.6283t_i)]$$

where Δx and Δy are the increments in x and y, respectively. The streaklines can be estimated by recursively applying the following relations:

$$x_{i+1} = x_i + \Delta x, \qquad y_{i+1} = y_i + \Delta y$$

From the given data, the original location is x_1 = 0 m and y_1 = 0 m, and for purposes of approximation, use Δt = 1 second. The results of the application are given in the following table.

x_1 (m)	y_1 (m)	t_1 (s)	t_2 (s)	Δx (m)	Δy (m)	x_2 (m)	y_2 (m)
0	0	0	1	1.30	2.40	1.30	2.40
1.30	2.40	1	2	1.80	1.87	3.10	4.27
3.10	4.27	2	3	1.98	1.00	5.08	5.27
5.08	5.27	3	4	1.80	0.13	6.88	5.40
6.88	5.40	4	5	1.30	−0.40	8.18	5.00
8.18	5.00	5	6	0.70	−0.40	8.88	4.60
8.88	4.60	6	7	0.20	0.13	9.08	4.73
9.08	4.73	7	8	0.02	1.00	9.10	5.73
9.10	5.73	8	9	0.20	1.87	9.30	7.60
9.30	7.60	9	10	0.70	2.40	10.00	10.00

A plot of the streakline is shown in Figure 3.4.

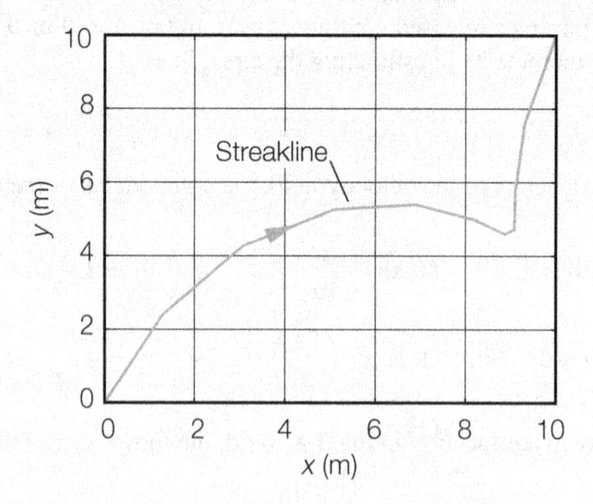

Figure 3.4: **Streakline**

It is apparent from the streakline that the released particle is moving with a pattern that is characteristic of periodic fluctuations in the velocity about a mean drift velocity. This pattern is frequently encountered with drift currents in tidal flows.

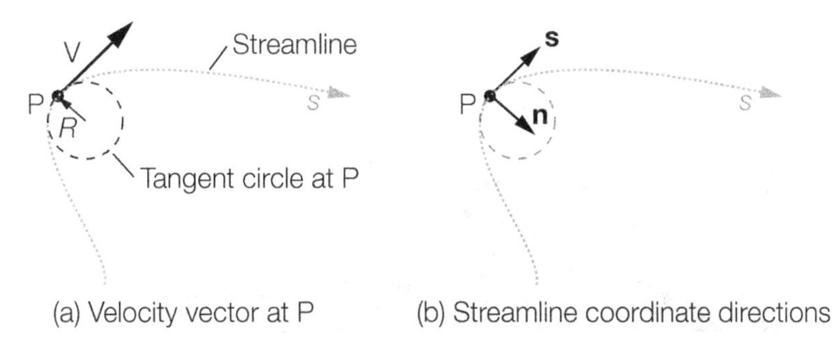

Figure 3.5: Unit vectors in a streamline coordinate system

Kinematics of fluid motion in a streamline coordinate system. Under steady-state conditions, streamlines, pathlines, and streaklines are coincident, and it is sometimes convenient to represent the motion of a fluid using a *streamline coordinate system.* Using this system, one coordinate direction is taken in the direction of flow; the other, in the direction perpendicular to the flow. The s-axis is usually denoted as the axis in the direction of the flow, and the n-axis is perpendicular to the flow. These coordinates are illustrated in Figure 3.5, where $\mathbf{s}$ and $\mathbf{n}$ are the unit vectors defining the positive s and n directions, respectively. The streamline coordinate description of flow is particularly useful for two-dimensional flow fields in which both $\mathbf{s}$ and $\mathbf{n}$ lie in the xy plane. Because the s-axis, by definition, coincides with the direction of flow, the velocity vector, $\mathbf{v}$, and acceleration, $\mathbf{a}$, at any point P on a streamline in steady two-dimensional flow can be represented as

$$\mathbf{v} = V\mathbf{s} \tag{3.21}$$

$$\mathbf{a} = V\frac{\partial V}{\partial s}\mathbf{s} + \frac{V^2}{R}\mathbf{n} \quad \text{(steady state)} \tag{3.22}$$

where V is the magnitude of the velocity and R is the radius of curvature of the streamline at Point P. Equation 3.22 explicitly expresses the acceleration of a fluid particle in terms of a convective acceleration in the direction of movement ($\mathbf{s}$) and a centripetal acceleration in the direction normal ($\mathbf{n}$) to the direction of fluid motion. Equations 3.21 and 3.22 apply to steady-state conditions. When the flow is unsteady along fixed streamlines, the velocity v is still given by Equation 3.21, and the acceleration $\mathbf{a}$ is given by

$$\mathbf{a} = \left(\left.\frac{\partial V}{\partial t}\right|_s + V\frac{\partial V}{\partial s}\right)\mathbf{s} + \left(\left.\frac{\partial V}{\partial t}\right|_n + \frac{V^2}{R}\right)\mathbf{n} \quad \text{(unsteady state)} \tag{3.23}$$

where the velocity $\partial V/\partial t|_s$ and $\partial V/\partial t|_n$ are the local accelerations in the $\mathbf{s}$ and $\mathbf{n}$ directions, respectively.

EXAMPLE 3.5

Consider the case in which a liquid flows over a vertical bump as shown in Figure 3.6. The radius of curvature at the apex of the bump is 1 m. To control flow separation over the apex, an engineer wants to limit the flow velocity so that the vertical component of the fluid acceleration over the apex of the bump is less than or equal to the acceleration due to gravity. Under steady-state conditions, what is the limiting flow velocity over the apex of the bump?

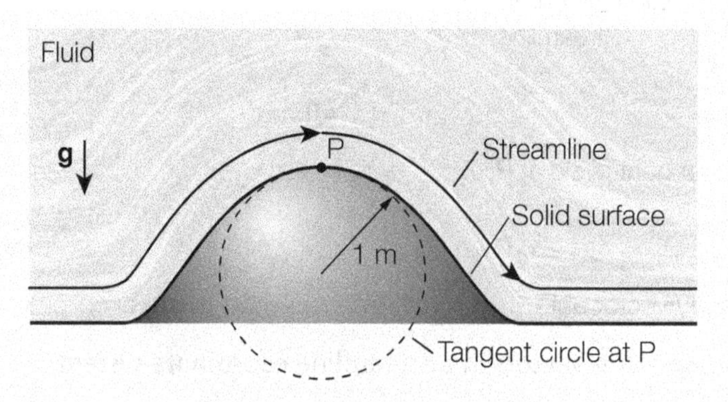

Figure 3.6: Flow over a bump

SOLUTION

From the given data: R = 1 m. The acceleration due to gravity is 9.81 m/s^2. The normal acceleration over the bump is equal to g, when

$$\frac{V^2}{R} = g \quad \rightarrow \quad \frac{V^2}{1} = 9.81 \quad \rightarrow \quad V = 3.13 \text{ m/s}$$

Therefore, separation between the liquid and the solid surface will be prevented if the fluid velocity over the apex is less than **3.13 m/s**. In reality, the limiting velocity for separation is slightly less than 3.13 m/s, because separation will begin to occur when the pressure near the interface is equal to the vapor pressure of the liquid. Theoretically, when V = 3.13 m/s, the pressure near the interface is equal to zero.

3.2.2 The Material Derivative

The *material derivative*, commonly denoted by $\mathrm{D}/\mathrm{D}t$, describes the rate of change of any fluid property that would be observed from the perspective of a person or an instrument moving with the fluid. In fact, the word "material" in the term "material derivative" is intended to reinforce the fact that the rate of change given by the material derivative is from the perspective of the moving "material," which in this case is the moving fluid. The moving perspective is commonly referred to as the *Lagrangian*[15] point of view, a contrast to the *Eulerian*[16] point of view, which is from the perspective of a person or an instrument located at a (fixed) point in space. The material derivative, which describes changes relative to a Lagrangian viewpoint, is sometimes called the *substantial derivative* and is defined by the relation

$$\frac{\mathrm{D}(\cdot)}{\mathrm{D}t} = \frac{\partial(\cdot)}{\partial t} + \frac{\partial(\cdot)}{\partial x}\frac{\mathrm{d}x}{\mathrm{d}t} + \frac{\partial(\cdot)}{\partial y}\frac{\mathrm{d}y}{\mathrm{d}t} + \frac{\partial(\cdot)}{\partial z}\frac{\mathrm{d}z}{\mathrm{d}t} \tag{3.24}$$

where $(\cdot)$ represents a designated property of a fluid particle that is being tracked, and Equation 3.24 is simply a direct application of the chain rule. Equation 3.24 is more commonly and conveniently expressed as

$$\frac{\mathrm{D}(\cdot)}{\mathrm{D}t} = \frac{\partial(\cdot)}{\partial t} + \frac{\partial(\cdot)}{\partial x}v_x + \frac{\partial(\cdot)}{\partial y}v_y + \frac{\partial(\cdot)}{\partial z}v_z \tag{3.25}$$

[15]Named in honor of the French mathematician and astronomer Joseph-Louis Lagrange (1736–1813).

[16]Named in honor of the Swiss mathematician and physicist Leonhard Euler (1707–1783).

which can also be expressed in vector notation as

$$\frac{\mathrm{D}(\cdot)}{\mathrm{D}t} = \frac{\partial(\cdot)}{\partial t} + (\mathbf{v} \cdot \boldsymbol{\nabla})(\cdot) \tag{3.26}$$

where $\mathbf{v}$ is the velocity vector with components v_x, v_y, and v_z. Comparing Equations 3.26 and 3.11, it is apparent that the acceleration of a fluid particle can be expressed in terms of the material derivative, where

$$\mathbf{a} = \frac{\mathrm{D}\mathbf{v}}{\mathrm{D}t} = \frac{\mathrm{d}\mathbf{v}}{\mathrm{d}t} \tag{3.27}$$

The acceleration of a fluid particle described by Equation 3.27 is commonly called the *material acceleration*. The material derivative, $\mathrm{D}(\cdot)/\mathrm{D}t$, can be used to represent changes in a variety of fluid-particle properties, such as temperature and pressure. For example, the rate of change of pressure, p, experienced by a moving fluid particle can be expressed as

$$\frac{\mathrm{D}p}{\mathrm{D}t} = \frac{\mathrm{d}p}{\mathrm{d}t} = \frac{\partial p}{\partial t} + (\mathbf{v} \cdot \boldsymbol{\nabla})p \tag{3.28}$$

where $p(\mathbf{r}, t)$ describes the pressure field within the fluid continuum. In Equation 3.28, the rate of pressure change from a Lagrangian perspective is $\mathrm{D}p/\mathrm{D}t$ (or $\mathrm{d}p/\mathrm{d}t$) and the rate of change of pressure from an Eulerian perspective is $\partial p/\partial t$. In general, the notations $\mathrm{D}(\cdot)/\mathrm{D}t$ and $\mathrm{d}(\cdot)/\mathrm{d}t$ are equivalent, with the capitalized "D" used to emphasize that the resulting derivative is from a Lagrangian perspective.

EXAMPLE 3.6

The velocity field, $\mathbf{v}$, and pressure field, p, in a two-dimensional converging duct of 3-meter width shown in Figure 3.7 are given by

$$\mathbf{v} = (5 + 2x)\mathbf{i} - 2y\mathbf{j} \text{ m/s}, \qquad p = 101 - 2.45\left(5x + x^2 + y^2\right) \text{ kPa}$$

where x [m] is the coordinate measured along the centerline of the duct and y [m] is the coordinate measured normal to the centerline. Find the material acceleration and rate of change of pressure experienced by a fluid particle moving through the duct. Do these quantities increase or decrease along the centerline of the duct?

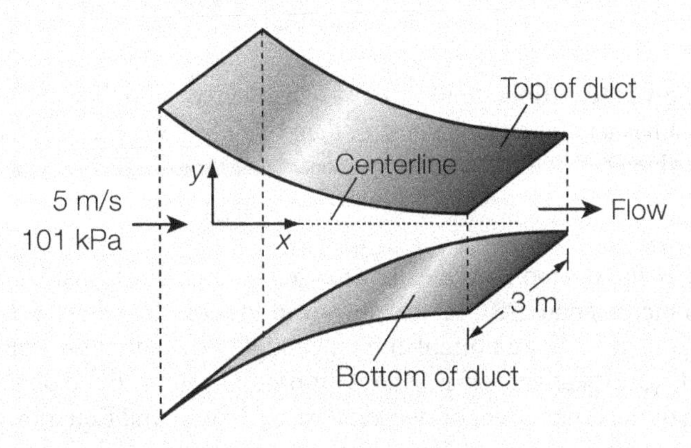

Figure 3.7: Duct flow

SOLUTION

The components of the material acceleration as defined by Equations 3.27 and 3.25 are given by

$$a_x = \frac{\partial v_x}{\partial t} + v_x\frac{\partial v_x}{\partial x} + v_y\frac{\partial v_x}{\partial y} + v_z\frac{\partial v_x}{\partial z} = 0 + (5+2x)(2) + (-2y)(0) + (0)(0) = 2(5+2x)\ \text{m/s}^2$$

$$a_y = \frac{\partial v_y}{\partial t} + v_x\frac{\partial v_y}{\partial x} + v_y\frac{\partial v_y}{\partial y} + v_z\frac{\partial v_y}{\partial z} = (0) + (5+2x)(0) + (-2y)(-2) + (0)(0) = 4y\ \text{m/s}^2$$

These results can be expressed compactly as

$$\mathbf{a} = \mathbf{2(5+2}\boldsymbol{x}\mathbf{)i + 4}\boldsymbol{y}\mathbf{j\ m/s^2}$$

It is apparent from this result that the x component of the acceleration increases along the centerline, whereas the y component of the acceleration remains constant and equal to zero along the centerline. Applying the material derivative, Equation 3.28, to the given pressure distribution yields

$$\frac{\mathrm{D}p}{\mathrm{D}t} = \frac{\partial p}{\partial t} + v_x\frac{\partial p}{\partial x} + v_y\frac{\partial p}{\partial y} + v_z\frac{\partial p}{\partial z} = (0) + (5+2x)(-2.45)(5+2x) + (-2y)(-2.45)(2y) + (0)(0)\ \text{kPa/s}$$

which simplifies to

$$\frac{\mathbf{D}\boldsymbol{p}}{\mathbf{D}\boldsymbol{t}} = \mathbf{9.8}\boldsymbol{y}^\mathbf{2} - \mathbf{2.45(5+2}\boldsymbol{x}\mathbf{)^2\ kPa/s}$$

It is apparent from this result that the rate of change of pressure ($\mathrm{D}p/\mathrm{D}t$) decreases with increasing distance (x) along the streamline.

3.2.3 Flow Rates

The quantity of fluid flowing per unit time across any designated area is called the *flow rate*, and this quantity can be expressed as a *volume flow rate* $[\mathrm{L^3T^{-1}}]$, a *mass flow rate* $[\mathrm{MT^{-1}}]$, or a *weight flow rate* $[\mathrm{FT^{-1}}]$.

Volume flow rate. If the fluid velocity at a given location is $\mathbf{v}$ and a planar area A is oriented such that it has a unit normal vector $\mathbf{n}$, then by definition, the volume flow rate, Q, across the area A is given by

$$Q = v_\mathrm{n} A \tag{3.29}$$

where v_n is the component of $\mathbf{v}$ normal to the area A. Consider the general case of flow across an incremental area, $\mathrm{d}A$, as shown in Figure 3.8, where the velocity, $\mathbf{v}$, is separated into components that are normal and tangential to $\mathrm{d}A$ and these components are represented by v_n and v_t, respectively. If $\mathbf{n}$ is the unit normal vector to the plane containing $\mathrm{d}A$, then by definition of the dot product of two vectors, $v_\mathrm{n} = \mathbf{v}\cdot\mathbf{n}$ and Equation 3.29 can be expressed as

$$\mathrm{d}Q = \mathbf{v}\cdot\mathbf{n}\,\mathrm{d}A \tag{3.30}$$

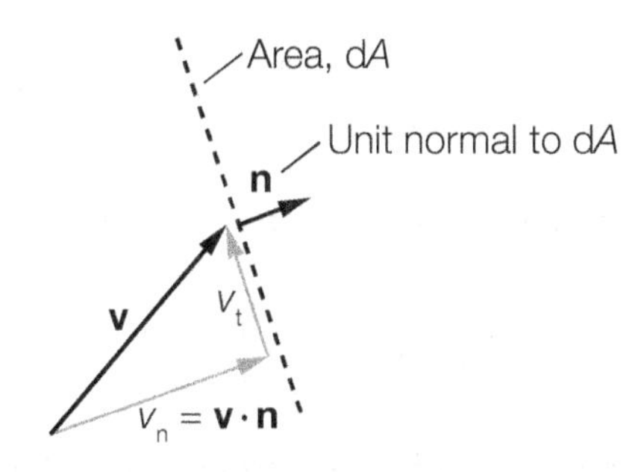

Figure 3.8: Volume flow rate across a given area

where $\mathrm{d}Q$ is the incremental volume flow rate across the incremental area $\mathrm{d}A$. If one wants to calculate the total volume flow rate, Q, across an area A, then Equation 3.30 can be integrated over A to yield

$$Q = \int_A \mathbf{v} \cdot \mathbf{n} \, \mathrm{d}A \quad \text{or} \quad Q = \int_A v_\mathrm{n} \, \mathrm{d}A \tag{3.31}$$

Application of Equation 3.31 takes into account any variability of the velocity **v** over the area A.

Average velocity. The average normal velocity, $\overline{V}_\mathrm{n}$, over an area A is defined as the total volume flow rate across A divided by the magnitude of A. The value of $\overline{V}_\mathrm{n}$ can be calculated analytically using the relation

$$\overline{V}_\mathrm{n} = \frac{Q}{A} = \frac{1}{A} \int_A v_\mathrm{n} \, \mathrm{d}A \tag{3.32}$$

It is also useful to remember that the average normal velocity, $\overline{V}_\mathrm{n}$, multiplied by the total area, A, gives the total volume flow rate across A, even if the velocity varies spatially over A. Hence, the volume flow rate, Q, can be expressed as

$$Q = A\overline{V}_\mathrm{n} \tag{3.33}$$

In many applications, the area, A, of interest is normal to the flow velocity, in which case it is commonplace to omit the subscript "n" in defining the velocity and average velocity, such that v_n is expressed as v and $\overline{V}_\mathrm{n}$ is expressed as $\overline{V}$.

Mass flow rate. By definition, the mass flow rate is equal to the volume flow rate $[\mathrm{L}^3\mathrm{T}^{-1}]$ multiplied by the density $[\mathrm{ML}^{-3}]$; therefore, the total mass flow rate, $\dot{m}$, across an area A can be calculated using any of the following relations:

$$\dot{m} = \rho \int_A v_\mathrm{n} \, \mathrm{d}A = \rho A \overline{V}_\mathrm{n} = \rho Q \tag{3.34}$$

Weight flow rate. The weight flow rate can be calculated by replacing the density, ρ, in Equation 3.34 with the specific weight, γ. The weight flow rate and the mass flow rate are more commonly used to quantify the flow rates of gases, in contrast to the volume flow rate, which is more commonly used to quantify the flow rates of liquids.

EXAMPLE 3.7

The velocity, $v(r)$, in a fluid at a distance r from the center of a pipe of radius R is given by

$$v(r) = V_{\text{max}}\left(1 - \frac{r^2}{R^2}\right)$$

where V_{max} is the maximum velocity in the pipe. Determine the volume flow rate, average velocity, and mass flow rate of the fluid in the pipe. If V_{max} remains constant, by what factors would the calculated quantities change if the diameter of the pipe were doubled?

SOLUTION

The volume flow rate, Q, is given by

$$Q = \int_A v\,\mathrm{d}A = \int_0^R V_{\text{max}}\left(1 - \frac{r^2}{R^2}\right) 2\pi r\,\mathrm{d}r$$

$$= V_{\text{max}} 2\pi \int_0^R \left(r - \frac{r^3}{R^2}\right) \mathrm{d}r = V_{\text{max}} 2\pi \left[\frac{r^2}{2} - \frac{r^4}{4R^2}\right]_0^R = V_{\text{max}} \frac{\pi R^2}{2}$$

$$\boldsymbol{Q = V_{\text{max}} \frac{\pi R^2}{2}}$$

The average velocity, $\overline{V}$, is given by

$$\overline{V} = \frac{Q}{A} = \frac{V_{\text{max}} \frac{\pi R^2}{2}}{\pi R^2} \quad \rightarrow \quad \boldsymbol{\overline{V} = \frac{V_{\text{max}}}{2}}$$

and the mass flow rate is given by

$$\dot{m} = \rho Q \quad \rightarrow \quad \boldsymbol{\dot{m} = \rho V_{\text{max}} \frac{\pi R^2}{2}}$$

where ρ is the density of the fluid that is flowing in the pipe. It is apparent that both the volume and mass flow rates increase as the square of the pipe radius. So **if V_{max} remains constant, doubling the diameter of the pipe doubles the radius and increases the volume and mass flow rates each by a factor of 4. The average velocity in the pipe remains unchanged.**

3.3 Dynamics of Flow along a Streamline

Consider the steady flow of a fluid element along a streamline as shown in Figure 3.9. The flow is assumed to be steady and inviscid. Under these conditions, the forces with components acting in the direction of flow include only the pressure forces on the upstream and downstream sides of the fluid element and the weight of the fluid element. For the fluid element shown in Figure 3.9, p and A are the pressure and area, respectively, on the upstream side of the fluid element, $\mathrm{d}p$ and $\mathrm{d}A$ are the changes in pressure and area, respectively, between the upstream and downstream sides of the fluid element, γ is the specific weight of the

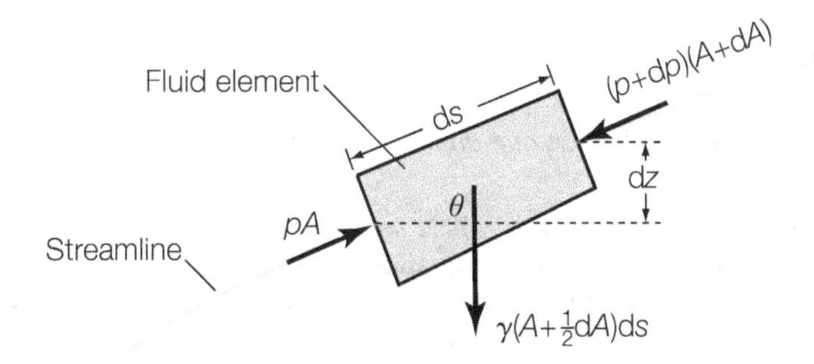

Figure 3.9: Motion of an inviscid fluid along a streamline

fluid, and θ is the angle the streamline makes with the horizontal. Applying Newton's second law to a fluid element moving along the streamline (s-direction) requires that

$$\sum F_s = ma_s \tag{3.35}$$

where $\sum F_s$ is the sum of the forces acting in the s-direction, m is the mass of the fluid element, and a_s is the component of the acceleration of the fluid element in the s-direction. Equation 3.35 can be expressed in terms of the variables shown in Figure 3.9 as

$$\underbrace{pA + (p + \tfrac{1}{2}\mathrm{d}p)\,\mathrm{d}A - (p + \mathrm{d}p)(A + \mathrm{d}A)}_{\text{net pressure force}} - \underbrace{\gamma\,\mathrm{d}s\,(A + \tfrac{1}{2}\mathrm{d}A)\sin\theta}_{\text{weight}} = \underbrace{\rho\,\mathrm{d}s\,(A + \tfrac{1}{2}\mathrm{d}A)\,a_s}_{\text{mass}\,\times\,\text{acceleration}} \tag{3.36}$$

where ρ is the density of the fluid. In Equation 3.36, the first three terms on the left-hand side account for the net pressure force exerted on the fluid element by the surrounding fluid, the fourth term is the component of the weight of the fluid element in the direction of flow, and the term on the right-hand side is equal to the mass of the fluid element times the acceleration of the fluid element. In physical terms, Equation 3.36 states that the rate of change of speed of a fluid element is a result of the combined effect of a pressure gradient in the direction of flow plus the component of the gravity force in the direction of flow. When simplifying Equation 3.36, the following equation is useful:

$$\sin\theta = \frac{\mathrm{d}z}{\mathrm{d}s} \tag{3.37}$$

where $\mathrm{d}z$ is the change in elevation, z, across the fluid element. Because the velocity of any fluid element on a streamline can be a function of both time and location along the streamline, the acceleration of a fluid element on a streamline, a_s, is given by

$$a_s = \frac{\mathrm{d}V}{\mathrm{d}t} = \frac{\partial V}{\partial t} + V\frac{\partial V}{\partial s} \quad \text{(general condition)} \tag{3.38}$$

Under steady-state conditions, V depends only on s, $\partial V/\partial t = 0$, and Equation 3.38 becomes

$$a_s = \frac{\mathrm{d}V}{\mathrm{d}t} = V\frac{\partial V}{\partial s} = V\frac{\mathrm{d}V}{\mathrm{d}s} \quad \text{(steady state)} \tag{3.39}$$

Combining Equations 3.36, 3.37, and 3.39, neglecting second-order terms (i.e., products of differentials), and simplifying yields

$$\frac{\mathrm{d}p}{\rho} + V\,\mathrm{d}V + g\,\mathrm{d}z = 0 \quad \text{(along a streamline)} \tag{3.40}$$

which can also be expressed as

$$\frac{dp}{\rho} + \frac{1}{2}d(V^2) + g\,dz = 0 \quad \text{(along a streamline)} \tag{3.41}$$

This equation is sometimes called the *one-dimensional Euler equation*.[17]

Pressure gradient along a streamline. It is sometimes convenient to express the one-dimensional Euler equation, as given by Equation 3.40 or 3.41, in terms of the pressure gradient in the direction, s, along a streamline. Multiplying Equation 3.40 by ρ/ds gives the following expression involving the pressure gradient, dp/ds, along a streamline:

$$\frac{dp}{ds} + \rho V\frac{dV}{ds} + \rho g\frac{dz}{ds} = 0 \tag{3.42}$$

where s is the coordinate measured along the streamline. Equation 3.42 is particularly useful in that it relates the pressure gradient to the velocity gradient and the rate of change of elevation along a streamline.

EXAMPLE 3.8

Consider the nozzle shown in Figure 3.10, where the diameter of the nozzle decreases linearly from 100 mm to 50 mm over a distance of 400 mm. If water at 20°C flows with a volume flow rate of 10 L/s through the nozzle, estimate the pressure gradient in the nozzle as a function of the distance, x, downstream of the nozzle entrance. Compare the pressure gradients at the entrance and exit of the nozzle.

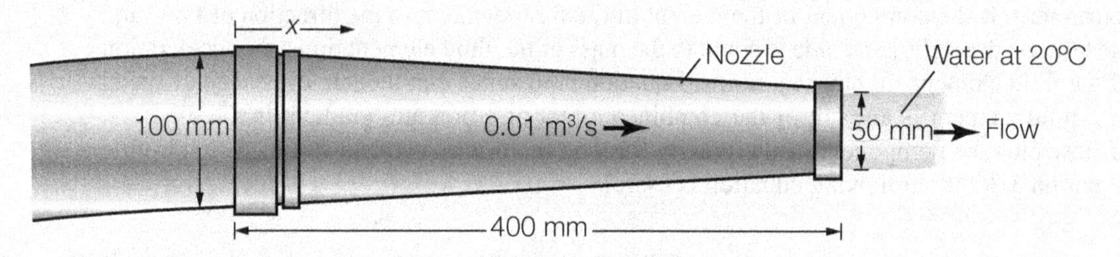

Figure 3.10: Flow through a nozzle

SOLUTION

From the given data: D_1 = 100 mm = 0.1 m, D_2 = 50 mm = 0.05 m, L = 400 mm = 0.4 m, and Q = 10 L/s = 0.01 m^3/s. For water at 20°C, ρ = 998 kg/m^3. The diameter of the nozzle as a function of x is given by

$$D(x) = D_1 + \frac{D_2 - D_1}{L}x = 0.1 + \frac{0.05 - 0.1}{0.4}x \quad \rightarrow \quad D(x) = 0.1 - 0.125x$$

Using this relationship, the cross-sectional area, A, velocity, V, and dV/dx, all as functions of x, are derived as follows:

$$A(x) = \frac{\pi}{4}D^2 = \frac{\pi}{4}(0.1 - 0.125x)^2$$

[17]Named in honor of the Swiss mathematician and physicist Leonhard Euler (1707–1783).

$$V(x) = \frac{Q}{A(x)} = \frac{(0.01)(4)}{\pi(0.1 - 0.125x)^2} = \frac{0.01273}{(0.1 - 0.125x)^2} \tag{3.43}$$

$$\frac{\mathrm{d}V}{\mathrm{d}x} = \frac{\mathrm{d}}{\mathrm{d}x}\left[\frac{0.01273}{(0.1 - 0.125x)^2}\right] = \frac{0.003183}{(0.1 - 0.125x)^3} \tag{3.44}$$

The relationship between the pressure gradient, $\mathrm{d}p/\mathrm{d}x$, and the velocity gradient, $\mathrm{d}V/\mathrm{d}x$, can be obtained by applying Equation 3.42 along the centerline of the nozzle, which yields

$$\frac{\mathrm{d}p}{\mathrm{d}x} + \rho V\frac{\mathrm{d}V}{\mathrm{d}x} + \rho g\frac{\mathrm{d}z}{\mathrm{d}x} = 0 \quad \rightarrow \quad \frac{\mathrm{d}p}{\mathrm{d}x} = -\rho V\frac{\mathrm{d}V}{\mathrm{d}x} \tag{3.45}$$

where the condition that $\mathrm{d}z/\mathrm{d}x = 0$ along the center streamline has been invoked. Substituting Equations 3.43 and 3.44 into Equation 3.45 yields the sought expression for the pressure gradient as follows:

$$\frac{\mathrm{d}p}{\mathrm{d}x} = -(998)\frac{0.01273}{(0.1 - 0.125x)^2} \cdot \frac{0.003183}{(0.1 - 0.125x)^3} \quad \rightarrow \quad \boldsymbol{\frac{\mathrm{d}p}{\mathrm{d}x} = -\frac{0.04044}{(0.1 - 0.125x)^5}\ \textbf{Pa/m}}$$

At the entrance to the nozzle section, $x = 0$ m and the pressure gradient is as follows:

$$\frac{\mathrm{d}p}{\mathrm{d}x} = -\frac{0.04044}{[0.1 - 0.125(0)]^5} = -4040\ \text{Pa/m} = \mathbf{-4.4\ kPa/m}$$

At the exit of the nozzle section, $x = 0.4$ m and the pressure gradient is as follows:

$$\frac{\mathrm{d}p}{\mathrm{d}x} = -\frac{0.04044}{[0.1 - 0.125(0.4)]^5} = -129000\ \text{Pa/m} = \mathbf{-129\ kPa/m}$$

Therefore, the (negative) pressure gradient increases by almost two orders of magnitude as the water moves through the nozzle section.

A particularly useful form of the one-dimensional momentum equation along a streamline can be obtained by integrating the one-dimensional Euler equation given by Equation 3.41. However, the preferred form of integrated momentum equation depends on whether the fluid is a liquid or a gas. One of the primary reasons for this difference in preferred form is that although gravity usually has a significant effect on the pressure distribution in liquids, gravity usually has a negligible effect on the pressure distribution in gases. Consequently, the preferred form of the integrated momentum equation for liquids explicitly includes g in two of the three terms, whereas the preferred form of the integrated equation for gases includes g in only one term, and that term is usually neglected. Even though there are different preferred forms of the integrated momentum equation for liquids and gases, each of the preferred forms is applicable to both liquids and gases.

Preferred form of integrated momentum equation for liquids. Dividing the one-dimensional Euler equation, Equation 3.41, by g and integrating yields

$$\int \frac{\mathrm{d}p}{\gamma} + \frac{V^2}{2g} + z = C \quad \text{(along a streamline)} \tag{3.46}$$

where γ is the specific weight of the fluid ($= \rho g$) and C is a constant of integration that can be evaluated by determining the value of the left-hand side of Equation 3.46 at any point on the streamline. Equation 3.46 applies to both compressible and incompressible flow. In the case of incompressible flow ($\gamma \approx$ constant), Equation 3.46 becomes

$$\frac{p}{\gamma} + \frac{V^2}{2g} + z = C \quad \text{or} \quad p + \tfrac{1}{2}\rho V^2 + \gamma z = C \quad \text{(along streamline)} \tag{3.47}$$

Because most liquids are incompressible, Equation 3.47 is the more useful equation form of the integrated momentum equation for inviscid and incompressible liquids under steady-state conditions. It is interesting to compare Equation 3.47 to the pressure distribution in a static fluid, where $p/\gamma + z$ = constant and $p + \gamma z$ = constant, which shows that as the velocity becomes small, the pressure distribution becomes hydrostatic.

Preferred form of integrated momentum equation for gases. Integrating Equation 3.41 directly yields the following form of the steady-state momentum equation for an inviscid fluid:

$$\int \frac{dp}{\rho} + \frac{V^2}{2} + gz = C \quad \text{(along a streamline)} \tag{3.48}$$

Equation 3.48 applies to both compressible and incompressible flow. In the case of an incompressible fluid (ρ = constant), Equation 3.48 yields

$$\frac{p}{\rho} + \frac{V^2}{2} + gz = C \quad \text{or} \quad p + \tfrac{1}{2}\rho V^2 + \gamma z = C \quad \text{(along streamline)} \tag{3.49}$$

Because gas flow can be either compressible or incompressible, depending on the magnitude of the pressure variations, both Equations 3.48 and 3.49 are useful under different frequently encountered conditions. For most gases and for most physical dimensions encountered in engineering applications, gravitational effects are negligible and the "gz" and "γz" terms in Equations 3.48 and 3.49 can be neglected.

The Bernoulli equation. The various forms of the integrated steady-state one-dimensional momentum equation for inviscid fluids given by Equations 3.46–3.49 are commonly referred to as the *Bernoulli equation*.[18] The constant of integration, C, on the right-hand side of these equations is called the *Bernoulli constant*. It is also quite common for the definition of the Bernoulli equation to be restricted to either Equation 3.47 or Equation 3.49, which are the forms of the equation corresponding to an incompressible fluid. The Bernoulli equation, in its various forms, is commonly used to explain the fact that (in a horizontal plane where z is constant) an increase in velocity along a streamline is accompanied by a decrease in pressure. In fact, such a relationship is expected on physical grounds, because the velocity increase is caused by the net pressure force and an accelerating fluid requires that the upstream pressure force be greater than the downstream pressure force. The same argument can be used to explain the pressure increase along a streamline in decelerating fluids.

Limitations of the Bernoulli equation. In applying the Bernoulli equation, great care must be taken to ensure that the assumptions implicit in the Bernoulli equation are respected. Otherwise, erroneous results will be obtained. Failure to limit the application of the Bernoulli equation to cases in which the assumptions are valid is a common error in engineering applications. Assumptions invoked in deriving the Bernoulli equation are listed below.

[18]Named in honor of the Swiss physicist and mathematician Daniel Bernoulli (1700–1782).

- The flow is steady and frictionless.
- The equation is applicable along a streamline.
- No energy is added to or removed from the fluid along the streamline.

The limitation of incompressible flow is also applicable when the Bernoulli equation is defined by Equation 3.47 or 3.49, which is commonly the case. A fluid that is both inviscid and incompressible is called an *ideal fluid*, and fluid flows in which viscosity and compressibility effects are negligible are called *ideal-fluid flows*. Hence, the Bernoulli equation is applicable to steady-state ideal-fluid flows. Although the assumptions of an ideal fluid and steady flow are nearly met in many practical applications, the Bernoulli equation is commonly misused by applying it to average conditions across a cross section of fluid bounded by solid surfaces (e.g., at cross sections of a pipe). This application of the Bernoulli equation violates the streamline assumption and neglects the relatively large energy losses associated with a fluid flowing over a (fixed) solid boundary. Fundamentally, the Bernoulli equation is applicable in regions of fluid flow where viscous forces on fluid elements are negligible, a condition that may occur even though a fluid is viscous. In so-called *inviscid regions of flow*, net viscous or frictional forces are negligibly small compared with other forces acting on fluid elements, and inviscid regions of flow typically occur where velocity gradients are negligible. Frictional effects are always important very close to solid walls (boundary layers) and directly downstream of solid bodies (in *wakes*). The various forms of the Bernoulli equation assume that the density is constant or is a function of pressure. In cases where the density varies due to changes in temperature caused by heat exchange between the fluid and its surroundings, the Bernoulli equation is not applicable. This condition violates the assumption that no energy is added to or removed from the fluid along the streamline. Such a violation would likely occur in the case of a gas flowing past a heating element located in the flow field. However, in cases where the temperature change is small and causes a negligible change in density, the (incompressible-flow) Bernoulli equation is approximately applicable; such cases are more likely to be found in liquids than in gases.

EXAMPLE 3.9

Water at 20°C flows along the streamlines shown in Figure 3.11. The (gauge) pressure, velocity, and elevation at Point 1 are 15.0 kPa, 1.0 m/s, and 5.02 m, respectively, and the velocity and elevation at Point 2 are 0.5 m/s and 5.21 m, respectively. Estimate the pressure at Point 2.

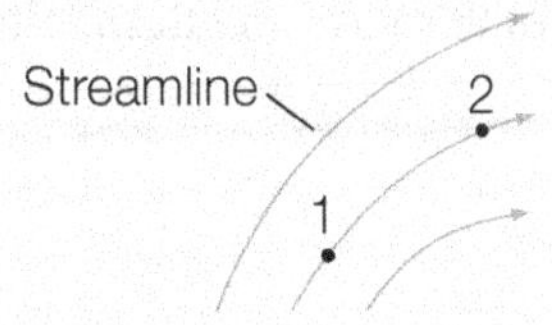

Figure 3.11: Streamline

SOLUTION

From the given data: $p_1 = 15.0$ kPa, $V_1 = 1.0$ m/s, $z_1 = 5.02$ m, $V_2 = 0.5$ m/s, and $z_2 = 5.21$ m. For water at 20°C, $\rho = 998.2$ kg/m^3; hence, $\gamma = \rho g = (998.2)(9.807) = 9789$ N/m^3 = 9.789 kN/m^3.

Substituting these data into the Bernoulli equation for an incompressible fluid (Equation 3.47) gives

$$\frac{p_1}{\gamma} + \frac{V_1^2}{2g} + z_1 = \frac{p_2}{\gamma} + \frac{V_2^2}{2g} + z_2$$

$$\frac{15.0}{9.789} + \frac{1.0^2}{2(9.807)} + 5.02 = \frac{p_2}{9.789} + \frac{0.5^2}{2(9.807)} + 5.21$$

which yields p_2 = 13.5 kPa. Hence, the pressure at Point 2 is estimated as **13.5 kPa**.

Energy components in the Bernoulli equation. The terms on the left-hand side of the Bernoulli equation, such as given by Equation 3.49, comprise the different forms of *mechanical energy*, where mechanical energy is defined as the form of energy that can be converted to mechanical work completely and directly by an ideal mechanical device such as an ideal turbine. The kinetic energy ($\frac{1}{2}mV^2$) and potential energy (mgz) of a mass, m, are familiar forms of mechanical energy. The pressure, p, is not a form of energy, but a pressure force acting on a fluid through a distance produces work, called *flow work* or *displacement work*; the quantity p/ρ is the work being done on a fluid element per unit mass, where ρ is the density of the fluid. The Bernoulli equation states that the work done on a fluid particle by the pressure force is equal to the change in kinetic plus potential energies; hence, the flow work appears in the Bernoulli equation. The mechanical energy per unit mass, e_{mech}, of a flowing fluid can therefore be expressed in the form

$$e_{\text{mech}} = \frac{p}{\rho} + \frac{V^2}{2} + gz \tag{3.50}$$

where $V^2/2$ and gz are the familiar kinetic and potential energies per unit mass. Comparing Equations 3.49 and 3.50, it is apparent that the Bernoulli equation states that mechanical energy is conserved along streamlines, provided the fluid is ideal, the flow is steady, and no energy is added to or removed from the fluid along the streamline.

Pressure components in the Bernoulli equation. A common application of the Bernoulli equation is where a fluid decelerates from a nonzero free-stream velocity to zero velocity along a streamline. An example of such a case is a fluid impinging on the nose of a submarine, a scenario that is usually analyzed relative to an (inertial) coordinate system that is attached to the submarine and moving with a constant velocity. Recall that a point of zero velocity is called a *stagnation point*. In this case, the Bernoulli equation requires that

$$\underbrace{p_{\text{s}}}_{\text{stagnation pressure}} = \underbrace{p_\infty}_{\text{static pressure}} + \underbrace{\tfrac{1}{2}\rho V_\infty^2}_{\text{dynamic pressure}} \tag{3.51}$$

where p_{s} is the pressure at the stagnation point, p_∞ is the upstream pressure, and V_∞ is the upstream velocity. The pressure p_{s} is commonly called the *stagnation pressure*, p_∞ is called the *static pressure*, and $\frac{1}{2}\rho V_\infty^2$ is called the *dynamic pressure*. The static pressure represents the actual thermodynamic pressure of the moving fluid (therefore, the term "static" can be misleading), and the stagnation pressure is the pressure at a point where the fluid is brought to a complete stop isentropically. It is apparent from Equation 3.51 that the stagnation pressure exceeds the static pressure by $\frac{1}{2}\rho V_\infty^2$. In cases where the compressibility of the fluid is taken into account, the stagnation pressure is given by

$$p_s = p_\infty + \frac{1}{2}\rho V_\infty^2 \left(1 + \frac{V_\infty^2}{4c^2} + \ldots\right) \tag{3.52}$$

where c is the speed of sound in the fluid. For air at 20°C, c = 345 m/s, and at V_∞ = 69 m/s, the error in using Equation 3.51 instead of Equation 3.52 is less than 1%.

EXAMPLE 3.10

A Los Angeles-class submarine is capable of cruising at a speed of 55 km/h when at a depth of 200 m. Estimate the static, dynamic, and stagnation pressures under these operating conditions. By what percentage is the stagnation pressure (at the front of the submarine) higher than the static pressure? Assume that seawater has a specific gravity of 1.03.

SOLUTION

Consider conditions relative to a coordinate system moving at the same speed as the submarine. From the given data: V_∞ = 55 km/h = 15 m/s, Δz = 200 m, and SG = 1.03. The density and specific weight of seawater are ρ_{sw} = (1.03)(1000) = 1030 kg/m^3 and γ_{sw} = (1030)(9.807) = 10100 N/m^3 = 10.10 kN/m^3, respectively. Assume standard atmospheric pressure of p_{atm} = 101 kPa. Using these data yields the following results:

$$\text{static pressure, } p_\infty = p_{atm} + \gamma_{sw}\Delta z = 101 + (10.10)(200) = \mathbf{2121\ kPa}$$

$$\text{dynamic pressure, } \frac{1}{2}\rho V_\infty^2 = \frac{1}{2}(1030)(15)^2 = 1.16 \times 10^5 \text{ Pa} = \mathbf{116\ kPa}$$

$$\text{stagnation pressure, } p_s = p_\infty + \frac{1}{2}\rho V_\infty^2 = 2121 + 116 = \mathbf{2237\ kPa}$$

Based on these results, the stagnation pressure is $116/2121 \times 100 = \mathbf{5.5\%}$ higher than the static pressure.

Approximate form of the Bernoulli equation applied to gases. Due to the small density of gases, it is a common circumstance that $p/\rho \gg gz$, in which case gz may be neglected and the Bernoulli equation is expressed as

$$p + \frac{1}{2}\rho V^2 = \text{constant} \tag{3.53}$$

which is the form of the Bernoulli equation most commonly used when dealing with gases.

Pressure distribution normal to streamlines: general relationships. In cases where streamlines are curved, the curvature of a streamline can be characterized by the *radius of curvature*, R, as shown in Figure 3.12. Consider the shaded fluid element between two closely spaced streamlines as shown in Figure 3.12. The momentum equation in the radial direction applied to the fluid element is given by

$$p\,\mathrm{d}A - (p + \mathrm{d}p)\,\mathrm{d}A + W\cos\theta = \rho\,\mathrm{d}A\,\mathrm{d}n\,a_n \tag{3.54}$$

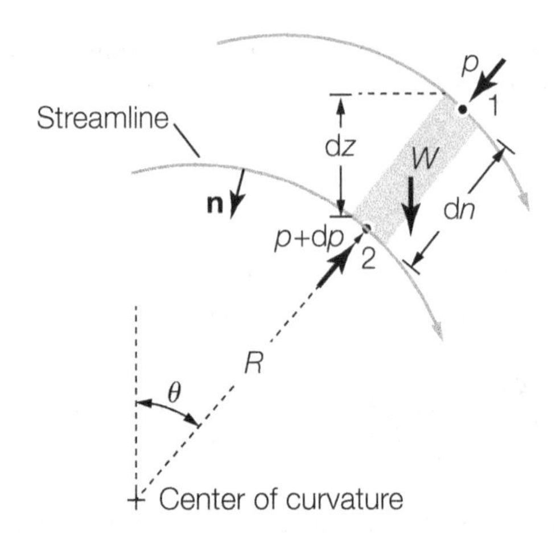

Figure 3.12: Pressure distribution normal to streamlines

where $(p + \mathrm{d}p)$ and $\mathrm{d}p$ are the pressures on the fluid element at the inner and outer bounding streamlines, respectively, $\mathrm{d}A$ is the surface area of the fluid element along the bounding streamlines, W is the weight of the fluid element, θ is the orientation of the fluid element relative to the vertical, $\mathrm{d}n$ is the distance between the bounding streamlines, and a_n is the acceleration of the fluid element in the direction normal to the bounding streamlines. The positive direction of the normal is taken as being inward toward the center of curvature. The following relationships are also applicable to the fluid element:

$$W = \gamma\,\mathrm{d}A\,\mathrm{d}n, \quad \mathrm{d}z = -\mathrm{d}n\,\cos\theta, \quad a_n = \frac{V^2}{R} \tag{3.55}$$

where $\mathrm{d}z$ is the difference in elevation between the inner and outer surfaces of the fluid element and V is the speed at which the fluid element is moving. Substituting the relationships in Equation 3.55 into Equation 3.54 yields

$$-\mathrm{d}p - \gamma\,\mathrm{d}z = \rho\frac{V^2}{R}\,\mathrm{d}n \tag{3.56}$$

which can be expressed in a form similar to that of the along-streamline Bernoulli equation as

$$\int\frac{\mathrm{d}p}{\rho} + \int\frac{V^2}{R}\,\mathrm{d}n + gz = \text{constant} \quad \text{(normal to streamline)} \tag{3.57}$$

To perform the integrations in Equation 3.57, it is necessary to know what the functional relationship is between ρ and p and how V and R vary as a function of n. The integrated form of Equation 3.56 can also be conveniently expressed as

$$\int_1^2 \mathrm{d}p + \int_1^2 \gamma\,\mathrm{d}z = -\int_1^2 \rho\frac{V^2}{R}\,\mathrm{d}n \quad \text{(normal to streamline)} \tag{3.58}$$

where Points 1 and 2 are located on the outer and inner streamlines bounding the fluid element, respectively.

Pressure distribution normal to streamlines: incompressible flow. Simplifying Equation 3.57 for the case of an incompressible fluid yields the following relationship describing the pressure distribution between the bounding streamlines:

$$p + \rho\int\frac{V^2}{R}\,\mathrm{d}n + \gamma z = \text{constant} \quad \text{(incompressible, normal to streamline)} \tag{3.59}$$

This pressure distribution can be compared with the pressure distribution in a static fluid given by $p + \gamma z$ = constant, which shows that as the radius of curvature, R, increases or the speed of the fluid, V, diminishes, the pressure distribution within the fluid becomes hydrostatic. An equivalent relationship to Equation 3.59 can be obtained by simplifying Equation 3.58 for the case of an incompressible fluid, which yields

$$\left(\frac{p_2}{\gamma} + z_2\right) - \left(\frac{p_1}{\gamma} + z_1\right) = -\frac{1}{g}\int_1^2 \frac{V^2}{R}\,\mathrm{d}n \quad \text{(incompressible, normal to streamline)} \tag{3.60}$$

Equations 3.59 and 3.60 both quantify the reality that when a fluid particle travels along a curved path, this must be accompanied by a net force directed toward the center of curvature, which is associated with a pressure variation normal to the streamlines. Equations 3.59 and 3.60 further indicate that the pressure, p, necessarily decreases in the direction toward the center of curvature. This is why central pressures in rotational flows, such as hurricanes and tornados, are at a minimum at the center of rotation. In cases where $z_2 \approx z_1$ (i.e., $\mathrm{d}z \approx 0$), Equation 3.56 can be approximated over small distance by

$$\frac{\Delta p}{\Delta n} = -\rho\frac{V^2}{R} \tag{3.61}$$

where Δp is the change in pressure over a distance Δn; recall that Δn is positive in the direction toward the center of curvature. It is further apparent from Equation 3.60 that as $R \to \infty$ (i.e., as the streamlines become straight), $p/\gamma + z$ becomes constant in the direction normal to the streamlines, which means that the pressure distribution normal to the streamlines becomes hydrostatic.

EXAMPLE 3.11

Water at 20°C flows around a two-dimensional bend as shown in Figure 3.13. The pressure measured at Point 1 (on the outside of the bend) is 300 kPa, and the velocity distribution of the fluid within the bend can be estimated by the relation

$$v(s) = 0.20\left(1 - \frac{s}{0.25}\right)^{\frac{1}{7}} \text{ m/s}$$

where $v(s)$ is the velocity at a distance s from the centerline of the bend. Assuming that the radius of curvature within the bend remains approximately constant, estimate the pressure at Point 2.

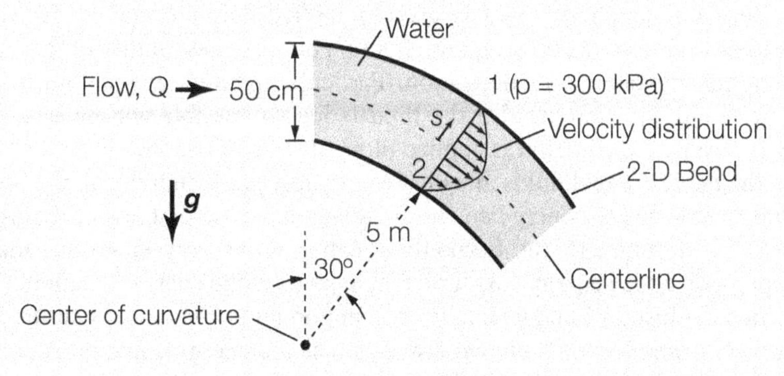

Figure 3.13: Flow around a two-dimensional bend

SOLUTION

From the given data: p_1 = 300 kPa and ρ = 998 kg/m^3. An average radius of curvature can be taken as $\overline{R}$ = 5 m + 0.5/2 m = 5.25 m. The momentum equation normal to the streamlines gives

$$\left(\frac{p_2}{\gamma}+z_2\right)-\left(\frac{p_1}{\gamma}+z_1\right)=-\frac{1}{g}\int_1^2\frac{V^2}{R}\,dn\approx\underbrace{-\frac{1}{g\overline{R}}\int_1^2 V^2\,dn}_{=I} \tag{3.62}$$

where the integral I can be evaluated as (taking g = 9.81 m/s^2), as follows:

$$\begin{aligned}
I &= -\frac{1}{(9.81)(5.25)}\left\{\int_0^{0.25}\left[0.20\left(1-\frac{0.25-n}{0.25}\right)^{\frac{1}{7}}\right]^2 dn+\int_{0.25}^{0.5}\left[0.20\left(1-\frac{n-0.25}{0.25}\right)^{\frac{1}{7}}\right]^2 dn\right\}\\
&= -\frac{(0.2)^2}{(9.81)(5.25)}\left\{\int_0^{0.25}\left(1-\frac{0.25-n}{0.25}\right)^{\frac{2}{7}} dn+\int_{0.25}^{0.5}\left(1-\frac{n-0.25}{0.25}\right)^{\frac{2}{7}} dn\right\}\\
&= -7.767\times10^{-4}\left\{\int_0^{0.25}(4n)^{\frac{2}{7}}\,dn+\int_{0.25}^{0.5}(2-4n)^{\frac{2}{7}}\,dn\right\}\\
&= -7.767\times10^{-4}\left\{\frac{7}{9}\cdot\frac{1}{4}(4n)^{\frac{9}{7}}\Big|_0^{0.25}+\frac{7}{9}\cdot\frac{1}{(-4)}(2-4n)^{\frac{9}{7}}\Big|_{0.25}^{0.5}\right\}\\
&= -7.767\times10^{-4}\,\{0.1944(1)-0.1944(-1)\}\\
&= -3.02\times10^{-4}\text{ m}
\end{aligned}$$

This small value of I indicates that the centripetal force contributes negligibly to the pressure distribution in the fluid. Substituting the value for I into Equation 3.62 gives

$$\left(\frac{300}{9.79}+5.5\cos 30^\circ\right)-\left(\frac{p_1}{9.79}+5\cos 30^\circ\right)=-3.02\times10^{-4}$$

which yields p_1 = **304 kPa**. This is approximately the same result that would be obtained by assuming a hydrostatic pressure distribution within the fluid.

3.4 Applications of the Bernoulli Equation

The Bernoulli equation is widely used in applied fluid mechanics. However, great caution should be exercised in applying this equation because the validity of the equation rests on the assumptions invoked in its derivation. Particular assumptions that must be met in applying the Bernoulli equation are that the fluid is incompressible and inviscid (or nearly so), the flow is steady, the equation is applied along a streamline, and there is no heat transfer to or from the fluid. For real fluids in which energy losses are significant, the Bernoulli equation is not applicable and the energy equation derived from the first law of thermodynamics is used instead. Unfortunately, the Bernoulli equation looks very much like the energy equation, except for an added energy-loss term that appears in the energy equation. (The two equations are often confused.) Fundamentally, it is important to remember that the Bernoulli equation is derived from Newton's second law, whereas the energy equation is derived from the first law of thermodynamics.

3.4.1 Flow through Orifices

An *orifice* is an opening through which fluid flows. The Bernoulli equation is widely used to estimate the velocity of flow through orifices, as well as to analyze flows originating from orifices. The basic principle used in these applications is that a streamline that passes through an orifice has a fixed amount of mechanical energy that can usually be determined from conditions upstream of the orifice. Applying the Bernoulli equation along the streamline that originates in a fluid reservoir and passes through an orifice gives

$$\frac{p_o}{\gamma} + \frac{V_o^2}{2g} + z_o = \frac{p_r}{\gamma} + \frac{V_r^2}{2g} + z_r \tag{3.63}$$

where the p_o, V_o, and z_o are the pressure, velocity, and elevation, respectively, at a point on the streamline as it passes through the orifice and p_r, V_r, and z_r are the pressure, velocity, and elevation, respectively, at a point in the reservoir that is located on the streamline that passes through the orifice.

Liquid flow through orifices. The case of liquid flow through an orifice is illustrated in Figure 3.14. In most cases of practical interest, the liquid reservoir is sufficiently large that the velocities within the reservoir can be neglected and the pressure distribution is assumed to be hydrostatic. This assumption is tantamount to assuming (quasi-)steady-state conditions in the liquid reservoir. Under this condition, the pressure distribution in the reservoir can be expressed as

$$\frac{p_r}{\gamma} + z_r = \frac{p_s}{\gamma} + z_s \tag{3.64}$$

where p_s and z_s are the pressure and elevation, respectively, at the liquid surface in the reservoir. Combining Equations 3.63 and 3.64 and using the large-reservoir assumption (i.e., $V_r \approx 0$) yields

$$\frac{p_o}{\gamma} + \frac{V_o^2}{2g} + z_o = \frac{p_s}{\gamma} + z_s$$

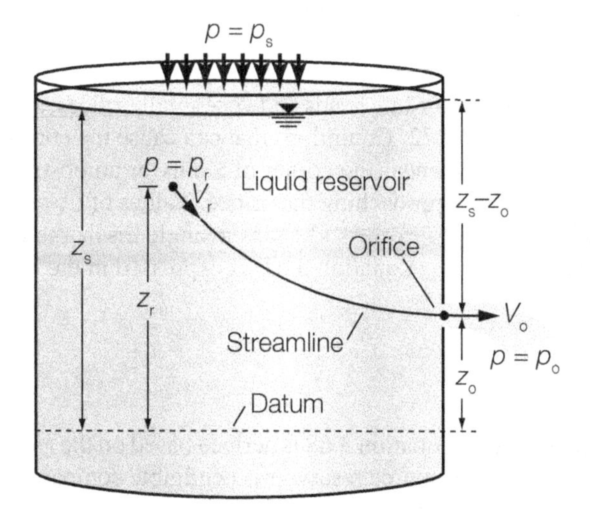

Figure 3.14: Liquid flow through an orifice

This is the form of the Bernoulli equation from which the theoretical flow of a liquid through an orifice is determined. Making V_o the subject of the formula yields the following explicit equation for the velocity of flow through the orifice:

$$V_o = \sqrt{2g\left(\frac{p_s - p_o}{\gamma} + z_s - z_o\right)} \tag{3.65}$$

In cases where an orifice discharges into the open atmosphere, $p_o = p_{atm}$ and Equation 3.65 indicates that the discharge velocity depends only on the gauge pressure at the reservoir surface (i.e., $p_s - p_o$) and the difference in elevation between the reservoir surface and the orifice opening (i.e., $z_s - z_o$). In applying Equation 3.65, the fact that z_o varies across the orifice opening is usually neglected and z_o is taken as the elevation of the centroid of the orifice. Equation 3.65 is sometimes referred to as Torricelli's theorem[19] and Equation 3.65 is commonly referred to as *Torricelli's equation* or *Torricelli's formula*. Because real fluid effects such as fluid friction and surface tension are neglected in deriving Equation 3.65, it is necessary to correct the theoretical velocity using a *correction coefficient for velocity*, C_v, such that the average velocity of the fluid passing through the orifice, V_o, is given by

$$V_o = C_v\sqrt{2g\left(\frac{p_s - p_o}{\gamma} + z_s - z_o\right)} \tag{3.66}$$

For well-made circular sharp-edged orifices producing free jets, values of C_v are typically in the range of 0.97–0.99, whereas for orifices that are not sharp-edged or of negligible thickness, the values of C_v can be much lower. Once the velocity through an orifice is determined, the discharge through the orifice can be estimated by multiplying this velocity by the area of the orifice. However, a complicating factor is that a jet of fluid exiting an orifice usually contracts (due to sharp corners at the orifice) such that the cross-sectional area of the jet stabilizes at a cross-sectional area that is less than the open area of the orifice. This contracted section is called the *vena contracta*,[20] although at low velocities, the downward deflection of the jet by gravity can lead to a poorly defined vena contracta. The ratio of the area at the vena contracta to the area of the orifice is called the *contraction coefficient*, which is commonly denoted by C_c, and taking the contraction coefficient into account, orifice discharges, Q, can be estimated by

$$Q = C_c C_v A_o\sqrt{2g\left(\frac{p_s - p_o}{\gamma} + z_s - z_o\right)} \tag{3.67}$$

where A_o is the orifice area. For circular sharp-edged orifices, values of C_c are typically in the range of 0.60–0.65, whereas for low heads and very small orifices, surface-tension effects might cause C_c to be as high as 0.72. Conditions that can cause the contraction coefficient to increase include the orifice being near the corner of a tank or an obstruction preventing the full convergence of streamlines approaching the orifice. Values of $C_c \approx 1$ can be attained for rounded orifices. It is common to express $C_c C_v$ as a single discharge coefficient, C_d, such that the orifice discharge equation (Equation 3.67) is expressed in the form

$$Q = C_d A_o\sqrt{2g\left(\frac{p_s - p_o}{\gamma} + z_s - z_o\right)} \tag{3.68}$$

It is important to remember that Equation 3.68 is derived based on the assumption that the rate of change of the fluid level in the source reservoir is negligible compared with the velocity of

[19]Named in honor of the Italian physicist and mathematician Evangelista Torricelli (1608–1647).
[20]Latin for "contracted vein."

the discharge through the orifice; in such cases, the flow is called *quasi-steady flow*. It should be apparent that steady flow is achievable only if liquid is being added to the source reservoir at the same rate it is being discharged through the orifice.

EXAMPLE 3.12

Compressed air is used to force water through a 25-mm-diameter orifice in a large tank as shown in Figure 3.15. The water level in the tank is kept stable by adding water at the same rate at which it is being discharged through the orifice. The orifice has rounded corners, and the manufacturer's literature indicates that the discharge coefficient is approximately equal to 0.94. Under a particular operating condition, the gauge pressure of the air above the water surface is 300 kPa, the water surface is 1.2 m above the centroid of the orifice, and the temperature of the water is 20°C. At what rate must water be supplied to the tank to maintain a steady-state condition?

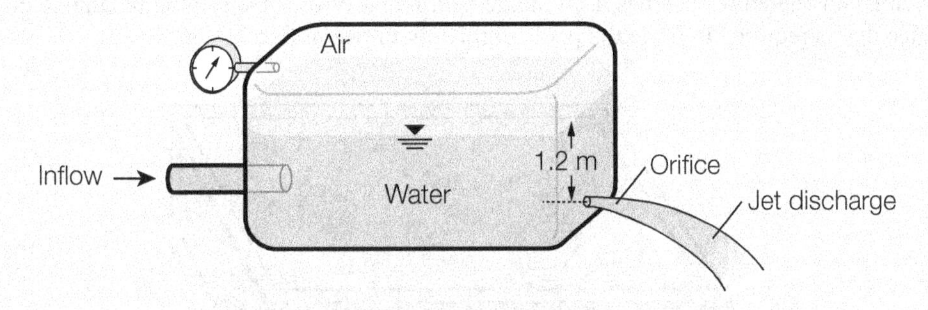

Figure 3.15: Flow through an orifice in a pressurized tank

SOLUTION

From the given data: D_o = 25 mm, C_d = 0.94, $p_\mathrm{s} - p_\mathrm{atm}$ = 300 kPa, and $z_\mathrm{s} - z_\mathrm{o}$ = 1.2 m. For water at 20°C, γ_w = 9.79 kN/m^3. The orifice area, A_o, is calculated as

$$A_\mathrm{o} = \frac{\pi}{4}D^2 = \frac{\pi}{4}(0.025)^2 = 4.909 \times 10^{-4}\ \mathrm{m}^2$$

Because the orifice discharges into the open atmosphere $p_\mathrm{o} = p_\mathrm{atm}$, and substituting the given and derived data into Equation 3.68 yields

$$\begin{aligned} Q &= C_\mathrm{d}A_\mathrm{o}\sqrt{2g\left(\frac{p_\mathrm{s}-p_\mathrm{o}}{\gamma_\mathrm{w}} + z_\mathrm{s} - z_\mathrm{o}\right)} \\ &= (0.94)(4.909 \times 10^{-4})\sqrt{2(9.81)\left(\frac{300}{9.79} + 1.2\right)} \\ &= 0.01153\ \mathrm{m}^3/\mathrm{s} = 11.5\ \mathrm{L/s} \end{aligned}$$

Therefore, water must be added to the tank at a rate of **11.5 L/s** to maintain a steady-state condition.

A frequently encountered case of orifice flow occurs when the reservoir supplying the orifice flow is open to the atmosphere, in which case $p_s = p_o$, and Equation 3.68 simplifies to

$$Q = C_d A_o \sqrt{2g\,(z_s - z_o)} \tag{3.69}$$

Under this circumstance, the flow velocity through the orifice depends only on the elevation of the reservoir surface relative to the elevation of the centroid of the orifice (i.e., $z_s - z_o$). The case of a reservoir surface open to the atmosphere occurs so often that Equation 3.69 is frequently cited as the orifice-flow formula, rather than Equation 3.68, which is the general form.

EXAMPLE 3.13

An orifice is to be used to drain a swimming pool as illustrated in Figure 3.16. The pool is 12 m long, 6 m wide, and 2.5 m deep. A commercially available drainage outlet (i.e., orifice) that is being considered for use has a discharge coefficient of 0.7. Determine the diameter of the orifice that is required to drain the pool completely in two hours.

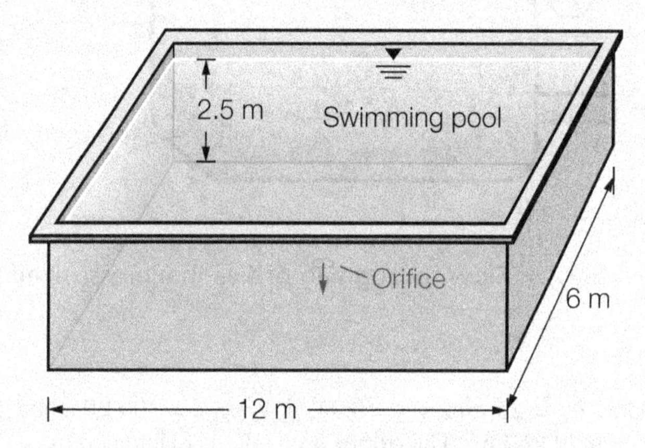

Figure 3.16: Swimming pool drain

SOLUTION

From the given data: L = 12 m, W = 6 m, H = 2.5 m, C_d = 0.7, and t_e = 2 h = 7200 s. Let h be the depth of water in the pool at any time t and let A_o be the area of the orifice. Conservation of (water) mass and the orifice discharge formula (Equation 3.69) requires that

$$-LW\frac{dh}{dt} = C_d A_o \sqrt{2gh} \quad \rightarrow \quad -\frac{dh}{\sqrt{h}} = \left[\frac{C_d A_o \sqrt{2g}}{LW}\right] dt$$

$$\rightarrow \quad -\int_H^0 \frac{dh}{\sqrt{h}} = \left[\frac{C_d A_o \sqrt{2g}}{LW}\right] \int_0^{t_e} dt$$

where t_e is the time to empty the pool. Performing the integration gives

$$2\sqrt{H} = \left[\frac{C_d A_o \sqrt{2g}}{LW}\right] t_e$$

Making A_o the subject of the formula and noting that $A_\mathrm{o} = \pi D_\mathrm{o}^2/4$, where D_o is the diameter of the orifice, gives

$$A_\mathrm{o} = \frac{\pi D_\mathrm{o}^2}{4} = \frac{2LW\sqrt{H}}{C_\mathrm{d}\sqrt{2g}t_\mathrm{e}} \quad \rightarrow \quad D_\mathrm{o} = \sqrt{\frac{8LW\sqrt{H}}{\pi C_\mathrm{d}\sqrt{2g}t_\mathrm{e}}}$$

Substituting known quantities yields

$$D_\mathrm{o} = \sqrt{\frac{8(12)(6)\sqrt{(2.5)}}{\pi(0.7)\sqrt{2(9.81)}(7200)}} = 0.114\ \mathrm{m} = 114\ \mathrm{mm}$$

Therefore, a drain with a diameter of **114 mm** will empty the pool in two hours. If a larger-diameter drain is used, the pool will empty in less than two hours.

Orifice with a submerged exit. An orifice with a submerged exit is defined as an orifice in which a fluid is discharged from one reservoir into another (discharge) reservoir containing the same fluid, where the orifice is located below the surface of the discharge reservoir. This scenario is illustrated in Figure 3.17. Applying the Bernoulli equation along the streamline that passes through the submerged orifice gives

$$\frac{p_\mathrm{o}}{\gamma} + \frac{V_\mathrm{o}^2}{2g} + z_\mathrm{o} = \frac{p_\mathrm{r1}}{\gamma} + \frac{V_\mathrm{r1}^2}{2g} + z_\mathrm{r1} \tag{3.70}$$

where p_o, V_o, and z_o are the pressure, velocity, and elevation, respectively, at a point on the streamline as it passes through the submerged orifice and p_r1, V_r1, and z_r1 are the pressure, velocity, and elevation, respectively, at a point in Reservoir 1 (= source reservoir) that is located on the streamline that passes through the submerged orifice. Assuming that the source reservoir is sufficiently large that the velocities within the reservoir can be neglected, the pressure distribution in the source reservoir can be assumed to be hydrostatic and expressed as

$$\frac{p_\mathrm{r1}}{\gamma} + z_\mathrm{r1} = \frac{p_\mathrm{s1}}{\gamma} + z_\mathrm{s1} \tag{3.71}$$

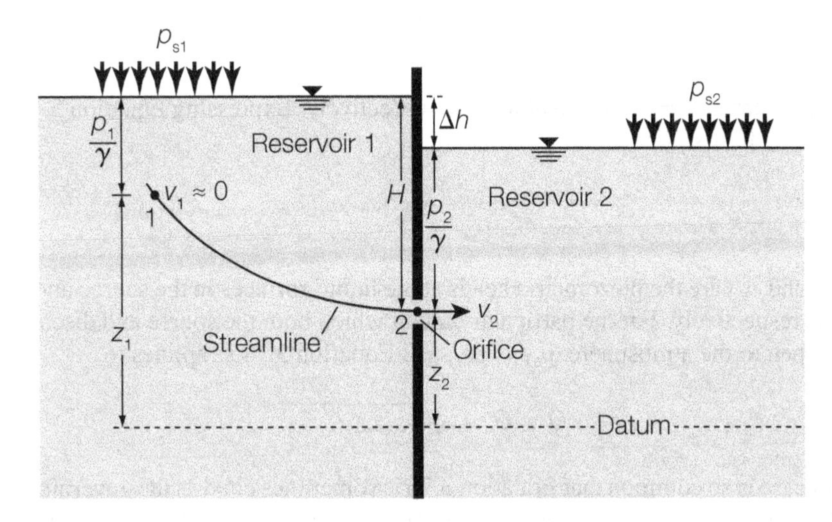

Figure 3.17: Flow through a submerged orifice

where p_{s1} and z_{s1} are the pressure and elevation, respectively, at the surface of the source reservoir. On the discharge side of the orifice, Reservoir 2 (= discharge reservoir) can also be assumed to be sufficiently large that velocities within the reservoir can be neglected, in which case the pressure distribution is hydrostatic and

$$\frac{p_o}{\gamma} + z_o = \frac{p_{s2}}{\gamma} + z_{s2} \tag{3.72}$$

where p_{s2} and z_{s2} are the pressure and elevation, respectively, at the surface of the discharge reservoir. Combining Equations 3.70–3.72 and using the large-reservoir assumptions yields

$$\frac{p_{s2}}{\gamma} + \frac{V_o^2}{2g} + z_{s2} = \frac{p_{s1}}{\gamma} + z_{s1}$$

Making V_o the subject of the formula yields the following explicit equation for the velocity of flow through the orifice:

$$V_o = \sqrt{2g\left(\frac{p_{s1} - p_{s2}}{\gamma} + z_{s1} - z_{s2}\right)} \tag{3.73}$$

As was done in the case of a free discharge, the volume flow rate through the submerged orifice can be derived from Equation 3.73 by applying a discharge coefficient, C_d, to account for the nonuniformity of the velocity distribution across the orifice and the contraction of the flow area downstream of the orifice opening (at the vena contracta). The volume flow rate from the submerged orifice is given by

$$Q = C_d A_o \sqrt{2g\left(\frac{p_{s1} - p_{s2}}{\gamma} + z_{s1} - z_{s2}\right)} \tag{3.74}$$

where A_o is the area of the submerged orifice. It is sometimes computationally convenient to express Equation 3.74 in terms of the piezometric heads in the source and receiving reservoirs. Recall that the piezometric head, ϕ, at any location in a fluid of specific weight γ is defined as

$$\phi = \frac{p}{\gamma} + z$$

where p and z are the pressure and elevation, respectively. Expressing Equation 3.74 in terms of piezometric heads gives

$$Q = C_d A_o \sqrt{2g\left(\phi_{s1} - \phi_{s2}\right)} \tag{3.75}$$

where ϕ_{s1} and ϕ_{s2} are the piezometric heads at the liquid surfaces in the source and discharge reservoirs, respectively. For the particular case in which both the source and discharge reservoirs are open to the atmosphere, $p_{s1} = p_{s2}$, and Equation 3.74 simplifies to

$$Q = C_d A_o \sqrt{2g\left(z_{s1} - z_{s2}\right)} \tag{3.76}$$

This latter case is so common that Equation 3.76 is sometimes cited as the governing equation for the discharge through submerged orifices, rather than the more general relationship given by Equations 3.74 and 3.75.

EXAMPLE 3.14

Two compartments of a pressurized tank are connected by a submerged 50-mm-diameter orifice as shown in Figure 3.18. In the source reservoir, the air pressure is 500 kPa and the depth of water above the centroid of the orifice is 1.7 m. In the discharge reservoir, the depth of water is 2.3 m. The tank contains inflow and outflow ports to maintain steady-state conditions in the two compartments, and the temperature of the water is 20°C. The discharge coefficient of the orifice is 0.87. What air pressure must be maintained in the discharge reservoir so that the volume flow rate through the tank is 10 L/s.

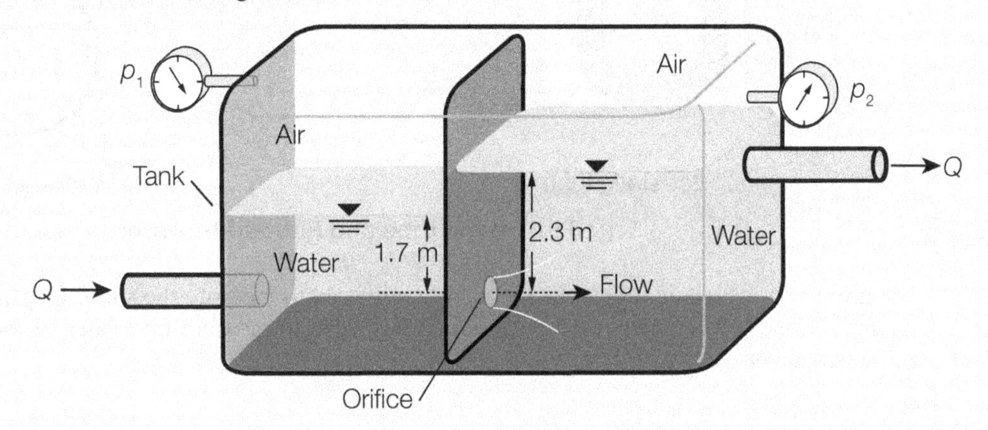

Figure 3.18: Submerged orifice in a pressurized tank

SOLUTION

From the given data: D_o = 50 mm, p_{s1} = 500 kPa, z_{s1} = 1.7 m, z_{s2} = 2.3 m, C_d = 0.87, and Q = 10 L/s = 0.01 m^3/s. For water at 20°C, γ_w = 9.79 kN/m^3. The area of the orifice, A_o, is given by

$$A_\text{o} = \frac{\pi}{4}D_\text{o}^2 = \frac{\pi}{4}(0.050)^2 = 0.001963 \text{ m}^2$$

The flow through the submerged orifice is given by Equation 3.74, and substituting the given and derived data yields

$$Q = C_\text{d}A_\text{o}\sqrt{2g\left(\frac{p_{\text{s}1} - p_{\text{s}2}}{\gamma_\text{w}} + z_{\text{s}1} - z_{\text{s}2}\right)}$$

$$10 = (0.87)(0.001963)\sqrt{2(9.81)\left(\frac{500 - p_{\text{s}2}}{9.79} + 1.7 - 2.3\right)} \quad [\times 10^3 \text{ L/m}^3] \quad \rightarrow \quad p_{\text{s}2} = 159 \text{ kPa}$$

Therefore, if the air pressure in the discharge compartment is maintained at **159 kPa**, the flow rate from the source to the discharge compartment will be 10 L/s. A lower pressure will yield a higher flow rate.

3.4.2 Flow Measurement

Devices that measure flow speeds and/or volume flow rates are generically called *flowmeters*, and the designs of such devices are generally based on known relationships between the speed of a fluid and easily measured flow properties such as pressure. Applications of the Bernoulli equation to many devices yield relationships between speed and pressure, in which cases the flow rates are estimated based on pressure measurements. Two such devices in common

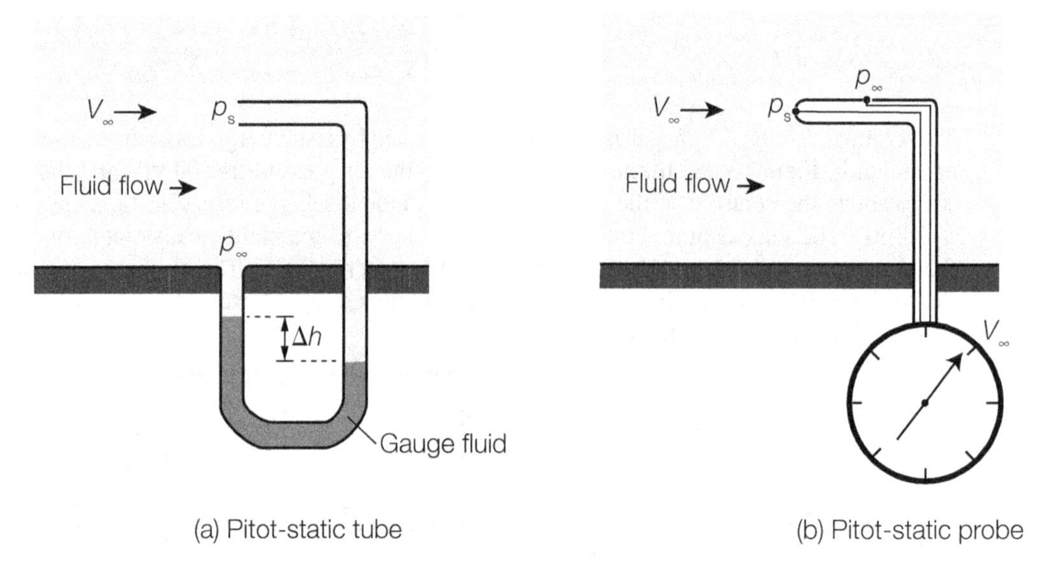

(a) Pitot-static tube (b) Pitot-static probe

Figure 3.19: Pitot-static tube and Pitot-static probe

use are the Pitot[21]-static tube, which is commonly used to measure the speed of gases such as air, and the Venturi meter, which is commonly used to measure flow rates of liquids in closed conduits.

3.4.2.1 Pitot-static tube

One of the most common applications of the Bernoulli equation is in the measurement of air velocity using a *Pitot-static tube*.[22] The Pitot-static tube, sometimes called simply a *Pitot tube*, is illustrated in Figure 3.19(a) and consists of a tube bent at right angles to the fluid flow and facing directly upstream. By pointing the tube directly upstream into the flow and measuring the difference between the (stagnation) pressure, p_s, and the (static) pressure of the surrounding airflow, p_∞, the Pitot-static tube gives an accurate estimate of the free-stream fluid velocity which, according to Equation 3.51, is given by

$$V_\infty = \sqrt{\frac{2(p_s - p_\infty)}{\rho}} \tag{3.77}$$

Commercially available Pitot-static tubes have tube diameters as small as 1.5 mm. In practical applications, Pitot-static tubes frequently use (pressure) transducers to measure the difference between the stagnation and static pressure ($p_s - p_\infty$). Such an application with the *Pitot-static probe* is shown in Figure 3.19(b), where a pressure transducer measuring $p_s - p_\infty$ is connected to an electronic gauge that converts this measurement into the free-stream velocity, V_∞, via Equation 3.77. A common application of the Pitot-static tube is to measure the flight speed of an aircraft relative to the air in which the aircraft is flying. In such applications, the Pitot-static tube is commonly mounted on the side of the aircraft near the nose. However, in some small aircraft, the Pitot-static tube is mounted on the underside of the wing (e.g., Cessna 172). Examples of Pitot-static tubes mounted on the side of military aircraft are shown in Figure 3.20, where Figure 3.20(a) shows a single Pitot-static tube on the side of an F-15A Eagle fighter aircraft and Figure 3.20(b) shows a triad of Pitot-static tubes on the side of a B-1b Lancer bomber.

Operational issues. Directional velocity fluctuations due to turbulence typically cause increased pressures to be measured by Pitot-static tubes, and the calculated velocities in these

[21] Pronounced "pea-toe."

[22] Named in honor of the French hydraulic engineer Henri de Pitot (1695–1771).

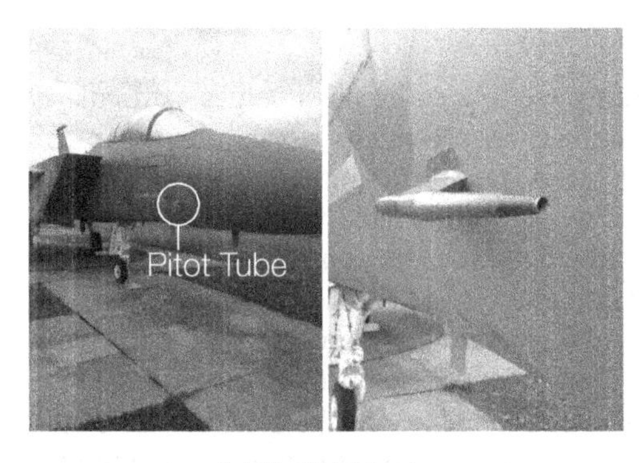

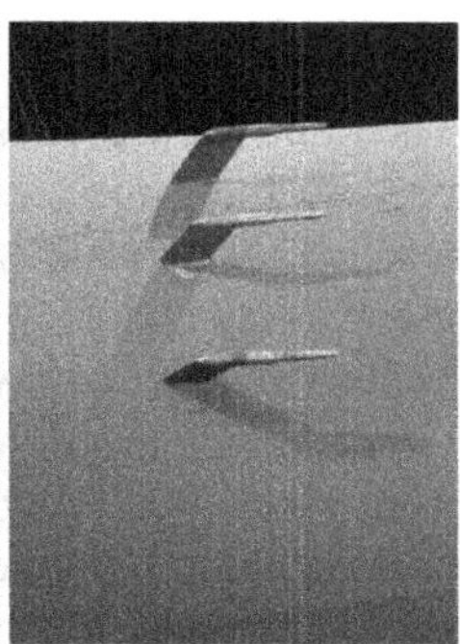

(a) F-15A Eagle (b) B1b Lancer

Figure 3.20: Pitot-static tubes on military aircraft
Source: Lt. Col. Dirk P. Yamamoto, United States Air Force.

cases (using Equation 3.77) are typically multiplied by factors in the range of 0.98–0.995. In addition to errors caused by velocity fluctuations and errors associated with imperfections in the construction of Pitot-static tubes, flow-measurement errors are caused by the Pitot-static tube not being properly aligned with the flow direction. However, misalignment of up to 12°–20° (depending on the tube design) gives results less than 1% in error from the perfectly aligned results. Because the Bernoulli equation is used as a basis for converting the pressure measurements in Pitot-static devices into an estimate of the velocity (via Equation 3.77), Pitot-static devices should only be used in cases where application of the Bernoulli equation is valid. Hence, Pitot-static devices should not be used is cases where viscous effects are significant or the flow is unsteady.

EXAMPLE 3.15

The Pitot-static tube shown in Figure 3.21 is used to measure the relative airspeed of a small aircraft. If the pressure measured at Point 1 is 105 kPa and the pressure at Point 2 is 101 kPa, estimate the relative airspeed.

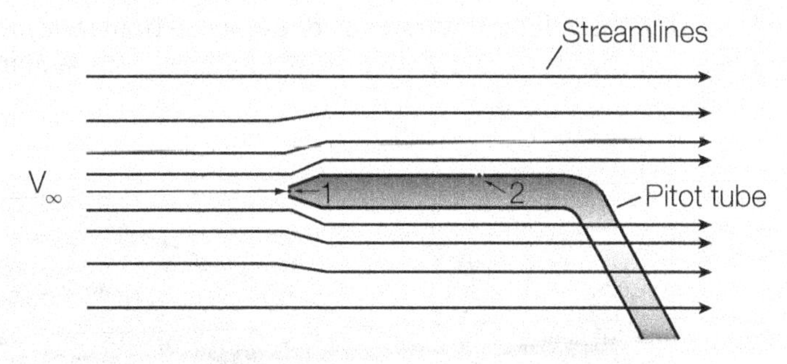

Figure 3.21: Airspeed measurement

SOLUTION

From the given data: p_∞ = 101 kPa and p_s = 105 kPa. Take the density of air as 1.204 kg/m^3 (at 20°C). Substituting into Equation 3.77 gives

$$V_\infty = \sqrt{\frac{2(p_\mathrm{s} - p_\infty)}{\rho}} = \sqrt{\frac{2(105 - 101)(1000)}{1.204}} = 81.5 \text{ m/s}$$

Therefore, the relative airspeed is **81.5 m/s**.

3.4.2.2 Venturi meter

The Pitot-static tube is not well suited to measuring the flow of liquids in pipes, particularly because the projection of the tube into the conduit is an obstruction to the flow than can lead to blockage and significant energy losses. The device more commonly used to measure the velocity and flow rate in pipes is the *Venturi meter*,[23] illustrated in Figure 3.22. In the Venturi meter, the pressure is measured at an upstream section (section 1) and at a contracted *throat section* (section 2); beyond the throat section, the pipe is gradually expanded back to its original size. The gradual expansion is intended to prevent flow separation from the walls of the conduit (and the associated energy losses) during expansion back to the original pipe size. However, more gradual expansions generally correspond to longer expansions and more losses due to wall friction. An expansion angle of around 6° is usually optimal to minimize overall head losses. Application of the Bernoulli equation to the central streamline between sections 1 and 2 yields

$$\frac{p_1}{\gamma} + \frac{V_1^2}{2g} + z_1 = \frac{p_2}{\gamma} + \frac{V_2^2}{2g} + z_2 \tag{3.78}$$

where p_1 and p_2 are the pressures at sections 1 and 2, respectively, V_1 and V_2 are the centerline velocities at sections 1 and 2, respectively, and z_1 and z_2 are the centerline elevations at sections 1 and 2, respectively. In accordance with the continuity equation, it is estimated that

$$Q = A_1\overline{V}_1 = A_2\overline{V}_2 \tag{3.79}$$

where A_1 and A_2 are the cross-sectional areas at sections 1 and 2, respectively. Because the velocity is never exactly uniform across the cross section of a pipe, V_1 and V_2 in Equation 3.78, which represent the centerline velocities at sections 1 and 2, respectively, are never exactly equal to $\overline{V}_1$ and $\overline{V}_2$ in Equation 3.79, which represent the average velocities across sections 1 and 2, respectively. However, assuming that $V_1 = \overline{V}_1$ and $V_2 = \overline{V}_2$ and combining Equations 3.78 and 3.79 yields the following expression for the volume flow rate, Q, in the pipe:

$$Q = A_2\sqrt{\frac{2[(p_1 - p_2) + \gamma(z_1 - z_2)]}{\rho\left[1 - \left(\frac{A_2}{A_1}\right)^2\right]}} \tag{3.80}$$

which indicates that the volume flow rate can be estimated from measurements of the pressure difference between sections 1 and 2 ($= p_1 - p_2$) for any given Venturi meter configuration

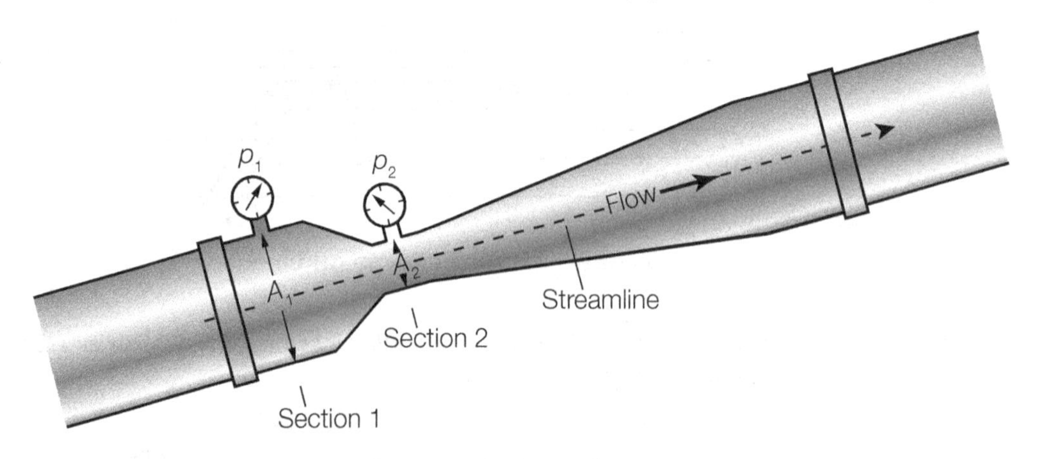

Figure 3.22: Venturi meter

[23]The Venturi meter is named after the Italian physicist Giovanni Battista Venturi (1746–1822).

in which A_1, A_2, z_1, and z_2 are fixed. For pipes of circular cross section, $A_2/A_1 = D_2^2/D_1^2$, where D_1 and D_2 are the diameters at the upstream and throat sections, respectively; typically, D_2/D_1 is around 0.5. In reality, the flow rate estimated using Equation 3.80 can deviate from the actual flow rate by amounts ranging from as low as 1% to as high as 40%, depending on the pipe geometry and the configuration of the Venturi meter. In practice, these deviations are taken into account by multiplying the theoretical flow rate, given by Equation 3.80, by a *discharge coefficient*, C_d, such that the actual flow rate is given by

$$Q = C_d A_2 \sqrt{\frac{2[(p_1 - p_2) + \gamma(z_1 - z_2)]}{\rho\left[1 - \left(\frac{A_2}{A_1}\right)^2\right]}} \tag{3.81}$$

Whenever a particular Venturi meter is used, the discharge coefficient of that meter should be determined by calibration or solicited from the manufacturer of the meter. The discharge coefficient for Venturi meters is typically in the range of 0.95–0.99, with the discharge coefficient typically increasing as the Reynolds number increases. In the absence of specific data, it is common to take $C_d = 0.98$. To ensure that the average pressure at each section is used in the discharge formula, Equation 3.81, the pressure at any cross section of a Venturi meter is typically taken as the average of several pressure measurements around an annular ring.

EXAMPLE 3.16

A horizontal Venturi meter with a 100-mm-diameter throat section is used to measure the volume flow rate of water in a 200-mm pipeline. If the pressure in the pipeline at the entrance to the Venturi meter is 480 kPa and the pressure in the throat of the Venturi meter is measured as 400 kPa, estimate the flow rate in the pipe in liters per second. Assume that the discharge coefficient of the Venturi meter is equal to 0.98 and that the water is at 20°C.

SOLUTION

From the given data: $z_1 = z_2$, D_1 = 200 mm, D_2 = 100 mm, p_1 = 480 kPa, p_2 = 400 kPa, and C_d = 0.98. At 20°C, ρ = 998 kg/m^3. Using the given data yields the following results:

$$A_2 = \frac{\pi}{4}D_2^2 = \frac{\pi}{4}(0.100)^2 = 0.007854 \text{ m}^2$$

$$Q = C_d A_2 \sqrt{\frac{2(p_1 - p_2)}{\rho\left[1 - \left(\frac{D_2^2}{D_1^2}\right)^2\right]}} = (0.98)(0.007854)\sqrt{\frac{2(480 - 400)10^3}{(998)\left[1 - \left(\frac{0.100}{0.200}\right)^4\right]}}$$

$$= 0.1007 \text{ m}^3\text{/s} = 100.7 \text{ L/s}$$

Therefore, the volume flow rate in the pipe (and through the Venturi meter) is approximately **101 L/s**.

3.4.3 Trajectory of a Liquid Jet

The Bernoulli equation can be used to estimate the trajectory of a liquid jet such as that shown in Figure 3.23. In this case, the path of the liquid jet is taken as a streamline, where air resistance is neglected and the pressure in the liquid jet is taken as constant and equal to atmospheric pressure along its entire trajectory. Under these circumstances, the Bernoulli equation gives

$$\frac{V^2}{2g} + z = \frac{V_0^2}{2g} \tag{3.82}$$

where V is the velocity along the trajectory, V_0 is the initial (discharge) velocity, z is the elevation relative to the discharge elevation, and g is gravity. In accordance with Newton's law for the trajectory of a fluid element, the x- and z- coordinates on the jet trajectory are

$$x = V_{x0}t \tag{3.83}$$

$$z = V_{z0}t - \frac{1}{2}gt^2 \tag{3.84}$$

where V_{x0} and V_{z0} are the components of the initial velocity, $\mathbf{V}_0$, in the x- and z-directions, respectively, and t is the time since the fluid element is discharged from its source. Combining Equations 3.83 and 3.84 by eliminating t yields the equation of the jet trajectory as

$$z = \frac{V_{z0}}{V_{x0}}\left(x - \frac{g}{2V_{x0}V_{z0}}x^2\right) \tag{3.85}$$

The maximum elevation of the jet can be determined by setting $\mathrm{d}z/\mathrm{d}x = 0$, which yields the following coordinates at the maximum elevation point:

$$x_{\text{max}} = \frac{V_{x0}V_{z0}}{g} \tag{3.86}$$

$$z_{\text{max}} = \frac{V_{z0}^2}{2g} \tag{3.87}$$

The corresponding velocity at the maximum height can be determined from the Bernoulli equation, Equation 3.82, by taking $z = z_{\text{max}}$.

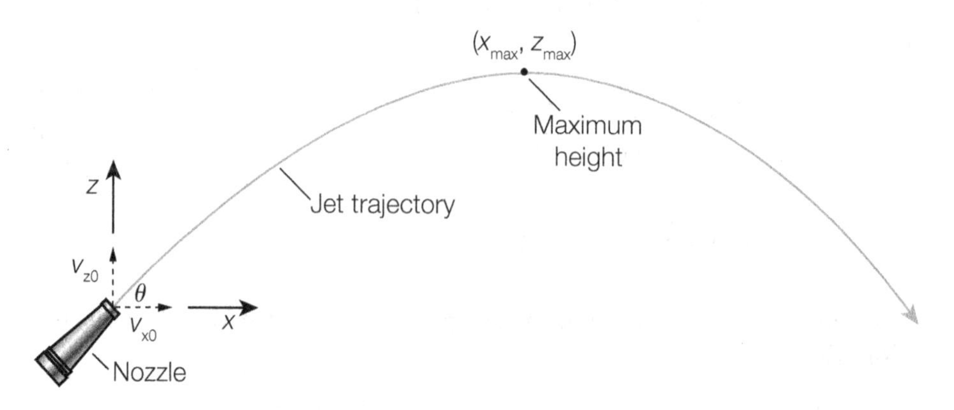

Figure 3.23: Jet trajectory

EXAMPLE 3.17

A fire hose discharges at ground level and is inclined at 45°. If water exits the fire hose at 15 m/s, determine the horizontal distance the water will travel to its ground level target. At what velocity will the water impinge on its target?

SOLUTION

From the given data: V_0 = 15 m/s and θ = 45°. Based on these data,

$$V_{x0} = 15 \cos 45^\circ = 10.61 \text{ m/s}, \qquad V_{z0} = 15 \sin 45^\circ = 10.61 \text{ m/s}$$

When the jet is at ground level, z = 0 m, and Equation 3.84 gives

$$t = \frac{2V_{z0}}{g}$$

Substituting this result into Equation 3.83 yields

$$x = V_{x0}t = \frac{2V_{x0}V_{z0}}{g} = \frac{2(10.61)(10.61)}{9.807} = 22.96 \text{ m}$$

Therefore, the jet will travel a distance of approximately **23.0 m**. The velocity at which the jet impinges at ground level can be estimated using Equation 3.82, where taking $z = 0$ m gives $V = V_0$. Hence, the jet impinges on the ground level target at the same velocity at which it leaves the fire hose (i.e., at a velocity of **15 m/s**).

Application. Vertical and near-vertical jets are commonly used in water fountains, which can be purely ornamental or serve practical purposes such as aerating a water body to elevate oxygen levels and improve water quality. An example of a water fountain in a lake is shown in Figure 3.24. In such applications, the maximum height, z_{max}, attained by a jet of water in the fountain is related to the vertical component of the jet exit velocity, V_{z0}, by Equation 3.87. For an individual jet in which the magnitude of the jet exit velocity is V_0, the energy per unit weight of the water exiting the jet is $V_0^2/2g$, and because most of this energy is dissipated by the jet, the power, P_{jet}, required to drive the jet can be estimated by

$$P_{\text{jet}} = \gamma Q \cdot \frac{V_0^2}{2g} \quad \rightarrow \quad P_{\text{jet}} = \gamma A_0 \frac{V_0^3}{2g} \tag{3.88}$$

where γ is the specific weight of water (≈ 9.79 kN/m^3), Q ($= A_0V_0$) is the volume discharge rate [L^{3}T^{-1}], and A_0 is the cross-sectional area of the jet at discharge. Therefore, when the number of jets (n_{jets}), the size of each jet discharge port (A_0), and the discharge velocity (V_0) of each jet required to attain a desired rise height (z_{max}) are known, the pumping power needed to achieve the desired effect can be estimated using Equation 3.88 as $n_{\text{jets}}P_{\text{jet}}$.

3.4.4 Compressibility Effects

The Bernoulli equation for both compressible and incompressible fluids is given by Equation 3.48, which is repeated as follows for easy reference:

$$\int \frac{dp}{\rho} + \frac{1}{2}V^2 + gz = C \quad \text{(along a streamline)} \tag{3.89}$$

In the case of compressible flows, the relationship between the density, ρ, of the fluid and the pressure, p, in the fluid must be taken into account. The relationship between density and pressure depends on the nature of the pressure (and corresponding density) changes. The word "incompressible" is commonly used to describe either a fluid property or a flow property. When used to describe a fluid property, an *incompressible fluid* is a fluid in which the density is a constant, and when used to describe a flow property, an *incompressible flow* occurs when pressure variations within the fluid flow cause negligible changes in the density of the fluid. A *Mach number criterion* is commonly used as a basis for determining whether a flow should be treated as compressible, and the supporting theory behind this criterion is described subsequently.

Isentropic flow. In the case of isentropic flow of a perfect gas, which is a reversible adiabatic process, the density and pressure are related by

$$\frac{p}{\rho^k} = C_1 \tag{3.90}$$

where k is the specific heat ratio and C_1 is a constant. Combining Equations 3.89 and 3.90 and integrating yields

$$\left(\frac{k}{k-1}\right)\frac{p_1}{\rho_1} + \frac{1}{2}V_1^2 + gz_1 = \left(\frac{k}{k-1}\right)\frac{p_2}{\rho_2} + \frac{1}{2}V_2^2 + gz_2 \quad \text{(along a streamline)} \tag{3.91}$$

Figure 3.24: Water fountain

which is the form of the Bernoulli equation that is applicable in cases where the fluid flow is compressible and isentropic between sections 1 and 2. Comparing Equation 3.91 with the Bernoulli equation for incompressible flows shows that the only differences are the presence of the factor $k/(k-1)$ and the fact that the densities at sections 1 and 2 are different.

Isentropic flow on a stagnation streamline. Applying Equation 3.91 to a stagnation streamline in which $z_1 = z_2$ and $V_2 = 0$, Equation 3.91 gives

$$\frac{p_2 - p_1}{p_1} = \left[\left(1 + \frac{k-1}{2}\mathrm{Ma}_1^2\right)^{\frac{k}{k-1}} - 1\right] \tag{3.92}$$

where Points 1 and 2 correspond to the free stream and the stagnation point, respectively, and Ma_1 is the *Mach number*[24] at Point 1, defined as

$$\mathrm{Ma}_1 = \frac{V_1}{c_1} \tag{3.93}$$

where c_1 is the speed of sound in the fluid at Point 1, given by

$$c_1 = \sqrt{kRT_1} \tag{3.94}$$

If the fluid was incompressible with density ρ, then the Bernoulli equation for flow toward a stagnation point would be given by

$$\tfrac{1}{2}\rho V_1^2 + p_1 = p_2 \tag{3.95}$$

and according to the ideal gas law,

$$p_1 = \rho RT_1 \tag{3.96}$$

Combining Equations 3.95 and 3.96 yields the following relationship for an incompressible fluid:

$$\frac{p_2 - p_1}{p_1} = \frac{k\,\mathrm{Ma}_1^2}{2} \tag{3.97}$$

The key (stagnation-flow) equations to be compared here are Equation 3.92 and Equation 3.97, which are applicable to compressible and incompressible fluids, respectively. Taking $k = 1.40$ (for air), it can be shown that the values of $(p_2 - p_1)/p_1$ given by these two equations differ by less than 2% for values of $\mathrm{Ma}_1 \leq 0.3$. This important result is the basis for the often-made statement that the flow of a perfect gas can be taken as incompressible for values of the Mach number less than about 0.3. Under standard conditions, T_1 = 15°C and $c_1 = \sqrt{kRT_1}$ = 340 m/s, which corresponds to $V_1 = \mathrm{Ma}_1 c_1 = 0.3(340\text{ m/s})$ = 104 m/s. Thus, high speeds are required for compressibility to be important. For gases moving at slower speeds (i.e., less than 100 m/s), compressibility need not be taken into account. Such would be the case in analyzing the flows surrounding cars, trains, light aircraft, most pipe flows, and turbomachinery at moderate rotational speeds.

[24]Named after the Austrian physicist Ernst Mach (1838–1916).

Isothermal flow. In the case of isothermal flow of a perfect gas, the density and pressure are related by

$$\frac{p}{\rho} = RT = \text{constant} \tag{3.98}$$

Combining Equations 3.98 and 3.89 and integrating yields

$$RT \ln p_1 + \tfrac{1}{2}V_1^2 + gz_1 = RT \ln p_2 + \tfrac{1}{2}V_2^2 + gz_2 \quad \text{(along a streamline)} \tag{3.99}$$

which is the form of the Bernoulli equation that is applicable in cases where the fluid flow is compressible and isothermal between sections 1 and 2. Isothermal compressible flows are less common than isentropic compressible flows, and most compressible flows are analyzed, at least partially, by assuming isentropic conditions.

3.4.5 Viscous Effects

The derivation of the Bernoulli equation can be modified by allowing for fluid friction. For the fluid element shown in Figure 3.25, the frictional force is given by

$$\text{frictional force} = \tau\left(P + \tfrac{1}{2}\mathrm{d}P\right)\mathrm{d}s \tag{3.100}$$

where τ is the shear stress on the surface of the fluid element, P is the perimeter of the upstream side of the fluid element, $\mathrm{d}P$ is the change in perimeter between the upstream and downstream sides of the fluid element, and $(P + \frac{1}{2}\mathrm{d}P)\,\mathrm{d}s$ is the approximate surface area over which the shear stress acts. Applying Newton's second law in the direction of motion of the fluid element, which is tangential to the streamline and designated as the s-direction, requires that

$$\underbrace{pA + (p + \tfrac{1}{2}\mathrm{d}p)\,\mathrm{d}A - (p + \mathrm{d}p)(A + \mathrm{d}A)}_{\text{net pressure force}} - \underbrace{\tau\left(P + \tfrac{1}{2}\mathrm{d}P\right)\mathrm{d}s}_{\text{shear force}}$$
$$- \underbrace{\gamma\,\mathrm{d}s\left(A + \tfrac{1}{2}\mathrm{d}A\right)\sin\theta}_{\text{weight}} = \underbrace{\rho\,\mathrm{d}s\left(A + \tfrac{1}{2}\mathrm{d}A\right)a_s}_{\text{mass}\,\times\,\text{acceleration}} \tag{3.101}$$

where p and A are the pressure and area, respectively, at the upstream side of the fluid element, $\mathrm{d}p$ and $\mathrm{d}A$ are the changes in pressure and area, respectively, between the upstream and downstream sides of the fluid element, γ is the specific weight of the fluid, and θ is the angle the streamline makes with the horizontal. Noting that $a_s = V\mathrm{d}V/\mathrm{d}s$ for steady-state flow, Equation 3.101 simplifies to

$$\frac{\mathrm{d}p}{\rho} + g\,\mathrm{d}z + V\,\mathrm{d}V = -\frac{\tau P}{\rho A}\,\mathrm{d}s \tag{3.102}$$

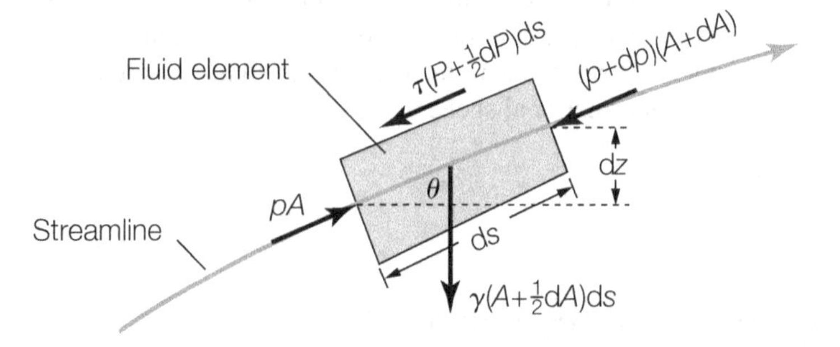

Figure 3.25: Motion of a viscous fluid along a streamline

This equation is similar to the Bernoulli equation for nonviscous (i.e., inviscid) fluids (Equation 3.40), except for the extra term $-\tau P\,\mathrm{d}s/(\rho A)$ that accounts for fluid friction. Equation 3.102 can be put in the form

$$\frac{\mathrm{d}p}{\gamma} + \mathrm{d}z + \mathrm{d}\frac{V^2}{2g} = -\frac{\tau P}{\gamma A}\,\mathrm{d}s \tag{3.103}$$

which applies to both compressible and incompressible fluids. In the case of an incompressible fluid, applying Equation 3.103 from Point 1 to Point 2 on the same streamline yields

$$\left(\frac{p_2}{\gamma} + z_2 + \frac{V_2^2}{2g}\right) - \left(\frac{p_1}{\gamma} + z_1 + \frac{V_1^2}{2g}\right) = -\frac{\tau PL}{\gamma A} \tag{3.104}$$

where L is the distance between Point 1 and Point 2, measured along the streamline. The assumptions associated with Equation 3.104 (which is applicable along a streamline) are (1) steady flow and (2) incompressible fluid.

Rectilinear viscous incompressible flow in closed conduits. In the case of incompressible rectilinear flow in closed conduits, $p/\gamma + z$ is constant across the cross section, because the pressure distribution is hydrostatic normal to rectilinear streamlines, even for viscous flow. Furthermore, for steady-state flow of an incompressible fluid in a prismatic conduit, the continuity equation requires the average velocity and the velocity distribution at all cross sections to be the same. Under these conditions, Equation 3.104 can be applied to a streamtube encompassing the entire cross-sectional area, A, of the conduit, which gives

$$\left(\frac{p_2}{\gamma} + z_2\right) - \left(\frac{p_1}{\gamma} + z_1\right) = -\frac{\tau PL}{\gamma A} \tag{3.105}$$

where $p_i/\gamma + z_i$ can be conveniently evaluated at the centroid of cross section i ($i = 1, 2$) and τ is the shear stress on the wall of the conduit, which can be represented by τ_0. Noting that the left-hand side of Equation 3.105 is the change in total head between the upstream and downstream sections of the conduit, the *wall-friction head loss*, h_f, is defined as

$$h_\mathrm{f} = \frac{\tau_0 PL}{\gamma A} \tag{3.106}$$

Combining Equations 3.105 and 3.106 yields

$$\left(\frac{p_1}{\gamma} + z_1\right) - \left(\frac{p_2}{\gamma} + z_2\right) = h_\mathrm{f} \tag{3.107}$$

Equation 3.107 is the integrated form of the momentum equation for steady rectilinear viscous flow of an incompressible fluid in a prismatic conduit. This equation is commonly applied in the analysis of flow in pipes. The combination of Equation 3.107 with the energy equation, as described in Chapter 4, yields the relationship between h_f and the thermodynamic properties of the flow, notably the relationship between head loss, temperature change of the fluid, and heat transfer from the fluid. Experimentally derived empirical expressions for h_f in terms of measurable physical quantities are used as bases for applying Equation 3.107 to estimate the pressure variations in closed conduits, and such applications are covered extensively in Chapter 7.

EXAMPLE 3.18

Water at 20°C flows at a rate of 40 L/s in a 200-mm-diameter pipeline where the pressure and elevation at a particular section are 480 kPa and 10 m, respectively. If the pressure and elevation at a section 50 m downstream are 400 kPa and 15 m, respectively, estimate the friction head loss and the average shear stress on the pipe between the two sections.

SOLUTION

From the given data: Q = 40 L/s = 0.040 m^3/s, D = 200 mm = 0.200 m, p_1 = 480 kPa, z_1 = 10 m, p_2 = 400 kPa, z_2 = 15 m, and L = 50 m. At 20°C, γ = 9.79 kN/m^3. Using these data gives

$$A = \frac{\pi}{4}D^2 = \frac{\pi}{4}(0.200)^2 = 0.03142 \text{ m}^2, \qquad P = \pi D = \pi(0.200) = 0.6283 \text{ m}$$

Applying Equation 3.107 yields

$$h_\text{f} = \left(\frac{p_1}{\gamma} + z_1\right) - \left(\frac{p_2}{\gamma} + z_2\right) = \left(\frac{480}{9.79} + 10\right) - \left(\frac{400}{9.79} + 15\right) = 3.171 \text{ m}$$

Therefore, the head loss in the pipeline section is approximately **3.17 m**. Using the definition of h_f given by Equation 3.106 gives

$$h_\text{f} = \frac{\tau_0 PL}{\gamma A} \quad \rightarrow \quad 3.171 = \frac{\tau_0(0.6283)(50)}{(9.79)(0.03142)} \quad \rightarrow \quad \tau_0 = 0.03105 \text{ kN/m}^2 = 31.05 \text{ Pa}$$

Therefore, the average shear stress on the pipe between the upstream and downstream sections is approximately **31.1 Pa**.

It is apparent from the above example that the application of Equation 3.107 does not depend on whether the flow causing the boundary shear stress, τ_0, is laminar or turbulent. As such, Equation 3.107 is valid under both flow conditions.

3.4.6 Branching Conduits

Application of the Bernoulli equation to flow in conduits commonly involves cases in which a single conduit branches into two or more conduits. This is illustrated in Figure 3.26, which shows a conduit that branches into two conduits. In this case, the upstream section is section 1 and the downstream sections are designated as sections 2 and 3. Application of the Bernoulli equation along the branching streamlines yields

$$\frac{p_1}{\gamma} + \frac{V_1^2}{2g} + z_1 = \frac{p_2}{\gamma} + \frac{V_2^2}{2g} + z_2 \tag{3.108}$$

$$\frac{p_{1'}}{\gamma} + \frac{V_{1'}^2}{2g} + z_{1'} = \frac{p_3}{\gamma} + \frac{V_3^2}{2g} + z_3 \tag{3.109}$$

where the subscripts denote the values of the variables at the respective sections. In prismatic conduits, the streamlines are rectilinear, so $p/\gamma + z$ is a constant across any given conduit section. Combining the requirement that $p/\gamma + z$ = constant with the assumption that the velocities on the section 1 streamlines are approximately the same gives

$$\frac{p_1}{\gamma} + z_1 = \frac{p_{1'}}{\gamma} + z_{1'}, \quad \text{and} \quad V_1 \approx V_{1'} \tag{3.110}$$

The velocity at section 1, V_1, is usually taken as being equal to the average velocity across the cross section. Applying the approximations in Equation 3.110 to Equations 3.108 and 3.109 yields

$$\frac{p_1}{\gamma} + \frac{V_1^2}{2g} + z_1 = \frac{p_2}{\gamma} + \frac{V_2^2}{2g} + z_2 = \frac{p_3}{\gamma} + \frac{V_3^2}{2g} + z_3 \tag{3.111}$$

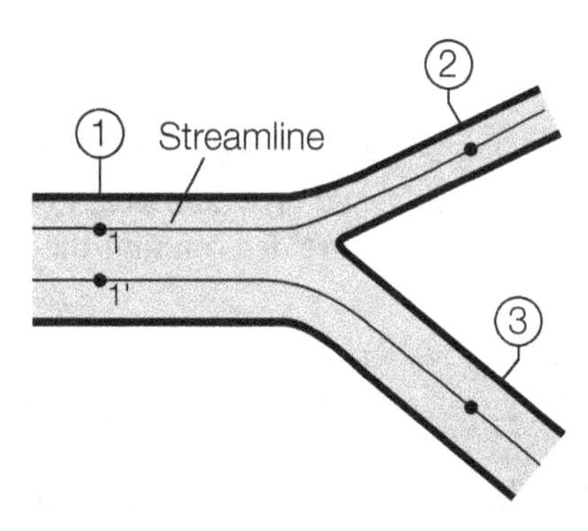

Figure 3.26: Branching conduit

In applying Equation 3.111, the pressure, p, and elevation, z, in each conduit are usually taken as centerline values, and the velocity, V, in each conduit is usually taken as the cross-sectionally averaged value. This latter approximation is tantamount to assuming a uniform velocity distribution across each flow section, which is generally appropriate in turbulent flows; many closed-conduit flows of engineering interest are turbulent. In cases where a single conduit branches into multiple conduits, $p/\gamma + z$ is still constant across each section, and one need only invoke the uniform-velocity assumption at each section, in which case the generalized form of Equation 3.111 is

$$\frac{p_1}{\gamma} + \frac{V_1^2}{2g} + z_1 = \frac{p_n}{\gamma} + \frac{V_n^2}{2g} + z_n \tag{3.112}$$

where an index n designates each downstream branch.

EXAMPLE 3.19

Consider a branching pipeline in the horizontal plane as shown in Figure 3.27, where a 500-mm-diameter pipe branches into three separate pipes with diameters of 100 mm, 250 mm, and 150 mm. The upstream pipe is designated as Pipe A, and the three branch pipes are designated as Pipes B, C, and D. When the flow rate of water through Pipe A is 450 L/s, the pressure in Pipe A is 350 kPa, the average velocity in Pipe C is 3 m/s, and the pressure in Pipe D is 250 kPa. (a) What is the volume flow rate in each of the three branch pipes? (b) What is the expected pressure in Pipes B and C?

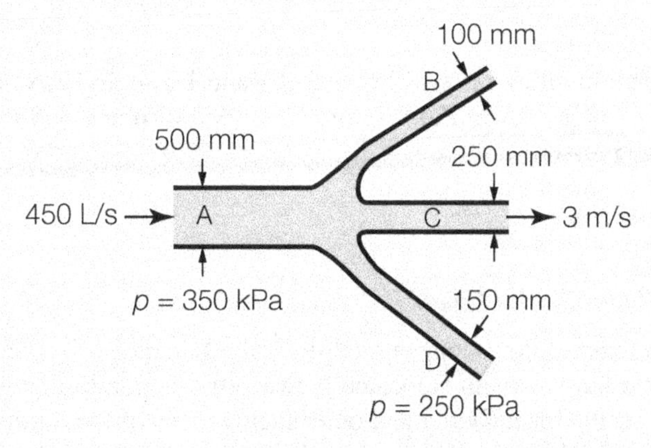

Figure 3.27: Flow in a branched pipeline

SOLUTION

From the given data: D_A = 500 mm, D_B = 100 mm, D_C = 250 mm, D_D = 150 mm, Q_A = 450 L/s = 0.45 m^3/s, p_A = 350 kPa, V_C = 3 m/s, and p_D = 250 kPa. For water, γ = 9.79 kN/m^3. Calculations in this problem are facilitated by first determining the values of the following quantities:

$$A_\mathrm{A} = \frac{\pi}{4}D_\mathrm{A}^2 = \frac{\pi}{4}(0.5)^2 = 0.1963\ \mathrm{m}^2, \qquad A_\mathrm{B} = \frac{\pi}{4}D_\mathrm{B}^2 = \frac{\pi}{4}(0.1)^2 = 0.007854\ \mathrm{m}^2$$

$$A_\mathrm{C} = \frac{\pi}{4}D_\mathrm{C}^2 = \frac{\pi}{4}(0.25)^2 = 0.04909\ \mathrm{m}^2, \qquad A_\mathrm{D} = \frac{\pi}{4}D_\mathrm{D}^2 = \frac{\pi}{4}(0.15)^2 = 0.01767\ \mathrm{m}^2$$

$$V_\mathrm{A} = \frac{Q_\mathrm{A}}{A_\mathrm{A}} = \frac{0.45}{0.1963} = 2.292\ \mathrm{m/s}, \qquad h_\mathrm{A} = \frac{p_\mathrm{A}}{\gamma} + \frac{V_\mathrm{A}^2}{2g} = \frac{350}{9.79} + \frac{2.292^2}{2(9.81)} = 36.02\ \mathrm{m}$$

(a) Applying the branched-pipe Bernoulli equation, Equation 3.112, between Pipe A and Pipe D gives

$$h_\mathrm{A} = \frac{p_\mathrm{D}}{\gamma} + \frac{V_\mathrm{D}^2}{2g} \quad\rightarrow\quad 36.02 = \frac{250}{9.79} + \frac{V_\mathrm{D}^2}{2(9.81)} \quad\rightarrow\quad V_\mathrm{D} = 14.34\ \mathrm{m/s}$$

With this result, the volume flow rates in the branch pipes can be calculated as

$$Q_\mathrm{D} = A_\mathrm{D}V_\mathrm{D} = (0.01767)(14.34) = 0.2534\ \mathrm{m^3/s} = 253\ \mathrm{L/s}$$

$$Q_\mathrm{C} = A_\mathrm{C}V_\mathrm{C} = (0.04909)(3) = 0.1473\ \mathrm{m^3/s} = 147\ \mathrm{L/s}$$

$$Q_\mathrm{B} = Q_\mathrm{A} - (Q_\mathrm{C} + Q_\mathrm{D}) = 450\ \mathrm{L/s} - (253\ \mathrm{L/s} + 147\ \mathrm{L/s}) = 50\ \mathrm{L/s}$$

(b) Based on the calculated volume flow rate in Pipe B, the corresponding velocity is $V_\mathrm{B} = Q_\mathrm{B}/A_\mathrm{B} = 6.279$ m/s. The pressures in Pipes B and C can be determined by applying branched-pipe Bernoulli equation, Equation 3.112, which gives

$$h_\mathrm{A} = \frac{p_\mathrm{B}}{\gamma} + \frac{V_\mathrm{B}^2}{2g} \quad\rightarrow\quad 36.02 = \frac{p_\mathrm{B}}{9.79} + \frac{6.279^2}{2(9.81)} \quad\rightarrow\quad p_\mathrm{B} = 333\ \mathrm{kPa}$$

$$h_\mathrm{A} = \frac{p_\mathrm{C}}{\gamma} + \frac{V_\mathrm{C}^2}{2g} \quad\rightarrow\quad 36.02 = \frac{p_\mathrm{C}}{9.79} + \frac{3^2}{2(9.81)} \quad\rightarrow\quad p_\mathrm{C} = 348\ \mathrm{kPa}$$

In summary, the volume flow rates in Pipes B, C, and D are **253 L/s**, **147 L/s**, and **50 L/s**, respectively, and the pressures in Pipes B and C are **333 kPa** and **348 kPa**, respectively. All pressures are gauge pressures.

3.5 Curved Flows and Vortices

Consider the case of flow along curved streamlines in the horizontal plane as shown in Figure 3.28, where a fluid element is located between two streamlines with radii of curvatures of r and $r + \mathrm{d}r$. If the inner-facing and outer-facing areas of the fluid element have an area $\mathrm{d}A$, then the mass of the fluid element is $\rho(\mathrm{d}A)(\mathrm{d}r)$ and the centripetal acceleration toward the center of curvature, O, is equal to V^2/r. If the pressure on the inner and outer sides of the

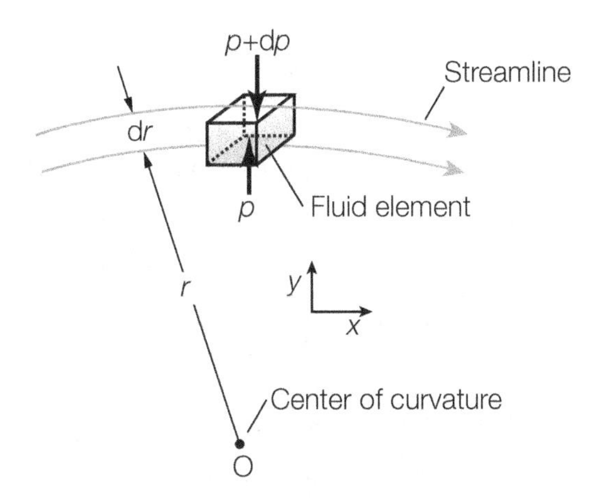

Figure 3.28: Flow along curved streamlines in the horizontal plane

fluid element are p and $p+\mathrm{d}p$ respectively, then applying Newton's law in the radial direction gives

$$(p+\mathrm{d}p)\mathrm{d}A - p\,\mathrm{d}A = \rho(\mathrm{d}A)(\mathrm{d}r)\frac{V^2}{r} \tag{3.113}$$

which simplifies to

$$\frac{\mathrm{d}p}{\mathrm{d}r} = \rho\frac{V^2}{r} \quad \text{(in horizontal plane)} \tag{3.114}$$

Equation 3.114 shows that the pressure increases with distance from the center of curvature, and the actual variation of p with distance from the center of curvature depends on how V varies with r. The relationship given by Equation 3.114 is a special case of the more general description of the pressure distribution normal to streamlines as derived in Section 3.3. The simplification here is that Equation 3.114 does not include a gravity component because it is derived for the horizontal plane, and gravity acts in the vertical direction. Several applications of Equation 3.114 are considered in the following sections.

3.5.1 Forced Vortices

In a forced vortex, the fluid rotates as a solid body, as in cases where the fluid container is rotated or the fluid is stirred. Examples of forced vortices are considered below.

3.5.1.1 Cylindrical forced vortex

In the case of a fluid contained in a cylinder rotating with an angular velocity, ω, the velocity, V, at a distance r from the center of the cylinder is given by

$$V = r\omega \tag{3.115}$$

Substituting Equation 3.115 into Equation 3.114 gives

$$\mathrm{d}p = \rho\omega^2 r\,\mathrm{d}r = \frac{\gamma}{g}\omega^2 r\,\mathrm{d}r \tag{3.116}$$

Integrating Equation 3.116 with $p = p_1$ at $r = r_1$ and $p = p_2$ at $r = r_2$ yields

$$\frac{p_2}{\gamma} - \frac{p_1}{\gamma} = \frac{\omega^2}{2g}(r_2^2 - r_1^2) \tag{3.117}$$

This equation describes the pressure variation in the radial direction in both closed and open tanks containing liquids. In cases where the tank is open, as shown in Figure 3.29, the liquid

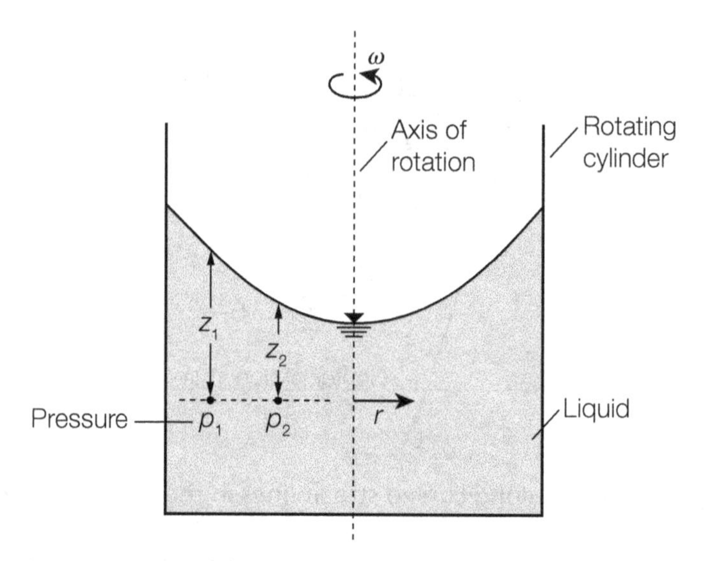

Figure 3.29: Cylindrical forced vortex in an open container

surface will adjust such that $z = p/\gamma$ and the liquid-surface profile can be derived from Equation 3.117 as

$$z_2 - z_1 = \frac{\omega^2}{2g}(r_2^2 - r_1^2) \tag{3.118}$$

Equations 3.117 and 3.118 apply to cases in which the container is rotated about a vertical axis. In the special case where the cylinder is closed and rotates about an inclined axis, the pressure varies with elevation, z, as well as radius. The pressure variation can be described by

$$\frac{p_2}{\gamma} - \frac{p_1}{\gamma} + z_2 - z_1 = \frac{\omega^2}{2g}(r_2^2 - r_1^2) \tag{3.119}$$

where (r_1, z_1) and (r_2, z_2) are the coordinates of two distinct points in the rotating fluid having pressures p_1 and p_2, respectively.

EXAMPLE 3.20

A liquid in a 15-cm-diameter cylindrical container is to be stirred such that the difference between the liquid level in the center of the container and the liquid level on the side of the container is not more than 3 cm. What is the maximum rate at which the liquid should be stirred?

SOLUTION

Assume that the liquid in the container rotates as a cylindrical forced vortex. From the given data: $r_1 = 0$ m, $r_2 = 7.5$ cm $= 0.075$ m, and $z_2 - z_1 = 3$ cm $= 0.03$ m. Substituting the given data into Equation 3.118 yields

$$z_2 - z_1 = \frac{\omega^2}{2g}(r_2^2 - r_1^2) \quad \rightarrow \quad 0.03 = \frac{\omega^2}{2(9.81)}(0.075^2 - 0^2) \quad \rightarrow \quad \omega = 10.23 \text{ rad/s} = 97.7 \text{ rpm}$$

Hence, the maximum rate at which the fluid should be stirred is **97.7 rpm.**

The fluid motion in a cylindrical forced vortex is an example of *rigid-body rotation* (also known as *solid-body rotation*) that was covered in Section 2.7. In such motions, the entire fluid rotates as a solid body. This type of flow can be generated by starting with a stationary container and then rotating the container with a constant angular velocity, ω. Under this circumstance, the viscosity of the fluid drags the fluid along with the rotating container until both are rotating as a solid body.

3.5.1.2 Spiral forced vortex

A *spiral forced vortex* occurs when an inward or outward flow is superimposed on a forced vortex, such as when a fluid is added to or removed from the center of a rotating cylindrical container. The radial pressure variation in a spiral forced vortex can be obtained (for an ideal fluid) by adding the pressure difference caused by the superimposed inward/outward flow to the pressure difference caused by the forced vortex flow. For an inward/outward flow, the pressure difference is given by

$$\frac{p_2}{\gamma} - \frac{p_1}{\gamma} = \frac{v_1^2}{2g} - \frac{v_2^2}{2g} \tag{3.120}$$

where v_1 and v_2 are the flow velocities of the superimposed (inward/outward) flows at locations where the pressures are p_1 and p_2, respectively. Combining Equations 3.117 and 3.120 yields the following equation for pressure variations in spiral forced vortices:

$$\frac{p_2}{\gamma} - \frac{p_1}{\gamma} = \frac{\omega^2}{2g}(r_2^2 - r_1^2) + \frac{v_1^2 - v_2^2}{2g} \tag{3.121}$$

A common case of a spiral forced vortex occurs in the rotor of a centrifugal pump as illustrated in Figure 3.30. In this case, flow enters the eye of the impeller and moves outward by the centrifugal effect of the rotating impeller. If there was no outward flow, pressure variations would be as described for a cylindrical forced vortex (Equation 3.117). When a liquid is supplied through the eye of the impeller at a flow rate Q, the relative flow velocities induced by Q at sections 1 and 2 are v_1 and v_2, respectively, such that

$$v_1 A_1 = v_2 A_2 \tag{3.122}$$

where A_1 and A_2 are the flow areas normal to the velocity at sections 1 and 2, respectively. It is apparent from Equation 3.121 that the pressure in the liquid leaving a centrifugal pump is higher than the pressure in the liquid entering the pump. This pressure increase is caused by the rotation of the impeller, and larger rotational speeds (ω) cause larger pressure increases. The decrease in the relative flow velocity ($v_2 < v_1$) as the fluid moves outward also contributes to the increased pressure. A similar analysis to the one presented here can be done for a turbine rotor, in which case the flow is inward through the rotor and the pressure of the

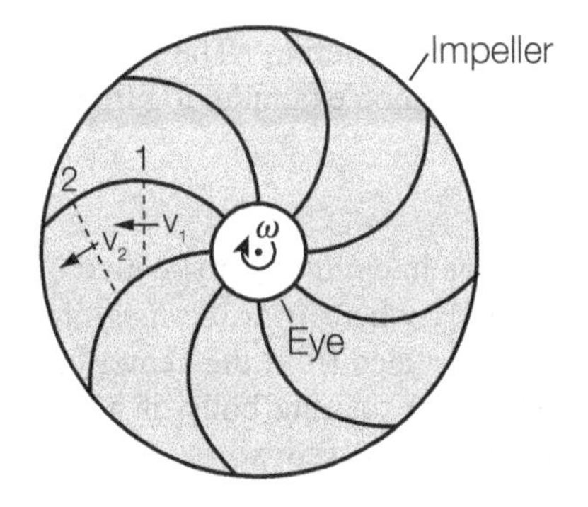

Figure 3.30: Flow through a pump impeller

outflow is less than the pressure of the inflow, with the difference depending on the angular speed of the rotor.

EXAMPLE 3.21

A small water pump has a 50-cm-diameter rotor where water enters the pump at a radial distance of 5 cm from the center of the rotor and exits the rotor at a distance of 25 cm from the center of the rotor. The width of the rotor is 20 cm, and when the rotor spins at 240 rpm, the pump delivers 100 L/s. Estimate the pressure difference between the inflow and outflow. For this preliminary analysis, neglect the thickness of the blades in the rotor and assume that the water is at 20°C.

SOLUTION

From the given data: r_1 = 0.05 m, r_2 = 0.25 m, W = 0.20 m, ω = 240 rpm = 25.14 rad/s, and Q = 100 L/s = 0.10 m^3/s. At 20°C, ρ = 998 kg/m^3. Using these data yields

$$A_1 = 2\pi r_1 W = 2\pi(0.05)(0.20) = 0.06283\ \text{m}^2, \qquad A_2 = 2\pi r_2 W = 2\pi(0.25)(0.20) = 0.3142\ \text{m}^2$$

$$v_1 = \frac{Q}{A_1} = \frac{0.1}{0.06283} = 1.592\ \text{m/s}, \qquad v_2 = \frac{Q}{A_2} = \frac{0.1}{0.3142} = 0.3183\ \text{m/s}$$

Using Equation 3.121 to calculate the pressure differential gives

$$\frac{p_2}{\gamma} - \frac{p_1}{\gamma} = \frac{\omega^2}{2g}(r_2^2 - r_1^2) + \frac{v_1^2 - v_2^2}{2g}$$

$$p_2 - p_1 = \frac{\rho\omega^2}{2}(r_2^2 - r_1^2) + \frac{\rho}{2}\left(v_1^2 - v_2^2\right)$$

$$p_2 - p_1 = \frac{(998)(25.14)^2}{2}(0.25^2 - 0.05^2) + \frac{998}{2}\left(1.592^2 - 0.3183^2\right) = 20140\ \text{Pa} = 20.14\ \text{kPa}$$

Therefore, it is estimated that the pressure of the outflow is **20.1 kPa** higher than the pressure at the inflow.

3.5.2 Free Vortices

In a *free vortex*, there is no external energy input and the fluid rotates either because of an initial rotation or because of an internal action. The rotary flow formed when a reservoir empties through a drain hole is a common example of a free vortex.

3.5.2.1 Cylindrical free vortex

In this case, the fluid elements move in circular paths with the angular momentum of a fluid element of mass, m, given by mVr, where V is the magnitude of the velocity (tangential to the circular path) and r is the distance from the center of rotation. In accordance with Newton's laws of motion applied to a rotating body, if no external force acts on the fluid element, then angular momentum is conserved; so

$$Vr = C = \text{constant} \tag{3.123}$$

Substituting Equation 3.123 into Equation 3.114 and integrating between distances r_1 and r_2 from the center of rotation yields

$$\frac{p_2}{\gamma} - \frac{p_1}{\gamma} = \frac{V_1^2}{2g}\left[1 - \left(\frac{r_1}{r_2}\right)^2\right] \tag{3.124}$$

where p_1 and p_2 are the pressures at locations r_1 and r_2, respectively, from the center of rotation, and V_1 is the magnitude of the velocity at a distance r_1 from the center of rotation. It is interesting to note that Equation 3.123 indicates that the velocity in a cylindrical free vortex increases without bound as $r \to 0$. However, in reality, high velocities induce high (viscous) dissipation of energy, which is proportional to the square of the velocity, and the core of the vortex tends to rotate as a solid body.

EXAMPLE 3.22

An atmospheric weather system circulates as a free vortex. The weather system has an outer radius of approximately 500 m, where the pressure is 101.0 kPa. Estimate the wind velocity 50 m from the center of the weather system, where the pressure is measured as 99.5 kPa. Assume that the system consists of air at 20°C.

SOLUTION

From the given data: r_1 = 50 m, r_2 = 500 m, p_1 = 99.5 kPa, and p_2 = 101.0 kPa. At 20°C, ρ = 1.204 kg/m^3. The pressure distribution is given by Equation 3.124, which can be put in the form

$$p_2 - p_1 = \frac{\rho V_1^2}{2}\left[1 - \left(\frac{r_1}{r_2}\right)^2\right]$$

$$(101.0 - 99.5)10^3 = \frac{(1.204)V_1^2}{2}\left[1 - \left(\frac{50}{500}\right)^2\right]$$

$$V_1 = 50.2 \text{ m/s}$$

Therefore, the estimated wind velocity is **50.2 m/s** at 50 m from the center of the system.

3.5.2.2 Spiral free vortex

A *spiral free vortex* results when a radial flow is superimposed on a cylindrical free vortex. The radial flow component can be induced by fluid being removed from the core of the vortex (e.g., what happens in the eye of a hurricane). If the flow is constrained to a height H and the radial volume flow rate is Q, then

$$Q = 2\pi r H V_\text{r} = \text{constant} \quad \to \quad rV_\text{r} = C_1 \tag{3.125}$$

where V_r is the radial component of the velocity and C_1 is a constant. Because a spiral free vortex is formed by the superposition of a radial flow with a free vortex (which has only a tangential velocity), the tangential component of the velocity in a spiral free vortex, V_t, is equal to the velocity of a free vortex, which can be expressed as

$$rV_\text{t} = C_2 \tag{3.126}$$

where C_2 is a constant. By combining Equations 3.125 and 3.126, the magnitude of the velocity, V, in a spiral free vortex can be expressed as

$$V = \sqrt{V_\text{r}^2 + V_\text{t}^2} = \frac{1}{r}\sqrt{C_1^2 + C_2^2} = \frac{C_3}{r} \quad \rightarrow \quad rV = C_3 \tag{3.127}$$

where C_3 is a constant. Comparing Equations 3.127 and 3.123 shows that the velocity variation with r is the same in both circular and radial flow. So Equation 3.124 is applicable for the combined flow, where V_1 now represents the magnitude of the combined radial and circumferential velocity components. Using this result, in the case of a spiral free vortex, the pressure variation can be expressed as

$$\frac{p_2}{\gamma} - \frac{p_1}{\gamma} = \frac{V_1^2}{2g} - \frac{V_2^2}{2g} \tag{3.128}$$

where V_1 and V_2 are the magnitudes of the combined velocities at locations 1 and 2, respectively. Equation 3.128 can be derived from Equation 3.124 by noting that $r_1^2V_1^2 = r_2^2V_2^2$, where V_1 and V_2 are the combined tangential and radial velocity components, and that $V^2 = V_\text{t}^2 + V_\text{r}^2$, where V_t and V_r are the tangential and radial components of the velocity, respectively, and V is the magnitude of the resultant.

EXAMPLE 3.23

The pressure and wind speed at the eye wall of a hurricane are measured as 97 kPa and 82 m/s, respectively. If the hurricane can be approximated by a spiral free vortex, estimate the wind speed where the pressure becomes equal to atmospheric pressure at 101 kPa. Assume air at 20°C.

SOLUTION

From the given data: p_1 = 97 kPa, V_1 = 82 m/s, and p_2 = 101 kPa. For air at 20°C, ρ = 1.204 kg/m^3. The relationship between pressure and velocity can be approximated by Equation 3.128, which can be expressed as

$$p_2 - p_1 = \frac{1}{2}\rho\left(V_1^2 - V_2^2\right) \quad \rightarrow \quad (101-97)10^3 = \frac{1}{2}(1.204)\left(82^2 - V_2^2\right) \quad \rightarrow \quad V_2 = 8.91 \text{ m/s}$$

Therefore, the estimated wind speed where the pressure becomes atmospheric is **8.91 m/s**.

Key Equations in Kinematics and Streamline Dynamics

The following list of equations are useful in solving problems related to the analysis of fluid flows involving kinematics and streamline dynamics. If one is able to recognize these equations and recall their appropriate use, then the learning objectives of this chapter have been met to a significant degree. Derivations of these equations, definitions of the variables, and detailed examples of usage can be found in the main text.

KINEMATICS

Position vector: $\mathbf{r} = x\mathbf{i} + y\mathbf{j} + z\mathbf{k}$

Velocity: $\mathbf{v} = \dfrac{d\mathbf{r}}{dt}$

Acceleration: $\mathbf{a} = \dfrac{d\mathbf{v}}{dt} = \dfrac{d^2\mathbf{r}}{dt}; \quad \mathbf{a} = \dfrac{\partial \mathbf{v}}{\partial t} + (\mathbf{v} \cdot \nabla)\mathbf{v}$

Pathline:

$$\mathbf{r} = \mathbf{r}_{\text{start}} + \int_{t_{\text{start}}}^{t} \mathbf{v}\,dt$$

$$\frac{dx}{dt} = v_x, \quad \frac{dy}{dt} = v_y, \quad \frac{dz}{dt} = v_z$$

Streamline: $\dfrac{dx}{v_x} = \dfrac{dy}{v_y} = \dfrac{dz}{v_z}$

Streakline: $\mathbf{r} = \mathbf{r}_{\text{inject}} + \displaystyle\int_{t_{\text{inject}}}^{t} \mathbf{v}\,dt$

Material derivative: $\dfrac{D(\cdot)}{Dt} = \dfrac{\partial(\cdot)}{\partial t} + \dfrac{\partial(\cdot)}{\partial x}v_x + \dfrac{\partial(\cdot)}{\partial y}v_y + \dfrac{\partial(\cdot)}{\partial z}v_z$

Volume flow rate: $Q = \displaystyle\int_A \mathbf{v} \cdot \mathbf{n}\,dA = \int_A v_n\,dA$

Average velocity: $\overline{V}_n = \dfrac{Q}{A} = \dfrac{1}{A}\displaystyle\int_A v_n\,dA$

Mass flow rate: $\dot{m} = \rho \displaystyle\int_A v_n\,dA = \rho A \overline{V}_n = \rho Q$

DYNAMICS OF FLOW ALONG A STREAMLINE

Bernoulli equation (BE), inviscid fluid: $\displaystyle\int \frac{dp}{\gamma} + \frac{V^2}{2g} + z = C; \quad \int \frac{dp}{\rho} + \frac{V^2}{2} + gz = C$

BE, incompressible: $\dfrac{p}{\gamma} + \dfrac{V^2}{2g} + z = C \quad \text{or} \quad p + \tfrac{1}{2}\rho V^2 + \gamma z = C$

BE, incompressible, stagnation, $z_1 = z_2$: $\dfrac{p_2 - p_1}{p_1} = \dfrac{k\,\mathrm{Ma}_1^2}{2}$

BE, isentropic: $\left(\dfrac{k}{k-1}\right)\dfrac{p_1}{\rho_1} + \dfrac{1}{2}V_1^2 + gz_1 = \left(\dfrac{k}{k-1}\right)\dfrac{p_2}{\rho_2} + \dfrac{1}{2}V_2^2 + gz_2$

BE, isentropic, stagnation: $\dfrac{p_2 - p_1}{p_1} = \left[\left(1 + \dfrac{k-1}{2}\mathrm{Ma}_1^2\right)^{\frac{k}{k-1}} - 1\right]$

BE, with viscous effect: $\left(\dfrac{p_2}{\gamma} + z_2 + \dfrac{V_2^2}{2g}\right) - \left(\dfrac{p_1}{\gamma} + z_1 + \dfrac{V_1^2}{2g}\right) = -\dfrac{\tau PL}{\gamma A}$

Mechanical energy per unit mass: $e_{\mathrm{mech}} = \dfrac{p}{\rho} + \dfrac{V^2}{2} + gz$

Stagnation pressure (compressible): $p_{\mathrm{s}} = p_\infty + \frac{1}{2}\rho V_\infty^2\left(1 + \dfrac{V_\infty^2}{4c^2} + \dots\right)$

Stagnation pressure (incompressible): $p_{\mathrm{s}} = p_\infty + \frac{1}{2}\,\rho V_\infty^2$

Approximation (for gases): $p + \frac{1}{2}\,\rho V^2 = \text{constant}$

Normal to streamline (compressible):

$$\int \frac{\mathrm{d}p}{\rho} + \int \frac{V^2}{R}\,\mathrm{d}n + gz = \text{constant}$$

$$\int_1^2 \mathrm{d}p + \int_1^2 \gamma\,\mathrm{d}z = -\int_1^2 \rho\frac{V^2}{R}\,\mathrm{d}n$$

Normal to streamline (incompressible):

$$p + \rho\int \frac{V^2}{R}\,\mathrm{d}n + \gamma z = \text{constant}$$

$$\left(\frac{p_2}{\gamma} + z_2\right) - \left(\frac{p_1}{\gamma} + z_1\right) = -\frac{1}{g}\int_1^2 \frac{V^2}{R}\,\mathrm{d}n$$

Normal to streamline (approximation): $\dfrac{\Delta p}{\Delta n} = -\rho\dfrac{V^2}{R}$

APPLICATIONS OF THE BERNOULLI EQUATION

Orifice: $Q = C_{\mathrm{d}}A_{\mathrm{o}}\sqrt{2g\left(\dfrac{p_{\mathrm{s}} - p_{\mathrm{o}}}{\gamma} + z_{\mathrm{s}} - z_{\mathrm{o}}\right)}$

Orifice (open to atmosphere): $Q = C_{\mathrm{d}}A_{\mathrm{o}}\sqrt{2g\,(z_{\mathrm{s}} - z_{\mathrm{o}})}$

Orifice (submerged):

$$Q = C_{\mathrm{d}}A_{\mathrm{o}}\sqrt{2g\left(\frac{p_{\mathrm{s1}} - p_{\mathrm{s2}}}{\gamma} + z_{\mathrm{s1}} - z_{\mathrm{s2}}\right)}$$

$$Q = C_{\mathrm{d}}A_{\mathrm{o}}\sqrt{2g\,(\phi_{\mathrm{s1}} - \phi_{\mathrm{s2}})}$$

Pitot-static tube: $V_\infty = \sqrt{\dfrac{2(p_{\mathrm{s}} - p_\infty)}{\rho}}$

Venturi meter: $$Q = C_d A_2 \sqrt{\frac{2[(p_1 - p_2) + \gamma(z_1 - z_2)]}{\rho\left[1 - \left(\frac{A_2}{A_1}\right)^2\right]}}$$

Jet trajectory: $$\frac{V^2}{2g} + z = \frac{V_0^2}{2g}; \quad z = \frac{V_{z0}}{V_{x0}}\left(x - \frac{g}{2V_{x0}V_{z0}}x^2\right)$$

Rectilinear viscous flow: $$\left(\frac{p_1}{\gamma} + z_1\right) - \left(\frac{p_2}{\gamma} + z_2\right) = h_f$$

Branching conduits: $$\frac{p_1}{\gamma} + \frac{V_1^2}{2g} + z_1 = \frac{p_n}{\gamma} + \frac{V_n^2}{2g} + z_n$$

CURVED FLOWS AND VORTICES

General equation: $$\frac{dp}{dr} = \rho\frac{V^2}{r}$$

Cylindrical forced vortex: $$\frac{p_2}{\gamma} - \frac{p_1}{\gamma} = \frac{\omega^2}{2g}(r_2^2 - r_1^2); \qquad z_2 - z_1 = \frac{\omega^2}{2g}(r_2^2 - r_1^2)$$

Spiral forced vortex: $$\frac{p_2}{\gamma} - \frac{p_1}{\gamma} = \frac{\omega^2}{2g}(r_2^2 - r_1^2) + \frac{v_2^2 - v_1^2}{2g}$$

Cylindrical free vortex: $$\frac{p_2}{\gamma} - \frac{p_1}{\gamma} = \frac{V_1^2}{2g}\left[1 - \left(\frac{r_1}{r_2}\right)^2\right]$$

Spiral free vortex: $$\frac{p_2}{\gamma} - \frac{p_1}{\gamma} = \frac{V_1^2}{2g} - \frac{V_2^2}{2g}$$

Applications of the above equations are provided in the text, and additional problems to practice using these equations can be found in the following section.

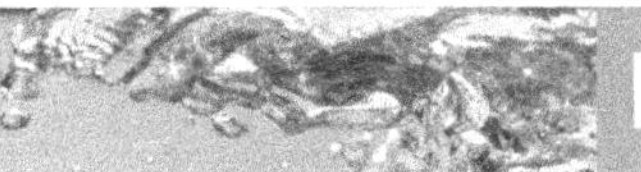

PROBLEMS

Section 3.2: Kinematics

3.1. A fluid flows radially toward a sink as shown in Figure 3.31. The velocity at any point P in the flow field is given by

$$\mathbf{v} = -\frac{1}{r}(\cos\theta\mathbf{i} + \sin\theta\mathbf{j}) \text{ m/s}$$

where r is the distance from the sink in meters and θ is the radial angle. Determine the velocity and acceleration at the location where $(x,y) = (2,1)$.

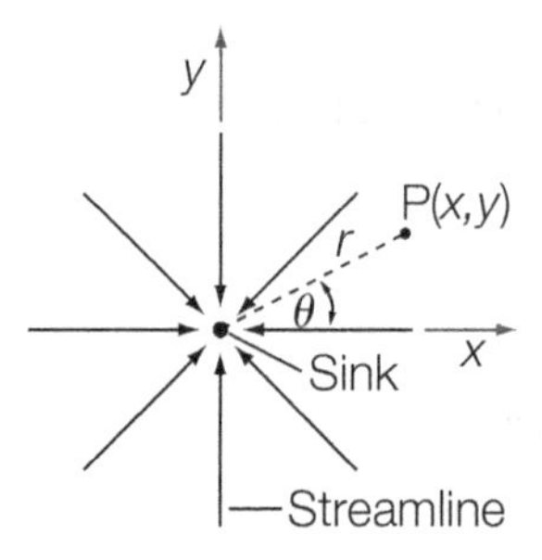

Figure 3.31: Radial flow toward a sink

3.2. The velocity field in a two-dimensional flow is given by

$$\mathbf{v} = (2 + 1.5x + 2.1y)\mathbf{i} + (1.8 - 3x + 4y)\mathbf{j}$$

(a) Calculate the acceleration field and (b) identify any stagnation points in the flow field.

3.3. In a uniform velocity field, the velocity does not change spatially. Consider the uniform unsteady velocity field, **v**, given by

$$\mathbf{v} = a\cos\omega t\mathbf{i} + b\sin\omega t\mathbf{j} \text{ m/s}$$

where a and b are the amplitudes of the velocity fluctuations and ω is the frequency of the fluctuations. In a particular case, a = 0.15 m/s, b = 0.2 m/s, $\omega = 2\pi/T$, and T = 30 h. Determine the acceleration as a function of time at all points within the velocity field. At what times, if any, are the acceleration equal to zero?

3.4. A steady, incompressible, two-dimensional velocity field is given by the following components in the xy plane:

$$v_x = 2 + 1.5x + 0.75y, \qquad v_y = 1 + 3x + 1y$$

What is the acceleration of the fluid at (x, y) = (2,4)?

3.5. A three-dimensional velocity field is given by

$$\mathbf{v} = (3z + 2y + 2)\mathbf{i} + (2x + z + 1)\mathbf{j} + (3x + y - 1)\mathbf{k}$$

Determine the following: (a) the magnitude of the velocity at the origin, (b) the acceleration field, (c) the location of the stagnation point, and (d) the location where the acceleration is equal to zero.

3.6. The radial coordinate system uses r and θ to identify the location of a point (P) in two-dimensional space as shown in Figure 3.32. When a radial coordinate system is used, the velocity field is commonly expressed in terms of the radial component, v_r, and the

θ component, v_θ, where v_r is in the direction of increasing r and v_θ is normal to v_r and in the direction of increasing θ. Consider the case where the velocity field is given by

$$v_r = 4r\cos\theta, \qquad v_\theta = -4r\sin 2\theta$$

Express this velocity field in terms of Cartesian coordinates, with the velocity vector expressed in terms of v_x and v_y. Are there any stagnation points in Quadrant I of the Cartesian domain, where $x, y > 0$?

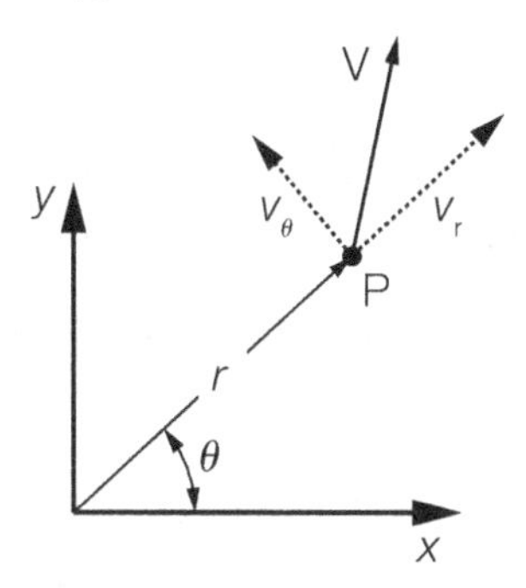

Figure 3.32: Velocity field in radial coordinates

3.7. A two-dimensional velocity field is given by

$$\mathbf{v} = 5x\mathbf{i} + 2y\mathbf{j} \text{ m/s}$$

where x and y are in meters. Determine the equation of the pathline that originates at the point (1,2). For what range of times since release does the released particle remain within the domain of $0 \leq x \leq 100$ m, $0 \leq y \leq 100$ m. Plot the pathline within this domain.

3.8. A fluid moves through a converging duct of width 3 m and length 5 m as shown in Figure 3.33. The velocity field, **v**, in the duct is given by

$$\mathbf{v} = (5 + 2x)\mathbf{i} - 2y\mathbf{j} \text{ m/s}$$

where x [m] is the coordinate measured along the centerline of the duct and y [m] is the coordinate measured normal to the centerline. (a) Determine the equation of the pathline of a fluid particle that originates at the center of the duct with coordinates $x = 0$ m, $y = 0$ m. (b) How long does it take the particle described in part (a) to exit the duct? (c) Determine the equation of the pathline followed by a fluid particle originating at the top of the duct, $x = 0$ m, $y = 2$ m, and compare the time it takes this particle to exit the duct with the time it takes the particle described in part (a) to exit the duct.

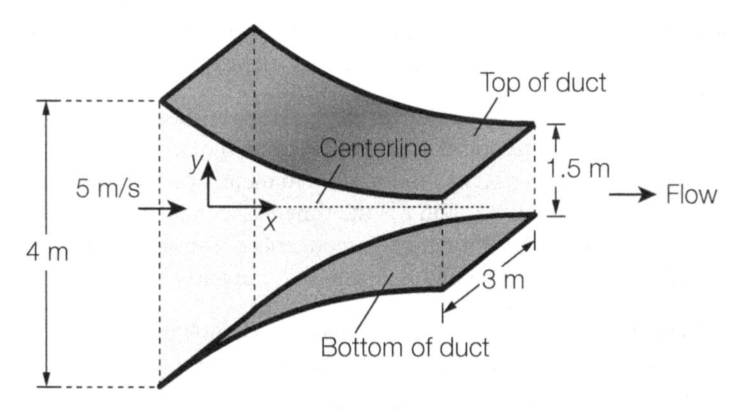

Figure 3.33: Flow through a duct

3.9. A two-dimensional velocity field within the domain $0 \leq x \leq 10$ m and $0 \leq y \leq 10$ m is given by

$$\mathbf{v} = 5x\mathbf{i} + 3xy\mathbf{j} \text{ cm/s}$$

where x and y are the Cartesian coordinates in meters. Find the equation of the streamline that passes through the point (1 m, 1 m). Plot the streamline.

3.10. A steady-state velocity field in Quadrant I of the Cartesian plane is given by

$$\mathbf{v} = (1 - 2y)\mathbf{i} + 2\mathbf{j} \text{ m/s}$$

where x and y are in meters and $x, y \geq 0$. Determine the equation of the streamlines and plot several streamlines for $0 \leq x \leq 10$ and $0 \leq x \leq 10$.

3.11. A velocity field, $\mathbf{v}$, is spatially uniform and varies with time according to the following relation:

$$\mathbf{v} = \begin{cases} 3\mathbf{i} + \mathbf{j} \text{ m/s}, & t \in [0 \text{ s}, 8 \text{ s}] \\ 5\mathbf{i} - 4\mathbf{j} \text{ m/s}, & t \in (8 \text{ s}, 15 \text{ s}] \end{cases}$$

If an injection point is located at the origin of a Cartesian coordinate system at (0 m, 0 m), sketch to scale the following at $t = 15$ s along with their key coordinates: (a) the pathline of a particle released at the injection point at $t = 0$ s, (b) the streakline of dye continuously released at the injection point starting at $t = 0$ s, and (c) the streamlines in the flow field at $t = 15$ s.

3.12. An unsteady-state two-dimensional velocity field in the xy plane is given by

$$\mathbf{v} = a \sin \omega t\mathbf{i} + b \cos \omega t\mathbf{j} \text{ cm/s}$$

where t is the time is hours, a and b are the amplitudes of the temporal velocity fluctuations, $\omega = 2\pi/T$ is the frequency of the velocity fluctuations, and T is the period of the fluctuations. Consider a particular case where a = 15 cm/s, b = 20 cm/s, and T = 12 h. Estimate the streakline for particles released continuously from (x, y) = (0 m, 0 m) starting at $t = 0$ h and ending at $t = 12$ h. Use 1-h time increments in estimating the streakline.

3.13. A circular streamline has a velocity of 2.5 m/s and a radius of 0.8 m. If the flow is steady, what are the normal and tangential components of the acceleration of a fluid particle located on the streamline?

3.14. A fluid particle follows a circular streamline that has a radius of 2 m. The magnitude of the velocity, V, of a fluid particle on the streamline varies with time according to the relation $V = (0.8 + 2.5t)$ m/s, where t is the time in seconds. What are the normal and tangential components of the acceleration of a fluid particle located on the streamline at $t = 3$ s?

3.15. The velocity along a circular pathline is given by the relation $V = s^5t^4$, where V is the velocity in the direction of fluid motion in meters per second, s is the coordinate along the pathline in meters, and t is the time is seconds. The radius of curvature of the pathline is 0.8 m. Determine the components of the acceleration in the directions tangential and normal to the pathline at $s = 2.5$ m and $t = 1.5$ s.

3.16. A liquid flows over a semicircular projection as shown in Figure 3.34, where the speed, V, of the liquid stream at any vertical distance, h, below the liquid surface is given by

$$V = \sqrt{2gh}$$

Consider the case where the liquid surface is held at 2 m above the projection and the radius of the semicircular projection is 3 m. (a) Plot the component of the acceleration normal to the projection as a function of θ. (b) How far below the liquid surface is the normal acceleration equal to 20 m/s^2? (c) Plot the streamwise component of the acceleration as a function of θ. (d) How far below the liquid surface is the streamwise acceleration equal to g?

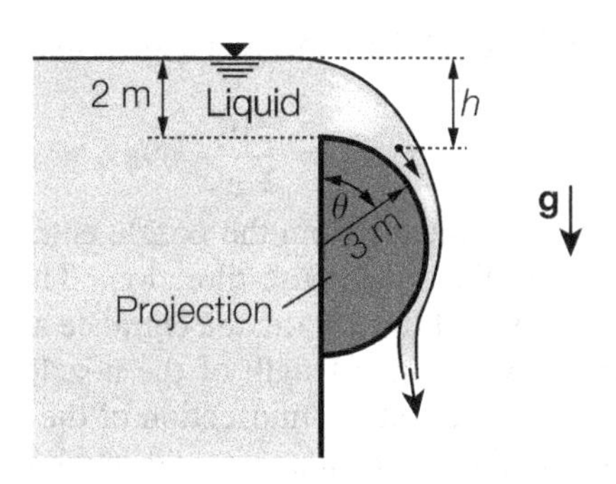

Figure 3.34: Flow over a projection

3.17. The velocity field as given in Problem 3.1 describes a case of radial flow toward a sink, where

$$\mathbf{v} = -\frac{1}{r}(\cos\theta\mathbf{i} + \sin\theta\mathbf{j}) \text{ m/s}$$

Determine the magnitudes of the streamwise and normal accelerations along any given streamline.

3.18. The velocity field downstream of a river discharge into an ocean is given by

$$v(r) = \frac{30}{r}\sin\left(\frac{\pi t}{6}\right)$$

where v is the velocity in cm/s, r is the distance from the outlet in meters, and t is the time in hours. If a person floats with the current starting at r = 1 m, estimate the time it will take to get 50 m from the starting point. Comment on whether you think the velocity field is realistic.

3.19. The velocity, V, and pressure, p, along a straight prismatic conduit are given by

$$V = 2.5 + 0.05x \text{ m/s}, \qquad \text{and} \qquad p = 6x^2 \text{ Pa}$$

where x is the distance along the conduit in meters. (a) Determine the rate of change of pressure with time experienced by a fluid particle as it passes the section where $x = 2$ m. (b) What is the rate of change of pressure with time that would be derived from measurements at the section where $x = 2$ m.

3.20. The temperature, T, in an 11-m-wide × 11-m-long × 4-m-high room varies according to the relation

$$T = 20 + 3x + 4y - 2z \text{ °C}$$

where x [m] and y [m] are the coordinates in the horizontal plane and z [m] is the vertical coordinate measured from the floor upward. If a 1.65-m-tall person walks through the room in the x- direction at 2 m/s, what is the rate of change of temperature at the top of the person's head?

3.21. The temperature, T, and the vertical component, w, of the wind velocity on the side of a very steep cliff are approximated by the relations

$$T(z,t) = 20(1 - 0.3z^2)\sin\left(\frac{\pi t}{6}\right)\ ^\circ\text{C}, \qquad w = 2.1(1 + 0.5z^2)\ \text{m/s}$$

where z is the elevation above sea level in km and t is the time in seconds. The horizontal components of the wind velocity are negligible along the cliff. Estimate the rate of change of temperature in the wind at $z = 1.2$ km and $t = 5400$ s (1.5 h).

3.22. The nozzle shown in Figure 3.35 is designed such that the average velocity, V, at any cross section is given by

$$V = \frac{10}{1 - x}\ \text{m/s}$$

where x is the distance in meters from the nozzle entrance to any cross section of the nozzle, measured in the streamwise direction. The velocity is approximately uniform across any section of the nozzle; the entrance and exit velocities are 10 m/s and 20 m/s, respectively; and the length of the nozzle is 0.5 m. (a) What is the acceleration of a fluid particle at the midsection of the nozzle, where $x = 0.25$ m? (b) How long does a fluid particle take to pass through the nozzle?

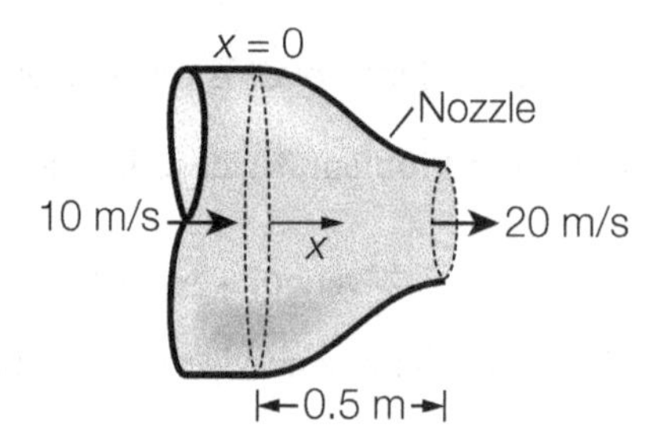

Figure 3.35: Flow through a nozzle

3.23. A reducer changes the diameter of a pipe linearly from D_1 to D_2 over a length L as shown in Figure 3.36. If the volume rate of fluid flow in the pipe is Q, determine a general expression for the average velocity and acceleration of the fluid as a function of the distance x from the beginning of the reducer.

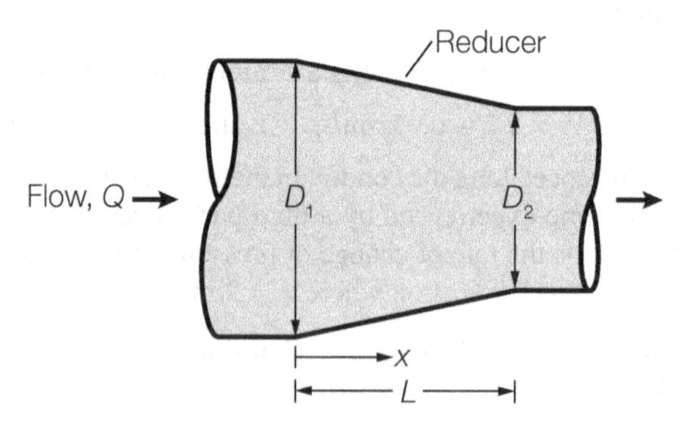

Figure 3.36: Linear reducer

3.24. A gas with a specific weight of 7.75 N/m^3 flows at a rate of 0.500 kg/s through a 350 mm × 510 mm duct. Estimate the average velocity in the duct.

3.25. A liquid with a specific gravity of 1.6 flows in a 210-mm-diameter pipe at an average velocity of 1.8 m/s. (a) What is the volume flow rate in L/min? (b) What is the mass flow rate in kg/s?

3.26. A fan draws air into a duct at a rate of 2.5 m^3/s from a room in which the temperature is 26°C and the pressure is 100 kPa. The diameter of the intake duct is 310 mm. Estimate the average velocity at which the air enters the duct and the mass flow rate into the duct.

3.27. A 300-mm-diameter pipe conveys carbon dioxide at a temperature of 5°C and a pressure of 200 kPa absolute. If the average velocity of the gas in the pipe is 12 m/s, what is the mass flow rate?

3.28. A single-family residence has an estimated indoor area of 350 m^2 and an average floor-to-ceiling height of 3 m. The local building code requires that residential houses be ventilated at a minimum air change rate of 0.3 h^{-1}, which means that 30% of the interior volume must be ventilated every hour. (a) If ventilation is to be done by an air conditioning (AC) system, determine the minimum capacity of the AC system in units of m^3/sec. (b) If the air intake is to be a single square vent and the maximum intake velocity to the vent is 4.5 m/s, what are the minimum dimensions of the vent?

3.29. A liquid drains at the rate of 30 L/min from a 27-mm-diameter opening in the bottom of a large tank as shown in Figure 3.37. Assume that the flow velocity toward the opening in the radial direction. (a) At what radial distances from the opening is the average approach velocity equal to 10%, 1%, and 0.1% of the exit velocity? (b) Estimate the convective acceleration of a fluid particle that is located at radial distance of 110 mm from the opening.

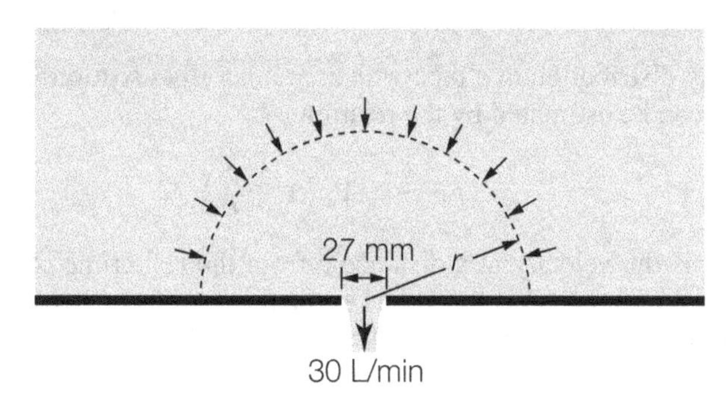

Figure 3.37: Radial flow toward a circular outlet

3.30. A thin layer of SAE 10 oil at 20°C flows in a rectangular channel of width 2 m. Under a particular flow condition, the depth of oil in the channel is 40 mm and the velocity distribution is given by $v = 110z$ m/s, where z is the distance from the bottom of the channel in meters. What is the volume flow rate of oil in the channel? What is the mass flow rate?

3.31. SAE 30 lubricating oil at 30°C is contained in the 30-mm space between two moving parallel flat surfaces as shown in Figure 3.38. The top surface is moving at a velocity of 3.5 m/s, and the bottom surface is moving at 1.5 m/s in the direction opposite that in which the top surface is moving. The velocity distribution is linear between the two surfaces, and the flat surfaces are approximately rectangular with a width of 0.75 m. Determine the volume flow rate and the mass flow rate of oil between the two surfaces.

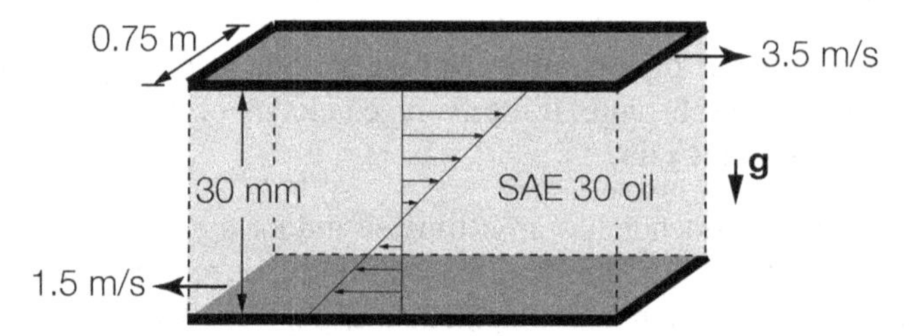

Figure 3.38: Oil between moving flat surfaces

3.32. The velocity distribution in a pipe with a circular cross section under laminar flow conditions can be approximated by the equation

$$v(r) = V_0 \left[1 - \left(\frac{r}{R}\right)^2\right]$$

where $v(r)$ is the velocity at a distance r from the centerline of the pipe, V_0 is the centerline velocity, and R is the radius of the pipe. (a) Calculate the average velocity and volume flow rate in the pipe in terms of V_0. (b) Based on the result in part (a), assess the extent to which the velocity can be assumed to be constant across the cross section.

3.33. The velocity distribution in a pipe with a circular cross section under turbulent flow conditions can be estimated by the relation

$$v(r) = V_0 \left(1 - \frac{r}{R}\right)^{\frac{1}{7}}$$

where $v(r)$ is the velocity at a distance r from the centerline of the pipe, V_0 is the centerline velocity, and R is the radius of the pipe. (a) Calculate the average velocity and the volume flow rate in the pipe in terms of V_0. Express your answers in rational form. (b) Based on the result in part (a), assess the extent to which the velocity can be assumed to be constant across the cross section.

3.34. The velocity distribution for laminar flow between two infinite parallel plates (called *Poiseuille flow*) is given by

$$u(y) = \frac{1}{2\mu}\frac{\mathrm{d}p}{\mathrm{d}x}(y^2 - hy)$$

where $u(y)$ is the velocity at a distance y from the bottom plate, h is the vertical distance between the top and bottom plates, μ is the dynamic viscosity, and $\mathrm{d}p/\mathrm{d}x$ is the constant pressure gradient driving the flow, where $\mathrm{d}p/\mathrm{d}x$ is negative. The density of the fluid is ρ. For a unit width perpendicular to the flow, determine (a) the volume flow rate, (b) the average flow velocity, and (c) the mass flow rate.

3.35. An empirical equation for the velocity distribution in an open channel is given by

$$u = u_{\text{max}} \left(\frac{y}{d}\right)^{\frac{1}{7}}$$

where u is the velocity at a distance y above the bottom of the channel, u_{max} is the maximum velocity, and d is the depth of flow. If $d = 2$ m and $u_{\text{max}} = 4$ m/s, what is the volume flow rate (i.e., discharge) in m^3/s per meter of width of channel? What is the average velocity in the channel?

Section 3.3: Dynamics of Flow along a Streamline

3.36. Consider the case in which an ideal fluid flows through a horizontal conduit. (a) Determine the acceleration of the fluid as a function of the pressure gradient and the density of the fluid. (b) If the fluid flowing in the conduit is water at 25°C and the pressure decreases at a rate of 1.5 kPa/m in the flow direction, at what rate is the fluid accelerating? (c) What pressure gradient is required to accelerate water at a rate of 6 m/s^2?

3.37. Water at 25°C flows into a conduit that is inclined upward at an angle of 35° to the horizontal. Near the entrance to the conduit, the flow along the centerline of the conduit can be assumed to be frictionless. If the fluid on the centerline of the conduit is accelerating at a rate of 5 m/s^2, what is the pressure gradient along the centerline of the conduit?

3.38. Consider the case in which an ideal fluid flows through a vertical conduit. (a) Determine the relationship between the pressure gradient and the acceleration and density of the fluid if the flow is upward. (b) Determine the relationship between the pressure gradient and the acceleration and density of the fluid if the flow is downward. (c) What pressure gradient is required to accelerate benzene vertically upward at a rate of 8 m/s^2? Assume benzene at 20°C.

3.39. Water at 25°C flows through a 1.5-m section of a converging conduit in which the velocity, V (in m/s), in the conduit increases linearly with distance, x (in m), from the entrance of the section according to the relation

$$V = 1.5(1 + 0.5x) \text{ m/s}$$

The flow can be approximated as being inviscid. (a) Determine the pressure gradient as a function of x and determine the pressure difference across the 1.5-m section of conduit by integrating the pressure gradient. (b) Determine the pressure difference across the conduit using the Bernoulli equation and compare your result with that obtained in part (a).

3.40. Water at 25°C enters a nozzle at a pressure of 560 kPa and a velocity of 2.1 m/s as shown in Figure 3.39. The outflow from the nozzle is into the open atmosphere where the pressure is 101 kPa. (a) What is the maximum velocity achievable at the exit of the nozzle? Why can't the velocity be any higher than the calculated value? (b) Derive an expression that relates the maximum outflow velocity to the pressure change across the nozzle in cases where the inflow velocity is negligible. Apply this relationship using the inflow and outflow pressures given in part (a) and assess the relative importance of the inflow velocity on the maximum outflow velocity. Assume a uniform velocity distribution at each cross section.

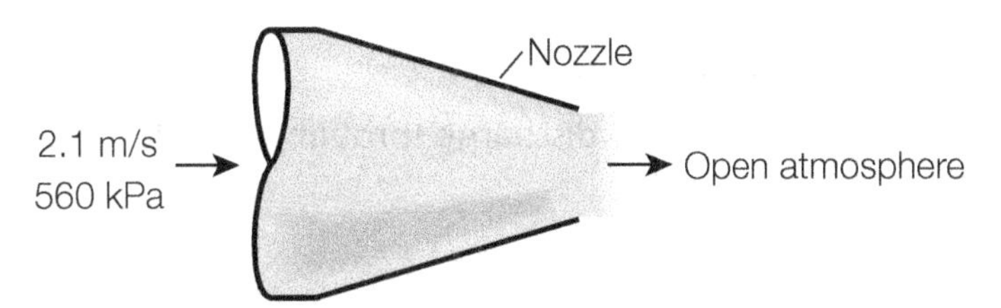

Figure 3.39: Flow through a nozzle

3.41. A stagnation streamline approaches a solid surface as shown in Figure 3.40, where the velocity, V, along the stagnation streamline is related to the distance from the solid surface, x, by

$$V = 5\left[1 - \frac{8}{(2+x)^3}\right]$$

where V is in m/s and x is in m. In a particular case, the fluid is air with a density of 1.2 kg/m^3 and the static pressure far upstream of the solid surface is 101 kPa. (a) What is the air pressure as a function of x along the stagnation streamline? (b) What is the difference between the stagnation pressure and the free-stream pressure. Assume that the elevation, z, remains constant along the stagnation streamline.

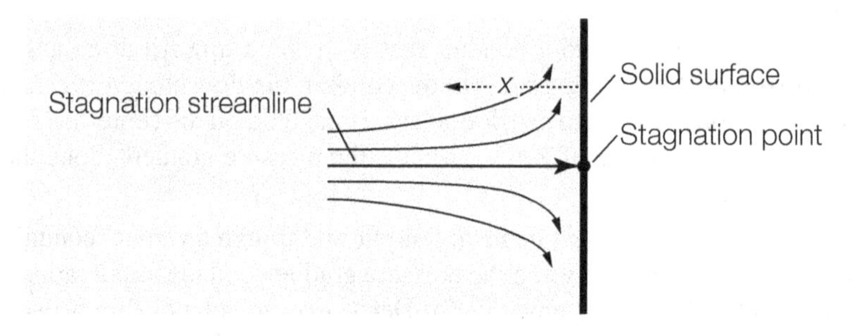

Figure 3.40: Flow toward a solid surface

3.42. Under severe conditions, the wind speed at a particular location is 35 m/s and the pressure is 101 kPa. At a downwind location, the wind flows between buildings, thereby increasing its velocity. Assuming that the density of air is constant as long as the pressure does not vary by more than 10%, determine the maximum between-building velocity for which the incompressible Bernoulli equation can be applied. Would you expect such velocities between buildings under hurricane conditions? Explain.

3.43. Compressed air discharges from a storage tank into the atmosphere through a nozzle as shown in Figure 3.41. The gauge pressure in the storage tank is 5 kPa, the temperature in the storage tank is 15°C, and atmospheric pressure is 101.3 kPa. The nozzle exit diameter is 20 mm, and an intermediate section of the nozzle has a diameter of 40 mm. (a) Estimate the volume flow rate at which air is being discharged. (b) Estimate the gauge pressure at the intermediate section of the nozzle. Assume that the flow is incompressible.

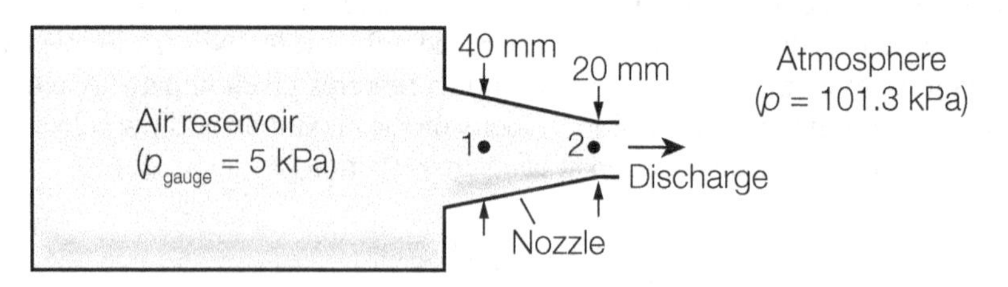

Figure 3.41: Air discharge through a nozzle

3.44. A pump is used to take water out of a large lake and deliver it to a farm to be used for irrigation. The pump location is illustrated in Figure 3.42, where the suction side of the pump (also called the *pump intake*) is 3.5 m above the water surface in the lake and the vertical intake pipe has a diameter of 130 mm. Assuming that viscous effects are negligible, what is the pressure on the suction side of the pump when the flow rate through the system is 120 L/s?

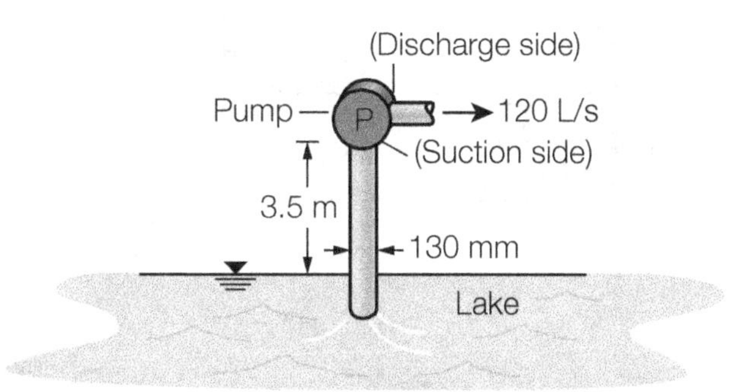

Figure 3.42: Pump withdrawal from a lake

3.45. A siphon tube is sometimes used to drain a liquid from a reservoir as shown in Figure 3.43. The driving force is gravity. (a) Determine the relationship between the volume flow rate, Q, the elevation difference, Δz_e, and the cross-sectional area, A, of the tube. (b) If the maximum allowable height of the tube above the exit of the tube, Δz_{max}, occurs when cavitation of water in the tube is incipient, determine Δz_{max}.

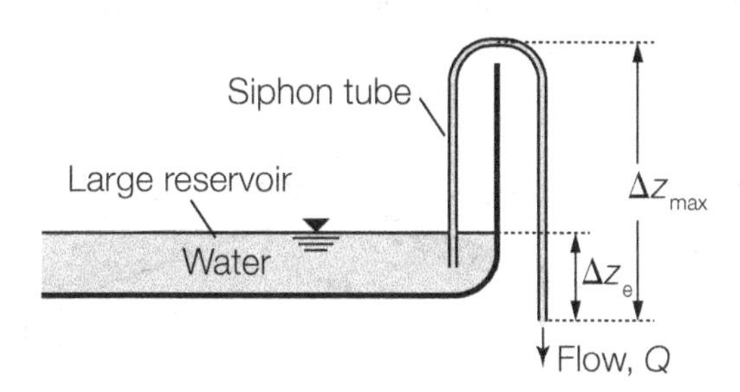

Figure 3.43: Siphon tube

3.46. The siphon tube shown in Figure 3.43 is made of a plastic material that will break if the pressure outside the tube is more than 50 kPa higher than the pressure inside the tube. Consider the case in which the liquid is water and the diameter of the tube is 25 mm. (a) Determine the maximum height, Δz_{max}, that can be used to prevent the siphon tube from breaking. (b) If the outlet of the siphon tube is fixed at an elevation 2 m below the surface elevation of the reservoir, provide an expression for the flow rate through the siphon tube as a function of Δz_{max} and determine the flow rate when the tube is at the breaking point? (c) How would the breaking-point flow rate calculated in part (b) change if a nozzle with an exit diameter of 10 mm was placed at the end of the 25-mm-diameter tube?

3.47. A 25-mm-diameter garden hose is used to empty a spa as shown in Figure 3.44. When completely full, the water depth in the spa is 1.5 m. The spa has a cross-sectional area of 10 m^2. To drain the spa, the hose entrance is placed on the bottom of the spa and the hose exit is placed 2 m below the bottom of the spa. Estimate how long it will take to drain the spa completely when it is initially completely full.

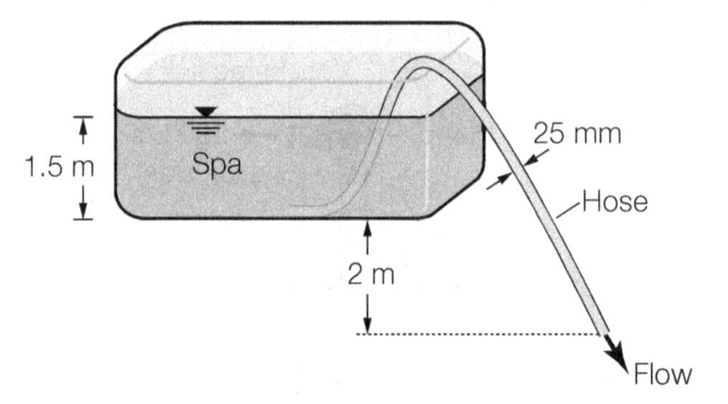

Figure 3.44: Garden hose emptying a spa

3.48. A strong wind blows over a house as illustrated in Figure 3.45. The approaching wind has a velocity of 60 km/h and a pressure of 100 kPa, and (due to the space taken up by the house) the wind velocity over the roof is higher that the approach velocity and is equal to 80 km/h. The pressure inside the house remains at 100 kPa. The house has two large windows each with an area of 1.3 m^2, one on the side of the house and one on the roof, which acts as a skylight. Estimate the net force on each window and the direction in which the net force acts. Based on these results, what is the failure mode of these windows? Assume a standard atmosphere at sea level.

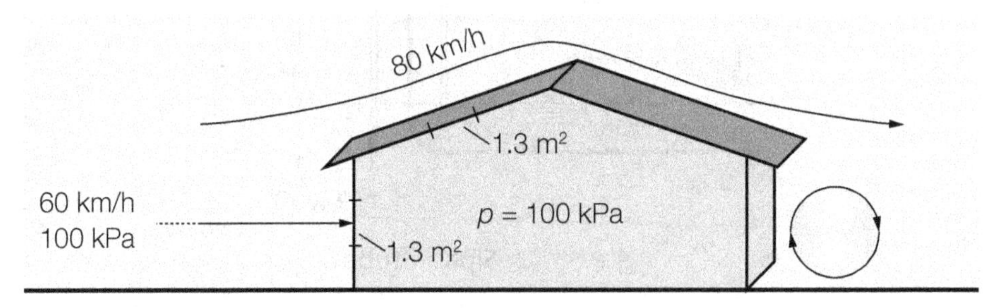

Figure 3.45: Wind flow over a house

3.49. An airplane operates at a low elevation where the pressure is 101 kPa and the temperature is 20°C. (a) Neglecting compressibility effects, estimate the flight speed at which the stagnation pressure is 15% higher than the static pressure. (b) Estimate by how much the result in part (a) would differ if compressibility effects were taken into account.

3.50. The airfoil shown in Figure 3.46 is designed such that when the airplane flies at a velocity V, the velocity of the air passing over the airfoil at its widest point (Point P) is 30% higher than the speed of the airplane. (a) Develop an expression for the airplane velocity V as a function of the static pressure, the pressure at Point P, and the air density at the flight elevation. (b) If the plane flies at an altitude of 5 km and measures a pressure of 48.5 kPa, what is the speed of the airplane? Assume a standard atmosphere.

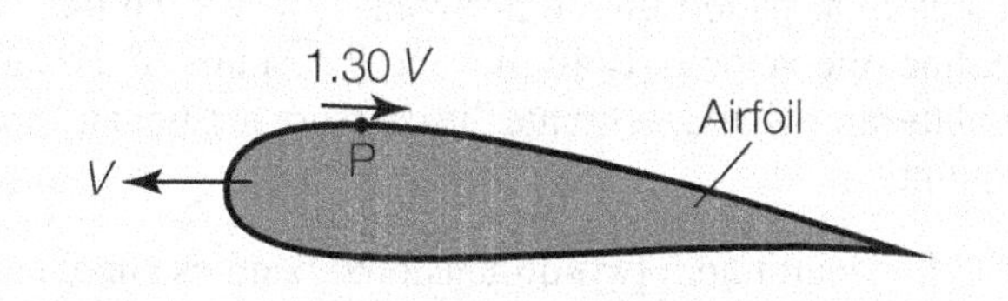

Figure 3.46: Pressure measurement on an airfoil to estimate airspeed

3.51. An airplane flies at 400 km/h at an altitude of 2700 m. The airflow over the wing is as illustrated in Figure 3.47. The stagnation streamline intersects the wing at Point 1, and the airspeed over the wing is faster than the speed of the aircraft such that the maximum airspeed relative to the aircraft occurs at Point 2. In the present case, the relative airspeed at Point 2 is 480 km/h. Estimate the pressures at Points 1 and 2. Assume a standard atmosphere.

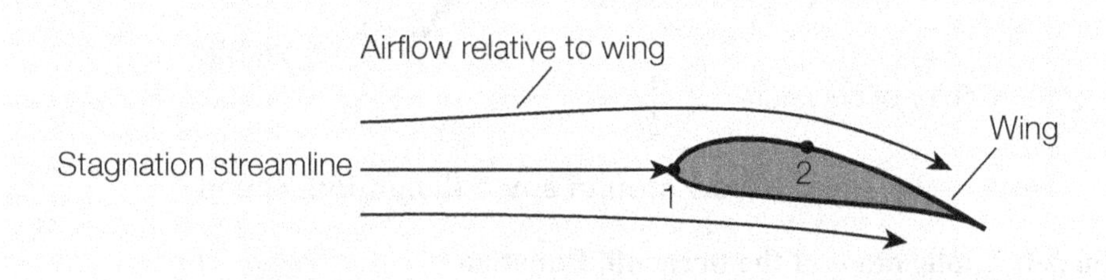

Figure 3.47: Airflow over a wing

3.52. The magnitudes of various pressures can be sensed by a person in a moving vehicle putting her hand out the window and orienting the palm of her hand toward the oncoming air. If the air temperature is 20°C, at what speed is the vehicle moving when she senses 1 Pa? What is the speed when she senses 1 kPa?

3.53. Consider the case in which a tropical cyclone generates wind speeds on the order of 100 km/h directly onto the front of a roadside advertising billboard. The billboard has dimensions of 6 m $\times$ 18 m and is mounted on poles 10 m high from the ground level to the base of the billboard. Estimate the wind force on the billboard and the total moment that must be supported by the poles at ground level. Assume a standard atmosphere at sea level.

3.54. Consider the case in which fluid of density ρ circulates around a central point such that the velocity, V, at any distance r from the central point is given by

$$V = Cr$$

where C is a constant. Note that the streamlines are circles and that V on each streamline is in the tangential direction. It is known that the pressure is equal to p_0 at a distance r_0 from the central point. Determine an expression for the pressure distribution in any horizontal plane in terms of r, where C, p_0, r_0 are parameters of the pressure distribution.

3.55. A fluid of density ρ circulates around a central point such that the velocity, V, at any distance r from the central point is given by

$$V = \frac{C}{r}$$

where C is a constant. The pressure is equal to p_0 at a distance r_0 from the central point. Determine an expression for the pressure distribution in any horizontal plane in terms of r, where C, p_0, r_0 are parameters of the pressure distribution.

3.56. Air flows in a circular pattern in the horizontal plane such that at a particular point on the streamline, the velocity is 40 m/s and the radius of curvature is 15 m. If the temperature of the air is 20°C, estimate the pressure gradient normal to the streamline at the given point.

3.57. Water at 20°C flows around a two-dimensional bend as shown in Figure 3.48. The velocity distribution across the bend is approximately uniform and equal to 40 cm/s. Determine the difference in pressure between the inside and outside of the bend. Perform your calculations with and without assuming that the radius of curvature is constant.

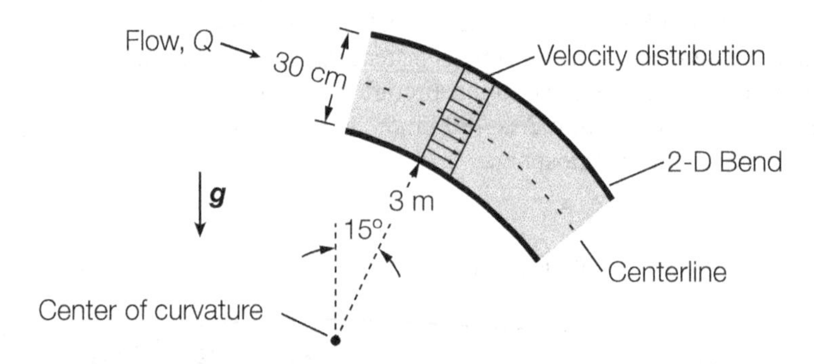

Figure 3.48: Flow around a two-dimensional bend

Section 3.4: Applications of the Bernoulli Equation

3.58. Water is discharged from a large pressurized tank through a 50-mm-diameter orifice into the open atmosphere as shown in Figure 3.49. The orifice is sharp and has a discharge coefficient equal to 0.62. Under a particular operating condition, the water surface is 2 m above the centroid of the orifice, a 2.9-m-thick layer of methanol overlays the water, and the air (gauge) pressure above the methanol is 250 kPa. All liquids are at 20°C, and the atmospheric pressure is 101.3 kPa. What is the volume flow rate through the orifice?

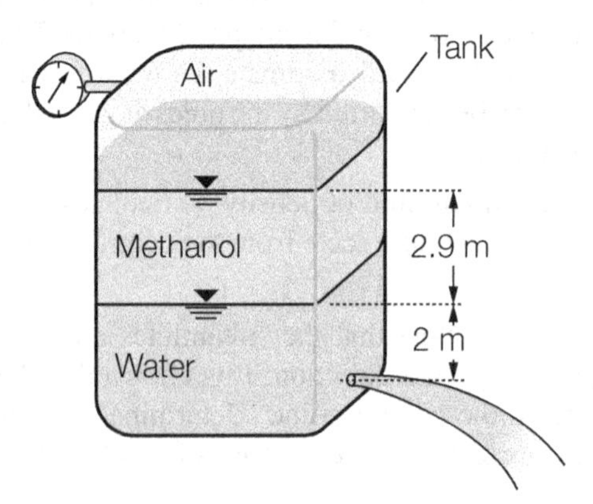

Figure 3.49: Orifice discharge from a pressurized tank

3.59. An orifice is located in the bottom of a reservoir that has a cross-sectional area of 144 m^2 and a depth of 4 m. The orifice has a discharge coefficient of 0.6 and a diameter of 50 mm. (a) Estimate how long it will take to drain the reservoir completely. (b) If the depth in the reservoir is to be maintained at 4 m, at what rate must liquid be added to the reservoir?

3.60. A 2-m-long tank with a trapezoidal cross section is drained from the bottom by a rectangular (slot) opening in the bottom center of the tank. The drain slot extends over the entire 2-m length of the tank and is 1 cm wide. The slot opening has been smoothed such that it has a discharge coefficient approximately equal to unity. The trapezoidal cross section of the tank has a bottom width of 1 m and side slopes that are at 60° to the horizontal. If the initial depth of liquid in the tank is 1 m, how long does it take to drain the tank completely? Comment on the influence of the density of the liquid on the drain time.

3.61. A seawall with a small slot opening is used to dampen the tidal influence in a coastal area. The seawall, which is shown in Figure 3.50, is 3 m long (in the direction perpendicular to the page), the mean sea level on the seaward side is 2.2 m above the centroid of the slot, and the mean sea level on the landward side of the wall is 1.4 m above the centroid of the slot. The seaward tide fluctuates ± 0.6 m, and the landward tide fluctuates ± 0.4 m. The slot extends the entire length of the wall (i.e., it is 3 m long) and is estimated to have a discharge coefficient of 0.80. If the volume of seawater moving from the seaward to the landward side of the seawall is to be no more than 12 000 m^3 in 24 hours, what is the maximum allowable height of the slot?

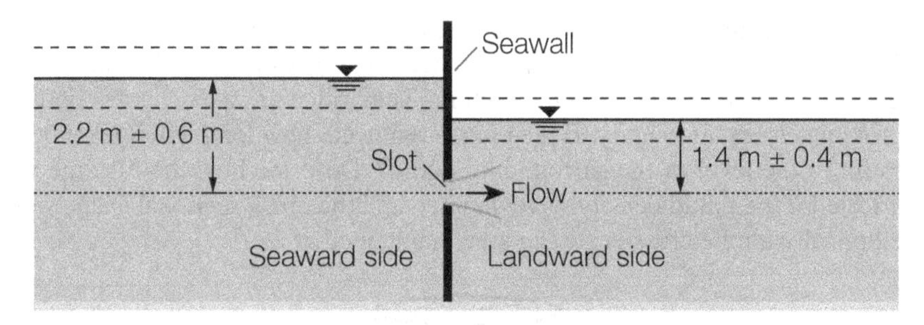

Figure 3.50: Seawall with a slot opening

3.62. A submerged 40-mm-diameter orifice joins two compartments in a pressurized tank as shown in Figure 3.51. In the first compartment, there is a 0.5-m depth of water overlaid by a 1.2-m depth of oil, and in the second compartment, there is a 1.3-m depth of water overlaid by a 0.6-m depth of oil. The water is at a temperature of 20°C, and the oil has a specific gravity of 0.88. The discharge coefficient of the orifice is 0.93. If the air pressure in the second compartment is 200 kPa and a volume flow rate through the system of 8 L/s is desired, what air pressure must be maintained in the first compartment?

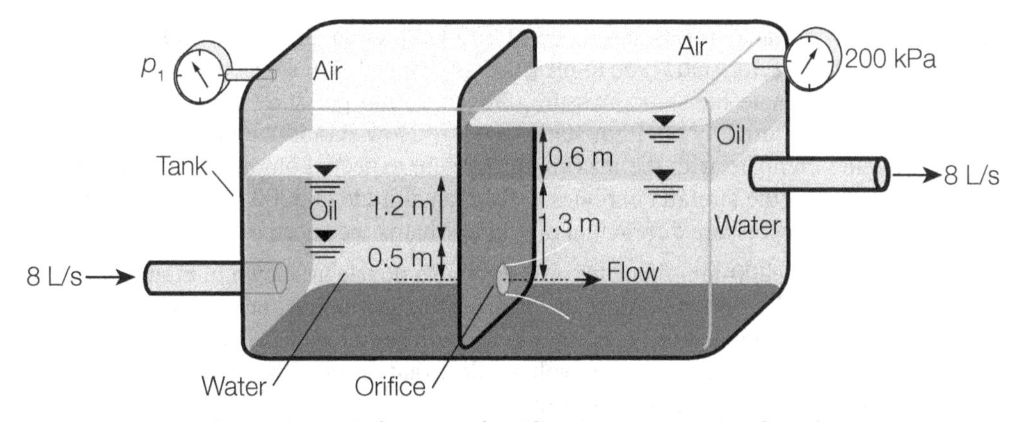

Figure 3.51: Submerged orifice in a pressurized tank

3.63. An orifice meter consists of an arrangement in which a plate containing a circular orifice at its center is placed normal to the flow in a conduit, and the flow rate in the conduit is related to the measured difference in pressure across the orifice. Consider the orifice meter shown in Figure 3.52, where the orifice has a diameter of 50 mm and a discharge coefficient of 0.62 and the conduit has a diameter of 100 mm. The fluid is water at 20°C, and the differential pressure gauge shows a pressure difference of 30 kPa. What is the volume flow rate in the conduit?

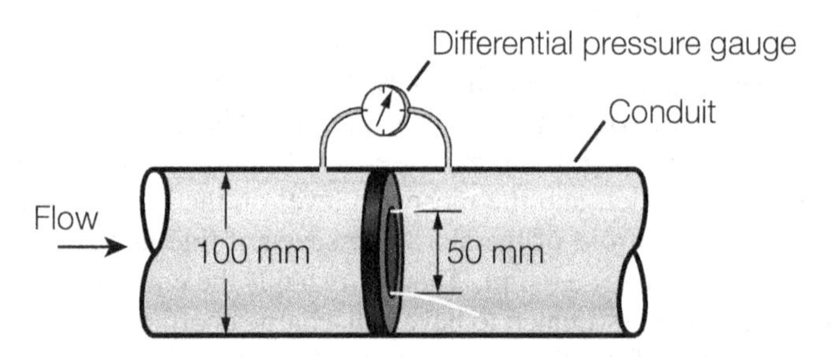

Figure 3.52: Flow through an orifice meter

3.64. A cylinder of internal cross-sectional area 2 m^2 is submerged in a liquid to a depth of 1 m as shown in Figure 3.53. The cylinder is initially empty and contains a plugged orifice of diameter 20 mm in the bottom. The orifice has a discharge coefficient approximately equal to unity. If the plug is removed, how long will it take the cylinder to fill to the level of the surrounding liquid? Describe how the fill time would be affected if the liquid density were doubled and the cross-sectional shape were made elliptical with the cross-sectional area maintained at 2 m^2.

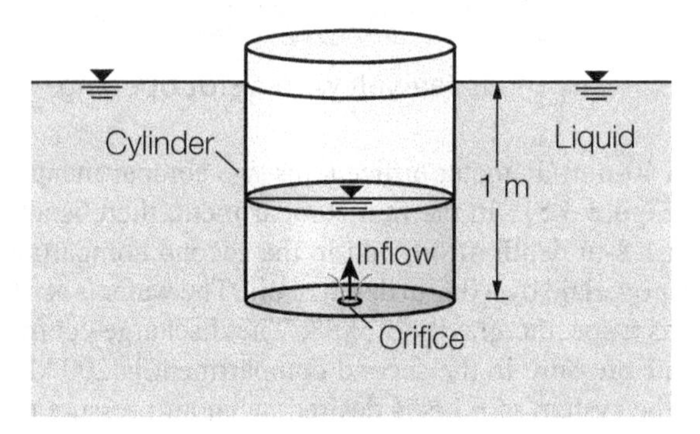

Figure 3.53: Submerged cylindrical tank

3.65. A small airplane cruises at a speed of 300 km/h at an altitude of 3000 m. The airplane uses a Pitot-static tube to measure its airspeed. (a) Neglecting compressibility effects, estimate the static pressure, the stagnation pressure, and the pressure difference measured by the Pitot-static tube. (b) How should the measured stagnation pressure compare with the maximum pressure expected on the fuselage of the airplane? (c) If the aircraft continues to fly at an altitude of 3000 m and the Pitot-static tube reads a pressure differential of 4 kPa, what is the speed of the aircraft?

3.66. Water exits a 200-mm-diameter pipe through a nozzle as shown in Figure 3.54. Under a particular flow condition, the pressure gauge mounted near the exit of the pipe measures 20 kPa and water in the Pitot tube inserted at the center of the discharge jet rises 2.5 m. (a) Estimate the volume flow rate through the pipe. (b) Estimate the exit velocity and exit diameter of the nozzle.

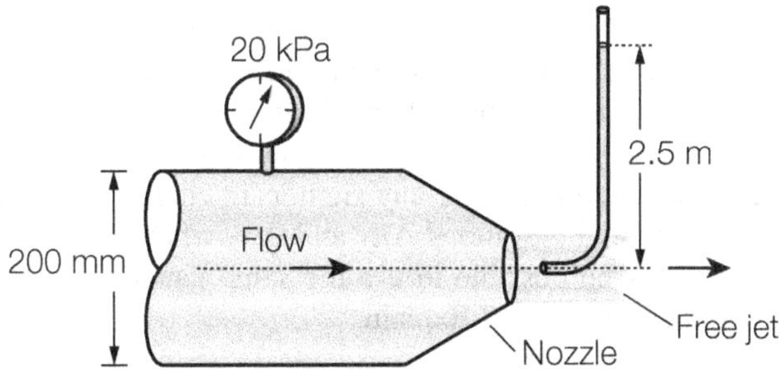

Figure 3.54: Flow measurement using a Pitot tube

3.67. A Pitot-static tube is used to measure the velocity of carbon dioxide (CO_2) in a 200-mm-diameter pipe. The temperature and (absolute) pressure of CO_2 in the pipe are 5°C and 250 kPa, respectively. If the Pitot-static tube indicates a differential pressure (i.e., stagnation minus static) of 0.4 kPa, estimate the volume flow rate in the pipe.

3.68. The air velocity in the duct of a heating system is measured by a Pitot-static probe inserted in the duct. If the differential height between water columns connected to the two ports of the probe is 2.4 cm, determine (a) the pressure rise (above static pressure) at the tip of the probe and (b) the flow velocity. The air temperature and pressure in the duct are 45°C and 98 kPa, respectively.

3.69. A 75-mm-diameter pipe is reduced to a 50-mm-diameter pipe as shown in Figure 3.55. A static tube is installed in the 75-mm pipe, and a Pitot tube is installed in the 50-mm pipe. Determine the difference, Δh, between the water levels in the static tube and the Pitot tube when the flow rate in the pipe is 10 L/s. How would Δh be different if a lighter (i.e., less dense) liquid were used instead of water. Assume the same volume flow rate in both cases.

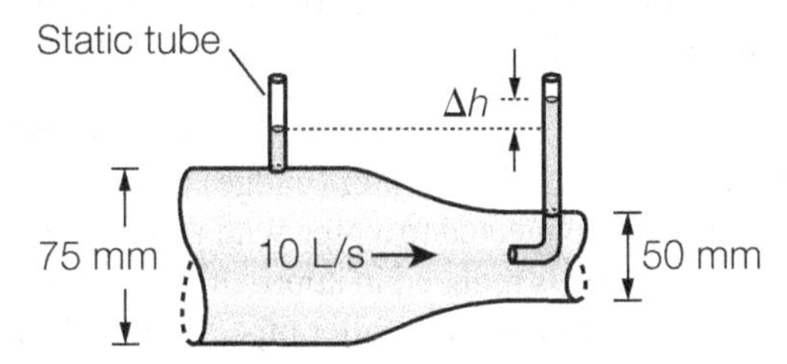

Figure 3.55: Pitot tube in a contracted pipe

3.70. A Pitot tube is placed in a river as shown in Figure 3.56. The submerged end of the tube is 0.50 m below the water surface, and the water in the tube rises to a level that is 40 mm above the water surface in the river. Estimate the velocity in the river upstream of the submerged end of the tube. Note that this is the original application of the Pitot tube used by Henri Pitot to measure the flow rate in the Seine River.

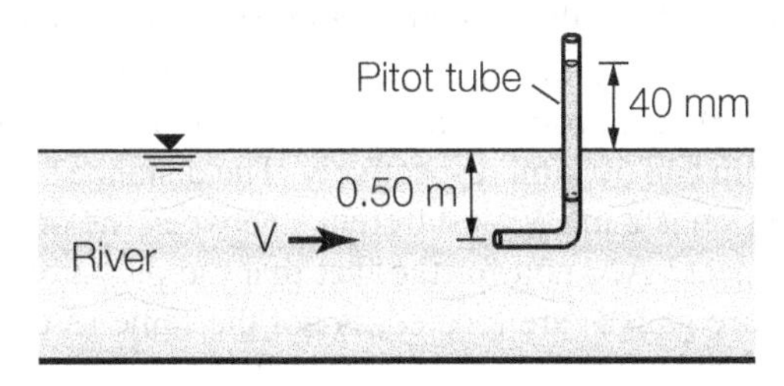

Figure 3.56: Pitot tube in a river

3.71. Water at 20°C flows in a 100-mm-diameter conduit containing two Pitot-static tubes as shown in Figure 3.57. In the upstream Pitot-static tube, the gauge fluid is above the conduit and has a density of 850 kg/m^3. In the downstream Pitot-static tube, the gauge fluid is below the conduit and has a density of 1500 kg/m^3. Assuming that the Pitot-static tubes do not cause any significant energy losses within the conduit, estimate the differential elevations Δh_1 and Δh_2 when the flow rate in the conduit is 10 L/s. Would it be possible to use the same gauge fluid in the upstream and downstream Pitot-static tubes? Explain.

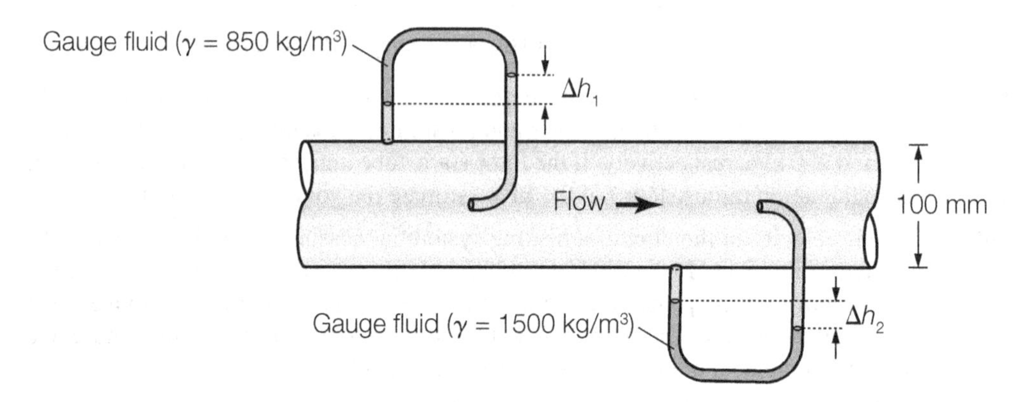

Figure 3.57: Conduit with two Pitot-static tubes

3.72. Carbon dioxide flows at a rate of 0.5 kg/s through a 200-mm-diameter conduit such that at a particular section, the absolute pressure is 300 kPa and the temperature is 15°C. Just downstream of this section, the diameter contracts to 150 mm. Estimate the difference in pressure between the uncontracted and contracted section. Assume that the flow is incompressible.

3.73. Water flows at 9.464×10^{-4} m^3/s in a pipe that has a diameter of 0.05080 m. The water temperature is 20°C, and the water (gauge) pressure is 206.9 kPa. The downstream end of this pipe is connected to a pipe with a much smaller diameter where the pressure in the smaller pipe is reduced in approximate accordance with the Bernoulli equation. To avoid cavitation in the smaller pipe, what is the minimum diameter that should be used for this pipe?

3.74. A pipe AB carries water and tapers uniformly from a diameter of 0.1 m at A to 0.2 m at B over a length of 2 m. Pressure gauges are installed at A, B, and C, the midpoint between A and B. If the pipe centerline slopes upward from A to B at an angle of 30° and the pressures recorded at A and B are 200 and 230 kPa, respectively, determine the flow through the pipe and pressure recorded at C, neglecting all losses.

3.75. Water at 20°C flows through an inclined circular conduit at a rate of 10 L/s as shown in Figure 3.58. The conduit is inclined at 40° and contracts from a larger to smaller diameter. The diameters upstream and downstream of the contraction, at sections A and B, are 100 mm and 50 mm, respectively. The distance between the upstream and downstream section is 2 m. What difference between the pressure readings at A and B would you expect? Account for the fact that the pressure on the side of the pipe, where the pressure gauges are located, is not the same as the pressure at the center of the pipe.

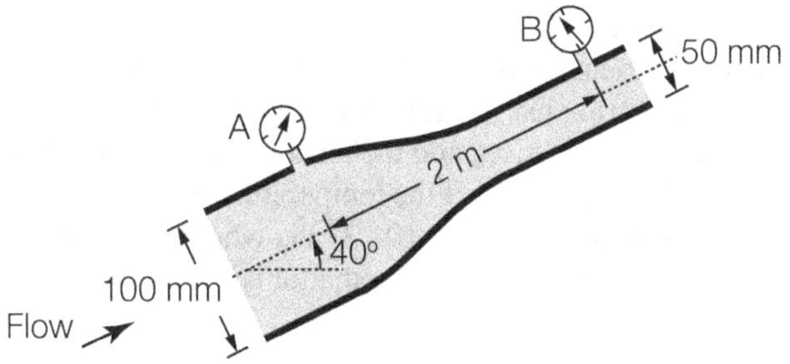

Figure 3.58: Inclined contraction

3.76. The Venturi meter shown in Figure 3.59 is to be used to measure flow rates in the range of 4–40 L/s. The liquid to be transported through the Venturi meter is gasoline at 20°C. The manufacturer claims that the meter has a discharge coefficient of 0.97. What range of pressure differences will be measured between the approach section and the throat section of the Venturi meter?

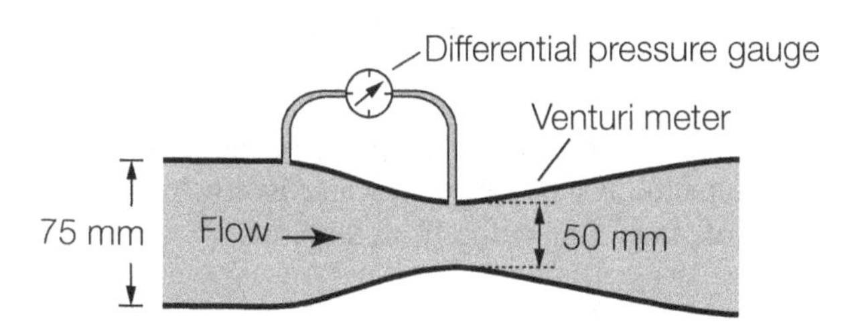

Figure 3.59: Venturi meter for measuring gasoline flows

3.77. The Venturi meter with attached manometer shown in Figure 3.60 is used to measure the flow rate of water. (a) Estimate the flow rate for the conditions shown in Figure 3.60. (b) Determine the maximum flow rate that can occur without the mercury being sucked out of the manometer.

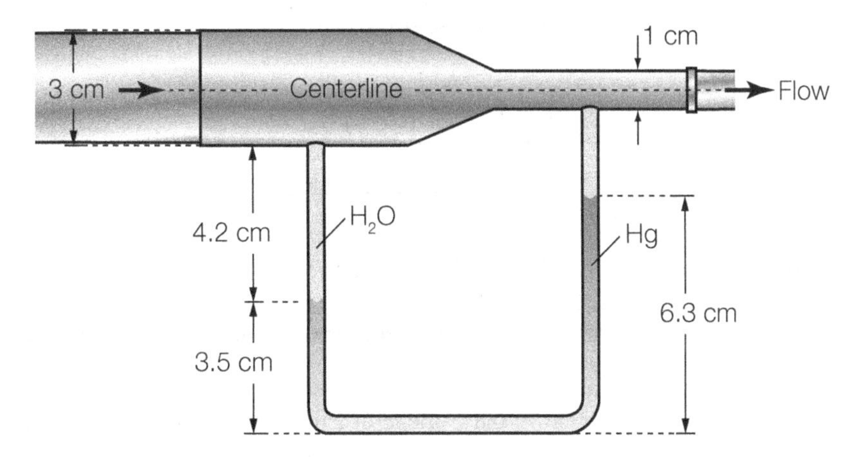

Figure 3.60: Venturi meter with a manometer

3.78. In the contracted conduit shown in Figure 3.61, liquid methanol is pulled up from an open reservoir into an airstream. The approach section has a diameter of 90 mm, and the contracted section has a diameter of 15 mm. The methanol reservoir is 0.5 m below the contracted air section, and the airflow rate is such that the (air) velocity in the contracted section is 30 m/s. If the temperature of the air in the approach section is 15°C and atmospheric pressure is 101.3 kPa, what is the maximum gauge pressure at the approach section so that liquid methanol is pulled into the contracted section?

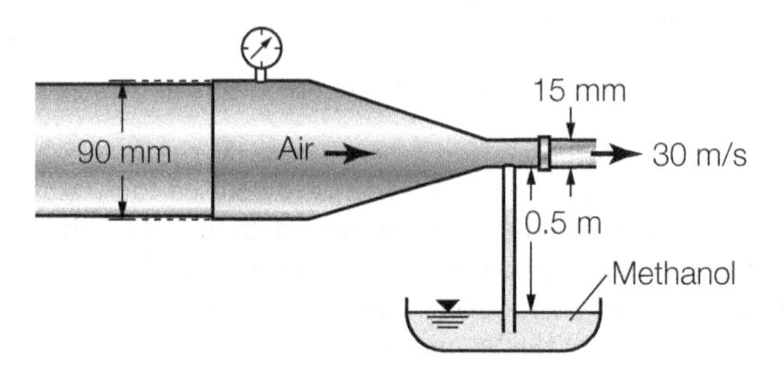

Figure 3.61: Methanol suction system

3.79. A small hole is punched in a water supply pipeline as shown in Figure 3.62. The pressure in the pipeline is 414 kPa, and water is flowing at 0.60 m/s. Estimate the height of the jet of water emanating from the hole.

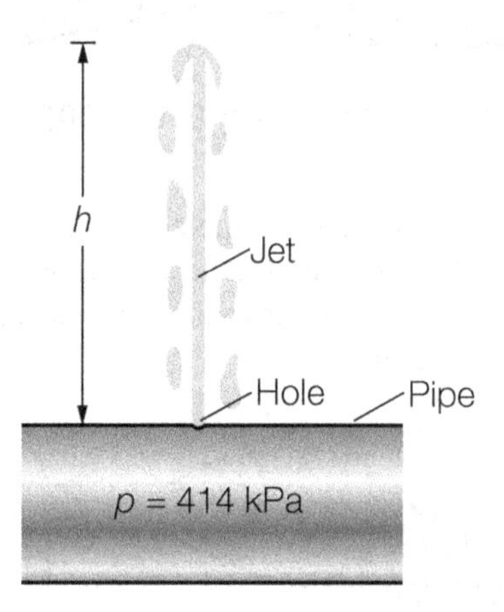

Figure 3.62: Pipe leak

3.80. A jet of water is initially 12 cm in diameter, and when directed vertically upward, it reaches a maximum height of 20 m. Assuming that the jet remains circular, determine the rate of water flowing and the diameter of the jet at a height of 10 m.

3.81. Several ornamental water fountains such as the one shown in Figure 3.63 can be found in the plaza adjacent to the College of Engineering on the University of Miami campus. These fountains typically rise to a height of around 40 cm, and the discharge port of each fountain has an approximate diameter of 19 mm. (a) Estimate the volume flow rate of water at each fountain. (b) Estimate the power requirement to operate each fountain.

Figure 3.63: Ornamental water fountain

3.82. A water fountain is to be constructed using the system shown in Figure 3.64. Water in the fountain is to come from a tank containing water and pressurized air. The depth of water in the tank is to be 6 m, and the nozzle of the water fountain is to be 1 m above the bottom of the tank. (a) Determine the relationship between the air pressure in the tank and the maximum height to which the water fountain can rise above the discharge nozzle. (b) If a water fountain height of 10 m is desired, what air pressure should be used? Assume water at 20°C.

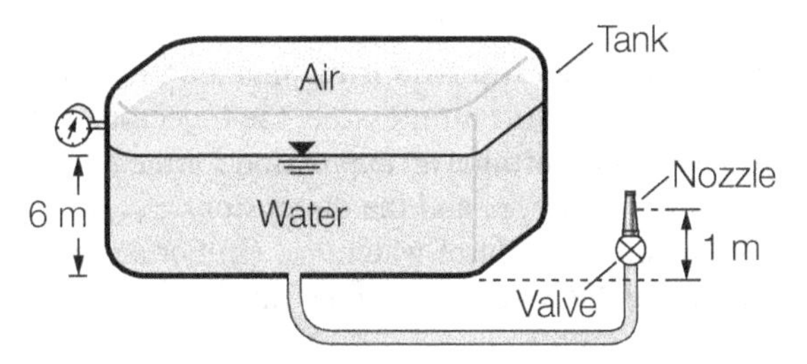

Figure 3.64: Water fountain system

3.83. Consider the illustration in Figure 3.65 where water is discharged vertically from a nozzle and rises to a height of 8 m. The exit diameter of the nozzle is 25 mm, and the diameter of the conduit leading to the nozzle is 100 mm. If a pressure gauge is located 1 m below the nozzle exit, what pressure is the gauge expected to show?

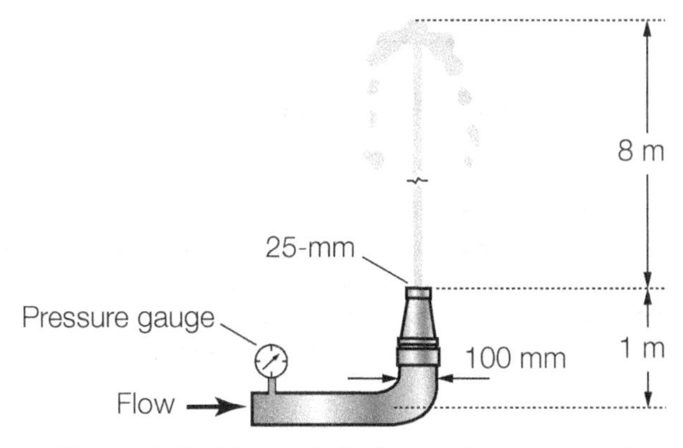

Figure 3.65: Vertical discharge from a nozzle

3.84. A liquid drains from a reservoir through a circular discharge conduit as shown in Figure 3.66. Under the influence of gravity, the diameter of the liquid stream decreases with distance below the drain. At a particular distance below the drain, the diameter of the stream is 30 mm, and 1 m below this location, the diameter is 20 mm. Estimate the volume flow rate in the liquid stream.

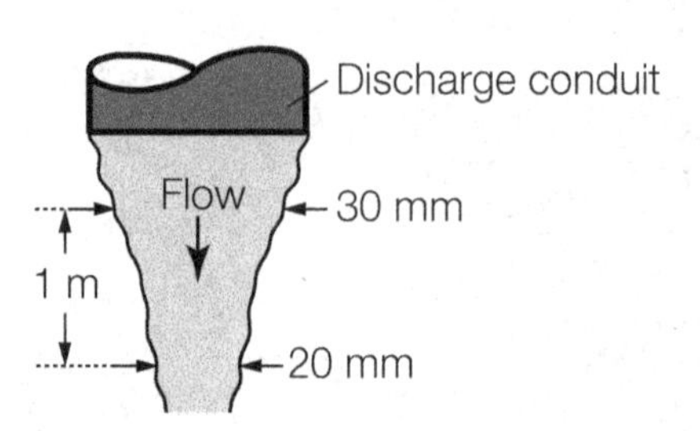

Figure 3.66: Liquid stream draining from a discharge conduit

3.85. A liquid drains from a storage tank into the atmosphere through a 20-mm-diameter opening at a rate of 0.4 L/s. The drained liquid strikes the ground, which is 0.5 m below the drain hole. (a) What is the diameter of the liquid stream when it strikes the ground. (b) Compare the velocity of the liquid when it leaves the drain with its velocity when it hits the ground.

3.86. A hose discharges at ground level and is inclined at 60°. If water exits the hose at 15 m/s, what is the maximum height the jet attains and what is the velocity at that height?

3.87. The pressure in a large water reservoir is maintained using pressurized air as shown in Figure 3.67. By changing the air pressure, the trajectory of the resulting water jet can be manipulated. (a) Determine an expression for the horizontal distance x_{max} as a function of the air pressure, p, and the dimensions z_{w}, z_0, and θ; this relationship also includes the specific weight of water, γ_{w}. (b) For a case in which p = 100 kPa, z_{w} = 3 m, z_0 = 5 m, and θ = 45°, determine the maximum distance, x_{max}, covered by the water jet.

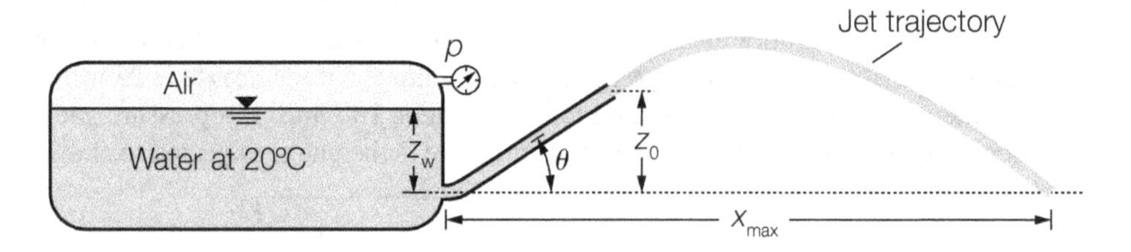

Figure 3.67: Water jet from a pressurized reservoir

3.88. A fire hose is to be used to direct a stream of water to a location on a building that is 15 m above ground level, and the building is 9 m away from the location of the nozzle. The hose has a diameter of 100 mm, and the nozzle exit has a diameter of 25 mm. What is the minimum pressure that must exist in the hose at the entrance to the nozzle such that the water stream will be able to reach the target location?

3.89. The nozzle shown in Figure 3.68 discharges water at a velocity of 40 m/s. The jet is intended to impact the building at point P that is located 25 m horizontally and 20 m vertically from the nozzle discharge. Determine the angle(s) θ for which the center of the jet will hit the building at Point P.

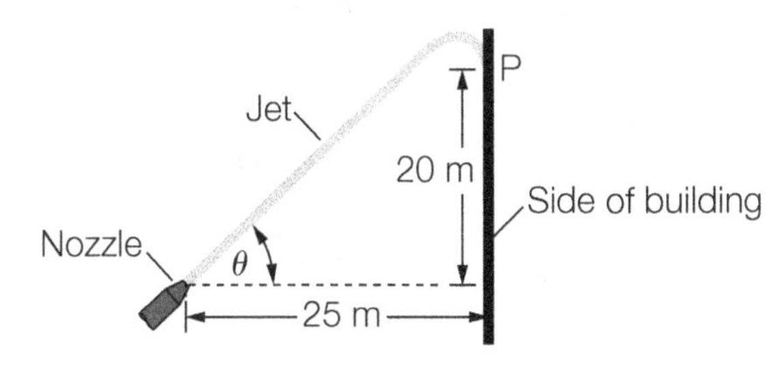

Figure 3.68: Jet directed toward the side of a building

3.90. A standard fire hose has a diameter of 75 mm, and the standard attached nozzle has a diameter of 30 mm. If a standard fire stream is expected to deliver 16 L/s, what is the minimum pressure that should be maintained in the fire hose?

3.91. A waterjet cutter that is capable of slicing metals and other materials uses a small nozzle to generate a 0.08-mm-diameter jet at a speed of 650 m/s. Assuming that the velocity head in the source conduit is negligible compared with the velocity head in the cutting jet, what pressure and water supply rate must be generated in the source conduit to operate the waterjet cutter?

3.92. Experiments indicate that the shear stress, τ_0, on the wall of a 200-mm pipe can be related to the flow in the pipe using the relation

$$\tau_0 = 0.04\rho V^2$$

where ρ and V are the density and velocity, respectively, of the fluid in the pipe. If water at 20°C flows in the pipe at a flow rate of 60 L/s and the pipe is horizontal, estimate the pressure drop per unit length along the pipe.

3.93. A 200-mm-diameter water pipe taps into a 300-mm pipe as shown in Figure 3.69. When the flow rate upstream of the junction is 400 L/s, the pressure in the 300-mm pipe upstream of the junction is 400 kPa and the pressure in the 300-mm pipe downstream of the junction is 410 kPa. The 200-mm pipe rises to an elevation 1 m above the 300-mm pipe. Estimate the flow rate and pressure in the 200-mm pipe.

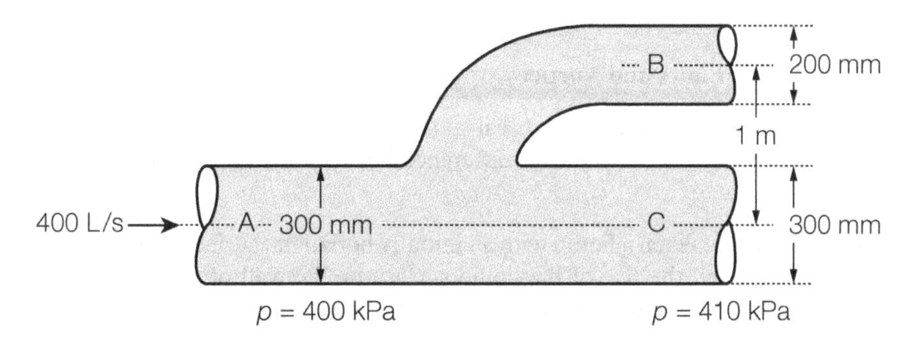

Figure 3.69: Tap into a horizontal pipe

3.94. Water flows into a pressurized tank at a rate of 4 m^3/s as shown in Figure 3.70. Water is discharged from the tank into the open atmosphere from two ports. The connection to the ports is 0.5 m below the water surface in the tank, the 250-mm-diameter port discharges 0.6 m above the connection, and the 400-mm-diameter port discharges 0.7 m below the connection. What air pressure is required to keep the system at steady state?

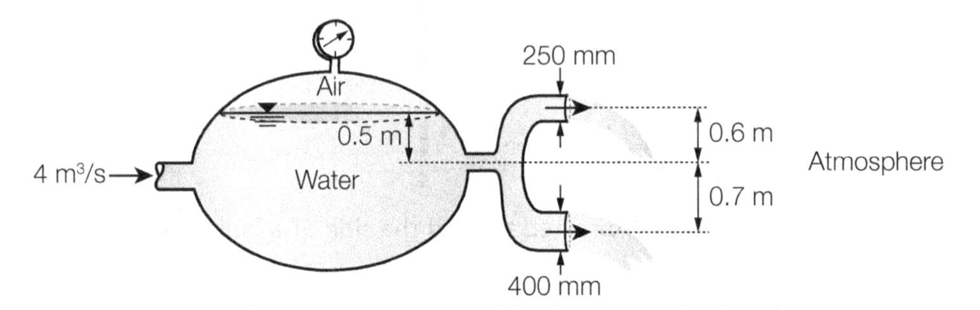

Figure 3.70: Port discharges from a pressurized cylinder

3.95. A cone can be used to estimate the airflow rate in a conduit by placing the cone in front of the exit of the conduit and measuring both the pressure near the pipe exit and the thickness of the airstream at the periphery of the cone, as shown in Figure 3.71. Consider a particular configuration where the diameter of the conduit is 200 mm and the diameter of the base of the cone is 300 mm. The flow conditions are such that the air is at 20°C, the gauge pressure near the exit is 30 kPa, atmospheric pressure is 101.3 kPa, and the airstream thickness at the periphery of the cone is 10 mm. What is the estimated volume flow rate of the air?

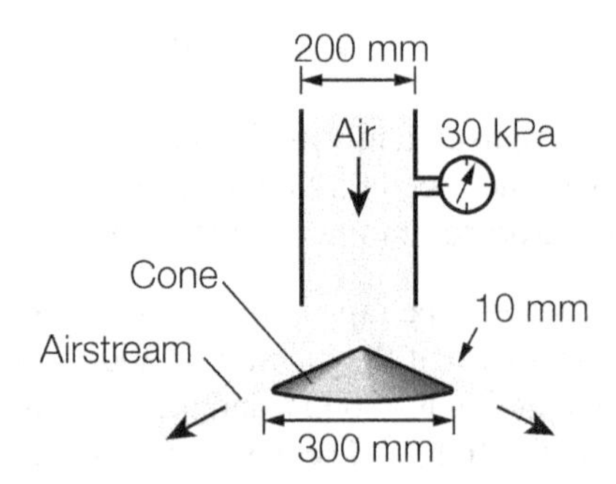

Figure 3.71: Airflow measured by a cone

Section 3.5: Curved Flows and Vortices

3.96. A closed 15-cm-diameter cylindrical tank is filled with water and is rotated at 600 rpm. What is the maximum pressure difference between the center and the wall of the tank?

3.97. It is known that when a liquid with an initially horizontal surface is rotated in a cylindrical container, the rise of the liquid surface on the wall of the container is equal to the fall of the liquid surface at the center of the container. Consider an 850-mm-diameter cylindrical container with an initial depth of water equal to 1.2 m. At what rotational speed does the water surface at the center of the container intersect the bottom of the container?

3.98. The pump rotor shown in Figure 3.72 has an inner diameter of 200 mm, an outer diameter of 600 mm, and a height of 75 mm. When the rate of angular rotation is 900 rpm, the flow through the rotor is 300 L/s. Estimate the difference in pressure between the inflow and the outflow side of the rotor. Assume water at 20°C and neglect head losses in the rotor.

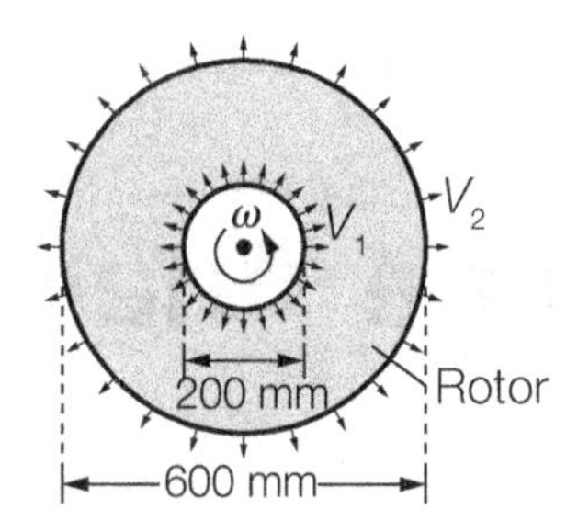

Figure 3.72: Flow through a pump rotor

3.99. A pump rotor has an inner radius of 15 cm, an outer radius of 40 cm, and a thickness of 0.80 m. If the speed of the rotor is 450 rpm, estimate the flow rate through the pump when the added pressure head is 35 m.

3.100. An atmospheric disturbance can be approximated by a circular rotation of air around a central location. If the pressure and velocity measured 110 m from the center are 95.0 kPa and 110 m/s, respectively, estimate the distance from the center to where the pressure is equal to the atmospheric pressure of 101 kPa. Assume air at 20°C.

3.101. Measurements in the outer regions of a hurricane indicate a pressure of 101 kPa and velocities of 6 m/s and 3 m/s in the radial and tangential directions, respectively. If the pressure at the center of the hurricane is estimated as 98 kPa, what wind speed do you expect at the center?

3.102. Airflow within the eye of a hurricane can be approximated as a spiral free vortex. The velocity at the center of the eye can be approximated as being equal to zero, and the velocity at the eye wall can be associated with the maximum wind velocity in the hurricane. Consider Hurricane Wilma, which attained pressure of 88.2 kPa at the center of the eye and a wind velocity of 295 km/h in the eye wall. (a) Estimate the difference in pressure between the center of the eye and the eye wall. What is the difference in sea-surface elevation between the center of the eye and eye wall caused by this difference in pressure? (b) If the pressure at the outer limit of the hurricane is 101 kPa with negligible wind velocities, estimate the difference in sea-surface elevation between the wall of the hurricane and the outer limit of the hurricane. What are the implications of this sea-surface difference to coastal communities? Assume a uniform atmospheric temperature of 25°C, assume a seawater temperature of 20°C, and neglect air density variations in the eye of the hurricane.

Chapter 4

Finite Control Volume Analysis

LEARNING OBJECTIVES

After reading this chapter and solving a representative sample of end-of-chapter problems, you will be able to:

- Understand and apply the fundamental laws of conservation of mass, linear momentum, and energy to fluid systems.
- Calculate hydrodynamic forces on structures.
- Apply the angular momentum principle to the analysis of fluid systems and turbomachines.
- Analyze fluid mechanical systems that include pumps, turbines, jet engines, and rockets.
- Understand the fundamental fluid mechanical principles involved in harnessing wind energy.

4.1 Introduction

The motion of a fluid can be described using a variety of reference frames, with different reference frames giving different perspectives of the flow. The two reference frames commonly used in fluid mechanics are (1) a reference frame moving with a fluid element that contains a defined mass of fluid and (2) a reference frame fixed in space. A reference frame moving with a fluid element containing a defined mass of fluid can be regarded as the fundamental reference frame, because the laws of motion and thermodynamics are directly applicable to a defined fluid mass as it moves within a fluid continuum. In contrast, a fixed reference frame describes the behavior of the fluid at a fixed point in space, which is the usual perspective from which fluid motions are observed and measured. The equations governing the behavior of fluids are typically formulated by first applying the fundamental laws to a defined fluid mass and then transforming the defined-fluid-mass equations into equivalent fixed-reference-frame equations to facilitate practical applications.

Lagrangian and Eulerian reference frames. A *Lagrangian reference frame* moves with a fluid element that contains a fixed mass of fluid, and fluid properties within the fluid element only change with time. For example, the velocity, $\mathbf{v}$, of a fluid element in a Lagrangian reference frame is expressed in the form $\mathbf{v}(t)$, where t is the time. An *Eulerian reference frame* is fixed in space, and changes in fluid properties are described at fixed locations. For example, the velocity field of a fluid in an Eulerian reference frame is expressed in the form $\mathbf{v}(\mathbf{x}, t)$, where $\mathbf{x}$ is the location and t is the time. From a practical viewpoint, we are usually interested in the behavior of fluids at particular locations in space, in which case an Eulerian reference frame is preferable. The complicating factor in working with Eulerian reference frames is that the fundamental equations of fluid motion and thermodynamics are all stated for fluid elements in Lagrangian reference frames. For example, Newton's second law states that the sum of the forces on any fluid element (as it moves within the fluid continuum) is equal to the mass of fluid within the fluid element multiplied by the acceleration of the fluid element. To transform the fundamental equations of fluid motion and thermodynamics into useful equations in Eulerian reference frames, it is necessary to understand the relationship between Lagrangian equations and Eulerian equations. The transformation of Lagrangian equations into Eulerian equations is generally done using Reynolds transport theorem.

4.2 Reynolds Transport Theorem

A *system* is defined as a collection of matter of fixed identity that may move, flow, and interact with its surroundings; a fluid system always contains the same set of atoms. A fluid system can be expected to change its size and shape continually. In contrast to a fluid system, which moves around within a fluid continuum, a *control volume* is a fixed volume in space through which fluid may flow in and out. A system is sometimes referred to as a *closed system*, and a control volume is sometimes referred to as an *open system*. The terms *closed system* and *open system* are more common in the field of thermodynamics, whereas the corresponding equivalent terms of *system* and *control volume* are more common in the field of fluid mechanics; in this text, the latter terminology is generally used. The behavior of a fluid within a system is described by Lagrangian equations; the behavior of a fluid within a control volume is described by Eulerian equations.

Derivation of Reynolds transport theorem. Consider the (fixed) control volumes shown in Figures 4.1(a) and (b). In Figure 4.1(a), the control volume is contained within a larger body of fluid that flows through the defined control volume. The control volume shown in Figure 4.1(b) differs from that in Figure 4.1(a) only in that part of the boundary of the defined control volume coincides with a solid physical boundary. The key features of control volumes in Figures 4.1(a) and (b) are that both control volumes are fixed volumes in space through which fluid flows. Because a fluid system corresponds to a fixed mass of fluid, we can define a particular fluid system as the fluid contained in the control volume at time t. In other words, at time t, the system coincides with the control volume. During a subsequent time interval,

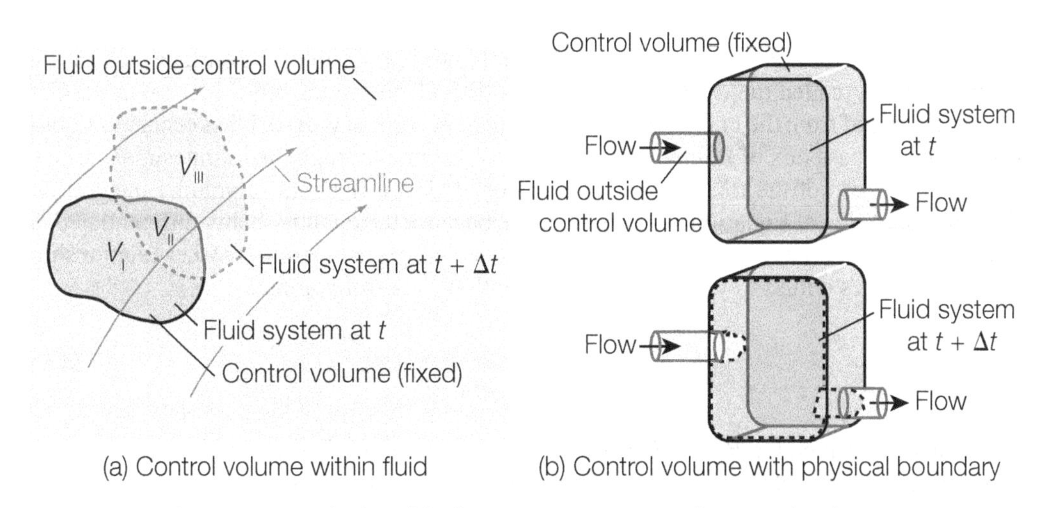

Figure 4.1: Relationship between system and control volume

Δt, a volume $V_{\rm I}$ of fluid enters the control volume, a volume $V_{\rm II}$ of fluid remains in the control volume, and a volume $V_{\rm III}$ of fluid leaves the control volume. Defining the amount of a fluid property within the system at time t as $B_{\rm sys}(t)$, the change in $B_{\rm sys}$ over the time interval Δt is given by

$$B_{\rm sys}(t+\Delta t) - B_{\rm sys}(t) = \left[\int_{V_{\rm II}} \rho\beta \, {\rm d}V + \int_{V_{\rm III}} \rho\beta \, {\rm d}V\right]_{t+\Delta t} - \left[\int_{V_{\rm II}} \rho\beta \, {\rm d}V\right]_t \tag{4.1}$$

where ρ is the density of the fluid and β is the fluid property per unit mass of the fluid in the system and is related to $B_{\rm sys}$ by

$$B_{\rm sys} = \int_{\rm sys} \rho\beta \, {\rm d}V \tag{4.2}$$

The quantity $B_{\rm sys}$ can represent fluid properties such as mass, momentum, and energy and is an *extensive property* because it depends on the amount of fluid in the system. In contrast, β is an *intensive property* because it is independent of the amount of fluid mass in the system. Adding and subtracting the quantity $(\int_{V_{\rm I}} \rho\beta \, {\rm d}V)_{t+\Delta t}$ to the right-hand side of Equation 4.1 and dividing by Δt yields

$$\frac{B_{\rm sys}(t+\Delta t) - B_{\rm sys}(t)}{\Delta t} = \frac{\left[\int_{V_{\rm II}} \rho\beta \, {\rm d}V + \int_{V_{\rm I}} \rho\beta \, {\rm d}V\right]_{t+\Delta t} - \left[\int_{V_{\rm II}} \rho\beta \, {\rm d}V\right]_t}{\Delta t} + \frac{\left[\int_{V_{\rm III}} \rho\beta \, {\rm d}V\right]_{t+\Delta t}}{\Delta t} - \frac{\left[\int_{V_{\rm I}} \rho\beta \, {\rm d}V\right]_{t+\Delta t}}{\Delta t} \tag{4.3}$$

Taking the limit as $\Delta t \rightarrow 0$, the term on the left-hand side of Equation 4.3 becomes ${\rm d}B_{\rm sys}/{\rm d}t$, the first term on the right-hand side is equal to the rate of change of the amount of fluid property in the control volume, and the last two terms are collectively equal to the net rate at which the fluid property leaves the control volume. Therefore, in the limit as $\Delta t \rightarrow 0$, Equation 4.3 can be expressed as

$$\underbrace{\frac{{\rm d}B_{\rm sys}}{{\rm d}t}}_{\text{rate of change of B in system}} = \underbrace{\frac{\rm d}{{\rm d}t}\int_{\rm cv} \beta\rho \, {\rm d}V}_{\text{rate of change of B in cv}} + \underbrace{\int_{\rm cs} \beta\rho(\mathbf{v}\cdot\mathbf{n}) \, {\rm d}A}_{\text{net flux of B out of cv}} \tag{4.4}$$

where "cv" denotes the volume of the control volume, "cs" denotes the surface area of the control volume (called the *control surface*), $\mathbf{v}$ is the velocity field, and $\mathbf{n}$ is the unit normal directed outward from the control surface. Note that the term $\rho(\mathbf{v}\cdot\mathbf{n})\,{\rm d}A$ that occurs in Equation 4.4 is the mass flux of fluid leaving the control volume through the infinitesimal surface area ${\rm d}A$. The use of the total derivative operator, ${\rm d}/{\rm d}t$, on the control volume integral on the right-hand side of Equation 4.4 is appropriate because the quantity being differentiated is only a function of time. Combining Equations 4.4 and 4.2 gives the following key relationship between changes within a system and changes within a control volume:

$$\frac{\rm d}{{\rm d}t}\int_{\rm sys} \rho\beta \, {\rm d}V = \frac{\rm d}{{\rm d}t}\int_{\rm cv} \beta\rho \, {\rm d}V + \int_{\rm cs} \beta\rho(\mathbf{v}\cdot\mathbf{n}) \, {\rm d}A \tag{4.5}$$

Equation 4.5 is called the *Reynolds*[25] *transport theorem.* It provides a fundamental relationship between the rate of change of a fluid property in a system and the rate of change of a fluid

[25]Named in honor of the British engineer Osborne Reynolds (1842–1912).

property in a control volume, where the fluid property per unit mass is given by β. A convenient verbal statement of the Reynolds transport theorem (Equation 4.5) is as follows: "The rate of change of a fluid property within a system at any instant of time is equal to the rate of change of that property in a control volume that coincides with the system at the given time plus the net rate of outflow of the property from the control volume." The Reynolds transport theorem, Equation 4.5, is one of the most useful relationships in fluid mechanics, because most of the fundamental laws governing the behavior of fluids are stated in terms of the properties of a fluid within a system, whereas most of the applications of these fundamental laws are concerned with fluid properties in a control volume. Viewed from another perspective, the Reynolds transport theorem is the finite-control-volume equivalent of the material derivative given by Equation 3.25, which is applicable to an infinitesimal control volume.

Control volume moving with a constant velocity. The Reynolds transport theorem given by Equation 4.5 was derived for a fixed control volume. In cases where the control volume is moving with a constant velocity, the Reynolds transport theorem can be easily adapted to this situation and stated as follows:

$$\frac{\mathrm{d}}{\mathrm{d}t}\int_{\mathrm{sys}} \rho\beta \,\mathrm{d}V = \frac{\mathrm{d}}{\mathrm{d}t}\int_{\mathrm{cv}} \beta\rho \,\mathrm{d}V + \int_{\mathrm{cs}} \beta\rho(\mathbf{v}_{\mathrm{r}}\cdot\mathbf{n})\,\mathrm{d}A \tag{4.6}$$

where $\mathbf{v}_{\mathrm{r}}$ is the fluid velocity relative to the moving control volume. A control volume that is stationary or is moving with a constant velocity is an *inertial frame of reference*, and all physical laws can be applied relative to such moving control volumes. This would not be the case for an accelerating control volume, which is a *noninertial frame of reference*.

4.3 Conservation of Mass

The *law of conservation of mass* states that the mass of fluid within any defined fluid system remains constant, and this law can be expressed quantitatively as

$$\frac{\mathrm{d}}{\mathrm{d}t}\int_{\mathrm{sys}} \rho \,\mathrm{d}V = 0 \tag{4.7}$$

where ρ is the density of the fluid and $\mathrm{d}V$ is an elemental volume of fluid within the system. The Reynolds transport theorem relates the rate of change of any fluid property within a system to the changes in that same fluid property within a control volume, such that

$$\frac{\mathrm{d}}{\mathrm{d}t}\int_{\mathrm{sys}} \rho\beta \,\mathrm{d}V = \frac{\mathrm{d}}{\mathrm{d}t}\int_{\mathrm{cv}} \beta\rho \,\mathrm{d}V + \int_{\mathrm{cs}} \beta\rho(\mathbf{v}\cdot\mathbf{n})\,\mathrm{d}A \tag{4.8}$$

where β is a fluid property per unit mass, "cv" denotes the volume of the control volume, "cs" denotes the surface area of the control volume, $\mathbf{v}$ is the velocity field, and $\mathbf{n}$ is the unit normal pointing out of the control volume. Taking $\beta = 1$ in Equation 4.8 gives

$$\frac{\mathrm{d}}{\mathrm{d}t}\int_{\mathrm{sys}} \rho \,\mathrm{d}V = \frac{\mathrm{d}}{\mathrm{d}t}\int_{\mathrm{cv}} \rho \,\mathrm{d}V + \int_{\mathrm{cs}} (\mathbf{v}\cdot\mathbf{n})\,\mathrm{d}A \tag{4.9}$$

Combining Equations 4.7 and 4.9 at the instant when the system coincides with the control volume yields the following expression for the law of conservation of mass that is applicable to any given control volume:

$$\frac{\mathrm{d}}{\mathrm{d}t}\int_{\mathrm{cv}} \rho \,\mathrm{d}V + \int_{\mathrm{cs}} \rho(\mathbf{v}\cdot\mathbf{n})\,\mathrm{d}A = 0 \tag{4.10}$$

This equation is sometimes referred to as the *continuity equation*, and it states that the rate of change of fluid mass within the control volume (first term) plus the net flux of fluid mass out of the control volume (second term) is equal to zero. Upon rearrangement, Equation 4.10 also states that the rate of change of fluid mass within a control volume is equal to the net influx of mass into the control volume.

Continuity equation in terms of average velocities. In practical applications, it is more convenient (and just as accurate) to express inflow and outflow velocities in terms of average velocities over subareas rather than as velocity distributions. Consider a control volume with multiple inlets and outlets as illustrated in Figure 4.2. At each inlet and outlet, the density, the average velocity, and the flow area are denoted by ρ, $\overline{V}$, and A, respectively, with the subscript being an index identifier of the opening in the control volume. Because $\mathbf{n}$ in Equation 4.10 is the unit normal pointing out of the control volume, $\mathbf{v} \cdot \mathbf{n} = v$ on outflow surfaces and $\mathbf{v} \cdot \mathbf{n} = -v$ on inflow surfaces, where v is the magnitude of $\mathbf{v}$. Assuming that the density of the fluid remains constant across each designed area,

$$\int_{\text{cs}} \rho(\mathbf{v} \cdot \mathbf{n})\,\mathrm{d}A = \int_{A_{\text{out}}} \rho v\,\mathrm{d}A - \int_{A_{\text{in}}} \rho v\,\mathrm{d}A = \sum_{i=1}^{N_{\text{out}}} \rho_i \overline{V}_i A_i - \sum_{j=1}^{N_{\text{in}}} \rho_j \overline{V}_j A_j \tag{4.11}$$

where A_{out} and A_{in} are the areas where the flow is outward and inward, respectively, N_{in} and N_{out} are the number of inlet and outlet areas, respectively, and ρ_i, $\overline{V}_i$, and A_i are the density, average velocity, and area at inlet (or outlet) i, respectively. Combining Equations 4.10 and 4.11 gives the following useful form of the continuity equation:

$$\frac{\mathrm{d}}{\mathrm{d}t}\int_{\text{cv}} \rho\,\mathrm{d}V = \sum_{j=1}^{N_{\text{in}}} \rho_j \overline{V}_j A_j - \sum_{i=1}^{N_{\text{out}}} \rho_i \overline{V}_i A_i \tag{4.12}$$

The left-hand side of Equation 4.12 is the rate at which fluid mass is accumulating within the control volume. The right-hand side of Equation 4.12 is equal to the rate at which fluid mass is entering the control volume minus the rate at which fluid mass is leaving the control volume.

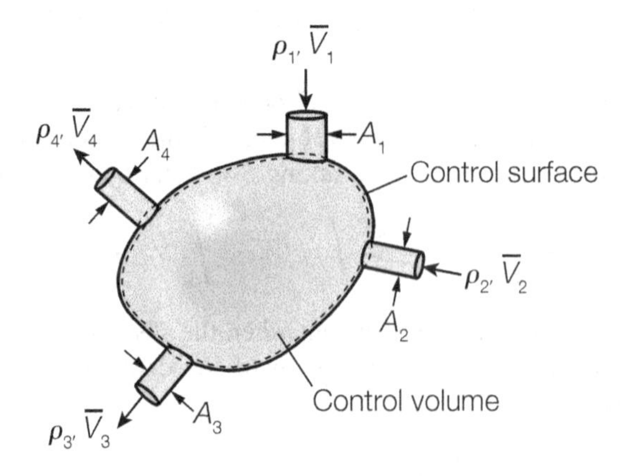

Figure 4.2: Control volume with multiple inlets and outlets

EXAMPLE 4.1

The velocity distribution, $v(r)$, in pipes with circular diameters can be approximated by an equation of the form

$$v(r) = V_{\text{max}}\left(1 - \frac{r^2}{R^2}\right)$$

where r is the radial distance from the centerline of the pipe, V_{max} is the maximum (centerline) velocity, and R is the radius of the pipe. If water at 20°C enters a reservoir through a 100-mm-diameter pipe with a maximum velocity of 1 m/s and leaves the reservoir through a 75-mm-diameter pipe with a maximum velocity of 0.8 m/s, at what rate is water accumulating in the reservoir?

SOLUTION

From the given data: $D_1 = 100$ mm, $R_1 = D_1/2 = 50$ mm, $V_{\text{max},1} = 1$ m/s, $D_2 = 75$ mm, $R_2 = D_2/2 = 37.5$ mm, and $V_{\text{max},2} = 0.8$ m/s. The subscripts 1 and 2 designate the inflow and outflow sections, respectively. For water at 20°C, $\rho = 998.2$ kg/m^3. The product of the average velocity and the flow area in the inflow and outflow pipes are calculated as follows:

$$\overline{V}_1 A_1 = \int_{A_1} v\,\text{d}A = \int_0^{R_1} V_{\text{max},1}\left(1 - \frac{r^2}{R_1^2}\right) 2\pi r\,\text{d}r = \frac{1}{2}\pi V_{\text{max},1} R_1^2 \tag{4.13}$$

$$\overline{V}_2 A_2 = \int_{A_2} v\,\text{d}A = \int_0^{R_2} V_{\text{max},2}\left(1 - \frac{r^2}{R_2^2}\right) 2\pi r\,\text{d}r = \frac{1}{2}\pi V_{\text{max},2} R_2^2 \tag{4.14}$$

The continuity equation, as given by Equation 4.12, requires that

$$\frac{\text{d}}{\text{d}t}\int_{\text{cv}} \rho\,\text{d}V = \rho_1 \overline{V}_1 A_1 - \rho_2 \overline{V}_2 A_2 \tag{4.15}$$

where the term on the left-hand side of this equation is the rate at which water is accumulating in the reservoir. Noting that $\rho_1 = \rho_2 = \rho$, Equations 4.13–4.15 combine to give

$$\frac{\text{d}}{\text{d}t}\int_{\text{cv}} \rho\,\text{d}V = \left[\frac{1}{2}\pi\rho V_{\text{max},1} R_1^2\right] - \left[\frac{1}{2}\pi\rho V_{\text{max},2} R_2^2\right] = \frac{1}{2}\pi\rho\left[V_{\text{max},1} R_1^2 - V_{\text{max},2} R_2^2\right]$$

Substituting the given values of the parameters yields

$$\frac{\text{d}}{\text{d}t}\int_{\text{cv}} \rho\,\text{d}V = \frac{1}{2}\pi(998.2)\left[(1)(0.05)^2 - (0.8)(0.0375)^2\right] = 2.16\ \text{kg/s}$$

Therefore, water is accumulating in the reservoir at the rate of **2.16 kg/s**. This is an unsteady-state condition in the flow system.

Continuity equation in terms of mass flow rates. The continuity equation as given by Equation 4.12 can also be expressed directly in terms of mass flow rates as

$$\frac{\text{d}}{\text{d}t}\int_{\text{cv}} \rho\,\text{d}V = \sum_{j=1}^{N_{\text{in}}} \dot{m}_j - \sum_{i=1}^{N_{\text{out}}} \dot{m}_i \tag{4.16}$$

where $\dot{m}_i$ is the mass flow rate through inlet (or outlet) i. In most applications of the continuity equation, the control volume is defined by the boundaries of a physical structure having inlets and outlets.

EXAMPLE 4.2

Air flows from multiple ports into a storage tank, and the air is discharged from a single port as shown in Figure 4.3. The port diameters and flow properties of the air are also shown in Figure 4.3. Estimate the rate at which air is accumulating in the storage tank.

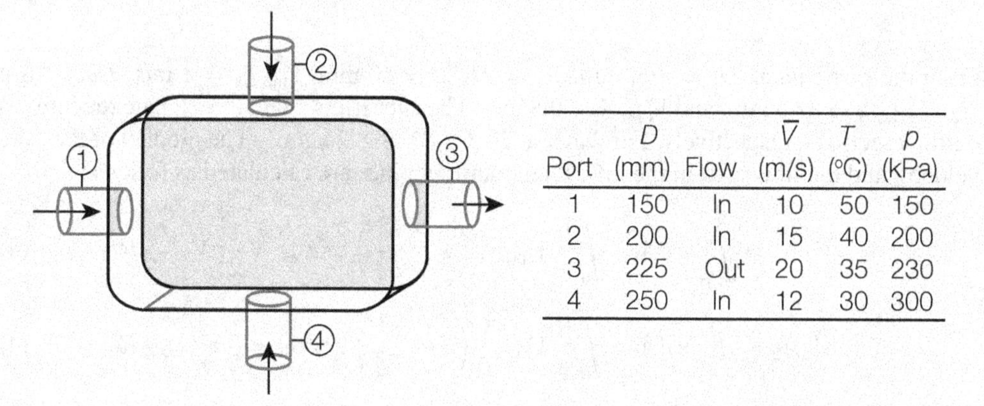

Port	D (mm)	Flow	$\overline{V}$ (m/s)	T (°C)	p (kPa)
1	150	In	10	50	150
2	200	In	15	40	200
3	225	Out	20	35	230
4	250	In	12	30	300

Figure 4.3: Inflow and outflow of air from a storage tank

SOLUTION

For air, $R = 287.1$ J/kg·K, and the following relationships can be used to determine the density, ρ, flow area, A, and mass flow rate, $\dot{m}$, across each port using the given data:

$$\rho = \frac{p}{RT}, \quad A = \frac{\pi D^2}{4}, \quad \dot{m} = \rho \overline{V} A$$

where T is the temperature in K. Applying these relationships to the given data yields the following results:

Port	T (K)	ρ (kg/m^3)	A (m^2)	$\dot{m}$ (kg/s)
1	323	1.618	0.0177	0.2859
2	313	2.226	0.0314	1.0489
3	308	2.601	0.0398	2.0685
4	303	3.449	0.0491	2.0315

Taking the storage tank as the control volume, the continuity equation, Equation 4.16, can be used to determine the rate at which air is accumulating in the storage tank as follows:

$$\frac{\mathrm{d}}{\mathrm{d}t}\int_{\mathrm{cv}} \rho\, \mathrm{d}V = (\dot{m}_1 + \dot{m}_2 + \dot{m}_4) - \dot{m}_3 = (0.2859 + 1.0489 + 2.0315) - 2.0685 = 1.298 \text{ kg/s}$$

Therefore, air is accumulating in the control volume at a rate of approximately **1.30 kg/s.**

4.3.1 Closed Conduits

Consider fluid flow through the section of a closed conduit (e.g., pipe or duct) shown in Figure 4.4, where the conduit section defines a control volume. The inflow section of the control volume has an area A_1, and the fluid passing this section has an average velocity $\overline{V}_1$ and density ρ_1; the fluid exits at the downstream section of the control volume that has an area A_2, with an average velocity $\overline{V}_2$ and density ρ_2. Hence, for the control volume shown in Figure 4.4, the applicable continuity equation can be expressed in either of the following simplified forms:

$$\frac{\mathrm{d}}{\mathrm{d}t}(\rho \mathcal{V}) = \dot{m}_1 - \dot{m}_2 \quad \text{or} \quad \frac{\mathrm{d}}{\mathrm{d}t}(\rho \mathcal{V}) = \rho_1 A_1 \overline{V}_1 - \rho_2 A_2 \overline{V}_2 \tag{4.17}$$

where $\mathcal{V}$ is the volume of the fluid in the conduit section (= control volume) and ρ is the average density of the fluid in the conduit section. Under steady-flow conditions, the continuity equation can be expressed as

$$\dot{m}_1 = \dot{m}_2 \quad \text{or} \quad \rho_1 A_1 \overline{V}_1 = \rho_2 A_2 \overline{V}_2 \tag{4.18}$$

In cases where the fluid is incompressible (i.e., same density at all sections of the conduit), the continuity equation further reduces to

$$Q_1 = Q_2 = Q \quad \text{or} \quad A_1 \overline{V}_1 = A_2 \overline{V}_2 \tag{4.19}$$

where Q is the *volume flow rate* through the conduit section. The relationships in Equation 4.19 simply state that the volume flow rate into the conduit section is equal to the volume flow rate out of the conduit section, provided the fluid is incompressible and flow conditions are steady. The volume flow rate, Q, is commonly referred to as the *flow rate* or the *discharge*.

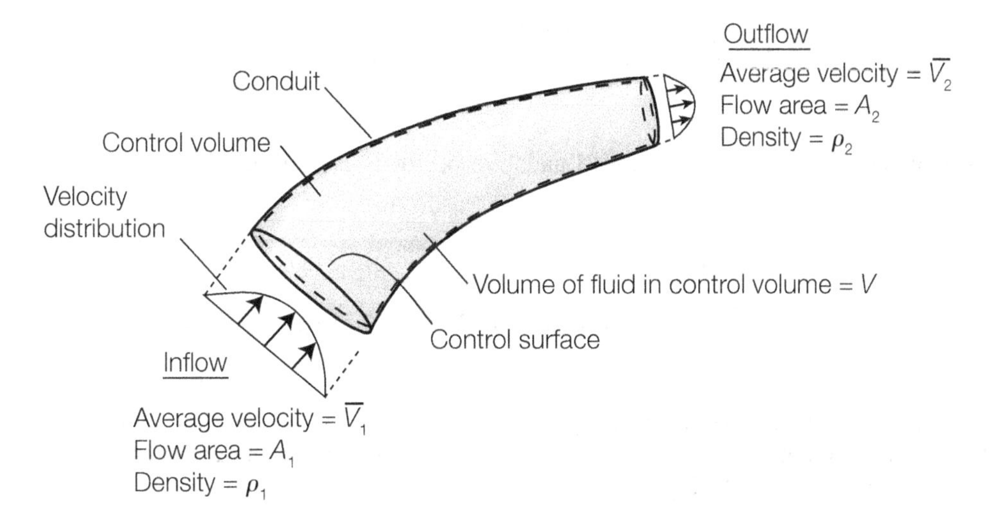

Figure 4.4: Flow through a conduit

EXAMPLE 4.3

Airflow through a conduit is commonly monitored for leaks by measuring the fluid properties at regular intervals along the conduit. Consider the case shown in Figure 4.5 where measurements at one section indicate a temperature, pressure, and average velocity of –23°C, 500 kPa, and 200 m/s, respectively, and measurements at a downstream section indicate a temperature, pressure, and average velocity of 17°C, 550 kPa, and 150 m/s, respectively. The conduit has a circular cross section and a diameter of 150 mm. Determine whether there is a leak in the conduit and estimate the magnitude of the leak in kilograms of air per second.

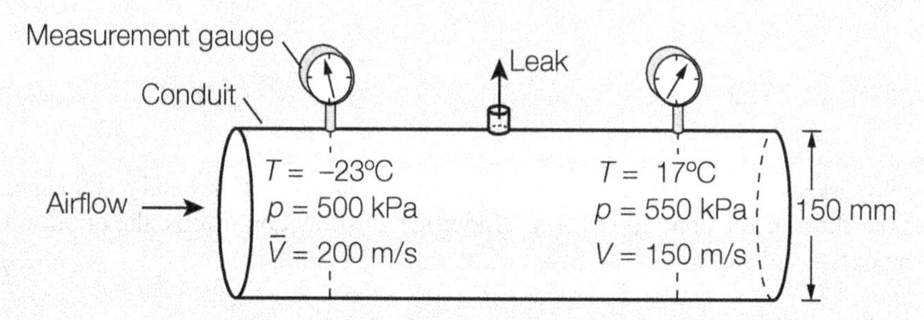

Figure 4.5: Airflow in a conduit

SOLUTION

From the given data: $T_1 = -23°\text{C} = 250$ K, $p_1 = 500$ kPa, $\overline{V}_1 = 200$ m/s, $T_2 = 17°\text{C} = 290$ K, $p_2 =$ 550 kPa, $\overline{V}_2 = 150$ m/s, $D = 150$ mm, and $A = \pi D^2/4 = 0.01767\ \text{m}^2$. For air, $R = 287.1$ J/kg·K. The air densities, ρ_1 and ρ_2, at sections 1 and 2, respectively, are as follows:

$$\rho_1 = \frac{p_1}{RT_1} = \frac{500 \times 10^3}{(287.1)(250)} = 6.967\ \text{kg/m}^3$$

$$\rho_2 = \frac{p_2}{RT_2} = \frac{550 \times 10^3}{(287.1)(290)} = 6.606\ \text{kg/m}^3$$

The mass flow rates, $\dot{m}_1$ and $\dot{m}_2$, at sections 1 and 2, respectively, are estimated as follows:

$$\dot{m}_1 = \rho_1 \overline{V}_1 A = (6.967)(200)(0.01767) = 24.62\ \text{kg/s}$$

$$\dot{m}_2 = \rho_2 \overline{V}_2 A = (6.606)(150)(0.01767) = 17.51\ \text{kg/s}$$

Because the mass flow rate at the downstream section (17.51 kg/s) is less than the mass flow rate at the upstream section (24.62 kg/s), **there is a leak**. The magnitude of the leak is estimated as 24.62 kg/s – 17.51 kg/s = **7.11 kg/s**.

4.3.2 Free Discharges from Reservoirs

Consider the case in which a liquid is contained within an open reservoir and is discharged through an outflow conduit as shown in Figure 4.6. In this case, the control volume includes the reservoir and the discharge port, and the continuity equation can be expressed as

$$\frac{\mathrm{d}}{\mathrm{d}t}\int_{V_\mathrm{c}} \rho \, \mathrm{d}V = -\dot{m}_\mathrm{out} \tag{4.20}$$

where V_c is the (fixed) volume of the control volume and $\dot{m}_\mathrm{out}$ is the mass flow rate out of the reservoir through the discharge port. Although the volume of the control volume is fixed, the liquid is only contained in a portion of the control volume. Let $\mathcal{V}$ be the volume of liquid contained in the control volume; then Equation 4.20 becomes

$$\frac{\mathrm{d}}{\mathrm{d}t}\left[\int_{\mathcal{V}} \rho \, \mathrm{d}V + \int_{V_\mathrm{c}-\mathcal{V}} \rho \, \mathrm{d}V\right] = -\dot{m}_\mathrm{out} \tag{4.21}$$

and because there is no liquid in the volume $V_\mathrm{c} - \mathcal{V}$,

$$\int_{V_\mathrm{c}-\mathcal{V}} \rho \, \mathrm{d}V = 0$$

and the continuity equation (Equation 4.20) can be stated simply as

$$\frac{\mathrm{d}}{\mathrm{d}t}\int_{\mathcal{V}} \rho \, \mathrm{d}V = -\dot{m}_\mathrm{out} \tag{4.22}$$

This equation forms a convenient starting point for the analysis of a reservoir being drained by one or more ports.

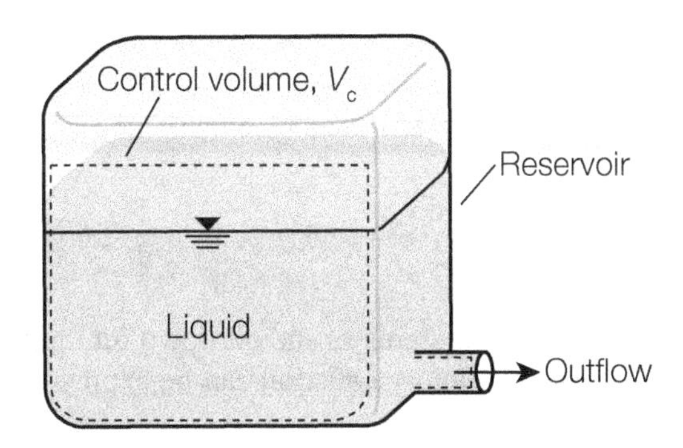

Figure 4.6: Flow from a reservoir

EXAMPLE 4.4

A cylindrical cistern shown in Figure 4.7 is 3 m high and 2 m in diameter. Water is released via a 10-mm-diameter circular orifice in the bottom of the cylinder, and it is known that the exit velocity, v, through the orifice is given by the relationship

$$v = \sqrt{2gh}$$

where h is the height of the water above the orifice. Determine how long a full cistern will take to empty.

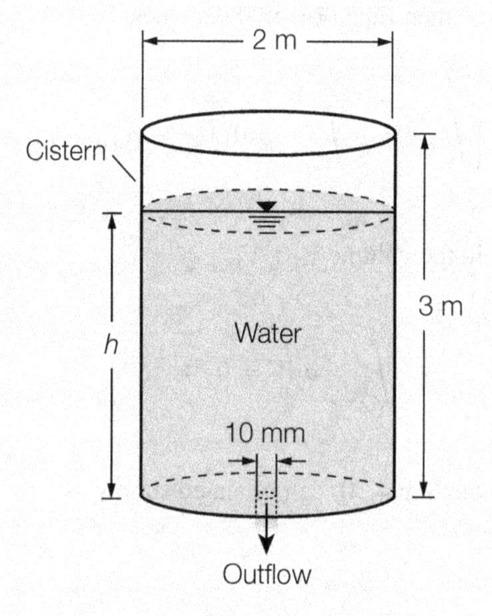

Figure 4.7: Water in a cistern

SOLUTION

From the given data: H = 3 m, D = 2 m, and D_o = 10 mm = 0.01 m. Noting that the density is constant and representing the volume of water in the cistern by $\mathcal{V}$, the continuity equation (Equation 4.22) requires that

$$\frac{\mathrm{d}}{\mathrm{d}t}\int_{\mathcal{V}} \rho \,\mathrm{d}V = -\dot{m}_{\mathrm{out}} \quad \rightarrow \quad \not{\rho}\frac{\mathrm{d}}{\mathrm{d}t}\int_{\mathcal{V}} \mathrm{d}V = \not{\rho} A\overline{V} \quad \rightarrow$$

$$\frac{\mathrm{d}}{\mathrm{d}t}\mathcal{V} = -A\overline{V} \quad \rightarrow \quad \frac{\mathrm{d}}{\mathrm{d}t}\left(\pi R^2 h\right) = -A_o\sqrt{2gh}$$

where R is the radius of the cistern, A_o is the area of the outflow port, and h is the depth of water in the cistern at any time t. The continuity equation can be expressed as the following differential equation:

$$\pi R^2 \frac{dh}{dt} = -A_o\sqrt{2g}\sqrt{h}$$

This equation can be solved by separating variables, integrating, and applying the initial condition that $h = H$ at $t = 0$, such that

$$\frac{\mathrm{d}h}{\sqrt{h}} = -\frac{A_\mathrm{o}\sqrt{2g}}{\pi R^2}\,\mathrm{d}t \quad\rightarrow\quad \int_H^0 \frac{\mathrm{d}h}{\sqrt{h}} = -\frac{A_\mathrm{o}\sqrt{2g}}{\pi R^2}\int_0^T \mathrm{d}t \quad\rightarrow\quad \left[2h^{\frac{1}{2}}\right]_H^0 = -\frac{A_\mathrm{o}\sqrt{2g}}{\pi R^2}\left[t\right]_0^T$$

which yields

$$T = \frac{2\sqrt{H}\pi R^2}{A_\mathrm{o}\sqrt{2g}}$$

where T is the time it takes to empty the cistern. Substituting the given values of the parameters yields

$$T = \frac{2\sqrt{3}\pi(1)^2}{\frac{\pi}{4}(0.01)^2\sqrt{2(9.81)}} = 31\,283\ \mathrm{s} = 8.69\ \mathrm{h}$$

Therefore, the cistern will take **8.69 hours** to empty. If desired, the emptying time could be controlled by adjusting the diameter of the discharge orifice in the bottom of the cistern.

4.3.3 Moving Control Volumes

The general continuity equation given by Equation 4.10 was derived for a control volume that is fixed in space. However, Equation 4.10 is also valid for a moving or deforming control volume provided the absolute velocity, $\mathbf{v}$, is replaced by the *relative velocity*, $\mathbf{v}_\mathrm{r}$, which is the fluid velocity relative to the velocity of the control volume and can be expressed as

$$\mathbf{v}_\mathrm{r} = \mathbf{v} - \mathbf{v}_\mathrm{cv} \tag{4.23}$$

where $\mathbf{v}_\mathrm{cv}$ is the velocity of the control volume. Hence, for a moving or deforming control volume, the continuity equation is given by

$$\frac{\mathrm{d}}{\mathrm{d}t}\int_\mathrm{cv} \rho\,\mathrm{d}V + \int_\mathrm{cs} \rho(\mathbf{v}_\mathrm{r}\cdot\mathbf{n})\,\mathrm{d}A = 0 \tag{4.24}$$

Moving control volumes of fixed shape differ from deforming control volumes in that for fixed-shape control volumes, $\mathbf{v}_\mathrm{cv}$ is constant on the surface of the control volume, whereas for deforming control volumes, $\mathbf{v}_\mathrm{cv}$ varies over the surface of the control volume. For a fixed control volume, where $\mathbf{v}_\mathrm{cv} = 0$, $\mathbf{v}_\mathrm{r} = \mathbf{v}$ and Equation 4.24 reverts back to Equation 4.10. In practical applications, it is convenient to work with the continuity equation in terms of average velocities over inflow and outflow areas of the control surface. The continuity equation for a control volume moving with a constant velocity, Equation 4.24, can be expressed as

$$\frac{\mathrm{d}}{\mathrm{d}t}\int_\mathrm{cv} \rho\,\mathrm{d}V = \sum_{j=1}^{N_\mathrm{in}} \rho_j \overline{V}_{j\mathrm{r}} A_j - \sum_{i=1}^{N_\mathrm{out}} \rho_i \overline{V}_{i\mathrm{r}} A_i \tag{4.25}$$

where N_in and N_out are the number of inlet and outlet areas, respectively, and ρ_i, $\overline{V}_{i\mathrm{r}}$, and A_i are the density, average velocity relative to the moving control volume, and area at inlet (or outlet) i, respectively. The continuity equation can also be expressed directly in terms of mass flow rates by the relation

$$\frac{\mathrm{d}}{\mathrm{d}t}\int_\mathrm{cv} \rho\,\mathrm{d}V = \sum_{j=1}^{N_\mathrm{in}} \dot{m}_{j\mathrm{r}} - \sum_{i=1}^{N_\mathrm{out}} \dot{m}_{i\mathrm{r}} \tag{4.26}$$

where $\dot{m}_{i\mathrm{r}}$ is the mass flow rate through inlet (or outlet) i relative to the moving control volume.

EXAMPLE 4.5

The jet engine shown in Figure 4.8 has an intake area of 0.9 m^2 and an exhaust area of 0.6 m^2. Performance data indicate that under normal operating conditions, the exhaust velocity is 20% higher than the intake velocity. If this jet engine is attached to an aircraft that is cruising at a speed of 800 km/h at an altitude of 10 km and the fuel consumption rate is 2.2 kg/s, estimate the density of the exhaust gas.

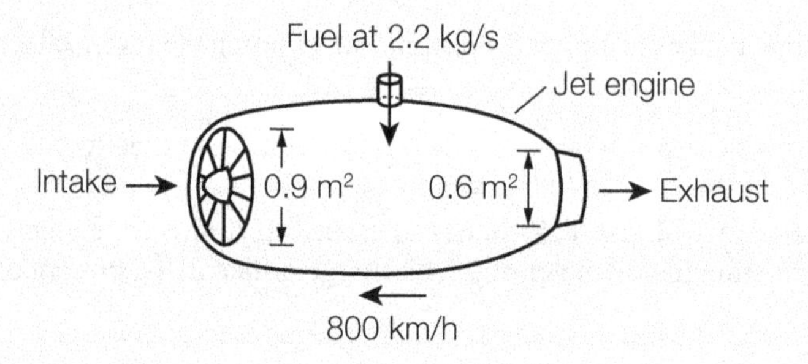

Figure 4.8: Jet engine

SOLUTION

From the given data: $A_1 = 0.9\,\text{m}^2$, $A_2 = 0.6\,\text{m}^2$, and $\dot{m}_{\text{fuel}} = 2.2$ kg/s. At an elevation of 10 km in a standard atmosphere, $\rho_{\text{air}} = \rho_1 = 0.4135$ kg/m^3. Expressing the intake and exhaust velocities relative to the jet engine, which is moving at a speed of 800 km/h, the relative velocities of the air at the intake and exhaust sections of the engine are $V_{1\text{r}}$ = 800 km/h = 222.2 m/s and $V_{2\text{r}}$ = 1.2(222.2) = 266.6 m/s. If the control volume is taken to include the entire jet engine and steady-state conditions are assumed, applying Equation 4.25 gives

$$\frac{\text{d}}{\text{d}t}\int_{\text{cv}} \rho\,\text{d}V = \underbrace{[\dot{m}_{\text{fuel}} + \rho_1 A_1 V_{1\text{r}}]}_{\text{mass inflow}} - \underbrace{[\rho_2 A_2 V_{2\text{r}}]}_{\text{mass outflow}}$$

and substituting the values of the known parameters gives

$$0 = [2.2 + (0.4135)(0.9)(222.2)] - [\rho_2\,(0.6)(266.6)] \quad \rightarrow \quad \rho_2 = 0.5307\ \text{kg/m}^3$$

Therefore, the density of the exhaust gas is estimated as **0.5307 kg/m^3**.

4.4 Conservation of Linear Momentum

The motion of any fluid system is governed by Newton's laws of motion, which are the same laws that govern the motion of solids. *Newton's second law of motion*, commonly referred to as the *law of conservation of linear momentum*, requires that the sum of the forces acting on the fluid within a system is equal to the rate of change of (linear) momentum of the system. Strictly speaking, use of the term "conservation of linear momentum" to describe Newton's second law is a misnomer, because the rate of change of the linear momentum of a system can be nonzero, and this rate of change is equal to the resultant force on the system. In other words, linear momentum can be created or destroyed by the application of a force. Consequently, linear momentum is not conserved in the same manner that mass is conserved as required by the law of conservation of mass. In spite of this anomalous terminology, "law of conservation of linear momentum" is widely used to refer to Newton's second law, and this convention will be followed here.

4.4.1 General Momentum Equations

Consider a fluid system containing a fixed mass of fluid. If the system is composed of elements of mass δm, then the force acting on each fluid element, $\delta \mathbf{F}$, is related to the momentum of the fluid element using Newton's second law (i.e., the law of conservation of linear momentum) which gives

$$\delta \mathbf{F} = \frac{\mathrm{d}}{\mathrm{d}t}(\delta m \, \mathbf{v}) \tag{4.27}$$

where $\mathbf{v}$ is the velocity of the fluid element. Integrating Equation 4.27 over a fluid system consisting of a continuum of fluid elements of infinitesimally small size gives

$$\int_{\text{sys}} \delta \mathbf{F} = \frac{\mathrm{d}}{\mathrm{d}t} \int_{\text{sys}} \mathbf{v} \, \delta m \tag{4.28}$$

Noting that the differential mass, δm, is equal to $\rho \, \mathrm{d}V$, where ρ is the density of the fluid and $\mathrm{d}V$ is an elemental volume within the fluid system, Equation 4.28 can be conveniently expressed as

$$\sum \mathbf{F}_{\text{sys}} = \frac{\mathrm{d}}{\mathrm{d}t} \int_{\text{sys}} \mathbf{v} \rho \, \mathrm{d}V \tag{4.29}$$

where $\sum \mathbf{F}_{\text{sys}}$ is the sum of the forces acting on the system. Recall that the Reynolds transport theorem relates the rate of change of any fluid property within a system to the changes in that same fluid property within a coincident control volume, such that

$$\frac{\mathrm{d}}{\mathrm{d}t} \int_{\text{sys}} \beta \rho \, \mathrm{d}V = \frac{\mathrm{d}}{\mathrm{d}t} \int_{\text{cv}} \beta \rho \, \mathrm{d}V + \int_{\text{cs}} \beta \rho (\mathbf{v} \cdot \mathbf{n}) \, \mathrm{d}A \tag{4.30}$$

where "cv" and "cs" denote the volume and surface area of the control volume, respectively, $\mathbf{n}$ is the unit normal pointing out of the control volume, and β is a fluid property per unit mass. Taking $\beta = \mathbf{v}$ in Equation 4.30 gives

$$\frac{\mathrm{d}}{\mathrm{d}t} \int_{\text{sys}} \mathbf{v} \rho \, \mathrm{d}V = \frac{\mathrm{d}}{\mathrm{d}t} \int_{\text{cv}} \mathbf{v} \rho \, \mathrm{d}V + \int_{\text{cs}} \mathbf{v} \rho (\mathbf{v} \cdot \mathbf{n}) \, \mathrm{d}A \tag{4.31}$$

This equation is a mathematical statement of the fact that at any given instant, the rate of change of momentum within a (moving) fluid system is equal to the rate of change of momentum within a (stationary) control volume that coincides with the fluid system, plus the net flux of momentum out of the control volume. Combining Equations 4.29 and 4.31 yields the following expression for the law of conservation of momentum:

$$\sum \mathbf{F}_{\text{cv}} = \frac{\mathrm{d}}{\mathrm{d}t} \int_{\text{cv}} \rho \mathbf{v} \, \mathrm{d}V + \int_{\text{cs}} \rho \mathbf{v} \, (\mathbf{v} \cdot \mathbf{n}) \, \mathrm{d}A \tag{4.32}$$

where $\sum \mathbf{F}_{\text{cv}}$ represents the sum of the forces on the control volume. Note that when the system coincides with the control volume, the sum of the forces on the control volume is equal to the sum of the forces on the system; that is,

$$\sum \mathbf{F}_{\text{sys}} = \sum \mathbf{F}_{\text{cv}} = \sum \mathbf{F} \tag{4.33}$$

where $\sum \mathbf{F}$ will be used in subsequent analyses to represent the sum of forces on a control volume. The momentum equation given by Equation 4.32 is sometimes referred to as the *linear momentum equation* or simply the *momentum equation* and can be written in component form as

$$\sum F_x = \frac{\mathrm{d}}{\mathrm{d}t}\int_{\mathrm{cv}} \rho v_x\,\mathrm{d}V + \int_{\mathrm{cs}} \rho v_x(\mathbf{v}\cdot\mathbf{n})\,\mathrm{d}A \tag{4.34}$$

$$\sum F_y = \frac{\mathrm{d}}{\mathrm{d}t}\int_{\mathrm{cv}} \rho v_y\,\mathrm{d}V + \int_{\mathrm{cs}} \rho v_y(\mathbf{v}\cdot\mathbf{n})\,\mathrm{d}A \tag{4.35}$$

$$\sum F_z = \frac{\mathrm{d}}{\mathrm{d}t}\int_{\mathrm{cv}} \rho v_z\,\mathrm{d}V + \int_{\mathrm{cs}} \rho v_z(\mathbf{v}\cdot\mathbf{n})\,\mathrm{d}A \tag{4.36}$$

where F_x, F_y, and F_z are the components of $\mathbf{F}_{\mathrm{cv}}$ (= $\mathbf{F}$) and v_x, v_y, and v_z are components of $\mathbf{v}$. The linear momentum equation is commonly applied to determine the forces on various hydraulic structures, such as deflectors and blades in turbomachines, pipe bends, and enlargements and contractions in pipeline systems. In most cases of practical interest, the reaction force of the hydraulic structure balances the rate of change of momentum as the fluid passes through the hydraulic structure. In calculating the forces on structures, the control volume is typically taken as the volume of fluid that is in contact with the structure of interest and bounded by identifiable inflow and outflow sections and/or structural (solid) boundaries. Selection of appropriate control volumes will be illustrated in subsequent applications and examples.

Momentum-flux correction factor. The various forms of the momentum equation all involve calculating the momentum flux across the surface area of a control volume (i.e., the control surface). The control surface is commonly selected such that the velocity is normal to the control surface, in which case the momentum flux across a control surface of area A is given by

$$\text{momentum flux across } A = \int_A \rho\mathbf{v}(\mathbf{v}\cdot\mathbf{n})\,\mathrm{d}A = \int_A \rho v_{\mathrm{n}}(v_{\mathrm{n}})\,\mathrm{d}A = \int_A \rho\, v_{\mathrm{n}}^2\,\mathrm{d}A$$

where v_{n} is the component of $\mathbf{v}$ that is normal to the control surface. It is convenient to represent the above integral expression in terms of the average velocity, $\overline{V}$, and mass flow rate, $\dot{m}$, across A such that

$$\int_A \rho\, v_{\mathrm{n}}^2\,\mathrm{d}A = \beta\rho\overline{V}^2 A = \beta\,\dot{m}\,\overline{V} \tag{4.37}$$

where β is a dimensionless factor that accounts for the variation of the velocity, v_{n}, over A. Note that the dimensionless factor β in Equation 4.37 is unrelated to the (dimensional) parameter β used in the Reynolds transport theorem to represent the amount of fluid property per unit mass of fluid. In conventional fluid mechanics applications, the symbol used for both of these quantities is β, and this convention will be followed here. It is apparent from Equation 4.37 that if the velocity is constant over A, then $\beta = 1$; if the velocity is not constant over A, then $\beta > 1$. The parameter β in Equation 4.37 is called the *momentum-flux correction factor*. For sections of a control surface over which the density does not vary significantly, the momentum-flux correction factor is commonly expressed as

$$\beta = \frac{1}{A}\int_A \left(\frac{v}{\overline{V}}\right)^2 \mathrm{d}A \tag{4.38}$$

where the velocity v is understood to be normal to the surface A. It is apparent from Equation 4.38 that when $v = \overline{V}$, $\beta = 1$, which again demonstrates that when the velocity is uniform across the cross section, $\beta = 1$. The extent to which β deviates from 1 can be used as a measure of the extent to which the velocity profile deviates from being uniform. For fully developed laminar flow in a circular pipe, $\beta = \frac{4}{3}$, and for turbulent flow in a circular pipe, β is typically in the range of 1.01–1.05. In the case of turbulent flow in circular pipes, it is commonly assumed that $\beta = 1$.

EXAMPLE 4.6

Estimate the momentum-flux correction factor, β, for the case of laminar flow in a circular pipe of radius R, where the velocity profile is radially symmetrical about the centerline of the pipe and is given by

$$v(r) = 2\overline{V}\left(1 - \frac{r^2}{R^2}\right)$$

where r is the radial distance from the center of the pipe and $\overline{V}$ is the average velocity in the pipe.

SOLUTION

Using the definition of β given by Equation 4.38,

$$\begin{aligned}\beta &= \frac{1}{A}\int_A \left(\frac{v}{\overline{V}}\right)^2 \mathrm{d}A = \frac{1}{\pi R^2}\int_0^R 4\left(1 - \frac{r^2}{R^2}\right)^2 2\pi r\,\mathrm{d}r = \frac{4 \cdot 2\cancel{\pi}}{\cancel{\pi}R^2}\int_0^R \left(r - \frac{2r^3}{R^2} + \frac{r^5}{R^4}\right)\mathrm{d}r \\ &= \frac{8}{R^2}\left[\frac{r^2}{2} - \frac{r^4}{2R^2} + \frac{r^6}{6R^4}\right]_0^R = \frac{8}{R^2}\left[\frac{R^2}{2} - \frac{R^2}{2} + \frac{R^2}{6}\right] = 8\left(\frac{1}{6}\right) = \frac{4}{3}\end{aligned}$$

Therefore, the momentum-flux correction factor is equal to **4/3**.

Momentum equation in terms of average velocities. In practical applications of the momentum equation, it is common to represent inflow and outflow velocities relative to the control volume by their average values rather than as velocity distributions. In such cases, the velocity distribution is accounted for by the momentum-flux correction factor. Because the momentum-flux correction factor can frequently be approximated by unity, the formulation of the momentum equation in terms of average velocities is further simplified. The momentum equation (Equation 4.32) expressed in terms of average velocities and momentum-flux correction factors is as follows:

$$\sum \mathbf{F} = \frac{\mathrm{d}}{\mathrm{d}t}\int_{\text{cv}} \rho\mathbf{v}\,\mathrm{d}V + \sum_{i=1}^{N_{\text{out}}} \beta_i \dot{m}_i \overline{\mathbf{V}}_i - \sum_{j=1}^{N_{\text{in}}} \beta_j \dot{m}_j \overline{\mathbf{V}}_j \tag{4.39}$$

where N_{in} and N_{out} are the number of inlets and outlets to the control volume, respectively, and β_i, $\dot{m}_i$, and $\overline{\mathbf{V}}_i$ are the momentum-flux correction factor, mass flow rate, and average velocity across inlet (or outlet) i, respectively. The momentum equation, as given by Equation 4.39, is frequently applied under steady-state conditions, in which case the following simplified form can be used directly:

$$\sum \mathbf{F} = \sum_{i=1}^{N_{\text{out}}} \beta_i \dot{m}_i \mathbf{V}_i - \sum_{j=1}^{N_{\text{in}}} \beta_j \dot{m}_j \mathbf{V}_j \quad \text{(steady state)} \tag{4.40}$$

This equation is particularly useful in finding the forces exerted by fluids on external structures when the inflow and outflow velocities and mass flow rates are known. The component forms of Equation 4.40 are more directly useful and are given by

$$\sum F_x = \sum_{i=1}^{N_{\text{out}}} \beta_i \dot{m}_i V_{x,i} - \sum_{j=1}^{N_{\text{in}}} \beta_j \dot{m}_j V_{x,j} \tag{4.41}$$

$$\sum F_y = \sum_{i=1}^{N_{\text{out}}} \beta_i \dot{m}_i V_{y,i} - \sum_{j=1}^{N_{\text{in}}} \beta_j \dot{m}_j V_{y,j} \tag{4.42}$$

$$\sum F_z = \sum_{i=1}^{N_{\text{out}}} \beta_i \dot{m}_i V_{z,i} - \sum_{j=1}^{N_{\text{in}}} \beta_j \dot{m}_j V_{z,j} \tag{4.43}$$

where the subscripts x, y, and z indicate vector components.

EXAMPLE 4.7

Water flowing at 30 m^3/s in a 10-m-wide horizontal rectangular channel passes under an open gate as shown in Figure 4.9(a). Assuming that the shear stress exerted on the water by the bottom of the channel can be neglected, calculate the force exerted by the water on the gate. How does this force on the gate compare with the force on the gate when it is closed and the water depth behind the gate is 3 m?

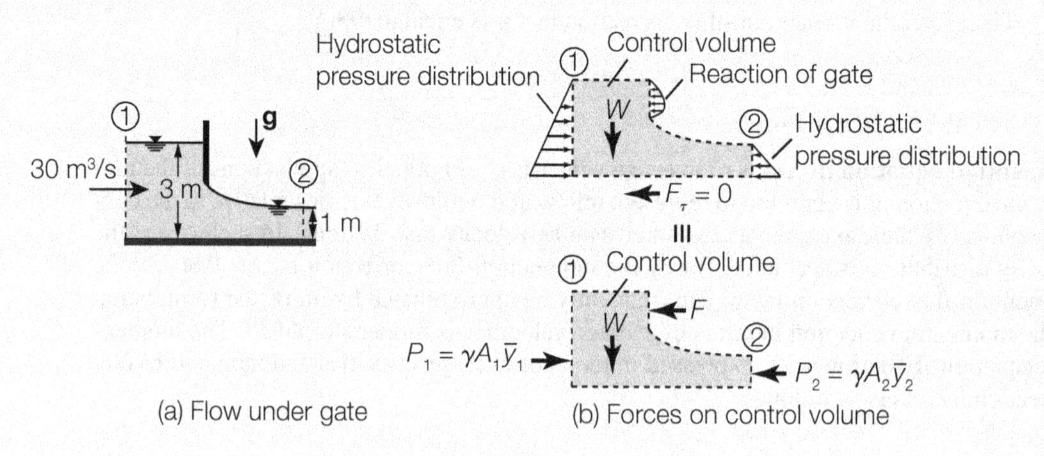

Figure 4.9: Flow under an open gate

SOLUTION

From the given data: Q = 30 m^3/s, b = 10 m, y_1 = 3 m, and y_2 = 1 m. Assume that $\beta = 1$, ρ = 998 kg/m^3, and γ = 9790 N/m^3 (at 20°C). A control volume between the upstream section (section 1) and the downstream section (section 2) is illustrated in Figure 4.9(b), where P_1 and P_2 are the resultant pressure forces on sections 1 and 2, respectively, and the resultant reaction of the gate to the force exerted by the water is F. The weight of the water within the control volume is W, and the shear force on the bottom of the control volume (exerted by the bottom of the channel) is taken as being negligible. The force on the gate can be calculated by applying the steady-state momentum equation, Equation 4.40, in the flow direction, which can be expressed as

$$P_1 - P_2 - F = \dot{m}(\overline{V}_2 - \overline{V}_1) \tag{4.44}$$

where V_2 and V_1 are the average inflow and outflow velocities, respectively. Using the given data,

$$\dot{m} = \rho Q = (998)(30) = 2.994 \times 10^4 \text{ kg/s}$$

$$A_1 = by_1 = (10)(3) = 30 \text{ m}^2, \qquad A_2 = by_2 = (10)(1) = 10 \text{ m}^2$$

$$\overline{V}_1 = \frac{Q}{A_1} = \frac{30}{30} = 1 \text{ m/s}, \qquad \overline{V}_2 = \frac{Q}{A_2} = \frac{30}{10} = 3 \text{ m/s}$$

$$P_1 = \gamma A_1 \bar{y}_1 = (9790)(30)\left(\frac{3}{2}\right) = 4.406 \times 10^5 \text{ N}$$

$$P_2 = \gamma A_2 \bar{y}_2 = (9790)(10)\left(\frac{1}{2}\right) = 4.895 \times 10^4 \text{ N}$$

Substituting the calculated parameters into Equation 4.44 yields

$$4.406 \times 10^5 - 4.895 \times 10^4 - F = (2.994 \times 10^4)(3 - 1)$$

which yields $F = 3.32 \times 10^5$ N = 332 kN. Hence, the hydrodynamic force on the gate is 332 kN. When the gate is closed, the hydrostatic force, F_h, on the gate is given by

$$F_\text{h} = \gamma A_1 \bar{y}_1 = (9.79)\,(30)\left(\frac{3}{2}\right) = 441 \text{ kN}$$

Therefore, opening the gate reduces the force on the gate from **441 kN** to **332 kN**, a reduction of approximately 25%.

4.4.2 Forces on Pressure Conduits

Conduits that transport fluids under pressure are commonplace in engineering applications. Components of these conduits that change the velocity and/or flow direction of a fluid must be designed with sufficient structural support to withstand dynamic forces. Two common components of pressure conduits are reducers and bends.

4.4.2.1 Forces on reducers

Consider the case of a fluid flowing through an inclined conduit with a reducer segment as shown in Figure 4.10(a). Upstream of the reducer, the cross-sectional area is A_1, the average velocity is V_1, the pressure is p_1, the corresponding variables downstream of the reducer are A_2, V_2, and p_2, respectively, the reducer is inclined at an angle θ to the horizontal, and gravity is assumed to act in the vertical direction. The objective here is to find the force acting on the reducer segment of the conduit, so it is logical to define a control volume that contains the volume of fluid within the reducer segment as shown in Figure 4.10(b). The forces acting on the surface of the control volume are shown in Figure 4.10(c). They consist of the pressure forces on the upstream and downstream (inflow and outflow, respectively) faces of the control volume, the reaction of the conduit wall to the pressure forces acting

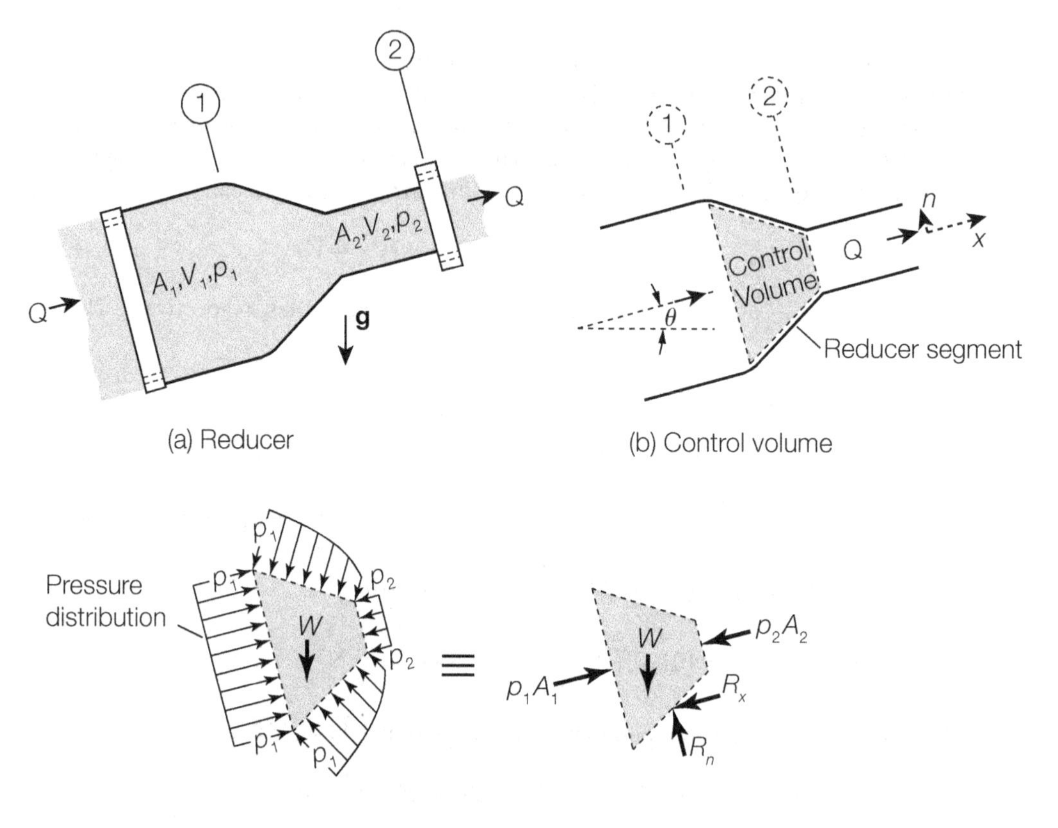

Figure 4.10: Force on a reducer

on the wall, and the weight of the fluid within the control volume. The resultant force on the upstream face is p_1A_1, the resultant force on the downstream face is p_2A_2, the resultant reaction of the conduit wall in the flow (x) direction is R_x, the resultant reaction of the conduit wall in the direction normal to the flow (in the n direction) is R_n, and the weight of the fluid in the control volume is W. Note that pressure forces always act inward to the control volume, even on surfaces where the flow is outward, because pressure forces represent the forces exerted by the surrounding fluid on the fluid contained within the control volume. The x component of the steady-state momentum equation is given by Equation 4.41, which yields

$$\sum F_x = \dot{m}(\beta_{\text{out}}\overline{V}_{x,\text{out}} - \beta_{\text{in}}\overline{V}_{x,\text{in}}) \tag{4.45}$$

$$p_1A_1 - p_2A_2 - R_x - W\sin\theta = \dot{m}(V_2 - V_1) \tag{4.46}$$

where it is assumed that $\beta_{\text{in}} \approx \beta_{\text{out}} \approx 1$. The reaction force R_x is therefore given by Equation 4.46 as

$$R_x = p_1A_1 - p_2A_2 - W\sin\theta - \dot{m}(V_2 - V_1) \tag{4.47}$$

The steady-state momentum equation can also be applied in the (n-)direction normal to the flow, which gives the normal reaction force, R_n, as

$$R_n = W\cos\theta \tag{4.48}$$

The surface-reaction forces, R_x and R_n, are equal and opposite to the forces exerted by the fluid on the reducer segment, and a support force equal to $\sqrt{R_x^2 + R_n^2}$ must be provided to keep the reducer in place.

EXAMPLE 4.8

A horizontal reducer is installed to connect a 100-mm-diameter pipe to a 50-mm-diameter pipe as shown in Figure 4.11. The pipe is to be used to transport water at 20°C and at a design flow rate of 20 L/s. At the design flow rate, the pressure upstream of the reducer is expected to equal 350 kPa. What force will be needed to support the reducer under the design condition?

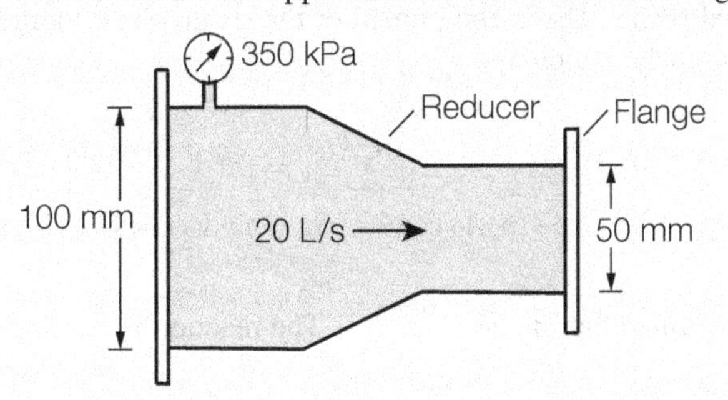

Figure 4.11: Reducer operating at the design condition

SOLUTION

From the given data: D_1 = 100 mm, D_2 = 50 mm, Q = 20 L/s = 0.02 m^3/s, and p_1 = 350 kPa. For water at 20°C, ρ = 998 kg/m^3 and γ = 9.79 kN/m^3. The following preliminary calculations are useful:

$$A_1 = \frac{\pi}{4}D_1^2 = \frac{\pi}{4}(0.1)^2 = 0.007854 \text{ m}^2, \qquad A_2 = \frac{\pi}{4}D_2^2 = \frac{\pi}{4}(0.05)^2 = 0.001963 \text{ m}^2$$

$$V_1 = \frac{Q}{A_1} = \frac{0.02}{0.007854} = 2.547 \text{ m/s} \qquad V_2 = \frac{Q}{A_2} = \frac{0.02}{0.001963} = 10.12 \text{ m/s}$$

The pressure at the downstream section can be estimated by applying the Bernoulli equation between the upstream section (section 1) and the downstream section (section 2), which yields

$$\frac{p_1}{\gamma} + \frac{V_1^2}{2g} + z_1 = \frac{p_2}{\gamma} + \frac{V_2^2}{2g} + z_2 \quad \rightarrow \quad \frac{350}{9.79} + \frac{2.547^2}{2(9.81)} + 0 = \frac{p_2}{9.79} + \frac{10.12^2}{2(9.81)} + 0$$

$$\rightarrow \quad p_2 = 301.5 \text{ kPa}$$

It is important to remember that this approximation of the downstream pressure is approximate in that it neglects any energy losses that occur in the reducer and does not take into account velocity variations across each section. However, these effects are likely to be relatively small, and p_2 = 301.5 kPa is expected to be a good approximation of the downstream pressure. The mass flow rate, $\dot{m}$, through the reducer is given by

$$\dot{m} = \rho Q = (998)(0.02) = 19.96 \text{ kg/s}$$

Applying the momentum equation given by Equation 4.47 yields

$$\begin{aligned} R_x &= p_1A_1 - p_2A_2 - \dot{m}(V_2 - V_1) \\ &= (350)(0.007854) - (301.5)(0.001963) - (19.96)(10.12 - 2.547)\ [\times 10^{-3} \text{ kN/N}] \\ &= 2.00 \text{ kN} \end{aligned}$$

Therefore, the force exerted by the water on the reducer is estimated as **2.00 kN**. The bolts in the upstream flange of the reducer should be capable of supporting this force.

4.4.2.2 Forces on bends

Consider a fluid flowing through the bend illustrated in Figure 4.12(a). Upstream of the bend, the cross-sectional area is A_1, the average velocity is V_1, the pressure is p_1, and the corresponding variables downstream of the bend are A_2, V_2, and p_2, respectively. A control volume within the bend is shown in Figure 4.12(b), where the forces acting on the fluid in the control volume are the reaction components, R_x and R_z, the pressure forces, p_1A_1 and p_2A_2, and the weight, W, of fluid within the control volume. This assumes that the bend is in the vertical plane. The x component of the steady-state momentum equation is given by Equation 4.41, which yields

$$\sum F_x = \dot{m}(\beta_{\text{out}}\overline{V}_{x,\text{out}} - \beta_{\text{in}}\overline{V}_{x,\text{in}}) \tag{4.49}$$

$$p_1A_1 - p_2A_2\cos\theta - R_x = \dot{m}(V_2\cos\theta - V_1) \tag{4.50}$$

where it is assumed that $\beta_{\text{in}} \approx \beta_{\text{out}} \approx 1$. The reaction force R_x is therefore given by Equation 4.50 as

$$R_x = p_1A_1 - p_2A_2\cos\theta - \dot{m}(V_2\cos\theta - V_1) \tag{4.51}$$

The z-component of the steady-state momentum equation is given by Equation 4.43, which yields

$$\sum F_z = \dot{m}(\beta_{\text{out}}\overline{V}_{z,\text{out}} - \beta_{\text{in}}\overline{V}_{z,\text{in}}) \tag{4.52}$$

$$R_z - p_2A_2\sin\theta - W = \dot{m}(V_2\sin\theta - 0) \tag{4.53}$$

where it is again assumed that $\beta_{\text{in}} \approx \beta_{\text{out}} \approx 1$. The reaction force R_z is therefore given by Equation 4.53 as

$$R_z = p_2A_2\sin\theta + W + \dot{m}V_2\sin\theta \tag{4.54}$$

As noted in the case of a reducer, the reaction force is equal and opposite to the force exerted on the bend by the fluid, and a supporting force equal to the force on the bend must be provided to keep the bend in place.

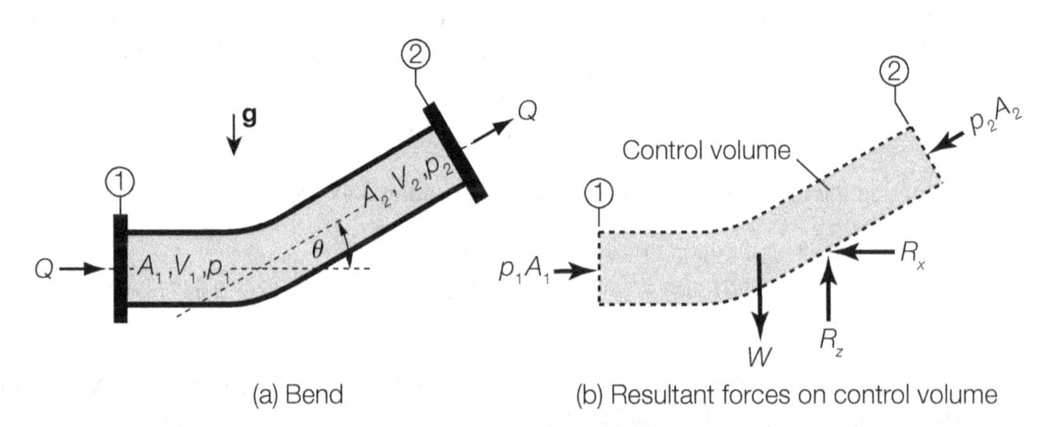

Figure 4.12: Force on a bend

EXAMPLE 4.9

Consider the 120° reducer-bend as shown in Figure 4.13. The entrance to the reducer bend has a diameter of 200 mm, and the exit has a diameter of 175 mm. When water at 20°C flows at a rate of 100 L/s through the reducer bend, the measured pressures at the entrance and exit are 350 kPa and 215 kPa, respectively. The reducer bend is in the horizontal plane. (a) What is the magnitude and direction of the force exerted by the water on the bend? (b) How does the pressure at the exit of the bend compare with that estimated using the Bernoulli equation? What is a possible reason for the discrepancy?

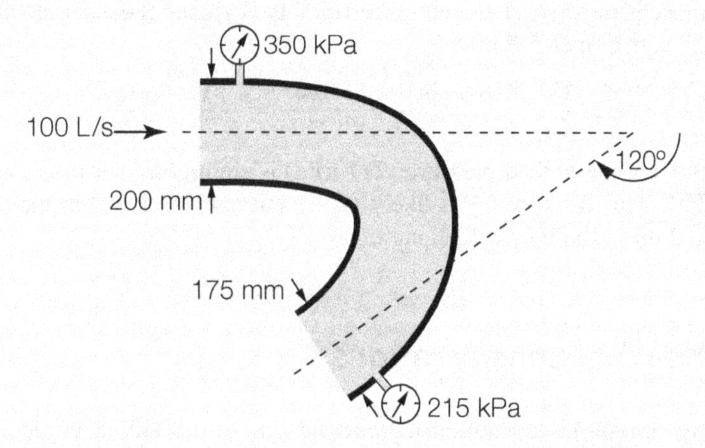

Figure 4.13: Flow through a reducer-bend

SOLUTION

From the given data: θ = 120°, D_1 = 200 mm, D_2 = 175 mm, Q = 100 L/s = 0.1 m^3/s, p_1 = 350 kPa, and p_2 = 215 kPa. For water at 20°C, γ = 9.79 kN/m^3 and ρ = 998 kg/m^3. The following preliminary calculations are useful:

$$A_1 = \frac{\pi}{4}D_1^2 = \frac{\pi}{4}(0.2)^2 = 0.03142 \text{ m}^2, \qquad A_2 = \frac{\pi}{4}D_2^2 = \frac{\pi}{4}(0.175)^2 = 0.006013 \text{ m}^2$$

$$V_1 = \frac{Q}{A_1} = \frac{0.1}{0.03142} = 3.183 \text{ m/s} \qquad V_2 = \frac{Q}{A_2} = \frac{0.1}{0.006013} = 16.63 \text{ m/s}$$

$$\dot{m} = \rho Q = (998)(0.1) = 99.8 \text{ kg/s}$$

(a) The component of the reaction in the inflow (x) direction, R_x, is given by Equation 4.51 as

$$\begin{aligned} R_x &= p_1A_1 - p_2A_2\cos\theta - \dot{m}[V_2\cos\theta - V_1] \\ &= (350)(0.03142) - (215)(0.006013)\cos 120^\circ \\ &\quad - (99.8)[(16.63)\cos 120^\circ - 3.183]\ [\times 10^{-3} \text{ kN/N}] \\ &= 12.79 \text{ kN} \end{aligned}$$

Similarly, the component of the reaction in the direction perpendicular to the inflow direction, R_y, is given by Equation 4.54, without the weight component, as

$$\begin{aligned} R_y &= p_2A_2\sin\theta + \dot{m}V_2\sin\theta \\ &= (215)(0.006013)\sin 120^\circ + (99.8)(16.63)\sin 120^\circ\ [\times 10^{-3} \text{ kN/N}] \\ &= 2.557 \text{ kN} \end{aligned}$$

The magnitude, R, and direction, α, of the resultant reaction are calculated as follows:

$$R = \sqrt{R_x^2 + R_y^2} = \sqrt{12.79^2 + 2.557^2} = 13.0 \text{ kN}$$

$$\alpha = 180^\circ + \tan^{-1}\left(\frac{R_y}{R_x}\right) = 180^\circ + \tan^{-1}\left(\frac{2.557}{12.79}\right) = 191.3^\circ$$

Because the force of the water on the reducer bend is equal and opposite to the reaction force, the force of the water on the bend has a magnitude of **13.0 kN** and makes an angle of 191.3° − 180° = **11.3°** with the inflow (x) axis.

(b) If the pressure at the exit of the reducer bend was not measured, it could be estimated from the entrance pressure (p_1), the entrance velocity (V_1), and the exit velocity (V_2) using the Bernoulli equation as follows:

$$\frac{p_1}{\gamma} + \frac{V_1^2}{2g} = \frac{p_2}{\gamma} + \frac{V_2^2}{2g} \quad \rightarrow \quad \frac{350}{9.79} + \frac{3.183^2}{2(9.81)} = \frac{p_2}{9.79} + \frac{16.63^2}{2(9.81)} \quad \rightarrow \quad p_2 = 217 \text{ kPa}$$

Therefore, the theoretical pressure (**217 kPa**) is slightly higher than the actual pressure (**215 kPa**). This discrepancy is likely due to energy losses within the bend that are not accounted for in the Bernoulli equation.

Practical Considerations

The structural supports required at pipe bends can be substantial, especially in cases where the fluid in the pipe is flowing at high speed and at high pressure. Public water supply pipelines typically transport water at fairly high pressures and at moderate speeds—usually at pressures greater than 240 kPa and at speeds of around 1 m/s. To resist the forces on the bends in water supply pipelines and to prevent the pipes from being pulled apart at bends, bolted support connections are commonly used and it is not unusual for some bends to be encased in concrete. An example of a bolted support at a bend is shown in Figure 4.14(a), and an example of a double-bend with concrete supports is shown in Figure 4.14(b). Because public water supply pipelines are generally buried, routine observation of these bend support features is not commonplace.

(a) Single bend

(b) Double bend

Figure 4.14: Water supply pipe bends

4.4.2.3 Forces on junctions

In the case of junctions, where multiple flow conduits intersect, it is usually easier to determine the reaction forces directly from the steady-state momentum equation rather than use generalized formulas. This approach is illustrated by the following example.

EXAMPLE 4.10

Several water pipes intersect at the junction box illustrated in Figure 4.15(a). Determine the force required to keep the junction box in place. Assume water at 20°C.

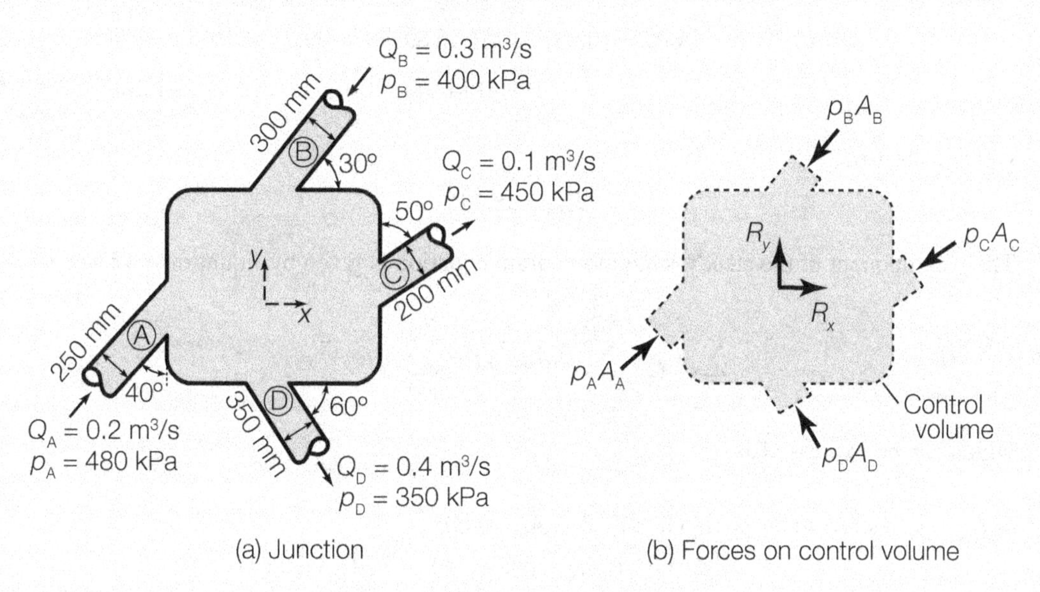

Figure 4.15: Force on a junction

SOLUTION

For water at 20°C, ρ = 998 kg/m^3. Consider the control volume that contains the fluid in the junction as shown in Figure 4.15(b). The x component of the steady-state momentum equation is given by Equation 4.41 as

$$\sum F_x = \sum_{i=1}^{N_\text{out}} \beta_i \dot{m}_i \overline{V}_{x,i} - \sum_{j=1}^{N_\text{in}} \beta_j \dot{m}_j \overline{V}_{x,j}$$

which can be expressed as

$$\begin{aligned} R_x + p_\text{A} A_\text{A} \sin 40^\circ - p_\text{B} A_\text{B} \cos 30^\circ - p_\text{C} A_\text{C} \sin 50^\circ \\ - p_\text{D} A_\text{D} \cos 60^\circ = \dot{m}_\text{C}(V_\text{C} \sin 50^\circ) + \dot{m}_\text{D}(V_\text{D} \cos 60^\circ) \\ - \dot{m}_\text{A}(V_\text{A} \sin 40^\circ) + \dot{m}_\text{B}(-V_\text{B} \cos 30^\circ) \quad (4.55) \end{aligned}$$

where the momentum-flux correction factors have been taken as unity. The velocity in each pipe is calculated using the relation $V = Q/A = Q/(\pi D^2/4)$, and the mass flow rate is calculated using the relation $\dot{m} = \rho Q$. Using the flow rates, Q, and diameters, D, given in Figure 4.15(a) yields

Pipe	Q (m^3/s)	D (m)	A (m^2)	V (m/s)	$\dot{m}$ (kg/s)
A	0.2	0.25	0.0491	4.07	199.6
B	0.3	0.30	0.0707	4.24	299.4
C	0.1	0.20	0.0314	3.18	99.8
D	0.4	0.35	0.0962	4.16	399.2

Substituting into Equation 4.55 gives

$$
\begin{aligned}
R_x + (480)(0.0491)\sin 40^\circ - (400)(0.0707)\cos 30^\circ - (450)(0.0314)\sin 50^\circ \\
- (350)(0.0962)\cos 60^\circ = (99.8)[(3.18)\sin 50^\circ] + (399.2)[(4.16)\cos 60^\circ] \\
- (199.6)[(4.07)\sin 40^\circ] - (299.4)[(-4.24)\cos 30^\circ]
\end{aligned}
$$

which leads to

$$R_x = 38.7 \text{ kN}$$

The y component of the steady-state momentum equation is given by Equation 4.42 as

$$\sum F_y = \sum_{i=1}^{N_{\text{out}}} \beta_i \dot{m}_i \overline{V}_{y,i} - \sum_{j=1}^{N_{\text{in}}} \beta_j \dot{m}_j \overline{V}_{y,j}$$

which can be expressed as

$$
\begin{aligned}
R_y + p_\text{A} A_\text{A} \cos 40^\circ - p_\text{B} A_\text{B} \sin 30^\circ - p_\text{C} A_\text{C} \cos 50^\circ \\
+ p_\text{D} A_\text{D} \sin 60^\circ = \dot{m}_\text{C}(V_\text{C} \cos 50^\circ) + \dot{m}_\text{D}(-V_\text{D} \sin 60^\circ) \\
- \dot{m}_\text{A}(V_\text{A} \cos 40^\circ) - \dot{m}_\text{B}(-V_\text{B} \sin 30^\circ) \quad (4.56)
\end{aligned}
$$

Substituting the given parameters into Equation 4.56 gives

$$
\begin{aligned}
R_y + (480)(0.0491)\cos 40^\circ - (400)(0.0707)\sin 30^\circ - (450)(0.0314)\cos 50^\circ \\
+ (350)(0.0962)\sin 60^\circ = (99.8)[(3.18)\cos 50^\circ] + (399.2)[(-4.16)\sin 60^\circ] \\
- (199.6)[(4.07)\cos 40^\circ] + (299.4)[(-4.24)\sin 30^\circ]
\end{aligned}
$$

which leads to

$$R_y = -25.3 \text{ kN}$$

The magnitude, R, of the force required to keep the junction in place is therefore given by

$$R = \sqrt{R_x^2 + R_y^2} = \sqrt{38.7^2 + 25.3^2} = \mathbf{46.2\ kN}$$

This support force could be provided by a brace that is reinforced in the direction that the junction force (R) acts.

4.4.3 Forces on Deflectors and Blades

Deflectors and blades are commonly used to adjust the flow direction (deflectors) and convert hydraulic energy into mechanical energy, and vice versa (blades). The forces exerted on deflectors and blades are of interest for a variety of reasons, such as to ensure that adequate structural support is provided, and to determine the mechanical energy that can be extracted from the flow. In cases where several blades are used to guide the flow, these blades are commonly referred to as *vanes*. Consider the case of a fluid jet deflected by a stationary blade or vane as illustrated in Figure 4.16(a). Such a fluid jet, in contact with the atmosphere, is called a *free jet*. Using the control volume shown in Figure 4.16(b), the x and y components of the steady-state momentum equation can be written as

$$\sum F_x = \dot{m}(\beta_2 \overline{V}_{x,2} - \beta_1 \overline{V}_{x,1}) \tag{4.57}$$

$$\sum F_y = \dot{m}(\beta_2 \overline{V}_{y,2} - \beta_1 \overline{V}_{y,1}) \tag{4.58}$$

Assuming that $\beta_1 \approx \beta_2 \approx 1$ and applying these equations directly to the control volume yields

$$-R_x = \dot{m}(V_2 \cos\theta - V_1) \tag{4.59}$$

$$-R_y = -\dot{m}V_2 \sin\theta \tag{4.60}$$

where R_x and R_y are the x and y components of the reaction force on the control volume, V_1 and V_2 are the inflow and outflow velocities, respectively, and θ is the deflection angle. External (gauge) pressures on the control volume are equal to zero because the fluid is surrounded by air at atmospheric pressure. Because R_x and R_y are the reactions of the blade, the components of the force exerted by the fluid on the blade are equal and opposite to R_x and R_y. In high-velocity flow, frictional effects of the blade can be significant, and these effects are usually accounted for by a reduction in the flow velocity between the entrance and exit of the blade.

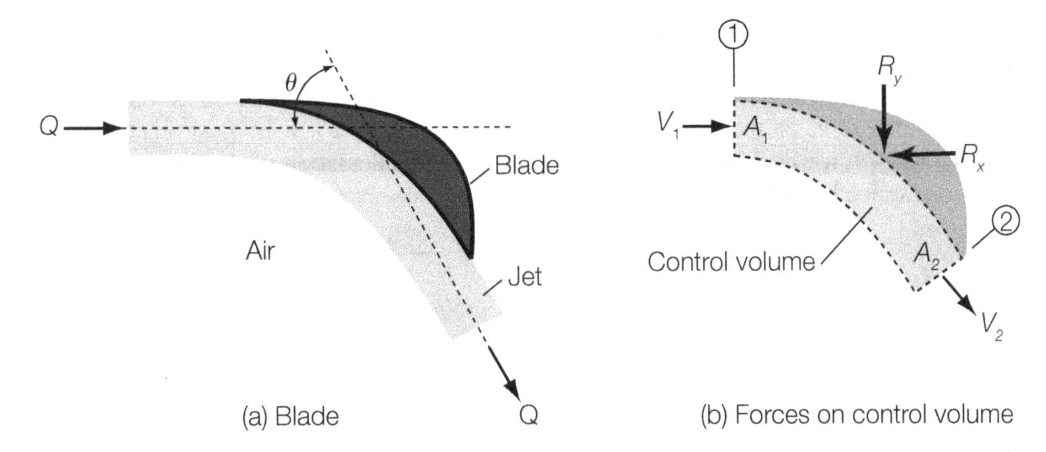

Figure 4.16: Force on a stationary blade

EXAMPLE 4.11

The rotor of a Pelton wheel turbine is driven by water jets impinging on vanes mounted on the periphery of the rotor. A particular case of a Pelton wheel vane is illustrated in Figure 4.17, where the incident water jet has a velocity of 30 m/s and a diameter of 150 mm. If the vane deflects the jet by 165°, what is the force exerted by the water on the vane? Assume water at 20°C.

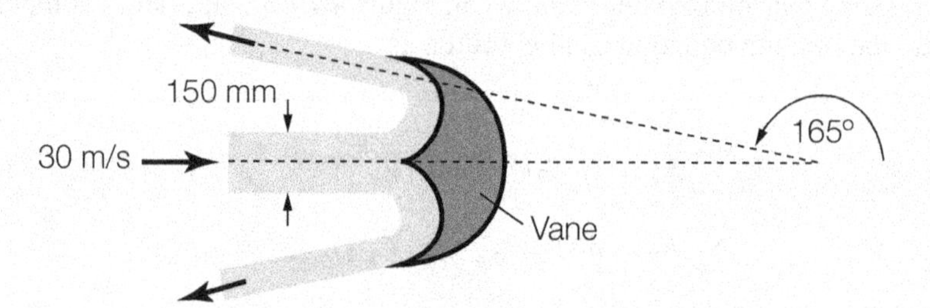

Figure 4.17: Force on Pelton wheel vane

SOLUTION

From the given data: V_j = 30 m/s, D_j = 150 mm, and θ = 165°. For water at 20°C, ρ = 998.2 kg/m^3. The following preliminary calculations are useful:

$$A_\mathrm{j} = \frac{\pi D_\mathrm{j}^2}{4} = \frac{\pi(0.15)^2}{4} = 1.767 \times 10^{-2}\ \mathrm{m}^2$$

$$\dot{m} = \rho V_\mathrm{j} A_\mathrm{j} = (998.2)(30)(1.767 \times 10^{-2}) = 529.1\ \mathrm{kg/s}$$

where A_j is the area of the incident jet, and $\dot{m}$ is the mass flow rate of water toward the vane. Because the vane is symmetrical about the centerline of the jet, one-half of the incident mass flow will go toward each side of the vane, and it can be further assumed that the flow velocity is the same in both the incident and deflected jets. With these assumptions, applying the momentum equation in the direction of the incident jet and denoting the reaction force of the vane by F gives

$$-F = 2\left(\frac{\dot{m}}{2}\right) V\cos\theta - \dot{m}V = \dot{m}V(\cos\theta - 1)$$

$$= (529.1)(30)(\cos 165^\circ - 1) = -3.121 \times 10^4\ \mathrm{N} \quad \rightarrow \quad F = 31.2\ \mathrm{kN}$$

Note that F is taken to be positive in the direction opposite the inflow direction. Based on the above calculation, the force on the vane is approximately equal to **31.2 kN**. In most practical applications of the Pelton wheel, the vane will be moving, and calculation of the force on a moving vane is described in the following section.

4.4.4 Forces on Moving Control Volumes

In some applications of the linear momentum equation, it is simpler to select a control volume that is moving at a constant velocity (e.g., in the cases of analyzing forces in rotodynamic machines or the thrust created by a jet engine in flight). In such cases, the fluid velocities can be expressed relative to the moving control volume and the appropriate form of the linear momentum equation is

$$\sum \mathbf{F} = \frac{\mathrm{d}}{\mathrm{d}t}\int_{\mathrm{cv}} \rho \mathbf{v}_\mathrm{r}\, \mathrm{d}V + \int_{\mathrm{cs}} \rho \mathbf{v}_\mathrm{r}\,(\mathbf{v}_\mathrm{r}\cdot\mathbf{n})\,\mathrm{d}A \tag{4.61}$$

where $\mathbf{v}_\mathrm{r}$ is the fluid velocity relative to the moving control volume. Equation 4.61 can be expressed more conveniently in terms of average velocities at the inflow and outflow sections of the control volume as

$$\sum \mathbf{F} = \frac{\mathrm{d}}{\mathrm{d}t}\int_{\mathrm{cv}} \rho \mathbf{v}_\mathrm{r}\, \mathrm{d}V + \sum_{i=1}^{N_\mathrm{out}} \beta_i \dot{m}_{i\mathrm{r}} \overline{\mathbf{V}}_{i\mathrm{r}} - \sum_{j=1}^{N_\mathrm{in}} \beta_j \dot{m}_{j\mathrm{r}} \overline{\mathbf{V}}_{j\mathrm{r}} \tag{4.62}$$

where $\dot{m}_{i\mathrm{r}}$ and $\dot{m}_{j\mathrm{r}}$ are the outflow and inflow mass flow rates, respectively, relative to the moving control volume and $\overline{\mathbf{V}}_{i\mathrm{r}}$ and $\overline{\mathbf{V}}_{j\mathrm{r}}$ are the average outflow and inflow velocities, respectively, relative to the moving control volume. In cases of steady flow, Equation 4.62 simplifies to

$$\sum \mathbf{F} = \sum_{i=1}^{N_\mathrm{out}} \beta_i \dot{m}_{i\mathrm{r}} \overline{\mathbf{V}}_{i\mathrm{r}} - \sum_{j=1}^{N_\mathrm{in}} \beta_j \dot{m}_{j\mathrm{r}} \overline{\mathbf{V}}_{j\mathrm{r}} \tag{4.63}$$

An *inertial reference frame* is any reference frame that is stationary or is moving with a constant velocity. An *inertial coordinate system* is any coordinate system fixed to an inertial reference frame. Equations 4.61–4.63 apply to any inertial reference frame. If a reference frame is accelerating, it is a *noninertial reference frame* and Equations 4.61–4.63 are not applicable.

EXAMPLE 4.12

A water jet with a diameter of 25 mm and velocity 20 m/s impacts a trolley that is moving at 3 m/s as shown in Figure 4.18(a). Estimate the force the water jet exerts on the trolley. Assume steady-state conditions and water at 20°C.

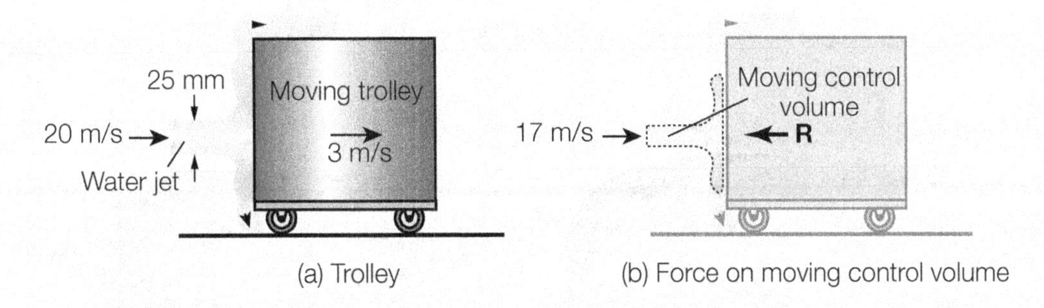

Figure 4.18: Water jet on a moving trolley

SOLUTION

From the given data: $D = 25$ mm, $v_\mathrm{j} = 20$ m/s, and $v_\mathrm{t} = 3$ m/s. For water at 20°C, $\rho = 998$ kg/m^3. Because the trolley is moving at a constant velocity, the control volume attached to the trolley, as shown in Figure 4.18(b), is inertial and the steady-state momentum equation (Equation 4.63) can be applied in the flow direction as follows:

$$-R = -\beta_{\text{in}}\dot{m}_{\text{r}}v_{\text{in,r}} \tag{4.64}$$

where R is the magnitude of the reaction force, β_{in} is the momentum-flux correction factor, $\dot{m}_{\text{r}}$ is the inflow (and outflow) mass flow rate relative to the moving control volume, and $v_{\text{in,r}}$ is the inflow velocity relative to the moving control volume. Assume that the inflow velocity is constant across the inflow surface of the control volume (i.e., $\beta_{\text{in}} \approx 1$). The relative inflow velocity, $v_{\text{in,r}}$, and the relative mass flow rate, $\dot{m}_{\text{r}}$, are given by

$$v_{\text{in,r}} = v_{\text{j}} - v_{\text{t}} = 17 \text{ m/s}, \qquad \dot{m}_{\text{r}} = \rho v_{\text{in,r}} A_{\text{in}} = (998)(17)\left[\frac{1}{4}\pi(0.025)^2\right] = 8.33 \text{ kg/s}$$

Substituting these parameter values into the momentum equation, Equation 4.64, gives

$$-R = -(1)(8.33)(17) = -142 \text{ N} \quad \rightarrow \quad R = 142 \text{ N}$$

Because the force exerted by the water jet on the trolley is equal and opposite to the reaction force, R, the force exerted by the jet on the trolley is equal to **142 N**.

Moving blade. Consider the case of a fluid jet deflected by a moving blade as illustrated in Figure 4.19(a), where the blade moves with a velocity V_{b}. As long as the blade moves with a constant velocity, the momentum equation is applicable to the moving reference frame. The control volume viewed relative to the moving blade is shown in Figure 4.19(b), and the x and y components of the momentum equation can be written as

$$-R_x = \dot{m}_{\text{r}}(V_{2\text{r}}\cos\theta - V_{1\text{r}}) \tag{4.65}$$

$$-R_y = -\dot{m}_{\text{r}}V_{2\text{r}}\sin\theta \tag{4.66}$$

where $V_{1\text{r}}$ and $V_{2\text{r}}$ are the relative fluid velocities across the inflow and outflow boundaries of the control volume, respectively, $\dot{m}_{\text{r}}$ is the relative mass flow rate into and out of the control volume, and θ is the deflection angle. In terms of the absolute velocity of the impinging jet, V_1, and the velocity of the blade, V_{b}, the relative velocities can be written as

$$V_{1\text{r}} = V_1 - V_{\text{b}}, \qquad V_{2\text{r}} = V_{1\text{r}}\frac{A_1}{A_2} \tag{4.67}$$

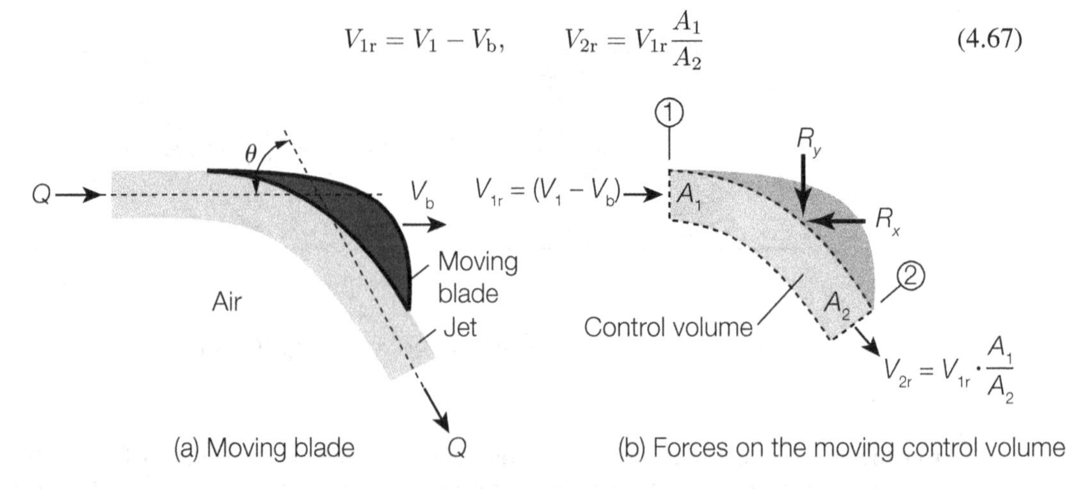

Figure 4.19: Force on a moving blade

where A_1 and A_2 are the inflow and outflow areas, respectively. Again, because R_x and R_y are the reactions of the blade, the force exerted by the fluid on the blade is equal and opposite to R_x and R_y.

EXAMPLE 4.13

A jet of water at 20°C having a velocity of 15 m/s impinges on a stationary vane whose section is in the form of a circular arc. The vane deflects the jet through an angle of 120°, as illustrated in Figure 4.20. (a) Find the magnitude and direction of the force on the vane when the jet discharge is 0.45 kg/s. (b) If the vane moves with a velocity of 6 m/s in the direction of the jet and the velocity relative to the moving vane remains constant, determine the power delivered by the jet to the moving blade.

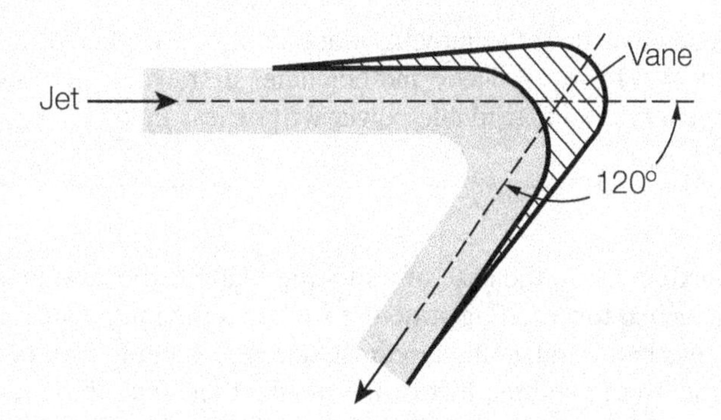

Figure 4.20: Force on a vane

SOLUTION

From the given data: V_1 = 15 m/s, $\theta = 120°$, $\dot{m}$ = 0.45 kg/s, and V_b = 6 m/s. For water at 20°C, ρ = 998 kg/s.

(a) In the case of a stationary vane, $V_{1r} = V_1$, $V_{2r} = V_2 = V_1$, and the x component of the reaction, R_x, is given by Equation 4.59 as

$$-R_x = \dot{m}(V_2\cos\theta - V_1) = (0.45)(15\cos 120° - 15) \quad \rightarrow \quad R_x = 10.1 \text{ N}$$

The y component of the reaction is given by Equation 4.60 as

$$-R_y = -\dot{m}V_2\sin\theta = -(0.45)(15)\sin 120° \quad \rightarrow \quad R_y = 5.85 \text{ N}$$

The force exerted by the water on the vane is equal and opposite to the reaction on the vane. The magnitude of the force on the vane, R, is given by

$$R = \sqrt{R_x^2 + R_y^2} = \sqrt{10.1^2 + 5.85^2} = \mathbf{11.7\ N}$$

and the angle with the horizontal, α, is given by

$$\alpha = \tan^{-1}\left(\frac{5.85}{10.1}\right) = \mathbf{30.1°}$$

(b) If the vane moves at 6 m/s, the x component of the reaction, R_x, is given by Equation 4.65 as

$$-R_x = \dot{m}_\mathrm{r}(V_{2\mathrm{r}}\cos\theta - V_{1\mathrm{r}})$$

where the mass flow rate and relative velocities over the moving vane are

$$\dot{m}_\mathrm{r} = \dot{m}\frac{V_1 - V_\mathrm{b}}{V_1} = 0.45 \times \frac{15-6}{15} = 0.27 \text{ kg/s},$$

$$V_{1\mathrm{r}} = 15 - 6 = 9 \text{ m/s}, \quad V_{2\mathrm{r}} = V_{1\mathrm{r}} = 9 \text{ m/s}$$

Substituting these values into the momentum equation gives

$$-R_x = 0.27(9\cos 120^\circ - 9) \quad \rightarrow \quad R_x = 3.65 \text{ N}$$

The power, P, delivered by the jet to the moving blade is given by

$$P = R_x V_\mathrm{b} = (3.65)(6) = 21.9 \text{ N·m} = \mathbf{21.9\ W}$$

A generator connected to the moving blade could be expected to deliver a maximum power of 21.9 W. However, due to inefficiencies in transferring power from the moving blade to the generator, the available power will be less that 21.9 W.

A common extension to the case of a moving blade is the case in which a series of moving blades are attached to a rotating structure such that the inflow jet constantly impinges on one blade after another. Under this circumstance, the volume flux of liquid impinging on the blades is equal to the volume flux of the incident jet, such that the components of the force on the structure with the rotating blades is given by

$$-R_x = \dot{m}(V_{2\mathrm{r}}\cos\theta - V_{1\mathrm{r}}) \tag{4.68}$$

$$-R_y = -\dot{m}V_{2\mathrm{r}}\sin\theta \tag{4.69}$$

where $\dot{m}$ is the absolute (not relative) mass flow rate of the jet discharging to the rotating blades.

Control volume moving with variable velocity. If a control volume is moving with a variable velocity (i.e., accelerating) relative to a fixed coordinate system, then the moving coordinate system is a *noninertial coordinate system*. If the acceleration of the control volume is rectilinear (i.e., straight-line motion), then it can be shown (e.g., Pritchard, 2011) that the linear momentum equation is given by

$$\sum \mathbf{F} = \frac{\mathrm{d}}{\mathrm{d}t}\int_\mathrm{cv} \rho\mathbf{v}_\mathrm{r}\,\mathrm{d}V + \int_\mathrm{cs} \rho\mathbf{v}_\mathrm{r}\,(\mathbf{v}_\mathrm{r}\cdot\mathbf{n})\,\mathrm{d}A + \int_\mathrm{cv} \rho\mathbf{a}\,\mathrm{d}V \quad \text{(for rectilinear motion)} \tag{4.70}$$

where $\mathbf{v}_\mathrm{r}$ is the fluid velocity relative to the moving control volume and $\mathbf{a}$ is the acceleration of the moving control volume relative to a fixed (inertial) coordinate system, where $\mathbf{a}$ can be expressed as $\mathbf{a} = \mathrm{d}\mathbf{V}_\mathrm{cv}/\mathrm{d}t$ and $\mathbf{V}_\mathrm{cv}$ is the velocity of the moving control volume. In most cases of practical interest, $\mathbf{a}$ is a function of time. In applying Equation 4.70, it is usually convenient to express the equation in terms of average inflow and outflow velocities to the control volume and the mass of substance, M, in the control volume. In terms of these variables, Equation 4.70 can be expressed as

$$\sum \mathbf{F} = \frac{\mathrm{d}}{\mathrm{d}t}\int_{\mathrm{cv}} \rho \mathbf{v}_{\mathrm{r}}\,\mathrm{d}V + \sum_{i=1}^{N_{\mathrm{out}}} \beta_i \dot{m}_{i\mathrm{r}} \mathbf{V}_{i\mathrm{r}} - \sum_{j=1}^{N_{\mathrm{in}}} \beta_j \dot{m}_{j\mathrm{r}} \mathbf{V}_{j\mathrm{r}} + M\mathbf{a} \tag{4.71}$$

In the (common) cases where the momentum of the substance within the control volume relative to the control volume remains fairly steady, such that its rate of change is small compared to the magnitude of the other terms, the first term in Equation 4.71 can be neglected and Equation 4.71 further simplifies into the following form:

$$\sum \mathbf{F} = \sum_{i=1}^{N_{\mathrm{out}}} \beta_i \dot{m}_{i\mathrm{r}} \mathbf{V}_{i\mathrm{r}} - \sum_{j=1}^{N_{\mathrm{in}}} \beta_j \dot{m}_{j\mathrm{r}} \mathbf{V}_{j\mathrm{r}} + M\mathbf{a} \tag{4.72}$$

In cases where the control volume is undergoing rectilinear acceleration, Equation 4.72 is usually the most useful form of the momentum equation. It is apparent that Equation 4.72 is similar to Newton's second law, where $\mathbf{F} = M\mathbf{a}$, with an additional force equal to the rate at which the fluid is gaining momentum as it passes through the control volume.

Problem-solving tip. For most problems that involve finding the acceleration of a moving body caused by fluid motion, it is preferable to define a control volume that encompasses the entire body that is accelerating in response to the moving fluid. This is in contrast to most other types of problems that involve finding the forces exerted on bodies by moving fluids, where it is usually more beneficial to define a control volume that encompasses only the fluid.

EXAMPLE 4.14

A rocket is to be tested by mounting it on level rails and igniting the solid fuel contained in the rocket. Consider the rocket shown in Figure 4.21(a), which has an initial mass of 500 kg, a fuel burn rate of 8 kg/s, and a nozzle exit velocity of 3000 m/s. The aerodynamic drag force, F_{D}, on the rocket depends on the velocity, v, of the rocket such that

$$F_{\mathrm{D}} = 4.50v^2$$

where F_{D} is in newtons and v is in meters per second. The coefficient of rolling friction between the wheels and the rail is approximately equal to 0.4. (a) Derive an equation that relates the velocity of the rocket to the time since ignition of the fuel. (b) Estimate the acceleration of the rocket when the fuel is first ignited.

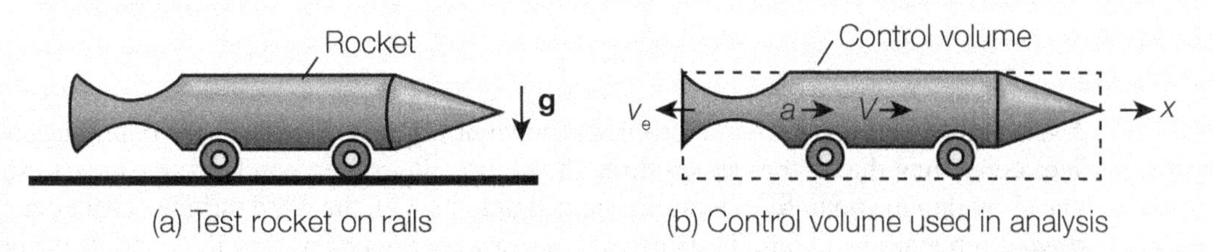

Figure 4.21: Test rocket

SOLUTION

From the given data: $M_0 = 500$ kg, $\dot{m} = 8$ kg/s, $v_e = 3000$ m/s, and $\mu = 0.4$. Let M be the mass of the rocket at any time, t, then, in terms of the given variables

$$M = M_0 - \dot{m}t$$

The rocket is undergoing rectilinear motion in the direction x as described by its velocity v and acceleration a.

(a) Utilizing the control volume shown in Figure 4.21(b), direct application of Equation 4.72 in the direction of motion (i.e., in the x-direction) gives

$$-F_D - F_f = \dot{m}(-v_e) + Ma \tag{4.73}$$

where F_f is the frictional force. It is assumed that the rate of change of the momentum of the control-volume contents relative to the moving control volume (i.e., the first term in Equation 4.71) is negligible compared to the magnitudes of the other terms in the momentum equation. Noting that $F_D = 4.50v^2$, $F_f = \mu Mg$, and $a = dv/dt$, Equation 4.73 can be expressed as

$$-4.50v^2 - \mu Mg = -\dot{m}v_e + M\frac{dv}{dt} \quad \rightarrow \quad \frac{dv}{dt} + \left(\frac{4.50}{M_0 - \dot{m}t}\right)v^2 = \frac{\dot{m}v_e}{M_0 - \dot{m}t} - \mu g$$

Substituting the given values of the parameters yields

$$\frac{dv}{dt} + \left(\frac{4.50}{500 - 8t}\right)v^2 = \frac{(8)(3000)}{500 - 8t} - (0.4)(9.81) \quad \rightarrow$$

$$\mathbf{\frac{dv}{dt} + \left(\frac{4.50}{500 - 8t}\right)v^2 = \frac{24\,000}{500 - 8t} - 3.92} \tag{4.74}$$

where t is in seconds and v is in meters per second. Equation 4.74 relates the velocity, v, of the rocket to the time, t, since ignition of the rocket fuel. Equation 4.74 is a nonlinear differential equation that could subsequently be solved (numerically) with the initial conditions: $v = 0$ when $t = 0$.

(b) The acceleration at the instant after the fuel is ignited is derived from Equation 4.74 by setting $t = 0$ and $v = 0$, which gives

$$a\big|_{t=0} = \left.\frac{dv}{dt}\right|_{t=0} = \left[-\left(\frac{4.5}{500 - 8t}\right)v^2 + \frac{24\,000}{500 - 8t} - 3.92\right]_{t=0,v=0} = 44.1 \text{ m/s}^2$$

Therefore, the initial acceleration of the rocket is **44.1 m/s²**, which is approximately 4.5 times the acceleration due to gravity.

4.4.5 Wind Turbines

A schematic diagram of a wind turbine is shown in Figure 4.22, where the bounding streamlines containing the air that passes through the propeller of the wind turbine are also shown. The air within the bounding streamlines that interacts with the wind turbine is called the *slipstream*, a bounding streamline is called a *slipstream boundary*, and the conduit formed by the bounding streamlines is a *streamtube*. At the upstream end of the streamtube (section 1), the free-stream wind velocity is V_1 and the free-stream pressure is p_∞. Considering a control

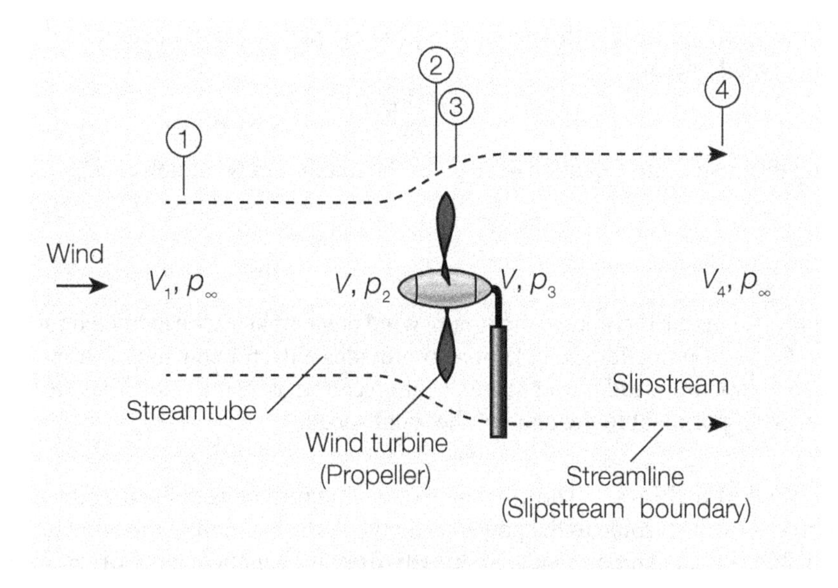

Figure 4.22: Wind turbine

volume within the streamtube bounded by sections 1 and 4 (where the pressure returns to the free-stream pressure), application of the momentum equation yields

$$-F = \dot{m}(V_4 - V_1) \tag{4.75}$$

where F is the force exerted by the wind on the propeller, $\dot{m}$ is the air mass flow rate through the propeller, and V_4 is the wind velocity at section 4. Applying the momentum equation between sections 2 and 3, which are just upstream and just downstream of the propeller, respectively, yields

$$-F + (p_2 - p_3)A = \dot{m}(V - V) = 0 \tag{4.76}$$

where p_2 and p_3 are the pressures at sections 2 and 3, respectively, and V is the velocity at which the air moves through the propeller, which is approximately equal to the velocities at sections 2 and 3, respectively. The mass flow rate of air, $\dot{m}$, within the streamtube is given by

$$\dot{m} = \rho AV \tag{4.77}$$

where ρ is the density of the air and A is the frontal area of the propeller. Applying the Bernoulli equation between sections 1 and 2 and between sections 3 and 4 yields

$$p_\infty + \frac{1}{2}\rho V_1^2 = p_2 + \frac{1}{2}\rho V^2 \tag{4.78}$$

$$p_3 + \frac{1}{2}\rho V^2 = p_\infty + \frac{1}{2}\rho V_4^2 \tag{4.79}$$

Combining Equations 4.75–4.79 yields

$$V = \frac{1}{2}(V_1 + V_4) \tag{4.80}$$

The power, P, extracted by the propeller can be determined by combining Equations 4.76 and 4.78–4.80, which yields

$$P = FV = \frac{1}{4}\rho A(V_1^2 - V_4^2)(V_1 + V_4) \tag{4.81}$$

For any given wind speed, V_1, the maximum possible power can be determined by setting $\mathrm{d}P/\mathrm{d}V_4 = 0$, which yields

$$V_4 = \frac{1}{3}V_1 \tag{4.82}$$

Substituting this result into Equation 4.81 gives the maximum wind power, P_{max}, as

$$P_{\text{max}} = \frac{8}{27}\rho A V_1^3 \tag{4.83}$$

This expression gives the theoretical maximum wind power that can be extracted for any given wind speed (V_1) and propeller size (A); it also provides a useful benchmark from which the feasibility for extracting wind power at any location can be assessed. Note that the maximum wind power is proportional to the cube of the wind speed.

Efficiency of wind turbines. The efficiency of a wind turbine is defined as the ratio of the power extracted by the turbine to the power available at the location of the wind turbine. The power available, P_{avail}, can be calculated directly from the kinetic energy of the wind and is given by

$$P_{\text{avail}} = \frac{1}{2}\rho(AV_1)V_1^2 \tag{4.84}$$

Combining Equations 4.81 and 4.84 gives the efficiency of the wind turbine, η, as

$$\eta = \frac{P}{P_{\text{avail}}} = \frac{1}{2}\left[1 - \left(\frac{V_4}{V_1}\right)^2\right]\left[1 + \left(\frac{V_4}{V_1}\right)\right] \tag{4.85}$$

The efficiency, η, has a maximum value when $V_4/V_1 = 1/3$, and this yields a maximum efficiency of $\eta = 0.593$. Hence, wind turbines generally have efficiencies less than 59.3%. This limitation is sometimes referred to as *Bette's law*, although this nomenclature is not widespread. Modern wind turbines constructed with aerofoil profiles can achieve efficiencies close to the maximum, in contrast to traditional European windmills constructed with large rotating sails that typically have efficiencies as low as 5% (Massey and Ward-Smith, 2012).

Wind farms. Wind turbines are commonly deployed in arrays as shown in Figure 4.23. Areas that contain multiple turbines for extracting wind energy are sometimes called *wind farms*, and the energy extracted in wind farms is commonly stored in various types of devices to regulate a steady output of energy. Turbines in wind farms typically have three blades that are adjusted to the wind direction by computer-controlled motors. The blades typically range in length from 20–40 m, rotate at 10–22 rpm, and have tip speeds of over 320 km/h. The blades are usually colored white for daytime visibility, especially by aircraft. The steel towers supporting wind turbines are typically 60–90 m tall. The typical spacing between wind turbines is on the order of 6–10 rotor diameters.

Feasibility of wind power. The feasibility of harnessing wind power at any location depends on the average annual wind speed at that location. The *wind power density* is defined as the mean annual power available per square meter of swept area of a wind turbine, and it is usually tabulated for different heights above the ground. In the United States, the wind power density is used as a basis for classifying locations according to their potential for generating wind power, with classifications varying from Class 1 to Class 7 as shown in Table 4.1; all classifications are based on average annual wind speeds at 50 m altitude. Commercial wind farms are typically located in Class 3 areas or higher (i.e., Class 3–7 areas). The average

Figure 4.23: Wind farm
Source: majeczka/Shutterstock.

annual wind speeds at 50 m above ground level in the 48 contiguous states in the United States are shown in Figure 4.24. It is apparent that the high plains states in the middle of the country have the highest potential for the development of wind power. The performance of wind turbines is commonly measured by their *capacity factor*, which is the ratio of the actual annual energy output to the rated annual energy output. Commercial wind turbines typically have capacity factors in the range of 0.3–0.4

Classification of wind turbines. Aside from the classification of locations by their wind power density, wind turbines have their own classification system. Wind turbines are classified by the average wind speed at which they are capable of functioning efficiently. Wind

Table 4.1: Classifications of Locations by Wind Power Density

Class	Wind Power Density (W/m^2)	Class	Wind Power Density (W/m^2)
1	0–100	5	250–300
2	100–150	6	300–400
3	150–200	7	400–1000
4	200–250		

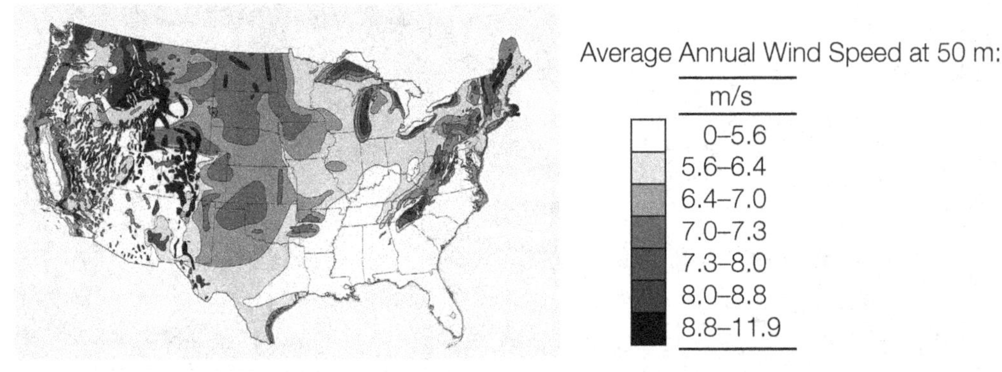

Figure 4.24: **Average annual wind speeds at 50 m in the United States**
Source: Courtesy of the Pacific Northwest National Laboratory, operated by Battelle for the U.S. Department of Energy.

turbine classifications are given in Table 4.2, where wind turbines are classified as Class I to Class IV, with A and B referring to the *turbulence ratio*. The turbulence ratio is the ratio of the root mean square of the turbulent velocity fluctuations to the mean velocity.

Table 4.2: Classifications of Wind Turbines

Turbine Class	Average Wind Speed (m/s)	Turbulence Ratio	Turbine Class	Average Wind Speed (m/s)	Turbulence Ratio
IA	10	0.18	IIIA	7.5	0.18
IB	10	0.16	IIIB	7.5	0.16
IIA	8.5	0.18	IVA	6	0.18
IIB	8.5	0.16	IVB	6	0.16

EXAMPLE 4.15

A wind turbine with a hub height of 50 m and rotor diameter of 60 m is being proposed for a wind farm in Amarillo, Texas. The average annual temperature in Amarillo is 13.9°C, and the corresponding average annual air density is 1.00 kg/m^3. (a) Estimate the wind power density and assess the feasibility of using wind as a source of energy in this area. (b) Estimate the wind power that could be developed per unit area of land.

SOLUTION

From the given data: D = 60 m and ρ_{air} = 1.00 kg/m^3. Locating Amarillo on the map in Figure 4.24 indicates that the average annual wind velocity at 50 m above ground is 7.0–7.3 m/s. So assume an average wind velocity of V = 7.15 m/s.

(a) The wind power density can be calculated using Equation 4.83, which gives

$$\text{wind power density} = \frac{P_{\text{max}}}{A} = \frac{8}{27}\rho_{\text{air}}V^3 = \frac{8}{27}(1.00)(7.15)^3 = \mathbf{108\ W/m^2}$$

Because the wind power density indicates a Class 2 location (100-150 W/m^2) which is less desirable than the minimally acceptable Class 3 location (150-200 W/m^2), the development of wind energy in Amarillo **does not appear to be feasible.**

(b) For a rotor diameter of 60 m, the area of the rotor, A, and the maximum power generation, P_{max}, are given by

$$A = \frac{\pi}{4}D^2 = \frac{\pi}{4}(60)^2 = 2827 \text{ m}^2$$

$$P_{\text{max}} = \frac{P_{\text{max}}}{A}A = (108)(2827) = 305\,000 \text{ W} = 305 \text{ kW}$$

Assume a spacing between units of 10 rotor diameters. So each turbine is in the center of a circle of radius 5 rotor diameters, which is 5 × 60 m = 300 m. Hence, the land area covered by each unit is $\pi(300)^2 = 2.83 \times 10^5 \text{ m}^2$ = 28.3 ha. Hence,

$$\text{power per unit area of land} = \frac{P_{\text{max}}}{\text{land area}} = \frac{305 \text{ kW}}{28.3 \text{ ha}} = 10.8 \text{ kW/ha}$$

Therefore, if the wind farm was developed, it could be expected to produce a maximum power output of around **10.8 kW/ha**.

4.4.6 Reaction of a Jet

The case of a liquid jet exiting a tank is illustrated in Figure 4.25(a). Assuming that the tank is large enough that the velocities within the tank can be neglected and assuming that the liquid is inviscid and incompressible, the Bernoulli equation can be applied to a streamline within the jet to give the jet velocity, V_2, as

$$V_2 = \sqrt{2gh} \tag{4.86}$$

where h is the height of the liquid surface above the centroid of the orifice from which the jet exits the tank. The force exerted by the exiting jet on the tank can be determined by considering the liquid within the tank as a control volume, as shown in Figure 4.25(b). Applying the momentum equation in the horizontal (x) direction yields

$$F_x = \dot{m}_2 V_2 = \rho Q_2 V_2 = \rho A_2 V_2^2 = \rho A_2(2gh) = 2\gamma h A_2 \tag{4.87}$$

where $\dot{m}_2$ is the mass flow rate of liquid leaving the tank, A_2 is the area of the orifice where the jet exits the tank, and γ is the specific weight of the liquid. The force F_x is the net force

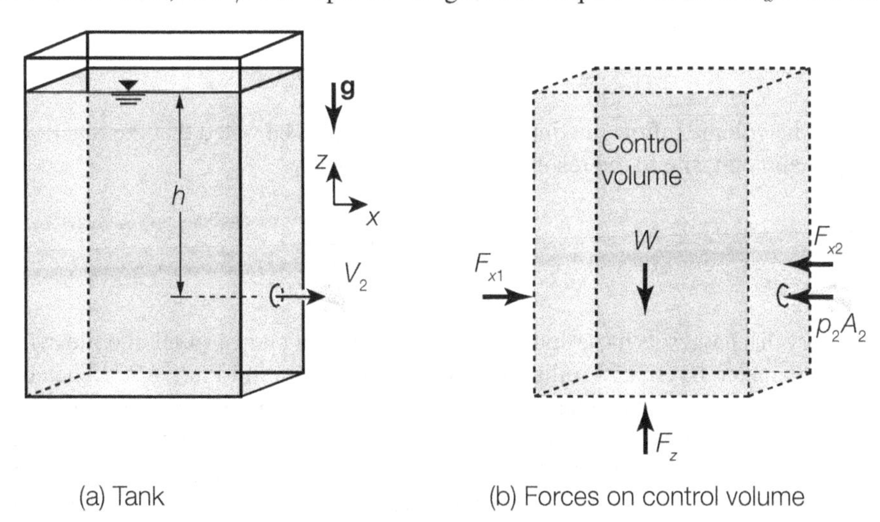

Figure 4.25: Jet exiting a tank

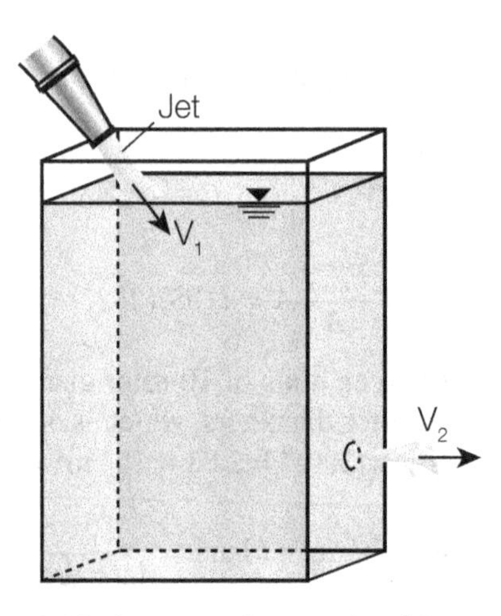

Figure 4.26: Jet entering and exiting a tank

of the tank on the liquid in the x direction; it acts toward the right and causes the velocity of the liquid to increase from zero to V_2. The net force F_x $(= \dot{m}_2 V_2)$ is equal to the difference in magnitude of the pressure forces on the two sides of the tank, F_{x1} and F_{x2}. On the left side of the tank, a typical hydrostatic pressure exists, whereas on the right side of the tank, the pressure is lower near the orifice because of the increase in velocity in that region. From the last term in Equation 4.87, it is apparent that the net jet reaction force is equal to twice the hydrostatic force on A_2. Equal and opposite to the force F_x given by Equation 4.87 is the force of the liquid on the tank, often referred to as the *jet reaction*. The force F_x will move the tank to the left if F_x is greater than the frictional force resisting the movement of the tank. The net force on the tank in the vertical direction, F_z, is equal to the weight, W, of liquid in the tank. The frictional force resisting the movement of the tank is equal to μF_z, where μ is the coefficient of friction between the tank and the surface on which the tank is resting.

Accounting for liquid inflow. Consider the case illustrated in Figure 4.26 where a jet of liquid of cross-sectional area A_1 discharges into a tank with a velocity V_1 and mass flow rate $\dot{m}_1$. In this case, the jet exerts a force $\mathbf{F} = \dot{m}_1 \mathbf{v}_1$ on the liquid, which in turn transmits the force to the tank. This force is called the *jet action*. In the case shown in Figure 4.26 where there are two jets, one entering at section 1 and the other leaving at section 2, the resultant force, $\mathbf{F}$, on the tank is given by

$$\mathbf{F} = \dot{m}_1 \mathbf{V}_1 - \dot{m}_2 \mathbf{V}_2 \quad \text{or} \quad \mathbf{F} = \rho Q_1 \mathbf{V}_1 - \rho Q_2 \mathbf{V}_2 \tag{4.88}$$

where Q_1 is the volume inflow rate into the tank. It can also be noted that the reaction of the tank is equal and opposite to the resultant force on the tank.

EXAMPLE 4.16

A research laboratory discharges its hazardous wastewater by putting it in a small, mobile cylindrical container of diameter 10 cm and rolling it to a sump as shown in Figure 4.27. The container is emptied through a 2-cm-diameter orifice whose center is 1.5 cm from the bottom of the container; the orifice geometry is such that the discharge coefficient is approximately equal to unity. The weight of the container is 10 N, and the coefficient of (rolling) friction between the container and the ground surface is 0.05. (a) Find the maximum depth of wastewater in the container so that it will not move when the outlet is opened. (b) If the container was filled to a depth of 40 cm, what support force would be required to keep it from moving? The density of the wastewater is approximately 998 kg/m^3.

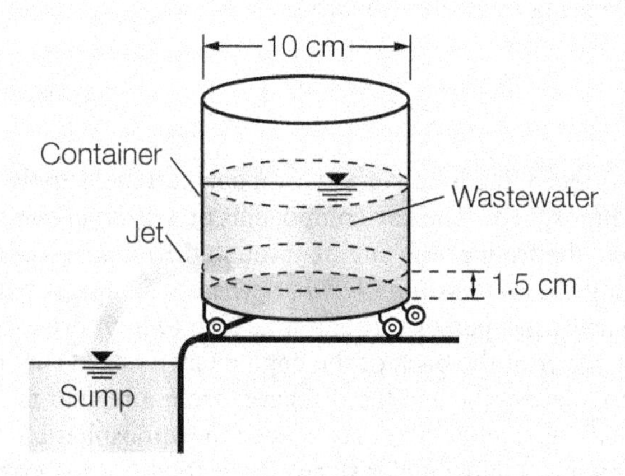

Figure 4.27: Container discharging wastewater

SOLUTION

From the given data: $D_c = 0.10$ m, $D_j = 0.02$ m, $C_d = 1$, $h_o = 0.015$ m, $W_c = 10$ N, $\mu = 0.05$, $\gamma_w = 9790$ N/m^3, and $\rho_w = 998$ kg/m^3. Using these given data and taking h as the depth of liquid in the container,

frictional force, $$F_f = \mu \cdot \text{Weight} = \mu \left[\gamma_w \frac{\pi D_c^2}{4} h + W_c\right]$$
$$= (0.05)\left[(9790)\frac{\pi(0.10)^2}{4} h + 10\right] = 3.845h + 0.5 \text{ N}$$

jet exit velocity, $$V_j = \sqrt{2g(h - 0.015)} = \sqrt{2(9.81)(h - 0.015)} = 4.429\sqrt{h - 0.015} \text{ m/s}$$

area of jet, $$A_j = \frac{\pi D_j^2}{4} = \frac{\pi(0.02)^2}{4} = 3.142 \times 10^{-4} \text{ m}^2$$

force on container, $$F_j = \dot{m}V_j = \rho_w Q V_j = \rho_w A_j V_j^2$$
$$= (998)(3.142 \times 10^{-4})(4.429\sqrt{h - 0.015})^2 = 6.151(h - 0.015) \text{ N}$$

(a) The container will move when the frictional force is equal to the force on the container caused by the jet, in which case

$$F_j = F_f \quad \rightarrow \quad 6.151(h - 0.015) = 3.845h + 0.5 \quad \rightarrow \quad h = 0.257 \text{ m} = 25.7 \text{ cm}$$

Hence, the container will move when the depth of wastewater in the container is more than **25.7 cm.**

(b) When the depth of wastewater in the container is 40 cm, $h = 0.40$ m and the net force on the container, F_{net}, is given by

$$F_{net} = F_j - F_f = [6.151(h - 0.015)] - [3.845h + 0.5]$$
$$= [6.151(0.40 - 0.015)] - [3.845(0.40) + 0.5] = 1.33 \text{ N}$$

Hence, the net force required to keep the container from moving is **1.33 N.**

4.4.7 Jet Engines and Rockets

The propulsion of jet engines and rockets is derived from the reaction caused by a very high-speed jet discharging from the structure.

Jet engine. A jet engine is connected to an onboard fuel supply and takes in air for combustion from the atmosphere. The key components of a *turbojet engine* are shown in Figure 4.28, where air enters the front of the engine through the *intake* (section 1), the air is compressed prior to entering the *combustion chamber* where it is mixed with fuel and ignited, and the high-speed exhaust gas passes through a turbine (which drives the compressor) on its way to the *exhaust nozzle* at the back of the engine (section 2). The main difference between the discharge from a jet engine and the discharge from a regular nozzle is that the exit pressure p_e of the gas leaving a jet engine may exceed the atmospheric pressure p_a. The *thrust force*, F, generated by a jet engine under steady-state conditions is given by

$$F = \begin{cases} \dot{m}_{2\mathrm{r}}v_{2\mathrm{r}} - \dot{m}_{1\mathrm{r}}v_{1\mathrm{r}} + (p_\mathrm{e} - p_\mathrm{a})A_2, & \text{(in terms of mass flow rates)} \\ \rho_2 A_2 v_{2\mathrm{r}}^2 - \rho_1 A_2 v_{1\mathrm{r}}^2 + (p_\mathrm{e} - p_\mathrm{a})A_2, & \text{(in terms of densities)} \end{cases} \tag{4.89}$$

where $\dot{m}_{1\mathrm{r}}$ and $\dot{m}_{2\mathrm{r}}$ are the (relative) rates at which fluid mass is entering and leaving the jet engine, respectively, $v_{1\mathrm{r}}$ is the relative velocity of the air entering the jet engine, $v_{2\mathrm{r}}$ is the relative velocity of the exhaust gas leaving the jet engine, A_2 is the exhaust area, ρ_1 is the density of the air entering the jet engine, and ρ_2 is the density of the exhaust gas. The velocity at which air enters the jet engine is usually equal to the velocity of flight. It is apparent from Equation 4.89 that the thrust force varies with the speed of flight, because the speed of flight is equal to $v_{1\mathrm{r}}$. In applying Equation 4.89, the rate at which fluid mass leaves the jet engine can usually be taken as being equal to the mass rate at which air enters the engine plus the rate at which fuel mass is being consumed within the combustion chamber. Also, it is not uncommon to assume that $p_\mathrm{e} \approx p_\mathrm{a}$ at the exit of a turbojet engine.

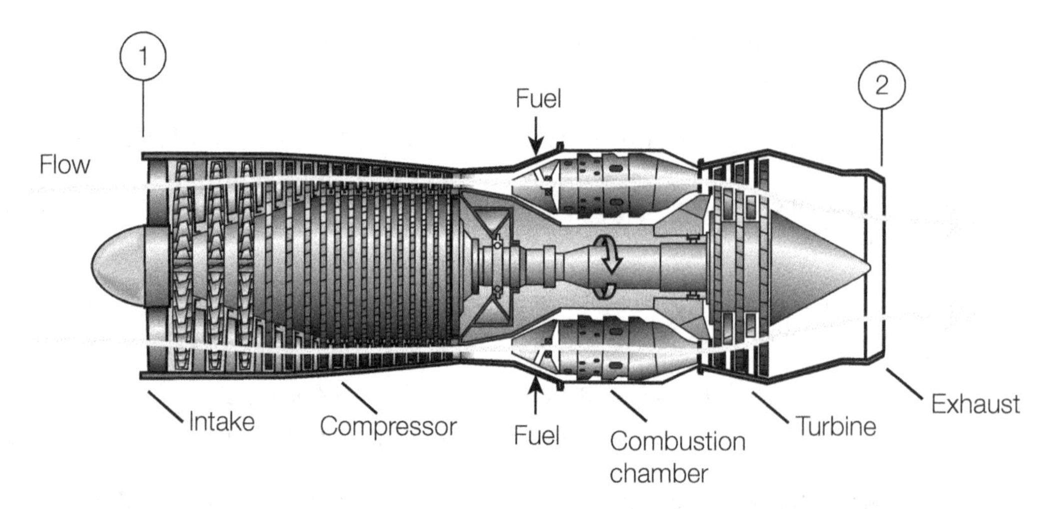

Figure 4.28: Jet engine

EXAMPLE 4.17

A commercial jetliner cruises at at 900 km/h at an elevation of 10 600 m where the air density is estimated to be 0.384 kg/m^3. The aircraft has four turbojet engines, it has a fuel capacity of 200 000 L, and fuel is consumed at the rate of 4 L/s. The density of the jet fuel is 804 kg/m^3. Each engine has an intake diameter of 2.19 m, and according to the manufacturer, each engine is capable of developing a thrust of 245 kN. Estimate the exhaust velocity.

SOLUTION

From the given data: V_eng = 900 km/h = 250 m/s, ρ_a = 0.384 kg/m^3, ρ_f = 804 kg/m^3, D_in = 2.19 m, and F = 245 kN = 245 000 N. Because there are four engines, the fuel consumption of each engine is Q_f = 4/4 = 1 L/s = 0.001 m^3/s. Assume that the exhaust pressure is equal to the intake pressure, so $p_\text{e} \approx p_\text{a}$. Using the given data gives the following derived parameters for each engine:

$$\text{intake area, } A_\text{in} = \frac{\pi}{4}D_\text{in}^2 = \frac{\pi}{4}(2.19)^2 = 3.767 \text{ m}^2$$

$$\text{air mass flow rate, } \dot{m}_\text{a} = \rho_\text{a} A_\text{in} V_\text{eng} = (0.384)(3.767)(250) = 361.6 \text{ kg/s}$$

$$\text{fuel mass flow rate, } \dot{m}_\text{f} = \rho_\text{f} Q_\text{f} = (804)(0.001) = 0.804 \text{ kg/s}$$

Applying the momentum equation, Equation 4.89, gives

$$F = \dot{m}_{2\text{r}} v_{2\text{r}} - \dot{m}_{1\text{r}} v_{1\text{r}} + (p_\text{e} - p_\text{a})A_2$$

$$245\,000 = (\dot{m}_\text{a} + \dot{m}_\text{f})v_{2\text{r}} - \dot{m}_\text{a} V_\text{eng} + 0$$

$$245\,000 = (361.6 + 0.804)v_{2\text{r}} - (361.6)(250) + 0 \quad \rightarrow \quad v_{2\text{r}} = 925 \text{ m/s}$$

Hence, the estimated relative exhaust velocity from the jet engine is **925 m/s**. Since the aircraft is traveling at **250 m/s**, the absolute velocity of the exhaust gas is **925 m/s − 250 m/s = 675 m/s**. Such a high exhaust velocity is theoretically necessary to provide sufficient thrust from the relatively small exhaust (nozzle) area of a jet engine.

Rocket. In the case of a rocket, both the fuel and the oxidant for combustion are contained within the rocket, which is analogous to the tank shown in Figure 4.25. Rockets typically do not require air, so they can operate in a vacuum. The rocket thrust, F, is given by

$$F = \begin{cases} \dot{m}_{2\text{r}} v_{2\text{r}} + (p_\text{e} - p_\text{a})A_2, & \text{(in terms of mass flow rates)} \\ \rho A_2 v_{2\text{r}}^2 + (p_\text{e} - p_\text{a})A_2, & \text{(in terms of densities)} \end{cases} \tag{4.90}$$

where $v_{2\text{r}}$ is the velocity of the exhaust gas relative to the moving rocket, A_2 is the exhaust area, and ρ is the density of the exhaust gas. It is apparent from Equation 4.90 that the thrust F is independent of the absolute speed of the rocket.

EXAMPLE 4.18

A streamlined 4000-kg rocket burns fuel at a rate of 20 kg/s, yielding an exhaust velocity of 2200 m/s relative to the moving rocket. Estimate the time elapsed between when the rocket lifts off and when it attains a velocity of 223 m/s. Neglect the drag force of the surrounding air and assume that the pressure of the exhaust gas is equal to the pressure of the surrounding atmosphere.

SOLUTION

From the given data: m_0 = 4000 kg, $\dot{m}_\mathrm{f}$ = $\dot{m}_{2\mathrm{r}}$ = 20 kg/s, and $v_{2\mathrm{r}}$ = 2200 m/s. If the rocket produces a thrust of F, then applying the momentum equation with $p_\mathrm{e} \approx p_\mathrm{a}$ gives

$$F - mg = ma \quad \rightarrow \quad \dot{m}_{2\mathrm{r}} v_{2\mathrm{r}} - mg = m\frac{\mathrm{d}V}{\mathrm{d}t} \tag{4.91}$$

where m is the mass of the rocket and V is the velocity of the rocket at any time t. The velocity of the rocket as a function of time is derived from Equation 4.91 as follows:

$$\frac{\mathrm{d}V}{\mathrm{d}t} = \frac{\dot{m}_{2\mathrm{r}} v_{2\mathrm{r}}}{m_0 - \dot{m}_\mathrm{f} t} - g$$

$$V(t) = \dot{m}_{2\mathrm{r}} v_{2\mathrm{r}} \int_0^t \frac{1}{m_0 - \dot{m}_\mathrm{f} t'}\,\mathrm{d}t' - gt$$

$$V(t) = -\frac{\cancel{\dot{m}_{2\mathrm{r}}} v_{2\mathrm{r}}}{\cancel{\dot{m}_\mathrm{f}}} \ln\left[\frac{m_0 - \dot{m}_\mathrm{f} t}{m_0}\right] - gt \quad \rightarrow \quad V(t) = v_{2\mathrm{r}} \ln\left[\frac{m_0}{m_0 - \dot{m}_\mathrm{f} t}\right] - gt$$

Let t^* be the time it takes the rocket to go from a speed of 0 m/s on the launch pad to a speed of 223 m/s; then

$$223 = (2200)\ln\left[\frac{4000}{4000 - 20\,t^*}\right] - 9.81\,t^* \quad \rightarrow \quad t^* = 25.4\ \mathrm{s}$$

Therefore, it takes approximately **25.4 seconds** to accelerate to a velocity of 223 m/s.

4.5 Angular Momentum Principle

The angular momentum principle states that the sum of the moments acting on any fluid system is equal to the rate of change of the angular momentum of the system. The angular momentum principle is derived from Newton's second law of motion by integrating Newton's second law over a finite-sized system. The angular momentum principle is particularly useful in determining the locations of resultant forces exerted by moving fluids.

Governing equation. Consider a fluid system containing a fixed mass of fluid. If the system is composed of elements of mass δm, then the force acting on each fluid element, $\delta\mathbf{F}$, can be expressed using Newton's second law (i.e., the law of conservation of linear momentum) as

$$\delta\mathbf{F} = \frac{\mathrm{d}(\delta m\,\mathbf{v})}{\mathrm{d}t} \tag{4.92}$$

where $\mathbf{v}$ is the velocity of the fluid element. The moment, $\delta\mathbf{M}$, of the force on the fluid element about a point P is given by

$$\delta\mathbf{M} = \mathbf{r} \times \delta\mathbf{F} \tag{4.93}$$

where $\mathbf{r}$ is the position vector that gives the location of the fluid element relative to P. Combining Equations 4.92 and 4.93 gives the moment on the fluid element as

$$\delta\mathbf{M} = \mathbf{r} \times \frac{\mathrm{d}(\delta m\,\mathbf{v})}{\mathrm{d}t} \tag{4.94}$$

The following identities can be used to transform and simplify Equation 4.94:

$$\frac{\mathrm{d}}{\mathrm{d}t}(\mathbf{r} \times \delta m\,\mathbf{v}) = \frac{\mathrm{d}\mathbf{r}}{\mathrm{d}t} \times \delta m\,\mathbf{v} + \mathbf{r} \times \frac{\mathrm{d}}{\mathrm{d}t}(\delta m\,\mathbf{v}) \tag{4.95}$$

$$\mathbf{v} = \frac{\mathrm{d}\mathbf{r}}{\mathrm{d}t} \tag{4.96}$$

$$\mathbf{v} \times \mathbf{v} = 0 \tag{4.97}$$

where Equation 4.95 applies the product rule for differentiating vectors, Equation 4.96 expresses the velocity, $\mathbf{v}$, in terms of the position vector, $\mathbf{r}$, and Equation 4.97 applies the rule that the cross product between two collinear vectors is equal to zero. Using the identities given by Equations 4.95–4.97, Equation 4.94 can be expressed as

$$\delta\mathbf{M} = \frac{\mathrm{d}}{\mathrm{d}t}(\mathbf{r} \times \delta m\,\mathbf{v}) \tag{4.98}$$

where the quantity in parentheses on the right-hand side of Equation 4.98 is the *moment of momentum* or *angular momentum* of the fluid element (of mass δm) relative to P. Integrating Equation 4.98 over a fluid system consisting of a continuum of fluid elements of infinitesimally small size gives

$$\int_{\text{sys}} \delta\mathbf{M} = \frac{\mathrm{d}}{\mathrm{d}t}\int_{\text{sys}} (\mathbf{r} \times \mathbf{v})\,\delta m \tag{4.99}$$

Because the differential mass, δm, is equal to $\rho\,\mathrm{d}V$, where ρ is the density of the fluid and $\mathrm{d}V$ is an elemental volume within the fluid system, Equation 4.99 can be conveniently expressed as

$$\sum \mathbf{M}_{\text{sys}} = \frac{\mathrm{d}}{\mathrm{d}t}\int_{\text{sys}} (\mathbf{r} \times \mathbf{v})\rho\,\mathrm{d}V \tag{4.100}$$

where $\sum \mathbf{M}_{\text{sys}}$ is the sum of the moments acting on the system. Equation 4.100 is commonly called the *moment of momentum equation* or *angular momentum equation*. The Reynolds transport theorem relates the rate of change of any fluid property within a system to the changes in that same fluid property within a coincident control volume, such that

$$\frac{\mathrm{d}}{\mathrm{d}t}\int_{\text{sys}} \beta\rho\,\mathrm{d}V = \frac{\mathrm{d}}{\mathrm{d}t}\int_{\text{cv}} \beta\rho\,\mathrm{d}V + \int_{\text{cs}} \beta\rho(\mathbf{v}\cdot\mathbf{n})\,\mathrm{d}A \tag{4.101}$$

where β is a fluid property per unit mass, "cv" and "cs" denote the volume and surface area of the control volume, respectively, and $\mathbf{n}$ is the unit normal pointing out of the control volume. Taking $\beta = \mathbf{r} \times \mathbf{v}$ in Equation 4.101 gives

$$\frac{\mathrm{d}}{\mathrm{d}t}\int_{\text{sys}} (\mathbf{r} \times \mathbf{v})\rho\,\mathrm{d}V = \frac{\mathrm{d}}{\mathrm{d}t}\int_{\text{cv}} (\mathbf{r} \times \mathbf{v})\rho\,\mathrm{d}V + \int_{\text{cs}} (\mathbf{r} \times \mathbf{v})\rho(\mathbf{v}\cdot\mathbf{n})\,\mathrm{d}A \tag{4.102}$$

Combining Equations 4.100 and 4.102 yields the following expression for the moment of momentum equation

$$\sum \mathbf{M}_{\rm cv} = \frac{\mathrm{d}}{\mathrm{d}t}\int_{\rm cv} \rho(\mathbf{r}\times\mathbf{v})\,\mathrm{d}V + \int_{\rm cs} \rho(\mathbf{r}\times\mathbf{v})\,(\mathbf{v}\cdot\mathbf{n})\,\mathrm{d}A \tag{4.103}$$

where $\sum \mathbf{M}_{\rm cv}$ represents the sum of the moments on the control volume. Note that when the system coincides with the control volume, the sum of the moments on the control volume is equal to the sum of the moments on the system, that is

$$\sum \mathbf{M}_{\rm sys} = \sum \mathbf{M}_{\rm cv} = \sum \mathbf{M} \tag{4.104}$$

where $\sum \mathbf{M}$ will be used in subsequent analyses to represent the sum of moments on a control volume. The term $\sum \mathbf{M}$ is commonly referred to as the *torque* on the fluid system; hence, Equation 4.103 relates the torque to the angular momentum of a fluid.

Moment of momentum equation in terms of average velocities. In most practical applications of the moment of momentum equation, it is more convenient to work with average velocities over inlet and outlet subareas rather than velocity distributions over the control surface. For a stationary control volume, Equation 4.103 can be expressed in the following approximate form that is more directly useful in practical applications:

$$\sum \mathbf{M} = \frac{\mathrm{d}}{\mathrm{d}t}\int_{\rm cv} \rho(\mathbf{r}\times\mathbf{v})\,\mathrm{d}V + \sum_{i=1}^{N_{\rm out}}(\bar{\mathbf{r}}_i \times \dot{m}_i\overline{\mathbf{V}}_i) - \sum_{j=1}^{N_{\rm in}}(\bar{\mathbf{r}}_j \times \dot{m}_j\overline{\mathbf{V}}_j) \tag{4.105}$$

where $\bar{\mathbf{r}}_i$ and $\overline{\mathbf{V}}_i$ refer to the average position vector and the average velocity, respectively, of the flow at inlet or outlet i. Where the flow is steady and all the forces and momentum flows are in the same plane, Equation 4.105 simplifies to

$$\sum M = \sum_{i=1}^{N_{\rm out}}(\bar{r}_i\dot{m}_i\overline{V}_{\theta i}) - \sum_{j=1}^{N_{\rm in}}(\bar{r}_j\dot{m}_j\overline{V}_{\theta j}) \quad \text{(steady state)} \tag{4.106}$$

where $\bar{r}_i$ and $\overline{V}_{\theta i}$ are the magnitude of the position vector and the θ component of the average velocity, respectively, at inlet or outlet i. Note that the θ component of any vector, $\mathbf{V}$, at any given location, as denoted by V_θ, is equal to the component of $\mathbf{V}$ normal to the position vector of that location. The positive direction of V_θ is the direction that creates a positive moment as described below.

Sign convention. The right-hand rule is used in specifying the positive directions of the coordinate axes, the moments about the coordinate axes, and the θ components of the velocities. Using the right-hand rule, curling the fingers of the right hand from the x- to y-axis points the thumb in the direction of the positive z-axis. Pointing the thumb of the right hand in the positive direction of any given axis, the fingers are curled in the positive direction of both the moment about that axis and the θ component of the velocity in the plane normal to that axis. Applying this rule, the direction of positive moments and the positive θ components of velocities in the yz, xy, and xz planes are shown in Figure 4.29. Note that the θ component of the velocity is, by definition, perpendicular to the position vector, which is shown in Figure 4.29 to have a magnitude of r. The positive direction sign conventions shown in Figure 4.29 must be used in applying the moment of momentum equation (Equation 4.106).

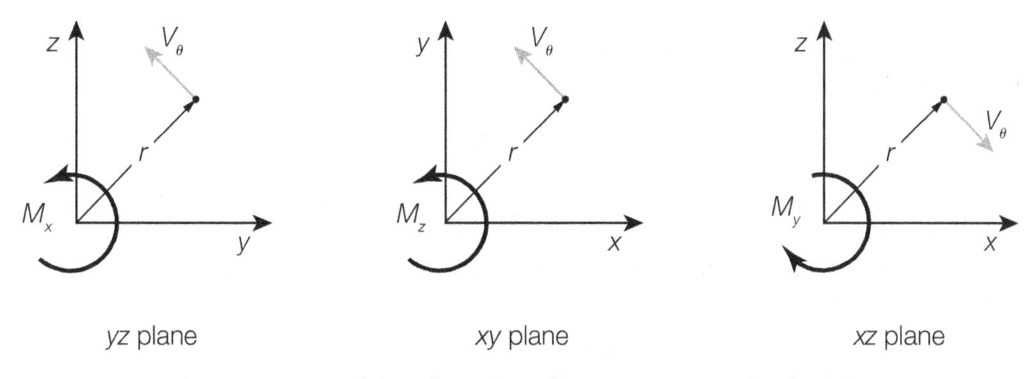

Figure 4.29: Positive directions for moments and velocities

EXAMPLE 4.19

Water flows at 8 L/s through the riser shown in Figure 4.30(a). If the weight of the pipe is 200 N/m, estimate the required restraining moment at P.

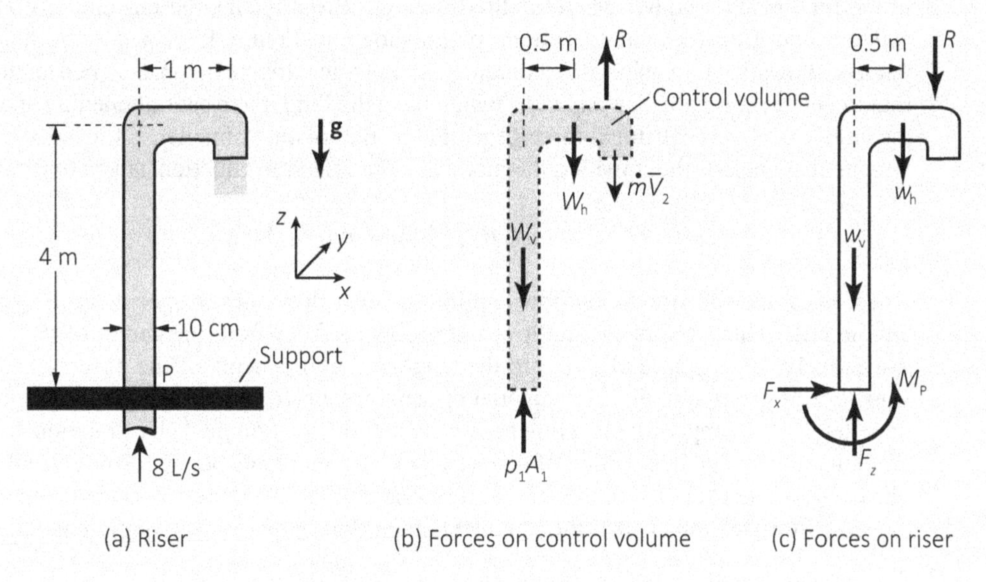

Figure 4.30: Riser

SOLUTION

The control volume containing the fluid in the riser is shown in Figure 4.30(b), and the forces on the riser are shown in Figure 4.30(c). From the given data, $\ell_1 = 4$ m, $\ell_2 = 1$ m, $D = 0.1$ m, $\gamma_{\mathrm{p}} = 200$ N/m, and $Q = 8$ L/s $= 0.008$ m^3/s. For water at 20°C, $\gamma_{\mathrm{w}} = 9789$ N/m^3 and $\rho_{\mathrm{w}} = 998.2$ kg/m^3. The following preliminary calculations are useful:

$$\dot{m} = \rho Q = (998.2)(0.008) = 7.986 \text{ kg/s}, \qquad \overline{V}_2 = \frac{Q}{A} = \frac{0.008}{\frac{\pi}{4}(0.10)^2} = 1.019 \text{ m/s}$$

$$W_{\mathrm{h}} = \gamma_{\mathrm{w}} \frac{\pi}{4} D^2 \ell_2 = (9789)\frac{\pi}{4}(0.1)^2(1) = 76.88 \text{ N}$$

For the control volume, where R is the resultant reaction force, and M_R is the moment of R about P, the moment-of-momentum equation gives

$$M_R - \left(\tfrac{1}{2}\ell_2\right) W_\mathrm{h} = -\ell_2 \dot{m} \overline{V}_2 \rightarrow M_R - \left(\tfrac{1}{2} \times 1\right)(76.88) = -(1)(7.986)(1.019) \qquad (4.107)$$

which gives $M_R = 30.30$ N·m. If M_P is the restraining moment on the riser at P, then M_P must support both the moment generated by the water flowing in the control volume ($= -M_R$) and the moment generated by the weight of the horizontal pipe. Hence, referring to Figure 4.30(c), for equilibrium of the riser:

$$M_\mathrm{P} - \gamma_\mathrm{P}(\tfrac{1}{2}\ell_2) + (-M_R) = 0 \rightarrow M_\mathrm{P} - (200)(\tfrac{1}{2} \times 1) + (-30.30) = 0 \rightarrow M_\mathrm{P} = 130.3 \text{ N·m}$$

Therefore, the restraining moment on the riser at P is approximately **130 N·m** in the counter-clockwise direction.

Pumps and fans. The moment of momentum equation is commonly applied in the analysis and design of rotating machines, and perhaps the most common application is to pumps. Once the application of the moment of momentum equation to pumps is understood, application to other rotating machines, such as turbines, is very similar. Pumps usually have the connotation of the fluid being a liquid; when the fluid is as gas (usually air), the machine is commonly called a *fan*. Consider the centrifugal pump shown in Figure 4.31, where the fluid being pumped flows into a casing that contains the impeller driven by an attached motor. The rotating element is called an *impeller*, which pulls the fluid into a central opening and pushes the fluid out by the centrifugal force generated by the rotating impeller. Using a fixed control volume that encases the impeller, the moment of momentum equation can be expressed as

$$T = \dot{m}_\mathrm{out} R_\mathrm{out} V_{\theta,\mathrm{out}} - \dot{m}_\mathrm{in} R_\mathrm{in} V_{\theta,\mathrm{in}} \qquad (4.108)$$

where $\dot{m}_\mathrm{out}$ and $\dot{m}_\mathrm{in}$ are the outflow and inflow mass flow rates, respectively, R_out and R_in are the radial distances of the outflow and inflow surfaces from the center of the impeller, respectively, and $V_{\theta,\mathrm{out}}$ and $V_{\theta,\mathrm{in}}$ are the tangential components of the outflow and inflow velocities, respectively. The key impeller dimensions are R_out and R_in, and the height of the impeller is H. In applying the moment of momentum equation given by Equation 4.108, the aforementioned sign convention of counterclockwise moments being positive should be applied. The moment acting on the fluid as it moves through the impeller is commonly referred to as the "torque" rather than the "moment," and this torque is commonly called the *shaft*

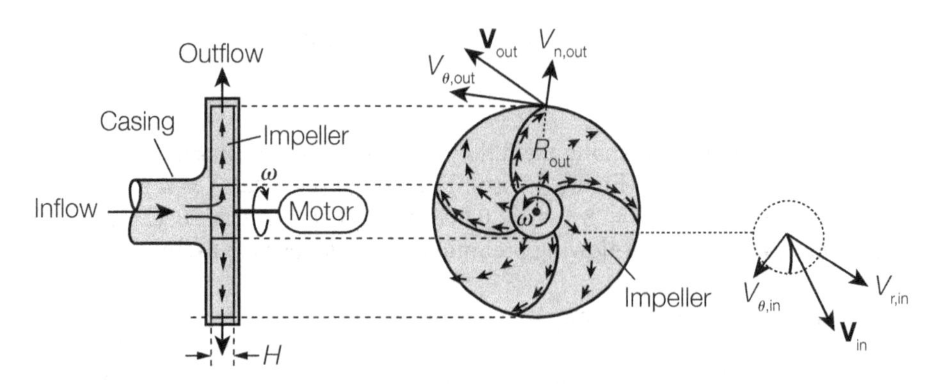

Figure 4.31: Centrifugal pump

torque, T_{shaft}. Noting that under steady-state conditions the mass flow rate into the impeller is equal to the mass flow rate out of the impeller, Equation 4.108 can be expressed as

$$T_{\text{shaft}} = \dot{m}(R_{\text{out}} V_{\theta,\text{out}} - R_{\text{in}} V_{\theta,\text{in}}) \tag{4.109}$$

Whenever a machine rotates at an angular velocity ω, while delivering a torque of T_{shaft}, the power delivered to the machine is called the *shaft power*, commonly denoted by $\dot{W}_{\text{shaft}}$ and given by

$$\dot{W}_{\text{shaft}} = T_{\text{shaft}}\omega \tag{4.110}$$

This equation is applicable to cases in which the fluid exerts a torque on the machine (e.g., turbines) as well as cases in which the machine exerts a torque on the fluid (e.g., pumps). Combining Equations 4.109 and 4.110 gives the following equivalent expressions for the shaft power:

$$\dot{W}_{\text{shaft}} = \dot{m}\omega(R_{\text{out}} V_{\theta,\text{out}} - R_{\text{in}} V_{\theta,\text{in}}) \quad \text{or} \quad \dot{W}_{\text{shaft}} = \dot{m}(U_{\text{out}} V_{\theta,\text{out}} - U_{\text{in}} V_{\theta,\text{in}}) \tag{4.111}$$

where U_{out} and U_{in} are the tangential velocities of the impeller tips at the inflow and outflow surfaces respectively, where

$$U_{\text{out}} = R_{\text{out}}\omega, \qquad U_{\text{in}} = R_{\text{in}}\omega \tag{4.112}$$

In some cases, the quantity of interest is the shaft work per unit mass of fluid being pumped, commonly denoted by $\dot{w}_{\text{shaft}}$. This quantity can be derived from Equation 4.111 by dividing both sides by $\dot{m}$, which gives

$$\dot{w}_{\text{shaft}} = \omega(R_{\text{out}} V_{\theta,\text{out}} - R_{\text{in}} V_{\theta,\text{in}}) \quad \text{or} \quad \dot{w}_{\text{shaft}} = (U_{\text{out}} V_{\theta,\text{out}} - U_{\text{in}} V_{\theta,\text{in}}) \tag{4.113}$$

The quantity $\dot{w}_{\text{shaft}}/g$ is equal to the energy added to the fluid per unit weight, which is commonly referred to as the head added by the pump, denoted by h_{p}.

EXAMPLE 4.20

The impeller of a water pump has an inner and outer diameter of 200 mm and 800 mm, respectively, and a height of 150 mm. Inflow to the impeller occurs in the radial direction, and at the exit of the impeller, the blade directs the flow at an angle of 35° to the outflow face. The pump motor rotates the impeller at 600 rpm. Estimate the power delivered to the water when the flow rate through the pump is 1000 L/min.

SOLUTION

From the given data: D_{in} = 200 mm, D_{out} = 800 mm, $R_{\text{out}} = D_{\text{out}}/2 = 400$ mm, H = 150 mm, θ = 35°, ω = 600 rpm = 62.83 rad/s, and Q = 1000 L/min = 0.01667 m^3/s. For water at 20°C, ρ = 998 kg/m^3. The mass flow rate through the pump, $\dot{m}$, the outflow area, A_{out}, and the component of the outflow velocity normal to the outflow face are calculated as follows:

$$\dot{m} = \rho Q = (998)(0.01667) = 16.63 \text{ kg/s}, \qquad A_{\text{out}} = \pi D_{\text{out}} H = \pi(0.8)(0.15) = 0.3770 \text{ m}^2$$

$$V_{\text{n,out}} = \frac{Q}{A_{\text{out}}} = \frac{0.01667}{0.3770} = 0.04421 \text{ m/s}$$

Calculation of the tangential component of the outflow velocity, $V_{\theta,\text{out}}$ requires consideration of the outflow velocity components as illustrated in Figure 4.32.

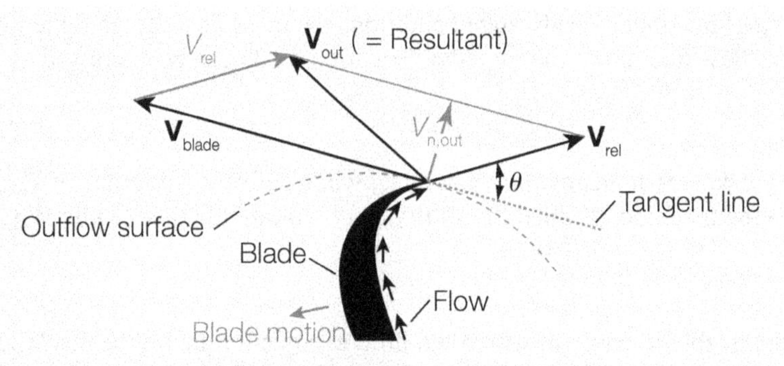

Figure 4.32: Components of impeller outflow velocity

The outflow velocity, $\mathbf{V}_{out}$, is equal to the vector sum of the outflow velocity relative to the moving blade, $\mathbf{V}_{rel}$, and the velocity of the blade at the exit, denoted by $\mathbf{V}_{blade}$. This relationship can be expressed analytically as follows:

$$\mathbf{V}_{out} = \mathbf{V}_{rel} + \mathbf{V}_{blade}$$

Based on this vectorial relationship and noting that $V_{blade} = \omega R_{out}$, the tangential component of the outflow velocity, $V_{\theta,out}$, can be derived as follows:

$$V_{\theta,out} = \omega R_{out} - V_{rel}\cos\theta \quad \rightarrow \quad V_{\theta,out} = \omega R_{out} - \frac{V_{n,out}}{\tan\theta}$$

$$\rightarrow \quad V_{\theta,out} = (62.83)(0.4) - \frac{0.04421}{\tan 35°} = 25.07 \text{ m/s}$$

The power input to the pumped water, $\dot{W}_{shaft}$, is given by Equation 4.111. Because the inflow velocity to the impeller is normal to the inflow surface, the radial component of the inflow velocity, $V_{\theta,in}$, is equal to zero, and Equation 4.111 gives

$$\dot{W}_{shaft} = \dot{m}\omega[R_{out}V_{\theta,out} - R_{in}V_{\theta,in}] = (16.63)(62.83)[(0.4)(25.07) - 0]$$
$$= 1.048 \times 10^4 \text{ W} = 10.48 \text{ kW}$$

Therefore, the power delivered to the pumped water is approximately **10.5 kW**. The power demand of the motor will be higher due to energy losses between the power input to the motor and the power input to the pumped water.

Turbines. The principal difference between a pump and a turbine is that a turbine extracts power from a fluid, whereas a pump adds power to a fluid. For both pumps and turbines, the shaft power is equal to shaft torque (= T_{shaft}) multiplied by the angular velocity (ω), as given by Equation 4.110. In the case of turbines, the impeller is called a *runner*, and the flow is typically from the periphery of the runner toward the center of the runner rather than from the center toward the periphery as in the case of pumps. A typical case of flow through a turbine runner is shown in Figure 4.33, where the inflow and outflow vectors are represented by $\mathbf{V}_{in}$ and $\mathbf{V}_{out}$, respectively, and the corresponding tangential components of these velocity vectors are $V_{\theta,in}$ and $V_{\theta,out}$, respectively. Applying the moment of momentum equation, the shaft torque, T_{shaft}, and the shaft power, $\dot{W}_{shaft}$, generated by the fluid are given by

$$T_{shaft} = \dot{m}(R_{out}V_{\theta,out} - R_{in}V_{\theta,in}) \tag{4.114}$$

$$\dot{W}_{shaft} = \dot{m}\omega(R_{out}V_{\theta,out} - R_{in}V_{\theta,in}) \tag{4.115}$$

where $\dot{m}$ is the mass flow rate through the runner, ω is the rotational speed, and R_{in} and R_{out} are the radii of the inflow and outflow surfaces, respectively.

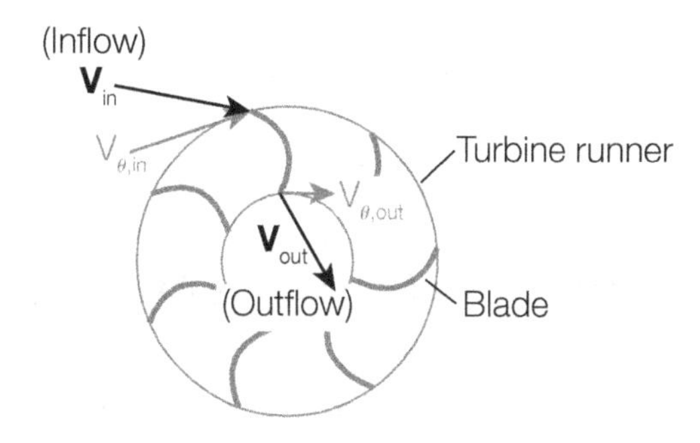

Figure 4.33: Flow through a turbine runner

EXAMPLE 4.21

A turbine runner has an outer diameter of 2 m and an inner diameter of 0.8 m, and it rotates at 60 rpm. On the (outer) inflow side, the blade angle is 50°, and on the (inner) outflow side, the blade angle is 110°. In addition, when the flow rate through the runner is 1 m³/s, the inflow velocity vector makes an angle of 30° with the inflow surface as shown in Figure 4.34(a). Determine the height of the runner required to ensure that the relative inflow velocity is in a direction that is tangent to the blade. What power is being generated by the runner?

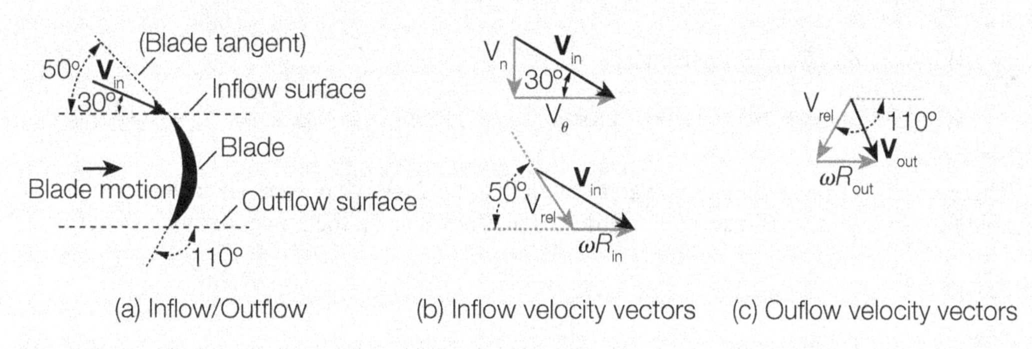

Figure 4.34: Velocity vectors for flow through a turbine

SOLUTION

Use subscripts 1 and 2 to denote the entrance and exit surfaces, respectively. From the given data: D_1 = 2 m, R_1 = 1 m, D_2 = 0.8 m, R_2 = 0.4 m, ω = 60 rpm = 6.283 rad/s, θ_{b1} = 50°, θ_{b2} = 110°, θ_{v1} = 30°, and Q = 1 m³/s. For water at 20°C, ρ = 998 kg/m³, and the mass flow rate through the runner is $\dot{m} = \rho Q = 998$ kg/s. Considering the inflow velocity components shown in Figure 4.34(b), the tangential component of the inflow velocity, $V_{\theta,1}$ is given by the following equations:

$$V_{\theta,1} = \frac{V_{\text{n},1}}{\tan 30^\circ}, \qquad V_{\theta,1} = \omega R_1 + \frac{V_{\text{n},1}}{\tan 50^\circ}$$

Combining these equations to eliminate $V_{\theta,1}$ and substituting the given data yields the following:

$$\frac{V_{\text{n},1}}{\tan 30^\circ} = \omega R_1 + \frac{V_{\text{n},1}}{\tan 50^\circ} \quad \rightarrow \quad \frac{V_{\text{n},1}}{\tan 30^\circ} = (6.283)(1) + \frac{V_{\text{n},1}}{\tan 50^\circ} \quad \rightarrow \quad V_{\text{n},1} = 7.036 \text{ m/s}$$

The corresponding tangential component of the inflow velocity, $V_{\theta,1}$, is given by

$$V_{\theta,1} = \frac{V_{n,1}}{\tan 30^\circ} = \frac{7.036}{\tan 30^\circ} = 12.19 \text{ m/s}$$

The required height, H, of the runner is derived from the given flow rate, Q, as follows:

$$V_{n,1} = \frac{Q}{A_1} = \frac{Q}{\pi D_1 H} \quad \rightarrow \quad H = \frac{Q}{\pi D_1 V_{n,1}} = \frac{1}{\pi(2)(7.036)} \quad \rightarrow \quad H = 0.02262 \text{ m} = 22.6 \text{ mm}$$

Therefore, using a runner height of approximately **23 mm** will ensure that the entering velocity is tangential to the blade. At the exit (inner) face of the runner, the velocity components are shown in Figure 4.34(c), which require that

$$V_{n,2} = \frac{Q}{A_2} = \frac{1}{\pi D_2 H} = \frac{1}{\pi(0.8)(0.02262)} = 17.59 \text{ m/s}$$

$$V_{\theta,2} = \omega R_2 + \frac{V_{n,2}}{\tan 110^\circ} = (6.283)(0.4) - \frac{17.59}{\tan(180^\circ - 110^\circ)} = -3.889 \text{ m/s}$$

Because the calculated value of $V_{\theta,2}$ is negative, $V_{\theta,2}$ is in the direction opposite that of ωR_2. The power, $\dot{W}_{\text{shaft}}$, added to the water as it moves through the runner is given by Equation 4.115. Respecting the directional sign conventions gives

$$\dot{W}_{\text{shaft}} = \dot{m}\omega[R_{\text{out}} V_{\theta,2} - R_{\text{in}} V_{\theta,1}]$$

$$\dot{W}_{\text{shaft}} = (998)(6.283)[(0.4 \times -3.889) - (1 \times 12.19)] = -8.62 \times 10^4 \text{ W} = -86.2 \text{ kW}$$

Therefore, it is expected that the water will deliver **86.2 kW** of power. The actual power available for use will be less than 86.2 kW due to inefficiencies in the power delivery system.

Control volume moving with constant velocity. In cases where the control volume is moving with a constant velocity, the velocity is expressed relative to the moving control volume and the appropriate form of the angular momentum equation is

$$\sum \mathbf{M} = \frac{\mathrm{d}}{\mathrm{d}t}\int_{\mathrm{cv}} \rho(\mathbf{r} \times \mathbf{v}_\mathrm{r})\,\mathrm{d}V + \int_{\mathrm{cs}} \rho(\mathbf{r} \times \mathbf{v}_\mathrm{r})\,(\mathbf{v}_\mathrm{r} \cdot \mathbf{n})\,\mathrm{d}A \tag{4.116}$$

where $\mathbf{v}_\mathrm{r}$ is the fluid velocity relative to the moving control volume. A commonly encountered circumstance is the case where the flow is steady and all the forces and flows are in the same plane. In this case, Equation 4.116 simplifies to

$$\sum M = \sum_{i=1}^{N_\text{out}} (\bar{r}_i \dot{m}_{\mathrm{r},i} \overline{V}_{\mathrm{r},\theta i}) - \sum_{j=1}^{N_\text{in}} (\bar{r}_j \dot{m}_{\mathrm{r},j} \overline{V}_{\mathrm{r},\theta j}) \tag{4.117}$$

where $\bar{r}_i$, $\dot{m}_{\mathrm{r},i}$, and $\overline{V}_{\mathrm{r},\theta i}$ are the magnitude of the position vector, average relative mass flow rate, and θ component of the average relative velocity, respectively, at inlet or outlet i.

Fixed systems with rotating components. In fixed systems with rotating components, the moment-of-momentum equation is applied using the absolute inflow and outflow velocities, along with the mass flow rate through the system. Turbomachines and sprinklers are examples of fixed systems with rotating components.

EXAMPLE 4.22

Water flows at 2 L/s into the rotating sprinkler shown in Figure 4.35. If the restraining torque exerted by friction is 4 N·m, determine the rate of rotation of the sprinkler.

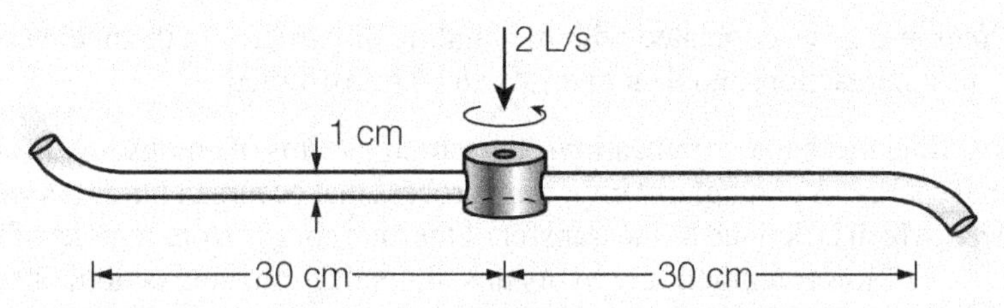

Figure 4.35: Sprinkler

SOLUTION

From the given data: $r = 0.30$ m, $D = 0.01$ m, $Q = 2$ L/s $= 0.002$ m^3/s, and $T = 4$ N·m. Assume that $\rho = 998$ kg/m^3. The moment-of-momentum equation requires that

$$\sum M = \sum_{\text{out}} r\dot{m}V \quad \rightarrow \quad T = 2r\dot{m}V \tag{4.118}$$

where T is the torque exerted on the flowing water and $\dot{m}$ is the mass flow rate out of each sprinkler arm, and V is the absolute velocity of the water exiting the sprinkler. Because one-half of the mass flow goes out of each sprinkler arm,

$$\dot{m} = \rho\frac{Q}{2} = (998)\frac{0.002}{2} = 0.998 \text{ kg/s}$$

Substituting into Equation 4.118 gives

$$4 = 2(0.30)(0.998)V \quad \rightarrow \quad V = 6.68 \text{ m/s}$$

The flow velocity within each rotating sprinkler, V_{flow}, and the velocity of nozzle rotation, V_{noz}, are given by

$$V_{\text{flow}} = \frac{Q/2}{A} = \frac{0.002/2}{\pi(0.01)^2/4} = 12.7 \text{ m/s}, \qquad V_{\text{noz}} = r\omega = 0.30\omega$$

The absolute velocity of the jet leaving the rotating sprinkler, V, is given by

$$V = V_{\text{flow}} - V_{\text{noz}} \quad \rightarrow \quad 6.68 \text{ m/s} = 12.7 - 0.30\omega \quad \rightarrow \quad \omega = 20.1 \text{ rad/s} = 192 \text{ rpm}$$

Therefore, with a restraining torque of 4 N·m, the sprinkler rotates at approximately **192 rpm**.

4.6 Conservation of Energy

Energy exists in many forms and can be conveniently categorized into the following forms: mechanical energy, thermal energy, chemical energy, electrical energy, and nuclear energy. These types of energy are defined as follows:

Mechanical energy is associated with motion and position in a field; kinetic energy and potential energy are forms of mechanical energy.

Thermal energy is associated with temperature changes and phase changes; energy required to raise the temperature of a substance is thermal energy. Thermal energy is associated with molecular-scale motion.

Chemical energy is the energy associated with chemical bonds between elements; such energy can be transformed during chemical reactions.

Electrical energy is associated with electrical charge; for example, capacitors can store electrical energy.

Nuclear energy is associated with the binding of particles in the nucleus of an atom; during fission reactions, nuclear energy can be transformed.

In most fluid mechanics applications, the variable forms of energy are mechanical energy and thermal energy. A particularly important process that occurs in fluids (as well as solids) is *heat transfer*, which is defined as the transfer of thermal energy from regions of higher temperature to regions of lower temperature by means of conduction, convection, or radiation.

4.6.1 The First Law of Thermodynamics

The *first law of thermodynamics* states that within any defined system, the heat added to the system, ΔQ_h, minus the work done by the system, ΔW, is equal to the change in internal energy within the system, ΔE. The first law of thermodynamics, also called the *law of conservation of energy*, can be put in the form

$$\Delta Q_\mathrm{h} - \Delta W = \Delta E \tag{4.119}$$

Dividing Equation 4.119 by a time interval, Δt, and taking the limit as Δt approaches zero leads to the differential form of the first law of thermodynamics,

$$\lim_{\Delta t\to 0} \frac{\Delta Q_\mathrm{h}}{\Delta t} - \lim_{\Delta t\to 0} \frac{\Delta W}{\Delta t} = \lim_{\Delta t\to 0} \frac{\Delta E}{\Delta t} \tag{4.120}$$

or

$$\frac{\mathrm{d}Q_\mathrm{h}}{\mathrm{d}t} - \frac{\mathrm{d}W}{\mathrm{d}t} = \frac{\mathrm{d}E_\mathrm{sys}}{\mathrm{d}t} \tag{4.121}$$

where E_sys is the system energy. Denoting the energy per unit mass within a fluid system as e, the law of conservation of energy, Equation 4.121, can be expressed as

$$\frac{\mathrm{d}}{\mathrm{d}t}\int_\mathrm{sys} e\rho\,\mathrm{d}V = \frac{\mathrm{d}Q_\mathrm{h}}{\mathrm{d}t} - \frac{\mathrm{d}W}{\mathrm{d}t} \tag{4.122}$$

where ρ is the density of the fluid and $\mathrm{d}V$ is an elemental volume within the fluid system. Recall that the Reynolds transport theorem relates the rate of change of any fluid property within a system to the changes in that same fluid property within a coincident control volume, such that

$$\frac{\mathrm{d}}{\mathrm{d}t}\int_\mathrm{sys} \beta\rho\,\mathrm{d}V = \frac{\mathrm{d}}{\mathrm{d}t}\int_\mathrm{cv} \beta\rho\,\mathrm{d}V + \int_\mathrm{cs} \beta\rho(\mathbf{v}\cdot\mathbf{n})\,\mathrm{d}A \tag{4.123}$$

where "cv" and "cs" denote the volume and surface area of the control volume, respectively, $\mathbf{v}$ is the fluid velocity, $\mathbf{n}$ is the unit normal pointing out of the control volume, and β is a fluid property per unit mass. Taking $\beta = e$ in Equation 4.123 gives

$$\frac{\mathrm{d}}{\mathrm{d}t}\int_\mathrm{sys} e\rho\,\mathrm{d}V = \frac{\mathrm{d}}{\mathrm{d}t}\int_\mathrm{cv} e\rho\,\mathrm{d}V + \int_\mathrm{cs} e\rho(\mathbf{v}\cdot\mathbf{n})\,\mathrm{d}A \tag{4.124}$$

Combining Equations 4.122 and 4.124 yields the following expression for the law of conservation of energy:

$$\frac{dQ_h}{dt} - \frac{dW}{dt} = \frac{d}{dt}\int_{cv} \rho e \, dV + \int_{cs} \rho e \, (\mathbf{v} \cdot \mathbf{n}) \, dA \tag{4.125}$$

4.6.2 Steady-State Energy Equation

Consider the general conditions between the inflow and outflow sections of a closed conduit as illustrated in Figure 4.36. Between the inflow and outflow sections of the conduit (sections 1 and 2) there can be an exchange of heat between the fluid and the conduit; there can also be a machine such as a pump or turbine that inputs or extracts energy from the flowing fluid. Under steady-state conditions (and considering the instant when the system and control volume coincide), Equation 4.125 gives

$$\frac{dQ_h}{dt} - \frac{dW}{dt} = \int_{cs} \rho e \, (\mathbf{v} \cdot \mathbf{n}) \, dA \quad \text{(steady state)} \tag{4.126}$$

where Q_h is the heat added to the fluid in the control volume [E[26]], W is the work done by the fluid in the control volume [E], "cs" denotes the surface area of the control volume [L^2], ρ is the density of the fluid in the control volume [ML^{-3}], and e is the energy per unit mass of fluid in the control volume [EM^{-1}] given by

$$e = gz + \frac{v^2}{2} + u \tag{4.127}$$

where z [L] is the elevation of a unit mass of fluid having a velocity v (= $|\mathbf{v}|$) [LT^{-1}] and internal energy u [E]. The internal energy, u, is the defined as the sum of all forms of energy at the molecular scale; therefore, Equation 4.127 represents a partitioning of energy between a macroscopic component ($= gz + v^2/2g$) and a microscopic component ($= u$). In fluid mechanics applications, u is associated only with thermal energy, which is the form of energy associated with molecular motion and molecular bonding forces and u is primarily a function of temperature. By convention, the heat added to a system and the work done by a system are positive quantities. The normal stresses on the inflow and outflow boundaries of the control volume are equal to the pressure, p, and the shear stresses on the inflow and outflow boundaries of the control volume are tangential to the control surface. As the fluid moves

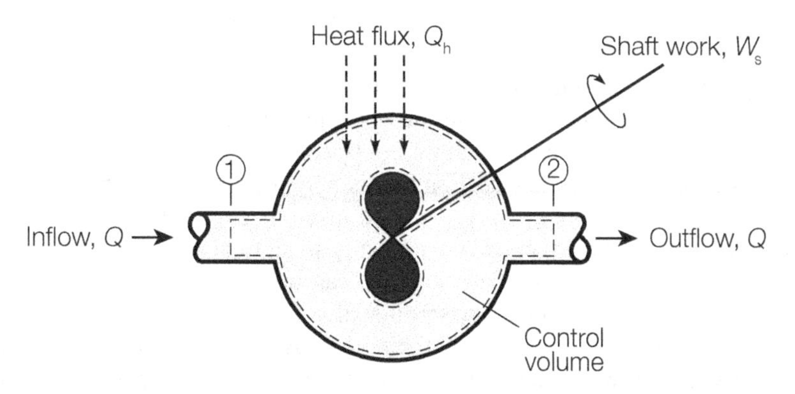

Figure 4.36: Energy balance in a closed conduit

[26] $E = ML^2T^{-2}$ denotes dimension of energy.

across the control surface with velocity $\mathbf{v}$, the power (= rate of doing work) expended by the fluid against the external pressure forces is given by

$$\frac{\mathrm{d}W_\mathrm{p}}{dt} = \int_\mathrm{cs} p(\mathbf{v}\cdot\mathbf{n})\,\mathrm{d}A \tag{4.128}$$

where W_p is the work done against external pressure forces [E]. The work done by a fluid in the control volume is typically separated into work done against external pressure forces, W_p, plus work done against rotating surfaces, W_s, commonly referred to as the *shaft work*. Examples of rotating surfaces include *rotors* in gas or steam turbines, *impellers* in pumps, and *runners* in hydraulic turbines. The rate at which work is done by a fluid system, $\mathrm{d}W/\mathrm{d}t$, can therefore be written as

$$\frac{\mathrm{d}W}{\mathrm{d}t} = \frac{\mathrm{d}W_\mathrm{p}}{\mathrm{d}t} + \frac{\mathrm{d}W_\mathrm{s}}{\mathrm{d}t} = \int_\mathrm{cs} p(\mathbf{v}\cdot\mathbf{n})\,\mathrm{d}A + \frac{\mathrm{d}W_\mathrm{s}}{\mathrm{d}t} \tag{4.129}$$

Combining Equation 4.129 with the steady-state energy equation (Equation 4.126) gives

$$\frac{\mathrm{d}Q_\mathrm{h}}{\mathrm{d}t} - \frac{\mathrm{d}W_\mathrm{s}}{\mathrm{d}t} = \int_\mathrm{cs} \rho\left(\frac{p}{\rho} + e\right)(\mathbf{v}\cdot\mathbf{n})\,\mathrm{d}A \tag{4.130}$$

Substituting the definition of the internal energy, e, given by Equation 4.127 into Equation 4.130 yields

$$\frac{\mathrm{d}Q_\mathrm{h}}{\mathrm{d}t} - \frac{\mathrm{d}W_\mathrm{s}}{\mathrm{d}t} = \int_\mathrm{cs} \rho\left(h + gz + \frac{v^2}{2}\right)(\mathbf{v}\cdot\mathbf{n})\,\mathrm{d}A \tag{4.131}$$

where h is the *enthalpy* of the fluid defined by

$$h = \frac{p}{\rho} + u \tag{4.132}$$

The enthalpy, h, is a convenient grouping of terms, particularly for gases, because for an ideal gas, h is only a function of temperature (i.e., $h = RT + u$ for an ideal gas). Denoting the rate at which heat is being added to the fluid system by $\dot{Q}$ and the rate at which work is being done against moving impervious boundaries (shaft work) by $\dot{W}_\mathrm{s}$, the energy equation can be written in the form

$$\dot{Q} - \dot{W}_\mathrm{s} = \int_\mathrm{cs} \rho\left(h + gz + \frac{v^2}{2}\right)(\mathbf{v}\cdot\mathbf{n})\,\mathrm{d}A \tag{4.133}$$

Consider the term $h + gz$, where

$$h + gz = \frac{p}{\rho} + u + gz = g\left(\frac{p}{\gamma} + z\right) + u \tag{4.134}$$

and γ is the specific weight of the fluid. Equation 4.134 indicates that $h + gz$ can be assumed to be constant across the inflow and outflow openings illustrated in Figure 4.36, provided $p/\gamma + z$ and u are constant across the section. For (incompressible) liquids, it has been shown that $p/\gamma + z$ is constant across any section that is perpendicular to the flow (under steady-state conditions), and for gases, $p/\gamma + z$ can be taken as approximately constant because pressure variations across flow sections are typically small (and $p/\gamma \gg z$ because γ for gases is typically very small). The internal energy, u, can be taken as a constant across a flow section if the temperature does not vary significantly across the cross section, because u is predominantly a function of temperature. Taking $h + gz$ as a constant across the inflow and outflow sections and knowing that $\mathbf{v}\cdot\mathbf{n}$ is equal to zero over the impervious boundaries in contact with the fluid system, Equation 4.133 can be integrated to yield

$$\dot{Q} - \dot{W}_\mathrm{s} = (h_1 + gz_1)\int_{A_1} \rho_1(\mathbf{v}\cdot\mathbf{n})\,\mathrm{d}A + \int_{A_1} \rho_1\frac{v^2}{2}(\mathbf{v}\cdot\mathbf{n})\,\mathrm{d}A$$

$$+ (h_2 + gz_2)\int_{A_2} \rho_2(\mathbf{v}\cdot\mathbf{n})\,\mathrm{d}A + \int_{A_2} \rho_2\frac{v^2}{2}(\mathbf{v}\cdot\mathbf{n})\,\mathrm{d}A$$

$$= -(h_1 + gz_1)\int_{A_1} \rho_1 v_1\,\mathrm{d}A - \int_{A_1} \rho_1\frac{v_1^3}{2}\,\mathrm{d}A$$

$$+ (h_2 + gz_2)\int_{A_2} \rho_2 v_2\,\mathrm{d}A + \int_{A_2} \rho_2\frac{v_2^3}{2}\,\mathrm{d}A \qquad (4.135)$$

where the subscripts 1 and 2 refer to the inflow and outflow boundaries, respectively, and the negative signs result from the fact that the unit normal points out of the control volume, causing $\mathbf{v}\cdot\mathbf{n}$ to be negative on the inflow boundary and positive on the outflow boundary. Equation 4.135 can be further simplified by noting that the assumption of steady flow requires that rate of mass inflow to the control volume be equal to the rate of mass outflow. Denoting the mass flow rate by $\dot{m}$, the continuity equation requires that

$$\dot{m} = \int_{A_1} \rho_1 v_1\,\mathrm{d}A = \int_{A_2} \rho_2 v_2\,\mathrm{d}A \qquad (4.136)$$

Furthermore, if the density variation across any given flow section is negligible, then the constants α_1 and α_2 can be defined by the equations

$$\int_{A_1} \rho_1\frac{v_1^3}{2}\,\mathrm{d}A = \alpha_1\rho_1\frac{V_1^3}{2}A_1 \quad\rightarrow\quad \alpha_1 = \frac{1}{A_1}\int_{A_1}\left(\frac{v_1}{V_1}\right)^3\mathrm{d}A \qquad (4.137)$$

$$\int_{A_2} \rho_2\frac{v_2^3}{2}\,\mathrm{d}A = \alpha_2\rho_2\frac{V_2^3}{2}A_2 \quad\rightarrow\quad \alpha_2 = \frac{1}{A_2}\int_{A_2}\left(\frac{v_2}{V_2}\right)^3\mathrm{d}A \qquad (4.138)$$

where A_1 and A_2 are the areas of the inflow and outflow boundaries, respectively, and V_1 and V_2 are the corresponding mean velocities across these boundaries. The constants α_1 and α_2 are determined by the velocity profile across the flow boundaries, and these constants are called *kinetic energy correction factors*. The general definition of the kinetic energy correction factor, α, for any flow section is

$$\alpha = \frac{1}{A}\int_A\left(\frac{v}{V}\right)^3\mathrm{d}A \qquad (4.139)$$

where A is the flow area, v is the velocity at a point within the flow area, and V is the average velocity over the flow area. It is apparent from Equation 4.139 that when the velocity is constant across a flow area, $v = V$ and $\alpha = 1$. If the velocity distribution deviates from being uniform, then $\alpha > 1$. Larger variations of velocity across a flow section will yield larger values of α. For fully developed laminar flows in circular conduits, it can be shown that $\alpha = 2$, whereas for turbulent flows in circular conduits, it can be shown that α is usually in the range of 1.04–1.11, although values as high as 1.15 have been reported. In most practical cases where the flow is turbulent, $\alpha = 1$ is assumed, and the resulting errors are usually negligible because the kinetic energy is typically a relatively small part of the energy balance. Also, in laminar flows, where $\alpha = 2$, the velocities are usually sufficiently low that the kinetic energy term in the energy equation is so small that taking $\alpha = 1$ or $\alpha = 2$ makes little difference. Combining Equations 4.135–4.138 leads to

$$\dot{Q} - \dot{W}_\mathrm{s} = -(h_1 + gz_1)\dot{m} - \alpha_1\rho_1\frac{V_1^3}{2}A_1 + (h_2 + gz_2)\dot{m} + \alpha_2\rho_2\frac{V_2^3}{2}A_2 \qquad (4.140)$$

Invoking the continuity equation requires that

$$\rho_1 V_1 A_1 = \rho_2 V_2 A_2 = \dot{m} \qquad (4.141)$$

and combining Equations 4.140 and 4.141 leads to

$$\dot{Q} - \dot{W}_s = \dot{m}\left[\left(h_2 + gz_2 + \alpha_2\frac{V_2^2}{2}\right) - \left(h_1 + gz_1 + \alpha_1\frac{V_1^2}{2}\right)\right] \quad (4.142)$$

which can be put in the form

$$\frac{\dot{Q}}{\dot{m}g} - \frac{\dot{W}_s}{\dot{m}g} = \left(\frac{p_2}{\gamma_2} + \frac{u_2}{g} + z_2 + \alpha_2\frac{V_2^2}{2g}\right) - \left(\frac{p_1}{\gamma_1} + \frac{u_1}{g} + z_1 + \alpha_1\frac{V_1^2}{2g}\right) \quad (4.143)$$

and can be further rearranged into the useful form

$$\underbrace{\left(\frac{p_1}{\gamma_1} + \alpha_1\frac{V_1^2}{2g} + z_1\right)}_{\text{inflow mechanical energy}} = \underbrace{\left(\frac{p_2}{\gamma_2} + \alpha_2\frac{V_2^2}{2g} + z_2\right)}_{\text{outflow mechanical energy}} + \underbrace{\left[\frac{1}{g}(u_2 - u_1) - \frac{\dot{Q}}{\dot{m}g}\right]}_{\text{head loss, } h_\ell} + \underbrace{\left[\frac{\dot{W}_s}{\dot{m}g}\right]}_{\text{shaft work, } h_s} \quad (4.144)$$

Equation 4.144 is a form of the steady-state energy equation that is widely applied in practice, particularly to describe the flow of liquids in closed conduits. Equation 4.144 is also applicable to gases, although other arrangements of this equation are more commonly used for gases. The steady-state energy equation expressed by Equation 4.144 indicates that the change in mechanical energy between the inflow and outflow from a control volume is equal to the gain in internal energy minus the heat added per unit weight, collectively called the *head loss*, h_ℓ, plus the (shaft) work done by the fluid per unit weight, h_s. Application of the steady-state energy equation generally requires characterization of h_ℓ and h_s.

Head loss. The energy loss per unit weight, commonly called the *head loss*, h_ℓ, is defined by the relation

$$h_\ell = \frac{1}{g}(u_2 - u_1) - \frac{\dot{Q}}{\dot{m}g} \quad (4.145)$$

which indicates that the head loss between the inflow and outflow sections of a control volume is equal to the gain in internal energy minus the heat added to the control volume. Equation 4.145 also indicates that the head loss is equal to the gain in internal energy plus the heat lost from the control volume. Thus, the head loss is the change in internal energy derived from sources internal to the control volume. Head loss can be regarded as lost mechanical energy that is converted to thermal energy through viscous action. In accordance with the second law of thermodynamics, h_ℓ is always positive (i.e., $h_\ell > 0$) because of fluid friction. The change in internal (thermal) energy per unit mass can be related to the change in temperature by the specific heat at constant volume, c_v, where

$$u_2 - u_1 = c_v(T_2 - T_1) \quad (4.146)$$

where the specific heat is defined as the energy required to generate a unit change in temperature of a unit mass of substance. For most liquids, the explicit condition of constant volume for the specific heat, c_v, is implicit in the fact that the liquid is incompressible; hence, for liquids, c_v is usually stated simply as the specific heat, c. Heat addition to a control volume, as quantified by $\dot{Q}$, includes all heat transfer processes generated by a temperature difference between the control volume and its surroundings, such as radiation, conduction, and convection. Processes in which there is no heat transfer between the system and its surroundings are called *adiabatic processes*. A process can be adiabatic in one of two ways: (1) the system is well insulated so that only a negligible amount of heat can pass through the system boundary or (2) both the system and surroundings are at the same temperature and therefore there is no driving force (temperature difference) for heat transfer. By combining the momentum equation and the energy equation, it can be shown that[27] the head loss in pipes of length L and diameter D can be expressed in the form

$$h_\ell = f\frac{L}{D}\frac{V^2}{2g} \tag{4.147}$$

Where f is a *friction factor* that depends on the pipe characteristics and flow conditions, and V is the average flow velocity in the pipe.

Shaft work. The (shaft) work done by the fluid per unit weight, h_s, is defined by the relation

$$h_s = \frac{\dot{W}_s}{\dot{m}g} \tag{4.148}$$

The shaft work, $\dot{W}_s$, is generally associated with the presence of a pump or a turbine. Turbines extract energy from a flowing fluid, in which case $\dot{W}_s > 0$ and $h_s > 0$, and pumps add energy to a flowing fluid, in which case $\dot{W}_s < 0$ and $h_s < 0$. In many applications, h_s is calculated from the energy equation to determine the pumping requirements (for pump) to achieve a desired objective or to determine the power that can be extracted from a flowing fluid (for turbine).

Features of the steady-state one-dimensional energy equation. The most common form of the steady-state one-dimensional energy equation as given by Equation 4.144 is

$$\left(\frac{p_1}{\gamma_1} + \alpha_1\frac{V_1^2}{2g} + z_1\right) = \left(\frac{p_2}{\gamma_2} + \alpha_2\frac{V_2^2}{2g} + z_2\right) + h_\ell + h_s \tag{4.149}$$

The energy equation states that the mechanical energy in a closed conduit is continuously being converted into thermal energy via head loss. Many practitioners incorrectly refer to the energy equation in the form of Equation 4.149 as the *Bernoulli equation*, which bears some resemblance to the Bernoulli equation but is fundamentally different in that the Bernoulli equation is derived from the momentum equation, which does not account for variations in thermal energy. Equation 4.149 applies to liquids, gases, and vapors; to viscous and inviscid fluids; and to compressible and incompressible flows. The only restriction is the requirement of steady flow. Each term in the energy equation (Equation 4.149) has units of length, and each term is commonly referred to as a "head." The term p/γ is called the *pressure head*, $V^2/2g$ is called the *velocity head*, z is called the *elevation head* or *potential head*, and the sum of these three head terms is called the *total head* or the *total mechanical energy*. The combined term $p/\gamma + z$ is called the *piezometric head*.

EXAMPLE 4.23

Water at 20°C flows vertically upward through a thermally insulated reducer with an entrance diameter of 15 cm and an exit diameter of 10 cm. The exit is at an elevation 1 m higher than the entrance. When the flow rate through the reducer is 18 L/s, pressure gauges at the entrance and exit of the reducer show 100 kPa and 67 kPa, respectively. Estimate the head loss and temperature difference between the entrance and exit of the reducer.

[27] See Chapter 7.

SOLUTION

From the given data: p_1 = 100 kPa, p_2 = 67 kPa, D_1 = 15 cm = 0.15 m, D_2 = 10 cm = 0.10 m, $z_2 - z_1$ = 1 m, and Q = 18 L/s = 0.018 m^3/s. For water at 20°C, γ = 9.79 kN/m^3. The following preliminary calculations are useful:

$$A_1 = \frac{\pi}{4}D_1^2 = \frac{\pi}{4}(0.15)^2 = 0.01767 \text{ m}^2, \qquad V_1 = \frac{Q}{A_1} = \frac{0.018}{0.01767} = 1.019 \text{ m/s}$$

$$A_2 = \frac{\pi}{4}D_2^2 = \frac{\pi}{4}(0.10)^2 = 0.007854 \text{ m}^2, \qquad V_2 = \frac{Q}{A_2} = \frac{0.018}{0.007854} = 2.292 \text{ m/s}$$

Because there is no shaft work between sections 1 and 2, the head loss, h_ℓ, is given by Equation 4.149 as

$$h_\ell = \left[\frac{p_1}{\gamma} + \alpha_1\frac{V_1^2}{2g} + z_1\right] - \left[\frac{p_2}{\gamma} + \alpha_2\frac{V_2^2}{2g} + z_2\right] \tag{4.150}$$

Substituting the given and derived data into Equation 4.150 and taking $\alpha_1 = \alpha_2 = 1.0$ gives

$$h_\ell = \left[\frac{100}{9.79} + \frac{1.019^2}{2(9.81)} + 0\right] - \left[\frac{67}{9.79} + \frac{2.292^2}{2(9.81)} + 1\right] = 2.156 \text{ m} \approx \mathbf{2.16\ m}$$

Because the reducer is insulated, $\dot{Q} = 0$, and the combination of Equations 4.145 and 4.146 gives

$$h_\ell = \frac{1}{g}(u_2 - u_1) - \frac{\dot{Q}}{\dot{m}g} = \frac{1}{g}c(T_2 - T_1) - 0 \quad \rightarrow \quad h_\ell = \frac{1}{g}c(T_2 - T_1) \tag{4.151}$$

where the specific heat of water, c, is 4187 J/(kg·K). Substituting h_ℓ = 2.156 m = 2.156 N·m/m into Equation 4.151 and rearranging yields

$$T_2 - T_1 = \frac{h_\ell g}{c} = \frac{(2.156)(9.81)}{4187} = 0.00505 \text{ K} = 0.00505°\text{C}$$

Therefore, the temperature change between sections 1 and 2 is a minuscule **0.00505°C**.

Forms of the energy equation useful for liquids. If a fluid is incompressible, which is the case usually encountered with liquids, then $\gamma_1 = \gamma_2 = \gamma$, and the energy equation, Equation 4.149, can be put in the form

$$\left(\frac{p_1}{\gamma} + \alpha_1\frac{V_1^2}{2g} + z_1\right) = \left(\frac{p_2}{\gamma} + \alpha_2\frac{V_2^2}{2g} + z_2\right) + h_\ell + h_s \quad \text{(incompressible fluid)} \tag{4.152}$$

It is usually convenient to separate h_s in Equation 4.152 into the head added by a pump, h_pump, and the head added by a turbine, h_turb, such that $h_s = h_\text{turb} - h_\text{pump}$, and Equation 4.152 becomes

$$\left(\frac{p_1}{\gamma} + \alpha_1\frac{V_1^2}{2g} + z_1\right) + h_\text{pump} = \left(\frac{p_2}{\gamma} + \alpha_2\frac{V_2^2}{2g} + z_2\right) + h_\text{turb} + h_\ell \quad \text{(incompressible fluid)} \tag{4.153}$$

In most practical cases where a machine is in the system, the machine will be either a pump or a turbine. Therefore, in applying Equation 4.153, either $h_\text{pump} = 0$ or $h_\text{turb} = 0$. Of course, if there is no machine, then $h_{\text{pump}} = 0$ and $h_{\text{turb}} = 0$. If the machine is a turbine, the power that can be delivered by the turbine, P_{turb}, can be expressed as

$$P_{\text{turb}} = \eta_{\text{turb}} \cdot \gamma Q h_{\text{turb}} \tag{4.154}$$

where η_{turb} is the efficiency of the turbine and Q is the volume flow rate through the turbine. If the machine is a pump, the power that must be delivered to the pump, P_{pump}, can be expressed as

$$P_{\text{pump}} = \frac{\gamma Q h_{\text{pump}}}{\eta_{\text{pump}}} \tag{4.155}$$

where η_{pump} is the efficiency of the pump. In cases where a liquid moves between two reservoirs in which the pressure on the surface of each reservoir is equal to atmospheric pressure and the velocity head in each reservoir is negligible, the pressure distribution in each reservoir can be assumed to be hydrostatic, and the energy equation given by Equation 4.149 can be expressed as

$$z_{s1} + h_{\text{pump}} = z_{s2} + h_{\text{turb}} + h_\ell \quad \text{(incompressible fluid)} \tag{4.156}$$

where z_{s1} and z_{s2} are the surface elevations of the inflow and outflow reservoirs, respectively.

EXAMPLE 4.24

Consider the system shown in Figure 4.37, where a pump is delivering water from a lower to an upper reservoir at a rate of 6.3 L/s. The efficiency of the pump in utilizing electricity is 80%, and the head loss in the delivery pipe, h_ℓ, varies with the volume flow rate, Q, according to the empirical relation

$$h_\ell = 10^5 Q^2$$

where h_ℓ is in meters and Q is in m^3/s. Determine the electricity demand of the pump in units of kilowatts.

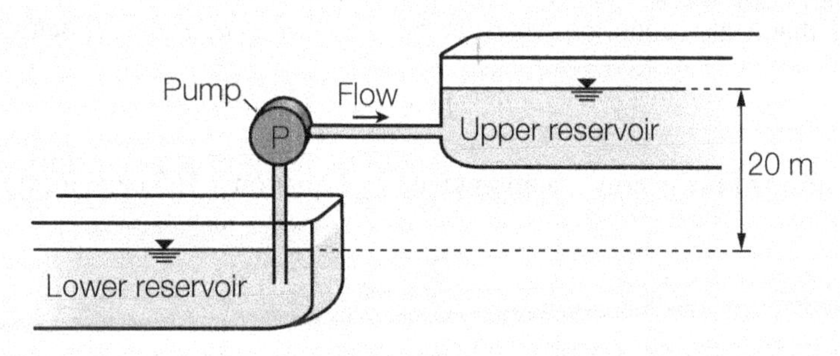

Figure 4.37: Pump system

SOLUTION

From the given data: Q = 6.3 L/s = 0.0063 m^3/s and η_{pump} = 80%. Assume water at 20°C, for which γ = 9.79 kN/m^3. The head loss in the delivery pipe, h_ℓ, can be calculated using the given empirical equations as

$$h_\ell = 10^5 (0.0063)^2 = 3.97 \text{ m}$$

Because the difference in water surface elevations in the reservoirs is 20 m, $z_{s2} - z_{s1} = 20$ m, and Equation 4.156 gives

$$z_{s1} + h_{pump} = z_{s2} + h_{turb} + h_\ell \quad \rightarrow$$
$$h_{pump} = (z_{s2} - z_{s1}) + h_{turb} + h_\ell = 20 \text{ m} + 0 + 3.97 \text{ m} = 23.97 \text{ m}$$

Therefore, the power, P_{pump}, that must be supplied to the pump to deliver the required flow (6.3 L/s) is given by Equation 4.155 as

$$P_{pump} = \frac{\gamma Q h_{pump}}{\eta_{pump}} = \frac{(9.79)(0.0063)(23.97)}{0.80} = \mathbf{1.85\ kW}$$

The theoretical power required to drive the pump is 1.85 kW. This requirement will likely need to be refined because a commercially available pump might not be able to deliver exactly 6.3 L/s when the head difference between the reservoirs is 20 m.

Forms of the energy equation useful for gases and vapors. For gases and vapors, heat exchange, temperature changes, and compressibility are usually processes of concern, and elevation changes usually have a negligible effect on the energy balance (due to the relatively small density of air). To facilitate analyses of gas and vapor flows, the energy equation, Equation 4.149, is commonly expressed in one of the following forms:

$$\frac{p_1}{\rho_1} + \frac{1}{2}V_1^2 + gz_1 = \frac{p_2}{\rho_2} + \frac{1}{2}V_2^2 + gz_2 + \left[\frac{\dot{W}_s}{\dot{m}}\right] + \left[(u_2 - u_1) - \frac{\dot{Q}}{\dot{m}}\right] \tag{4.157}$$

or

$$h_1 + \frac{1}{2}V_1^2 + gz_1 = h_2 + \frac{1}{2}V_2^2 + gz_2 + \left[\frac{\dot{W}_s}{\dot{m}}\right] - \left[\frac{\dot{Q}}{\dot{m}}\right] \tag{4.158}$$

where h_i is the enthalpy at section i as defined by Equation 4.132. The energy equation in the forms given by Equations 4.157 and 4.158 are valid for any gas or vapor and for any process. A term that is also commonly used is the *stagnation enthalpy*, H, defined as

$$H = h + \frac{1}{2}V^2 + gz \tag{4.159}$$

So the form of the energy equation given by Equation 4.158 can be expressed as

$$H_2 - H_1 = \frac{\dot{Q}}{\dot{m}} - \frac{\dot{W}_s}{\dot{m}} \tag{4.160}$$

where H_1 and H_2 are the stagnation enthalpies at sections 1 and 2, respectively. Therefore, the energy equation in the form given by Equation 4.160 states that the change in stagnation enthalpy is equal to the heat transfer into the control volume minus the shaft work done by the fluid in the control volume. Some knowledge of thermodynamics is necessary to evaluate enthalpies, and in the case of vapors, vapor tables or charts are commonly required because vapor properties cannot be expressed by the simple ideal gas law equation. By using the enthalpy instead of the internal energy to represent the energy of a flowing fluid, the (pressure) energy associated with pushing the fluid is automatically taken into account by the enthalpy, which is one of the main reasons for defining enthalpy as a property.

EXAMPLE 4.25

A turbine is used to extract energy from steam as illustrated in Figure 4.38, where the inflow conduit has a diameter of 300 mm and the outflow conduit has a diameter of 500 mm. On the inflow side, steam enters at a velocity of 35 m/s, with a density of 0.6 kg/m^3 and an enthalpy of 5000 kJ/kg. At the outflow side, steam exits with a density of 0.1 kg/m^3 and an enthalpy of 3000 kJ/kg. The centerlines of the inflow and outflow conduits are at approximately the same elevation, and the system is sufficiently well insulated that adiabatic conditions can be assumed. Estimate the power produced by the turbine.

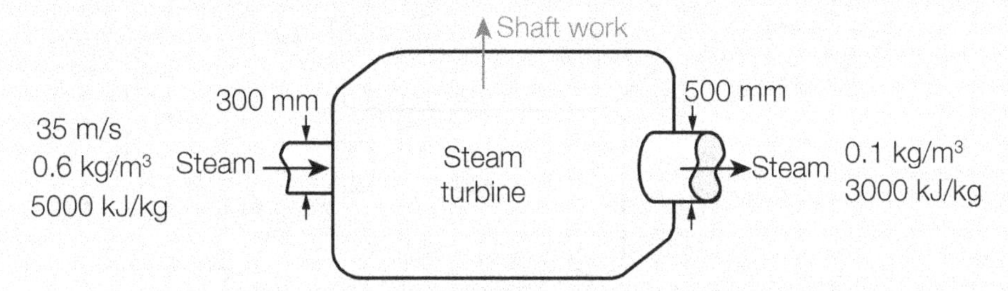

Figure 4.38: Power generation by a steam turbine

SOLUTION

From the given data: $D_1 = 300$ mm, $D_2 = 500$ mm, $V_1 = 35$ m/s, $\rho_1 = 0.6$ kg/m^3, $h_1 = 5000$ kJ/kg, $\rho_2 = 0.1$ kg/m^3, $h_2 = 3000$ kJ/kg, $z_1 = z_2$, and $\dot{Q} = 0$. The inflow area, A_1, and the mass flow rate through the system, $\dot{m}$, can be calculated from the given data as follows:

$$A_1 = \frac{\pi}{4}D_1^2 = \frac{\pi}{4}(0.3)^2 = 0.07069\,\text{m}^2, \qquad \dot{m} = \rho_1 V_1 A_1 = (0.6)(35)(0.07069) = 1.484\,\text{kg/s}$$

The outflow velocity can be estimated by applying the steady-state conservation of mass equation, which requires that

$$\dot{m} = \rho_1 V_1 A_1 = \rho_2 V_2 A_2 \quad \rightarrow \quad V_2 = \left(\frac{\rho_1}{\rho_2}\right)\left(\frac{D_1}{D_2}\right)^2 V_1 = \left(\frac{0.6}{0.1}\right)\left(\frac{300}{500}\right)^2 (35) = 75.6\ \text{m/s}$$

The inflow and outflow stagnation enthalpies, H_1 and H_2, respectively, are given by Equation 4.159 as follows:

$$H_1 = h_1 + \frac{1}{2}V_1^2 + gz_1 = 5000 + \frac{1}{2}(35)^2\ [\times 10^{-3}\ \text{kJ/J}] + 0 = 5000\ \text{kJ/kg}$$

$$H_2 = h_2 + \frac{1}{2}V_2^2 + gz_2 = 3000 + \frac{1}{2}(75.6)^2\ [\times 10^{-3}\ \text{kJ/J}] + 0 = 3000\ \text{kJ/kg}$$

It is apparent from these results that the kinetic energy contributes a negligible amount to the stagnation enthalpy. Applying the energy equation in the form of Equation!4.160 gives

$$H_2 - H_1 = \frac{\dot{Q}}{\dot{m}} - \frac{\dot{W}_s}{\dot{m}} \quad \rightarrow \quad 3000 - 5000 = 0 - \frac{\dot{W}_s}{1.484} \quad \rightarrow \quad \dot{W}_s = 2.97 \times 10^3\ \text{kW} = 2.97\ \text{MW}$$

The theoretical power output generated by the steam turbine is **2.97 MW**. The usable power output will be less than 2.97 MW due to inefficiencies in power transmission and generator operation.

Forms of the energy equation useful for ideal gases. In many applications, gases are sufficiently far removed from the liquid state that they can be treated as ideal gases. Differential and finite changes in the internal energy and enthalpy of an ideal gas can be expressed in terms of specific heats as

$$du = c_v\,dT, \qquad dh = c_p\,dT \tag{4.161}$$

where c_v and c_p are the constant volume and constant pressure specific heats of an ideal gas. Assuming that specific heats remain constant within the temperature range encountered by a gas, the energy equation in the forms of Equations 4.157 and 4.158 can be conveniently expressed in terms of temperature changes as

$$\frac{p_1}{\rho_1} + \frac{1}{2}V_1^2 + gz_1 = \frac{p_2}{\rho_2} + \frac{1}{2}V_2^2 + gz_2 + c_v(T_2 - T_1) + \left[\frac{\dot{W}_s}{\dot{m}}\right] - \left[\frac{\dot{Q}}{\dot{m}}\right] \tag{4.162}$$

or

$$\frac{1}{2}V_1^2 + gz_1 = \frac{1}{2}V_2^2 + gz_2 + c_p(T_2 - T_1) + \left[\frac{\dot{W}_s}{\dot{m}}\right] - \left[\frac{\dot{Q}}{\dot{m}}\right] \tag{4.163}$$

EXAMPLE 4.26

Air flows into a compressor through a 200-mm-diameter tube and out of the compressor through a 100-mm-diameter tube. The inflow air has a velocity of 60 m/s, a temperature of 25°C, and a pressure of 101 kPa. The outflow air has a temperature of 80°C and a pressure of 250 kPa. Cooling water within the compressor assembly removes heat at a rate of 20 kJ per kg of air that passes through the compressor. Estimate the power consumption of the compressor.

SOLUTION

From the given data: D_1 = 200 mm, D_2 = 100 mm, V_1 = 60 m/s, T_1 = 25°C = 298.15 K, p_1 = 101 kPa, T_2 = 80°C = 353.15 K, p_2 = 250 kPa, and $\dot{Q}$ = −20 kJ/kg. For standard air, R = 287.1 J/kg·K and c_p = 1003 J/kg·K. The following preliminary calculations are useful:

$$A_1 = \frac{\pi D_1^2}{4} = \frac{\pi(0.2)^2}{4} = 3.142 \times 10^{-2}\ \text{m}^2, \qquad A_2 = \frac{\pi D_2^2}{4} = \frac{\pi(0.1)^2}{4} = 7.854 \times 10^{-3}\ \text{m}^2$$

$$\rho_1 = \frac{p_1}{RT_1} = \frac{101 \times 10^3}{(287.1)(298.15)} = 1.180\ \text{kg/m}^3, \quad \rho_2 = \frac{p_2}{RT_2} = \frac{250 \times 10^3}{(287.1)(353.15)} = 2.466\ \text{kg/m}^3$$

$$\dot{m} = \rho_1 V_1 A_1 = (1.180)(60)(3.142 \times 10^{-2}) = 2.224\ \text{kg/s},$$

$$V_2 = \frac{\dot{m}}{\rho_2 A_2} = \frac{2.224}{(2.466)(7.854 \times 10^{-3})} = 114.8\ \text{m/s}$$

The energy equation in the form of Equation 4.163 can be applied directly. Gravitational effects associated with the difference in elevation (if any) between the inflow and outflow tubes can be assumed negligible in comparison with the magnitude of the other terms in the energy equation. Applying Equation 4.163 gives

$$\frac{1}{2}V_1^2 + \cancel{gz_1} = \frac{1}{2}V_2^2 + \cancel{gz_2} + c_p(T_2 - T_1) + \left[\frac{\dot{W}_s}{\dot{m}}\right] - \left[\frac{\dot{Q}}{\dot{m}}\right]$$

$$\frac{1}{2}(60)^2 = \frac{1}{2}(114.8)^2 + 1003(353.15 - 298.15) + \left[\frac{\dot{W}_s}{2.224}\right] - \left[\frac{-20 \times 10^3}{2.224}\right] \rightarrow$$

$$\dot{W}_s = -1.354 \times 10^5 \text{ W}$$

Therefore, the power consumption of the compressor is approximately **135 kW**.

Practical application. The energy equation is commonly applied in the analysis of airflow through ducts that are used to distribute air in heating, ventilating, and air conditioning (HVAC) systems. Circulating air in these ducts is usually driven by fans that do shaft work on the air in the duct.

EXAMPLE 4.27

A 1-kW intake fan is used to pull in air from a room and push it through a 1 m × 0.5 m duct as shown in Figure 4.39(a). The air pressure on the downstream side of the fan is approximately equal to the room air pressure, with the energy input of the fan going toward increasing the velocity of the air. (a) Estimate the air velocity in the duct if the fan is 100% efficient in transferring energy to the air. (b) What would be the air velocity in the duct if the energy transfer was 70% efficient?

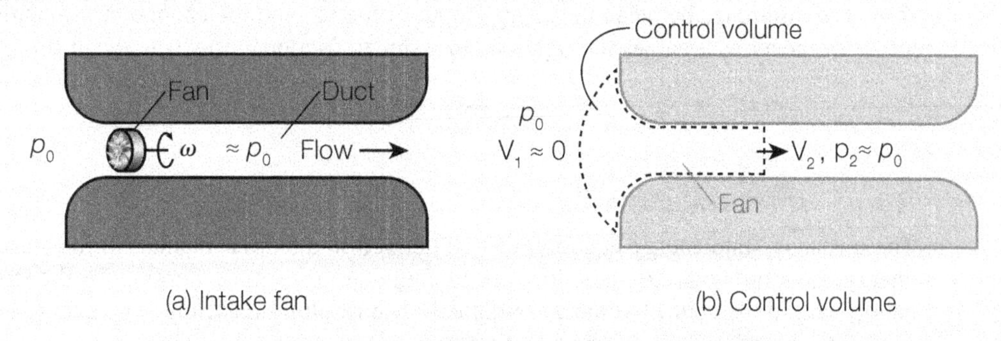

Figure 4.39: Intake fan for a ventilating system

SOLUTION

From the given data: $\dot{W}_s$ = −1 kW = −10^3 W and A = (1)(0.5) = 0.5 m^2. For standard air, ρ = 1.225 kg/m^3.

(a) If the fan is 100% efficient in transferring energy to the air, then the energy loss is equal to zero and the energy equation (Equation 4.157) gives

$$\frac{p_1}{\rho_1} + \frac{1}{2}V_1^2 + gz_1 = \frac{p_2}{\rho_2} + \frac{1}{2}V_2^2 + gz_2 + \left[\frac{\dot{W}_s}{\dot{m}}\right] \tag{4.164}$$

where sections 1 and 2 are the upstream and downstream sections, respectively, of the control volume as shown in Figure 4.39(b). From the given conditions, $p_1 \approx p_2$, $V_1 \approx 0$, and elevation changes can be neglected such that $z_1 \approx z_2$. It can be further assumed that the flow is incompressible, such that $\rho_1 \approx \rho_2 \approx \rho$. Substituting the given data and approximations into Equation 4.164 gives

$$\cancel{\frac{p_1}{\rho_1}}+0+0 = \cancel{\frac{p_2}{\rho_2}}+\frac{1}{2}V_2^2+0+\left[\frac{-10^3}{\rho V_2 A}\right] \;\rightarrow\; 0 = \frac{1}{2}V_2^2+\left[\frac{-10^3}{(1.225)V_2(0.5)}\right] \;\rightarrow\; V_2 = \mathbf{14.8\,m/s}$$

(b) If the fan is not 100% efficient in transferring energy to the air, then the energy equation (Equation 4.157) can be expressed as

$$\frac{p_1}{\rho_1} + \frac{1}{2}V_1^2 + gz_1 = \frac{p_2}{\rho_2} + \frac{1}{2}V_2^2 + gz_2 - E_s + E_\ell \tag{4.165}$$

where E_s is the shaft work done by the fan on the air per unit mass of air and E_ℓ is the energy loss per unit mass of air. Representing the efficiency as η and taking $\eta = 0.7$ gives

$$\eta = \frac{E_s - E_\ell}{E_s} \quad\rightarrow\quad 0.7 = \frac{E_s - E_\ell}{E_s} \quad\rightarrow\quad E_\ell = 0.3E_s \tag{4.166}$$

Substituting Equation 4.166 into Equation 4.165 and implementing the aforementioned flow approximations gives

$$0 = \frac{1}{2}V_2^2 - (1-0.3)E_s \;\rightarrow\; 0 = \frac{1}{2}V_2^2 - (1-0.3)\left[\frac{10^3}{(1.225)V_2(0.5)}\right] \;\rightarrow\; V_2 = \mathbf{13.2\ m/s}$$

Hence, when the efficiency of the fan is reduced from 100% to 70%, the air velocity induced by the fan is reduced from 14.8 m/s to 13.2 m/s, a reduction of approximately 11%. The difference between the energy input of the fan and the change in mechanical energy of the air as it passes through the fan is due to frictional effects in the airflow.

4.6.3 Unsteady-State Energy Equation

The unsteady-state energy equation can be derived in a similar manner to the steady-state energy equation, with the main difference being that the term accounting for the storage of energy within the control volume is retained. The resulting unsteady-state energy equation can be expressed in the following useful form:

$$\dot{Q} - \dot{W}_s = \frac{d}{dt}\int_{cv} e\rho\, d\mathcal{V} + \dot{m}_2\left(h_2 + gz_2 + \alpha_2\frac{V_2^2}{2}\right) - \dot{m}_1\left(h_1 + gz_1 + \alpha_1\frac{V_1^2}{2}\right) \tag{4.167}$$

where e is the energy unit mass as defined by Equation 4.127, "cv" represents the control volume, and $d\mathcal{V}$ is a volume element within the control volume.

EXAMPLE 4.28

The 0.2-m^3 insulated tank shown in Figure 4.40 is to be filled with air from a high-pressure supply line. The tank initially contains air at a temperature of 20°C and an absolute pressure of 120 kPa. The supply line maintains air at a temperature of 15°C and an absolute pressure of 1.8 MPa. At the instant the valve connecting the supply line to the tank is opened, a flow meter measures an airflow rate of 5.08×10^{-4} m^3/s. Estimate the rate of change of temperature in the tank. The velocities in the supply line and the tank are sufficiently small that they may be neglected in the analysis.

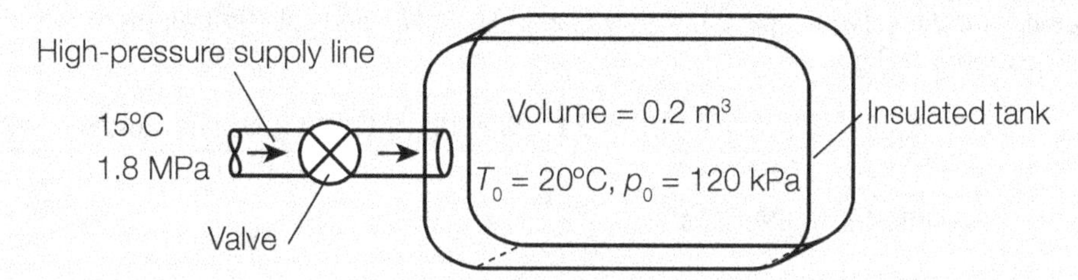

Figure 4.40: Tank connected to a supply line

SOLUTION

From the given data: $V = 0.2$ m^3, $T_0 = 20°\text{C} = 293$ K, $p_0 = 120$ kPa, $T_1 = 15°\text{C} = 288$ K, $p_1 = 1.8$ MPa, and $Q_0 = 5.08 \times 10^{-4}$ m^3/s. For standard air: $R = 287.1$ J/kg·K and $c_v = 716$ J/kg·K. Applying the energy equation, Equation 4.167, with the tank as the control volume and recalling that $e = V^2/2 + gz + u$ gives

$$\underbrace{\dot{Q}}_{=0} - \underbrace{\dot{W}_\text{s}}_{=0} = \frac{\text{d}}{\text{d}t}\int_\text{cv}\Big(\underbrace{\frac{1}{2}V^2}_{\approx 0} + \underbrace{gz}_{=0} + u\Big)\rho\,\text{d}V$$
$$+ \underbrace{\dot{m}_2}_{=0}\Big(h_2 + gz_2 + \alpha_2\frac{1}{2}V_2^2\Big) - \dot{m}_1\Big(h_1 + \underbrace{gz_1}_{=0} + \underbrace{\alpha_1\frac{1}{2}V_1^2}_{\approx 0}\Big)$$

where the conditions of an insulated tank ($\dot{Q} = 0$), no shaft work on the air in the tank ($\dot{W}_\text{s} = 0$), negligible velocities in the tank and the supply line ($\frac{1}{2}V^2$ and $\frac{1}{2}V_1^2 \approx 0$), zero datum and negligible gravitational effect (gz and $gz_1 = 0$), and no outflow ($\dot{m}_2 = 0$) are indicated. Taking $h_1 = p_1/\rho_1 + u_1$, the energy equation simplifies to

$$0 = \frac{\text{d}}{\text{d}t}\int_\text{cv} u\rho\,\text{d}V - \dot{m}\Big(\frac{p_1}{\rho_1} + u_1\Big) \tag{4.168}$$

where $\dot{m}$ is the mass flow rate into the tank, previously represented by $\dot{m}_1$. Representing the mass of air in the tank as M and noting that $p_1/\rho_1 = RT_1$ and $\text{d}u = c_v\,\text{d}T$, Equation 4.168 gives

$$0 = \frac{\text{d}}{\text{d}t}(uM) - \dot{m}(RT_1 + u_1) \quad \rightarrow \quad 0 = u\frac{\text{d}M}{\text{d}t} + M\frac{\text{d}u}{\text{d}t} - \dot{m}(RT_1 + u_1)$$
$$\rightarrow \quad 0 = u\dot{m} + Mc_v\frac{\text{d}T}{\text{d}t} - \dot{m}(RT_1 + u_1) \quad \rightarrow \quad \frac{\text{d}T}{\text{d}t} = \frac{\dot{m}[RT_1 + u_1 - u]}{Mc_v}$$
$$\rightarrow \quad \frac{\text{d}T}{\text{d}t} = \frac{\dot{m}[RT_1 + c_v(T_1 - T)]}{Mc_v} \tag{4.169}$$

Equation 4.169 can be used as a basis for determining the rate of change of the temperature (dT/dt) in the tank. To calculate the initial rate of change of temperature when the valve is first opened, the following calculated variables are used:

$$\rho_0 = \frac{p_0}{RT_0} = \frac{120 \times 10^3}{(287.1)(293)} = 1.427 \text{ kg/m}^3, \qquad \rho_1 = \frac{p_1}{RT_1} = \frac{1.8 \times 10^6}{(287.1)(288)} = 21.77 \text{ kg/m}^3$$

$$M_0 = \rho_0 V = (1.427)(0.2) = 0.2854 \text{ kg}, \qquad \dot{m}_0 = \rho_1 Q_0 = (21.77)(5.08 \times 10^{-4})$$

$$= 1.106 \times 10^{-2} \text{ kg/s}$$

At the instant the valve is opened, $\dot{m} = \dot{m}_0$ and $M = M_0$, and substituting these initial values into Equation 4.169 gives

$$\left.\frac{dT}{dt}\right|_{t=0} = \frac{\dot{m}_0[RT_1 + c_v(T_1 - T_0)]}{M_0 c_v} = \frac{(1.106 \times 10^{-2})[(287.1)(288) + 716(288 - 293)]}{(0.2854)(716)}$$

$$= 4.28\text{K/s} = 4.28\ ^\circ\text{C/s}$$

Therefore, it is expected that the temperature of the air in the tank will initially increase at the rate of **4.28°C/s**.

Key Equations in Finite Control Volume Analysis

The following list of equations is particularly useful in solving problems related to the analysis of fluid flows using finite control volumes. If one is able to recognize these equations and recall their appropriate use, then the learning objectives of this chapter have been met to a significant degree. Derivations of these equations, definitions of the variables, and detailed examples of usage can be found in the main text.

REYNOLDS TRANSPORT THEOREM

Fixed control volume:

$$\frac{\mathrm{d}}{\mathrm{d}t}\int_{\text{sys}} \rho\beta\,\mathrm{d}V = \frac{\mathrm{d}}{\mathrm{d}t}\int_{\text{cv}} \beta\rho\,\mathrm{d}V + \int_{\text{cs}} \beta\rho(\mathbf{v}\cdot\mathbf{n})\,\mathrm{d}A$$

Moving (inertial) control volume:

$$\frac{\mathrm{d}}{\mathrm{d}t}\int_{\text{sys}} \rho\beta\,\mathrm{d}V = \frac{\mathrm{d}}{\mathrm{d}t}\int_{\text{cv}} \beta\rho\,\mathrm{d}V + \int_{\text{cs}} \beta\rho(\mathbf{v}_\text{r}\cdot\mathbf{n})\,\mathrm{d}A$$

CONSERVATION OF MASS

Fixed control volume:

$$\frac{\mathrm{d}}{\mathrm{d}t}\int_{\text{cv}} \rho\,\mathrm{d}V + \int_{\text{cs}} \rho(\mathbf{v}\cdot\mathbf{n})\,\mathrm{d}A = 0$$

Moving control volume:

$$\frac{\mathrm{d}}{\mathrm{d}t}\int_{\text{cv}} \rho\,\mathrm{d}V + \int_{\text{cs}} \rho(\mathbf{v}_\text{r}\cdot\mathbf{n})\,\mathrm{d}A = 0$$

Steady state (fixed control volume):

$$\int_{\text{cs}} \rho(\mathbf{v}\cdot\mathbf{n})\,\mathrm{d}A = 0$$

Steady state (moving control volume):

$$\int_{\text{cs}} \rho(\mathbf{v}_\text{r}\cdot\mathbf{n})\,\mathrm{d}A = 0$$

Reservoir with multiple inlets/outlets:

$$\frac{\mathrm{d}}{\mathrm{d}t}\int_{\text{cv}} \rho\,\mathrm{d}V = \sum_{j=1}^{N_\text{in}} \dot{m}_j - \sum_{i=1}^{N_\text{out}} \dot{m}_i$$

Reservoir with single inlet and outlet:

$$\frac{\mathrm{d}}{\mathrm{d}t}(\rho\mathcal{V}) = \dot{m}_1 - \dot{m}_2 = \rho_1 A_1 \overline{V}_1 - \rho_2 A_2 \overline{V}_2$$

Flow in a conduit (steady):

$$\dot{m}_1 = \dot{m}_2; \quad \rho_1 A_1 \overline{V}_1 - \rho_2 A_2 \overline{V}_2$$

Incompressible flow in a conduit:

$$Q_1 = Q_2 = Q; \quad A_1 \overline{V}_1 = A_2 \overline{V}_2$$

CONSERVATION OF LINEAR MOMENTUM

Fixed control volume:

$$\sum \mathbf{F}_\text{cv} = \frac{\mathrm{d}}{\mathrm{d}t}\int_{\text{cv}} \rho\mathbf{v}\,\mathrm{d}V + \int_{\text{cs}} \rho\mathbf{v}\,(\mathbf{v}\cdot\mathbf{n})\,\mathrm{d}A$$

Moving (inertial) control volume:

$$\sum \mathbf{F}_\text{cv} = \frac{\mathrm{d}}{\mathrm{d}t}\int_{\text{cv}} \rho\mathbf{v}_r\,\mathrm{d}V + \int_{\text{cs}} \rho\mathbf{v}_\text{r}\,(\mathbf{v}_\text{r}\cdot\mathbf{n})\,\mathrm{d}A$$

Steady state (fixed control volume): $$\sum \mathbf{F}_{\text{cv}} = \int_{\text{cs}} \rho \mathbf{v} \, (\mathbf{v} \cdot \mathbf{n}) \, \mathrm{d}A$$

Steady state (moving control volume): $$\sum \mathbf{F}_{\text{cv}} = \int_{\text{cs}} \rho \mathbf{v}_{\text{r}} \, (\mathbf{v}_{\text{r}} \cdot \mathbf{n}) \, \mathrm{d}A$$

Momentum-flux correction factor: $$\beta = \frac{1}{A} \int_A \left(\frac{v}{\overline{V}} \right)^2 \mathrm{d}A$$

In terms of average velocity (unsteady): $$\sum \mathbf{F} = \frac{\mathrm{d}}{\mathrm{d}t} \int_{\text{cv}} \rho \mathbf{v} \, \mathrm{d}V + \sum_{i=1}^{N_{\text{out}}} \beta_i \dot{m}_i \overline{\mathbf{V}}_i - \sum_{j=1}^{N_{\text{in}}} \beta_j \dot{m}_j \overline{\mathbf{V}}_j$$

In terms of average velocity (steady): $$\sum \mathbf{F} = \sum_{i=1}^{N_{\text{out}}} \beta_i \dot{m}_i \overline{\mathbf{V}}_i - \sum_{j=1}^{N_{\text{in}}} \beta_j \dot{m}_j \overline{\mathbf{V}}_j$$

Reducers: $$R_x = p_1 A_1 - p_2 A_2 - W \sin\theta - \dot{m}(V_2 - V_1)$$ $$R_n = W \cos\theta$$

Bends: $$R_x = p_1 A_1 - p_2 A_2 \cos\theta - \dot{m}(V_2 \cos\theta - V_1)$$ $$R_z = p_2 A_2 \sin\theta + W + \dot{m} V_2 \sin\theta$$

Power extracted by propeller: $$P = \frac{1}{4} \rho A (V_1^2 - V_4^2)(V_1 + V_4)$$

Maximum wind power: $$P_{\max} = \frac{8}{27} \rho A V_1^3$$

Efficiency of wind turbine: $$\eta = \frac{P}{P_{\text{avail}}} = \frac{(V_1^2 - V_4^2)(V_1 + V_4)}{2V_1^3}$$

Force of liquid on reservoir: $$\mathbf{F} = \dot{m}_1 \mathbf{V}_1 - \dot{m}_2 \mathbf{V}_2, \qquad \mathbf{F} = \rho Q_1 \mathbf{V}_1 - \rho Q_2 \mathbf{V}_2$$

Thrust of a jet engine: $$F = \dot{m}_{2\text{r}} v_{2\text{r}} - \dot{m}_{1\text{r}} v_{1\text{r}} + (p_{\text{e}} - p_{\text{a}}) A_2$$ $$F = \rho_2 A_2 v_{2\text{r}}^2 - \rho_1 A_2 v_{1\text{r}}^2 + (p_{\text{e}} - p_{\text{a}}) A_2$$

Thrust of a rocket engine: $$F = \dot{m}_{2\text{r}} v_{2\text{r}} + (p_{\text{e}} - p_{\text{a}}) A_2$$ $$F = \rho A_2 v_{2\text{r}}^2 + (p_{\text{e}} - p_{\text{a}}) A_2$$

ANGULAR MOMENTUM PRINCIPLE

Fixed control volume: $$\sum \mathbf{M}_{\text{cv}} = \frac{\mathrm{d}}{\mathrm{d}t} \int_{\text{cv}} \rho (\mathbf{r} \times \mathbf{v}) \, \mathrm{d}V + \int_{\text{cs}} \rho (\mathbf{r} \times \mathbf{v}) \, (\mathbf{v} \cdot \mathbf{n}) \, \mathrm{d}A$$

In terms of average velocity (steady): $\sum \mathbf{M}_{\text{cv}} = \sum_{i=1}^{N_{\text{out}}} (\bar{\mathbf{r}}_i \times \dot{m}_i \overline{\mathbf{V}}_i) - \sum_{j=1}^{N_{\text{in}}} (\bar{\mathbf{r}}_j \times \dot{m}_j \overline{\mathbf{V}}_j)$

In terms of average velocity (planar, fixed cv): $\sum M = \sum_{i=1}^{N_{\text{out}}} (\bar{r}_i \dot{m}_i \overline{V}_{\theta i}) - \sum_{j=1}^{N_{\text{in}}} (\bar{r}_j \dot{m}_j \overline{V}_{\theta j})$

Shaft torque: $T_{\text{shaft}} = \dot{m}(R_{\text{out}} V_{\theta,\text{out}} - R_{\text{in}} V_{\theta,\text{in}})$

Shaft power:

$$\dot{W}_{\text{shaft}} = T_{\text{shaft}}\omega$$
$$\dot{W}_{\text{shaft}} = \dot{m}\omega(R_{\text{out}} V_{\theta,\text{out}} - R_{\text{in}} V_{\theta,\text{in}})$$
$$\dot{W}_{\text{shaft}} = \dot{m}(U_{\text{out}} V_{\theta,\text{out}} - U_{\text{in}} V_{\theta,\text{in}})$$

In terms of average velocity (planar, moving cv): $\sum M = \sum_{i=1}^{N_{\text{out}}} (\bar{r}_i \dot{m}_{\text{r},i} \overline{V}_{\text{r},\theta i}) - \sum_{j=1}^{N_{\text{in}}} (\bar{r}_j \dot{m}_{\text{r},j} \overline{V}_{\text{r},\theta j})$

Shaft power: $\dot{W}_{\text{shaft}} = T_{\text{shaft}}\omega$

CONSERVATION OF ENERGY

Steady state, general:

$$\left(\frac{p_1}{\gamma_1} + \alpha_1 \frac{V_1^2}{2g} + z_1\right) = \left(\frac{p_2}{\gamma_2} + \alpha_2 \frac{V_2^2}{2g} + z_2\right) + \left[\frac{1}{g}(u_2 - u_1) - \frac{\dot{Q}}{\dot{m}g}\right] + \left[\frac{\dot{W}_{\text{s}}}{\dot{m}g}\right]$$

Steady state, useful for gases:

$$\frac{p_1}{\rho_1} + \frac{1}{2}V_1^2 + gz_1 = \frac{p_2}{\rho_2} + \frac{1}{2}V_2^2 + gz_2 + \left[\frac{\dot{W}_{\text{s}}}{\dot{m}}\right] + \left[(u_2 - u_1) - \frac{\dot{Q}}{\dot{m}}\right]$$

Steady state, useful for gases:

$$h_1 + \frac{1}{2}V_1^2 + gz_1 = h_2 + \frac{1}{2}V_2^2 + gz_2 + \left[\frac{\dot{W}_{\text{s}}}{\dot{m}}\right] - \left[\frac{\dot{Q}}{\dot{m}}\right]$$

Steady state, useful for gases:

$$H_2 - H_1 = \frac{\dot{Q}}{\dot{m}} - \frac{\dot{W}_{\text{s}}}{\dot{m}}$$

Head loss:

$$h_\ell = \frac{1}{g}(u_2 - u_1) - \frac{\dot{Q}}{\dot{m}g}$$

Shaft work (head):

$$h_{\text{s}} = \frac{\dot{W}_{\text{s}}}{\dot{m}g}$$

Incompressible fluid:

$$\left(\frac{p_1}{\gamma} + \alpha_1 \frac{V_1^2}{2g} + z_1\right) + h_{\text{pump}} =$$

$$\left(\frac{p_2}{\gamma} + \alpha_2 \frac{V_2^2}{2g} + z_2\right) + h_{\text{turb}} + h_\ell$$

Power delivered by turbine: $$P_{\text{turb}} = \eta_{\text{turb}} \cdot \gamma Q h_{\text{turb}}$$

Power supplied to pump: $$P_{\text{pump}} = \frac{\gamma Q h_{\text{pump}}}{\eta_{\text{pump}}}$$

Flow from upper to lower reservoir: $$z_{\text{s1}} + h_{\text{pump}} = z_{\text{s2}} + h_{\text{turb}} + h_\ell$$

Applications of the above equations are provided in the text, and additional problems to practice using these equations can be found in the following section.

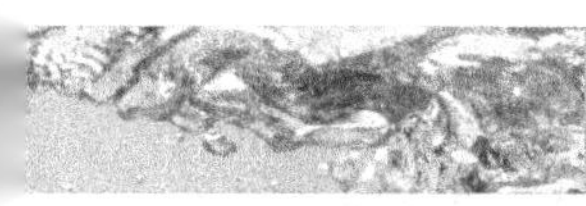

PROBLEMS

Section 4.3: Conservation of Mass

4.1. The volume flow rate on the intake side of a pump is equal to Q as illustrated in Figure 4.41. If the power delivered by the pump to the fluid is 150 kW, how does the volume flow rate on the discharge side of the pump compare with Q?

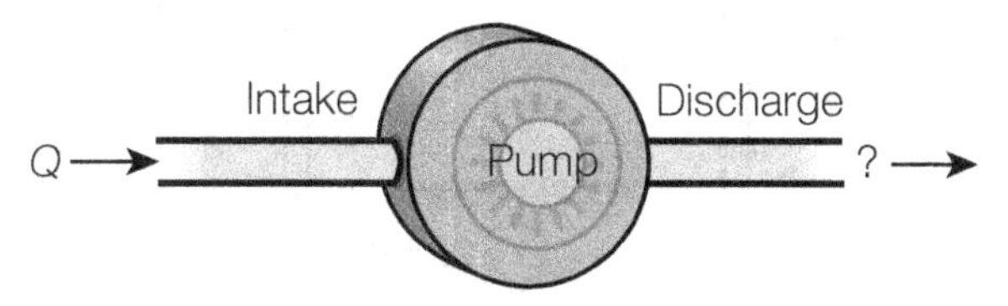

Figure 4.41: Flow through a pump

4.2. In the United States, flow rates through showerheads are regulated to be no greater than 9.5 L/min under any water pressure condition likely to be encountered in a home. Water pressures in homes are typically less than 550 kPa. A practical showerhead delivers water at a velocity of at least 5 m/s. If nozzles in a showerhead can be manufactured with diameters of 0.75 mm, what is the maximum number of nozzles required to make a practical showerhead?

4.3. An air conditioner delivers air into a 250-m^3 room at a rate of 4 m^3/min. Air from the room is removed by a 0.55 m $\times$ 0.7 m duct. Assuming steady-state conditions and incompressible flow, what is the average velocity of air in the exhaust duct?

4.4. At a particular section in a 100-mm-diameter pipe, the velocity is measured as 1200 m/s. Downstream of this section, the flow is expanded to a 300-mm pipe in which the velocity and density of the airflow is measured as 700 m/s and 1.1 kg/m^3, respectively. If flow conditions are steady, estimate the density of the air at the upstream (100-mm-diameter) section.

4.5. Air enters a heat exchanger at a rate of 200 kg/h and exits at a rate of 195 kg/h as illustrated in Figure 4.42. The purpose of the heat exchanger is to dehumidify the air. Estimate the rate at which liquid water drains from the system.

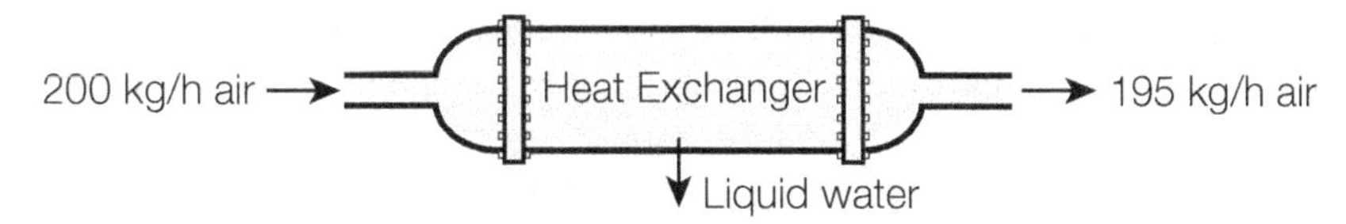

Figure 4.42: Flow through a heat exchanger

4.6. A piston moving at 15 mm/s displaces air from a 110-mm-diameter cylinder into a 20-mm discharge line as shown in Figure 4.43. Assuming that compressibility effects can be neglected, what is the volume flow rate and velocity in the discharge line?

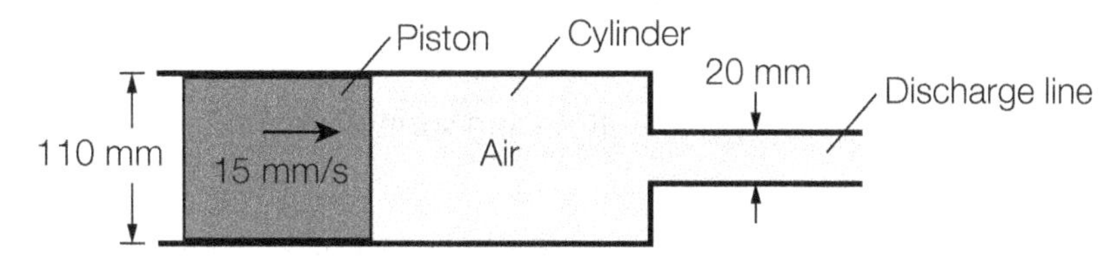

Figure 4.43: Airflow driven by a piston

4.7. Air enters a 1.2-m^3 storage tank at a rate of 0.5 m^3/s and at a temperature of 20°C and pressure of 101 kPa as shown in Figure 4.44. When air is released from the tank at a rate of 0.3 m^3/s, the pressure of the released air is measured as 150 kPa, and a weight scale indicates that the mass of the air in the tank is increasing at a rate of 0.1 kg/s. Estimate the temperature and the density of the air being released from the tank.

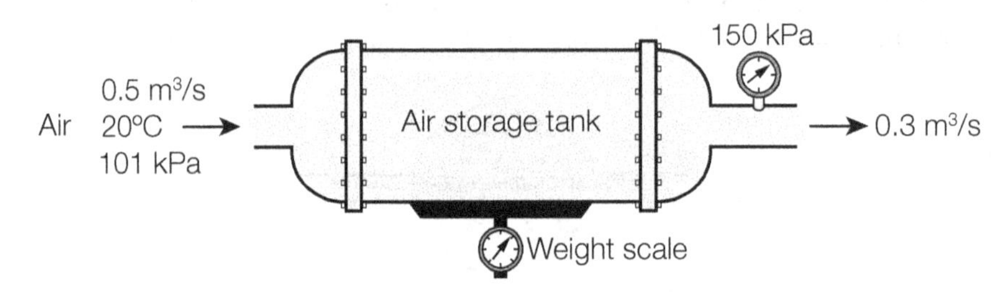

Figure 4.44: Air storage in a tank

4.8. Pipelines containing gasoline, water, and methanol merge to create a mixture as shown in Figure 4.45. All fluids are at 20°C, and the known diameters and fluid velocities in the merging pipes are also shown in Figure 4.45. (a) What is the volume flow rate of the mixture? (b) What is the density of the mixture?

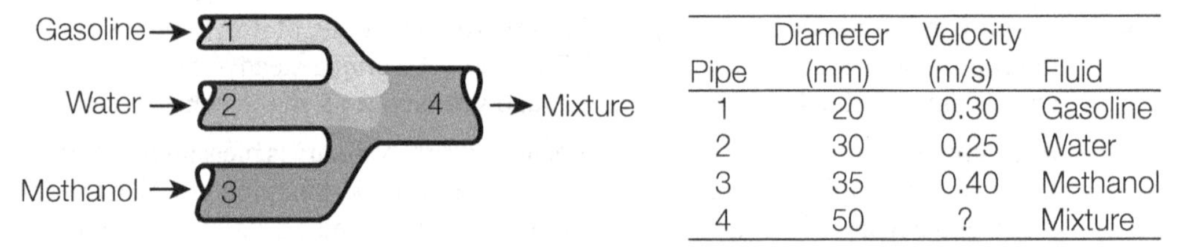

Pipe	Diameter (mm)	Velocity (m/s)	Fluid
1	20	0.30	Gasoline
2	30	0.25	Water
3	35	0.40	Methanol
4	50	?	Mixture

Figure 4.45: Mixing in pipes

4.9. Water flows from a garden hose into a sprinkler at a volume flow rate, Q, of 30 L/min. The flow exits the sprinkler via four rotating arms as shown Figure 4.46. The radius, R, of the sprinkler is 0.25 m, and the diameter of each discharge port is 12 mm. At what rate, ω, must the sprinkler rotate such that the absolute velocity of the discharged water is equal to zero. Describe how this would look to a stationary observer.

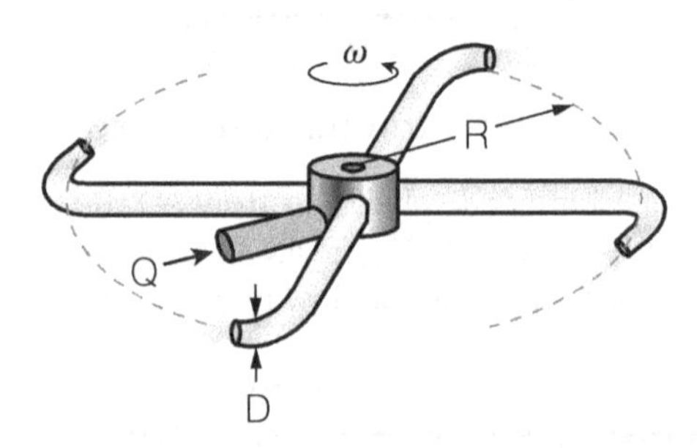

Figure 4.46: Rotating sprinkler

4.10. The flow in a pipe is divided as shown in Figure 4.47. The diameter of the pipe at sections 1, 2, and 3 are 200 mm, 150 mm, and 100 mm, respectively, and the volume flow rate at section 1 is 20 L/s. Calculate the volume flow rate and velocity at section 2.

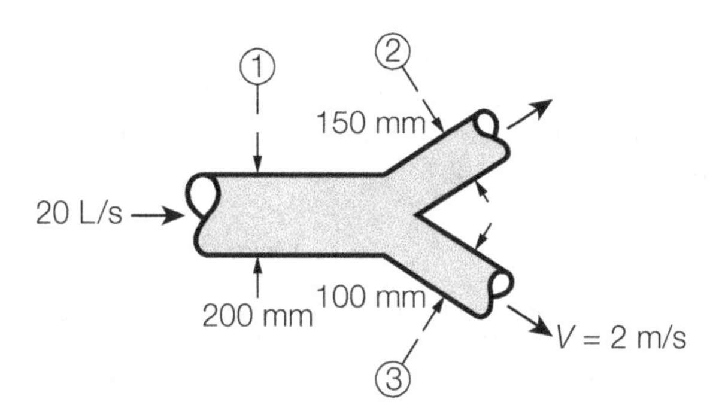

Figure 4.47: Divided flow

4.11. Water at 20°C is flowing in a 100-mm-diameter pipe at an average velocity of 2 m/s. If the diameter of the pipe is suddenly expanded to 150 mm, what is the velocity in the expanded pipe? What are the volume and mass flow rates in the pipe?

4.12. A 400-mm-diameter pipe divides into two smaller pipes, each of diameter 200 mm. If the flow divides equally between the two smaller pipes and the velocity in the 400-mm pipe is 2 m/s, calculate the velocity and flow rate in each of the smaller pipes.

4.13. Water flows steadily through the pipe shown in Figure 4.48. The pipe has a circular cross section, the entrance velocity is constant, $u = U_0$, and the exit velocity has a distribution given by

$$u(r) = u_{\rm max}\left(1 - \frac{r}{R}\right)^{\frac{1}{7}} \tag{4.170}$$

(a) Use the velocity distribution given by Equation 4.170 to determine the volume flow rate in the pipe in terms of $u_{\rm max}$ and R. (b) Determine the ratio $U_0/u_{\rm max}$.

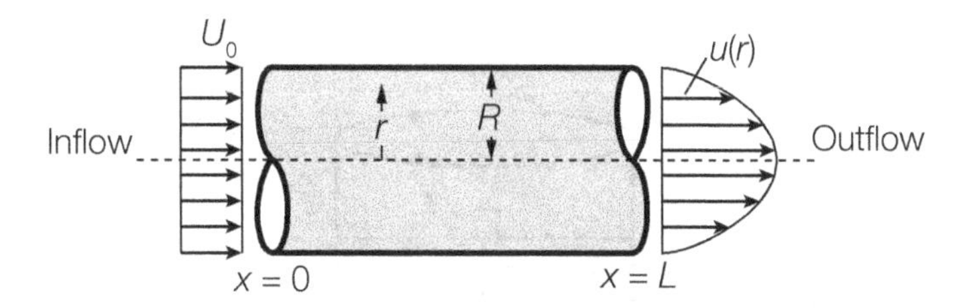

Figure 4.48: Flow through a pipe

4.14. Consider the 0.5-m^3 storage tank shown in Figure 4.49. Initially, the density of the liquid in the tank is 980 kg/m^3. At a particular instant, the inlet valve is opened and a fluid of density 800 kg/m^3 flows into the tank at a rate of 90 L/min. At this same instant, the outlet valve is opened and adjusted such that the liquid level in the tank is maintained at a constant elevation. A mixer is also turned on to ensure that the liquid in the tank is well mixed. After the valves are open and the mixer is turned on, how long will it take the density of the liquid in the tank to decline from 980 kg/m^3 to 880 kg/m^3? If the system is left operating long enough, what will be the ultimate density of the liquid in the tank?

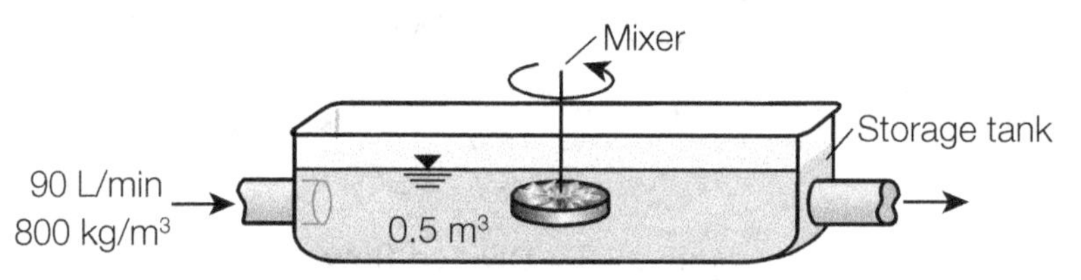

Figure 4.49: Flow through a tank

4.15. Water drains from a funnel that has a top diameter of 20 cm, a bottom (outlet) diameter of 0.5 cm, and a height of 25 cm. The exit velocity, V, is related to the depth of water, h, in the funnel by the relation

$$V = \sqrt{2gh}$$

Determine the relationship between the rate of change of depth (dh/dt) and the depth (h). What is the rate of change of depth when the funnel is half full?

4.16. Water enters the cylindrical reservoir shown in Figure 4.50 through a pipe at Point A at a rate of 2 L/s and exits through a 3-cm-diameter orifice at B. The diameter of the cylindrical reservoir is 50 cm, and the velocity, v, of water leaving the orifice is given by

$$v = \sqrt{2gh}$$

where h is the height of the water surface (in the reservoir) above the centroid of the orifice. How long will it take the water surface in the reservoir to drop from $h = 3$ m to $h = 2$ m? (*Hint*: You might need to use integral tables.)

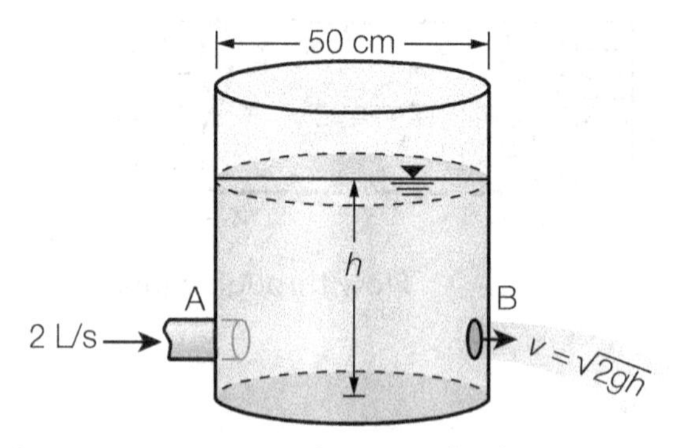

Figure 4.50: Water exiting a cylindrical reservoir

4.17. Water leaks out of a 5-mm-diameter hole in the side of a large cup as shown in Figure 4.51. Estimate how long it takes the cup to go from being full to being half full. (You can assume that the velocity at the outlet is $\sqrt{2gh}$ where h is the height of the water above the centroid of the outlet. Also, you might need to know that the volume of a cone is $\pi r^2 h/3$.)

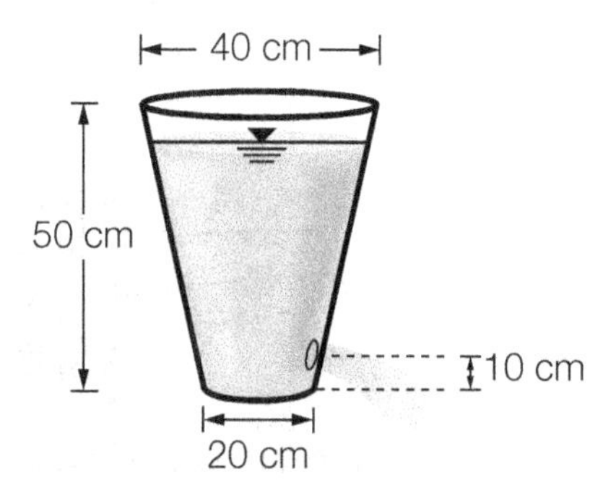

Figure 4.51: Leaking cup

4.18. A 2-m-high cone with a 1-m-diameter base is turned upside down to form a reservoir and is filled with water. A drain hole in the side of the reservoir is opened to release water, and the release rate, Q, is given by

$$Q = 7.8\sqrt{h - 0.02}$$

where Q is in L/s and h is the height of the water surface above the bottom of the reservoir in meters. Estimate how long it will take the reservoir to empty half of its volume. (The volume of a cone is $\pi r^2 h/3$.)

4.19. A storage tank with a volume of 0.2 m^3 contains compressed air. Air is released from the tank by opening a valve connected to a 12-mm-diameter tube. At the instant the valve is opened, air exits at a velocity of 150 m/s. Find the density and the rate of change of density of the air in the tank the instant the valve is opened under the following conditions: (a) The density of the air exiting the tank is 6 kg/m^3 and (b) the air exiting the tank has a temperature of -10°C and an absolute pressure of 400 kPa.

Section 4.4: Conservation of Linear Momentum

4.20. Calculate the momentum-flux correction factor, β, for the following velocity distribution:

$$v(r) = V_0 \left[1 - \left(\frac{r}{R}\right)^2\right]$$

where $v(r)$ is the velocity at a radial distance r from the centerline of a pipe of radius R.

4.21. Water flows over a 0.2-m-high step in a 5-m-wide channel as illustrated in Figure 4.52. If the flow rate in the channel is 15 m^3/s and the upstream and downstream depths are 3.00 m and 2.79 m, respectively, calculate the force on the step.

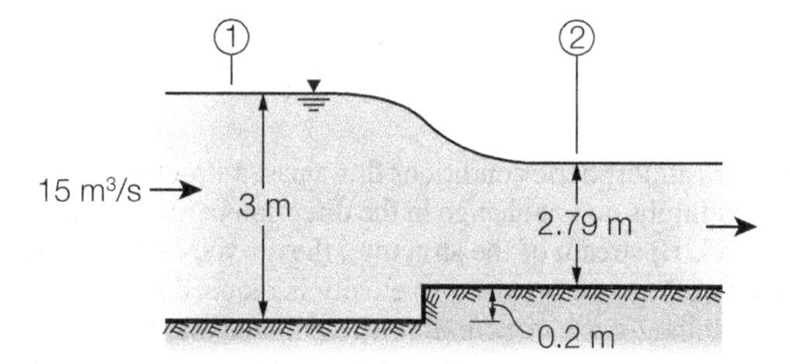

Figure 4.52: Flow over a step

4.22. Water at 20°C flows at 30 m^3/s in a rectangular channel that is 10 m wide. The flowing water encounters three piers as shown in Figure 4.53, and the flow depths upstream and downstream of the piers are 3 m and 2.5 m, respectively. Estimate the force on each pier.

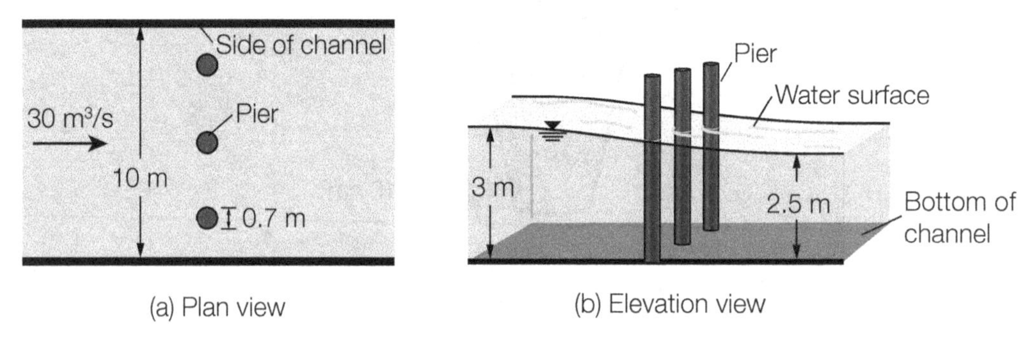

Figure 4.53: Piers in a stream

4.23. The 3-m-high and 7-m-wide spillway shown in Figure 4.54 is designed to accommodate a flow of 35 m^3/s. Under design conditions, the upstream and downstream flow depths are 6 m and 1.5 m, respectively. The base length of the spillway is 10 m. (a) Estimate the force exerted by the water on the spillway and the shear force exerted by the spillway on the underlying soil. (b) If the shear force on the underlying soil is resisted by piles and each pile can resist a shear force of 100 kN, how may piles are necessary.

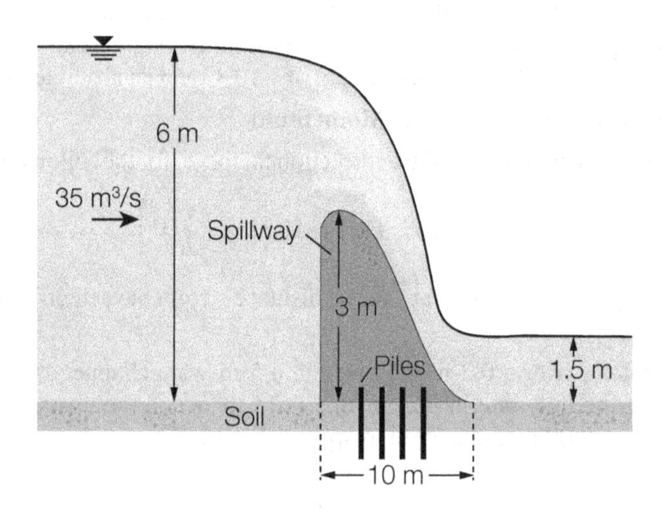

Figure 4.54: Flow over a spillway

4.24. Air at standard atmospheric conditions flows past a structure as shown in Figure 4.55. The flow conditions do not change in the direction perpendicular to the plane shown in Figure 4.55. Upstream of the structure, the air velocity is uniform at 25 m/s, and downstream of the structure the air velocity is reduced linearly behind the structure as shown in Figure 4.55. A bounding streamline expands by an amount δ between locations upstream and downstream of the structure, and within the bounding streamline, the volume flow rate remains constant. (a) Estimate the value of δ. (b) Estimate the force of the air on the structure per unit length perpendicular to the plane of the airflow.

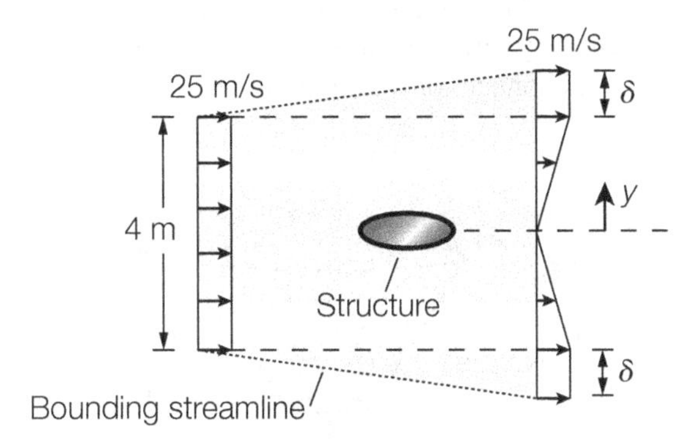

Figure 4.55: Airflow past a structure

4.25. A jet pump such as that illustrated in Figure 4.56 is used in a variety of practical applications, such as in extracting water from groundwater wells. Consider a particular case in which a jet of diameter 160 mm with a velocity of 50 m/s is used to drive the flow of water in a 400-mm-diameter pipe where the velocity of flow in the pipe at the location of the jet is 6 m/s. At the location of the jet, section 1, the pressure is approximately the same across the entire cross section, and at the downstream location, section 2, the jet momentum is dissipated and the velocity is the same across the entire cross section. Estimate the increase in the water pressure between section 1 and section 2. Assume water at 20°C.

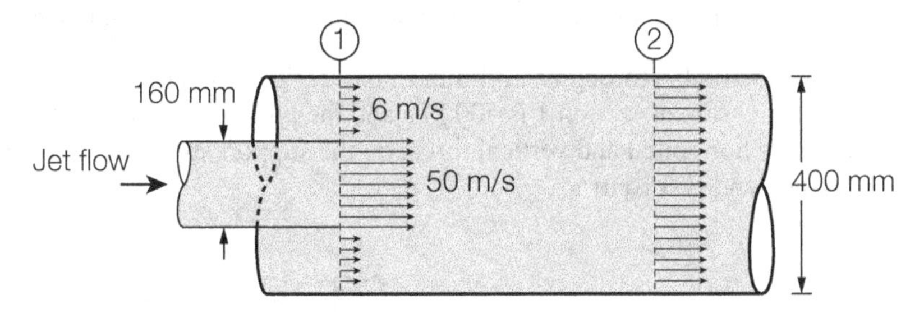

Figure 4.56: Jet pump

4.26. A reducer is to be used to attach a 400-mm-diameter pipe to a 300-mm-diameter pipe. For any flow rate, the pressures in the pipes upstream and downstream of the reducer are expected to be related by

$$\frac{p_1}{\gamma} + \frac{V_1^2}{2g} = \frac{p_2}{\gamma} + \frac{V_2^2}{2g}$$

where p_1 and p_2 are the upstream and downstream pressures, respectively, and V_1 and V_2 are the upstream and downstream velocities, respectively. Estimate the force on the reducer when water flows through the reducer at 200 L/s and the upstream pressure is 400 kPa.

4.27. The nozzle shown in Figure 4.57 is attached to a pipe via a thread connection. The diameter of the source pipe is 30 mm, the diameter of the nozzle exit is 12 mm, the length of the nozzle is 110 mm, and the mass of the nozzle is 0.25 kg. The nozzle is oriented vertically downward and discharges water at 20°C. What force will be exerted on the thread connection when the flow through the nozzle is 25 L/min?

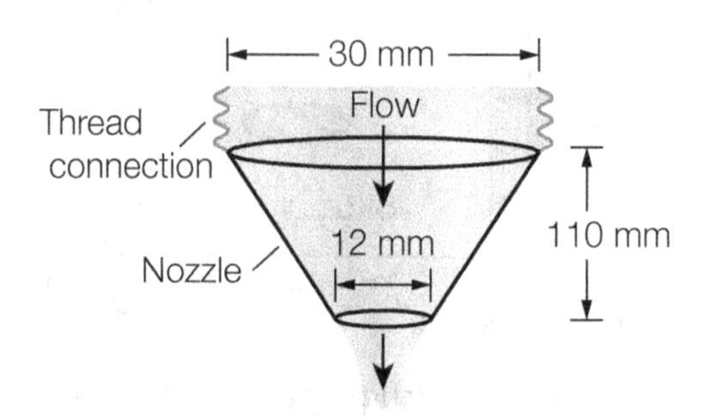

Figure 4.57: Flow through a vertical nozzle

4.28. Air flows in a 150-mm-diameter duct, and at a particular section, the velocity, pressure, and temperature are measured as 70 m/s, 600 kPa, and −23°C, respectively. At a downstream section, the pressure and temperature are measured as 150 kPa and −73°C, respectively. Estimate (a) the average velocity at the downstream section and (b) the wall friction force between the upstream and downstream sections.

4.29. Water under a pressure of 500 kPa flows with a velocity of 5 m/s through a 90° bend in the horizontal plane. If the bend has a uniform diameter of 200 mm and assuming no drop in pressure, calculate the force required to keep the bend in place.

4.30. Water flows at 100 L/s through a 200-mm-diameter vertical bend as shown in Figure 4.58. If the pressure at section 1 is 500 kPa and the pressure at section 2 is 450 kPa, determine the horizontal and vertical thrust on the support structure. The volume of water in the bend is 0.16 m^3.

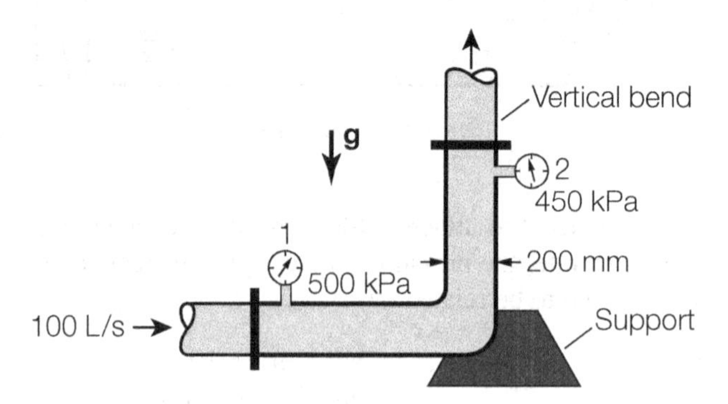

Figure 4.58: Flow through a vertical bend

4.31. The bend shown in Figure 4.59 discharges water into the atmosphere. Determine the force components at the flange required to hold the bend in place. The bend lies in the horizontal plane, the interior volume of the bend is 0.25 m^3, and the mass of the bend material is 250 kg.

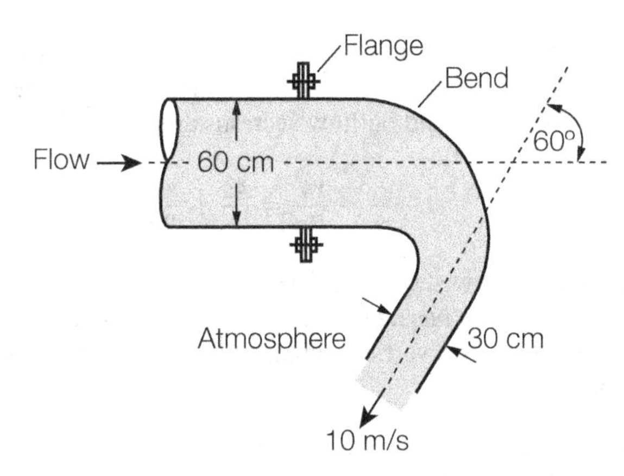

Figure 4.59: Bend

4.32. The fire hose nozzle shown in Figure 4.60 is to be held by two firefighters who (working together) can support a force of 2 kN. (a) Determine the maximum flow rate in the fire hose that can be supported by the firefighters. Give your answer in liters per minute. If the firefighters were to let go of the nozzle, in what direction would it move? (b) If the nozzle is pointed vertically upward, how high will the water jet rise?

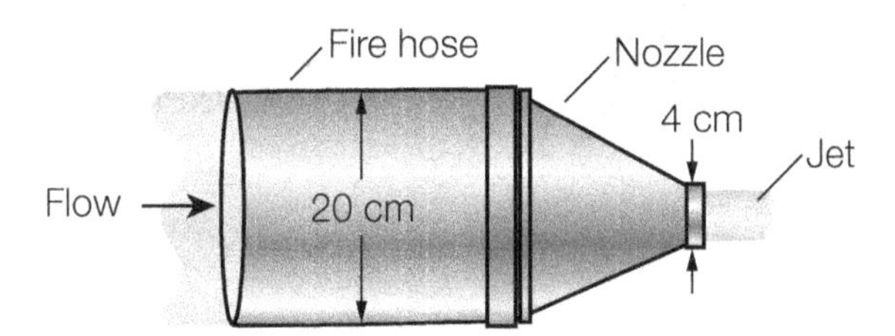

Figure 4.60: Nozzle on a fire hose

4.33. Water at 20°C flows through a 300-mm-diameter pipe that discharges into the atmosphere as shown in Figure 4.61. When a partially open valve extends 200 mm from the top of the pipe, the flow rate in the pipe is 0.2 m^3/s and the (gauge) pressure just upstream of the valve is 600 kPa. Estimate the force exerted by the water on the partially open valve.

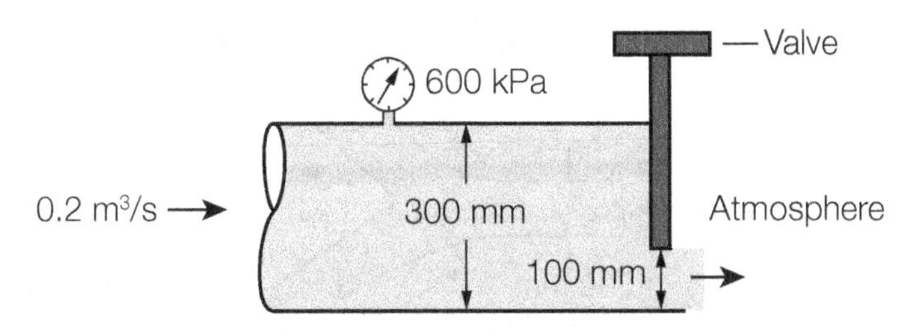

Figure 4.61: Flow past a partially open valve

4.34. A 30° bend connects a 250-mm-diameter pipe (inflow) to a 400-mm-diameter pipe (outflow). The volume of the bend is 0.2 m^3, the weight of the bend is 400 N, and the pressures at the inflow and outflow sections are related by

$$\frac{p_1}{\gamma} + \frac{V_1^2}{2g} = \frac{p_2}{\gamma} + \frac{V_2^2}{2g}$$

The pressure at the inflow section is 500 kPa, the bend is in the vertical plane, and the bend support structure has a maximum allowable load of 18 kN in the horizontal direction and 40 kN in the vertical direction. Determine the maximum allowable flow rate in the bend.

4.35. Determine the force required to restrain the pipe junction shown in Figure 4.62. Assume that the junction is in the horizontal plane and explain how your answer would differ if the junction was in the vertical plane.

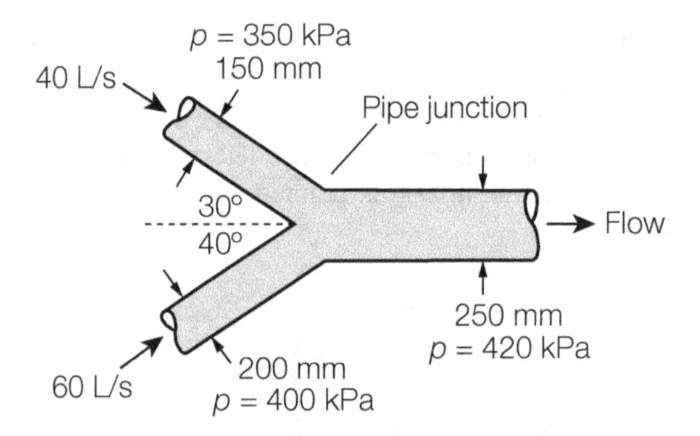

Figure 4.62: Pipe junction

4.36. The pipe connection shown in Figure 4.63 is used to split the flow from one pipe into two pipes. The pipe connection weighs 890 N, the pressure at the incoming section (section 1) is 400 kPa, the flow at section 1 is 0.04 m^3/s, the flow divides such that 60% goes to section 2 and 40% goes to section 3, and the temperature of the water is 20°C. Determine the force required to support the connection.

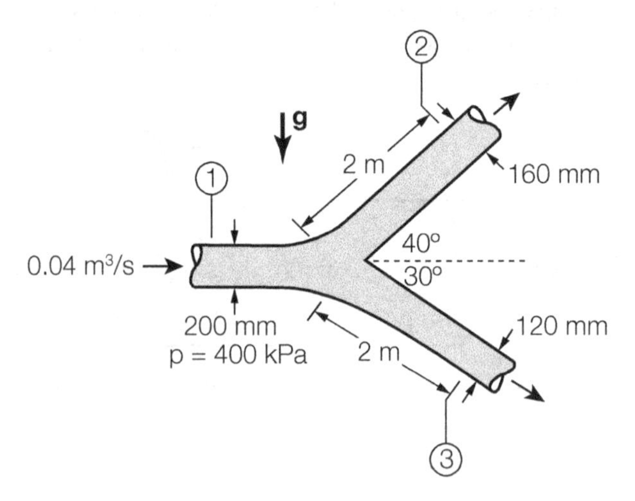

Figure 4.63: Pipe connection

4.37. Find the x- and y-force components on the horizontal T-section shown in Figure 4.64. Neglect viscous effects.

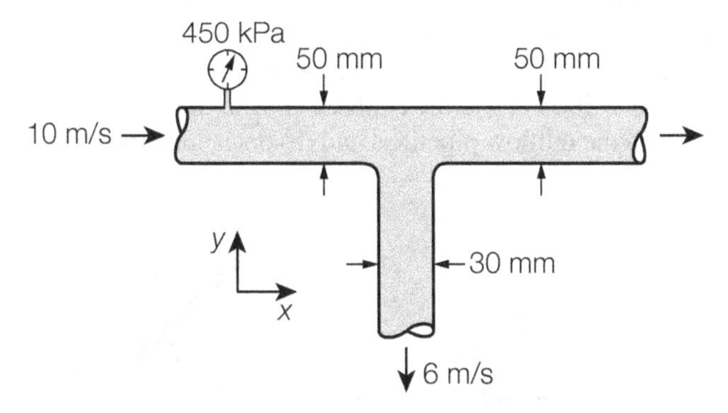

Figure 4.64: T-section of a pipe

4.38. Water is flowing into and discharging from a pipe U-section as shown in Figure 4.65. At flange 1, the absolute pressure is 200 kPa and 30 kg/s flows into the pipe. At flange 2, the absolute pressure is 150 kPa and 22 kg/s flows out of the pipe. At location 3, 8 kg/s of water discharges to the atmosphere, which is at an absolute pressure of 100 kPa. The center of the pipe at flange 2 is located 4.0 m above the center of the pipe at flange 1. The weight of the water in the bend is 280 N, and the weight of the bend is 200 N. Determine the total x- and z-forces (including their directions) to be supported by the two flanges of the pipe U-section. Assume that the momentum-flux correction factor is equal to 1.03.

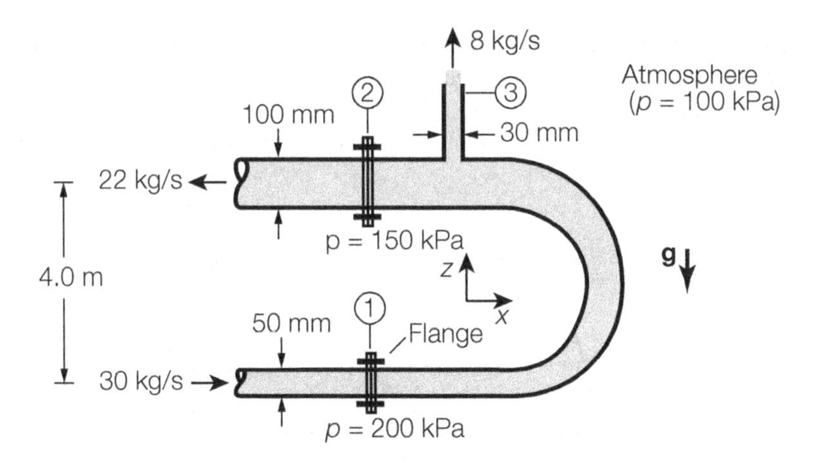

Figure 4.65: Pipe U-section

4.39. Water flowing at 20 L/s in a 150-mm-diameter pipeline is delivered to two separate pipelines via the connection shown in Figure 4.66. At the connection, one outflow pipeline has a diameter of 100 mm, makes an angle of 45° with the inflow pipeline, and carries 8 L/s, whereas the other outflow pipeline has a diameter of 120 mm, makes an angle of 60° with the inflow pipeline, and carries 12 L/s. The pressure in the inflow pipe is 500 Pa, and the connection is in the horizontal plane. (a) Estimate the pressures in the outflow pipelines and (b) determine the force on the connection.

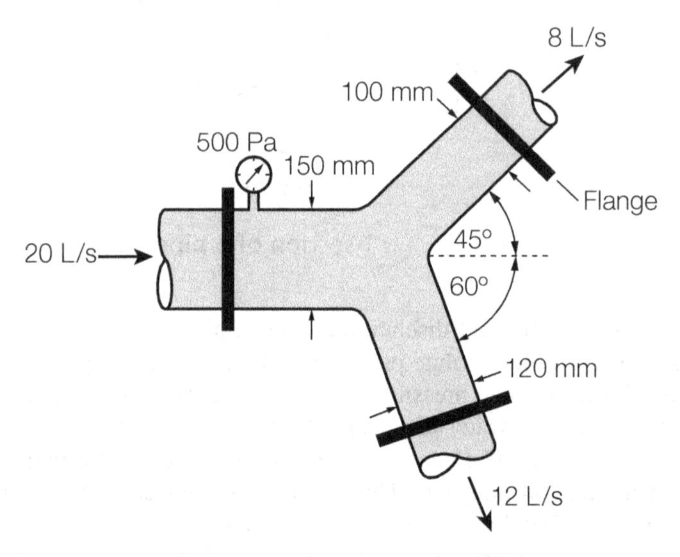

Figure 4.66: Pipe junction

4.40. Consider a 50-mm-diameter thin-plate orifice centrally located in a 110 mm-diameter pipe as shown in Figure 4.67. When water at 20°C flows at a rate of 0.04 m^3/s through the pipe, a pressure drop of 150 kPa between sections 1 and 2 is measured. If the wall friction is negligible, estimate the force of the water on the orifice plate.

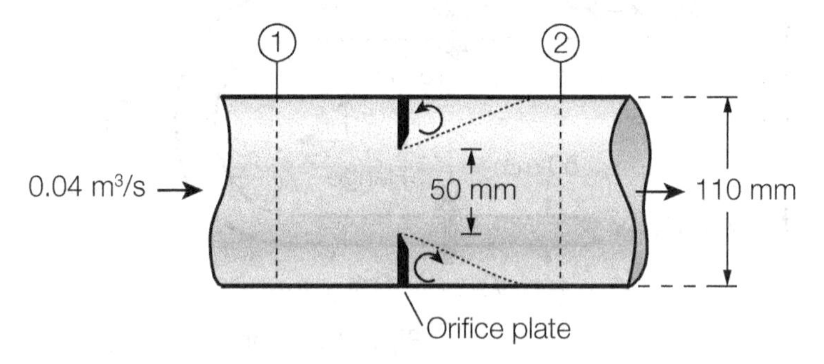

Figure 4.67: Thin-plate orifice

4.41. A nozzle with a quarter-circle shape discharges water at 4 m/s as shown in Figure 4.68. The radius of the quarter circle is 300 mm, and the height of the nozzle is 25 mm. Estimate the force required to hold the nozzle in a fixed position.

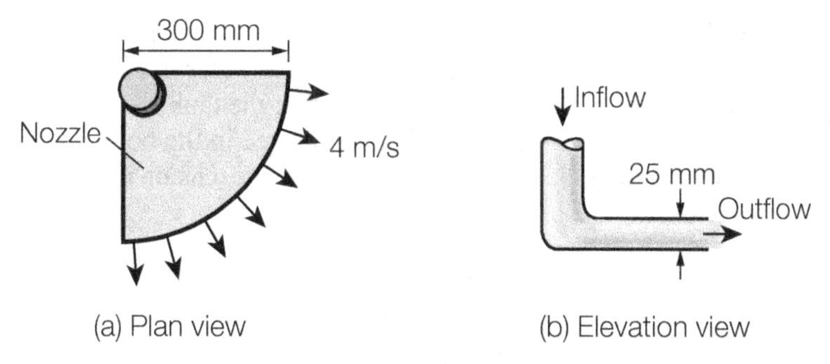

Figure 4.68: Quarter-circle nozzle

4.42. A jet of water at 20°C exits a nozzle into the open atmosphere and strikes a stagnation tube as shown in Figure 4.69. If head losses are neglected, estimate the mass flow rate in kg/s and the height h of the water in the stagnation tube.

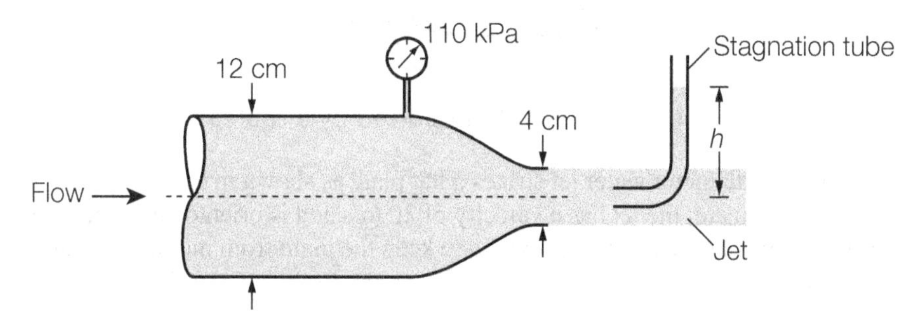

Figure 4.69: Water striking a stagnation tube

4.43. A jet of water at 20 m^3/s impinges on a stationary deflector that changes the flow direction of the jet by 60°. The velocity of the impinging jet is 20 m/s, and the velocity of the deflected jet is 18 m/s. What force is required to keep the deflector in place?

4.44. The concrete structure shown in Figure 4.70 is to be overturned by a jet of water striking the center, P, of the vertical panel. The structure is expected to overturn along the edge QR. The diameter of the jet is 200 mm, and it can be assumed that the density of concrete is 2300 kg/m^3. Estimate the velocity of the jet required to overturn the structure. The panel is made of concrete.

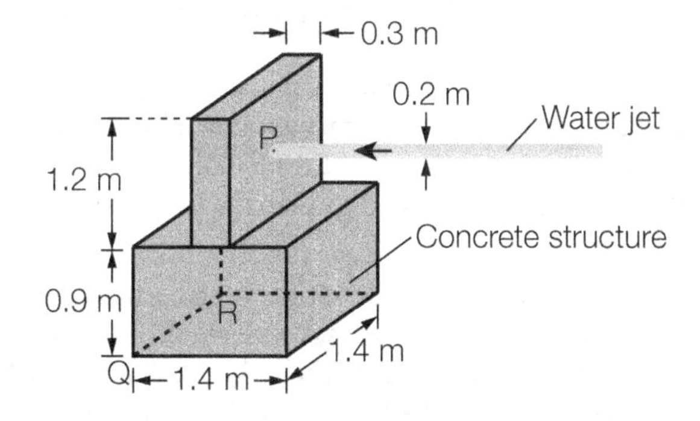

Figure 4.70: Jet impinging on a concrete structure

4.45. Water at 20°C is contained in a pressurized tank that is supported by a cable as shown in Figure 4.71. The (gauge) air pressure above the water in the tank is 600 kPa, the depth of water in the tank is 1 m, the diameter of the tank is 1 m, the weight of the tank is 1 kN, and there is a 50-mm-diameter orifice in the bottom of the tank. The discharge coefficient of the orifice is 0.8. Estimate the tension in the cable supporting the tank.

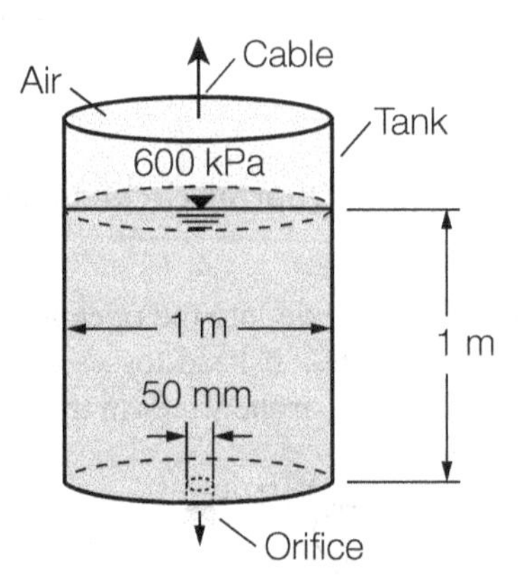

Figure 4.71: Tank supported by a cable

4.46. A 100-mm-diameter water jet strikes a flat plate as shown in Figure 4.72(a). Prior to striking the plate, the jet has a velocity of 20 m/s and is oriented normal to the plate. What support force must be provided to keep the plate from moving? Assume water at 20°C.

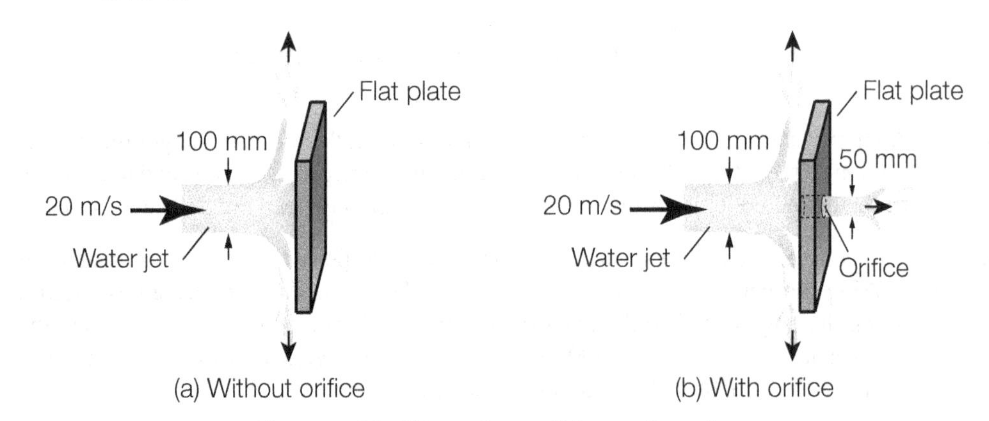

Figure 4.72: Water jet striking a flat plate

4.47. A 100-mm-diameter water jet strikes a flat plate with a 50-mm-diameter orifice whose center is aligned with the water jet as shown in Figure 4.72(b). Prior to striking the plate, the jet has a velocity of 20 m/s and is oriented normal to the plate. What support force must be provided to keep the plate from moving? Assume water at 20°C.

4.48. Water-jet cutters are used to cut a wide variety of materials using a very high-pressure water jet. A particular water-jet cutter uses water at 20°C at a flow rate of 5 L/min and a jet diameter of 0.20 mm. If this jet is used to cut a flat plate where the plate is oriented normal to the jet, what is the force per unit area exerted by the jet on the plate?

4.49. Air at 20°C and 101.3 kPa flows in a 25-cm-diameter duct at 15 m/s as shown in Figure 4.73; the exit is choked by a 90° cone. Estimate the force of the airflow on the cone.

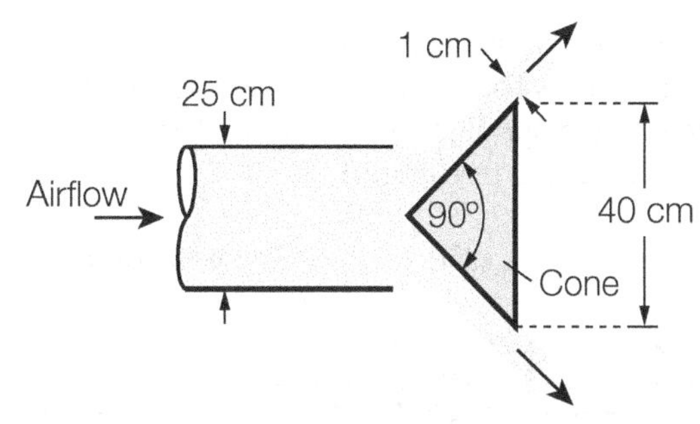

Figure 4.73: Airflow onto a cone

4.50. An air tank hovers above a solid surface supported only by the air being released through a 15-mm-diameter orifice that is located immediately below the center of gravity of the tank as shown in Figure 4.74. The temperature of the air in the tank is 23°C, and the weight of the tank plus air in the tank is 100 N. The flow of air through the orifice can be assumed to be incompressible and frictionless, and atmospheric pressure can be taken as 101 kPa. Estimate the air pressure in the tank.

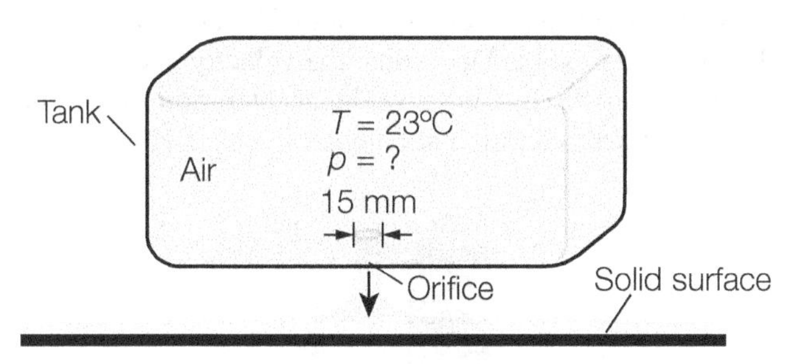

Figure 4.74: Tank supported by an air jet

4.51. A blade attached to a turbine rotor is driven by a stream of water that has a velocity of 20 m/s. The blade moves with a velocity of 10 m/s and deflects the stream of water through an angle of 80°; the entrance and exit flow areas are each equal to 2 m^2. Estimate the force on the moving blade and the power transferred to the turbine rotor.

4.52. A circular jet of water at 20°C impinges on the vane shown in Figure 4.75. The incident jet has a velocity of 25 m/s and a diameter of 200 mm. The vane divides the incident jet equally, and the jet exits the vane as two jets, each making an angle of 20° with the incoming jet. (a) Determine the force on the vane when the vane is moving at a speed of 15 m/s in the same direction as that of the incident jet. (b) Determine the force on the vane when the vane is moving at a speed of 15 m/s in the opposite direction to that of the incident jet. Assume water at 20°C.

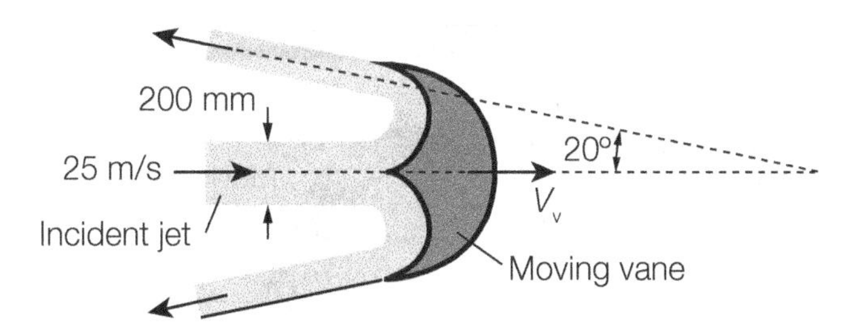

Figure 4.75: Force on a moving vane

4.53. Water flows out the 6-mm-diameter slots in a sprinkler rotor as shown in Figure 4.76. Assuming that the rotor is frictionless, estimate the angular velocity (ω) if 20 kg/s of water is delivered to the sprinkler rotor. (Note: The velocities of the exit jets might not be uniform as shown in the figure.)

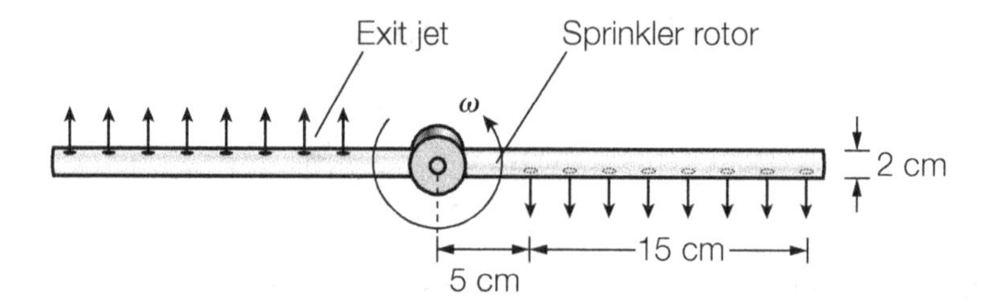

Figure 4.76: Sprinkler

4.54. The rocket shown in Figure 4.77 is launched vertically from high in the atmosphere where aerodynamic drag on the rocket is negligible. The rocket has an initial mass of 600 kg, it burns fuel at a rate of 7 kg/s, and the velocity of the exhaust gas relative to the rocket is 2800 m/s. (a) Determine the velocity and acceleration of the rocket 10 seconds after launch. (b) Compare the acceleration of the rocket at 10 seconds after launch with the acceleration at launch.

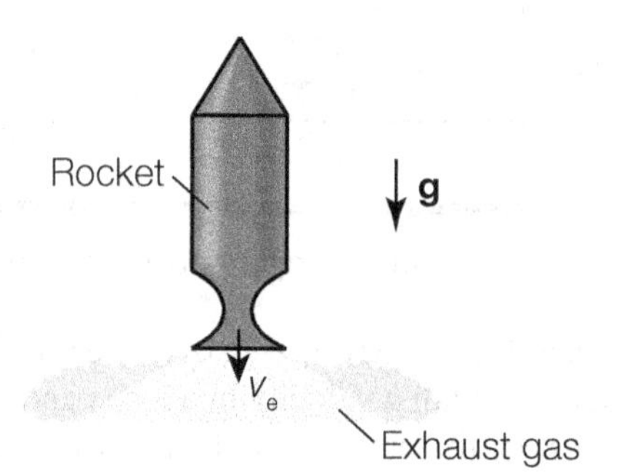

Figure 4.77: Rocket with a vertical trajectory

4.55. A 50-mm-diameter jet of water impacts a vane at a velocity of 20 m/s as shown in Figure 4.78. The vane has a mass of 90 kg, and it deflects the jet through a 55° angle. The vane is constrained to move in the direction of the incoming jet, and there is no force opposing the motion of the vane. Determine the acceleration of the vane when it is first released. Assume water at 20°C.

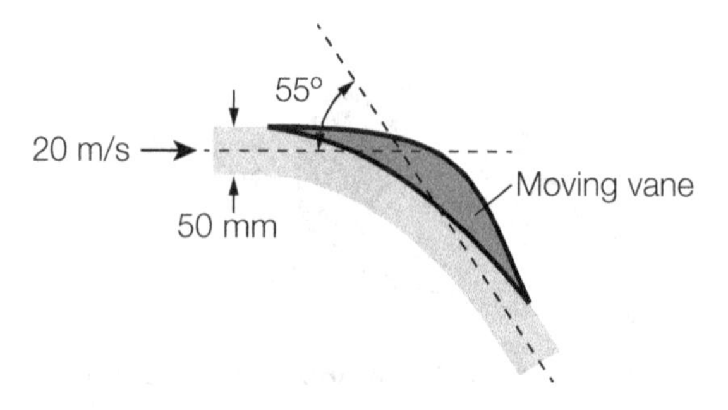

Figure 4.78: Motion of a vane caused by a water jet

4.56. A wind turbine is being proposed for a commercial development in Miami, Florida. The proposed turbine will have a hub height of 50 m and a rotor diameter of 70 m. Miami is approximately at sea level. Estimate the wind power density and assess the feasibility of using wind as a source of energy in Miami.

4.57. Estimate the maximum wind power density available at any location within the land area of the 48 contiguous United States. For such a location, provide a rough layout of a 259-hectare wind farm, where each wind turbine would have a 90-m-diameter rotor. What is the maximum power that could be produced from this wind farm? Assume standard temperature and pressure.

4.58. A 3-cm-diameter orifice is 1 m below the water surface in a 1.5-m-diameter barrel containing water at 20°C. Estimate the force on the barrel when water is flowing freely out of the orifice.

4.59. A vertical jet discharges water onto the surface of a reservoir, and water is discharged from the same reservoir via an opening that is inclined at 45° to the horizontal as shown in Figure 4.79. The inflow jet has a diameter of 25 mm, the outflow jet has a diameter of 35 mm, and conditions are at steady state. (a) What is the net hydrodynamic force on the reservoir container? (b) How high will the discharge jet rise above its discharge elevation?

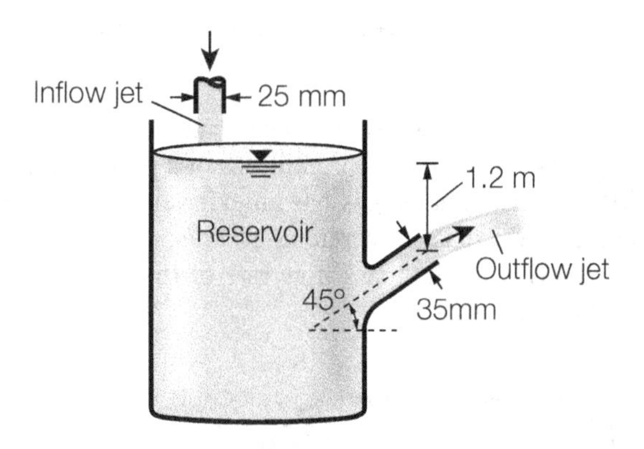

Figure 4.79: Inflow and outflow jets in a reservoir

4.60. A jet engine is mounted on an aircraft that is cruising at an altitude of 10 km in a standard atmosphere, and the speed of the aircraft is 300 m/s. The intake area of the engine is 2 m^2, the fuel consumption rate is 30 kg/s, and the exhaust gas exits at a speed of 250 m/s relative to the moving aircraft. The pressure of the exhaust gas is approximately equal to the ambient atmospheric pressure. Under these conditions, what thrust is expected from the engine?

4.61. The specifications of a jet engine indicate that it has an intake diameter of 2.70 m, consumes fuel at a rate of 1.8 L/s, and produces an exhaust jet with a velocity of 900 m/s. If the jet engine is mounted on an airplane that is cruising at 850 km/h where the air density is 0.380 kg/m^3, estimate the thrust produced by the engine. Assume that the jet fuel has a density of 810 kg/m^3.

4.62. Jet engines are usually tested on a static thrust stand as shown in Figure 4.80. In such tests, the pressures of the inflow and outflow gases are usually expressed as gauge pressures. In a particular test of an engine with an inflow area of 1.2 m^2 and an outflow area of 0.5 m^2, the inflow air has a velocity of 250 m/s, a (gauge) pressure of 50 kPa, and a temperature of -50°C. The exhaust gas has a velocity of 550 m/s and is at atmospheric pressure, which is 101 kPa (absolute). (a) Estimate the thrust on the test stand, assuming that the mass flow rate of the exhaust gas is equal to the mass flow rate of the inflow air. (b) Estimate the thrust on the test stand, taking into account that fuel is supplied to the engine at a mass flow rate equal to 2% of the air mass flow rate. Which estimated thrust is likely to be more accurate?

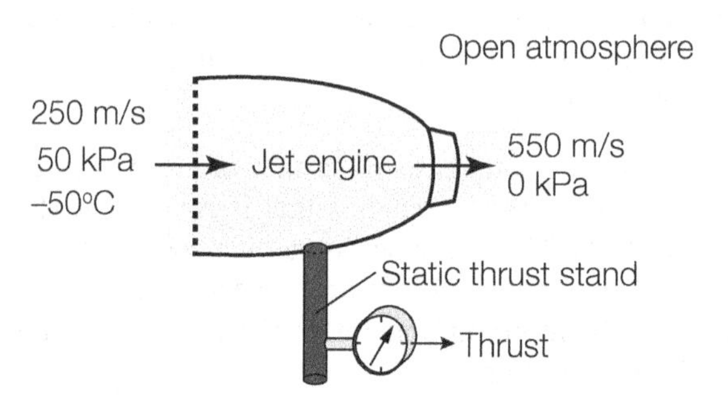

Figure 4.80: Jet engine on a test stand

4.63. The performance of a rocket with a 220-mm-diameter nozzle exit is tested on a stand as shown in Figure 4.81. Under the conditions of a particular test, atmospheric pressure is 101 kPa and the exhaust gas has an exit velocity of 1550 m/s, a pressure of 120 kPa absolute, and a mass flow rate of 15 kg/s. (a) Estimate the thrust generated by the rocket. (b) How is the mass flow rate of the exhaust gas related to the fuel consumption rate of the rocket?

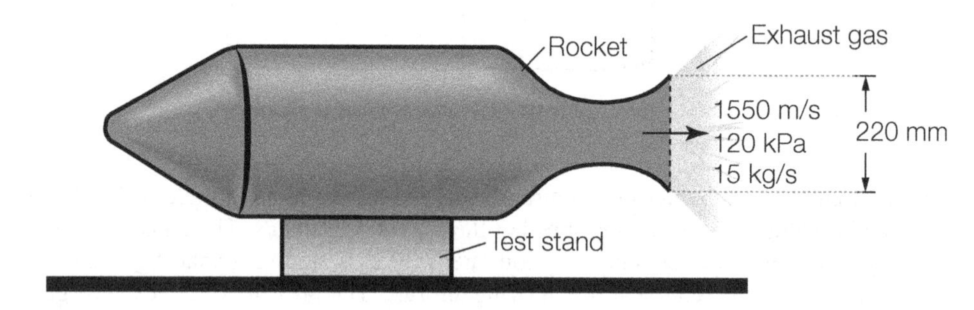

Figure 4.81: Rocket on a test stand

4.64. A rocket weighs 6000 kg, burns fuel at a rate of 40 kg/s, and has an exhaust velocity of 3000 m/s. Estimate the initial acceleration of the rocket and the velocity after 10 seconds. Neglect the drag force of the surrounding air and assume that the pressure of the exhaust gas is equal to the pressure of the surrounding atmosphere.

Section 4.5: Angular Momentum Principle

4.65. Water flows through a 12-cm-diameter pipe that consists of a 3-m-long riser and a 2-m-long horizontal section with a 90° elbow to force the water to be discharged downward. Water discharges to the atmosphere at a velocity of 4 m/s, and the weight of the pipe section is 175 N per meter length. (a) Determine the restraining moment at the riser support (i.e., Point S in Figure 4.82). (b) What would be the support moment if the flow was discharged upward instead of downward? (c) Which discharge orientation creates the greatest stress on the support?

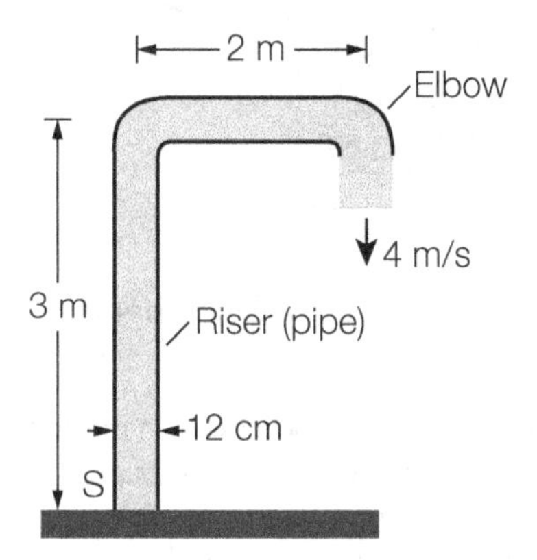

Figure 4.82: Pipe system

4.66. Water flows through the pipe bend and nozzle arrangement shown in Figure 4.83, which lies in the horizontal plane. The water issues from the nozzle into the atmosphere as a parallel jet, and friction may be neglected. Find the moment of the resultant force due to the water on this arrangement about a vertical axis through the point X.

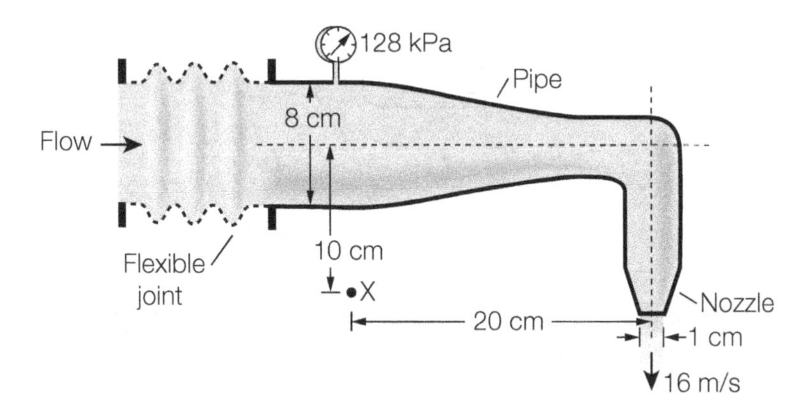

Figure 4.83: Force on a nozzle

4.67. Water at 20°C flows through the pipe bend in Figure 4.84 at 0.2524 m^3/s, where the bend is in the horizontal plane. (a) Compute the torque required at Point B to hold the bend stationary and (b) determine the location of the line of action of the resultant force.

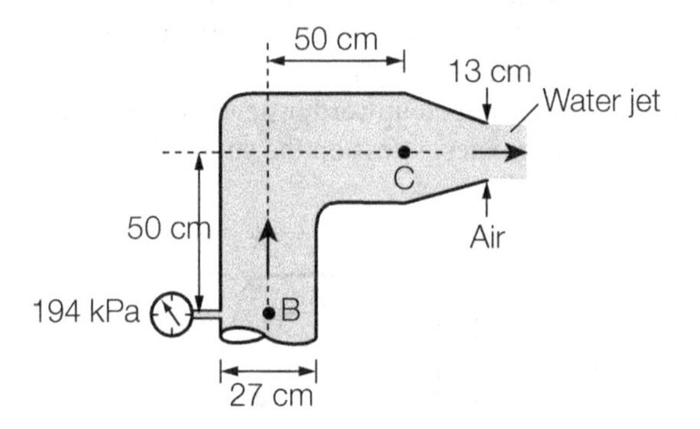

Figure 4.84: Pipe bend

4.68. The portable 0.05 m diameter pipe bend shown in Figure 4.85 is to be held in place by a 68.03 kg person. The entrance to the bend is at A, and the water discharges freely (underwater) at B. If the person holding the bend can resist a 444.82 N force being exerted on the bend, what is the maximum flow rate that should be used? Determine the support location, C, where the person would experience zero torque. Assume that the head loss in the bend is $2V^2/2g$, where V is the velocity in the bend.

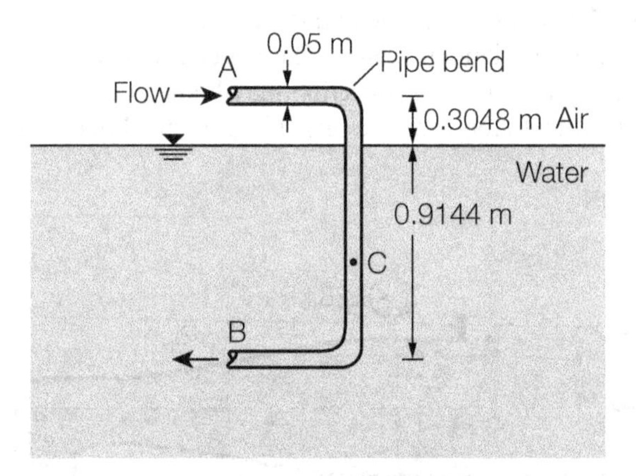

Figure 4.85: Flow in a bend

4.69. The pipe double-bend shown in Figure 4.86 is commonly used in pipeline systems to avoid obstructions. In a particular application, the pipe has a diameter of 200 mm and carries water at a flow rate of 28 L/s and at a temperature of 15°C. The pressure at the entrance to the bend is 25 kPa, and the head loss within the bend can be neglected. If the bend is supported at P, estimate the magnitude of the force and the moment on the support. What is the location of the resultant force exerted by the flowing water on the bend?

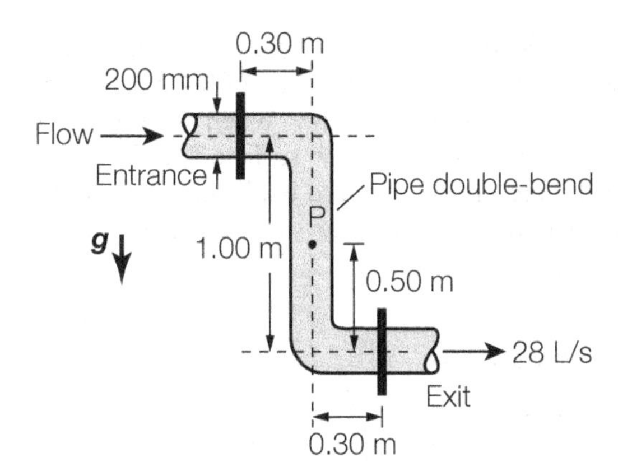

Figure 4.86: Pipe double-bend

4.70. Water at 20°C flows through the 100-mm-diameter pipe-bend section that is supported at S as shown in Figure 4.87. The velocity in the pipe is 5 m/s, the inflow pressure is 550 kPa, and the outflow pressure is 450 kPa. The centerline of the inflow pipe is 80 mm below the support, and the centerline of the outflow pipe is 300 mm below the support. The pipe bend is in the horizontal plane. Determine the horizontal-plane moment that is exerted on the support by the flowing water.

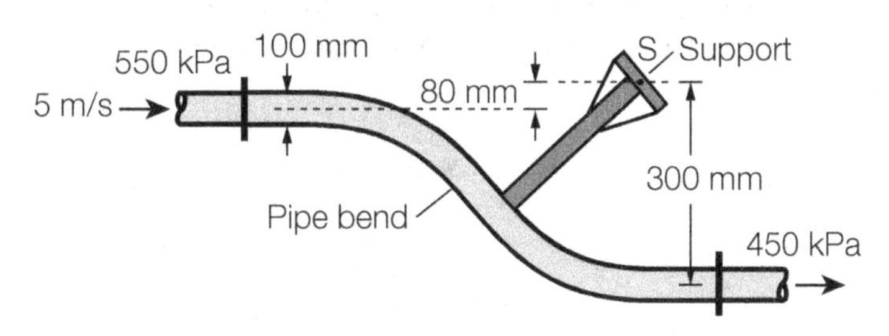

Figure 4.87: Pipe bend in a horizontal plane

4.71. Water at 20°C is discharged through a slot in a 225-mm-diameter pipe as shown in Figure 4.88. The slot has a length of 0.8 m and a width of 30 mm, and the velocity through the slot increases linearly from 6 m/s to 18 m/s over the length of the slot. The (gauge) pressure in the pipe just upstream of the slot is 80 kPa, and the slot discharges to the open atmosphere. Estimate the component forces and moment that are exerted on the support section.

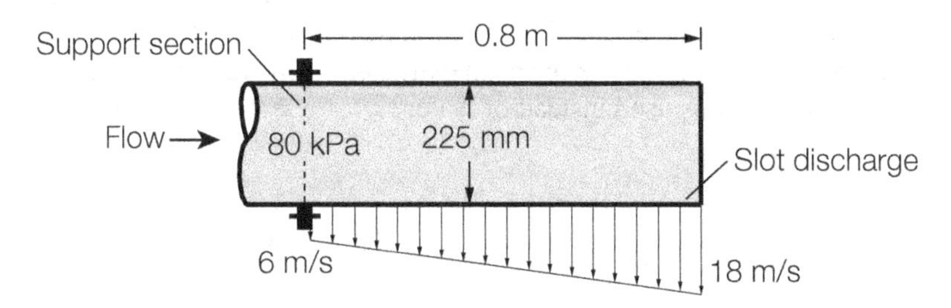

Figure 4.88: Discharge through a slot

4.72. A centrifugal fan blows air using the rotor shown in Figure 4.89. The rotor has an inner diameter of 0.2 m, an outer diameter of 0.4 m, a height of 30 mm, and a blade angle of 25° with the outflow surface. When the rotor turns at 1800 rpm, the inflow velocity is normal to the inflow surface and the flow rate through the fan is 10 m^3/min. Estimate (a) the required blade angle at the inflow surface so that the inflow velocity is normal to the inflow surface and (b) the power that the fan delivers to the air. Assume standard air.

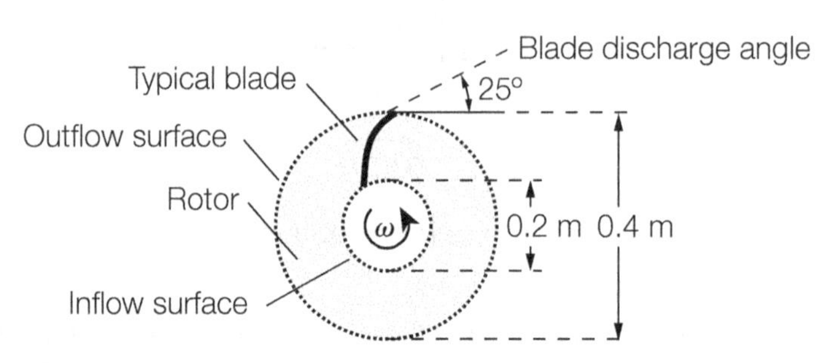

Figure 4.89: Airflow through a fan

4.73. A water pump impeller has an inner diameter of 0.1 m, an outer diameter of 0.4 m, and a height of 0.08 m. When the impeller is rotating at 1800 rpm, the flow rate through the impeller is 1000 L/s, the inflow velocity is normal to the inflow surface, and the outflow velocity is 18 m/s as illustrated in Figure 4.90. Estimate the head added by the pump.

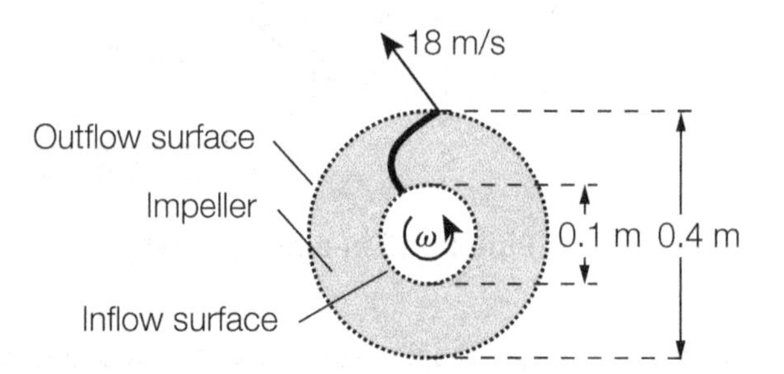

Figure 4.90: Flow through a pump

4.74. The water turbine runner shown in Figure 4.91 is driven by an inflow velocity that makes an angle of 50° with the inflow surface. The outflow velocity is normal to the outflow surface and has a magnitude of 15 m/s. The outer diameter of the runner is 4 m, the inner diameter is 1.5 m, and the runner rotates at 60 rpm in the counterclockwise direction. Estimate the head extracted by the turbine. Would your answer be different if air was used instead of water?

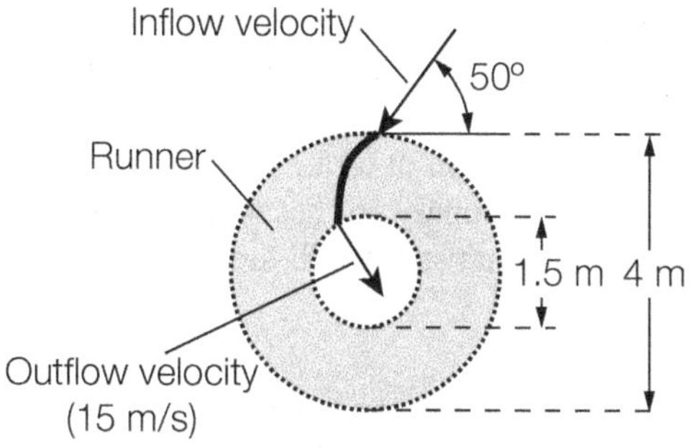

Figure 4.91: Flow through a turbine

4.75. Water enters a turbine runner that has an inner diameter of 0.8 m and an outer diameter of 2 m as shown in Figure 4.92. Water crosses the inflow surface at an angle of 25° with an absolute flow velocity of 15 m/s. The outflow velocity is normal to the outflow surface. The height of the runner is 0.4 m, and the runner is rotating at a rate of 180 rpm. What power is being generated by the runner?

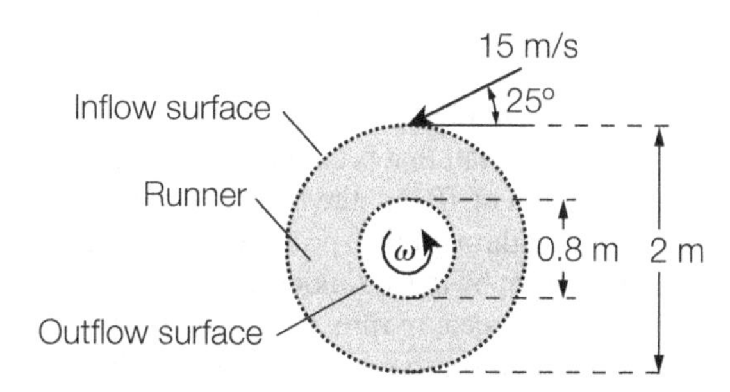

Figure 4.92: Flow through a turbine runner

4.76. A three-arm sprinkler is used to water a garden by rotating in a horizontal plane. Water enters the sprinkler along the axis of rotation at a rate of 40 L/s and leaves the 1.2-cm-diameter nozzles in the tangential direction. The central sprinkler bearing applies a retarding torque of 50 N·m due to friction at the anticipated operating speeds. For a normal distance of 40 cm between the axis of rotation and the center of the nozzles, determine the rate of rotation (in rpm) of the sprinkler.

4.77. The two-arm water sprinkler shown in Figure 4.93 is connected to a hose that delivers 10 L/min to the sprinkler. The radius of the sprinkler is 175 mm, and the nozzles of the sprinkler have a diameter of 5 mm and are directed outward at an angle of 15° to the tangent of the circle of rotation. What is the torque exerted on the sprinkler support?

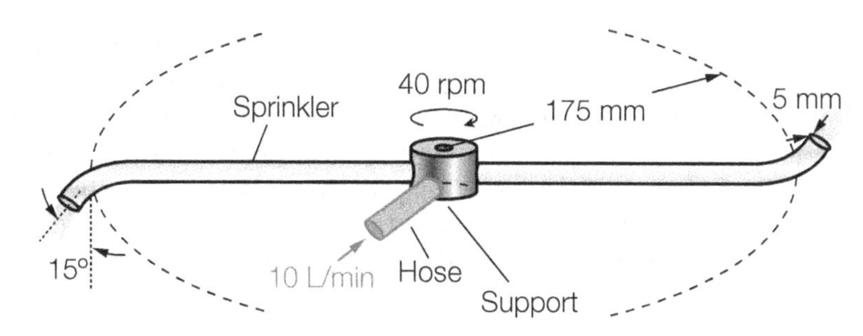

Figure 4.93: Two-arm sprinkler

4.78. The four-arm water sprinkler shown in Figure 4.94 has a radius-of-rotation of 0.7 m. The nozzles at the four outlets of the sprinkler each have an area of 30 mm^2 and are oriented at an angle of of 40° to the circle-of-rotation. Consider the case where water is supplied to the sprinkler at a rate of 8 L/s. (a) What torque is necessary to hold the sprinkler rotors stationary so that they do not rotate? (b) If the restraining torque is equal to zero, at what speed will the sprinkler rotate?

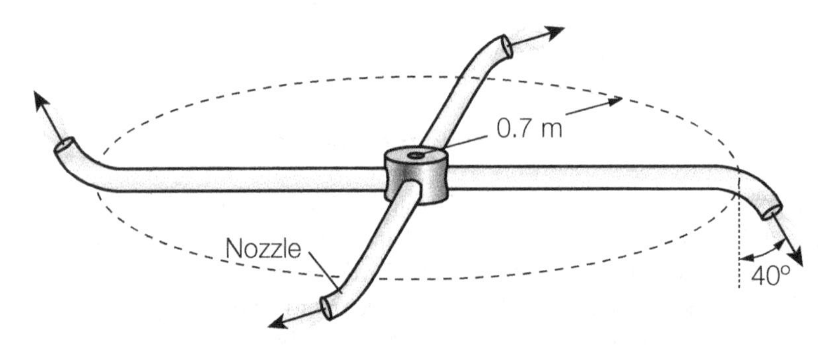

Figure 4.94: Four-arm sprinkler

4.79. Consider the system shown in Figure 4.95 where a water sprinkler is used to generate electric power by rotating a shaft that is connected to an electric generator. Each arm of the sprinkler has a length of 10 cm, the nozzle at the end of each arm has a diameter of 1 cm, the total flow through the sprinkler is 8 L/s, and the rotational speed of the sprinkler rotor can be set by the sprinkler operator using a device called a governor. (a) At what rotational speed, in rpm, would the maximum power be generated? (b) What is the maximum power that can be generated by the sprinkler-generator?

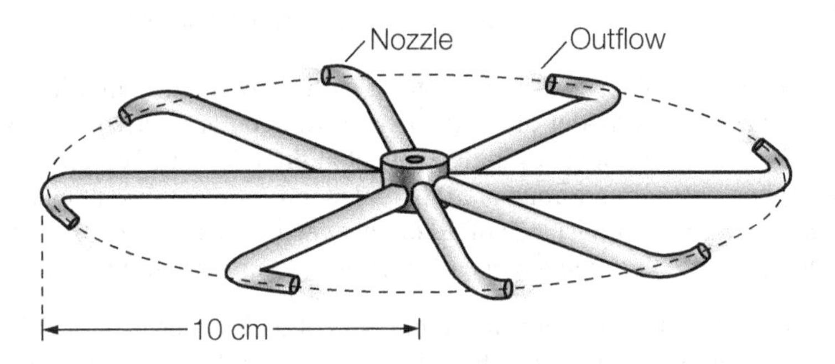

Figure 4.95: Sprinkler for electricity generation

Section 4.6: Conservation of Energy

4.80. (a) Air flows at 0.05 kg/s through an insulated circular duct such that the (gauge) pressure immediately downstream of a blower is 10 kPa. If the pressure at a downstream location within the duct is measured as 8 kPa, estimate the change in the temperature of the air that would be expected between upstream and downstream locations. You may neglect the impact of elevation change in your analysis. (b) If a blower was used to raise the pressure in the duct described in part (a) by 2 kPa at a single location, estimate the power required to drive the blower if it has an efficiency of 90%.

4.81. An incompressible fluid flows with an average velocity of 1.2 m/s in a horizontal 300-mm-diameter pipe. If the pressure drop between two cross sections is 15 kPa, what is the rate at which energy is being lost between these two sections?

4.82. Water flows along the straight 300-mm-diameter conduit shown in Figure 4.96, where the pressure at the lower-elevation section of the conduit is 560 kPa and the pressure at the higher-elevation section is 400 kPa. The conduit makes an angle of 20° with the horizonal, and the distance between the gauged sections is 12 m. If the fluid is incompressible, confirm the assumed direction of flow shown in Figure 4.96 and estimate the head loss between the two sections.

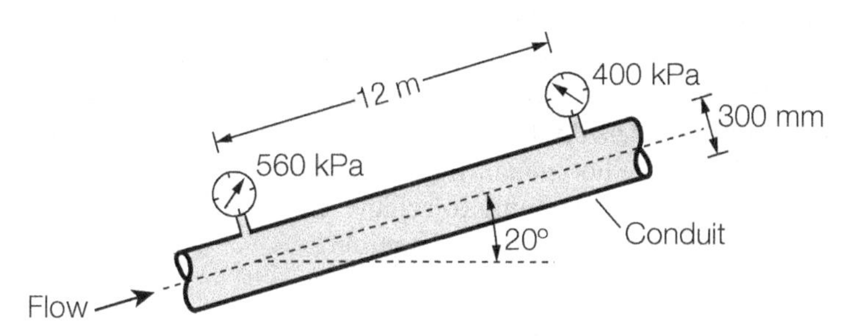

Figure 4.96: Flow in an inclined conduit

4.83. A valve connects two pipes, where the upstream pipe has a diameter of 300 mm and the downstream pipe has a diameter of 450 mm. When water flows through the valve and the velocity in the upstream pipe is 2.5 m/s, the difference in pressure between the inflow pipe and the outflow pipe is measured as 60 kPa. If the inflow and outflow pipes are at the same elevation, what is the head loss through the valve? What is the power loss across the valve?

4.84. Calculate the rate of heat loss between the upstream and downstream sections in Figure 4.52. Assume that the temperature of the water does not change between the upstream and downstream sections.

4.85. Water flows at a velocity of 1.8 m/s in a 150-mm-diameter pipeline as shown in Figure 4.97. At a particular section, the (gauge) pressure is equal to 150 kPa, and downstream of this section, the pipe elevation increases by 1.2 m and the water is discharged through a 50-mm nozzle. Estimate the head loss between the section where the pressure is 150 kPa and the section is just downstream of the nozzle discharge.

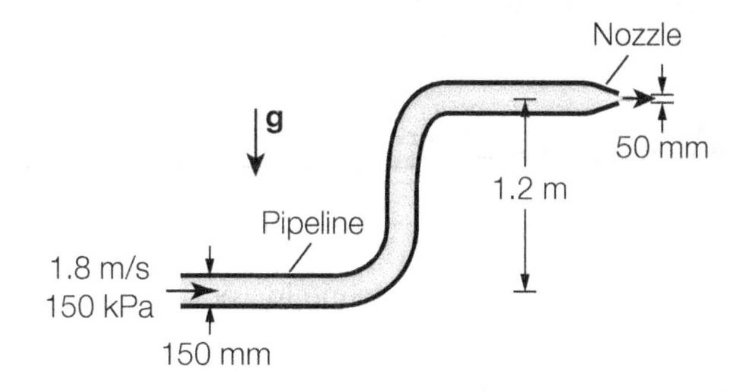

Figure 4.97: Pipeline with a nozzle discharge

4.86. Water at 20°C flows through a pump in which the inflow pipe has a diameter of 150 mm and the outflow pipe has a diameter of 75 mm. The centerlines of the inflow and outflow pipes are aligned so that there is no difference in elevation. When the flow rate through the pump is 30 L/s, the inflow and outflow pressures are 150 kPa and 500 kPa, respectively, and measurements indicate that the temperature of the water increases by 0.1°C between the inflow and the outflow. Estimate the rate at which energy is being delivered by the pump to the water. Approximately what percentage of the energy input by the pump goes toward raising the temperature?

4.87. Calculate the rate of energy loss in the reducer described in Problem 4.26.

4.88. A pump is attached to a fire hydrant as shown in Figure 4.98, where it is desired to produce a jet that rises to a maximum height of 40 m and delivers water at a rate of 100 L/s. At the attachment location, the diameter is 100 mm and the pressure is estimated as 150 kPa. The discharge nozzle has a diameter of 80 mm and is inclined at an angle of 50° to the horizontal. If friction losses are neglected, estimate the head that must be added by the pump to achieve the desired objective. Approximately how much power will it take to run the pump?

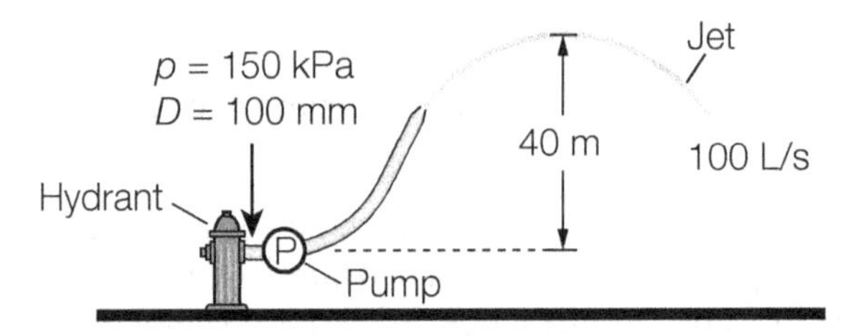

Figure 4.98: Pump attached to a fire hydrant

4.89. Calculate the rate of energy loss in the junction described in Problem 4.35.

4.90. Water at 20°C in a 10-m-diameter, 2-m-deep aboveground swimming pool is to be emptied by unplugging a 3-cm-diameter, 25-m-long horizontal plastic drain pipe attached to the bottom of the pool. Determine the time it will take to empty the swimming pool completely, assuming that the entrance to the drain pipe is well-rounded with negligible loss. Assume the friction factor of the pipe to be 0.022.

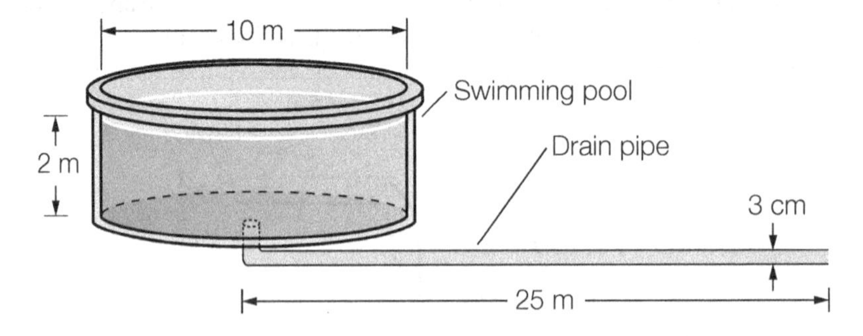

Figure 4.99: Swimming pool

4.91. Water flows at a rate of 0.025 m^3/s in a horizontal pipe whose diameter increases from 6 to 11 cm by an enlargement section. If the head loss across the enlargement section is 0.45 m and the kinetic energy correction factor at both the inlet and the outlet is 1.05, determine the pressure change.

4.92. The suction and discharge pipes of a pump both have diameters of 150 mm and are at the same elevation. Under a particular operating condition, the pump delivers 1600 L/min, the pressures in the suction and discharge lines are 30 kPa and 300 kPa, respectively, and the power consumption is 8 kW. Estimate the efficiency of the pump under this operating condition. Assume water at 20°C.

4.93. Water is to be be pumped from a lake to irrigate a tree as shown in Figure 4.100. The head loss in the 50-cm-diameter pipeline, h_ℓ [m], is a function of the flow rate, Q [m^3/s], according to the relation

$$h_\ell = 1.05Q^2$$

The desired flow rate is 1 m^3/s, and the pump efficiency is assumed to be 80%. (a) What size pump would you select for the job? (b) If a nozzle of diameter 25 cm is attached to the end of the pipeline, will the same size pump be adequate? If not, calculate the pump size that is required in this case.

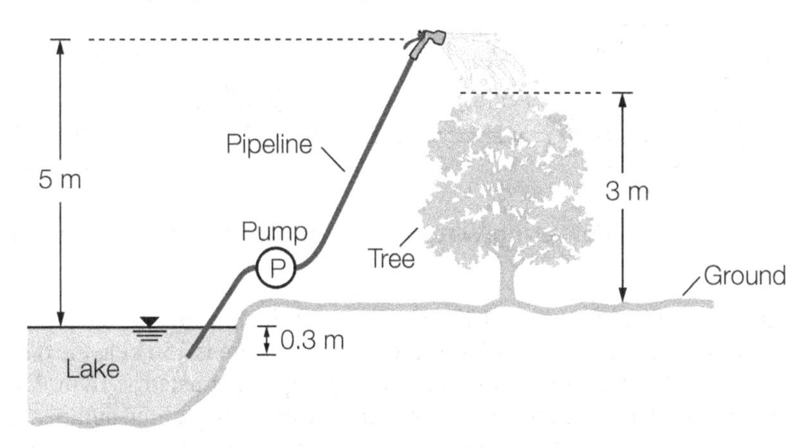

Figure 4.100: Tree irrigation

4.94. Water is pumped from a lower reservoir to a higher reservoir using a pump that delivers 30 kW of power. The elevation difference between the upper and lower reservoirs is 15 m, and the head loss in the pipeline is estimated as 6 m. Find the flow rate in the pipeline, the rate at which energy is lost from the system, and the fraction of input power that is lost.

4.95. Water is pumped from an open lower reservoir to a pressurized upper reservoir as shown in Figure 4.101. Under design conditions, the water level in the upper reservoir is 30 m above the water level in the lower reservoir, and it is desired to have a minimum delivery rate of 20 L/s from the lower to the upper reservoir. If a 10-kW pump is used and friction losses in the pipeline are neglected, what is the maximum allowable air pressure in the upper reservoir?

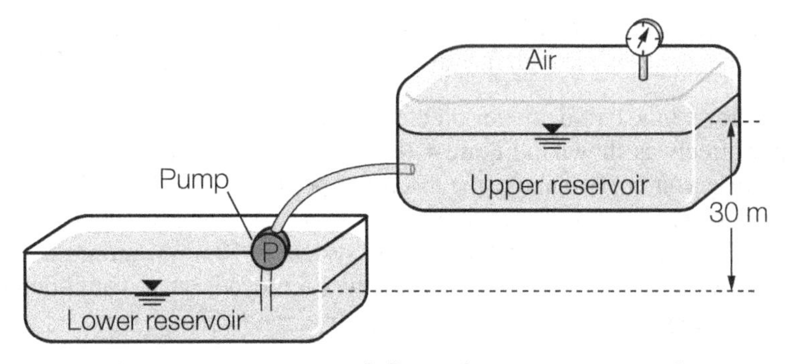

Figure 4.101: Water delivery between reservoirs

4.96. A pump delivers water from a source reservoir to a terminal reservoir through a 200-mm-diameter pipe as shown in Figure 4.102. The water surface elevations in the reservoirs differ by 2 m, and the delivery pipe discharges at an elevation 5 m below the water surface elevation of the source reservoir. Both the frictional head loss, h_ℓ, and the head added by the pump, h_p, are functions of the flow rate, Q, between reservoirs, and these functional relationships are given by

$$h_\ell = 3\frac{V^2}{2g}, \qquad h_p = 18(1 - 100Q^2)$$

where h_ℓ is in m, V is the velocity in the pipeline in m/s, h_p is in m, and Q is in m^3/s. (a) Estimate the flow rate between the reservoirs. (b) If the terminal reservoir was empty such that the pipeline discharged into the open atmosphere, what would be the flow rate under this condition?

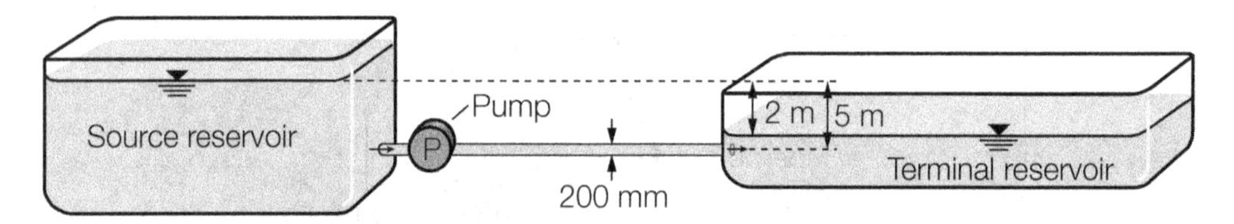

Figure 4.102: Pump delivering water from a source to a terminal reservoir

4.97. A pump delivers water from a 125-mm-diameter pipeline to a 75-mm pipeline. When the flow through the pump is 40 L/s, the inflow pressure is 150 kPa, the outflow pressure is 450 kPa, and the head loss within the pump is estimated as 18 m. Estimate the power required to drive the pump motor and the efficiency of the pump in converting electrical energy to mechanical energy of the flow. Assume that the pump is 90% efficient in transforming electrical energy to the energy of the rotating impeller.

4.98. The inflow and outflow pipes to a turbine have diameters of 400 mm and 500 mm, respectively, and are located at approximately the same elevation. Under a particular operating condition, the flow rate of water through the turbine is 1 m^3/s and the power extracted from the water by the turbine is 100 kW. Estimate the change in water pressure across the turbine. Assume water at 20°C.

4.99. A hydroelectric facility has an upstream reservoir elevation of 110 m and a downstream river elevation of 85 m. The river can be taken as having a velocity of 1 m/s. The head loss, h_ℓ, through the tunnel leading to and from the turbine is given by

$$h_\ell = 0.0826Q^2$$

where h_ℓ is in meters and Q is the flow in m^3/s. The efficiency of the turbine is 80%. Determine the flow through the turbine that is required to generate 1 MW of power.

4.100. Water flows in a 1-m-diameter pipeline from a reservoir through a turbine and discharges freely as shown in Figure 4.103. Under design conditions, the water surface in the reservoir is 30 m above the discharge point. The head loss through the system can be estimated using the relation $h_\ell = 0.8V^2/2g$, where V is the velocity in the pipeline. (a) If the flow rate through the system is 10 m^3/s, estimate the power extracted by the turbine. (b) If the turbine was removed, what would be the flow rate through the system?

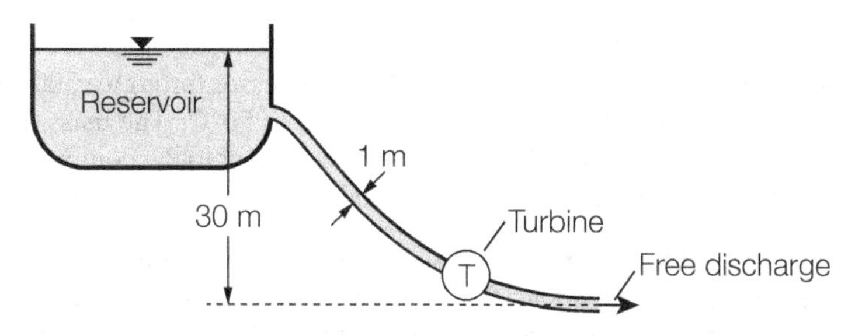

Figure 4.103: Flow through a turbine with free discharge

4.101. Water flows through a turbine as shown in Figure 4.104, where the pipeline upstream of the turbine has a diameter of 1.2 m and the downstream pipeline has a diameter of 1.7 m. The inflow section of the upstream pipeline is 4 m above the outflow section of the downstream pipeline. When the flow through the turbine is 6 m^3/s, the pressure at the inflow section is 450 kPa, the pressure at the outflow section is −30 kPa, and the head loss between these two sections is estimated as 20 m. What is the power being delivered to the turbine?

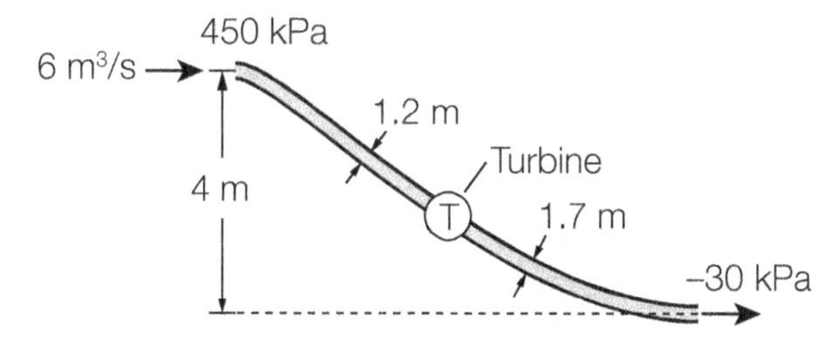

Figure 4.104: Flow through a turbine

4.102. Angel Falls in Venezuela is the second-highest waterfall in the world, with a vertical drop of approximately 979 m. If adiabatic conditions were to exist along the entire length of the waterfall, estimate the difference in temperature in an element of water in the upstream pool and this same element of water in the downstream pool after the kinetic energy acquired during the fall has been dissipated.

4.103. A room has dimensions 10 m × 10 m × 3 m, and air from the room is vented outside through a 100-mm-diameter vent. The head loss in this type of vent can be estimated as $h_\ell = 0.1\,V^2/2g$, where V is the flow velocity through the vent. If the pressure inside the room is 2 kPa higher than the outside pressure, estimate the volume flow rate of air through the vent and the time it takes to fully exchange the air in the room. Assume standard air.

4.104. Air flows from a large tank through a nozzle such that the change in enthalpy between the nozzle inflow and outflow sections is 200 kJ/kg. The flow through the nozzle is approximately adiabatic, and the inflow velocity is approximately equal to zero. Estimate the nozzle exit velocity.

4.105. Air flows through a turbomachine at a steady mass flow rate of 1 kg/s. The turbomachine is well insulated to minimize heat exchange with the surroundings. The inflow and outflow ducts are of the same size; the temperature and pressure on the inflow side are approximately equal to standard atmospheric (sea-level) conditions; and the temperature and pressure on the discharge side are 150°C and 600 kPa, respectively. The discharge velocity is 120 m/s. Estimate the power input to the turbomachine.

4.106. Air flows into a compressor through a large intake duct at a pressure of 101 kPa and a temperature of 25°C. Air flows out of the compressor through a 300-mm-diameter pipe at a pressure of 400 kPa and a temperature of 50°C. The mass flow rate of air through the compressor is 10 kg/s, and the power input to the compressor is 500 kW. Estimate the rate at which heat is transferred from the compressor.

4.107. A 0.8-m-diameter fan is driven by a 1-kW motor. The fan pulls in air from a large room, and air leaves the fan at 10 m/s. Estimate the efficiency of the fan in converting electrical energy to the energy added to the air by the fan (i.e., the efficiency of the fan-motor system). Assume standard air.

4.108. A 200-L insulated storage tank contains compressed air at a pressure of 600 kPa and a temperature of 18°C. A valve is opened to release air from the tank, and at the instant the valve is opened, the mass flow rate of air out of the tank is 0.02 kg/s. Estimate the rate of change of temperature in the tank at the instant the valve is opened. You may neglect the macroscopic kinetic energy of the outflow air.

Chapter 5

Differential Analysis

LEARNING OBJECTIVES

After reading this chapter and solving a representative sample of end-of-chapter problems, you will be able to:

- Describe the kinematics of fluid motion in terms of translation, rotation, and deformation.
- Understand and apply the differential forms of the continuity and momentum equations.
- Calculate stream and potential functions for incompressible irrotational flows.
- Apply potential-flow theory to the solution of practical problems.
- Understand turbulent flow and the analysis of turbulent flow.

5.1 Introduction

Fundamental equations that govern the flow of fluids are the conservation equations of mass, momentum, and energy. In the previous chapter, these equations were put in forms that are applicable to finite control volumes. The finite-control-volume forms of the governing equations are particularly useful when we are interested in the overall features of a flow, such as the mass flow rate into and out of a control volume or the net force exerted on a structure in contact with a flowing fluid. Although the control-volume forms of the governing equations are exact and useful in many cases, in some cases, we are interested in the detailed spatial variations in fluid flow that are not resolved by the spatially coarse finite-control-volume formulation. In those cases where finer detail is required, it is preferable to utilize the governing flow equations in forms of differential equations that are applicable at each point in the flow field. This approach is called the *differential analysis of fluid flow*, and the differential equations that govern fluid flow can be derived from the equations for finite control volumes by taking the limit as the size of the control volume becomes infinitesimally small. In mathematical terms, utilization of the control-volume forms of the governing equations constitute an *integral approach* to the analysis of fluid flow, whereas utilization of the differential equation forms of the governing equations constitute a *differential approach* to the analysis of fluid flow.

Applications of the material in this chapter. In many engineering applications, finite-control-volume analyses are sufficient to achieve the desired objectives and the application of differential analysis is not required. For such cases, the material contained in this chapter is not essential to the analysis and no loss of accuracy is associated with using control-volume formulations. In those cases where detailed spatial and temporal variations in the flow field are of interest, solution of the flow equations in differential form is required. The material contained in this chapter is particularly useful in describing those few flow fields that can be described exactly by analytic expressions. In many commonly encountered flow fields, viscous and compressibility effects are negligible. Under these circumstances, the fluid can be approximated as an *ideal fluid*, which is defined as a fluid that is inviscid and incompressible. Analytic expressions for several commonly encountered ideal-fluid flows are covered in this chapter.

5.2 Kinematics

The field of *kinematics* deals with the geometry of motion. Within a fluid continuum, infinitesimal fluid elements undergo motion that can be described in terms of the following four components: (1) translation, (2) rotation, (3) linear deformation, and (4) angular deformation. *Translation* describes the movement of the fluid element from one point to another, *rotation* describes the spin of the element about an axis, *linear deformation* describes the expansion or contraction of the element without changing shape, and *angular deformation* describes the change in the shape of the element (as measured by the change in the angle between adjacent sides of the element). Linear deformation and angular deformation can also be grouped together and collectively classified as *strain*, with linear deformation called *extensional strain* and angular deformation called *shear strain*. Complete characterization of the kinematics of fluid motion requires specification of the translation, rotation, linear deformation, and angular deformation of fluid elements undergoing motion. Quantification of these kinematic parameters are described in the following sections.

5.2.1 Translation

The translation of a fluid element within a fluid continuum is described by the velocity field, $\mathbf{V}$, which can be expressed as

$$\mathbf{V} = u\mathbf{i} + v\mathbf{j} + w\mathbf{k} \tag{5.1}$$

where u, v, and w are the components of the velocity vector, $\mathbf{V}$. The velocity within a fluid domain is usually expressed as a function of location and time (i.e., using an Eulerian reference frame), in which case the velocity can be expressed in functional form as $\mathbf{V}(\mathbf{r}, t)$, where t is time and $\mathbf{r}$ is the position vector given by

$$\mathbf{r} = x\mathbf{i} + y\mathbf{j} + z\mathbf{k} \tag{5.2}$$

where x, y, and z are the Cartesian coordinates at the location where the velocity is $\mathbf{V}$. The acceleration of a fluid element, $\mathbf{a}$, is derived from the velocity field using the relation

$$\mathbf{a} = \frac{\mathrm{D}\mathbf{V}}{\mathrm{D}t} = \frac{\partial \mathbf{V}}{\partial t} + u\frac{\partial \mathbf{V}}{\partial x} + v\frac{\partial \mathbf{V}}{\partial y} + w\frac{\partial \mathbf{V}}{\partial z} \tag{5.3}$$

where $\mathrm{D}\mathbf{V}/\mathrm{D}t$ is the *material derivative* of the velocity, which is the rate of change in $\mathbf{V}$ when viewed from a Lagrangian reference frame moving with the fluid element. The material derivative $\mathrm{D}\mathbf{V}/\mathrm{D}t$ is exactly the same as the total derivative $\mathrm{d}\mathbf{V}/\mathrm{d}t$, with the notation of the material derivative used to emphasize the Lagrangian nature of the motion. The terms in Equation 5.3 express the acceleration as the sum of the acceleration caused by variation in the velocity at a fixed location ($= \partial\mathbf{V}/\partial t$) and the acceleration caused by spatial variations in the velocity ($= u\partial\mathbf{V}/\partial x + v\partial\mathbf{V}/\partial y + w\partial\mathbf{V}/\partial z$). These components of the acceleration are called the *local acceleration* and the *convective acceleration*, respectively; hence,

$$\mathbf{a} = \underbrace{\frac{\partial\mathbf{V}}{\partial t}}_{\text{local acceleration}} + \underbrace{u\frac{\partial\mathbf{V}}{\partial x} + v\frac{\partial\mathbf{V}}{\partial y} + w\frac{\partial\mathbf{V}}{\partial z}}_{\text{convective acceleration}}$$

where it is apparent that the local acceleration is a result of temporal variations in velocity. Under steady-state conditions, the acceleration of a fluid element is associated entirely with spatial variations in the velocity. This further emphasizes that the acceleration of a fluid element is always described from a Lagrangian perspective.

EXAMPLE 5.1

Consider a velocity field that is described by the analytic expression

$$\mathbf{V} = (x^2 + z^2)t\mathbf{i} + (y^2 - z)t\mathbf{j} + (x - y)\mathbf{k}\ \text{cm/s}$$

where (x, y, z) are the Cartesian coordinates in meters and t is the time in seconds. Determine the analytic expression for the acceleration field and the velocity and acceleration at location (1 m, 4 m, 2 m) at t = 1 sec.

SOLUTION

The components of the velocity vector are given as

$$u = (x^2 + z^2)t, \quad v = (y^2 - z)t, \quad w = (x - y) \tag{5.4}$$

and the partial derivatives of the velocity vector are

$$\frac{\partial\mathbf{V}}{\partial t} = (x^2 + z^2)\mathbf{i} + (y^2 - z)\mathbf{j}, \quad \frac{\partial\mathbf{V}}{\partial x} = 2xt\mathbf{i} + \mathbf{j}, \quad \frac{\partial\mathbf{V}}{\partial y} = 2yt\mathbf{j} - \mathbf{k}, \quad \frac{\partial\mathbf{V}}{\partial z} = 2z\mathbf{i} - \mathbf{j} \tag{5.5}$$

Combining Equations 5.4 and 5.5 as defined by Equation 5.3 yields the acceleration field

$$\mathbf{a} = \left[(x^2 + z^2)\mathbf{i} + (y^2 - z)\mathbf{j}\right] + \left[(x^2 + z^2)t\right][2xt\mathbf{i} + \mathbf{j}] + \left[(y^2 - z)t\right][2yt\mathbf{j} - \mathbf{k}] + [(x - y)][2z\mathbf{i} - \mathbf{j}]$$

which can be expressed in component form as

$$\mathbf{a} = \left[(x^2 + z^2)\left(1 + 2xt^2\right) + 2(x - y)z\right]\mathbf{i} + \left[(y^2 - z)\left(1 + 2yt^2\right) + (x^2 + z^2)\,t - (x - y)\right]\mathbf{j} - \left[(y^2 - z)t\right]\mathbf{k}$$

Using the given and derived expressions, the velocity and acceleration at t = 1 and (x, y, z) = (1,4,2) are

$$\mathbf{V} = 5\mathbf{i} + 14\mathbf{j} - 3\mathbf{k}\ \text{cm/s}, \quad \text{and} \quad \mathbf{a} = 3\mathbf{i} + 134\mathbf{j} - 14\mathbf{k}\ \text{cm/s}^2$$

The units of **V** and **a** have been included in the final result to emphasize that the above approach is valid even though the given velocity equation is not dimensionally homogeneous.

5.2.2 Rotation

The rate at which a fluid element rotates within a fluid continuum can be expressed in terms of the velocity field. Consider the fluid element ABCD of dimensions $\Delta x \times \Delta y$ in the xy plane as shown in Figure 5.1. At an initial time, t_0, the diagonal of the fluid element (AC) makes an angle θ_0 with the x-axis. After a time interval Δt, the sides AB and AD rotate by θ_1 and θ_2, respectively, which changes the angle of the diagonal (AC) by $\Delta\theta$ where

$$\Delta\theta = \frac{1}{2}(\theta_1 - \theta_2) \tag{5.6}$$

Denoting the rate of angular rotation of the diagonal by ω_z [rad T^{-1}] and taking the derivative of Equation 5.6 with respect to time yields

$$\omega_z = \frac{\mathrm{d}}{\mathrm{d}t}(\Delta\theta) = \frac{1}{2}\left(\frac{\mathrm{d}\theta_1}{\mathrm{d}t} - \frac{\mathrm{d}\theta_2}{\mathrm{d}t}\right) \tag{5.7}$$

In terms of the velocity field, it can be readily shown that

$$\frac{\mathrm{d}\theta_1}{\mathrm{d}t} = \frac{\partial v}{\partial x} \quad \text{and} \quad \frac{\mathrm{d}\theta_2}{\mathrm{d}t} = \frac{\partial u}{\partial y} \tag{5.8}$$

where u and v are x and y components of the velocity, respectively. Combining Equations 5.7 and 5.8 yields the following expression for the rate of rotation of the fluid about the z-axis:

$$\omega_z = \frac{1}{2}\left(\frac{\partial v}{\partial x} - \frac{\partial u}{\partial y}\right) \tag{5.9}$$

Repeating the above analysis for fluid elements in the yz and xz planes yields the rates of angular rotation as

$$\omega_x = \frac{1}{2}\left(\frac{\partial w}{\partial y} - \frac{\partial v}{\partial z}\right) \quad \text{and} \quad \omega_y = \frac{1}{2}\left(\frac{\partial u}{\partial z} - \frac{\partial w}{\partial x}\right) \tag{5.10}$$

where ω_x and ω_y are the rates of rotation about the x- and y-axes, respectively. Combining Equations 5.9 and 5.10 gives the following expression for the *rate-of-rotation vector* or *angular-velocity vector*, $\boldsymbol{\omega}$, as

$$\boldsymbol{\omega} = \frac{1}{2}\left(\frac{\partial w}{\partial y} - \frac{\partial v}{\partial z}\right)\mathbf{i} + \frac{1}{2}\left(\frac{\partial u}{\partial z} - \frac{\partial w}{\partial x}\right)\mathbf{j} + \frac{1}{2}\left(\frac{\partial v}{\partial x} - \frac{\partial u}{\partial y}\right)\mathbf{k} \tag{5.11}$$

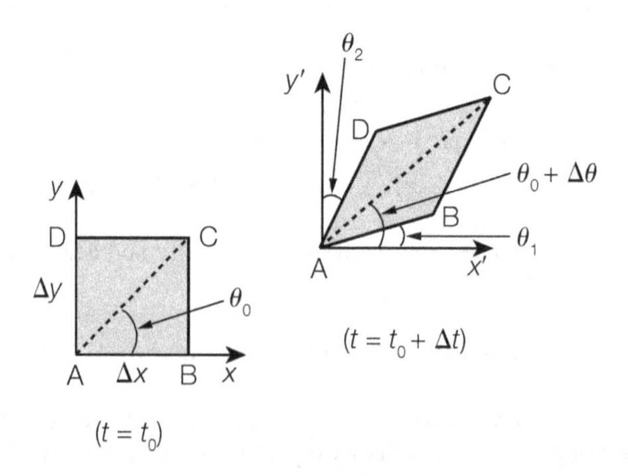

Figure 5.1: Rotation of a fluid element

Based on the aforementioned analysis, $\boldsymbol{\omega}$ can be taken as either the average rate of rotation of two perpendicular lines in a fluid element or the rate of rotation of the bisector of these lines. Equation 5.11 can be expressed in vector notation as

$$\boldsymbol{\omega} = \frac{1}{2}(\nabla \times \mathbf{V}) \tag{5.12}$$

where ∇ is the gradient operator given by

$$\nabla f = \frac{\partial f}{\partial x}\mathbf{i} + \frac{\partial f}{\partial y}\mathbf{j} + \frac{\partial f}{\partial z}\mathbf{k} \tag{5.13}$$

where $f(x, y, z, t)$ is any scalar function. Equation 5.12 can be conveniently expressed and evaluated using the following determinant relationship:

$$\boldsymbol{\omega} = \frac{1}{2}(\nabla \times \mathbf{V}) = \frac{1}{2}\begin{vmatrix} \mathbf{i} & \mathbf{j} & \mathbf{k} \\ \frac{\partial}{\partial x} & \frac{\partial}{\partial y} & \frac{\partial}{\partial z} \\ u & v & w \end{vmatrix} \tag{5.14}$$

In lieu of measuring the rotational characteristics of a fluid by ω, it is slightly more convenient to measure rotationality using the *vorticity*, $\boldsymbol{\zeta}$, which is defined as

$$\boldsymbol{\zeta} = \nabla \times \mathbf{V} = 2\boldsymbol{\omega} \tag{5.15}$$

Using the vorticity, $\boldsymbol{\zeta}$, instead of the angular velocity, $\boldsymbol{\omega}$, removes the need to carry the additional factor of $\frac{1}{2}$. If the vorticity at a point in a flow field is nonzero, the fluid element that occupies that point in space is rotating, and the flow in that region is characterized as *rotational flow*. Conversely, if the vorticity in a region of flow is zero, the flow in that region is characterized as *irrotational flow*.

EXAMPLE 5.2

Consider the following steady-state three-dimensional velocity field:

$$\mathbf{V} = (5xyz + 2x^2)\mathbf{i} + (xz^2)\mathbf{j} + (2z^2)\mathbf{k}$$

Determine the vorticity field and assess whether the flow is irrotational at any point.

SOLUTION

For the given velocity field: $u = 5xyz + 2x^2$, $v = xz^2$ and $w = 2z^2$. In accordance with Equations 5.15 and 5.11, the vorticity vector, $\boldsymbol{\zeta}$, is given by

$$\boldsymbol{\zeta} = \left(\frac{\partial w}{\partial y} - \frac{\partial v}{\partial z}\right)\mathbf{i} + \left(\frac{\partial u}{\partial z} - \frac{\partial w}{\partial x}\right)\mathbf{j} + \left(\frac{\partial v}{\partial x} - \frac{\partial u}{\partial y}\right)\mathbf{k}$$

$$= (0 - 2xz)\mathbf{i} + (5xy - 0)\mathbf{j} + (z^2 - 5xz)\mathbf{k}$$

$$\rightarrow \quad \boldsymbol{\zeta} = -2xz\mathbf{i} + 5xy\mathbf{j} + (z^2 - 5xk)\mathbf{k}$$

The flow is rotational when $\boldsymbol{\zeta} \neq 0$. It is apparent that the vorticity varies spatially, and **the flow is irrotational at x = (0,0,0).**

Fluid elements originating in irrotational flow can only acquire rotation if they experience surface shear stresses, because body forces and (normal) pressure forces cannot cause rotation. As a consequence, rotational flow always occurs in viscous flows. Rotational flows generally occur in wakes, boundary layers, and flows with heat transfer.

Vorticity vector in cylindrical polar coordinates. In cases where there is radial or cylindrical polar symmetry, it is usually advantageous to work in cylindrical polar coordinates, because the dimensionality of the problem can be reduced in such a coordinate system. The expression for the vorticity, ζ, in cylindrical polar coordinates is given by

$$\zeta = \left(\frac{1}{r}\frac{\partial v_r}{\partial \theta} - \frac{\partial v_\theta}{\partial z}\right)\mathbf{e}_r + \left(\frac{\partial v_r}{\partial z} - \frac{\partial v_z}{\partial r}\right)\mathbf{e}_\theta + \frac{1}{r}\left(\frac{\partial (r v_\theta)}{\partial r} - \frac{\partial v_r}{\partial \theta}\right)\mathbf{e}_z \tag{5.16}$$

where v_r, v_θ, and v_z are the r, θ, and z components, respectively, of the velocity vector and $\mathbf{e}_r$, $\mathbf{e}_\theta$, and $\mathbf{e}_z$ are the unit vectors in the r, θ, and z directions, respectively.

EXAMPLE 5.3

A liquid is contained in the rotating cylinder as shown in Figure 5.2. Under steady-state conditions, the liquid undergoes rigid-body motion, and the velocity distribution, $\mathbf{V}$, is given by

$$\mathbf{V} = r\omega\,\mathbf{e}_\theta$$

where r is the radial distance from the axis of rotation and ω is the rate of rotation of the cylinder. If the cylinder is rotating at a rate of 120 rpm, determine the vorticity of the liquid in the cylinder.

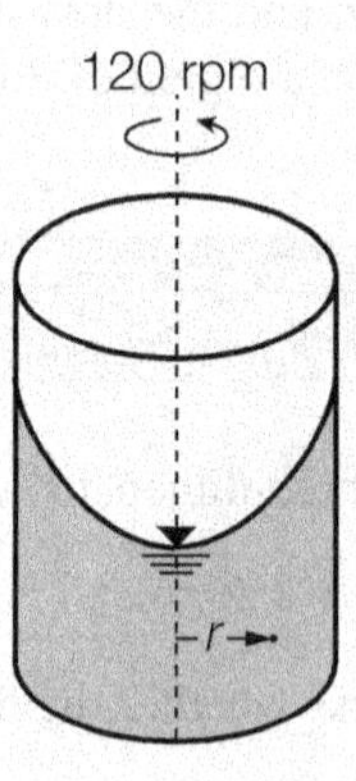

Figure 5.2: Fluid in a rotating cylinder

SOLUTION

From the given data: $\omega = 120$ rpm $= 12.57$ rad/s, $v_\theta = r\omega = 12.57r$, $v_r = 0$, and $v_z = 0$. Because motion is in the $r\theta$ plane, the only nonzero component of the vorticity is in the z-direction. The z component of the vorticity is given by Equation 5.16 as

$$\zeta_z = \frac{1}{r}\left\{\frac{\partial (r v_\theta)}{\partial r} - \frac{\partial v_r}{\partial \theta}\right\} = \frac{1}{r}\left\{\frac{\partial}{\partial r}[r(12.57r)] - 0\right\} = \frac{1}{r}(25.13r) = 25.13 \text{ rad/s}$$

Therefore, the vorticity of the liquid in the cylinder is equal to **25.13 rad/s**, which is twice the rate of rotation of the cylinder.

5.2.3 Angular Deformation

Angular deformation refers to the change in the shape of a fluid element. The rate of angular deformation of a fluid element is measured by the rate of change in the angle between adjacent sides of the element that are originally perpendicular to each other. In a positive angular deformation, the angle between the sides of a fluid element is decreasing. From Figure 5.1, it is apparent that the rate of change of the (90°) angle between the adjacent sides AB and AD in the xy plane is given by $\mathrm{d}\theta_1/\mathrm{d}t + \mathrm{d}\theta_2/\mathrm{d}t$ which, according to Equation 5.8, can be expressed in terms of the velocity field as

$$\Omega_{xy} = \frac{\mathrm{d}\theta_1}{\mathrm{d}t} + \frac{\mathrm{d}\theta_2}{\mathrm{d}t} = \frac{\partial v}{\partial x} + \frac{\partial u}{\partial y} \tag{5.17}$$

where Ω_{xy} denotes the angular deformation rate in the xy plane. Similarly, the angular deformation rates in the other planes are given by

$$\Omega_{xz} = \frac{\partial w}{\partial x} + \frac{\partial u}{\partial z}, \quad \text{and} \quad \Omega_{yz} = \frac{\partial w}{\partial y} + \frac{\partial v}{\partial z} \tag{5.18}$$

Angular deformation rates are typically caused by shear stresses in viscous fluids. In fact, by definition, Newtonian fluids obey the following relationship in the xy plane:

$$\tau_{xy} = \mu \left(\frac{\partial v}{\partial x} + \frac{\partial u}{\partial y} \right)$$

where τ_{xy} is the shear stress on the fluid element and μ is viscosity of the fluid. Hence, in a Newtonian fluid, the shear stress on a fluid element is linearly proportional to the rate of angular deformation.

EXAMPLE 5.4

Consider the two-dimensional velocity field given by

$$\mathbf{V} = 6xy\mathbf{i} - 4(x^2 - y^2)\mathbf{j}$$

Determine the rate of angular deformation of a fluid element located at $(x, y) = (1,2)$ and identify any locations in the xy plane where the rate of angular deformation is equal to zero.

SOLUTION

For the given velocity field: $u = 6xy$ and $v = -4(x^2 - y^2)$. The angular deformation in the xy plane is given by

$$\Omega_{xy} = \frac{\partial v}{\partial x} + \frac{\partial u}{\partial y} = \frac{\partial}{\partial x}[-4(x^2 - y^2)] + \frac{\partial}{\partial y}[6xy] = (-8x) + (6x) = -2x$$

Therefore, at $(x, y) = (1,2)$, $\Omega_{xy} = -2(1) =$ **-2 rad/s**. Because Ω_{xy} does not depend on y and $\Omega_{xy} = 0$ at $x = 0$, **$\boldsymbol{\Omega_{xy} = 0}$ at any coordinate location $\boldsymbol{(0, y)}$.**

5.2.4 Linear Deformation

Linear deformation of a fluid element is commonly called *dilatation*, which is defined as the process by which a fluid element expands as it moves. Consider the fluid element shown in Figure 5.3, where the velocity at Point P is (u,v,w) and the velocities in the x-, y-, and z-directions at distances Δx, Δy, and Δz from P, respectively, are $u+\partial u/\partial x\,\Delta x$, $v+\partial v/\partial y\,\Delta y$, and $w+\partial w/\partial z\,\Delta z$, respectively. Therefore, if a fluid volume initially has a volume, V, equal to $\Delta x\Delta y\Delta z$ at time t_0, then

$$V(t_0) = \Delta x\Delta y\Delta z \tag{5.19}$$

$$V(t_0+\Delta t) = \Delta x\Delta y\Delta z + \left(\frac{\partial u}{\partial x}\Delta x\cdot\Delta y\Delta z + \frac{\partial v}{\partial y}\Delta y\cdot\Delta x\Delta z + \frac{\partial w}{\partial z}\Delta z\cdot\Delta x\Delta y\right)\Delta t \tag{5.20}$$

Defining the *volumetric dilatation rate* as the rate of change of the fluid volume per unit volume, in accordance with Equations 5.19 and 5.20,

$$\text{volumetric dilatation rate} = \lim_{\Delta t\to 0}\left[\frac{V(t_0+\Delta t)-V(t_0)}{\Delta t}\cdot\frac{1}{V(t_0)}\right] = \frac{\partial u}{\partial x}+\frac{\partial v}{\partial y}+\frac{\partial w}{\partial z} \tag{5.21}$$

which can be expressed in vector notation as

$$\text{volumetric dilatation rate} = \nabla\cdot\mathbf{V} \tag{5.22}$$

The volumetric dilatation rate as defined by Equation 5.22 is also referred to as the *volumetric strain rate*, *bulk strain rate*, *rate of volumetric dilatation*, or simply the *dilatation rate*. Because the fluid element depicted in Figure 5.3 contains a fixed mass of fluid, if the fluid is incompressible (i.e., ρ = constant), the volumetric dilatation rate must necessarily be equal to zero.

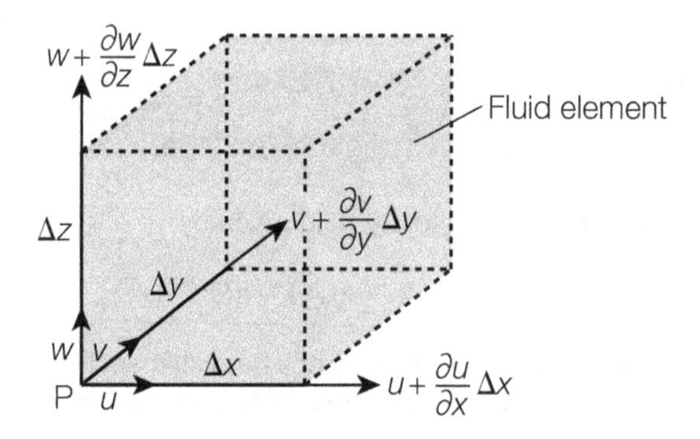

Figure 5.3: Dilatation of a fluid element

EXAMPLE 5.5

Consider the velocity field given by

$$\mathbf{V} = 6xyz\mathbf{i} + 2(x^2+y^2)\mathbf{j} + 3yz\mathbf{k}\ \text{m/s}$$

where x, y, and z are Cartesian-coordinate locations in meters. Find the volumetric dilatation rate of the fluid at (x, y, z) = (1 m, 1 m, 3 m) and assess whether the fluid is being compressed or expanded at this location.

SOLUTION

For the given velocity field: $u = 6xyz$, $v = 2(x^2+y^2)$, and $w = 3yz$. The dilatation rate $(\nabla\cdot\mathbf{V})$ is given by Equation 5.21 as

$$\text{dilatation rate} = \frac{\partial u}{\partial x} + \frac{\partial v}{\partial y} + \frac{\partial w}{\partial z} = \frac{\partial}{\partial x}[6xyz] + \frac{\partial}{\partial y}[2(x^2+y^2)] + \frac{\partial}{\partial z}[3yz]$$
$$= 6yz + 4y + 3y = 6yz + 7y\ \text{s}^{-1}$$

Hence, at (x, y, z) = (1 m, 1 m, 3 m), the dilatation rate is $6(1)(3)+7(1) = \mathbf{25\ s^{-1}}$. Because the dilatation rate is greater than zero, the flow is compressible and **a fluid element is expanding when located at (1 m, 1 m, 3 m).**

5.3 Conservation of Mass

The law of conservation of mass states that the total mass of a given substance is conserved. In a Lagrangian reference frame this can be stated as follows: The mass of fluid in a defined system is constant. In an Eulerian reference frame, the law of conservation of mass can be stated as follows: The net rate at which fluid mass is entering a control volume is equal to the rate at which mass is accumulating within the control volume. Differential analysis of fluid flow typically uses an Eulerian reference frame, and the general equation for the law of conservation of mass is based on a control volume analysis.

5.3.1 Continuity Equation

Consider the control volume of dimensions $\mathrm{d}x \times \mathrm{d}y \times \mathrm{d}z$ as shown in Figure 5.4. The control volume is located within a fluid continuum, where P is the point in the center of the control volume, ρ is the fluid density at P, and u, v, and w are the velocity components in the x-, y-, and z-directions, respectively, at P. In formulating differential equations using infinitesimal control volumes, it is usually necessary to express the value of a variable on the surface of the control volume in terms of the value of the variable at the center of the control volume. For this purpose, it is common to use the Taylor expansion,[28] which can be expressed as

$$f(x_i + \Delta x_i) = f(x_i) + \left.\frac{\partial f}{\partial x_i}\right|_{x_i} \Delta x_i + \frac{1}{2!}\left.\frac{\partial^2 f}{\partial x_i^2}\right|_{x_i} (\Delta x_i)^2 + \ldots \tag{5.23}$$

where f is the variable of interest and x_i is the coordinate location. Applying the first terms of the Taylor expansion, the mass flow rate through each face of the control volume is shown in Figure 5.5. Using the mass flow rates shown in Figure 5.5, the net mass inflow rate, Δm_{in},

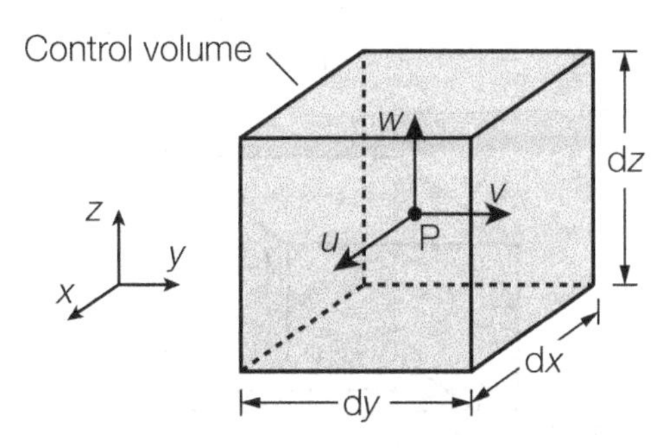

Figure 5.4: Control volume in fluid flow

[28]Named in honor of the English mathematician Brook Taylor (1685–1731).

and net mass outflow rate, Δm_{out}, from the control volume of dimensions $\mathrm{d}x \times \mathrm{d}y \times \mathrm{d}z$ are given by

$$\Delta m_{\text{in}} = \left[\rho u - \frac{\partial(\rho u)}{\partial x}\frac{\mathrm{d}x}{2}\right]\mathrm{d}y\,\mathrm{d}z + \left[\rho v - \frac{\partial(\rho v)}{\partial y}\frac{\mathrm{d}y}{2}\right]\mathrm{d}x\,\mathrm{d}z + \left[\rho w - \frac{\partial(\rho w)}{\partial z}\frac{\mathrm{d}z}{2}\right]\mathrm{d}x\,\mathrm{d}y \tag{5.24}$$

$$\Delta m_{\text{out}} = \left[\rho u + \frac{\partial(\rho u)}{\partial x}\frac{\mathrm{d}x}{2}\right]\mathrm{d}y\,\mathrm{d}z + \left[\rho v + \frac{\partial(\rho v)}{\partial y}\frac{\mathrm{d}y}{2}\right]\mathrm{d}x\,\mathrm{d}z + \left[\rho w + \frac{\partial(\rho w)}{\partial z}\frac{\mathrm{d}z}{2}\right]\mathrm{d}x\,\mathrm{d}y \tag{5.25}$$

The law of conservation of mass can be expressed analytically as

$$\Delta m_{\text{in}} - \Delta m_{\text{out}} = \frac{\partial \rho}{\partial t}\,\mathrm{d}x\,\mathrm{d}y\,\mathrm{d}z \tag{5.26}$$

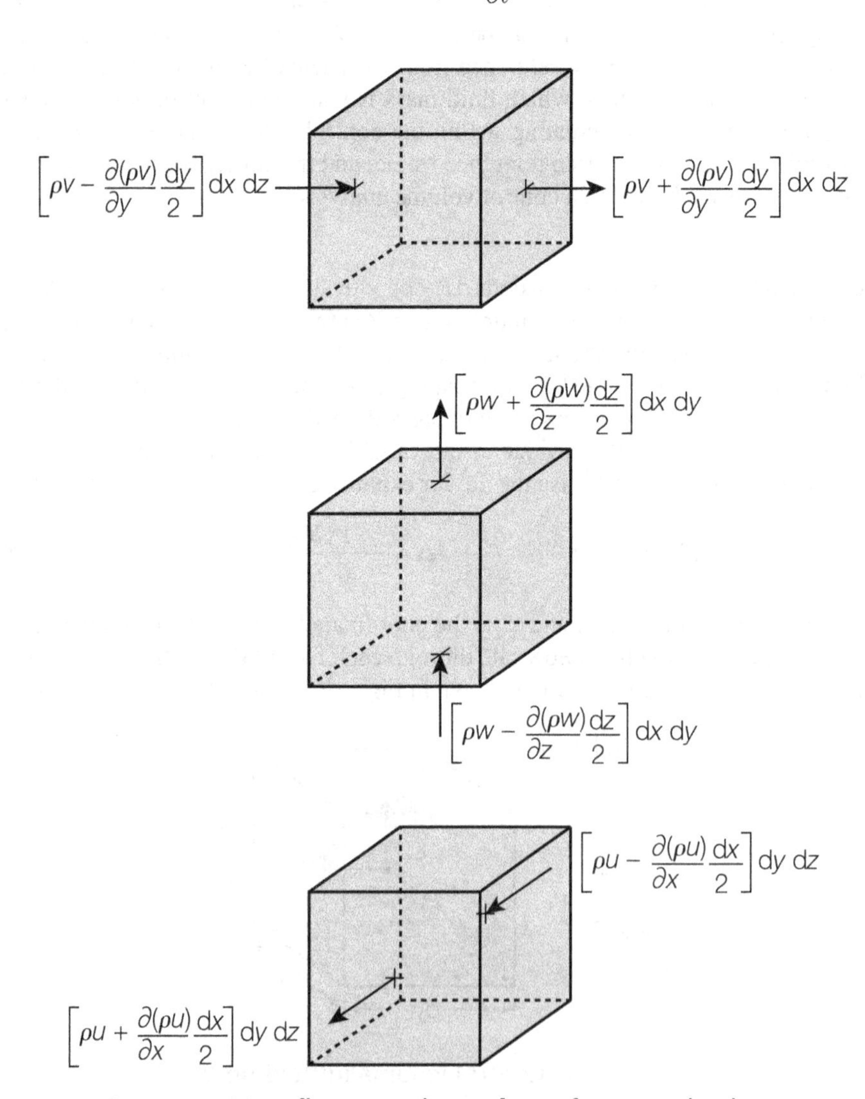

Figure 5.5: Mass fluxes on the surface of a control volume

which states that the mass inflow rate into the control volume minus the mass outflow rate from the control volume is equal to the rate at which mass is accumulating within the control volume. Substituting Equations 5.24 and 5.25 into Equation 5.26 gives

$$\left[-\frac{\partial(\rho u)}{\partial x} - \frac{\partial(\rho v)}{\partial y} - \frac{\partial(\rho w)}{\partial z}\right] \mathrm{d}x\,\mathrm{d}y\,\mathrm{d}z = \frac{\partial \rho}{\partial t}\,\mathrm{d}x\,\mathrm{d}y\,\mathrm{d}z \tag{5.27}$$

which simplifies to the following differential form of the law of conservation of mass:

$$\frac{\partial \rho}{\partial t} + \frac{\partial(\rho u)}{\partial x} + \frac{\partial(\rho v)}{\partial y} + \frac{\partial(\rho w)}{\partial z} = 0 \tag{5.28}$$

This equation is commonly referred to as the *continuity equation*, which can be expressed in vector notation as

$$\frac{\partial \rho}{\partial t} + \nabla \cdot (\rho \mathbf{V}) = 0 \tag{5.29}$$

Interestingly, the continuity equation, Equation 5.29, can be expressed in terms of the material derivative of ρ, because

$$\frac{\partial \rho}{\partial t} + \nabla \cdot (\rho \mathbf{V}) = \underbrace{\frac{\partial \rho}{\partial t} + \mathbf{V} \cdot \nabla \rho}_{=\mathrm{D}\rho/\mathrm{D}t} + \rho \nabla \cdot \mathbf{V} = 0$$

which gives the following form of the continuity equation:

$$\frac{1}{\rho}\frac{\mathrm{D}\rho}{\mathrm{D}t} + \nabla \cdot \mathbf{V} = 0 \tag{5.30}$$

Based on Equation 5.30, it is apparent that the rate of change of fluid density within any fluid system ($\mathrm{D}\rho/\mathrm{D}t$) is equal to $-\nabla \cdot \mathbf{V}$. If a fluid is incompressible, then

$$\frac{\mathrm{D}\rho}{\mathrm{D}t} = \frac{\partial \rho}{\partial t} + u\frac{\partial \rho}{\partial x} + v\frac{\partial \rho}{\partial y} + w\frac{\partial \rho}{\partial z} = 0 \tag{5.31}$$

By combining Equations 5.31 and 5.30, it is apparent that an incompressible fluid requires that

$$\nabla \cdot \mathbf{V} = 0 \quad \text{(incompressible fluid)} \tag{5.32}$$

Comparing Equations 5.30 and 5.32, it is apparent that a fluid flow can be approximated as incompressible when $\rho \nabla \cdot \mathbf{V} \gg \mathrm{D}\rho/\mathrm{D}t$.

Applications of the continuity equation. Incompressibility of a fluid does not preclude spatial variations in density, such as those that occur when the density of a liquid varies spatially due to spatial variations in temperature or when the density varies spatially due to spatial variations in the concentration of a dissolved substance. In such cases, $\mathrm{D}\rho/\mathrm{D}t = 0$, but each of the terms in $\mathrm{D}\rho/\mathrm{D}t$, as given by Equation 5.31, need not necessarily be zero. Physically, this means that in an incompressible fluid, the density of a fluid element does not change as it moves, so $\mathrm{D}\rho/\mathrm{D}t = 0$. However, the local component terms, such as $\partial\rho/\partial t$ and $u\partial\rho/\partial x$, could be nonzero due to temporal and spatial variations in the density. In contrast to variable-density fluids, in constant-density fluids, all of the local component terms of $\mathrm{D}\rho/\mathrm{D}t$ must be zero. Compressible fluids behave like incompressible fluids when pressure variations within the flow field are insufficient to cause significant variations in density. It was demonstrated previously (see Section 3.4.4) that a fluid can be approximated as being incompressible when the Mach number, Ma, is less than 0.3, which can be stated analytically as

$$\mathrm{Ma} = \frac{V}{c} < 0.3 \quad \text{(for approximate incompressibility)} \tag{5.33}$$

where V is the magnitude of the velocity and c is the sonic velocity in the fluid. Commonly, compressibility effects are associated with gases, and the sonic velocity in an ideal gas is given by

$$c = \sqrt{RTk} \tag{5.34}$$

where R is the gas constant, T is the absolute temperature, and k is the specific heat ratio. For standard air, Equation 5.34 gives $c = 340$ m/s and Ma = 0.3 corresponds to $V = 104$ m/s. Airflow through ducts in building air conditioning systems is an example of an engineering application involving the incompressible flow of a compressible fluid (i.e., air). In such applications, the Mach number is generally much less than 0.3, and the flow can generally be assumed to be incompressible.

EXAMPLE 5.6

A duct network is being designed to transport carbon dioxide at a temperature of 20°C. Estimate the maximum allowable gas-flow speed in the duct network for which the flow can be assumed incompressible.

SOLUTION

From the given data: $T = 20°\text{C} = 293$ K. For carbon dioxide (CO_2), $c_p = 858$ J/kg·K, $c_v = 670$ J/kg·K, and $R = 189$ J/kg·K (from Appendix B.5). From these data, $k = c_p/c_v = 1.28$. Substituting in Equation 5.34 gives

$$c = \sqrt{RTk} = \sqrt{(189)(293)(1.28)} = 266 \text{ m/s}$$

In accordance with Equation 5.33, the flow can be assumed incompressible when Ma < 0.3, which requires that

$$V < 0.3c = 0.3(266) \quad \rightarrow \quad V < 79.9 \text{ m/s}$$

Therefore, as long as the flow speed of CO_2 in the duct network is less than **79.9 m/s**, the flow can be assumed incompressible.

Alternative derivation of the continuity equation. The continuity equation given by Equation 5.29 was derived by applying the law of conservation of mass to the fixed infinitesimal control volume shown in Figure 5.4. To gain a different perspective, the continuity equation can alternatively be derived by applying the finite-control-volume continuity equation given by

$$\frac{\mathrm{d}}{\mathrm{d}t}\int_{\mathrm{cv}} \rho \, \mathrm{d}V + \int_{\mathrm{cs}} \rho(\mathbf{V}\cdot\mathbf{n})\,\mathrm{d}A = 0 \tag{5.35}$$

where "cv" and "cs" denote the volume and surface area, respectively, of any control volume within the fluid and $\mathbf{n}$ is a unit normal vector pointing outward from the control volume.

Because the control volume is fixed, the time derivative in Equation 5.35 can be put inside the integral such that

$$\int_{\text{cv}} \frac{\partial \rho}{\partial t}\, \mathrm{d}V + \int_{\text{cs}} \rho(\mathbf{V} \cdot \mathbf{n})\, \mathrm{d}A = 0 \tag{5.36}$$

where the partial derivative is used inside the integral (compared with the total derivative used outside the integral), because the density of the fluid can be spatially variable. The vector identity known as the *Gauss divergence theorem* (also called the *divergence theorem* or *Gauss's theorem*) states that for any vector function **F**, there exists the following relationship between a surface integral and a volume integral:

$$\int_A \mathbf{F} \cdot \mathbf{n}\, \mathrm{d}A = \int_V \boldsymbol{\nabla} \cdot \mathbf{F}\, \mathrm{d}V \tag{5.37}$$

where A is the surface area of the volume V and $\mathbf{n}$ is the unit vector pointing outward. Taking $\mathbf{F} = \rho\mathbf{V}$, Equation 5.37 can be expressed as

$$\int_{\text{cs}} \rho\mathbf{V} \cdot \mathbf{n}\, \mathrm{d}A = \int_{\text{cv}} \boldsymbol{\nabla} \cdot (\rho\mathbf{V})\, \mathrm{d}V \tag{5.38}$$

Combining Equations 5.36 and 5.38 and putting both integrals under one volume integration gives

$$\int_{\text{cv}} \left[\frac{\partial \rho}{\partial t} + \boldsymbol{\nabla} \cdot (\rho\mathbf{V})\right] \mathrm{d}V = 0 \tag{5.39}$$

Because this relationship is valid regardless of the location and extent of the control volume, "cv," the integrand must necessarily be equal to zero everywhere within the fluid, which means that

$$\frac{\partial \rho}{\partial t} + \boldsymbol{\nabla} \cdot (\rho\mathbf{V}) = 0 \tag{5.40}$$

This continuity equation for a fluid continuum is identical to Equation 5.29 that was derived previously using an alternative procedure based on the specification of an infinitesimal control volume.

Special flow conditions. Special flow conditions that are commonly encountered in engineering applications are steady flow and incompressible flow. In the case of steady flow, all time derivatives are equal to zero, $\partial\rho/\partial t = 0$, and the continuity equation (in Cartesian coordinates) simplifies to

$$\frac{\partial(\rho u)}{\partial x} + \frac{\partial(\rho v)}{\partial y} + \frac{\partial(\rho w)}{\partial z} = 0 \quad \text{(steady flow)} \tag{5.41}$$

For incompressible flow, ρ = constant, and the continuity equation simplifies to

$$\frac{\partial u}{\partial x} + \frac{\partial v}{\partial y} + \frac{\partial w}{\partial z} = 0 \quad \text{(incompressible flow)} \tag{5.42}$$

which is the most common form of the continuity equation encountered in practice. Equation 5.42 can also be written in vector form:

$$\boldsymbol{\nabla} \cdot \mathbf{V} = 0 \quad \text{(incompressible flow)} \tag{5.43}$$

Comparing Equations 5.43 and 5.22, it is apparent that for incompressible flows, the volumetric dilatation rate is equal to zero. This is consistent with our physical interpretation of dilatation, because an incompressible fluid cannot deform in response to pressure (which is defined as a normal stress).

EXAMPLE 5.7

Velocity variations within an incompressible fluid in the xy plane are described by

$$u = x^2 - 2y, \quad v = 2xy$$

The presence of a solid boundary on the $z = 0$ plane requires that $w = 0$ at all locations on the surface of this plane. Find an expression for the vertical component of the velocity within the flow field.

SOLUTION

From the given velocity field,

$$\frac{\partial u}{\partial x} = \frac{\partial}{\partial x}(x^2 - 2y) = 2x, \quad \text{and} \quad \frac{\partial v}{\partial y} = \frac{\partial}{\partial y}(2xy) = 2x$$

To satisfy the law of conservation of mass for an incompressible fluid,

$$\frac{\partial u}{\partial x} + \frac{\partial v}{\partial y} + \frac{\partial w}{\partial z} = 0 \quad \rightarrow \quad \frac{\partial w}{\partial z} = -\frac{\partial u}{\partial x} - \frac{\partial v}{\partial y} = -2x - 2x = -4x$$

Integrating this result and applying the boundary condition that $w = 0$ when $z = 0$ yields

$$\frac{\partial w}{\partial z} = -4x \quad \rightarrow \quad w = -4xz + f(x, y) \quad \rightarrow$$
$$0 = -4x(0) + f(x, y) \quad \rightarrow \quad f(x, y) = 0 \quad \rightarrow \quad w = -4xz$$

Therefore, the vertical velocity component in the flow field is described by $\boldsymbol{w = -4xz}$.

Continuity equation in cylindrical polar coordinates. The continuity equation, as given in the scalar form of Equation 5.28, can also be expressed in cylindrical polar coordinates (r, θ, z) as

$$\frac{\partial \rho}{\partial t} + \frac{1}{r}\frac{\partial(r\rho v_r)}{\partial r} + \frac{1}{r}\frac{\partial(\rho v_\theta)}{\partial \theta} + \frac{\partial(\rho v_z)}{\partial z} = 0 \tag{5.44}$$

where v_r, v_θ, and v_z are the r, θ, and z components of the velocity, respectively. When cylindrical polar coordinates are used, the velocity vector is given by

$$\mathbf{V} = v_r\mathbf{e}_r + v_\theta\mathbf{e}_\theta + v_z\mathbf{e}_z \tag{5.45}$$

where $\mathbf{e}_r$, $\mathbf{e}_\theta$, and $\mathbf{e}_z$ are unit vectors in the r-, θ-, and z- directions, respectively. You might recall that the relationship between cylindrical and Cartesian coordinates is given by

$$r = \sqrt{x^2 + y^2}, \quad \theta = \tan^{-1}\left(\frac{y}{x}\right), \quad z = z$$

In cases where the flow is steady, Equation 5.44 can be simplified by setting $\partial\rho/\partial t = 0$. In cases where the fluid is incompressible, $\rho =$ constant, in which case the derivatives of ρ in time and space are all equal to zero; hence, ρ disappears from the continuity equation, which becomes

$$\frac{1}{r}\frac{\partial(rv_r)}{\partial r} + \frac{1}{r}\frac{\partial v_\theta}{\partial \theta} + \frac{\partial v_z}{\partial z} = 0 \quad \text{(incompressible flow)} \tag{5.46}$$

Utilization of cylindrical polar coordinates is particularly useful in cases where the flow is symmetrical relative to one of the coordinate axes (i.e., in cases of *axisymmetric flow*); in such cases, the dimensionality of the continuity equation can be reduced by using cylindrical polar coordinates. For example, in cases where the flow is radially symmetric and thus the velocity does not depend on θ, the velocity expressed in cylindrical polar coordinates is dependent on three variables (r, z, and t) rather than four variables (x, y, z, and t) that would be necessary to describe the flow in Cartesian coordinates. Practical examples of axisymmetric flows include flows around spheres and torpedoes.

EXAMPLE 5.8

An incompressible liquid is injected steadily at a volume flow rate Q [$\mathrm{L^3T^{-1}}$] uniformly over the depth H between parallel plates, and both an elevation and a plan view of this injection process is illustrated in Figure 5.6. (a) Determine the velocity field between the plates in cylindrical polar coordinates. (b) Show that this velocity field satisfies the continuity equation expressed in cylindrical polar coordinates.

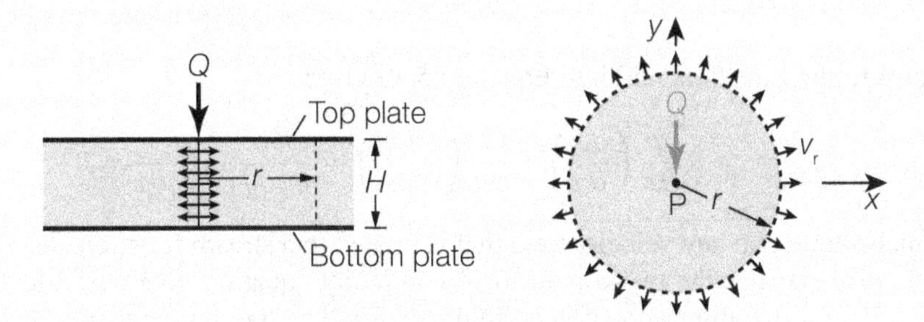

Figure 5.6: Injection of a liquid in the xy plane

SOLUTION

(a) Because the flow is at steady state, the volume flow rate across the surface of any imaginary cylinder centered at the injection point must be equal to Q. Therefore, for any imaginary cylinder of radius r and height H,

$$Q = 2\pi r H v_r \quad \rightarrow \quad v_r = \frac{Q}{2\pi r H} \tag{5.47}$$

where v_r is the radial component of the velocity at a distance r from the injection point. Because the induced flow is in the radial direction, the θ and z components of the velocity are equal to zero; hence, $v_\theta = 0$ and $v_z = 0$. The velocity field in cylindrical polar coordinates between the parallel plates is therefore given by

$$\mathbf{v} = \frac{\mathbf{Q}}{\mathbf{2\pi r H}} \mathbf{e}_r$$

(b) For an incompressible flow in which $v_\theta = 0$ and $v_z = 0$, the continuity equation given by Equation 5.46 becomes

$$\frac{1}{r}\frac{\partial (r v_r)}{\partial r} = 0$$

Substituting the derived expression for v_r given by Equation 5.47 into the continuity equation yields

$$\frac{1}{r}\underbrace{\frac{\partial}{\partial r}\left[\frac{Q}{2\pi H}\right]}_{=0} = 0 \tag{5.48}$$

Note that the quantity $Q/(2\pi H)$ is a constant, so its derivative with respect to r is equal to zero. It is apparent from Equation 5.48 that the derived velocity field satisfies the continuity equation expressed in cylindrical polar coordinates.

5.3.2 The Stream Function

Consider the case of an incompressible two-dimensional flow, where the continuity equation is given by

$$\frac{\partial u}{\partial x} + \frac{\partial v}{\partial y} = 0 \tag{5.49}$$

The *stream function*, ψ, is defined as a function that is related to the velocity field by

$$u = \frac{\partial \psi}{\partial y}, \quad v = -\frac{\partial \psi}{\partial x} \tag{5.50}$$

Substituting Equation 5.50 into Equation 5.49 gives

$$\frac{\partial}{\partial x}\left(\frac{\partial \psi}{\partial y}\right) + \frac{\partial}{\partial y}\left(-\frac{\partial \psi}{\partial x}\right) = \frac{\partial^2 \psi}{\partial x \partial y} - \frac{\partial^2 \psi}{\partial y \partial x} = 0 \tag{5.51}$$

which shows that any velocity field that is related to a stream function, $\psi(x, y)$, by Equation 5.50 also satisfies the two-dimensional continuity equation given by Equation 5.49. Conversely, if a two-dimensional flow field satisfies the continuity equation given by Equation 5.49, then a stream function can be found to describe the flow field. A key benefit to these results is that the flow field can be represented by a single function, $\psi(x, y)$, rather than two functions, $u(x, y)$ and $v(x, y)$.

Stream-function contour lines. To give physical meaning to stream-function contours (i.e., lines on which ψ = constant) in two-dimensional space, consider a streamline in the xy plane as shown in Figure 5.7. It is clear that along each streamline,

$$\frac{dy}{dx} = \frac{v}{u} \tag{5.52}$$

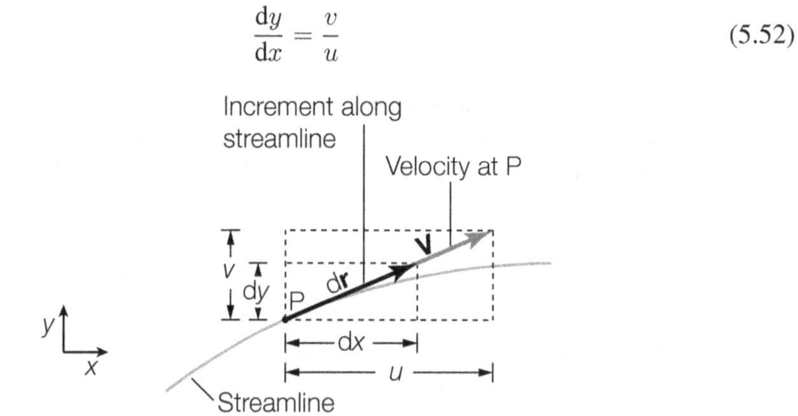

Figure 5.7: Streamline in xy plane

which combined with Equation 5.50 and applying the chain rule yields

$$-v\,\mathrm{d}x + u\,\mathrm{d}y = 0 \tag{5.53}$$

$$\frac{\partial \psi}{\partial x}\mathrm{d}x + \frac{\partial \psi}{\partial y}\mathrm{d}y = 0 \tag{5.54}$$

$$\mathrm{d}\psi = 0 \tag{5.55}$$

which shows that the value of ψ does not change along a streamline or, in other words, ψ = constant defines a streamline. If the volume flow rate between streamlines ψ and $\psi + \mathrm{d}\psi$ is equal to $\mathrm{d}q$ as illustrated in Figure 5.8, then

$$\mathrm{d}q = u\ \mathrm{d}y - v\ \mathrm{d}x \tag{5.56}$$

where u and v are the x and y components of the velocity where the fluid crosses an incremental surface between the streamlines with coordinate increments $\mathrm{d}x$ and $\mathrm{d}y$. Combining Equations 5.50 and 5.56 yields

$$\mathrm{d}q = \frac{\partial \psi}{\partial y}\mathrm{d}y + \frac{\partial \psi}{\partial x}\mathrm{d}x = \mathrm{d}\psi \tag{5.57}$$

Hence, the incremental volume flow rate between the streamlines is equal to the increment in the value of the stream function. Equation 5.57 can also be expressed in the integrated form

$$q = \int_{\psi_1}^{\psi_2} \mathrm{d}\psi = \psi_2 - \psi_1 \tag{5.58}$$

where q is the volume flow rate between streamlines $\psi(x, y) = \psi_1$ and $\psi(x, y) = \psi_2$. It can further be inferred that the sign convention used in Equation 5.50 guarantees that the flow is from left to right as ψ increases in the y-direction. This implies that the value of ψ increases to the left of the direction of flow in the xy plane. It is noteworthy that because the flow rate between any two streamlines is constant, converging streamlines indicate increases in velocity and diverging streamlines indicate decreases in velocity. Although stream-function properties can be very useful in practice, single-stream-function applications cannot be extended to general three-dimensional flows.

Stream function in polar coordinates. In polar coordinates (r, θ), the continuity equation for two-dimensional (planar) incompressible flow is given by

$$\frac{\partial (r v_r)}{\partial r} + \frac{\partial v_\theta}{\partial \theta} = 0 \tag{5.59}$$

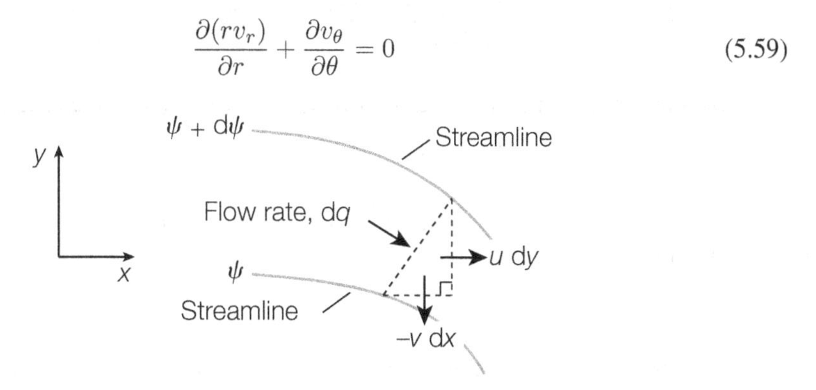

Figure 5.8: Flow between streamlines

where v_r and v_θ are the r and θ components of the velocity, respectively. If the stream function is related to the velocity components v_r and v_θ by

$$v_r = \frac{1}{r}\frac{\partial \psi}{\partial \theta}, \qquad v_\theta = -\frac{\partial \psi}{\partial r} \tag{5.60}$$

then it can also be shown that the continuity equation (Equation 5.59) is identically satisfied, lines of constant ψ are streamlines, and the volume flow rate between adjacent streamlines is equal to the difference in ψ values between the streamlines. Formulating the stream function in terms of polar coordinates is particularly useful is analyzing the flows in vortices, line-sources, and line-sinks, because the symmetries of these configurations commonly allow ψ to be expressed in terms of a single coordinate, usually r.

Stream function in axisymmetric flow. A three-dimensional flow represented in cylindrical polar coordinates (r, θ, z) is called *axisymmetric* when there is no circumferential variations in flow characteristics, in which case $\partial v_\theta/\partial\theta = 0$. Under this condition, the continuity equation for incompressible flow is given by

$$\frac{1}{r}\frac{\partial(r v_r)}{\partial r} + \frac{\partial v_z}{\partial z} = 0 \tag{5.61}$$

where v_r and v_z are the r and z components of the velocity, respectively. If the stream function is related to the velocity components v_r and v_z by

$$v_r = -\frac{1}{r}\frac{\partial \psi}{\partial z}, \qquad v_z = \frac{1}{r}\frac{\partial \psi}{\partial r} \tag{5.62}$$

then it can also be shown that the continuity equation (Equation 5.61) is identically satisfied, lines of constant ψ are streamlines, and the volume flow rate between adjacent streamlines is equal to 2π times the difference in ψ values between the streamlines.

Existence of stream functions. In general, a stream function can be found for any incompressible flow field in which the velocity components are functions of only two independent variables. This is evidenced in the aforementioned cases, where the velocities can be expressed in terms of (x, y), (r, θ), or (r, z) coordinates.

EXAMPLE 5.9

Consider a two-dimensional incompressible flow field in which the velocity is given by

$$\mathbf{V} = 2\mathbf{i} + 4x\mathbf{j}$$

Determine the location(s) of any stagnation points and derive the expression for the stream function. Sketch the streamlines of this flow.

SOLUTION

For the given velocity field: u = 2 and v = $4x$. A stagnation point is a location where the flow velocity is equal to zero. Because u = 2 (= non-zero constant), **there are no stagnation**

points in this flow field. Applying the definition of the stream function for two-dimensional incompressible flow in a Cartesian coordinate system,

$$\frac{\partial \psi}{\partial y} = u = 2 \quad \rightarrow \quad \psi = 2y + f_1(x)$$

$$\frac{\partial \psi}{\partial x} = -v = -4x \quad \rightarrow \quad \psi = -2x^2 + f_2(y)$$

which requires that

$$\psi = 2y + f_1(x) = -2x^2 + f_2(y) \quad \rightarrow \quad f_1(x) = -2x^2 + C_1 \quad \text{and} \quad f_2(y) = 2y + C_1$$

Hence, the stream function is given by

$$\boldsymbol{\psi = -2x^2 + 2y + C_1}$$

where C_1 is a constant. Each streamline is described by $\psi =$ constant, or $-2x^2+2y+C_1 = C_2$, where C_2 is a constant. Hence, the equation of each streamline is given by

$$-2x^2 + 2y = C$$

where C $(= C_2 - C_1)$ is a constant that is different for each streamline. To maintain an equal volume flow rate between streamlines, equal increments in C should be used in plotting the streamlines. The plotted streamlines, described by $y = x^2 + C/2$ with C values in increments of 2, are shown in Figure 5.9. It is apparent that the flow field has parabolic streamlines that are concave upward.

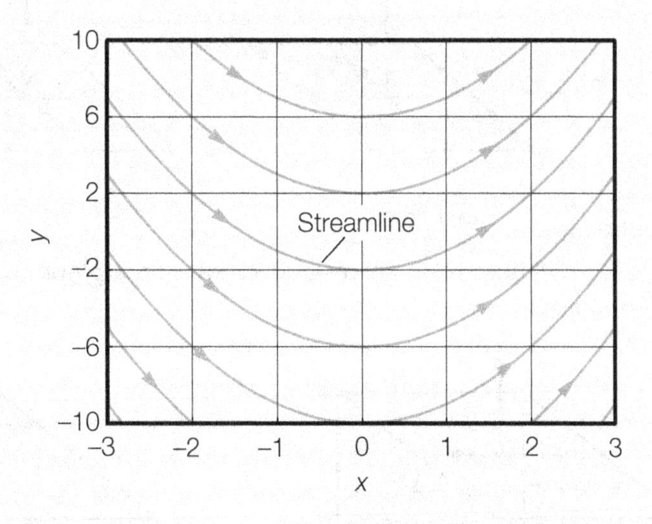

Figure 5.9: Streamlines

5.4 Conservation of Momentum

Newton's second law of motion states that the sum of the forces on a fluid element is equal to the mass of the element multiplied by its acceleration. Based on this requirement, an equation for the motion of a fluid can be derived by applying Newton's second law to a fluid element and taking into account forces due to gravity, viscosity, and pressure.

5.4.1 General Equation

Consider the force acting on a small area, $\mathrm{d}A$, of a fluid element, where this force consists of normal and tangential components as shown in Figure 5.10. The components of the force on $\mathrm{d}A$ can be expressed in terms of a normal stress, denoted by σ, and two orthogonal tangential (shear) stresses, denoted by τ_1 and τ_2. The normal component of the force F_n and tangential components of the force F_t1 and F_t2 acting on $\mathrm{d}A$ are related to the surface stresses by

$$F_\mathrm{n} = \sigma\,\mathrm{d}A, \quad F_\mathrm{t1} = \tau_1\,\mathrm{d}A, \quad F_\mathrm{t2} = \tau_2\,\mathrm{d}A \tag{5.63}$$

For fluids in motion (and for solids), the components of stress on any plane surface depends on the orientation of the surface, and it can be shown that the stress components on any plane surface passing through a given point can be expressed in terms of the stress components on three orthogonal planes passing through that point. Stresses on orthogonal planes are illustrated in Figure 5.11, which also shows the double-subscript notation commonly used to represent stresses, as well as the directions in which these stresses are taken as positive. All normal stresses are denoted by σ, and all tangential stresses are denoted by τ; the first subscript indicates the coordinate axis normal to the plane surface, and the second subscript indicates the coordinate axis along which the stress acts. The sign convention is as follows: All outward (tensile) normal stresses are taken as positive; in cases where the outward normal stress is in the positive coordinate direction, the shear stresses in the other positive coordinate directions are taken as positive; and in cases where the outward normal stress is in the negative coordinate direction, the shear stresses in the other negative coordinate directions are taken as positive.

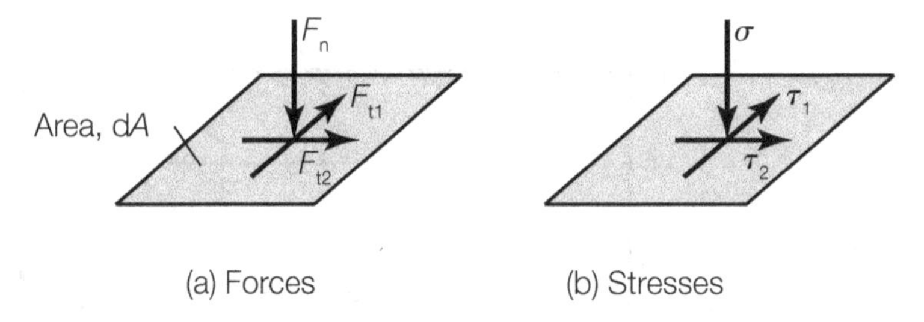

Figure 5.10: Forces and stresses on a plane surface

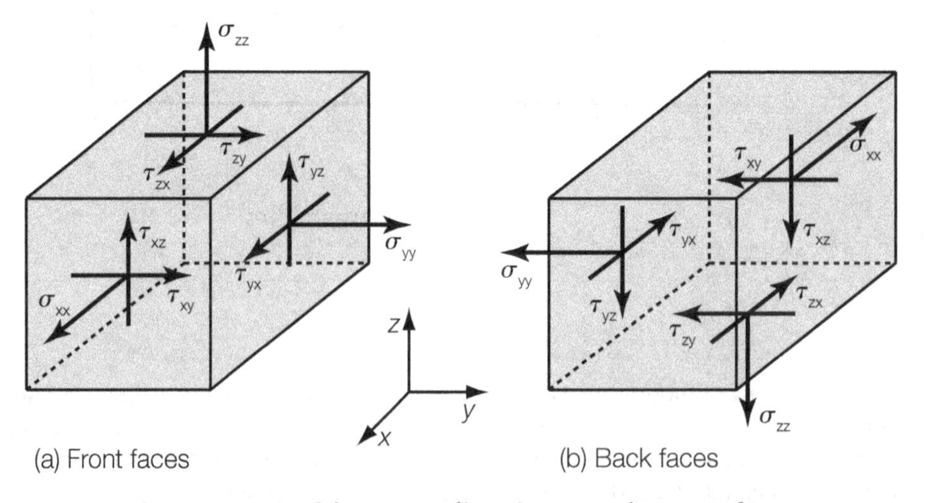

Figure 5.11: Positive stress directions on planar surfaces

Derivation of the momentum equation. Consider the fluid element of dimensions $\Delta x \times \Delta y \times \Delta z$ shown in Figure 5.12, where the surface forces in the z-direction and the body force due to gravity in the z-direction are also shown. The body force is equal to $\rho g_z \Delta x \Delta y \Delta z$, where ρ is the density of the fluid and g_z is the component of the acceleration due to gravity in the z-direction. Other body forces besides gravity could also be considered, such as electromagnetic forces that develop in the flow of electrically conducting fluids in electric or magnetic fields. In the present case, the gravity force is the only body force being considered, which covers the vast majority of cases likely to be encountered in engineering applications. Summing the forces in the z-direction yields the total force in the z-direction as

$$\sum F_z = \left(\rho g_z + \frac{\partial \sigma_{zz}}{\partial z} + \frac{\partial \tau_{yz}}{\partial z} + \frac{\partial \tau_{xz}}{\partial z} \right) \Delta x \Delta y \Delta z \tag{5.64}$$

If the x and y components of the forces on the fluid element are similarly summed, the total force in the x- and y-directions are given by

$$\sum F_x = \left(\rho g_x + \frac{\partial \sigma_{xx}}{\partial x} + \frac{\partial \tau_{yx}}{\partial x} + \frac{\partial \tau_{zx}}{\partial x} \right) \Delta x \Delta y \Delta z \tag{5.65}$$

$$\sum F_y = \left(\rho g_y + \frac{\partial \sigma_{yy}}{\partial y} + \frac{\partial \tau_{xy}}{\partial y} + \frac{\partial \tau_{zy}}{\partial y} \right) \Delta x \Delta y \Delta z \tag{5.66}$$

Applying the momentum equation to the fluid element and assuming that the fluid is incompressible requires that

$$\sum \mathbf{F} = m\mathbf{a} = \rho \Delta x \Delta y \Delta z \left[\frac{\partial \mathbf{V}}{\partial t} + (\mathbf{V} \cdot \nabla)\mathbf{V} \right] \tag{5.67}$$

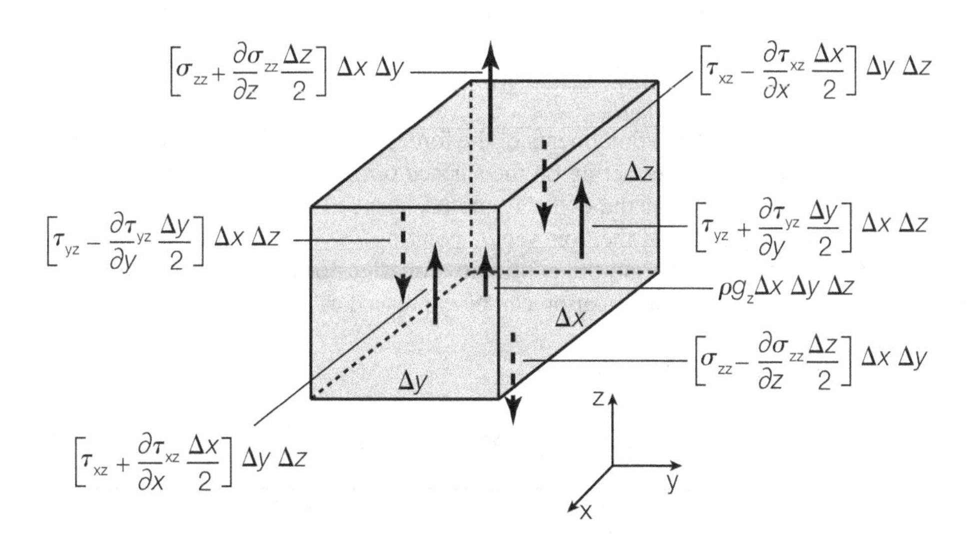

Figure 5.12: Forces in the z-direction on a fluid element

Combining Equations 5.64–5.67 yields the following components of the momentum equation:

$$\rho g_x + \frac{\partial \sigma_{xx}}{\partial x} + \frac{\partial \tau_{yx}}{\partial x} + \frac{\partial \tau_{zx}}{\partial x} = \rho\left(\frac{\partial u}{\partial t} + u\frac{\partial u}{\partial x} + v\frac{\partial u}{\partial y} + w\frac{\partial u}{\partial z}\right) \tag{5.68}$$

$$\rho g_y + \frac{\partial \sigma_{yy}}{\partial y} + \frac{\partial \tau_{xy}}{\partial y} + \frac{\partial \tau_{zy}}{\partial y} = \rho\left(\frac{\partial v}{\partial t} + u\frac{\partial v}{\partial x} + v\frac{\partial v}{\partial y} + w\frac{\partial v}{\partial z}\right) \tag{5.69}$$

$$\rho g_z + \frac{\partial \sigma_{zz}}{\partial z} + \frac{\partial \tau_{yz}}{\partial z} + \frac{\partial \tau_{xz}}{\partial z} = \rho\left(\frac{\partial w}{\partial t} + u\frac{\partial w}{\partial x} + v\frac{\partial w}{\partial y} + w\frac{\partial w}{\partial z}\right) \tag{5.70}$$

Equations 5.68–5.70 can be expressed compactly using vector notation as

$$\rho \mathbf{g} + \nabla \cdot \sigma_{ij} = \rho\left[\frac{\partial \mathbf{V}}{\partial t} + (\mathbf{V}\cdot\nabla)\mathbf{V}\right] \tag{5.71}$$

where $\boldsymbol{\sigma}_{ij}$ are the components of the stress tensor given by

$$\boldsymbol{\sigma}_{ij} = \begin{bmatrix} \sigma_{xx} & \tau_{yx} & \tau_{zx} \\ \tau_{xy} & \sigma_{yy} & \tau_{zy} \\ \tau_{xz} & \tau_{yz} & \sigma_{zz} \end{bmatrix} \tag{5.72}$$

The momentum equation given in vector form by Equation 5.71 and in component form by Equations 5.68–5.70 is called *Cauchy's*[29] *equation*, and the components of Cauchy's (momentum) equation relate the forces on a fluid element to the rate of change of momentum of the fluid element. Cauchy's (momentum) equation involves 13 unknowns: ρ, u, v, w, σ_{xx}, σ_{yy}, σ_{zz}, τ_{xy}, τ_{xz}, τ_{yz}, τ_{yx}, τ_{zx}, τ_{zy}. Applications of the momentum equation to particular cases typically require the utilization of case-specific constitutive relationships between the stresses (σ_{ij}, τ_{ij})[30] and the velocity field, $\mathbf{V}$, as well as a constitutive relationship describing the behavior of the density, ρ.

Alternative derivation of the momentum equation. Using index notation to represent the Cartesian coordinate axes such that the subscripts 1, 2, and 3 refer to the x-, y-, and z-coordinates, respectively, then using Newton's second law with the Reynolds transport theorem, the i component of the momentum equation for a finite control volume can be expressed as

$$\sum F_i = \int_{\text{cv}} \frac{\partial(\rho v_i)}{\partial t}\,\mathrm{d}V + \int_{\text{cs}} (\rho v_i)(\mathbf{V}\cdot\mathbf{n})\,\mathrm{d}A \tag{5.73}$$

where $\sum F_i$ is the sum of the i components of the forces acting on the control volume (which includes body forces and forces acting on the surface of the system), "cv" and "cs" refer to the volume and surface area of the control volume, respectively, v_i is the i component of the velocity vector $\mathbf{V}$, and $\mathbf{n}$ is the unit vector pointing out of the control volume. Using the Gauss divergence theorem and the continuity equation for an incompressible fluid, the left-hand side of the momentum equation can be expressed as

$$\int_{\text{cv}} \frac{\partial(\rho v_i)}{\partial t}\,\mathrm{d}V + \int_{\text{cs}} (\rho v_i)(\mathbf{V}\cdot\mathbf{n})\,\mathrm{d}A = \underbrace{\int_{\text{cv}} \left[\frac{\partial(\rho v_i)}{\partial t} + \nabla\cdot(\rho v_i \mathbf{V})\right]\mathrm{d}V}_{\text{using the divergence theorem}} = \underbrace{\int_{\text{cv}} \rho\left[\frac{\partial v_i}{\partial t} + \mathbf{V}\cdot\nabla v_i\right]\mathrm{d}V}_{\text{for an incompressible fluid}} \tag{5.74}$$

[29]Named in honor of the French engineer and mathematician Augustin-Louis Cauchy (1789–1857).

[30]The index subscripts are such the 1 = x, 2 = y, and 3 = z.

Combining Equations 5.73 and 5.74 with expressions for the gravity force and surficial stresses on the control volume gives

$$\int_{\text{cv}} \rho g_i \, \mathrm{d}V + \int_{\text{cs}} \boldsymbol{\sigma}_{ij} \cdot \mathbf{n} \, \mathrm{d}A = \int_{\text{cv}} \rho \left[\frac{\partial v_i}{\partial t} + \mathbf{V} \cdot \boldsymbol{\nabla} v_i \right] \mathrm{d}V \tag{5.75}$$

where g_i is the i component of the acceleration due to gravity $\mathbf{g}$. Applying the Gauss divergence theorem to the integral over the control surface gives

$$\int_{\text{cs}} \boldsymbol{\sigma}_{ij} \cdot \mathbf{n} \, \mathrm{d}A = \int_{\text{cv}} \boldsymbol{\nabla} \cdot \boldsymbol{\sigma}_{ij} \, \mathrm{d}V \tag{5.76}$$

Substituting Equation 5.76 into Equation 5.75 and combining integrands into a single term yields

$$\int_{\text{cv}} \left\{ \rho g_i + \boldsymbol{\nabla} \cdot \boldsymbol{\sigma}_{ij} - \rho \left[\frac{\partial v_i}{\partial t} - \mathbf{V} \cdot \boldsymbol{\nabla} v_i \right] \right\} \mathrm{d}V = 0 \tag{5.77}$$

Because this relationship is valid regardless of the location and extent of the control volume, "cv," the integrand must necessarily be equal to zero throughout the fluid, which means that

$$\rho g_i + \boldsymbol{\nabla} \cdot \boldsymbol{\sigma}_{ij} = \rho \left[\frac{\partial v_i}{\partial t} + \mathbf{V} \cdot \boldsymbol{\nabla} v_i \right] \tag{5.78}$$

This is the same expression for the momentum equation (i.e., Cauchy's equation) that was derived previously as Equation 5.71 using an infinitesimal control volume.

5.4.2 Navier–Stokes Equation

For incompressible Newtonian fluids, the relationships between stress and strain rate, with viscosity as a parameter, are given by

$$\sigma_{xx} = -p + 2\mu \frac{\partial u}{\partial x} \tag{5.79}$$

$$\sigma_{yy} = -p + 2\mu \frac{\partial v}{\partial y} \tag{5.80}$$

$$\sigma_{zz} = -p + 2\mu \frac{\partial w}{\partial z} \tag{5.81}$$

$$\tau_{xy} = \tau_{yx} = \mu \left(\frac{\partial u}{\partial y} + \frac{\partial v}{\partial x} \right) \tag{5.82}$$

$$\tau_{yz} = \tau_{zy} = \mu \left(\frac{\partial v}{\partial z} + \frac{\partial w}{\partial y} \right) \tag{5.83}$$

$$\tau_{zx} = \tau_{xz} = \mu \left(\frac{\partial w}{\partial x} + \frac{\partial u}{\partial z} \right) \tag{5.84}$$

where the *mechanical pressure*, p, is related to the normal stresses by σ_{xx}, σ_{yy}, and σ_{zz} by

$$p = -\frac{1}{3}(\sigma_{xx} + \sigma_{yy} + \sigma_{zz}) \tag{5.85}$$

Equations 5.79–5.84 are called *constitutive equations*, because they provide empirical relationships between flow variables with a fluid property (i.e., viscosity) as a parameter. When a fluid is at rest, the normal stresses on the faces of any fluid element are equal to the mechanical pressure, p. However, in moving fluids, viscous stresses are added in accordance with Equations 5.79–5.81. Keep in mind that Equations 5.79–5.84 apply to incompressible fluids. In the case of compressible fluids, the term $\lambda \boldsymbol{\nabla} \cdot \mathbf{V}$ is added to the right-hand sides of Equations 5.79–5.81, where λ is called the *second coefficient of viscosity*. The compressible-flow

Navier–Stokes equations are seldom used, and subsequent analyses will deal exclusively with the incompressible-flow form of the Navier–Stokes equation. Substituting Equations 5.79–5.85 into Equations 5.68–5.70, assuming that the viscosity, μ, is constant and that the density, ρ, is constant (which requires that Equation 5.42 be satisfied) gives the differential form of the components of the momentum equation as

$$\rho\left[\frac{\partial u}{\partial t}+u\frac{\partial u}{\partial x}+v\frac{\partial u}{\partial y}+w\frac{\partial u}{\partial z}\right]=-\frac{\partial p}{\partial x}+\rho g_x+\mu\left(\frac{\partial^2 u}{\partial x^2}+\frac{\partial^2 u}{\partial y^2}+\frac{\partial^2 u}{\partial z^2}\right) \tag{5.86}$$

$$\rho\left[\frac{\partial v}{\partial t}+u\frac{\partial v}{\partial x}+v\frac{\partial v}{\partial y}+w\frac{\partial v}{\partial z}\right]=-\frac{\partial p}{\partial y}+\rho g_y+\mu\left(\frac{\partial^2 v}{\partial x^2}+\frac{\partial^2 v}{\partial y^2}+\frac{\partial^2 v}{\partial z^2}\right) \tag{5.87}$$

$$\rho\left[\frac{\partial w}{\partial t}+u\frac{\partial w}{\partial x}+v\frac{\partial w}{\partial y}+w\frac{\partial w}{\partial z}\right]=-\frac{\partial p}{\partial z}+\rho g_z+\mu\left(\frac{\partial^2 w}{\partial x^2}+\frac{\partial^2 w}{\partial y^2}+\frac{\partial^2 w}{\partial z^2}\right) \tag{5.88}$$

These are the three components of a single vector equation called the *Navier–Stokes equation*.[31] A basic assumption in the Navier–Stokes equation is that the fluid is homogeneous, which means that the fluid properties μ and ρ are constant within the flow field. The assumption that the fluid properties are constant corresponds to assuming that the fluid is incompressible (or at least the flow is incompressible) and that temperature variations, if any, within the fluid have negligible effects on the fluid properties. In cases where temperature effects are significant, the (differential) energy equation must be solved simultaneously with the Navier–Stokes equation, where the solution of the energy equation gives the temperature distribution within the flow field. The differential momentum equation can also be derived for the more general case in which the fluid properties are not constant, but vary in both space and time. The momentum equation in this general case is also referred to as the Navier–Stokes equation. The assumption of constant μ and ρ is common and usually justified in engineering applications.

Navier–Stokes equation in vector notation. It is usually more convenient and compact to state the Navier–Stokes equation in vector notation. The Navier–Stokes equation for constant μ and ρ as given by Equations 5.86–5.88 can be expressed using vector notation as

$$\rho\left(\frac{\partial \mathbf{V}}{\partial t}+\mathbf{V}\cdot\nabla\mathbf{V}\right)=-\nabla p+\rho\mathbf{g}+\mu\nabla^2\mathbf{V} \tag{5.89}$$

or terms of the material derivative as

$$\rho\frac{\mathrm{D}\mathbf{V}}{\mathrm{D}t}=-\nabla p+\rho\mathbf{g}+\mu\nabla^2\mathbf{V} \tag{5.90}$$

or in component form as

$$\rho\left(\frac{\partial V_i}{\partial t}+\mathbf{V}\cdot\nabla V_i\right)=-\nabla p+\rho g_i+\mu\nabla^2 V_i \tag{5.91}$$

The flow of fluids having a constant temperature is described by the combined Navier–Stokes and continuity equations, which combine to form four equations in four unknowns: u, v, w,

[31]Named in honor of the French engineer Claude-Louis Marie Henri Navier (1785–1836) and the English mathematician and physicist Sir George Gabriel Stokes (1819–1903), both of whom (independently) derived these equations.

and p. Notably, for nonmoving (static) fluids, the velocities in Equation 5.89 are equal to zero and Equation 5.89 simplifies to

$$\nabla p = \rho \mathbf{g} \tag{5.92}$$

which was derived previously in Chapter 2 for the case of static fluids.

Nomenclature. In practice, the Navier–Stokes equation is sometimes referred to in plural form as the "Navier–Stokes equations," with the plural form usually referring to the three components of the single vector Navier–Stokes equation. The singular term "Navier–Stokes equation" is preferable, because this terminology is consistent with fact that the Navier–Stokes equation is derived from the single fundamental (vector) momentum equation, which states that the sum of the force vectors on a fluid element must be equal to the mass times the acceleration vector of the fluid element.

Navier–Stokes equation in cylindrical polar coordinates. The components of the Navier–Stokes equation given by Equations 5.86–5.88 can also be expressed in cylindrical polar coordinates as

$$\begin{aligned}\rho\left(\frac{\partial v_r}{\partial t} + v_r\frac{\partial v_r}{\partial r} + \frac{v_\theta}{r}\frac{\partial v_r}{\partial \theta} - \frac{v_\theta^2}{r} + v_z\frac{\partial v_r}{\partial z}\right) \\ = -\frac{\partial p}{\partial r} + \rho g_r + \mu\left[\frac{1}{r}\frac{\partial}{\partial r}\left(r\frac{\partial v_r}{\partial r}\right) - \frac{v_r}{r^2} + \frac{1}{r^2}\frac{\partial^2 v_r}{\partial \theta^2} - \frac{2}{r^2}\frac{\partial v_\theta}{\partial \theta} + \frac{\partial^2 v_r}{\partial z^2}\right]\end{aligned} \tag{5.93}$$

$$\begin{aligned}\rho\left(\frac{\partial v_\theta}{\partial t} + v_r\frac{\partial v_\theta}{\partial r} + \frac{v_\theta}{r}\frac{\partial v_\theta}{\partial \theta} - \frac{v_r v_\theta}{r} + v_z\frac{\partial v_\theta}{\partial z}\right) \\ = -\frac{1}{r}\frac{\partial p}{\partial \theta} + \rho g_\theta + \mu\left[\frac{1}{r}\frac{\partial}{\partial r}\left(r\frac{\partial v_\theta}{\partial r}\right) - \frac{v_\theta}{r^2} + \frac{1}{r^2}\frac{\partial^2 v_\theta}{\partial \theta^2} + \frac{2}{r^2}\frac{\partial v_r}{\partial \theta} + \frac{\partial^2 v_\theta}{\partial z^2}\right]\end{aligned} \tag{5.94}$$

$$\begin{aligned}\rho\left(\frac{\partial v_z}{\partial t} + v_r\frac{\partial v_z}{\partial r} + \frac{v_\theta}{r}\frac{\partial v_z}{\partial \theta} + v_z\frac{\partial v_z}{\partial z}\right) \\ = -\frac{\partial p}{\partial z} + \rho g_z + \mu\left[\frac{1}{r}\frac{\partial}{\partial r}\left(r\frac{\partial v_z}{\partial r}\right) + \frac{1}{r^2}\frac{\partial^2 v_z}{\partial \theta^2} + \frac{\partial^2 v_z}{\partial z^2}\right]\end{aligned} \tag{5.95}$$

5.4.3 Nondimensional Navier–Stokes Equation

The Navier–Stokes equation can be nondimensionalized in terms of reference values of variables that are characteristic of the environment in which fluid flow is occurring. The following fixed reference values (constants) can be defined: a length, L [L]; a time, [T], or frequency, ω [T^{-1}]; a speed, V [LT^{-1}]; a pressure, p_0 [FL^{-2}]; and gravity, g [LT^{-2}]. Using these reference values, leads to the nondimensional variables

$$\mathbf{x}^* = \frac{\mathbf{x}}{L}, \quad t^* = \omega t, \quad \mathbf{V}^* = \frac{\mathbf{V}}{V}, \quad p^* = \frac{p - p_\infty}{p_0 - p_\infty}, \quad \mathbf{g}^* = \frac{\mathbf{g}}{g}, \quad \text{and} \quad \nabla^* = L\nabla \tag{5.96}$$

where p_∞ is the pressure under undisturbed conditions, and the nondimensional gradient operator, ∇^*, is defined by

$$\nabla^*(\cdot) = \frac{\partial(\cdot)}{\partial x^*}\mathbf{i} + \frac{\partial(\cdot)}{\partial y^*}\mathbf{j} + \frac{\partial(\cdot)}{\partial z^*}\mathbf{k} \tag{5.97}$$

Table 5.1: Dimensionless Groups in Navier–Stokes Equation

Group	Ratio	Name	Acronym
$\dfrac{\omega L}{V}$	$\dfrac{\text{advection time scale}}{\text{oscillation time scale}}$	Strouhal number	St
$\dfrac{p_0 - p_\infty}{\rho V^2}$	$\dfrac{\text{pressure force}}{\text{inertial force}}$	Euler number	Eu
$\dfrac{V}{\sqrt{gL}}$	$\dfrac{\text{inertial force}}{\text{gravity force}}$	Froude number	Fr
$\dfrac{\rho V L}{\mu}$	$\dfrac{\text{inertial force}}{\text{viscous force}}$	Reynolds number	Re

Expressing the Navier–Stokes equation (Equation 5.89) in terms of the nondimensional quantities defined by Equations 5.96 and 5.97 gives

$$\left[\frac{\omega L}{V}\right]\frac{\partial \mathbf{V}^*}{\partial t^*} + \mathbf{V}^* \cdot \nabla^* \mathbf{V}^* = -\left[\frac{p_0 - p_\infty}{\rho V^2}\right]\nabla p^* + \left[\frac{gL}{V^2}\right]\mathbf{g} + \left[\frac{\mu}{\rho V L}\right]\nabla^{*2}\mathbf{V}^* \tag{5.98}$$

The groups of constants in square brackets are all dimensionless quantities (as are all of the terms in Equation 5.98), and these dimensionless groups are widely used in fluid mechanics to measure the ratios of time and force scales as given in Table 5.1. Substituting the conventional acronyms for the dimensionless groups into Equation 5.98 gives the following nondimensional form of the Navier–Stokes equation:

$$(\mathrm{St})\frac{\partial \mathbf{V}^*}{\partial t^*} + \mathbf{V}^* \cdot \nabla^* \mathbf{V}^* = -(\mathrm{Eu})\nabla p^* + \left(\frac{1}{\mathrm{Fr}^2}\right)\mathbf{g}^* + \left(\frac{1}{\mathrm{Re}}\right)\nabla^{*2}\mathbf{V}^* \tag{5.99}$$

where St is the *Strouhal number*, Eu is the *Euler number*, Fr is the *Froude number*, and Re is the *Reynolds number*. The form of the Navier–Stokes equation given by Equation 5.99 is particularly useful in cases where the reference variables (L, ω^{-1}, V, and p_0) are on the same order of magnitude as the corresponding flow variables ($\mathbf{x}$, t, $\mathbf{V}$, and p), in which case the nondimensional variables ($\mathbf{x}^*$, t^*, $\mathbf{V}^*$, and p^*) in Equation 5.99 along with their derivatives are all on the order of unity (i.e., order 1). As a consequence, the relative magnitudes of each of the terms in Equation 5.99 can be determined from the relative magnitudes of the nondimensional parameters (St, Eu, Fr, and Re). This property of Equation 5.99 is particularly useful, because it facilitates the identification of flow environments in which some terms in the Navier–Stokes are more important than others, thereby providing justification for neglecting some terms and providing a simpler equation to solve. It is worth noting that the Navier–Stokes equation must generally be solved simultaneously with the continuity equation, and the continuity equation can be expressed in terms of nondimensional variables as

$$\nabla^* \cdot \mathbf{V}^* = 0 \tag{5.100}$$

The continuity equation (Equation 5.100) does not involve any nondimensional parameters. Consequently, all terms in the continuity equation are of equal importance and simplification of the continuity equation cannot be obtained based on the magnitudes of the characteristic nondimensional groups.

Special case: steady flow. When the flow is steady, the derivative with respect to time in Equation 5.99 (i.e., the first term on the left-hand side) is equal to zero; therefore, the magnitude of the Strouhal number, St, has no influence on the flow. One might attribute this result to the reference frequency, ω, being infinite under steady-state flow conditions. However, even if a finite reference frequency was selected, the derivative with respect to time would still be equal to zero under steady-state conditions. Hence, under steady-state conditions, the Navier–Stokes equation is given by

$$\mathbf{V}^* \cdot \nabla^* \mathbf{V}^* = -(\text{Eu})\nabla^* p^* + \left(\frac{1}{\text{Fr}^2}\right)\mathbf{g}^* + \left(\frac{1}{\text{Re}}\right)\nabla^{*2}\mathbf{V}^* \quad \text{(steady state)} \tag{5.101}$$

Special case: no free surface. When a fluid does not have a free surface, which means that the boundaries of the fluid are not free to move, the pressure generated by the flowing fluid can be taken as additive to the hydrostatic pressure under nonflowing condition. Consequently, a modified pressure, p', can be defined that is equal to the actual pressure in the flowing fluid, p, minus the pressure under the static (nonflowing) condition such that

$$\nabla p' = \nabla p - \rho \mathbf{g} \tag{5.102}$$

Substituting Equation 5.102 into the dimensional Navier–Stokes equation (Equation 5.89) and nondimensionalizing yields

$$(\text{St})\frac{\partial \mathbf{V}^*}{\partial t^*} + \mathbf{V}^* \cdot \nabla^* \mathbf{V}^* = -(\text{Eu}')\nabla^* p'^* + \left(\frac{1}{\text{Re}}\right)\nabla^{*2}\mathbf{V}^* \quad \text{(no free surface)} \tag{5.103}$$

where Eu$'$ is the Euler number based on the reference pressure p_0' that is on the same order of magnitude as p'. In principle, after the solution for p' is obtained from Equation 5.103, the gravity effect can be added back in via Equation 5.102. The useful result here is that the Froude number, Fr, is not a relevant parameter in the solution of flow problems without a free surface.

Special case: viscous effects are negligible. Perhaps the most commonly invoked special case occurs when viscous effects are negligible. In this case, $\mu \to 0$, Re $\to \infty$, and the nondimensional Navier–Stokes equation (Equation 5.99) becomes

$$(\text{St})\frac{\partial \mathbf{V}^*}{\partial t^*} + \mathbf{V}^* \cdot \nabla^* \mathbf{V}^* = -(\text{Eu})\nabla p^* + \left(\frac{1}{\text{Fr}^2}\right)\mathbf{g}^* \tag{5.104}$$

The remarkable result here is that the second-order derivatives in the Navier–Stokes equation have disappeared, thereby reducing the equation to a lower-order equation and consequently reducing the number of boundary conditions that can be specified in solving the equation. This is precisely the approximation that precludes specifying the no-slip boundary condition in inviscid flows. This mathematical limitation is a reflection of the reality that viscous effects become important (and ultimately dominant) in the region of flow immediately adjacent to a solid boundary, because it is the viscosity of the fluid that causes the fluid velocity to become equal to the boundary velocity at the fluid-boundary interface. As a consequence of the importance of viscosity in the boundary region, the reduced Navier–Stokes equation (Equation 5.104) is not valid near the boundary.

EXAMPLE 5.10

Consider the flow of water at 20°C through a 100-mm-diameter, 1-km-long pipe segment as shown in Figure 5.13. The average gauge pressure in the pipe section is on the order of 400 kPa, the average velocity is on the order of 0.3 m/s, and flow conditions in the pipe fluctuate approximately periodically with a period of 12 hours. What simplified form of the Navier–Stokes equation would be applicable in this situation?

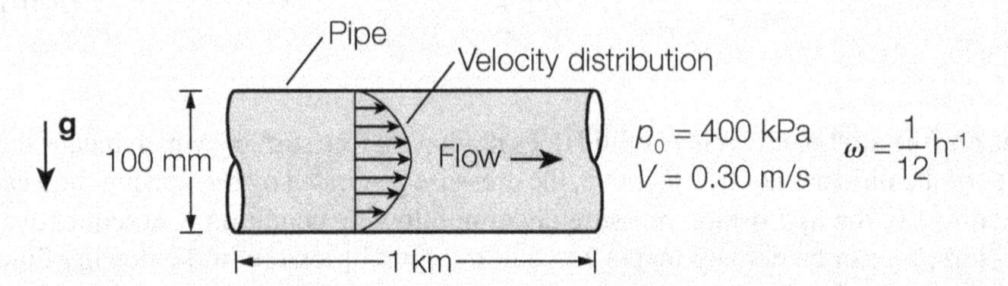

Figure 5.13: Flow through a pipe

SOLUTION

From the given data, the reference parameters are p_0 = 400 kPa = 400 × 10^3 Pa, V = 0.3 m/s, L_1 = 100 mm = 0.1 m, L_2 = 1 km = 1000 m, and ω = 1/12 h^{-1}. Because p_0 is a gauge pressure, take p_∞ = 0 kPa. For water at 20°C, ρ = 998 kg/m^3 and ν = 10^{-6} m^2/s. The nondimensional groups in the Navier–Stokes equation are

$$\text{St} = \frac{\omega L_2}{V} = \frac{(\frac{1}{12})1000}{0.3} = 278$$

$$\text{Eu} = \frac{p_0 - p_\infty}{\rho V^2} = \frac{400 \times 10^3 - 0}{(998)(0.3)^2} = 4435$$

$$\text{Eu}' = \frac{p_0 - \rho g L_1 - p_\infty}{\rho V^2} = \frac{400 \times 10^3 - (998)(9.81)(0.1) - 0}{(998)(0.3)^2} = 4442$$

$$\text{Fr} = \frac{V}{\sqrt{gL_1}} = \frac{0.3}{\sqrt{(9.81)(0.1)}} = 0.30$$

$$\text{Re} = \frac{VL_1}{\nu} = \frac{(0.3)(0.1)}{10^{-6}} = 3 \times 10^4$$

Because St $\gg$ 1, transient effects are likely to be significant; because Eu $\gg$ 1, pressure effects are likely to be significant; because there is no free surface, gravity effects can be included with the pressure (and gravity makes a small contribution because Eu$'$ $\simeq$ Eu); and because 1/Re $\ll$ 1, viscous effects are not likely to be significant except very close to the wall of the pipe (i.e., within the boundary layer). Based on this order-of-magnitude analysis, the approximate form of the Navier–Stokes equation that can be used to describe the flow outside the boundary layer is

$$(\text{St})\frac{\partial \mathbf{V}^*}{\partial t^*} + \mathbf{V}^* \cdot \nabla^* \mathbf{V}^* = -(\text{Eu}')\nabla^* p'^* \quad \rightarrow \quad 278\frac{\partial \mathbf{V}^*}{\partial t^*} + \mathbf{V}^* \cdot \nabla^* \mathbf{V}^* = -4442\nabla^* p'^*$$

where the pressure p'^* would subsequently be adjusted to account for the gravitational effect.

5.5 Solutions of the Navier–Stokes Equation

Although the Navier–Stokes equation applies to both laminar and turbulent flow, the random velocity fluctuations that are characteristic of turbulent flow make the solution of the Navier–Stokes equation for turbulent flow intractable. There are a limited number of analytic solutions to the combined Navier–Stokes and continuity equations and their associated boundary conditions; therefore, numerical solutions are commonly sought. The field of fluid mechanics dealing with numerical solutions to the equations of fluid mechanics is called *computational fluid dynamics* (CFD). A few analytic solutions can be found for laminar flow, and some of the more useful laminar-flow solutions are described in the following sections.

5.5.1 Steady Laminar Flow Between Stationary Parallel Plates

Consider the case of laminar flow between two infinite parallel plates as shown in Figure 5.14. In this case, both the top and bottom plates are fixed and the fluid moves between the plates in response to a pressure gradient in the direction of flow. Because the x-axis is in the direction of flow, $u \neq 0$, $v = 0$, and $w = 0$, and the continuity equation gives

$$\frac{\partial u}{\partial x} + \frac{\partial v}{\partial y} + \frac{\partial w}{\partial z} = 0 \quad \rightarrow \quad \frac{\partial u}{\partial x} = 0 \tag{5.105}$$

which means that the velocity does not change in the x-direction for any given y- and z-coordinate. It can also be inferred that the velocity does not change in the y-direction because the flow is two-dimensional (in the xz plane). Based on these considerations and the fact that the flow is at steady state so that u is not a function of time, t, the fluid velocity u is only a function of z. Using these characteristics of the velocity field, along with the fact that the gravitational force is in the negative z-direction, the components of the Navier–Stokes equation become

$$0 = -\frac{\partial p}{\partial x} + \mu\left(\frac{\partial^2 u}{\partial z^2}\right) \tag{5.106}$$

$$0 = -\frac{\partial p}{\partial y} \tag{5.107}$$

$$0 = -\frac{\partial p}{\partial z} - \rho g \tag{5.108}$$

Combining Equations 5.107 and 5.108 and integrating yields the following expression for the pressure distribution:

$$p = -\rho g z + f(x) \tag{5.109}$$

where $f(x)$ is any function of x. Equation 5.109 indicates that at any x location, the pressure has a hydrostatic distribution in the z-direction. Because $\partial p/\partial x$ is a constant, which is a

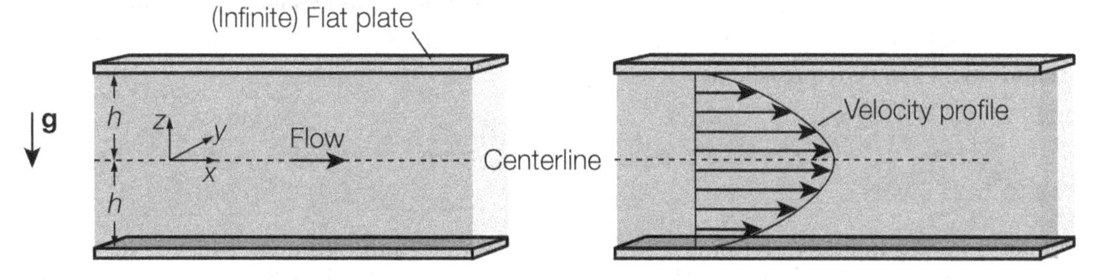

Figure 5.14: Laminar flow between infinite stationary parallel plates

requirement for the flow to be uniform in the x-direction, the pressure distribution given by Equation 5.109 can be expressed as

$$p = -\rho g z + \left(\frac{\partial p}{\partial x}\right) x + p_0 \tag{5.110}$$

where p_0 is the pressure at the location where $x = 0$ and $z = 0$. To determine the velocity distribution, $u(z)$, Equation 5.106 can be integrated twice with the boundary conditions that $u = 0$ at $z = \pm h$, yielding the following velocity distribution:

$$u = \frac{1}{2\mu}\left(\frac{\partial p}{\partial x}\right)(z^2 - h^2) \tag{5.111}$$

The pressure gradient, $\partial p/\partial x$, is generally a negative quantity because a decreasing pressure in the direction of flow is necessary to "drive" the flow. The parabolic velocity distribution given by Equation 5.111, which is sometimes called a *Poiseuille parabola*, is illustrated in Figure 5.14. The volume flow rate per unit width, q, between the plates can be determined from Equation 5.111 using the relation

$$q = \int_{-h}^{h} u\,\mathrm{d}z \rightarrow q = -\frac{2h^3}{3\mu}\frac{\partial p}{\partial x} \tag{5.112}$$

It is apparent that the volume flow rate is proportional to h^3, and hence, the flow rate between the plates increases as the cube of the spacing between the plates. The average velocity, V, and the maximum velocity, u_{max}, are derived as from Equations 5.111 and 5.112 as follows:

$$V = \frac{q}{2h} \rightarrow V = -\frac{h^2}{3\mu}\frac{\partial p}{\partial x} \quad \text{and} \quad u_{\text{max}} = u(0) \rightarrow u_{\text{max}} = -\frac{h^2}{2\mu}\frac{\partial p}{\partial x} = \frac{3}{2}V \tag{5.113}$$

where it is apparent that the average velocity is proportional to the square of the spacing between the plates. Steady flow between stationary parallel plates is commonly called *Hele-Shaw flow*.

Application. This type of flow is frequently encountered in the study of *microchannel flows*. Flows in the small annular spaces between bounding plates of large radii of curvature are commonly approximated as the flow between parallel plates.

Laminar-flow restriction. The equations derived in this section apply only to laminar flow, which usually occurs at values of the Reynolds number (Re) less than around 1400. This requirement can be expressed as

$$\text{Re} = \frac{\rho V(2h)}{\mu} < 1400 \tag{5.114}$$

In cases where Equation 5.114 is violated, the flow is usually turbulent and the velocity is three-dimensional and unsteady. Under these conditions, the laminar-flow equations derived in this section are not applicable.

EXAMPLE 5.11

A lubrication system calls for SAE 30 oil at 20°C to flow at a rate of 1.5 L/s between 10 m × 2 m parallel plates spaced 5 mm apart as illustrated in Figure 5.15. (a) What is the required pressure difference between the inflow and the outflow sections? (b) What force must be used to keep each plate from moving? (c) What is the maximum velocity between the plates?

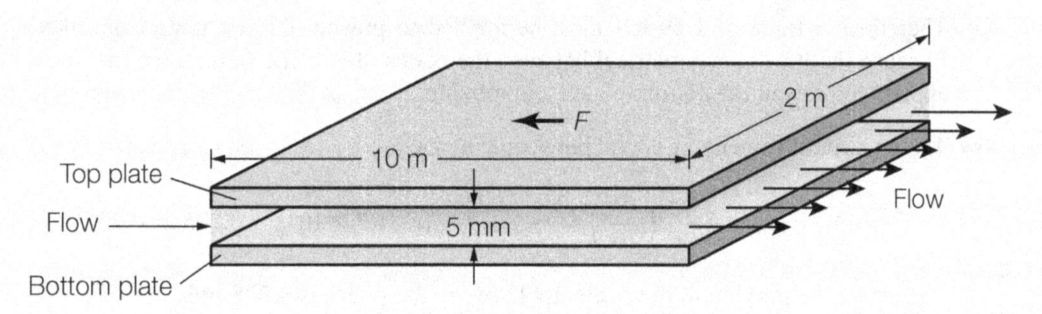

Figure 5.15: Flow between parallel plates

SOLUTION

From the given data: h = 5 mm/2 = 2.5 mm = 0.0025 m, L = 10 m, W = 2 m, and Q = 1.5 L/s = 0.0015 m^3/s. At 20°C, the properties of SAE 30 oil are ρ = 918 kg/m^3 and μ = 440 mPa·s = 0.440 Pa·s (from Appendix B). Check the Reynolds number to determine whether the flow is laminar.

$$V = \frac{Q}{W(2h)} = \frac{0.0015}{(2)(0.005)} = 0.15 \text{ m/s}$$

$$\text{Re} = \frac{\rho V(2h)}{\mu} = \frac{(918)(0.15)(0.005)}{0.440} = 1.56$$

Because Re $\ll$ 1400, the flow is laminar and the laminar-flow equations can be used.

(a) The theoretical flow rate per unit width, q, is given by Equation 5.112; hence,

$$Q = qW = -\frac{2h^3W}{3\mu}\frac{\partial p}{\partial x} \quad \rightarrow \quad 0.0015 = -\frac{2(0.0025)^3(2)}{3(0.440)}\frac{\partial p}{\partial x}$$

$$\rightarrow \quad \frac{\partial p}{\partial x} = -31\,680 \text{ Pa/m} = -31.7 \text{ kPa/m}$$

Because the required pressure gradient is -31.7 kPa/m and there is a 10-m distance between the inflow and outflow sections, the required pressure differential is $(-31.7)(10) = -317$ kPa. The pressure at the inflow section must be **317 kPa** higher than the pressure at the outflow section.

(b) The shear stress, τ, at any distance z from the midpoint between the plates can be derived from the velocity distribution as follows:

$$\tau = \mu\frac{\text{d}u}{\text{d}z} = \mu\frac{\text{d}}{\text{d}z}\left[\frac{1}{2\mu}\left(\frac{\partial p}{\partial x}\right)(z^2 - h^2)\right] = \frac{1}{2}\left(\frac{\partial p}{\partial x}\right)\frac{\text{d}}{\text{d}z}(z^2 - h^2) = z\frac{\partial p}{\partial x}$$

On the surface of the top plate, $z = h$, so the shear stress on the top plate, τ_{top}, is given by

$$\tau_{\text{top}} = h\frac{\partial p}{\partial x} = (0.0025)(-31.7) = -0.0793 \text{ kPa}$$

and the shear force on the top plate, F, is given by

$$F = |\tau_{\text{top}}|\, LW = (0.0793)(10)(2) = 1.59 \text{ kN}$$

Therefore, a force of **1.59 kN** must be applied to prevent the top plate from moving. Because the flow is symmetrical between the plates, this is the same force that must be applied to prevent the bottom plate from moving.

(c) The maximum velocity of the oil between the plates, u_{max}, is given by Equation 5.113 as

$$u_{\text{max}} = -\frac{h^2}{2\mu}\left(\frac{\partial p}{\partial x}\right) = -\frac{0.0025^2}{2(0.440)}(-31.7 \times 10^3) = 0.225 \text{ m/s}$$

Therefore, the maximum flow velocity between the plates is **0.225 m/s.**

5.5.2 Steady Laminar Flow Between Moving Parallel Plates

In the previous analysis, both the top and bottom plates were stationary relative to the moving fluid. However, a solution to the Navier–Stokes equation can also be obtained for the case where one plate is fixed and the other plate is moving with a constant velocity. This scenario is illustrated in Figure 5.16, where the top plate is moving with velocity U and the two plates are spaced a distance b apart. In this case, Equations 5.106–5.108 are still applicable, the velocity in the x-direction is only a function of z, and the pressure distribution is also given by Equation 5.110. However, the boundary conditions to be applied in integrating Equation 5.106 to determine the velocity distribution are

$$u(0) = 0, \quad \text{and} \quad u(b) = U \tag{5.115}$$

Integrating Equation 5.106 twice with boundary conditions given by Equation 5.115 gives the velocity distribution

$$u = U\frac{z}{b} + \frac{1}{2\mu}\left(\frac{\partial p}{\partial x}\right)(z^2 - bz) \tag{5.116}$$

This velocity distribution is illustrated in Figure 5.16 for the case of $\partial p/\partial x < 0$. It is noteworthy that according to Equation 5.116, the velocity distribution is linear in z when $\partial p/\partial x = 0$, and a portion of the velocity distribution is directed in the negative x-direction

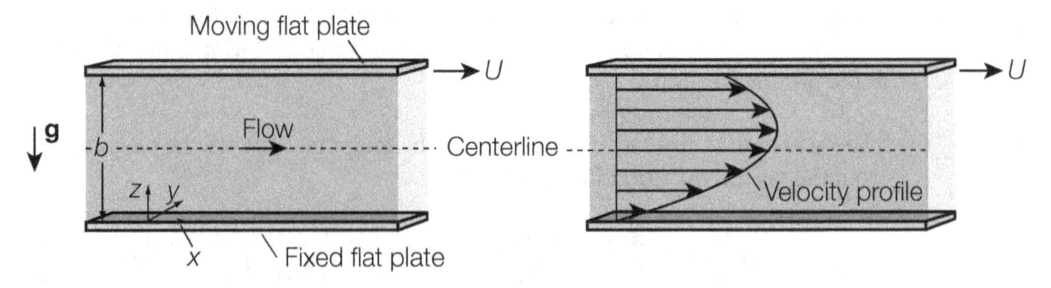

Figure 5.16: Laminar flow between infinite moving parallel plates

when $\partial p/\partial x > 0$. The volume flow rate per unit width, q, between the plates can be determined from Equation 5.116 using the relation

$$q = \int_0^b u\,\mathrm{d}z \rightarrow q = -\left(\frac{\partial p}{\partial x}\right)\frac{b^3}{12\mu} + \frac{Ub}{2} \tag{5.117}$$

The flow field represented by Equation 5.116 is called *Couette flow*.[32] It is characterized by one parallel plate moving with a constant velocity relative to the other parallel plate.

Application. Couette flow commonly occurs within a fluid bounded by concentric cylinders, where the inner cylinder is fixed and the outer cylinder is rotating with a constant angular velocity.

Flow restriction. For Couette flow, laminar conditions usually occur at values of the Reynolds number (Re) less than 1500, and this requirement can be expressed as

$$\text{Re} = \frac{\rho U b}{\nu} < 1500 \tag{5.118}$$

which applies to conditions in which $\partial p/\partial x = 0$. Not much information on the limiting Reynolds number is available for cases where $\partial p/\partial x \neq 0$. When the flow is not laminar, the velocity distribution derived in this section is not applicable.

EXAMPLE 5.12

SAE 30 oil at 20°C flows between parallel plates separated by 40 mm. The plates have a length of 1 m in the flow direction and a width of 0.50 m; the flow is laminar between the plates. Compare the velocity distributions for the following cases: (a) the bottom plate is stationary, the top plate moves at 5 cm/s, and the pressure gradient is −500 Pa/m; (b) the bottom plate is stationary, the top plate moves at 5 cm/s, and the pressure gradient is +500 Pa/m; and (c) the bottom plate moves at −3 cm/s, the top plate moves at +2 cm/s, and the pressure gradient is −500 Pa/m. (d) For the conditions described in part (a), what force must be applied to move the top plate?

SOLUTION

From the given data: L = 1 m, W = 0.5 m, and b = 40 mm = 0.040 m. At 20°C, the properties of SAE 30 oil are ρ = 918 kg/m^3 and μ = 440 mPa·s = 0.440 Pa·s (from Appendix B).

(a) In this case, U = 5 cm/s = 0.05 m/s, $\partial p/\partial x$ = −500 Pa/m, and Equation 5.116 gives the velocity distribution as

$$u = U\frac{z}{b} + \frac{1}{2\mu}\left(\frac{\partial p}{\partial x}\right)(z^2 - bz) = (0.05)\frac{z}{0.04} + \frac{1}{2(0.440)}(-500)\,(z^2 - 0.04z)$$

which yields

$$\boldsymbol{u = 24.0z(1 - 23.7z)}$$

This velocity distribution is shown in Figure 5.17(a).

[32]Named in honor of the French physicist Maurice Marie Alfred Couette (1858–1943).

(b) In this case, U = 5 cm/s = 0.05 m/s, $\partial p/\partial x$ = +500 Pa/m, and Equation 5.116 gives the velocity distribution as

$$u = U\frac{z}{b} + \frac{1}{2\mu}\left(\frac{\partial p}{\partial x}\right)(z^2 - bz) = (0.05)\frac{z}{0.04} + \frac{1}{2(0.440)}(+500)(z^2 - 0.04z)$$

which yields

$$\boldsymbol{u = 21.5z(26.4z - 1)}$$

This velocity distribution is shown in Figure 5.17(b).

(c) In this case, U_{bot} = −3 cm/s = −0.03 m/s, U_{top} = +2 cm/s = +0.02 m/s, and $\partial p/\partial x$ = −500 Pa/m. In terms of relative velocities, this condition is the same as the bottom plate being stationary and the top plate moving at 0.05 m/s, which is the condition in part (a). Hence, the (absolute) velocity distribution is derived from the part (a) distribution by subtracting 0.03 m/s, which yields

$$\boldsymbol{u = 24.0z(1 - 23.7z) - 0.03}$$

This velocity distribution is shown in Figure 5.17(c).

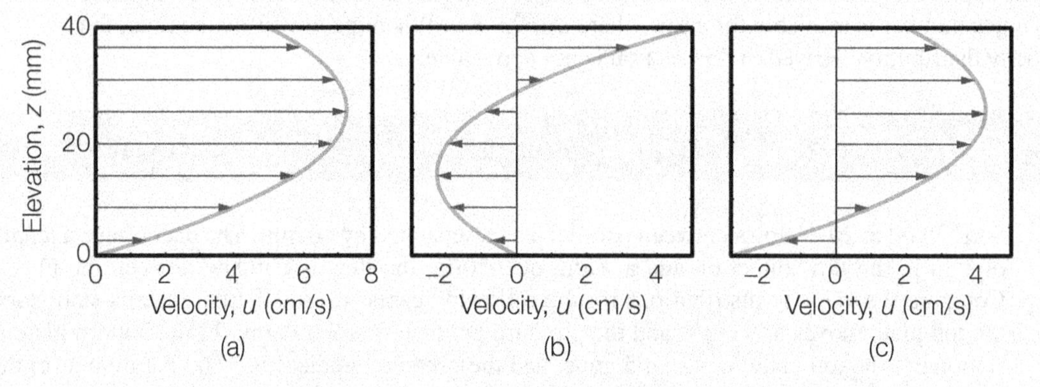

Figure 5.17: Velocity profiles

(d) The shear stress, τ, as a function of z is derived from the velocity distributions as follows:

$$\tau = \mu\frac{\mathrm{d}u}{\mathrm{d}z} = \mu\frac{\mathrm{d}}{\mathrm{d}z}\left[U\frac{z}{b} + \frac{1}{2\mu}\left(\frac{\partial p}{\partial x}\right)(z^2 - bz)\right] = \mu\frac{U}{b} + \frac{1}{2}\left(\frac{\partial p}{\partial x}\right)(2z - b)$$

Taking $z = 0.04$ m gives the shear stress on the top plate, τ_{top}, as

$$\tau_{\text{top}} = (0.440)\frac{0.05}{0.04} + \frac{1}{2}(-500)[2(0.04) - (0.04)] = -9.45 \text{ Pa}$$

Hence, the force, F, required to move the top plate is

$$F = |\tau_{\text{top}}|\, LW = (9.45)(1)(0.5) = \mathbf{4.73\ N}$$

Note that the force required to restrain the bottom plate from moving can be calculated similarly, although due to lack of symmetry in the velocity distribution, this force will be different from that required to move the top plate.

5.5.3 Steady Laminar Flow Adjacent to Moving Vertical Plate

An interesting variation of Couette flow is shown in Figure 5.18, where a plate or similar bounding surface moves vertically with velocity W through a fluid reservoir. This condition causes the fluid to be pulled upward along the plate due to the no-slip requirement on the surface of the plate and the viscosity of the fluid. Assuming that the flow within the layer of fluid adjacent to the plate is laminar, steady, and fully developed, within this layer, it is apparent that $u = 0$, $v = 0$, and $w = w(x)$, where u, v, and w are the components of the fluid velocity in the x-, y-, and z-coordinate directions, respectively. Combining these conditions with the three components of the Navier–Stokes equation (Equations 5.86–5.88) yields the following component equations:

$$\frac{\partial p}{\partial x} = 0 \tag{5.119}$$

$$\frac{\partial p}{\partial y} = 0 \tag{5.120}$$

$$0 = -\underbrace{\frac{\partial p}{\partial z}}_{\approx 0} - \rho g + \mu\frac{\partial^2 w}{\partial x^2} \quad \rightarrow \quad \frac{\mathrm{d}^2 w}{\mathrm{d}x^2} = \frac{\rho g}{\mu} \tag{5.121}$$

The requirement that $\partial p/\partial z = 0$ in Equation 5.121 stems from the fact that the pressure on the surface of the layer of fluid is equal to atmospheric pressure that is assumed to vary negligibly with elevation. Combining this with requirements given by Equations 5.119 and 5.120 that the pressure within the fluid does not vary in any horizontal plane, then vertical variation of pressure within the fluid must necessarily be negligible; hence, $\partial p/\partial z \approx 0$. The resulting ordinary differential equation describing the velocity distribution $w(x)$ given by Equation 5.121 can be integrated subject to the following boundary conditions:

$$\tau_{\text{surface}} = \mu \left.\frac{\mathrm{d}w}{\mathrm{d}x}\right|_{x=h} \quad \rightarrow \quad \mu \left.\frac{\mathrm{d}w}{\mathrm{d}x}\right|_{x=h} = 0 \tag{5.122}$$

$$w(0) = W \tag{5.123}$$

where Equation 5.122 represents the requirement that the shear stress on the surface of the fluid layer (at $x = h$) is equal to zero (which assumes that the air exerts negligible shear

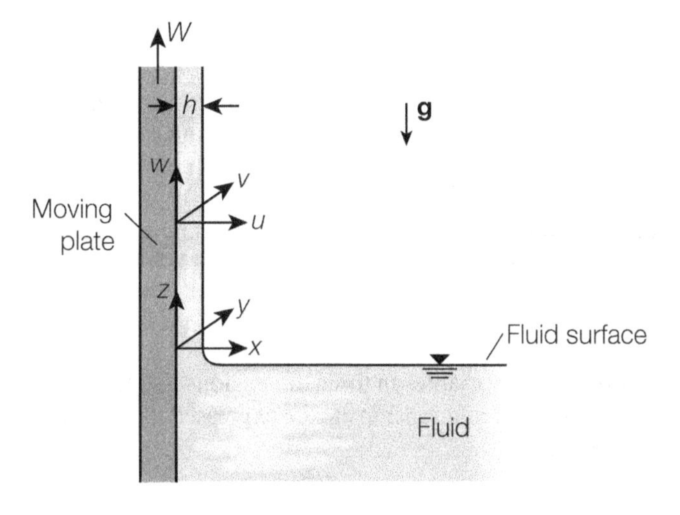

Figure 5.18: Vertical moving plate

stress on the fluid) and Equation 5.123 represents the no-slip requirement on the surface of the moving plate. Solving Equation 5.121 subject to the boundary conditions given by Equations 5.122 and 5.123 yields the following velocity distribution in the fluid layer adjacent to the moving plate:

$$w = \frac{\gamma}{2\mu}x^2 - \frac{\gamma h}{\mu}x + W \tag{5.124}$$

where $\gamma = \rho g$. The volume flow rate per unit width of the plate, q, and the average velocity within the fluid layer, V, are derived from the velocity distribution as follows:

$$q = \int_0^h w\,\mathrm{d}x \rightarrow q = Wh - \frac{\gamma h^3}{3\mu} \quad \text{and} \quad V = \frac{q}{h} \rightarrow V = W - \frac{\gamma h^2}{3\mu} \tag{5.125}$$

It is interesting to note that according to Equation 5.124, it is possible for $w(x) < 0$ for some outer values of x, which means that the fluid adjacent to the plate is flowing upward and the fluid in an outer region of the fluid layer is flowing downward. This condition is more likely to occur in low-viscosity (i.e., small μ) fluids and in instances when W is relatively small. Related to this phenomenon is the result that the net flow is downward (i.e., $q \leq 0$) when $W \leq \gamma h^2/3\mu$. Although the results presented in this section are applied to a moving plate, the results are equally applicable when the moving plate is replaced by a moving belt.

EXAMPLE 5.13

Show that $\gamma h^2/\mu W$ is the single dimensionless group that determines the velocity distribution in a fluid layer of thickness h adjacent to a vertical flat plate, where the plate moves with velocity W through a reservoir of fluid having a specific weight γ and dynamic viscosity μ. Verify that your result is consistent with the result that you would get using the Buckingham pi theorem. Compare the theoretical velocity distributions for values of $\gamma h^2/\mu W$ equal to 2 and 4.

SOLUTION

The velocity distribution in the fluid layer adjacent to the plate is given by Equation 5.124; dividing this equation by W and rearranging yields

$$\frac{w}{W} = \frac{1}{2}\left(\frac{\gamma h^2}{\mu W}\right)\left(\frac{x}{h}\right)^2 - \left(\frac{\gamma h^2}{\mu W}\right)\left(\frac{x}{h}\right) + 1 \tag{5.126}$$

Hence, the distribution of the normalized velocity w/W, with respect to the normalized distance from the surface of the plate, x/h, depends only on the value of the nondimensional group $\gamma h^2/\mu W$.

If the velocity distribution in the fluid layer was unknown, it could nevertheless be represented by the following functional relationship:

$$w = f_1(W, x, h, \gamma, \mu)$$

Because this relationship involves six variables in three dimensions (M, L, T), according to the Buckingham pi theorem the functional relationship can be expressed as a relationship between three dimensionless groups. Hence,

$$\frac{w}{W} = f_2\left(\frac{x}{h}, \frac{\gamma h^2}{\mu W}\right) \tag{5.127}$$

It is apparent that Equation 5.127 is a functional expression of Equation 5.126; therefore, the theoretical result is consistent with the Buckingham pi theorem.

The velocity distribution for $\gamma h^2/\mu W = 2$ is derived from Equation 5.126 and is shown in Figure 5.19(a); the velocity distribution for $\gamma h^2/\mu W = 4$ is shown in Figure 5.19(b).

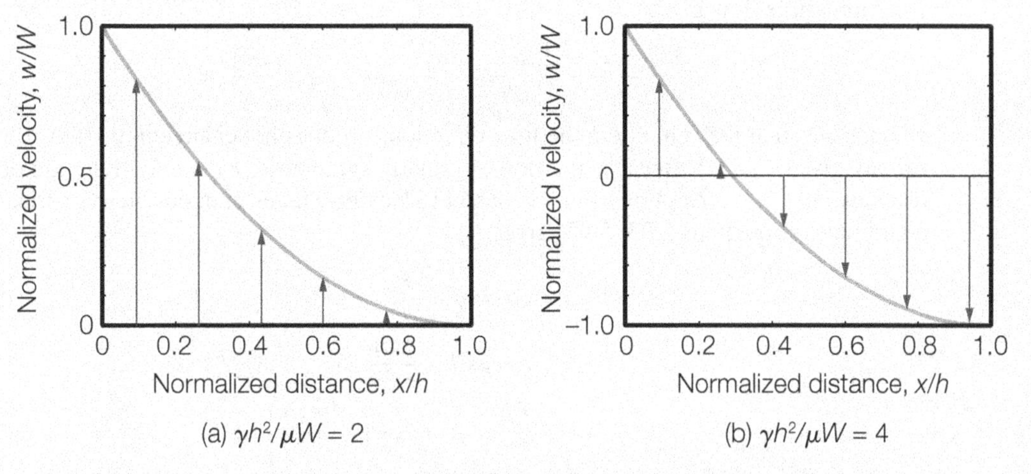

Figure 5.19: Velocity distribution adjacent to a vertical plate

It is apparent that the flow is generally in the same direction as the moving plate when $\gamma h^2/\mu W = 2$, whereas when $\gamma h^2/\mu W = 4$, the flow is upward near the plate and downward for x/h greater that around 0.29. These results are consistent with the theoretical result that the net flow is downward when $\gamma h^2/\mu W > 3$. It is also apparent from Figures 5.19(a) and (b) that $\mathrm{d}w/\mathrm{d}x = 0$ at $x/h = 1$, which is consistent with the included boundary condition that the shear stress is equal to zero at this interface with the surrounding air.

Flow down an inclined plane. A case that is closely related to that of a viscous liquid flowing adjacent to a moving vertical plane surface is that of a viscous liquid flowing down a fixed inclined plane surface under steady-state conditions as shown in Figure 5.20. An exact solution of this problem can be derived by direct application of the Navier–Stokes equation and is given as a practical exercise in Problem 5.46. In this case, the velocity distribution and the volume flow rate down the incline are given by

$$u(y) = \frac{\gamma \sin\theta}{2\mu}(2h - y)y \quad \text{and} \quad q = \frac{\gamma h^3 \sin\theta}{3\mu} \tag{5.128}$$

where $u(y)$ is the velocity at a distance y from the surface of the inclined plane (measured normal to the plane surface), θ is the angle of inclination of the surface relative to the horizontal, and h is the depth of flow of the liquid.

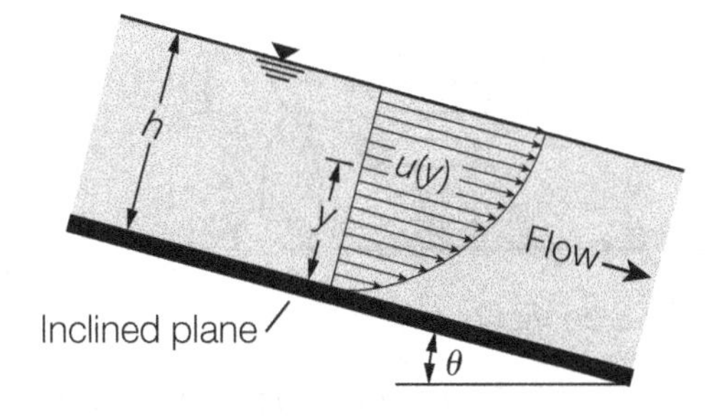

Figure 5.20: Flow down an inclined plane

5.5.4 Steady Laminar Flow Through a Circular Tube

Consider the case of steady laminar flow through a horizontal tube of circular cross section and of radius R as shown in Figure 5.21. In this case, because of the cylindrical symmetry, it is convenient to work with cylindrical polar coordinates (r, θ, x) and with velocity components v_r, v_θ, and v_x. Because $v_r = 0$ and $v_\theta = 0$, the continuity equation, Equation 5.44, for steady incompressible flow gives

$$\frac{1}{r}\frac{\partial(rv_r)}{\partial r} + \frac{1}{r}\frac{\partial v_\theta}{\partial \theta} + \frac{\partial v_x}{\partial x} = 0 \quad \rightarrow \quad \frac{\partial v_x}{\partial x} = 0 \tag{5.129}$$

which states that the velocity in the flow direction, v_x, does not change in the flow direction for any given r and θ. Because the flow is radially symmetric, v_x is not a function of θ and depends only on r. Applying this result to the Navier–Stokes equation in cylindrical polar coordinates (Equations 5.93–5.95) gives

$$0 = -\rho g \sin\theta - \frac{\partial p}{\partial r} \tag{5.130}$$

$$0 = -\rho g \cos\theta - \frac{1}{r}\frac{\partial p}{\partial \theta} \tag{5.131}$$

$$0 = -\frac{\partial p}{\partial x} + \mu\left[\frac{1}{r}\frac{\partial}{\partial r}\left(r\frac{\partial v_x}{\partial r}\right)\right] \tag{5.132}$$

Integrating and combining Equations 5.130 and 5.131 gives

$$p = -\rho g(r\sin\theta) + f(x) \tag{5.133}$$

which indicates that the pressure is distributed hydrostatically perpendicular to the flow direction and that the pressure gradient, $\partial p/\partial x$, is not a function of r or θ. The pressure gradient, $\partial p/\partial x$, must necessarily be negative because it is the pressure that drives the flow against the frictional resistance. So the pressure must necessarily decrease in the downstream direction. Also, for uniform flow in the x-direction, $\partial p/\partial x$ must necessarily be a constant that is independent of x. Integrating Equation 5.132 twice and using the boundary conditions that v_x is finite at $r = 0$ and $v_x = 0$ at $r = R$, yields the velocity distribution

$$v_x = \frac{1}{4\mu}\left(\frac{\partial p}{\partial x}\right)(r^2 - R^2) \tag{5.134}$$

This parabolic velocity distribution is illustrated in Figure 5.21. The volume flow rate, Q, through the circular tube can be determined from Equation 5.134 using the relation

$$Q = \int_0^R v_x 2\pi r\,\mathrm{d}r \rightarrow Q = -\frac{\pi R^4}{8\mu}\left(\frac{\partial p}{\partial x}\right) \rightarrow Q = -\frac{\pi D^4}{128\mu}\left(\frac{\partial p}{\partial x}\right) \tag{5.135}$$

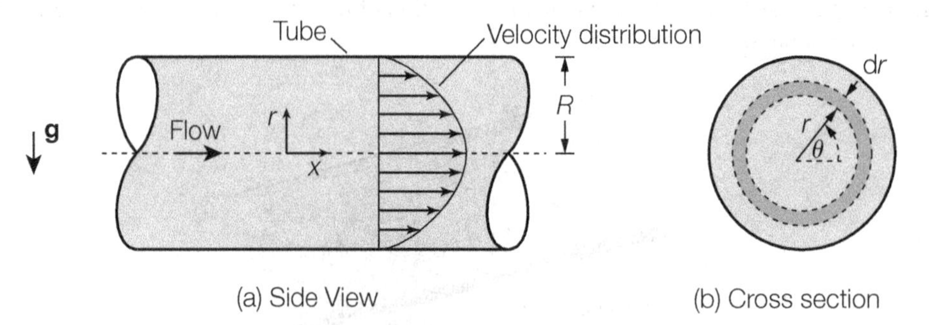

Figure 5.21: Laminar flow in a tube

where the flow rate has been preferably expressed in terms of the tube diameter, $D = 2R$, because it is more common to describe the size of a tube in terms of its diameter rather than its radius. It is apparent from Equation 5.135 that the volume flow rate increases as the fourth power of the diameter of the tube. The average velocity, V, and the maximum velocity in the tube, v_{max}, can be derived directly from Equations 5.134 and 5.135 as

$$V = \frac{Q}{A} = \frac{Q}{\pi D^2/4} \rightarrow V = -\frac{D^2}{32\mu}\left(\frac{\partial p}{\partial x}\right) \quad \text{and} \quad v_{\text{max}} = v_x(0) \rightarrow v_{\text{max}} = -\frac{D^2}{16\mu}\left(\frac{\partial p}{\partial x}\right) \tag{5.136}$$

where it is apparent that the average velocity is proportional to the second power of the diameter of the tube and the maximum velocity is equal to twice the average velocity.

Hagen-Poiseuille flow. The flow field represented by Equation 5.134 is called *Hagen-Poiseuille flow*[33] or *Poiseuille flow*, and Equation 5.135 is known as the *Hagen-Poiseuille formula* or *Poiseuille's formula*. Interestingly, neither Hagen nor Poiseuille derived Equation 5.135, but they both laid the groundwork for its development. The *Hagen-Poiseuille formula* has been widely validated by experiments, and it is this agreement with experiments that provides the primary validation that a fluid continuum does not slip past a solid boundary. The *Hagen-Poiseuille formula* was derived for laminar flow in a straight tube and is not applicable to curved tubes, although the formula yields approximate results when the curvature of the tube is slight (i.e., when the radius of curvature is large).

Laminar-flow restriction. The equations derived in this section apply only to laminar flow, which is estimated to occur at values of the Reynolds number (Re) less than 2100. This requirement can be expressed as

$$\text{Re} = \frac{\rho V D}{\mu} < 2100 \tag{5.137}$$

In cases where Equation 5.137 is violated, the flow is turbulent and the velocity is three-dimensional and unsteady. Under these conditions, the laminar-flow equations derived in this section are not applicable.

EXAMPLE 5.14

Gasoline at 20°C flows through the 5-mm-diameter tube shown in Figure 5.22, where upstream and downstream pressure measurements taken at gauges 5 m apart are 120 Pa and 70 Pa, respectively. Determine the volume flow rate through the tube and verify that the flow is laminar. What is the range of velocity across any cross section of the tube?

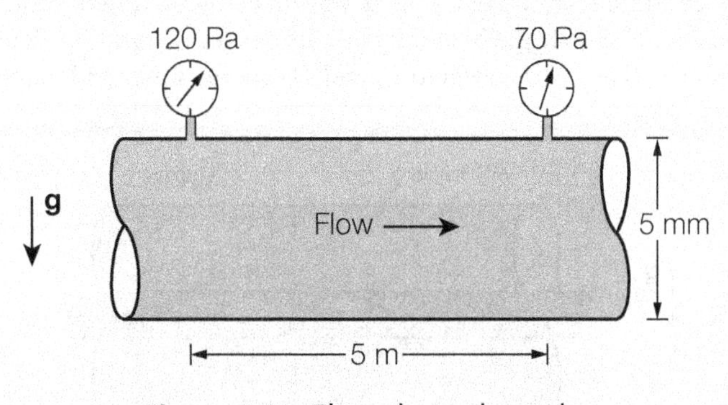

Figure 5.22: Flow through a tube

[33]Named in honor of the German hydraulic engineer Gotthilf H. L. Hagen (1797–1884) and the French physician Jean Louis Marie Poiseuille (1799–1869).

SOLUTION

From the given data: p_1 = 120 Pa, p_2 = 70 Pa, Δx = 5 m, and D = 5 mm = 0.005 m. At 20°C, the properties of gasoline are ρ = 680 kg/m^3 and μ = 0.29 mPa·s = 2.9 × 10^{-4} Pa·s (from Appendix B). The derived parameters are as follows:

$$\frac{\partial p}{\partial x} = \frac{p_2 - p_1}{\Delta x} = \frac{70 - 120}{5} = -10 \text{ Pa/m}$$

$$Q = -\frac{\pi D^4}{128\mu}\left(\frac{\partial p}{\partial x}\right) = -\frac{\pi(0.005)^4}{128(2.9 \times 10^{-4})}(-10) = 5.29 \times 10^{-7} \text{ m}^3\text{/s} = 0.529 \text{ mL/s}$$

$$A = \frac{\pi}{4}D^2 = \frac{\pi}{4}(0.005)^2 = 1.96 \times 10^{-5} \text{ m}^2$$

$$V = \frac{Q}{A} = \frac{5.29 \times 10^{-7}}{1.96 \times 10^{-5}} = 0.0270 \text{ m/s} = 27.0 \text{ mm/s}$$

$$v_\text{max} = 2V = 2(0.0270) = 0.054 \text{ m/s} = 54.0 \text{ mm/s}$$

$$\text{Re} = \frac{\rho V D}{\mu} = \frac{(680)(0.054)(0.005)}{2.9 \times 10^{-4}} = 633$$

Based on these results, the volume flow rate through the tube is **0.529 mL/s** and **the flow is laminar** because Re $=$ 633 $<$ 2100. The confirmation of laminar flow validates using the laminar-flow theoretical expression for Q. The velocities across any section of the tube vary in the range of 0–v_max, which in this case is **0–54 mm/s**.

5.5.5 Steady Laminar Flow Through an Annulus

Consider the case of steady flow through an annulus as shown in Figure 5.23. The simplified continuity and momentum equations describing flow in a tube, Equations 5.129–5.132, are still applicable. However the boundary conditions for integrating Equation 5.132 are v_x = 0 at $r = R_\text{i}$ and v_x = 0 at $r = R_\text{o}$, where R_i and R_o are the inner and outer radii, respectively. Ap-

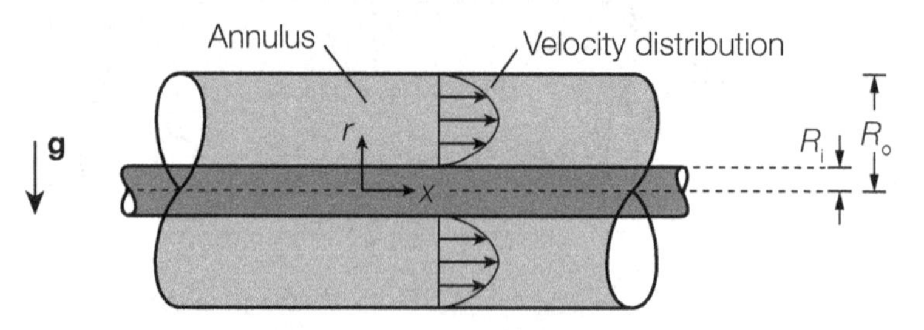

Figure 5.23: Laminar flow through an annulus

plying these boundary conditions yields the following velocity distribution within the annular region:

$$v_x = \frac{1}{4\mu}\left(\frac{\partial p}{\partial x}\right)\left[r^2 - R_\mathrm{o}^2 + \frac{R_\mathrm{i}^2 - R_\mathrm{o}^2}{\ln(R_\mathrm{o}/R_\mathrm{i})}\ln\frac{r}{R_\mathrm{o}}\right] \tag{5.138}$$

This parabolic velocity distribution is illustrated in Figure 5.23. The radial location, R_max, at which the maximum velocity occurs can be found by setting $\mathrm{d}v_x/\mathrm{d}r$ equal to zero and solving for r (= R_max), which yields

$$R_\mathrm{max} = \left[\frac{R_\mathrm{o}^2 - R_\mathrm{i}^2}{2\ln(R_\mathrm{o}/R_\mathrm{i})}\right]^{\frac{1}{2}} \tag{5.139}$$

This result indicates that the maximum velocity does not occur at the midpoint of the annular space, but occurs nearer the inner cylinder with the exact location depending on R_i and R_o. The maximum velocity can be determined by substituting $r = R_\mathrm{max}$ into Equation 5.138. The volume flow rate, Q, can be determined from Equation 5.138 using the relation

$$Q = \int_{R_\mathrm{i}}^{R_\mathrm{o}} v_x 2\pi r\,\mathrm{d}r \rightarrow Q = -\frac{\pi}{8\mu}\left(\frac{\partial p}{\partial x}\right)\left[R_\mathrm{o}^4 - R_\mathrm{i}^4 - \frac{(R_\mathrm{o}^2 - R_\mathrm{i}^2)^2}{\ln(R_\mathrm{o}/R_\mathrm{i})}\right] \tag{5.140}$$

The average velocity, V, in the annulus can be obtained by dividing the volume flow rate, Q, by the cross-sectional area, $\pi(R_\mathrm{o}^2 - R_\mathrm{i}^2)$, which gives

$$V = -\frac{1}{8\mu}\left(\frac{\partial p}{\partial x}\right)\left[R_\mathrm{o}^2 + R_\mathrm{i}^2 - \frac{R_\mathrm{o}^2 - R_\mathrm{i}^2}{\ln(R_\mathrm{o}/R_\mathrm{i})}\right] \tag{5.141}$$

It is noteworthy that as $R_\mathrm{i} \rightarrow 0$, the analytic expressions derived here for flow through an annulus become the same as those derived previously for flow through a circular tube.

Laminar-flow restriction. The equations derived in this section apply only to laminar flow, which usually occurs at Reynolds numbers (Re) less than 2100. This requirement can be expressed as

$$\mathrm{Re} = \frac{\rho V D_\mathrm{h}}{\mu} < 2100 \tag{5.142}$$

where D_h is the *hydraulic diameter*, which is defined as the ratio of four times the flow area, A, to the perimeter, P, of the conduit that is in contact with the fluid. Hence, for flow in an annulus, D_h is given by

$$D_\mathrm{h} = \frac{4A}{P} = \frac{4\pi(R_\mathrm{o}^2 - R_\mathrm{i}^2)}{2\pi(R_\mathrm{i} + R_\mathrm{o})} = 2(R_\mathrm{o} - R_\mathrm{i}) \tag{5.143}$$

In general, D_h is used to describe the characteristic size of noncircular conduits because it is equal to the actual diameter D when the conduit is circular. In cases where the Reynolds number of the flow, Re, as defined by Equation 5.142, exceeds 2100, the flow is turbulent and the laminar-flow equations derived in this section are not applicable.

EXAMPLE 5.15

Two reservoirs are connected by a 2-m-long annulus with an inner diameter of 2 mm and an outer diameter of 5 mm as shown in Figure 5.24. The fluid in the system has a density of 920 kg/m^3 and a dynamic viscosity of 0.300 Pa·s. (a) Determine the pressure difference between the reservoirs that would be required to obtain a flow rate of 10 mL/s between the reservoirs. (b) What force must be applied to support the inner (solid) cylinder? (c) If the inner cylinder was removed and the pressure differential maintained, what percentage increase in flow rate would be obtained?

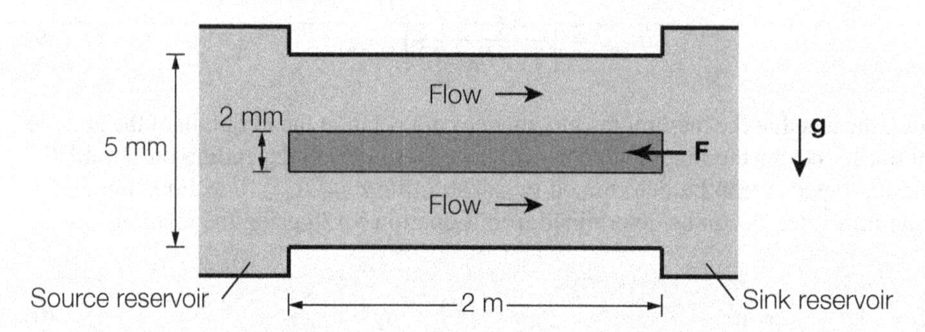

Figure 5.24: Flow in an annulus between reservoirs

SOLUTION

From the given data: L = 2 m, D_i = 2 mm, D_o = 5 mm, ρ = 920 kg/m^3, and μ = 0.300 Pa·s. The inner and outer radii of the annulus are $R_\text{i} = D_\text{i}/2 = 0.001$ m and $R_\text{o} = D_\text{o}/2 = 0.0025$ m, respectively.

(a) For Q = 10 mL/s = 10^{-5} m^3/s, check whether the flow is laminar.

$$A = \pi(R_\text{o}^2 - R_\text{i}^2) = \pi(0.0025^2 - 0.001^2) = 1.649 \times 10^{-5} \text{ m}^2$$

$$D_\text{h} = 2(R_\text{o} - R_\text{i}) = 2(0.0025 - 0.001) = 0.0030 \text{ m}$$

$$V = \frac{Q}{A} = \frac{10^{-5}}{1.649 \times 10^{-5}} = 0.606 \text{ m/s}$$

$$\text{Re} = \frac{\rho V D_\text{h}}{\mu} = \frac{(920)(0.606)(0.0030)}{0.300} = 11$$

Because Re = 11 < 2100, the flow is laminar and the theoretical equations for laminar flow can be applied. For a flow rate of 10^{-5} m^3/s,

$$Q = -\frac{\pi}{8\mu}\left(\frac{\partial p}{\partial x}\right)\left[R_\text{o}^4 - R_\text{i}^4 - \frac{(R_\text{o}^2 - R_\text{i}^2)^2}{\ln(R_\text{o}/R_\text{i})}\right]$$

$$10^{-5} = -\frac{\pi}{8(0.300)}\left(\frac{\partial p}{\partial x}\right)\left[0.0025^4 - 0.001^4 - \frac{(0.0025^2 - 0.001^2)^2}{\ln(0.0025/0.001)}\right]$$

$$\rightarrow \left(\frac{\partial p}{\partial x}\right) = -4.73 \text{ Pa/m}$$

Hence, the required pressure gradient is −4.73 Pa/m. Because the distance between reservoirs is 2 m, the required pressure differential is (4.73)(2)= **9.46 Pa**.

(b) A general expression for the shear stress, τ_i, on the inner (solid) cylinder can be derived as follows:

$$\tau_\mathrm{i} = \mu \left.\frac{\mathrm{d}v_x}{\mathrm{d}r}\right|_{r=R_\mathrm{i}} = \mu \frac{\mathrm{d}}{\mathrm{d}r}\left\{\frac{1}{4\mu}\left(\frac{\partial p}{\partial x}\right)\left[r^2 - R_\mathrm{o}^2 + \frac{R_\mathrm{i}^2 - R_\mathrm{o}^2}{\ln(R_\mathrm{o}/R_\mathrm{i})}\ln\frac{r}{R_\mathrm{o}}\right]\right\}_{r=R_\mathrm{i}}$$

$$= \mu\frac{1}{4\mu}\left(\frac{\partial p}{\partial x}\right)\left[2r + \frac{R_\mathrm{i}^2 - R_\mathrm{o}^2}{\ln(R_\mathrm{o}/R_\mathrm{i})}\frac{1}{r}\right]_{r=R_\mathrm{i}} = \frac{1}{4}\left[2R_\mathrm{i} + \frac{R_\mathrm{i}^2 - R_\mathrm{o}^2}{\ln(R_\mathrm{o}/R_\mathrm{i})}\frac{1}{R_\mathrm{i}}\right]$$

Therefore, in this particular case, the shear stress and the resulting force, F, on the surface of the inner cylinder are given by

$$\tau_\mathrm{i} = \frac{1}{4}\left[2(0.001) + \frac{0.001^2 - 0.0025^2}{\ln(0.0025/0.001)}\frac{1}{0.001}\right] = -9.32 \times 10^{-4}\ \mathrm{Pa}$$

$$F = |\tau_\mathrm{i}|\,(\pi D_\mathrm{i} L) = (9.32 \times 10^{-4})\pi(0.002)(2) = 1.175 \times 10^{-5}\ \mathrm{N}$$

Hence, the resulting force on the inner cylinder is approximately **11.8 μN**.

(c) If the inner cylinder is removed, then purely tube flow is obtained, which can be described by taking $R_\mathrm{i} = 0$, and yields

$$Q = -\frac{\pi}{8\mu}\left(\frac{\partial p}{\partial x}\right)R_\mathrm{o}^4 = -\frac{\pi}{8(0.300)}(-9.46)(0.0025)^4 = 1.29 \times 10^{-5}\ \mathrm{m^3/s} = 12.9\ \mathrm{mL/s}$$

Therefore, removing the inner cylinder increases the flow rate between reservoirs by $(12.9 - 10)/10 \times 100$ = **12.9%**.

5.5.6 Steady Laminar Flow Between Rotating Cylinders

Consider the case of a fluid contained in the annulus between a rotating cylinder on the inside and a stationary cylinder on the outside. This situation commonly occurs in the field of lubrication, where the fluid is usually oil and the inner cylinder is a rotating shaft. This condition is illustrated in Figure 5.25, where R_i and R_o are the radii of the inner and outer cylinders, respectively, ω is the rate of rotation of the inner cylinder, and r is the distance from the center of the inner cylinder. The flow induced by the rotating cylinder is in the (r, θ) plane,

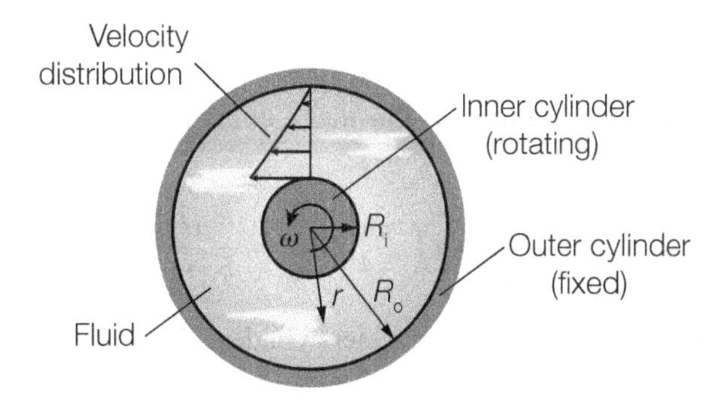

Figure 5.25: Laminar flow between rotating cylinders

the z-coordinate direction is along the axes of the inner and outer cylinders (perpendicular to the page), and $v_z = 0$ because there is no axial motion. The continuity equation is therefore given by

$$\frac{1}{r}\frac{\partial(rv_r)}{\partial r} + \frac{1}{r}\frac{\partial v_\theta}{\partial \theta} = 0 \tag{5.144}$$

Because the induced flow will have circular symmetry, $\partial v_\theta/\partial\theta = 0$, and Equation 5.144 simplifies to

$$\frac{1}{r}\frac{\mathrm{d}(rv_r)}{\mathrm{d}r} = 0 \quad \rightarrow \quad rv_r = \text{constant} \tag{5.145}$$

Because v_r must equal zero at both $r = R_\mathrm{i}$ and $r = R_\mathrm{o}$, the "constant" in Equation 5.145 must be equal to zero, which means that $v_r = 0$ everywhere between the cylinders. This leaves v_θ as the only nonzero component of the velocity, and v_θ must necessarily be only a function of r. The θ component of the momentum equation is given by Equation 5.94 as

$$\rho\left(\frac{\partial v_\theta}{\partial t} + v_r\frac{\partial v_\theta}{\partial r} + \frac{v_\theta}{r}\frac{\partial v_\theta}{\partial \theta} - \frac{v_r v_\theta}{r} + v_z\frac{\partial v_\theta}{\partial z}\right) = \tag{5.146}$$

$$-\frac{1}{r}\frac{\partial p}{\partial \theta} + \rho g_\theta + \mu\left[\frac{1}{r}\frac{\partial}{\partial r}\left(r\frac{\partial v_\theta}{\partial r}\right) - \frac{v_\theta}{r^2} + \frac{1}{r^2}\frac{\partial^2 v_\theta}{\partial \theta^2} + \frac{2}{r^2}\frac{\partial v_r}{\partial \theta} + \frac{\partial^2 v_\theta}{\partial z^2}\right]$$

which in this case simplifies to

$$0 = \frac{1}{r}\frac{\mathrm{d}}{\mathrm{d}r}\left(r\frac{\mathrm{d}v_\theta}{\mathrm{d}r}\right) - \frac{v_\theta}{r^2} \tag{5.147}$$

where the partial derivatives have been replaced by total derivatives because v_θ is only a function of r. The solution of Equation 5.147 is

$$v_\theta = ar + \frac{b}{r} \tag{5.148}$$

where a and b are constants. The boundary conditions to be satisfied by Equation 5.148 are

$$v_\theta(R_\mathrm{i}) = R_\mathrm{i}\omega \tag{5.149}$$

$$v_\theta(R_\mathrm{o}) = 0 \tag{5.150}$$

Using Equations 5.149 and 5.150 to determine the constants a and b in Equation 5.148 yields the velocity distribution

$$v_\theta = \frac{R_\mathrm{o}/r - r/R_\mathrm{o}}{R_\mathrm{o}/R_\mathrm{i} - R_\mathrm{i}/R_\mathrm{o}} R_\mathrm{i}\omega \tag{5.151}$$

where the shape of this velocity profile is illustrated in Figure 5.25. The shear stress on the inner cylinder, τ_i, is given by

$$\tau_\mathrm{i} = -\left[\mu r\frac{\mathrm{d}}{\mathrm{d}r}\left(\frac{v_\theta}{r}\right)\right]_{r=R_\mathrm{i}} = \frac{2\mu R_\mathrm{o}^2\omega}{R_\mathrm{o}^2 - R_\mathrm{i}^2} \tag{5.152}$$

Hence, the torque, T, required to rotate the inner cylinder is given by

$$T = \tau_\mathrm{i}A_\mathrm{i}R_\mathrm{i} = \frac{4\pi\mu R_\mathrm{i}^2 R_\mathrm{o}^2 L\omega}{R_\mathrm{o}^2 - R_\mathrm{i}^2} \tag{5.153}$$

where A_i and L are the surface area and length of the inner cylinder, respectively. The power, P, required to rotate the inner shaft is given by

$$P = T\omega = \frac{4\pi\mu R_\mathrm{i}^2 R_\mathrm{o}^2 L\omega^2}{R_\mathrm{o}^2 - R_\mathrm{i}^2} \tag{5.154}$$

In accordance with the first law of thermodynamics, the power, P, given by Equation 5.154 causes an increase in internal energy of the fluid, which is manifested in an increase in temperature. A cooling mechanism is usually necessary to moderate the temperature of the fluid.

Flow restriction. Although the velocity distribution given by Equation 5.151 is mathematically exact, experiments have shown that this flow becomes unstable when

$$\frac{R_\mathrm{i}(R_\mathrm{o} - R_\mathrm{i})^3\omega^2}{\nu^2} \approx 1700 \tag{5.155}$$

where ν is the kinematic viscosity of the fluid. Whenever the flow becomes unstable, the plane flow described by Equation 5.151 becomes three dimensional in a pattern consisting of rows of nearly square toroidal vortices, sometimes called *Taylor vortices* after G. I. Taylor, who first identified them (Taylor, 1923). This type of flow is sometimes referred to as *Taylor-Couette flow*. The quantity on the left-hand side of Equation 5.155 is called the *Taylor number*, commonly denoted by "Ta."

EXAMPLE 5.16

A shaft of length 1.5 m and diameter of 50 mm rotates inside an outer cylinder of diameter 75 mm, with the volume between the rotating shaft and the outer cylinder filled with a lubricating fluid of density 950 kg/m^3 and dynamic viscosity 0.25 Pa·s. At what rotational speed will the flow of the lubricant become unstable, and what is the power required to drive the shaft at this rotational speed?

SOLUTION

From the given data: L = 1.5 m, D_i = 50 mm, D_o = 75 mm, ρ = 950 kg/m^3, and μ = 0.25 Pa·s. Hence, $R_\mathrm{i} = D_\mathrm{i}/2$ = 25 mm = 0.025 m, $R_\mathrm{o} = D_\mathrm{o}/2$ = 37.5 mm = 0.0375 m, and $\nu = \mu/\rho = 0.25/950 = 2.63 \times 10^{-4}$ m^2/s. The flow of the lubricant becomes unstable when Ta $\approx$ 1700, and according to Equation 5.155, this occurs (approximately) when

$$\omega = \sqrt{\frac{1700\nu^2}{R_\mathrm{i}(R_\mathrm{o} - R_\mathrm{i})^3}} = \sqrt{\frac{1700(2.63 \times 10^{-4})^2}{0.025(0.0375 - 0.025)^3}} = 49.1 \text{ rad/s} = 469 \text{ rpm}$$

The power, P, required to rotate the shaft at 469 rpm (= 49.1 rad/s) is given by Equation 5.154 as

$$P = \frac{4\pi\mu R_\mathrm{i}^2 R_\mathrm{o}^2 L\omega^2}{R_\mathrm{o}^2 - R_\mathrm{i}^2} = \frac{4\pi(0.25)(0.025)^2(0.0375)^2(1.5)(49.1)^2}{0.0375^2 - 0.025^2} = 12.8 \text{ W}$$

Therefore, the flow of the lubricant will become unstable at a rotational speed of **469 rpm**, and the power required to rotate the shaft at this speed is **12.8 W**.

5.6 Inviscid Flow

The word "inviscid," means "not viscous," and is a term that is commonly used to describe either a fluid property or a flow property. When used to describe a fluid property, an *inviscid fluid* is a fluid in which the viscosity is negligibly small, and when used to describe a flow property, *inviscid flow* occurs when viscous forces are negligible compared to other driving forces such as pressure and gravity. Inviscid flows are generally characterized by high Reynolds numbers. The class of flows where the inviscid flow approximation is most widely applicable is *external flows*, where the fluid flow is external to a solid boundary. In these types of flows, viscous effects are usually restricted to regions that are very close to the solid boundary. Under inviscid flow conditions, viscous forces are negligible compared to inertial forces, Re $\gg$ 1, and in accordance with Equation 5.99, the Navier–Stokes equation becomes

$$\rho\left[\frac{\partial \mathbf{V}}{\partial t} + (\mathbf{V}\cdot\nabla)\mathbf{V}\right] = -\nabla p + \rho\mathbf{g} \tag{5.156}$$

which is called the *Euler equation*. The Euler equation given by Equation 5.156 can also be conveniently represented using the material derivative as

$$\rho\frac{\mathrm{D}\mathbf{V}}{\mathrm{D}t} = -\nabla p + \rho\mathbf{g} \tag{5.157}$$

where, for a unit volume of a fluid element, the left-hand side of Equation 5.157 is equal to the mass times the acceleration of the fluid element and the right-hand side of Equation 5.157 is equal to the sum of the net pressure and the gravitational force acting on the fluid element. Typically, the Euler equation is used in high-Reynolds-number regions of the flow, where viscous forces are negligible, and far away from walls and wakes. The Euler equation is typically used as a first approximation to the governing equations in computational fluid dynamics (CFD) codes. Because the Euler equation is so widely used in a variety of applications, the forms of the Euler equation in both Cartesian and cylindrical polar coordinates are given below.

Euler equation in Cartesian coordinates. The Euler equation in Cartesian coordinates is given by

$$\rho\left(\frac{\partial u}{\partial t} + u\frac{\partial u}{\partial x} + v\frac{\partial u}{\partial y} + w\frac{\partial u}{\partial z}\right) = -\frac{\partial p}{\partial x} + \rho g_x \tag{5.158}$$

$$\rho\left(\frac{\partial v}{\partial t} + u\frac{\partial v}{\partial x} + v\frac{\partial v}{\partial y} + w\frac{\partial v}{\partial z}\right) = -\frac{\partial p}{\partial y} + \rho g_y \tag{5.159}$$

$$\rho\left(\frac{\partial w}{\partial t} + u\frac{\partial w}{\partial x} + v\frac{\partial w}{\partial y} + w\frac{\partial w}{\partial z}\right) = -\frac{\partial p}{\partial z} + \rho g_z \tag{5.160}$$

Euler equation in cylindrical polar coordinates. The Euler equation in cylindrical polar coordinates is given by

$$\rho\left(\frac{\partial v_r}{\partial t} + v_r\frac{\partial v_r}{\partial r} + \frac{v_\theta}{r}\frac{\partial v_r}{\partial \theta} + v_z\frac{\partial v_r}{\partial z} - \frac{v_\theta^2}{r}\right) = -\frac{\partial p}{\partial r} + \rho g_r \tag{5.161}$$

$$\rho\left(\frac{\partial v_\theta}{\partial t} + v_r\frac{\partial v_\theta}{\partial r} + \frac{v_\theta}{r}\frac{\partial v_\theta}{\partial \theta} + v_z\frac{\partial v_\theta}{\partial z} + \frac{v_r v_\theta}{r}\right) = -\frac{1}{r}\frac{\partial p}{\partial \theta} + \rho g_\theta \tag{5.162}$$

$$\rho\left(\frac{\partial v_z}{\partial t} + v_r\frac{\partial v_z}{\partial r} + \frac{v_\theta}{r}\frac{\partial v_z}{\partial \theta} + v_z\frac{\partial v_z}{\partial z}\right) = -\frac{\partial p}{\partial z} + \rho g_z \tag{5.163}$$

Solution of the Euler equations. Solution of the Euler equations to describe flows that are approximately inviscid generally requires the specification of boundary conditions describing the particular problem at hand. An essential boundary condition at an impermeable surface is that the component of the fluid velocity normal to the surface is equal to zero and the component of the fluid velocity tangential to the surface is equal to the velocity surface; this latter boundary condition is called the *no-slip condition.* Whereas the normal-flow boundary condition can generally be enforced when using the Euler equation, because of the mathematical form of the Euler equation, it is not possible to require that the *no-slip condition* at a solid boundary be satisfied. As a consequence, the flows near solid boundaries calculated using the Euler equation are not accurately estimated, thereby limiting the utility of the Euler equation near solid boundaries.

EXAMPLE 5.17

The velocity components in a two-dimensional inviscid incompressible flow are $u = (-4x + 2y)t$ m/s and $v = (4y + 2x)t$ m/s, where x and y are the Cartesian coordinates in meters and t is the time in seconds. The gravity force acts in the negative z-direction, and the density of the fluid is equal to 1.20 kg/m^3. Determine the components of the pressure gradient at $x = 2$ m, $y = 1$ m, and $t = 3$ seconds.

SOLUTION

The given velocity field is $u = (-4x + 2y)t$ m/s and $v = (4y + 2x)t$ m/s; the density of the fluid is $\rho = 1.20$ kg/m^3. The following derived relationships at $x = 2$ m, $y = 1$ m, and $t = 3$ are relevant to this problem:

$$\frac{\partial u}{\partial t} = -4x + 2y \quad \rightarrow \quad \frac{\partial u}{\partial t} = -6 \text{ m/s}^2 \quad (\text{at } x = 2 \text{ m}, y = 1 \text{ m}, t = 3 \text{ s})$$

$$u\frac{\partial u}{\partial x} = [(-4x + 2y)t][-4t] \quad \rightarrow \quad u\frac{\partial u}{\partial x} = 216 \text{ m/s}^2 \quad (\text{at } x = 2 \text{ m}, y = 1 \text{ m}, t = 3 \text{ s})$$

$$v\frac{\partial u}{\partial y} = [(4y + 2x)t][2t] \quad \rightarrow \quad v\frac{\partial u}{\partial y} = 144 \text{ m/s}^2 \quad (\text{at } x = 2 \text{ m}, y = 1 \text{ m}, t = 3 \text{ s})$$

$$\frac{\partial v}{\partial t} = 4y + 2x \quad \rightarrow \quad \frac{\partial v}{\partial t} = 8 \text{ m/s}^2 \quad (\text{at } x = 2 \text{ m}, y = 1 \text{ m}, t = 3 \text{ s})$$

$$u\frac{\partial v}{\partial x} = [(-4x + 2y)t][2t] \quad \rightarrow \quad u\frac{\partial v}{\partial x} = -108 \text{ m/s}^2 \quad (\text{at } x = 2 \text{ m}, y = 1 \text{ m}, t = 3 \text{ s})$$

$$v\frac{\partial v}{\partial y} = [(4y + 2x)t][2t] \quad \rightarrow \quad v\frac{\partial v}{\partial y} = 144 \text{ m/s}^2 \quad (\text{at } x = 2 \text{ m}, y = 1 \text{ m}, t = 3 \text{ s})$$

Noting that $g_x = g_y = 0$ and $g_z = -9.807$ m/s^2, substituting the above results into the Euler equations (Equations 5.158–5.160) gives

$$x\text{-direction: } \rho\left(\frac{\partial u}{\partial t} + u\frac{\partial u}{\partial x} + v\frac{\partial u}{\partial y} + w\frac{\partial u}{\partial z}\right) = -\frac{\partial p}{\partial x} + \rho g_x$$

$$\rightarrow \quad (1.20)\,(-6 + 216 + 144 + 0) = -\frac{\partial p}{\partial x} + 0 \quad \rightarrow \quad \frac{\partial p}{\partial x} = -425 \text{ Pa/m}$$

$$y\text{-direction: } \rho\left(\frac{\partial v}{\partial t}+u\frac{\partial v}{\partial x}+v\frac{\partial v}{\partial y}+w\frac{\partial v}{\partial z}\right)=-\frac{\partial p}{\partial y}+\rho g_y$$

$$\rightarrow \quad (1.20)\,(8-108+144+0)=-\frac{\partial p}{\partial y}+0 \quad \rightarrow \quad \frac{\partial p}{\partial y}=-52.8\ \text{Pa/m}$$

$$z\text{-direction: } \rho\left(\frac{\partial w}{\partial t}+u\frac{\partial w}{\partial x}+v\frac{\partial w}{\partial y}+w\frac{\partial w}{\partial z}\right)=-\frac{\partial p}{\partial z}+\rho g_z$$

$$\rightarrow \quad (1.20)\,(0+0+0+0)=-\frac{\partial p}{\partial z}+(1.20)(-9.807) \quad \rightarrow \quad \frac{\partial p}{\partial z}=-11.8\ \text{Pa/m}$$

Combining the derived results gives the pressure gradient, ∇p, as

$$\boldsymbol{\nabla p=-425\,\mathbf{i}-52.8\,\mathbf{j}-11.8\,\mathbf{k}\ \text{Pa/m}}$$

5.6.1 Bernoulli Equation for Steady Inviscid Flow

Consider the case of steady inviscid flow, in which case the Euler equation as given in vector form by Equation 5.156 becomes

$$\rho(\mathbf{V}\cdot\boldsymbol{\nabla})\mathbf{V}=-\boldsymbol{\nabla}p+\rho\mathbf{g} \tag{5.164}$$

Taking the z-axis as vertical, the z-coordinate as positive upward, and the gravity force as acting vertically downward

$$\mathbf{g}=-g\boldsymbol{\nabla}z=-g\mathbf{k} \tag{5.165}$$

To simplify Equation 5.164, the following vector identity is useful:

$$(\mathbf{V}\cdot\boldsymbol{\nabla})\mathbf{V}=\frac{1}{2}\boldsymbol{\nabla}(V^2)-\mathbf{V}\times(\boldsymbol{\nabla}\times\mathbf{V}) \tag{5.166}$$

where V is the magnitude of $\mathbf{V}$. Substituting Equations 5.165 and 5.166 into Equation 5.164 and rearranging yields

$$\frac{\boldsymbol{\nabla}p}{\rho}+\frac{1}{2}\boldsymbol{\nabla}(V^2)+g\boldsymbol{\nabla}z=\mathbf{V}\times(\boldsymbol{\nabla}\times\mathbf{V}) \tag{5.167}$$

Defining a differential length vector **ds** that is in the same direction as the velocity vector, which also means that **ds** is parallel to a streamline, and taking the dot product of **ds** with each of the terms in Equation 5.167 yields

$$\frac{\boldsymbol{\nabla}p}{\rho}\cdot\mathbf{ds}+\frac{1}{2}\boldsymbol{\nabla}(V^2)\cdot\mathbf{ds}+g\boldsymbol{\nabla}z\cdot\mathbf{ds}=\mathbf{V}\times(\boldsymbol{\nabla}\times\mathbf{V})\cdot\mathbf{ds} \tag{5.168}$$

Because the cross product of two vectors yields a vector that is normal to both vectors, $\mathbf{V}\times(\boldsymbol{\nabla}\times\mathbf{V})$ is a vector normal to $\mathbf{V}$, and because **ds** is defined to be in the direction of $\mathbf{V}$ and the dot product of two orthogonal vectors is equal to zero

$$\mathbf{V}\times(\boldsymbol{\nabla}\times\mathbf{V})\cdot\mathbf{ds}=0 \tag{5.169}$$

It is also useful to note that for any scalar function f and differential length vector **ds**,

$$\boldsymbol{\nabla}f\cdot\mathbf{ds}=\mathrm{d}f \tag{5.170}$$

where $\mathrm{d}f$ is the change in f over the differential length vector **ds**. Applying Equations 5.169 and 5.170 to Equation 5.168 yields

$$\frac{\mathrm{d}p}{\rho}+\frac{1}{2}\mathrm{d}(V^2)+g\,\mathrm{d}z=0 \quad \text{(along a streamline)} \tag{5.171}$$

where Equation 5.171 is applicable along a streamline because **ds** is defined to be in the direction of a streamline (i.e., parallel to **V**). Equation 5.171 is sometimes used to determine the pressure gradient along the streamline, $\mathrm{d}p/\mathrm{d}s$, in which case Equation 5.171 is expressed in the form

$$\frac{\mathrm{d}p}{\mathrm{d}s} + \frac{1}{2}\rho\frac{\mathrm{d}(V^2)}{\mathrm{d}s} + \rho g\frac{\mathrm{d}z}{\mathrm{d}s} = 0 \tag{5.172}$$

where Equation 5.172 is derived from Equation 5.171 by multiplying both sides of Equation 5.171 by $\rho/\mathrm{d}s$. Integrating Equation 5.171 along a streamline yields

$$\int\frac{\mathrm{d}p}{\rho} + \frac{V^2}{2} + gz = \text{constant} \quad \text{(along a streamline)} \tag{5.173}$$

This is the classical Bernoulli equation that is applicable to both compressible and incompressible fluids in which the flow is steady and frictional forces are negligible. For incompressible fluids, ρ = constant, and Equation 5.173 becomes

$$\frac{p}{\rho} + \frac{V^2}{2} + gz = \text{constant} \quad \text{(along a streamline)} \tag{5.174}$$

These same forms of the Bernoulli equation were previously derived in Chapter 3 (Equations 3.41 and 3.47) by a simpler control-volume analysis in which the fluid was a priori assumed to be inviscid.

EXAMPLE 5.18

A fluid that is approximately incompressible and inviscid flows through a contracting pipe as shown in Figure 5.26. The pipe diameter decreases linearly from 300 mm to 100 mm over a distance of 2 m, and the pipe is inclined at an angle of 30° to the horizontal. The density of the fluid is 1000 kg/m^3, and under particular steady-state flow conditions, the velocity and pressure at the entrance of the contraction are 1.2 m/s and 300 kPa, respectively. (a) Determine the pressure as a function of the distance from the entrance of the contraction. (b) Determine the pressure gradient as a function of the distance from the entrance of the contraction. (c) Plot the pressure and pressure-gradient functions.

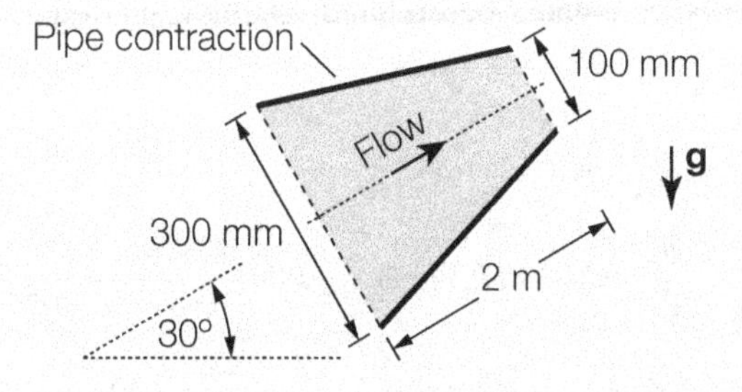

Figure 5.26: Flow through an inclined pipe

SOLUTION

From the given data: $D_1 = 300$ mm, $D_2 = 100$ mm, $L = 2$ m, $\theta = 30°$, $\rho = 1000$ kg/m^3, $V_1 = 1.2$ m/s, and $p_1 = 300$ kPa. Taking x as the distance from the entrance of the contraction, the following preliminary calculations are useful:

$$A_1 = \frac{\pi}{4}D_1^2 = \frac{\pi}{4}(0.3)^2 = 7.069 \times 10^{-2}\ \text{m}^2$$

$$Q = V_1 A_1 = (1.2)(7.069 \times 10^{-2}) = 8.482 \times 10^{-2}\ \text{m}^3/\text{s}$$

$$D(x) = 0.3 - 0.1x\ \text{m}, \qquad A(x) = \frac{\pi}{4}D(x)^2 = 0.7854(0.3 - 0.1x)^2\ \text{m}^2$$

$$V(x) = \frac{Q}{A(x)} = \frac{0.108}{(0.3 - 0.1x)^2}\ \text{m/s}$$

$$V(x)^2 = \frac{1.166 \times 10^{-2}}{(0.3 - 0.1x)^4}\ \text{m}^2/\text{s}^2, \qquad \frac{\text{d}V(x)^2}{\text{d}x} = \frac{4.664 \times 10^{-3}}{(0.3 - 0.1x)^5}\ \text{m/s}^2$$

$$z(x) = x \sin 30° = 0.5x\ \text{m}, \qquad \frac{\text{d}z}{\text{d}x} = 0.5$$

(a) Because the flow is steady, incompressible, and inviscid, the Bernoulli equation as given by Equation 5.174 can be applied. In this case, the centerline streamline is selected, and it is assumed that the given velocity and pressure are characteristic of the fluid properties on the centerline streamline at the entrance of the contraction. Determine the Bernoulli constant, C, by applying Equation 5.174 at the entrance of the contraction.

$$C = \frac{p_1}{\rho} + \frac{V_1^2}{2} + gz_1 = \frac{300 \times 10^3}{1000} + \frac{1.2^2}{2} + 0 = 300.7\ \text{m}^2/\text{s}^2$$

The pressure, $p(x)$ [kPa], within the contraction is determined by applying the Bernoulli equation along the centerline streamline as follows:

$$\frac{p(x)}{\rho} + \frac{V(x)^2}{2} + gz(x) = 300.7\ \text{m}^2/\text{s}^2$$

$$\rightarrow \frac{p(x)\,[\times 10^3\ \text{Pa/kPa}]}{1000} + \frac{1.166 \times 10^{-2}}{2(0.3 - 0.1x)^4} + (9.807)(0.5x) = 300.7$$

$$\rightarrow \boldsymbol{p(x) = 300.7 - 4.904x - \frac{5.832 \times 10^{-3}}{(0.3 - 0.1x)^4}\ \text{kPa}}$$

(b) The pressure gradient, $\text{d}p/\text{d}x$ [kPa/m], can be obtained by differentiating $p(x)$ or by using Equation 5.172. Normally, the advantage of using Equation 5.172 is that it does not require that $p(x)$ be calculated beforehand. Using Equation 5.172 with $\text{d}p/\text{d}x$ in kPa/m gives

$$\frac{\text{d}p}{\text{d}x} + \frac{1}{2}\rho\frac{\text{d}(V^2)}{\text{d}x} + \rho g\frac{\text{d}z}{\text{d}x} = 0$$

$$\rightarrow \frac{\text{d}p}{\text{d}x}[\times 10^3\ \text{Pa/kPa}] + \frac{1}{2}(1000)\frac{4.664 \times 10^{-3}}{(0.3 - 0.1x)^5} + (1000)(9.807)(0.5) = 0$$

$$\rightarrow \boldsymbol{\frac{\text{d}p}{\text{d}x} = -4.904 - \frac{2.332 \times 10^{-7}}{(0.3 - 0.1x)^5}\ \text{kPa/m}}$$

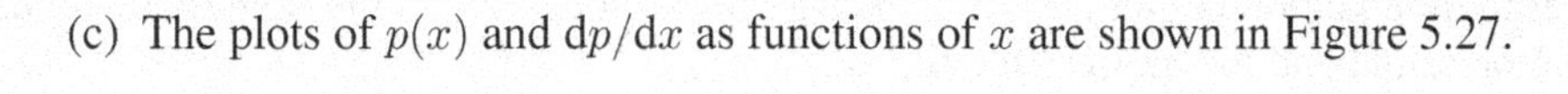

(c) The plots of $p(x)$ and $\mathrm{d}p/\mathrm{d}x$ as functions of x are shown in Figure 5.27.

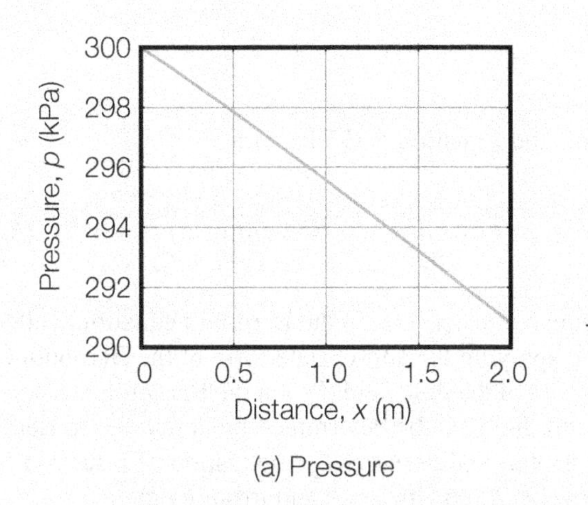

(a) Pressure

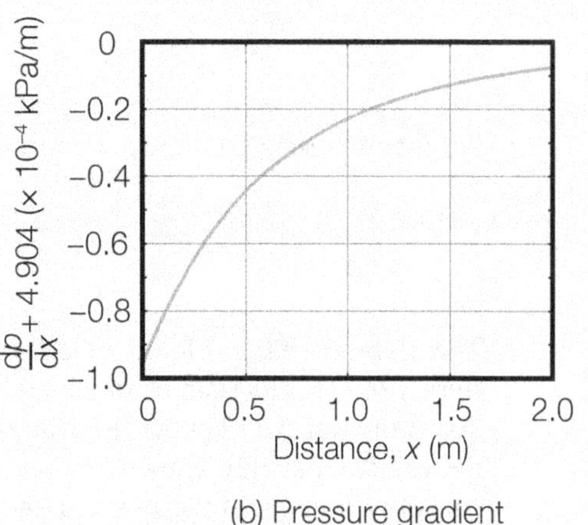

(b) Pressure gradient

Figure 5.27: **Pressure and pressure gradient in a contracting pipeline**

It is apparent from Figure 5.27(a) that the pressure, $p(x)$, decreases approximately linearly in the contraction, where $\mathrm{d}p/\mathrm{d}x$ remains approximately constant and equal to 4.904 kPa/m. To clarify the deviation of the slope from -4.904 Pa/m, the pressure gradient is plotted as $\mathrm{d}p/\mathrm{d}x + 4.904$. It is apparent from Figure 5.27(b) that deviations of the slope from -4.904 Pa/m are on the order of 10^{-4} kPa/m.

5.6.2 Bernoulli Equation for Steady Irrotational Inviscid Flow

It has been shown that the rate of rotation of a fluid element is measured by the vorticity vector, ζ, defined by the relation

$$\zeta = \nabla \times \mathbf{V} \tag{5.175}$$

where the vector components of the vorticity are equal to twice the angular velocity of the fluid element about the component axes. When a fluid is inviscid, the only forces acting on a fluid element are the weight of the element and pressure forces that are normal to the faces of the fluid element. Because none of these forces can cause the fluid element to rotate, it can be asserted that if some part of a flow field is irrotational, then all fluid elements emanating from this region will be irrotational. In other words, the rotation of a fluid element, can only be caused by shear forces on the surface of the fluid element and such forces are impossible in an inviscid fluid. For irrotational flows, the vorticity (and angular rotation) is equal to zero; hence,

$$\nabla \times \mathbf{V} = 0 \tag{5.176}$$

The vorticity of a fluid element cannot change except through the action of viscosity, nonuniform heating (temperature gradients), or other nonuniform phenomena. Rotation of fluid elements is associated with wakes, boundary layers, flow through turbomachinery (such as pumps, fans, and turbines), and flow with heat transfer.

Bernoulli equation in irrotational flow. Under irrotational-flow conditions, Equation 5.176 is true. Thus, Equation 5.169 must also be true regardless of the direction of **ds**, which was previously required to be a vector parallel to **V**. In other words, Equation 5.169 is true if **ds** is parallel to a streamline or the flow is irrotational. Therefore, if the flow is irrotational, then the Bernoulli equation (Equation 5.173) is true for all locations in the flow field and hence, for compressible and incompressible fluids.

$$\int \frac{dp}{\rho} + \frac{V^2}{2} + gz = \text{constant} \quad \text{(throughout the flow field)} \tag{5.177}$$

For incompressible fluids, ρ = constant, and Equation 5.177 becomes

$$\frac{p}{\rho} + \frac{V^2}{2} + gz = \text{constant} \quad \text{(throughout the flow field)} \tag{5.178}$$

Therefore, once the velocity field is known, the constant in the Bernoulli equation (called the *Bernoulli constant*) can be obtained by applying the appropriate form of the Bernoulli equation (Equation 5.177 or 5.178) to any point in the flow field. When the Bernoulli constant and the velocity field are known, the pressure field can be determined by applying the Bernoulli equation throughout the inviscid and irrotational flow field. Application of Equation 5.177 or 5.178 generally requires that the flow be steady, inviscid, and irrotational.

EXAMPLE 5.19

A two-dimensional flow field in the xy plane is given by $u = 2x + y$ m/s and $v = x - 2y$ m/s, where x and y are the Cartesian coordinates in meters. The gravity force acts in the negative z-direction, the density of the fluid is 1000 kg/m^3, and the pressure at the location (2 m, 1 m) is known to be equal to 105 kPa. (a) Determine an analytic expression for the pressure field. (b) What is the pressure at the point (4 m, 3 m)?

SOLUTION

The given velocity distribution is $u = 2x + y$ m/s and $v = x - 2y$ m/s. For the fluid, $\rho =$ 1000 kg/m^3, and at (x_0,y_0) = (2 m, 1 m), $p_0 = 105$ kPa. From the given velocity field,

$$\frac{\partial u}{\partial x} = 2, \qquad \frac{\partial v}{\partial y} = -2 \quad \rightarrow \quad \frac{\partial u}{\partial x} + \frac{\partial v}{\partial y} = 2 + (-2) = 0 \qquad (\therefore \text{ incompressible})$$

$$\frac{\partial v}{\partial x} = 1, \qquad \frac{\partial u}{\partial y} = 1 \quad \rightarrow \quad \omega_z = \frac{1}{2}\left(\frac{\partial v}{\partial x} - \frac{\partial u}{\partial y}\right) = 1 - 1 = 0 \quad (\therefore \text{ irrotational})$$

Because the flow is incompressible and irrotational, the Bernoulli equation as given by Equation 5.178 is applicable.

(a) Because the z-coordinate is constant in the xy plane, Equation 5.178 can be expressed as

$$\frac{p}{\rho} + \frac{V^2}{2} = C \tag{5.179}$$

where C is the Bernoulli constant for the flow. At the point (2 m, 1 m), $u = 2(2) + 1 =$ 5 m/s, $v = 2 - 2(1) = 0$ m/s, and $V = \sqrt{u^2 + v^2} = 5$ m/s. Because Equation 5.179 must be satisfied at this location,

$$\frac{105 \times 10^3}{1000} + \frac{5^2}{2} = C \quad \rightarrow \quad C = 117.5 \text{ m}^2/\text{s}^2$$

Using $C = 117.5$ m²/s², the pressure distribution in the xy plane can be derived using Equation 5.179, which yields

$$p = \rho\left[C - \frac{V^2}{2}\right] = (1000)\left[117.5 - \frac{(2x+y)^2 + (x-2y)^2}{2} Big\right][\times 10^{-3}\ \text{kPa/Pa}]$$

$$\rightarrow \quad \mathbf{p = 117.5 - 2.5(x^2 + y^2)\ kPa}$$

(b) At the location (4 m, 3 m), the pressure is calculated as

$$p = 117.5 - 2.5(x^2 + y^2) = 117.5 - 2.5(4^2 + 3^2) = \mathbf{55\ kPa}$$

5.6.3 Velocity Potential

For any scalar function such as $\phi(\mathbf{x})$, the quantity $\nabla\phi$ is called the gradient of ϕ and can be expressed in the form

$$\nabla\phi = \frac{\partial\phi}{\partial x}\mathbf{i} + \frac{\partial\phi}{\partial y}\mathbf{j} + \frac{\partial\phi}{\partial x}\mathbf{k} \tag{5.180}$$

Hence, the gradient of any scalar function is a vector given by Equation 5.180. A useful vector identity is that the curl of the gradient of any scalar function is equal to zero. So in general,

$$\nabla \times \nabla\phi = 0 \tag{5.181}$$

for any scalar function ϕ. The utility of this relationship is that according to Equations 5.176 and 5.181, whenever the flow is irrotational; the velocity vector can be expressed as the gradient of a scalar function; hence,

$$\mathbf{V} = \nabla\phi \tag{5.182}$$

and ϕ is commonly referred to as the *velocity potential function*, the *velocity potential*, or the *potential function*. A minus sign is sometimes used in the definition of ϕ given by Equation 5.182 such that $\mathbf{V} = -\nabla\phi$. This is done to ensure that the velocity potential decreases in the direction of flow. Flow fields in which Equation 5.182 is applicable are called *potential flows*, and regions of irrotational flow are commonly called *regions of potential flow*. Equation 5.182 can be expressed in the scalar form

$$u = \frac{\partial\phi}{\partial x}, \qquad v = \frac{\partial\phi}{\partial y}, \qquad w = \frac{\partial\phi}{\partial z} \tag{5.183}$$

Recall the continuity equation for incompressible flows:

$$\frac{\partial u}{\partial x} + \frac{\partial v}{\partial y} + \frac{\partial w}{\partial z} = 0 \tag{5.184}$$

Combining Equations 5.184 and 5.183 yields the following equation for ϕ in inviscid incompressible flows:

$$\frac{\partial^2\phi}{\partial x^2} + \frac{\partial^2\phi}{\partial y^2} + \frac{\partial^2\phi}{\partial z^2} = 0 \tag{5.185}$$

which can be written in vector notation as

$$\nabla^2\phi = 0 \tag{5.186}$$

which is known as the *Laplace equation*. The vector notation, $\nabla^2(\cdot) = \nabla \cdot \nabla(\cdot)$, is called the *Laplacian operator*, which can be expressed as

$$\nabla^2 f = \frac{\partial^2 f}{\partial x^2} + \frac{\partial^2 f}{\partial y^2} + \frac{\partial^2 f}{\partial z^2} \tag{5.187}$$

where $f(x, y, z)$ is any scalar function. The value of Equation 5.186 in describing incompressible irrotational inviscid flow is that the entire flow field can be determined by the solution of one scalar equation for ϕ, from which we can determine the velocity components using Equation 5.182. Interestingly, Equation 5.186 contains no physical parameters; any such parameters are contained in the imposed boundary conditions on ϕ. As a consequence, solutions to the Laplace equation (Equation 5.186) are purely geometric, depending only on the shape of the boundaries and the free-stream conditions. Once the velocity field is derived from the velocity potential, the pressure field, $p(\mathbf{x})$, can be obtained by substituting the velocity field into the Bernoulli equation, which requires that

$$\frac{p}{\rho} + \frac{V^2}{2} + gz = \text{constant} \tag{5.188}$$

where ρ is the fluid density, V is the magnitude of the velocity (= $|\nabla\phi|$), g is gravity, and z is the vertical coordinate.

Potential flows in engineering applications. All flows in which the velocity vector can be expressed as the gradient of a scalar (potential) function are called *potential flows*, and contour lines of constant ϕ are called *equipotential lines*. It is noteworthy that any function ϕ that satisfies Laplace's equation (Equation 5.186) is the velocity potential of a possible irrotational flow field. Potential flows are frequently encountered in high-Reynolds-number external flows, where fluids flow around solid objects and viscous effects are limited to a region very close to the surface of the object, called the *boundary layer*. Flow around airfoils is a case in point. Another common application of potential flows is in laminar flow through porous media, where the seepage velocity is proportional to the gradient in the piezometric head ($= p/\gamma + z$); thus such flows are potential flows. Flow through aquifers is an example of flow through porous media, and groundwater hydrologists have developed many useful relationships based on potential-flow theory.

Velocity potential in cylindrical polar coordinates. It is sometimes convenient to use cylindrical polar coordinates to analyze potential flows, particularly in cases where the flow is symmetric relative to one to the coordinate axes. In cylindrical polar coordinates, the velocity vector is given by

$$\mathbf{V} = v_r \mathbf{e}_r + v_\theta \mathbf{e}_\theta + v_z \mathbf{e}_z \tag{5.189}$$

where $\mathbf{e}_r$, $\mathbf{e}_\theta$, and $\mathbf{e}_z$ are unit vectors in the r-, θ-, and z-directions, respectively. If the velocity is equal to the gradient of the potential, ϕ, as given by Equation 5.182, then

$$v_r = \frac{\partial \phi}{\partial r}, \qquad v_\theta = \frac{1}{r}\frac{\partial \phi}{\partial \theta}, \qquad v_z = \frac{\partial \phi}{\partial z} \tag{5.190}$$

and combining Equation 5.190 with the continuity equation yields the Laplace equation in cylindrical polar coordinates as

$$\frac{1}{r}\frac{\partial}{\partial r}\left(r\frac{\partial \phi}{\partial r}\right) + \frac{1}{r^2}\frac{\partial^2 \phi}{\partial \theta^2} + \frac{\partial^2 \phi}{\partial z^2} = 0 \tag{5.191}$$

Interestingly, the Laplace equation for the velocity potential (Equation 5.191 or 5.185) is valid for inviscid incompressible irrotational flows under both steady and unsteady conditions, because the assumption of steadiness was not invoked during its derivation.

5.6.4 Two-Dimensional Potential Flows

In general, potential flows can be one dimensional, two dimensional, or three dimensional and all flows that can be expressed in terms of a potential function are necessarily irrotational. The analysis of irrotational flows in two dimensions is further enhanced by the fact that two-dimensional flows can also be described by a stream function whenever the flow is incompressible. Hence, in two-dimensional flows, the potential function exists by virtue of irrotationality and the stream function exists by virtue of incompressibility.

Stream-function governing equation. Section 5.3.2 showed that for incompressible flows in two dimensions, the continuity equation is satisfied by any velocity field that can be expressed in terms of the stream function $\psi(x, y)$, where

$$u = \frac{\partial \psi}{\partial y}, \qquad v = -\frac{\partial \psi}{\partial x} \tag{5.192}$$

If the flow field also is irrotational in the xy plane, then the z-component of the vorticity vector, ζ, is equal to zero, which can be expressed as

$$\zeta_z = \frac{\partial v}{\partial x} - \frac{\partial u}{\partial y} = 0 \tag{5.193}$$

Combining Equations 5.193 and 5.192 yields

$$\nabla^2 \psi = 0 \tag{5.194}$$

Hence, in cases of two-dimensional, incompressible, and irrotational flows, the stream function satisfies the Laplace equation.

Joint use of the potential and stream function. Both the stream function, ψ, and the potential function, ϕ, satisfy the Laplace equation in incompressible, irrotational flows in the xy plane. The velocity field can be determined by solving the Laplace equation for either ϕ or ψ. It is often more convenient to use ψ, because boundary conditions on ψ are usually easier to specify. The functions $\phi(\mathbf{x})$ and $\psi(\mathbf{x})$ are called *harmonic functions*, and ϕ and ψ are called *harmonic conjugates*. It is important to remember that the stream function is defined for two-dimensional incompressible flows and only satisfies the Laplace equation when the flows are also irrotational. Conversely, the potential function is defined for irrotational flows and only satisfies the Laplace equation when the flows are incompressible. Therefore, the stream function and the potential function are used together only in flow fields that are two-dimensional, incompressible, and irrotational. In engineering applications, many practical flow fields meet these requirements. It is apparent from the definitions of ϕ and ψ that when both functions exist, they are related by the equations

$$\frac{\partial \phi}{\partial x} = \frac{\partial \psi}{\partial y}, \qquad \frac{\partial \phi}{\partial y} = -\frac{\partial \psi}{\partial x} \tag{5.195}$$

These equations, called the *Cauchy-Riemann*[34] *equations*, are widely used in the theory of complex variables. In fact, the theory of complex variables is intimately related to the description of potential flows of incompressible fluids, because all analytic complex functions can be expressed in the form $\phi(x, y) + i\psi(x, y)$, where $\phi(x, y)$ and $\psi(x, y)$ satisfy the Cauchy-Riemann equations. Hence, all analytic complex functions represent potential flows. The utility of using corresponding stream and potential functions in the analysis of fluid flows is a powerful tool that has withstood the test of time; these functions were first introduced into the analysis of fluid flows in 1781 by the Italian mathematician Joseph-Louis Lagrange (1736–1813).

[34]Named in honor of the French mathematician Augustin-Louis Cauchy (1789–1857) and German mathematician Georg Friedrich Bernhard Riemann (1826–1866).

Orthogonality condition. Curves of constant ψ define streamlines, and curves of constant ϕ define *equipotential lines*. Along any given streamline,

$$d\psi = \frac{\partial \psi}{\partial x}dx + \frac{\partial \psi}{\partial y}dy = 0$$

which gives the slope, dy/dx, of the streamline $\psi = \psi_0$ as

$$\left.\frac{dy}{dx}\right|_{\psi=\psi_0} = -\frac{\partial\psi/\partial x}{\partial\psi/\partial y} = -\frac{-v}{u} = \frac{v}{u} \tag{5.196}$$

Similarly, along any given equipotential line,

$$d\phi = \frac{\partial \phi}{\partial x}dx + \frac{\partial \phi}{\partial y}dy = 0$$

which gives the slope, dy/dx, of the equipotential line $\phi = \phi_0$ as

$$\left.\frac{dy}{dx}\right|_{\phi=\phi_0} = -\frac{\partial\phi/\partial x}{\partial\phi/\partial y} = -\frac{u}{v} = -\frac{u}{v} \tag{5.197}$$

Combining Equations 5.196 and 5.197 gives the product of the slopes of intersecting streamlines and equipotential lines as

$$\left.\frac{dy}{dx}\right|_{\psi=\psi_0} \cdot \left.\frac{dy}{dx}\right|_{\phi=\phi_0} = \left(\frac{v}{u}\right)\left(-\frac{u}{v}\right) = -1 \tag{5.198}$$

It is known that the product of the slopes of orthogonal lines is equal to -1. Hence, Equation 5.198 proves that streamlines intersect equipotential lines at right angles, indicating that these lines are mutually orthogonal.

Planar potential flows in polar coordinates. In planar flows, the flow field is the same in all parallel planes, and these types of flows can be conveniently described by either (Cartesian) xy coordinates or (polar) $r\theta$ coordinates. When polar coordinates are used, the velocity components and the stream function are related by

$$v_r = \frac{1}{r}\frac{\partial \psi}{\partial \theta}, \quad v_\theta = -\frac{\partial \psi}{\partial r} \tag{5.199}$$

where v_r and v_θ are the r and θ components of the velocity vector, respectively. Substituting Equation 5.199 into the requirement for irrotationality ($\nabla \times \mathbf{V} = 0$) shows that the Laplace equation (Equation 5.194) also describes the stream function when polar coordinates are used to describe planar potential flows.

Axisymmetric potential flows in cylindrical polar coordinates. In axisymmetric flows, the flow field is symmetric with respect to one of the coordinate axes. Commonly encountered axisymmetric flows are ones in which symmetry is about the z-axis and the flow variables depend only on z and r (i.e., they are independent of θ). The incompressible continuity equation ($\nabla \cdot \mathbf{V} = 0$) for flow fields that depend on only r and z is given by

$$\frac{1}{r}\frac{\partial}{\partial r}(rv_r) + \frac{\partial v_z}{\partial z} = 0 \tag{5.200}$$

where v_r and v_z are the r and z components of the velocity vector, respectively. For a stream function, ψ, to satisfy the continuity equation (Equation 5.200), it must be related to the velocity components by

$$v_r = -\frac{1}{r}\frac{\partial \psi}{\partial z}, \qquad v_z = \frac{1}{r}\frac{\partial \psi}{\partial r} \tag{5.201}$$

Substituting the relationships given in Equation 5.201 into the requirement for irrotationality ($\nabla \times \mathbf{V} = 0$) yields the following equation for determining the stream function:

$$r\frac{\partial}{\partial r}\left(\frac{1}{r}\frac{\partial \psi}{\partial r}\right) + \frac{\partial^2 \psi}{\partial z^2} = 0 \tag{5.202}$$

The notable result here is that Equation 5.202 is not the Laplace equation (in rz-coordinates). Thus, in axisymmetric irrotational inviscid flows, the stream function does not obey the Laplace equation. It can be shown that the potential function, ϕ, still obeys the Laplace equation under axisymmetric flow conditions. A consequence of ψ not satisfying the Laplace equation is that contours of constant ψ and ϕ are not mutually orthogonal as they are in planar incompressible irrotational flows.

EXAMPLE 5.20

The stream function, ψ, for a two-dimensional flow is given in terms of Cartesian coordinates x and y by

$$\psi = 3x^2y - y^3$$

(a) If possible, determine the potential function describing this flow. (b) Sketch the stream-function contours for x and y in the range of ± 100 and for ψ values of 0, 5×10^4, 20×10^4, 50×10^4, and 100×10^4. What type of flow is described by these streamlines. Plot contours of the potential function to show that they appear orthogonal to the streamlines. (c) Find the pressure distribution in the xy plane in terms of the pressure and velocity at a given location in the plane.

SOLUTION

(a) Using the relationships between the stream function, ψ, the potential function, ϕ, and the x and y components of the velocity, u and v, respectively, the following relationships can be derived:

$$\psi = 3x^2y - y^3$$

$$u = \frac{\partial \psi}{\partial y} = \frac{\partial}{\partial y}(3x^2y - y^3) = 3x^2 - 3y^2$$

$$v = -\frac{\partial \psi}{\partial x} = -\frac{\partial}{\partial x}(3x^2y - y^3) = -6xy$$

$$u = \frac{\partial \phi}{\partial x} = 3x^2 - 3y^2 \rightarrow \phi = x^3 - 3xy^2 + f_1(y) \tag{5.203}$$

$$v = \frac{\partial \phi}{\partial y} = -6xy \rightarrow \phi = -3xy^2 + f_2(x) \tag{5.204}$$

Equating the expressions for ϕ given by Equations 5.203 and 5.204 yields

$$x^3 - 3xy^2 + f_1(y) = -3xy^2 + f_2(x)$$

which requires that

$$f_1(y) = C, \quad \text{and} \quad f_2(x) = x^3 + C$$

where C is an arbitrary constant, which is usually taken as zero for convenience. Using the derived expressions for $f_1(x)$ and $f_2(x)$ in the equations for ϕ (and taking $C = 0$), yields the following velocity potential:

$$\boldsymbol{\phi = x^3 - 3xy^2}$$

It should be noted that in this case, we are able to find a functional expression for ϕ only because the flow is irrotational. If that were not the case, ϕ would not exist.

(b) Contours of the stream function $\psi = 3x^2y - y^3$ for $\psi = 0,\ 5 \times 10^4,\ 20 \times 10^4,\ 50 \times 10^4$, and 100×10^4 are shown in Figure 5.28

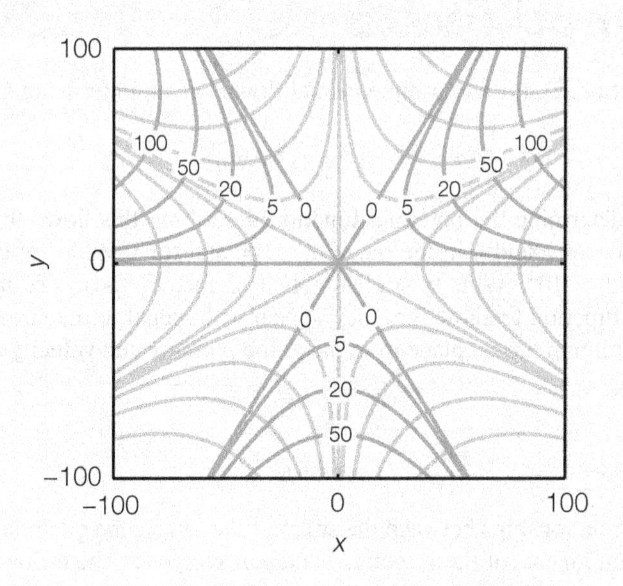

Figure 5.28: Streamlines and equipotential lines

Because stream-function contours represent streamlines, it is apparent from Figure 5.28 that flow described by these streamlines represent **identical flows in multiple corners.** A possible scenario is that we are really only interested in the flow in one corner, and the (identical) flows in the other corners are superfluous. Contours of the potential function $\phi = x^3 - 3xy^2$ for $\phi = 0,\ \pm 5 \times 10^4,\ \pm 20 \times 10^4,\ \pm 50 \times 10^4$, and $\pm 100 \times 10^4$ are shown in the background of Figure 5.28. Any constant ϕ values could have been chosen, but these particular values were chosen to provide contrast with the stream-function contours. Comparing the streamlines and equipotential lines, it is apparent that these lines are orthogonal.

(c) Because the flow is steady, incompressible, and irrotational and z is a constant in the xy plane, the following form of the Bernoulli equation must be satisfied:

$$\frac{p}{\rho} + \frac{V^2}{2} = \text{constant} \rightarrow \frac{p}{\rho} + \frac{u^2 + v^2}{2} = \text{constant}$$

Using the previously derived expressions for the velocity gives the following equation for the pressure distribution in the xy plane

$$\frac{p}{\rho} + \frac{(3x^2 - 3y^2)^2 + (-6xy)^2}{2} = \frac{p_0}{\rho} + \frac{u_0^2 + v_0^2}{2}$$

$$\rightarrow \frac{p}{\rho} = \left[\frac{p_0}{\rho} + \frac{u_0^2 + v_0^2}{2}\right] - \frac{9}{2}\left(x^2 - y^2\right)^2 - 18x^2y^2$$

where p_0, u_0, and v_0 are the known pressure and velocity components, respectively, at a given location within the flow field.

5.7 Fundamental and Composite Potential Flows

Incompressible flows that can be characterized by potential functions are commonly referred to as *potential flows*. Two-dimensional potential flows can also be characterized by stream functions, where stream-function contours are streamlines. An attractive feature of potential-flow formulations is that they can be applied to describe flows around solid bodies, because the flow normal to a streamline is equal to zero and hence, any streamline can be taken as coinciding with the surface of a solid body.

5.7.1 Principle of Superposition

The *principle of superposition* states that if there are multiple solutions to a differential equation that is both linear and homogeneous, then any linear combination of the solutions is also a solution to the given equation. The principle of superposition can be demonstrated for the particular case of the (linear homogeneous) Laplace equation by noting that if we can identify N solutions to the Laplace equation, denoted by $f_1, f_2, \ldots, f_N$, such that

$$\nabla^2 f_1 = 0, \quad \nabla^2 f_2 = 0, \quad \ldots \nabla^2 f_N = 0$$

then equating the sum of the left-hand sides of these equations to the sum of the right-hand sides of these equations (= 0) and using the linearity property of the Laplace equation gives

$$\nabla^2 f_1 + \nabla^2 f_2 + \ldots + \nabla^2 f_N = 0 \quad \rightarrow \quad \nabla^2 (f_1 + f_2 + \ldots + f_N) = 0$$

The principle of superposition is particularly useful in dealing with potential flows because both the stream function and associated potential function that characterize any potential flow satisfy the Laplace equation. Therefore, any linear combination of potential-flow fields is also a potential-flow field; this is the basis of generating composite potential-flow fields from fundamental potential-flow fields. It is also relevant to note that superposition of potential functions is tantamount to superposition of velocity fields. Hence, the velocity at any point in a combined flow is equal to the sum of the velocities at that point in each of the superimposed flow fields.

Practical Considerations

In applying the principle of superposition to linear homogeneous equations (such as the Laplace equation), it is important to remember that the boundary conditions corresponding to the superimposed solutions are also superimposed. Therefore, the combined solution satisfies the combined boundary condition. In rotational (nonpotential) flow fields, the governing equations for ϕ are not linear; consequently, rotational flow fields cannot be superimposed on irrotational flow fields to yield valid results. Based on the principle of superposition, various elementary potential flow fields can be used as building blocks to form other flow fields, with the objective being to form flow fields that meet desired boundary conditions. The most commonly used elementary flow fields are uniform flow, source/sink flow, vortex flow, and doublet flow.

EXAMPLE 5.21

The stream function, ψ_1, for a two-dimensional uniform flow field and the stream function, ψ_2, for a flow field induced by a line sink are as follows:

$$\psi_1 = 2x - 5y, \qquad \psi_2 = 3\tan^{-1}(y/x)$$

where the line sink is located at the origin of the xy-coordinate axes. The flow fields described by ψ_1 and ψ_2 are both known to be irrotational. (a) What is the stream function of the flow field caused by activating a line sink in a uniform flow? (b) Show that this combined flow field is also irrotational.

SOLUTION

(a) Because the flow fields described by ψ_1 and ψ_2 are both irrotational, they satisfy the Laplace equation and hence can be superimposed (because the Laplace equation is linear and homogeneous). The stream function of the combined (i.e., superimposed) flow, ψ, is equal to the sum of the stream functions of the individual flows; hence,

$$\boldsymbol{\psi = 3x - 5y + 3\tan^{-1}\left(\frac{y}{x}\right)}$$

This is the flow field that would exist if a line sink were activated in a uniform flow.

(b) For the purpose of this analysis, the stream function of the combined flow is more conveniently represented in polar coordinates; hence,

$$\psi = 3r\cos\theta - 5r\sin\theta + 3\theta \tag{5.205}$$

If ψ given by Equation 5.205 was irrotational, then ψ would satisfy the Laplace equation which, in polar coordinates, can be expressed as

$$\nabla^2\psi = \frac{1}{r}\frac{\partial}{\partial r}\left(r\frac{\partial\psi}{\partial r}\right) + \frac{1}{r^2}\frac{\partial^2\psi}{\partial\theta^2} = 0 \tag{5.206}$$

Using ψ of the combined flow given by Equation 5.205, the following derivatives can be obtained:

$$r\frac{\partial\psi}{\partial r} = r\frac{\partial}{\partial r}(3r\cos\theta - 5r\sin\theta + 3\theta) = r(3\cos\theta - 5\sin\theta)$$

$$\frac{1}{r}\frac{\partial}{\partial r}\left(r\frac{\partial\psi}{\partial r}\right) = \frac{1}{r}\frac{\partial}{\partial r}[r(3\cos\theta - 5\sin\theta)] = \frac{1}{r}(3\cos\theta - 5\sin\theta) \tag{5.207}$$

$$\frac{\partial\psi}{\partial\theta} = \frac{\partial}{\partial\theta}(3r\cos\theta - 5r\sin\theta + 3\theta) = r(-3\sin\theta - 5\cos\theta) + 3$$

$$\frac{1}{r^2}\frac{\partial^2\psi}{\partial\theta^2} = \frac{1}{r^2}\frac{\partial}{\partial\theta}\left(\frac{\partial\psi}{\partial\theta}\right) = \frac{1}{r^2}\frac{\partial}{\partial\theta}[r(-3\sin\ theta - 5\cos\theta) + 3] = \frac{1}{r}(-3\cos\theta + 5\sin\theta) \tag{5.208}$$

Substituting Equations 5.207 and 5.208 into Equation 5.206 yields

$$\nabla^2\psi = \frac{1}{r}(3\cos\theta - 5\sin\theta) + \frac{1}{r}(-3\cos\ theta + 5\sin\theta) = 0 \quad\rightarrow\quad \nabla^2\psi = 0$$

Because the combined flow satisfies the Laplace equation, **the combined flow is irrotational**, just like the individual flows that were combined. This result is predicted by the principle of superposition.

5.7.2 Uniform Flow

Uniform flow or *rectilinear flow* is a condition in which a constant velocity, **V**, characterizes the entire flow field. If the positive x-axis coincides with the direction of **V**, then

$$\frac{\partial\phi}{\partial x} = V, \quad \frac{\partial\phi}{\partial y} = 0 \tag{5.209}$$

where V is the magnitude of **V**. Integrating Equation 5.209 yields

$$\phi = Vx + C \tag{5.210}$$

where C is a constant. With the requirement that $\phi = 0$ at $x = 0$, we can take $C = 0$ and thus represent the potential function for uniform flow as

$$\phi = Vx \tag{5.211}$$

Expressing the velocity field in terms of the stream function $\psi(x, y)$ requires that

$$\frac{\partial\psi}{\partial y} = V, \quad \frac{\partial\psi}{\partial x} = 0 \tag{5.212}$$

Integrating Equation 5.212 and imposing the condition that $\psi = 0$ at $x = 0$ gives the stream function for uniform flow as

$$\psi = Vy \tag{5.213}$$

Both Equations 5.211 and 5.213 were derived by taking the positive x-direction to coincide with the direction of **V**. In cases where **V** is at an angle of θ to the positive x-axis, the potential and stream functions are given by the general relations

$$\phi = V(x\cos\theta + y\sin\theta) \tag{5.214}$$

$$\psi = V(y\cos\theta - x\sin\theta) \tag{5.215}$$

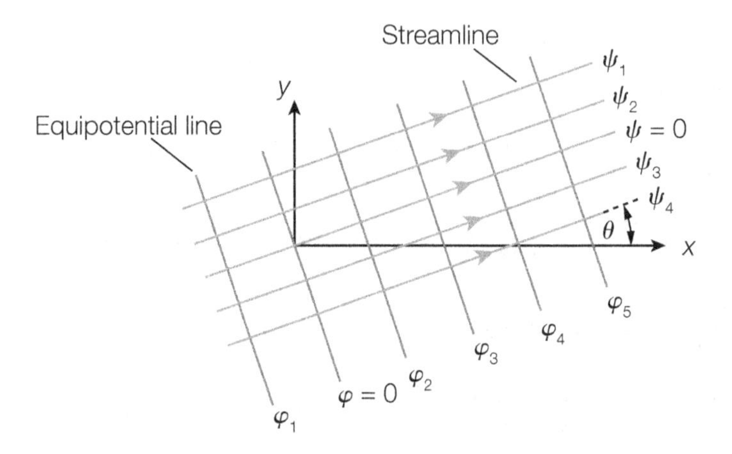

Figure 5.29: **Streamlines and equipotential lines in uniform flow**

The streamlines (ψ = constant) and equipotential lines (ϕ = constant) under uniform-flow conditions are illustrated in Figure 5.29.

EXAMPLE 5.22

Consider the case of a uniform two-dimensional flow in the xy plane, where the Cartesian coordinates x and y are in meters and the velocity distribution is given by

$$\mathbf{V} = 2\mathbf{i} + 3\mathbf{j} \text{ m/s}$$

Find the stream and potential functions for this flow field.

SOLUTION

From the given data, the components of the velocity, u and v, are

$$u = 2 \text{ m/s}, \quad \text{and} \quad v = 3 \text{ m/s}$$

Because $u = V\cos\theta$ and $v = V\sin\theta$, the potential and stream functions given by Equations 5.214 and 5.215 can be expressed as

$$\phi = V(x\cos\theta + y\sin\theta) = ux + vy \quad \rightarrow \quad \boldsymbol{\phi = 2x + 3y}$$
$$\psi = V(y\cos\theta - x\sin\theta) = uy - vx \quad \rightarrow \quad \boldsymbol{\psi = 2y - 3x}$$

Both ϕ and ψ must have units of m^2/s to be consistent with the given units of (x, y) in m and the velocity in m/s.

5.7.3 Line Source/Sink Flow

Line source/sink flow is a condition in which the direction of the velocity vector is either radially outward (line source) or radially inward (line sink) from a line that is perpendicular to the flow domain. Physically, this corresponds to fluid being continuously injected (source) or continuously extracted (sink) from an infinitely long line that is discharging fluid uniformly along its length; such a system is representative of a thin *pipe manifold*. The flow condition caused by such a line source can be conveniently expressed in polar coordinates as

$$(2\pi r)v_r = q \tag{5.216}$$

where r is the distance from the source [L], v_r is the radial component of the velocity [LT^{-1}], and q is the volume flow rate per unit length of the line source [L^{3}T^{-1}L^{-1}]. Rearranging Equation 5.216, the velocity components are given by

$$v_r = \frac{q}{2\pi r}, \quad v_\theta = 0 \tag{5.217}$$

Hence, in accordance with the requirement of potential flow, the velocity field is related to the potential function, ϕ, by the relations

$$\frac{\partial \phi}{\partial r} = \frac{q}{2\pi r}, \quad \frac{1}{r}\frac{\partial \phi}{\partial \theta} = 0 \tag{5.218}$$

which integrates to yield

$$\phi = \frac{q}{2\pi} \ln r \tag{5.219}$$

It is apparent from Equation 5.219 that ϕ is undefined at $r = 0$, where the radial component of the velocity given by Equation 5.217 is infinite. Consequently, $r = 0$ is a *singular point* (also known as a *singularity*) where the derived equations for sources and sinks are not applicable. Expressing the velocity field in terms of the stream function, ψ, requires that

$$\frac{1}{r}\frac{\partial \psi}{\partial \theta} = \frac{q}{2\pi r}, \quad -\frac{\partial \psi}{\partial r} = 0 \tag{5.220}$$

and integrating Equation 5.220 yields

$$\psi = \frac{q}{2\pi}\theta \tag{5.221}$$

By plotting Equation 5.219, it is apparent that lines of equal potential (ϕ = constant) are concentric circles and streamlines (ψ = constant) are radial lines. Equations 5.219 and 5.221 were derived for the case in which the line source/sink is located at the origin of the coordinate system and can be expressed in terms of Cartesian coordinates as

$$\phi = \frac{q}{2\pi} \ln \sqrt{x^2 + y^2} \tag{5.222}$$

$$\psi = \frac{q}{2\pi} \tan^{-1}\left(\frac{y}{x}\right) \tag{5.223}$$

If the line source/sink is located at (x_0,y_0) instead of at the origin, then ϕ and ψ are defined by

$$\phi = \frac{q}{2\pi} \ln \sqrt{(x - x_0)^2 + (y - y_0)^2} \tag{5.224}$$

$$\psi = \frac{q}{2\pi} \tan^{-1}\left(\frac{y - y_0}{x - x_0}\right) \tag{5.225}$$

The streamlines (ψ = constant) and equipotential lines (ϕ = constant) for a line source are illustrated in Figure 5.30. Streamlines and equipotential lines for a line sink are the same as illustrated in Figure 5.30, with the exception that the flows along streamlines are radially inward rather than outward.

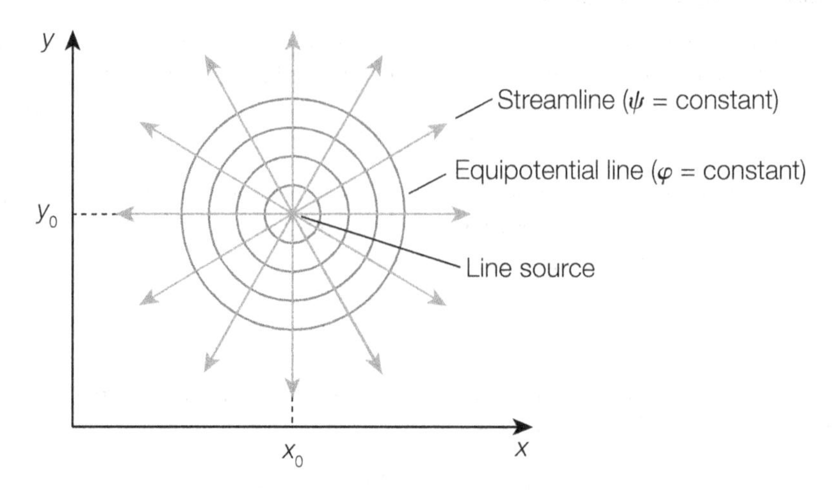

Figure 5.30: Streamlines and equipotential lines for a line source

EXAMPLE 5.23

An inviscid liquid is discharged at a constant rate through a line source embedded in a wall as shown in Figure 5.31. The line source is sufficiently long that it creates a two-dimensional flow field in the interior of the flow domain, and liquid is discharged at the rate of 2 m^3/s per m of source length. The density of the liquid is 1000 kg/m^3. (a) Estimate the stream function and potential function in the flow field. (b) If the pressure at a radial distance of 2 m from the source is measured as 50 Pa, derive an expression for the pressure at any point P in the flow field.

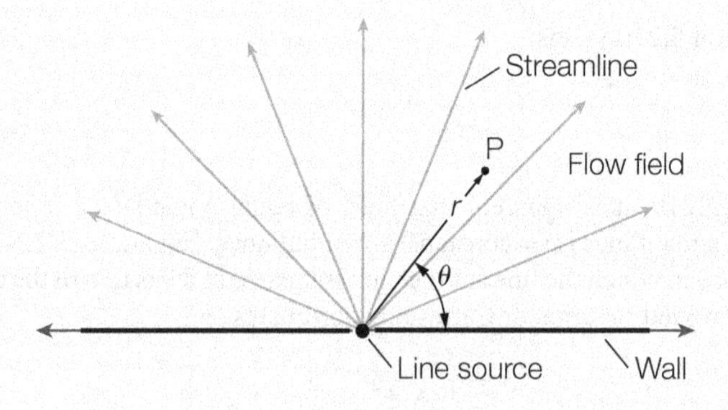

Figure 5.31: Source of flow in a wall

SOLUTION

(a) From the given data: q = 2 m^3/s/m = 2 m^2/s. Neglecting the frictional effects associated with the solid wall and noting that the liquid flow is steady, the stream function, ψ, and potential function, ϕ, in the half-plane shown in Figure 5.31 are given by

$$\psi = \frac{(2q)}{2\pi}\theta = \frac{(2\times 2)}{2\pi}\theta \quad \rightarrow \quad \boldsymbol{\psi = 0.6366\,\theta}$$

$$\phi = \frac{(2q)}{2\pi}\ln r = \frac{(2\times 2)}{2\pi}\ln r \quad \rightarrow \quad \boldsymbol{\phi = 0.6366\ln r}$$

Note that the intake rate in the derived (conventional) formulas for ψ and ϕ has been taken as $2q$ to account for the condition that q is the half-plane flow.

(b) At a radial distance of 2 m from the source, the radial component of the velocity, $v_{r,2}$ must satisfy the following continuity relation:

$$\pi \cdot (2\ \text{m}) \cdot v_{r,2} = q \quad \rightarrow \quad \pi \cdot (2\ \text{m}) \cdot v_{r,2} = 2 \quad \rightarrow \quad v_{r,2} = 0.3183\ \text{m/s}$$

Because the flow is radial, the θ-component of the velocity is equal to zero; and so the magnitude of the velocity at $r = 2$ m, denoted by V_2, is $V_2 = v_{r,2} = 0.3183$ m/s. Similarly, the velocity, V, at any distance r from the source is given by

$$\pi \cdot r \cdot V = q \quad \rightarrow \quad V = \frac{q}{\pi r} \quad \rightarrow \quad V = \frac{2}{\pi r} \quad \rightarrow \quad V = \frac{0.6366}{r}$$

The pressure at $r = 2$ m is given as $p_2 = 50$ Pa. Because the flow is steady and irrotational, the pressure distribution in the flow field is given by

$$\frac{p}{\rho} + \frac{V^2}{2} = \frac{p_2}{\rho} + \frac{V_2^2}{2} \quad \rightarrow \quad \frac{p}{1000} + \frac{0.6366^2}{2r^2} = \frac{50}{1000} + \frac{0.3183^2}{2}$$

$$\rightarrow \quad \mathbf{p = 50.06 - \frac{0.2026}{r^2}\ \textbf{Pa}}$$

5.7.4 Line Vortex Flow

In *line vortex flow*, the streamlines in any given plane consist of concentric circles, and because equipotential lines are orthogonal to streamlines, the equipotential lines are radial lines emanating from the center of the vortex. A line vortex is sometimes called a *plane circular vortex*. In the analysis of source/sink flow described in Section 5.7.3, a stream function and potential function were identified that satisfy the Laplace equation and consist of concentric circles and radial lines, respectively. Consequently, this result can be applied to the analysis of vortex flow, with the only adjustments being that the streamlines in vortex flow correspond to potential lines in source/sink flow, potential lines in vortex flow correspond to streamlines in source/sink flow, and a negative sign is introduced in the expression for the stream function to account for this switch. Based on these conditions, the potential function and stream function for line vortex flow are given by

$$\phi = C\theta \tag{5.226}$$

$$\psi = -C \ln r \tag{5.227}$$

where C is a constant that depends on the strength of the vortex. The components of the velocity vector can be derived from ϕ and ψ as follows:

$$v_r = \frac{\partial \phi}{\partial r} = \frac{1}{r}\frac{\partial \psi}{\partial \theta} = 0 \tag{5.228}$$

$$v_\theta = \frac{1}{r}\frac{\partial \phi}{\partial \theta} = -\frac{\partial \psi}{\partial r} = \frac{C}{r} \tag{5.229}$$

which indicate that the velocity component normal to the concentric streamlines (v_r) is equal to zero and the velocity component tangential to the streamlines (v_θ) decreases in inverse proportion to the distance from the center of the line vortex (r). Physically, this particular velocity distribution is necessary to keep individual fluid elements from rotating while the fluid moves in a circular motion. This type of motion exists in a *free vortex*, which is also called an *irrotational vortex*. In cases where the fluid rotates as a solid body, the vortex is called a *forced vortex*, which is not irrotational; thus, the results developed here are not applicable to a forced vortex. A notable difference between a free vortex and a forced vortex is that in a free vortex, the velocity decreases inversely with distance from the center of circulation, whereas in a forced vortex, the velocity increases linearly with the distance from the center. A third type of vortex is a *combined vortex*, which has an inner region that is a forced vortex, and an outer region with a velocity distribution that is characteristic of a free vortex.

Circulation. The strength of a vortex is commonly expressed in terms of the *circulation*, Γ, which is defined for any flow field as

$$\Gamma = \oint_{\mathrm{P}} \mathbf{V} \cdot \mathbf{ds} \tag{5.230}$$

where the integral is taken around a closed path, P, in the counterclockwise direction, and **ds** is a differential length along the curve. In cases where the flow is irrotational,

$$\Gamma = \oint_{\mathrm{P}} \mathbf{V} \cdot \mathbf{ds} = \oint_{\mathrm{P}} \nabla\phi \cdot \mathbf{ds} = \oint_{\mathrm{P}} \mathrm{d}\phi = 0 \tag{5.231}$$

which indicates that the circulation, Γ, is equal to zero for irrotational flows. However, in cases where the path P encloses a singularity (i.e., where the velocity is undefined or infinite) within the irrotational flow field, the circulation is not necessarily equal to zero. In the case of a free vortex, there is a singularity at $r = 0$ because $v_\theta = C/r$, and the circulation around a circular path that includes this singularity is given by

$$\Gamma = \int_0^{2\pi} \frac{C}{r}(r\,\mathrm{d}\theta) = 2\pi C \rightarrow C = \frac{\Gamma}{2\pi} \tag{5.232}$$

In the context of vortices, the circulation, Γ, is sometimes called the *vortex strength*; a positive value of Γ represents a counterclockwise vortex, and a negative value of Γ represents a clockwise vortex.

Potential and stream functions in terms of circulation. Combining Equation 5.232 with Equations 5.226 and 5.227 yields the potential and stream functions in terms of the circulation as

$$\phi = \frac{\Gamma}{2\pi}\theta \tag{5.233}$$

$$\psi = -\frac{\Gamma}{2\pi}\ln r \tag{5.234}$$

Equations 5.233 and 5.234 were derived for the case in which the vortex is located at the origin of the coordinate system and can be expressed in terms of Cartesian coordinates as

$$\phi = \frac{\Gamma}{2\pi}\tan^{-1}\left(\frac{y}{x}\right) \tag{5.235}$$

$$\psi = -\frac{\Gamma}{2\pi}\ln\sqrt{x^2 + y^2} \tag{5.236}$$

If the vortex is located at (x_0,y_0) instead of the origin, then ϕ and ψ are defined by

$$\phi = \frac{\Gamma}{2\pi}\tan^{-1}\left(\frac{y - y_0}{x - x_0}\right) \tag{5.237}$$

$$\psi = -\frac{\Gamma}{2\pi}\ln\sqrt{(x - x_0)^2 + (y - y_0)^2} \tag{5.238}$$

The streamlines (ψ = constant) and equipotential lines (ϕ = constant) for a line vortex are illustrated in Figure 5.32. A real-life example of a free vortex occurs (approximately) when water drains slowly through the outlet of a sink or bathtub. A real-life example of a forced vortex occurs when a fluid is enclosed in a rotating container. Tornados, hurricanes, and tropical storms are examples of a combined vortex, which has an inner region behaving like a forced vortex and an outer region behaving like a free vortex. The vortex structure of a typhoon (hurricane) is shown in Figure 5.33. The image of Typhoon Neoguri shown in Figure 5.33 was captured as it was approaching Okinawa, Japan, using the Visible Infrared Imaging Radiometer Suite (VIIRS) on the satellite Suomi-NPP. At the time the image was taken, maximum sustained wind speeds in the typhoon were around 240 km/h, with significant wave heights of 12 m generated by the typhoon.

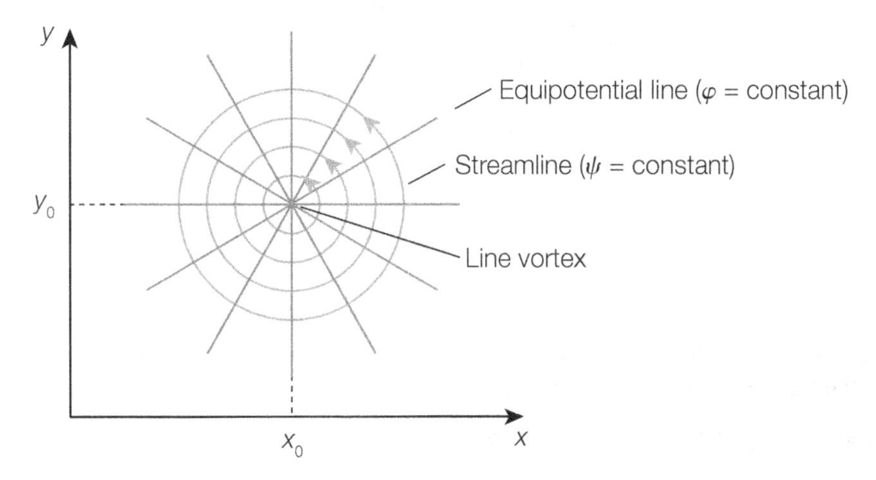

Figure 5.32: Streamlines and equipotential lines for a line vortex

Figure 5.33: Typhoon Neoguri
Source: NASA.

EXAMPLE 5.24

The wind speed in a tornado is measured at 160 km/h at a distance of 30 m from the center of the tornado, as illustrated in Figure 5.34. The rotational core of the tornado has a radius less than 30 m. Estimate the following: (a) the circulation, (b) the velocity at a distance 50 m from the center of the tornado, and (c) the potential function and stream function for the velocity field outside the rotational core.

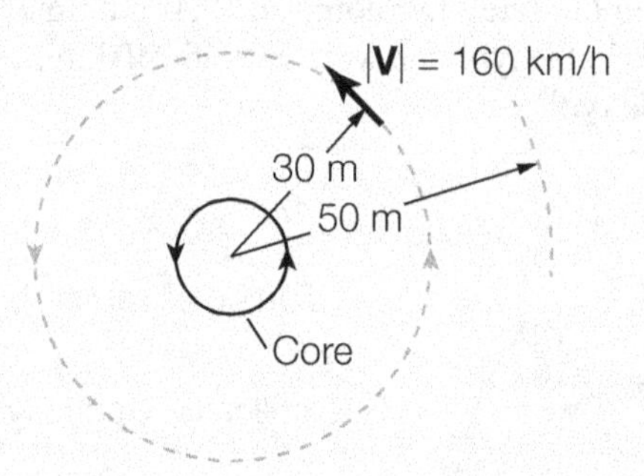

Figure 5.34: Flow in a tornado

SOLUTION

From the given data: v_θ = 160 km/h = 44.4 m/s at r = 30 m. It can be assumed that a free vortex exists outside the (rotational) core of the tornado.

(a) The velocity field outside the core is given by

$$v_\theta = \frac{1}{r}\frac{\partial\phi}{\partial\theta} = \frac{1}{r}\frac{\partial}{\partial\theta}\left(\frac{\Gamma}{2\pi}\,theta\right) = \frac{\Gamma}{2\pi r}$$

This equation is satisfied by the measured velocity of v_θ = 44.4 m/s at r = 30 m; hence,

$$44.4 = \frac{\Gamma}{2\pi(30)} \quad \rightarrow \quad \mathbf{\Gamma = 8380\ m^2/s}$$

(b) The velocity at a distance r = 50 m from the center of the tornado, $v_{\theta,50}$, is given by

$$v_{\theta,50} = \frac{\Gamma}{2\pi(50)} = \frac{8380}{2\pi(50)} = 26.7\ \text{m/s} = \mathbf{96.0\ km/h}$$

(c) Because the flow is assumed to be an irrotational free vortex, the potential function, ϕ, and stream function, ψ, are given by

$$\phi = \frac{\Gamma}{2\pi}\theta = \frac{8380}{2\pi}\theta \quad \rightarrow \quad \boldsymbol{\phi = 1330\,\theta\ \mathbf{m^2/s}}$$

$$\psi = -\frac{\Gamma}{2\pi}\ln r = -\frac{8380}{2\pi}\ln r \quad \rightarrow \quad \boldsymbol{\psi = -1330\ln r\ \mathbf{m^2/s}}$$

5.7.5 Spiral Flow Toward a Sink

Irrotational flow toward a sink, as is routinely observed in drains, can be described by the superposition of a line sink and a line vortex, yielding what is sometimes called a *spiral vortex*. The velocity potential, ϕ, and stream function, ψ, of this combined flow are given by

$$\phi = \frac{q}{2\pi}\ln r + \frac{\Gamma}{2\pi}\theta \tag{5.239}$$

$$\psi = \frac{q}{2\pi}\theta - \frac{\Gamma}{2\pi}\ln r \tag{5.240}$$

Rearranging Equation 5.240 gives the equation of the streamline $\psi = \psi_0$ as

$$r = \exp\left(\frac{q\theta - 2\pi\psi_0}{\Gamma}\right) \tag{5.241}$$

A typical spiral pathline described by Equation 5.241 is shown in Figure 5.35. In this type of flow, the radial and tangential components of the velocity, v_r and v_θ, are given by

$$v_r = \frac{1}{r}\frac{\partial\psi}{\partial\theta} = \frac{q}{2\pi r}, \qquad v_\theta = -\frac{\partial\psi}{\partial r} = \frac{\Gamma}{2\pi r} \tag{5.242}$$

It is apparent that v_r is due entirely to the sink and v_θ is due entirely to the vortex, as would be expected from the principle of superposition. Airflow in a tornado and the flow induced by a rapidly draining bathtub are closely approximated by spiral-vortex flow. However, instead of having an infinite velocity at the center of the vortex, real flows approximate solid-body rotation at the center.

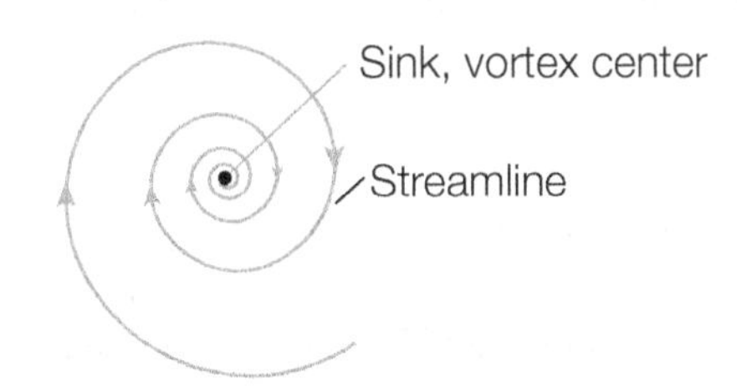

Figure 5.35: Spiral flow towards a sink

EXAMPLE 5.25

The wind velocity is measured at a point P located 25 m from the center of a tornado as shown in Figure 5.36. The measured velocity has a magnitude of 95 km/h and is directed at an angle of 55° west of north. If the tornado can be represented as a spiral vortex, what are the strengths of the sink and vortex?

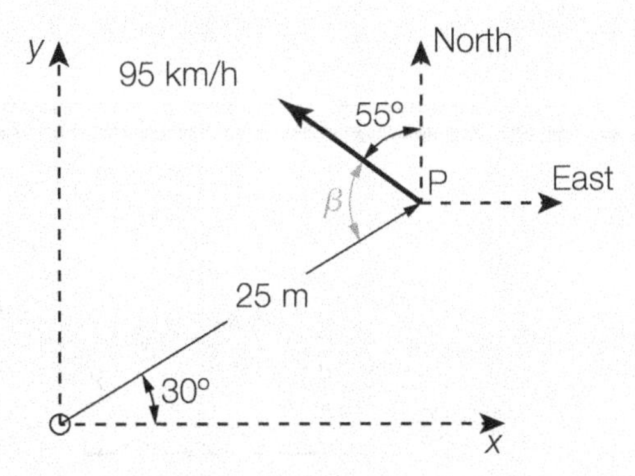

Figure 5.36: Measured wind velocity in a tornado

SOLUTION

From the given data: r = 25 m, θ = 30°, V = 95 km/h = 26.4 m/s, and α = 55°. The angle, β, that the velocity vector makes with the (radial) position vector is given by

$$\beta = (90° - 55°) + 30° = 65°$$

The r and θ components of the measured velocity vector, v_r and v_θ, are calculated as follows:

$$v_r = -V\cos\beta = -26.4\cos(65°) = -11.2 \text{ m/s}$$
$$v_\theta = V\sin\beta = 26.4\sin(65°) = 23.9 \text{ m/s}$$

These components of the velocity are related directly to the strength of the line sink, q, and the strength of the line vortex, Γ, that constitute the spiral vortex; hence,

$$v_r = \frac{q}{2\pi r} \quad\rightarrow\quad -11.2 = \frac{q}{2\pi(25)} \quad\rightarrow\quad q = \mathbf{-1760\ m^2/s}$$
$$v_\theta = \frac{\Gamma}{2\pi r} \quad\rightarrow\quad 23.9 = \frac{\Gamma}{2\pi(25)} \quad\rightarrow\quad \Gamma = \mathbf{3750\ m^2/s}$$

Based on these results, it can be inferred that the tornado is taking in air at a rate of 1750 m^3/s per m of (vertical) height and has a rotational strength of 3750 m^2/s.

5.7.6 Doublet Flow

Doublet flow occurs in the limit as a line source and a line sink of equal strength approach each other and become coincident. The doublet configuration of a source and a sink is sometimes called a *dipole*. Consider the line source and line sink as shown in Figure 5.37. In accordance with the principle of superposition, the stream function of a doublet flow field is obtained by adding the stream function of a source to the stream function of a sink, which gives the combined stream function, ψ, as

$$\psi = -\frac{q}{2\pi}(\theta_1 - \theta_2) \tag{5.243}$$

Using the geometric properties of the triangles shown in Figure 5.37, Equation 5.243 can be expressed in the form

$$\psi = -\frac{q}{2\pi}\tan^{-1}\left(\frac{2ar\sin\theta}{r^2 - a^2}\right) \tag{5.244}$$

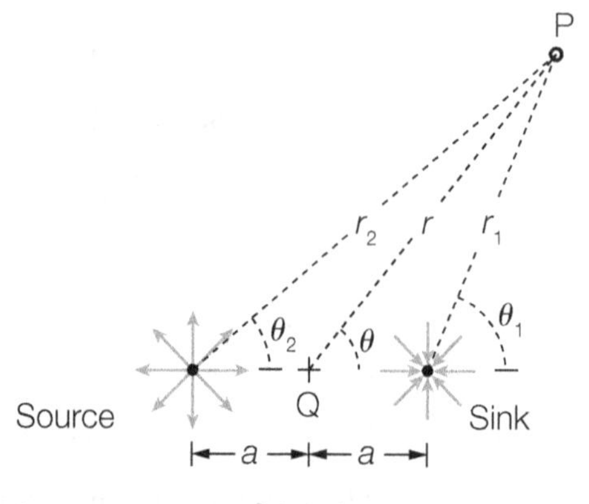

Figure 5.37: Combined source and sink

where r is the distance of a point P from the origin Q and a source and sink are each located a distance a from the origin Q. Taking the limit as $a \to 0$ and noting that $\tan^{-1}(\cdot) \to (\cdot)$ as $(\cdot) \to 0$, Equation 5.244 becomes

$$\psi = -\frac{q}{2\pi}\frac{2ar\sin\theta}{r^2 - a^2} = -\frac{qar\sin\theta}{\pi(r^2 - a^2)} \tag{5.245}$$

In mathematical terms, a doublet flow is obtained as $a \to 0$ (i.e., the source and sink become coincident) and $q \to \infty$ (i.e., the strengths of the source and sink become infinite) such that qa = constant. Under these circumstances, the stream function for a doublet as given by Equation 5.245 becomes

$$\psi = -\frac{K\sin\theta}{r} \tag{5.246}$$

where K is the *doublet strength* defined as

$$K = \frac{qa}{\pi} \tag{5.247}$$

The potential function, ϕ, of a doublet can be derived similarly and is given by

$$\phi = \frac{K\cos\theta}{r} \tag{5.248}$$

Streamlines and equipotential lines in doublet flow are illustrated in Figure 5.38, which indicates that the streamlines are circles that are tangent to the x-axis at the point Q, which is the location of the merged source and sink. The equation of a streamline corresponding to a given value of the stream function, ψ_0, can be derived directly from Equation 5.246, which yields, in Cartesian coordinates,

$$x^2 + \left(y + \frac{K}{2\psi_0}\right)^2 = \left(\frac{K}{2\psi_0}\right)^2 \tag{5.249}$$

Equipotential lines are circles tangent to the y-axis, and the equation of an equipotential line corresponding to a given value of the potential, ϕ_0, can be derived directly from Equation 5.248, which yields, in Cartesian coordinates,

$$\left(x - \frac{K}{2\phi_0}\right)^2 + y^2 = \left(\frac{K}{2\phi_0}\right)^2 \tag{5.250}$$

The streamlines and equipotential lines intersect at 90° everywhere except at the origin, which is at the location of the merged source and sink and is also a singular point.

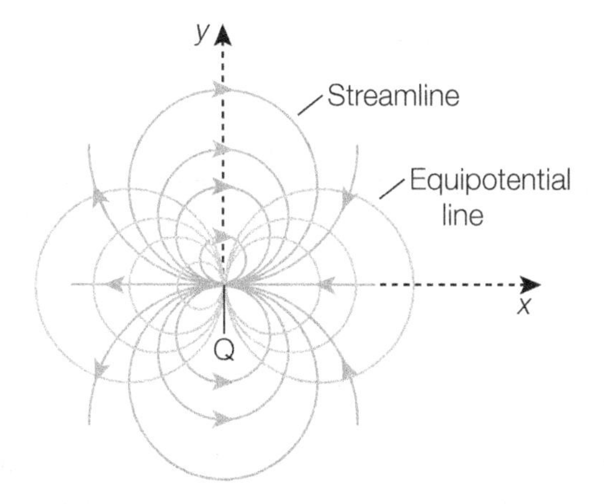

Figure 5.38: Streamlines and equipotential lines in doublet flow

Application. Although a doublet is not a realistic flow, just as pure line sources and line sinks are not realistic flows, the value of defining a doublet flow is that it can be combined with other potential flows to represent realistic flow fields, such as the flow around a circular cylinder.

EXAMPLE 5.26

A doublet of strength 10 m^3/s is located at the origin of a Cartesian coordinate system. Determine the velocity at (x, y) = (3 m, 2 m).

SOLUTION

From the given data: K = 10 m^3/s. At (x_0, y_0) = (3 m, 2 m),

$$r_0 = \sqrt{x_0^2 + y_0^2} = \sqrt{3^2 + 2^2} = 3.61 \text{ m}$$

$$\theta_0 = \tan^{-1}\left(\frac{y_0}{x_0}\right) = \tan^{-1}\left(\frac{2}{3}\right) = 0.588 \text{ rad}$$

The components of the velocity at any location (r, θ) are given by

$$v_r = \frac{\partial \phi}{\partial r} = \frac{\partial}{\partial r}\left(\frac{K\cos\theta}{r}\right) = -\frac{K\cos\theta}{r^2}$$

$$v_\theta = \frac{1}{r}\frac{\partial \phi}{\partial \theta} = \frac{1}{r}\frac{\partial}{\partial \theta}\left(\frac{K\cos\theta}{r}\right) = -\frac{K\sin\theta}{r^2}$$

Therefore, the velocity components at (r, θ) = (3.61 m, 0.588 rad) are

$$v_{r,0} = -\frac{10\cos(0.588)}{3.61^2} = -0.638 \text{ m/s}$$

$$v_{\theta,0} = -\frac{10\sin(0.588)}{3.61^2} = -0.426 \text{ m/s}$$

The velocity, **V**, at (x, y) = (3 m, 2 m) is therefore given by

$$\mathbf{V} = \mathbf{-0.638\,e_\mathit{r} - 0.426\,e_\theta\ m/s}$$

5.7.7 Flow Around a Half-Body

Consider the two-dimensional flow field obtained by superimposing a uniform flow field with the flow field generated by a (vertical) line source as shown in Figure 5.39(a). In accordance with the principle of superposition, the velocity potential, ϕ, and stream function, ψ, of this combined flow are given by

$$\phi = Vr\cos\theta + \frac{q}{2\pi}\ln r \tag{5.251}$$

$$\psi = Vr\sin\theta + \frac{q}{2\pi}\theta \tag{5.252}$$

where V is the uniform-flow velocity and q is the volume flow rate emanating from the line source. The streamlines described by Equation 5.252 are shown in Figure 5.39(b), where it is apparent that this (combined) flow field can be used to described the flow around a blunt

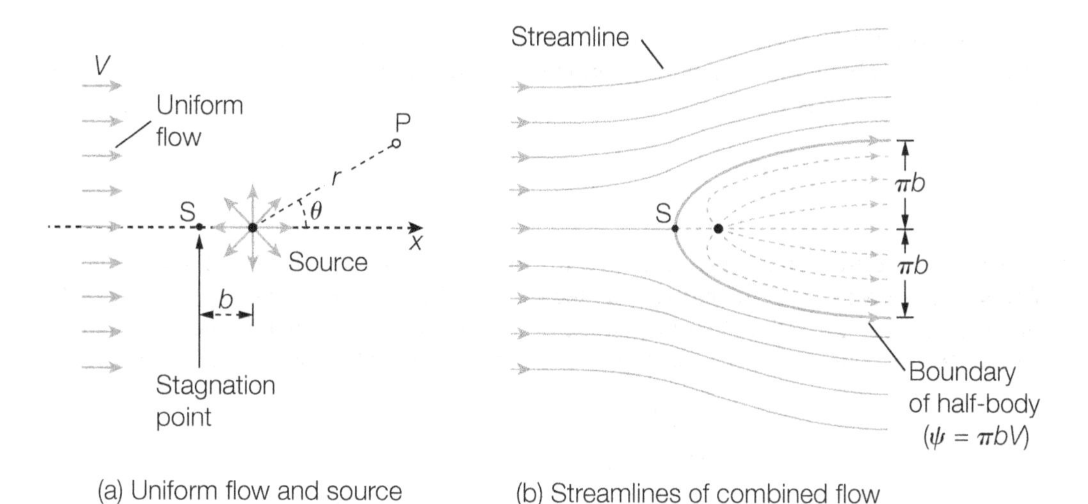

Figure 5.39: Flow around a half-body

object, where the surface of the blunt object coincides with the streamline shown in bold in Figure 5.39(b). This blunt body is referred to as a *half-body* (or *Rankine half-body*),[35] because it is open at the downstream end. To reflect reality, the surface of the half-body must coincide with the streamline on which the fluid velocity is equal to zero at the point where the free-stream velocity directly impinges on the body; this is Point S in Figure 5.39(b). A location at which the velocity is equal to zero in any flow field is called a *stagnation point.* The location of the stagnation point in the case of half-body flow can be determined by considering the radial component of the velocity, v_r, given by

$$v_r = \frac{\partial \phi}{\partial r} = V \cos\theta + \frac{q}{2\pi r} \tag{5.253}$$

Because the stagnation point occurs when $v_r = 0$, $r = b$, and $\theta = \pi$, Equation 5.253 gives the distance of the stagnation point from the location of the source, b, as

$$b = \frac{q}{2\pi V} \quad \text{or} \quad \frac{q}{2\pi} = bV \tag{5.254}$$

At $r = b$ and $\theta = \pi$, the value of the stream function, ψ_0, given by Equation 5.252 is

$$\psi_0 = Vb \sin\pi + \left(\frac{q}{2\pi}\right)(\pi) = 0 + (bV)(\pi) = \pi bV \tag{5.255}$$

Combining Equations 5.255 and 5.252 gives the equation of the streamline bounding the half-body as

$$Vr_\text{s}\sin\theta_\text{s} + \left(\frac{q}{2\pi}\right)\theta_\text{s} = \psi_0 \quad \rightarrow \quad Vr_\text{s}\sin\theta_\text{s} + (bV)\theta_\text{s} = \pi bV$$

where $(r_\text{s}, \theta_\text{s})$ are the polar coordinates of points on the surface of the half-body. The equation of the surface of the half-body is therefore given by

$$r_\text{s} = \frac{b(\pi - \theta_\text{s})}{\sin\theta_\text{s}} \tag{5.256}$$

Because the width of the half-body is equal to $r_\text{s}\sin\theta_\text{s}$, it is apparent from Equation 5.256 that the half-width is equal to $b(\pi - \theta_\text{s})$, which asymptotically approaches πb as $r_\text{s} \rightarrow \infty$ and $\theta_\text{s} \rightarrow 0$.

[35] Named in honor of the Scottish engineer W. J. M. Rankine (1820–1872).

Pressure distribution. The pressure distribution on the surface of a half-body can be determined by applying the Bernoulli equation to the bounding streamline, in which case the pressure distribution is given by

$$p_0 + \frac{1}{2}\rho V^2 = p_s + \frac{1}{2}\rho v_s^2 \tag{5.257}$$

where p_0 is the free-stream pressure (i.e., the pressure in the approaching fluid), V is the free-stream velocity, p_s is the pressure on the surface of the half-body, and v_s is the velocity on the bounding streamline. The magnitude of the velocity at any point in the flow field, v, can be derived from the stream function of the combined flow (Equation 5.252) such that

$$\begin{aligned} v^2 = v_r^2 + v_\theta^2 &= \left(\frac{1}{r}\frac{\partial \psi}{\partial \theta}\right)^2 + \left(-\frac{\partial \psi}{\partial r}\right)^2 \\ &= \left(V\cos\theta + \frac{q}{2\pi r}\right)^2 + (-V\sin\theta)^2 = \left(V\cos\theta + \frac{bV}{r}\right)^2 + (V\sin\theta)^2 \\ \to v^2 &= V^2\left(1 + 2\frac{b}{r}\cos\theta + \frac{b^2}{r^2}\right) \end{aligned} \tag{5.258}$$

Using this relationship, the velocity, v_s on the surface of the half-body can be obtained using the relation

$$v_s^2 = V^2\left(1 + 2\frac{b}{r_s}\cos\theta_s + \frac{b^2}{r_s^2}\right) \tag{5.259}$$

where (r_s, θ_s) are the polar coordinates of any given location on the surface of the half-body having a velocity v_s. Combining Equations 5.259 and 5.256 gives the following expression for v_s as a function of θ_s:

$$v_s^2 = V^2\left[1 + \frac{\sin 2\theta_s}{\pi - \theta_s} + \frac{\sin^2\theta_s}{(\pi - \theta_s)^2}\right] \tag{5.260}$$

The combination of Equations 5.260 and 5.257 gives the pressure distribution around the surface of the half-body, which can be integrated to yield the net pressure force on the body. Caution should be taken in applying this net pressure force because this net force does not include the friction force associated with fluid motion relative to a stationary body, which can be considered as additive. A related consequence of neglecting fluid friction in this analysis is that the velocity, v_s, on the surface of the body in not equal to zero, which violates the no-slip boundary condition that must be met at fluid-solid interfaces. However, when frictional (i.e., viscous) effects are limited to a very thin boundary layer on the solid surface, both v_s and p_s closely approximate the conditions at the outer end of the boundary layer. Furthermore, because the pressure does not typically vary significantly across a boundary layer, the net pressure force that is calculated using potential theory usually provides a fairly accurate approximation to the actual net pressure force on a body.

Point of minimum pressure. Using Equation 5.260, it can be shown that the maximum value of v_s is equal to $1.26V$ and occurs at $\theta_s = 63°$ which, according to the Bernoulli equation, is also the point of minimum pressure. Downstream of this maximum-velocity/minimum-pressure point, the flow over the half-body decelerates, which corresponds to an adverse pressure gradient that can cause flow separation, in which case the potential-flow approximation is no longer valid.

Practical Considerations

Flow around a half-body is most commonly used in studying the flow at the upstream end of a symmetrical body that is long relative to its width. Realistic flow conditions that can be represented by flow around a half-body include the flow around bridge piers, struts, and other columnar structures. The upper part of a half-body can also be taken to represent the flow of a liquid over a rise in the bottom of a channel that is transporting the liquid, the flow of wind over a hillside, or the flow past a side contraction in a wide channel.

EXAMPLE 5.27

Air flows at a velocity of 0.5 m/s toward a columnar structure that is symmetric about its centerline, with each side of the structure having the shape of a Rankine half-body. The total width of the column is 3 m as illustrated in Figure 5.40. (a) Determine the source strength that parameterizes each of the two half-bodies that make up the column and derive the equation describing the shape of each side of the column. (b) Compare the shape of the half-body column with a column whose front has a semicircular shape. (c) Plot the velocity distribution and pressure distribution on the surface of each side of the half-body column.

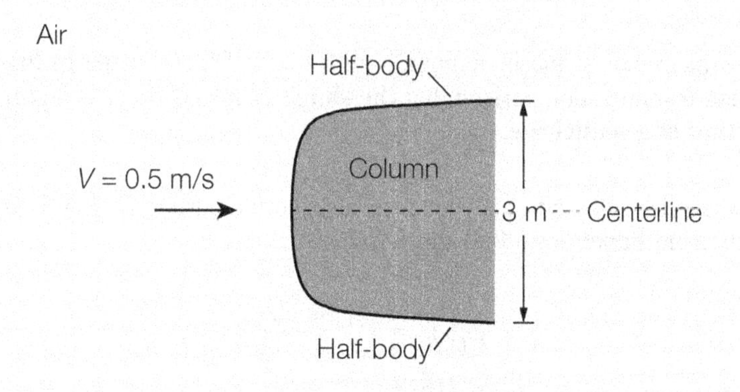

Figure 5.40: Flow toward a half-body column

SOLUTION

From the given data: $V = 0.5$ m/s and $W = 3$ m. The flow and pressure distribution is the same on each side of the column, so consider only one side of the column and apply the results to the other side.

(a) From the general analysis of half-bodies presented in this section, b is the distance from the front stagnation point to the origin of the polar coordinate system, where the origin also coincides with the location of the sink. Because the asymptotic half-width is πb

$$\pi b = \frac{W}{2} = \frac{3\ \text{m}}{2} \quad \rightarrow \quad b = 0.477\ \text{m}$$

The source strength, q, required to form a half-body with a characteristic dimension of $b = 0.477$ m is given by Equation 5.254 as

$$\frac{q}{2\pi} = bV \quad \rightarrow \quad \frac{q}{2\pi} = (0.477)(0.5) \quad \rightarrow \quad q = \mathbf{1.50\ m^2/s}$$

Therefore, the required source strength is 1.50 m^2/s. The surface of the half-body is described by Equation 5.256, which gives

$$r_s = \frac{b(\pi - \theta_s)}{\sin\theta_s} \quad \rightarrow \quad \boldsymbol{r_s = \frac{0.477(\pi - \theta_s)}{\sin\theta_s}}$$

where θ_s is in radians and r_s is in meters.

(b) The location of the stagnation point at the front of the half-body column is given in Cartesian coordinates by $(x, y) = (-0.477\ \text{m}, 0\ \text{m})$. If the front face of the column was semicircular, the column would have a radius of 1.5 m, the surface of the column would pass through the stagnation point, and the center of the semicircle would be on the centerline of the column. Therefore, the center of the semicircle would be at $(x_0, y_0) = (1.5\ \text{m} - 0.477\ \text{m}, 0) = (1.023\ \text{m}, 0)$. The equation of the semicircle is therefore given by

$$(x - 1.023)^2 - y^2 = 1.5^2$$

This semicircular shape is compared with the half-body shape in Figure 5.41(a). It is apparent from this comparison that **the shape of a half-body is significantly different from that of a semicircle.**

(c) The normalized velocity and pressure distributions around a half-body can be derived directly from Equations 5.260 and 5.257, which yield

$$\left(\frac{v_s}{V}\right)^2 = 1 + \frac{\sin 2\theta_s}{\pi - \theta_s} + \frac{\sin^2\theta_s}{(\pi - \theta_s)^2}$$

$$\frac{p_s - p_0}{\frac{1}{2}\rho V^2} = 1 - \left(\frac{v_s}{V}\right)^2$$

These distributions are plotted in Figures 5.41(b) and 5.41(c) in terms of the x-coordinates of points on the surface of the half-body, where x and θ_s are related using Equation 5.256, where

$$x = r_s\cos\theta_s = \frac{b(\pi - \theta_s)}{\tan\theta_s} = \frac{0.477(\pi - \theta_s)}{\tan\theta_s}\ \text{m}$$

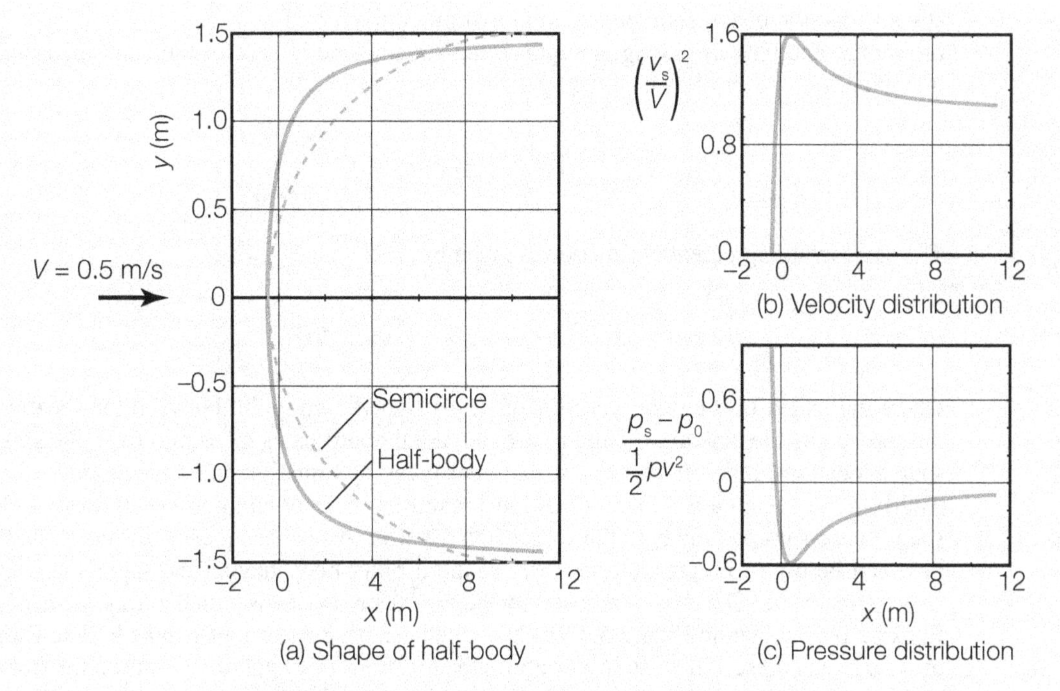

Figure 5.41: **Flow around a half-body column**

As predicted by theory, the velocity is a maximum at $x = 0.496$ m, which corresponds to $\theta_s = 63°$. Because the pressure varies inversely with the velocity, the location of minimum pressure coincides with the location of maximum velocity. Both the velocity and the pressure asymptote to their upstream values as $x \to \infty$.

5.7.8 Rankine Oval

Consider the flow field obtained by superimposing a uniform flow, V, with a source and a sink as shown in Figure 5.42(a). Invoking the principle of superposition, noting that the

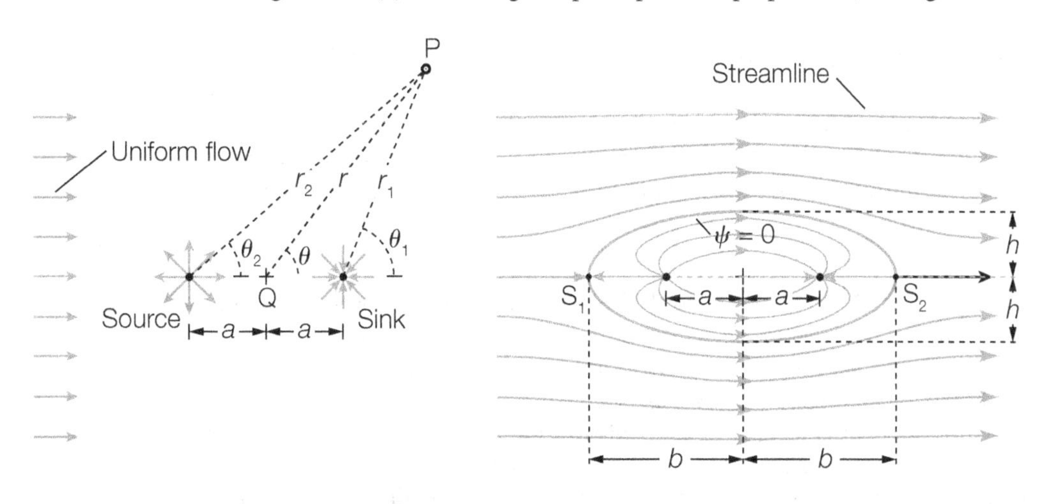

(a) Combined source, sink, uniform flow

(b) Flow around a Rankine oval

Figure 5.42: **Flow around a Rankine oval**

stream function of the source/sink combination is given by Equation 5.244, and specifying the source and sink as having strength q, the stream function, ψ, of the combined uniform flow plus source/sink flow is given by

$$\psi = Vr\sin\theta - \frac{q}{2\pi}\tan^{-1}\left(\frac{2ar\sin\theta}{r^2 - a^2}\right) \tag{5.261}$$

The corresponding potential function is given by

$$\phi = Vr\cos\theta + \frac{q}{2\pi}(\ln r_1 - \ln r_2) \tag{5.262}$$

where the origin of the polar coordinates (r, θ) is at the midpoint between the locations of the source and the sink and r_1 and r_2 are the radial coordinates measured from the sink and source locations, respectively. The streamlines given by Equation 5.261 are plotted in Figure 5.42(b), from which it is apparent that the streamline corresponding to $\psi = 0$ forms a closed streamline encompassing both the source and the sink. As a consequence, the combined flow field can be used to represent the flow around a blunt body having the same shape as the $\psi = 0$ streamline. The shape enclosed by the $\psi = 0$ streamline is called a *Rankine oval*. The Rankine oval shown in Figure 5.42(b) has a half-length b and a half-height h. The Rankine oval is not an ellipse. The distance b defines the locations of stagnation points on the front and back of the Rankine oval, identified by S_1 and S_2, respectively, in Figure 5.42(b). Equating the velocity of the combined flow field to zero at the stagnation points gives

$$\frac{b}{a} = \sqrt{\frac{q}{\pi Va} + 1} \tag{5.263}$$

The value of h in the Rankine oval can be determined from Equation 5.261 by specifying that $r = h$ when $\psi = 0$ and $\theta = \pi/2$, which yields the following implicit equation:

$$\frac{h}{a} = \frac{1}{2}\left[\left(\frac{h}{a}\right)^2 - 1\right]\tan\left[2\left(\frac{\pi Va}{q}\right)\frac{h}{a}\right] \tag{5.264}$$

It is apparent from Equations 5.263 and 5.264 that the shape of the Rankine oval depends on the single dimensionless parameter, Va/q, and by varying this parameter, various lengths and heights can be obtained for the Rankine oval. As the value of Va/q increases, the Rankine oval changes from a blunt object to a long slender body. Specifically, as Va/q increases from zero to infinity, the Rankine oval goes from being a large circular cylinder to being a flat plate of length $2a$.

Pressure distribution. In accordance with the Bernoulli equation, on the front side of the body, the pressure decreases as the fluid moves over the body (called a *favorable pressure gradient*), whereas on the back side of the body, the pressure decreases as the fluid moves past the body (called an *adverse pressure gradient*). Under turbulent flow conditions, an adverse pressure gradient is usually associated with the thin viscous boundary layer on the surface of a body separating from the body, under which condition the downstream fluid motion within the wake region becomes rotational and the potential flow solution is not applicable. Boundary-layer separation is not predicted by potential-flow theory. Most realistic Rankine ovals have large adverse pressure gradients on the leeward side, resulting in boundary-layer separation and the existence of a wake region.

Maximum velocity. The maximum velocity, V_{max}, on the surface of the Rankine oval is given by

$$\frac{V_{\text{max}}}{V} = 1 + \left(\frac{q}{\pi V a}\right) \frac{1}{1 + h^2/a^2} \tag{5.265}$$

where V_{max} occurs at the highest point on the Rankine oval or, equivalently, the point where the Rankine oval is widest.

Practical Considerations

Although the streamlines within the Ranking oval, as shown in Figure 5.42, are typically not of great concern to fluid mechanicians, these within-oval streamlines are of particular concern to groundwater hydrologists in cases where contaminated groundwater is to be remediated by a pump-and-treat system. In this application, the pumping well acts as the sink, the injection well acts as the source, and the regional flow in the aquifer acts as the uniform flow field. Within the Rankine oval formed by this system, the contaminated groundwater is isolated from the regional flow, thereby preventing the contaminated groundwater from moving downstream.

EXAMPLE 5.28

A fluid flows around an elliptical body that is 8 m long and 4 m wide as shown in Figure 5.43. The fluid approaches the ellipse along its major axis at a speed of 1 m/s. Assuming that the flow around the ellipse can be approximated by the flow around a Rankine oval, (a) estimate the parameters describing the Rankine-oval flow field, (b) compare the shape of the Rankine oval to the shape of the ellipse, and (c) estimate the maximum velocity on the surface of the Rankine oval.

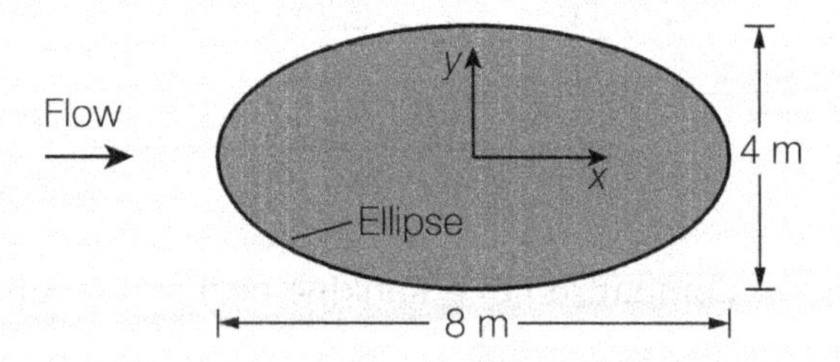

Figure 5.43: Flow around an ellipse

SOLUTION

From the given data: L = 8 m, W = 4 m, and V = 1 m/s. A Rankine oval has length $2b$ and height $2h$, so take $b = L/2 = 4$ m and $h = W/2 = 2$ m.

(a) The parameters of the Rankine-oval flow field are q and a, which can be determined by the simultaneous solution of Equations 5.263 and 5.264 for the given values of h, b, and V. Combining Equations 5.263 and 5.264 to eliminate $\pi Va/q$ yields

$$\frac{h}{a} = \frac{1}{2}\left[\left(\frac{h}{a}\right)^2 - 1\right]\tan\left\{2\left[\left(\frac{b}{a}\right)^2 - 1\right]^{-1}\frac{h}{a}\right\}$$

$$\rightarrow \frac{2}{a} = \frac{1}{2}\left[\left(\frac{2}{a}\right)^2 - 1\right]\tan\left\{2\left[\left(\frac{4}{a}\right)^2 - 1\right]^{-1}\frac{2}{a}\right\}$$

which yields a = 3.125 m. Using this result in Equation 5.263 requires that

$$\frac{b}{a} = \sqrt{\frac{q}{\pi Va} + 1} \quad\rightarrow\quad \frac{4}{3.125} = \sqrt{\frac{q}{\pi(1)(3.125)} + 1} \quad\rightarrow\quad q = 6.277\ \text{m}^2\text{/s}$$

Therefore, the parameters of the Rankine oval are **a = 3.125 m** and **q = 6.277 m²/s**.

(b) The equation of the Rankine oval that has the same length and width as the ellipse is given by Equation 5.261 by taking ψ = 0, which yields

$$\psi = Vr\sin\theta - \frac{q}{2\pi}\tan^{-1}\left(\frac{2ar\sin\theta}{r^2 - a^2}\right)$$

$$\rightarrow 0 = (1)r\sin\theta - \frac{6.277}{2\pi}\tan^{-1}\left[\frac{2(3.125)r\sin\theta}{r^2 - (3.125)^2}\right]$$

$$\rightarrow r = \frac{0.999}{\sin\theta}\tan^{-1}\left[\frac{6.25\,r\sin\theta}{r^2 - 9.766}\right]$$

This equation is plotted and compared to the given ellipse in Figure 5.44. **It is apparent from this comparison that the Rankine oval closely approximates the shape of the ellipse.**

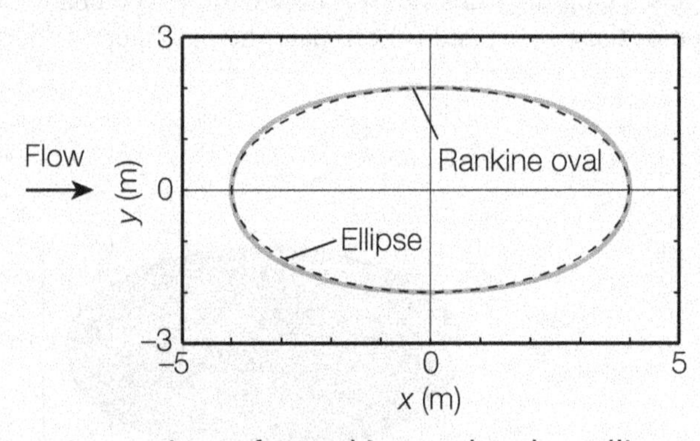

Figure 5.44: Comparison of a Rankine oval and an ellipse

(c) The maximum velocity, V_{max}, on the surface of the Rankine oval is given by Equation 5.265, where

$$\frac{V_{max}}{V} = 1 + \left[\frac{q}{\pi Va}\right]\frac{1}{1 + h^2/a^2} \quad\rightarrow\quad \frac{V_{max}}{1} = 1 + \left[\frac{6.277}{\pi(1)(3.125)}\right]\frac{1}{1 + 2^2/3.125^2}$$

which yields V_{max} = **1.45 m/s**. Hence, the maximum velocity is about 45% higher than the approach velocity, V, of 1 m/s.

5.7.9 Flow Around a Circular Cylinder

In the limit as the spacing between the source and the sink approaches zero in the Rankine-oval flow, the Rankine oval becomes a circle, thus, the flow becomes representative of the flow around a cylinder with a circular cross section. Recalling that a coincident source and sink is a doublet, the stream function and potential function of the combined uniform flow and doublet flow are given by

$$\psi = Vr\sin\theta - \frac{K\sin\theta}{r} \tag{5.266}$$

$$\phi = Vr\cos\theta + \frac{K\cos\theta}{r} \tag{5.267}$$

Requiring that the surface of the cylinder, defined by $\psi = 0$, be at $r = R$, where R is the radius of the cylinder, gives $K = VR^2$; hence, the stream and potential functions can be represented as

$$\psi = Vr\left(1 - \frac{R^2}{r^2}\right)\sin\theta \tag{5.268}$$

$$\phi = Vr\left(1 + \frac{R^2}{r^2}\right)\cos\theta \tag{5.269}$$

The streamlines and flow characteristics described by Equation 5.268 are shown in Figure 5.45(a). The components of the flow velocity within the flow field, v_r and v_θ, can be derived from the potential function (Equation 5.269) and are given by

$$v_r = \frac{\partial\phi}{\partial r} = V\left(1 - \frac{R^2}{r^2}\right)\cos\theta \tag{5.270}$$

$$v_\theta = \frac{1}{r}\frac{\partial\phi}{\partial\theta} = -V\left(1 + \frac{R^2}{r^2}\right)\sin\theta \tag{5.271}$$

On the surface of the cylinder, $r = R$, and Equations 5.270 and 5.271 yield

$$v_{r,\text{s}} = 0, \quad v_{\theta,\text{s}} = -2V\sin\theta \tag{5.272}$$

where the subscript "s" indicates that the velocities are those on the surface of the cylinder. It is apparent from Equation 5.272 that the velocity normal to the cylinder is equal to zero (as expected) and that the tangential velocity is a maximum at the extremities of the cylinder where $\theta = \pm\pi/2$; the maximum velocity at this location is equal to twice the free-stream velocity.

Pressure distribution. The pressure distribution on the circumference of the cylinder, $p_\text{s}(\theta)$, can be calculated using the Bernoulli equation, which can be expressed as

$$p_\text{s} + \frac{1}{2}\rho v_{\theta,\text{s}}^2 = p_0 + \frac{1}{2}\rho V^2 \tag{5.273}$$

where p_0 is the free-stream (i.e., static) pressure. Combining Equations 5.273 and 5.272 gives the pressure distribution on the surface of the cylinder as

$$p_\text{s} = p_0 + \frac{1}{2}\rho V^2(1 - 4\sin^2\theta) \tag{5.274}$$

As was the case for the Rankine oval, the pressure distribution derived from a potential-flow analysis is less likely to be realistic on the back side of the cylinder (where $\theta < \pi/2$) due

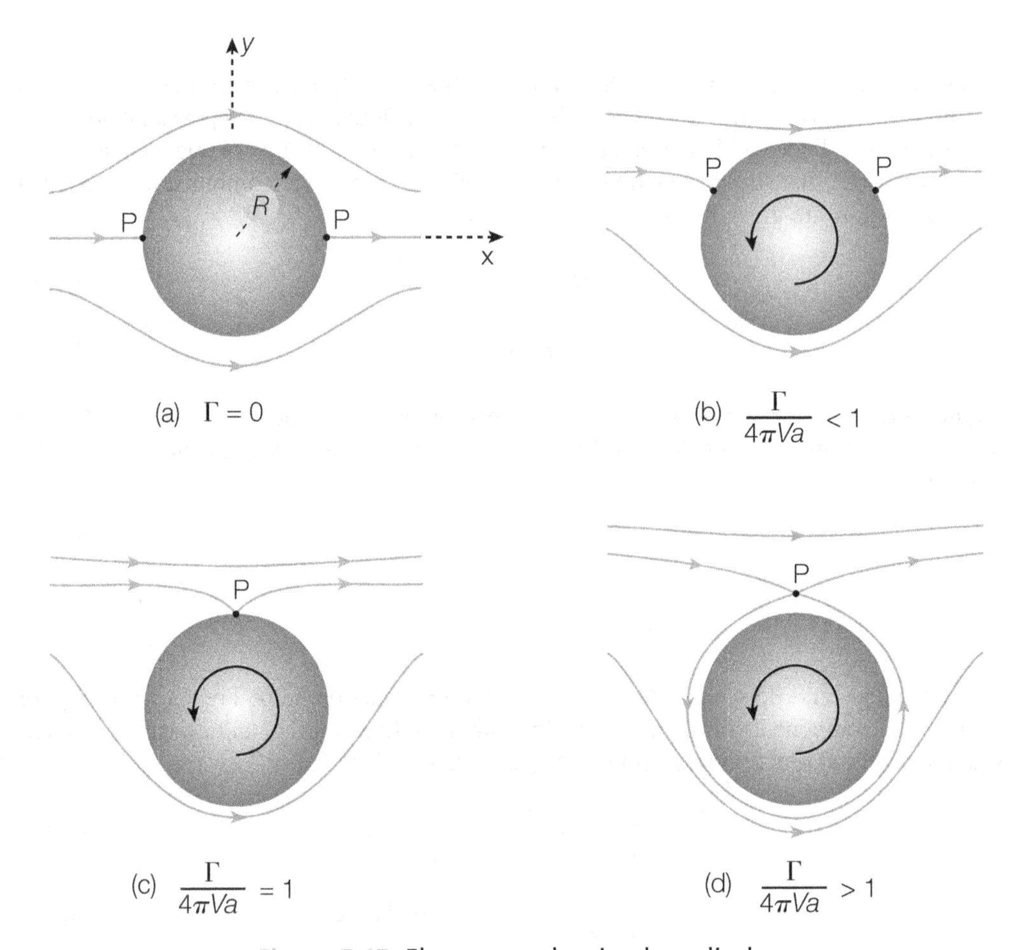

Figure 5.45: Flow around a circular cylinder

to the adverse pressure gradient and the resulting separated boundary layer that is likely to exist in this region. An interesting corollary to Equation 5.274 is that the second term on the right-hand side of Equation 5.274 disappears when

$$1 - 4\sin^2\theta = 0 \quad \rightarrow \quad \theta = \sin^{-1}\left(\pm\frac{1}{2}\right) = 30^\circ, 150^\circ, 210^\circ, 330^\circ$$

which means that at these four locations on the surface of the cylinder, Equation 5.274 gives $p_s = p_0$ regardless of the free-stream velocity. These points are called *zero-pressure points*. Interestingly, the location of zero-pressure points on the front sides of fish has been reported to coincide with the locations of their eyes, which enables them not to have blurred vision due to changes in pressure as they swim at various speeds. Another result derived from Equation 5.274 is that the pressure on the surface of a cylinder is a minimum when

$$\sin^2\theta = 1 \quad \rightarrow \quad \theta = \sin^{-1}(\pm 1) = 90^\circ, 270^\circ$$

These points identify the *aerodynamic shoulder* of the body, which for any body is defined as the location(s) of minimum pressure. Returning to the case of the fish, it has also been reported that gills of fish are located near their aerodynamic shoulder, which makes it easier for them to exhale. In fact, inhaling at the location of maximum pressure and exhaling at the location of minimum pressure is ideal.

Drag and lift forces. The total pressure forces in the x- and y-directions exerted on a circular cylinder by a flowing fluid are calculated from the surface pressure distribution, p_s, using the relations

$$F_x = F_D = -L\int_0^{2\pi} p_s \cos\theta R \, d\theta \tag{5.275}$$

$$F_y = F_L = -L\int_0^{2\pi} p_s \sin\theta R \, d\theta \tag{5.276}$$

where L is the length of the cylinder. The net pressure force in the flow direction, commonly represented as F_x or F_D, is called the *drag force*, and the net pressure force normal to the flow direction, commonly represented as F_y or F_L, is called the *lift force*. Substituting Equation 5.274 into Equations 5.275 and 5.276 yields $F_D = 0$ and $F_L = 0$, which is a consequence of the symmetry of the flow about the x- and y-axes. In reality, the drag force on a circular cylinder caused by a flowing fluid is primarily due to viscous effects, which cause a shear force on the solid boundary and boundary-layer separation, none of which are accounted for in the potential-flow assumption.

Rotating cylinder. An interesting case of flow past a cylinder occurs when the cylinder is rotating. In this case, the fluid adjacent to the cylinder rotates at the same tangential velocity as the rotating cylinder, whereas the normal velocity on the surface of the cylinder is equal to zero. The irrotational flow field external to the boundary layer that corresponds to this scenario can be approximated by adding a free vortex to the source/sink and uniform-flow combination that corresponds to flow around a cylinder. The stream function, ψ, and potential function, ϕ, are then given by

$$\psi = Vr\left(1 - \frac{R^2}{r^2}\right)\sin\theta - \frac{\Gamma}{2\pi}\ln r \tag{5.277}$$

$$\phi = Vr\left(1 + \frac{R^2}{r^2}\right)\cos\theta + \frac{\Gamma}{2\pi}\theta \tag{5.278}$$

where Γ is the circulation of the vortex. Following a similar derivation to that described previously (see Equations 5.270–5.272), the radial and tangential components of the fluid velocity, $v_{r,s}$ and $v_{\theta,s}$, respectively, on the surface of the cylinder are

$$v_{r,s} = 0, \quad v_{\theta,s} = -2V\sin\theta + \frac{\Gamma}{2\pi R} \tag{5.279}$$

Using the expression for $v_{\theta,s}$, the location of the stagnation points can be identified where $v_{\theta,s} = 0$, and hence, stagnation points occur where

$$\sin\theta_s = \frac{\Gamma}{4\pi VR} \tag{5.280}$$

where θ_s is the value of θ at which stagnation conditions occur. Equation 5.280 yields two solutions for θ_s as long as $|\Gamma|/(4\pi VR) < 1$ as shown in Figure 5.45(b), and a single solution for θ_s when $|\Gamma|/(4\pi VR) = 1$ as shown in Figure 5.45(c). In cases where $|\Gamma|/(4\pi VR) > 1$, there is no stagnation point on the surface of the cylinder, with the stagnation point located away from the cylinder as shown in Figure 5.45(d). Applying the Bernoulli equation to the streamline around the circular cylinder and using the surface velocity given by Equation 5.279 yields

$$p_0 + \frac{1}{2}\rho V^2 = p_s + \frac{1}{2}\rho\left(-2V\sin\theta + \frac{\Gamma}{2\pi R}\right)^2$$

which gives the pressure distribution on the surface of the rotating cylinder as

$$p_s = p_0 + \frac{1}{2}\rho V^2 \left(1 - 4\sin^2\theta + \frac{2\Gamma\sin\theta}{\pi R V} - \frac{\Gamma^2}{4\pi^2 R^2 V^2}\right) \tag{5.281}$$

Substituting the pressure distribution given by Equation 5.281 into Equations 5.275 and 5.276 gives the drag force, F_D, and the lift force, F_L, per unit length of the cylinder as

$$F_\mathrm{D} = 0, \quad F_\mathrm{L} = -\rho V \Gamma \tag{5.282}$$

Therefore, a rotating cylinder generates a lift force, in contrast to a nonrotating cylinder that does not generate a lift force. The negative sign in the expression for F_L in Equation 5.282 indicates that if V is positive (i.e, directed in the positive x-direction) and Γ is positive (i.e., a counterclockwise rotation), then F_L is directed downward. The expression for F_L in Equation 5.282 is called the *Kutta-Joukowski theorem*,[36] or the *Kutta-Joukowski lift theorem*, and it has been shown to apply to cylinders of any shape, including airfoils. The development of lift on rotating bodies, called the *Magnus effect*,[37] is largely responsible for the curved trajectory of spinning golf balls and baseballs.

Concluding notes. Although inviscid-flow theory can be used to describe the flow around solid bodies, where viscosity effects are limited to a very small region adjacent to the surface of the body, the calculated flow fields predict no viscous drag (because the no-slip condition on the surface of the body cannot be satisfied when viscosity is neglected). The irrotational-flow assumption also does not account for the separation of the boundary layer behind bodies and generally yields a net pressure drag of zero. This is embodied in *d'Alembert's paradox*,[38] which states that the irrotational flow approximation leads to zero aerodynamic drag force on any nonlifting body of any shape immersed in a uniform stream. Potential-flow theory can provide acceptable approximations in cases where fluids of low viscosity are moving at relatively high velocities in regions of accelerating flow. This section has described the superposition of fundamental flows to yield flow fields around some regular body shapes. In cases where body shapes are irregular, advanced techniques are available to determine the corresponding superpositions of fundamental flows that would be required to simulate the flow field.

EXAMPLE 5.29

Air flows at 96 km/h toward a 4-m-long × 0.5-m-wide beam with a semicircular cross section as shown in Figure 5.46. The air, which is at a temperature of 20°C, separates from the beam at its widest point such that the pressure behind the beam is equal to the pressure at the separation point. Estimate the drag force on the beam.

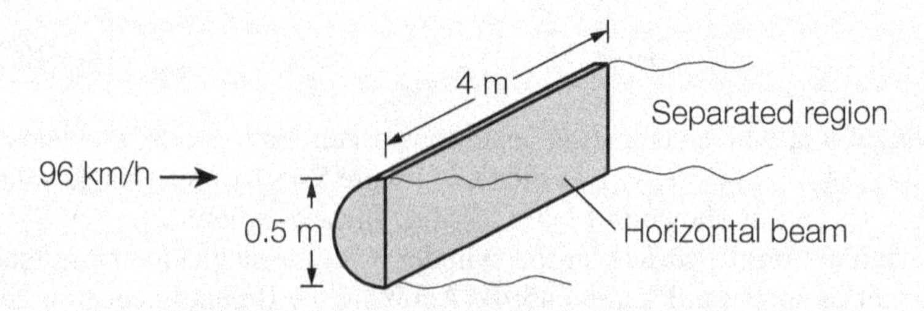

Figure 5.46: Flow around a semicircular beam

[36]Named in honor of the German mathematician Martin Wilhelm Kutta (1867–1944) and the Russian mathematician Nikolai E. Joukowski (1847–1921).

[37]Named in honor of the German scientist Heinrich G. Magnus (1802–1870).

[38]Named in honor of the French mathematician and philosopher Jean-le-Rond d'Alembert (1717–1783).

SOLUTION

From the given data: V = 96 km/h = 26.7 m/s, L = 4 m, D = 0.5 m, and $R = D/2$ = 0.25 m. At 20°C, for air, ρ = 1.204 kg/m^3. From the pressure distribution given by Equation 5.274, the pressure force on the front side of the beam, F_{front}, is given by

$$\begin{aligned} F_{\text{front}} &= -L\int_{\frac{\pi}{2}}^{\frac{3\pi}{2}} p_s \cos\theta\, R\, d\theta = -L\int_{\frac{\pi}{2}}^{\frac{3\pi}{2}}\left[p_0 + \frac{1}{2}\rho V^2(1-4\sin^2\theta)\right]\cos\theta\, R\, d\theta \\ &= -\left(p_0 + \frac{1}{2}\rho V^2\right)RL\int_{\frac{\pi}{2}}^{\frac{3\pi}{2}}\cos\theta\, d\theta + 2\rho V^2 RL\int_{\frac{\pi}{2}}^{\frac{3\pi}{2}}\sin^2\theta\cos\theta\, d\theta \\ &= -\left(p_0 + \frac{1}{2}\rho V^2\right)RL\Big[\sin\theta\Big]_{\frac{\pi}{2}}^{\frac{3\pi}{2}} + \frac{2}{3}\rho V^2 RL\Big[\sin^3\theta\Big]_{\frac{\pi}{2}}^{\frac{3\pi}{2}} \\ &= \left(2p_0 - \frac{4}{3}\rho V^2\right)RL \end{aligned}$$

The minimum pressure on the surface of the beam occurs at $\theta = \pi/2$, and using Equation 5.274, the minimum pressure, p_{min}, is given by

$$p_{\text{min}} = p_0 + \frac{1}{2}\rho V^2\left[1 - 4\sin^2\left(\frac{\pi}{2}\right)\right] = p_0 - \frac{3}{2}\rho V^2$$

Because the pressure on the back of the beam is equal to p_{min}, and the area of the back of the beam is $2RL$, the pressure force on the back of the beam, F_{back}, is given by

$$F_{\text{back}} = (p_0 - \frac{3}{2}\rho V^2)(2RL) = (2p_0 - 3\rho V^2)RL$$

The net pressure force, F_{net}, is therefore given by

$$F_{\text{net}} = F_{\text{front}} - F_{\text{back}} = \left(2p_0 - \frac{4}{3}\rho V^2\right)RL - (2p_0 - 3\rho V^2)RL = \frac{5}{3}\rho V^2 RL$$

Using the given values for ρ, V, R, and L, the drag force (= net pressure force) is calculated as follows:

$$F_{\text{net}} = \frac{5}{3}(1.204)(26.7)^2(0.25)(4) = 1430\ \text{N} = 1.43\ \text{kN}$$

Therefore, the drag force on the beam is **1.43 kN**.

5.8 Turbulent Flow

Turbulent flow is characterized by persistent random instabilities in the flow field called *turbulence*. Turbulence in fluids is a result of spatial variations in velocity called *velocity shear*, where the fluid viscosity causes small eddies to form in the fluid as a result of velocity shear. If the inertia of the fluid is sufficiently high, then some of these eddies become unstable and grow to form large-scale disturbances that propagate within the flow field. This view of turbulence in terms of the stability of small-scale eddies is known as the *Tollmien-Schlichting theory*. If the inertia of the fluid is relatively low, then the small-scale eddies caused by viscous effects do not become unstable, and large-scale random perturbations in the flow field are not present. Under these circumstances, the flow is called *laminar flow*. The word "laminar" comes from the English word "lamina," which is defined as a thin plate, sheet, or layer. In the context of fluid mechanics, under "laminar flow" conditions, fluid elements do not mix significantly with neighboring fluid elements (except by molecular diffusion); so the fluid elements tend to stay within their own "lamina." This behavior is in contrast to turbulent flow, where there is rapid mixing between fluid laminae. This distinction is illustrated in Figure 5.47, which shows a tracer being mixed into adjacent layers of a fluid by eddies. Under some circumstances, fluid flow can be intermittently laminar and turbulent; this type of flow is called *intermittent flow*.

Eddies in turbulent flow. An *eddy* is defined as a flow structure in which the flow direction differs from that of the mean flow of the fluid. Turbulent flow is characterized by the dominant presence of eddies in the flow field, and generally speaking, the more eddies there are in a flow field, the more turbulent the flow. In the *Kolmogorov*[39] *theory* of an energy cascade in turbulent flow, energy is input to the turbulent flow field at the scale of the largest eddies and energy is dissipated at the scale of the smallest eddies, where energy dissipation (into waste heat) is due to friction associated with the viscosity of the fluid. The length scale of the largest eddies in a turbulent flow field is commonly called the *integral length scale*, and the length scale of the smallest eddies in a turbulent flow field is called the *Kolmogorov length scale*. The size of the integral length scale typically depends on the physical dimensions of the flow field and/or the scale at which energy is input to the fluid. In contrast, the size of the Kolmogorov length scale, η, is typically independent of the physical dimensions of the flow field and depends only on the kinematic viscosity of the fluid, ν, and the energy dissipation rate, ϵ, according to the relation

$$\eta = \left(\frac{\nu^3}{\epsilon}\right)^{\frac{1}{4}} \tag{5.283}$$

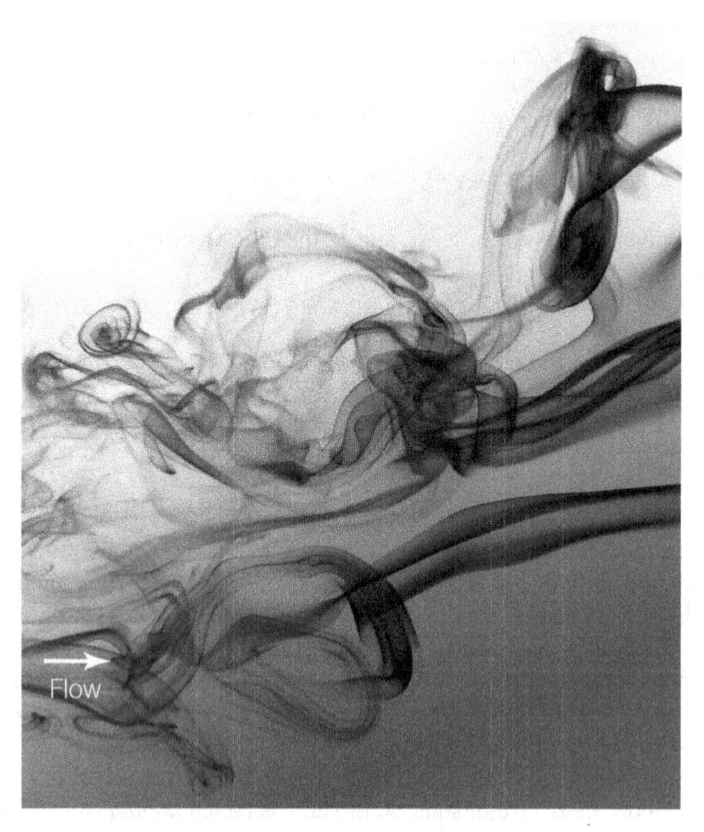

Figure 5.47: Mixing in turbulent flow
Source: Luis Francisco Corde/Fotolia.

[39]Named in honor of the Russian mathematician Andrey Nikolaevich Kolmogorov (1903–1987).

Eddies can be generated by a variety of mechanisms. For example, turbulent flow past a body of characteristic size L will have eddies with an integral scale on the order of L, and turbulent flow in circular conduits of diameter D will have eddies with an integral scale on the order of D. If the fluid and the energy dissipation rate were the same in these two systems, they would have the same Kolmogorov length scale.

5.8.1 Occurrence of Turbulence

The tendency of small-scale eddies to cause turbulence depends on the relative magnitudes of the inertial and viscous forces. The inertial force, F_I, is proportional to the mass times acceleration of a fluid and can be approximated (to within an order of magnitude) by

$$F_\mathrm{I} = ma \sim \rho L^3 \frac{V^2}{L} = \rho V^2 L^2 \tag{5.284}$$

where m is the mass of fluid within a fluid element, ρ is the density of the fluid, L is the length scale of the fluid element, and V is a measure of the velocity of the fluid element. The viscous force on the fluid element, F_V, can be estimated from the definition of the viscosity (Equation 1.44) as

$$F_\mathrm{V} \sim \mu \frac{V}{L} L^2 = \mu V L \tag{5.285}$$

where μ is the dynamic viscosity of the fluid. The ratio of the inertial to the viscous force is called the *Reynolds number*,[40] Re, and is given by

$$\mathrm{Re} = \frac{F_\mathrm{I}}{F_\mathrm{V}} = \frac{\rho V L}{\mu} \tag{5.286}$$

For higher values of the Reynolds number, inertial forces are much larger than viscous forces and turbulence tends to occur within the fluid. On the other hand, small values of the Reynolds number indicate that viscous forces are comparable to or greater than inertial forces, indicating that turbulence is less likely to occur within the fluid. Turbulence has a significant influence on energy losses within a flowing fluid, and the values of Re at which turbulent flow occurs depends on the geometry of the flow and the characteristic length scale. In the case of flow through a pipe, where the diameter of the pipe is the characteristic length scale, for $\mathrm{Re} \leq 2000$, the flow is laminar, for $2000 < \mathrm{Re} < 4000$, there is a gradual change to turbulent flow with random oscillations between laminar and turbulent conditions, and for $\mathrm{Re} > 4000$, the flow is turbulent. In the case of open-channel flow, such as in canals and rivers, the depth of flow is used as the characteristic length scale, and open-channel flows are typically turbulent for $\mathrm{Re} \geq 1000$. In most engineering applications involving pipe and open-channel flow, the Reynolds number limits are far exceeded and the flows are turbulent.

Steady turbulent flow. Because turbulent flow is characterized by random perturbations in the flow velocity, strictly speaking, turbulent flow cannot be characterized as steady flow. However, in cases where the flow velocity fluctuates randomly about a steady mean velocity, the flow field is still regarded as steady. As an example, steady turbulent flow in open channels such as rivers is commonly assumed.

5.8.2 Turbulent Shear Stress

Consider the typical turbulent-flow scenario shown in Figure 5.48, where the x and y components of the turbulent velocity fluctuations are u' and v', respectively, the mean velocity, $\bar{u}$, is in the x-direction and varies spatially, and the density of the fluid is ρ. In accordance with the momentum equation, if a fluid with a velocity fluctuation u' moves across a plane area A,

[40] Named in honor of the English physicist Osborne Reynolds (1842–1912).

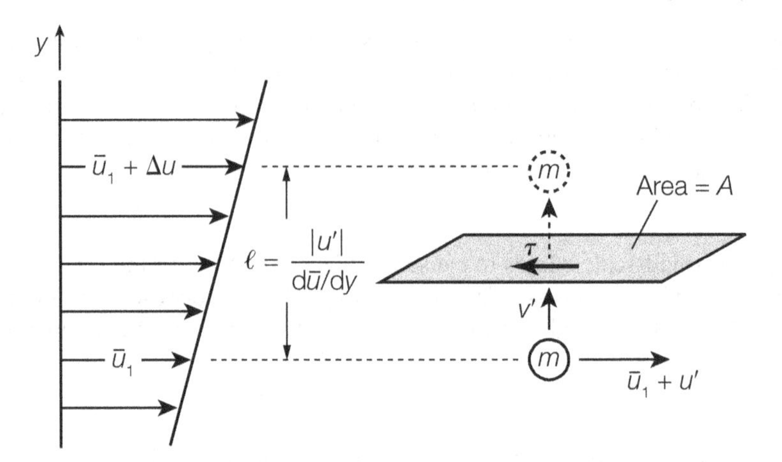

Figure 5.48: Shear stress caused by turbulence

the mass flux is $\rho A v'$ and the change in velocity is u', so the force, F, and shear stress, τ_{turb}, exerted on the fluid above the plane area A can be estimated as

$$F = -(\rho A v')u' \tag{5.287}$$

$$\tau_{\text{turb}} = \frac{F}{A} = -\rho v' u' \tag{5.288}$$

where the negative sign originates from the expectation that for fluid masses moving from a lower-velocity environment to a higher-velocity environment $u' < 0$ when $v' > 0$. Similarly, for fluid masses moving from a higher-velocity environment to a lower-velocity environment it is expected that $u' > 0$ when $v' < 0$. Hence, on average, the turbulent shear stress expressed by Equation 5.288 is expected to be a positive quantity. Mixing of fluid elements between adjacent fluid layers can be assumed to occur over a *mixing length*, ℓ, which is sometimes taken to be analogous to mean free path in molecular theory or, alternatively, the average distance over which the velocity fluctuation of a fluid element is dissipated. Taking the temporal average of Equation 5.288 (denoted by an overbar) yields

$$\tau_{\text{turb}} = -\rho \overline{u'v'} \tag{5.289}$$

where the quantity $-\rho\overline{u'v'}$ is called the *turbulent shear stress*, *apparent shear stress*, or *Reynolds stress*. In accordance with the definition of the mixing length, ℓ, and assuming that u' and v' are on the same order of magnitude, the following order-of-magnitude estimates are appropriate:

$$|u'| \sim \ell \frac{\mathrm{d}\bar{u}}{\mathrm{d}y} \tag{5.290}$$

$$|v'| \sim |u'| \tag{5.291}$$

Combining Equations 5.289–5.291 gives

$$\tau_{\text{turb}} = \rho \ell^2 \left(\frac{\mathrm{d}\bar{u}}{\mathrm{d}y}\right)^2 \tag{5.292}$$

The concept of using a mixing length in the analysis of turbulent flows was originally suggested by Ludwig Prandtl (1875–1953), a German engineer. In utilizing the mixing-length concept, the mixing length, ℓ, is not generally a constant within a flow field but is typically a variable that is commonly assumed to be proportional to the distance from the flow boundary. In fact, the mixing length approaches zero as the flow boundary is approached. Equation

5.292 can be related to the Newtonian definition of viscosity, where viscosity is the proportionality constant between the shear stress and the velocity gradient. Thus, Equation 5.292 can be expressed as

$$\tau_{\text{turb}} = \eta \frac{\mathrm{d}\bar{u}}{\mathrm{d}y} \tag{5.293}$$

where η is called the *eddy viscosity* or *turbulent viscosity*, in contrast to the dynamic viscosity, μ, which is a property of the fluid. Comparing Equations 5.289, 5.292, and 5.293, shows that the eddy viscosity, η, is given by

$$\eta = -\rho\overline{u'v'}\left(\frac{\mathrm{d}\bar{u}}{\mathrm{d}y}\right)^{-1} = \rho\ell^2\frac{\mathrm{d}\bar{u}}{\mathrm{d}y} \tag{5.294}$$

It is apparent from Equation 5.294 that the eddy viscosity depends on both the fluid and the flow conditions and is not only a property of the fluid. It is also apparent from Equation 5.294 that the eddy viscosity, η, necessarily approaches zero at the boundary of the flow because the mixing length approaches zero at the boundary. Because the eddy viscosity depends on the flow conditions, to model turbulent flows, it is usually necessary to express the eddy viscosity, η, as a function of averaged flow velocities. The need for such a relationship is called the *closure problem*, and the relationship used is sometimes called the *eddy viscosity closure*.

Total shear stress. In general, the total shear stress in turbulent flow is the sum of the laminar shear stress and turbulent shear stress, in which case the actual (total) shear stress, τ, is given by

$$\tau = \mu\frac{\mathrm{d}\bar{u}}{\mathrm{d}y} + \eta\frac{\mathrm{d}\bar{u}}{\mathrm{d}y} = \rho(\nu + \epsilon)\frac{\mathrm{d}\bar{u}}{\mathrm{d}y} \tag{5.295}$$

where ϵ is called the *kinematic eddy viscosity*, *kinematic turbulent viscosity*, or *eddy diffusivity of momentum* and is equal to η/ρ. In turbulent flow, the viscosity due to turbulence is typically much greater than the (laminar-flow) fluid viscosity.

Turbulence intensity. The degree of turbulence is sometimes characterized by the *turbulence intensity*, I, which is defined as the root mean square of the random fluctuations divided by the mean velocity; hence,

$$I = \frac{\sqrt{\overline{(u')^2}}}{\bar{u}} \tag{5.296}$$

and in some instances $\overline{(u')^2}$ is referred to as the turbulence intensity. Typically, u' is only a small fraction of $\bar{u}$, with larger turbulence intensities corresponding to larger velocity fluctuations. Values of $I > 0.1$ are typically found in atmospheric and river flows.

5.8.3 Mean Steady Turbulent Flow

Turbulent flow is by definition unsteady because it involves random perturbations in velocity. The velocity, $\mathbf{v}$, at any point in a turbulent flow field can be expressed in terms of deviations, $\mathbf{v}'$, from an averaged velocity, $\bar{\mathbf{v}}$, such that the velocity at any time t is given by

$$\mathbf{v}(t) = \bar{\mathbf{v}}(t) + \mathbf{v}' \tag{5.297}$$

If the averaging time for $\bar{\mathbf{v}}(t)$ is Δt, then Equation 5.297 can be more appropriately expressed as

$$\mathbf{v}(t) = \bar{\mathbf{v}}_{\Delta t}(t) + \mathbf{v}' \tag{5.298}$$

where $\bar{\mathbf{v}}_{\Delta t}$ is the averaged velocity over time Δt, which is calculated as

$$\bar{\mathbf{v}}_{\Delta t}(t) = \frac{1}{\Delta t}\int_{t-\frac{\Delta t}{2}}^{t+\frac{\Delta t}{2}} \mathbf{v}\,\mathrm{d}t \tag{5.299}$$

Typical values of Δt used in averaging depend on the time scales of interest in a particular application. Mean steady turbulent flow occurs in cases where $\bar{\mathbf{v}}_{\Delta t}$ remains constant at any given location and independent of t such that the velocity at any given location can be expressed as

$$\mathbf{v}(t) = \bar{\mathbf{v}}_{\Delta t} + \mathbf{v}' \tag{5.300}$$

Such mean steady turbulent flows occur widely in engineering applications and are normally treated as a mean flow with the velocity field described by spatial variations in $\bar{\mathbf{v}}_{\Delta t}$, with random perturbations accounted for in the eddy viscosity.

5.9 Conservation of Energy[41]

The *conservation of energy equation*, or simply the *energy equation*, is an analytic expression of the first law of thermodynamics. In fluid-mechanics applications, the energy equation is mostly used in the control-volume form; the derivation of the control-volume form of the energy equation is covered extensively in Chapter 4. This section presents an abbreviated derivation of the differential form of the energy equation, with the objectives of familiarizing the reader with what the differential form of the energy equation looks like and of elucidating the mechanisms by which energy is lost and transferred to different forms in fluid flows. The differential energy equation is derived by combining the first law of thermodynamics, Fourier's law of heat conduction, the continuity equation, and the linear momentum equation. These component equations are briefly discussed below and then are combined to form the differential energy equation.

The first law of thermodynamics. The first law of thermodynamics states that the heat added to a system minus the work done by the system is equal to the change in internal energy of the system. The control-volume form of the energy equation was derived previously and is given by Equation 4.125, which can be expressed as

$$\dot{Q} - \dot{W} = \frac{\mathrm{d}}{\mathrm{d}t}\int_{\mathrm{cv}} \rho e\,\mathrm{d}V + \int_{\mathrm{cs}} \rho e\,(\mathbf{v}\cdot\mathbf{n})\,\mathrm{d}A \tag{5.301}$$

where $\dot{Q}$ is the rate at which heat is added to a system, $\dot{W}$ is the rate at which work is done by the system, "cv" denotes the control volume, "cs" denotes the surface of the control volume, ρ is the density of the fluid, e is the energy per unit mass of fluid, $\mathbf{v}$ is the velocity vector, and $\mathbf{n}$ is the unit normal pointing outward from the control surface. The energy per unit mass, e, is defined as

$$e = u + \tfrac{1}{2}|\mathbf{v}|^2 + gz$$

where u is the internal energy per unit mass, $|\mathbf{v}|$ is the magnitude of the velocity, g is the gravity constant, and z is the elevation. If the control volume is taken as a rectangular prism with sides $\mathrm{d}x$, $\mathrm{d}y$, and $\mathrm{d}z$, then Equation 5.301 can be expressed in the following differential form:

$$\dot{Q} - \dot{W} = \left[\rho\frac{\mathrm{d}e}{\mathrm{d}t} + \frac{\partial}{\partial x}(\rho u\zeta) + \frac{\partial}{\partial y}(\rho v\zeta) + \frac{\partial}{\partial z}(\rho w\zeta)\right]\mathrm{d}x\,\mathrm{d}y\,\mathrm{d}z \tag{5.302}$$

where u, v, and w are the components of $\mathbf{v}$ in the (Cartesian) coordinate directions and ζ is the energy plus flow work defined by

$$\zeta = e + \frac{p}{\rho}$$

[41]This section can be omitted without loss of continuity.

The differential form of the continuity equation was derived previously as Equation 5.28 and is given by

$$\frac{\partial \rho}{\partial t} + \frac{\partial(\rho u)}{\partial x} + \frac{\partial(\rho v)}{\partial y} + \frac{\partial(\rho w)}{\partial z} = 0 \tag{5.303}$$

Combining the form of the first law of thermodynamics given by Equation 5.302 with the continuity equation given by Equation 5.303 yields

$$\dot{Q} - \dot{W} = \left[\rho\frac{\mathrm{d}e}{\mathrm{d}t} + \mathbf{v}\cdot\nabla p + p\nabla\cdot\mathbf{v}\right]\mathrm{d}x\,\mathrm{d}y\,\mathrm{d}z \tag{5.304}$$

This analytic expression of the first law of thermodynamics will be used to derive the final form of the differential energy equation.

Fourier's law. *Fourier's law of heat conduction*,[42] also known as *Fourier's law of heat transfer*, states that the heat flux in any direction is proportional to the temperature gradient in that direction. Fourier's law is usually expressed in the form

$$\mathbf{q} = -k\nabla T \tag{5.305}$$

where k is a proportionality constant called the *thermal conductivity*, which can be a function of temperature and pressure, particularly for gases. The thermal conductivity measures the ease with which heat is conducted within the fluid. If heat loss due to radiation is neglected, then the rate, $\dot{Q}$, at which heat is added to any control volume of dimensions $\mathrm{d}x \times \mathrm{d}y \times \mathrm{d}z$ can be expressed by the relation

$$\dot{Q} = -\nabla\cdot\mathbf{q}\,\mathrm{d}x\,\mathrm{d}y\,\mathrm{d}z \tag{5.306}$$

This equation is derived using the property that the divergence of a flux of a quantity is equal to the rate of decrease of that quantity per unit volume. Combining Equations 5.305 and 5.306 gives the following expression for $\dot{Q}$ in terms of the temperature field:

$$\dot{Q} = \nabla\cdot(k\nabla T)\,\mathrm{d}x\,\mathrm{d}y\,\mathrm{d}z \tag{5.307}$$

Rate of doing work. The rate at which work is done by any force acting on a fluid system is equal to the product of the force and the velocity at which the fluid moves against the force. Consider the stress components illustrated in Figure 5.11, with the slight change of notation that $\tau_{xx} = \sigma_{xx}$, $\tau_{yy} = \sigma_{yy}$, and $\tau_{zz} = \sigma_{zz}$. There are three force components on each surface (= stress component $\times$ surface area). Taking u, v, and w as the fluid velocity components, the rate that work is done by each force on the surface of the control volume is equal to the product of the force component and the velocity component in the direction opposing the force. Summing the work done by all forces on the rectangular prism gives the rate at which work is done as

$$\dot{W} = -\left[\frac{\partial}{\partial x}(u\tau_{xx} + v\tau_{xy} + w\tau_{xz}) + \frac{\partial}{\partial y}(u\tau_{yx} + v\tau_{yy} + w\tau_{yz}) + \frac{\partial}{\partial z}(u\tau_{zx} + v\tau_{zy} + w\tau_{zz})\right]\mathrm{d}x\,\mathrm{d}y\,\mathrm{d}z$$

$$\rightarrow \quad \dot{W} = -\nabla\cdot(\mathbf{v}\cdot\boldsymbol{\tau}_{ij})\,\mathrm{d}x\,\mathrm{d}y\,\mathrm{d}z \quad \rightarrow \quad \dot{W} = -\left[\mathbf{v}\cdot(\nabla\cdot\boldsymbol{\tau}_{ij}) + \Phi\right]\mathrm{d}x\,\mathrm{d}y\,\mathrm{d}z \tag{5.308}$$

where, for Newtonian fluids, Φ is a collection of terms only involving velocity gradients and is given by

$$\Phi = \mu\left[2\left(\frac{\partial u}{\partial x}\right)^2 + 2\left(\frac{\partial v}{\partial y}\right)^2 + 2\left(\frac{\partial w}{\partial z}\right)^2 + \left(\frac{\partial v}{\partial x} + \frac{\partial u}{\partial y}\right)^2 + \left(\frac{\partial w}{\partial y} + \frac{\partial v}{\partial z}\right)^2 + \left(\frac{\partial u}{\partial z} + \frac{\partial w}{\partial x}\right)^2\right] \tag{5.309}$$

[42] Named in honor of the French mathematician and physicist Jean Baptiste Joseph Fourier (1768–1830).

where Φ is necessarily positive because it involves only quadratic terms. The quantity Φ is called the *viscous dissipation function*. Also, because shear stress, τ_{ij}, within a fluid is entirely due to the viscosity of the fluid, $\dot{W}$ as defined by Equation 5.308 is commonly referred to as the *viscous work*.

Linear momentum equation. The linear momentum equation, derived previously as Equation 5.78, is used to eliminate the shear stress tensor τ_{ij} in the viscous work given by Equation 5.308. The useful form of the linear momentum equation for this purpose is

$$\rho \mathrm{g} - \nabla p + \nabla \cdot \tau_{ij} = \rho \frac{\mathrm{d}\mathrm{v}}{\mathrm{d}t} \tag{5.310}$$

where g is the gravity vector and p is the mechanical pressure that exists in addition to the shear stresses caused by the viscosity of the fluid.

Combined energy equation. The combined form of the energy equation that is most used in practice is obtained by combining Equations 5.304, 5.307, 5.308, and 5.310 to give

$$\rho \frac{\mathrm{d}u}{\mathrm{d}t} + p(\nabla \cdot \mathrm{v}) = \nabla \cdot (k \nabla T) + \Phi \tag{5.311}$$

This form of the energy equation is applicable to Newtonian fluids in cases where radiation heat transfer and internal sources of heat are not present. Equation 5.311 can be applied to unsteady and compressible flows. In cases where the density (ρ), specific heat at constant volume (c_v), viscosity (μ), and thermal conductivity (k) can be assumed constant, Equation 5.311 simplifies to

$$\rho c_v \frac{\mathrm{d}T}{\mathrm{d}t} = k \nabla^2 T + \Phi \qquad (\text{for } \rho,\, c_v,\, \mu, \text{ and } k \text{ constant}) \tag{5.312}$$

where the relations $\nabla \cdot \mathrm{v} = 0$ and $\mathrm{d}u = c_v\,\mathrm{d}T$ have been applied in deriving Equation 5.312 from Equation 5.311. For fluids at rest, $\Phi = 0$ and c_v is replaced by c_p in Equation 5.312. Also note that $c_v \approx c_p$ in liquids. Numerical methods are usually used to obtain solutions to these equations for given boundary and initial conditions.

Practical Considerations

The differential form of the energy equation should generally be used in cases where spatial variations in temperature affect the variables in the continuity and momentum equations. This is generally the case in compressible flows where the pressure and density are related to temperature by an equation of state. In addition, the flows of incompressible fluids can be significantly influenced by temperature variations. For example, in free convection, the relationship between density and temperature is of central importance, and in lubrication flows, the relationship between temperature and viscosity can significantly influence the flow.

Key Equations in Differential Analysis of Fluid Flows

The following list of equations is particularly useful in solving problems related to the differential analysis of fluid flows. If one is able to recognize these equations and recall their appropriate use, then the learning objectives of this chapter have been met to a significant degree. Derivations of these equations, definitions of the variables, and detailed examples of usage can be found in the main text.

KINEMATICS

Velocity:

$$\mathbf{V} = u\mathbf{i} + v\mathbf{j} + w\mathbf{k}$$

Acceleration:

$$\mathbf{a} = \frac{\mathrm{D}\mathbf{V}}{\mathrm{D}t} = \frac{\partial \mathbf{V}}{\partial t} + u\frac{\partial \mathbf{V}}{\partial x} + v\frac{\partial \mathbf{V}}{\partial y} + w\frac{\partial \mathbf{V}}{\partial z}$$

Angular velocity:

$$\boldsymbol{\omega} = \frac{1}{2}\left(\frac{\partial w}{\partial y} - \frac{\partial v}{\partial z}\right)\mathbf{i} + \frac{1}{2}\left(\frac{\partial u}{\partial z} - \frac{\partial w}{\partial x}\right)\mathbf{j} + \frac{1}{2}\left(\frac{\partial v}{\partial x} - \frac{\partial u}{\partial y}\right)\mathbf{k}$$

$$\boldsymbol{\omega} = \frac{1}{2}(\boldsymbol{\nabla} \times \mathbf{V}) = \frac{1}{2}\begin{vmatrix} \mathbf{i} & \mathbf{j} & \mathbf{k} \\ \dfrac{\partial}{\partial x} & \dfrac{\partial}{\partial y} & \dfrac{\partial}{\partial z} \\ u & v & w \end{vmatrix}$$

Vorticity:

$$\boldsymbol{\zeta} = \boldsymbol{\nabla} \times \mathbf{V} = 2\boldsymbol{\omega}$$

$$\boldsymbol{\zeta} = \left(\frac{1}{r}\frac{\partial v_z}{\partial \theta} - \frac{\partial v_\theta}{\partial z}\right)\mathbf{e}_r + \left(\frac{\partial v_r}{\partial z} - \frac{\partial v_z}{\partial r}\right)\mathbf{e}_\theta + \frac{1}{r}\left(\frac{\partial (r v_\theta)}{\partial r} - \frac{\partial v_r}{\partial \theta}\right)\mathbf{e}_z$$

Angular deformation rate:

$$\Omega_{xy} = \frac{\partial v}{\partial x} + \frac{\partial u}{\partial y}, \quad \Omega_{xz} = \frac{\partial w}{\partial x} + \frac{\partial u}{\partial z}, \quad \Omega_{yz} = \frac{\partial w}{\partial y} + \frac{\partial v}{\partial z}$$

Volumetric dilatation rate:

$$\boldsymbol{\nabla} \cdot \mathbf{V}$$

CONSERVATION OF MASS

Continuity equation:

$$\frac{\partial \rho}{\partial t} + \frac{\partial (\rho u)}{\partial x} + \frac{\partial (\rho v)}{\partial y} + \frac{\partial (\rho w)}{\partial z} = 0$$

$$\frac{\partial \rho}{\partial t} + \frac{1}{r}\frac{\partial (r\rho v_r)}{\partial r} + \frac{1}{r}\frac{\partial (\rho v_\theta)}{\partial \theta} + \frac{\partial (\rho v_z)}{\partial z} = 0$$

$$\frac{\partial \rho}{\partial t} + \boldsymbol{\nabla} \cdot (\rho \mathbf{V}) = 0$$

$$\frac{1}{\rho}\frac{\mathrm{D}\rho}{\mathrm{D}t} + \boldsymbol{\nabla} \cdot \mathbf{V} = 0$$

$$\nabla \cdot \mathbf{V} = 0 \quad \text{(incompressible fluid/flow)}$$

$$\frac{\partial(\rho u)}{\partial x} + \frac{\partial(\rho v)}{\partial y} + \frac{\partial(\rho w)}{\partial z} = 0 \quad \text{(steady flow)}$$

$$\frac{\partial u}{\partial x} + \frac{\partial v}{\partial y} + \frac{\partial w}{\partial z} = 0 \quad \text{(incompressible flow)}$$

$$\frac{1}{r}\frac{\partial(r v_r)}{\partial r} + \frac{1}{r}\frac{\partial(v_\theta)}{\partial \theta} + \frac{\partial(v_z)}{\partial z} = 0 \quad \text{(incompressible flow)}$$

Mach criterion:

$$\mathrm{Ma} = \frac{V}{c} < 0.3 \quad \text{(for approximate incompressibility)}$$

Stream function:

$$u = \frac{\partial \psi}{\partial y}, \quad v = -\frac{\partial \psi}{\partial x}$$

$$v_r = \frac{1}{r}\frac{\partial \psi}{\partial \theta}, \quad v_\theta = -\frac{\partial \psi}{\partial r}$$

$$v_r = -\frac{1}{r}\frac{\partial \psi}{\partial z}, \quad v_z = \frac{1}{r}\frac{\partial \psi}{\partial r}$$

Flow between streamlines:

$$q = \int_{\psi_1}^{\psi_2} \mathrm{d}\psi = \psi_2 - \psi_1$$

CONSERVATION OF MOMENTUM

Momentum equation:

$$\rho \mathbf{g} + \nabla \cdot \boldsymbol{\sigma}_{ij} = \rho \left[\frac{\partial \mathbf{V}}{\partial t} + (\mathbf{V} \cdot \nabla)\mathbf{V}\right]$$

$$\rho g_i + \nabla \cdot \boldsymbol{\sigma}_{ij} = \rho \left[\frac{\partial v_i}{\partial t} + \mathbf{V} \cdot \nabla v_i\right]$$

Navier–Stokes equation:

$$\rho\left(\frac{\partial \mathbf{V}}{\partial t} + \mathbf{V} \cdot \nabla \mathbf{V}\right) = -\nabla p + \rho \mathbf{g} + \mu \nabla^2 \mathbf{V}$$

$$\rho \frac{\mathrm{D}\mathbf{V}}{\mathrm{D}t} = -\nabla p + \rho \mathbf{g} + \mu \nabla^2 \mathbf{V}$$

$$\rho\left(\frac{\partial V_i}{\partial t} + \mathbf{V} \cdot \nabla V_i\right) = -\nabla p + \rho g_i + \mu \nabla^2 V_i$$

$$(\mathrm{St})\frac{\partial \mathbf{V}^*}{\partial t^*} + \mathbf{V}^* \cdot \nabla^* \mathbf{V}^* = -(\mathrm{Eu})\nabla p^* + \left(\frac{1}{\mathrm{Fr}^2}\right)\mathbf{g}^* + \left(\frac{1}{\mathrm{Re}}\right)\nabla^{*2}\mathbf{V}^*$$

$$\mathbf{V}^* \cdot \nabla^* \mathbf{V}^* = -(\mathrm{Eu})\nabla^* p^* + \left(\frac{1}{\mathrm{Fr}^2}\right)\mathbf{g}^* + \left(\frac{1}{\mathrm{Re}}\right)\nabla^{*2}\mathbf{V}^* \quad \text{(steady state)}$$

$$(\text{St})\frac{\partial \mathbf{V}^*}{\partial t^*} + \mathbf{V}^* \cdot \boldsymbol{\nabla}^* \mathbf{V}^* = -(\text{Eu}')\boldsymbol{\nabla}^* p'^* + \left(\frac{1}{\text{Re}}\right) \nabla^{*2}\mathbf{V}^* \quad \text{(no free surface)}$$

$$(\text{St})\frac{\partial \mathbf{V}^*}{\partial t^*} + \mathbf{V}^* \cdot \boldsymbol{\nabla}^* \mathbf{V}^* = -(\text{Eu})\boldsymbol{\nabla} p^* + \left(\frac{1}{\text{Fr}^2}\right) \mathbf{g}^* \quad \text{(inviscid flow)}$$

Continuity equation: $\boldsymbol{\nabla}^* \cdot \mathbf{V}^* = 0$ (incompressible flow)

SOLUTIONS OF THE NAVIER–STOKES EQUATION

Fixed plates (Hele-Shaw):

$$p = -\rho g z + \left(\frac{\partial p}{\partial x}\right) x + p_0, \qquad u = \frac{1}{2\mu}\left(\frac{\partial p}{\partial x}\right)(z^2 - h^2)$$

$$q = -\frac{2h^3}{3\mu}\frac{\partial p}{\partial x}, \qquad V = \frac{h^2}{3\mu}\frac{\partial p}{\partial x}, \qquad u_{\max} = -\frac{h^2}{2\mu}\frac{\partial p}{\partial x} = \frac{3}{2}V$$

Moving plate (Couette flow):

$$u = U\frac{z}{b} + \frac{1}{2\mu}\left(\frac{\partial p}{\partial x}\right)(z^2 - bz), \qquad q = -\left(\frac{\partial p}{\partial x}\right)\frac{b^3}{12\mu} + \frac{Ub}{2}$$

Moving vertical plate:

$$w = \frac{\gamma}{2\mu}x^2 - \frac{\gamma h}{\mu}x + W, \quad q = Wh - \frac{\gamma h^3}{3\mu}, \quad V = W - \frac{\gamma h^2}{3\mu}$$

Flow down inclined plane:

$$u(y) = \frac{\rho g \sin\theta}{2\mu}(2h - y)y, \qquad q = \frac{\rho g h^3 \sin\theta}{3\mu}$$

Flow through circular tube:

$$p = -\rho g(r\sin\theta) + f(x), \quad v_x = \frac{1}{4\mu}\left(\frac{\partial p}{\partial x}\right)(r^2 - R^2)$$

$$Q = -\frac{\pi D^4}{128\mu}\left(\frac{\partial p}{\partial x}\right), \quad V = -\frac{D^2}{32\mu}\left(\frac{\partial p}{\partial x}\right), \quad v_{\max} = -\frac{D^2}{16\mu}\left(\frac{\partial p}{\partial x}\right)$$

Flow through annulus:

$$v_x = \frac{1}{4\mu}\left(\frac{\partial p}{\partial x}\right)\left[r^2 - R_\text{o}^2 + \frac{R_\text{i}^2 - R_\text{o}^2}{\ln(R_\text{o}/R_\text{i})}\ln\frac{r}{R_\text{o}}\right]$$

$$Q = -\frac{\pi}{8\mu}\left(\frac{\partial p}{\partial x}\right)\left[R_\text{o}^4 - R_\text{i}^4 - \frac{(R_\text{o}^2 - R_\text{i}^2)^2}{\ln(R_\text{o}/R_\text{i})}\right]$$

$$V = -\frac{1}{8\mu}\left(\frac{\partial p}{\partial x}\right)\left[R_\text{o}^2 + R_\text{i}^2 - \frac{R_\text{o}^2 - R_\text{i}^2}{\ln(R_\text{o}/R_\text{i})}\right]$$

Flow between rotating cylinders:

$$v_\theta = \frac{R_\text{o}/r - r/R_\text{o}}{R_\text{o}/R_\text{i} - R_\text{i}/R_\text{o}}R_\text{i}\omega, \qquad T = \frac{4\pi\mu R_\text{i}^2 R_\text{o}^2 L\omega}{R_\text{o}^2 - R_\text{i}^2}$$

$$P = T\omega = \frac{4\pi\mu R_\text{i}^2 R_\text{o}^2 L\omega^2}{R_\text{o}^2 - R_\text{i}^2}$$

INVISCID FLOW

Navier–Stokes equation (inviscid): $\rho\left[\frac{\partial \mathbf{V}}{\partial t} + (\mathbf{V}\cdot\nabla)\mathbf{V}\right] = -\nabla p + \rho \mathbf{g}$

Euler equation: $\rho\frac{\mathrm{D}\mathbf{V}}{\mathrm{D}t} = -\nabla p + \rho\mathbf{g}$

Pressure gradient (streamline): $\frac{\mathrm{d}p}{\mathrm{d}s} + \frac{1}{2}\rho\frac{\mathrm{d}(V^2)}{\mathrm{d}s} + \rho g\frac{\mathrm{d}z}{\mathrm{d}s} = 0$

Along streamline: $\int\frac{\mathrm{d}p}{\rho} + \frac{V^2}{2} + gz = \text{constant}, \qquad \frac{p}{\rho} + \frac{V^2}{2} + gz = \text{constant}$

In irrotational flow field: $\frac{p}{\rho} + \frac{V^2}{2} + gz = \text{constant}$

Velocity potential: $\mathbf{V} = \nabla\phi, \qquad \nabla^2\phi = 0$

Bernoulli equation: $\frac{p}{\rho} + \frac{V^2}{2} + gz = \text{constant}$

Continuity equation: $\frac{1}{r}\frac{\partial}{\partial r}\left(r\frac{\partial\phi}{\partial r}\right) + \frac{1}{r^2}\frac{\partial^2\phi}{\partial\theta^2} + \frac{\partial^2\phi}{\partial z^2} = 0$

Stream function: $\nabla^2\psi = 0$

Potential and stream function: $\frac{\partial\phi}{\partial x} = \frac{\partial\psi}{\partial y}, \qquad \frac{\partial\phi}{\partial y} = -\frac{\partial\psi}{\partial x}$

Plane potential flow: $v_r = \frac{1}{r}\frac{\partial\psi}{\partial\theta}, \quad v_\theta = -\frac{\partial\psi}{\partial r}$

Axisymmetric potential flow: $v_r = -\frac{1}{r}\frac{\partial\psi}{\partial z}, \quad v_z = \frac{1}{r}\frac{\partial\psi}{\partial r}, \qquad r\frac{\partial}{\partial r}\left(\frac{1}{r}\frac{\partial\psi}{\partial r}\right) + \frac{\partial^2\psi}{\partial z^2} = 0$

FUNDAMENTAL AND COMPOSITE POTENTIAL FLOWS

Uniform flow: $\phi = Vx, \qquad \phi = V(x\cos\theta + y\sin\theta)$

$\psi = Vy, \qquad \psi = V(y\cos\theta - x\sin\theta)$

Line source/sink flow: $\phi = \frac{q}{2\pi}\ln r, \qquad \phi = \frac{q}{2\pi}\ln\sqrt{(x-x_0)^2 + (y-y_0)^2}$

$\psi = \frac{q}{2\pi}\theta, \qquad \psi = \frac{q}{2\pi}\tan^{-1}\left(\frac{y-y_0}{x-x_0}\right)$

Line vortex flow: $\phi = \frac{\Gamma}{2\pi}\theta, \qquad \phi = \frac{\Gamma}{2\pi}\tan^{-1}\left(\frac{y-y_0}{x-x_0}\right)$

$\psi = -\frac{\Gamma}{2\pi}\ln r, \qquad \psi = -\frac{\Gamma}{2\pi}\ln\sqrt{(x-x_0)^2 + (y-y_0)^2}$

Spiral flow: $\phi = \frac{q}{2\pi}\ln r + \frac{\Gamma}{2\pi}\theta, \qquad \psi = \frac{q}{2\pi}\theta - \frac{\Gamma}{2\pi}\ln r$

Doublet: $\psi = -\frac{K\sin\theta}{r}, \qquad \phi = \frac{K\cos\theta}{r}$

$$x^2 + \left(y + \frac{K}{2\psi_0}\right)^2 = \left(\frac{K}{2\psi_0}\right)^2 \quad \text{(streamline)}$$

$$\left(x - \frac{K}{2\phi_0}\right)^2 + y^2 = \left(\frac{K}{2\phi_0}\right)^2 \quad \text{(equipotential)}$$

Flow around half-body: $\phi = Vr\cos\theta + \frac{q}{2\pi}\ln r, \qquad \psi = Vr\sin\theta + \frac{q}{2\pi}\theta$

$$b = \frac{q}{2\pi V}, \qquad r_{\mathrm{s}} = \frac{b(\pi - \theta_{\mathrm{s}})}{\sin\theta_{\mathrm{s}}}, \qquad p_0 + \frac{1}{2}\rho V^2 = p_{\mathrm{s}} + \frac{1}{2}\rho v_{\mathrm{s}}^2$$

$$v_{\mathrm{s}}^2 = V^2\left[1 + \frac{\sin 2\theta_{\mathrm{s}}}{\pi - \theta_{\mathrm{s}}} + \frac{\sin^2\theta_{\mathrm{s}}}{(\pi - \theta_{\mathrm{s}})^2}\right]$$

Rankine oval: $\psi = Vr\sin\theta - \frac{q}{2\pi}\tan^{-1}\left(\frac{2ar\sin\theta}{r^2 - a^2}\right)$

$$\phi = Vr\cos\theta + \frac{q}{2\pi}(\ln r_1 - \ln r_2), \qquad \frac{b}{a} = \sqrt{\frac{q}{\pi V a} + 1}$$

$$\frac{h}{a} = \frac{1}{2}\left[\left(\frac{h}{a}\right)^2 - 1\right]\tan\left[2\left(\frac{\pi V a}{q}\right)\frac{h}{a}\right]$$

$$\frac{V_{\mathrm{max}}}{V} = 1 + \left(\frac{q}{\pi V a}\right)\frac{1}{1 + h^2/a^2}$$

Flow around circular cylinder: $\psi = Vr\sin\theta - \frac{K\sin\theta}{r}, \qquad \psi = Vr\left(1 - \frac{R^2}{r^2}\right)\sin\theta$

$$\phi = Vr\cos\theta + \frac{K\cos\theta}{r}, \qquad \phi = Vr\left(1 + \frac{R^2}{r^2}\right)\cos\theta$$

$$p_{\mathrm{s}} = p_0 + \frac{1}{2}\rho V^2(1 - 4\sin^2\theta), \quad F_x = F_{\mathrm{D}} = -L\int_0^{2\pi} p_{\mathrm{s}}\cos\theta R\,\mathrm{d}\theta$$

$$F_y = F_{\mathrm{L}} = -L\int_0^{2\pi} p_{\mathrm{s}}\sin\theta R\,\mathrm{d}\theta$$

Rotating cylinder: $\psi = Vr\left(1 - \frac{R^2}{r^2}\right)\sin\theta - \frac{\Gamma}{2\pi}\ln r,$

$$\phi = Vr\left(1 + \frac{R^2}{r^2}\right)\cos\theta + \frac{\Gamma}{2\pi}\theta$$

$$p_{\mathrm{s}} = p_0 + \frac{1}{2}\rho V^2\left(1 - 4\sin^2\theta + \frac{2\Gamma\sin\theta}{\pi R V} - \frac{\Gamma^2}{4\pi^2 R^2 V^2}\right)$$

$$F_{\mathrm{D}} = 0, \qquad F_{\mathrm{L}} = -\rho V \Gamma$$

TURBULENT FLOW

Turbulent shear stress: $\tau_{\text{turb}} = -\rho\overline{u'v'}, \qquad \tau_{\text{turb}} = \eta \dfrac{\mathrm{d}\bar{u}}{\mathrm{d}y}$

Eddy viscosity: $\eta = -\rho\overline{u'v'}\left(\dfrac{\mathrm{d}\bar{u}}{\mathrm{d}y}\right)^{-1} = \rho\ell^2 \dfrac{\mathrm{d}\bar{u}}{\mathrm{d}y}$

Total shear stress: $\tau = \mu\dfrac{\mathrm{d}\bar{u}}{\mathrm{d}y} + \eta\dfrac{\mathrm{d}\bar{u}}{\mathrm{d}y} = \rho(\nu + \epsilon)\ frac\mathrm{d}\bar{u}\mathrm{d}y$

Turbulence intensity: $I = \dfrac{\sqrt{\overline{(u')^2}}}{\bar{u}}$

Turbulent velocity: $\mathbf{v}(t) = \bar{\mathbf{v}}_{\Delta t} + \mathbf{v}'$

CONSERVATION OF ENERGY

Fourier's law: $\mathbf{q} = -k\boldsymbol{\nabla}T$

Energy equation: $\rho\dfrac{\mathrm{d}u}{\mathrm{d}t} + p\,(\boldsymbol{\nabla}\cdot\mathbf{v}) = \boldsymbol{\nabla}\cdot(k\boldsymbol{\nabla}T) + \Phi$

$$\rho c_v \frac{\mathrm{d}T}{\mathrm{d}t} = k\nabla^2 T + \Phi \qquad \text{(for } \rho,\ c_v,\ \mu, \text{ and } k \text{ constant)}$$

PROBLEMS

Section 5.2: Kinematics

5.1. Velocity measurements in a steady-state two-dimensional incompressible flow field are taken at the four points shown in Figure 5.49. The x and y components of the velocity, designated as u and v, respectively, are as follows:

Point	u (m/s)	v (m/s)
1	3.30	0.52
2	4.30	0.65
3	5.50	0.33
4	3.20	–

(a) Estimate the x component of the fluid acceleration at Point 2. (b) Inadvertently, the velocity v was not measured at Point 4; estimate this missing value based on the available measurements and the continuity equation.

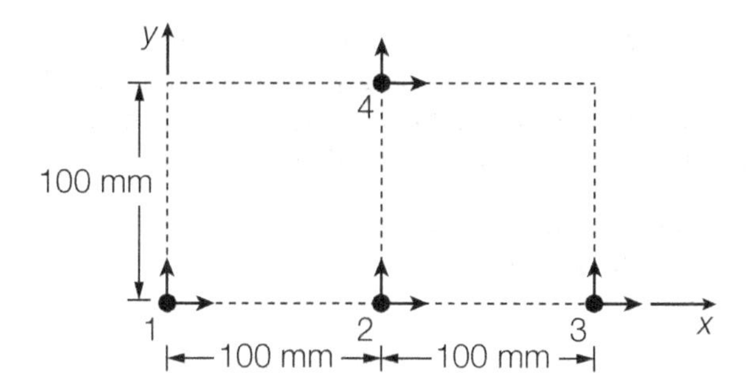

Figure 5.49: Velocity measurements in a two-dimensional flow field

5.2. The steady two-dimensional flow field of a compressible fluid is described by the velocity components $u = 5(x^2 + y^2)$ and $v = 5xy$. At the coordinate location $(x, y) = (2, 3)$, the density of the fluid is 1.5 kg/m^3. Determine the rate of change of the density of the fluid as it passes through this point.

5.3. Consider a flow with the following steady-state velocity field:

$$\mathbf{V} = 6xy\mathbf{i} + 3yz\mathbf{j} + 4xz^2\mathbf{k}$$

Determine the acceleration field. Determine the velocity and acceleration at (x, y, z) = (3,4,2).

5.4. The nozzle shown in Figure 5.50 is designed such that the velocity, $\mathbf{v}$, at any point near the centerline of the nozzle under steady-state conditions can be estimated by

$$\mathbf{v} = \frac{4}{0.8 - x}\mathbf{i} \text{ m/s}$$

where x is the distance in meters from the center of the nozzle entrance, measured along the centerline of the nozzle. The centerline velocities at the entrance and exit of the nozzle are 4 m/s and 20 m/s, respectively, and the length of the nozzle is 0.6 m. What is the acceleration of a fluid element at the center of the midsection of the nozzle, where $x = 0.2$ m?

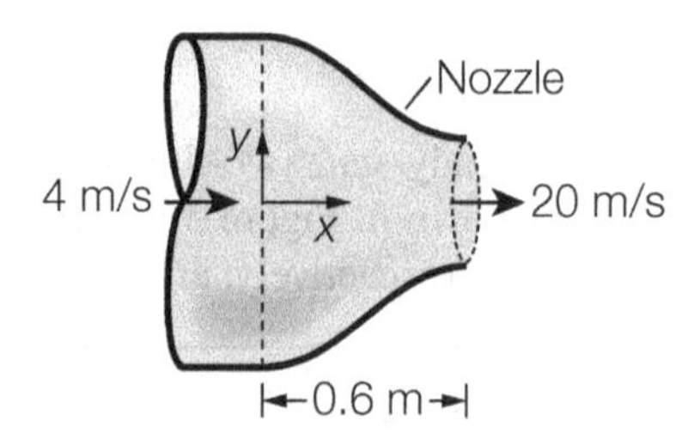

Figure 5.50: Flow through a nozzle

5.5. Prove Equation 5.8.

5.6. The velocity field in a two-dimensional flow is given by

$$\mathbf{V} = (2 + 8x + 4y)\mathbf{i} + (1 + 4x - 6y)\mathbf{j}$$

Determine the vorticity field and assess the rotationality of the flow.

5.7. The velocity components in a two-dimensional velocity field in the yz plane are $u = 2y^2$ m/s and $v = -2yz$ m/s, where y and z are in meters. Determine the rate of rotation of a fluid element about the point (1 m, 1 m). Indicate whether the rotation is in the clockwise or counterclockwise direction.

5.8. The velocity field, $\mathbf{V}$ [m/s], in the two-dimensional converging duct of width 3 m shown in Figure 5.51 is given by

$$\mathbf{V} = 8x\mathbf{i} + (3 - 4y)\mathbf{j}$$

where x [m] is the coordinate measured along the centerline of the duct and y [m] is the coordinate measured normal to the centerline. Assess the rotationality of the flow within the duct.

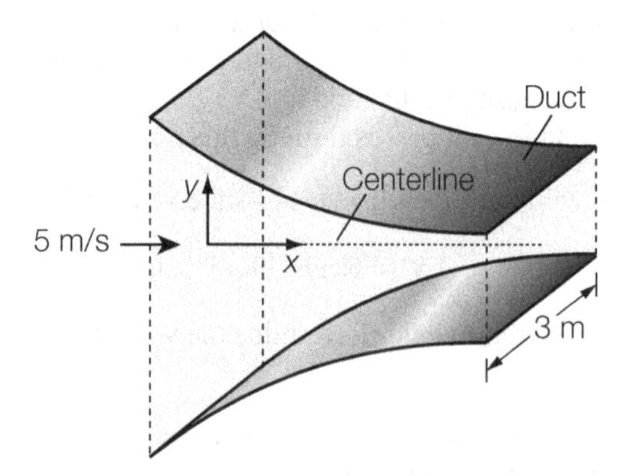

Figure 5.51: Flow through a duct

5.9. A two-dimensional velocity field is given by

$$\mathbf{V} = 3(x^2 + y^2)\mathbf{i} + 4(x^2 - y^2)\mathbf{j}$$

What is the rate of angular deformation at location (x, y) = (2,3). Identify any locations where the rate of angular deformation is equal to zero.

5.10. Consider the following steady three-dimensional velocity field:

$$\mathbf{V} = (5.0 + 3.0x - 2.0y)\mathbf{i} + (3.0x - 6.0y)\mathbf{j} + (2.0xy)\mathbf{k}$$

Determine the angular velocity in the xy plane at (x, y) = (1,2).

5.11. A viscous fluid is contained between two parallel plates spaced 30 mm apart as shown in Figure 5.52. The bottom plate remains stationary, the top plate moves at 2 m/s, and the velocity of the fluid varies linearly between the bottom and top plate. (a) Determine the rate of angular rotation and the vorticity of a fluid element contained between the plates and (b) determine the rate of angular deformation of the fluid element.

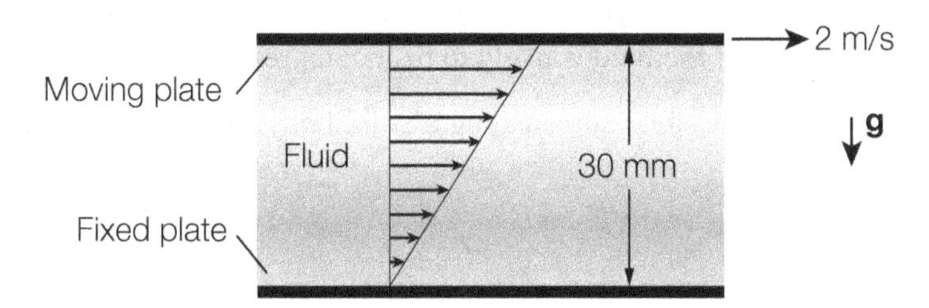

Figure 5.52: Flow between parallel plates

5.12. Measurements in a flow field indicate that the velocity components are $u = 6$ m/s and $v = 2$ m/s at a location where $x = 3$ m and $y = 4$ m. Express the given location in polar coordinates (r, θ) and determine the r and θ components of the velocity, which are commonly represented by v_r and v_θ.

5.13. Measurements in a flow field indicate that the velocity components are $v_r = 3$ m/s and $v_\theta = -2$ m/s at a location where $r = 2.5$ m and $\theta = 60°$. Express the given location in Cartesian coordinates (x, y) and determine the x and y components of the velocity.

5.14. The velocity profile, $u(r)$, in a circular conduit under laminar flow conditions is shown in Figure 5.53 and can be expressed analytically as

$$u(r) = U_0 \left[1 - 4\left(\frac{r}{D}\right)^2\right]$$

where U_0 is the centerline velocity, r is the radial distance from the centerline, and D is the diameter of the conduit. Determine functional expressions for the components of the angular deformation rate of a fluid element as a function of location in the conduit.

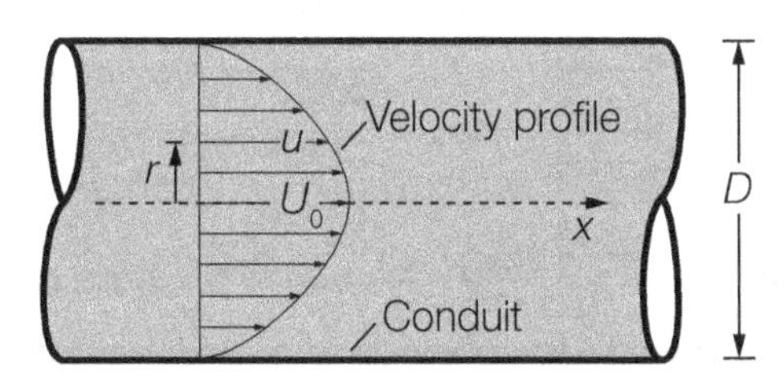

Figure 5.53: Flow in a circular conduit

5.15. Consider the velocity field, $\mathbf{V}$, given by

$$\mathbf{V} = xz\mathbf{i} - 2yz\mathbf{j} + 3xy\mathbf{k} \text{ m/s}$$

where x, y, and z, are the coordinate locations in meters. Find the dilatation rate of the fluid at (x, y, z) = (2 m,3 m,1 m) and assess whether the fluid is being compressed or expanded.

Section 5.3: Conservation of Mass

5.16. A duct is being used to transport methane at a temperature of 25°C. Estimate the maximum allowable speed in the duct for which the flow can be assumed incompressible.

5.17. Observations on the spatial variations in velocity within a fluid indicate that the velocity components can be estimated by $u = 0$, $v = z(z^2 - 4y^2)$, and $w = y(4z^2 - y^2)$. Determine whether the fluid is likely to be incompressible.

5.18. The theoretical flow field surrounding a solid body called a Rankine half-body is given by

$$v_r = V\cos\theta + \frac{q}{2\pi}\frac{1}{r}, \qquad v_\theta = -V\sin\theta$$

where V and q are parameters of the flow field, and r and θ are polar coordinates. Show that the fluid is incompressible.

5.19. When a fluid flow approaches a flat solid surface as shown in Figure 5.54, the velocity of the fluid in contact with the solid surface must necessarily be equal to zero. In addition, the viscosity of the fluid causes the velocity of the fluid away from the solid surface to be retarded. Retardation of the fluid is significant within a region called the boundary layer, which can be taken to have a thickness, δ, that depends on the distance, x, from the front of the solid surface. Observations indicate that the x-component of the velocity, u, can be expressed in normalized form by

$$\frac{u}{V} = 2\left(\frac{y}{\delta}\right) - \left(\frac{y}{\delta}\right)^2, \qquad \text{where: } \frac{\delta}{x} = \frac{4.91}{\text{Re}^{\frac{1}{2}}}, \ \text{Re} = \frac{Vx}{\nu}$$

where V is the free-stream velocity, y is the distance normal to the solid surface, and ν is the kinematic viscosity of the fluid. (a) Assuming that the fluid is incompressible, determine an expression for the normalized y component of the velocity, v/V, as a function of x, y, and δ. (b) Boundary layer analyses typically assume that v is negligible compared with u. Assess the justification of this assumption by finding the range of values of Re for which $v/u \leq 0.1$ at the outer limit of the boundary layer.

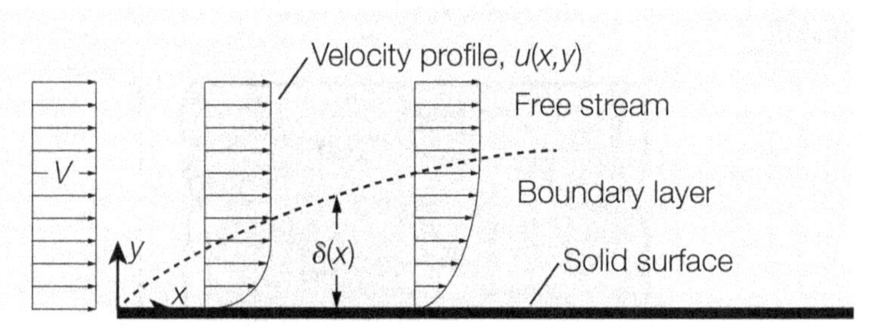

Figure 5.54: Flow in a boundary layer

5.20. In a two-dimensional incompressible flow field, the y component of the velocity, v, is given by $v = 5x^2$. The x component of the velocity, $u(x, y)$, is unknown, but it is known that $u(x, y)$ must satisfy the boundary condition that $u(0, y) = 0$. Determine $u(x, y)$.

5.21. The velocity components in a three-dimensional velocity field are given by $u = ax^2z$, $v = bxz^2$, and $w = cxz^2 + d$, where a, b, c, and d are constants. Determine the relationship between the constants that would be required for the flow to be incompressible.

5.22. The x and z components of the velocity in a three-dimensional flow field are given by: $u = ayz - bxy^2$ and $w = 2az + bxz + cx^2$. If the fluid is incompressible, determine the y component of the velocity, v, as a function of x, y, and z.

5.23. The velocity field of an incompressible fluid is given by

$$\mathbf{V} = 7xy^2\mathbf{i} + 4xz\mathbf{j} + xzf(x, y)\mathbf{k}$$

where $f(x, y)$ is an unknown function of x and y. Determine the required functional form of $f(x, y)$.

5.24. Flow of an incompressible fluid is in the xy plane, and the y component of the velocity, v, is given by $v = 2y$. (a) Determine the required functional form of the velocity field. (b) Give a particular velocity field that satisfies the required functional form.

5.25. Two-dimensional unsteady flow of an incompressible fluid occurs in the xy plane such that the x-component of the velocity, u, is given by $u = 3xy^2 + 4y$. Determine the required functional form of the y component of the velocity.

5.26. Using cylindrical polar coordinates, a three-dimensional flow field has velocity components that can be approximated by

$$v_r = \frac{r}{2}\sin\theta, \qquad v_\theta = 4r\cos\theta, \qquad v_z = 3z\sin\theta$$

(a) Assess whether this flow field can be classified as incompressible. (b) If the flow field can be classified as incompressible, does this mean that the fluid is incompressible? Explain.

5.27. Two-dimensional unsteady flow of an incompressible fluid occurs in the xy plane. When polar coordinates are used, the θ-component of the velocity is given by

$$v_\theta = -\frac{C\sin\theta}{r^2}$$

where C is a constant. Determine the required functional form of the r-component of the velocity.

5.28. Two-dimensional unsteady flow of an incompressible fluid occurs in the xy plane. When polar coordinates are used, the r component of the velocity is given by

$$v_r = r^2\sin\theta + r\cos\theta$$

Determine the required functional form of the θ-component of the velocity.

5.29. A two-dimensional steady flow in the $r\theta$ plane has a θ-component of the velocity given by

$$v_\theta = 10\left(1 + \frac{1}{r^2}\right)\cos\theta - \frac{15}{r}$$

A boundary condition that must be satisfied by the flow field is $v_r(1, \theta) = 0$. Determine the r-component of the velocity as a function of r and θ.

5.30. Air flows at a steady-state through a straight pipe as shown in Figure 5.55. Measurements at section A indicate that the temperature is 15°C, the pressure is 400 kPa absolute, and the velocity is 100 m/s. At section B, which is 500 mm downstream of section A, the velocity is measured as 140 m/s. Estimate the density of the air at sections A and B. The flow can be assumed to be one-dimensional in the x-direction.

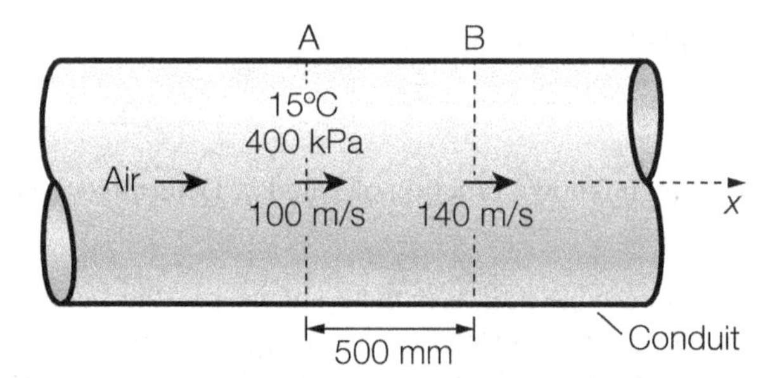

Figure 5.55: Airflow through a pipe

5.31. Air flows through a 1-m-long two-dimensional converging duct as shown in Figure 5.56. Under steady-state conditions, the inflow and outflow velocities are 50 m/s and 100 m/s, respectively, and the inflow and outflow densities are 1.3 kg/m^3 and 0.8 kg/m^3, respectively. Locations within the duct are identified relative to Cartesian coordinate axes with the origin at the point where the centerline of the duct intersects the inflow section. The x component of velocity, u, varies nonlinearly between the entrance and the exit according to the relation $u = a + bx^2$, and the density, ρ, varies linearly between the entrance and exit according to the relation $\rho = c + dx$, where a, b, c, and d are constants. It is known that the y component of the velocity is equal to zero along the centerline of the duct but is nonzero elsewhere. Determine v as a function of x and y.

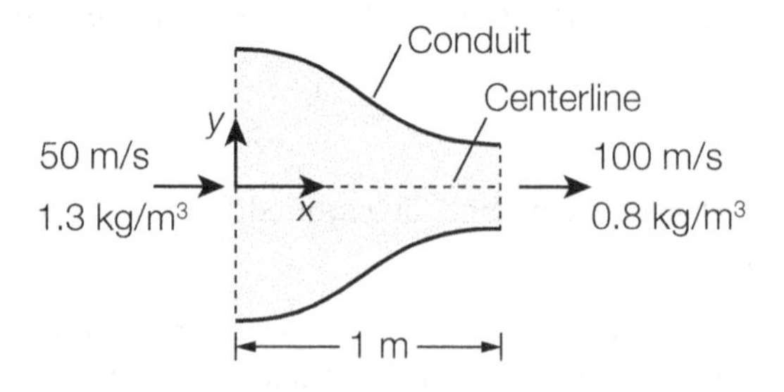

Figure 5.56: Two-dimensional flow through a converging duct

5.32. Standard air is contained within an insulated chamber with a fixed side and a moving side as shown in Figure 5.57. The moving side oscillates such that within a time interval of 0.1 seconds, the velocity, u, in the direction of oscillation is given by

$$u = \frac{10x}{0.12 + 10t} \text{ m/s}$$

where x is the distance from the fixed side in meters and t is the time from the beginning of the time interval in seconds. Within this time interval, the density, ρ, is spatially uniform within the chamber but varies with time. The density of the air at the beginning of the time interval is 3 kg/m^3. What is the air density within the chamber at the end of the time interval?

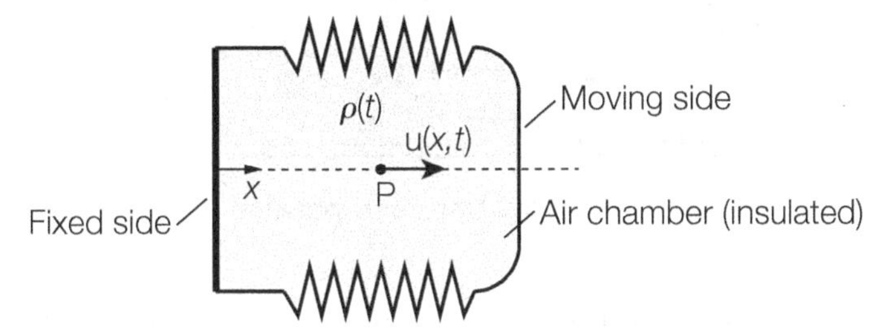

Figure 5.57: Expanding insulated air chamber

5.33. The two-dimensional velocity field of an incompressible fluid is given by $\mathbf{V} = 3x\mathbf{i} - 3y\mathbf{j}$. Determine the location(s) of the stagnation points and the expression for the stream function.

5.34. The two-dimensional velocity field of an incompressible fluid is given by $u = 3y(4x + 1)$, $v = -6y^2 + x^2$. Determine the analytic expression for the stream function and describe how you would use this stream function to plot the streamlines of the flow.

5.35. The stream function in a two-dimensional flow field is given by $\psi = x^2 - y^2$ m^2/s, where x and y are the Cartesian coordinates in meters. Determine the rate of angular rotation of a fluid element located at (5 m, 5 m). How does this rate of rotation compare with the rate of rotation at other points in the flow field?

5.36. A fluid flows over a flat plate such that the velocity increases linearly with distance from the surface of the plate as shown in Figure 5.58. (a) Determine the stream function. (b) Use the stream function to calculate the volume flow rate of the fluid between the plate surface and a distance 2 m from the surface. (c) Determine distance beyond 2 m for the volume flow rate to be the same as that between the plate surface and 2 m.

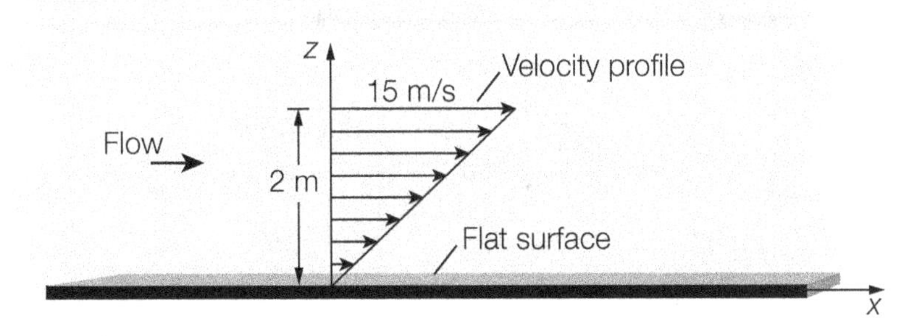

Figure 5.58: Flow adjacent to flat plate

5.37. The two-dimensional fluid flow between two flat plates is illustrated in Figure 5.59. The velocity distribution, $u(z)$, between the plates can be expressed as $u(z) = a(h^2 - z^2)$, where a is a constant, z is the vertical coordinate measured from the midplane between the plates, and $2h$ is the distance from the bottom plate to the top plate. (a) Determine the stream function that can be used to describe the flow. (b) Determine the flow between the plates using the derived stream function. (c) Verify the result in part (b) by integrating the velocity distribution over the interval between the plates.

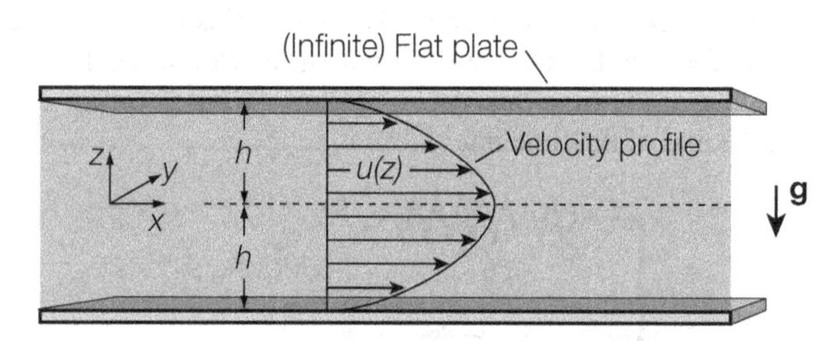

Figure 5.59: Flow between parallel plates

5.38. The stream function of a particular two-dimensional flow field is given by $\psi = 0.2xy$ m^2/s, where x and y are the Cartesian coordinates in meters. Determine the volume flow rate between the points (1 m, 1 m) and (2 m, 3 m).

5.39. The two-dimensional flow field of an incompressible fluid is described in polar coordinates as $v_r = 2/r$, $v_\theta = 4/r^2$. Determine the analytic expression for the stream function.

5.40. A particular flow field experienced by an incompressible fluid is described in polar coordinates by the stream function $\psi(r, \theta) = -3r \sin\theta + 2\theta$ m^2/s, where r is in meters and θ is in radians. (a) Determine the r and θ components of the velocity vector as a function of r and θ. (b) At the point $r = 2$ m and $\theta = \pi/6\ rad = 30°$, determine the r and θ components of the velocity and the x and y components of the velocity, where x and y are the Cartesian coordinates.

5.41. A free vortex is an irrotational flow in which there are no external forces acting on the fluid. Streamlines of a particular free vortex are shown in Figure 5.60, where the tangential component of the velocity, v_θ, is given by $v_\theta = 3.5/r$, where r is the radial distance from the center of the vortex. The radial component of the velocity is equal to zero. (a) Determine the stream function. (b) Explain how you can deduce from the result in part (a) that the streamlines are circles. (c) Use the stream function to estimate the volume flow rate between the streamlines at $r = 1.2$ m and $r = 1.4$ m.

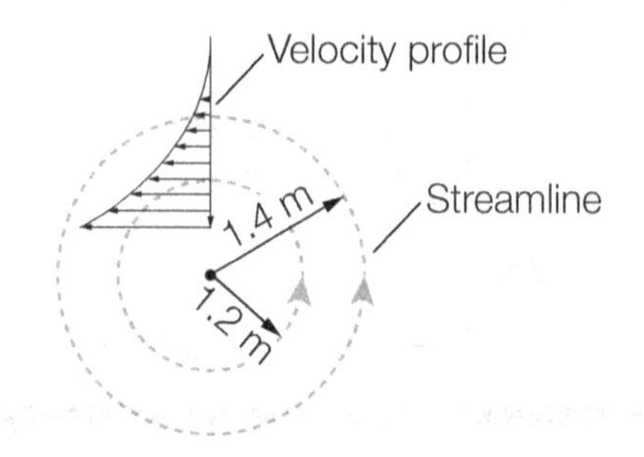

Figure 5.60: Flow in a free vortex

Section 5.4: Conservation of Momentum

5.42. A velocity field can be represented by the velocity components $u = 8z$, $v = 0$, and $w = 2x$, where gravity has a magnitude g and acts in the negative z direction. Within the flow field, the viscosity is constant, and the pressure and density are equal to p_0 and ρ_0, respectively, at the location (x_0, y_0, z_0). Use the Navier–Stokes equation to determine the pressure distribution in terms of the given parameters.

5.43. Observations of a fluid flow indicate that a velocity field can be represented by the velocity components $u = 8z$, $v = 6x$, and $w = 2x$, where gravity has a magnitude g and acts in the negative z-direction. Within the flow field, the viscosity is presumed constant. Assess whether this given velocity field satisfies the Navier–Stokes equation. If not, speculate on what may be the reason. If so, find an expression for the pressure distribution.

5.44. A liquid with a density ρ and viscosity μ flows between parallel plates as shown in Figure 5.61. The parallel plates are inclined at angle θ with the horizontal, and the flow is driven by gravity. Flow conditions are steady and do not change in the dimension perpendicular to the page. The x-coordinate is measured along the incline, the z-coordinate is measured perpendicular to the incline, and the magnitude of the acceleration due to gravity is g. Write the simplified form of the Navier–Stokes equation in Cartesian coordinates that applies to this scenario.

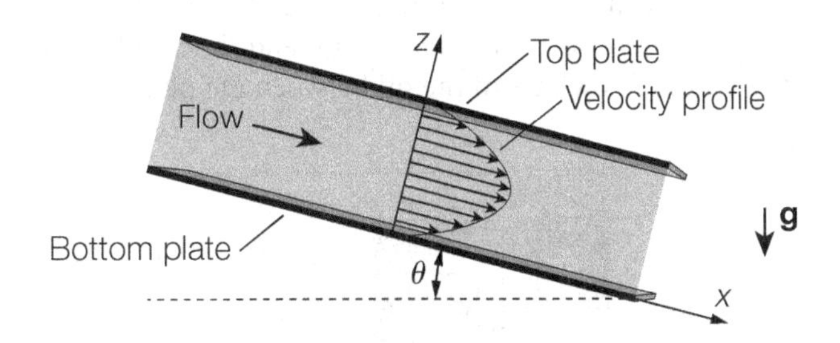

Figure 5.61: Flow between inclined parallel plates

5.45. A viscous liquid flows down an inclined plane as shown in Figure 5.62. The flow is two-dimensional in the xy plane, and the x-axis is oriented in the flow direction. Apply the Navier–Stokes equation to this problem and write the components of the Navier–Stokes equation in their most simplified forms.

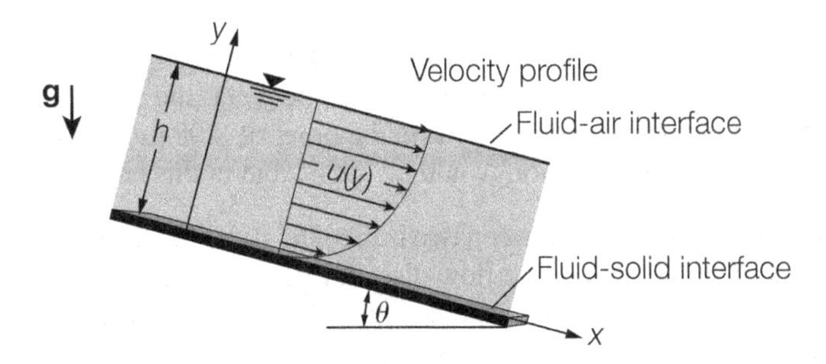

Figure 5.62: Flow down an inclined plane

Section 5.5: Solutions of the Navier–Stokes Equation

5.46. Consider the case of a viscous fluid flowing down an inclined plane as shown in Figure 5.62. The only nonzero component of the velocity is $u(y)$, and there is no pressure gradient in the x-direction. (a) Apply the Navier–Stokes equation in the x-direction to determine the longitudinal velocity distribution $u(y)$ and the volume flow rate down the incline. In your derivation, use the no-slip condition and the condition that the shear stress is negligible at the air-liquid interface. (b) If the fluid is SAE 30 oil at 20°C, the depth of flow is 10 mm, and the angle of incline is 20°, determine the volume flow rate.

5.47. SAE 50 oil at 20°C flows between two horizontal parallel plates spaced 7 mm apart. The plates are 3 m long and 2 m wide, and the oil flow is driven by a pressure differential between the inflow and outflow sections. The inflow pressure is 200 kPa gauge, and the outflow pressure is atmospheric. (a) Estimate the volume flow rate of the oil between the plates. (b) Estimate the average flow velocity between the plates. (c) Validate your results using a Reynolds number analysis.

5.48. A fluid flows freely between two vertical parallel plates that are of width W and spaced a distance h apart as shown in Figure 5.63. The force driving the flow is gravity, and it can be reasonably approximated that the pressure remains constant throughout the fluid. (a) Determine an expression for the volume flow rate of the fluid as a function of the fluid properties, W, and h. (b) If the fluid is SAE 30 oil at 20°C, the spacing between plates is 5 mm, and the width of the plates is 2.5 m, determine the volume flow rate of the oil between the plates.

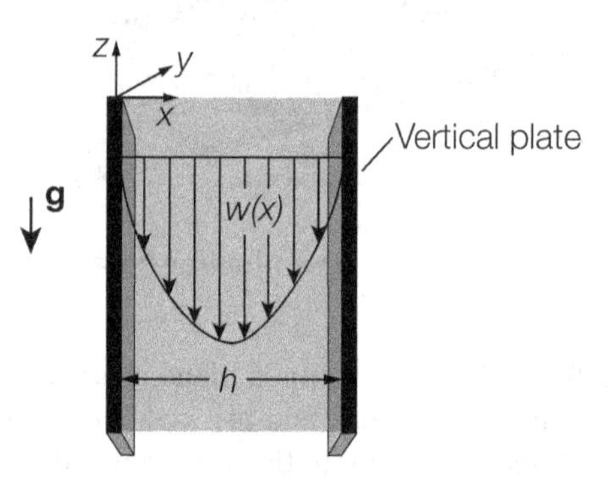

Figure 5.63: Flow between vertical parallel plates

5.49. Consider the steady flow of SAE 10 oil at 20°C between two horizontal flat plates spaced 20 mm apart. The average pressure between the plates is on the order of 100 kPa, and the average velocity is on the order of 0.08 m/s. What simplified form of the Navier–Stokes equation would be applicable in this situation?

5.50. SAE 10 oil at 20°C flows between two horizontal parallel plates spaced 20 mm apart. The length of the plates in the flow direction is 1 m, and the width of the plates is 90 cm. Find the volume flow rate between the plates and the force exerted by the fluid on each of the plates if a pressure gradient of 100 Pa/m is applied to drive the flow.

5.51. SAE 30 oil at 20°C flows between two horizontal 0.50-m-wide parallel plates separated by 30 mm. The length of the top and bottom plates in the direction of flow is 2 m, the bottom plate is stationary, the top plate moves at 3 cm/s, an adverse pressure gradient of 700 Pa/m is applied between the plates, and the flow is laminar between the plates. Determine the flow rate between the plates and the force that must be applied to move the top plate.

5.52. Consider the case of a vertically oriented plate moving with a (vertical) velocity W through a fluid reservoir. Show that the distance, x_0, from the plate surface to the location where the velocity in the fluid is equal to zero is given by

$$\frac{x_0}{h} = 1 - \sqrt{1 - \frac{2\mu W}{\gamma h^2}}$$

where h is the thickness of the fluid layer and γ and μ are the specific weight and dynamic viscosity of the fluid, respectively.

5.53. SAE 50 oil at 20°C is introduced uniformly at a volume flow rate of q per unit width at the top of a plane wall as shown in Figure 5.64. The oil ultimately attains a steady velocity profile and a uniform thickness, h. If the thickness of the oil layer is 8 mm, at what rate is oil being added at the top of the wall?

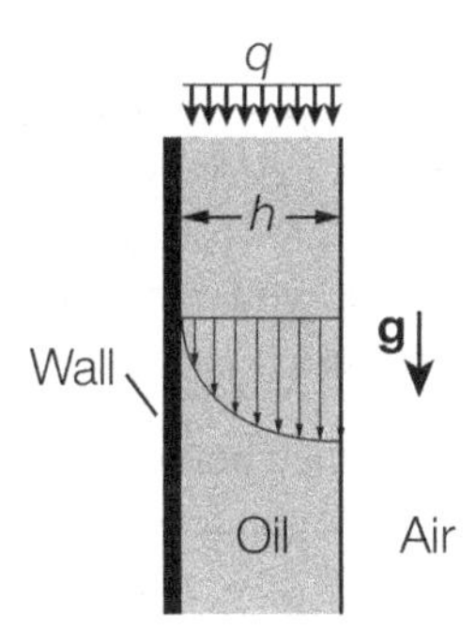

Figure 5.64: Oil flow down a wall

5.54. Show that the shear stress on the wall of a tube of diameter D in which laminar flow is occurring is given by

$$\tau_{\text{wall}} = -\frac{D}{4}\left(\frac{\partial p}{\partial x}\right)$$

where $\partial p/\partial x$ is the pressure gradient in the flow direction.

5.55. SAE 30 oil at 20°C flows in a 15-m-long, 225-mm-diameter pipe. The pipe is horizontal, and under a particular flow condition, the volume flow rate is 20 L/min. (a) Verify that Hagen-Poiseuille flow can be assumed in the pipe. (b) What is the difference in pressure between the entrance and exit of the pipe?

5.56. Gasoline at 20°C is to be delivered in a 25-m-long tube where the difference in pressure between the entrance and exit of the tube is 5 kPa. If laminar flow conditions are to be maintained in the tube, (a) what is the maximum tube diameter than can be used and (b) what is the volume flow rate through the tube when using this maximum diameter?

5.57. Ethylene glycol at 20°C is to be delivered between two pressurized reservoirs that have a pressure difference of 5 kPa. Delivery is to be through multiple tubes, with each tube having a length of 1.2 m and a diameter of 2 mm. If the desired volume flow rate between the reservoirs is 1.5 L/min, how many tubes are required?

5.58. A fluid flows under the influence of gravity in a vertical tube of diameter D as shown in Figure 5.65. It can be reasonably approximated that the pressure remains constant throughout the fluid. (a) Determine an expression for the volume flow rate as a function of the fluid properties and D. (b) If the fluid is SAE 30 oil at 20°C and the diameter of the tube is 50 mm, determine the volume flow rate in the tube.

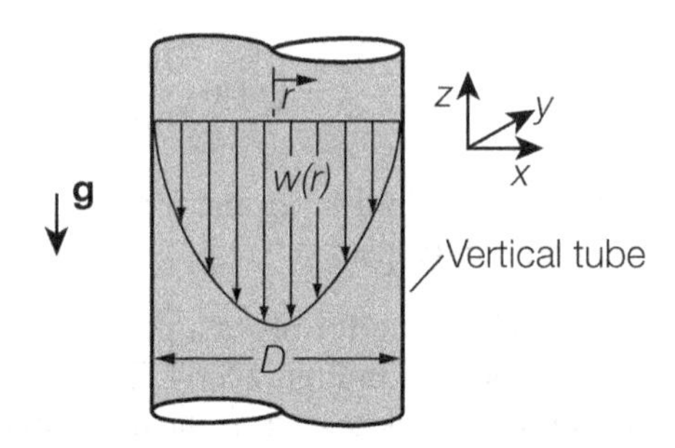

Figure 5.65: Flow in a vertical tube

5.59. A fluid flows through a glass annulus with an inner diameter of 5 mm and an outer diameter of 10 mm. The fluid has a density of 930 kg/m^3 and a dynamic viscosity of 0.200 Pa·s. If a pressure gradient of -100 Pa/m is applied, what flow rate is expected through the annulus? How would you expect this flow rate to change if the annulus was made of steel rather than glass?

5.60. An incompressible fluid is contained between two cylinders as shown in Figure 5.66. The radii of the inner and outer cylinders are R_i and R_o, respectively, and the viscosity of the fluid is μ. The fluid flow is caused by a pressure gradient in the flow direction and by movement of the inner cylinder at a velocity V in the direction of flow. (a) Determine an expression for the steady-state velocity distribution in the annular region as a function of r, R_i, R_o, μ, V, and the pressure gradient. (b) If the fluid is SAE 10 oil at 20°C, the inner and outer radii are 10 mm and 20 mm, respectively, the pressure gradient is -80 Pa/m, and the inner cylinder moves at 50 mm/s, determine the fluid velocity midway between the inner and outer cylinders.

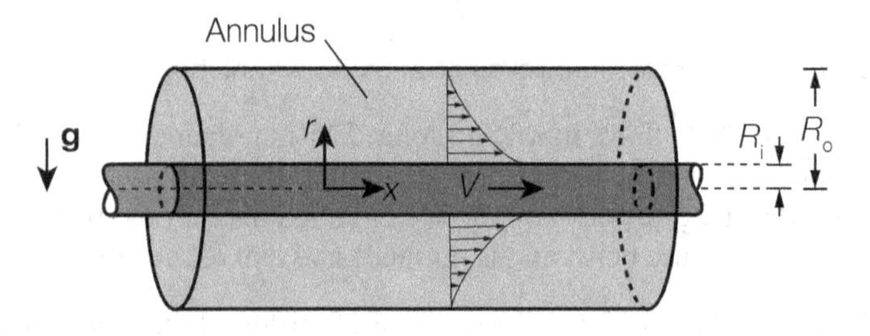

Figure 5.66: Flow between cylinders

5.61. A liquid with a density ρ and viscosity μ is contained in the annular space between two cylinders as shown in Figure 5.67. The inner cylinder rotates at a constant angular speed of ω, the outer cylinder is stationary, flow conditions in the fluid are at steady state and uniform in the z-dimension (perpendicular to the page), and gravity acts in the z-direction. Write the simplified form of the Navier–Stokes equation in cylindrical coordinates.

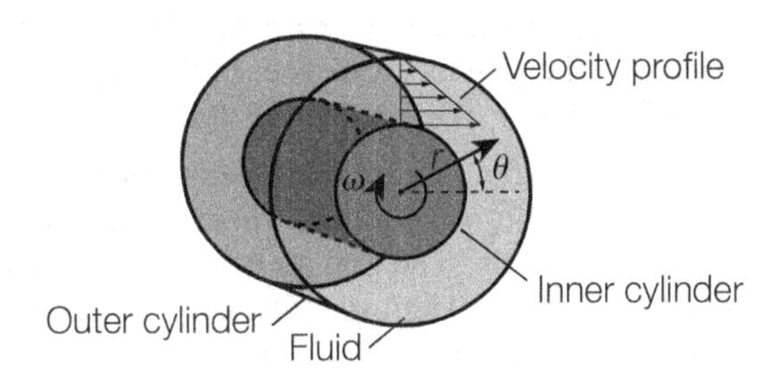

Figure 5.67: Flow in the annular space between cylinders

5.62. A viscous liquid is contained between two cylinders as shown in Figure 5.68. The inner cylinder has a radius $R_{\rm i}$, the outer cylinder has a radius $R_{\rm o}$, and the outer cylinder rotates at an angular speed ω. Use the θ-component of the Navier–Stokes equation along with the appropriate inner and outer boundary conditions to determine the velocity distribution in the fluid contained between the two cylinders.

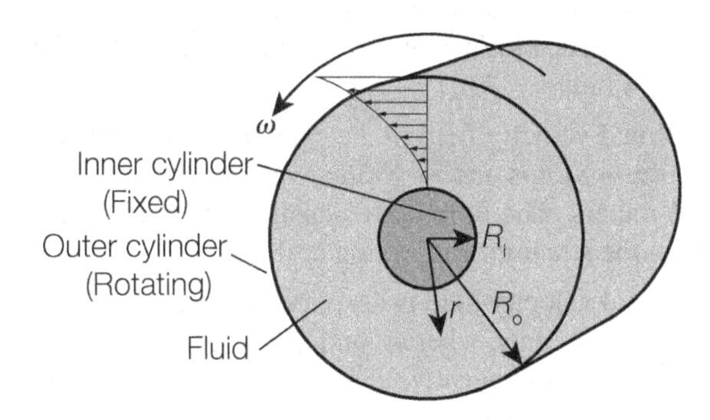

Figure 5.68: Fluid between cylinders

5.63. A lubricant is located in the annular region between a 1.0-m-long rotating shaft of diameter 40 mm and an outer cylinder of diameter 50 mm. The density and viscosity of the lubricant are 900 kg/m^3 and 0.35 Pa·s, respectively. What is the power required to rotate the shaft at 50 rpm? Is the flow of the lubricant stable? Explain.

Section 5.6: Inviscid Flow

5.64. A two-dimensional velocity field in the xy plane is described by the following velocity components:

$$u = \frac{10x}{x^2 + y^2}, \qquad v = \frac{10y}{x^2 + y^2}$$

Does the velocity field represent a possible incompressible flow? If so, find the pressure gradient ∇p assuming a frictionless flow with negligible body forces.

5.65. A two-dimensional velocity field in the xz plane is described by the velocity components $u = 2(x^2 - z^2) - 6x$ m/s and $w = 4xz - 6z$ m/s, where x and z are the Cartesian coordinates in meters. The gravity force acts in the negative z-direction, and the fluid has a density of 998 kg/m^3. Determine the pressure gradients in the x- and z-directions at the point $x = 2$ m, $z = 1$ m.

5.66. A two-dimensional velocity field in the xy plane is described by the velocity components $u = -3x + 9y^2$ m/s and $v = 3y$ m/s, where x and y are the Cartesian coordinates in meters. The gravity force acts in the negative z-direction, and the fluid has a density of 1.2 kg/m^3. Determine ∇p at the point $x = 2$ m, $z = 1$ m.

5.67. A two-dimensional velocity field in the $r\theta$ plane is described by the velocity components $v_r = -6/r$ m/s and $v_\theta = 3/r$ m/s, where r and θ are polar coordinates in meters and radians, respectively. The gravity force acts in the negative z-direction, and the fluid has a density of 1.20 kg/m^3. Calculate the pressure gradients in the r-, θ-, and z-directions at $r = 2$ m and $\theta = \pi/4$ rad.

5.68. A fundamental two-dimensional flow field that is used as a building block for constructing more complex ideal-fluid flows is that of a doublet, which has a flow field described by the velocity components

$$v_r = -\frac{K\cos\theta}{r^2}, \qquad v_\theta = -\frac{K\sin\theta}{r^2}$$

where K is a constant. The gravity force acts in the negative z-direction, the acceleration due to gravity is g, and the density of the fluid is ρ. (a) Determine a functional expression for the pressure gradient, ∇p, in terms of r, θ, z, ρ, and g. (b) If $K = 3$ m^3/s, $\rho = 1000$ kg/m^3, and $g = 9.81$ m/s^2, what is the pressure gradient at $r = 2$ m, $\theta = \pi/4$ rad, $z = 5$ m?

5.69. A two-dimensional velocity field in the xz plane is described by the velocity components $u = 5(x + z)$ m/s and $w = 5(x - z)$ m/s, where x and z are the Cartesian coordinates in meters. The fluid has a density of 998.2 kg/m^3. Determine the pressure gradient in the streamline direction at the point $x = 2$ m, $z = 1$ m.

5.70. A two-dimensional velocity field is described by the velocity components $u = 4x - 10$ m/s and $v = 6 - 4y$ m/s, where x and y are the Cartesian coordinates in meters. The gravity force is in the negative z-direction, and the density of the fluid is equal to 1.5 kg/m^3. (a) Show that the flow field is incompressible and irrotational. (c) Estimate the difference in pressure between the points (1 m, 1 m) and (2 m, 2 m).

5.71. A flow field in which the radial component of the velocity is everywhere equal to zero and in which the tangential component of the velocity depends only on the radial distance, r, from a center point is called a vortex. Consider the vortex illustrated in Figure 5.69, where Point P is the center of the vortex, Point 1 is on one streamline, and Point 2 is on another streamline. For which of the following velocity fields can the Bernoulli equation be applied between Points 1 and 2: $v_\theta = ar$ or $v_\theta = a/r$, where a is any constant?

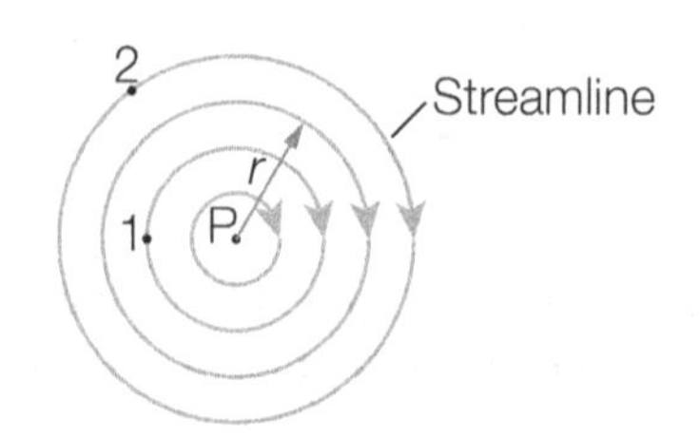

Figure 5.69: Streamlines in a vortex

5.72. The hydrofoil shown in Figure 5.70 was tested, and the minimum pressure on the surface of the hydrofoil was found to be 70 kPa when the hydrofoil was submerged 1.83 m and towed at a speed of 8 m/s. At the same depth, at what speed will cavitation first occur? Assume irrotational flow for both cases and a water temperature of 10°C.

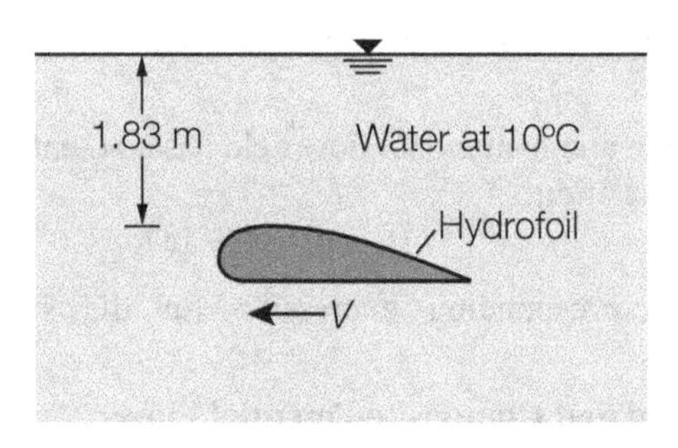

Figure 5.70: Flow around a hydrofoil

5.73. A fluid that is approximately inviscid and incompressible flows through a 3-m-long horizontal diffuser with a cross-sectional area, A, that increases according to the relation $A(x) = 0.1(1 + 0.5x^2)$ m^2, where x is the distance from the entrance of the diffuser in meters. The fluid has a density of 1100 kg/m^3, and under a particular steady-state condition, the pressure and velocity of the fluid at the entrance of the diffuser are 240 kPa and 2 m/s, respectively. (a) Determine an expression for the pressure in the diffuser as a function of x. (b) Determine an expression for the pressure gradient in the diffuser as a function of x.

5.74. The stream function of a flow field is given as $\psi = x^2 - y^2$. (a) Is the flow irrotational throughout the flow field? (b) If the flow is irrotational, determine the potential function of the flow field.

5.75. The velocity components in a two-dimensional flow field are $u = x+2$ and $v = 3-y$. (a) Is the flow irrotational throughout the flow field? (b) If the flow is irrotational, determine the potential function of the flow field.

5.76. Consider a two-dimensional flow described by the following stream function:

$$\psi = 2r\sin\theta - 4\theta \text{ m}^2/\text{s}$$

where r is in meters and θ is in radians. The density of the fluid is 1000 kg/m^3. (a) If possible, find the potential function for this flow. (b) Find the difference in pressure between two points with (x, y) locations given by (1 m,1 m) and (3 m,2 m).

5.77. The stream function of two irrotational flows are given by

$$\psi_1 = 15x^2y - 5y^3, \qquad \psi_2 = -2(x - y)$$

(a) What is the stream function of the flow field generated by combining these two flows? (b) Show that the combined flow field is irrotational.

5.78. The stream function, ψ, for a particular flow field is given by

$$\psi = 4x + 3y$$

(a) Describe the velocity field. (b) If possible, find the potential function for the velocity field. (c) Determine the velocity at $(x, y) = (1,2)$.

5.79. An incompressible and irrotational flow field has a potential function, ϕ, that can be described by the relation

$$\phi = y^6 - 15x^2y^4 + 15x^4y^2 - x^6$$

Determine an expression for the stream function that can be used in plotting the streamlines.

5.80. An incompressible and irrotational flow field has a potential function, ϕ, that can be described by the relation

$$\phi = r\cos\theta + \ln r$$

Determine the corresponding expression for the stream function in polar coordinates.

Section 5.7: Fundamental and Composite Potential Flows

5.81. Consider the line sink embedded in a solid boundary as shown in Figure 5.71. The sink withdraws fluid at a rate of 3 L/s per m length. The fluid is inviscid, and the flow is at steady state. Determine (a) an analytic expression for the velocity field and (b) an analytic expressions for the stream and potential functions.

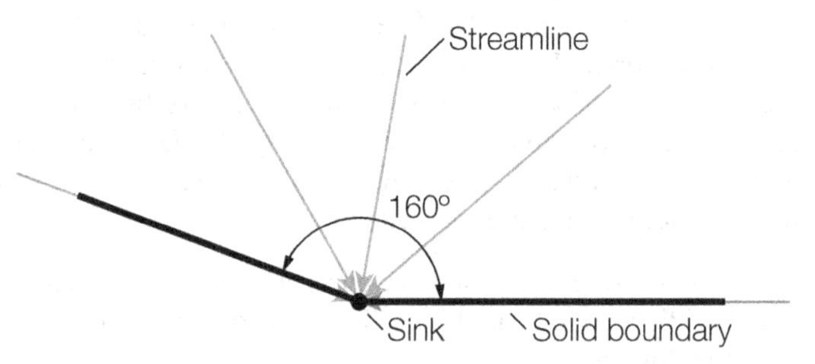

Figure 5.71: Sink embedded in a solid boundary

5.82. A two-dimensional flow field is generated in the xy plane by injecting a liquid uniformly at a rate of 2 m^3/s over the 3-m depth as shown in Figure 5.72. In the absence of the injection, the liquid has a uniform velocity of 0.015 m/s. If the origin of the coordinate axes is located on the injection line, determine the (x, y) coordinates where the fluid velocity is equal to zero (i.e., the location of the stagnation point).

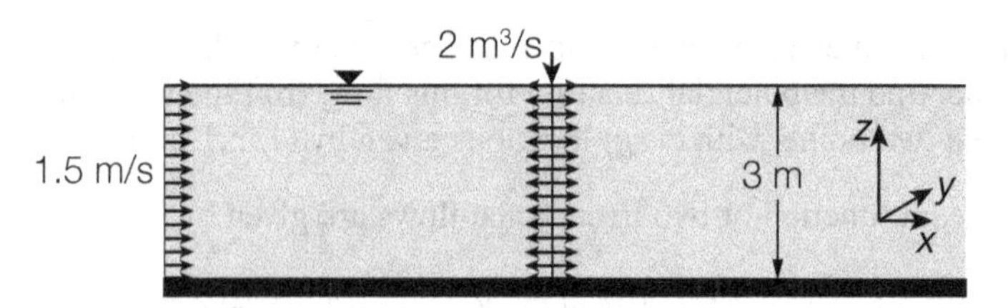

Figure 5.72: Injection into a two-dimensional flow field

5.83. The streamlines resulting from the superposition of a sink and uniform flow are illustrated in Figure 5.73. This composite flow is commonly used by groundwater hydrologists to represent the flow field surrounding a pumped well (sink) in a regional flow (uniform flow). Locations in the flow field are referenced to coordinate axes centered at the sink location, the magnitude of the uniform flow is V, and the strength of the sink is q. (a) Determine the coordinate location of the stagnation point, P, in terms of V and q. (b) A portion of the upstream flow is intercepted by the sink, and a portion of the upstream flow bypasses the sink. Determine the width, W, of the upstream flow that is intercepted by the sink. (c) If a pumped well extracts 70 L/s over a vertical interval of 5 m in an aquifer that has a regional velocity of 1 m/day, determine the distance of the stagnation point from the well and the width of the regional flow intercepted by the well.

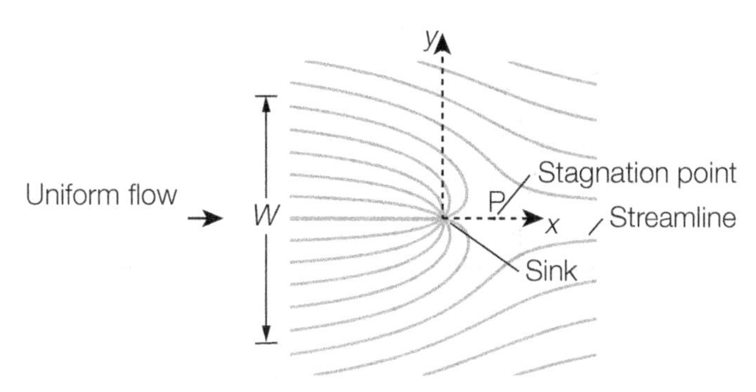

Figure 5.73: Sink in uniform flow

5.84. A two-dimensional velocity field has the velocity components $u = -2x$ and $v = 2y$. (a) Verify that this flow field can be described by a potential function. (b) Determine the circulation, Γ, around a closed path of straight lines connecting the following points: (1,1), (2,1), (2,4), (1,4), and (1,1). Explain the significance of your result.

5.85. A flow field closely approximates that of a free line vortex with a vortex strength Γ. The pressure far from the vortex center is p_∞, and the density of the fluid is ρ. (a) Determine the pressure at any distance r from the vortex center in terms of r, p_∞, ρ, and Γ. (b) If the fluid is air with a density of 1.225 kg/m^3, the pressure far from the influence of the vortex is 101 kPa, and the strength of the vortex is 3000 m^2/s, determine the radial distance from the vortex center to where the pressure is 10% less than the undisturbed pressure of 101 kPa.

5.86. The velocity field in a hurricane can be approximated by a forced vortex between the center and a radial distance of R from the center and as a free vortex beyond the radial distance R. In such an approximation, the θ-component of the velocity can be approximated by

$$v_\theta = \begin{cases} r\omega, & 0 \leq r \leq R \\ \dfrac{\Gamma}{2\pi r}, & r \geq R \end{cases}$$

where ω is the rotational speed of the forced vortex, Γ is the circulation of the free vortex, and R is the match point of the two velocity distributions. The radial component of the velocity, v_r, is equal to zero. (a) Explain why R is where the wind speed is a maximum. (b) If the undisturbed pressure outside the hurricane is p_0 and the maximum wind speed is V_{max}, determine an expression for the pressure at the match point in terms of the given variables and the relevant properties of the air. (c) For an intense Category 4 hurricane, the maximum velocity is 251 km/h, the match-point radius is 15 km, and conditions outside the hurricane are standard sea-level conditions. Determine the angular speed, ω, and the circulation, Γ, of the hurricane.

5.87. A liquid forms a free vortex around a drain as shown in Figure 5.74. If the depth of liquid at a distance, r_0, from the drain is h_0, determine an expression for the depth, h, of the liquid as a function of the distance, r, from the drain. Your final result should be in terms of the strength Γ of the vortex. Assume steady flow.

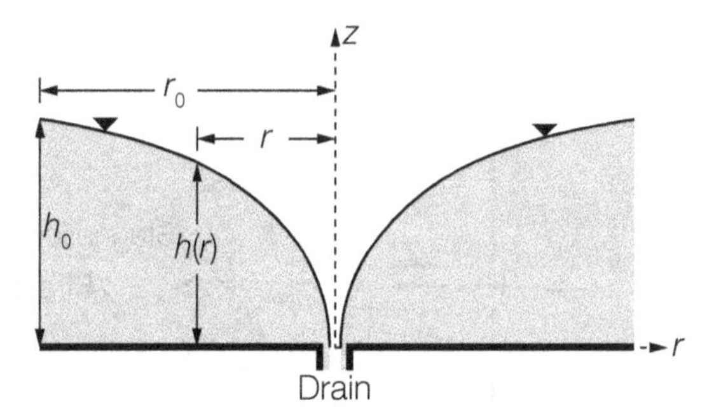

Figure 5.74: Vortex surrounding a drain

5.88. A two-dimensional flow field can be approximated by the superposition of two line sources and a line vortex in the xy plane. The first line source is located at the point (1 m, 1 m) and has a strength of 2 m^2/s, the second line source is located at the point (2 m, 2 m) and has a strength of 3 m^2/s, and the line vortex is located at the point (3 m, 3 m) and has a strength of 2.5 m^2/s. Determine the x and y components of the velocity at the point (4 m, 4 m).

5.89. Consider a flow field that can be approximated as a spiral free vortex, with sink strength q and vortex strength Γ. If locations within the flow field are specified in polar coordinates r and θ, express the pressure distribution within a spiral free vortex in terms of r, θ, q, and Γ.

5.90. The airflow velocity is measured 35 m from the center of a tornado as shown in Figure 5.75. The measured velocity has a magnitude of 110 km/h and is directed at an angle of 60° south of west. Assuming that the tornado can be represented as a spiral free vortex, find the equation of the streamline that passes through the measurement location.

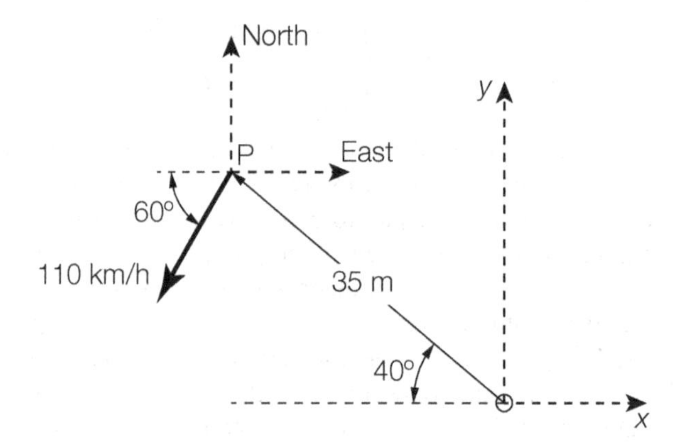

Figure 5.75: Measured wind velocity

5.91. Consider a tornado that is approximated as a spiral free vortex. In a particular tornado the circulation is estimated as 8000 m^2/s, and at a distance of 50 m from the center of the tornado the pressure is measured as 2 kPa below the pressure outside the tornado. Conditions outside the tornado are approximately equal to standard atmospheric conditions at sea level. (a) Estimate the sink strength of the tornado. (b) If a circle of radius 50 m is drawn around the center of the tornado, at what angle do the velocity vectors cross this circle?

5.92. A flow field is generated by two doublets, each of strength 5 m^3/s. If one doublet is located at (x, y) = (−1 m,0 m) and the other doublet is located at (x, y) = (1 m, 0 m), determine the resulting flow velocity at (x, y) = (2 m, 2 m).

5.93. Air flows at 10 m/s towards a structure whose side can be approximated by a Rankine half-body as shown in Figure 5.76. The geometry of the structure is referenced to Cartesian axes that are centered at the "toe" of the structure. A point of interest on the surface of the structure has coordinates (4 m, 4 m). (a) Determine the coordinates of Point C that represents the location of the imaginary source used in constructing the Rankine half-body. (b) What is the equation of the surface of the structure in terms of polar coordinates centered at C? (c) Determine the equation of the stream function that can be used to generate the streamline of airflow over the structure.

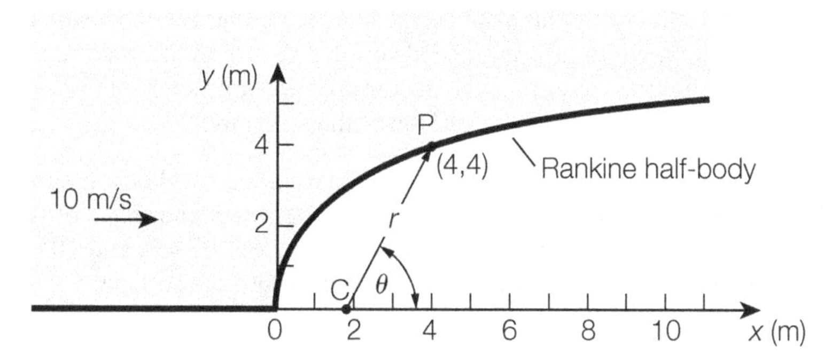

Figure 5.76: Airflow over a Rankine half-body

5.94. Water flowing with a velocity of 0.4 m/s and a depth of 3 m in a channel encounters a hump whose shape can be approximated as a Rankine half-body as shown in Figure 5.77. The flow in the channel can be approximated as irrotational and two dimensional in the xz plane. (a) Estimate the velocity and gauge pressure at the point P that is 3 m downstream and 1.5 m above the toe of the hump. (b) Compare the velocity and gauge pressure at Point P calculated in part (a) with the velocity and gauge pressure at Point P that would be obtained if the flow was assumed to be normal to the flow section at P. Comment on any discrepancies.

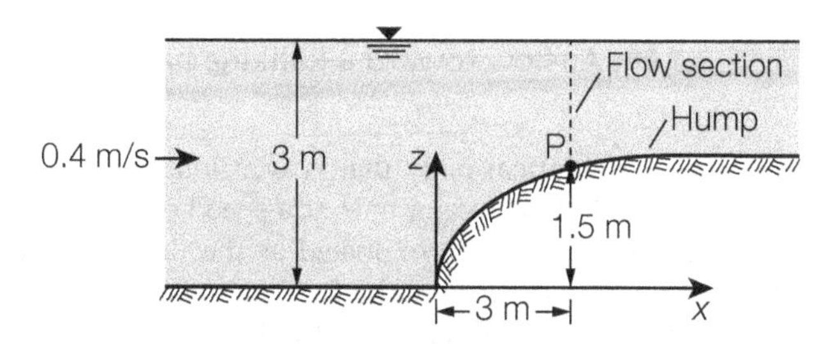

Figure 5.77: Water flow over a Rankine half-body

5.95. Air flows toward a steep hill at 60 km/h, where the shape of the hill can be approximated as a Rankine half-body of height 100 m as illustrated in Figure 5.78. The air is at a temperature of 20°C, and atmospheric pressure at the base of the hill is 101 kPa. Estimate the maximum velocity and the minimum pressure on the face of the hill. Neglect any change in temperature as the air flows up the hill.

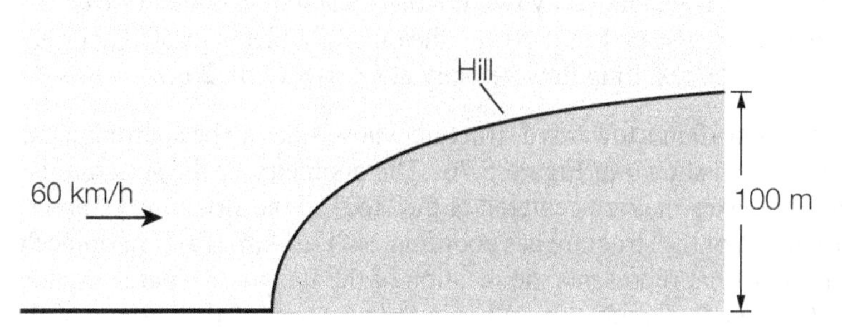

Figure 5.78: Airflow over a hill

5.96. Water at 20°C flows around a submerged structure that has the shape of a Rankine half-body with a width of 3 m. The absolute pressure in the water upstream of the structure is 130 kPa. Determine the maximum allowable approach velocity such that cavitation does not occur on the surface of the structure.

5.97. Air flows toward a surface with the shape of a Rankine half-body as shown in Figure 5.79. The point P is the source location that determines the shape of the half-body. Pressure taps at Points 1 and 2 measure pressures of 207 kPa and 205 kPa, respectively. Estimate the velocity of the air flowing toward the surface.

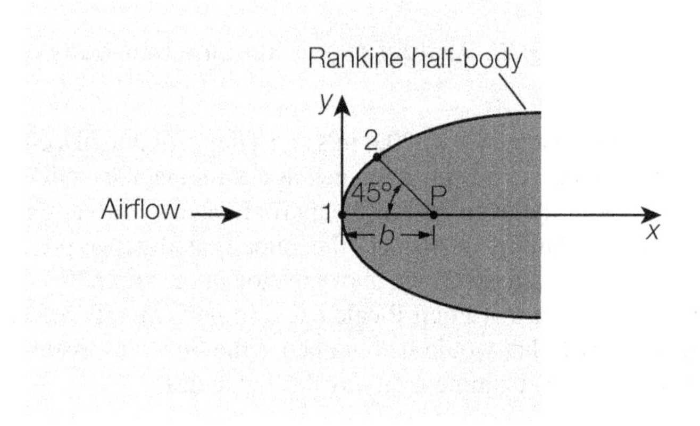

Figure 5.79: Airflow toward a Rankine half-body

5.98. Fluid flows around an elliptical body that is 8 m long and 7 m wide. The fluid approaches the ellipse along its major axis at a speed of 1 m/s. Assuming that the flow around the ellipse can be approximated as the flow around a Rankine oval, (a) estimate the parameters describing the Rankine-oval flow field and (b) estimate the maximum velocity on the surface of the Rankine oval.

5.99. Water at 20°C flows toward a structure that has the shape of a Rankine oval as shown in Figure 5.80. The structure has a length of 3 m and a width of 1.5 m. The absolute pressure in the water approaching the structure is equal to 125 kPa. Estimate the approach velocity, V, of the water that will initiate cavitation on the structure.

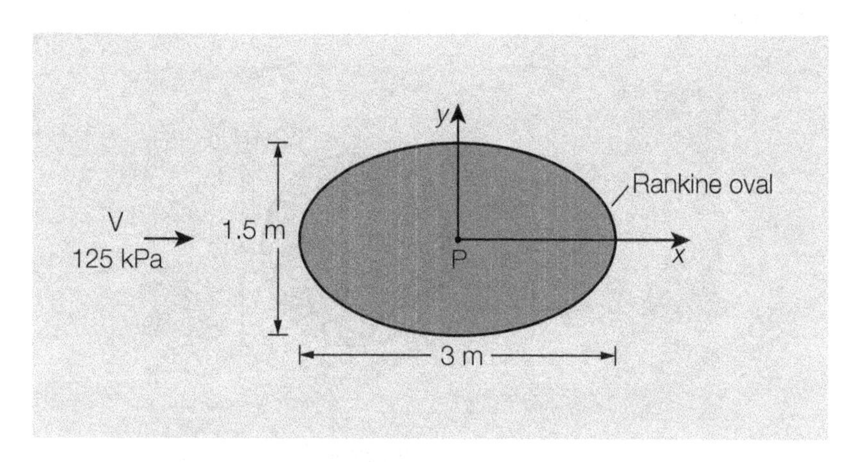

Figure 5.80: Water flow toward a Rankine oval

5.100. A storage bunker has the shape of a half-cylinder of radius 5 m as shown in Figure 5.81. The bunker structural supports are to be designed to withstand the lift force generated by a 96-km/h wind when both the static atmospheric pressure and the air pressure inside the bunker are 101 kPa and the air temperature is 15°C. What is the lift force on the bunker?

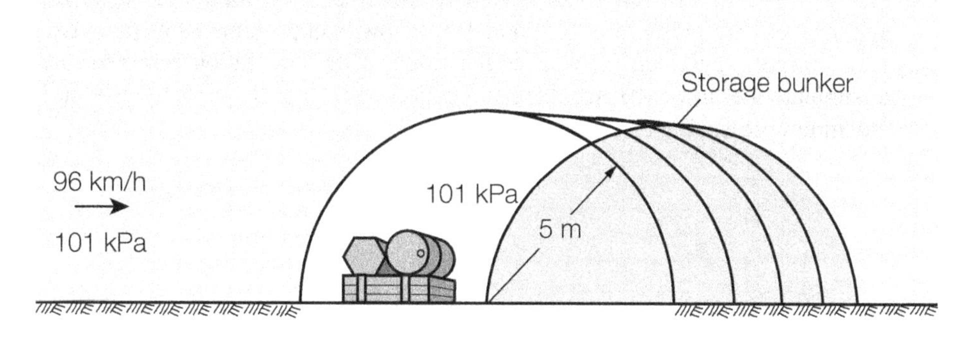

Figure 5.81: Airflow over a storage bunker

5.101. Air under conditions of standard temperature and pressure flows at 15 m/s toward a vertical cylindrical flagpole with a diameter of 0.5 m. (a) Estimate the maximum and minimum speed of the air adjacent to the flagpole. (b) Estimate the maximum and minimum pressure deviation from standard pressure by the air adjacent to the flagpole.

5.102. Water at 20°C flows toward a 1-m-diameter cylindrical column as shown in Figure 5.82. Pressure measurements at Point P indicate that the pressure at that point is approximately 12.5 kPa less than the pressure in the flow approaching the column. (a) Estimate the approach velocity. (b) Determine the doublet strength that would be used with potential theory to simulate the flow field.

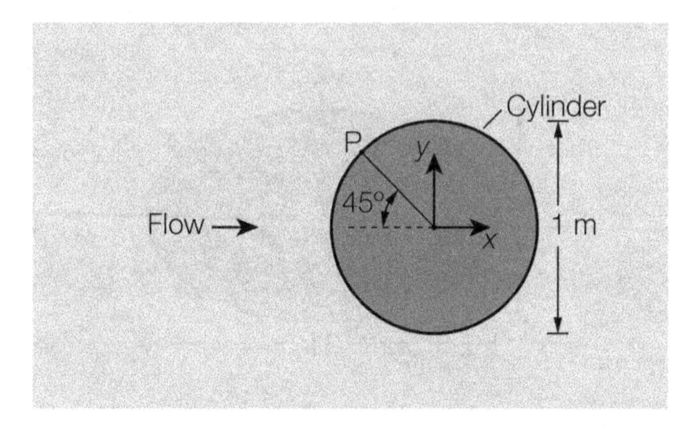

Figure 5.82: Water flow toward a cylinder

5.103. Consider the flow toward the cylinder as shown in Figure 5.82. If the approach velocity is equal to 5 m/s, estimate the rate of rotation of the cylinder that would make the point P a stagnation point. What direction of rotation would be required: clockwise or counterclockwise?

5.104. Air flows toward a rotating cylindrical cable at 100 km/h, where the diameter of the cable is 150 mm and the length of the cable is 10 m. Force measurements indicate that the lift on the rotating cable is 1.00 kN when the temperature of the air is 20°C and atmospheric pressure is 101 kPa. If the flow is approximated as potential flow past a rotating cylinder, (a) what is the value of the circulation parameter needed to simulate the flow? (b) Are there stagnation points on the cable? (c) What is the maximum pressure at any point on the cable?

Chapter 6

Dimensional Analysis and Similitude

LEARNING OBJECTIVES

After reading this chapter and solving a representative sample of end-of-chapter problems, you will be able to:

- Understand the roles of dimensions and units in the analysis of fluid flow problems.
- Apply the principles of dimensional analysis to determine functional relationships.
- Identify dimensionless groups commonly used in fluid mechanics.
- Understand and apply the theory of similitude.
- Design model studies to simulate the performance of full-scale systems.

6.1 Introduction

The primary objective of *dimensional analysis* is to transform functional relationships between dimensional variables into corresponding functional relationships between dimensionless groups. The advantage of using dimensional analysis is that there is generally a smaller number of dimensionless groups than variables that are required to describe any given problem. *Similitude* describes the relationship between similar systems, with the primary application of similitude in fluid mechanics being to quantify how the behavior of one system can be inferred from the behavior of another (similar) system. The rules of similitude can be derived using dimensional analysis, thereby facilitating the design of experiments to study the behavior of full-scale systems using laboratory-scale models. Such model studies and similitude applications are particularly important in the design of large and expensive structures, where safety and economic concerns require increased certainty in how the structure will perform once it is constructed. Although the focus of this book is on the applications of dimensional analysis in the field of fluid mechanics, dimensional analysis is a mathematical technique that is applicable in any field where interrelated variables can be identified.

6.2 Dimensions in Equations

A *dimension* is a measure of a physical quantity, and the seven fundamental dimensions in nature are mass, length, time, temperature, electric current, amount of light, and amount of matter. In contrast to a dimension, a *unit* is a particular convention used to measure a dimension. For example, centimeters, meters, and kilometers are all units of length. Equations that do not depend on the units used, provided the units are consistent, are called *dimensionally homogeneous equations.* For example, suppose the SI system of units is used in a dimensionally homogeneous equation, such that length, mass, and time are expressed in terms of meters, kilograms, and seconds, respectively. Because the equation is dimensionally homogeneous, any calculated quantity will have length, mass, and time units that are in meters, kilograms, and seconds, respectively. If the units in the same equation were changed to another system of units (such as U.S. customary units using feet, slugs, and seconds), the magnitude of the calculated quantity would be the same as before, with the only differences being the numerical value and the system of units in which the calculated quantity was represented. Dimensionally homogeneous equations have the property that if they involve sums (or differences), then each term being summed has the same units. All equations derived from fundamental physical laws are dimensionally homogeneous, and the *principle of dimensional homogeneity* is sometimes stated as follows: If an equation expresses the proper relationship between variables in a physical process, it must necessarily be dimensionally homogeneous.

Dimensions used in fluid mechanics applications. In fluid mechanics, the most commonly encountered dimensions are mass, length, time, and temperature and the symbols commonly used to represent these dimensions are M, L, T, and Θ, respectively. In some cases, the dimension of mass (M) is replaced by the dimension of force (F). However, in dimensional analysis applications, utilization of MLTΘ dimensions is usually preferable to using FLTΘ dimensions.

EXAMPLE 6.1

The force, F, on a sluice gate is given by the equation

$$F = \dot{m}(V_1 - V_2) + \frac{W}{2}\rho g(y_1^2 - y_2^2)$$

where $\dot{m}$ is the mass flow rate under the gate, V_1 and V_2 are the average flow velocities upstream and downstream of the gate, respectively, W is the width of the gate, ρ is the density of the fluid, g is gravity, and y_1 and y_2 are the flow depths upstream and downstream of the gate, respectively. Determine whether this equation is dimensionally homogeneous.

SOLUTION

Express each variable in terms of its fundamental dimensions.

$$F = \dot{m}(V_1 - V_2) + \frac{W}{2}\rho g(y_1^2 - y_2^2)$$

$$\mathrm{M}\left(\frac{\mathrm{L}}{\mathrm{T}^2}\right) = \left(\frac{\mathrm{M}}{\mathrm{T}}\right)\left(\frac{\mathrm{L}}{\mathrm{T}} - \frac{\mathrm{L}}{\mathrm{T}}\right) + (\mathrm{L})\left(\frac{\mathrm{M}}{\mathrm{L}^3}\right)\left(\frac{\mathrm{L}}{\mathrm{T}^2}\right)(\mathrm{L}^2 - \mathrm{L}^2)$$

$$\frac{\mathrm{ML}}{\mathrm{T}^2} = \frac{\mathrm{M}}{\mathrm{T}}\cdot\frac{\mathrm{L}}{\mathrm{T}} + \frac{\mathrm{M}}{\mathrm{LT}^2}\cdot\mathrm{L}^2$$

$$\frac{\mathrm{ML}}{\mathrm{T}^2} = \frac{\mathrm{ML}}{\mathrm{T}^2} + \frac{\mathrm{ML}}{\mathrm{T}^2}$$

Therefore, **the equation is dimensionally homogeneous** because each term in the equation has the same dimensions, which also means that the terms on each side of the equation have the same units.

Nondimensional equations. An equation is classified as a *nondimensional equation* when each term in the equation is dimensionless. The units used in a nondimensional equation are immaterial as long as a consistent set of units is used. Any dimensionally homogeneous equation can be made nondimensional by dividing each term in the equation by a common combination of variables that makes each term dimensionless. A group of variables in which the dimensions of the variables cancel out such that the group is dimensionless is called a *nondimensional group* or *dimensionless group* (of variables). The benefit of nondimensionalization is that the number of relevant variables is reduced by considering the dimensionless groups as variables rather than considering the raw variables individually.

EXAMPLE 6.2

The force, F, on a sluice gate that is used to regulate the flow of a liquid in an open channel is given by the equation

$$F = \rho W y_1 V_1 (V_1 - V_2) + \frac{W}{2} \rho g (y_1^2 - y_2^2)$$

where ρ is the density of the liquid, W is the width of the gate, y_1 and y_2 are the flow depths upstream and downstream of the gate, respectively, V_1 and V_2 are the average flow velocities upstream and downstream of the gate, respectively, and g is gravity. Express the given equation in nondimensional form.

SOLUTION

The given equation expresses the force, F, in terms of the parameters ρ, V_1, V_2, W, g, y_1, and y_2, and the dimensions involved in these variables are M, L, and T. The conventional approach used to nondimensionalize an equation is to select a fixed set of variables, or combinations of variables, to cancel out each dimension in each term. The following parameters can be selected:

$$\text{mass, M} : \rho y_1^3 \qquad \text{length, L} : y_1 \qquad \text{time, T} : \frac{y_1}{V_1}$$

The dimension of each term in the given equation is MLT^{-2}; therefore, the combination of reference variables that has this dimension is determined as follows:

$$\text{MLT}^{-2} = \underbrace{(\rho y_1^3)}_{\text{M}} \cdot \underbrace{(y_1)}_{\text{L}} \cdot \underbrace{(y_1/V_1)^{-2}}_{\text{T}^{-2}} = \rho V^2 y_1^2$$

Dividing each term in the given equation by $\rho V^2 y_1^2$ gives the following equation in which each term is dimensionless:

$$\frac{F}{\rho V^2 y_1^2} = \frac{\rho W y_1 V_1 (V_1 - V_2)}{\rho V^2 y_1^2} + \frac{W \rho g (y_1^2 - y_2^2)}{2(\rho V^2 y_1^2)}$$

Some of the terms in this equation can be further simplified by expressing them in terms of ratios of variables, which are easier to visualize. Such a simplification yields

$$\frac{F}{\rho y_1^2 V_1^2} = \left(\frac{W}{y_1}\right)\left(1 - \frac{V_2}{V_1}\right) + \frac{1}{2}\left(\frac{g y_1}{V_1^2}\right)\left(\frac{W}{y_1}\right)\left[1 - \left(\frac{y_2}{y_1}\right)^2\right]$$

Each of the terms in parentheses in this equation are dimensionless, and these dimensionless terms all involve ratios of dimensional variables.

Normalized equations. In cases where the dimensionless variables in a nondimensional equation are on order 1, the nondimensional equation is sometimes referred to as a *normalized equation*. Normalized equations are usually obtained by introducing fixed (constant) *scaling parameters* that are of the same magnitude as the values of the variables in the equation. Characteristic (constant) fluid properties and characteristic length and time scales are used as scaling parameters. Transformation of a dimensional equation into a normalized equation is typically done using the following two-step procedure: (1) each variable is normalized (i.e., made dimensionless) by a combination of the scaling parameters that has the same dimensions as that variable and (2) the dimensional variables in the original equation are expressed in terms of the normalized variables. The resulting equation is a normalized nondimensional equation.

EXAMPLE 6.3

A sluice gate is commonly used to regulate the flow of a liquid in an open channel. The force, F, exerted by the liquid on a sluice gate is given by the equation

$$F = \rho W y_1 V_1 (V_1 - V_2) + \frac{W}{2} \rho g (y_1^2 - y_2^2)$$

where ρ is the density of the liquid, W is the width of the gate, y_1 and y_2 are the flow depths upstream and downstream of the gate, respectively, V_1 and V_2 are the average flow velocities upstream and downstream of the gate, respectively, and g is gravity. Use the characteristic length L_0 and the characteristic velocity V_0 to normalize the given equation.

SOLUTION

The scaling parameters and their dimensions are ρ [ML^{-3}], L_0 [L], and V_0 [LT^{-1}]. The (dimensional) variables are F [MLT^{-2}], W [L], y_1 [L], V_1 [LT^{-1}], y_2 [L], and V_2 [LT^{-1}]; and g [LT^{-2}] is a dimensional constant.

Step 1: Normalize each variable. Denoting the normalized variables with asterisks, the normalized variables are

$$F^* = \frac{F}{\rho V_0^2 L_0^2}, \quad W^* = \frac{W}{L_0} \quad y_1^* = \frac{y_1}{L_0}, \quad V_1^* = \frac{V_1}{V_0}, \quad y_2^* = \frac{y_2}{L_0}, \quad V_2^* = \frac{V_2}{V_0}$$

Step 2: Express the dimensional variables in the given equation in terms of the normalized variables, which gives

$$F = \rho W y_1 V_1 (V_1 - V_2) + \frac{W}{2} \rho g (y_1^2 - y_2^2)$$

$$F^* (\rho V_0^2 L_0^2) = \rho (W^* L_0) (y_1^* L_0) (V_1^* L_0) L_0 (V_1^* - V_2^*) + \frac{W^* L_0}{2} \rho g L_0^2 (y_1^{*2} - y_2^{*2})$$

which simplifies to

$$F^* = W^* (V_1^* - V_2^*) + \frac{1}{2} \left(\frac{g L_0}{V_0^2} \right) W^* \left[(y_1^*)^2 - (y_2^*)^2 \right]$$

The dimensionless group $V/\sqrt{gL}$ occurs frequently in fluid mechanics applications and is called the *Froude number*, usually represented as Fr. The normalized equation with the nondimensional parameter Fr is given by

$$\boldsymbol{F^* = W^*\left(V_1^* - V_2^*\right) + \frac{1}{2}\left(\frac{1}{\mathrm{Fr_0}^2}\right)W^*\left[(y_1^*)^2 - (y_2^*)^2\right]}$$

where Fr_0 is the Froude number in terms of the scaling parameters.

Because all of the normalized variables are on the order of unity, each term in the above equation that involves only normalized variables is also on the order of unity.

6.3 Dimensional Analysis

Dimensional analysis is the process by which functional relationships are formulated in terms of dimensionless groups. For example, consider the case where there is an unknown relationship between the N variables $x_1, x_2, \ldots, x_N$, which can be written in the functional form

$$f_1(x_1, x_2, \ldots, x_N) = 0 \tag{6.1}$$

where f_1 is an unknown function. In cases where the relationship given by Equation 6.1 must be determined by experiment, performance of these experiments can become a daunting task as N becomes large, because all possible combinations of values of the variables must be considered. Fortunately, the complexity of this problem can be reduced by combining the variables into a smaller number of dimensionless groups, thereby reducing the number of combinations of variables that must be considered in designing the experiments. This process of transforming a functional relationship between N variables into a functional relationship between a smaller number of dimensionless groups is called *dimensional analysis*. The foundation of dimensional analysis is the *Buckingham pi* (Π) *theorem*[43] (Buckingham, 1915), which can be stated as follows:

Theorem 6.1. *If there are N dimensional variables in a dimensionally homogeneous equation, described by m fundamental dimensions, then this equation can be transformed into a relationship between $N - m$ dimensionless groups.*

The SI system of units has seven fundamental dimensions. These dimensions and their corresponding units are as follows: (1) length (meter), (2) mass (kilogram), (3) time (second), (4) temperature (kelvin), (5) electric current (ampere), (6) luminous intensity (candela), and (7) amount of substance (mole). To illustrate the application of the Buckingham pi theorem, consider again the functional relationship between the N variables given in Equation 6.1 and assume that m fundamental dimensions are involved in the units of these variables. Then according to the Buckingham pi theorem, Equation 6.1 can also be written in the form

$$f_2(\Pi_1, \Pi_2, \ldots, \Pi_{N-m}) = 0 \tag{6.2}$$

where $\Pi_1, \Pi_2, \ldots, \Pi_{N-m}$ are independent dimensionless groups of the original dimensional variables $x_1, x_2, \ldots, x_N$. A benefit of this reformulation is that the number of experiments required to determine an empirical relationship between the $N - m$ dimensionless groups is less than the number of experiments required to determine the relationship between the N dimensional variables; the time and cost required for the requisite experiments are also (typically) reduced.

[43]Named in honor of the American physicist Edgar Buckingham (1867–1940).

Notation. Using the Greek variable Π to represent a dimensionless group of variables is particularly appropriate because in mathematical notation Π is conventionally used to denote the product of variables. Dimensionless groups are formed by the products of dimensional variables, hence the justification for the Π notation.

EXAMPLE 6.4

The motion of an object falling in a vacuum under the influence of gravity is described by the relation

$$s = u_0 t + \tfrac{1}{2} g t^2 \tag{6.3}$$

where s is the distance fallen in time t, u_0 is the initial velocity, and g is the acceleration due to gravity. Demonstrate the validity of the Buckingham pi theorem.

SOLUTION

The equation describing the motion of the falling object is given by

$$s - u_0 t - \tfrac{1}{2} g t^2 = 0$$

which can be put in the functional form

$$f_1(s, u_0, t, g) = 0 \tag{6.4}$$

where the dimensions of the variables are

Variable	Dimension
s	[L]
u_0	[LT^{-1}]
t	[T]
g	[LT^{-2}]

In this case, there are four variables ($N = 4$) and two dimensions ($m = 2$). The Buckingham pi theorem states that Equation 6.4 can be expressed as a relation between $N - m = 2$ nondimensional groups. Defining the dimensionless groups, Π_1 and Π_2, as

$$\Pi_1 = \frac{s}{u_0 t} \quad \text{and} \quad \Pi_2 = \frac{gt}{u_0}$$

then according to the Buckingham pi theorem, Equation 6.4 can be written as

$$f_2\left(\frac{s}{u_0 t}, \frac{gt}{u_0}\right) = 0 \tag{6.5}$$

This can be verified by rearranging Equation 6.3 as

$$\left(\frac{s}{u_0 t}\right) - 1 - \frac{1}{2}\left(\frac{gt}{u_0}\right) = 0$$

which is the actual functional relation between the dimensionless groups and **verifies the result** (Equation 6.5) derived using the Buckingham pi theorem.

Note: In this example, the analytic relationship between variables is known and it is therefore possible to verify the result of the dimensional analysis (given by Equation 6.5). In the more usual case, the analytic relationship between variables is unknown and the result of a dimensional analysis is the functional relationship between dimensionless groups. The actual relationship is then determined by experimentation.

The three key steps in applying dimensional analysis to any problem are (1) selection of (dimensional) variables, (2) identification of fundamental dimensions, and (3) formulation of dimensionless groups. Guidelines for performing these steps are given below.

Step 1: Selection of dimensional variables. The selection of a complete set of relevant dimensional variables is required for a correct dimensional analysis. Selected variables must describe (a) the geometry of the flow system, (b) the properties of the fluid, (c) the external effects driving the fluid flow, and (d) the internal property of the fluid flow that is of interest. The dimensional variables selected must be independent of each other, which means that none of the variables can be obtained by combining the other variables.

Step 2: Identification of fundamental dimensions. After selecting the N dimensional variables that describe the behavior of a system, the next step is to identify the m fundamental dimensions. In most cases, m is equal to the number of conventional fundamental dimensions used to describe the variables, of which there are seven in nature; the four conventional fundamental dimensions usually encountered in fluid mechanics are mass (M), length (L), time (T), and temperature (Θ). In some unusual cases, conventional dimensions might occur in the same combination in all the variables. For example, in a set of variables involving M, L, and T, the dimensions L and T might always occur as LT. Under such a circumstance, the *combined conventional dimension* (LT in this case) is taken as a single fundamental dimension. Hence, in the aforementioned example, the number of fundamental dimensions, m, is equal to two (M and LT), not three (M, L, and T). Remember that the occurrence of a combined conventional dimension is rare and that in most cases, m is equal to the number of conventional fundamental dimensions. It is a basic prerequisite that for there to be a possible relationship between a given set of variables, a fundamental dimension must occur in at least two dimensional variables or not at all. If a fundamental dimension was to occur in only one variable, that variable would be precluded from being expressed in terms of the other variables.

Step 3: Formulation of dimensionless groups. After selecting the N variables and identifying the m fundamental dimensions, the next step is to form the $N - m$ dimensionless groups. A variety of methodologies for formulating dimensionless groups are available, and the valid groups that result from these methods are not necessarily unique. The exception is in cases where there is only one dimensionless group, and that group is certainly unique. Regardless of the method used to form the dimensionless groups, the total number of dimensionless groups remains fixed by the number of variables and the number of fundamental dimensions involved; any combination of dimensionless groups can be converted to any other combination of dimensionless groups by multiplying and/or dividing the dimensionless groups by each other. The ideal method of formulating dimensionless groups yields groups to which physical significance can be attached. Methods of formulating dimensionless groups are described in the following sections.

6.3.1 Conventional Method of Repeating Variables

A popular method of formulating dimensionless groups is the *method of repeating variables*, commonly attributed to Edgar Buckingham, although the method was first published by the Russian scientist Dimitri Riabouchinsky in 1911. The six steps that comprise the method of repeating variables are as follows:

Step 1: List the relevant variables (including dimensional constants such as gravity). These variables should include those necessary to describe the geometry of the fluid system, the fluid properties, the external forces causing the fluid flow, and the dependent variable that is of interest. Let N be the total number of variables in the problem.

Step 2: List the primary dimensions for each of the N variables. If any of the primary dimensions always occur in the same combination, then this combination is taken as a primary dimension in lieu of the primary dimensions that make up the combination. A primary dimension must occur in at least two variables; otherwise, the list of variables must be revised.

Step 3: Set m equal to the number of primary dimensions determined in Step 2. In accordance with the Buckingham pi theorem, the number of dimensionless groups is equal to $N - m$.

Step 4: Choose m repeating variables that will be used to construct each dimensionless group. These repeating variables should be selected such that each repeating variable contains one primary dimension that is not contained in the other repeating variables. When properly selected, the repeating variables cannot be combined to form a dimensionless group. Additional guidelines for choosing repeating variables are (1) never pick the dependent variable; (2) never pick variables that are already dimensionless; (3) whenever possible, choose dimensional constants (such as gravity) over dimensional variables; and (4) pick simple variables over complex variables whenever possible. In general, viscosity (μ) should appear in only one dimensionless group; therefore, μ should not be chosen as a repeating variable. When given the choice, it usually is best to choose density, ρ [ML^{-3}], speed, V [LT^{-1}], and a characteristic length, ℓ [L], as the repeating variables.

Step 5: Generate the dimensionless groups one at a time by grouping the m repeating variables with one of the remaining variables, forcing the product to be dimensionless. Follow this approach to construct all $N - m$ dimensionless groups.

Step 6: Check that all groups are dimensionless and express the dimensionless group containing the dependent variable as a function of the other dimensionless groups.

Keep in mind that variables that are already dimensionless become dimensionless groups by themselves. For example, an angle (which is derived from a ratio of sides) is already a dimensionless variable and no further nondimensionalization is required. The functional relationship derived by dimensional analysis is not unique in that it depends on the choice of repeating variables. Judicious choices for the repeating variables are key to identifying dimensionless groups that are closely aligned with the physics of the problem. If only one dimensionless group, Π, results from the analysis, then the resulting dimensionless group must necessarily be equal to a constant because the functional relation

$$f(\Pi) = 0$$

could theoretically be solved for Π to yield Π = constant. This constant could then be determined by a single experiment in which all of the relevant variables are measured.

Limitations and applications of dimensional analysis. Dimensional analysis can only determine the functional relationship between dimensionless groups; it cannot determine the form of the function or the exact equation relating these groups. Consequently, the next step after dimensional analysis is usually experimentation to determine an empirical relationship between the dimensionless groups. Another follow-on application is in the area of physical modeling, which is based on the fact that if two geometrically similar systems of different size have the same values for each of the independent dimensionless groups, then the value of the dependent dimensionless group will be the same in both systems. The application of dimensional analysis to physical modeling is covered in detail in Section 6.6.

EXAMPLE 6.5

The drag force on a submarine of a given shape is a function of the size of the submarine, the velocity of the submarine relative to the surrounding fluid, and the density and viscosity of the fluid. Derive a nondimensional expression relating the drag force to the other variables.

SOLUTION

Step 1: The relevant variables are the drag force on the submarine (F_D), the size of the submarine (L), the velocity of the submarine (V), the density of the fluid (ρ), and the viscosity of the fluid (μ). The problem variables can be expressed in the functional form

$$F_D = f_1(\rho, \mu, V, L)$$

There are five variables, so $N = 5$.

Step 2: The variables and their dimensions are as follows: F_D [MLT^{-2}], ρ [ML^{-3}], μ [ML^{-1}T^{-1}], V [LT^{-1}], and L [L].

Step 3: There are three primary dimensions [M, L, and T]. So m = 3 and the number of dimensionless groups is $N - m = 5 - 3 = 2$.

Step 4: Choose ρ, L, and V as the three repeating variables ($m = 3$), which contain the units of mass, length, and time.

Step 5: Form the dimensionless groups with each of the two remaining variables (F_D and μ) as follows:

Combination	Units	Combination	Units
F_D	MLT^{-2}	μ	$ML^{-1}T^{-1}$
$\frac{F_D}{\rho}$	$\frac{MLT^{-2}}{ML^{-3}} = L^4T^{-2}$	$\frac{\mu}{\rho}$	$\frac{ML^{-1}T^{-1}}{ML^{-3}} = L^2T^{-1}$
$\frac{F_D}{\rho V^2}$	$\frac{L^4T^{-2}}{L^2T^{-2}} = L^2$	$\frac{\mu}{\rho V}$	$\frac{L^2T^{-1}}{LT^{-1}} = L$
$\frac{F_D}{\rho V^2 L^2}$	$\frac{L^2}{L^2}$ = dimensionless	$\frac{\mu}{\rho V L}$	$\frac{L}{L}$ = dimensionless

Therefore, based on these results, the two dimensionless groups can be taken as

$$\frac{F_D}{\rho V^2 L^2} \quad \text{and} \quad \frac{\rho V L}{\mu}$$

The group $\rho V L/\mu$ is recognized as the Reynolds number (Re), which gives a measure of the ratio of the inertial to the viscous force.

Step 6: The functional relation between dimensionless groups can therefore be expressed in the form

$$\frac{F_D}{\rho V^2 L^2} = f_2\left(\frac{\rho V L}{\mu}\right)$$

which more commonly would be written as

$$\boldsymbol{\frac{F_D}{\rho V^2 L^2} = f_2\,(\mathrm{Re})}$$

6.3.2 Alternative Method of Repeating Variables

An alternative method of repeating variables was proposed by Ipsen (1960). The Ipsen method of repeating variables differs from the conventional method of repeating variables in that it is not necessary to quantify the number of primary dimensions, hence obviating the need to identify any combined conventional dimensions. The steps to be followed in applying the Ipsen method of repeating variables are as follows:

Step 1: List the relevant variables using the same guidelines as in the conventional method of repeating variables (i.e., same as Step 1 for conventional method).

Step 2: Choose a repeating variable containing a conventional primary dimension (e.g., M, L, T, or Θ) and then use this repeating variable to remove the selected primary dimension from the other variables. After completing this step, discard the repeating variable (because it is no longer a separate variable). Repeat this step until all variable groups are dimensionless.

Step 3: Express the dimensionless group containing the dependent variable as a function of the other dimensionless groups.

This approach can be both illustrated and contrasted with the conventional method of repeating variables by redoing the previous problem using the Ipsen method.

EXAMPLE 6.6

The drag force on a submarine of a given shape is a function of the size of the submarine, the velocity of the submarine relative to the surrounding fluid, and the density and viscosity of the fluid. Derive a nondimensional expression relating the drag force to the other variables.

SOLUTION

Step 1: The functional relationship between the variables can be expressed in the form

$$F_\text{D} = f_1(\rho, \mu, V, L)$$

The variables and their dimensions are F_D [MLT^{-2}], ρ [ML^{-3}], μ [ML^{-1}T^{-1}], V [LT^{-1}], and L [L].

Step 2.1: Using ρ as the repeating variable to remove the mass (M) dimension yields

$$\frac{F_\text{D}}{\rho} = f_2\left(\frac{\mu}{\rho}, V, L\right)$$

Step 2.2: Using L as the repeating variable to remove the length (L) dimension yields

$$\frac{F_\text{D}}{\rho L^4} = f_3\left(\frac{\mu}{\rho L^2}, \frac{V}{L}\right)$$

Step 2.3: Using V/L as the repeating variable to remove the time (T) dimension yields

$$\frac{F_\text{D}}{\rho (V/L)^2 L^4} = f_4\left(\frac{\mu}{\rho (V/L) L^2}\right) \quad \rightarrow \quad \frac{F_\text{D}}{\rho V^2 L^2} = f_4\left(\frac{\mu}{\rho V L}\right)$$

Step 3: The functional expression derived in Step 2.3 more commonly would be written as

$$\boldsymbol{\frac{F_\text{D}}{\rho V^2 L^2} = f_4\,(\text{Re})}$$

where $\text{Re} = \rho V L/\mu$. This is the same result derived previously using the conventional method of repeating variables.

6.3.3 Method of Inspection

In contrast to the more formal methods based on repeating variables, the less formal method of inspection identifies dimensionless groups of variables based on prior knowledge of dimensionless groups that have associated physical meanings. Specifically, for any given functional relationship between N variables in m dimensions, such as

$$f_1(x_1, x_2, \ldots, x_N) = 0$$

by inspection, we pick out $N - m$ dimensionless groups such that

$$\Pi_1 = f_2(\Pi_2, \ldots, \Pi_{N-m})$$

where Π_1 is the dimensionless group containing the dependent variable, and $\Pi_2, \ldots, \Pi_{N-m}$ are dimensionless groups that contain the independent variables. The most commonly used dimensionless groups have physical meanings and are named after engineers and scientists who pioneered their use. These commonly used dimensionless groups are listed in Table 6.1. Therefore, if some of the variables in a dimensional analysis problem (e.g., $x_i, \ldots, x_j$) include variables that constitute a particular named dimensionless group in Table 6.1, that named group would be chosen because it is already known to be dimensionless, and it also has physical meaning. Besides choosing dimensionless groups with physical meaning, such as those shown in Table 6.1, other guidelines in applying the method of inspection are as follows:

- Normalize pressures and stresses by ρV^2 or $\frac{1}{2}\rho V^2$.
- Normalize forces by $\rho V^2 L^2$ or $\frac{1}{2}\rho V^2 L^2$.
- In cases where there are multiple length scales in the list of variables, select one length scale, L, as the reference length scale and use the ratios of the other length scales to the reference length scale as dimensionless groups.

Application of the method of inspection can be contrasted with the method of repeating variables by comparing the following example with the previous example.

Table 6.1: Common Dimensionless Groups in Fluid Mechanics

Name	Symbol	Formula	Physical Meaning	Named After
Cauchy number*	Ca	$\dfrac{\rho V^2}{E_\text{v}}$	$\dfrac{\text{inertial force}}{\text{elastic force}}$	Augustin Cauchy (1789–1857)
Euler number	Eu	$\dfrac{\Delta p}{\rho V^2}$ or $\dfrac{\Delta p}{\frac{1}{2}\rho V^2}$	$\dfrac{\text{pressure force}}{\text{inertial force}}$	Leonhard Euler (1707–1783)
Froude number	Fr	$\dfrac{V}{\sqrt{gL}}$	$\dfrac{\text{inertial force}}{\text{gravitational force}}$	William Froude (1810–1879)
Mach number*	Ma	$\dfrac{V}{\sqrt{E_\text{v}/\rho}}$	$\dfrac{\text{inertial force}}{\text{elastic force}}$	Ernst Mach (1838–1916)
Prandtl number	Pr	$\dfrac{\mu c_p}{k}$	$\dfrac{\text{viscous diffusion rate}}{\text{thermal diffusion rate}}$	Ludwig Prandtl (1875–1953)
Reynolds number	Re	$\dfrac{\rho V L}{\mu}$	$\dfrac{\text{inertial force}}{\text{viscous force}}$	Osborne Reynolds (1842–1912)
Richardson number	Ri	$\dfrac{\left(\dfrac{\Delta\rho}{\rho}g\right)L}{V}$	$\dfrac{\text{buoyancy force}}{\text{inertial force}}$	Lewis Richardson (1881–1953)
Strouhal number	St	$\dfrac{\omega L}{V}$	$\dfrac{\text{local inertial force}}{\text{convective inertial force}}$	Vincenz Strouhal (1850–1922)
Weber number	We	$\dfrac{\rho V^2 L}{\sigma}$	$\dfrac{\text{inertial force}}{\text{surface tension force}}$	Moritz Weber (1871–1951)

*The Cauchy number and Mach number are related by Ca = Ma^2.

EXAMPLE 6.7

The drag force on a submarine of a given shape is a function of the size of the submarine, the velocity of the submarine relative to the surrounding fluid, and the density and viscosity of the fluid. Derive a nondimensional expression relating the drag force to the other variables.

SOLUTION

The relevant variables can be expressed in the functional form

$$F_{\mathrm{D}} = f_1(\rho, \mu, V, L)$$

This function involves five variables in three dimensions, so according to the Buckingham pi theorem, the relationship between variables can be expressed as a relationship between two dimensionless groups. Upon inspection, it can be seen that the variables (ρ, V, L, μ) are included in the Reynolds number, Re, and the force, F_{D}, can be normalized by $\rho V^2 L$; hence, by inspection,

$$\frac{F_{\mathrm{D}}}{\rho V^2 L^2} = f_2\left(\frac{\rho V L}{\mu}\right) \quad \text{or} \quad \frac{F_{\mathrm{D}}}{\rho V^2 L^2} = f_2\,(\mathrm{Re})$$

This is the same result obtained previously using the method of repeating variables. However, using the method of inspection requires much less effort.

6.4 Dimensionless Groups as Force Ratios

The forces acting on fluid elements can originate from several sources, such as an applied pressure force, a frictional force due to the viscosity of the fluid, a gravity force due to the weight of the fluid, a surface tension force due to the surface tension of the fluid, and an elastic force due to the compressibility of the fluid. The sum of all the forces acting on a fluid element is commonly called the *inertial force*, which is equal to the mass times the acceleration of the fluid element. From an alternative perspective, the inertial force can be regarded as the force required to reduce the acceleration of a fluid element to zero. This is so because if the net force on a fluid element is F and the mass times acceleration of the fluid element is ma, then applying an opposing force equal to ma gives a net force on the element of $F - ma = 0$, which reduces the acceleration of the fluid element to zero.

Inertial force. Denoting the inertial force by F_{I}, the relationship between F_{I} and the length and velocity scales, L and V, is as follows:

$$F_{\mathrm{I}} = ma \propto (\rho L^3)\left(\frac{L}{T^2}\right) = \rho L^2 \left(\frac{L}{T}\right)^2 = \rho L^2 V^2 \tag{6.6}$$

where m and a are the mass and acceleration of the fluid element, respectively, ρ is the density of the fluid, and $T = L/V$ is the time scale for the particular system being considered. Equation 6.6 states that the inertial force is proportional to the product of the density times the length scale squared times the velocity scale squared, which is expressed as

$$F_{\mathrm{I}} \propto \rho L^2 V^2 \tag{6.7}$$

Force ratios. The inertial force, F_I, is a measure of the resultant force on a fluid element, and a dimensionless group with physical meaning is typically formed as a ratio of a particular force to the inertial force. A dimensionless ratio formed in this manner provides a measure of the relative influence of a particular force on the fluid motion. Several dimensionless groups with physical meaning are listed in Table 6.1 along with their corresponding force ratios. The derivation of the gravity-to-inertial (= Froude number^{-1}), viscous-to-inertial (= Reynolds number^{-1}), and pressure-to-inertial (Euler number) force ratios are given below as illustrations; derivations of the expressions for other force ratios can be performed similarly. Other force ratios that are commonly used include the surface tension-to-inertial (Weber number) and elastic-to-inertial (Mach number) force ratios.

Froude number. The gravitational force, F_G, on a fluid element that has a length scale L is measured by the weight of the fluid element and is given by

$$F_G = mg \propto \rho L^3 g \tag{6.8}$$

The ratio of the inertial force to the gravitational force can be obtained by combining Equations 6.7 and 6.8, which yields

$$\frac{F_I}{F_G} \propto \frac{\rho L^2 V^2}{\rho L^3 g} = \frac{V^2}{gL}$$

This dimensionless ratio is related to the *Froude number*,[44] Fr, which is defined as

$$\text{Fr} = \frac{V}{\sqrt{gL}}$$

Thus,

$$\text{Fr}^2 = \frac{V^2}{gL} \propto \frac{F_I}{F_G} = \frac{\text{inertial force}}{\text{gravitational force}}$$

Therefore, the Froude number, Fr, is a measure of the ratio of the inertial to the gravitational force on a fluid element having a velocity scale V and length scale L. Gravitational forces are typically dominant in free-surface flows such as the flows around ships, waves generated by moving ships, flows in open channels such as rivers and streams, flows over weirs and spillways, flows involving two fluids separated by an interface, and flows over submerged objects where cavitation is occurring. Two interesting properties of the Froude number are that (1) it does not involve any fluid properties and (2) no mass dimension is involved in the defining variables. These properties of the Froude number reflect the fact that the Froude number is basically a kinematic parameter that fixes the relation between length and time in cases where gravity forces dominate.

Reynolds number. The viscous force, F_V, on a fluid element that has a length scale L is measured by the shear force on the fluid element, which is due to the viscosity of the fluid and is given by

$$F_V = \mu \frac{dV}{dy} A \propto \mu \left(\frac{V}{L}\right) L^2 = \mu V L \tag{6.9}$$

The ratio of the inertial force to the viscous force can be obtained by combining Equations 6.7 and 6.9, which yields

$$\frac{F_I}{F_V} \propto \frac{\rho L^2 V^2}{\mu V L} = \frac{\rho V L}{\mu}$$

[44]Named in honor of the British naval architect William Froude (1810–1879).

This dimensionless ratio is related to the *Reynolds number*,[45] Re, which is defined as

$$\text{Re} = \frac{\rho V L}{\mu}$$

Then

$$\text{Re} = \frac{\rho V L}{\mu} \propto \frac{F_\text{I}}{F_\text{V}} = \frac{\text{inertial force}}{\text{viscous force}}$$

Therefore, the Reynolds number, Re, is a measure of the ratio of the inertial to the viscous force on a fluid element having a velocity scale V, length scale L, and fluid properties ρ and μ. Viscous forces are typically dominant in cases of high-viscosity fluids (e.g., oils) flowing in small-diameter conduits. Fluid flows are laminar at low Reynolds numbers and turbulent at high Reynolds numbers, and the Reynolds number is typically used as an indicator to distinguish between laminar and turbulent flows.

Euler number. Pressure forces that contribute to fluid motion are associated with pressure differences, because there is no net pressure force on a fluid element unless there is a pressure difference. The pressure force, F_P, on a fluid element acted on by a pressure difference Δp is given by

$$F_\text{P} = \Delta p A \propto \Delta p L^2 \tag{6.10}$$

where A is the area over which the net pressure acts. The ratio of the net pressure force to the inertial force can be obtained by combining Equations 6.7 and 6.10, which yields

$$\frac{F_\text{P}}{F_\text{I}} \propto \frac{\Delta p L^2}{\rho L^2 V^2} = \frac{\Delta p}{\rho V^2}$$

This dimensionless ratio is related to the *Euler number*,[46] Eu, which is defined as

$$\text{Eu} = \frac{\Delta p}{\frac{1}{2}\rho V^2}$$

Thus,

$$\text{Eu} = \frac{\Delta p}{\frac{1}{2}\rho V^2} \propto \frac{F_\text{P}}{F_\text{I}} = \frac{\text{pressure force}}{\text{inertial force}}$$

Therefore, the Euler number, Eu, is a measure of the ratio of the net pressure force to the inertial force on a fluid element having a pressure-difference scale Δp, velocity scale V, and density ρ. The Euler number, Eu, is sometimes called the *pressure coefficient*. In aerodynamic applications, it is convenient to work with the difference between the local pressure on the surface of an aircraft and the free-stream pressure of the approaching air. This formulation is particularly appealing because it contrasts the local pressure difference (Δp) with the dynamic pressure ($\frac{1}{2}\rho V^2$), which is the maximum pressure difference along a stagnation streamline. In hydrodynamic applications where cavitation of a liquid is a concern, the difference between the local pressure, p, and the saturation vapor pressure, p_v, is of interest, because the larger this difference, the less likely that cavitation occurs. In such applications, the Euler number is typically used in the form

$$\text{Eu} = \frac{p - p_\text{v}}{\frac{1}{2}\rho V^2} \quad (= \text{Ca})$$

where the Euler number is commonly called the *cavitation number*, Ca, to indicate the phenomenon (cavitation) that is being measured by this dimensionless number. Note that "Ca" is used to represent both the cavitation number and the *Cauchy number* ($= \rho V^2/E_\text{v}$), which are unrelated.

[45] Named in honor of the British engineer Osborne Reynolds (1842–1912).

[46] Named in honor of the Swiss mathematician and physicist Leonhard Euler (1707–1783).

Application of dimensionless force ratios. The dimensionless groups listed in Table 6.1 are mostly (with the exception of the Prandtl number) measures of the ratios of the magnitudes of various forces to the magnitude of the inertial force, and the physical meanings of these dimensionless groups are widely used as bases for either including or neglecting variables in any given problem. For example, high values of the Reynolds number would indicate that viscous forces are small relative to inertial forces; therefore, the viscosity of the fluid could be neglected as an influential variable. Conversely, low values of a dimensionless group could indicate that the associated fluid property exerts a significant effect on the flow and should not be neglected. The Prandtl number (Pr) gives a measure of the relative rate at which heat moves by (viscous) momentum transfer compared with the rate at which heat moves by conduction within a fluid. The Prandtl number can also be defined as the ratio of the viscous diffusion rate to the thermal diffusion rate. In cases where heat is transferred from a boundary into a moving fluid, the Prandtl number gives a measure of the relative thicknesses of the momentum and thermal boundary layers. Unlike the force ratios, the Prandtl number is a property of the fluid only, with $\text{Pr} = 7$ for water at 20°C and $\text{Pr} = 0.7 - 0.8$ for air and many other gases.

EXAMPLE 6.8

The pressure drop Δp between any two sections in a pipe that are located a distance L apart can be assumed to be a function of the pipe diameter, D, the roughness height of the pipe material, ϵ, the fluid density, ρ, the fluid viscosity, μ, and the velocity, V, of the fluid. (a) Perform a dimensional analysis to formulate this functional relationship in terms of dimensionless groups; the dependent variable of interest is the pressure drop. (b) Experiments are performed in a laboratory to determine an empirical formula that can be used to describe the relationship derived in part (a). A 100-mm-diameter steel pipe carrying water at 20°C at a flow rate of 100 L/s is used in the experiments, and the following data were measured:

L (m)	Δp (kPa)
1	0.093
10	1.032
50	4.439
75	8.511
100	9.002

Viscous effects were negligible for the flow conditions in the experiments. Use the measured data to derive an empirical relationship that can be used to estimate the pressure drop for any fluid flowing in any pipe that has the same ratio of roughness height to pipe diameter as the steel pipe used in the experiments, provided viscous effects are negligible.

SOLUTION

From the given data, the seven interrelated variables and their dimensions are pressure drop, Δp [FL^{-2}], distance, L [L], diameter, D [L], roughness height, ϵ [L], density, ρ [ML^{-3}], viscosity, μ [FL^{-2}T], and velocity, V [LT^{-1}].

(a) Because the seven variables involve three fundamental dimensions (MLT or FLT), the functional relationship between these variables can be transformed into a functional relationship between $7 - 3 = 4$ dimensionless groups. Using the method of inspection and selecting conventional dimensionless groups gives

$$f_1(\Delta p, L, D, \epsilon, \rho, \mu, V) = 0 \quad \rightarrow \quad \frac{\Delta p}{\frac{1}{2}\rho V^2} = f_2\left(\frac{L}{D}, \frac{\epsilon}{D}, \frac{\rho V D}{\mu}\right)$$

Recognizable dimensionless groups in this relationship are the Euler number, Eu; the Reynolds number, Re; and ratios of length scales. The relationship between the dimensionless groups can be expressed as:

$$\mathbf{Eu} = f_2\left(\frac{L}{D}, \frac{\epsilon}{D}, \mathbf{Re}\right)$$

(b) In cases where the ratio of the roughness height to the pipe diameter, ϵ/D, is fixed and viscous effects are negligible, the functional relationship between the variable dimensionless groups becomes

$$\frac{\Delta p}{\frac{1}{2}\rho V^2} = f_3\left(\frac{L}{D}\right)$$

For the experimental conditions, D = 100 mm, Q = 100 L/s, $A = \pi D^2/4 = 0.007854\ \text{m}^2$, and $V = Q/A$ = 1.273 m/s. For water at 20°C, ρ = 998.2 kg/m^3. Using these conditions, the given experimental data can be expressed as follows:

$\frac{L}{D}$	$\frac{\Delta p}{\frac{1}{2}\rho V^2}$
10	0.12
100	1.28
500	5.49
750	10.52
1000	11.13

These data are plotted in Figure 6.1, where it is apparent that an approximate linear relationship can be fit to the data.

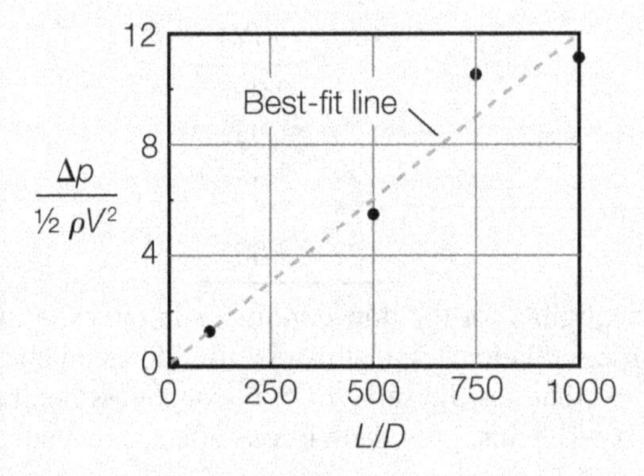

Figure 6.1: Experimental data and best-fit line

The best-fit line shown in Figure 6.1 was determined from a least-squares analysis and is given by

$$\frac{\Delta p}{\frac{1}{2}\rho V^2} = 0.012\frac{L}{D}$$

This relationship can be used to estimate the pressure drop in pipes for fluids of any density, provided the pipes of interest have the same ratio of roughness height to diameter as that used in the experiments, and viscous effects are negligible.

6.5 Dimensionless Groups in Other Applications

By definition, a dimensionless group is any combination of variables that is dimensionless. Because a variety of different variables are used in various engineering applications, the number of possible dimensionless groups is very large, if not infinite. The common dimensionless groups listed in Table 6.1 are mostly force ratios, so these dimensionless groups occur mostly in applications related to the motion of a fluid and the forces exerted on moving fluids. There are many additional dimensionless groups that commonly occur in cognate areas of fluid mechanics—in particular, applications related to heat transfer, mass transfer, and energy. Other dimensionless groups involving force ratios also appear when heat and mass transfer occur simultaneously with momentum transfer. A list of several of the more prominent dimensionless groups used in cognate areas of fluid mechanics is given in Table 6.2. The key variables involved in the dimensionless groups listed in Table 6.2 are c_p = specific heat at constant pressure, D = mass diffusion coefficient, g = gravitational constant, h = convective heat transfer coefficient, k = thermal conductivity, L = length scale, L_v = latent heat of vaporization, t = time, V = velocity scale, α = thermal diffusivity, β = volumetric thermal expansion coefficient, Θ = temperature, Θ_s = surface temperature, Θ_∞ = bulk temperature, Θ_sat = saturation temperature, and ν = kinematic viscosity. Aside from fluid mechanics and its cognate areas, specialized dimensionless groups have been developed in diverse areas such as astrophysics, botany, chemistry, economics, mathematics, medicine, meteorology, and several areas of social science.

Table 6.2: Dimensionless Groups in Various Applications

Name	Symbol	Formula	Physical Meaning	Named After
Biot number	Bi	$\dfrac{hL}{k}$	$\dfrac{\text{surface thermal resistance}}{\text{internal thermal resistance}}$	Jean-Baptiste Biot (1774–1862)
Eckert number	Ec	$\dfrac{V^2}{c_p\Theta}$	$\dfrac{\text{kinetic energy}}{\text{enthalpy}}$	Ernst Eckert (1904–2004)
Fourier number	Fo	$\dfrac{\alpha t}{L^2}$	$\dfrac{\text{physical time}}{\text{thermal diffusion time}}$	Jean Baptiste Joseph Fourier (1768–1830)
Grashof number	Gr	$\dfrac{g\beta(\Theta_\text{s}-\Theta_\infty)L^3}{\nu^2}$	$\dfrac{\text{buoyancy force}}{\text{viscous force}}$	Franz Grashof (1826–1893)
Jakob number	Ja	$\dfrac{c_p(\Theta-\Theta_\text{sat})}{L_\text{v}}$	$\dfrac{\text{sensible energy}}{\text{latent energy}}$	Max Jakob (1879–1955)
Lewis number	Le	$\dfrac{\alpha}{D}$	$\dfrac{\text{convection heat transfer}}{\text{conduction heat transfer}}$	Warren Lewis (1882–1975)
Nusselt number	Nu	$\dfrac{Lh}{k}$	$\dfrac{\text{buoyancy force}}{\text{inertial force}}$	Wilhelm Nusselt (1882–1957)
Peclet number	Pe	$\dfrac{VL}{\alpha}$	$\dfrac{\text{advection heat transfer}}{\text{conduction heat transfer}}$	Jean Peclet (1793–1857)
Rayleigh number	Ra	$\dfrac{g\beta(\Theta_\text{s}-\Theta_\infty)L^3}{\nu\alpha}$	$\dfrac{\text{buoyancy force}}{\text{viscous force}}$	Lord Rayleigh (1842–1919)
Schmidt number	Sc	$\dfrac{\nu}{D}$	$\dfrac{\text{viscous diffusion}}{\text{mass diffusion}}$	Ernst Schmidt (1892–1975)
Sherwood number	Sh	$\dfrac{VL}{D}$	$\dfrac{\text{advection mass transfer}}{\text{diffusion mass transfer}}$	Thomas Sherwood (1903–1976)
Stanton number	St	$\dfrac{h}{\rho c_p V}$	$\dfrac{\text{heat transfer}}{\text{thermal capacity}}$	Thomas Stanton (1865–1931)

Variations. Variations in the names of the dimensionless groups listed in Table 6.2 also occur, with some dimensionless numbers having different names depending on the application. For example, in mass-transfer applications in environmental engineering, the Peclet number is generally used to measure the ratio of advection mass transfer to diffusion mass transfer. This makes the Peclet number the same as the Sherwood number, with the Peclet number being the more commonly used name in environmental engineering applications. Further, in such applications, the name Peclet is usually accented as "Péclet" to more accurately represent the originator's name and its pronunciation.

6.6 Modeling and Similitude

In cases where a flow system is to be either constructed or modified, it is sometimes prudent to study the expected behavior of the planned system (called the *prototype*) by constructing a laboratory-scale *model* of the prototype and inferring the behavior of the prototype from observations on the model. The model can be smaller or larger than the prototype, although smaller-scale models are more common. The more descriptive term "physical model" is sometimes used to differentiate this type of model from other types of models, such as numerical models and analytical models. Smaller-scale physical models are typically used to study the performance of fluid-driven machinery such as hydraulic turbines and pumps; the performance of large structures such as aircraft, ships, spillways of large dams, and buildings; and processes such as the action of waves and tides on beaches, soil erosion, and sediment transport. In building applications, small-scale models are used particularly for studying the effects of wind loads on tall buildings and the airflow patterns generated by the construction of new buildings. Larger-scale models are used to study the performance of small structures such as biomedical devices, injection nozzles, and carburetors.

Example application. A 1:100 scale hydraulic model of a reservoir, a spillway, and the downstream topography of a proposed prototype design is shown in Figure 6.2. To give an idea of scale, the model dam is approximately 1 m high, which corresponds to a prototype dam height of approximately 100 m. Although model tests provide valuable insight into the design of prototype systems, model tests should be used to supplement but not replace the theoretical knowledge, good judgment, and experience of the design engineer.

Figure 6.2: Hydraulic model

Theory of modeling. A fundamental question in interpreting data measured in models is how the magnitudes of various parameters in a model are related to the magnitudes of the same parameters in the prototype. The answer to this question is derived from a dimensional analysis of the flow system, the results of which can generally be written in the form

$$\Pi_0 = f(\Pi_1, \Pi_2, \ldots, \Pi_N) \tag{6.11}$$

where Π_0 is a dimensionless group containing the variable of interest and $\Pi_1, \Pi_2, \ldots, \Pi_N$ are dimensionless groups containing other variables that affect the variable of interest. Based on Equation 6.11, it is clear that if both the model and prototype represent the same physical system and if the model is constructed and operated such that $\Pi_1, \Pi_2, \ldots, \Pi_N$ are the same in both the model and prototype, then Π_0 also must be the same in both the model and prototype. Therefore, by measuring Π_0 in the model, the parameter of interest in the prototype can be inferred from the relation

$$(\Pi_0)_\mathrm{p} = (\Pi_0)_\mathrm{m} \tag{6.12}$$

because

$$(\Pi_1)_\mathrm{p} = (\Pi_1)_\mathrm{m}\,; \quad (\Pi_2)_\mathrm{p} = (\Pi_2)_\mathrm{m}\,; \ldots \text{ and } (\Pi_N)_\mathrm{p} = (\Pi_N)_\mathrm{m} \tag{6.13}$$

where the subscript "m" denotes the value of a dimensionless group in the model and the subscript "p" denotes the value of a dimensionless group in the prototype. Equation 6.12 is called the *prediction equation*, and the relationships in Equation 6.13 are called the *modeling laws* or *similarity requirements*. In the course of meeting the similarity requirements given by Equation 6.13, the model and prototype must necessarily be geometrically, kinematically, and dynamically similar.

Geometric similarity. Usually several of the independent Π groups on the right-hand side of Equation 6.11 involve the ratio of length scales, and equality of these Π groups in the model and prototype require that the model be *geometrically similar* to the prototype. Geometric similarity requires that the model and prototype have the same shape and differ only in size. Under such circumstances, the model is characterized by the *length-scale ratio* or *scale factor*, L_r, given by

$$L_\mathrm{r} = \frac{L_\mathrm{p}}{L_\mathrm{m}}$$

where L_p is a prototype dimension (length) and L_m is the corresponding model dimension (length). For each point in the prototype, there is a corresponding point in the model, and these corresponding points are called *homologous points*. Values of variables at homologous points are related quantitatively by similarity rules derived from dimensional analysis. Complete geometric similarity is not always possible to obtain; for example, it might not be possible to reduce the surface roughness in a small model to maintain complete geometric similarity. Also, in the study of sediment transport, it might not be possible to scale down the bed material in the model without having bed material so fine as to be impractical. Fine powder, because of the cohesive forces between particles, does not simulate the behavior of sand.

Kinematic similarity. *Kinematic similarity* requires that length and time scales be similar between the model and the prototype, implying that velocities at corresponding points are similar. When fluid motions are kinematically similar, the patterns formed by streamlines are geometrically similar, and because impermeable boundaries also represent streamlines, kinematically similar flows are possible only past geometrically similar boundaries. Therefore, geometric similarity is a necessary, but not sufficient, condition for kinematic similarity. A common application of kinematic similarity is in turbomachines, where the length scale is measured by the size of the rotor, the time scale is measured by the inverse of the rotational speed, and the same ratio of length scale to time scale in a model and prototype results in kinematic similarity. In the case of turbomachines, kinematic similarity is usually more influential than dynamic similarity in achieving similitude between a model and a prototype.

Dynamic similarity. Dynamic similarity requires the similarity of all forces acting on fluid elements within the model and prototype systems. Therefore, dynamic similarity is achieved when

$$\left(\frac{F_{\rm I}}{F_i}\right)_{\rm p} = \left(\frac{F_{\rm I}}{F_i}\right)_{\rm m} \quad \text{or} \quad \left(\frac{F_i}{F_{\rm I}}\right)_{\rm p} = \left(\frac{F_i}{F_{\rm I}}\right)_{\rm m} \tag{6.14}$$

where F_i the $i^{\rm th}$ force on a fluid element and $F_{\rm I}$ is the inertial force of the fluid element. Equation 6.14 must apply to all relevant forces in the model and prototype systems. For example, if pressure differences, viscosity, and gravity influence the fluid motion in both the model and prototype, then dynamic similarity requires that

$$\left[\frac{\Delta p}{\frac{1}{2}\rho V^2}\right]_{\rm p} = \left[\frac{\Delta p}{\frac{1}{2}\rho V^2}\right]_{\rm m}, \qquad \left[\frac{\rho V L}{\mu}\right]_{\rm p} = \left[\frac{\rho V L}{\mu}\right]_{\rm m}, \qquad \left[\frac{V}{\sqrt{gL}}\right]_{\rm p} = \left[\frac{V}{\sqrt{gL}}\right]_{\rm m}$$

or equivalently, these force ratios can be expressed in terms of conventional dimensionless force ratios as

$$\Big[\text{Eu}\Big]_{\rm p} = \Big[\text{Eu}\Big]_{\rm m}, \qquad \Big[\text{Re}\Big]_{\rm p} = \Big[\text{Re}\Big]_{\rm m}, \qquad \Big[\text{Fr}\Big]_{\rm p} = \Big[\text{Fr}\Big]_{\rm m}$$

where Eu, Re, and Fr are the Euler number, Reynolds number, and Froude number, respectively. Several of the independent Π groups on the right-hand side of Equation 6.11, resulting from dimensional analysis of the relevant variables, typically involve the conventional ratios of forces, and equality of these Π groups in the model and prototype typically enforce *dynamic similarity* between the model and the prototype. In most models, dynamic similarity is only required for the dominant force(s). For example, in most open-channel models, the dominant force is gravity, and similitude is only enforced with respect to gravity forces while neglecting the effects of viscosity, surface tension, and other forces. Surface-tension effects in models are usually minimized by using model flow depths greater than 5 cm, and viscous effects are usually minimized by requiring that the flow be turbulent in the model, assuming that the flow is turbulent in the prototype. Distortions in model performance that result from neglecting certain similarity requirements are called *scale effects.*

EXAMPLE 6.9

The impact of major channel modifications on the water surface profile in a river are to be studied in the laboratory using a scale model. Dimensional analysis shows that the water surface profile in the river can be written in the functional form

$$\frac{y}{L} = f\left(\frac{x}{L}, \frac{V}{\sqrt{gL}}\right), \quad \text{or} \quad \frac{y}{L} = f\left(\frac{x}{L}, \text{Fr}\right)$$

where y is the depth of flow in the river at a distance x downstream from a reference location, L is the depth at the reference location, V is the velocity at the reference location, and Fr is the Froude number. The model is to be designed based on Froude number similarity, which is required because the dominant force is gravity. The model scale is 1:20, and the flow rate in the prototype is 50 m^3/s. (a) What flow rate should be used in the model? (b) If the reference depth, L, in the prototype is 5 m, what is the corresponding reference depth in the model? (c) If the measured water surface profile in the model is given in the following table, tabulate the water surface profile in the prototype.

x/L	0	1	10	100	1000
y/L	1.00	0.96	0.90	0.71	0.83

SOLUTION

Because the model and prototype are geometrically similar, the theory of models requires that when x/L and $V/\sqrt{gL}$ are the same in the model and prototype, then y/L is also the same in the model and the prototype. That is, using subscripts "m" and "p" to denote the model and prototype, respectively

$$\left(\frac{V}{\sqrt{gL}}\right)_{\text{m}} = \left(\frac{V}{\sqrt{gL}}\right)_{\text{p}}, \quad \text{and} \quad \left(\frac{x}{L}\right)_{\text{m}} = \left(\frac{x}{L}\right)_{\text{p}}$$

requires that

$$\left(\frac{y}{L}\right)_{\text{m}} = \left(\frac{y}{L}\right)_{\text{p}}$$

Also, because

$$Q \propto VL^2$$

the Froude similarity requirement can be written as

$$\left(\frac{VL^2}{\sqrt{gL^5}}\right)_{\text{m}} = \left(\frac{VL^2}{\sqrt{gL^5}}\right)_{\text{p}} \rightarrow \left(\frac{Q}{\sqrt{gL^5}}\right)_{\text{m}} = \left(\frac{Q}{\sqrt{gL^5}}\right)_{\text{p}}$$

which leads to

$$\frac{Q_{\text{m}}}{Q_{\text{p}}} = \left(\frac{L_{\text{m}}}{L_{\text{p}}}\right)^{\frac{5}{2}}$$

(a) Because $L_{\text{m}}/L_{\text{p}} = 1/20$ and $Q_{\text{p}} = 50$ m^3/s, the required flow rate in the model, Q_{m}, is given by

$$Q_{\text{m}} = Q_{\text{p}}\left(\frac{L_{\text{m}}}{L_{\text{p}}}\right)^{\frac{5}{2}} = 50\left(\frac{1}{20}\right)^{\frac{5}{2}} = 0.028 \text{ m}^3\text{/s} = \textbf{28 L/s}$$

(b) Because the reference depth, L_{p}, in the prototype is 5 m, the corresponding reference depth in the model, L_{m}, is given by

$$L_{\text{m}} = L_{\text{p}}\left(\frac{1}{20}\right) = 5\left(\frac{1}{20}\right) = 0.25 \text{ m} = \textbf{25 cm}$$

(c) Because the Froude number is the same in the model and the prototype, the nondimensional water surface profile is the same in both the model and prototype. This can be seen from the nondimensional functional relationship between y/L, x/L, and Fr given in the problem. Therefore, the water surface profile in the prototype can be calculated using the relations

$$x_{\text{p}} = \left(\frac{x}{L}\right)_{\text{m}} L_{\text{p}} = 5\left(\frac{x}{L}\right)_{\text{m}}, \quad \text{and} \quad y_{\text{p}} = \left(\frac{y}{L}\right)_{\text{m}} L_{\text{p}} = 5\left(\frac{y}{L}\right)_{\text{m}}$$

The following water surface profile in the prototype can be derived from the given water surface profile in the model:

x (m)	0	5	50	500	5,000
y (m)	5	4.8	4.5	3.6	4.2

Figure 6.3: Wind tunnel test of a model aircraft
Source: Jeff Caplan/NASA.

Applications. Wind tunnels are used to test the responses of various structures to high-velocity airflow around the structures. Applications of wind tunnels include studying, developing, and designing new aircraft and studying the behavior of buildings and combinations of buildings under hurricane- and tornado-force winds. Wind tunnels are also used to test the performance of submarines, with air replacing water as the test fluid and nondimensional similarity relationships used to relate the performance of a scale model in air to the performance of a full-size prototype in seawater. Notably, air traveling at a high speed can simulate the drag effect of a liquid at a lower speed. A typical wind tunnel being used to test a model of an aircraft under development is shown in Figure 6.3. The model aircraft shown in Figure 6.3 is 8.5% of the size of the (planned) prototype and is particularly novel in that it does not have a tail. Model testing is typically necessary to validate the stability and controllability of the aircraft.

EXAMPLE 6.10

The wind force on a particular type of building depends on the wind speed, the size of the building, and the density and viscosity of the air. (a) Express the relationship between these variables in terms of dimensionless groups. (b) A scale model is to be used to assess the forces on the building caused by 30-meter-per-hour winds on the full-scale building. If a one-half scale model of the building is tested in a wind tunnel, what airspeed must be used in the model? Using the required airspeed, the measured force on a particular component of the model is 100 N, what is the magnitude of the corresponding force in the full-scale building?

SOLUTION

(a) The relevant variables are force on building (F), wind speed (V), size of building (L), density of air (ρ), and viscosity of air (μ). The functional relationship can be expressed as

$$F = f(V, L, \rho, \mu)$$

This functional relationship involves five variables in three dimensions, which leads to two dimensionless groups. Using recognizable dimensionless groups gives the following functional relationship:

$$\frac{F}{\rho V^2 L^2} = f_1\left(\frac{\rho V L}{\mu}\right) \tag{6.15}$$

(b) The relationship between the prototype and the model is illustrated in Figure 6.4, where it should be clear that Equation 6.15 applies to both the model and the prototype.

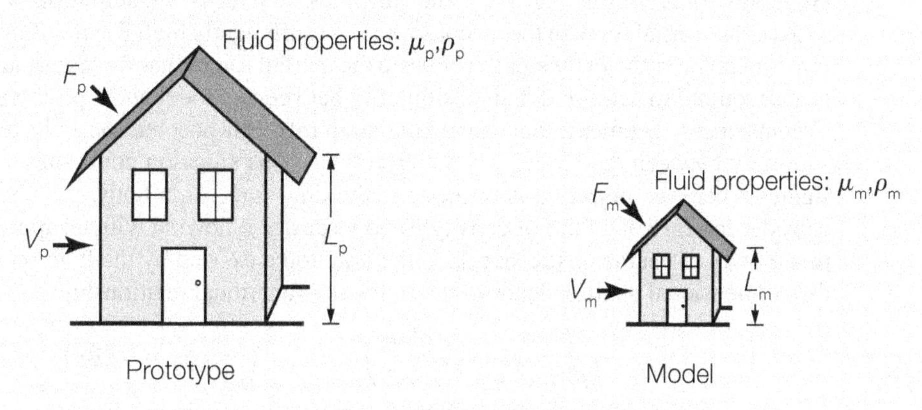

Figure 6.4: **Model building**

Based on Equation 6.15, the theory of modeling requires that if $\rho VL/\mu$ (= Reynolds number) is the same in the model and prototype, then $F/\rho V^2 L^2$ must also be the same in the model and prototype. So for model-prototype similarity, it is required that

$$\left[\frac{\rho V L}{\mu}\right]_{\mathrm{m}} = \left[\frac{\rho V L}{\mu}\right]_{\mathrm{p}} \quad \rightarrow \quad \frac{\rho_{\mathrm{m}} V_{\mathrm{m}} L_{\mathrm{m}}}{\mu_{\mathrm{m}}} = \frac{\rho_{\mathrm{p}} V_{\mathrm{p}} L_{\mathrm{p}}}{\mu_{\mathrm{p}}}$$

where the subscripts "m" and "p" refer to model and prototype, respectively. Rearranging this scaling equation gives

$$V_{\mathrm{m}} = \left[\frac{\rho_{\mathrm{p}}}{\rho_{\mathrm{m}}}\right]\left[\frac{L_{\mathrm{p}}}{L_{\mathrm{m}}}\right]\left[\frac{\mu_{\mathrm{m}}}{\mu_{\mathrm{p}}}\right] V_{\mathrm{p}}$$

In this case, $\rho_{\mathrm{p}} = \rho_{\mathrm{m}}$, $\mu_{\mathrm{p}} = \mu_{\mathrm{m}}$, $L_{\mathrm{p}} = 2L_{\mathrm{m}}$, and V_{p} = 30 m/s, so the scaling equation requires that

$$V_{\mathrm{m}} = (1)(2)(1)30 = \mathbf{60\ m/s}$$

Based on the required Reynolds number similarity, the airspeed in the wind tunnel containing the model must be 60 m/s. As a consequence,

$$\left[\frac{F}{\rho V^2 L^2}\right]_{\mathrm{m}} = \left[\frac{F}{\rho V^2 L^2}\right]_{\mathrm{p}}$$

which gives (with F_{m} = 100 N),

$$F_{\mathrm{p}} = \left(\frac{\rho_{\mathrm{p}}}{\rho_{\mathrm{m}}}\right)\left(\frac{V_{\mathrm{p}}}{V_{\mathrm{m}}}\right)^2\left(\frac{L_{\mathrm{p}}}{L_{\mathrm{m}}}\right)^2 F_{\mathrm{m}} \quad \rightarrow \quad F_{\mathrm{p}} = (1)\left(\frac{30}{60}\right)^2\left(\frac{2}{1}\right)^2(100) = 100\ \mathrm{N}$$

Therefore, the magnitude of the force on the prototype is **100 N**, which is the same as that measured in the model.

Model validation. When possible, predictions of prototype behavior derived from model studies should be validated by measurements on the prototype, if the prototype is constructed. Such validation data are sometimes available when model studies are being done to predict the performance of modifications to existing structures, in which case the satisfactory performance of the (unmodified) model of the existing structure serves to support the accuracy of the modified model. In some unusual cases, the scaling laws used in constructing and operating a model are tested by constructing multiple models and validating the scaling laws by comparing data collected from models at different scales.

High Reynolds number flows. The Reynolds number is, by definition, a measure of the ratio of the inertial force to the viscous force. At sufficiently high Reynolds numbers, viscous forces become sufficiently small relative to the inertial forces that Reynolds number similarity is not required to achieve dynamic similarity between a model and a prototype. Under these circumstances, geometric and kinematic similarity can become the only requirements for similarity between the model and the prototype. This situation commonly occurs in model studies where the objective is to measure the drag force on a body. As a typical example, consider the case of a fluid of density ρ and viscosity μ flowing with an approach velocity V past a body of characteristic size L. If the drag force exerted by the fluid on the body is F_{D}, then dimensional analysis leads to the following functional relationship:

$$\frac{F_{\mathrm{D}}}{\frac{1}{2}\rho V^2 L^2} = f\left(\frac{\rho V L}{\mu}\right) \quad \rightarrow \quad \frac{F_{\mathrm{D}}}{\frac{1}{2}\rho V^2 L^2} = f\,(\mathrm{Re}) \tag{6.16}$$

If the Reynolds number, Re, is sufficiently high, then the function on the right-hand side of Equation 6.16 must necessarily become independent of the Reynolds number, leading to the requirement that

$$\frac{F_{\mathrm{D}}}{\frac{1}{2}\rho V^2 L^2} = \text{constant} \tag{6.17}$$

The consequence of the high-Reynolds-number result given by Equation 6.17 is that the constant on the left-hand side of Equation 6.17 must be the same in both the model and the prototype; hence, the relationship between the drag force in the model and the drag force in the prototype under the high-Reynolds-number and geometric similarity conditions is given by

$$\left[\frac{F_{\mathrm{D}}}{\frac{1}{2}\rho V^2 L^2}\right]_{\mathrm{m}} = \left[\frac{F_{\mathrm{D}}}{\frac{1}{2}\rho V^2 L^2}\right]_{\mathrm{p}} \qquad (= C_{\mathrm{D}}) \tag{6.18}$$

where the subscripts "m" and "p" refer to the model and prototype, respectively. The ratio $F_{\mathrm{D}}/\frac{1}{2}\rho V^2 L^2$ is called the *drag coefficient*, and it is commonly represented by C_{D}. At low Reynolds numbers, C_{D} is dependent on Re as given by Equation 6.16, and if Re exceeds some critical value (e.g., $\mathrm{Re}_{\mathrm{crit}}$), then C_{D} is independent of Re, as given by Equation 6.17. The value of $\mathrm{Re}_{\mathrm{crit}}$ generally depends on the shape of the body.

EXAMPLE 6.11

The drag force on a particular body is to be estimated using a model study. Prototype conditions consist of air at 15°C flowing at a velocity, V, of 50 km/h toward a body with a characteristic size, L, of 2.25 m. The drag force on a body with this geometry becomes independent of the Reynolds number when the Reynolds number is greater than 10^5. The Reynolds number, Re, is defined as $\mathrm{Re} = VL/\nu$, where ν is the kinematic viscosity of the fluid. A geometrically similar model body with a characteristic size of 0.15 m is tested in air at 15°C with an approach velocity of 50 km/h. A drag force of 5 N is measured in the model. What is the corresponding drag force in the prototype?

SOLUTION

From the given data: $V_\text{p} = V_\text{m} = 50$ km/h $= 13.89$ m/s, $L_\text{p} = 2.25$ m, $\text{Re}_\text{crit} = 10^5$, $L_\text{m} = 0.15$ m, and $F_\text{m} = 5$ N. For air at 15°C, $\nu = 1.461 \times 10^{-5}$ m²/s (from Appendix B.2). Check the Reynolds number of the flow in both the prototype and the model.

$$\text{Re}_\text{p} = \frac{V_\text{p}L_\text{p}}{\nu_\text{p}} = \frac{(13.89)(2.25)}{1.461 \times 10^{-5}} = 2.14 \times 10^6, \qquad \text{Re}_\text{m} = \frac{V_\text{m}L_\text{m}}{\nu_\text{m}} = \frac{(13.89)(0.15)}{1.461 \times 10^{-5}} = 1.43 \times 10^6$$

Because both Re_p and Re_m are greater than 10^5, viscous effects are not important in the model design, and hence, the relationship between the drag forces in the model and prototype are given by Equation 6.18 as

$$\left[\frac{F_\text{D}}{\frac{1}{2}\rho V^2 L^2}\right]_\text{m} = \left[\frac{F_\text{D}}{\frac{1}{2}\rho V^2 L^2}\right]_\text{p} \quad \rightarrow \quad \left[\frac{F_\text{D}}{V^2 L^2}\right]_\text{m} = \left[\frac{F_\text{D}}{V^2 L^2}\right]_\text{p}$$

Substituting the known values of the parameters gives

$$\left[\frac{5}{(13.89)^2(0.15)^2}\right]_\text{m} = \left[\frac{F_\text{D}}{(13.89)^2(2.25)^2}\right]_\text{p} \quad \rightarrow \quad F_{\text{Dp}} = 1.125 \times 10^3 \text{ N} \approx 1.13 \text{ kN}$$

Therefore, the drag force on the prototype corresponding to a drag force of 5 N on the model is estimated as **1.13 kN**.

Testing of model vehicles. Models of automobiles are typically tested in wind tunnels at $\frac{3}{8}$ scale. At this scale, a typical model automobile has a frontal area of around 0.3 m² and can be tested in wind tunnels with a test section area of 6 m² or larger. Using a $\frac{3}{8}$ scale model, a wind speed of around 240 km/h is needed to model a prototype automobile moving at the typical highway speed limit. At this speed, compressibility effects are negligible in both the model and the prototype and Reynolds similarity is an adequate basis for model and experimental design. Trucks and buses are typically tested at $\frac{1}{8}$ scale, which typically requires model wind speeds of up to 700 km/h to achieve Reynolds similarity. At such wind speeds, compressibility effects are significant in the model but not in the prototype. Fortunately, the drag forces on bluff bodies such as trucks and buses are independent of the Reynolds number once a minimum Reynolds number is exceeded; therefore, lower wind speeds can be used in the model to determine the drag characteristics in both the model and the prototype.

Simultaneous Reynolds and Mach similarity. In aerodynamic applications where both viscous and compressibility effects are important, simultaneous Reynolds and Mach similarity is required to achieve dynamic similarity between the prototype and the model. Such cases usually occur in model testing of high-speed aircraft in wind tunnels and is a particular issue when small-scale models are used. Reynolds and Mach similarity require that

$$\text{Re}_\text{m} = \text{Re}_\text{p} : \quad \frac{V_\text{m}L_\text{m}}{\nu_\text{m}} = \frac{V_\text{p}L_\text{p}}{\nu_\text{p}} \quad \rightarrow \quad \frac{V_\text{m}}{V_\text{p}} = \frac{\nu_\text{m}}{\nu_\text{p}}\frac{L_\text{p}}{L_\text{m}} \tag{6.19}$$

$$\text{Ma}_\text{m} = \text{Ma}_\text{p} : \quad \frac{V_\text{m}}{c_\text{m}} = \frac{V_\text{p}}{c_\text{p}} \quad \rightarrow \quad \frac{V_\text{m}}{V_\text{p}} = \frac{c_\text{m}}{c_\text{p}} \tag{6.20}$$

where c_m and c_p are the sonic speeds in the model and prototype fluids, respectively. Comparing Equations 6.19 and 6.20 shows that to achieve Reynolds number and Mach number similarity, the kinematic viscosities in the model and prototype must be related such that

$$\frac{\nu_\text{m}}{\nu_\text{p}} = \left(\frac{c_\text{m}}{c_\text{p}}\right)\left(\frac{L_\text{m}}{L_\text{p}}\right) \tag{6.21}$$

Therefore, as the model scale deviates from the prototype scale, it is usually desirable to adjust the test fluid such that sonic speed in the test fluid increases relative to the sonic speed in the prototype fluid (usually air) to compensate. Commonly, air is the working fluid in both the model and the prototype, and cooling and pressurizing of the air in the wind tunnel are used to increase $c_{\rm m}$ relative to $c_{\rm p}$. For models with scales of 1:10 and smaller, it is usually impossible to satisfy Equation 6.21. Under such circumstances, where compressibility is important in both the model and the prototype, Mach similarity is usually preferentially chosen and viscous effects are estimated by some other means. It might seem that using a different fluid (e.g., a different gas) in the model might be a reasonable approach to satisfying Equation 6.21. However, a different gas will likely have a different specific heat ratio, k, which would cause the flows not to be dynamically similar, even if the Reynolds and Mach numbers were the same. In cases where Ma ≤ 0.3 in the prototype and model, Mach similarity is not required. However, if Reynolds similarity requires that Ma > 0.3 in the model while Ma ≤ 0.3 in the prototype, then the properties of the working fluid in the model will need to be adjusted such that Ma ≤ 0.3 in the model. If this cannot be achieved, the model and prototype will not be dynamically similar, even if Reynolds similarity is enforced.

Simultaneous Reynolds and Froude similarity. An issue that might be confronted in modeling a liquid flow that has a free surface is how to construct a model of a free-surface system in which both viscous and gravity forces are important. An example of where this situation occurs is the design of physical models to study the drag forces on ships, where viscous forces cause *skin friction* on the hull of the ship and gravity forces cause *surface-wave resistance* as the ship moves through water. Fundamentally, surface-wave resistance is generated because a ship must generate the force that is required to lift the water as it moves, and this "lifted" water results in surface waves. Because both skin friction and surface-wave resistance are present as a ship moves through water, the Reynolds number and the Froude number must be the same in the model and the prototype. These requirements can be expressed as

$$\mathrm{Re}_{\rm m} = \mathrm{Re}_{\rm p}: \quad \frac{V_{\rm m}L_{\rm m}}{\nu_{\rm m}} = \frac{V_{\rm p}L_{\rm p}}{\nu_{\rm p}} \quad \rightarrow \quad \frac{V_{\rm m}}{V_{\rm p}} = \frac{\nu_{\rm m}}{\nu_{\rm p}}\frac{L_{\rm p}}{L_{\rm m}} \tag{6.22}$$

$$\mathrm{Fr}_{\rm m} = \mathrm{Fr}_{\rm p}: \quad \frac{V_{\rm m}}{(gL_{\rm m})^{\frac{1}{2}}} = \frac{V_{\rm p}}{(gL_{\rm p})^{\frac{1}{2}}} \quad \rightarrow \quad \frac{V_{\rm m}}{V_{\rm p}} = \left(\frac{L_{\rm m}}{L_{\rm p}}\right)^{\frac{1}{2}} \tag{6.23}$$

Comparing Equations 6.22 and 6.23 shows that to achieve both Reynolds number and Froude number similarity, the kinematic viscosities in the model and prototype must be such that

$$\frac{\nu_{\rm m}}{\nu_{\rm p}} = \left(\frac{L_{\rm m}}{L_{\rm p}}\right)^{\frac{3}{2}} \tag{6.24}$$

Therefore, for anything other than a full-size model (i.e., $L_{\rm m} \neq L_{\rm p}$), different fluids must be used in the model and prototype, with a fluid of lesser viscosity required in the model when $L_{\rm m} < L_{\rm p}$. For typical length-scale ratios, finding a fluid to use in the model that satisfies Equation 6.24 is usually impractical or impossible. As a consequence, such models are usually constructed using either Reynolds or Froude similarity depending on which is deemed more important relative to the phenomenon being studied. Specialized analytic techniques are then used to take into account the effects that are not being modeled properly. In the case of modeling the drag on ships, it is commonplace to design model tests using Froude similarity so that surface-wave resistance can be modeled accurately; a theoretical (numerical) model is then used to remove the effect of skin friction in the model and predict the amount of skin friction in the prototype. Length-scale ratios of 1:100 are typically used in the testing of ship models.

Distorted models. Models that do not maintain a constant length-scale ratio in all dimensions are called *distorted models*. Distorted models are usually necessary when a single force (e.g., gravity) is dominant at the prototype scale, but a geometrically similar model sized using the appropriate scaling law (e.g., Froude number) results in multiple forces becoming important in the model (e.g., gravity, viscosity, surface tension), thereby leading to incomplete similarity. In models of open-channel flow, a common circumstance is where Froude number similarity is the appropriate scaling law but the resulting laboratory-scale models are sufficiently small that both the Froude number and Reynolds number are important in the model. To address this limitation, models of rivers, harbors, estuaries, and reservoirs are usually distorted and have horizontal-scale ratios that range from 1:2000 to 1:100 and vertical-scale ratios that range from 1:150 to 1:50. In such cases, model flow ways are deeper relative to their width than in the prototype. Under such conditions, bottom friction effects are smaller in the model than in the prototype, thereby leading to *incomplete similarity*. To partially account for this discrepancy, flow ways in the model are sometimes given increased roughness to create increased frictional effects that are more comparable to the actual shallower flow ways. In addition to such physical adjustments, empirical corrections and correlations are sometimes required to scale up model data properly in distorted models.

Example of a distorted model. A portion of the U.S. Army Corps of Engineers distorted model of San Francisco Bay (locally called the *Bay Model*) is shown in Figure 6.5. The model is a three-dimensional hydraulic model of San Francisco Bay and Delta areas that is capable of simulating tides and currents. The model area shown in Figure 6.5 is the main portion of San Francisco Bay, the far water area at top of image is the Pacific Ocean, the bridge in the center of the image is the Golden Gate Bridge (the "Bay Bridge"), and to the left of the Bay Bridge is the deep water channel leading to the Port of Oakland. The Bay Model is over 4000 m^2 in size, the horizontal scale of the model is 1:1000, and the vertical scale

Figure 6.5: Distorted physical model of San Francisco Bay
Source: Larry Quintana, U.S. Army Corps of Engineers.

is 1:100; hence the model is distorted by a factor of ten between the horizontal and vertical scales. The effects of model distortion are corrected by the use of copper strips throughout the model, with the number of copper strips being adjusted to calibrate the model.

Periodic flows. In modeling prototype conditions with periodic flows, the periodicity of a flow is typically characterized by its frequency, ω. Dimensional analysis of the relevant variables usually results in a similarity variable being the dimensionless group $\omega L/V$, where L and V are the characteristic length and velocity scales, respectively. The dimensionless group $\omega L/V$ is called the *Strouhal number*.

EXAMPLE 6.12

A 10-m-diameter rotor for a wind turbine is to be designed for use in an area where the characteristic wind speed is 40 km/h. The rotor will be restricted to rotating at 8 rpm, and the turbine is expected to deliver 550 kW under design conditions. For the rotor shape being considered, viscous effects are negligible for Reynolds numbers greater than 10^5. A 1:20 scale model of the rotor is tested in a wind tunnel with an airspeed that is the same as in the prototype. (a) What rotational speed should be used in the model? (b) What is the power output expected in the model? Assume standard air at sea level.

SOLUTION

From the given data: $L_\text{p} = 10$ m, $V_\text{p} = V_\text{m} = 40$ km/h $= 11.11$ m/s, $\omega_\text{p} = 8$ rpm, $P_\text{p} = 550$ kW, $L_\text{r} = 20$, and $\text{Re}_\text{crit} = 10^5$. For standard air at sea level, $\nu = 1.461 \times 10^{-5}$ m^2/s. A dimensional analysis of the relevant variables yields the following functional relationship:

$$\frac{P}{\frac{1}{2}\rho V^3 L^2} = f\left(\frac{\omega L}{V}, \frac{VL}{\nu}\right) \quad \rightarrow \quad \frac{P}{\frac{1}{2}\rho V^3 L^2} = f_1\,(\text{St}, \text{Re})$$

Check the Reynolds numbers in the prototype and model to determine whether viscous effects can be neglected.

$$\text{Re}_\text{p} = \frac{V_\text{p} L_\text{p}}{\nu} = \frac{(11.11)(10)}{1.461 \times 10^{-5}} = 7.60 \times 10^6,$$

$$\text{Re}_\text{m} = \frac{V_\text{m} L_\text{m}}{\nu} = \frac{(11.11)(10/20)}{1.461 \times 10^{-5}} = 3.80 \times 10^5$$

Because $\text{Re}_\text{p} > 10^5$ and $\text{Re}_\text{m} > 10^5$, viscous effects can be neglected in the model and prototype. The relevant functional relationship for model-prototype similarity becomes

$$\frac{P}{\frac{1}{2}\rho V^3 L^2} = f_2\,(\text{St})$$

Therefore, the model tests should be designed in accordance with Strouhal number similarity.

(a) Strouhal number similarity requires that

$$\left[\frac{\omega L}{V}\right]_\text{p} = \left[\frac{\omega L}{V}\right]_\text{m} \quad \rightarrow \quad \left[\frac{(8)(10)}{(40)}\right]_\text{p} = \left[\frac{\omega(10/20)}{40}\right]_\text{m} \quad \rightarrow \quad \omega_\text{m} = 160 \text{ rpm}$$

Therefore, the rotor speed in the model should be set at **160 rpm**.

(b) If Strouhal number similarity is enforced, then

$$\left[\frac{P}{\frac{1}{2}\rho V^3 L^2}\right]_{\rm p} = \left[\frac{P}{\frac{1}{2}\rho V^3 L^2}\right]_{\rm m} \quad \rightarrow$$

$$\left[\frac{550}{(40)^3(10)^2}\right]_{\rm p} = \left[\frac{P}{(40)^3(10/20)^2}\right]_{\rm m} \quad \rightarrow \quad P_{\rm m} = 1.38\ \text{kW}$$

Therefore, the power output from the model is expected to be **1.38 kW**.

Key Equations for Dimensional Analysis and Similitude

The following list of equations is particularly useful in solving problems related to dimensional analysis and similitude. If one is able to recognize these equations and recall their appropriate use, then the learning objectives of this chapter have been met to a significant degree. Derivations of these equations, definitions of the variables, and detailed examples of usage can be found in the main text.

MODELING AND SIMILITUDE

Dimensional analysis: $\Pi_0 = f(\Pi_1, \Pi_2, \ldots, \Pi_N)$

Similitude: $(\Pi_0)_\mathrm{p} = (\Pi_0)_\mathrm{m}$

Reynolds and Mach similarity: $\dfrac{\nu_\mathrm{m}}{\nu_\mathrm{p}} = \left(\dfrac{c_\mathrm{m}}{c_\mathrm{p}}\right)\left(\dfrac{L_\mathrm{m}}{L_\mathrm{p}}\right)$

Reynolds and Froude similarity: $\dfrac{\nu_\mathrm{m}}{\nu_\mathrm{p}} = \left(\dfrac{L_\mathrm{m}}{L_\mathrm{p}}\right)^{\frac{3}{2}}$

Section 6.2: Dimensions in Equations

6.1. It is known from engineering analysis that for a liquid flowing upward in a vertical pipe, the gauge pressure, p, in the a pipe at a height z above the liquid surface in the source reservoir is given by

$$p = -\gamma\left(1 + 0.24\frac{Q^2}{gD^5}\right)z \tag{6.25}$$

where γ is the specific weight of the liquid, Q is the volume flow rate ($\mathrm{L^3T^{-1}}$), and D is the diameter of the pipe. Show that Equation 6.25 is dimensionally homogeneous.

6.2. When a liquid of depth h flows in a wide inclined channel that makes an angle θ with the horizontal, the velocity u in the flow direction at a distance y from the bottom of the channel can be estimated by the relation

$$u(y) = \frac{\rho g}{\mu}\sin\theta\left(hy - \frac{y^2}{2}\right) \tag{6.26}$$

where ρ and μ are the density and viscosity of the liquid, respectively, and g is the gravity constant. Determine whether Equation 6.26 is dimensionally homogeneous.

6.3. The displacement x [L] of a vibrating particle and a function of time t [T] can sometimes be described by the equation

$$\frac{\mathrm{d}^2x}{\mathrm{d}t^2} + a\frac{\mathrm{d}x}{\mathrm{d}t} + bx = c$$

What are the dimensions of a, b, and c that are required for this equation to be dimensionally homogeneous?

6.4. If a heated object of mass m is placed in a liquid that is maintained at a temperature T_ℓ, the temperature, T, of the object as a function of time, t, can be estimated using the relation

$$\frac{\mathrm{d}T}{\mathrm{d}t} = \frac{h}{mc}(T - T_\ell)$$

where c is the specific heat of the object and h is the heat transfer coefficient. In a typical application, the units of the variables are as follows: T (°C), t (s), m (kg), c (kJ/kg·°C), and T_ℓ (°C). In what units should h be expressed?

6.5. Consider the energy equation for flow of a fluid in a closed conduit, which can be expressed in the form

$$\frac{p_1}{\gamma} + \frac{V_1^2}{2g} + z_1 = \frac{p_2}{\gamma} + \frac{V_2^2}{2g} + z_2 + h_\mathrm{f} \tag{6.27}$$

where p_1, V_1, and z_1 are the pressure, average velocity, and centerline elevation at an upstream section of the conduit, p_2, V_2, and z_2 are the corresponding variables at a downstream section, h_f is the head loss between the two sections, γ is the specific weigh of the fluid, and g is the gravity constant. State the SI unit of energy and state the SI unit(s) of the terms (not the individual variables) in Equation 6.27. Explain why each of the terms in Equation 6.27 represents energy and express Equation 6.27 as a nondimensional equation.

6.6. The Reynolds number, Re, of the flow at any cross section of a closed conduit is defined by the relation

$$\mathrm{Re} = \frac{\rho \overline{V} D_\mathrm{h}}{\mu}$$

where ρ is the density of the fluid, $\overline{V}$ is the average velocity, D_h is the hydraulic diameter, and μ is the dynamic viscosity. (a) Determine the dimensions of Re. (b) Determine the Reynolds number at a section of pipe with a hydraulic diameter of 100 mm where methane is flowing at a velocity of 25 m/s. The temperature of the methane is 20°C, and the pressure of the methane is standard sea-level pressure of 101 kPa.

6.7. The Froude number, Fr, at any cross section of an open channel is defined by the relation

$$\mathrm{Fr} = \frac{\overline{V}}{\sqrt{g D_\mathrm{h}}}$$

where $\overline{V}$ is the average velocity, g is the acceleration due to gravity, and D_h is the hydraulic depth. The hydraulic depth is defined as A/T, where A is the flow area and T is the top width of the flow area. (a) Show that Fr is dimensionless. (b) Determine the value of Fr in a trapezoidal channel that has a bottom width of 3 m, side slopes 2.5:1 (H:V), an average velocity of 0.4 m/s, and a flow depth of 1.5 m.

6.8. One-dimensional flow in a horizontal conduit is illustrated in Figure 6.6, where the velocity profile is symmetric about the centerline of the conduit and varies with time. The differential equation describing the velocity profile, $u(r,t)$, is given by

$$\rho \frac{\partial u}{\partial t} = \mu \left(\frac{1}{r}\frac{\partial u}{\partial r} + \frac{\partial^2 u}{\partial r^2} \right)$$

where ρ is the density of the fluid, t is time, μ is the dynamic viscosity of the fluid, and r is the radial distance from the conduit centerline. Consider the case where R is the radius of the conduit and the velocity fluctuates such that the maximum velocity is U and the temporal frequency of the oscillation is ω. Using R, ω, and U as reference variables, express the governing equation in normalized form.

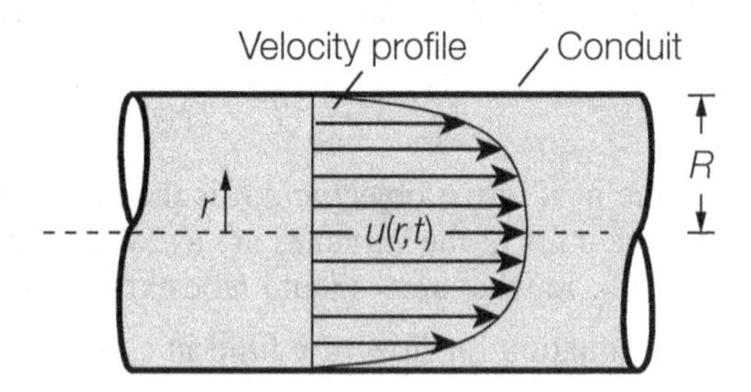

Figure 6.6: One-dimensional flow in a conduit

6.9. A two-dimensional flow field in the xy plane has velocity components u and v in the x- and y-directions, respectively, and the density, ρ, of the fluid varies in both space (x, y) and time, t. The continuity equation for this two-dimensional compressible flow is given by

$$\frac{\partial \rho}{\partial t} + \frac{\partial}{\partial x}(\rho u) + \frac{\partial}{\partial y}(\rho v) = 0 \tag{6.28}$$

The velocity, length, time, and density scales in the flow field are V, L, T, and ρ_0, respectively. It is common practice to represent the time scale, T, in terms of a frequency scale, ω, where $\omega = 2\pi/T$; in such cases, the time scale becomes ω^{-1}. (a) Express Equation 6.28 in normalized form. Use the Strouhal number, St, defined as $\text{St} = \omega L/V$, in the final expression. (b) Give the asymptotic form of the normalized continuity equation as the Strouhal number becomes small and state a physical scenario in which the Strouhal number might become negligibly small.

6.10. The momentum equation that describes the motion of a nonviscous incompressible fluid in a three-dimensional (xyz) flow field can be expressed in the form

$$\rho\left[\frac{\partial \mathbf{V}}{\partial t} + u\frac{\partial \mathbf{V}}{\partial x} + v\frac{\partial \mathbf{V}}{\partial y} + w\frac{\partial \mathbf{V}}{\partial z}\right] = -\boldsymbol{\nabla} p - \rho g\mathbf{k} \tag{6.29}$$

where ρ is the density of the fluid, $\mathbf{V}$ is the velocity vector with components u, v, and w, p is the pressure, and g is the gravity constant. Consider the case in which the only relevant scales are the length scale, L, and the velocity scale, V. Note that time can be normalized by L/V and pressure can be normalized by ρV^2. Express Equation 6.29 in normalized form, using the Froude number, Fr, defined as $\text{Fr} = V/\sqrt{gL}$, in the final expression. What happens to the effect of gravity on the flow as the Froude number becomes large?

6.11. In a particular two-dimensional flow field of an incompressible fluid in the xz plane, the z component of the momentum equation is given by

$$\rho u\frac{\partial w}{\partial x} = \mu\left(\frac{\partial^2 w}{\partial x^2} + \frac{\partial^2 w}{\partial z^2}\right) - \rho g \tag{6.30}$$

where u and w are the x and z components of the velocity, respectively, ρ and μ are the density and dynamic viscosity of the fluid, respectively, and g is the gravity constant. The relevant scales are the length scale, L, and the velocity scale, V. Express Equation 6.30 in normalized form, using the Reynolds number, Re, defined as $\text{Re} = \rho VL/\mu$, and the Froude number, Fr, defined as $\text{Fr} = V/\sqrt{gL}$, in the final expression. What is the asymptotic form of the governing equation as the Reynolds number becomes large?

6.12. The temperature distribution, $T(x, y)$, within a two-dimensional flow field of an incompressible and nonviscous fluid is governed by the following (energy) equation:

$$\rho c_p\left[u\frac{\partial T}{\partial x} + v\frac{\partial T}{\partial y}\right] = k\left(\frac{\partial^2 T}{\partial x^2} + \frac{\partial^2 T}{\partial y^2}\right) \tag{6.31}$$

where ρ and c_p are the density and specific heat of the fluid, respectively, u and v are the velocity components in the x- and y-directions, respectively, and k is the thermal conductivity of the fluid. The relevant scales are the length scale, L; the velocity scale, V; and the temperature scale, T_0. Express Equation 6.31 in normalized form, using the Prandtl number, Pr, defined as $\text{Pr} = \mu c_p/k$, and the Reynolds number, Re, defined as $\text{Re} = \rho VL/\mu$, where μ is the dynamic viscosity of the fluid.

Section 6.3: Dimensional Analysis

6.13. The velocity distribution, $V(r)$, in a pipe with fully developed laminar flow is given by

$$V(r) = \frac{\Delta p}{4\mu L}\left(\frac{D^2}{4} - r^2\right)$$

where Δp is the pressure drop over a distance L, μ is the dynamic viscosity, D is the diameter of the pipe, and r is the radial distance from the pipe centerline. Derive this functional relationship using the Buckingham pi theorem.

6.14. Under laminar flow conditions, the average velocity, V, in a pipe of diameter D depends on the pressure gradient in the direction of flow, $\partial p/\partial x$, and the dynamic viscosity, μ, of the fluid. Use dimensional analysis to determine the functional relationship between V and the influencing variables.

6.15. The force F on a reducer is related to the flow conditions and the cross-sectional areas of the inflow and outflow conduits by

$$F = p_1 A_1 - p_2 A_2 - \rho A_1 V_1 (V_2 - V_1)$$

where p_1 and p_2 are the inflow and outflow pressure, respectively, A_1 and A_2 are the corresponding flow areas, and V_1 and V_2 are the corresponding velocities. Derive the given functional relationship using the Buckingham pi theorem.

6.16. When a liquid flows at a high velocity through a pipe, sudden closure of a valve in the pipe can cause the generation of a pressure wave of sufficient magnitude to damage the pipe. This phenomenon is called water hammer. The maximum water hammer pressure, p_{max}, depends on the pipe diameter, D, the velocity of flow in the pipe prior to valve closure, V, the density of the fluid prior to valve closure, ρ, and the bulk modulus of elasticity of the liquid, E_v. Use dimensional analysis to determine the relationship between p_{max} and the influencing variables.

6.17. A pump impeller of diameter D contains a mass m of fluid and rotates at an angular velocity of ω. Use dimensional analysis to obtain a functional expression for the centrifugal force F on the fluid in terms of D, m, and ω.

6.18. The energy per unit mass (of fluid), e, added by a pump of a given shape depends on the pump size, D, volume flow rate, Q, speed of the rotor, ω, density of the fluid, ρ, and dynamic viscosity of the fluid, μ. This functional relation can be stated as

$$e = f(D, Q, \omega, \rho, \mu)$$

Express this as a relationship between dimensionless groups. What is gained by expressing the pump performance as an empirical relationship between dimensionless groups versus expressing the pump performance as a relationship between the given dimensional variables?

6.19. The power required to drive a rotary hydraulic machine typically depends on the speed of rotation, the size of the rotor, and the density and viscosity of the working fluid. Determine the functional relationship between the power requirement and the influencing variables, expressed in terms of dimensionless groups. Identify any named conventional dimensionless groups.

6.20. An orifice plate is a flat plate with a central opening that is sometimes used to measure the volume flow rate in a pipe. The pressure drop across an orifice plate can be assumed to be a function of the diameter of the opening in the plate, the diameter of the pipe, the velocity of flow in the pipe, and the density and viscosity of the fluid. Use dimensional analysis to determine the functional relationship between the pressure drop across the plate and the influencing variables. Express this functional relationship in both a dimensional and nondimensional form. Identify any named conventional dimensionless groups that appear in the nondimensional relationship.

6.21. When freewheeling, the angular velocity, Ω, of a windmill is assumed to be a function of the windmill diameter, D, the wind velocity, V, the air density, ρ, air viscosity, μ, and the windmill height, H, as compared to the thickness of the atmospheric boundary layer, L, and the number of blades, N. This functional relationship can be expressed analytically as follows:

$$\Omega = f\left(D, V, \rho, \frac{H}{L}, N\right)$$

If viscosity effects are negligible, find the appropriate dimensionless groups and rewrite the given function in dimensionless form. Identify any conventional dimensionless groups.

6.22. The airflow in large wind tunnels is driven by large fans that exert a thrust force on their support structure. The thrust force, T, on a support structure can be expected to be a function of the density and viscosity of the air, the size of the fan, and the rate at which the fan rotates. Determine a nondimensional functional expression that relates the thrust generated by a fan to the influencing variables.

6.23. A model differential equation for chemical reaction dynamics in a plug-flow reactor is as follows:

$$u\frac{\partial C}{\partial x} = D\frac{\partial^2 C}{\partial x^2} - kC - \frac{\partial C}{\partial t}$$

where u is the velocity, D is the diffusion coefficient, k is the reaction-rate constant, x is the distance along the reactor, and C is the (dimensionless) concentration for a given chemical in the reactor. Determine the appropriate dimensions of D and k. Using a characteristic length scale L, average velocity V, and background concentration C_0 as parameters, rewrite the given equation in normalized form and comment on any dimensionless groups that appear.

6.24. A fluid of density ρ and dynamic viscosity μ flows with a velocity V toward a rectangular plate of width W, height H, and thickness T. The approaching fluid flow makes an angle θ with the direction normal to the plate. Determine the functional relationship between dimensionless groups that would be appropriate for studying the relationship between the drag force on the plate and the given independent variables.

6.25. For the flow condition shown in Figure 6.7, the velocity, u, at any distance y from a fixed bottom plate depends on the velocity V of the top plate, the spacing h between the plates, the dynamic viscosity μ of the fluid, and the pressure gradient $\mathrm{d}p/\mathrm{d}x$. (a) Use the Buckingham pi theorem to obtain a functional relationship between dimensionless groups. (b) A theoretical analysis postulates that the actual relationship between the variables in this problem is given by

$$u = \frac{Vy}{h} + \frac{1}{2\mu}\frac{\mathrm{d}p}{\mathrm{d}x}(y^2 - hy)$$

Nondimensionalize this equation and show that the resulting nondimensional equation validates your analysis in part (a).

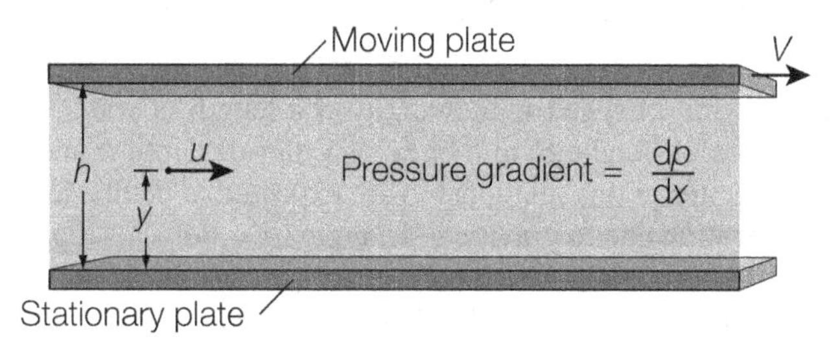

Figure 6.7: Flow between parallel plates

6.26. A cable of length L and diameter D is strung tightly between two poles. A fluid of density ρ and viscosity μ flows at a velocity V past the cable, producing a deflection δ. The modulus of elasticity of cable material is E, and the cable is sufficiently long that the geometry of the end poles does not affect the cable deflection. Determine a functional expression relating dimensionless groups that would be appropriate for studying the relationship between the cable deflection and the given independent variables.

6.27. Wind blowing at a steady velocity V toward a vertical column of diameter, D, generates velocity fluctuations behind the column that have a frequency ω. The frequency of the generated velocity fluctuations can also depend on the density and viscosity of the air. Determine a nondimensional functional relationship that relates the frequency of the velocity fluctuations to the influencing variables. Identify any named conventional dimensionless groups.

6.28. Standard air blows at 144 km/h past a 330-mm-diameter flagpole during an intense hurricane. (a) Estimate the magnitude of the viscous force relative to the inertial force of the wind. (b) Estimate the magnitude of the gravity force relative to the inertial force of the wind. (c) Partially based on the results in parts (a) and (b), estimate the relative importance of fluid viscosity and gravity in determining the force of the wind on the flagpole.

6.29. Air flowing past an object typically exerts a drag force on the object. For the given shape of the object, the drag force is commonly assumed to depend on the speed at which the air approaches the object, the size of the object, and the density and viscosity of the air. At sufficiently high velocities, the drag force asymptotically becomes independent of the viscosity of the fluid. Use dimensional analysis to express the asymptotic functional behavior of the drag force and state the functional relationship between the drag force and the influencing variables under the asymptotic condition.

6.30. A sphere of diameter D that falls in a fluid ultimately attains a constant velocity, called the terminal velocity, when the drag force exerted by the fluid is equal to the net weight of the sphere in the fluid. The net weight of a sphere can be represented by $\Delta\gamma\mathcal{V}$, where $\Delta\gamma$ is the difference between the specific weight of the sphere and the specific weight of the fluid and $\mathcal{V}$ is the volume of the sphere. In cases where the sphere falls slowly in a viscous fluid, the terminal velocity, V, depends only on the size of the sphere, D, the specific-weight difference $\Delta\gamma$, and the viscosity of the fluid, μ. Use dimensional analysis to determine the functional relationship between the terminal velocity and the other relevant variables.

6.31. When an object of a given shape falls very slowly in a fluid, the steady-state velocity of the object is observed to depend only on the size of the object and the density and viscosity of the fluid. Use dimensional analysis to determine the functional relationship between the steady-state velocity and the influencing variables.

6.32. The force, F, on a satellite in Earth's upper atmosphere depends on the mean path length of molecules, λ, the density, ρ, the diameter of the body, D, and the molecular speed, c. Express this functional relationship in terms of dimensionless groups.

6.33. Surface waves are generated by wind blowing over large bodies of water. It is postulated that a wind of speed V blowing over a length of water, F, called a "fetch," generates waves of height H in a body of water of depth h and density ρ_w. Other important variables in this relationship are expected to be the density of the air, ρ_air, and the acceleration due to gravity, g. Using h, ρ_w, and V as repeating variables, determine an appropriate relationship between dimensionless groups that can be used to concisely express the relationship between the variables.

6.34. The propagation speed, V, of small ripples in a shallow depth, h, of a liquid depends on the surface tension, σ, at the air-liquid interface, the density of the liquid, ρ, and the gravity constant, g. Determine the functional relationship between these variables in terms of dimensionless groups. If possible, also express this relationship in terms of named conventional dimensionless groups.

6.35. Observations on the International Space Station indicate that if a small spherical liquid droplet suspended in space is deformed slightly, then its surface will oscillate with a frequency that depends on the diameter of the droplet and the density and viscosity of the liquid. Use dimensional analysis to estimate the functional relationship between the oscillation frequency of the droplet surface and the influencing variables.

6.36. A 5-mm-deep film of SAE 30 oil flows steadily down an inclined plane at a velocity of 8 cm/s. (a) Estimate the magnitude of the viscous force relative to the inertial force. (b) Estimate the magnitude of the gravity force relative to the inertial force. (c) Estimate the magnitude of the surface-tension force relative to the inertial force. (d) Partially based on the results in parts (a), (b), and (c), estimate the relative importance of fluid viscosity, gravity, and surface tension on the flow rate of the oil.

6.37. The height of rise, h, of a fluid in a capillary tube depends on the specific weight, γ, the surface tension, σ, of the fluid, and the diameter, D, of the capillary tube. In other words,

$$h = f(\gamma, \sigma, D)$$

Express this relationship in terms of dimensionless groups. How does your result compare with the following theoretical expression?

$$h = \frac{4\sigma}{\gamma D}$$

6.38. The characteristic diameter of the liquid droplets delivered into the air from a spray bottle depends on the diameter of the jet in the spray bottle, the velocity of the spray jet, and the density, viscosity, and surface tension of the liquid. Determine a functional relationship, in terms of dimensionless groups, that relates the characteristic diameter of the liquid droplets to the influencing variables. Identify any named conventional dimensionless groups involved in this relationship.

6.39. It is well known that tiny objects can be supported on the surface of a liquid by surface-tension forces. For an object of a particular shape, the weight of the object that can be supported is a function of the size of the object, the surface tension and density of the fluid, and the gravity constant. Determine a nondimensional functional relationship that relates the weight of the supported object to the influencing variables.

6.40. An orifice in the side of an open storage tank discharges the stored liquid at an average velocity of V when the depth of liquid above the orifice is h. The liquid has a density and viscosity of ρ and μ, respectively, and the acceleration due to gravity is g. (a) Show by dimensional reasoning that if viscous effects are negligible, the density of the fluid does not influence the average discharge velocity. (b) If viscous effects are negligible, use dimensional analysis to determine the functional relationship between V and the influencing variables.

6.41. An orifice of diameter D in the side of an open storage tank discharges liquid at an average velocity of V when the depth of liquid above the orifice is h. The liquid has a density and viscosity of ρ and μ, respectively, and the acceleration due to gravity is g. If experiments are to be done to determine the relationship between these variables, identify the dimensionless groups that should be used in analyzing the experimental data.

6.42. The volume flow rate over a sharp-crested weir, Q, is commonly expressed in the form

$$Q = f\left(\text{We}, \text{Re}, \frac{H}{H_\text{w}}\right) b\sqrt{g}H^{\frac{3}{2}} \tag{6.32}$$

where We is the Weber number defined by

$$\text{We} = \frac{\rho V^2 H}{\sigma}$$

and V is the flow velocity over the weir, σ is the surface tension of the fluid, ρ is the density of the fluid, H is the depth of fluid over the crest of the weir, and Re is the Reynolds number defined by

$$\text{Re} = \frac{\rho V H}{\mu}$$

where μ is the viscosity of the fluid, H_w is the height of the weir, and b is the length of the weir crest. Derive Equation 6.32 using dimensional analysis.

6.43. A hydraulic jump is a phenomenon associated with high-velocity flow of a liquid in an open channel in which the flow depth of the liquid suddenly changes from a lower depth, h_1, to a higher depth, h_2. When the flow is occurring in a horizontal rectangular channel, the relationship between h_1 and h_2 depends only on the lower-depth velocity of flow, V_1, and the gravity constant, g. Use dimensional analysis to determine the functional relationship between h_2 and the influencing variables, expressed in terms of dimensionless groups. Identify any named conventional dimensionless groups that occur in this relationship.

6.44. A ship of length L moves at a cruising speed of V in a calm ocean where the density and viscosity of the seawater are ρ and μ, respectively. Synoptic experimental measurements of the drag force on the ship and the influencing variables are available. Determine the dimensionless groups that should be used to organize these data to determine an empirical relation between the drag force and the influencing variables. Why is it important to include gravity as a variable?

6.45. The thrust, T, generated by a ship's propeller is generally thought to be a function of the diameter, D, forward speed, V, and rate of rotation, ω, of the propeller, the density, ρ, and viscosity, μ, of the liquid in which the propeller is submerged, the pressure, p, at the level of the propeller, and the gravity constant, g. Determine a nondimensional functional relationship between the thrust, T, and its influencing variables. Identify any named conventional dimensionless groups that appear in this relationship.

6.46. The lift force on an airfoil depends on the forward speed of the airfoil, the angle the airfoil makes with the forward direction, the length of the airfoil, the thickness of the airfoil, and the density of the air. Determine a nondimensional functional relationship that relates the normalized lift force to the influencing variables.

6.47. The drag force on a supersonic aircraft of a particular shape depends on its size, its speed, the density of the air, and the sonic speed in the air. Determine a nondimensional functional relationship between the drag force and its influencing variables. Identify any named conventional dimensionless groups.

6.48. The length, L_w, of the wake behind a particular model aircraft depends on the speed of the aircraft, the size of the aircraft, L, and the density and viscosity of the ambient air. Develop a nondimensional functional relationship that relates the size of the wake to its influencing variables.

6.49. The power, P, required to drive the propeller on a turboprop aircraft is thought to be a function of the diameter, D, forward speed, V, and rate of rotation, ω, of the propeller, the density, ρ, and viscosity, μ, of the air, and the sonic speed, c, in the air. Determine a nondimensional functional relationship between the power, P, and its influencing variables. Identify any named conventional dimensionless groups that appear in this relationship.

Section 6.4: Dimensionless Groups as Force Ratios

6.50. Show that the ratio of the inertial force to the net pressure force is measured by the Euler number.

6.51. Show that the ratio of the inertial force to the surface tension force is measured by the Weber number.

6.52. Show that the ratio of the inertial force to the compressibility force is measured by the Cauchy number.

6.53. An atomizer is a common term used to describe a spray nozzle that produces small droplets of a liquid that is drawn into the nozzle by the low pressure that exists in the nozzle. Consider a nozzle of diameter D that generates droplets of diameter d when the velocity at the nozzle exit is V. The relevant liquid properties are the density, ρ, the viscosity, μ, and the surface tension, σ. With the objective of predicting the size of the droplets generated by an atomizer, express the relationship between the relevant variables in nondimensional form where, to the extent possible, dimensionless groups representing force ratios are used with each of the fluid properties.

Dimensional analysis of laboratory data:

6.54. The performance of the expansion shown in Figure 6.8 was studied in the laboratory to determine the relationship between the pressure rise, Δp, across the expansion and the diameter ratio D_1/D_2, where D_1 and D_2 are the upstream and downstream diameters, respectively. The fluid is water at 20°C. (a) Assuming that the flow is fully turbulent, perform a dimensional analysis to determine the functional relationship between dimensionless groups to be investigated in the laboratory. (b) A laboratory investigation of the expansion is performed using a fixed upstream diameter of 150 mm and a constant flow rate of 20 L/s. The downstream diameter was varied, and the corresponding increases in pressure between the upstream and downstream sections were measured. The results are as follows:

D_2 (mm)	750	375	250	188
Δp (Pa)	326.5	275.2	174.6	81.8

Viscous effects were negligible for all flow conditions. Plot the measured data in nondimensional form and derive an empirical relationship that can be used to estimate the change in pressure for a given diameter ratio.

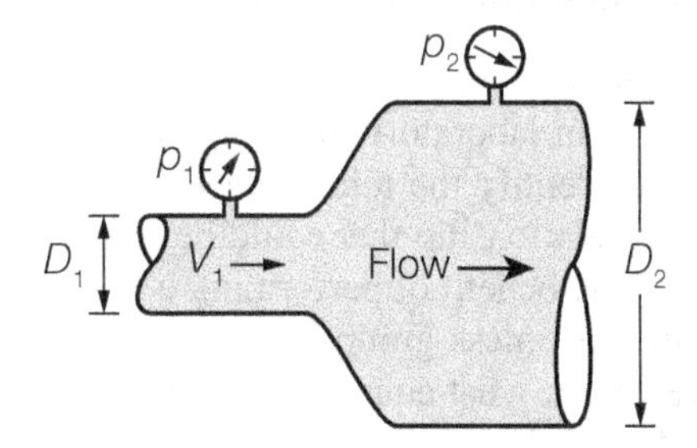

Figure 6.8: Performance of an expansion

6.55. The performance of the sudden contraction shown in Figure 6.9 was studied in the laboratory. The goal of the study was to determine the head loss coefficient, K, of the contraction, where the head loss, h_ℓ, in the contraction can be expressed as

$$h_\ell = K\frac{V_2^2}{2g}$$

where V_2 is the velocity in the section downstream of the contraction. The laboratory study was conducted by fixing the diameter, D_2, of the downstream section at 150 mm, using water at 20°C as the fluid, and fixing the flow rate at 42.41 L/s. The diameter, D_1, of the upstream section was varied between experiments, and the corresponding pressure differences, $p_1 - p_2$, between the upstream and downstream sections were measured. The measured laboratory data were as follows:

D_1 (mm)	$p_1 - p_2$ (Pa)	D_1 (mm)	$p_1 - p_2$ (Pa)
150	0	330	1630
165	796	375	1651
180	1287	450	1631
210	1637	600	1569
240	1688	750	1547
270	1652	1500	1523
300	1660	—	—

(a) Perform a dimensional analysis on the variables and present the laboratory results in an appropriate nondimensional form that relates the pressure change to the characteristics of the contraction, fluid, and flow. Plot the laboratory results in dimensionless form. (b) Use the results of the laboratory study to plot the head loss coefficient, K, as a function of the diameter ratio D_1/D_2. Estimate an empirical equation that relates K to D_1/D_2.

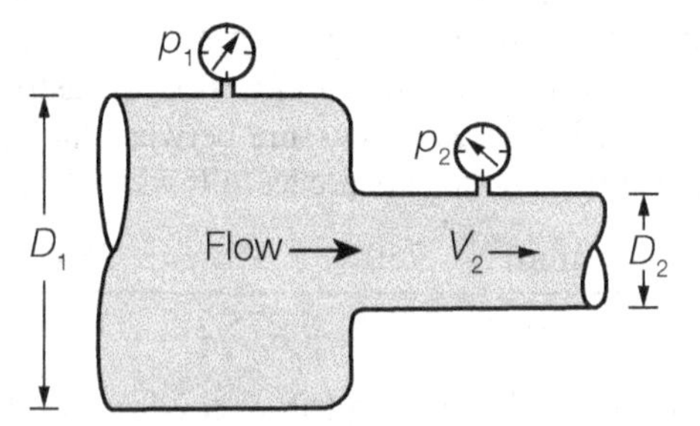

Figure 6.9: Performance of sudden contraction

Section 6.6: Modeling and Similitude

6.56. A rectangular duct is to be used to transport a very viscous fluid, and you have been commissioned to perform laboratory experiments to determine the shear stress on the walls of the duct. Identify the relevant variables and use dimensional analysis to derive a functional relationship that could be used as a basis for designing your experiments. Explain the benefit of performing a dimensional analysis and identify any conventional dimensionless groups that occur in your derived functional relationship. For a given fluid, what equation would be used to relate the fluid velocity in the model to the fluid velocity in the prototype to achieve dynamic similarity between the model and the prototype?

6.57. Show that if Reynolds number similarity is used in a model study and the same fluid is used in the model and prototype, the force on a particular component of the model will always be equal to the force on the corresponding component of the prototype.

6.58. The shear stress, τ_0, exerted on a pipe of diameter, D, by a fluid of density, ρ, and viscosity, μ, moving with a velocity, V, is given by the following dimensionless relation:

$$\frac{\tau_0}{\rho V^2} = f\left(\frac{\rho V D}{\mu}, \frac{\epsilon}{D}\right)$$

where ϵ is the height of the roughness elements on the surface of the pipe. In a prototype pipe, the velocity of flow is 2 m/s, the diameter is 3 m, and the height of the roughness elements is 2 mm. If a model of the pipe is to be constructed based on Reynolds number similarity (because the viscous force is important), a model scale of 1:20 is used, and the same fluid is used in both the model and prototype, what velocity and roughness height should be used in the model? If the shear stress on the pipe surface in the model is 2.25 kPa, what is the corresponding shear stress in the prototype?

6.59. The performance of a valve used to regulate the flow of water in a pipeline is to be tested using a $\frac{1}{2}$-scale model. At the prototype scale, the characteristic dimension of the valve is 220 mm and the design flow rate through the valve is 325 L/s. What scaling law should be used to design the model test? What is the model flow rate corresponding to the design flow rate in the prototype?

6.60. Kerosene at 20°C flows through a section of 170-mm-diameter pipeline with a velocity of 2.5 m/s. The pressure loss in the section of pipeline is to be studied using a scale model that has a diameter of 30 mm, and water at 20°C is to be used in the model study. What water velocity should be used in the model? If a pressure loss of 15 kPa is measured in the model, what is the corresponding pressure drop in the actual pipeline section?

6.61. Water flows out of a fire hose at 20 L/s, where the hose has a diameter of 20 cm and the nozzle at the end of the hose has an exit diameter of 5 cm. (a) What force must a firefighter exert to hold the nozzle in place? (b) If a scale model of the nozzle is constructed to have the same Froude number, with a model flow rate of 0.03 L/s, what must be the dimension of the model nozzle? What restraining force would you expect in the model nozzle?

6.62. Water with a kinematic viscosity of 10^{-6} m^2/s flows through a 15-cm-diameter pipe. Assuming that the viscosity of the fluid significantly affects the dynamics of the flow, what would the velocity of water have to be for the water flow to be dynamically similar to oil ($\nu = 10^{-5}$ m^2/s) flowing through the same pipe at a velocity of 3 m/s? What is meant by "dynamic similarity"?

6.63. The speed of sound, c, in a fluid that is contained within a conduit is given by

$$c = \sqrt{\frac{E_\mathrm{v}/\rho}{1 + \left(\dfrac{E_\mathrm{v}D}{eE}\right)}} \tag{6.33}$$

where E_v is the bulk modulus of elasticity of the fluid, ρ is the density of the fluid, D is the diameter of the conduit, e is the thickness of the conduit wall, and E is the modulus of elasticity of the pipe wall material. (a) Express Equation 6.33 as a relationship between dimensionless groups. (b) Perform a dimensional analysis on the variables in Equation 6.33 to determine a functional relationship between dimensionless groups. Identify any commonly used dimensionless groups. (c) Discuss whether your result in part (b) is confirmed by your result in part (a). (d) A model of a steel pipe is to be constructed to study the propagation of a pressure wave. The scale model is to have a diameter that is 1/20 the diameter of the prototype, and the model will use benzene at 20°C instead of water at 20°C in the prototype. If the pipe wall thickness in the prototype is 20 mm, what pipe wall thickness would you use in the model to yield dynamic similarity between pressure wave propagation in the model and in the prototype?

6.64. A performance of a prototype water pump is to be tested using a scale model. The prototype will have an impeller of diameter 260 mm, and when delivering 990 L/s, the difference in water pressure between the inlet and the outlet of the pump is expected to be 350 kPa. The impeller diameter in the model is to be 50 mm, and the model tests are to use water at the same temperature as that used in the prototype. (a) What dimensionless functional relationship would describe the performance of the pump? Express this relationship in terms of conventional dimensionless groups. (b) What flow rate should be used in the model to achieve dynamic similarity? (c) What would be the expected pressure rise between the inlet and the outlet in the model?

6.65. A prototype pump increases the pressure by 515 kPa when the mass flow rate through the pump is 640 kg/s. The performance of this pump is to be studied using a 1:7 scale model. (a) State the nondimensional relationship between the relevant variables that forms the basis of the model study. (b) If water at the same temperature is used in both the prototype and the model, what mass flow rate should be used in the model? (c) What pressure is expected to be added by the model pump?

6.66. A prototype water pump with an impeller diameter of 470 mm is designed to operate at a rotational speed of 950 rpm, and at this speed, the pump delivers a volume flow rate of 1.7 m^3/s. The temperature of the water in the prototype pump is 20°C. The performance of a 1:6 scale model of the pump is tested using standard air as the fluid. The model pump has a rotational speed of 1750 rpm, and the power required to drive the model pump is 95 W. The Reynolds number in the prototype and the model are both sufficiently high that Reynolds similarity is not a requirement. (a) Determine the volume flow rate of air in the model that corresponds to the design volume flow rate of water in the prototype. (b) Determine the power requirement of the prototype that corresponds to the measured power requirement in the model.

6.67. A manufacturer makes several fans in different sizes, but all have the same shape (i.e., they are geometrically similar). Such a series of models is called a homologous series. Two fans in the series are to be operated under dynamically similar conditions. The first fan has a diameter of 260 mm, has a rotational speed of 2500 rpm, and produces an airflow rate of 0.9 m^3/s. The second fan is to have a rotational speed of

1500 rpm and produce a flow rate of 3 m^3/s. What diameter should be selected for the second fan?

6.68. A prototype turbine is to be designed to deliver a power of 270 kW when the flow rate through the turbine is 3 m^3/s. The performance of this turbine is to be studied using a 1:9 scale model. (a) State the nondimensional relationship between the relevant variables that forms the basis of this investigation and identify the scaling relationship to be used in the model study. (b) If water at the same temperature is used in both the prototype and the model, what flow rate should be used in the model? (c) What power is expected to be produced by the model?

6.69. A new boomerang is being proposed that has a characteristic size of 215 mm, travels at an average velocity of 18 m/s, and rotates at 168 rpm. The performance of a model of this boomerang is to be tested in a wind tunnel, where the model is to be four times the size of the prototype and Reynolds number and Strouhal similarity laws are to be used in designing the experiments. What wind speed and boomerang rotation rates should be used in the laboratory tests?

6.70. In open-channel flow in a microchannel, both gravity and surface tension forces are important. A 1:15 scale model is to be constructed to study microchannel flow, where the fluid used in the model is to have the same surface tension as the fluid in the prototype. What similarity laws should be used in designing the model study? What is the required ratio of the density of the fluid in the model to the density of the fluid in the prototype to achieve dynamic similarity between the model and the prototype?

6.71. A 1:12 model of a spillway is tested under a particular upstream condition. The measured velocity and flow rate over the model spillway are 0.68 m/s and 0.12 m^3/s, respectively. What are the corresponding velocity and flow rate in the actual spillway?

6.72. A hydraulic model is to be constructed to study the performance of a spillway. The prototype spillway has a length of 26 m and has a design overflow rate of 144 m^3/s. The model scale is to be 1:18. (a) What scaling law should be used in designing the model study? (b) What flow rate should be used in the model? (c) If the force on a component of the spillway is measured in the model as 86 N, what is the force on the corresponding component in the prototype spillway?

6.73. A sluice gate is to be installed to control the flow in a coastal canal. The volume flow rate through the gate, Q, depends on the upstream water depth, h_1, the downstream water depth, h_2, the gate opening, s, the width of the gate, b, and the acceleration due to gravity, g. Use dimensional analysis to determine a functional relationship between dimensionless groups that can be used to guide laboratory model experiments on the gate. Identify the physical significance of each dimensionless group. A 1:7 scale model having a width of 1 m is constructed in the laboratory, and the following operating condition is observed in the model:

h_1 (m)	h_2 (m)	s (m)	Q (m^3/s)
0.57	0.50	0.16	0.14

Find the corresponding operating condition in the prototype. How would you justify neglecting the viscosity of the water in your analysis?

6.74. Water flows at 27 m^3/s in a 15-m-wide rectangular channel at a depth of 2.5 m. A 1:17 scale model of the channel is to be built in the laboratory. (a) What flow rate should be used in the model to achieve Reynolds similarity? (b) What flow rate should be used in the model to achieve Froude similarity? (c) Find the viscosity of the model fluid required to achieve both Reynolds and Froude similarity. Does such a fluid exist? (d) Explain the circumstances under which a Froude model would be preferable to a Reynolds model. (e) A force of 2 N is measured on a hydraulic structure placed in a Froude model. What is the corresponding force in the prototype?

6.75. A simple flow measurement device for streams and channels is a notch, of angle α, cut into the face of a dam as shown in Figure 6.10. The volume flow rate, Q, depends only on α, the acceleration due to gravity, g, and the height, δ, of the upstream water surface above the notch vertex. Tests of a model notch, of angle α = 55°, yield the following flow rate data:

δ (cm)	10	20	30	40
Q (m^3/h)	8	47	126	263

(a) Find a dimensionless correlation for the data. (b) Use the model data to predict the flow rate of a prototype notch, also of angle α = 55°, when the upstream height δ is 3.2 m.

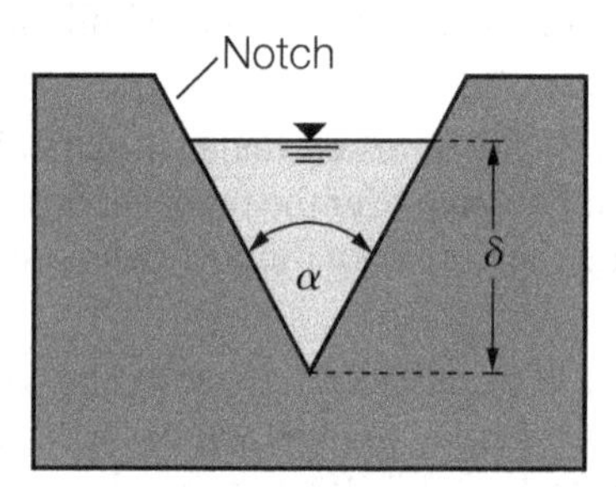

Figure 6.10: Flow over a V-notch weir

6.76. Consider the case where a particle of diameter d and density ρ_s rests on the bottom of an open channel where a liquid with density ρ and viscosity μ flows over the particle. The particle begins to move when the fluid velocity is V_c, and in this analysis, gravity, g, is a relevant variable. (a) Determine a dimensionless group of the given variables that measures the ratio of the fluid force tending to move the particle to the weight of the particle. Take into account that the particle does not weigh the same when submerged in a liquid as it does when not submerged. (b) Using this result, perform a dimensional analysis on the relevant variables to determine a functional relationship between dimensionless groups that can be used to analyze measurements in a scale model to measure the critical fluid velocity, V_c, to move a particle of a given size, d. In your dimensional analysis, incorporate the fact that ρ_s and g can be combined into the single variable $(\rho_s - \rho)g$. (c) If a scale model is to be built that is twice the size of the actual system, what dimensionless group(s) would you make the same in the model and the prototype? If model measurements using a 3-mm-diameter particle show a critical velocity of 25 cm/s, what is the corresponding critical velocity and particle size in the prototype?

6.77. The performance of a low-flying aircraft at a speed of 146 km/h is to be investigated using a model in a wind tunnel. Standard air is representative of flying conditions, and the wind tunnel will also use standard air. If the maximum airspeed that can be achieved in the wind tunnel is 92 m/s, what scale ratio corresponds to the smallest-size model aircraft that can be used in the study?

6.78. A supersonic aircraft has a design speed of 1650 km/h when operating at an elevation of 15 km. If the aircraft is operated at an elevation of 9 km, what speed would cause the same compressibility effects as when the aircraft operated at the higher altitude?

6.79. A light aircraft cruises at 370 km/h at an elevation of 4 km, and the performance of this aircraft is to be studied in a wind tunnel using a 1:12 scale model. If standard air is to be used in the wind tunnel tests, what airspeed should be used in the model study? Comment on the practicality of using the calculated airspeed in the model.

6.80. A small plane is designed to cruise at 350 km/h at an elevation of 1 km, where the typical atmospheric pressure is 90 kPa. The drag characteristics of the aircraft are to be studied in a pressurized wind tunnel, where the air in the wind tunnel is the same temperature as the air in the prototype. The model is to be at 1:6 scale, and the airspeed in the wind tunnel is to be 260 km/h. (a) What scaling law should be used in designing the model study? (b) What pressure should be used in the wind tunnel? (c) If the drag on a component of the model aircraft is measured as 15 N, what is the drag on the corresponding component in the prototype?

6.81. A model of a new airplane is tested in a wind tunnel. Measurements taken during the model tests indicate that when the approach airspeed is 925 km/h, the stagnation pressure on the nose of the airplane is 45 kPa higher than the free-stream pressure. The temperature of the air in the wind tunnel is 25°C. The prototype airplane is designed to cruise at an altitude of 10.6 km. What are the airspeed and the relative stagnation pressure in the prototype that correspond to the model scenario?

6.82. Model aircraft are sometimes tested in a water tunnel instead of a wind tunnel. For a 1:7 scale model of an aircraft, estimate the ratio of the required airspeed in a wind tunnel to the required water speed in a water tunnel. For the wind tunnel tests, assume standard air, and for the water tunnel tests, assume water at 20°C.

6.83. The performance of an aircraft is to be tested using a scale model. The condition of interest in the prototype is at a speed of 180 km/h. Ambient conditions in the prototype can be taken as standard air at sea level. A 1:12 scale model is to be used in the study. (a) If the performance of the aircraft is to be tested in a wind tunnel, determine the required airspeed in the wind tunnel, which will also use standard air at sea level. Comment on the practicality of using the required speed. (b) If the performance of the aircraft is tested in a water tunnel, determine the required speed of the water. Assume water at 20°C and comment on the practicality of the required speed.

6.84. The lift force generated by a new wing (airfoil) design is to be tested in a pressurized wind tunnel. Under design conditions, the aircraft will have a speed of 216 km/h and operate at an elevation where the temperature is 10°C and the absolute pressure is 92.5 kPa. The model airfoil will have a scale of 1:10, and the temperature of the air in the wind tunnel will be maintained at 10°C. Determine the airspeed and pressure that should be maintained in the wind tunnel to achieve dynamic similarity and minimize compressibility effects.

6.85. The performance of a new blimp is to be tested using a scale model in a wind tunnel. The full-scale blimp is designed to move at 54 km/h in standard air. If a 1:12 model is used in a wind tunnel with air at 15°C and an airspeed of 80 m/s, what must the air pressure in the wind tunnel be to achieve dynamic similarity? If dynamic similarity is achieved and a drag force of 200 N is measured in the model, what is the corresponding drag force in the prototype?

6.86. When a fluid flows around a cylinder, under some circumstances, transient oscillating vortices are formed behind the cylinder as illustrated in Figure 6.11. The path of these vortices is called the von Kármán vortex street. Consider the case in which a fluid of density ρ and viscosity μ approaches a cylinder of diameter D and generates vortices with a frequency ω. A model is to be constructed to study this phenomenon in the laboratory. (a) What scaling law should be used to design the model? (b) In a prototype, standard air flows past a cylinder of diameter 300 mm at a velocity of 70 km/h. If the model cylinder is to have a diameter of 50 mm and the model test is to be conducted in a water tunnel using water at 20°C, what approach velocity should be used in the model? (c) If a shedding frequency of 32 Hz is observed in the model, what shedding frequency is expected in the prototype?

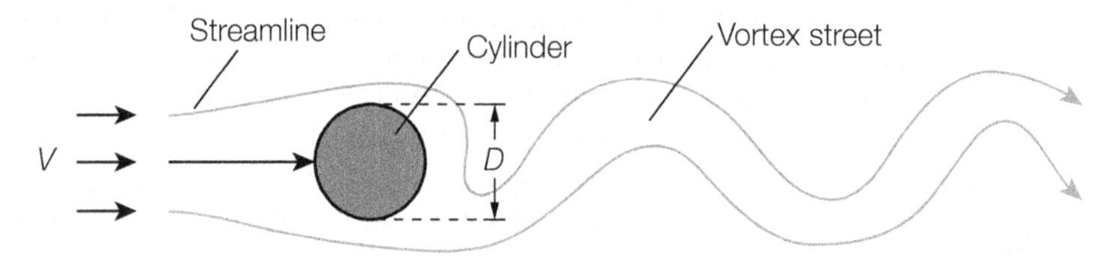

Figure 6.11: Vortex street behind a cylinder

6.87. A 1:60 scale model of a ship is used in a water tank to simulate a ship speed of 10 m/s. What should be the model speed? If a towing force of 10 N is measured in the model, what force is expected on the prototype? Neglect viscous effects.

6.88. A 1:50 scale model of a ship is to be tested to determine the wave drag. The geometry and surface properties of the ship are such that viscous drag is negligible. In the model test, the model is moved at a velocity of 2 m/s and the measure drag on the model is 20 N. Water at the same temperature is used in both the model and the prototype. What are the corresponding velocity and drag force in the prototype?

6.89. The performance of a ship is to be tested using a 1:50 scale model. In this particular ship, under normal operating conditions, both gravity and viscous effects have a significant effect on the drag. Estimate the dynamic viscosity of the fluid that must be used in the model test to achieve dynamic similarity with the prototype. Assume seawater at 20°C for the prototype.

6.90. A boat of length L is designed to move through water at a velocity V while generating bow waves of height H. To study the range of wave heights that will be generated by the boat, a scale model of the boat is constructed and tested in a hydraulics laboratory. Viscous effects might be important at the laboratory scale. (a) What nondimensional relationship will you use to design your laboratory study? (b) How will you use this relationship? (c) If your model scale is 1:10, the model length is 60 cm, and you measure a wave height of 5 cm when the model boat is moving at 30 cm/s, what are the corresponding conditions in the prototype? (d) How does the ratio of wave height to boat length in the model compare with the corresponding ratio in the prototype?

6.91. A prototype ship 35 m long is designed to cruise at 11 m/s. Its drag is to be simulated by a 1-m-long model pulled in a tow tank. For Froude scaling, determine the tow speed, the ratio of prototype to model drag, and the ratio of prototype to model power.

6.92. The performance of a new hydrofoil is tested in a laboratory open (water) channel using a 1:15 scale model. The geometry of the hydrofoil is sufficiently aerodynamic that viscous effects have a negligible effect on the flow around the hydrofoil. Laboratory measurements show that when the speed of the model hydrofoil is 10 m/s, the force on a critical component of the model hydrofoil is 10 N. The water properties in the model and prototype can be assumed to be approximately the same. Determine the speed and force in the prototype hydrofoil corresponding to the model measurements.

6.93. A 1:15 scale model of a hydrofoil is to be tested using Froude number similarity. Under design conditions, the prototype hydrofoil will operate in fresh water at 10°C, the free-stream pressure at the hydrofoil level is 150 kPa, and the hydrofoil has a cruising speed of 100 km/h. To duplicate the prototype cavitation conditions in a model test, the cavitation number must be the same in both the model and the prototype. To facilitate achieving similarity, the temperature of the water used in the model tests is 50°C. Determine the free-stream pressure at the level of the model hydrofoil that is required to achieve similarity in cavitation conditions.

6.94. A particular submarine is capable of cruising at 20 km/h at the surface of the ocean and far below the surface of the ocean. Tests of both of these conditions are to be performed using a 1:40 scale model of the submarine. Determine the required model speed and the ratio of prototype drag to model drag for the following model conditions: (a) a model of cruising on the surface and (b) a model of cruising far below the surface. Assume that the prototype ocean water properties can be duplicated in the model tests.

6.95. The performance of a new submarine propeller is to be tested using a 1:15 scale model. Under design conditions, the prototype propeller is expected to rotate at 1500 rpm. When the submarine is near the surface, gravitational effects are more important than viscous effects, and when the submarine is far below the surface, viscous effects are more important than gravity effects. Determine the theoretical model rotational speed that should be used to simulate each of these conditions. Comment on the practicality of the model rotational speeds.

6.96. A particular torpedo is designed to travel underwater at a velocity of 50 km/h, and the performance of this torpedo is to be studied using a 1:8 scale model. Options suggested for the model tests are to use standard air in a wind tunnel or to use water at 20°C in a water tunnel. Determine the required test speed for each option and suggest which option is more feasible.

6.97. The performance of a torpedo is to be tested using a model in a wind tunnel. The prototype torpedo is designed to move at 30 m/s in water at 20°C. If a 1:4 model is used in a wind tunnel with air at 20°C and an airspeed of 110 m/s, what must the air pressure in the wind tunnel be to achieve dynamic similarity? If dynamic similarity is achieved and a drag force of 600 N is measured in the model, what is the corresponding drag force in the prototype?

6.98. The aerodynamic drag on a car is to be predicted at a speed of 96.6 km/h at an air temperature of 25°C. Automotive engineers build a 1:4 scale model of the car to test in a wind tunnel. Determine the required airspeed in the wind tunnel. If the aerodynamic drag on the model is measured as 160 N, estimate the drag at full scale.

6.99. A small wind tunnel has a test section that is 50.8 cm × 50.8 cm in cross section, is 1.22 m long, and can sustain an airspeed of 48.77 m/s. A model of an 18-wheeler truck is to be built to study the aerodynamic drag on the truck. A full-size tractor-trailer truck is 15.85 m long, 2.54 m wide, and 3.66 m high. Both the air in the wind tunnel and the air flowing over the prototype is at 27°C and at atmospheric pressure. What is the largest scale model that can be accommodated in the wind tunnel that will still stay within the guidelines for blockage? What are the dimensions of the model truck in centimeters? What is the maximum model truck Reynolds number that can be simulated in the wind tunnel. Can Reynolds number independence be achieved? What is meant by Reynolds number independence?

6.100. The drag force on a truck at a speed of 88 km/h is to be measured in a wind tunnel using a 1:10 scale model. Standard air is to be used as a basis for assessment in both the prototype and the model. (a) What airspeed should be used in the model tests? (b) Determine whether compressibility will be a complicating factor in the model tests.

6.101. A prototype vehicle has a design speed of 88 km/h at an ambient air temperature of 20°C. The drag on this vehicle is to be investigated using a 1:8 scale model in a wind tunnel. The temperature of the air in the wind tunnel is 5°C. Both the model and prototype air are at the same pressure. (a) Estimate the airspeed at which the model tests should be run in the wind tunnel. (b) Determine whether compressibility will preclude dynamic similarity at the wind speed calculated in part (a).

Chapter 7

Flow in Closed Conduits

LEARNING OBJECTIVES

After reading this chapter and solving a representative sample of end-of-chapter problems, you will be able to:

- Understand the hydraulic classification of smooth and rough pipes.
- Apply the energy equation to solve practical problems related to flow in closed conduits.
- Calculate water hammer pressures and critical closure times for valves.
- Calculate flow distributions in pipe networks.
- Analyze and design building water supply systems.

7.1 Introduction

A flow in which the fluid completely fills a conduit is classified as a *flow in a closed conduit.* Gases generally fill the conduit in which they are being transported. In contrast to gases, liquids only undergo closed-conduit flow when there is no free surface—in other words, only when the liquid completely fills the conduit. The cross sections of closed conduits can be of any shape or size, and conduits can be made of any material. For the most part, the same general equations can be used to describe the flows of both liquids and gases in closed conduits, with the only difference being the fluid properties. In cases where the closed conduit has a circular cross section, the conduit is commonly referred to as a *pipe*, and in cases where the cross section is not circular, the conduit is commonly referred to as a *duct*, particularly when the fluid being transported is a gas. Small-diameter pipes are commonly referred to as *tubes*.

Laminar versus turbulent flow. Flows in closed conduits can be laminar or turbulent, with each type of flow having such sufficient distinguishing features that it is worthwhile to consider them separately. The type of flow (laminar or turbulent) is usually indicated by the Reynolds number, Re, which, for pipes of circular cross section, is generally defined by

$$\text{Re} = \frac{VD}{\nu} \tag{7.1}$$

where V is the average flow velocity, D is the pipe diameter, and ν is the kinematic viscosity of the fluid being transported by the pipe. Laminar flows in pipes usually occur for $\text{Re} < 2000$, and turbulent flows usually occur for $\text{Re} > 4000$, with either laminar or turbulent flows usually occurring in the range $2000 < \text{Re} < 4000$.

It is also common to cite the range of laminar flow as Re < 2300 instead of Re < 2000, with the former condition more widely used in the United States and the latter condition more widely used in the United Kingdom. To accommodate both conditions, it is safe to assume that laminar flow generally occurs for Re < 2000. In exceptional and highly controlled cases, laminar flows have been maintained up to Re = 100,000. Turbulent flows commonly occur in applications involving the flow of water in large-diameter pipes, such as in water distribution networks and household plumbing systems. Laminar flows commonly occur in applications involving the transport of viscous fluids in small-diameter pipes, such as in hydraulic and lubrication systems.

7.2 Steady Incompressible Flow

Steady-state incompressible flow is perhaps the most widely encountered flow condition in engineering practice, and this type of flow typically occurs in pipelines and other closed conduits. In practical applications, engineers are usually interested in relating pressure changes in a conduit to the flow rate through the conduit, the fluid properties, the shape of the conduit, and the surface properties of the conduit material. The steady-state energy equation when applied to a conduit control volume between an inflow section (section 1) and an outflow section (section 2) can be expressed in the form

$$\left(\frac{p_1}{\gamma} + \frac{v_1^2}{2g} + z_1\right) - \left(\frac{p_2}{\gamma} + \frac{v_2^2}{2g} + z_2\right) = h_\ell \quad (= h_\mathrm{f}) \tag{7.2}$$

where p_1, v_1, and z_1 are the pressure, velocity, and (center-)elevation, respectively, at the inflow section, p_2, v_2, and z_2 are the corresponding quantities at the outflow section, and h_ℓ is the head loss between the upstream and downstream sections. In cases where the head loss, h_ℓ, is due only to friction between the flowing fluid and the surface of the conduit, the lead loss h_ℓ is commonly represented by h_f, where the subscript "f" indicates that friction between the flowing fluid and the conduit surface is the cause of the head loss. Once h_ℓ or h_f is known or estimated, pressure changes along the conduit can be determined directly using Equation 7.2. The quantity h_f is commonly called the *head loss due to friction* or the *friction loss*, and methods used to estimate h_f are described below.

Relationship between friction loss and shear stress. Consider the cylindrical fluid element within a closed conduit as shown in Figure 7.1. Under steady-state conditions, the momentum equation applied to the fluid element requires that the sum of the forces on the fluid element be equal to zero. Equating the sum of the forces on the fluid element in the flow direction to zero yields the following relation:

$$p_1 A - p_2 A - \tau P L - \gamma A L \sin\theta = 0 \tag{7.3}$$

where p_1 and p_2 are the pressures on the upstream and downstream ends of the fluid element, respectively, A is the cross-sectional area of the fluid element, τ is the shear stress on the side of the fluid element, P and L are the perimeter and length of the fluid element, respectively, γ is the specific weight of the fluid, and θ is the angle of inclination of the conduit containing the fluid element. The inclination of the conduit is related to the center-elevations of the upstream and downstream ends of the fluid element by

$$\sin\theta = \frac{z_2 - z_1}{L} \tag{7.4}$$

where z_1 and z_2 are the (center-)elevations at the upstream and downstream ends of the fluid element, respectively. Combining Equations 7.3 and 7.4 and rearranging gives the momentum equation in the following form:

$$\left(\frac{p_1}{\gamma} + z_1\right) - \left(\frac{p_2}{\gamma} + z_2\right) = \frac{\tau P L}{\gamma A} \tag{7.5}$$

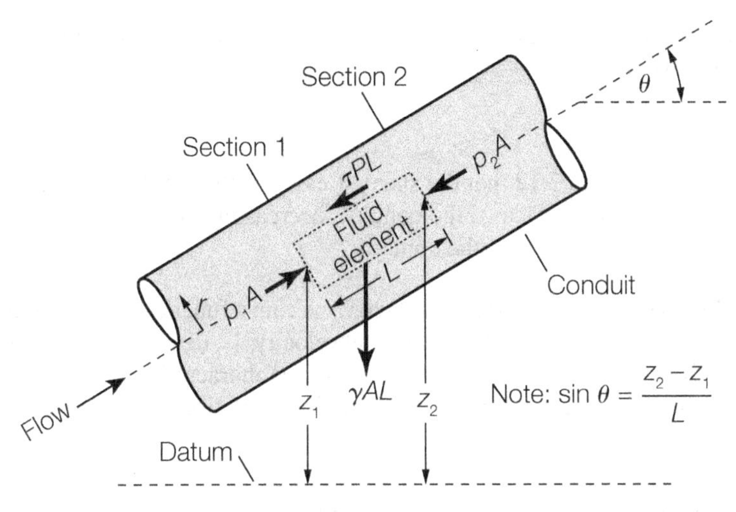

Figure 7.1: Forces on a fluid element in a conduit

If the fluid element moving along a streamline has a radius r, then the perimeter, P, and area, A, of the fluid element are given by

$$P = 2\pi r, \qquad A = \pi r^2 \tag{7.6}$$

and combining Equations 7.5 and 7.6 yields

$$\left(\frac{p_1}{\gamma} + z_1\right) - \left(\frac{p_2}{\gamma} + z_2\right) = \frac{2\tau L}{r\gamma} \tag{7.7}$$

Equation 7.7 is the form of the momentum equation that is most useful in this analysis. When applied to the streamtube containing the fluid element shown in Figure 7.1, the energy equation (Equation 7.2) can be expressed in the form

$$\left(\frac{p_1}{\gamma} + \frac{v_1^2}{2g} + z_1\right) - \left(\frac{p_2}{\gamma} + \frac{v_2^2}{2g} + z_2\right) = h_\mathrm{f} \tag{7.8}$$

where the velocities v_1 and v_2 at the upstream and downstream ends of the fluid element, respectively, are equal because the conduit is of constant diameter; hence, the velocity distribution does not change along the conduit. The simplified form of the energy equation as given by Equation 7.8 is

$$\left(\frac{p_1}{\gamma} + z_1\right) - \left(\frac{p_2}{\gamma} + z_2\right) = h_\mathrm{f} \tag{7.9}$$

Combining the momentum and energy equations (Equations 7.7 and 7.9) yields

$$\frac{2\tau L}{r\gamma} = h_\mathrm{f} \tag{7.10}$$

For a given change in piezometric head $(p/\gamma + z)$ along the center of the conduit, h_f will be the same regardless of the radius, r, of the fluid element. Hence, if the surface of the fluid element is extended to the (inner) surface of a straight circular conduit with diameter D, it follows that

$$h_\mathrm{f} = \frac{2\tau L}{r\gamma} = \frac{4\tau_0 L}{D\gamma} = \text{constant} \tag{7.11}$$

where τ_0 is the shear stress exerted on the fluid element by the surface of the conduit. The quantity τ_0 is commonly called the *boundary shear stress*, because it is the shear stress exerted on the boundary of the flowing fluid. Equation 7.11 shows that the distribution of shear stress across the conduit is given by

$$\tau = \tau_0 \frac{2r}{D} \tag{7.12}$$

It is apparent from Equation 7.12 that the shear stress, τ, between fluid layers varies linearly with the distance from the surface of the conduit, decreasing from $\tau = \tau_0$ on the surface of the conduit to $\tau = 0$ at the center of the conduit.

Relationship between shear stress and velocity. From a functional viewpoint, the boundary shear stress, τ_0, is a function of the mean flow velocity, V, density of the fluid, ρ, dynamic viscosity of the fluid, μ, diameter of the conduit, D, and characteristic size of roughness projections on the surface of the conduit, ϵ. This functional relationship can be expressed as

$$\tau_0 = f_1(V, \rho, \mu, D, \epsilon) \tag{7.13}$$

According to the Buckingham pi theorem, this relationship between six variables in three fundamental dimensions (M, L, and T) can also be expressed as a relationship between three dimensionless groups. The following relationship between dimensionless groups can be used:

$$\frac{\tau_0}{\frac{1}{8}\rho V^2} = f\left(\text{Re}, \frac{\epsilon}{D}\right) \tag{7.14}$$

where Re is the Reynolds number defined by

$$\text{Re} = \frac{\rho V D}{\mu} \tag{7.15}$$

The function $f(\text{Re}, \epsilon/D)$ is, at this point, unknown and cannot be determined simply by dimensional analysis. However, the contribution of dimensional analysis is that the ratio of the boundary shear stress, τ_0, to $\frac{1}{8}\rho V^2$ is equal to a quantity f that depends on Re and ϵ/D. This exact relationship will subsequently be determined separately for laminar and turbulent flow. It is interesting to note that the boundary shear stress, τ_0, in Equation 7.14 is normalized by $\frac{1}{8}\rho V^2$ rather than the more standard normalization of shear stress by the dynamic pressure $\frac{1}{2}\rho V^2$. The latter normalization is the norm in the United Kingdom (e.g., Massey and Ward-Smith, 2012). However, the former normalization is universally used in the United States and will be used exclusively in this text. To avoid confusion, f_B can be used to represent the normalization of the shear stress by $\frac{1}{2}\rho V^2$ such that

$$\frac{\tau_0}{\frac{1}{2}\rho V^2} = f_\text{B}\left(\text{Re}, \frac{\epsilon}{D}\right) \tag{7.16}$$

Subsequent expressions will be formulated in terms of f, and to convert these expressions in terms of f_B for British usage, simply use the relation $f = 4f_\text{B}$.

The Darcy–Weisbach equation. Equations 7.11 and 7.14 are independent equations that apply generally to steady incompressible flows in closed conduits. These equations can be expressed as

$$h_\text{f} = \frac{4\tau_0 L}{D\gamma} \quad \text{and} \quad \frac{\tau_0}{\frac{1}{8}\rho V^2} = f$$

Combining these equations by eliminating τ_0 gives the following relationship between the head loss, h_f, and the average velocity, V, in any closed conduit with a circular cross section:

$$h_\text{f} = \frac{fL}{D}\frac{V^2}{2g} \tag{7.17}$$

This expression is particularly useful because it provides a direct relationship between the head loss and measurable quantities involved in closed-conduit flow, having eliminated τ_0, which is not an (easily) measured quantity. Equation 7.17 is called the *Darcy–Weisbach equation*,[47] and the factor f is called the *friction factor* or the *Darcy–Weisbach friction factor*. Equation 7.17 is sometimes referred to simply as the Darcy equation; however, this is inappropriate, because it was Weisbach who first proposed the exact form of Equation 7.17 in 1845, with Darcy's contribution being the functional dependence of f on V and D in 1857 (Rouse and Ince, 1957; Brown, 2002). Differences in head loss between laminar and turbulent flow were later quantified by Reynolds[48] in 1883 (Reynolds, 1883).

Relationship between pressure gradient and head loss. A quantity that is commonly of interest in flow through closed conduits is the pressure gradient along the conduit, commonly denoted by $\mathrm{d}p/\mathrm{d}x$, where p is the pressure and x is the coordinate measured along the conduit. For prismatic conduits, the pressure gradient can be derived directly from the energy equation (Equation 7.9), which yields

$$\frac{\mathrm{d}p}{\mathrm{d}x} = -\gamma\left(\frac{h_\mathrm{f}}{L} + \sin\theta\right) \tag{7.18}$$

where γ is the specific weight of the fluid, h_f is the head loss over a distance L along the conduit, and θ is the inclination of the conduit, which is related to the change in elevation of the conduit by Equation 7.4. For steady flow of incompressible fluids in prismatic conduits, the pressure gradient remains constant along the conduit.

EXAMPLE 7.1

SAE 30 oil at 20°C flows at a rate of 3 L/min through a straight 50-mm-diameter pipeline inclined at 15° to the horizontal. It is known that for this (laminar) flow condition, the friction factor, f, is related to the Reynolds number, Re, by the relation $f = 64/\text{Re}$. Determine the pressure gradient and the change in pressure over a 5-m section of the pipeline.

SOLUTION

From the given data: $Q = 3$ L/min $= 5.00 \times 10^{-5}$ m³/s, $D = 50$ mm, and $\theta = 15°$. For SAE 30 oil at 20°C, $\rho = 918$ kg/m³, $\mu = 440$ mPa·s, and $\gamma = 9.003$ kN/m³ (from Appendix B.4). The following derived parameters can be calculated from the given data:

$$A = \frac{\pi D^2}{4} = \frac{\pi(0.050)^2}{4} = 1.963 \times 10^{-3}, \qquad V = \frac{Q}{A} = \frac{5.00 \times 10^{-5}}{1.963 \times 10^{-3}} = 0.02546 \text{ m/s}$$

$$\text{Re} = \frac{\rho V D}{\mu} = \frac{(918)(0.02546)(0.050)}{440 \times 10^{-3}} = 2.656, \quad f = \frac{64}{\text{Re}} = \frac{64}{2.656} = 24.09$$

Using the values of the given and derived parameters in the Darcy–Weisbach equation gives

$$\frac{h_\mathrm{f}}{L} = \frac{f}{D}\frac{V^2}{2g} = \frac{24.09}{0.050}\frac{(0.02546)^2}{2(9.807)} = 0.01592$$

[47] Named in honor of the French engineer Henry Darcy (1803–1858) and the German engineer Julius Weisbach (1806–1871).

[48] Osborne Reynolds (1842–1912) was a British engineer who made seminal contributions in the area of fluid dynamics.

The pressure gradient along the pipeline is given by Equation 7.18 as

$$\frac{dp}{dx} = -\gamma\left(\frac{h_f}{L} + \sin\theta\right) = -(9.003)\,(0.01592 + \sin 15^\circ) = \mathbf{-2.47\ kPa/m}$$

Over a distance of 5 m, the pressure change, Δp, is calculated as

$$\Delta p = \frac{dp}{dx} \cdot L = (-2.47)(5) = -12.4\ \text{kPa}$$

Therefore, the pressure drop along any 5-m segment of the pipeline is estimated as **12.4 kPa**. These calculations assume that the flow is fully developed within the pipeline.

Transition to fully developed flow. *Fully developed flow* refers to a flow condition that does not change along a conduit, and fully developed flow within a conduit does not occur until the frictional effect of the conduit surface propagates to the center of the conduit. Regions of flow that are not fully developed are typically found immediately downstream of the entrances to closed conduits. The region of flow affected by the conduit surface is called the *boundary layer*; therefore, fully developed flow does not occur until the boundary layer extends to the center of the conduit. As the boundary layer is extending toward the center of the conduit, the velocity in the conduit beyond the boundary-layer region must necessarily be increasing from its value at the conduit entrance to account for the retardation of the flow within the boundary layer. The region between the conduit inlet and the location where fully developed flow occurs is sometimes called the *hydrodynamic entry region*, and the length of conduit containing the hydrodynamic entry region is called the *hydrodynamic entry length* or, simply, the *entrance length*. For laminar flow entering a circular conduit, the entrance length, L_e, is usually estimated by the relation

$$\frac{L_e}{D} = 0.058\,\text{Re} \quad \text{(laminar flow)} \tag{7.19}$$

where Re is the Reynolds number of the fully developed flow and D is the conduit diameter. The coefficient in Equation 7.19 is sometimes taken as 0.065 (e.g., Potter et al., 2012) or 0.06 (e.g., White, 2011); however, a coefficient of 0.058 is more common. The entrance length for turbulent flow is considerably shorter than the entrance length for laminar flow. Under turbulent flow conditions, the entrance length can be estimated by the relation

$$\frac{L_e}{D} = 4.4\,\text{Re}^{\frac{1}{6}} \quad \text{(turbulent flow)} \tag{7.20}$$

Equations 7.19 and 7.20 can be contrasted by noting that for $10^5 < \text{Re} < 10^6$ (turbulent flow), Equation 7.20 gives $20 < L_e/D < 30$, whereas for $\text{Re} = 2000$ (laminar flow), Equation 7.19 gives $L_e/D = 116$. Equations for fully developed flow in circular conduits are applicable in regions beyond the entrance length. In most practical applications, the total length, L, of the pipeline is such that $L/D > 1000$; consequently, the irregular frictional characteristics in the entrance region can usually be neglected. For shorter conduits (e.g., $L/D \leq 100$), conditions within the entrance length might need to be taken into account, particularly for laminar flows. In some applications, the length of the conduit is purposely kept short to prevent boundary effects from significantly affecting the flow in the central portion of the conduit. This is the case in designing wind tunnels that are used to test the performance of (model) aircraft, where in-flight conditions are typically unaffected by boundaries. Within the hydrodynamic entry region, the boundary shear stress is highest when the fluid first enters the conduit, because the boundary layer is thinnest at that location. As the thickness of the boundary layer grows, the boundary shear stress (which is proportional to the velocity gradient adjacent to the conduit surface) decreases as fully developed flow is approached. As a consequence, the friction factor is higher in the entrance region than in the region of fully developed flow.

Flow in noncircular conduits. In cases where a prismatic conduit does not have a circular cross section, Equation 7.5 can be applied directly to a fluid element that encompasses the entire cross section of the conduit. Combining Equation 7.5 with the energy equation (Equation 7.2) gives the following expression for the head loss between two sections of the conduit that are a distance L apart:

$$h_\mathrm{f} = \frac{\tau_0 PL}{\gamma A} \tag{7.21}$$

The ratio of the cross-sectional area, A, of the conduit to the perimeter, P, of the conduit can be expressed in terms of either the *hydraulic radius*, R_h, or the *hydraulic diameter*, D_h, of the conduit, where

$$R_\mathrm{h} = \frac{A}{P}, \quad \text{and} \quad D_\mathrm{h} = 4\frac{A}{P} \tag{7.22}$$

Hence, the head loss can be related to the boundary shear stress in terms of either the hydraulic radius or hydraulic diameter as

$$h_\mathrm{f} = \frac{\tau_0 L}{\gamma R_\mathrm{h}}, \quad \text{or} \quad h_\mathrm{f} = \frac{4\tau_0 L}{\gamma D_\mathrm{h}} \tag{7.23}$$

For circular conduits, $R_\mathrm{h} = D/4$ and $D_\mathrm{h} = D$, and substituting these relationships into Equation 7.23 yields Equation 7.11. Therefore, either form of Equation 7.23 can be considered as a general relationship for relating the head loss to the boundary shear stress for flow in closed conduits. Although both R_h and D_h are widely used, preference should be given to using D_h because it leads to a head-loss expression that has the same form in both circular and noncircular conduits. Using Equation 7.23 to relate the head loss to the boundary shear stress, the frictional head losses, h_f, in noncircular conduits can be estimated using the following form of the Darcy–Weisbach equation

$$h_\mathrm{f} = \frac{fL}{D_\mathrm{h}}\frac{V^2}{2g} \tag{7.24}$$

where the friction factor, f, is calculated using a (hydraulic) Reynolds number, Re_h, defined by

$$\mathrm{Re}_\mathrm{h} = \frac{\rho V D_\mathrm{h}}{\mu} \tag{7.25}$$

It is important to remember that characterizing a noncircular conduit by the hydraulic diameter, D_h, is necessarily approximate, because the geometric characteristics of noncircular conduits cannot be described with a single parameter (e.g., D_h). Using the hydraulic diameter as a basis for calculating frictional head losses in noncircular conduits is usually accurate to within 15% for turbulent flow and is much less accurate for laminar flows, where the accuracy is on the order of $\pm$40%. Characterization of noncircular conduits by the hydraulic diameter can be used for rectangular conduits where the ratio of sides, called the *aspect ratio*, does not exceed about 8:1 (Olson and Wright, 1990), although some engineers state that aspect ratios must not exceed 4:1 (e.g., Potter and Wiggert, 2012; Pritchard, 2011; Mott, 2006).

7.3 Friction Effects in Laminar Flow

Laminar flow occurs in closed conduits when Re $\leq$ 2000. In cases where the flow of a Newtonian fluid is laminar, the volume flow rate and velocity profile can be related analytically to the head loss along the conduit, and the friction factor can be related analytically to the Reynolds number of the flow.

Volume flow rate and velocity profile. For laminar flow, the shear stress, τ, between fluid layers is related to the velocity distribution within the fluid by Newton's law of viscosity, which requires that

$$\tau = \mu \frac{\mathrm{d}u}{\mathrm{d}y} \tag{7.26}$$

where μ is the dynamic viscosity, u is the magnitude of the velocity, y is the coordinate in the direction normal to the direction of the velocity, and $\mathrm{d}u/\mathrm{d}y$ is the gradient of u in the y-direction. Combining Equations 7.10 and 7.26, where $y = D/2 - r$, $\mathrm{d}y = -\mathrm{d}r$, and $u(r)$ is the velocity at a distance r from the pipe centerline, gives

$$\mathrm{d}u = -\frac{h_\mathrm{f}\gamma}{2\mu L} r \, \mathrm{d}r \tag{7.27}$$

At the surface of the pipe, the no-slip boundary condition requires that $u = 0$ at $r = D/2$, and combining this boundary condition with Equation 7.27 yields the integrated result

$$u(r) = \frac{h_\mathrm{f}\gamma}{16\mu L}(D^2 - 4r^2) \tag{7.28}$$

This velocity distribution is illustrated in Figure 7.2, where V_c is the centerline velocity (at $r = 0$) and is given by Equation 7.28 as

$$V_\mathrm{c} = \frac{h_\mathrm{f}\gamma}{16\mu L} D^2 \tag{7.29}$$

The average velocity, V, in the pipe can be calculated from the velocity distribution using the relation

$$V = \frac{1}{A}\int_A u \, \mathrm{d}A = \frac{1}{\pi D^2/4}\int_0^{D/2} u \, 2\pi r \, \mathrm{d}r \tag{7.30}$$

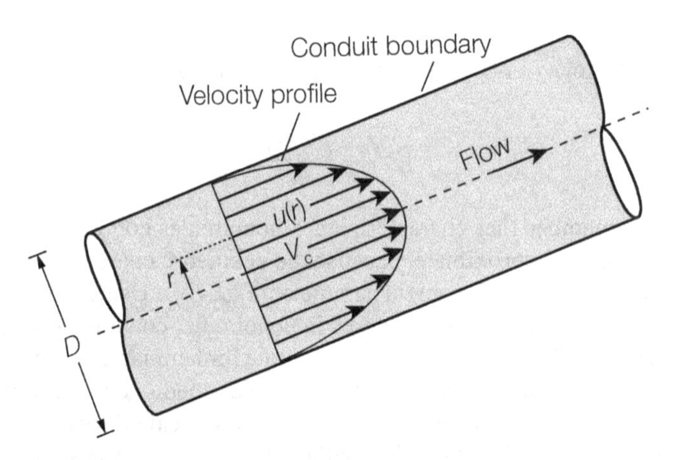

Figure 7.2: Laminar flow in a pipe

and substituting Equation 7.28 for u into Equation 7.30 yields

$$V = \frac{h_f \gamma}{32\mu L} D^2 \tag{7.31}$$

The volume flow rate, Q, through the pipe can be determined directly from Equation 7.31 by multiplying the average velocity, V, by the cross-sectional area, $A = \pi D^2/4$, which yields

$$Q = \frac{\pi h_f \gamma}{128\mu L} D^4 \tag{7.32}$$

It is useful to note that the volume flow rate, Q, is proportional to D^4, which indicates that the flow rate is very sensitive to the diameter of the pipe. For example, doubling the diameter of the pipe will increase the volume flow rate, Q, by a factor of 16. Equations 7.28 and 7.31 can be combined to express the velocity distribution as

$$u(r) = \frac{2V}{D^2}\left(D^2 - 4r^2\right) = 2V\left(1 - 4\frac{r^2}{D^2}\right) \tag{7.33}$$

and taking $R = D/2$, where R is the radius of the pipe, yields

$$u(r) = 2V\left(1 - \frac{r^2}{R^2}\right) \tag{7.34}$$

which demonstrates that under laminar-flow conditions, the velocity distribution in the pipeline is parabolic. The energy and momentum correction factors, α and β, respectively, derived from the velocity profile given by Equation 7.34 are $\alpha = 2$ and $\beta = \frac{4}{3}$. Hence, under laminar-flow conditions, the commonly used approximations that $\alpha = 1$ and $\beta = 1$ might not be appropriate.

The friction factor in laminar flow. In many practical applications, the volume flow rate, Q, and the cross-sectional area, A, of the pipe are known or given quantities, the average velocity of flow, V $(= Q/A)$, can be derived from Q and A, and the objective is to estimate the head loss, h_f, along the pipe. In these cases, Equation 7.31 can be used in the following rearranged form:

$$h_f = 32\nu\frac{L}{gD^2}V, \qquad \text{or} \qquad h_f = \left(\frac{64}{\text{Re}}\right)\frac{L}{D}\frac{V^2}{2g} \tag{7.35}$$

where the Reynolds number, Re, is as defined by Equation 7.15. Comparing Equation 7.35 with Equation 7.17, it is apparent that Equation 7.35 is in the form of the Darcy–Weisbach equation, where the friction factor, f, is given by

$$f = \frac{64}{\text{Re}} \tag{7.36}$$

Laminar flow in noncircular conduits. Friction factors have been determined experimentally for many conduit shapes, and the derived friction factors in circular, rectangular, elliptical, and triangular conduits under laminar-flow conditions are compared in Table 7.1, where the Reynolds number, Re_h, is based on the hydraulic diameter, D_h. It is apparent from Table 7.1 that the friction factors under laminar-flow conditions can differ significantly between conduits with the same value of Re_h.

Table 7.1: Laminar-Flow Friction Factors

Shape	Illustration	Geometric Parameter	$f \times \text{Re}_\text{h}$
Circle		D	64
Rectangle		a/b	
		1	56.92
		2	62.20
		3	68.36
		4	72.92
		6	78.80
		8	82.32
		∞	96.00
Ellipse		a/b	
		1	64.00
		2	67.28
		4	72.96
		8	76.60
		16	78.16
Triangle		θ	
		10°	50.80
		30°	52.28
		60°	53.32
		90°	52.60
		120°	50.96

Source: Çengel and Cimbala, 2014.

Practical Considerations

Laminar flow in a pipe is sometimes called *Hagen–Poiseuille flow*,[49] the associated equation for head loss (Equation 7.35) is sometimes called the *Hagen–Poiseuille law*, and the corresponding flow equation (Equation 7.32) is sometimes called *Poiseuille's law*. Typically, laminar flow in a pipe occurs when Re $\leq$ 2000, and under these conditions, Equation 7.35 can be used to determine the head loss over a distance L of pipe. In engineering applications, transition from laminar to turbulent flow typically occurs for Re in the range of 2000–4000. As a reference point, for water at 15°C, for diameters greater than 25 mm, and for velocities greater than 9 cm/s, the flow will be turbulent, assuming a transition Re of 2000.

[49]Named in honor of the German engineer Gotthilf Heinrich Ludwig Hagen (1797–1884) and the French scientist Jean Léonard Marie Poiseuille (1799–1869) who both published this result independently between 1839 (Hagen) and 1840 (Poiseuille).

EXAMPLE 7.2

SAE 10W oil at 20°C flows through a 100-mm-diameter pipe inclined at 30° to the horizontal. The pressure and elevation at a particular pipe section are 500 kPa and 5.00 m, respectively. If the flow rate in the pipe is 9.39 L/s, assess the validity of assuming Poiseuille flow and determine the head loss over 10 m of pipe and the pressure 10 m downstream of the given section. What is the average shear stress on the pipe between the two sections?

SOLUTION

From the given data: D = 100 mm = 0.1 m, θ = 30°, p_1 = 500 kPa, z_1 = 5.00 m, Q = 9.39 L/s = 0.00939 m^3/s, and L = 10 m. For SAE 10W oil at 20°C: $\mu = 0.104$ Pa·s and ρ = 870 kg/m^3, which gives $\nu = \mu/\rho = 1.19 \times 10^{-4}$ m^2/s and $\gamma = \rho g = 8530$ N/m^3 $= 8.53$ kN/m^3. Using these data,

$$A = \frac{\pi}{4}D^2 = \frac{\pi}{4}(0.1)^2 = 0.00785 \text{ m}^2, \qquad V = \frac{Q}{A} = \frac{0.00939}{0.00785} = 1.20 \text{ m/s}$$

The validity of assuming laminar flow can be checked by calculating the Reynolds number, Re, which yields

$$\text{Re} = \frac{\rho V D}{\mu} = \frac{(870)(1.20)(0.1)}{(0.104)} = 1003$$

Because Re $<$ 2000, the flow in the pipe is laminar and the assumption of **Poiseuille flow is validated**. The head loss between two sections 10 m apart is given by Equation 7.35 as

$$h_\text{f} = 32\nu\frac{L}{gD^2}V = 32(1.19 \times 10^{-4})\frac{10}{(9.81)(0.10)^2}(1.20) = 0.466 \text{ m}$$

Therefore, the head loss between the two sections is **0.466 m**. Applying the definition of h_f, the pressure 10 m downstream of the given section is calculated as follows:

$$h_\text{f} = \left(\frac{p_1}{\gamma} + z_1\right) - \left(\frac{p_2}{\gamma} + z_2\right) \quad \rightarrow$$

$$0.466 = \left(\frac{500}{8.53} + 0\right) - \left(\frac{p_2}{8.53} + 10\sin 30^\circ\right) \quad \rightarrow \quad p_2 = 453 \text{ kPa}$$

Therefore, the pressure 10 m downstream of the given section is **453 kPa**. The change in pressure between sections is mostly due to the change in elevation. The average shear stress, τ_0, on the pipe surface between the two sections is given by Equation 7.11 as

$$h_\text{f} = \frac{4\tau_0 L}{D\gamma} \quad \rightarrow \quad 0.466 = \frac{4\tau_0(10)}{(0.1)(8530)} \quad \rightarrow \quad \tau_0 = 9.94 \text{ Pa}$$

Therefore, the average shear stress on the boundary of the pipeline between the two sections is **9.94 Pa**.

7.4 Friction Effects in Turbulent Flow

The head loss due to friction in a fluid flowing under any regime (laminar or turbulent) can be estimated using the Darcy–Weisbach equation, which is given by Equation 7.17 and is repeated here for easy reference as

$$h_f = \frac{fL}{D}\frac{V^2}{2g} \tag{7.37}$$

where h_f is the head loss between conduit sections spaced L apart, f is the friction factor, D is the conduit diameter, and V is the average velocity in the conduit. Application of the Darcy–Weisbach equation to estimate friction losses in laminar and turbulent flows differ only in the formulations used to estimate the friction factor, f. In the case of laminar flow, f can be determined analytically and is given by Equation 7.36. However, in the case of turbulent flow, f is determined by empirical relations derived from experimental data.

Smooth pipes versus rough pipes. Under turbulent flow conditions, a pipe behaves like a *smooth pipe* when the friction factor, f, does not depend on the height of the roughness projections on the pipe surface. In contrast, in *rough pipes*, the friction factor is influenced by the height of the roughness projections. This difference in behavior can be explained by the presence of a viscous boundary layer that develops adjacent to the pipe surface. A viscous layer occurs as a result of the very small fluid velocities that must necessarily occur near the pipe surface, because the fluid velocity must reduce to zero at the pipe surface. The thickness, δ_v, of the viscous boundary layer can be determined by a combination of dimensional analysis and experimentation and is commonly taken as

$$\delta_v = 5\frac{\nu}{u_*} \tag{7.38}$$

where ν is the kinematic viscosity of the fluid and u_* (pronounced "you star") is the *friction velocity*, which is defined in terms of the boundary shear stress, τ_0, as

$$u_* = \sqrt{\frac{\tau_0}{\rho}} \quad \left(= \sqrt{\frac{f}{8}}V\right) \tag{7.39}$$

where ρ is the density of the fluid. The term in parentheses is derived from the definition of the friction factor (Equation 7.14), which relates τ_0 to V. Equations 7.38 and 7.39 collectively demonstrate that the thickness of the viscous layer adjacent to the pipe surface can be estimated from the friction factor, f, and the mean velocity, V. For any given pipe, the thickness of the viscous boundary layer decreases as the mean velocity increases, and at sufficiently high mean velocities, the roughness projections protrude outside the viscous boundary layer. Therefore, a pipe is considered smooth or rough based not solely on the magnitude of the roughness projections, but on the relative magnitude of the roughness projections to the thickness of the viscous boundary layer. Taking the characteristic roughness height on the pipe surface as ϵ, it is common practice to infer smooth pipe conditions when $\epsilon < \delta_v$, rough pipe conditions when $\epsilon > 14\delta_v$, and transition conditions when $\delta_v \leq \epsilon \leq 14\delta_v$. Keep in mind that the definition of the viscous boundary layer is only applicable to turbulent flows; in laminar flows, the viscous layer does not exist because the entire flow field is influenced by viscous effects.

EXAMPLE 7.3

Pressure measurements along a pipe that is carrying water at 10°C indicate that the friction factor remains constant at 0.0321 as long as the average velocity of the water exceeds 1.53 m/s. The pipe is observed to be very rough with encrustations caused by rust. Based on this information, estimate the height of the roughness projections on the pipe surface. (Hint: Under rough pipe conditions the friction factor depends only on the relative roughness and is independent of the flow velocity.)

SOLUTION

From the given data: $f = 0.0321$ and $V_{\text{min}} = 1.53$ m/s. For water at 10°C, $\rho = 999.1$ kg/m^3, $\mu = 1.139$ mPa·s, and $\nu = \mu/\rho = 1.140 \times 10^{-6}$ m^2/s (from Appendix B.1). Rough pipe conditions begin when the shear velocity, u_*, is given by Equation 7.39 as

$$u_* = \sqrt{\frac{f}{8}}V_{\text{min}} = \sqrt{\frac{0.0321}{8}}(1.53) = 0.00969 \text{ m/s}$$

At the beginning of rough pipe conditions, it can be estimated that the roughness height, ϵ, is given by

$$\epsilon = 14\,\delta_{\text{v}} = 14\left(5\frac{\nu}{u_*}\right) = 14\left(5\frac{1.140 \times 10^{-6}}{0.00969}\right) = 0.00824 \text{ m} = 8.24 \text{ mm}$$

Therefore, the roughness height is estimated as approximately **8 mm**, which is consistent with the surface of the pipe characterized as being "very rough with encrustations."

Empirical estimates of the friction factor. Nikuradse (1932, 1933) performed landmark experiments to study the friction factor in pipes that were artificially roughened by sticking sand grains of uniform size onto the surface of smooth pipes. Based on Nikuradse's experimental results, the following empirical formulas were developed for estimating the friction factor in turbulent pipe flows:

$$\frac{1}{\sqrt{f}} = \begin{cases} -2\log\left(\dfrac{2.51}{\text{Re}\sqrt{f}}\right), & \text{for smooth pipe} \\ -2\log\left(\dfrac{k_s/D}{3.7}\right), & \text{for rough pipe} \end{cases} \tag{7.40}$$

where k_s is the size of the sand grains on the surface of the pipe. The relationships given in Equation 7.40 show that f is independent of k_s/D and depends only on Re under smooth pipe conditions, whereas f depends only on k_s/D under rough pipe conditions. Because the flow is independent of Re under rough pipe conditions, this type of flow is commonly referred to as *fully turbulent flow*. Actually, the flow is turbulent under both smooth pipe and rough pipe conditions. Between the smooth and rough pipe conditions, there is a transition region in which the friction factor depends on both the Reynolds number and the relative roughness.

Friction factor for commercial pipes: the Colebrook equation. A limitation of the Nikuradse experiments on sand-roughened pipes was that the roughness heights were of uniform size, whereas in commercial pipes, the roughness heights are generally nonuniform in size. Colebrook and White (1939) performed experiments on commercial pipes and showed that friction factors in commercial pipes could still be characterized asymptotically by the relations in Equation 7.40; however, the transition region where f depends on both Re and k_s/D has a different behavior in commercial pipes than in Nikuradse's sand-roughened pipes. Colebrook (1939) identified the *equivalent sand roughness* of several commercial pipe materials that would give the same head losses as sand-roughened pipes under fully turbulent conditions and developed the following relationship that adequately describes the friction factor of commercial pipes under turbulent flow conditions:

$$\frac{1}{\sqrt{f}} = -2\log\left(\frac{k_s/D}{3.7} + \frac{2.51}{\mathrm{Re}\sqrt{f}}\right) \tag{7.41}$$

This equation is commonly referred to as the *Colebrook equation* or the *Colebrook–White* equation. Values of the friction factor, f, predicted by the Colebrook equation (Equation 7.41) are generally accurate to within 10%–15% of experimental data and the Colebrook equation asymptotes to the relations in Equation 7.40. The accuracy of the Colebrook equation deteriorates significantly for small pipe diameters, and it is recommended that this equation not be used for pipes with diameters smaller than 2.5 mm (Yoo and Singh, 2005). The Colebrook equation is valid only for turbulent flow. In cases where the flow is in the laminar-turbulent transition regime (typically taken as $2000 < \mathrm{Re} < 4000$), alternative approximations should be used (e.g., Cheng, 2008); however, such conditions are rare in engineering applications.

Estimation of equivalent sand roughness. The equivalent sand roughness, k_s, of several commercial pipe materials is given in Table 7.2. These values of k_s apply to clean new pipe only. Water supply pipes that have been in service for a long time usually experience corrosion or scale buildup that results in values of k_s orders of magnitude larger than the values given in Table 7.2 (Echávez, 1997; Butler and Davies, 2011). An example of the large roughness heights that can occur from corrosion in a public water supply pipe is shown in Figure 7.3. It is apparent from the figure that the increased roughness is of sufficient magnitude also to reduce the effective diameter of the pipe. In general, the rate of increase of the equivalent sand roughness, k_s, in water supply pipelines with time depends primarily on the quality of the water being transported. The roughness coefficients for older water mains are usually determined through field testing.

The Moody diagram. The expression for the friction factor derived by Colebrook (Equation 7.41) was plotted by Moody (1944) in what is commonly referred to as the *Moody diagram*[50] or *Moody chart*, reproduced in Figure 7.4. The Moody diagram indicates that for

Figure 7.3: Corroded pipe
Source: Courtesy of Mark Reutter, Baltimore Brew.

[50]Named in honor of the American engineer Lewis F. Moody (1880–1953). This type of diagram was originally suggested by Blasius in 1913 and Stanton in 1914 (Stanton and Pannell, 1914). The Moody diagram is sometimes called the *Stanton diagram*.

Table 7.2: Typical Equivalent Sand Roughness for Various Materials

Material	Equivalent sand roughness, k_s (mm)
Asbestos cement:	
Coated	0.038
Uncoated	0.076
Brass	0.0015–0.003
Brick	0.6–6.0
Concrete:	
General	0.3–3.0
Steel forms	0.18
Wooden forms	0.6
Centrifugally spun	0.13–0.36
Clay	0.03–0.15
Copper	0.0015–0.003
Corrugated metal	45
Glass	0.0015–0.003
Iron:	
Cast iron	
General	0.2–5.5
Lined with asphalt	0.1–2.1
Uncoated	0.226
Coated	0.102
Ductile iron	0.26
Lined with bitumen	0.12
Lined with spun concrete	0.030–0.038
Galvanized iron	0.102–4.6
Wrought iron	0.046–2.4
Lead	0.0015
Plastic	0.0015–0.06
PVC	0.0015
Rubber, smoothed	0.01
Steel	
Coal-tar enamel	0.0048
Commercial	0.045
New unlined	0.028–0.076
Riveted	0.9–9.0
Stainless	0.002
Wood stave	0.18–0.5

Sources: Butler and Davies (2011); Çengel and Cimbala (2014); Haestad et al., (2004); Haestad Methods, Inc. (2002); Moody (1944); Sanks (1998).

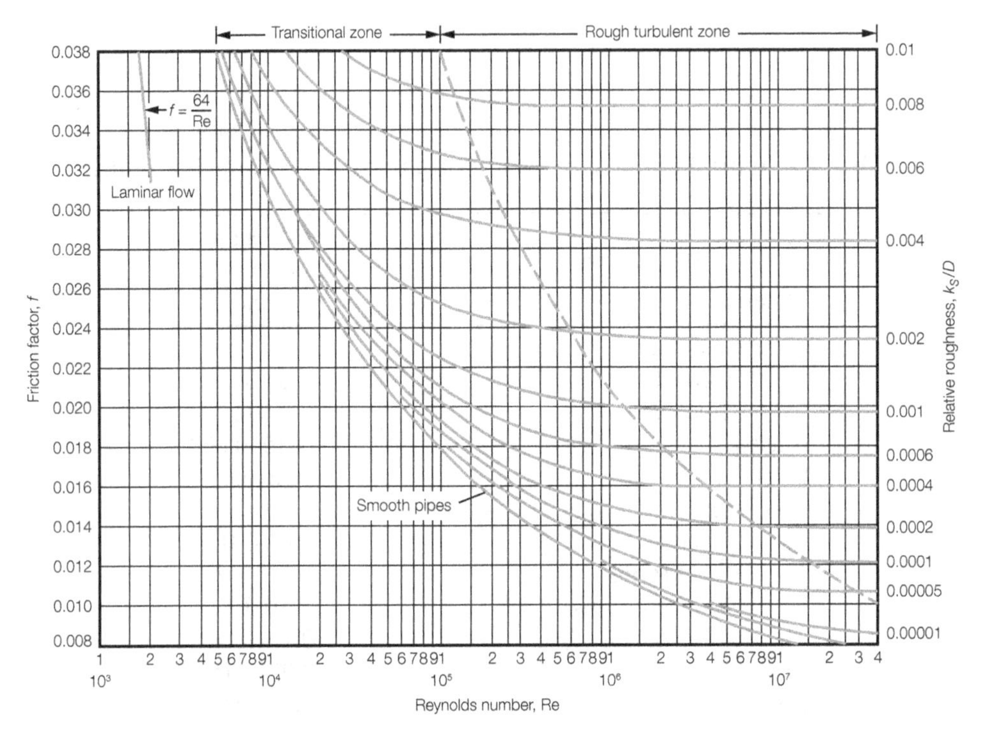

Figure 7.4: Moody diagram
Source: Moody (1944).

Re $\leq$ 2000, the flow is laminar and the friction factor is given by

$$f = \frac{64}{\text{Re}} \tag{7.42}$$

which is commonly known as the *Hagen–Poiseuille formula* and was derived theoretically in Section 7.3 based on the assumption of laminar flow of a Newtonian fluid. For 2000 < Re $\leq$ 4000, there is no fixed relationship between the friction factor and the Reynolds number or relative roughness, and flow conditions are generally uncertain. Beyond a Reynolds number of 4000, the flow is usually turbulent and the friction factor is controlled by the thickness of the viscous boundary layer relative to the height of the roughness projections on the surface of the pipe. The dashed line in Figure 7.4 indicates the boundary between the fully turbulent flow regime, where f is independent of Re, and the transition regime, where f depends on both Re and the relative roughness, k_s/D. The equation of this dashed line is given by (Mott, 2006)

$$\frac{1}{\sqrt{f}} = \frac{\text{Re}}{200(D/k_s)} \tag{7.43}$$

The line in the Moody diagram corresponding to a relative roughness of zero describes the friction factor for pipes that are *hydraulically smooth*.

Swamee–Jain equation. Although the Colebrook equation (Equation 7.41) can be used to calculate the friction factor, this equation has the drawback that it is an implicit equation for the friction factor and must be solved iteratively. This minor inconvenience was circumvented by Swamee and Jain (1976), who suggested the following explicit equation for the friction factor:

$$\frac{1}{\sqrt{f}} = -2\log\left(\frac{k_s/D}{3.7} + \frac{5.74}{\text{Re}^{0.9}}\right), \quad 10^{-6} \leq \frac{k_s}{D} \leq 10^{-2},\ 5000 \leq \text{Re} \leq 10^8 \tag{7.44}$$

where, according to Swamee and Jain (1976), Equation 7.44 deviates by less than 1% from the Colebrook equation within the entire turbulent flow regime, provided the restrictions on k_s/D and Re are honored. The *Swamee–Jain equation* (Equation 7.44) can be more conveniently written as

$$f = \frac{0.25}{\left[\log\left(\frac{k_s/D}{3.7} + \frac{5.74}{\mathrm{Re}^{0.9}}\right)\right]^2} \tag{7.45}$$

In addition to the Swamee–Jain equation, several other explicit equations have been proposed that approximate, to within about 1%, the friction factor given by the Colebrook equation (e.g., Haaland, 1983; Sonnad and Goudar, 2006). Uncertainties in the relative roughness and in the data used to produce the Colebrook equation make the use of accuracy to several places in pipe flow problems unjustified. As a rule of thumb, an accuracy of 10% in calculating the friction factor from either the Colebrook equation or close approximations is to be expected (Gerhart et al., 1992; Munson et al., 2013).

Velocity profile. The velocity profile in both smooth and rough pipes of circular cross section under turbulent flow conditions can be estimated by the semiempirical equation

$$v(r) = \left[(1 + 1.326\sqrt{f}) - 2.04\sqrt{f}\log\left(\frac{R}{R-r}\right)\right]V \tag{7.46}$$

where $v(r)$ is the velocity at a radial distance r from the centerline of the pipe, R is the radius of the pipe, f is the friction factor, and V is the average velocity across the pipe. The velocity distribution given by Equation 7.46 agrees well with velocity measurements in both smooth and rough pipes. This equation, however, is not applicable within the small region close to the centerline of the pipe and is not applicable in the small region close to the pipe surface. This is apparent because at the centerline of the pipe, $\mathrm{d}v/\mathrm{d}r$ must be equal to zero, but Equation 7.46 does not have a zero slope at $r = 0$. At the pipe surface, v must be equal to zero, but Equation 7.46 gives a velocity of zero at a small distance from the pipe surface, with a velocity of $-\infty$ at $r = R$. The energy and momentum correction factors, α and β, respectively, derived from the velocity profile given by Equation 7.46 are (Moody, 1950)

$$\alpha = 1 + 2.7f \tag{7.47}$$

$$\beta = 1 + 0.98f \tag{7.48}$$

Another commonly used equation to describe the velocity distribution in turbulent pipe flow is the empirical *power law* equation given by

$$v(r) = V_0\left(1 - \frac{r}{R}\right)^{\frac{1}{n}} \tag{7.49}$$

where V_0 is the centerline velocity. Values of n are typically in the range of 5–10. For any given value of f, the power-law distribution (Equation 7.49) gives the same mean velocity (V) and the same centerline velocity (V_0) as given by Equation 7.46 when the following relationship is satisfied:

$$\left[\frac{n}{n+1} - \frac{n}{2n+1}\right] = \frac{1}{2(1 + 1.326\sqrt{f})} \tag{7.50}$$

The power law distribution is not applicable within $0.04R$ of the pipe surface, and the power law gives an unrealistic infinite velocity gradient at the pipe surface. Although the power-law profile fits measured velocities close to the centerline of the pipe, it does not give zero slope at the centerline. The energy and momentum correction factors, α and β, respectively, derived from the power-law velocity distribution are

$$\alpha = \frac{(n+1)^3(2n+1)^3}{4n^4(n+3)(2n+3)} \tag{7.51}$$

$$\beta = \frac{(n+1)^2(2n+1)^2}{2n^2(n+2)(2n+2)} \tag{7.52}$$

For n between 5 and 10, α varies from 1.11–1.03 and β varies from 1.04–1.01. In most engineering applications, it is assumed that $\alpha = 1$ and $\beta = 1$ because errors introduced by these assumptions are usually negligible. It is worthwhile to remember that the above results are applicable only for turbulent flow.

EXAMPLE 7.4

Water at 20°C flows through a 300-mm-diameter pipe at a rate of 150 L/s. The equivalent sand roughness of the pipe is 0.5 mm. (a) Compare the velocity distributions given by Equations 7.46 and 7.49 based on matching the maximum velocity and using an n value given by Equation 7.50. (b) Compare the energy and momentum correction factors for each of the velocity distributions considered in part (a) and assess the common assumption that these factors can be taken as approximately equal to unity.

SOLUTION

From the given data: $D = 300$ mm, $Q = 150$ L/s, and $k_\mathrm{s} = 0.5$ mm. For water at 20°C, $\nu = 1.00 \times 10^{-6}$ m²/s. The following parameters can be calculated from the given data:

$$A = \frac{\pi D^2}{4} = \frac{\pi(0.300)^2}{4} = 7.069 \times 10^{-2}\ \mathrm{m}^2, \qquad V = \frac{Q}{A} = \frac{150 \times 10^{-3}}{7.069 \times 10^{-2}} = 2.122\ \mathrm{m/s}$$

$$\mathrm{Re} = \frac{VD}{\nu} = \frac{(2.122)(0.300)}{1.00 \times 10^{-6}} = 6.366 \times 10^5, \qquad \frac{k_\mathrm{s}}{D} = \frac{0.5}{300} = 1.667 \times 10^{-3}$$

The friction factor, f, can be determined by substituting Re and k_s/D into the Colebrook equation (Equation 7.41), which gives

$$\frac{1}{\sqrt{f}} = -2\log\left[\frac{k_\mathrm{s}/D}{3.7} + \frac{2.51}{\mathrm{Re}\sqrt{f}}\right] \quad \rightarrow$$

$$\frac{1}{\sqrt{f}} = -2\log\left[\frac{1.667 \times 10^{-3}}{3.7} + \frac{2.51}{6.366 \times 10^5\sqrt{f}}\right] \quad \rightarrow \quad f = 0.0227$$

(a) The centerline velocity, V_0, given by the logarithmic velocity profile (Equation 7.46) occurs at $r = 0$. Therefore,

$$V_0 = \left[(1 + 1.326\sqrt{f}) - 2.04\sqrt{f}\log\left(\frac{R}{R-0}\right)\right]V \quad \rightarrow \quad V_0 = (1 + 1.326\sqrt{f})V$$

where V is the average velocity. Normalizing the logarithmic velocity distribution by V_0 and using the calculated value for f gives

$$\frac{v}{V_0} = 1 - \frac{2.04\sqrt{f}}{1 + 1.326\sqrt{f}}\log\left[\frac{1}{1-(r/R)}\right] \quad \rightarrow \quad \frac{v}{V_0} = 1 - 0.265\log\left[\frac{1}{1-(r/R)}\right] \tag{7.53}$$

Considering the power-law velocity distribution, the exponent, n, as given by Equation 7.50 is

$$\left[\frac{n}{n+1} - \frac{n}{2n+1}\right] = \frac{1}{2(1+1.326\sqrt{0.0227})} \quad \rightarrow \quad n = 7.84$$

Thus, the normalized power-law velocity distribution given by Equation 7.49 can be expressed as

$$\frac{v}{V_0} = \left(1 - \frac{r}{R}\right)^{\frac{1}{n}} \quad \rightarrow \quad \frac{v}{V_0} = \left(1 - \frac{r}{R}\right)^{0.128} \tag{7.54}$$

The logarithmic and power-law velocity distributions are plotted and compared in Figure 7.5.

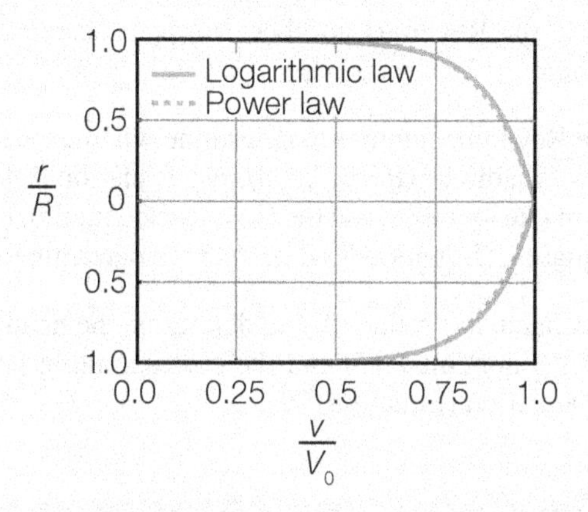

Figure 7.5: Comparison of turbulent velocity distributions

It is apparent from Figure 7.5 that the logarithmic and power-law velocity distributions are in very close agreement. However, the discrepancy that the velocity gradient does not asymptote to zero at the centerline of the pipe gives a sense of unreality to the computed velocity distribution near the centerline.

(b) The energy and momentum correction factors corresponding to the logarithmic velocity distribution are given by Equations 7.47 and 7.48, which yield

$$\alpha = 1 + 2.7f = 1 + 2.7(0.0227) = \mathbf{1.06}, \qquad \beta = 1 + 0.98f = 1 + 0.98(0.0227) = \mathbf{1.02}$$

The energy and momentum correction factors corresponding to the power-law velocity distribution are given by Equations 7.51 and 7.52, which yield

$$\alpha = \frac{(n+1)^3(2n+1)^3}{4n^4(n+3)(2n+3)} = \frac{(7.84+1)^3[2(7.84)+1]^3}{4(7.84)^4(7.84+3)[2(7.84)+3]} = \mathbf{1.05}$$

$$\beta = \frac{(n+1)^2(2n+1)^2}{2n^2(n+2)(2n+2)} = \frac{(7.84+1)^2[2(7.84)+1]^2}{2(7.84)^2(7.84+2)[2(7.84)+2]} = \mathbf{1.02}$$

Comparing the derived values of α and β for the logarithmic and power-law velocity distributions shows that with the retention of three significant digits, the values of α differ by 0.01 and the values of β are the same. For this case, the common assumption that $\alpha \approx 1$ and $\beta \approx 1$ seems reasonable.

7.5 Practical Applications

Practical applications of the energy equation are encountered in the analysis and design of pipeline systems. The analysis and design of such systems can include a variety of objectives, such as sizing pipelines to ensure adequate pressures and flow rates, estimating flow rates for given head differences, and determining the pumping requirements to achieve desired pressures and flow rates. Several of these applications are illustrated in the following sections.

7.5.1 Estimation of Pressure Changes

A case that is frequently encountered is where the volume flow rate through a pipe is known and the objective is to calculate the corresponding pressure change over a given length of the pipe. The following step-by-step approach can be used in solving such problems:

Step 1: Calculate the Reynolds number, Re, and the relative roughness, k_s/D, from the given data.

Step 2: Use the Reynolds number to determine whether the flow is laminar or turbulent. If the flow is laminar (Re $\leq$ 2000), use Equation 7.36 to determine f; if the flow is turbulent (Re $\geq$ 4000), use the Colebrook equation (Equation 7.41) or the Swamee–Jain equation (Equation 7.44 or 7.45) to determine f.

Step 3: Use the calculated value of f to determine the head loss from the Darcy–Weisbach equation (Equation 7.17) and the corresponding pressure change by applying the energy equation (Equation 7.9).

EXAMPLE 7.5

Water from a treatment plant is pumped into a distribution system at a rate of 4.38 m^3/s, a pressure of 480 kPa, and a temperature of 20°C. The pipe has a diameter of 750 mm and is made of ductile iron. (a) Estimate the pressure 200 m downstream of the treatment plant if the pipeline remains horizontal. (b) Compare the friction factor estimated using the Colebrook equation with the friction factor estimated using the Swamee–Jain equation. (c) After 20 years in operation, scale buildup is expected to cause the equivalent sand roughness of the pipe to increase by a factor of 10. Determine the effect on the water pressure 200 m downstream of the treatment plant.

SOLUTION

From the given data: $Q = 4.38$ m^3/s, $p_1 = 480$ kPa, $D = 750$ mm, $L = 200$ m, and $z_1 = z_2$. For ductile iron pipe, $k_s = 0.26$ mm. For water at 20°C, $\nu = 1.00 \times 10^{-6}$ m^2/s and $\gamma = 9.79$ kN/m^3. The following preliminary calculations are useful:

$$A = \frac{\pi}{4}D^2 = \frac{\pi}{4}(0.75)^2 = 0.442 \text{ m}^2, \qquad V = \frac{Q}{A} = \frac{4.38}{0.442} = 9.91 \text{ m/s}$$

(a) The Reynolds number, Re, and the relative roughness, k_s/D, are given by

$$\text{Re} = \frac{VD}{\nu} = \frac{(9.91)(0.75)}{1.00 \times 10^{-6}} = 7.43 \times 10^6, \qquad \frac{k_s}{D} = \frac{0.26}{750} = 3.47 \times 10^{-4}$$

Because Re $\gg$ 4000, the flow is turbulent and it is appropriate to use the Colebrook equation to estimate the friction factor. The Colebrook equation gives

$$\frac{1}{\sqrt{f}} = -2\log\left[\frac{k_s/D}{3.7} + \frac{2.51}{\text{Re}\sqrt{f}}\right]$$
$$= -2\log\left[\frac{3.47\times10^{-4}}{3.7} + \frac{2.51}{7.43\times10^{6}\sqrt{f}}\right] \quad\rightarrow\quad f = 0.016$$

The head loss, h_f, between the upstream and downstream sections is given by the Darcy–Weisbach equation:

$$h_f = \frac{fL}{D}\frac{V^2}{2g} = \frac{(0.016)(200)}{0.75}\frac{(9.91)^2}{(2)(9.81)} = 21.4 \text{ m}$$

Using the definition of head loss, h_f, and noting that the conduit is horizontal (i.e., $z_1 = z_2$),

$$h_f = \left(\frac{p_1}{\gamma} + z_1\right) - \left(\frac{p_2}{\gamma} + z_2\right) \quad\rightarrow\quad 21.4 = \frac{480}{9.79} - \frac{p_2}{9.79} \quad\rightarrow\quad p_2 = 270 \text{ kPa}$$

Therefore, the pressure 200 m downstream of the treatment plant is **270 kPa**.

(b) Using the Swamee–Jain approximation for f gives

$$\frac{1}{\sqrt{f}} = -2\log\left[\frac{k_s/D}{3.7} + \frac{5.74}{\text{Re}^{0.9}}\right]$$
$$= -2\log\left[\frac{3.47\times10^{-4}}{3.7} + \frac{5.74}{(7.43\times10^{6})^{0.9}}\right] \quad\rightarrow\quad f = \mathbf{0.016}$$

This is the same friction factor obtained using the Colebrook equation within an accuracy of two significant digits.

(c) After 20 years, the equivalent sand roughness, k_s, of the pipe is 2.6 mm, the (previously calculated) Reynolds number is 7.43×10^6, and the Colebrook equation gives

$$\frac{1}{\sqrt{f}} = -2\log\left[\frac{2.6/750}{3.7} + \frac{2.51}{7.43\times10^{6}\sqrt{f}}\right] \quad\rightarrow\quad f = 0.027$$

The head loss, h_f, between the upstream and downstream sections is given by the Darcy–Weisbach equation:

$$h_f = \frac{fL}{D}\frac{V^2}{2g} = \frac{(0.027)(200)}{0.75}\frac{(9.91)^2}{(2)(9.81)} = 36.0 \text{ m}$$

Hence, the pressure, p_2, 200 m downstream of the treatment plant can be derived from the energy equation,

$$h_f = \frac{p_1}{\gamma} - \frac{p_2}{\gamma} \quad\rightarrow\quad 36.0 = \frac{480}{9.79} - \frac{p_2}{9.79} \quad\rightarrow\quad p_2 = 128 \text{ kPa}$$

Therefore, pipe aging over 20 years will cause the pressure 200 m downstream of the treatment plant to **decrease from 270 kPa to 128 kPa**. This is a significant drop and shows why velocities as high as 9.91 m/s are not used in these types of pipelines, even for short lengths of pipe.

7.5.2 Estimation of Flow Rate for a Given Head Loss

In many cases, the volume flow rate through a pipe is not controlled, but attains a level that matches the available pressure drop. For example, the flow rate through a wide-open faucet in home plumbing is determined by the pressure drop in the service line between the pressurized water main at one end and atmospheric pressure at the other end. To derive an exact analytic solution to this problem, consider the following rearranged forms of the Darcy–Weisbach, Colebrook, and volume flow rate equations:

$$\text{Re}\sqrt{f} = \left(\frac{2gh_\text{f}D^3}{\nu^2 L}\right)^{\frac{1}{2}}, \qquad \text{(Darcy-Weisbach equation)} \qquad (7.55)$$

$$\text{Re} = -2.0(\text{Re}\sqrt{f})\log\left(\frac{k_\text{s}/D}{3.7} + \frac{2.51}{\text{Re}\sqrt{f}}\right), \qquad \text{(Colebrook equation)} \qquad (7.56)$$

$$Q = \frac{1}{4}\pi D^2 V = \frac{1}{4}\pi D\nu\text{Re}, \qquad \text{(Volume flow rate equation)} \qquad (7.57)$$

Equations 7.55–7.57 can be combined to yield the following expression for the volume flow rate in terms of known parameters:

$$Q = -0.965D^2\sqrt{\frac{gDh_\text{f}}{L}}\ln\left(\frac{k_\text{s}/D}{3.7} + \frac{1.784\nu}{D\sqrt{gDh_\text{f}/L}}\right) \qquad (7.58)$$

This equation was originally derived by Fay (1994), is dimensionally homogeneous, and is particularly useful because the flow rate, Q, can be explicitly calculated for given values of D, h_f, L, and k_s. The use of Equation 7.58 must necessarily be validated by verifying that $\text{Re} > 4000$, which is required for application of the Colebrook equation.

EXAMPLE 7.6

A 50-mm-diameter galvanized iron service pipe is connected to a water main in which the pressure is 450 kPa gauge. If the length of the service pipe to a faucet is 40 m and the faucet is 1.2 m above the main, estimate the flow rate when the faucet is fully open. Neglect any head loss in the faucet.

SOLUTION

From the given data: D = 50 mm, p_1 = 450 kPa, L = 40 m, and $z_2 - z_1$ = 1.2 m. Assuming that the faucet is open to the atmosphere, p_2 = 0 kPa. For galvanized iron, k_s = 0.15 mm (from Table 7.2). For water at 20°C, γ = 9.79 kN/m^3 and $\nu = 1.00 \times 10^{-6}$ m^2/s. The head loss, h_f, in the pipe is estimated by

$$h_\text{f} = \left(\frac{p_1}{\gamma} + z_1\right) - \left(\frac{p_2}{\gamma} + z_2\right) = \left(\frac{450}{9.79} + 0\right) - \left(0 + 1.2\right) = 44.8 \text{ m}$$

The flow rate equation given by Equation 7.58 yields

$$Q = -0.965D^2\sqrt{\frac{gDh_\text{f}}{L}}\ln\left(\frac{k_\text{s}/D}{3.7} + \frac{1.784\nu}{D\sqrt{gDh_\text{f}/L}}\right)$$

$$= -0.965(0.05)^2\sqrt{\frac{(9.81)(0.05)(44.8)}{40}}\ln\left[\frac{0.15/50}{3.7} + \frac{1.784(1.00 \times 10^{-6})}{(0.05)\sqrt{(9.81)(0.05)(44.8)/40}}\right]$$

$$= 0.0126 \text{ m}^3\text{/s} = 12.6 \text{ L/s}$$

Therefore, the faucet can be expected to deliver **12.6 L/s** when fully open. From this result, Re = 3.21×10^5, which means that the flow is turbulent (i.e., Re ≥ 4000) and use of Equation 7.58 is validated. Such validation is commonly taken for granted when the flow of water in pipes is being considered.

7.5.3 Estimation of Diameter for a Given Flow Rate and Head Loss

In many cases, the design engineer must select a pipe diameter to provide a given level of service. For example, the minimum required flow rate for an available head loss might be specified for a water delivery pipe and the design engineer wants to calculate the minimum pipe diameter that will satisfy these design constraints. This is the usual design problem in gravity-driven pipeline systems (e.g., Jones, 2011). Solution of this design problem necessarily requires a numerical procedure and can be easily accomplished by using a programmable calculator, a spreadsheet with solver capability, or a numerical computing environment such as MATLAB.

EXAMPLE 7.7

A galvanized iron service pipe from a water main is required to deliver 200 L/s during a fire. If the length of the service pipe is 35 m and the head loss in the pipe is not to exceed 50 m, calculate the minimum pipe diameter that can be used.

SOLUTION

From the given data: $Q = 200$ L/s $= 0.200$ m^3/s, $L = 35$ m, and $h_\mathrm{f} = 50$ m. For water at 20°C, $\nu = 10^{-6}$ m^2/s. For galvanized iron, $k_\mathrm{s} = 0.15$ mm (from Table 7.2). Substituting these data into Equation 7.58 gives

$$Q = -0.965D^2\sqrt{\frac{gDh_\mathrm{f}}{L}}\ln\left[\frac{k_\mathrm{s}/D}{3.7} + \frac{1.784\nu}{D\sqrt{gDh_\mathrm{f}/L}}\right]$$

$$0.2 = -0.965D^2\sqrt{\frac{(9.81)D(50)}{(35)}}\ln\left[\frac{0.00015}{3.7D} + \frac{1.784(10^{-6})}{D\sqrt{(9.81)D(50)/(35)}}\right] \quad \rightarrow \quad D = 0.136 \text{ m}$$

Therefore, the minimum pipe diameter that will deliver 200 L/s is **136 mm**. Validating use of Equation 7.58 requires showing that Re ≥ 4000. In this case, such a validation can be taken for granted.

Most numerical procedures that could be used for obtaining D in the previous example converge quickly and do not pose any computational difficulties. In lieu of determining an exact numerical solution, Swamee and Jain (1976) suggested the following explicit formula for calculating the pipe diameter, D:

$$D = 0.66\left[k_\mathrm{s}^{1.25}\left(\frac{LQ^2}{gh_\mathrm{f}}\right)^{4.75} + \nu Q^{9.4}\left(\frac{L}{gh_\mathrm{f}}\right)^{5.2}\right]^{0.04} \tag{7.59}$$

Equation 7.59 is dimensionally homogeneous, is valid for $4000 \leq \text{Re} \leq 3 \times 10^8$ and $10^{-6} \leq k_\mathrm{s}/D \leq 2 \times 10^{-2}$, and yields a D within 5% of the value obtained by an exact solution of Equation 7.58. Use of Equation 7.59 is illustrated by repeating the previous example.

EXAMPLE 7.8

A galvanized iron service pipe from a water main is required to deliver 200 L/s during a fire. If the length of the service pipe is 35 m and the head loss in the pipe is not to exceed 50 m, use the Swamee–Jain approximation given by Equation 7.59 to calculate the minimum pipe diameter that can be used.

SOLUTION

Because $k_s = 0.15$ mm, $L = 35$ m, $Q = 0.2$ m^3/s, $h_f = 50$ m, and $\nu = 1.00 \times 10^{-6}$ m^2/s, the Swamee–Jain approximation (Equation 7.59) gives

$$D = 0.66\left[k_s^{1.25}\left(\frac{LQ^2}{gh_f}\right)^{4.75} + \nu Q^{9.4}\left(\frac{L}{gh_f}\right)^{5.2}\right]^{0.04}$$

$$= 0.66\left\{(0.00015)^{1.25}\left[\frac{(35)(0.2)^2}{(9.81)(50)}\right]^{4.75} + (1.00 \times 10^{-6})(0.2)^{9.4}\left[\frac{35}{(9.81)(50)}\right]^{5.2}\right\}^{0.04}$$

$$= 0.140 \text{ m}$$

The calculated pipe diameter (**140 mm**) is about 3% higher than that calculated by Equation 7.58 (i.e., 136 mm).

7.5.4 Head Losses in Noncircular Conduits

In cases where a pipe has a noncircular cross section, estimation of head losses is still done using the combination of Darcy–Weisbach and Colebrook equations for circular pipes, with the exception that the diameter, D, in these equations is replaced by the hydraulic diameter, D_h, as defined by Equation 7.22.

EXAMPLE 7.9

Water flows through a rectangular concrete culvert of width 2 m and depth 1 m. If the length of the culvert is 10 m and the flow rate is 6 m^3/s, estimate the head loss through the culvert. Assume that the culvert flows full.

SOLUTION

From the given data: $w = 2$ m, $d = 1$ m, $L = 10$ m, and $Q = 6$ m^3/s. A median equivalent sand roughness for concrete is $k_s = 1.6$ mm (Table 7.2). For water at 20°C, $\nu = 1.00 \times 10^{-6}$ m^2/s. The following preliminary calculations are useful:

$$A = wd = (2)(1) = 2 \text{ m}^2, \qquad P = 2(w + d) = 2(2 + 1) = 6 \text{ m}$$

$$D_h = \frac{4A}{P} = \frac{4(2)}{6} = 1.333 \text{ m}, \qquad V = \frac{Q}{A} = \frac{6}{2} = 3 \text{ m/s}$$

$$\text{Re} = \frac{VD_h}{\nu} = \frac{(3)(1.333)}{1.00 \times 10^{-6}} = 4.00 \times 10^6, \qquad \frac{k_s}{D_h} = \frac{1.6 \times 10^{-3}}{1.333} = 0.00120$$

Substituting Re and k_s/D_h into the Swamee–Jain equation (Equation 7.45) for the friction factor gives

$$f = \frac{0.25}{\left[\log\left(\frac{k_s/D_h}{3.7} + \frac{5.74}{\mathrm{Re}^{0.9}}\right)\right]^2} = \frac{0.25}{\left[\log\left(\frac{0.00120}{3.7} + \frac{5.74}{(4.00 \times 10^6)^{0.9}}\right)\right]^2} = 0.0206$$

Therefore, the frictional head loss in the rectangular concrete culvert, h_f, is given by the Darcy–Weisbach equation as

$$h_f = \frac{fL}{D_h}\frac{V^2}{2g} = \frac{(0.0206)(10)}{1.333}\frac{3^2}{2(9.81)} = 0.0709 \text{ m}$$

The head loss in the culvert can therefore be estimated as approximately **7.1 cm**.

7.5.5 Empirical Friction Loss Formulas

Friction loss in pipelines should generally be calculated using the Darcy–Weisbach equation. However, a minor inconvenience in using the Darcy–Weisbach equation to relate the friction loss to the flow velocity results from the possible dependence of the friction factor on the flow velocity; in which case the Darcy–Weisbach equation must be solved simultaneously with the Colebrook equation. In modern engineering practice, computers make this a minor inconvenience. In earlier years, however, this was considered a real problem, and various empirical head loss formulas were developed to relate the head loss directly to the flow velocity. Those most commonly used are the *Hazen–Williams formula* and the *Manning formula*.

Hazen–Williams formula. The Hazen–Williams formula (Williams and Hazen, 1920) is applicable only to the flow of water in pipes and is given by

$$V = 0.849 C_H R^{0.63} S_f^{0.54} \tag{7.60}$$

where V is the flow velocity [m/s], C_H is the Hazen–Williams roughness coefficient [dimensionless], R is the hydraulic radius [m], and S_f is the slope of the energy grade line [dimensionless], defined by

$$S_f = \frac{h_f}{L} \tag{7.61}$$

where h_f [L] is the head loss due to friction over a length L [L] of pipe. Values of C_H for a variety of commonly used pipe materials are given in Table 7.3, where the value of C_H typically varies in the range of 70–150 depending on the pipe material, diameter, and age. Combining Equations 7.60 and 7.61 yields the following expression for the frictional head loss:

$$h_f = 6.82 \frac{L}{D^{1.17}} \left(\frac{V}{C_H}\right)^{1.85} \tag{7.62}$$

where D is the diameter of the pipe [m]. The Hazen–Williams equation is commonly assumed to be applicable to the flow of water at 16°C in pipes with diameters in the range of 50–1850 mm and flow velocities less than 3 m/s (e.g., Mott, 2006). Street and colleagues (1996) and Liou (1998) have shown that the Hazen–Williams coefficient has a strong Reynolds number dependence and is mostly applicable when the pipe is relatively smooth and is in the early part of its transition to rough flow. Furthermore, Jain and colleagues (1978) have shown that an error of up to 39% can be expected in the estimation of the velocity by the Hazen–Williams formula over a wide range of diameters and slopes. In spite of

Table 7.3: Pipe Roughness Coefficients

Pipe material	C_H Range	C_H Typical	n Range	n Typical
Ductile and cast iron:				
New, unlined	120–140	130	—	0.013
Old, unlined	40–100	80	—	0.025
Cement lined and seal coated	100–140	120	0.011–0.015	0.013
Steel:				
Welded and seamless	80–150	120	—	0.012
Riveted	100–140	110	0.012–0.018	0.015
Mortar lining	120–145	130	—	—
Asbestos cement		140	—	0.011
Concrete	100–140	120	0.011–0.015	0.012
Vitrified clay pipe (VCP)	110–140	110	0.012–0.014	—
Polyvinyl chloride (PVC)	135–150	140	0.007–0.011	0.009
Corrugated metal pipe (CMP)	—	—	—	0.025

Sources: Cruise et al.(2007); Velon and Johnson (1993); Wurbs and James (2002).

these cautionary notes, the Hazen–Williams formula is frequently used in the United States for the design of large water supply pipes and building plumbing systems without regard to its limited range of applicability, practices that can have detrimental effects on pipe design and may lead to litigation (e.g., Bombardelli and García, 2003). In some cases, engineers have calculated correction factors for the Hazen–Williams roughness coefficient to account for these errors (e.g., Valiantzas, 2005). The attractiveness of the Hazen–Williams formula to practicing engineers is likely related to the relatively large database for C_H values compared with the relatively small database for equivalent sand roughnesses, k_s, required for application of the (preferred) Darcy–Weisbach equation. This reality can be partially addressed by determining the k_s values corresponding to measured C_H values under given experimental conditions; the k_s values remain constant under all flows and pipe sizes, but the C_H values do not. The relationship between k_s and C_H can be approximated by

$$k_s = D(3.320 - 0.021C_H D^{0.01})^{2.173} \exp(-0.04125C_H D^{0.01}) \tag{7.63}$$

where k_s is the equivalent sand roughness [m] and D is the diameter of the pipe [m] used in determining the Hazen–Williams coefficient C_H by experimentation; Equation 7.63 is approximately valid for $100 \leq C_H D^{0.01} \leq 155$ (Travis and Mays, 2007). In cases where D is unknown, a reasonable relationship between k_s and C_H can be determined by assuming that D is equal to 300 mm.

Manning formula. A second empirical formula that is sometimes used to describe flow in pipes is the Manning formula, which is given by

$$V = \frac{1}{n} R^{\frac{2}{3}} S_f^{\frac{1}{2}} \tag{7.64}$$

where V, R, and S_f have the same meaning and units as in the Hazen–Williams formula, and n is the Manning roughness coefficient. Values of n for a variety of commonly used pipe materials are given in Table 7.3. Combining Equations 7.64 and 7.61 yields the following expression for the frictional head loss:

$$h_f = 6.35 \frac{n^2 L V^2}{D^{\frac{4}{3}}} \tag{7.65}$$

The Manning formula applies only to fully turbulent flows, where the frictional head losses are controlled by the relative roughness. Such conditions are delineated by Equation 7.43. Practical application of the Manning equation to pipe flow is encountered in the design of storm sewers.

EXAMPLE 7.10

Water flows at a velocity of 1 m/s in a 150-mm-diameter new ductile iron pipe. Estimate the head loss over 500 m using (a) the Hazen–Williams formula, (b) the Manning formula, and (c) the Darcy–Weisbach equation. Compare your results and assess the validity of each head loss equation.

SOLUTION

(a) The Hazen–Williams roughness coefficient, C_H, can be taken as 130 (Table 7.3), L = 500 m, D = 0.150 m, and V = 1 m/s. Therefore the head loss, h_f, is given by Equation 7.62 as

$$h_\text{f} = 6.82\frac{L}{D^{1.17}}\left(\frac{V}{C_\text{H}}\right)^{1.85} = 6.82\frac{500}{(0.15)^{1.17}}\left(\frac{1}{130}\right)^{1.85} = \mathbf{3.85\ m}$$

(b) The Manning roughness coefficient, n, can be taken as 0.013 (approximation from Table 7.3). Therefore the head loss, h_f, is given by Equation 7.65 as

$$h_\text{f} = 6.35\frac{n^2LV^2}{D^{\frac{4}{3}}} = 6.35\frac{(0.013)^2(500)(1)^2}{(0.15)^{\frac{4}{3}}} = \mathbf{6.73\ m}$$

(c) The equivalent sand roughness, k_s, can be taken as 0.26 mm (Table 7.2), and the Reynolds number, Re, is given by

$$\text{Re} = \frac{VD}{\nu} = \frac{(1)(0.15)}{1.00\times10^{-6}} = 1.5\times10^5$$

where $\nu = 1.00\times10^{-6}$ m²/s at 20°C. Substituting k_s, D, and Re into the Colebrook equation yields the friction factor, f, where

$$\frac{1}{\sqrt{f}} = -2\log\left[\frac{k_\text{s}/D}{3.7} + \frac{2.51}{\text{Re}\sqrt{f}}\right]$$

$$= -2\log\left[\frac{0.26/150}{3.7} + \frac{2.51}{1.5\times10^5\sqrt{f}}\right] \quad\rightarrow\quad f = 0.0238$$

The head loss, h_f, is therefore given by the Darcy–Weisbach equation as

$$h_\text{f} = f\frac{L}{D}\frac{V^2}{2g} = 0.0238\frac{500}{0.15}\frac{1^2}{2(9.81)} = \mathbf{4.04\ m}$$

It is reasonable to assume that the Darcy–Weisbach equation yields the most accurate estimate of the head loss. In this case, the Hazen–Williams formula gives a head loss that is 5% less than the Darcy–Weisbach equation and the Manning formula yields a head loss that is 67% higher than the Darcy–Weisbach equation.

From the given data, Re = 1.5×10^5, $D/k_s = 150/0.26 = 577$, and Equation 7.43 gives the limit of fully turbulent flow as

$$\frac{1}{\sqrt{f}} = \frac{\text{Re}}{200(D/k_s)} = \frac{1.5 \times 10^5}{200(577)} \quad \rightarrow \quad f = 0.591$$

Because the actual friction factor (= 0.0238) is much less than the minimum friction factor for fully turbulent flow (= 0.591), the flow is not fully turbulent and the **Manning equation is not valid**. Because the pipe diameter (= 150 mm) is between 50 mm and 1850 mm and the velocity (= 1 m/s) is less than 3 m/s, the **Hazen–Williams formula can be taken as valid**. The **Darcy–Weisbach equation is unconditionally valid.** Given these results, it is not surprising that the Darcy–Weisbach and Hazen–Williams formulas are in close agreement, with the Manning equation giving a significantly different result. These results indicate why application of the Manning equation to closed-conduit flows is strongly discouraged.

Practical note. In spite of the fact that the combined Darcy–Weisbach and Colebrook equations provide the most accurate description of flow in pipes, approximations continue to be developed to circumvent the relatively minor computational inconveniences in using these equations.

7.5.6 Local Head Losses

Head losses in straight pipes of constant diameter are primarily caused by friction between the moving fluid and the (stationary) pipe and are estimated using the Darcy–Weisbach equation. Flow through pipe fittings, around bends, and through changes in pipeline geometry cause additional head losses that are typically quantified by an equation of the form

$$h_{\text{local}} = K\frac{V^2}{2g} \tag{7.66}$$

where h_{local} is the *local head loss* [L], K is a *local loss coefficient* [dimensionless], and V is the average velocity [LT^{-1}] at a defined location within a transition or fitting. In practical applications, K is commonly called the *head loss coefficient*. The head loss coefficient, K, is technically a function of both the geometry of a fitting and a characteristic Reynolds number (Re) of the flow through the fitting. However, Re is usually assumed to be sufficiently high that K is taken to be a function of the geometry of the fitting only. If inflow and outflow sections of a fitting have the same cross-sectional area and elevation, then the steady-state energy equation requires that

$$h_{\text{local}} = -\frac{\Delta p}{\gamma} \tag{7.67}$$

where Δp is the difference between the outflow pressure and the inflow pressure and γ is the specific weight of the fluid. Combining Equations 7.66 and 7.67 gives the following useful relationship between the head loss coefficient and the pressure change across a fitting:

$$K = -\frac{\Delta p}{\frac{1}{2}\rho V^2} \tag{7.68}$$

It is important to remember that the pressure generally decreases between the inflow and outflow sections of a fitting, so values of Δp are generally negative. In applications involving head loss coefficients, Equation 7.66 is the relationship that is most commonly used in accounting for local head losses. Head losses in transitions and fittings are called *local head losses* or *minor head losses*; however, the latter term should be avoided because in some cases these head losses are a significant portion of the total head loss in a pipe. Commonly encountered transitions include bends, tees, and changes in diameter, whereas commonly encountered fittings include various types of valves, such as gate valves that are used to open and close pipelines carrying liquids, and globe valves that are used to regulate the flow of liquids in pipelines and are used as faucets at the end of pipes in household plumbing. The head loss coefficients for several transitions and fittings are shown in Figures 7.6 and 7.7, where the locations of the velocities to be used in Equation 7.66 to calculate the local head loss are also shown.

General guidelines. There is considerable variability in the head loss coefficients given in Figures 7.6 and 7.7, because head loss coefficients generally vary with surface roughness, Reynolds number, and geometric details of the design. The head loss coefficients of two seemingly identical valves from two different manufacturers can differ by a factor of 2 or more. Also, whether a fitting is screw-connected or flange-connected can significantly affect the head loss coefficient of the fitting. In cases of identical fitting geometry with the only difference being a screw connection versus a flange connection, the fitting with the screw connection typically has a significantly higher head loss coefficient. However, differences in the geometry of the fitting can mask differences in the type of connection. For purposes of brevity, the ranges of head loss coefficients given in Figures 7.6 and 7.7 include both screw and flange connections and reflect the range of designs and materials that are used in various transitions and fittings. Based on these considerations, recent and fitting-specific manufacturers' data should be consulted in the final design of piping systems rather than relying on the representative values in textbooks and handbooks.

Head losses at inlets and outlets. Head losses at submerged (in liquid) pipe inlets are influenced significantly by the geometry of the inlet, with the head loss coefficient, K, varying from 0.03 for a well-rounded inlet to 0.80 for a reentrant (i.e., protruding) inlet. It is apparent that minimization of inlet losses can be achieved by rounding inlets. At submerged outlets, pipe flows typically lose all their velocity head, so K at a submerged pipe outlet is equal to the kinetic energy correction coefficient, α, regardless of the geometry of the outlet. For fully developed turbulent flows in pipes, $\alpha \approx 1$, so $K = 1$ is widely used to estimate the head loss at a submerged outlet. However, recognizing that $\alpha = 2$ for fully developed laminar flow in pipes, it is prudent to use the general relation $K = \alpha$ in calculating local head losses at submerged outlets. Because the head loss at a submerged outlet is independent of the outlet geometry, there is no advantage to rounding sharp edges in pipe outlets.

Head losses in expansions and contractions. Within pipeline systems, changes in diameter can occur suddenly or gradually within a fitting, with a greater head loss occurring for a sudden change in diameter. For both expansions and contractions, the velocity in the smaller pipe is used in Equation 7.66 to calculate the local head loss, and head losses in expansions are typically much higher than head losses in comparable contractions, primarily due to the flow separation that occurs in expansions. The head loss coefficient in a sudden contraction, K_{sc}, can be estimated by

$$K_{sc} = \begin{cases} 0.42\left(1 - \dfrac{D_2^2}{D_1^2}\right), & \text{for } D_2/D_1 \leq 0.76 \\[2ex] \left(1 - \dfrac{D_2^2}{D_1^2}\right)^2, & \text{for } D_2/D_1 > 0.76 \end{cases} \tag{7.69}$$

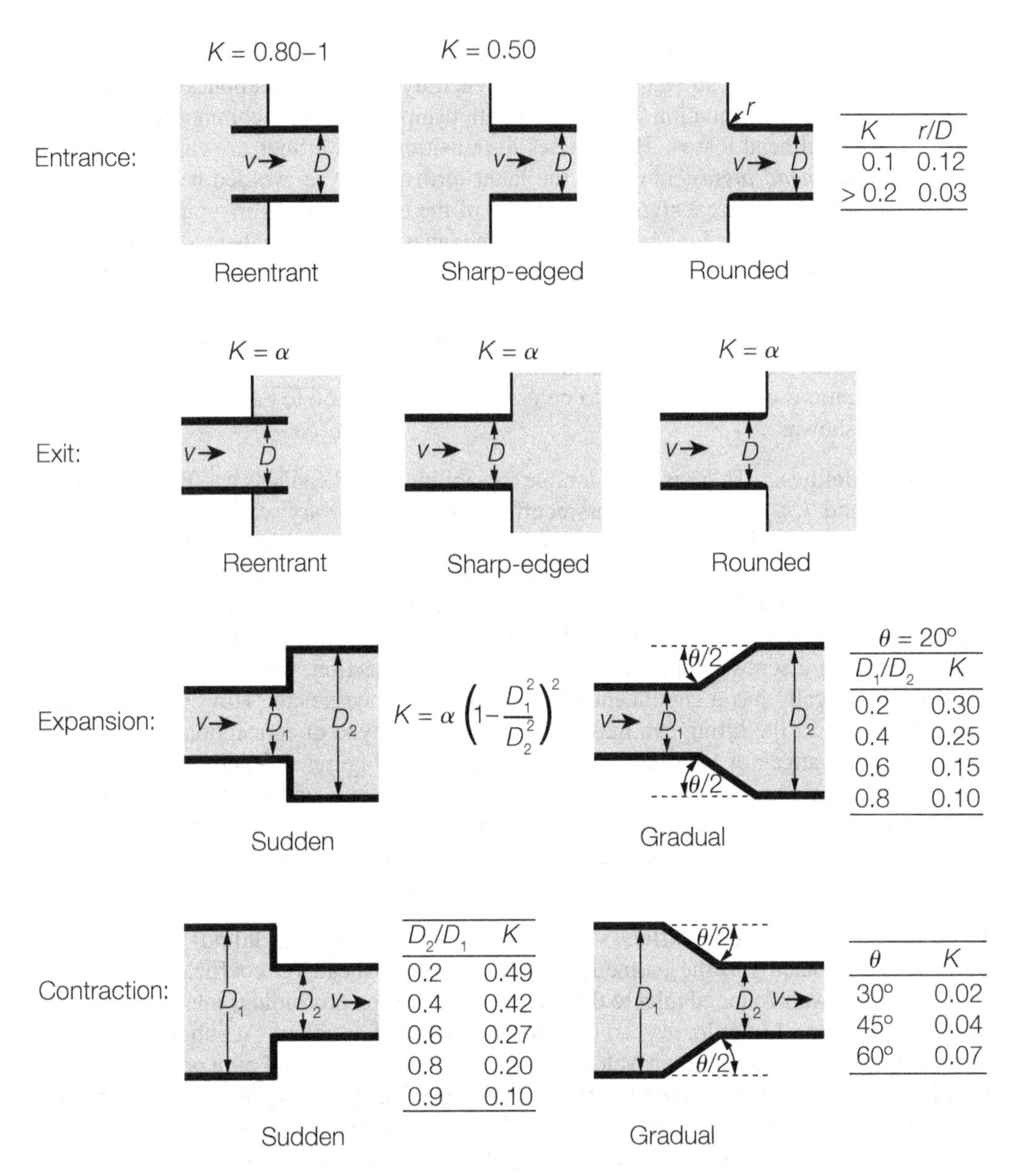

Figure 7.6: **Head loss coefficients for pipe entrances, exits, expansions, and contractions**

where D_1 and D_2 are the upstream and downstream diameters, respectively. The head loss coefficient in a sudden expansion, K_{se}, can be estimated by

$$K_{se} = \left(1 - \frac{D_1^2}{D_2^2}\right)^2 \tag{7.70}$$

where D_1 and D_2 are the upstream and downstream diameters, respectively. Equation 7.70 can be derived theoretically (e.g., Massey and Ward-Smith, 2012), and the head loss calculated using this coefficient is sometimes called the *Borda–Carnot head loss*.[51] A gradual expansion is commonly called a *diffuser*, a term that is sometimes used generically to refer to an expanding conduit. Head losses due to flow separation are of particular concern in the

[51] Named in honor of the French scientist Jean-Charles de Borda (1733–1799) and the French engineer Lazare Nicolas Marguérite Carnot (1753–1823).

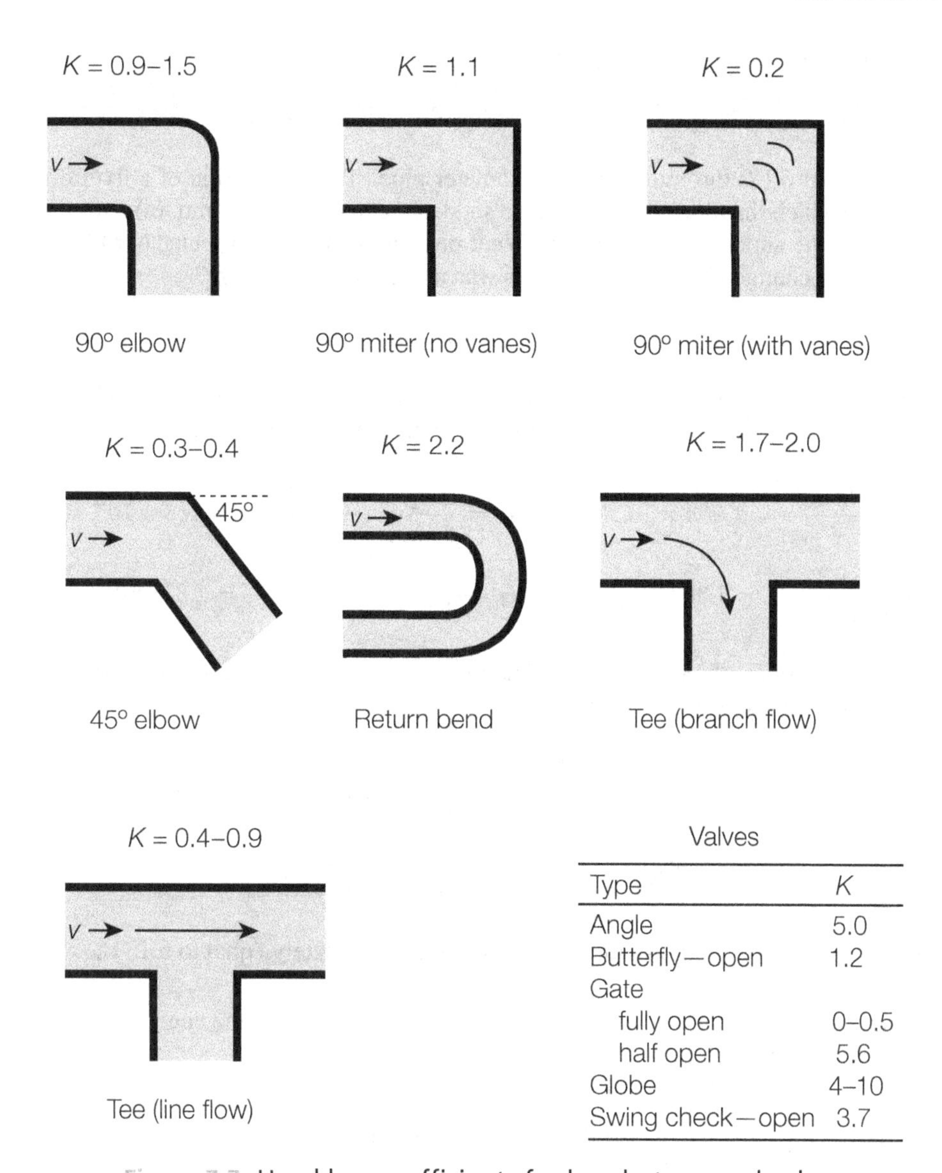

Valves

Type	K
Angle	5.0
Butterfly—open	1.2
Gate	
fully open	0–0.5
half open	5.6
Globe	4–10
Swing check—open	3.7

Figure 7.7: Head loss coefficients for bends, tees, and valves

design of diffusers, and such losses can be controlled by proper selection of the expansion angle, θ, illustrated in Figure 7.6. The head loss coefficient in a diffuser or gradual expansion, K_{ge}, can be estimated by the relation

$$K_{\text{ge}} = 2.61\sin\left(\frac{\theta}{2}\right)\left(1 - \frac{D_1^2}{D_2^2}\right)^2 + \frac{f_{\text{ge}}L_{\text{ge}}}{\overline{D}_{\text{ge}}}, \quad \text{for } \theta \leq 45^\circ \tag{7.71}$$

where f_{ge}, L_{ge}, and $\overline{D}_{\text{ge}}$ are the friction factor, length, and mean diameter, respectively, of the diffuser. For large expansion angles ($\theta > 45^\circ$), the coefficient of $2.61\sin(\theta/2)$ can be taken as 1.0, and Equation 7.71 gives the same loss coefficient as a sudden expansion, plus the friction loss. Minimum losses in diffusers and gradual expansions occur when $5^\circ < \theta < 15^\circ$, and the optimal value of θ is commonly cited as 6°. For $\theta < 5^\circ$, the diffuser is too long and has too much friction, and for $\theta > 15^\circ$, flow separation results in poor pressure recovery. The head loss in diverging flow is always greater than that in converging flow. In fact, in contrast to the head loss in a gradual expansion, a gradual contraction without sharp corners causes such a small head loss that it can usually be neglected.

EXAMPLE 7.11

Water at 20°C flows at 6 L/s through a 50-mm-diameter pipe. The installation of a fixture is being considered in which the pipe diameter will be suddenly expanded to 75 mm. (a) Estimate the head loss due to the sudden expansion. (b) What pressure change is expected? (c) What would be the pressure change if the local head loss was not taken into account?

SOLUTION

From the given data: $Q = 6$ L/s $= 0.006$ m^3/s, $D_1 = 50$ mm, and $D_2 = 75$ mm. For water at 20°C, $\rho = 998$ kg/m^3 and $\gamma = 9790$ N/m^3. The following preliminary calculations are useful:

$$A_1 = \frac{\pi D_1^2}{4} = \frac{\pi(0.050)^2}{4} = 1.963 \times 10^{-3}\ \text{m}^2, \quad A_2 = \frac{\pi D_2^2}{4} = \frac{\pi(0.075)^2}{4} = 4.418 \times 10^{-3}\ \text{m}^2$$

$$V_1 = \frac{Q}{A_1} = \frac{0.006}{1.963 \times 10^{-3}} = 3.056\ \text{m/s}, \qquad V_2 = \frac{Q}{A_2} = \frac{0.006}{4.418 \times 10^{-3}} = 1.358\ \text{m/s}$$

(a) Using the head loss coefficient given by Equation 7.70, the head loss, h_ℓ, at the sudden expansion is calculated as follows:

$$K = \left(1 - \frac{D_1^2}{D_2^2}\right)^2 = \left(1 - \frac{50^2}{75^2}\right)^2 = 0.3086$$

$$h_\ell = K\frac{V_1^2}{2g} = (0.3086)\frac{3.056^2}{2(9.807)} = 0.1469\ \text{m}$$

Therefore, the head loss at the sudden expansion is approximately equal to **0.15 m**.

(b) The pressure change, Δp, at the expansion is calculated by applying the energy equation across the expansion, such that

$$-\left(\frac{\Delta p}{\gamma} + \frac{\Delta V^2}{2g}\right) = h_\ell \quad \rightarrow \quad \Delta p = -\left(\gamma h_\ell + \tfrac{1}{2}\rho\Delta V^2\right) \qquad (7.72)$$

Substituting the given and derived data into Equation 7.72 gives

$$\Delta p = -\left[(9790)(0.1469) + \tfrac{1}{2}(998)(1.358^2 - 3.056^2)\right]\ [\times 10^{-3}\ \text{kPa/Pa}] = -1.25\ \text{kPa}$$

Therefore, it is expected that the pressure will decrease by **1.25 kPa** across the sudden expansion.

(c) If the local head loss is not taken into account, then $h_\ell = 0$, and Equation 7.72 gives

$$\Delta p = -\left[0 + \tfrac{1}{2}(998)(1.358^2 - 3.056^2)\right]\ [\times 10^{-3}\ \text{kPa/Pa}] = 0.1961\ \text{kPa}$$

Therefore, if the local head loss was neglected, the pressure would be predicted to increase by approximately **0.20 kPa**, rather than the more likely outcome that it would decrease by 1.25 kPa.

Head losses in bends. Local head losses in bends are caused primarily by flow separation on the curved walls and the swirling secondary flow caused by centrifugal acceleration as the fluid rounds the bend. Head losses in bends typically depend on the angle of the bend and on the relative radius of curvature R_c/D, where R_c is the radius of curvature of the centerline of the bend and D is the diameter of the conduit. Local head losses in bends and elbows can be reduced by using smooth rounded turns, rather than sharp turns as occur in miter bends. In cases where sharp turns are necessary, such as when space is limited, losses within the bend can be minimized by using guide vanes, a configuration that is sometimes called a *cascade*.

Head losses in valves. Valves are used in pipeline systems to regulate the flow, and they do so by creating increased local head losses as they are closed, with $K \to \infty$ as a valve is closed. The most common types of valves are (1) *gate valves*, which slide down across the open section; (2) *globe valves*, which close a hole in the valve; (3) *angle valves*, which are similar to globe valves but have a 90° turn; (4) *swing check valves*, which allow only one-way flow; and (5) *butterfly valves*, which operate by rotating a disk about a central axis in the pipe, where the valve is open when the face of the disk is perpendicular to the flow and the valve is closed when the face of the disk is normal to the flow. It is usually desirable that K for a valve be very small when it is fully open so as not to interfere with the flow; however, $K = 0$ for a fully open valve is seldom achievable.

Downstream influences. Although local head losses are commonly assumed to occur within a fitting or transition, these appurtenances typically influence the flow for several pipe diameters downstream. This is why most flow meter manufacturers recommend installing flow meters at least 10–20 pipe diameters downstream of any elbows or valves. This allows the swirling turbulent eddies generated by the elbow or valve to largely disappear and the velocity profile to become fully developed before entering the flow meter.

Head losses in nozzles. Head loss coefficients in nozzles are typically small due to the relative efficiency of accelerating a fluid. Typical head loss coefficients are in the range of K = 0.02–0.07, and the exit velocity of the nozzle is used in the calculation of the local head loss.

Hydrodynamic entrance length. The *hydrodynamic entrance length* of a pipe is defined as the distance from a pipe entrance to where the boundary shear stress (and thus the friction factor) reaches within 2% of the fully developed value. In many pipe flows of practical interest, the hydrodynamic entrance length is approximately equal to 10 pipe diameters. Over distances on the order of the hydrodynamic entrance length, frictional losses are greater than predicted by assuming fully developed flow from the pipe entrance; however, over distances much longer than the hydrodynamic entrance length, entrance effects are negligible.

Local head loss in terms of equivalent length. The *equivalent length* of conduit associated with a local head loss is defined as the length of the conduit that will produce a head loss equal to the local head loss. This definition can be expressed as

$$K\frac{V^2}{2g} = f\frac{L_{eq}}{D}\frac{V^2}{2g} \quad \to \quad L_{eq} = \frac{K}{f}D \tag{7.73}$$

where L_{eq} is the equivalent length, D is the conduit diameter, f is the friction factor, and K is the local head loss coefficient. Using Equation 7.73, a local head loss can be accounted for by adding a length L_{eq} to the conduit. The utilization of equivalent lengths to account for local head losses is commonplace in representing various fixtures in building plumbing systems.

EXAMPLE 7.12

Water at 15°C flows at a rate of 60 L/min in a 25-mm-diameter galvanized iron pipe that has an estimated equivalent sand roughness of 0.1 mm. The pipeline has a total length of 14 m and contains five threaded bend fittings, each with a head loss coefficient of 1.5. (a) What is the equivalent length of each bend? (b) What is the equivalent length of the entire pipeline? (c) Assess whether the bends contribute significantly to the head loss in the pipeline.

SOLUTION

From the given data: $Q = 60$ L/min $= 1.00 \times 10^{-3}$ m^3/s, $D = 25$ mm, $k_s = 0.1$ mm, $L = 14$ m, and $K_b = 1.5$. For water at 15°C, $\rho = 999.1$ kg/m^3, $\mu = 1.139$ mPa·s, and $\nu = \mu/\rho = 1.140 \times 10^{-6}$ m^2/s. The following preliminary calculations are useful:

$$A = \frac{\pi D^2}{4} = \frac{\pi(0.025)^2}{4} = 4.909 \times 10^{-4}\ \text{m}^2, \qquad V = \frac{Q}{A} = \frac{1.00 \times 10^{-3}}{4.909 \times 10^{-4}} = 2.037\ \text{m/s}$$

$$\text{Re} = \frac{VD}{\nu} = \frac{(2.037)(0.025)}{1.140 \times 10^{-6}} = 4.468 \times 10^4, \qquad \frac{k_s}{D} = \frac{0.1}{25} = 0.0040$$

Using the Colebrook equation to determine the friction factor, f,

$$\frac{1}{\sqrt{f}} = -2\log\left[\frac{k_s/D}{3.7} + \frac{2.51}{\text{Re}\sqrt{f}}\right] \rightarrow$$

$$\frac{1}{\sqrt{f}} = -2\log\left[\frac{0.0040}{3.7} + \frac{2.51}{4.468 \times 10^4\sqrt{f}}\right] \quad \rightarrow \quad f = 0.0307$$

(a) The equivalent length, L_b, of each bend is given by Equation 7.73 as

$$L_b = \frac{K_b}{f}D = \frac{1.5}{0.0307}(0.025) = \mathbf{1.22\ m}$$

(b) Because there are five bends in the entire system, the equivalent length of the entire system, L_e, is given by

$$L_e = L + 5L_b = 14 + 5(1.22) = \mathbf{20.1\ m}$$

(c) The fraction, p_b, of the total head loss in the system that is attributed to the bends is given by

$$p_b = \frac{5\dfrac{fL_b}{D}}{\dfrac{fL}{D} + 5\dfrac{fL_b}{D}} = \frac{5L_b}{L + 5L_b} = \frac{5(1.22)}{14 + 5(1.22)} = 0.304$$

Therefore, approximately 30.4% of the total head loss in the pipeline is contributed by the bends. The bends certainly have a **significant influence on the head loss** in the pipeline.

7.5.7 Pipelines with Pumps or Turbines

Pipelines that contain either pumps or turbines are typically analyzed using the energy equation in the form

$$E_0 - \sum_{i=1}^{\mathrm{NP}} \frac{f_i L_i}{D_i} \frac{V_i^2}{2g} - \sum_{j=1}^{\mathrm{NL}} K_j \frac{V_j^2}{2g} + h_\mathrm{p} - h_\mathrm{t} = E_1 \tag{7.74}$$

where E_0 and E_1 are the energy per unit weight (i.e., head) at the beginning and end of the pipeline, respectively, NP is the number of distinct pipeline segments in which segment i has a friction factor, f_i, length L_i, diameter (or hydraulic diameter) D_i, and velocity V_i, NL is the number of local-loss locations in which location j has a head loss coefficient K_j and reference velocity V_j, h_p is the total head added by the pump(s), and h_t is the total head extracted by the turbine(s). In cases where a pipeline begins and ends at large reservoirs, E_0 and E_1 are equal to the surface elevations of the starting and ending reservoirs, respectively. If the outlet of the pipeline is open to the atmosphere, then E_1 is equal to the elevation of the outlet plus the velocity head ($V^2/2g$) of the flow immediately after it exits the pipeline. In general, both h_p and h_t are functions of the volume flow rate, Q, through these units; therefore, values of h_p and h_t derived for a single flow rate are not applicable at other flow rates.

EXAMPLE 7.13

A pump is to be selected that will pump water from a well into a storage reservoir. To fill the reservoir in a timely manner, the pump is required to deliver 5 L/s when the water level in the reservoir is 5 m above the water level in the well. Find the head that must be added by the pump. The pipeline is shown in Figure 7.8. Assume that the local loss coefficient for each of the bends is equal to 0.25 and that the temperature of the water is 20°C.

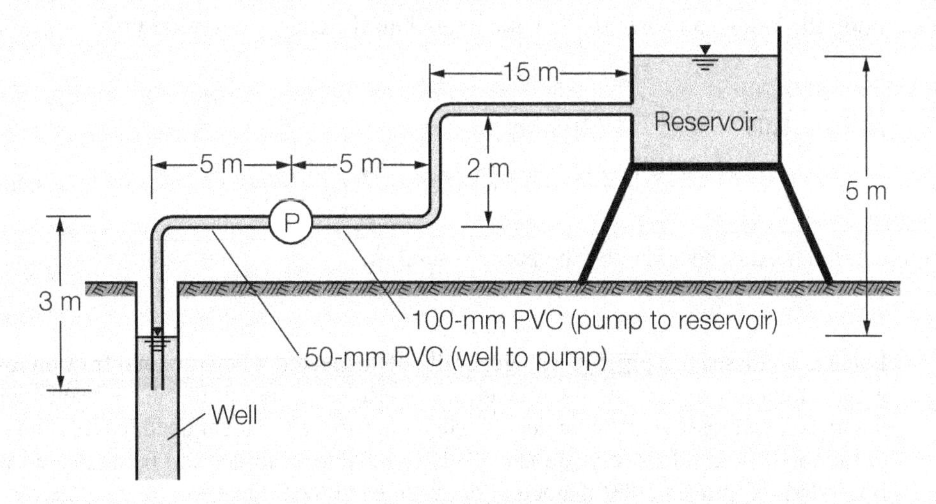

Figure 7.8: Pipeline system

SOLUTION

From the given data: Q = 5 L/s, D_1 = 50 mm, D_2 = 100 mm, and Δz = 5 m. PVC pipe can be considered smooth, so $k_\mathrm{s} \approx 0$. For water at 20°C, $\nu = 1.00 \times 10^{-6}$ m²/s. Taking the elevation of the water surface in the well (z_1) to be equal to 0 m and proceeding from the well to the storage reservoir (where the head is equal to 5 m), the energy equation can be written as

$$0 - \frac{V_1^2}{2g} - \frac{f_1 L_1}{D_1}\frac{V_1^2}{2g} - K_1\frac{V_1^2}{2g} + h_\text{p} - \frac{f_2 L_2}{D_2}\frac{V_2^2}{2g} - (K_2 + K_3)\frac{V_2^2}{2g} - \frac{V_2^2}{2g} = 5$$

where V_1 and V_2 are the velocities in the 50-mm ($= D_1$) and 100-mm ($= D_2$) pipes, respectively, L_1 and L_2 are the corresponding pipe lengths, f_1 and f_2 are the corresponding friction factors, K_1, K_2, and K_3 are the loss coefficients for each of the three bends, and h_p is the head added by the pump. The following preliminary calculations are useful:

$$A_1 = \frac{\pi}{4}D_1^2 = \frac{\pi}{4}(0.05)^2 = 0.001963 \text{ m}^2, \qquad V_1 = \frac{Q}{A_1} = \frac{0.005}{0.001963} = 2.54 \text{ m/s}$$

$$A_2 = \frac{\pi}{4}D_2^2 = \frac{\pi}{4}(0.10)^2 = 0.007854 \text{ m}^2, \qquad V_2 = \frac{Q}{A_2} = \frac{0.005}{0.007854} = 0.637 \text{ m/s}$$

$$\text{Re}_1 = \frac{V_1 D_1}{\nu} = \frac{(2.54)(0.05)}{1.00 \times 10^{-6}} = 1.27 \times 10^5, \ \text{Re}_2 = \frac{V_2 D_2}{\nu} = \frac{(0.637)(0.10)}{1.00 \times 10^{-6}} = 6.37 \times 10^4$$

The friction factor, f, can be estimated using the Swamee–Jain equation as

$$f = \frac{0.25}{\left[\log_{10}\dfrac{5.74}{\text{Re}^{0.9}}\right]^2}$$

Applying the Swamee–Jain equation to the 50-mm and 100-mm pipes,

$$f_1 = \frac{0.25}{\left[\log_{10}\dfrac{5.74}{(1.27 \times 10^5)^{0.9}}\right]^2} = 0.0170, \qquad f_2 = \frac{0.25}{\left[\log_{10}\dfrac{5.74}{(6.37 \times 10^4)^{0.9}}\right]^2} = 0.0197$$

Substituting the known and calculated parameters into the energy equation yields

$$0 - \left[1 + \frac{(0.0170)(8)}{0.05} + 0.25\right]\frac{2.54^2}{(2)(9.81)} + h_\text{p} - \left[\frac{(0.0197)(22)}{0.10} + 0.25 + 0.25 + 1\right]\frac{0.637^2}{(2)(9.81)} = 5 \quad \rightarrow \quad h_\text{p} = 6.43 \text{ m}$$

Therefore, the head to be added by the pump is **6.43 m.**

Local head losses in a pipeline can generally be neglected when the total friction loss is much greater than the sum of the local head losses. Rules of thumb that have been suggested are that local head losses can be neglected when the sum of the local head losses is less than 10% of the total friction loss (e.g., Jones, 2011) or local head losses can be neglected when there is a length of at least 1000 diameters between each local head loss (e.g., Potter et al., 2012; Streeter et al., 1998). Both of these rules have merit.

7.6 Water Hammer

The sudden closure of a valve in a pipeline transporting a liquid generates a pressure wave that can sometimes cause severe structural damage to the pipeline. This process is called *water hammer*. The high pressures generated by water hammer are generally of concern in the design of hydropower pipelines, in power plant piping, and in long water and oil pipeline delivery systems.

Governing equations. Consider the situation illustrated in Figure 7.9, where the flow of a fluid in a pipe is halted suddenly by the rapid closure of a valve and the pipe walls are *rigid* such that they do not flex in response to pressure changes. Before the valve is closed, the velocity in the pipe is V, the fluid pressure is p_0, and the density of the fluid is ρ_0. As the valve is closed suddenly, the fluid adjacent to the valve is halted immediately and the effect of valve closure is propagated upstream by a pressure wave that moves with a velocity c. Behind the pressure wave, the velocity is equal to zero, the fluid pressure is $p_0 + \Delta p$, and the fluid density is $\rho_0 + \Delta\rho$, whereas in front of the pressure wave, the velocity is V, the pressure is p_0, and the density is ρ_0. For the control volume illustrated in Figure 7.9, the momentum equation in the flow direction can be expressed in the conventional form as

$$\sum F_x = \frac{\mathrm{d}}{\mathrm{d}t}\int_{\mathcal{V}} \rho v_x \,\mathrm{d}\mathcal{V} + \int_{\mathcal{A}} \rho v_x (\mathbf{v}\cdot\mathbf{n})\,\mathrm{d}\mathcal{A} \tag{7.75}$$

where x is the coordinate in the flow direction, $\sum F_x$ is the sum of the forces in the x-direction that are acting on the fluid in the control volume, v_x is the x component of the fluid velocity, $\mathbf{v}$ is the fluid velocity vector, $\mathbf{n}$ is the unit normal pointing out of the control volume, and $\mathcal{V}$ and $\mathcal{A}$ represent the volume and surface area of the control volume, respectively. Neglecting the shear resistance on the pipe surface, Equation 7.75 can be written as

$$p_0 A - (p_0 + \Delta p)A = \frac{\mathrm{d}}{\mathrm{d}t}[\rho_0 V(L - ct)A] + \rho_0 V(-VA) \tag{7.76}$$

where A is the cross-sectional area of the pipe and L is the length of the control volume. Simplifying Equation 7.76 gives

$$(-\Delta p)A = \frac{\mathrm{d}}{\mathrm{d}t}(\rho_0 LAV - \rho_0 ctAV) - \rho_0 V^2 A \quad \rightarrow \quad \Delta pA = \rho_0 cAV + \rho_0 V^2 A$$

which gives

$$\Delta p = \rho_0 cV + \rho_0 V^2 \tag{7.77}$$

This equation relates the pressure increase, Δp, caused by sudden valve closure to the fluid and flow properties. The pressure change Δp is commonly called the *water hammer pressure* or the *Joukowsky pressure rise*.[52] In most cases, the velocity of the pressure wave, c, is much greater than the fluid velocity, V, and Equation 7.77 can be approximated as

$$\Delta p = \rho_0 cV \tag{7.78}$$

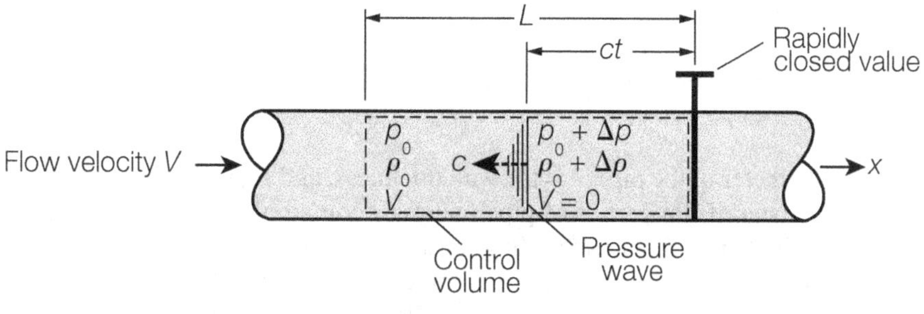

Figure 7.9: Pressure wave

[52] Named in honor of the Russian scientist Nicolai E. Joukowsky (1847–1921). Surname also spelled *Zhukovsky* in some publications (e.g., Simon and Korom, 1997).

This equation is commonly known as the *Joukowsky equation*, but it is sometimes called the *Joukowsky–Frizell* or *Allievi equation* (e.g., Tijsseling and Anderson, 2007). Applying the continuity equation to the control volume shown in Figure 7.9 requires that

$$\frac{\mathrm{d}}{\mathrm{d}t}\int_{\mathcal{V}} \rho \, \mathrm{d}\mathcal{V} + \int_{\mathcal{A}} \rho(\mathbf{v}\cdot\mathbf{n})\,\mathrm{d}\mathcal{A} = 0 \tag{7.79}$$

Equation 7.79 can be written as

$$\frac{\mathrm{d}}{\mathrm{d}t}[\rho_0(L-ct)A + (\rho_0+\Delta\rho)(ct)A] + \rho_0(-V)A = 0 \quad \rightarrow$$

$$\frac{\mathrm{d}}{\mathrm{d}t}[\rho_0 LA + \Delta\rho ctA] - \rho_0 VA = 0$$

which gives

$$\Delta\rho cA - \rho_0 VA = 0 \quad \rightarrow \quad \frac{\Delta\rho}{\rho_0} = \frac{V}{c} \tag{7.80}$$

Recalling the definition of the bulk modulus of elasticity, E_v, as

$$E_\mathrm{v} = \frac{\Delta p}{\Delta\rho/\rho_0} \tag{7.81}$$

Equations 7.80 and 7.81 combine to give

$$c = \frac{VE_\mathrm{v}}{\Delta p} \tag{7.82}$$

and substituting Equation 7.78 into Equation 7.82 yields wave speed, c, in terms of the fluid properties as

$$c = \sqrt{\frac{E_\mathrm{v}}{\rho_0}} \tag{7.83}$$

The wave speed, c, is also equal to the speed of sound in the fluid, because sound waves are pressure waves. In the case of water at 20°C, $E_\mathrm{v} = 2.15 \times 10^6$ kPa and $\rho_0 = 998$ kg/m^3; therefore, $c = 1470$ m/s. Clearly, the approximation that $V \ll c$ is reasonable for water in all practical cases.

Deformable pipe walls. The foregoing analysis has assumed that the pipe walls are rigid. If the pipe walls are slightly deformable, the wave speed, c, can be shown to be given by (Roberson et al., 1998)

$$c = \sqrt{\frac{E_\mathrm{v}/\rho_0}{1+(E_\mathrm{v}D/eE_\mathrm{p})}} \tag{7.84}$$

where D is the diameter of the pipe, e is the wall thickness, and E_p is the modulus of elasticity of the pipe wall material. Properties of pipe materials commonly used in engineering applications are given in Appendix D. According to Equation 7.84, in pipes that are not rigid, the wave speed, c, decreases as the pipe diameter increases and the wall thickness decreases. In general, the wave speed in a non-rigid pipe (e.g., PVC) is less than the wave speed in a rigid pipe (e.g., steel) with the same diameter and wall thickness. For normal pipe dimensions, the speed, c, of a pressure wave in a water pipe is usually in the range of 600–1200 m/s, but is always less than 1470 m/s, which is the speed of a pressure wave in free water.

Water hammer caused by sudden valve closure. The propagation of a pressure wave generated by sudden valve closure is illustrated in Figure 7.10, where a fluid (e.g., water) source that maintains an approximately constant pressure p_0 is located at a distance L upstream of the valve. The initial condition, at the instant just before valve closure, is shown in Figure 7.10(a), where the pressure head in the pipeline decreases with distance from the source due to frictional effects. If the valve is closed at time $t = 0$, during the time interval $0 < t < L/c$, the pressure wave propagates upstream, as illustrated in Figure 7.10(b), and at $t = L/c$, the pressure wave reaches the source. At this instant, the velocity in the pipe is zero, the pressure in the pipe is $p_0 + \Delta p$, and the pressure at the source is p_0. This abrupt pressure change at the source is the same as what occurred at the valve when the valve was suddenly closed. To equalize the pressure difference between the source and the pipe, the fluid flows with a velocity, V, from the pipe into the source. This causes a pressure wave to reflect back in the direction of the valve ($L/c < t < 2L/c$), as illustrated in Figure 7.10(c). At $t = 2L/c$, the wave arrives at the valve, where the velocity must equal zero. Because the velocity, V, in the pipe is directed toward the source, this causes a sudden pressure drop of Δp, creating another pressure wave where the pressure on the valve side of the wave is Δp less than the pressure on the source side of the wave. The pressure drop, Δp, has the same magnitude as the pressure rise when the valve was closed, because the abrupt change in flow velocity, V, is the same in both cases. During time $2L/c < t < 3L/c$, this wave

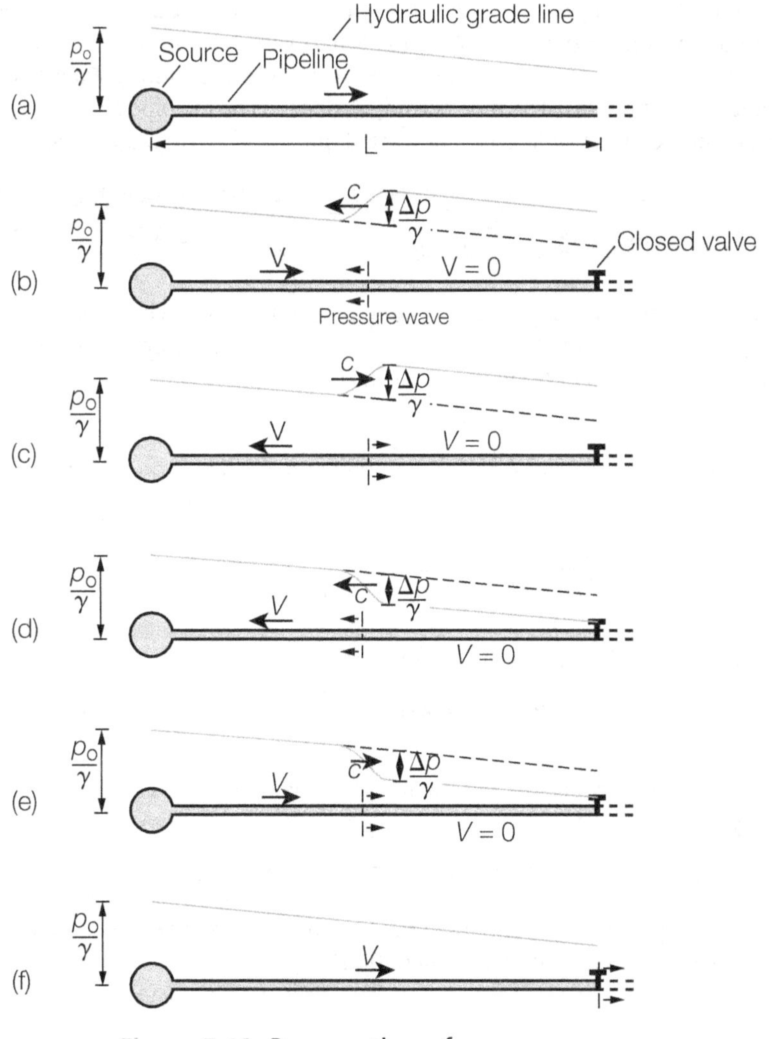

Figure 7.10: Propagation of a pressure wave

travels toward the source, as illustrated in Figure 7.10(d). At $t = 3L/c$, the pressure wave reaches the source, at which time the pressure in the pipe is Δp less than the pressure in the source. This pressure difference is the same as what occurred at the valve when the valve was closed. To equalize the pressures, fluid enters the pipe with velocity V. During the time interval $3L/c < t < 4L/c$, a pressure wave propagates back in the direction of the closed valve, as illustrated in Figure 7.10(e). At $t = 4L/c$, the pressure wave arrives back at the valve, Figure 7.10(f), and conditions are now exactly the same as immediately after the valve was closed. The cycle then repeats itself until the pressure wave is dissipated by frictional effects.

Critical time of closure. If the valve is closed gradually rather than instantaneously, the pressure increase also occurs gradually. As long as the valve is completely closed in a time less than $2L/c$, however, the maximum pressure occurring as a result of valve closure is still equal to Δp. The *critical time of closure* or *pipe period*, t_c, is therefore equal to

$$t_c = \frac{2L}{c} \tag{7.85}$$

If the valve is closed in a time longer than t_c, then any pressure increases generated in the pipe are damped by the open valve. Valve closures that take less than the pipe period are frequently referred to as *rapid valve closures*, whereas valve closures that take longer than the pipe period are called *slow valve closures*. Rapid valve closure is sometimes defined as closure times less than $10t_c$, and slow closure defined as closure times greater than $10t_c$ (e.g., Chadwick and Morfett, 1998). The pipe period, t_c, depends on the pipe elasticity via Equation 7.84; hence, compared with rigid pipes, non-rigid pipes will have longer pipe periods.

EXAMPLE 7.14

Estimate the maximum water hammer pressure generated in a rigid pipe where the initial water velocity is 2.5 m/s, the pipe is 5 km long, and a valve at the downstream end of the pipe is closed in 4 seconds. Assume a water temperature of 25°C.

SOLUTION

From the given data: $V = 2.5$ m/s, and $L = 5$ km $= 5000$ m. At 25°C, $E_v = 2.22 \times 10^6$ kPa and $\rho_0 = 997$ kg/m^3 (Table B.1, Appendix B). The speed of the pressure wave, c, is given by Equation 7.83 as

$$c = \sqrt{\frac{E_v}{\rho_0}} = \sqrt{\frac{2.22 \times 10^9}{997}} = 1490 \text{ m/s}$$

The critical time of closure, t_c, is given by Equation 7.85 as

$$t_c = \frac{2L}{c} = \frac{2(5000)}{1490} = 6.71 \text{ s}$$

Because the closure time of 4 seconds is less than t_c, the maximum pressure increase in the pipe, Δp, is given by Equation 7.78 as

$$\Delta p = \rho_0 c V = (997)(1490)(2.5) = 3.71 \times 10^6 \text{ Pa} = \mathbf{3710\ kPa}$$

This pressure increase is significantly higher than the pressures normally encountered in water distribution systems, which are typically less than 620 kPa.

Practical Considerations

Pipelines in water distribution systems experience transient pressures when valves are suddenly closed and when pumps are suddenly stopped. The possible occurrence of damaging water hammer pressures are controlled to some extent by keeping the pipeline velocities low. A variety of approaches can be taken to mitigate the effects of water hammer in pipeline systems. In some cases, valves that prevent rapid closure can be used, whereas in other cases, *pressure relief valves*, *surge tanks*, or *air chambers* are more practical. Pressure relief valves, which are also called *surge relief valves*, are designed to open when the pressure in the pipeline at the valve exceeds a specified value. Surge tanks are large open tanks directly connected to the pipeline and are commonly used in hydropower systems. An air chamber is a hydro-pneumatic tank (containing both water and air) that is connected to the pipeline, often at the discharge side of pumps. Although high water hammer pressures are usually of concern, the pressure wave generated by sudden valve closure also includes pressure reductions that are equal to the pressure increases. If the reduced pressure in the pipe falls below the saturation vapor pressure of the liquid, then cavitation will occur with the consequent vaporization of the liquid.

7.7 Pipe Networks

Pipe networks are commonly encountered in the context of water distribution systems. The performance criteria of these systems are typically specified in terms of minimum flow rates and minimum pressures that must be maintained at the specified points in the network. Analyses of pipe networks are usually within the context of (1) designing a new network, (2) designing a modification to an existing network, and/or (3) evaluating the reliability of an existing or proposed network. The procedure for analyzing a pipe network usually aims at finding the flow distribution within the network corresponding to given boundary conditions, with the pressure distribution being derived from the calculated flow distribution using the energy equation. A typical pipe network is illustrated in Figure 7.11, where the boundary conditions consist of inflows, outflows, and constant-head boundaries such as storage reservoirs. Inflows are typically from water treatment facilities, and outflows are usually from consumer withdrawals or for firefighting. Outflows are typically assumed to occur at network junctions.

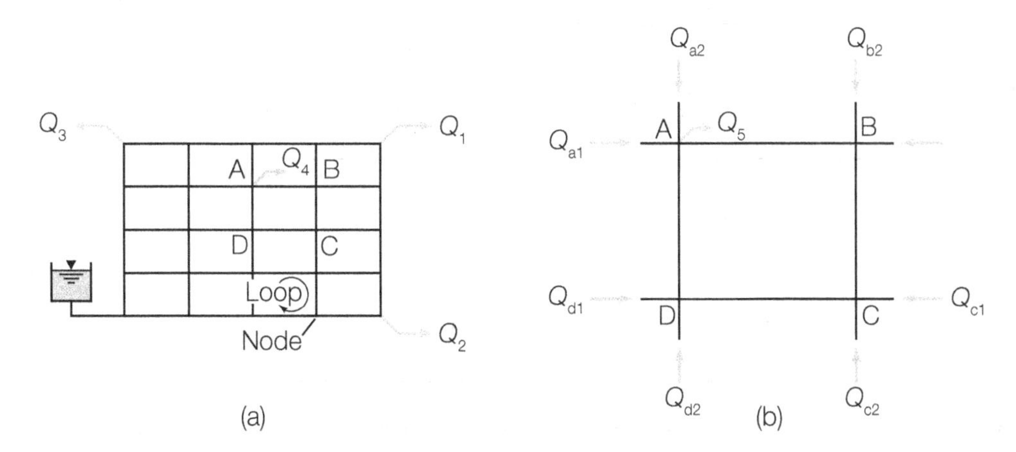

Figure 7.11: Typical pipe network

Governing equations. The basic equations to be satisfied in pipe networks are the continuity and energy equations. The continuity equation requires that at each junction in the network, the sum of the outflows is equal to the sum of the inflows. This requirement is expressed by the relation

$$\sum_{i=1}^{NP(j)} Q_{ij} - F_j = 0, \qquad j = 1, \ldots, NJ \tag{7.86}$$

where NP(j) is the number of pipes meeting at junction j, Q_{ij} is the flow rate in pipe i at junction j (inflows positive), F_j is the external flow rate (outflows positive) at junction j, and NJ is the total number of junctions in the network. The energy equation requires that the heads at each of the junctions in the pipe network be consistent with the head losses in the pipelines connecting the junctions. There are two principal methods of calculating the flows in pipe networks: the nodal method and the loop method. In the nodal method, the energy equation is expressed in terms of the heads at the network nodes (= junctions), and in the loop method, the energy equation is expressed in terms of the flows in closed loops within the pipe network.

7.7.1 Nodal Method

In the nodal method, the energy equation is written for each pipeline in the network as

$$h_2 = h_1 - \left(\frac{fL}{D} + \sum K_m\right)\frac{Q|Q|}{2gA^2} + \frac{Q}{|Q|}h_p \tag{7.87}$$

where h_1 and h_2 are the heads at the upstream and downstream ends of a pipe, respectively, the terms in parentheses are the friction loss and local losses, respectively, and h_p is the head added by pumps in the pipeline. The energy equation given by Equation 7.87 has been modified to account for the fact that the flow direction is commonly unknown, in which case a positive flow direction in each pipeline must be assumed and a consistent set of energy equations must be stated for the entire network. Equation 7.87 assumes that the positive flow direction is from node 1 to node 2. Application of the nodal method is usually limited to relatively simple networks.

EXAMPLE 7.15

The high-pressure ductile iron pipeline shown in Figure 7.12 becomes divided at Point B and rejoins at Point C. The pipeline characteristics are given in the following tables.

Pipe	Diameter (mm)	Length (m)
1	750	500
2	400	600
3	500	650
4	700	400

Location	Elevation (m)
A	5.0
B	4.5
C	4.0
D	3.5

If the flow rate in Pipe 1 is 2 m^3/s and the pressure at Point A is 900 kPa, calculate the pressure at Point D. Assume that the flows are fully turbulent in all pipes.

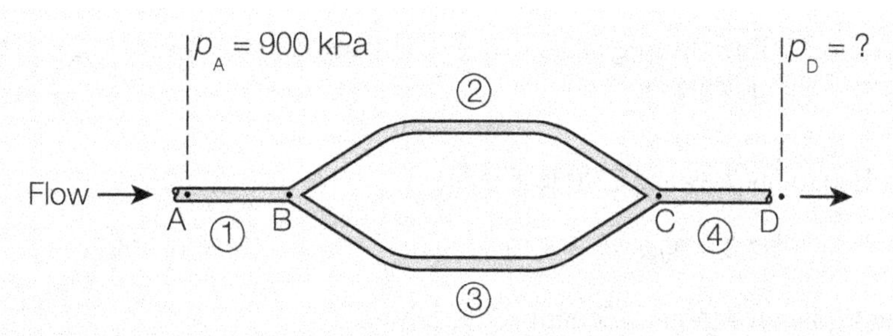

Figure 7.12: Pipe network

SOLUTION

The equivalent sand roughness, k_s, of ductile iron pipe is 0.26 mm, and the pipe and flow characteristics are as follows:

Pipe	Area (m^2)	Velocity (m/s)	k_s/D	f
1	0.442	4.53	0.000347	0.0154
2	0.126	—	0.000650	0.0177
3	0.196	—	0.000520	0.0168
4	0.385	5.20	0.000371	0.0156

where it has been assumed that all flows are fully turbulent. Taking $\gamma = 9.79$ kN/m^3, the head at Point A, h_A, is given by

$$h_A = \frac{p_A}{\gamma} + \frac{V_1^2}{2g} + z_A = \frac{900}{9.79} + \frac{4.53^2}{(2)(9.81)} + 5.0 = 98.0 \text{ m}$$

and applying the energy equation to each pipe gives the following relationships:

$$\text{Pipe 1:} \quad h_B = h_A - \frac{f_1 L_1}{D_1} \frac{V_1^2}{2g} = 98.0 - \frac{(0.0154)(500)}{0.75} \frac{4.53^2}{(2)(9.81)} \rightarrow h_B = 87.3 \text{ m} \tag{7.88}$$

$$\text{Pipe 2:} \quad h_C = h_B - \frac{f_2 L_2}{D_2} \frac{Q_2^2}{2gA_2^2} = 87.3 - \frac{(0.0177)(600)}{0.40} \frac{Q_2^2}{(2)(9.81)(0.126)^2} \rightarrow h_C = 87.3 - 85.2Q_2^2 \tag{7.89}$$

$$\text{Pipe 3:} \quad h_C = h_B - \frac{f_3 L_3}{D_3} \frac{Q_3^2}{2gA_3^2} = 87.3 - \frac{(0.0168)(650)}{0.50} \frac{Q_3^2}{(2)(9.81)(0.196)^2} \rightarrow h_C = 87.3 - 29.0Q_3^2 \tag{7.90}$$

$$\text{Pipe 4:} \quad h_D = h_C - \frac{f_4 L_4}{D_4} \frac{Q_4^2}{2gA_4^2} = h_C - \frac{(0.0156)(400)}{0.70} \frac{Q_4^2}{(2)(9.81)(0.385)^2} \rightarrow h_D = h_C - 3.07Q_4^2 \tag{7.91}$$

The continuity equations at the two pipe junctions are

$$\text{Junction B:} \quad Q_2 + Q_3 = 2 \text{ m}^3\text{/s} \tag{7.92}$$

$$\text{Junction C:} \quad Q_2 + Q_3 = Q_4 \tag{7.93}$$

Equations 7.89–7.93 are five equations in five unknowns: h_C, h_D, Q_2, Q_3, and Q_4. Equations 7.92 and 7.93 indicate that

$$Q_4 = 2 \text{ m}^3/\text{s}$$

Combining Equations 7.89 and 7.90 leads to

$$87.3 - 85.2Q_2^2 = 87.3 - 29.0Q_3^2 \quad \rightarrow \quad Q_2 = 0.583Q_3 \tag{7.94}$$

Substituting Equation 7.94 into Equation 7.92 gives

$$2 = (0.583 + 1)Q_3 \quad \rightarrow \quad Q_3 = 1.26 \text{ m}^3/\text{s}$$

and from Equation 7.94

$$Q_2 = 0.74 \text{ m}^3/\text{s}$$

According to Equation 7.90

$$h_C = 87.3 - 29.0Q_3^2 = 87.3 - 29.0(1.26)^2 = 41.3 \text{ m}$$

and Equation 7.91 gives

$$h_D = h_C - 3.07Q_4^2 = 41.3 - 3.07(2)^2 = 29.0 \text{ m}$$

Therefore, because the total head at D, h_D, is equal to 29.0 m

$$29.0 = \frac{p_D}{\gamma} + \frac{V_4^2}{2g} + z_D = \frac{p_D}{9.79} + \frac{5.20^2}{(2)(9.81)} + 3.5 \quad \rightarrow \quad p_D = 236 \text{ kPa}$$

Therefore, the pressure at location D is **236 kPa**. This problem has been solved by assuming that the flows in all pipes are fully turbulent. This is generally not known a priori; therefore, a complete solution requires that the Colebrook equation (with the Reynolds number term included) also be satisfied in all pipes.

7.7.2 Loop Method

In the loop method, the energy equation is written for each loop of the network, in which case the algebraic sum of the head losses within each loop is equal to zero. This requirement is expressed by the relation

$$\sum_{j=1}^{\text{NP}(i)} (h_{\text{L},ij} - h_{\text{p},ij}) = 0, \qquad i = 1, \ldots, \text{NL} \tag{7.95}$$

where NP(i) is the number of pipes in loop i, $h_{\text{L},ij}$ is the head loss in pipe j of loop i, $h_{\text{p},ij}$ is the head added by any pumps that may exist in pipe j of loop i, and NL is the number of loops in the network. Combining Equations 7.86 and 7.95 with an expression for calculating the head losses in pipes (such as the Darcy–Weisbach equation) and the pump characteristic curves, which relate the head added by the pump to the flow rate through the pump, yields a complete mathematical description of the flow problem. Solution of this system of flow equations is complicated by the fact that the equations are nonlinear, and numerical methods must be used to solve for the flow distribution in the pipe network.

Hardy Cross method. The Hardy Cross method (Cross, 1936) is a simple iterative technique used for manual solution of the loop system of equations governing flow in pipe networks. This iterative method was developed before the advent of computers, and much more efficient algorithms are now used for numerical computations. However, the Hardy Cross method is presented here to illustrate the manual iterative solution of the loop equations in pipe networks, which is sometimes desirable. The Hardy Cross method assumes that the head loss, h_ℓ, in each pipe is proportional to the volume flow rate in the pipe, Q, raised to some power, n, in which case

$$h_\ell = rQ^n \tag{7.96}$$

where typical values of n range from 1 to 2, where $n = 1$ corresponds to laminar flow and $n = 2$ to fully turbulent flow. The proportionality constant, r, depends on the head loss equation that is used and the types of losses in the pipe. If all head losses are due to friction and the Darcy–Weisbach equation is used to calculate the head losses, then

$$r = \frac{fL}{2gA^2D}, \quad n = 2 \tag{7.97}$$

If the flow in each pipe is approximated as $\hat{Q}$ and ΔQ is the error in this approximation, then the actual flow rate, Q, is related to $\hat{Q}$ and ΔQ by

$$Q = \hat{Q} + \Delta Q \tag{7.98}$$

and the head loss in each pipe is given by

$$\begin{aligned} h_\ell &= rQ^n = r(\hat{Q} + \Delta Q)^n \\ &= r\left[\hat{Q}^n + n\hat{Q}^{n-1}\Delta Q + \frac{n(n-1)}{2}\hat{Q}^{n-2}(\Delta Q)^2 + \cdots + (\Delta Q)^n\right] \end{aligned} \tag{7.99}$$

If the error in the flow estimate, ΔQ, is small, then the higher-order terms in ΔQ can be neglected and the head loss in each pipe can be approximated by

$$h_\ell \approx r\hat{Q}^n + rn\hat{Q}^{n-1}\Delta Q \tag{7.100}$$

This relation approximates the head loss in the flow direction. However, in working with pipe networks, the algebraic sum of the head losses in any loop of the network (see Figure 7.11) must be equal to zero. Therefore, we must define a positive flow direction (such as clockwise) and count head losses as positive in pipes when the flow is in the positive direction and negative when the flow is opposite the selected positive direction. Under these circumstances, the sign of the head loss must be the same as the sign of the flow direction. Further, when the flow is in the positive direction, positive values of ΔQ require a positive correction to the head loss; when the flow is in the negative direction, positive values in ΔQ also require a positive correction to the calculated head loss. To preserve the algebraic relation among head loss, flow direction, and flow error (ΔQ), Equation 7.100 for each pipe can be written as

$$h_\ell = r\hat{Q}|\hat{Q}|^{n-1} + rn|\hat{Q}|^{n-1}\Delta Q \tag{7.101}$$

where the approximation has been replaced by an equal sign. On the basis of Equation 7.101, the requirement that the algebraic sum of the head losses around each loop must be equal to zero can be written as

$$\sum_{j=1}^{\mathrm{NP}(i)} r_{ij}Q_j|Q_j|^{n-1} + \Delta Q_i \sum_{j=1}^{\mathrm{NP}(i)} r_{ij}n|Q_j|^{n-1} = 0, \qquad i = 1, \ldots, \mathrm{NL} \tag{7.102}$$

where $\mathrm{NP}(i)$ is the number of pipes in loop i, r_{ij} is the head loss coefficient in pipe j (in loop i), Q_j is the estimated flow in pipe j, ΔQ_i is the flow correction for the pipes in loop i, and NL is the number of loops in the entire network. The approximation given by Equation 7.102 assumes that there are no pumps in the loop and that the flow correction, ΔQ_i, is the same for each pipe in each loop. Solving Equation 7.102 for ΔQ_i leads to

$$\Delta Q_i = -\frac{\sum_{j=1}^{NP(i)} r_{ij} Q_j |Q_j|^{n-1}}{\sum_{j=1}^{NP(i)} n r_{ij} |Q_j|^{n-1}} \tag{7.103}$$

This equation forms the basis of the Hardy Cross method.

Calculation of the flow distribution. The steps in using the Hardy Cross method to calculate the flow distribution in pipe networks are as follows:

Step 1: Assume a reasonable distribution of flows in the pipe network. This assumed flow distribution must satisfy continuity.

Step 2: For each loop, i, in the network, calculate the quantities $r_{ij}Q_j|Q_j|^{n-1}$ and $nr_{ij}|Q_j|^{n-1}$ for each pipe in the loop. Calculate the flow correction, ΔQ_i, using Equation 7.103. Add the correction algebraically to the estimated flow in each pipe. (*Note*: Values of r_{ij} occur in both the numerator and denominator of Equation 7.103; therefore, values proportional to the actual r_{ij} may be used to calculate ΔQ_i.)

Step 3: Repeat Step 2 until the corrections (ΔQ_i) are acceptably small. Values of ΔQ_i are usually deemed to be acceptably small when the flow corrections do not change the calculated flows for the number of significant digits being used in the calculations (typically 3 significant digits).

The application of the Hardy Cross method is best demonstrated by an example.

EXAMPLE 7.16

Compute the distribution of flows in the pipe network shown in Figure 7.13(a), where the head loss in each pipe is given by

$$h_\ell = rQ^2$$

and the relative values of r are shown in Figure 7.13(a). The flows are taken as dimensionless for the sake of illustration.

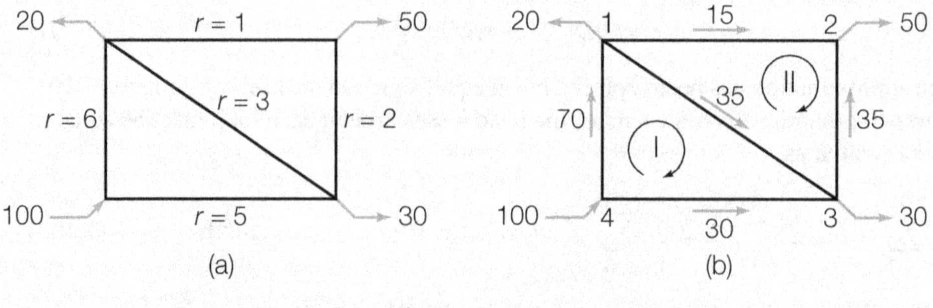

Figure 7.13: Flows in a pipe network

SOLUTION

The first step is to assume a distribution of flows in the pipe network that satisfies continuity. The assumed distribution of flows is shown in Figure 7.13(b), along with the positive flow directions in each of the two loops. The flow correction for each loop is calculated using Equation 7.103. Because $n = 2$ in this case, the flow correction formula becomes

$$\Delta Q_i = -\frac{\sum_{j=1}^{\mathrm{NP}(i)} r_{ij} Q_j |Q_j|}{\sum_{j=1}^{\mathrm{NP}(i)} 2r_{ij} |Q_j|}$$

The calculation of the numerator and denominator of this flow correction formula for loop I is tabulated as follows:

Loop	Pipe	Q	$rQ\|Q\|$	$2r\|Q\|$
I	4–1	70	29 400	840
	1–3	35	3675	210
	3–4	−30	−4500	300
			28 575	1350

The flow correction for loop I, ΔQ_{I}, is therefore given by

$$\Delta Q_{\mathrm{I}} = -\frac{28\,575}{1350} = -21.2$$

and the corrected flows are

Loop	Pipe	Q
I	4–1	48.8
	1–3	13.8
	3–4	−51.2

Moving to loop II, the calculation of the numerator and denominator of the flow correction formula for loop II is given by

Loop	Pipe	Q	$rQ\|Q\|$	$2r\|Q\|$
II	1–2	15	225	30
	2–3	−35	−2450	140
	3–1	−13.8	−574	83
			−2799	253

The flow correction for loop II, ΔQ_{II}, is therefore given by

$$\Delta Q_{\mathrm{II}} = -\frac{-2799}{253} = 11.1$$

and the corrected flows are

Loop	Pipe	Q
II	1–2	26.1
	2–3	−23.9
	3–1	−2.7

This procedure is repeated in the following table until the calculated flow corrections do not affect the calculated flows, to the level of significant digits retained in the calculations.

Iteration	Loop	Pipe	Q	$rQ\|Q\|$	$2r\|Q\|$	ΔQ	Corrected Q
2	I	4–1	48.8	14 289	586		47.7
		1–3	2.7	22	16		1.6
		3–4	−51.2	−13 107	512		−52.3
				1204	1114	−1.1	
	II	1–2	26.1	681	52		29.1
		2–3	−23.9	−1142	96		−20.9
		3–1	−1.6	−8	10		1.4
				−469	157	3.0	
3	I	4–1	47.7	13 663	573		47.7
		1–3	1.4	6	8		1.4
		3–4	−52.3	−13 666	523		−52.3
				3	1104	0.0	
	II	1–2	29.1	847	58		29.2
		2–3	−20.9	−874	84		−20.8
		3–1	1.4	6	8		1.5
				−21	150	0.1	
4	I	4–1	47.7	13 662	573		47.7
		1–3	1.5	7	9		1.5
		3–4	−52.3	−13 668	523		−52.3
				1	1104	0.0	
	II	1–2	29.2	853	58		29.2
		2–3	−20.8	−865	83		−20.8
		3–1	1.5	7	9		1.5
				−5	150	0.0	

The final flow distribution, after four iterations, is given by

Pipe	Q
1–2	**29.2**
2–3	**−20.8**
3–4	**−52.3**
4–1	**47.7**
1–3	**−1.5**

It is clear that the final results are fairly close to the flow estimates after only one iteration.

As the above example illustrates, complex pipe networks can generally be treated as a combination of simple loops, with each loop balanced in turn until compatible flow conditions exist in all loops. In applying the Hardy Cross method to each loop, there is some evidence that convergence can be accelerated by applying a correction of $0.6\Delta Q$ instead of ΔQ at each iteration (Elger et al., 2013); however, this is not part of the conventional approach. Typically, after the flows have been computed for all pipes in a network, the elevation of the hydraulic grade line and the pressure are computed for each junction node. These pressures are then assessed relative to acceptable operating pressures.

7.8 Building Water Supply Systems

Water supply to individual buildings is typically provided by a service line that is connected to the water main that abuts the building. The service line is typically made of plastic, copper, or cast iron. Copper is commonly regarded as the standard pipe material for service lines. However, plastic service lines are viewed by some as being just as durable and less expensive than copper and are sometimes preferred because of potential copper contamination of tap water if the water is corrosive. Cast iron is typically used for service lines larger than 50 mm in diameter.

Typical residential system. An isometric view of a typical residential water supply system is shown in Figure 7.14. In this system, the service line first connects to a water meter that measures the flow entering the building, and a backflow preventer is typically provided to prevent backflows from the building into the water main. Backflows of possibly contaminated water can be caused by in-building sources such as pumps, boilers, and heat-exchange equipment. Water heaters are used to heat a portion of the water supply, and branches from the hot and cold water lines provide water to different parts of the building. The different parts of the building illustrated in Figure 7.14 correspond to delivery points A, B, and C. Fixtures such as toilets, sinks, showers, and washing machines are located near these delivery points.

Design problem. The basic hydraulic design problem is to size the pipes to ensure that water is delivered to all water supply fixtures in the building with an adequate flow rate and at an adequate pressure. Guidelines for minimum flow rates and pressures at various fixtures are typically stated in the applicable local plumbing code. A typical design procedure for building water supply systems is as follows:

Step 1: Estimate the minimum daily pressure at the water supply main. This pressure is best determined from direct measurements of pressure in the existing water main, and such measurements are usually available from the utility that serves the area. Alternately, a hydraulic analysis of the existing water distribution system might

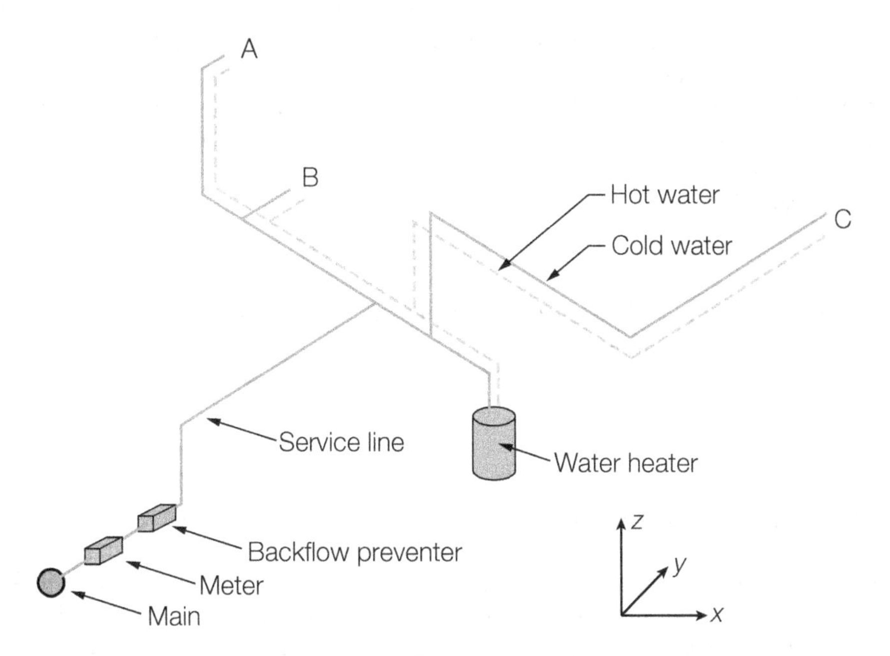

Figure 7.14: Isometric view of a residential water supply system

be necessary to estimate the water main pressure at the building location. Typical minimum daily pressures in water mains are on the order of 350 kPa.

Step 2: Estimate the design flow in the service line and the principal pipeline branches in the building.

Step 3: Specify the minimum required pressures at the fixtures connected to the principal branches.

Step 4: Determine the minimum pipe diameter in each branch that will provide adequate pressures at all fixtures connected to the branch.

The procedures to be followed in Steps 2–4 are given in more detail in the following sections.

7.8.1 Specification of Design Flows

Each principal branch in a building water supply system has its own design flow based on the number and types of fixtures to be connected to the branch. The principal branches are laid out to support the architectural design and function of the building, and the locations and types of fixtures connected to each of the branches are identified. The water demand at each fixture is expressed in terms of *water supply fixture units* (WSFU), which is an abstract number that takes into account both the flow rate to be delivered to the fixture and the frequency of use of the fixture. A commonly used relationship between the type of fixture and the number of fixture units assigned to the fixture is shown in Table 7.4 (Uniform Plumbing Code, 2009). For each branch in the building, the total number of fixture units to be supplied by the branch is determined by summing the fixture units connected to the branch. Using the fixture unit approach is generally preferred over summing the flow rates to be delivered at each fixture because the more fixtures there are, the less likelihood that all the fixtures are in operation at the same time. Once the number of fixture units to be supported by each branch is determined, the fixture units are converted into design flow rates using a curve similar to that shown in Figure 7.15. Curves relating design flow rates to fixture units, such as those shown in Figure 7.15, are called *Hunter curves* after Roy B. Hunter, who first suggested this relationship (Hunter, 1940). Hunter curves are included in most local plumbing codes, although some research has indicated that the peak flows estimated from fixture units provide conservative estimates of peak flows (AWWA, 2004).

Typical design flow rate. As a point of reference, for a typical house having two bathrooms, a full laundry, a kitchen, and one or two hose bibs, the maximum instantaneous flow rate to the residence is about 60 L/min.

7.8.2 Specification of Minimum Pressures

For each principal branch in a building water supply system, the minimum pressure required at the locations where the fixtures are connected must be specified. Attainment of these pressures at the fixture connections is necessary to yield the required minimum flow rates through the fixtures, and these minimum flow rates are essential for keeping the fixtures clean and sanitary. Recommended minimum pressures at various types of fixtures and their associated flow rates are given in Table 7.5.

Table 7.4: Water Supply Fixture Units (WSFU)

Fixture	Occupancy	Supply control	Load values in WSFU		
			Cold	Hot	Total
Bathroom group	Private	Flush tank	2.7	1.5	3.6
Bathroom group	Private	Flush valve	6	3	8
Bathtub	Private	Faucet	1	1	1.4
Bathtub	Public	Faucet	3	3	4
Bidet	Private	Faucet	1.5	1.5	2
Combination fixture	Private	Faucet	2.25	2.25	3
Dishwashing machine	Private	Automatic	—	1.4	1.4
Drinking fountain	Offices, etc.	9.5-mm valve	0.25	—	0.25
Kitchen sink	Private	Faucet	1	1	1.4
Kitchen sink	Hotel, restaurant	Faucet	3	3	4
Laundry trays (1 to 3)	Private	Faucet	1	1	1.4
Lavatory	Private	Faucet	0.5	0.5	0.7
Lavatory	Public	Faucet	1.5	1.5	2
Service sink	Offices, etc.	Faucet	2.25	2.25	3
Shower head	Public	Mixing valve	3	3	4
Shower head	Private	Mixing valve	1	1	1.4
Urinal	Public	25-mm flush valve	10	—	10
Urinal	Public	19-mm flush valve	5	—	5
Urinal	Public	Flush tank	3	—	3
Washing machine, 3.6 kg	Private	Automatic	1	1	1.4
Washing machine, 3.6 kg	Public	Automatic	2.25	2.25	3
Washing machine, 6.8 kg	Public	Automatic	3	3	4
Water closet	Private	Flush valve	6	—	6
Water closet	Private	Flush tank	2.2	—	2.2
Water closet	Public	Flush valve	10	—	10
Water closet	Public	Flush tank	5	—	5
Water closet	Public or private	Flushometer tank	2	—	2

Source: International Plumbing Code (2012).

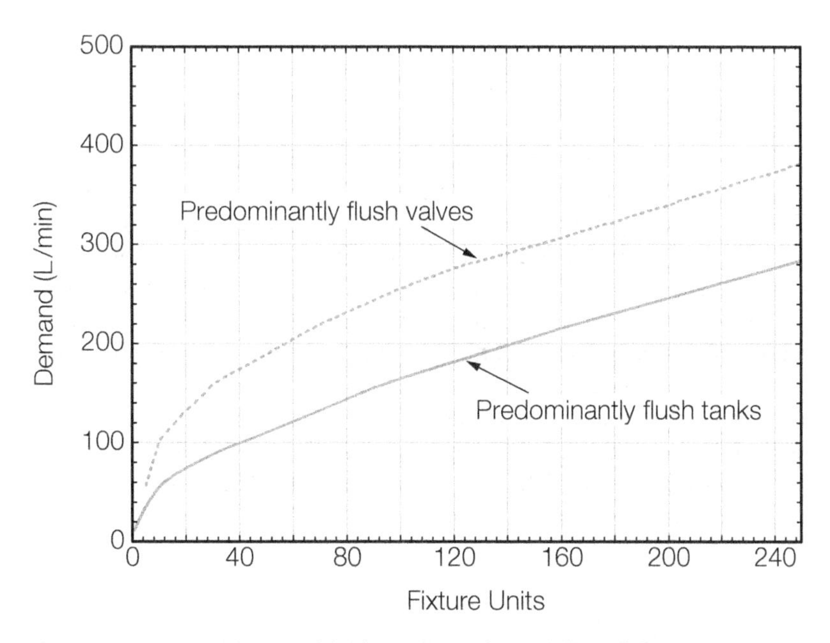

Figure 7.15: Relationship between demand and fixture units

Table 7.5: Typical Minimum Flow Rates and Pressures in Building Fixtures

Fixture	Flow rate (L/min)	Pressure (kPa)
Bathtub	15	138
Bidet	8	138
Combination fixture	15	55
Dishwasher, residential	10.4	55
Drinking fountain	2.8	55
Laundry tray	15	55
Lavatory, private	8	55
Lavatory, public	8	55
Shower head	11	55
Shower head, temperature controlled	11	138
Sink, residential	9.5	55
Sink, service	11	55
Urinal, valve	45	172
Water closet, blow out, flushometer valve	95	310
Water closet, siphonic, flushometer valve	95	241
Water closet, tank, close coupled	11	138
Water closet, tank, one piece	23	138

Source: International Plumbing Code (2012).

7.8.3 Determination of Pipe Diameters

Pipe diameters in each branch of a building water supply system are selected so that the required pressure heads at the fixture connections are attained. To determine the required pipe diameters, the energy equation is applied between the water main and the delivery point in each branch pipe such that

$$\frac{p_0}{\gamma} - \left[\frac{p_1}{\gamma} + \frac{V_1^2}{2g} + \Delta z\right] = \sum h_f + \sum h_{local} \tag{7.104}$$

where p_0 is the pressure in the water main [FL^{-2}], γ is the specific weight of water [FL^{-3}], p_1 is the pressure at the delivery point of the branch [FL^{-2}], V_1 is the flow velocity in the branch [LT^{-1}], g is gravity [LT^{-2}], Δz is the height of the delivery point above the water main [L], $\sum h_f$ is the sum of the friction losses in the pipes between the water main and the delivery point [L], and $\sum h_{local}$ is the sum of the local head losses between the water main and the delivery point [L]. For each branch, the design fixture is usually the fixture with the greatest head loss from the main and/or the fixture with the greatest elevation difference (Δz) from the main and/or the fixture that requires the largest pressure. It is usually assumed that the total flow to be accommodated by a branch occurs up to the location of the design fixture for that branch. In some cases, several design fixtures might need to be tried to determine the one(s) that control specification of the branch pipe diameter. The fixture requiring the largest branch pipe diameter is called the *critical fixture*.

Estimation of friction head losses. The head loss due to friction in each pipeline is best described by the Darcy–Weisbach equation; however, it is also common practice in the United

States to use the Hazen–Williams equation to calculate friction losses in building water supply pipes. These head loss equations can be put in the following convenient forms:

$$h_f = \begin{cases} 0.230\left(\dfrac{fL}{D^5}\right)Q^2, & \text{Darcy–Weisbach equation} \\ 85.4\left(\dfrac{L}{C_H^{1.85}D^{1.87}}\right)Q^{1.85}, & \text{Hazen–Williams equation} \end{cases} \tag{7.105}$$

where h_f is the lead loss due to friction [m], f is the friction factor [dimensionless], L is the length of the pipeline [m], D is the diameter of the pipeline [cm], Q is the flow rate in the pipeline [L/min], and C_H is the Hazen–Williams coefficient [dimensionless]. Design codes commonly incorporate graphical plots of these equations; however, if the parameters of friction loss (f or C_H) can be estimated from available data, it is generally preferable to use the analytic head loss relationships directly rather than read friction head losses from plots of these equations. A typical roughness height used in the Darcy–Weisbach equation for copper tubing and PVC pipe is 0.0015 mm (1.5 μm). Typical values of C_H are 130 for copper tubing and 140 for PVC pipe.

Estimation of local head losses. Local head losses caused by valves and fittings in building water supply systems can account for a significant portion of the total head losses. Local head losses include losses at the water meter, backflow preventer, transitions (e.g., tees and ells), and valves. Even when valves are fully open, they can cause significant head losses. It is common practice to express local head losses in terms of an equivalent pipe length that would cause the same head loss due to friction as the local head loss. Such relationships are shown in Table 7.6 for standard fittings of various sizes. It is apparent from Table 7.6 that the magnitude of the local head loss depends on the size of the fitting, with globe valves generally causing the highest local head losses.

Table 7.6: Head Loss in Standard Fittings in Terms of Equivalent Pipe Lengths

Fitting Diameter	90° Ell	45° Ell	90° Tee	Gate valve	Globe valve	Angle valve
(mm)	(m)	(m)	(m)	(m)	(m)	(m)
9.5	0.305	0.183	0.457	0.061	2.438	1.219
12.7	0.610	0.366	0.914	0.122	4.572	2.438
19.1	0.762	0.457	1.219	0.152	6.096	3.658
25.4	0.914	0.549	1.524	0.183	7.620	4.572
32	1.219	0.732	1.829	0.244	10.668	5.486
38	1.524	0.914	2.134	0.305	13.716	6.706
51	2.134	1.219	3.048	0.396	16.764	8.534
64	2.438	1.524	3.658	0.488	19.812	10.363
76	3.048	1.829	4.572	0.610	24.384	12.192
102	4.267	2.438	6.401	0.823	38.100	16.764
127	5.182	3.048	7.620	1.006	42.672	21.336
152	6.096	3.658	9.144	1.219	50.292	24.384

Source: Uniform Plumbing Code (2012).

Typical pressure requirement. A pressure of 100 kPa is adequate to operate most household fixtures. Therefore, if 70 kPa is allowed for the static lift from a basement to a second floor, adequate pressure to operate household fixtures will correspond to having a pressure of 170 kPa on the customer side of the water meter for a typical maximum residential flow of 60 L/min. For a typical residential service consisting of 12 m of 19-mm copper pipe and a 16-mm disk meter, the friction loss for simultaneous fixture use with a flow rate of 23–45 L/min is 35–130 kPa (Hammer and Hammer, 2010). Consequently, if 170 kPa is to be available to operate the household fixtures satisfactorily, the pressure in the water main must be in the range of 200–300 kPa.

Typical pipe diameters. The minimum allowable diameter of service lines is typically 19 mm, and diameters of service lines in residential and small commercial buildings rarely exceed 50 mm. The diameters of pipelines in buildings are generally less than or equal to the diameter of the service line. The standard commercial diameters of building water supply lines are shown in Table 7.6. Minimum allowable pipe sizes in buildings are typically 13 mm or 19 mm, depending on the types of fixtures served by the connected pipe.

Constraints on flow velocities. After pipe sizes are selected, the flow velocities under design conditions should be checked to ensure that they are in an acceptable range. Pipe flows with velocities above 3 m/s are usually too noisy, and flows with velocities above 1.8 m/s might be too noisy for acoustical-critical locations (Stein et al., 2006).

Illustration of design procedure. The recommended design procedure is illustrated in the following example.

EXAMPLE 7.17

The water supply system for a two-story factory building is shown in Figure 7.16, where Type L copper pipe is to be used for the service pipe and all pipelines in the building. The minimum daily pressure in the water main is 380 kPa, and the diameter of the service line is 64 mm. Based on manufacturers' data, the tap into the water main is expected to cause a pressure drop of 10 kPa, the meter is expected to cause a pressure drop of 76 kPa, and the backflow preventer is expected to cause a pressure drop of 62 kPa. The distribution of demand in the building is as follows:

Type	Pipe	Fixture units (WSFU)	Flow (L/min)
Service	AB	288	409
	BC	264	396
Cold	CF	132	291
	CE	132	291
	B'C'	24	144
Hot	C'F'	12	45
	C'E'	12	45

The critical fixtures on both floors of the building require a minimum allowable pressure of 103 kPa. All valves are expected to perform as gate valves, and all ells and tees are at 90°. Determine the required pipe diameters for the cold water lines.

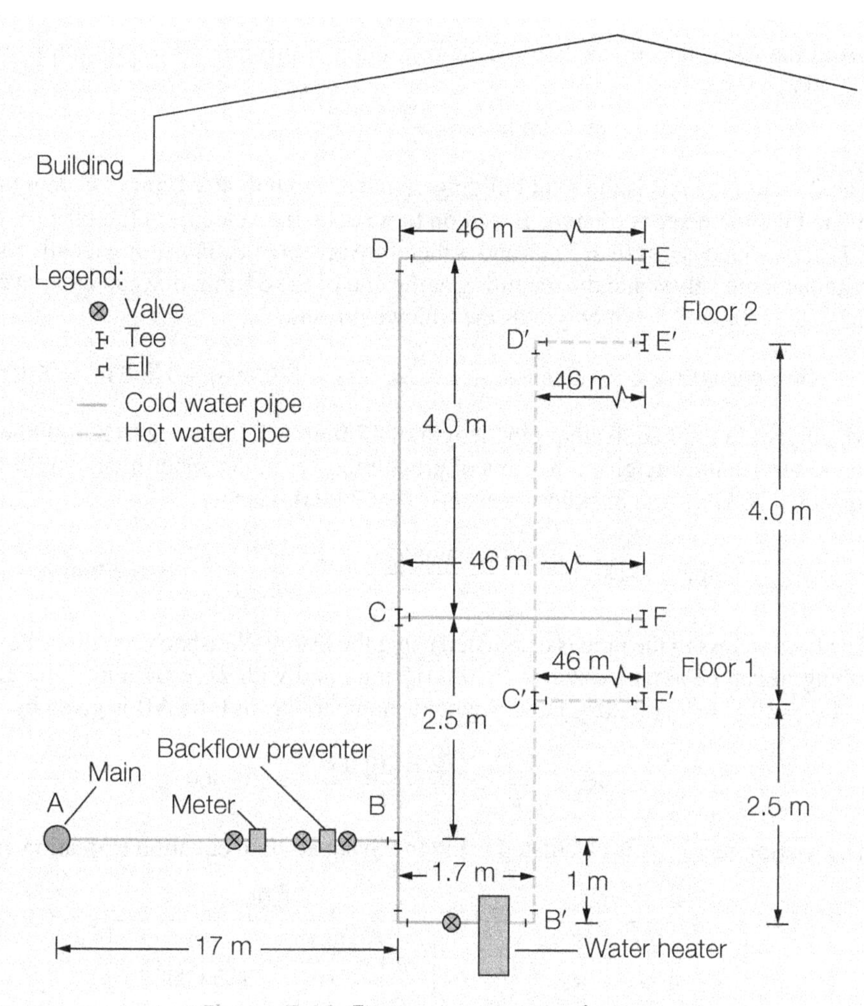

Figure 7.16: Factory water supply system

SOLUTION

The calculation to determine the pressures at the critical fixtures is an iterative process in which different pipe sizes are tried until the calculated pressures at all critical fixtures are greater than their minimum required values. The results of the design calculations are summarized in Table 7.7. The calculation procedures for lines AB and BC are typical and are described in detail below.

Table 7.7: Results of Design Calculations

Pipe	Starting head (m)	Flow (L/min)	Len. (m)	Diam. (mm)	Velocity (m/s)	Fitting length (m)	Total length (m)	Friction loss (m)	Other losses (m)	Elev. diff. (m)	Terminal head (m)	Terminal pressure (kPa)
AB	38.82	409	17.0	**64**	2.12	5.12	22.12	1.37	15.12	0	22.33	216
BC	22.33	396	2.5	**64**	2.05	3.66	6.16	0.36	0	2.5	21.97	188
CF	21.97	291	46.0	**51**	2.37	3.05	49.05	4.92	0	0	17.05	164
CD	21.97	291	4.0	**51**	2.37	3.05	7.05	0.71	0	4.0	21.26	166
DE	21.26	291	46.0	**51**	2.37	3.05	49.05	4.92	0	0	16.34	157

Line AB: Line AB is the service line. At the beginning of this line, the (main) pressure, p_0, is 380 kPa, so

$$\text{starting head} = \frac{p_0}{\gamma} = \frac{380}{9.79} = 38.82 \text{ m}$$

The flow in Pipe AB is the total building demand, equal to 409 L/min; the length is 17.0 m; and the diameter is 64 mm. Based on these data, the velocity in the pipe ($= Q/A$) is 2.12 m/s, which is less than 3 m/s and is therefore acceptable. The fitting length associated with the three valves and the tee fitting at the end of the 64-mm line can be derived using the data in Table 7.6, which yields the following result:

$$\text{fitting length} = 3 \times \text{valve loss} + 1 \times \text{tee loss} = 3(0.488) + 1(3.658) = 5.12 \text{ m}$$

which gives a total equivalent pipe length of 17.0 m + 5.12 m = 22.12 m. Other losses in the service line are given in terms of pressure losses at the water main tap (= 10 kPa), meter (= 76 kPa), and backflow preventer (= 62 kPa). Hence,

$$\text{other losses} = \frac{\text{total pressure loss}}{\gamma} = \frac{10 + 76 + 62}{9.79} = 15.12 \text{ m}$$

The friction loss in the pipe is calculated using the Darcy–Weisbach equation. For copper tubing, it can be assumed that $k_s = 0.0015$ mm, and with $D = 64$ mm, $V = 2.12$ m/s, and $\nu = 10^{-6}$ m²/s (at 20°C), the Reynolds number, Re, in Pipe AB is given by

$$\text{Re} = \frac{VD}{\nu} = \frac{(2.12)(0.064)}{10^{-6}} = 135\,600$$

The friction factor, f, is calculated using the Swamee–Jain equation (Equation 7.45) as

$$f = \frac{0.25}{\left[\log\left(\frac{k_s}{3.7D} + \frac{5.74}{\text{Re}^{0.9}}\right)\right]^2} = \frac{0.25}{\left[\log\left(\frac{0.0015}{3.7(64)} + \frac{5.74}{135\,600^{0.9}}\right)\right]^2} = 0.0173$$

and the head loss due to friction, h_f, is given by the Darcy–Weisbach equation as

$$h_f = \frac{fL}{D}\frac{V^2}{2g} = \frac{(0.0173)(22.12)}{(0.064)}\frac{(2.12)^2}{2(9.81)} = 1.37 \text{ m}$$

The terminal head in Pipe AB is then given by

$$\begin{aligned} \text{terminal head} &= \text{starting head} - \text{friction loss} - \text{other losses} \\ &= 38.82 \text{ m} - 1.37 \text{ m} - 15.12 \text{ m} = 22.33 \text{ m} \end{aligned}$$

The terminal pressure, p, is related to the terminal head by the relationship

$$p = \gamma\left[\text{terminal head} - \Delta z - \frac{V^2}{2g}\right] = (9.79)\left[22.33 - 0 - \frac{2.12^2}{2(9.81)}\right] = 216 \text{ kPa}$$

Line BC: Line BC originates at the end of line AB. The flow in Pipe BC is 396 L/min, and the length is 2.5 m. Taking the diameter as 51 mm yields a velocity ($= Q/A$) of 3.23 m/s, which is greater than 3 m/s and is therefore unacceptable. Taking the diameter as 64 mm yields a velocity of 2.05 m/s, which is less than 3 m/s and therefore acceptable. The starting head in line BC is equal to the terminal head in Pipe AB, which was calculated previously as 22.33 m. The fitting length associated with the tee at the end of the 64-mm pipe is derived from Table 7.6, which gives

$$\text{fitting length} = \text{tee loss} = 3.658 \text{ m}$$

So the total equivalent pipe length of BC is 2.5 m + 3.658 m = 6.158 m ≈ 6.16 m. The Reynolds number, Re, in Pipe BC is

$$\text{Re} = \frac{VD}{\nu} = \frac{(2.05)(0.064)}{10^{-6}} = 13\,100$$

The friction factor, f, is

$$f = \frac{0.25}{\left[\log\left(\frac{k_s}{3.7D} + \frac{5.74}{\text{Re}^{0.9}}\right)\right]^2} = \frac{0.25}{\left[\log\left(\frac{0.0015}{3.7(64)} + \frac{5.74}{13\,100^{0.9}}\right)\right]^2} = 0.0174$$

The head loss due to friction, h_f, is

$$h_f = \frac{fL}{D}\frac{V^2}{2g} = \frac{(0.0174)(6.16)}{(0.064)}\frac{(2.05)^2}{2(9.81)} = 0.36 \text{ m}$$

The terminal head in Pipe BC is given by

$$\text{terminal head} = \text{starting head} - \text{friction loss} = 22.33 \text{ m} - 0.36 \text{ m} = 21.97 \text{ m}$$

Because the change in elevation, Δz, between B and C is 2.5 m, the terminal pressure, p, in line BC is given by

$$p = \gamma\left[\text{terminal head} - \Delta z - \frac{V^2}{2g}\right] = (9.79)\left[21.97 - 2.5 - \frac{2.05^2}{2(9.81)}\right] = 188 \text{ kPa}$$

Calculations to determine the pressures at the ends of the other cold water lines are done in the same manner as for lines AB and BC. Because the objective of the design is to achieve a minimum pressure of 103 kPa at terminal locations E and F, diameter adjustments show that it is acceptable to use 51-mm pipe for all lines except AB and BC, where 64-mm lines should be used. With this water supply system, the expected pressures at the critical fixtures are 164 kPa on the first floor and 157 kPa on the second floor, both of which exceed the minimum required pressure of 103 kPa.

Supplemental components of water supply systems. In cases where the water main pressure is inadequate to provide adequate pressures throughout a building, supplementary components to the building water supply system are used. Common supplementary building systems for augmenting water pressure are gravity tanks, direct-feed booster pumps, and hydro-pneumatic systems. These are described briefly as follows:

Gravity Tanks. Gravity tank systems are widely used throughout the world. This system uses a roof tank to provide stable pressures and storage to supplement the available water supply. One or two pumps are used to pump water to the roof tank, which is typically sized to cover the peak building demands during the day. Water level controls mounted in the tank control the on/off pump operation to maintain the water level in the tank. A gravity tank system is recommended for buildings with long periods of low water demands and is ideal for many commercial and residential applications.

Direct-Feed Booster Pumps. Booster pump systems are the most widely used type of supplementary system in the United States. In these systems, pumps are connected directly to the building water supply for the sole purpose of increasing the water pressure in the building. A booster pump system is usually recommended for buildings that have a continuous water demand, such as hotels and hospitals.

Hydro-Pneumatic Systems. Hydro-pneumatic systems use pressurized air and stored water contained in the same tank. A pump supplies water to the tank, with the water being contained in an expandable bladder surrounded by compressed air. As the volume of water stored in the bladder changes, the volume of air in the pressurized tank adjusts, thereby maintaining the water pressure in the pipes that are connected to the bladder.

Key Equations for Flow in Closed Conduits

The following list of equations is particularly useful in solving problems related to flow in closed conduits. If one is able to recognize these equations and recall their appropriate use, then the learning objectives of this chapter have been met to a significant degree. Derivations of these equations, definitions of the variables, and detailed examples of usage can be found in the main text.

STEADY INCOMPRESSIBLE FLOW

Steady-state energy equation: $$\left(\frac{p_1}{\gamma}+\frac{v_1^2}{2g}+z_1\right)-\left(\frac{p_2}{\gamma}+\frac{v_2^2}{2g}+z_2\right)=h_\ell\ (=h_\mathrm{f})$$

Energy equation, constant diameter: $$\left(\frac{p_1}{\gamma}+z_1\right)-\left(\frac{p_2}{\gamma}+z_2\right)=h_\mathrm{f}$$

Combined energy and momentum equation: $$h_\mathrm{f}=\frac{2\tau L}{r\gamma}=\frac{4\tau_0 L}{D\gamma}=\text{constant}$$

Shear stress distribution: $$\tau=\tau_0\frac{2r}{D}$$

Dimensional analysis: $$\frac{\tau_0}{\frac{1}{8}\rho V^2}=f\left(\mathrm{Re},\frac{\epsilon}{D}\right),\qquad \frac{\tau_0}{\frac{1}{2}\rho V^2}=f_\mathrm{B}\left(\mathrm{Re},\frac{\epsilon}{D}\right)$$

Darcy–Weisbach equation: $$h_\mathrm{f}=\frac{fL}{D}\frac{V^2}{2g}$$

Pressure gradient: $$\frac{\mathrm{d}p}{\mathrm{d}x}=-\gamma\left(\frac{h_\mathrm{f}}{L}+\sin\theta\right)$$

Entrance length: $$\frac{L_\mathrm{e}}{D}=\begin{cases}0.058\,\mathrm{Re} & \text{(laminar flow)}\\ 4.4\,\mathrm{Re}^{\frac{1}{6}} & \text{(turbulent flow)}\end{cases}$$

Hydraulic radius: $$R_\mathrm{h}=\frac{A}{P}$$

Hydraulic diameter: $$D_\mathrm{h}=4\frac{A}{P}$$

Head loss, noncircular conduits: $$h_\mathrm{f}=\frac{\tau_0 L}{\gamma R_\mathrm{h}},\qquad h_\mathrm{f}=\frac{4\tau_0 L}{\gamma D_\mathrm{h}},\qquad h_\mathrm{f}=\frac{fL}{D_\mathrm{h}}\frac{V^2}{2g}$$

FRICTION EFFECTS IN LAMINAR FLOW

Velocity distribution: $$u(r)=\frac{h_\mathrm{f}\gamma}{16\mu L}(D^2-4r^2),\qquad u(r)=2V\left(1-\frac{r^2}{R^2}\right)$$

Centerline velocity: $$V_\mathrm{c}=\frac{h_\mathrm{f}\gamma}{16\mu L}D^2$$

Average velocity: $$V = \frac{h_f \gamma}{32\mu L} D^2$$

Volume flow rate: $$Q = \frac{\pi h_f \gamma}{128\mu L} D^4$$

Friction factor, laminar flow: $$f = \frac{64}{\text{Re}}$$

FRICTION EFFECTS IN TURBULENT FLOW

Darcy–Weisbach equation: $$h_f = \frac{f L}{D} \frac{V^2}{2g}$$

Thickness of viscous layer: $$\delta_v = 5\frac{\nu}{u_*}$$

Friction factor, turbulent flow: $$\frac{1}{\sqrt{f}} = \begin{cases} -2\log\left(\dfrac{2.51}{\text{Re}\sqrt{f}}\right), & \text{for smooth pipe} \\ -2\log\left(\dfrac{k_s/D}{3.7}\right), & \text{for rough pipe} \end{cases}$$

$$\frac{1}{\sqrt{f}} = -2\log\left(\frac{k_s/D}{3.7} + \frac{2.51}{\text{Re}\sqrt{f}}\right)$$

(Colebrook equation)

$$f = \frac{0.25}{\left[\log\left(\dfrac{k_s/D}{3.7} + \dfrac{5.74}{\text{Re}^{0.9}}\right)\right]^2}$$

(Swamee–Jain equation)

Limit of fully turbulent regime: $$\frac{1}{\sqrt{f}} = \frac{\text{Re}}{200(D/k_s)}$$

Velocity distribution: $$v(r) = \left[(1 + 1.326\sqrt{f}) - 2.04\sqrt{f}\log\left(\frac{R}{R-r}\right)\right] V$$

$$v(r) = V_0\left(1 - \frac{r}{R}\right)^{\frac{1}{n}}$$

Energy correction factor: $$\alpha = 1 + 2.7f$$

Momentum correction factor: $$\beta = 1 + 0.98f$$

PRACTICAL APPLICATIONS

Volume flow rate (Darcy–Weisbach): $$Q = -0.965D^2\sqrt{\frac{gDh_f}{L}}\ln\left(\frac{k_s/D}{3.7} + \frac{1.784\nu}{D\sqrt{gDh_f/L}}\right)$$

Pipe diameter: $$D = 0.66\left[k_s^{1.25}\left(\frac{LQ^2}{gh_f}\right)^{4.75} + \nu Q^{9.4}\left(\frac{L}{gh_f}\right)^{5.2}\right]^{0.04}$$

Hazen–Williams formula: $$V = 0.849 C_H R^{0.63} S_f^{0.54}$$

Head loss, Hazen–Williams formula: $$h_f = 6.82\frac{L}{D^{1.17}}\left(\frac{V}{C_H}\right)^{1.85}$$

Manning formula: $$V = \frac{1}{n}R^{\frac{2}{3}}S_f^{\frac{1}{2}}$$

Head loss, Manning formula: $$h_f = 6.35\,\frac{n^2LV^2}{D^{\frac{4}{3}}}$$

Local head loss: $$h_{\text{local}} = K\frac{V^2}{2g}$$

Head loss coefficient (general): $$K = -\frac{\Delta p}{\frac{1}{2}\rho V^2}$$

Head loss coefficient:
(sudden contraction)
$$K_{sc} = \begin{cases} 0.42\left(1 - \dfrac{D_2^2}{D_1^2}\right), & \text{for } D_2/D_1 \leq 0.76 \\ \left(1 - \dfrac{D_2^2}{D_1^2}\right)^2, & \text{for } D_2/D_1 > 0.76 \end{cases}$$

Head loss coefficient:
(sudden expansion)
$$K_{se} = \left(1 - \frac{D_1^2}{D_2^2}\right)^2$$

Head loss coefficient:
(gradual expansion)
$$K_{ge} = 2.61\sin\left(\frac{\theta}{2}\right)\left(1 - \frac{D_1^2}{D_2^2}\right)^2 + \frac{f_{ge}L_{ge}}{\overline{D}_{ge}}, \quad \text{for } \theta \leq 45^\circ$$

Equivalent length: $$L_{eq} = \frac{K}{f}D$$

Energy equation: $$E_0 - \sum_{i=1}^{NP}\frac{f_iL_i}{D_i}\frac{V_i^2}{2g} - \sum_{j=1}^{NL}K_j\frac{V_j^2}{2g} + h_p - h_t = E_1$$

WATER HAMMER

Water hammer pressure: $$\Delta p = \rho_0 c V$$

Speed of pressure wave:
$$c = \begin{cases} \sqrt{\dfrac{E_v}{\rho_0}}, & \text{(rigid walls)} \\ \sqrt{\dfrac{E_v/\rho_0}{1 + (E_vD/eE_p)}}, & \text{(flexible walls)} \end{cases}$$

Critical valve closure time: $$t_c = \frac{2L}{c}$$

PIPE NETWORKS

Continuity equation:

$$\sum_{i=1}^{\mathrm{NP}(j)} Q_{ij} - F_j = 0, \qquad j = 1, \ldots, \mathrm{NJ}$$

Energy equation:

$$h_2 = h_1 - \left(\frac{fL}{D} + \sum K_\mathrm{m}\right)\frac{Q|Q|}{2gA^2} + \frac{Q}{|Q|}h_\mathrm{p}$$

(nodal method)

$$\sum_{j=1}^{\mathrm{NP}(i)} (h_{\mathrm{L},ij} - h_{\mathrm{p},ij}) = 0, \qquad i = 1, \ldots, \mathrm{NL}$$

(loop method)

Head loss parameters:
(Darcy–Weisbach equation)

$$r = \frac{fL}{2gA^2D}, \qquad n = 2$$

Hardy Cross equation:

$$\Delta Q_i = -\frac{\sum_{j=1}^{\mathrm{NP}(i)} r_{ij}Q_j|Q_j|^{n-1}}{\sum_{j=1}^{\mathrm{NP}(i)} nr_{ij}|Q_j|^{n-1}}$$

BUILDING WATER SUPPLY SYSTEMS

Energy equation:

$$\frac{p_0}{\gamma} - \left[\frac{p_1}{\gamma} + \frac{V_1^2}{2g} + \Delta z\right] = \sum h_\mathrm{f} + \sum h_\mathrm{local}$$

Head loss:

$$h_\mathrm{f} = \begin{cases} 0.230\left(\dfrac{fL}{D^5}\right)Q^2, & \text{Darcy–Weisbach equation} \\ 85.4\left(\dfrac{L}{C_\mathrm{H}^{1.85}D^{4.87}}\right)Q^{1.85}, & \text{Hazen–Williams equation} \end{cases}$$

Section 7.1: Introduction

7.1. (a) If water at 15°C flows with a velocity of 3 m/s in a pipeline, what is the maximum diameter for which the flow is likely to be laminar and what is the minimum diameter for which the flow is likely to be turbulent? (b) If standard air flows with a velocity of 3 m/s in a pipeline, what is the maximum diameter for which the flow is likely to be laminar and what is the minimum diameter for which the flow is likely to be turbulent? (c) What is the maximum air velocity for which the flow is likely to be laminar in a 0.85 mm diameter pipeline? (d) What is the minimum air velocity for which the flow is likely to be turbulent in a 0.85 mm diameter pipeline?

7.2. Air at 80°C and standard pressure is to be carried in a pipeline with a weight flow rate of 0.8 N/s. For what range of pipeline diameters will the flow be laminar?

7.3. Methane at 20°C and 600 kPa (absolute) is to be transported via pipeline at a flow rate of 0.3 N/s. (a) Calculate the density of methane under these conditions and perform an online search to determine the dynamic viscosity of methane at the given temperature and pressure. (b) For what range of pipe diameters will the flow be turbulent?

7.4. Carbon dioxide at 20°C and 300 kPa (absolute) is to be transported via a 30-mm-diameter pipeline. (a) Calculate the density of carbon dioxide under these conditions and perform an online search to determine the dynamic viscosity of carbon dioxide at the given temperature and pressure. (b) For what range of volume flow rates will the flow be laminar?

Section 7.2: Steady Incompressible Flow

7.5. A fluid flows at a rate of 6 L/min in a straight 20-mm-diameter pipe. The fluid has a density of 918 kg/m^3 and a viscosity of 440 mPa·s. Pressure measurements at two locations 12 m apart show an upstream pressure of 400 kPa and a downstream pressure of 340 kPa. The downstream section is 1.75 m higher than the upstream section. (a) Estimate the average shear stress on the surface of the pipe. (b) Estimate the friction factor for the flow. (c) Estimate the distance from the pipe entrance required to establish fully developed flow. (d) Would your answers to part (a) and part (b) be different if the flow between the upstream and downstream sections was not fully developed?

7.6. A 125-mm-diameter duct admits air from a chamber in which the air temperature is 80°C and the air pressure is 101.3 kPa (absolute). (a) Determine the minimum volume inflow rate required to ensure that the flow in the duct is turbulent. (b) At the limiting condition determined in part (a), what is the hydrodynamic entrance length required for fully developed flow in the duct?

7.7. A fluid with a specific gravity of 0.87 and kinematic viscosity of 1.20×10^{-4} m^2/s flows in a straight 100-mm-diameter pipe that is 100 m long and inclined (upward) at an angle of 8° to the horizontal. If the pressure at the downstream (higher-elevation) end of the pipe is 150 kPa and the maximum allowable shear stress on the pipe surface is 180 Pa, what is the maximum allowable pressure at the upstream end of the pipe?

7.8. Standard air from a large room enters a 250-mm-diameter duct at a flow rate of 125 L/s. Estimate the distance from the duct entrance to where the flow is fully developed.

Section 7.3: Friction Effects in Laminar Flow

7.9. A liquid is forced through a small horizontal tube to determine its viscosity. The diameter of the tube is 0.5 mm; the length of the tube is 1.2 m; and when a pressure difference of 0.8 MPa is applied across the tube, the volume flow rate is 800 mm^3/s. Estimate the viscosity of the liquid.

7.10. SAE 30 oil at 20°C flows with an average velocity of 0.3 m/s in a 20-mm-diameter pipe that is inclined upward at 10° to the horizontal. (a) Determine the maximum velocity and the volume flow rate in the pipeline. (b) What longitudinal pressure gradient is causing this flow? (c) What is the shear stress on the pipe surface?

7.11. Consider the case of laminar flow through a pipeline of diameter D, where h_f is the head loss between two sections a distance L apart. (a) Explain how you would use pressure gauges and a tape measure to measure h_f. (b) For a given fluid and for measured values of h_f and L, determine (analytically) the relationship between the percentage error in the volume flow rate, Q, and the percentage error in estimating D.

7.12. A fluid with a density and viscosity of 850 kg/m^3 and 0.5 Pa·s, respectively, flows in a straight inclined pipe of diameter 25 mm at a rate of 3 L/min. Verify the assumption of laminar flow. Two sections of interest are spaced 10 m apart. (a) What is the head loss between the two sections? (b) What pipe inclination would cause the pressure at the two sections to be the same? (c) What is the average shear stress on the pipe surface between the two sections?

7.13. For the general case of laminar flow in a pipe of diameter D, at what distance from the pipe centerline is the velocity equal to the average velocity? Give your answer as a fraction of the pipe diameter.

7.14. Engine oil (SAE 50) at 20°C flows in a 2-cm-diameter pipe. The pressure at one location in the pipe is measured as 30 kPa, and at a location 30 m downstream, the pipe elevation is 1 m higher and the measured pressure is 5 kPa. If the dynamic viscosity of the oil is 0.86 Pa·s and the density is 902 kg/m^3, determine the following: (a) the average velocity in the pipe; (b) the velocity distribution in the pipe; (c) the Reynolds number, also verifying that the flow is laminar according to the Reynolds number criterion; and (d) the friction factor.

7.15. Water at 20°C flows in a straight 1-mm-diameter pipe that is inclined downward at an angle of 5°. (a) What is the maximum allowable pressure gradient along the pipe that will maintain laminar flow? (b) What is the volume flow rate in the pipe under the limiting laminar-flow condition?

7.16. Blood at 37°C flows in a straight 50-mm-diameter tube inclined upward at 80° to the horizontal. If the pressure gradient along the tube is measured as −20 kPa/m, estimate the volume flow rate in the tube.

Section 7.4: Friction Effects in Turbulent Flow

7.17. A fluid is to be carried in a 125-mm-diameter copper pipe such that the Reynolds number of the flow is equal to 5×10^4. The pipe is highly polished, and it can be assumed that $k_s \approx 0$. Using the given data, estimate the thickness of the viscous boundary layer adjacent to the pipe surface. Is the flow turbulent? Is the flow fully turbulent?

7.18. A fluid with a specific gravity and kinematic viscosity of 0.87 and 1.20×10^{-4} m^2/s, respectively, flows in a straight 75-mm-diameter PVC pipe. The roughness height of the PVC pipe is sufficiently small so that when the flow is turbulent, it can be assumed that the flow is hydraulically smooth. If the maximum allowable shear stress on the pipe surface is 125 Pa, what is the maximum allowable volume flow rate in the pipe?

7.19. Water at 20°C is to be transported in a 250-mm-diameter pipe at a volume flow rate of 50 L/s. The objective is to use pipes with sufficiently smooth surfaces so that the flow will be in the hydraulically smooth regime. What range of equivalent sand roughnesses can be used to achieve this objective?

7.20. Water is flowing in a horizontal 100-mm-diameter pipe at a rate of 0.06 m^3/s, and the pressures at sections 50 m apart are equal to 500 kPa at the upstream section and 400 kPa at the downstream section. Estimate the average shear stress on the pipe surface and the friction factor.

7.21. Water at 20°C flows through a 125-mm-diameter steel pipe at a rate of 10 L/s. The equivalent sand roughness of the pipe is estimated as 0.3 mm. (a) Estimate the friction factor of the flow. (b) Estimate the thickness of the viscous layer adjacent to the pipe surface. (c) Compare the thickness of the viscous layer to the sand roughness to estimate whether the flow is in the regime of smooth pipe, transition, or rough pipe.

7.22. Standard air flows with a velocity of 18 m/s in a 6-mm-diameter copper tube. (a) Confirm that the flow is turbulent and calculate the head loss per unit length of tube. (b) If great care is taken to maintain laminar-flow conditions in the tube, determine the head loss per unit length under the laminar-flow condition and the percentage change in head loss per unit length that occurs when the flow changes from the laminar to the turbulent flow condition.

7.23. Water at 20°C flows at a velocity of 1 m/s in a straight 250-mm-diameter horizontal ductile iron pipe. (a) Compare the friction factors derived from the Moody diagram, the Colebrook equation, and the Swamee–Jain equation. State whether the flow is fully turbulent. (b) Estimate the change in pressure over 100 m of pipeline. (c) How would the friction factor and pressure change be affected if the pipe was not horizontal, but was 1 m lower at the downstream section?

7.24. A straight pipe has a diameter of 50 mm, a roughness height of 0.1 mm, and is inclined upward at an angle of 10°. The pipe transports water at 20°C. (a) If the pressure at a given section of the pipe is to be maintained at 550 kPa, determine the pressure at a section 120 m downstream for flows of 2 L/min and 20 L/min (two answers are required). (b) For a flow of 20 L/min, compare the value of the friction factor calculated using the Colebrook equation with the value of the friction factor calculated using the Swamee–Jain equation. Does your result support Swamee–Jain's assertion that the results are within 15% of each other?

7.25. Show that the Colebrook equation can be written in the (slightly) more convenient form

$$f = \frac{0.25}{\left\{\log\left[\left(\frac{k_s/D}{3.7}\right) + \frac{2.51}{\mathrm{Re}\sqrt{f}}\right]\right\}^2}$$

Why is this form of the Colebrook equation termed "(slightly) more convenient"?

7.26. If you had your choice of estimating the friction factor from the Moody diagram or from the Colebrook equation, which one would you pick? Explain your reasons.

7.27. A frequently cited empirical equation for determining the friction factor in pipes under turbulent flow conditions is the Haaland equation, which is given by

$$\frac{1}{\sqrt{f}} = -1.8 \log\left[\left(\frac{k_s/D}{3.7}\right)^{1.11} + \frac{6.9}{\text{Re}}\right]$$

Consider flows in which $k_s/D = 1 \times 10^{-4}$ and Re is in the range of 10^4–10^8. (a) Determine the maximum percentage error introduced by using the Haaland equation instead of the Colebrook equation. (b) Compare the maximum percentage error determined in part (a) with the maximum percentage error introduced by using the Swamee–Jain equation instead of the Colebrook equation for the same range of k_s/D and Re. (c) State which equation (Haaland or Swamee-Jain) would be preferable for the given ranges of parameters.

7.28. Estimate the kinetic energy correction factor, α, for pipe flow with a velocity distribution given by

$$v(r) = V_0\left[1 - \left(\frac{r}{R}\right)^2\right]$$

where $v(r)$ is the velocity at a radial distance r from the centerline of the pipe, V_0 is the centerline velocity, and R is the radius of the pipe.

7.29. The velocity profile, $v(r)$, for turbulent flow in smooth pipes is sometimes estimated by the seventh-root law, originally proposed by Blasius (1913) as

$$v(r) = V_0\left(1 - \frac{r}{R}\right)^{\frac{1}{7}}$$

where V_0 is the centerline velocity and R is the radius of the pipe. Estimate the energy and momentum correction factors corresponding to the seventh-root law.

7.30. Show that the kinetic energy correction factor, α, corresponding to the power-law velocity profile is given by Equation 7.51. Use this result to confirm your answer to Problem 7.29.

7.31. The logarithmic velocity distribution and the power-law velocity distribution are both used to describe turbulent flow in pipes, and these equations are given by Equations 7.46 and 7.49, respectively. The parameter accounting for pipe friction in the logarithmic equation is the friction factor, f, and the parameter accounting for pipe friction in the power-law equation is the exponent, n. (a) Derive the relationship between f and n when both velocity distributions give the same average velocity and the same maximum velocity. (b) If $f = 0.02$, what is the corresponding value of n when the relationship derived in part (a) is used? (c) For $f = 0.02$, plot and compare the normalized logarithmic and power-law velocity distributions. Note that in the normalized distributions, the velocity is normalized by the maximum velocity and the radial distance is normalized by the radius of the conduit.

Section 7.5: Practical Applications

7.32. Water at 20°C is pumped at a rate of 15 L/s into a reservoir as shown in Figure 7.17. Between the pump and the reservoir is 150 m of 100-mm-diameter PVC pipe, and the water surface elevation in the reservoir is 8 m above the centerline of the inflow pipe. The roughness height in the PVC pipe can be assumed to be negligibly small. Estimate the gauge pressure on the downstream side of the pump.

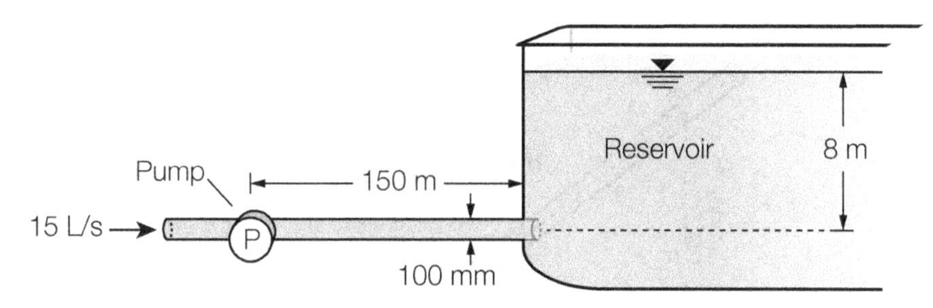

Figure 7.17: Pumped flow into a reservoir

7.33. Crude oil at 20°C is transported at a rate of 500 L/s through a 750-mm-diameter steel pipe that has an estimated equivalent sand roughness of 1.5 mm. The pipeline is 3 km long, and the downstream end of the pipeline is at an elevation that is 1.5 m higher than the elevation at the beginning of the pipeline. (a) What change in pressure is to be expected over the length of the pipeline. (b) At what rate is energy being consumed to overcome friction? (c) If a smooth lining installed in the pipe such that the roughness height is reduced by 70%, what is the percentage change in the quantities calculated in parts (a) and (b)?

7.34. An oil pipeline has a diameter of 1200 mm, an estimated roughness height of 0.15 mm, and an upward slope of 5%. This pipeline is to be used to transport crude oil that has an average density of 860 kg/m^3 and an average viscosity of 7.2 mPa·s. The design flow rate of oil is 3 m^3/s, and pumps are to be placed at fixed intervals to maintain the flow rate. If the maximum allowable pressure in the pipeline is 8 MPa and the minimum allowable pressure is 350 kPa, estimate the required spacing between pumps.

7.35. Water leaves a treatment plant in a 500-mm-diameter ductile iron pipeline at a pressure of 600 kPa and at a flow rate of 0.50 m^3/s. If the elevation of the pipeline at the treatment plant is 120 m, estimate the pressure in the pipeline 1 km downstream where the elevation is 100 m. Assess whether the pressure in the pipeline would be sufficient to serve the top floor of a ten-story building (approximately 30 m high).

7.36. Water flows at a rate of 100 L/s in a 100-m section of 200-mm-diameter ductile iron pipeline, where the end of the pipeline has an elevation that is 0.8 m higher than the beginning of the pipeline. Under the given flow condition, the pressure at the end of the pipeline is measured to be 90 kPa less than the pressure at the beginning of the pipeline. (a) Estimate the friction factor of the pipeline. (b) Assess the condition of the pipeline in terms of likely deterioration of the interior surface of the pipeline.

7.37. Methane at a temperature of 5°C and a pressure of 500 kPa (absolute) flows in a straight 50-mm-diameter duct at the rate of 5 L/s. The duct is inclined upward at 10° to the horizontal, and the pressure drop over a 20-m segment of the duct is measured as 300 Pa. A literature search for the properties of methane at the given temperature and pressure indicates that the dynamic viscosity of the gas is 0.01054 mPa·s under these conditions. Estimate the friction factor and equivalent sand roughness of the duct.

7.38. SAE 30 oil at 20°C flows in the 3-cm-diameter pipe shown in Figure 7.18. The pipe slopes downward at 37°. For the pressure measurements shown, determine the flow rate in m^3/h.

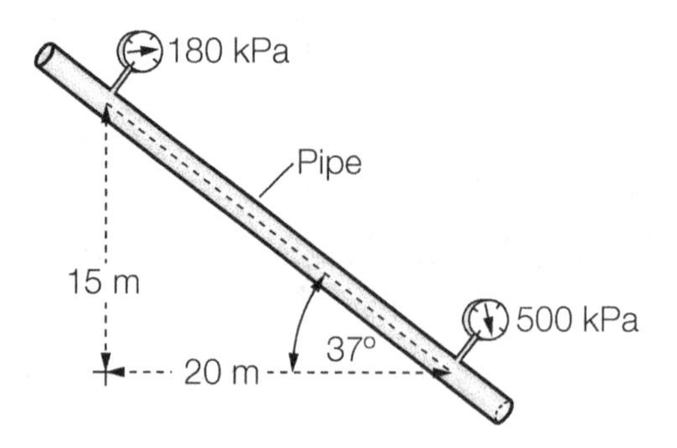

Figure 7.18: Flow in a pipe

7.39. A 25-mm-diameter galvanized iron service pipe is connected to a water main in which the pressure is 400 kPa. If the length of the service pipe to a faucet is 20 m and the faucet is 2.0 m above the main, estimate the flow rate when the faucet is fully open. Neglect any local head loss in the faucet.

7.40. A galvanized iron service pipe from a water main is required to deliver 300 L/s during a fire. If the length of the service pipe is 40 m and the head loss in the pipe is not to exceed 45 m, calculate the minimum pipe diameter that can be used. Use the Colebrook equation in your calculations.

7.41. Repeat Problem 7.40 using the Swamee–Jain approximation (Equation 7.59).

7.42. Air at standard atmospheric pressure and at a temperature of 20°C flows at a rate of 60 L/s through a 20-m-long duct that has a circular cross section. The duct material has an estimated roughness height of 0.5 mm. Design limitations require that the pressure change in the air as it passes through the duct be less than 1.5 kPa. What is the minimum acceptable diameter of the duct?

7.43. Air is to be moved between two chambers where the difference in pressure is 350 kPa. The material to be used in the connecting conduit has an estimated roughness height of 1.5 mm and a length of 100 m. If the desired air transfer rate is 500 L/s, what duct diameter should be used? Use the Colebrook equation in your analysis.

7.44. Methane is delivered to a 200-mm-diameter distribution line at a rate of 50 N/s and at a temperature and pressure of 10°C and 450 kPa, respectively. At the given temperature and pressure (at the entrance to the pipeline), the dynamic viscosity of methane is estimated as 0.01073 mPa·s. The equivalent sand roughness of the pipeline is estimated as 1 mm. Elevation changes along the distribution pipeline are negligible, and the temperature remains approximately constant along the entire length of the pipeline. (a) Assuming incompressible flow, estimate the change in pressure per kilometer of pipeline. (b) If the total length of the pipeline is 15 km, compare the densities at the beginning and end of the pipeline and reassess the assumption of incompressible flow.

7.45. Water flows at 5 m^3/s in a 1 m × 2 m rectangular concrete pipe. Calculate the head loss over a length of 100 m.

7.46. Water flows at 10 m^3/s in a 2 m × 2 m square reinforced concrete pipe. If the pipe is laid on a (downward) slope of 0.002, what is the change in pressure in the pipe over a distance of 500 m?

7.47. Air flows at 10 m/s in a 400 mm × 400 mm square duct that has an estimated equivalent sand roughness of 0.003 mm. The temperature of the air is 70°C, and the air is at standard atmospheric pressure. (a) Determine the head loss per unit length in the square duct. (b) If the square duct is changed to a circular duct made of the same material, what diameter should be used for the circular duct so that the head loss per unit length remains unchanged?

7.48. A fan circulates standard air at 200 L/s in a horizontal 200 mm × 400 mm duct. The length of the duct is 60 m, and the equivalent sand roughness of the duct material is estimated as 0.5 mm. What is the rate of energy loss in the air that can be attributed to friction between the moving air and the (stationary) duct? Is this loss rate approximately equal to the power that must be supplied by the fan to move the air through the duct? Explain.

7.49. Standard air flows at a rate of 500 L/s in an existing duct that is 250 mm wide and 500 mm high. The estimated equivalent sand roughness of the duct material is 0.1 mm. Ductwork renovations call for replacement of the existing duct by another rectangular duct having the same cross-sectional area and the same length but a different aspect ratio. The aspect ratio is defined as the height divided by the width. Aspect ratios in the range of 0.1–3 are being considered for the replacement duct. (a) Plot the ratio of the head loss in the replacement duct to the head loss in the existing duct as a function of the aspect ratio. (b) What aspect ratio in the replacement duct would give the least head loss?

7.50. Derive the Hazen–Williams head loss relation, Equation 7.62, starting from Equation 7.60.

7.51. Compare the Hazen–Williams formula for head loss with the Darcy–Weisbach equation for head loss to determine the expression for the friction factor that is assumed in the Hazen–Williams formula. Based on your result, identify the type of flow condition (rough, smooth, or transition) incorporated in the Hazen–Williams formula.

7.52. Derive the Manning head loss relation, Equation 7.65.

7.53. Compare the Manning formula for head loss with the Darcy–Weisbach equation for head loss to determine the expression for the friction factor that is assumed in the Manning formula. Based on your result, identify the type of flow condition (rough, smooth, or transition) incorporated in the Manning formula.

7.54. Determine the relationship between the Hazen–Williams roughness coefficient and the Manning roughness coefficient.

7.55. Given a choice between using the Darcy–Weisbach, Hazen–Williams, or Manning equation to estimate the friction losses in a pipeline, which equation would you choose? Why?

7.56. Water flows at a velocity of 2 m/s in a 300-mm-diameter new ductile iron pipe. Estimate the head loss over 500 m using (a) the Hazen–Williams formula, (b) the Manning formula, and (c) the Darcy–Weisbach equation. (d) Compare your results. (e) Calculate the Hazen–Williams roughness coefficient that should be used to obtain the same head loss as the Darcy–Weisbach equation. (f) Calculate the Manning coefficient that should be used to obtain the same head loss as the Darcy–Weisbach equation.

7.57. Water leaves a reservoir at 0.06 m^3/s through a 200-mm-diameter riveted steel pipeline that protrudes into the reservoir and then immediately makes a 90° bend with a local (minor) loss coefficient equal to 0.3. Estimate the length of pipeline required for the friction losses to account for 90% of the total losses, which includes both friction losses and so-called "minor losses." Would it be fair to say that for pipe lengths shorter than the length calculated in this problem, the word "minor" should not be used? Explain.

7.58. Floodwater from a residential neighborhood is discharged into a river through a 200-m-long, 100-mm-diameter pipe that has an estimated roughness height of 0.5 mm. The discharge end of the pipe is open to the atmosphere and is at an elevation that is 1.2 m below the entrance to the pipe. Appurtenances within the pipe combine to give a total local loss coefficient of 8.7. Under design conditions, the pressure at the entrance to the pipe is 300 kPa. Estimate the discharge through the pipe under design conditions. Assume water at 20°C.

7.59. Water at 20°C is to be delivered at a rate of 100 L/s through a 200-m-long pipe with an estimated roughness height of 0.1 mm and an estimated local loss coefficient of 3.4. The downstream end of the pipe has an elevation that is 1.8 m higher than the upstream end of the pipe. The pressure at the upstream end of the pipe can be maintained at 450 kPa using a pump, and the minimum allowable pressure at the downstream end of the pipe is 200 kPa. What is the minimum diameter pipe that can be used?

7.60. The 1-km-long tunnel shown in Figure 7.19 is to be sized such that water can be moved from the upper reservoir to the lower reservoir at a rate of 2 m^3/s when all control gates are fully open. The tunnel will be lined with a material that has an estimated roughness height of 1.5 mm, and it is expected that the water will be at approximately 20°C. (a) What tunnel diameter is required to achieve the desired flow rate if local losses are neglected? (b) If local losses were lumped into a single loss coefficient equal to 15.3, what tunnel diameter would be required? What can you conclude about the effect of local losses in this system?

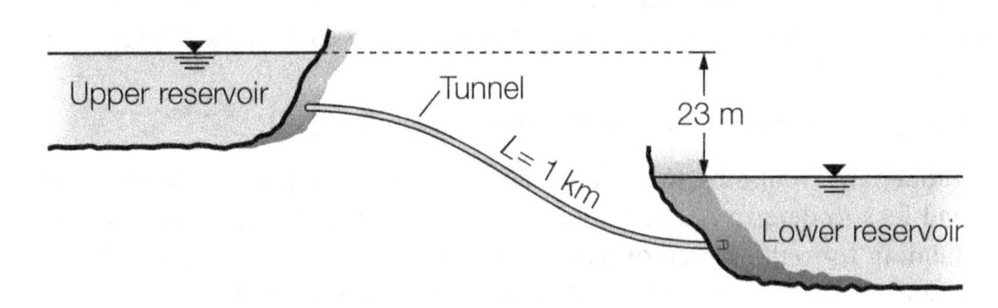

Figure 7.19: Flow through a tunnel

7.61. A large-capacity faucet is to be connected to a water main by 20-mm-diameter copper tubing as shown in Figure 7.20. The copper tubing has an estimated equivalent sand roughness of 0.003 mm, and local head losses are created by the corporation stop (C-stop), four bends, and the faucet itself. The faucet exit has a diameter of 12 mm. What minimum pressure in the water main is required to obtain a desired flow rate of 50 L/min when the faucet is wide open?

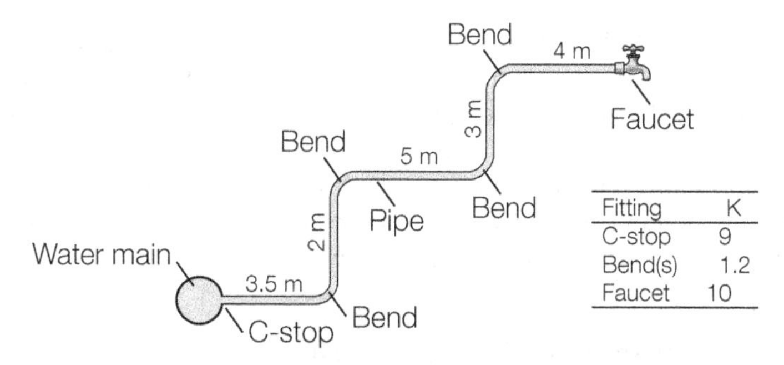

Fitting	K
C-stop	9
Bend(s)	1.2
Faucet	10

Figure 7.20: Faucet connected to a water main

7.62. Standard air flows through a 200-mm-diameter duct at a rate of 100 L/s. The roughness height of the duct material is estimated as 1 mm. The air duct system contains three bends, each with a head loss coefficient of 1.0, and a single valve with a head loss coefficient of 5.0. (a) Estimate the length of duct that will have the same head loss at each bend. (b) Estimate the length of duct that will have the same head loss as the valve. (c) If the total length of the duct system is 20 m, estimate what percentage of the system head loss is caused by the bends and valve combined.

7.63. Air at a temperature and pressure of 40°C and 150 kPa, respectively, flows in a 100-mm-diameter duct at a volume flow rate of 10 L/s. A fixture is inserted in the duct that causes the pressure to decrease by 20 Pa. What is the head loss coefficient of the fixture?

7.64. Water at 20°C flows at 6 L/s through a 75-mm-diameter pipe, and a fixture is installed such that the diameter is suddenly reduced to 50 mm. (a) Estimate the head loss due to the sudden contraction. (b) What pressure change is expected at the contraction? (c) What would be the pressure change if the local head loss was not taken into account?

7.65. A 41-mm-diameter Schedule 40 galvanized steel pipe is connected to a water main as shown in Figure 7.21. The pressure in the water main is 350 kPa, the flow in the pipeline is regulated by a globe valve, and flow exits the pipeline through a nozzle with an exit diameter of 30 mm. A globe valve is to be selected such that a water fountain as high as 5 m can be generated. The 90° bend between the globe valve and the pipe exit is estimated to have a head loss coefficient equal to 1.2, the nozzle is estimated to have a loss coefficient of 0.05, and the galvanized steel pipe is estimated to have an equivalent sand roughness of 0.1 mm. Estimate the maximum allowable head loss coefficient for the open globe valve. Assume water at 20°C.

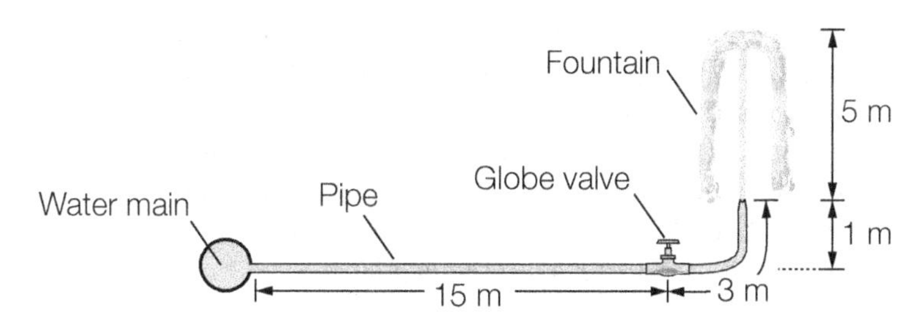

Figure 7.21: Water fountain

7.66. A small playground is to be served by a galvanized iron supply line connected to the water main. The pressure in the water main is 400 kPa, and the length of the supply line to the playground spigot is 50 m. It is estimated that the equivalent sand roughness of the supply line is 0.1 mm and that the sum of the local head loss coefficients of the pipeline appurtenances is equal to 30 when the spigot is fully open. If a maximum flow rate of 100 L/min is to be delivered to the playground, what is the minimum pipe diameter that should be used? Use the Colebrook equation. Assume water at 15°C.

7.67. The storage tank shown in Figure 7.22 is to be used to provide water to a community such that the pressure at Point B is always in the range of 330–440 kPa. The water demand of the community is expected to vary in the range of 0.3–1 m^3/s. The bottom of the storage tank is 42 m above ground level, and the supply pipe is 1.5 m below ground level. The length of steel pipe between the bottom of the storage tank (Point A) and Point B is 1 km, the diameter of the pipe is 600 mm, and the pipe contains five 90° bends. The equivalent sand roughness of the pipe is estimated as 1 mm, and the head loss coefficient of each bend is estimated as 0.8. An auxiliary pumping system will ensure that the tank never becomes empty. Determine the range of water depth (h) in the tank that will meet the design pressure requirements at B. At the minimum flow rate, what will be the minimum pressure? Assume water at 20°C.

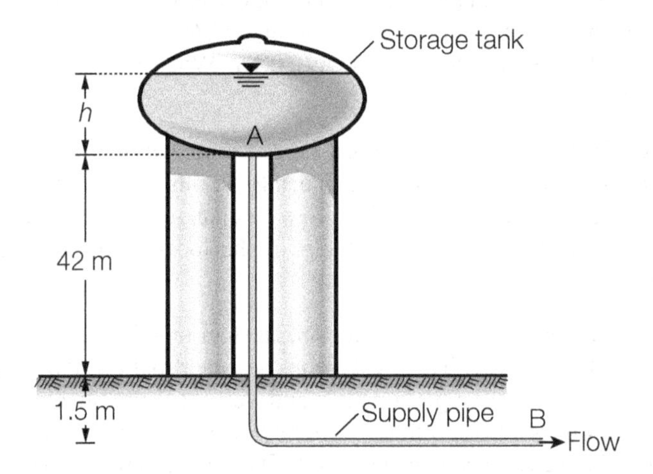

Figure 7.22: Storage tank for a water supply

7.68. The effect of a nozzle on the flow rate out of a garden hose is to be investigated. The 25-m-long, 19-mm-diameter hose is connected to a spigot as shown in Figure 7.23, where the spigot maintains a (gauge) pressure of 200 kPa and the hose exit is 0.9 m above the spigot. The equivalent sand roughness of the hose material is estimated as 0.1 mm. The nozzle has an exit diameter of 10 mm and is estimated to have a head loss coefficient of 0.05 based on the exit velocity. The water is at 15°C. (a) What is the exit velocity and volume flow rate out of the hose without the nozzle? (b) What is the exit velocity and volume flow rate out of the hose with the nozzle attached? (c) Identify a benefit of using a nozzle on a hose.

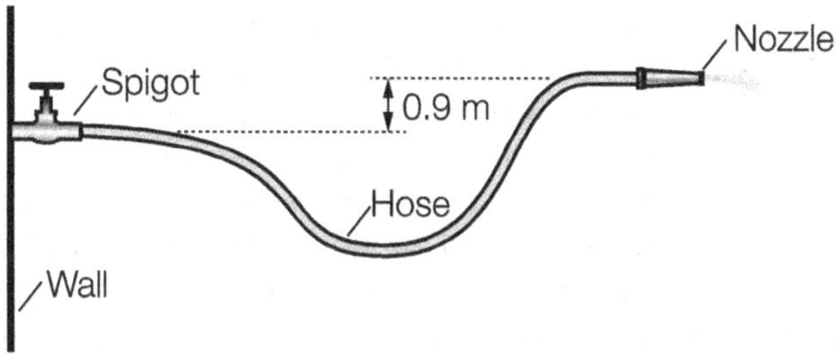

Figure 7.23: Hose discharging water from a spigot

7.69. A pump creates a water pressure of 1 MPa in the supply line to a 65-mm-diameter, 300-m-long hose with a roughness height of 0.5 mm. The end of the hose is at an elevation 2.8 m higher than the beginning of the hose. The objective is to place a nozzle at the end of the hose that will create a jet of water that rises 30 m into the air. If the local head loss coefficient of the nozzle is 0.04, what nozzle diameter should be used? Assume water at 20°C.

7.70. The pressurized water source shown in Figure 7.24 is connected to a 200-mm-diameter, 1-km-long pipeline that discharges into the atmosphere at an elevation 7 m below the connection point. When the pressure at the water source is 140 kPa, the pipeline discharges 60 L/s. A new connection midway along the existing pipeline is being contemplated such that the new connection will deliver 20 L/s while the outlet of the existing pipeline continues to deliver 60 L/s. What is the required pressure at the water source to accomplish this design objective? Assume water at 20°C.

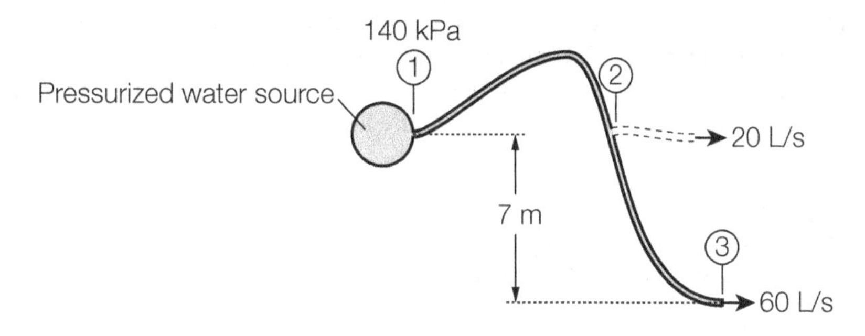

Figure 7.24: Existing pipeline with a new connection

7.71. Water enters and leaves a pump in pipelines of the same diameter and approximately the same elevation. If the pressure on the suction side of the pump is 30 kPa and a pressure of 500 kPa is desired on the discharge side of the pump, what is the head that must be added by the pump? What is the power delivered to the water?

7.72. Water at 20°C is pumped from a lower reservoir to an upper reservoir as shown in Figure 7.25. The riveted steel pipeline has a length of 1 km and a diameter of 500 mm, and the difference between the water surface elevations in the upper and lower reservoirs is 20 m. An antiquated gate valve controls the flow in the pipeline, and when it is fully open, it has a head loss coefficient of 45. The currently installed pump delivers 250 kW of power to the water when it is turned on and produces a flow of 500 L/s when the gate valve is fully open. (a) Estimate the equivalent sand roughness of the pipeline. (b) If the gate valve is replaced by a new one such that the head loss coefficient is reduced by 70%, what pumping power is needed to generate the same flow rate?

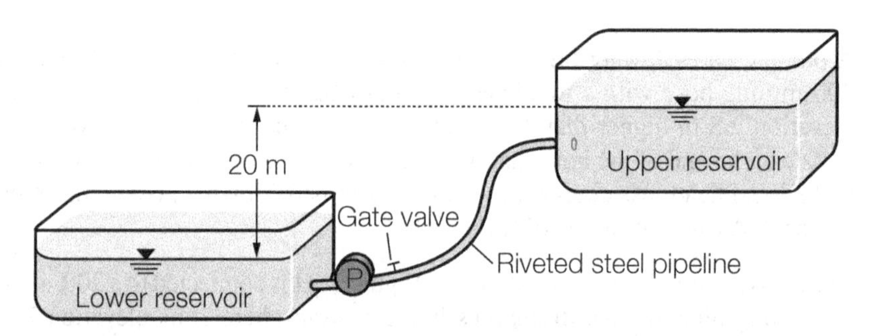

Figure 7.25: Pumping from a lower to upper reservoir

7.73. Water is pumped from a supply reservoir through a ductile iron water transmission line as shown in Figure 7.26. The high point of the transmission line is at Point A, 1 km downstream of the supply reservoir, and the low point of the transmission line is at Point B, 1 km downstream of A. If the flow rate through the pipeline is 1 m^3/s, the diameter of the pipe is 750 mm, and the pressure at A is to be 350 kPa, (a) estimate the head that must be added by the pump, (b) estimate the power supplied by the pump, and (c) calculate the water pressure at B.

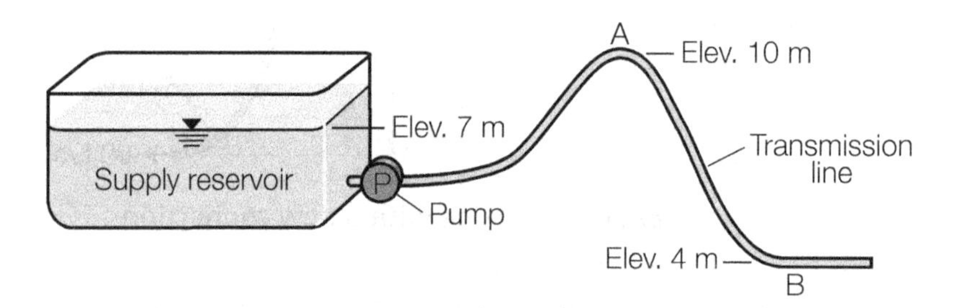

Figure 7.26: Water pumped into a transmission line

7.74. Water is to be pumped at a rate of 0.7 m^3/s from a supply reservoir into a 600-mm steel water supply pipeline as shown in Figure 7.27. The equivalent sand roughness of the steel pipe is estimated as 1 mm. The length of the pipeline from the reservoir to Point A is 1 km, from A to B is 1.5 km, and from B to C is 1.8 km. Point A is at the lowest elevation of the pipeline, Point B is at the highest elevation, and Point C is at the delivery point to the distribution network where the pressure is 400 kPa under the given flow condition. (a) Estimate the range of pressures in the pipeline downstream of the pump? (b) If the maximum pressure in the pipeline is to be no more than 600 kPa, what is the minimum diameter that should be used for the pipeline? Assume water at 20°C.

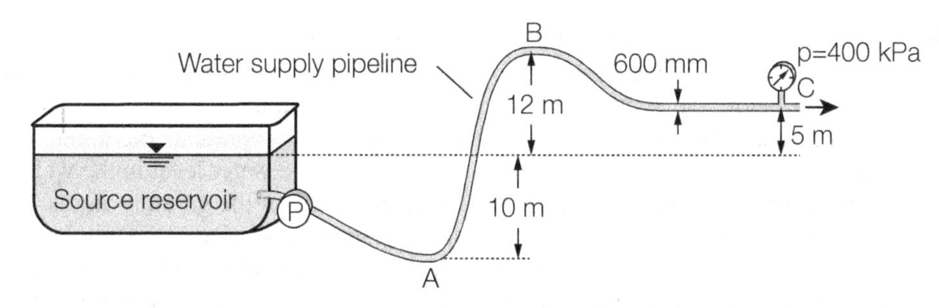

Figure 7.27: Water delivery from a supply reservoir

7.75. Water is to be pumped into a storage reservoir through a 230-m-long galvanized pipeline that has an estimated roughness height of 0.2 mm. Under design conditions, the water level in the reservoir is 18 m above the centerline of the pump discharge. When the pump is operating, the desired water pressure on the discharge side of the pump is 400 kPa and the desired flow rate is 50 L/s. What pipe diameter is required to attain these operating conditions? Assume water at 20°C.

7.76. A pump is being used to fill a storage tank with water as shown in Figure 7.28. The bottom of the storage tank is 4 m above the center of the discharge line of the pump, and the depth of water in the tank is represented by h. The plan area of the tank is 10 m^2, and the pipeline between the pump and the tank has a diameter of 75 mm, a length of 65 m, and an estimated roughness height of 0.5 mm. The pressure at the discharge side of the pump is maintained at 550 kPa. Estimate the time it takes to fill the tank from a depth of 0.5 m to a depth of 3 m. Assume fully turbulent flow and water at 20°C.

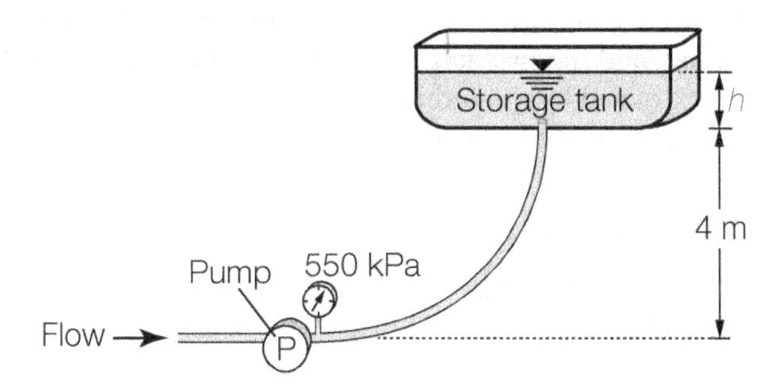

Figure 7.28: Filling a storage tank using a pump

7.77. A pipeline is to be run from a water treatment plant to a major suburban development 3 km away. The average daily demand for water at the development is 0.0175 m^3/s, and the peak demand is 0.578 m^3/s. Determine the required diameter of ductile iron pipe such that the flow velocity during peak demand is 2.5 m/s. Round the pipe diameter upward to the nearest 25 mm (i.e., 25 mm, 50 mm, 75 mm, and so on). The water pressure at the development is to be at least 340 kPa during average demand conditions and 140 kPa during peak demand. If the water at the treatment plant is stored in a ground-level reservoir where the level of the water is 10.00 m above the mean sea level and the ground elevation at the suburban development is 8.80 m above the mean sea level, determine the pump power (in kilowatts) that must be available to meet the average daily and peak demands.

7.78. Water is pumped at a rate of 240 L/min from an open well as shown in Figure 7.29. Groundwater in the well enters the vertical 50-mm-diameter intake pipe at a depth of 15 m below the suction side of the pump. The equivalent sand roughness of the intake pipe is estimated as 0.5 mm. The distance h represents the difference in elevation between the suction side of the pump and the water level in the well. Atmospheric pressure is 101.3 kPa, and the temperature of the groundwater is 15°C. (a) For the given pumping rate and intake characteristics, what is the maximum allowable value of h that will prevent the occurrence of cavitation on the suction side of the pump? (b) Estimate the maximum allowable value of h regardless of the pumping rate. (c) What general statement can you make about the maximum elevation of pumps relative to their source reservoirs?

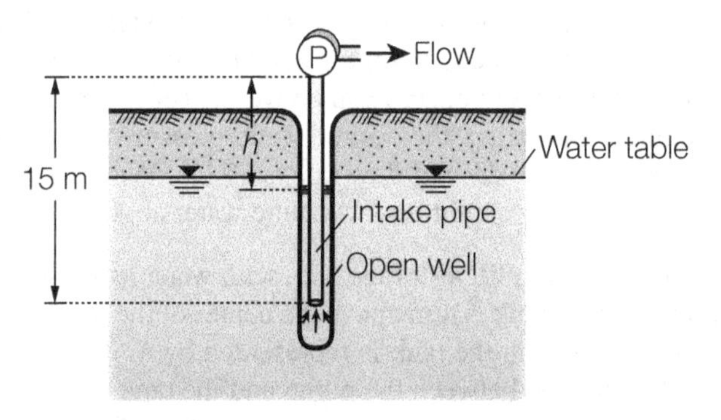

Figure 7.29: Pumping from a well

7.79. Water exits a reservoir through a 125-m-long, 5-cm-diameter horizontal cast iron pipe as shown in Figure 7.30. The pipe entrance is sharp-edged, the water flows through a turbine, and the discharge is to the atmosphere. If the flow rate is 4 L/s, what power is extracted by the turbine? Describe a practical situation in which you might encounter this type of problem.

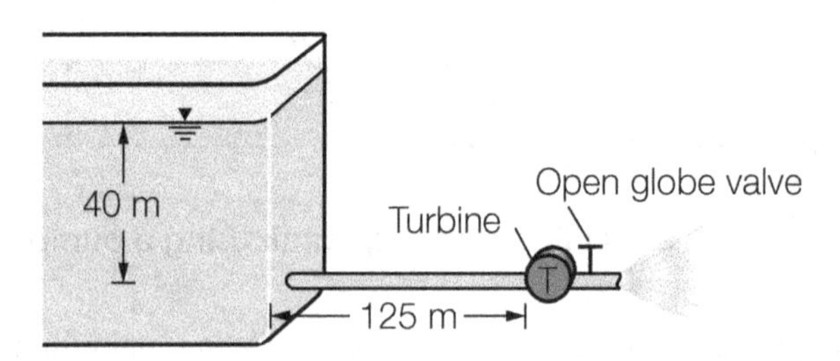

Figure 7.30: Flow out of a reservoir

7.80. Water flows from an upper reservoir to a lower reservoir through a 150-m-long, 350-mm-diameter tunnel that feeds a turbine as shown in Figure 7.31. The tunnel has an estimated roughness height of 1 mm, and the turbine is expected to extract energy from the water at a minimum rate of 60 kW. The entrance and exit loss coefficients are 0.8 and 1.0, respectively. What is the minimum flow rate through the tunnel that is required to accomplish this objective?

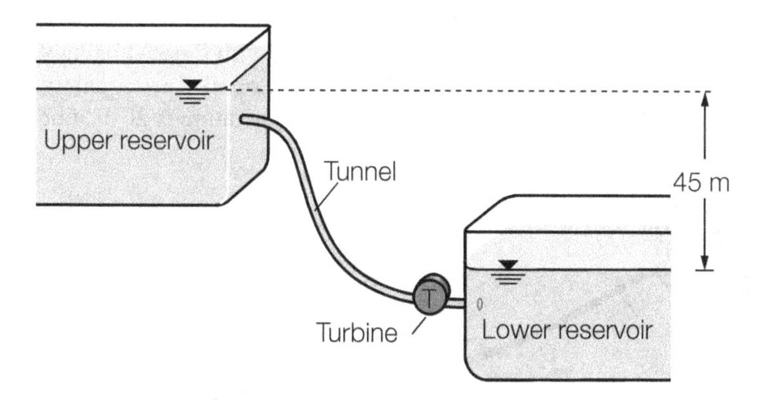

Figure 7.31: Flow between two reservoirs

Section 7.6: Water Hammer

7.81. Plot the water hammer pressure versus time at the midpoint of the pipeline shown in Figure 7.10. Assume that the valve is closed instantaneously.

7.82. Water flows in a 100-m-long pipe at 3 m/s. If the water temperature is 20°C, determine the minimum valve closure time to avoid creating maximum water hammer pressures. What is the maximum water hammer pressure that can occur? How is this pressure affected if the water temperature drops to 10°C?

7.83. Based on your result in Problem 7.82, can water hammer be a serious problem in household plumbing? Explain.

7.84. Water at 20°C flows in a 150-m-long, 50-mm-diameter ductile iron pipe at 4 m/s. The thickness of the pipe wall is 1.5 mm, and the modulus of elasticity of ductile iron is 1.655×10^5 MN/m^2. What is the maximum water hammer pressure that can occur?

7.85. The pipeline described in Problem 7.84 is replaced by 50-mm-diameter PVC pipe with a wall thickness of 2 mm and a modulus of elasticity of 1.7×10^4 MN/m^2. How will this affect the maximum water hammer pressure?

Section 7.7: Pipe Networks

7.86. A water supply pipeline is to deliver 0.6 m^3/s from a water treatment plant through a 1000-mm-diameter, 3.2-km-long ductile iron pipeline. The water surface elevation in the source storage reservoir is at 10.06 m above the mean sea level, and the water is to be pumped from the storage reservoir into the delivery pipe. The pump is located 1.5 m below the water surface in the source reservoir, and the end of the delivery pipe is at elevation 11.52 m above the mean sea level. (a) Estimate the pump requirement in kW so that the pressure at the delivery point is 350 kPa. (b) What is the pressure in the pipeline immediately downstream of the pump? (c) Recalculate the answer to part (a) if the pipeline is divided into two 600-mm-diameter, 3.2-km-long ductile iron pipelines; the pipes are separated just downstream of the pump and rejoined at the downstream location.

7.87. Reservoirs A, B, and C are connected as shown in Figure 7.32. The water elevations in reservoirs A, B, and C are 100 m, 80 m, and 60 m, respectively. The three pipes connecting the reservoirs meet at the junction J, with Pipe AJ being 900 m long, BJ 800 m long, and CJ 700 m long and the diameter of all pipes equal to 850 mm. If all pipes are made of ductile iron and the water temperature is 20°C, find the flow in or out of each reservoir.

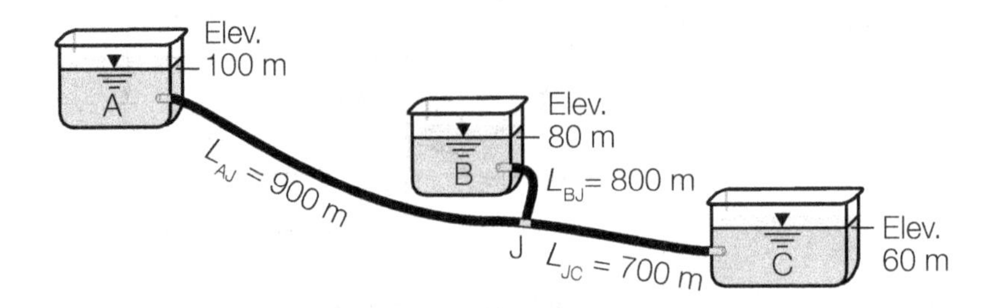

Figure 7.32: Connected reservoirs

7.88. A 3-km-long, 150-mm-diameter existing riveted steel pipeline connects two reservoirs as shown in Figure 7.33. The existing pipeline extends from location A to C and has an estimated roughness height of 1 mm. To increase the flow rate between reservoirs, it is proposed to add a new 200-mm-diameter pipeline such that the new pipeline is connected to the existing pipeline at location B that is 1.5 km from A. The new pipeline will have a length of 1.6 km and an estimated roughness height of 0.5 mm. Under design conditions, the difference in elevation between the water surfaces in the two reservoirs is 8 m. For this analysis, you may assume that the flows in all pipes are fully turbulent and that local losses are negligible compared to friction losses. (a) Compare the existing and new flows between the reservoirs. (b) What percentage change in flow will occur in the existing pipe? (c) Using the calculated flows in the new system, review your assumption of fully turbulent flow and state how you would modify your analysis based on your assessment results. Assume water at 20°C.

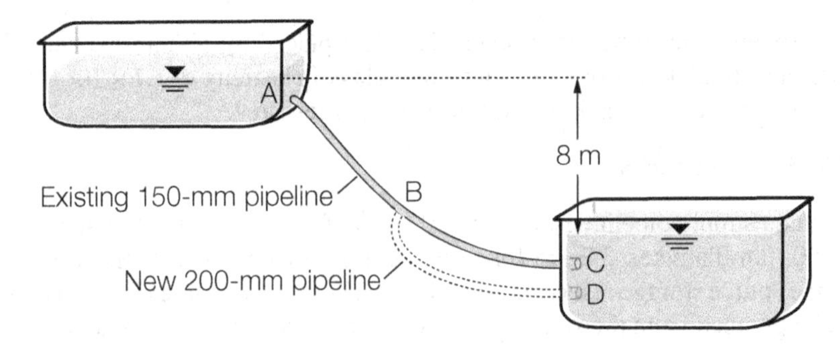

Figure 7.33: Existing and modified pipeline system

7.89. The water main shown in Figure 7.34 diverges at Point A and serves locations at Points B, C, and D. The lengths and diameters of the pipeline segments are tabulated in Figure 7.34, along with the elevations of Points A, B, C, and D. All pipes are made of ductile iron with an estimated roughness height of 0.2 mm. Under normal operating conditions, the synoptic pressures at the key points in the system are tabulated in the figure. A 1.8-km-long, 150-mm-diameter ductile iron pipe is being considered for installation that will connect Points B and C. (a) Estimate the flows Q_B, Q_C, and Q_D for the existing system. (b) Estimate the flows Q_B, Q_C, and Q_D with the new pipeline connecting B to C in operation. Assume water at 20°C. Do not assume that the flow is fully turbulent.

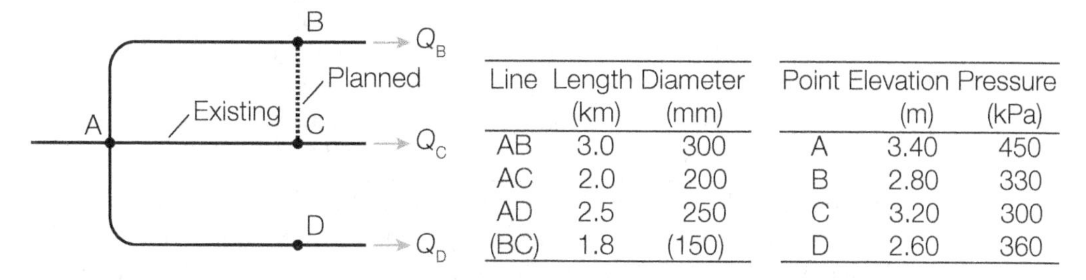

Line	Length (km)	Diameter (mm)
AB	3.0	300
AC	2.0	200
AD	2.5	250
(BC)	1.8	(150)

Point	Elevation (m)	Pressure (kPa)
A	3.40	450
B	2.80	330
C	3.20	300
D	2.60	360

Figure 7.34: Flows in a pipeline system

7.90. The water supply network shown in Figure 7.35 has constant-head elevated storage tanks at A and B, with inflows and withdrawals at C and D. The network is on flat terrain, and the pipeline characteristics are as follows:

Pipe	L (km)	D (mm)
AD	1.0	400
BC	0.8	300
BD	1.2	350
AC	0.7	250

If all pipes are made of ductile iron, calculate the inflows/outflows from the storage tanks. Assume that the flows in all pipes are fully turbulent.

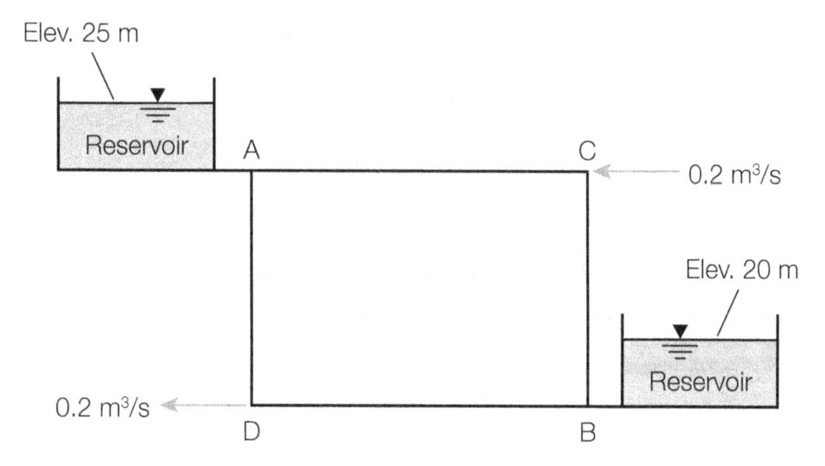

Figure 7.35: Water supply network

7.91. Water is handled by a system of pipes as shown in Figure 7.36, and the pipe characteristics are as follows:

Pipe	Length (m)	Diameter (m)	f
AC, BC	100	0.50	0.0055
CD	300	0.75	0.0050
DE	500	0.30	0.0060
DF	400	0.25	0.0060
DG	500	0.30	0.0060

The elevation of outlets E, F, and G is 100 m above the elevation of inlets A and B. All outlets and inlets are at atmospheric pressure. If the mean velocity in the pipes AC and BC is 2.5 m/s, calculate the flow rate through the pump P, the pressure difference across the pump, and the power consumed by the pump. Assume that the pump efficiency is 76%.

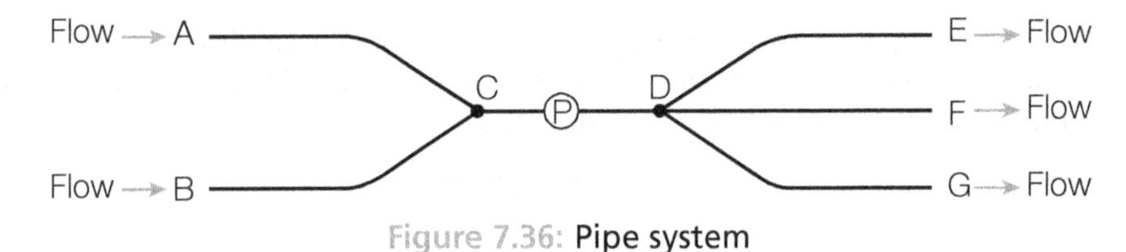

Figure 7.36: Pipe system

7.92. Consider the pipe network shown in Figure 7.37. The Hardy Cross method can be used to calculate the pressure distribution in the system, where the friction loss, h_f, is estimated using the equation

$$h_\mathrm{f} = rQ^n$$

and all pipes are made of ductile iron. What values of r and n would you use for each pipe in the system? The pipeline characteristics are as follows:

Pipe	L (m)	D (mm)
AB	1000	300
BC	750	325
CD	800	200
DE	700	250
EF	900	300
FA	900	250
BE	950	350

Assume that the flow in each pipe is hydraulically rough.

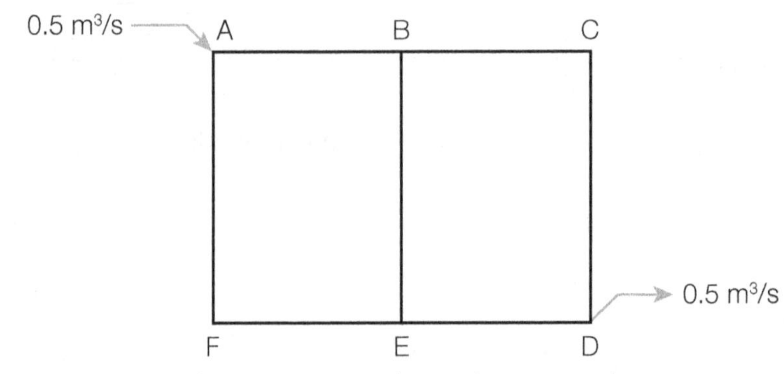

Figure 7.37: Two-loop pipe network

7.93. A portion of a municipal water distribution network is shown in Figure 7.38, where all pipes are made of ductile iron and have diameters of 300 mm. Use the Hardy Cross method to find the flow rate in each pipe. If the pressure at Point P is 500 kPa and the distribution network is on flat terrain, determine the water pressures at each pipe intersection.

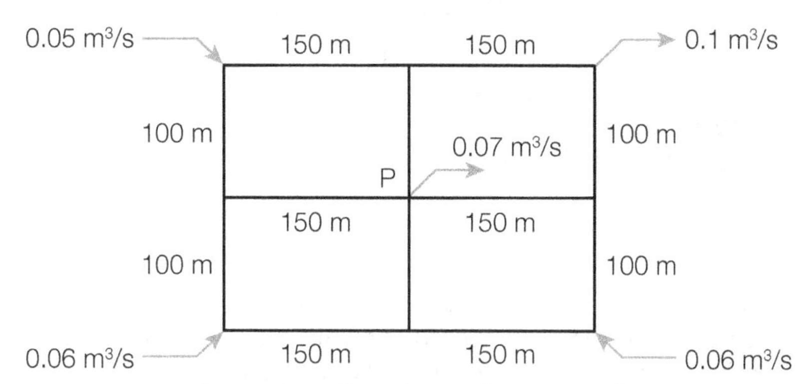

Figure 7.38: Four-loop pipe network

7.94. Water is delivered at a rate of 48 000 m^3/d at node C of the distribution system shown in Figure 7.39, and the water is withdrawn at a rate of 8000 m^3/d at each of the nodes. All pipes are made of steel with roughness heights of 0.05 mm and are of diameter 1120 mm. The pipe lengths are as follows: CD = 5 km, DE = 12 km, EF = 9 km, FC = 6 km, FG = 7 km, GH = 8 km, and HC = 10 km. Determine the flow distribution in the pipes.

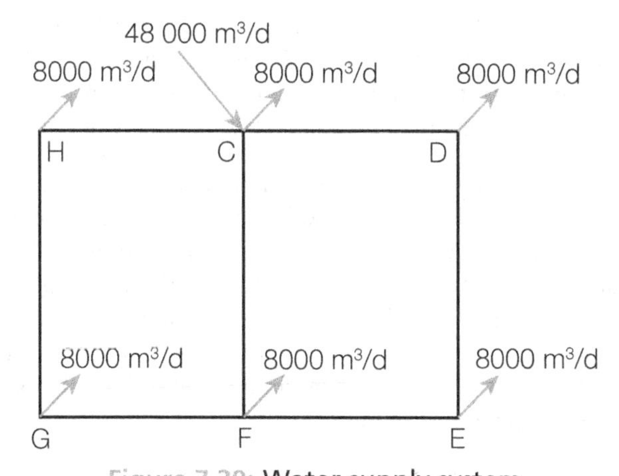

Figure 7.39: Water supply system

Section 7.8: Building Water Supply Systems

7.95. Water in a household plumbing system originates at the neighborhood water main where the pressure is 480 kPa, the velocity is 5 m/s, and the elevation is 2.44 m. A 19-mm copper service line supplies water to a two-story residence where the faucet in the master bedroom is 40 m (of pipe) away from the main and at an elevation of 7.62 m. If the sum of the minor-loss coefficients is 3.5, estimate the maximum (open faucet) flow. Is this flow rate typical of a bathroom faucet? How would this flow rate be affected by the operation of other faucets in the house?

7.96. A 50-mm-diameter pipeline made of PVC is connected directly to the water main as shown in Figure 7.40.

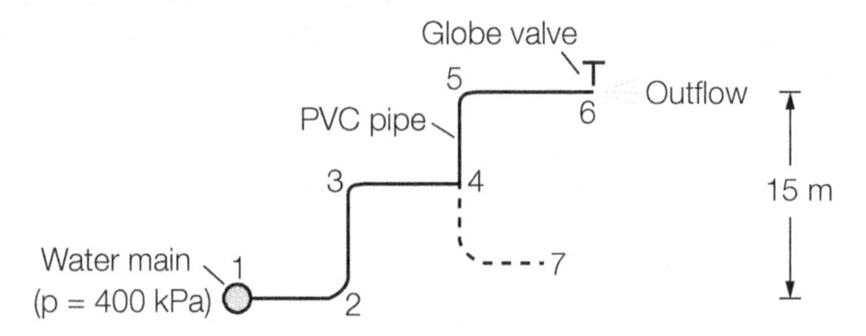

Figure 7.40: Pipeline

The pressure in the water main at Point 1 is 400 kPa, and the faucet of interest is at Point 6, which is 15.0 m vertically above the water main connection. The length of pipeline from the water main to the faucet is 30.0 m, and the pipeline has the following characteristics:

Location	Type of Fixture
1	sharp-edged entrance
2, 3, 5	90° elbow
4	tee
6	globe valve

There is no flow in the line from Point 4 to Point 7. According to the manufacturer of the faucet (= globe valve), the local loss coefficient is equal to 4.0 when the faucet is fully open. In this case, the copper tubing is so smooth that the roughness height can be taken as zero. Estimate the flow rate in the pipeline when the faucet is fully open.

7.97. The top floor of an office building is 40 m above street level and is to be supplied with water from a municipal pipeline buried 1.5 m below street level. The water pressure in the municipal pipeline is 450 kPa, the sum of the local loss coefficients in the building pipes is 10.0, and the flow is to be delivered to the top floor at 20 L/s through a 150-mm-diameter PVC pipe. The length of the pipeline in the building is 60 m, the water temperature is 20°C, and the water pressure on the top floor must be at least 150 kPa. Will a booster pump be required for the building? If so, what power must be supplied by the pump?

7.98. Design flow rates in household plumbing are based on the number and types of plumbing fixtures as measured by "fixture units." The Hunter curve is used to relate fixture units to flow rate. For a particular office building, the total fixture units are determined to be 120 and the corresponding flow rate is 4.67 L/s. The pressure at the water main is estimated to be 380 kPa, the maximum velocity in the plumbing is not to exceed 2.4 m/s, copper lines are to be used, the length of the line to the most remote fixture is 110 m, the minimum allowable pressure in the pipe (after accounting for head losses) is 240 kPa, and the elevation difference between the water main connection and the most remote fixture in the building is 3 m. Copper pipe is available in diameters starting at 12.5 mm and increasing in increments of 6.25 mm. Determine the minimum diameter that could be used for the plumbing line. Neglect local losses.

7.99. Design the hot water pipes for the building given in Example 7.17.

7.100. Water is to be delivered from a public water supply line to a two-story building. Under design conditions, each floor of the building is to be simultaneously supplied with water at a rate of 200 L/min. The pipes in the building plumbing system are to be made of galvanized iron. The length of pipe from the public water supply line to the delivery point on the first floor is 20 m, the length of pipe from the delivery point on the first floor to the delivery point on the second floor is 5 m, the water delivery point on the first floor is 2 m above the water main connection, and the delivery point on the second floor is 3 m above the delivery point on the first floor. If the water pressure at the water main is 380 kPa, what is the minimum diameter pipe in the building plumbing system to ensure that the pressure is at least 240 kPa on the second floor? Neglect minor losses and consider pipe diameters in increments of $\frac{1}{4}$ cm, with the smallest allowable diameter being $\frac{1}{2}$ cm. For the selected diameter under design conditions, what is the water pressure on the first floor?

Chapter

8

Turbomachines

LEARNING OBJECTIVES

After reading this chapter and solving a representative sample of end-of-chapter problems, you will be able to:

- Apply the fundamental equations of fluid mechanics to analyze the performance of pumps and turbines.
- Apply the affinity laws to relate the performances of homologous pumps and turbines.
- Analyze pumps in pipeline systems and choose the appropriate pump to meet various design objectives.
- Calculate limits on pump and turbine location to avoid cavitation effects.
- Analyze multi-pump and multi-turbine systems.
- Analyze hydropower installations that use impulse turbines or reaction turbines and assess hydropower generation potential.

8.1 Introduction

Machines used in fluid flow systems can be broadly categorized as either *turbomachines* or *positive-displacement machines*. Turbomachines add or extract energy from a fluid by means of a rotating component. The prefix "turbo" generally means "spinning," so a turbomachine is literally a "spinning machine." Machines with rotating components are also called *rotodynamic machines*. In contrast to turbomachines, positive displacement machines move fluids by forcing a fluid into and out of a chamber. The main focus of this chapter is on turbomachines, which are widely used in engineering applications. Turbomachines that add mechanical energy to flowing fluids are generically classified as *turbopumps*, whereas turbomachines that convert the energy of a flowing fluid into the mechanical energy are classified as *turbines*. A change in pressure is usually the dominant component of the change in mechanical energy between the inlet and outlet of a turbomachine, with pressure generally increasing across pumps and decreasing across turbines. In common usage, the term "pump" is usually applied to machines that move liquids, whereas machines that move gases are variously called *fans*, *blowers*, and *compressors*, depending on their applications. Typically, if the pressure rise is very small, a gas pump is called a *fan*; if the pressure rise is on the order of 101.3 kPa, the gas pump is called a *blower*; and for pressure rises much greater than 101.3 kPa, the gas pump is called a *compressor*. The primary use of compressors is to increase the pressure of a gas, whereas the primary use of fans and blowers is to move a gas. Turbines also have a variety of uses; for example, *hydraulic turbines* are used to extract energy from flowing water in hydroelectric facilities, *wind turbines* and *windmills* are used to extract energy from wind, and *gas turbines* and *steam turbines* are used to extract energy from various flowing gases.

Characteristics of turbomachines. All turbomachines include one or more rotating components, called *rotors*, between the inflow and outflow sections of the turbomachine. In the case of turbopumps, the rotor is commonly called an *impeller* and is powered by an external motor, and the impeller does work on the fluid. In the case of turbines, the rotor is commonly called a *runner*. The fluid does work on the runner, which transmits mechanical energy to an external generator. The flow of a fluid toward the rotor of a turbine is typically controlled by fixed or adjustable *vanes* or *blades*. For both turbopumps and turbines, the assembly of rotor and vanes is usually contained within a *casing*, which is sometimes called a *housing*. Turbomachines can be classified by the direction of fluid flow relative to the rotor, with the three classifications being *axial flow*, *mixed flow*, and *radial flow*. In axial-flow machines, the predominant flow direction is along the axis of rotation of the rotor; in radial-flow machines, the predominant flow direction is normal to the axis of rotation of the rotor; and in mixed-flow machines, the predominant flow direction has components in both the axial and radial directions.

Positive displacement machines. In contrast to turbomachines, which have rotating components, positive displacement machines operate by displacing fluids within a contained volume. Examples of positive displacement pumps include a *reciprocating pump* in which a piston moves up and down in a cylinder, a *diaphragm pump* in which the change in volume is caused by the deformation of a flexible boundary surface (such as in the human heart), and a *gear pump* in which two rotors mesh within a close-fitting housing. Positive displacement pumps are characterized by their ability to generate high pressures at low flow rates, in contrast to rotodynamic pumps that generate moderate pressures at high flow rates. For example, a positive displacement pump might be capable of generating a pressure on the order of 30.4 MPa with a flow rate of 6 L/s, whereas a rotodynamic pump might be capable of generating a pressure on the order of 0.51 MPa with a flow rate of 20 000 L/s.

8.2 Mechanics of Turbomachines

A turbomachine is characterized by the passage of a fluid over rotating blades, and such turbomachines include both turbopumps and turbines. Although a pump can be classified as either a turbopump or a positive displacement pump, in the present context, the focus is on turbomachines and a turbopump will be referred to simply as a pump. In pumps, the blades that are in contact with the fluid are attached to a central shaft that is turned by an attached motor. The rotating blades are inclined at such an angle that they force the fluid to flow. In the case of turbines, the fluid is directed to flow past blades that are attached to a central shaft, which is itself attached to a generator. The blades are inclined at such an angle that the fluid causes the blades and shaft to rotate, which causes the attached generator to produce electricity.

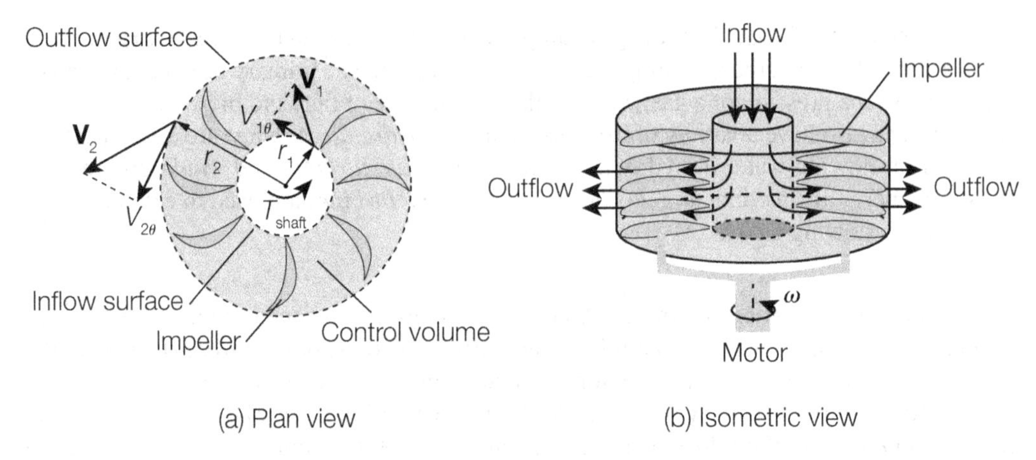

Figure 8.1: Flow through a pump

Mechanics of pumps. Consider a control volume bounded by concentric cylinders and containing the pump impeller (= rotor) as shown in Figure 8.1(a). The fluid enters from the inner surface at velocity $\mathbf{V}_1$ and exits through the outer surface at velocity $\mathbf{V}_2$. Applying the moment of momentum equation (Equation 4.106) yields the expression for the *shaft torque*, T_{shaft} [FL], exerted by the rotor on the fluid in terms of the velocities of the fluid entering and exiting the control volume containing the rotor,

$$T_{\text{shaft}} = \dot{m}\,(r_2 V_{2\theta} - r_1 V_{1\theta}) \tag{8.1}$$

where $\dot{m}$ is the mass flow rate of fluid entering and leaving the control volume [MT^{-1}], r_1 and r_2 are the radii of the inner and outer surfaces of the control volume [L], respectively, and $V_{1\theta}$ and $V_{2\theta}$ are the tangential (θ-)components of the fluid velocity at the entrance and exit of the control volume [LT^{-1}], respectively. The θ direction at the exit of the rotor is sometimes called the *whirl direction*, and the component of the entrance velocity, $V_{1\theta}$, in the whirl direction is called the *velocity of whirl*. (Usage of these latter terms involving "whirl" is particularly common in the United Kingdom.) Equation 8.1 is commonly called the *Euler turbomachine equation*. The pump impeller is connected to a rotating shaft as illustrated in Figure 8.1(b). If the shaft rotates at a rate ω [rad/s], the *shaft power*, $\dot{W}_{\text{shaft}}$ [ET^{-1}][53] can be derived from the shaft torque and is given by

$$\dot{W}_{\text{shaft}} = T_{\text{shaft}}\omega \tag{8.2}$$

and combining Equations 8.1 and 8.2 yields

$$\dot{W}_{\text{shaft}} = \dot{m}\,\omega\,(r_2 V_{2\theta} - r_1 V_{1\theta}) \tag{8.3}$$

The energy added to the fluid per unit mass of fluid, w_{shaft} [EM^{-1}], can be derived from Equation 8.3 by dividing both sides by $\dot{m}$, which yields

$$w_{\text{shaft}} = \frac{\dot{W}_{\text{shaft}}}{\dot{m}} \rightarrow w_{\text{shaft}} = \omega\,(r_2 V_{2\theta} - r_1 V_{1\theta}) \tag{8.4}$$

The relationships given by Equations 8.1 and 8.4 apply to centrifugal pumps, where the flow enters from the center of the rotor, is driven by an impeller (attached to an external motor) and the flow exits from the periphery of the rotor. The relationships developed in this section are also applicable to centrifugal fans and blowers.

[53] "E" denotes a dimension of energy.

EXAMPLE 8.1

The pump impeller shown in Figure 8.2 has a diameter of 800 mm, a blade width of 60 mm, and an exit blade angle of 30°, and it rotates at 1200 rpm. Under design conditions, water at 20°C flows through the pump at a flow rate of 200 L/s. Flow enters the impeller in a direction normal to the inflow surface. Estimate the required shaft power at the design flow rate.

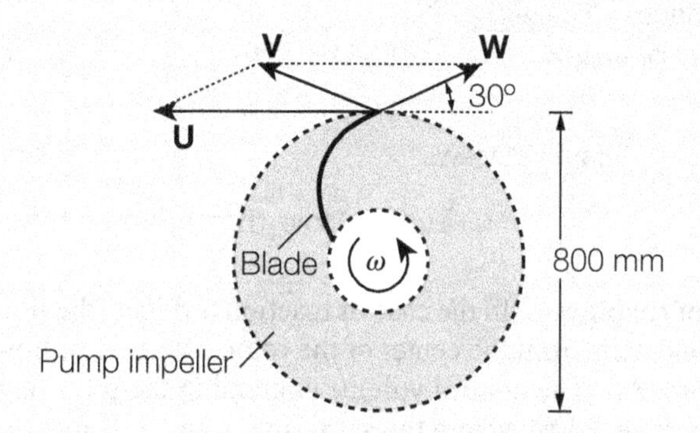

Figure 8.2: Flow through an impeller

SOLUTION

From the given data: $D = 800$ mm, $r_2 = D/2 = 400$ mm, $H = 60$ mm, $\beta = 30°$, $\omega =$ 1200 rpm $= 125.7$ rad/s, $Q = 200$ L/s $= 0.2$ m^3/s, and $V_{1\theta} = 0$ m/s. For water at 20°C, $\rho = 998.2$ kg/m^3. The velocity components at the exit of the blade are shown in Figure 8.2, where **U** is the peripheral velocity of the impeller, **W** is the velocity of the exiting fluid relative to a blade of the impeller, and **V** is the absolute velocity of the fluid exiting the impeller. The aforementioned vectors are related by $\mathbf{V} = \mathbf{U} + \mathbf{W}$. From the given data and velocity relationships,

$$\dot{m} = \rho Q = (998.2)(0.2) = 199.6 \text{ kg/s}, \quad V_{2r} = \frac{Q}{\pi DH} = \frac{0.3}{\pi(0.8)(0.06)} = 1.326 \text{ m/s}$$

$$W = \frac{V_{2r}}{\sin\beta} = \frac{1.326}{\sin 30°} = 2.653 \text{ m/s}, \qquad U = r_2\omega = (0.4)(125.7) = 50.27 \text{ m/s}$$

$$V_{2\theta} = U - W\cos\beta = 50.27 - (2.653)\cos 30° = 47.97 \text{ m/s}$$

where V_{2r} is the radial component of the exit velocity, **V**. Using the given and derived parameters in Equation 8.3 yields

$$\dot{W}_{\text{shaft}} = \dot{m}\,\omega\,[r_2V_{2\theta} - r_1V_{1\theta}] = (199.6)(125.7)\,[(0.4)(47.97) - 0] = 4.814 \times 10^5 \text{ W}$$

Therefore, the theoretical shaft power required to drive the pump at the design flow rate is approximately **481 kW**.

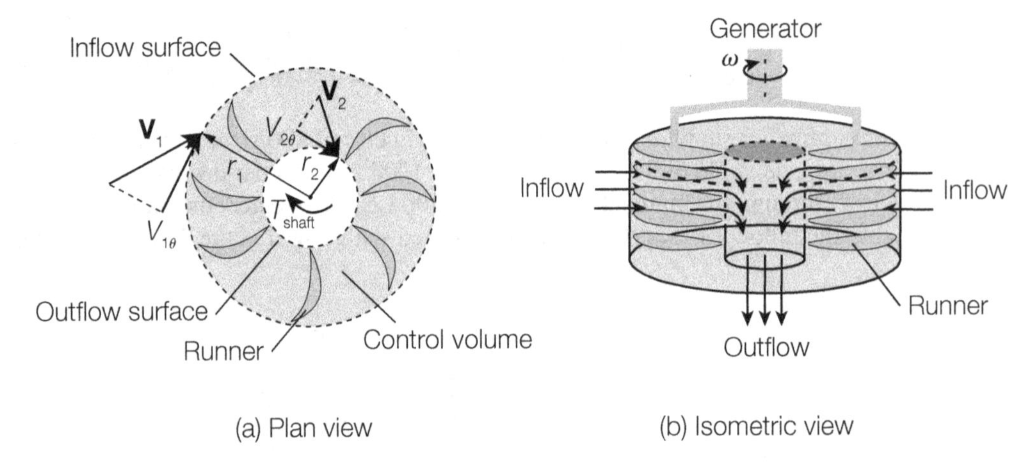

Figure 8.3: Flow through a turbine

Mechanics of turbines. In the case of reaction turbines, the flow enters from the periphery of the rotor and exits from the center of the rotor. The case of flow through a turbine is illustrated in Figure 8.3. The control volume containing the rotor (commonly called the runner) is shown in Figure 8.3(a), where the subscripts 1 and 2 denote the entering and exiting conditions, respectively. With this notation, Equation 8.1 can be used to determine the torque, T, exerted by the fluid on the rotor and Equation 8.4 can be used to determine the energy per unit mass extracted from the fluid.

EXAMPLE 8.2

A fluid enters a turbine runner at an angle of 35° to the inflow surface of the runner and exits at 8 m/s in a direction normal to the outflow surface of the runner as shown in Figure 8.4. The diameters of the inflow and outflow surfaces of the turbine runner are 600 mm and 200 mm, respectively, and the width of the runner is 80 mm. The turbine is designed to operate at a rotational speed of 160 rpm. (a) Determine the power extracted by the turbine runner if the fluid is water at 20°C. (b) Determine the power extracted if the fluid is standard air.

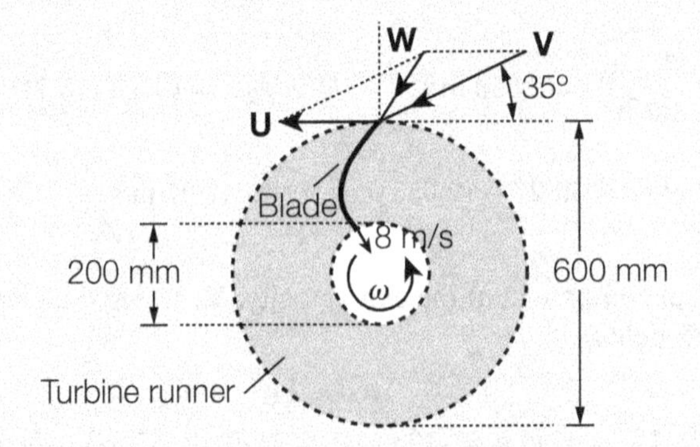

Figure 8.4: Flow through a turbine runner

SOLUTION

From the given data: $\beta = 35°$, $V_{2r} = 8$ m/s, $D_1 = 600$ mm, $r_1 = D_1/2 = 300$ mm, $D_2 = 200$ mm, $r_2 = D_2/2 = 100$ mm, $H = 80$ mm, and $\omega = 160$ rpm $= 16.76$ rad/s. The subscripts

1 and 2 are used to denote the inflow and outflow sections, respectively. The inflow and outflow surface areas, A_1 and A_2, are derived from the given data as follows:

$$A_1 = 2\pi r_1 H = 2\pi(0.3)(0.08) = 0.1508 \text{ m}^2, \qquad A_2 = 2\pi r_2 H = 2\pi(0.1)(0.08) = 0.05027 \text{ m}^2$$

The normal component, V_{1r}, of the (absolute) inflow velocity can be determined from the continuity equation as follows:

$$\dot{m} = \rho V_{1r} A_1 = \rho V_{2r} A_2 \quad \rightarrow \quad V_{1r} = \frac{A_2}{A_1} V_{2r} = \frac{0.05027}{0.1508}(8) = 2.667 \text{ m/s}$$

Because the inflow velocity makes an angle of 35° with the tangent to the inflow surface, the tangential component of the inflow velocity, $V_{1\theta}$ is given by

$$\tan 35° = \frac{V_{1r}}{V_{1\theta}} \quad \rightarrow \quad \tan 35° = \frac{2.667}{V_{1\theta}} \quad \rightarrow \quad V_{1\theta} = 3.809 \text{ m/s}$$

(a) If the fluid is water at 20°C, then $\rho = 998$ kg/m^3 and the mass flow rate, $\dot{m}$, through the turbine and shaft power, $\dot{W}_{\text{shaft}}$, extracted from the fluid are given by

$$\dot{m} = \rho V_{2r} A_2 = (998)(8)(0.05027) = 401.3 \text{ kg/s}$$

$$\dot{W}_{\text{shaft}} = \dot{m}\omega[r_2 V_{2\theta} - r_1 V_{1\theta}] = (401.3)(16.76)[0 - (0.3)(3.809)] = -7686 \text{ W}$$

Therefore, when the fluid is water, the turbine runner extracts a power of **7.69 kW**.

(b) If the fluid is standard air, then $\rho = 1.225$ kg/m^3 and the shaft power extracted from the fluid is calculated as follows:

$$\dot{m} = \rho V_{2r} A_2 = (1.225)(8)(0.05027) = 0.4926 \text{ kg/s}$$

$$\dot{W}_{\text{shaft}} = \dot{m}\omega[r_2 V_{2\theta} - r_1 V_{1\theta}] = (0.4926)(16.76)[0 - (0.3)(3.809)] = -9.44 \text{ W}$$

Therefore, when the fluid is standard air, the turbine runner extracts a power of **9.44 W**.

It is apparent from this analysis that the power extracted by the turbine runner is directly proportional to the density of the fluid that flows through the turbine.

Commonalities between pumps and turbines. It is apparent from Equation 8.4 that the energy delivered by a pump or the energy extracted by a turbine is determined by the inner and outer radius (r_1,r_2), the inner and outer tangential velocities ($V_{1\theta}$,$V_{2\theta}$), and the rotational speed (ω) of the rotor. It is further apparent from Equation 8.4 that for pumps, the maximum energy input per unit mass occurs when the inflow is normal to the inner surface, in which case $V_{1\theta} = 0$ and w_{shaft} is a maximum and is equal to $r_2\omega V_{2\theta}$. This condition produces the maximum pressure rise through the pump. For turbines, the maximum energy extraction occurs when the outflow from the runner is normal to the inner surface. Although the pressure varies between the inlet and outlet of a turbomachine, a pressure term does not appear in the moment of momentum equation because the inner and outer control surfaces are concentric cylinders where the (normal) pressure forces pass through the central axis of the cylindrical control surfaces and hence produce no moment about the central axis.

8.3 Hydraulic Pumps and Pumped Systems

The three types of rotodynamic pumps commonly encountered in engineering practice are radial-flow pumps (commonly referred to as centrifugal pumps), axial-flow pumps, and mixed-flow pumps. These three types of pumps are described below.

Centrifugal pumps. The centrifugal pump is the most widely used type of pump. In centrifugal pumps, where the rotor is commonly referred to as the *impeller*, the flow enters the pump chamber along the axis of the impeller, called the *eye* of the impeller, and the flow is discharged radially by centrifugal action into a casing that encloses the impeller, as illustrated in Figure 8.5(a). The snail-shell shaped casing enclosing the impeller is sometimes referred to as the *housing* or *volute* of the pump, or simply the *scroll*. The casing is shaped like a diffuser that decelerates the fast-moving fluid that exits the impeller, converts the velocity head to pressure head, and directs the flow to the outlet of the pump. Impellers can be either *single suction* or *double suction*. In a single-suction impeller, the fluid enters from only one side of the impeller, and in a double-suction impeller, the fluid enters from both sides of the impeller. The double-suction arrangement has the advantage of reducing the end thrust on the shaft and of reducing the inlet velocity; however, in cases where these effects are not significant, a single suction impeller is adequate. Impellers can be either *open impellers* or *closed impellers*, and they are illustrated in Figure 8.6. In open impellers, the blades are mounted on a *hub* or *backing plate* that is open on one side, as shown in Figure 8.6(a); the open side is sometimes referred to as the *casing side* or the *shroud side*. In closed impellers, which are sometimes called *shrouded impellers*, the blades are contained between a front plate and a back plate, as shown in Figure 8.6(b). A schematic diagram of a typical centrifugal pump installation is illustrated in Figure 8.7(a). Key components of the system shown in Figure 8.7(a) are the *foot valve* installed in the suction pipe to prevent the liquid draining from the pump when the pump is stopped and a *check valve* in the discharge pipe to prevent backflow through the pump if there is a power failure. If the suction line is empty prior to starting the pump, then the suction line must be *primed* (filled) prior to start-up. Unless the liquid is known to be very clean, a strainer should be installed at the inlet to the suction piping. Adequate submergence of the suction pipe is necessary to prevent air entrainment. The pipe size of the suction line should never be smaller than the inlet connection on the pump; if a reducer is required, it should be of the eccentric type, because concentric reducers place part of the supply pipe above the pump inlet where an air pocket can form. The discharge line from the pump should contain a valve close to the pump to allow service or pump replacement. A typical centrifugal pump is shown in Figure 8.8(a).

Axial-flow pumps. In axial-flow pumps, the flow enters and leaves the pump chamber along the axis of the impeller, as shown in Figures 8.5(b), 8.7(b), and 8.8(b). The impeller blades of axial-flow pumps are designed to maintain a constant axial-flow velocity as the fluid moves through the impeller. Some axial-flow pumps have adjustable blades and/or fixed diffuser vanes, called *stator vanes*, on the downstream side of the impeller to remove the swirl component of the velocity as it exits the impeller. Axial-flow pumps, which are sometimes called *propeller pumps*, are commonly used to deliver high flows with little added head and are typically found in drainage and irrigation applications.

Mixed-flow pumps. In mixed-flow pumps, outflows have both radial and axial components. Mixed-flow pumps perform optimally at a state that is intermediate between the high-head/low-flow state that is optimal for centrifugal pumps and the low-head/high-flow state that is optimal for axial flow pumps. Mixed flow pumps are commonly used in irrigation applications.

The pumps illustrated in Figure 8.7 are *single-stage* pumps, which means that they have only one impeller. In *multistage* pumps, two or more impellers are arranged in series in such a way that the discharge from one impeller enters the eye of the next impeller. If a pump has

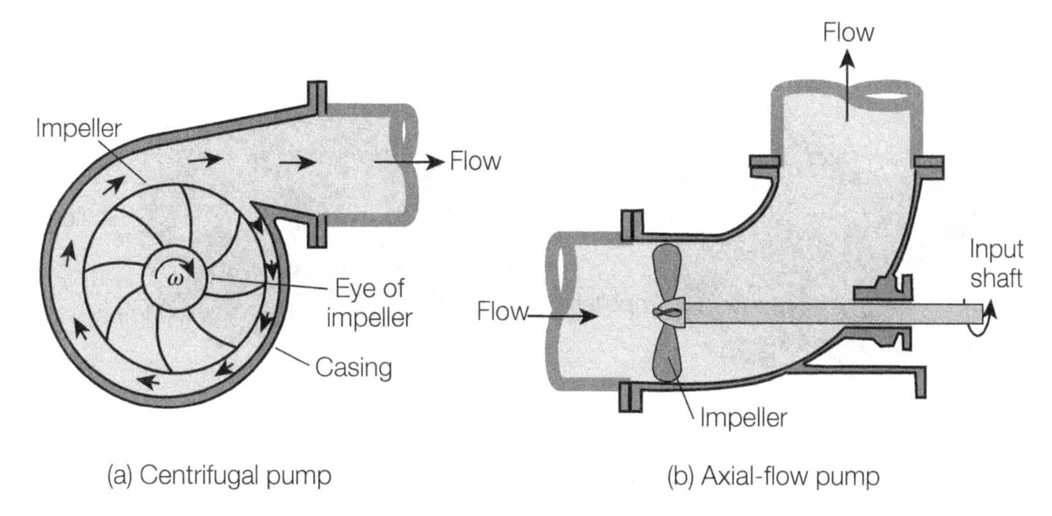

Figure 8.5: Types of pumps

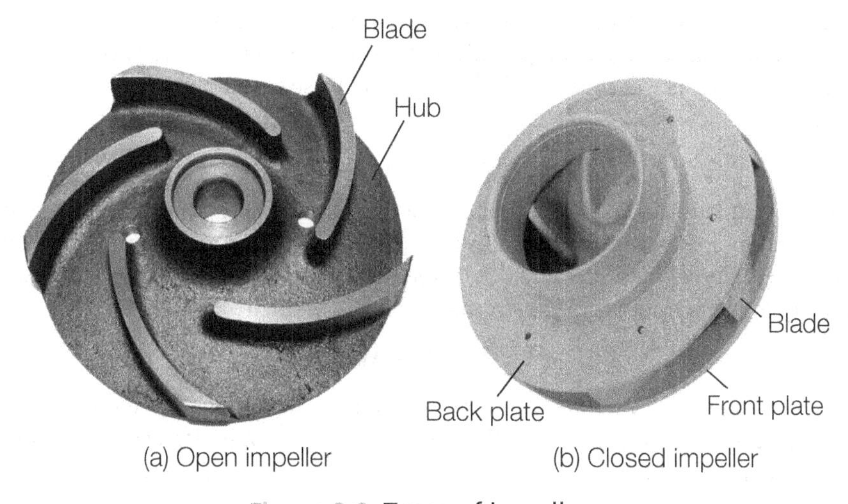

Figure 8.6: Types of impellers

Source: (a) marek_usz/Fotolia, (b) Marek Uszynski/Alamy.

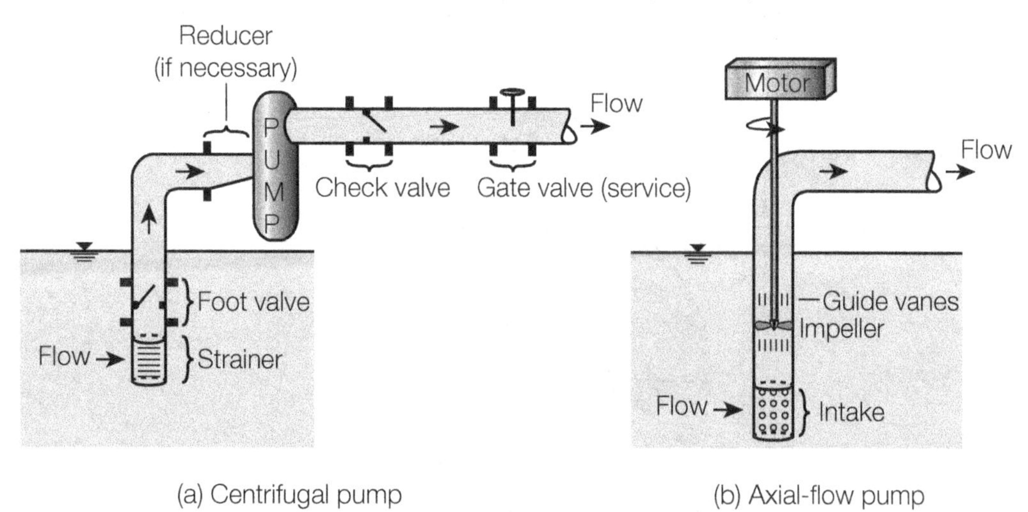

Figure 8.7: Schematic illustrations of centrifugal and axial-flow pump installations

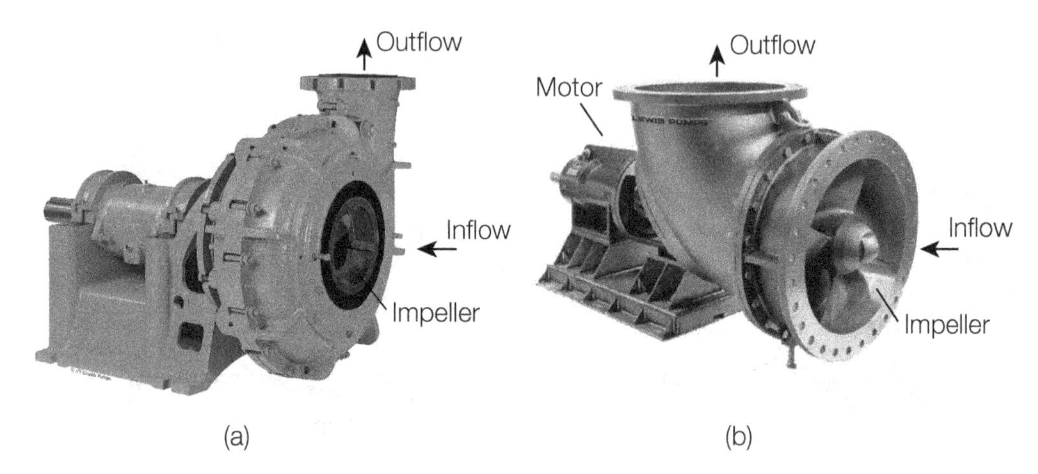

Figure 8.8: (a) Centrifugal pump and (b) axial-flow pump

Sources: (a) ITT Goulds Pumps, (b) Clark D. Hall, Weir Minerals - Lewis Pumps.

two impellers in series, it is called a two-stage pump, and if the pump has three impellers in series, it is called a three-stage pump, and so on. Multistage pumps are typically used when large pumping heads are required and are commonly used in the extraction of water from deep underground sources.

8.3.1 Flow Through Centrifugal Pumps

Application of Equations 8.1 and 8.4 to the rotor of a centrifugal pump is illustrated in Figure 8.9, where the fluid velocities at the inflow and outflow sections are equal to the vector sum of the rotor velocity, **U**, and the velocity of the fluid relative to the rotor, **W**. Hence, if the fluid velocity at the entrance and exit from the rotor is **V**, then

$$\mathbf{V} = \mathbf{U} + \mathbf{W} \tag{8.5}$$

and the θ components of **V**, represented by $V_{1\theta}$ and $V_{2\theta}$, are equal to the components of **V** that are tangential to the inflow and outflow surfaces, respectively. Referring to Figure 8.9, it is sometimes convenient to represent the magnitude of the rotor velocity, U, by $r\omega$ such that

$$U_1 = r_1\omega, \quad \text{and} \quad U_2 = r_2\omega \tag{8.6}$$

Then the energy per unit mass added by the turbomachine, Equation 8.4, can be expressed as

$$w_{\text{shaft}} = U_2V_{2\theta} - U_1V_{1\theta} \tag{8.7}$$

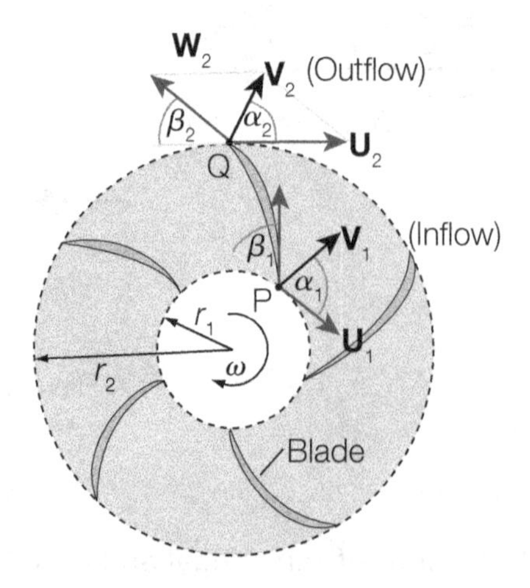

Figure 8.9: Rotor inflow and outflow

The head, $h_{\rm p}$, added to the fluid by a pump is equal to the difference in mechanical energy between the entrance and exit of the pump and is expressed as

$$h_{\rm p} = \left(\frac{p}{\gamma} + \frac{V^2}{2g} + z\right)_2 - \left(\frac{p}{\gamma} + \frac{V^2}{2g} + z\right)_1 \tag{8.8}$$

where subscripts 1 and 2 refer to conditions at the inlet and outlet of the pump, respectively, and p, V, and z are the pressure, average velocity, and centerline elevation of the inflow/outflow conduits, respectively. It is apparent from Equation 8.8 that when the inflow and outflow conduits have the same diameter and are at the same elevation, the head added by the pump, $h_{\rm p}$, is equal to the increase in pressure head (p/γ) between the inlet and outlet sections. The rate at which energy is added to the fluid by the pump, $P_{\rm f}$, is given by

$$P_{\rm f} = \gamma Q h_{\rm p} \tag{8.9}$$

The power, $P_{\rm f}$, given by Equation 8.9 is sometimes referred to as the *water horsepower*, which is an unfortunate term because $P_{\rm f}$ as defined by Equation 8.9 is neither limited to use with water nor requires the use of the unit of horsepower. The primary reason for using the name "water horsepower" in the context of a pump is to distinguish this power from the *brake horsepower*, which is the power delivered by an external motor to the shaft of the pump; the water horsepower, $P_{\rm f}$, delivered to the fluid is generally less than the external power delivered to the rotating shaft. The brake horsepower, BHP, can be expressed as

$$\text{BHP} = T_{\rm shaft}\,\omega \tag{8.10}$$

where $T_{\rm shaft}$ is the torque delivered to the pump rotor that spins at an angular velocity ω.

Energy equation applied to a pump. The energy equation applied to the control volume containing the pump rotor, shown in Figure 8.9, requires that the energy input per unit weight of fluid be equal to the head added to the fluid, $h_{\rm p}$, plus the head loss within the pump, h_ℓ; hence,

$$\frac{w_{\rm shaft}}{g} = h_{\rm p} + h_\ell$$

which can be combined with Equation 8.7 to yield

$$h_{\rm p} = \frac{U_2V_{2\theta} - U_1V_{1\theta}}{g} - h_\ell \tag{8.11}$$

Normal inflow condition. If the fluid enters the pump rotor with an absolute velocity V_1 that is normal to the control surface, then

$$V_{1\theta} = 0 \tag{8.12}$$

Referring to Figure 8.9, the following relationships are generally valid:

$$\cot\beta_2 = \frac{U_2 - V_{2\theta}}{V_{2r}}, \qquad Q = 2\pi r_2 b_2 V_{2r} \tag{8.13}$$

where β_2 is the angle the blade makes with the control surface at the exit, V_{2r} is the radial component of the velocity at the exit, Q is the volume flow rate through the rotor, and b_2 is the width of the rotor at the exit. Substituting Equations 8.12 and 8.13 into Equation 8.11 yields

$$h_{\rm p} = \frac{U_2^2}{g} - \frac{U_2\cot\beta_2}{2\pi r_2 b_2 g}Q - h_\ell(Q) \tag{8.14}$$

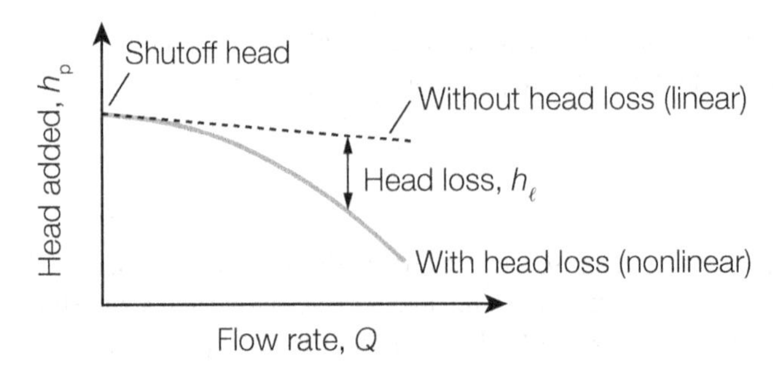

Figure 8.10: Pump performance curve

where h_ℓ is expressed as $h_\ell(Q)$ to emphasize the dependence of the head loss on the flow rate, Q. Equation 8.14 demonstrates that for a given rotor geometry and rate of rotation, the head added by a pump, h_p, is linearly proportional to the flow rate, Q, plus a head loss, h_ℓ, that depends on the flow rate. The relationship between the head added by a pump, h_p, and the volume flow rate through the pump, Q, is called the *pump performance curve*. The theoretical performance curve given by Equation 8.14 can be expressed as

$$h_\text{p} = h_0 - f(Q) \tag{8.15}$$

where h_0 is called the *shutoff head*, which is equal to the head added by the pump when the flow rate through the pump is equal to zero, as occurs when the valve on the discharge side of the pump is closed. Comparing Equations 8.15 and 8.14 gives

$$h_0 = \frac{U_2^2}{g} = \frac{r_2^2\omega^2}{g}, \qquad f(Q) = \frac{r_2\omega\cot\beta_2}{2\pi r_2 b_2 g}Q + h_\ell(Q)$$

The functional relationship given by Equation 8.15 is illustrated in Figure 8.10, where it is apparent that the head loss, h_ℓ, increases nonlinearly as the flow rate, Q, increases. Centrifugal pumps generally have performance curves of the form shown in Figure 8.10. However, a notable difference is that the shutoff head, h_0, in actual pumps is generally less than the theoretical shutoff head of $r_2^2\omega^2/g$, typically about 60% less.

Typical blade configurations. Typical values of the blade angle, β_2, are in the range of 15°–35°, and typical values of β_1 are in the range 15°–50°. Blades with $\beta_2 < 90°$ are called *backward-curved blades*, and blades with $\beta_2 > 90°$ are called *forward-curved blades*, which are not normally used because they create unstable flow conditions.

EXAMPLE 8.3

Water at 20°C flows through a centrifugal pump at the rate of 4000 L/min. The impeller has an inner radius of 50 mm, an outer radius of 100 mm, and a width of 75 mm, and it rotates at 1800 rpm. The blades of the impeller have a blade angle of 30° at the inflow surface and a blade angle of 20° at the outflow surface. (a) What is the head delivered to the fluid? (b) What would be the difference in pressure between the outflow and inflow if head losses were neglected and the inflow and outflow pipes were at the same elevation and of the same diameter? (c) What inflow blade angle would cause the inflow velocity to be normal to the inflow surface? (d) Under the normal inflow condition, what would be the difference in pressure between the outflow and inflow if head losses were neglected?

SOLUTION

From the given data: $Q = 4000$ L/min $= 0.06667$ m^3/s, $r_1 = 50$ mm, $r_2 = 100$ mm, $b = 75$ mm, $\omega = 1800$ rpm $= 188.5$ rad/s, $\beta_1 = 30°$, and $\beta_2 = 20°$. For water at 20°C, $\rho = 998$ kg/m^3. The inflow and outflow velocity triangles are illustrated in Figure 8.11.

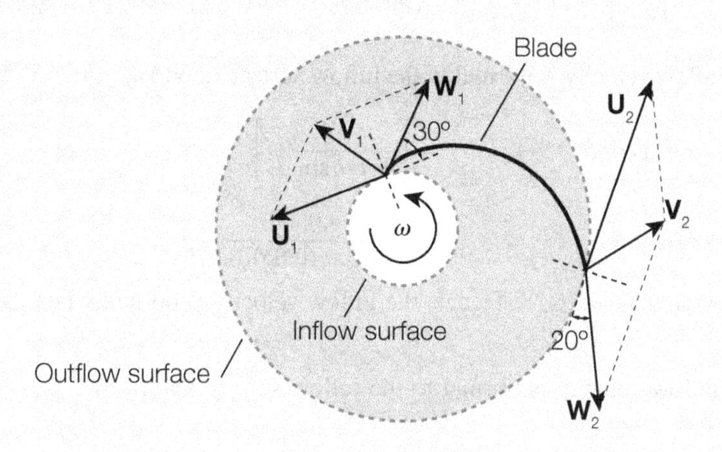

Figure 8.11: Inflow and outflow velocities

The quantities $\mathbf{U}_1$, $\mathbf{W}_1$, and $\mathbf{V}_1$ represent the blade tip speed, the fluid velocity relative blade at the inflow surface, and the absolute velocity of the fluid at the inflow surface, respectively. The quantities $\mathbf{U}_2$, $\mathbf{W}_2$, and $\mathbf{V}_2$ are the corresponding quantities at the outflow surface. The inflow and outflow tip speeds are given by

$$U_1 = r_1\omega = (0.05)(188.5) = 9.425 \text{ m/s}, \qquad U_2 = r_2\omega = (0.1)(188.5) = 18.85 \text{ m/s}$$

Applying the continuity equation to the inflow and outflow surfaces gives

$$W_1 \sin\beta_1 = \frac{Q}{2\pi r_1 b} \quad \rightarrow \quad W_1 \sin 30° = \frac{0.06667}{2\pi(0.05)(0.075)} \quad \rightarrow \quad W_1 = 8.273 \text{ m/s}$$

$$W_2 \sin\beta_2 = \frac{Q}{2\pi r_2 b} \quad \rightarrow \quad W_2 \sin 20° = \frac{0.06667}{2\pi(0.1)(0.075)} \quad \rightarrow \quad W_2 = 2.829 \text{ m/s}$$

Using the velocity triangles shown in Figure 8.11, the θ-component of the inflow and outflow velocities are given by

$$V_{1\theta} = U_1 - W_1 \cos\beta_1 = 9.425 - (8.273)\cos 30° = 1.651 \text{ m/s}$$

$$V_{2\theta} = U_2 - W_2 \cos\beta_2 = 18.85 - (2.829)\cos 20° = 16.40 \text{ m/s}$$

(a) The head delivered to the fluid is derived from Equation 8.7 as

$$\begin{aligned}\text{energy added to fluid} = w_\text{shaft} &= U_2V_{2\theta} - U_1V_{1\theta} \\ &= (18.85)(16.40) - (9.425)(1.651) = 293.6 \text{ J/kg}\end{aligned}$$

$$\text{head delivered to fluid} = \frac{w_\text{shaft}}{g} = \frac{293.6}{9.807} = \mathbf{29.9\ m}$$

(b) If the head loss in the pump is neglected and the inflow and outflow pipes are at the same elevation and are of the same diameter, then

$$h_\text{p} = \frac{w_\text{shaft}}{g} = \frac{p_2 - p_1}{\gamma} \quad \rightarrow$$

$$p_2 - p_1 = \rho(w_\text{shaft}) = (998)(293.6) = 2.930 \times 10^5 \text{ Pa} = \mathbf{293\ kPa}$$

(c) If the inflow velocity is normal to the inflow surface, then $V_{\theta 1} = 0$, which requires that

$$U_1 = W_1 \cos\beta_1 \quad \rightarrow \quad U_1 = \left[\frac{Q}{2\pi r_1 b \sin\beta_1}\right] \cos\beta_1 \quad \rightarrow \quad \tan\beta_1 = \frac{Q}{U_1 2\pi r_1 b}$$

$$= \frac{0.06667}{(9.425)2\pi(0.05)(0.075)}$$

which yields $\beta_1 = 16.7°$. Hence, the inflow velocity is normal when the blade angle is **16.7°**.

(d) If the inflow velocity is normal to the inflow surface and head losses in the pump are negligible, then

$$\text{energy added to fluid} = w_\text{shaft} = U_2 V_{2\theta} - U_1 V_{1\theta}$$
$$= (18.85)(16.40) - (9.425)(0) = 309.1 \text{ J/kg}$$

$$p_2 - p_1 = \rho(w_\text{shaft}) = (998)(309.1) = 3.085 \times 10^5 \text{ Pa} = 309 \text{ kPa}$$

Hence, the pressure added to the fluid is **309 kPa**, which is approximately 5% higher than the case in which the inflow velocity is not normal to the inflow surface.

Guidelines for problem solving. Solving problems involving the flow through pump impellers is best done in a step-by-step manner with an explicit consideration of velocity triangles at the entrance and exit of the impeller, as shown in the previous problem. After acquiring a "feel" for the mechanics of the flow, direct application of the following key formulas in terms of the base variables will expedite the solution of most problems:

$$\text{outflow surface: (a) } V_{2\theta} = r_2\omega - \frac{Q}{2\pi r_2 b_2 \tan\beta_2}, \qquad \text{(b) } V_{2r} = \frac{Q}{2\pi r_2 b_2} \tag{8.16}$$

$$\text{inflow surface: (a) } V_{1\theta} = r_1\omega - \frac{Q}{2\pi r_1 b_1 \tan\beta_1}, \qquad \text{(b) } V_{1r} = \frac{Q}{2\pi r_1 b_1} \tag{8.17}$$

$$\text{impeller: (a) } \dot{W}_\text{shaft} = \eta_\text{m} P_\text{m} = \rho Q\omega[r_2 V_{2\theta} - r_1 V_{1\theta}], \text{ (b) } h_\text{p} = \frac{\eta_\text{p} \dot{W}_\text{shaft}}{\rho g Q} \tag{8.18}$$

where r_1 and r_2 are the radii of the inflow and outflow surfaces, respectively, b_1 and b_2 are the widths of the inflow and outflow surfaces, respectively, β_1 and β_2 are the blade angles at the inflow and outflow surfaces, respectively (as shown in Figure 8.9), ω is the angular speed of the impeller, Q is the volume flow rate through the pump, $\dot{W}_\text{shaft}$ is the power input at the shaft of the pump, P_m is the power input to the pump motor, η_m is the efficiency of the motor, η_p is the efficiency of the pump, and h_p is the head added to the fluid as it flows through the pump. Equations 8.16–8.18 can be used to estimate the optimal performance of a pump of given geometry and rotational speed. The optimal condition is associated with the inflow velocity being normal to the inflow surface and the relative inflow velocity being tangential to the blade surface at the inflow section, a condition that is commonly called a *shockless entry*. Under these conditions, $V_{1\theta} = 0$ and the power, $\eta_\text{p}\dot{W}_\text{shaft}$, delivered to the fluid is maximized in accordance with Equation 8.18. Combined variables that can be used in Equations 8.16–8.18 are the mass flow rate, $\dot{m} = \rho Q$, entrance tip speed, $U_1 = r_1\omega$, and exit tip speed, $U_2 = r_2\omega$. For many problems, only a subset of Equations 8.16–8.18 are required to obtain a solution.

EXAMPLE 8.4

A pump impeller has an outflow surface with a radius of 200 mm and a width of 15 mm, and the impeller rotates at an angular speed of 1200 rpm. When the flow rate of water through the pump is 70 L/s, the inflow velocity is normal to the inflow surface, the power delivered to the pump motor is 50 kW, and the efficiency of the motor is 70%. Determine the blade angle at the outflow surface. Assume water at 20°C.

SOLUTION

From the given data: r_2 = 200 mm, b_2 = 15 mm, ω = 1200 rpm = 125.7 rad/s, Q = 70 L/s, P_m = 50 kW, and $\eta_\text{m} = 0.70$. For water at 20°C, $\rho = 998.2$ kg/m^3. Because the inflow velocity is normal to the inflow surface, $V_{1\theta} = 0$, and Equation 8.18(a) gives

$$\eta_\text{m}P_\text{m} = \rho Q\omega[r_2V_{2\theta}] \;\rightarrow\; V_{2\theta} = \frac{\eta_\text{m}P_\text{m}}{\rho Q\omega r_2} = \frac{(0.70)(50\times 10^3)}{(998.2)(0.070)(125.7)(0.2)} \;\rightarrow\; V_{2\theta} = 19.92 \text{ m/s}$$

Using this result in Equation 8.16(a) gives

$$V_{2\theta} = r_2\omega - \frac{Q}{2\pi r_2 b_2\tan\beta_2} \;\rightarrow\; 19.92 = (0.2)(125.7) - \frac{0.070}{2\pi(0.2)(0.015)\tan\beta_2} \;\rightarrow\; \beta_2 = 35.5°$$

Therefore, for the given conditions, the blade angle at outflow surface must be **35.5°**.

8.3.2 Efficiency

The performance of a pump is measured by the head added by the pump and the efficiency of the pump. The head added by a pump, h_p, is equal to the difference between the total head (i.e., total mechanical energy) on the discharge side of the pump and the total head on the suction side of the pump. The head added by a pump (h_p) is sometimes referred to as the *total dynamic head* (TDH). The efficiency of a pump, η_p, is defined by

$$\eta_\text{p} = \frac{\text{power delivered to the fluid}}{\text{power supplied to the shaft}} = \frac{\gamma Q h_\text{p}}{\text{BHP}} = \frac{\gamma Q h_\text{p}}{\dot{W}_\text{shaft}} = \frac{\gamma Q h_\text{p}}{T_\text{shaft}\omega} \tag{8.19}$$

In British practice, the quantity h_p is commonly called the *manometric head*, because it is the difference in head that occurs between manometers placed on the suction and discharge sides of a pump. In this context, the efficiency, η_p, given by Equation 8.19 is called the *manometric efficiency*. Pumps are inefficient for a variety of reasons, such as frictional losses as the fluid moves over the solid surfaces, viscous dissipation within the fluid, separation losses, leakage of fluid between the impeller and the casing, mechanical losses in the bearings and sealing glands of the pump, and *shock losses* due to the inlet flow angle not matching the blade angle.

A pump unit is typically considered as being separate from the motor that drives the pump, and the efficiency of a motor, $\eta_{\rm m}$, is defined as

$$\eta_{\rm m} = \frac{\text{power supplied to shaft}}{\text{power delivered to motor}} = \frac{\text{BHP}}{P_{\rm m}} = \frac{T_{\rm shaft}\omega}{P_{\rm m}} \tag{8.20}$$

where $P_{\rm m}$ is the power delivered to the motor, usually given in units of kilowatts. Combining Equations 8.20 and 8.19 gives the overall efficiency, $\eta_{\rm overall}$, of a motor/pump combination as

$$\eta_{\rm overall} = \eta_{\rm p} \cdot \eta_{\rm m} \tag{8.21}$$

In most formulations describing the performance of a pump, the subscript "p" in $\eta_{\rm p}$ is dropped, in which case the efficiency of the pump is represented simply by η. Care should be taken to ensure that the appropriate efficiency is used in calculations.

8.3.3 Dimensional Analysis

The performance of any given pump is characterized by (1) the relationship between the head added to the fluid and the flow rate through the pump and (2) the relationship between the power input to the pump and the flow rate through the pump. The head added to the fluid is denoted by $h_{\rm p}$ [L], and the power input to the pump is the *brake horsepower*, BHP [FLT^{-1}]. The pump performance parameters, $h_{\rm p}$ and BHP, are functions of the flow rate through the pump, the properties of the pumped fluid, and the physical characteristics of the pump. These functional relationships can be expressed as

$$gh_{\rm p} = f_1(\rho, \mu, D, \omega, Q) \tag{8.22}$$

$$\text{BHP} = f_2(\rho, \mu, D, \omega, Q) \tag{8.23}$$

where the energy added per unit mass of fluid, $gh_{\rm p}$, is used instead of $h_{\rm p}$ (to remove the effect of gravity), f_1 and f_2 are unknown functions, ρ and μ are the density and dynamic viscosity of the fluid, respectively, Q is the volume flow rate through the pump, D is a characteristic dimension of the pump (usually the inlet or outlet diameter), and ω is the angular speed of the pump impeller. In accordance with Equation 8.19, the efficiency of the pump, η, can be derived from Equations 8.22 and 8.23 to yield the following functional relationship:

$$\eta = \frac{\gamma Q h_{\rm p}}{\text{BHP}} = f_3(\rho, \mu, D, \omega, Q) \tag{8.24}$$

where f_3 is an unknown function. Equations 8.22–8.24 are functional relationships between six variables in three dimensions. According to the Buckingham pi theorem, these relationships can be expressed as relationships between three dimensionless groups as follows:

$$\frac{gh_{\rm p}}{\omega^2 D^2} = f_4\left(\frac{Q}{\omega D^3}, \frac{\rho\omega D^2}{\mu}\right) \tag{8.25}$$

$$\frac{\text{BHP}}{\rho\omega^3 D^5} = f_5\left(\frac{Q}{\omega D^3}, \frac{\rho\omega D^2}{\mu}\right) \tag{8.26}$$

$$\eta = f_6\left(\frac{Q}{\omega D^3}, \frac{\rho\omega D^2}{\mu}\right) \tag{8.27}$$

where $gh_{\rm p}/\omega^2 D^2$ is called the *head coefficient* or *head rise coefficient*, BHP$/\rho\omega^3 D^5$ is called the *power coefficient*, and $Q/\omega D^3$ is called the *flow coefficient*, *discharge coefficient*, or *capacity coefficient*. In most cases, the flow through the pump is fully turbulent and viscous forces are negligible relative to the inertial forces. Under these circumstances, the viscosity of the fluid can be neglected, and Equations 8.25–8.27 become

$$\frac{gh_p}{\omega^2 D^2} = f_7\left(\frac{Q}{\omega D^3}\right) \tag{8.28}$$

$$\frac{\text{BHP}}{\rho\omega^3 D^5} = f_8\left(\frac{Q}{\omega D^3}\right) \tag{8.29}$$

$$\eta = f_9\left(\frac{Q}{\omega D^3}\right) \tag{8.30}$$

These relationships describe the performance of a set of geometrically similar pumps in which viscous effects are negligible. Interestingly, the similarity variable $Q/\omega D^3$ represents the ratio of the time scale of rotation to the time scale of the flow through the pump. Hence, similarity of pump behavior relies on geometric and kinematic similarity; dynamic similarity (via $\rho\omega D^2/\mu$) is not a significant contributing factor.

Homologous pumps. A set of pumps having the same shape (but different sizes) are expected to have the same functional relationships between $Q/(\omega D^3)$ and the three performance characteristics: $gh_p/(\omega^2 D^2)$, $\text{BHP}/\rho\omega^3 D^5$, and η. A set of pumps that have the same shape (i.e., are geometrically similar) is called a *homologous series* of pumps, and the performance characteristics of a homologous series of pumps are described by curves such as those in Figure 8.12. Pumps are usually selected such that they will operate at or near the point of maximum efficiency, indicated by Point P in Figure 8.12.

Affinity laws. In accordance with the functional relationships given in Equations 8.28–8.30, it is apparent that if two geometrically similar pumps have the same value of $Q/\omega D^3$, then they must necessarily have the same values of $gh_p/\omega^2 D^2$, $\text{BHP}/\rho\omega^3 D^5$, and η. This result can be expressed in the form

$$\left[\frac{Q}{\omega D^3}\right]_1 = \left[\frac{Q}{\omega D^3}\right]_2 \quad \rightarrow \quad \begin{cases} \left[\dfrac{gh_p}{\omega^2 D^2}\right]_1 = \left[\dfrac{gh_p}{\omega^2 D^2}\right]_2 \\[2ex] \left[\dfrac{\text{BHP}}{\rho\omega^3 D^5}\right]_1 = \left[\dfrac{\text{BHP}}{\rho\omega^3 D^5}\right]_2 \\[2ex] \eta_1 = \eta_2 \end{cases} \tag{8.31}$$

where the subscripts 1 and 2 are corresponding operational states in geometrically similar pumps. The relationships given in Equation 8.31 are sometimes called the *affinity laws* or

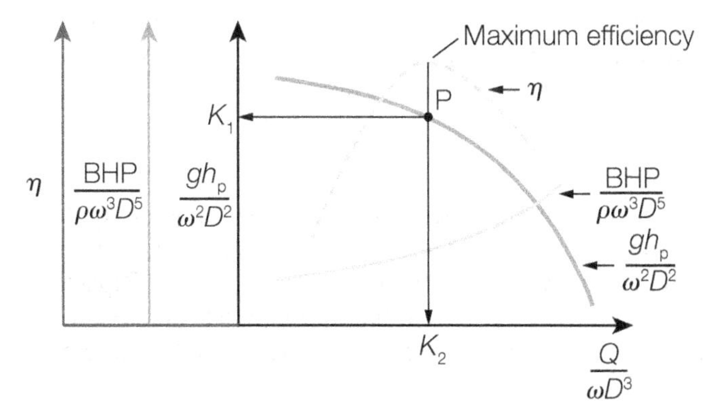

Figure 8.12: Performance curves of a homologous series of pumps

similarity rules for homologous pumps. In practice, it is common to represent the flow coefficient by C_Q, the head coefficient by C_H, and the power coefficient by C_P, such that

$$C_Q = \frac{Q}{\omega D^3}, \qquad C_H = \frac{gh_p}{\omega^2 D^2}, \qquad C_P = \frac{\text{BHP}}{\rho\omega^3 D^5} \tag{8.32}$$

where C_Q, C_H, and C_P are nondimensional parameters. In terms of these parameters, the affinity laws given by Equation 8.31 can be expressed as

$$[C_Q]_1 = [C_Q]_2 \quad \rightarrow \quad \begin{cases} [C_H]_1 = [C_H]_2 \\ [C_P]_1 = [C_P]_2 \\ \eta_1 = \eta_2 \end{cases} \tag{8.33}$$

It is noteworthy that manufacturers commonly put impellers of different diameters in the same pump casing while still classifying these different pumps as belonging to the same homologous series. In such cases, geometric similarity is not maintained because the width of the pump casing remains fixed while the diameter of the impeller changes. To accommodate this irregularity, the affinity laws given by Equation 8.31 must be modified such that the flow rate, Q, is normalized relative to D^2 rather than D^3 and the input power, BHP, which is the product of the flow rate and the head, must be normalized relative to D^4 rather than D^5. The modified affinity laws can be used to predict the performance of a pump with a changed impeller diameter based on the known performance curve of a given diameter in the same casing. However, it is generally preferable to perform separate tests on modified pumps to more accurately determine their performance.

Viscous effects. In accordance with the functional relationships describing pump performance, Equations 8.25–8.27, the affinity laws for scaling pump performance within any homologous series are valid as long as viscous effects are negligible and complete geometric similarity in all dimensions is maintained. The effect of viscosity is measured by the Reynolds number, Re, defined by

$$\text{Re} = \frac{\rho\omega D^2}{\mu} \tag{8.34}$$

Deviations from the affinity laws are negligible when $\text{Re} > 3 \times 10^5$ (Gerhart et al., 1992).

Scale effects. Deviations from exact geometric similarity between pumps in a homologous series typically occur due to inexact scaling of roughness heights and tip clearances. These aforementioned deviations typically result in larger pumps being more efficient than smaller pumps under the same operating conditions. This reality is commonly referred to as the *scale effect*, which is not accounted for in the affinity laws. The scale effect on the maximum efficiency of a pump can be approximated by (Stepanoff, 1957; Moody and Zowski, 1989)

$$\frac{1-\eta_2}{1-\eta_1} = \left(\frac{D_1}{D_2}\right)^{\frac{1}{4}} \tag{8.35}$$

where η_1 and η_2 are the maximum efficiencies of homologous pumps of diameters D_1 and D_2, respectively. The exponent of $\frac{1}{4}$ in Equation 8.35 has been found to vary between manufacturers, with reported alternative exponents being $\frac{1}{5}$, $\frac{1}{6.5}$, and $\frac{1}{7.5}$ (Ziparro and Hasen, 1993). The effect of changes in the rated flow rate (i.e., the flow rate at maximum efficiency) on the maximum efficiency is sometimes estimated using the relation (Anderson, 1980)

$$\frac{0.94-\eta_2}{0.94-\eta_1} = \left(\frac{Q_1}{Q_2}\right)^{0.32} \tag{8.36}$$

where Q_1 and Q_2 are corresponding rated flow rates. Equation 8.36 assumes a maximum possible pump efficiency of 94%. Equations 8.35 and 8.36 are applicable to peak efficiencies and should be used with great caution in predicting scale effects on off-peak efficiencies.

EXAMPLE 8.5

A particular pump in a homologous series has an impeller diameter of 700 mm. Under a particular operating condition, this pump delivers 3.2 m³/s with an added head of 140 m at an efficiency of 73%. The working fluid is water at 20°C. Another pump in this series has an impeller diameter of 175 mm, and both pumps have motors with a speed of 900 rpm. (a) Neglecting viscous and scale effects, what is the head added and the corresponding discharge in the smaller pump when operating at an efficiency of 73%? (b) Are viscous effects significant in any of the two pumps? Explain. (c) Assess whether scale effects are likely to have a significant effect on the performance of the smaller pump.

SOLUTION

From the given data: D_1 = 700 mm, Q_1 = 3.2 m³/s, h_1 = 140 m, $\eta_1 = \eta_2$ = 73%, D_2 = 175 mm, and $\omega_1 = \omega_2$ = 900 rpm = 94.2 rad/s. For water at 20°C, $\nu = 1.004 \times 10^{-6}$ m²/s.

(a) Assume that viscous and scale effects are negligible. Because both pumps are geometrically similar and have the same efficiency, Equation 8.31 requires that

$$\left[\frac{Q}{\omega D^3}\right]_1 = \left[\frac{Q}{\omega D^3}\right]_2 \quad \rightarrow \quad Q_2 = \left(\frac{D_2}{D_1}\right)^3 \left(\frac{\omega_2}{\omega_1}\right) Q_1$$

$$= \left(\frac{175}{700}\right)^3 (1)(3.2) = 0.050 \text{ m}^3/\text{s} = 50 \text{ L/s}$$

$$\left[\frac{gh_\text{p}}{\omega^2 D^2}\right]_1 = \left[\frac{gh_\text{p}}{\omega^2 D^2}\right]_2 \quad \rightarrow \quad h_2 = \left(\frac{D_2}{D_1}\right)^2 \left(\frac{\omega_2}{\omega_1}\right)^2 h_1$$

$$= \left(\frac{175}{700}\right)^2 (1)^2(140) = 8.75 \text{ m}$$

Therefore, neglecting scale effects, when operating with an efficiency of 73%, the smaller pump delivers **50 L/s** with an added head of **8.75 m**.

(b) The Reynolds numbers of the smaller and larger pumps, Re_1 and Re_2, respectively, are given by

$$\text{Re}_1 = \left[\frac{\omega D^2}{\nu}\right]_1 = \frac{(94.2)(0.7)^2}{1.004 \times 10^{-6}} = 4.60 \times 10^7$$

$$\text{Re}_2 = \left[\frac{\omega D^2}{\nu}\right]_2 = \frac{(94.2)(0.175)^2}{1.004 \times 10^{-6}} = 2.87 \times 10^6$$

Because $\text{Re}_1 > 3 \times 10^5$ and $\text{Re}_2 > 3 \times 10^5$, it is likely that **viscous effects are negligible** in both pumps. Hence, the affinity (i.e., scaling) laws applied in part (a) are valid.

(c) Assuming that the given efficiency of the larger pump is the maximum (rated) efficiency, then the corresponding maximum efficiency of the smaller pump can be estimated using Equation 8.35, which gives

$$\frac{1-\eta_2}{1-\eta_1} = \left(\frac{D_1}{D_2}\right)^{\frac{1}{4}} \quad \rightarrow \quad \frac{1-\eta_2}{1-0.73} = \left(\frac{700}{175}\right)^{\frac{1}{4}} \quad \rightarrow \quad \eta_2 = 0.62$$

The efficiency can also be estimated by Equation 8.36, which gives

$$\frac{0.94-\eta_2}{0.94-\eta_1} = \left(\frac{Q_1}{Q_2}\right)^{0.32} \quad \rightarrow \quad \frac{0.94-\eta_2}{0.94-0.73} = \left(\frac{3.2}{0.05}\right)^{0.32} \quad \rightarrow \quad \eta_2 = 0.15$$

Based on these results, it is estimated that the 175-mm pump will have a maximum efficiency somewhere in the range of 15%–62%. Based on this wide range of uncertainty and the fact that the lower limit of this range is much less that the rated efficiency of the larger pump, it can be inferred that **scale effects will be significant** and the smaller pump will likely be significantly less efficient than the larger pump.

In selecting a pump for any application, consideration should generally be given to the required pumping rate, commercially available pumps, characteristics of the system in which the pump operates, and the physical limitations associated with pumping water.

8.3.4 Specific Speed

The point of maximum efficiency is commonly called the *best efficiency point* (BEP) or sometimes the *nameplate* or *design point*. Maintaining operation near the BEP allows a pump to function for years with little maintenance, and as the operating point moves away from the BEP, pump thrust and radial loads increase, which increases the wear on the pump bearings and shaft. For these reasons, it is generally recommended that pump operation should be maintained between 70% and 130% of the BEP flow rate (Lansey and El-Shorbagy, 2001). At the BEP in Figure 8.12,

$$\frac{gh_\mathrm{p}}{\omega^2 D^2} = K_1 \quad \text{and} \quad \frac{Q}{\omega D^3} = K_2 \tag{8.37}$$

where K_1 and K_2 are the (constant) values of $gh_\mathrm{p}/\omega^2 D^2$ and $Q/\omega D^3$, respectively, at the most efficient operating point for the homologous pump series. Eliminating D from the expressions in Equation 8.37 yields

$$\frac{\omega Q^{\frac{1}{2}}}{(gh_\mathrm{p})^{\frac{3}{4}}} = \sqrt{\frac{K_2}{K_1^{\frac{3}{2}}}} \quad (= \text{constant}) \tag{8.38}$$

The term on the right-hand side of this equation is a constant for a homologous series of pumps and is denoted by the *specific speed*, n_s, defined by

$$n_\mathrm{s} = \frac{\omega Q^{\frac{1}{2}}}{(gh_\mathrm{p})^{\frac{3}{4}}} \tag{8.39}$$

Typical SI units used in calculating the specific speed, n_s, are ω in rad/s, Q in m^3/s, g in m/s^2, and h_p in m; however, because n_s is dimensionless, any consistent set of units can be used.

Because the specific speed, n_s, is based on the parameters at the most efficient operating point of a homologous series, n_s is the basis for selecting the appropriate homologous series for any desired set of operating conditions. The required pump operating point is defined by the flow rate, Q, and head, h_p, required from the pump. The specific speed calculated from the required pump operating point, along with consideration of the rotational speeds of available motors, is the basis for selecting the appropriate pump. The nomenclature of calling n_s the specific speed is somewhat unfortunate, because n_s is dimensionless and hence does not have units of speed. The specific speed, n_s, is also called the *shape number* (Hwang and Houghtalen, 1996; Wurbs and James, 2002) or the *type number* (Douglas et al., 2001), which may be preferable terminology in that n_s is more often used to select the shape or type of the required pump rather than the speed of the pump.

Dimensional specific speed. In lieu of defining the *nondimensional specific speed*, n_s, by Equation 8.39, it is also common practice to define a *dimensional specific speed*, N_s, as

$$N_s = \frac{\omega Q^{\frac{1}{2}}}{h_p^{\frac{3}{4}}} \tag{8.40}$$

where N_s is not dimensionless. Thus, when Equation 8.40 is used to define the specific speed, the units of ω, Q, and h_p must be explicitly specified. Be cautious. Although N_s has dimensions, the units are seldom stated in practice.

Relationship between specific speed and type of pump. Because the specific speed is independent of the size of a pump and all homologous pumps (of varying sizes) have the same specific speed, the calculated specific speed at the desired operating point indicates the type of pump that must be selected to ensure optimal efficiency. The types of pumps that give the maximum efficiency for given specific speeds, n_s, are listed in Table 8.1, along with typical flow rates delivered by the pumps. Table 8.1 indicates that centrifugal pumps are most efficient at low specific speeds, $n_s < 1.5$; mixed-flow pumps are most efficient at medium specific speeds, $1.5 < n_s < 3.7$; and axial-flow pumps are most efficient at high specific speeds, $n_s > 3.7$. This indicates that centrifugal pumps are most efficient at delivering low flows at high heads, whereas axial-flow pumps are most efficient at delivering high flows at low heads. The efficiencies of centrifugal pumps increase with increasing specific speed, whereas the efficiencies of mixed-flow and axial-flow pumps decrease with increasing specific speed. Pumps with specific speeds less than 0.3 tend to be inefficient. Pump capacities generally increase with increasing specific speed and, for a given specific speed, scale effects cause larger pumps to be more efficient than smaller ones.

Table 8.1: Pump Selection Guidelines

Type of pump	Range of specific speeds, n_s (dimensionless)	Typical flow rates (L/s)	Typical efficiencies (%)
Centrifugal	0.15–1.5	<60	70–94
Mixed-flow	1.5–3.7	60–300	90–94
Axial-flow	3.7–5.5	>300	84–90

Figure 8.13: Axial-flow pumps operating in a canal
Source: Used with permission from the South Florida Water Management District.

Practical application. Because axial-flow pumps are most efficient at delivering high flows at low heads, this type of pump is commonly used to move large volumes of water through major canals. An example of this application is shown in Figure 8.13, where three axial-flow pumps are operating in parallel and these pumps are driven by motors housed in the pump station.

Synchronous speeds and rated speeds. Most pumps are driven by standard electric motors that use an alternating current (AC). The standard speed of AC synchronous induction motors at 220–440 volts is given by

$$\text{Synchronous speed (rpm)} = \frac{60\,\Omega}{\text{no. of pairs of poles}} \tag{8.41}$$

where Ω is the frequency of the alternating current. The synchronous speed of a motor is attained when there is no resistance to shaft rotation. Actual motor speeds are less than synchronous speeds due to resistance to shaft rotation. The actual speed at a rated load is called the *rated speed*, and the difference between the synchronous speed and the rated speed is called the *slip*. The slip is typically 3%–5% of the synchronous speed for the kinds of motors typically used to drive centrifugal pumps. Typical synchronous and rated speeds are given in Table 8.2 for AC frequencies of 60 Hz and 50 Hz; an AC frequency of 60 Hz is typical of the United States, and 50 Hz is typically used in Europe. As a general rule, motors with higher rotational speeds make more noise than motors with lower rotational speeds. Pumps usually use motors with a synchronous speed of 3600 rpm, with a rated speed of around

Table 8.2: Typical Rated and Synchronous Motor Speeds

No. of Pairs of Poles	60 Hz		50 Hz	
	Synchronous	Rated	Synchronous	Rated
1	3600	3450	3000	2850
2	1800	1725	1500	1425
3	1200	1140	1000	950
4	900	850	750	700

3450 rpm. These speeds are typically found in sump pumps, swimming pool pumps, and water recirculating equipment. Motors with synchronous speeds of 1800 rpm are typically found in blowers, fans, air-handling equipment, compressors, and some conveyors. Motors with synchronous speeds of 1200 rpm and 900 rpm are typically found in window fans, furnace blowers, room air conditioners, and heat pumps; in such applications, less noise is an important factor in motor selection. It is important to remember that motor speeds other than more common synchronous speeds are possible, particularly when variable-speed motors are used. In some special cases, motor speeds higher than those available from standard electrical motors are required, particularly in cases where pumps are delivering fluids at very high heads and moderate flow rates. These pumps are sometimes driven by steam engines or gas turbines.

Matching specific speed to available pumps. A common problem is that for the motor speed chosen, the calculated specific speed does not exactly equal the specific speed of available pumps. In these cases, it is recommended to choose a pump with a specific speed that is close to and greater than the required specific speed or to use a variable-speed motor in the pump. The efficiencies of variable-speed motors can be significantly lower for lower speeds, which may in turn significantly affect the overall efficiency of the pump (Ulanicki et al., 2008). In rare cases, a new pump might be designed to meet the exact design conditions; however, this is usually very costly and only justified for very large pumps.

EXAMPLE 8.6

The most popular series of homologous pumps from a particular manufacturer has a specific speed of 0.45, and these pumps are typically used in countries with alternating current (AC) frequencies of 60 Hz. The pumps are typically sold with motors having a shaft speed of 1200 rpm. (a) If a particular model from this pump series is to be used in a water delivery system where the head added by the pump is 60 m, what will be the flow rate through the pump when it is operating in its most efficient state? (b) What type of pumps are included in this series? (c) How may pairs of poles are in the typical pump motor?

SOLUTION

From the given data: $n_\mathrm{s} = 0.45$, $\omega = 1200$ rpm $= 125.7$ rad/s, and $h_\mathrm{p} = 60$ m. The AC frequency is 60 Hz.

(a) The specific speed is derived from the head and discharge when the pump is operating at its most efficient state. Using the specific speed defined by Equation 8.39 gives

$$n_\mathrm{s} = \frac{\omega Q^{\frac{1}{2}}}{[gh_\mathrm{p}]^{\frac{3}{4}}} \quad \rightarrow \quad 0.45 = \frac{(125.7)Q^{\frac{1}{2}}}{[(9.807)(60)]^{\frac{3}{4}}} \quad \rightarrow \quad Q = 0.183\ \mathrm{m^3/s} = 183\ \mathrm{L/s}$$

Therefore, the flow rate through the pump when it is operating at its most efficient state is approximately **183 L/s.**

(b) The type of pump can be inferred from Table 8.1. Because the specific speed of the pump series is in the range of $0.15 \leq n_\mathrm{s} \leq 1.5$, the pumps in this series are likely to be **centrifugal pumps**.

(c) The relationship between the synchronous speed of a motor and the number of pairs of poles for AC frequencies of 60 Hz is given by Equation 8.41, which yields

$$\omega \text{ rpm} = \frac{3600}{\#\text{ pairs of poles}} \rightarrow 1200 = \frac{3600}{\#\text{ pairs of poles}} \rightarrow \#\text{ pairs of poles} = 3$$

Therefore, it is estimated that **three pairs of poles** are in the motor.

8.3.5 Performance Curves

Specification of a pump generally requires selection of a manufacturer, model, size (D), and rotational speed (ω). For each model, size, and rotational speed, pump manufacturers usually provide a *pump performance curve* or *pump characteristic curve* that shows the relationship between the head added by the pump, h_p, and the flow rate, Q, through the pump. A typical set of pump performance curves (h_p versus Q) provided by a manufacturer for a given pump model is shown in Figure 8.14. In this case, the model (Model 3409) can be fitted with impeller diameters in the range of 307–445 mm with a rated rotational speed of 885 revolutions per minute. Presentation of several impeller diameters on a single graph and under a single model number usually means that any one of the different sizes of impellers can be placed in the same pump casing and hence constitutes an interchangeable part. Allowing for several impeller sizes in the same pump casing is done

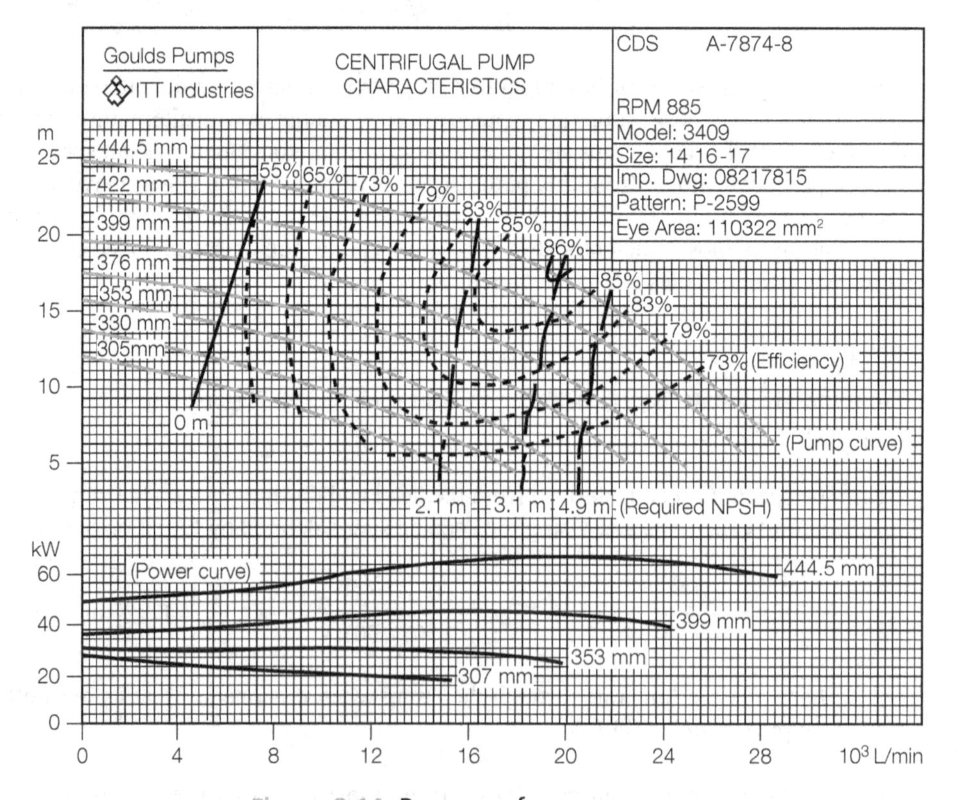

Figure 8.14: Pump performance curve

Source: Goulds Pumps (www.gouldspumps.com).

for various reasons, which include saving on manufacturing costs, facilitating impeller replacement to improve pump performance and increase capacity, standardizing installation mountings, and enabling reuse of support equipment for different applications. Superimposed on the performance curves for the various impeller sizes are lines of constant efficiency for efficiencies ranging from 55%–86%. It is apparent from Figure 8.14 that for a given pump casing, the larger the impeller, the higher the maximum achievable efficiency. Also superimposed on the pump curves are (dashed) isolines of *required net positive suction head*, NPSH_R, which is defined as the minimum allowable difference between the head on the suction side of the pump and the pressure head at which water vaporizes (i.e., the saturation vapor pressure of water). In Figure 8.14, NPSH_R ranges from 4.9 m for higher flow rates to approximately zero, which is indicated by a bold line that meets the 55% efficiency contour. Also shown in Figure 8.14, below the performance curves, is the power delivered to the pump (in kW) for various flow rates and impeller diameters. This power input to the pump shaft is the brake horsepower (BHP); it is related to the head added by the pump, h_p, by

$$\text{BHP} = \frac{\gamma Q h_\text{p}}{\eta} \tag{8.42}$$

where, typically, BHP is in kW, γ is the specific weight of water in kN/m^3, Q is the discharge rate in m^3/s, h_p is the head added by the pump in m, and η is the pump efficiency, which is dimensionless. Equation 8.42 is dimensionally homogeneous, so any consistent set of units can be used.

Shutoff head and rated capacity. Two key parameters of any pump performance curve are the *shutoff head*, which is the head added by the pump when the discharge is equal to zero, and the *rated capacity*, which is the discharge when the pump is operating its best efficiency point. The shutoff condition and best efficiency condition on a typical pump performance curve are illustrated in Figure 8.15. The shutoff head, h_0, is the maximum head that can be added by a pump; it also represents the pressure head developed across the pump when the discharge valve is closed. Operation of a pump with the discharge valve closed is not desirable due to overheating and the large mechanical stresses developed under this condition. The shutoff head and rated capacity are the key points that must be matched when the pump performance curve is expressed as a parabolic equation of the form $h_\text{p} = a - bQ^2$, where a and b are constants. The point (Q, h_p) on the performance curve at the shutoff head is $(0, h_0)$, and at

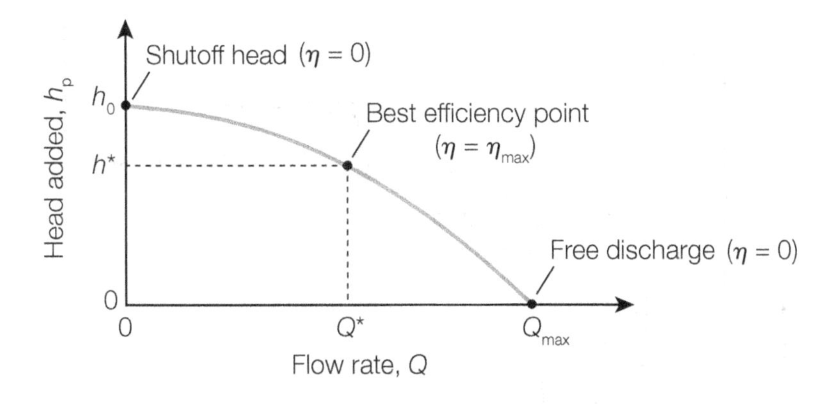

Figure 8.15: Key parameters on a pump performance curve

the rated capacity, the point on the curve is (Q^*, h^*). Requiring that the parabolic performance curve pass through these two points yields the curve

$$h_p = h_0 - \left(\frac{h_0 - h^*}{Q^{*2}}\right) Q^2 \tag{8.43}$$

Because this is an empirical equation relating the variables h_p and Q, care must be taken to state the applicable units for these variables. A condition that is sometimes referred to but seldom shown on practical performance curves is the *free discharge condition*, which is the point on the performance curve where the added head is equal to zero. The flow at the free discharge condition, Q_{max}, is sometimes referred to as the *free discharge* and is the maximum flow that can be handled by the pump. Typically, the rated capacity, Q^*, is around 60% of the free discharge.

Effect of fluid properties. As long as the flow through a pump is fully turbulent, a pump performance curve is independent of the properties of the fluid being pumped. The rationale for this assertion can be seen from the generalized nondimensional performance curves for pumps given by Equation 8.25 and repeated here for convenience as

$$\frac{gh_p}{\omega^2 D^2} = f\left(\frac{Q}{\omega D^3}, \frac{\rho\omega D^2}{\mu}\right) \tag{8.44}$$

It is apparent from this relationship that for fully turbulent flow, $\rho\omega D^2/\mu$ need not be considered. Therefore, the performance of the pump is determined by the functional relationship between $gh_p/\omega^2 D^2$ and $Q/\omega D^3$; none of these terms involve a fluid property.

8.3.6 System Characteristics

The goal in pump selection is to choose a pump that operates at a point of maximum efficiency and with a net positive suction head that exceeds the minimum allowable value. Pumps are placed in pipeline systems such as that illustrated in Figure 8.16, in which case the energy equation for the pipeline system requires that the head, h_p, added by the pump be given by

$$h_p = \Delta z + Q^2\left[\sum \frac{fL}{2gA^2D} + \sum \frac{K_{local}}{2gA^2}\right] \tag{8.45}$$

where Δz is the difference in elevation between the water surfaces of the source and destination reservoirs, the first term in the square brackets is the sum of the head losses due to

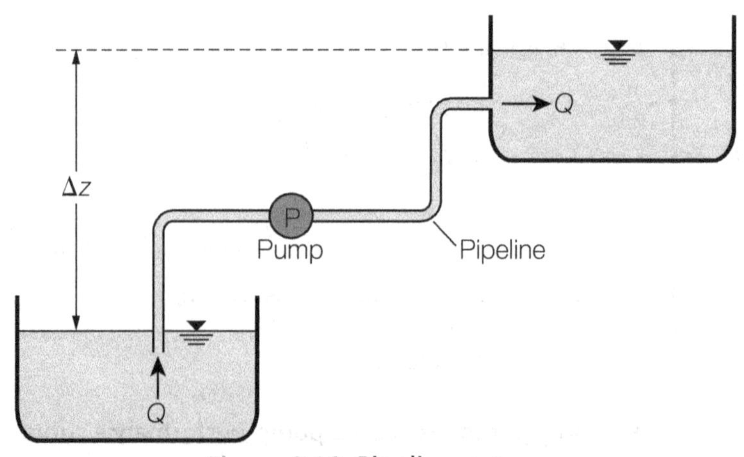

Figure 8.16: Pipeline system

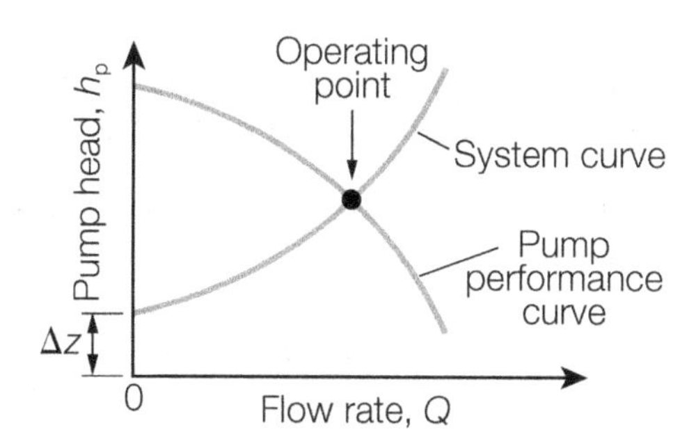

Figure 8.17: Operating point in a pipeline system

friction, and the second term is the sum of the local head losses. Equation 8.45 gives the required relationship between h_p and Q for the pipeline system, and this relationship is commonly called the *system curve*. Because the flow rate and head added by the pump must satisfy both the system curve and the pump performance curve, Q and h_p are determined by simultaneous solution of Equation 8.45 and the pump performance curve. The values of Q and h_p that satisfy both the system curve and the pump performance curve define the *operating point* or *duty point* of the pump. The location of the operating point on the performance curve is illustrated in Figure 8.17. Ideally, a pump should be selected such that the operating point is as close as possible to the best efficiency point of the pump.

EXAMPLE 8.7

A pump is to be used to transfer water from a lower reservoir to an upper reservoir through a 190-m-long, 100-mm-diameter ductile iron pipe with an estimated roughness height of 0.3 mm. The water surface in the upper reservoir is at an elevation 5 m higher than the water surface in the lower reservoir. The sum of the local head loss coefficients in the pipeline system is estimated as 5.4. The pump being considered for installation has a performance curve described by

$$h_\mathrm{p} = 40 - 0.00245Q^2$$

where h_p is the head added by the pump in m and Q is the flow rate through the pump in liters per second. (a) Plot the system and performance curves to show the operating point. (b) What is the expected flow rate through the system? Assume water at 20°C.

SOLUTION

From the given data: $L = 190$ m, $D = 100$ mm, $k_\mathrm{s} = 0.3$ mm, $\Delta z = 5$ m, and $K = 5.4$. For water at 20°C, $\gamma = 9.789$ kN/m^3 and $\nu = 1.00 \times 10^{-6}$ m^2/s. Using the given data,

$$A = \frac{\pi D^2}{4} = \frac{\pi(0.1)^2}{4} = 7.854 \times 10^{-3}\ \mathrm{m}^2, \qquad \frac{k_\mathrm{s}}{D} = \frac{0.3}{100} = 0.003$$

The independent variable is Q, so it is also convenient to define the following functions of Q, where Q is in L/s:

$$V(Q) = \frac{Q}{A} = \frac{Q \times 10^{-3}}{7.854 \times 10^{-3}}, \qquad \mathrm{Re}(Q) = \frac{V(Q)D}{\nu} = \frac{V(Q)(0.1)}{1.00 \times 10^{-6}}$$

$$f(Q) = f_\mathrm{CE}\left(\mathrm{Re}(Q), \frac{k_\mathrm{s}}{D}\right), \qquad h_\mathrm{v}(Q) = \frac{V(Q)^2}{2g} = \frac{V(Q)^2}{2(9.807)}$$

where $f_{CE}(\text{Re}, k_s/D)$ denotes the friction factor obtained by solution of the Colebrook equation, and $h_v(Q)$ is the velocity head corresponding to a flow rate of Q.

(a) Applying the energy equation between the lower and upper reservoirs and using the above-defined functions gives

$$0 - \left[\frac{f(Q)L}{D} + K\right] h_v(Q) + h_p = \Delta z \quad \rightarrow \quad h_p = \Delta z + \left[\frac{f(Q)L}{D} + K\right] h_v(Q)$$

This is the equation for the system curve. Substituting the values of the given and derived parameters into the system curve equation gives

$$h_p = 5 + \left[\frac{f(Q)(190)}{0.1} + 5.4\right] h_v(Q) \quad \rightarrow \quad h_p = 5 + \left[1900f(Q) + 5.4\right] h_v(Q)$$

This system curve is plotted along with the given pump curve in Figure 8.18.

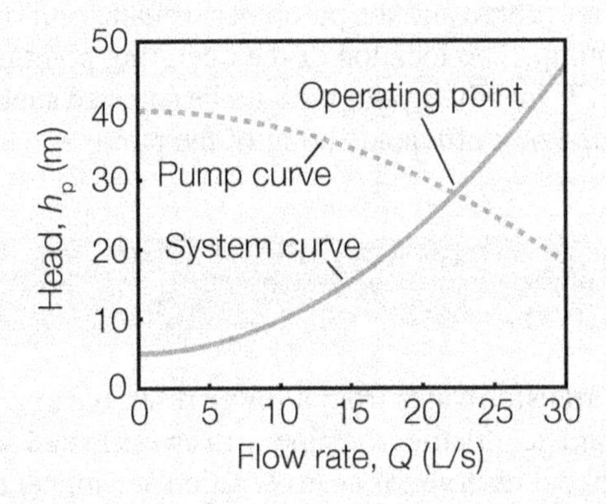

Figure 8.18: System and pump curves for a pump pipeline system

It is apparent that the system curve and pump curve intersect at a flow rate in the range of 20–25 L/s. The intersection point is the operating point of the pump pipeline system.

(b) The operating point of the pump pipeline system is determined by simultaneous solution of the system curve and the pump curve. Both equations are satisfied when

$$5 + \left[1900f(Q) + 5.4\right] h_v(Q) = 40 - 0.00245Q^2 \quad \rightarrow \quad Q = 22.2 \text{ L/s}$$

Therefore, the flow rate in the pipeline at the operating point is approximately **22.2 L/s**. This flow rate could be controlled by valves within the pipeline. Using valves to control the flow rate in the pipeline would be reflected analytically by changing the local head loss coefficient.

Note that this analysis is complicated by the fact that the friction factor, f, is assumed to depend on the flow rate, Q, via the Colebrook equation. This dependence must be taken into account for smooth and transitional flow conditions. Under rough flow conditions, the friction factor is a constant that is independent of Q, and this problem could then be solved analytically.

8.3.7 Limits on Pump Location

If the absolute pressure on the suction side of a pump falls below the saturation vapor pressure of the liquid being pumped, the liquid will begin to vaporize. This process of vaporization within a liquid is called *cavitation.* Cavitation is a transient phenomenon that occurs as a liquid enters the low-pressure suction side of a pump then, as the liquid containing vapor cavities moves toward the high-pressure environment of the discharge side of the pump, the vapor cavities are compressed and ultimately implode, creating small, localized high-velocity jets that can cause considerable damage to the pump machinery. Collapsing vapor cavities have been associated with jet velocities on the order of 110 m/s and pressures of up to 800 MPa when the jets strike a solid wall (Knapp et al., 1970; Finnemore and Franzini, 2002). The damage caused by collapsing vapor cavities usually manifests itself as *pitting* of the metal casing and impeller, reduced pump efficiency, and excessive vibration of the pump. The noise generated by imploding vapor cavities resembles the sound of gravel going through a centrifugal pump. Because the saturation vapor pressure increases with temperature, a system that operates satisfactorily without cavitation during the winter may have problems with cavitation during the summer.

Available net positive suction head. The potential for cavitation is measured by the *available net positive suction head*, NPSH_A, defined as the difference between the head on the suction side of the pump (at the inlet to the pump) and the head when cavitation begins; hence,

$$\text{NPSH}_\text{A} = \left(\frac{p_\text{s}}{\gamma} + \frac{V_\text{s}^2}{2g} + z_\text{s}\right) - \left(\frac{p_\text{v}}{\gamma} + z_\text{s}\right) = \frac{p_\text{s}}{\gamma} + \frac{V_\text{s}^2}{2g} - \frac{p_\text{v}}{\gamma} \tag{8.46}$$

where p_s, V_s, and z_s are the pressure, velocity, and elevation, respectively, of the liquid at the suction side of the pump and p_v is the saturation vapor pressure of the liquid. The NPSH_A can also be equivalently defined as the difference between the stagnation pressure ($p_\text{s}+\frac{1}{2}\rho V_\text{s}^2$) and the vapor pressure (p_v) on the suction side of the pump, where both pressures are expressed as heads of the flowing fluid. In cases where the liquid is being pumped from a reservoir by a pump that is located above the reservoir, the NPSH_A can be calculated by applying the energy equation between the reservoir and the suction side of the pump; in this case, the available net positive suction head is given by

$$\text{NPSH}_\text{A} = \frac{p_0}{\gamma} - \Delta z_\text{s} - h_\ell - \frac{p_\text{v}}{\gamma} \tag{8.47}$$

where p_0 is the pressure at the surface of the reservoir (usually atmospheric), Δz_s is the difference in elevation between the suction side of the pump and the liquid surface in the source reservoir (called the *suction lift* or *static suction head* or *static head*), h_ℓ is the head loss in the pipeline between the source reservoir and suction side of the pump (including local losses), and p_v is the saturation vapor pressure of the liquid. In applying either Equation 8.46 or 8.47 to calculate NPSH_A, care must be taken to use a consistent measure of pressure, using either gauge pressure or absolute pressure. Absolute pressures are usually more convenient, because the vapor pressure is typically expressed as an absolute pressure.

Required net positive suction head. A pump requires a minimum NPSH_A to prevent the onset of cavitation within the pump, and this minimum NPSH_A is called the *required net positive suction head*, NPSH_R, which is generally a function of the head, h_p, added by the pump. This relationship is commonly expressed in terms of the *critical cavitation parameter* or *critical cavitation number*, σ_c, where

$$\text{NPSH}_\text{R} = \sigma_\text{c} h_\text{p} \tag{8.48}$$

Pump manufacturers either present curves showing the relationship between NPSH_R and h_p or provide values of σ_c for each pump. For pumps to operate without cavitation problems, it is required that $\text{NPSH}_\text{A} \geq \text{NPSH}_\text{R}$. Many engineers use a safety factor, SF, to ensure that $\text{NPSH}_\text{A} \geq \text{NPSH}_\text{R}$, in which case the system is designed such that $\text{NPSH}_\text{A} = \text{SF} \cdot \text{NPSH}_\text{R}$. Safety factors of 1.1 and higher are common. The maximum elevation, z_max, of a pump above a source reservoir occurs when $\text{NPSH}_\text{A} = \text{NPSH}_\text{R}$, in which case Equation 8.47 can be conveniently expressed as

$$z_\text{max} = \frac{p_0 - p_\text{v}}{\gamma} - h_\ell - \text{NPSH}_\text{R} \tag{8.49}$$

In some cases, the NPSH_R requirement could be sufficiently large to require that a pump be located below the source reservoir.

Scaling of required net positive suction head. Required net positive suction heads in homologous pumps with the same flow coefficient ($Q/\omega D^3$) can be scaled according to the relationship

$$\left[\frac{g(\text{NPSH}_\text{R})}{\omega^2 D^2}\right]_1 = \left[\frac{g(\text{NPSH}_\text{R})}{\omega^2 D^2}\right]_2 \quad \text{when} \quad \left[\frac{Q}{\omega D^3}\right]_1 = \left[\frac{Q}{\omega D^3}\right]_2 \tag{8.50}$$

where the subscripts 1 and 2 indicate homologous operating points in two homologous pumps. The quantity $g(\text{NPSH}_\text{R})/\omega^2 D^2$ is commonly called the *suction head coefficient*. Eliminating D in Equation 8.50 yields the quantity, S, defined as

$$S = \frac{\omega Q^{\frac{1}{2}}}{(g\,\text{NPSH}_\text{R})^{\frac{3}{4}}} \tag{8.51}$$

where S is called the *suction specific speed*, and it defines homologous points with the same cavitation characteristics. In particular, two geometrically similar pumps have the same value of S when they have the same value of the flow coefficient ($= Q/\omega D^3$).

EXAMPLE 8.8

Water at 20°C is being pumped from a lower to an upper reservoir through a 200-mm-diameter pipe in the system shown in Figure 8.16. The water surface elevations in the source and destination reservoirs differ by 5.2 m, and the length of the steel pipe ($k_\text{s} = 0.046$ mm) connecting the reservoirs is 21.3 m. The pump is to be located 1.5 m above the water surface in the source reservoir, and the length of the pipeline between the source reservoir and the suction side of the pump is 3.5 m. The performance curves of the 885-rpm homologous series of pumps being considered for this system are given in Figure 8.14. If the desired flow rate in the system is 315 L/s, what size and specific speed pump should be selected? Assess the adequacy of the pump location based on a consideration of the available net positive suction head. Assume that the pipe intake loss coefficient is 0.1.

SOLUTION

For the system pipeline: $L = 21.3$ m, $D = 200$ mm $= 0.2$ m, $k_s = 0.046$ mm, and (neglecting local head losses) the energy equation for the system is given by

$$h_p = 5.2 + \frac{fL}{2gA^2D}Q^2 \tag{8.52}$$

where h_p is the head added by the pump (in m), f is the friction factor (dimensionless), Q is the flow rate through the system (in m^3/s), and A is the cross-sectional area of the pipe (in m^2) given by

$$A = \frac{\pi}{4}D^2 = \frac{\pi}{4}(0.2)^2 = 0.03142 \text{ m}^2 \tag{8.53}$$

The friction factor, f, can be calculated using the Swamee–Jain formula (Equation 7.45)

$$f = \frac{0.25}{\left[\log\left(\frac{k_s}{3.7D} + \frac{5.74}{\text{Re}^{0.9}}\right)\right]^2} \tag{8.54}$$

where Re is the Reynolds number given by

$$\text{Re} = \frac{VD}{\nu} = \frac{QD}{A\nu} \tag{8.55}$$

and $\nu = 1.00 \times 10^{-6}$ m^2/s at 20°C. Combining Equations 8.54 and 8.55 with the given data yields

$$f = \frac{0.25}{\left[\log\left(\frac{4.6 \times 10^{-5}}{3.7(0.2)} + \frac{5.74}{\left(\frac{Q(0.2)}{(0.03142)(1.00 \times 10^{-6})}\right)^{0.9}}\right)\right]^2} \tag{8.56}$$

which simplifies to

$$f = \frac{1}{4[\log(6.216 \times 10^{-5} + 4.32 \times 10^{-6}Q^{-0.9})]^2} \tag{8.57}$$

Combining Equations 8.52 and 8.57 gives the following relation:

$$\begin{aligned} h_p &= 5.2 + \frac{21.3}{2(9.81)(0.03142)^2(0.2)(4)[\log(6.216 \times 10^{-5} + 4.32 \times 10^{-6}Q^{-0.9})]^2}Q^2 \\ &= 5.2 + \frac{1375Q^2}{[\log(6.216 \times 10^{-5} + 4.32 \times 10^{-6}Q^{-0.9})]^2} \end{aligned} \tag{8.58}$$

This relation is applicable for h_p in m and Q in m^3/s.

Equation 8.58 is the "system curve," which relates the head added by the pump to the flow rate through the system, as required by the energy equation. Because the pump performance curve must also be satisfied, the operating point of the pump is at the intersection of the system curve (Equation 8.58) and the pump performance curve given in Figure 8.14. The system curve and the pump performance curves are both plotted in Figure 8.19, and the operating points for the various pump sizes are listed in the following table.

Pump Size (mm)	Operating Point (10^3 L/min)
307	11
330	12.8
353	14.7
376	16.6
399	18.3
422	20.6
444.5	22.1

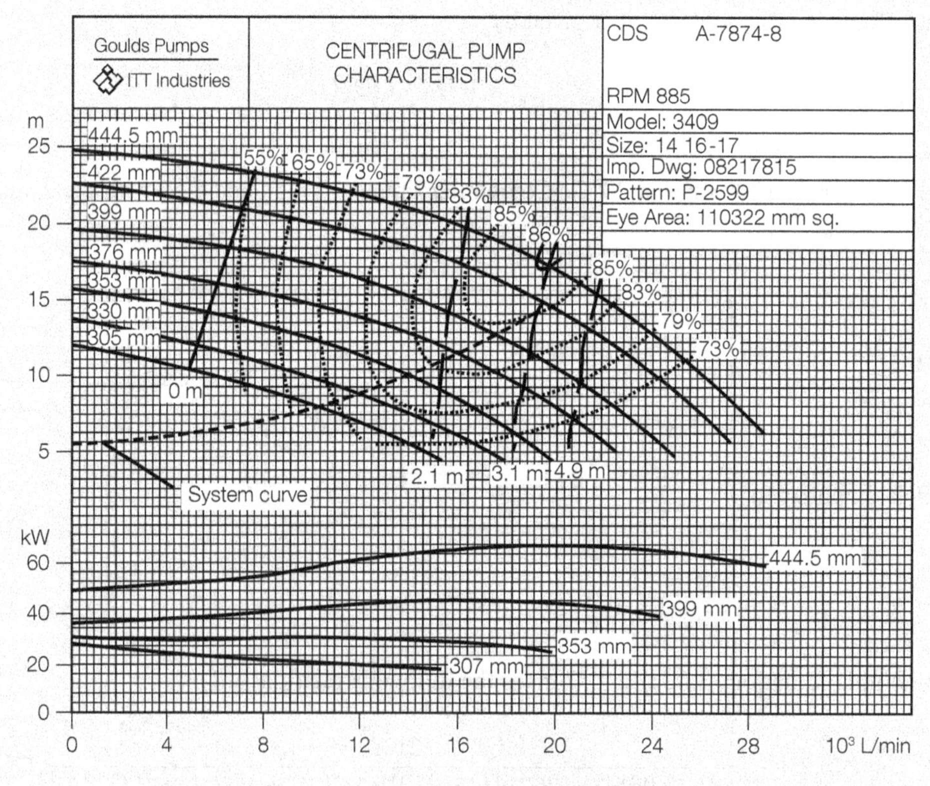

Figure 8.19: Pump operating points

Because the desired flow rate in the system is 315 L/s, the **42.2-cm pump** should be selected. This 42.2-cm pump will deliver 344 L/s when all system valves are open, and it can be throttled down to deliver 315 L/s as required. If a closer match between the desired flow rate and the operating point is desired for the given system, then an alternative series of homologous pumps should be considered. For the selected 42.2-cm pump, the maximum efficiency point is at $Q = 315$ L/s $= 0.315$ m^3/s, $h_\text{p} = 15.8$ m, and $\omega = 885$ rpm = 92.68 rad/s; hence, the specific speed, N_s, of the selected pump is given by

$$N_\text{s} = \frac{\omega Q^{\frac{1}{2}}}{h_\text{p}^{\frac{3}{4}}} = \frac{(92.68)(0.315)^{\frac{1}{2}}}{(15.8)^{\frac{3}{4}}} = \mathbf{6.56}$$

This dimensional specific speed corresponds to a dimensionless specific speed of n_s = 1.18. Comparing this result with the pump selection guidelines in Table 8.1 confirms that the pump being considered must be a centrifugal pump.

The available net positive suction head, NPSH_A, is defined by Equation 8.47 as

$$\text{NPSH}_\text{A} = \frac{p_0}{\gamma} - \Delta z_\text{s} - h_\ell - \frac{p_\text{v}}{\gamma} \tag{8.59}$$

Atmospheric pressure, p_0, can be taken as 101 kPa; the specific weight of water, γ, is 9.79 kN/m^3; the suction lift, Δz_s, is 1.5 m; and at 20°C, the saturated vapor pressure of water, p_v, is 2.34 kPa. The head loss, h_ℓ, is estimated as

$$h_\ell = \left(0.1 + \frac{fL}{D}\right)\frac{V^2}{2g} \tag{8.60}$$

where the entrance loss at the pump intake is $0.1\ V^2/2g$. For a flow rate, Q, equal to 344 L/s, Equation 8.57 gives the friction factor, f, as

$$\begin{aligned} f &= \frac{1}{4[\log(6.216 \times 10^{-5} + 4.32 \times 10^{-6}Q^{-0.9})]^2} \\ &= \frac{1}{4[\log(6.216 \times 10^{-5} + 4.32 \times 10^{-6}(0.344)^{-0.9})]^2} \\ &= 0.00366 \end{aligned}$$

and the average velocity of flow in the pipe, V, is given by

$$V = \frac{Q}{A} = \frac{0.344}{0.03142} = 10.9 \text{ m/s}$$

Substituting $f = 0.00366$, $L = 3.5$ m, $D = 0.2$ m, and $V = 10.9$ m/s into Equation 8.60 yields

$$h_\ell = \left[0.1 + \frac{(0.00366)(3.5)}{0.2}\right]\frac{10.9^2}{2(9.81)} = 3.44 \text{ m}$$

Thus, the available net positive suction head, NPSH_A (Equation 8.59), is

$$\text{NPSH}_\text{A} = \frac{101}{9.79} - 1.5 - 3.44 - \frac{2.34}{9.79} = 5.13 \text{ m}$$

According to the pump properties given in Figure 8.19, the required net positive suction head, NPSH_R, for the 42.2-cm pump at the operating point is 3.66. Because the available net positive suction head (5.13 m) is greater than the required net positive suction head (3.66 m), the pump location relative to the intake reservoir is adequate and **cavitation problems are not expected**.

8.3.8 Multiple Pump Systems

In cases where a single pump is inadequate to achieve a desired operating condition, multiple pumps can be used. Combinations of pumps are referred to as *pump systems*, and the pumps within these systems are typically arranged in series or in parallel. The performance curve of a pump system is determined by the arrangement of pumps.

Pumps in series. Consider the case of two identical pumps in series, illustrated in Figure 8.20(a). The flow through each pump is equal to Q, the head added by each pump is h_{p}, and the head added by the two-pump system is equal to $2h_{\text{p}}$. Consequently, the performance curve of the two-pump (in series) system is related to the performance curve of a single pump in that for any flow, Q, the head added by the two-pump system is twice the head added by a single pump. The relationship between the single-pump performance curve and the two-pump performance curve is illustrated in Figure 8.20(b). Also shown in Figure 8.20(b) is the operating condition when one pump is used (Point 1) and the operating condition when two pumps are used (Point 2). It is apparent that adding a second pump in series does not double the head that is added to the flow in the pump pipeline system. This analysis can be extended to cases where the pump system contains n identical pumps in series, in which case the n-pump performance curve is derived from the single-pump performance curve by multiplying the ordinate of the single-pump performance curve (h_{p}) by n. Pump systems that include multiple pumps in series are called *multistage pump systems*, and pumps that include multiple smaller pumps within a single housing are called *multistage pumps*. Multistage pump systems are commonly used in applications involving unusually high heads, such as in pumping water from deep boreholes. It is generally not appropriate simply to connect two or more pumps in series, because significant additional losses would be created if all pumps were not being operated simultaneously. Pumps and piping should be arranged so that individual pumps can be taken out of service without disrupting the flow through the remainder of the operating pumps.

Pumps in parallel. The case of two identical pumps arranged in parallel is illustrated in Figure 8.21. In this case, the flow through each pump is Q and the head added is h_{p}; therefore, the flow through the two-pump system is equal to $2Q$ and the head added is h_{p}. Consequently, the performance curve of the two-pump system is derived from the performance curve of a single pump by multiplying the abscissa (Q) by two. This is illustrated in Figure 8.21(b), which also shows the operating condition when one pump is used (Point 1) and the operating condition when two pumps are used (Point 2). It is apparent that adding a second pump in parallel will not double the flow in the pump pipeline system. In a similar manner to the two-pump example, the performance curves of systems containing n identical pumps in parallel can be derived from the single-pump performance curve by multiplying the abscissa (Q) by n. Pumps in parallel are used in cases where the desired flow rate is beyond the range of a single pump and to provide flexibility in pump operations, because some pumps in the system can be shut down during low-demand conditions or for service. This arrangement is common

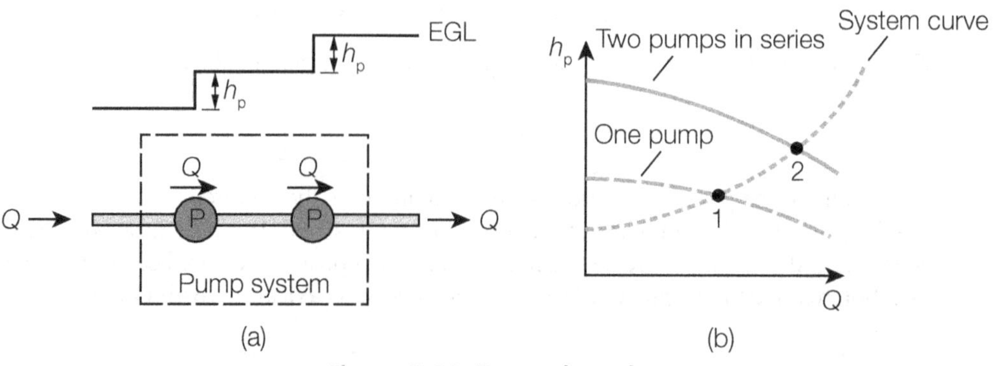

Figure 8.20: Pumps in series

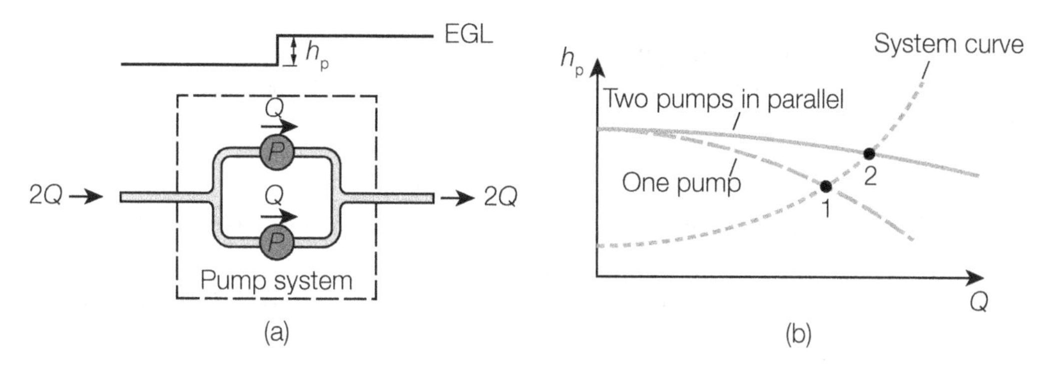

Figure 8.21: Pumps in parallel

in sanitary-sewer pump stations and in public water-supply systems, where flow rates vary significantly during the course of a day. Care should be taken in arranging parallel pumps and piping such that backflow through pumps that are not operating is prevented.

Guidelines for multiple pump systems. When pumps are placed in series or in parallel, it is usually desirable that these pumps be identical; otherwise, the pumps will be loaded unequally and the efficiency of the pump system will be less than optimal. In cases where nonidentical pumps are placed in series, the performance curve of the pump system is obtained by summing the heads added by the individual pumps for a given flow rate. In cases where nonidentical pumps are placed in parallel, the performance curve of the pump system is obtained by summing the flow rates through the individual pumps for a given head. In any case where dissimilar pumps are used in a pump system, care should be taken to ensure that each pump operates at or near its best efficiency point.

EXAMPLE 8.9

A pump has a performance curve described by the relation

$$h_\mathrm{p} = 12 - 0.1Q^2$$

where h_p is in m and Q is in L/s. (a) What is the performance curve for a system having three of these pumps in series? (b) What is the performance curve for a system having three of these pumps in parallel?

SOLUTION

In the pump system, denote the head added by the system as H_sys, the flow rate through the system as Q_sys, the head added by an individual pump within the system as h_p, and the flow rate through an individual pump in the system as Q.

(a) For a system with three pumps in series, the same flow, Q, goes through each pump, and each pump adds one-third of the head, H_sys, added by the pump system. Therefore,

$$H_\mathrm{sys} = 3h_\mathrm{p}, \qquad Q_\mathrm{sys} = Q$$

Substituting these relations in the given pump curve yields the system curve as follows:

$$h_\mathrm{p} = 12 - 0.1Q^2 \quad \rightarrow \quad \frac{H_\mathrm{sys}}{3} = 12 - 0.1Q_\mathrm{sys}^2 \quad \rightarrow \quad \mathbf{H_{sys} = 36 - 0.3\,Q_{sys}^2}$$

(b) For a system consisting of three pumps in parallel, one-third of the system flow goes through each pump, and the head added by each pump is the same as the total head added by the pump system. Therefore,

$$H_{sys} = h_p, \qquad Q_{sys} = 3Q$$

Substituting these relations in the given pump curve yields the system curve as follows:

$$h_p = 12 - 0.1Q^2 \quad \rightarrow \quad H_{sys} = 12 - 0.1\left(\frac{Q_{sys}}{3}\right)^2 \quad \rightarrow \quad \mathbf{H_{sys} = 12 - 0.0111\,Q_{sys}^2}$$

8.3.9 Variable-Speed Pumps

In variable-speed pumps, the rotational speed can be adjusted, in contrast to having only on-off modes. For any homologous (i.e., geometrically similar) series of pumps, the performance curve of any pump within the series is given by

$$\frac{gh_p}{\omega^2 D^2} = f\left(\frac{Q}{\omega D^3}\right)$$

where $f(Q/\omega D^3)$ is a function that is unique to the homologous series and is the same for all pumps within the series. For a fixed pump size, D, for two different motor speeds, ω_1 and ω_2, there will be different points on the performance curve corresponding to the different speeds, and these points can be expressed as

$$\frac{h_1}{\omega_1^2 D^2} = f\left(\frac{Q_1}{\omega_1 D^3}\right) \quad \text{and} \quad \frac{h_2}{\omega_2^2 D^2} = f\left(\frac{Q_2}{\omega_2 D^3}\right)$$

where h_1 and Q_1 are the corresponding head and flow rate when the rotational speed is ω_1 and h_2 and Q_2 are corresponding head and flow rate when the rotational speed is ω_2. Because the pumps are part of a homologous series the function f is fixed, and for a fixed D,

$$\frac{Q_1}{\omega_1} = \frac{Q_2}{\omega_2} \tag{8.61}$$

requires that

$$\frac{h_1}{\omega_1^2} = \frac{h_2}{\omega_2^2} \tag{8.62}$$

These relationships are used to relate the performance curve for any pump operating at ω_1 to the performance curve of the same pump operating at ω_2 as shown in Figure 8.22. P_1 is

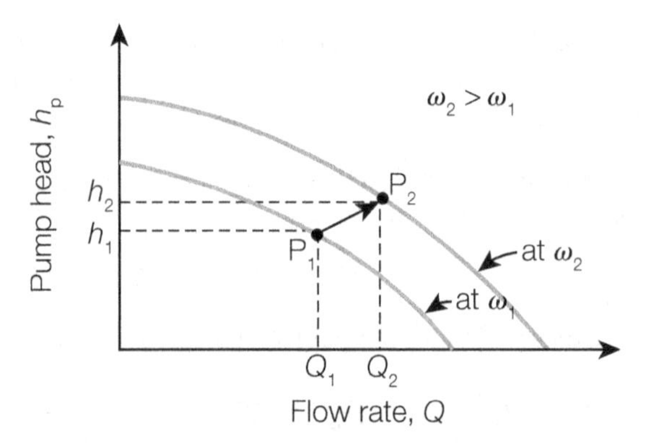

Figure 8.22: Pump curves for a variable-speed motor

a point on the ω_1 operating curve with coordinates (Q_1,h_1), and P_2 is a point on the ω_2 operating curve with coordinates (Q_2,h_2). Then according to Equations 8.61 and 8.62, these coordinates are related by

$$Q_2 = Q_1\left(\frac{\omega_2}{\omega_1}\right) \quad \text{and} \quad h_2 = h_1\left(\frac{\omega_2}{\omega_1}\right)^2 \tag{8.63}$$

It should be noted from these results that increased flows will generally be attained with increased rotational speed and that variable-speed pumps provide a viable option for adjusting the operating point in pump pipeline systems.

Scaling of NPSH$_\mathrm{R}$. Because the required net positive suction head (NPSH$_\mathrm{R}$) of a pump depends on the head added by the pump, NPSH$_\mathrm{R}$ must also be adjusted to account for the change in speed. The adjustment relationship is given by

$$\mathrm{NPSH}_{\mathrm{R}_2} = \left(\frac{\omega_2}{\omega_1}\right)^2 \mathrm{NPSH}_{\mathrm{R}_1} \tag{8.64}$$

where the numerical subscripts refer to conditions at two different motor speeds. It is apparent from Equation 8.64 that NPSH$_\mathrm{R}$ increases with increased rotational speed.

Scaling of pump efficiency. The affinity laws indicate that the efficiencies at homologous operating points are the same. Hence, for operating points (h_1, Q_1) and (h_2, Q_2) as given by Equation 8.63, $\eta_1 = \eta_2$. This approximation has been shown to be reasonably accurate for larger pumps ($\sim$ 500 kW). However, for smaller pumps ($\sim$ 5 kW) operating at reduced speeds, the efficiencies tend to be less than that predicted by the affinity laws (Simpson and Marchi, 2013). Taking η_1 and η_2 as the rated efficiencies, a relationship that accounts for the effect of changing the rotational speed on the rated efficiency is

$$\eta_2 = 1 - (1 - \eta_1)\left(\frac{\omega_1}{\omega_2}\right)^{0.1} \tag{8.65}$$

This equation, originally proposed by Sarbu and Borza (1988), produces good results (better than the affinity laws) as long as the pump speed is not reduced below 70% of the rated speed.

EXAMPLE 8.10

A pump with a 1200-rpm motor has a performance curve of

$$h_\mathrm{p} = 12 - 0.1Q^2$$

where h_p is in meters and Q is in cubic meters per minute. If the speed of the motor is changed to 2400 rpm, estimate the new performance curve.

SOLUTION

From the given data: $\omega_1 = 1200$ rpm, $\omega_2 = 2400$ rpm, and the affinity laws (Equation 8.63) state that

$$Q_1 = \frac{\omega_1}{\omega_2}Q_2 = \frac{1200}{2400}Q_2 = 0.5Q_2, \qquad h_1 = \frac{\omega_1^2}{\omega_2^2}h_2 = \frac{1200^2}{2400^2}h_2 = 0.25h_2$$

Because the performance curve of the pump at speed ω_1 is given by

$$h_1 = 12 - 0.1Q_1^2$$

the performance curve at speed ω_2 is given by

$$0.25h_2 = 12 - 0.1(0.5Q_2)^2 \quad \rightarrow \quad h_2 = 48 - 0.1Q_2^2$$

The performance curve of the pump with a 2400-rpm motor is therefore given by

$$\mathbf{h_p = 48 - 0.1Q^2}$$

Comparing the performance curve of the two pumps, the shutoff head is considerably higher in the 2400-rpm pump (48 m) compared with the shutoff head in the 1200 rpm pump (12 m).

Practical Considerations

Although variable-speed motors are more expensive than fixed-speed motors, in some hydraulic systems, using variable-speed pumps can be significantly more cost effective than using fixed-speed pumps, because the pump operation can be adjusted to be most efficient under a variety of operating conditions (e.g., Bene and Hös, 2012). Variable-speed pumps can also have undesirable effects if the operating speed is too low. For example, at low flows, check valves might not fully open and solids in slurries might settle out; in addition, lower speeds might impair lubrication and cooling of the pump. Speeds higher than normal can place excessive loads on couplings and other drive components.

8.4 Fans

Fans are used to move air, gases, and vapors. A characteristic of fans is that changes in pressure as a gas moves through a fan are insufficient to cause significant changes in density; hence, the analysis of flow through fans generally assumes incompressible flow. A typical centrifugal fan is illustrated in Figure 8.23. Typically, density changes in gases flowing through fans are less than 10%; pressure changes are also less than 10%. In contrast to fans, machines that move air and cause significant changes in density and pressure are typically classified as *compressors*. Fans are typically constructed of relatively thin sheet metal and can serve various functions and be of varying sizes. For example, fans can pull air into ducts, push air out of ducts, move air within ducts, or serve to move large volumes of air in and out of large spaces in buildings.

8.4.1 Performance Characteristics of Fans

Because fans are characterized by a rotating component that moves an incompressible fluid, the nondimensional functional relationships that are the bases for describing the performance of pumps are also applicable to fans. These nondimensional relationships developed for pumps are given by Equations 8.28–8.30. However, in the case of fans, the head added, h_p, is expressed in terms of the pressure increase, Δp, and the break horsepower, BHP, is simply referred to as the shaft power, P. Applying these changes to Equations 8.28–8.30

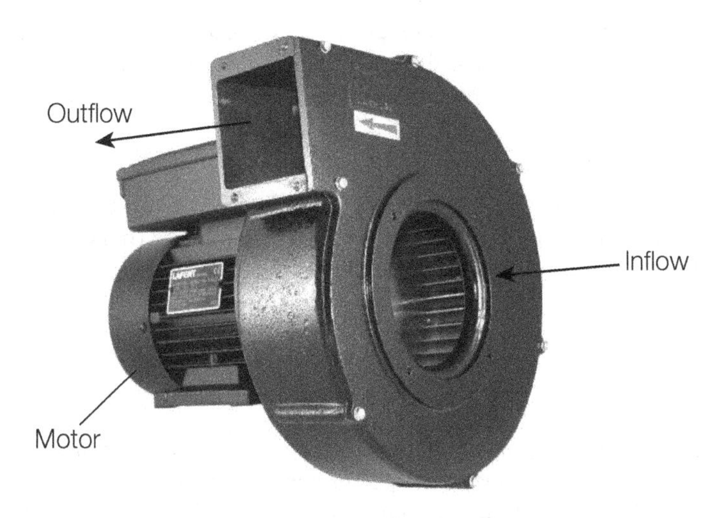

Figure 8.23: Typical fan
Source: Air Control Industries.

yields the following nondimensional functional relations that are the bases for describing the performance of fans:

$$\frac{\Delta p}{\rho\omega^2 D^2} = \Phi_1\left(\frac{Q}{\omega D^3}\right) \tag{8.66}$$

$$\frac{P}{\rho\omega^3 D^5} = \Phi_2\left(\frac{Q}{\omega D^3}\right) \tag{8.67}$$

$$\eta = \Phi_3\left(\frac{Q}{\omega D^3}\right) \tag{8.68}$$

where Φ_i represents a functional relationship, different values of i indicate different functional relationships, ρ is the density of the gas being moved by the fan, ω is the rotational speed of the fan, D is the diameter of the fan impeller, and Q is the volume flow rate at which the gas moves through the fan. The nondimensional functional relationships given by Equations 8.66–8.68 have similar shapes to those of pumps. Another similarity to pumps is that fans can be classified as being centrifugal, mixed-flow, or axial-flow, which have similar flow characteristics to their counterparts in pumps.

8.4.2 Affinity Laws of Fans

The affinity laws, which follow directly from the nondimensional functional relationships given by Equations 8.66–8.68, can be expressed as

$$\left[\frac{Q}{\omega D^3}\right]_1 = \left[\frac{Q}{\omega D^3}\right]_2 \quad \rightarrow \quad \begin{cases} \left[\dfrac{\Delta p}{\rho\omega^2 D^2}\right]_1 = \left[\dfrac{\Delta p}{\rho\omega^2 D^2}\right]_2 \\ \left[\dfrac{P}{\rho\omega^3 D^5}\right]_1 = \left[\dfrac{P}{\rho\omega^3 D^5}\right]_2 \\ \eta_1 = \eta_2 \end{cases} \tag{8.69}$$

where the subscripts 1 and 2 are corresponding operational states in geometrically similar fans, and η represents the efficiency with which the shaft power is transferred to the gas moving through the fan. The relationships given in Equation 8.69 are sometimes called the *fan laws* and can be used to translate the performance characteristics of a given fan to the performance characteristics of a geometrically similar (i.e., homologous) fan.

8.4.3 Specific Speed

As in the case of pumps, the specific speed of a fan is used to identify the combination of performance variables at the best efficiency operating point. The (nondimensional) specific speed of homologous series of fans, n_s, is defined by

$$n_s = \frac{\omega Q^{\frac{1}{2}} \rho^{\frac{3}{4}}}{(\Delta p)^{\frac{3}{4}}} \tag{8.70}$$

In selecting a fan for a given application, the usual approach is to calculate a desired specific speed based on the requisite values of the performance variables and then attempt to find a commercially available fan with the desired specific speed that is capable of delivering the required flow rate.

EXAMPLE 8.11

A particular centrifugal fan has an impeller diameter of 900 mm with a motor that rotates at 580 rpm. At the best efficiency point, the fan is capable of delivering 14.16 m^3/s with a pressure increment of 0.8 kPa and a power requirement of 16 kW. These test data were obtained using air at a temperature of 10°C. A fan from the same homologous series has a diameter of 1100 mm, has a motor that operates at 1140 rpm, and is intended to move air at a temperature of 20°C. (a) Estimate the flow rate, pressure increment, and power requirement of the 1100-mm fan operating at its best efficiency point. (b) Compare the specific speeds of the 900-mm and 1100-mm fans.

SOLUTION

From the given data: $D_1 = 900$ mm, $\omega_1 = 580$ rpm, $Q_1 = 14.16$ m^3/s, $\Delta p_1 = 0.8$ kPa, $P_1 = 16$ kW, $T_1 = 10$°C, $D_2 = 1100$ mm, $\omega_2 = 1140$ rpm, and $T_2 = 20$°C. For air at 10°C, $\rho_1 = 1.246$ kg/m^3, and for air at 20°C, $\rho_2 = 1.204$ kg/m^3.

(a) Because the fans are geometrically similar, applying the affinity laws gives

$$\frac{Q_1}{\omega_1 D_1^3} = \frac{Q_2}{\omega_2 D_2^3} \quad \rightarrow \quad \frac{14.16}{(580)(900)^3} = \frac{Q_2}{(1140)(1100)^3}$$

$$\rightarrow \quad Q_2 = \mathbf{50.80\ m^3/s}$$

$$\frac{\Delta p_1}{\rho_1 \omega_1^2 D_1^2} = \frac{\Delta p_2}{\rho_2 \omega_2^2 D_2^2} \quad \rightarrow \quad \frac{0.8}{(1.246)(580)^2(900)^2} = \frac{\Delta p_2}{(1.204)(1140)^2(1100)^2}$$

$$\rightarrow \quad \Delta p_2 = \mathbf{4.46\ kPa}$$

$$\frac{P_1}{\rho_1 \omega_1^3 D_1^5} = \frac{P_2}{\rho_2 \omega_2^3 D_2^5} \quad \rightarrow \quad \frac{16}{(1.246)(580)^3(900)^5} = \frac{P_2}{(1.204)(1140)^3(1100)^5}$$

$$\rightarrow \quad P_2 = \mathbf{320\ kW}$$

(b) The specific speed of each fan is calculated using Equation 8.70. The specific speeds of these fans are the same since they are from the same homologous series. This can be verified by calculating the specific speeds separately. The following values have to be used to calculate specific speeds: $\omega_1 = 60.74$ rad/s, $Q_1 = 14.16$ m^3/s, $\omega_2 = 119.4$ rad/s, and $Q_2 = 50.80$ m^3/s. The specific speeds are then calculated as follows:

$$\text{900-mm fan:} \quad n_\text{s} = \frac{(60.74)(14.16)^{\frac{1}{2}}(1.246)^{\frac{3}{4}}}{(800)^{\frac{3}{4}}} = 1.79$$

$$\text{1100-mm fan:} \quad n_\text{s} = \frac{(119.4)(50.80)^{\frac{1}{2}}(1.204)^{\frac{3}{4}}}{(4460)^{\frac{3}{4}}} = 1.79$$

Therefore, the specific speed of the fans is **1.79**.

Application. An interesting application of fans is in firefighting, where large fans are used to blow air into a building that is on fire. Such fans are called portable positive ventilation (PPV) fans, and a typical PPV fan is shown in Figure 8.24. When applied to tall structures, PPV fans, if used correctly, can limit the amount of smoke and heat entering the stairway and push smoke and deadly gases out of the structure.

Figure 8.24: Application of a fan in firefighting

Source: Daniel N. Madrzykowski, Fire Protection Engineer, National Institute of Standards and Technology (NIST).

8.5 Hydraulic Turbines and Hydropower

Turbines are used to extract energy from flowing fluids and are used with both liquids and gases. Turbines that are used to extract energy from flowing water are the central components of all hydropower facilities, and such turbines are commonly called *hydraulic turbines* or *hydroturbines*. Turbines used to extract energy from flowing gases are called *gas turbines*, and both compressibility and thermodynamic considerations are usually important in gas turbines. *Wind turbines* and *steam turbines* are types of gas turbines used to extract energy from wind and steam, respectively. Fluid mechanics and thermodynamics principles are intertwined in explaining the behavior of steam and gas turbines, and the functioning of these types of turbines is usually covered extensively in courses related to thermodynamics. The focus of the following sections is on hydraulic turbines, in which thermodynamic considerations are relatively simple and the performance of hydraulic turbines is based mostly on the principles of fluid mechanics. Hydraulic turbines used in engineering applications have a wide range of capacities. At the lower end of the spectrum are units that generate around 5 kW of power, whereas some very large units are capable of generating on the order of 400 MW of power. Hydraulic turbines can be broadly classified into two categories: impulse turbines and reaction turbines. These types of turbines are discussed separately in the following sections.

8.5.1 Impulse Turbines

In an impulse turbine, a free jet of water impinges on a rotating component, called a *runner*, which is exposed to atmospheric pressure. The runner in an impulse turbine is sometimes called an *impulse wheel* or *Pelton wheel*.[54] A typical impulse turbine is shown in Figure 8.25, where the runner has a series of split buckets located around its periphery. When the waterjet strikes the dividing ridge of the bucket, it is split into two parts that discharge at both sides of the bucket. Only one jet is used on small turbines, but two or more jets impinging at different points around the runner are often used on large units. The jet flows are usually controlled by a needle nozzle, and some jet velocities exceed 150 m/s. The generator rotor is usually mounted on a horizontal shaft between two bearings with the runner installed on the projecting end of the shaft; this is called a *single-overhung* installation. In some cases, runners are installed on both sides of the generator to equalize bearing loads, and this is called a *double-overhung* installation. The diameters of runners range up to about 5 m.

Figure 8.25: Impulse turbine
Source: Becki Rudig/Canyon Industries, Inc.

[54]Named in honor of the American mining engineer Lester A. Pelton (1829–1908).

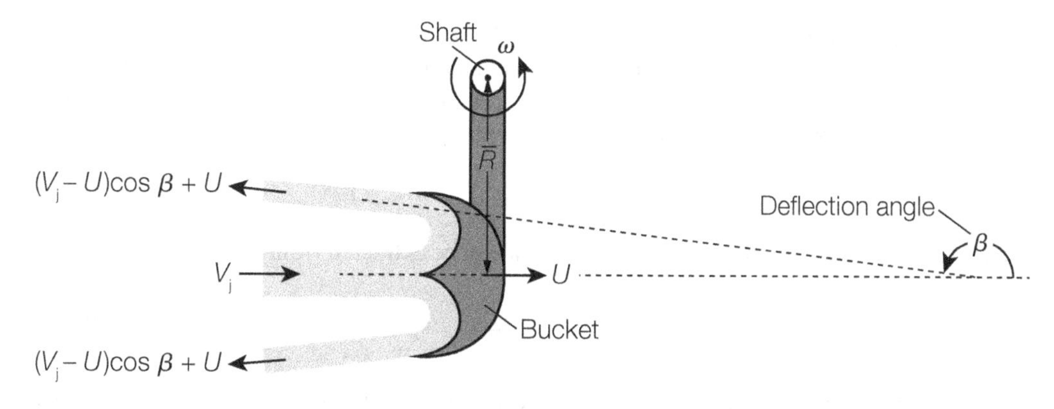

Figure 8.26: Mechanics of an impulse turbine

Mechanics. The mechanics of an impulse turbine are illustrated in Figure 8.26, where V_j is the velocity at which a waterjet strikes a bucket of the impulse turbine, U is the velocity of the bucket, and β is the deflection angle of the incoming jet. Power is generated when the force exerted on the bucket by the jet is applied at an average distance $\overline{R}$ from the shaft, which is rotating with an angular velocity ω. A generator attached to the rotating shaft then produces electricity. Application of the moment of momentum equation to the fluid entering and exiting the bucket gives the torque on the shaft, T_shaft, that can be approximated as

$$T_\mathrm{shaft} = \dot{m}\overline{R}(V_{2\theta} - V_{1\theta}) \tag{8.71}$$

where $\dot{m}$ is the mass inflow rate delivered by the jet and $V_{1\theta}$ and $V_{2\theta}$ are the θ components of the absolute velocity at the inflow and outflow sections of the bucket, respectively. The θ components of the velocity are defined as those components that are perpendicular to the direction in which $\overline{R}$ is measured, with positive directions of $V_{1\theta}$ and $V_{2\theta}$ creating positive moments about the shaft. This means that the positive directions of $V_{1\theta}$ and $V_{2\theta}$ are directed toward the right in Figure 8.26. The shaft power, $\dot{W}_\mathrm{shaft}$, can be derived from the shaft torque, T_shaft, given by Equation 8.71 as follows:

$$\dot{W}_\mathrm{shaft} = T_\mathrm{shaft}\omega = \dot{m}\,\overline{R}\,(V_{2\theta} - V_{1\theta})\,\omega \quad \rightarrow \quad \dot{W}_\mathrm{shaft} = \dot{m}\,U(V_{2\theta} - V_{1\theta}) \tag{8.72}$$

where the relationship between the bucket speed, U, and the rate of angular rotation, ω, is utilized (i.e., $U = \overline{R}\omega$). Because the inflow to the bucket shown in Figure 8.26 is normal to the bucket,

$$V_{1\theta} = V_\mathrm{j} \tag{8.73}$$

The inflow velocity relative to the moving bucket is $V_\mathrm{j} - U$, and this inflow relative velocity can be assumed to be equal to the outflow relative velocity. Thus, the θ component of the outflow relative velocity is $(V_\mathrm{j} - U)\cos\beta$. The absolute velocity at the exit is obtained by adding the bucket velocity, U, which gives

$$V_{2\theta} = (V_\mathrm{j} - U)\cos\beta + U \tag{8.74}$$

Substituting the values of $V_{1\theta}$ and $V_{2\theta}$ from Equations 8.73 and 8.74 into Equation 8.72 and simplifying gives

$$\dot{W}_\mathrm{shaft} = \dot{m}\,U(V_\mathrm{j} - U)(\cos\beta - 1) \tag{8.75}$$

This key equation shows that the power generated by an impulse turbine is fundamentally dependent on the mass flow rate of the fluid, $\dot{m}$, the speed of the bucket, U, the speed of the jet, V_j, and the bucket angle, β. Values of $\dot{W}_\mathrm{shaft}$ given by Equation 8.75 are generally negative, because the outflow momentum flux is less than the inflow momentum flux and hence the turbine extracts energy from the fluid. In design applications, it is useful to utilize a bucket speed that will maximize the power generated by the turbine for given values of $\dot{m}$, V_j, and β. Taking the partial derivative of $\dot{W}_\mathrm{shaft}$, as given by Equation 8.75, with respect to U and setting this partial derivative equal to zero yields

$$U = \tfrac{1}{2}V_\mathrm{j} \tag{8.76}$$

which gives the theoretical maximum power that can be extracted from an impulse turbine as

$$\dot{W}_\mathrm{shaft}\Big|_\mathrm{max} = \tfrac{1}{4}\dot{m}V_\mathrm{j}^2(\cos\beta - 1) \tag{8.77}$$

The optimal bucket speed given by Equation 8.76 shows that the maximum power will be generated when the bucket speed is equal to one-half the jet speed. The ratio U/V_j is commonly referred to as the *speed ratio*.

Practical Considerations

Although Equation 8.76 indicates that maximum power is achieved when the speed ratio is equal to 0.5, in practical applications, increased energy losses at high wheel rotation speeds usually result in an optimal speed ratio of around 0.46 (Massey and Ward-Smith, 2012). It is also apparent from Equation 8.75 that maximum power is generated when $\beta = 180°$; however, it is usually necessary to use lesser values of β to avoid the outflow jet interfering with the inflow jet. Consequently, $\beta < 180°$ is the norm, with the upper limit of β being around 165° to avoid interference between incoming and exiting jets. In contrast to the wheel speed that gives the maximum efficiency (Equation 8.76), the wheel speed that gives the minimum efficiency is $U = V_\mathrm{j}$, which yields $\dot{W}_\mathrm{shaft}$ equal to zero and hence an efficiency of zero. The wheel speed $U = V_\mathrm{j}$ is called the *runaway speed*, and when the wheel speed is equal to the runaway speed, the impulse turbine is said to be *freewheeling*.

Measures of efficiency. The *wheel efficiency*, η_w, of an impulse turbine is defined as the ratio of the shaft power output, $\dot{W}_\mathrm{shaft}$, to the power of the jet driving the turbine wheel; $\frac{1}{2}\dot{m}V_\mathrm{j}^2$; hence,

$$\text{wheel efficiency, } \eta_\mathrm{w} = \frac{|\dot{W}_\mathrm{shaft}|}{\frac{1}{2}\dot{m}V_\mathrm{j}^2} \tag{8.78}$$

If the theoretical shaft power given by Equation 8.75 is used to calculate η_w, then the theoretical wheel efficiency is given by

$$\eta_\mathrm{w,theory} = \frac{2U(V_\mathrm{j} - U)(1 - \cos\beta)}{V_\mathrm{j}^2} \tag{8.79}$$

The theoretical wheel efficiency accounts only for the deviation of U from its theoretically optimum value of $\frac{1}{2}V_\mathrm{j}$ and for the deviation of β from its theoretically optimum value of 180°. So if $U = \frac{1}{2}V_\mathrm{j}$ and $\beta = 180°$, Equation 8.79 gives $\eta_\mathrm{w} = 1$. The actual wheel efficiency, η_w, as defined by Equation 8.78, is generally less than the theoretical wheel efficiency, $\eta_\mathrm{w,theory}$, defined by Equation 8.79, due to additional influences such as mechanical friction, aerodynamic drag on the buckets, friction drag on the surface of the buckets, misalignment of the jet and bucket, and backsplashing. Typically, for Pelton wheels operating under optimal conditions, $\eta_\mathrm{w} \approx 0.9$. The wheel efficiency defined by Equation 8.78 takes the energy input to the turbine as being equal to the kinetic energy of the jet driving the wheel of the turbine. Usual practice is to include the jet nozzle as part of the turbine assembly and define the energy input to the turbine as the head just upstream of the nozzle. In such cases, the additional head loss in the nozzle must be accounted for in the input energy. The efficiency of the turbine, η_t, is then given by

$$\text{turbine efficiency, } \eta_t = \frac{|\dot{W}_{\text{shaft}}|}{\gamma Q h_e} \tag{8.80}$$

where h_e is *effective head*, which is equal to the head just upstream of the nozzle, which is related to the jet velocity, V_j, by the relation

$$V_j = C_v\sqrt{2gh_e} \tag{8.81}$$

where C_v is a velocity coefficient that accounts for head losses in the nozzle, typically in the range of 0.92–0.98. Combining Equations 8.75, 8.80, and 8.81, it can be shown analytically that η_t is maximized when $U = \frac{1}{2}C_v\sqrt{2gh_e}$. The ratio $U/\sqrt{2gh_e}$ is called the *peripheral velocity factor*, ϕ, such that

$$\phi = \frac{U}{\sqrt{2gh_e}} \tag{8.82}$$

Hence, under (theoretically) optimum conditions, $\phi = \frac{1}{2}C_v$. The theoretical turbine efficiency, η_t, can be expressed in terms of ϕ and C_v by

$$\eta_t = \frac{|\dot{W}_{\text{shaft}}|}{\gamma Q h_e} = \frac{\rho Q U(V_j - U)(1 - \cos\beta)}{\gamma Q h_e} = 2\phi(C_v - \phi)(1 - \cos\beta) \tag{8.83}$$

Under the optimal condition that $\phi = \frac{1}{2}C_v$, the maximum efficiency of an impulse turbine is given by Equation 8.83 as

$$\eta_t\Big|_{\text{max}} = \frac{1}{2}C_v^2(1 - \cos\beta) \tag{8.84}$$

It is important to remember that even with nozzle losses taken into account, the theoretical maximum power is still obtained when $U = \frac{1}{2}V_j$. This relation can be combined with the requirement that $U = \frac{1}{2}C_v\sqrt{2gh_e}$ to give the optimum jet velocity for a given effective head.

EXAMPLE 8.12

Consider the impulse turbine shown in Figure 8.27 where water at 20°C is delivered at a rate of 300 L/s through a 300-mm-diameter supply pipe and the nozzle at the end of the supply pipe has a diameter of 100 mm. The impulse turbine has a diameter of 2 m, and the buckets mounted on the periphery of the turbine have a deflection angle of 160°. (a) What is the maximum power that can be delivered by the turbine? (b) When operating at the maximum power condition, what is the rotational speed of the turbine runner?

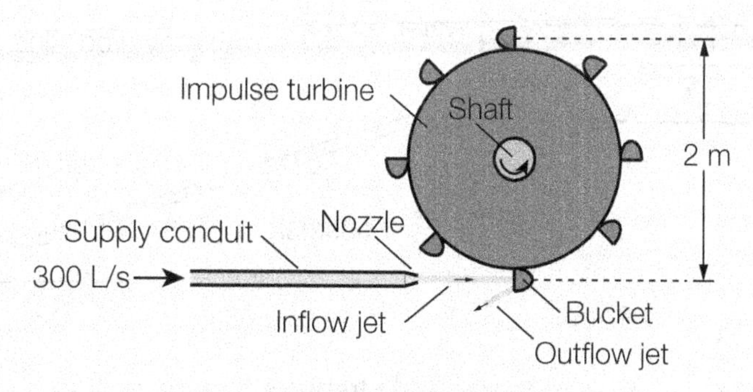

Figure 8.27: Impulse turbine

SOLUTION

From the given data: $Q = 300$ L/s $= 0.3$ m^3/s, $D_\text{p} = 300$ mm, $D_\text{j} = 100$ mm, $\overline{R} = 2\,\text{m}/2 = 1$ m, and $\beta = 160°$. For water at 20°C, $\rho = 998.2$ kg/m^3. The following preliminary calculations are useful:

$$\dot{m} = \rho Q = (998.2)(0.3) = 299.4 \text{ kg/s}, \qquad A_\text{j} = \frac{\pi D_\text{j}^2}{4} = \frac{\pi (0.1)^2}{4} = 7.854 \times 10^{-3} \text{ m}^2$$

$$V_\text{j} = \frac{Q}{A_\text{j}} = \frac{0.3}{7.854 \times 10^{-3}} = 38.20 \text{ m/s}$$

(a) Under maximum power operating conditions, the bucket speed, U, and the corresponding power are given by Equations 8.76 and 8.77 as

$$U = \tfrac{1}{2} V_\text{j} = \tfrac{1}{2}(38.20) = 19.10 \text{ m/s}$$

$$\dot{W}_\text{shaft}\Big|_\text{max} = \tfrac{1}{4}\dot{m}V_\text{j}^2(\cos\beta - 1) = \tfrac{1}{4}(299.4)(38.20)^2(\cos 160° - 1)$$

$$= -2.118 \times 10^5 \text{ W} = -212 \text{ kW}$$

Therefore, the theoretical maximum power that can be extracted from the given impulse turbine system is estimated as **212 kW**.

(b) The rotational speed, ω, of the runner under the maximum power condition is given by

$$\omega = \frac{U}{\overline{R}} = \frac{19.10}{1} = 19.10 \text{ rad/s} = 182 \text{ rpm}$$

Therefore, the rotational speed of the runner is estimated as **182 rpm**.

Energy transformations. The schematic layout of an impulse turbine installation is shown in Figure 8.28. Water is delivered from the upstream reservoir to the turbine through a large-diameter pipe called a *penstock*. The head loss in the penstock between the reservoir and the nozzle entrance is h_ℓ, and the static head, H, is equal to the difference between the water elevation in the reservoir and the elevation of the nozzle. The nozzle is considered to be an

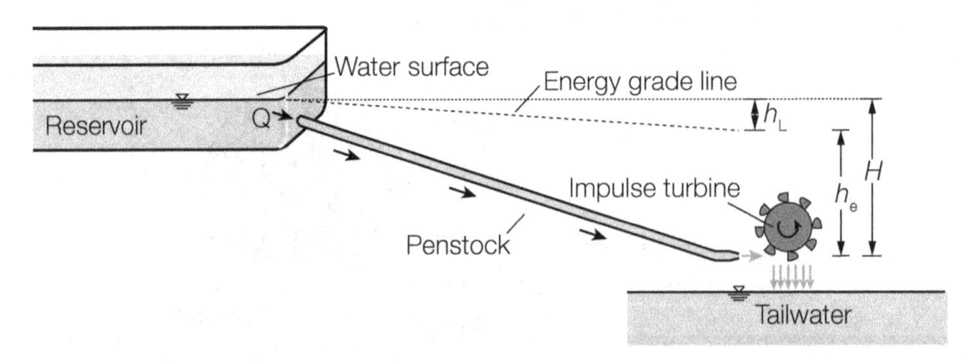

Figure 8.28: Impulse turbine system

integral part of the impulse turbine, so the *net head* or *effective head*, h_e, on the turbine is equal to the static head minus the head loss, h_ℓ, and is given by

$$h_e = H - h_\ell \tag{8.85}$$

The head at the entrance to the nozzle is expended in four ways: (1) energy is lost in fluid friction in the nozzle, known as *nozzle loss*; (2) energy is lost in fluid friction over the buckets of the turbine runner; (3) kinetic energy is carried away in the water discharged from the buckets; and (4) energy is used in rotating the buckets. Therefore, the head directly available to the runner, h_t, is given by

$$h_t = h_e - k_j \frac{V_j^2}{2g} - k \frac{v_2^2}{2g} - \frac{V_2^2}{2g} \tag{8.86}$$

where k_j is a nozzle head loss coefficient that depends on the geometry of the nozzle, V_j is the jet velocity, k is the bucket friction loss coefficient, v_2 is the velocity of water relative to the bucket at the exit from the bucket, and V_2 is the absolute velocity of the water leaving the bucket. Typical values of k_j are in the range of 0.01–0.03, and typical values of k are in the range of 0.2–0.6. The (theoretical) power extracted by the turbine, P_t, is therefore given by

$$P_t = \gamma Q h_t \tag{8.87}$$

where γ is the specific weight of water. Interestingly, for a given pipeline, there is a unique jet diameter that will deliver maximum power to a jet. This is apparent by noting that the power of the jet, P_j, issuing from the nozzle is given by

$$P_j = \gamma Q \frac{V_j^2}{2g} \tag{8.88}$$

As the size of the nozzle opening is increased, the flow rate, Q, increases while the jet velocity, V_j, decreases; hence, some intermediate size of nozzle opening will provide maximum power to the jet.

Efficiency. The *hydraulic efficiency*, η_t, of an impulse turbine is the ratio of the power delivered to the turbine buckets to the power in the flow at the entrance to the nozzle. Thus, for impulse turbines,

$$\eta_t = \frac{\gamma Q h_t}{\gamma Q h_e} = \frac{h_t}{h_e} \tag{8.89}$$

The overall efficiency, η, of an impulse turbine is less than the hydraulic efficiency, η_t, because part of the energy delivered to the buckets is lost in the mechanical friction in the bearings and in the air resistance associated with the spinning runner. The overall efficiency of an impulse turbine, η, is given by

$$\eta = \frac{\text{output (shaft) power}}{\text{input power}} = \frac{T_{\text{shaft}}\omega}{\gamma Q h_e} \tag{8.90}$$

where T_{shaft} is the torque delivered to the shaft by the turbine and ω is the angular speed of the runner. In actual installations, the peak efficiency occurs when the wheel speed is slightly less than one-half the jet speed, and this condition is usually used to fix the wheel speed once the jet speed is determined. The efficiencies, η, of well-designed Pelton wheels can approach 90%.

EXAMPLE 8.13

It is proposed to install a small Pelton wheel turbine system where the static head above the nozzle is 50 m and the delivery line is 455-mm-diameter concrete pipe with a length of 150 m and a roughness height of 1 mm. The discharge nozzle has a diameter of 100 mm and a head loss coefficient of 0.5. Model tests on the Pelton wheel show that the bucket friction loss coefficient is 0.3, the velocity of water relative to the bucket at the exit from the bucket is 1 m/s, and the absolute velocity of the water leaving the bucket is 5 m/s. Estimate the power that could be derived from the system and the hydraulic efficiency of the turbine.

SOLUTION

From the given data: $H = 50$ m, $D = 455$ mm, $L = 150$ m, $k_s = 1$ mm, $D_j = 100$ mm, $k_j = 0.5$, $k = 0.3$, $v_2 = 1$ m/s, and $V_2 = 5$ m/s. The flow rate, Q, is determined by application of the energy equation between the upstream reservoir and the exit of the nozzle, which requires that

$$H - \frac{fL}{D}\frac{Q^2}{2gA^2} - k_j\frac{Q^2}{2gA_j^2} = \frac{Q^2}{2gA_j^2} \tag{8.91}$$

where f is the friction factor of the concrete pipe and A and A_j are the cross-sectional areas of the pipe and jet, respectively. Using the Swamee–Jain equation (Equation 7.45) to relate f to Q, Equation 8.91 yields $Q = 0.200$ m^3/s. The corresponding velocities in the delivery pipeline and nozzle jet are $V = 1.23$ m/s and $V_j = 25.4$ m/s, and the friction factor is $f = 0.0244$. Using these derived data and taking the specific weight of water as $\gamma = 9.79$ kN/m^3 yields the following results:

$$h_e = H - h_\ell = H - \frac{fL}{D}\frac{V^2}{2g} = 50 - \frac{(0.0244)(150)}{(0.455)}\frac{1.23^2}{2(9.81)} = 49.4 \text{ m}$$

$$h_t = h_e - k_j\frac{V_j^2}{2g} - k\frac{v_2^2}{2g} - \frac{V_2^2}{2g} = 49.4 - 0.5\frac{25.4^2}{2(9.81)} - 0.3\frac{1^2}{2(9.81)} - \frac{5^2}{2(9.81)} = 31.6 \text{ m}$$

$$P_t = \gamma Q h_t = (9.79)(0.200)(31.6) = 61.9 \text{ kW}$$

$$\eta_t = \frac{h_t}{h_e} = \frac{31.6}{49.4} = 0.64$$

For the given configuration, the maximum power output from the system is **61.9 kW** with a hydraulic efficiency of **64%**.

8.5.2 Reaction Turbines

In a reaction turbine, flow takes place in a closed chamber under pressure, and the flow through a reaction turbine may be radially inward, axial, or mixed. Two types of reaction turbines are in common use: the *Francis turbine*[55] and the *axial-flow turbine*, also called a *propeller turbine*.

Francis turbines. All inward-flow turbines are called Francis turbines. In the conventional Francis turbine, water enters a *scroll case* and moves into the runner through a series of *guide vanes* with contracting passages that convert pressure head to velocity head. The guide vanes typically consist of a set of fixed vanes called *stay vanes* that direct the flow into a set of adjustable vanes called *wicket gates*. The adjustable wicket gates allow the quantity and direction of flow into the runner to be controlled. A Francis turbine runner is shown in Figure 8.29. When the runner is installed, flow enters at an oblique angle at the top of the runner as shown in Figure 8.29, and flow exits through the underside of the runner; such

[55]Named in honor of the American engineer James B. Francis (1815–1892).

turbines are therefore mixed-flow turbines. A scroll case that surrounds the runner is designed to decrease the cross-sectional area in proportion to the decreasing flow rate passing a given section of the casing. Constant rotative speed of the runner under varying load is achieved by a governor that actuates a mechanism that regulates the gate openings. A relief valve or surge tank is generally necessary to prevent serious water hammer pressures.

Propeller turbines. The propeller turbine is an axial-flow machine with its runner confined in a closed conduit. The usual runner has four to eight blades mounted on a hub, with very little clearance between the blades and the conduit wall. A *Kaplan turbine*[56] is a propeller turbine with movable blades whose pitch can be adjusted to suit existing operating conditions. A typical Kaplan turbine runner is shown in Figure 8.30. Other types of propeller turbines include the *Deriaz turbine*, an adjustable-blade, diagonal-flow turbine where the flow is directed inward as it passes through the blades, and the *tube turbine*, an inclined-axis type turbine that is particularly well adapted to low-head installations, because the water passages can be formed directly in the concrete structure of the dam.

Figure 8.29: Francis turbine runner
Source: Andritz Hydro GmBH.

Figure 8.30: Kaplan turbine runner
Source: U.S. Army Corps of Engineers

[56]Named in honor of the Austrian engineer Viktor Kaplan (1876–1934).

Mechanics. The control volume of a typical reaction turbine is illustrated in Figure 8.31, where the fluid enters the outer surface of the turbine runner with the absolute velocity $\mathbf{V}_1$ and exits through the inner surface of the runner with an absolute velocity $\mathbf{V}_2$. Applying the moment of momentum equation to the control volume containing the runner yields the following expression for the shaft torque, T_{shaft}, exerted by the rotor on the fluid in terms of the velocities of the fluid entering and exiting the control volume containing the rotor:

$$T_{\text{shaft}} = \dot{m}(r_2V_{2\theta} - r_1V_{1\theta}) \tag{8.92}$$

where $\dot{m}$ is the mass flow rate of fluid flowing through the runner, r_1 and r_2 are the radii of the outer and inner surfaces of the runner, respectively, and $V_{1\theta}$ and $V_{2\theta}$ are the tangential (θ-) components of the fluid velocity at the entrance and exit of the control volume, respectively. If the shaft rotates at a rate ω, the shaft power, $\dot{W}_{\text{shaft}}$ can be derived from the shaft torque as follows:

$$\dot{W}_{\text{shaft}} = T_{\text{shaft}}\omega = \dot{m}\omega(r_2V_{2\theta} - r_1V_{1\theta}) \tag{8.93}$$

The shaft power given by Equation 8.93 is generally less than the difference in mechanical energy between the inflow and outflow from the turbine. The difference in head between the inflow and the outflow is called the *available head.* Denoting the available head by h_a, the *hydraulic efficiency* of the turbine, η_h, can be estimated by

$$\eta_h = \frac{|\dot{W}_{\text{shaft}}|}{\dot{m}gh_a} \quad \rightarrow \quad \eta_h = \frac{\omega(r_1V_{1\theta} - r_2V_{2\theta})}{gh_a} \tag{8.94}$$

The power delivered to the runner (i.e., $\dot{W}_{\text{shaft}}$) is generally less than the power delivered by the entire turbine unit due to additional losses such as those caused by the bearings in the shaft of the runner. The hydraulic efficiency, η_h, given by Equation 8.94 is generally greater than the overall efficiency, η, of the turbine unit, where the overall efficiency is defined by the relation

$$\eta = \frac{P}{\dot{m}gh_a} \tag{8.95}$$

where P is the power output of the turbine unit.

Operational characteristics. To operate properly, reaction turbines must have a submerged discharge. After passing through the runner, the water enters a *draft tube*, which is an expanding conduit that directs the water to the discharge location in the downstream channel called the *tailrace*. Care must be taken to ensure that the pressure at the entrance to the draft tube

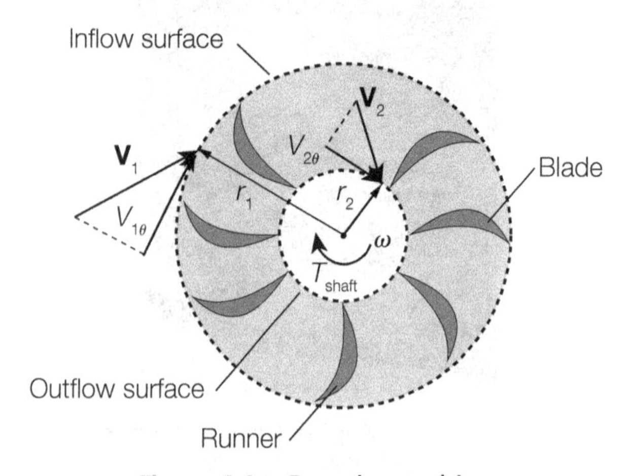

Figure 8.31: Reaction turbine

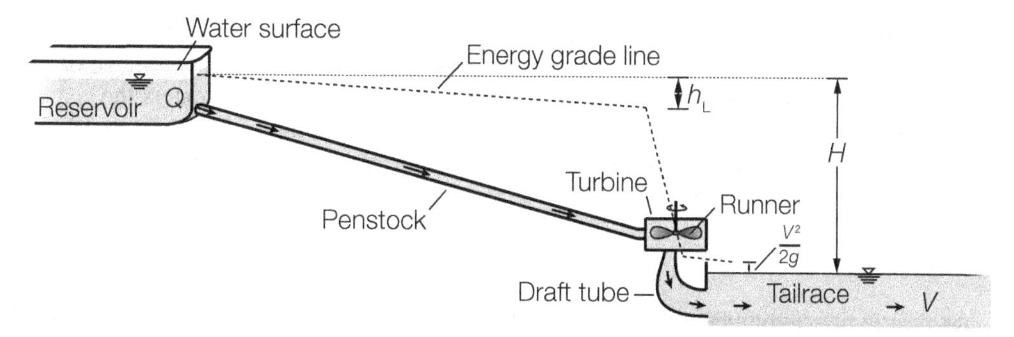

Figure 8.32: Reaction turbine system

is greater than the saturation vapor pressure of water so that cavitation does not occur; this generally requires that the runner be less than 10 m above the tailrace. A schematic diagram of a reaction turbine system is shown in Figure 8.32. Applying the energy equation between the upstream reservoir and the tailrace, the available head, h_a, is given by

$$h_a = H - h_\ell - h_{td} - \frac{V^2}{2g} \tag{8.96}$$

where H is the height of the water surface in the upstream reservoir above the water surface in the tailrace, h_ℓ is the head loss between the reservoir and the turbine inlet, h_{td} is the sum of the head losses in the turbine and the draft tube, and V is the velocity in the tailrace. In most cases, $V^2/2g$ is relatively small and may be neglected. The maximum (theoretical) power, P_{max}, that can be extracted by the turbine, can be estimated as $P_{max} = \gamma Q h_a$.

EXAMPLE 8.14

A hydropower facility is being planned in which the water surface in the tailrace is 75 m below the water surface of the supply reservoir. The 1.70-m-diameter concrete-lined tunnel leading from the reservoir to the Francis turbine intake (i.e., the penstock) is 300 m long and has an estimated roughness height of 10 mm. Model studies indicate that when the flow rate through the system is 16 m³/s, the combined head loss in the turbine and draft tube is 2.0 m and the average velocity in the tailrace is 0.50 m/s. Estimate the maximum power that can be extracted from the system.

SOLUTION

From the given data: $H = 75$ m, $D = 1.70$ m, $L = 300$ m, $k_s = 10$ mm, $Q = 16$ m³/s, $h_{td} = 2.0$ m, and $V = 0.5$ m/s. The velocity ($= Q/A$) in the penstock is calculated as $V_p = 7.05$ m/s. Using the given values of Q, D, and k_s and assuming that the kinematic viscosity of water, ν, is 1.00×10^{-6} m²/s (at 20°C), the friction factor, f, of the penstock is calculated using the Swamee–Jain equation (Equation 7.45) as $f = 0.0319$. The available head, h_a, is given by Equation 8.96 as

$$h_a = H - \frac{fL}{D}\frac{V_p^2}{2g} - h_{td} - \frac{V^2}{2g} = 75 - \frac{(0.0319)(300)}{1.70}\frac{7.05^2}{2(9.81)} - 2.0 - \frac{0.5^2}{2(9.81)} = 58.72 \text{ m}$$

which gives

$$P_{max} = \gamma Q h_a = (9.79)(16)(58.72) = 9200 \text{ kW} = 9.2 \text{ MW}$$

Therefore, the system will extract a maximum of **9.2 MW** of power from the water flowing through the turbine. Because of frictional losses within the turbine and transmission inefficiencies, the actual amount of electrical power available for distribution will be less.

Pump turbines. *Pump turbines*, which are used at pumped-storage facilities, are very similar in design and construction to the Francis turbine. When water enters the rotor at the periphery and flows inward, the machine acts as a turbine. When water enters the center (or eye) and flows outward, the machine acts as a pump. The pump turbine is connected to a motor generator, which acts as a motor or a generator, depending on whether the pump or turbine mode is being used.

8.5.3 Practical Considerations

Aside from the operational characteristics of turbines covered in the previous sections, there are several practical considerations in selecting the appropriate turbine to use in any particular circumstance.

Turbine performance curves. The functional forms of the nondimensional performance relationships derived for pumps also apply to reaction turbines. Therefore, the performance of a homologous series of reaction turbines can be described by the following functional relationships:

$$\frac{gh_a}{\omega^2 D^2} = \Phi_1\left(\frac{Q}{\omega D^3}\right), \quad \frac{\dot{W}_{shaft}}{\rho\omega^3 D^5} = \Phi_2\left(\frac{Q}{\omega D^3}\right), \quad \eta = \Phi_3\left(\frac{Q}{\omega D^3}\right) \tag{8.97}$$

where Φ_i denotes functional relationships, h_a is the available head, ω is the angular speed of the runner, D is the characteristic size of the runner, Q is the volume flow rate through the turbine, g is the acceleration due to gravity, $\dot{W}_{shaft}$ is the shaft power delivered to the turbine, and η is the efficiency of the turbine, which is defined as

$$\eta = \frac{\dot{W}_{shaft}}{\rho g Q h_a} \tag{8.98}$$

A notable difference between pumps and turbines is that the efficiency of a turbine is defined as $\dot{W}_{shaft}/\rho g Q h_a$, compared with $\rho g Q h_p/\dot{W}_{shaft}$ for pumps.

Affinity laws and scale effects. In accordance with the functional relationships in Equation 8.97, if two geometrically similar turbines have the same value of $Q/\omega D^3$, then they must necessarily have the same values of $gh_a/\omega^2 D^2$, $\dot{W}_{shaft}/\rho\omega^3 D^5$, and η. This result is expressed in the following affinity laws:

$$\left[\frac{Q}{\omega D^3}\right]_1 = \left[\frac{Q}{\omega D^3}\right]_2 \rightarrow \begin{cases} \left[\dfrac{gh_a}{\omega^2 D^2}\right]_1 = \left[\dfrac{gh_a}{\omega^2 D^2}\right]_2 \\ \left[\dfrac{\dot{W}_{shaft}}{\rho\omega^3 D^5}\right]_1 = \left[\dfrac{\dot{W}_{shaft}}{\rho\omega^3 D^5}\right]_2 \\ \eta_1 = \eta_2 \end{cases} \tag{8.99}$$

where the subscripts 1 and 2 are corresponding operational states in geometrically similar turbines. Corresponding operational states are commonly assumed to have the same efficiency. However, in a similar manner to pumps, scale effects are also present in homologous turbines due primarily to viscous effects and inexact geometric scaling of roughness heights and tip clearances. As a consequence, larger turbines in a homologous series tend be more efficient than smaller turbines within the series. The scale effect on the rated efficiency can be approximated by (Stepanoff, 1957; Moody and Zowski, 1989)

$$\frac{1-\eta_2}{1-\eta_1} = \left(\frac{D_1}{D_2}\right)^{\frac{1}{5}} \tag{8.100}$$

where η_1 and η_2 are the rated efficiencies of homologous turbines of diameters D_1 and D_2, respectively. The exponent in Equation 8.100 is somewhat uncertain, with an exponent of $\frac{1}{4}$ sometimes being used. The effect of changes in rated flow rate on the rated efficiency can also be estimated using the relation (Anderson, 1980)

$$\frac{0.95-\eta_2}{0.95-\eta_1} = \left(\frac{Q_1}{Q_2}\right)^{0.32} \tag{8.101}$$

where Q_1 and Q_2 are corresponding rated flow rates. Equation 8.101 assumes a maximum possible turbine efficiency of 95%. Equations 8.100 and 8.101 are applicable to scaling rated efficiencies only. Also, Equations 8.100 and 8.101 are not applicable to impulse turbines because the efficiencies of impulse turbines are independent of size.

Specific speed and turbine selection. In a similar fashion to pumps, the operating point of greatest efficiency is defined by the dimensionless specific speed, n_s, where

$$n_s = \frac{\omega Q^{\frac{1}{2}}}{(gh_a)^{\frac{3}{4}}} \tag{8.102}$$

and any consistent set of units can be used. The specific speed of turbines is commonly expressed in the alternative form

$$n_s' = \frac{\omega\sqrt{\dot{W}_{shaft}/\rho}}{(gh_a)^{\frac{5}{4}}} \tag{8.103}$$

The specific speeds given by Equations 8.102 and 8.103 are related by

$$n_s' = \eta_h^{\frac{1}{2}} n_s \tag{8.104}$$

where η_h is the hydraulic efficiency of the turbine unit. In cases where the output power, P_{out}, of the turbine unit is used instead of $\dot{W}_{shaft}$ in Equation 8.103, the specific speed is called the *power specific speed*, n_P, such that

$$n_P = \frac{\omega\sqrt{P_{out}/\rho}}{(gh_a)^{\frac{5}{4}}} \tag{8.105}$$

For $n_s' < 0.2$, impulse turbines are typically most efficient; for $0.2 \leq n_s' < 2$, Francis turbines are most efficient; and for $2 \leq n_s' \leq 5$, propeller turbines are most efficient. These efficiencies indicate that high-head, low-flow installations work best with impulse turbines and that low-head, high-flow installations work best with propeller-type turbines. In cases where impulse turbines have multiple jets, the specific speed, n_s', is calculated using a shaft power equal to $\dot{W}_{shaft}/N_{jets}$, where $\dot{W}_{shaft}$ is the total power generated by the turbine and N_{jets} is the number of jets. Impulse turbines are commonly used for heads greater than 150–300 m, whereas the upper limit for using a Francis turbine is on the order of 450 m because of possible cavitation and the difficulty of building casings to withstand such high pressures. Run-of-river hydroelectric power plants typically use propeller-type turbines for power generation.

Peripheral velocity factor. The *peripheral velocity factor*, ϕ, can be defined for any type of turbine as

$$\phi = \frac{U}{\sqrt{2gh_a}} \tag{8.106}$$

where U is the tip speed of the runner and h_a is the available head. Although values of ϕ vary with the operating state of a turbine, the value of ϕ at the best efficiency point, denoted by ϕ^*, falls within a fairly narrow range for each type of turbine. Noting that $U = (D/2)\omega$, at the best efficiency point, Equation 8.106 can be rearranged and put in the convenient form

$$D = \frac{2\phi^*\sqrt{2gh_a}}{\omega} \tag{8.107}$$

where D is the diameter of the runner. Typical ranges of ϕ^* for impulse, Francis, and propeller turbines are as follows:

Type	ϕ^*
Impulse turbine	0.43–0.48
Francis turbine	0.70–0.80
Propeller turbine	1.4–2.0

Using the typical values of ϕ^* in combination with Equation 8.107 is particularly useful in estimating the required size of a turbine runner based on the available head and the rotary speed of the runner.

EXAMPLE 8.15

A small run-of-river hydropower installation has an available head of 7.5 m, and the flow that can be consistently channeled through the turbine is 0.8 m^3/s. A generator with a shaft rotation speed of 720 rpm is feasible at this facility, and it is anticipated that the optimum turbine will have a maximum efficiency of 94%. The average water temperature throughout the year is 10°C. What type of turbine should be used? Estimate the size of the turbine runner required.

SOLUTION

From the given data: $h_a = 7.5$ m, $Q = 0.8$ m^3/s, $\omega = 720$ rpm $= 75.40$ rad/s, and $\eta = 0.94$. For water at 10°C, $\rho = 999.7$ kg/m^3 and $\gamma = 9804$ N/m^3. Using the given data, the specific speed of the turbine, n_s', is derived as follows:

$$\dot{W}_{shaft} = \eta\gamma Q h_a = (0.94)(9804)(0.8)(7.5) \quad \rightarrow \quad \dot{W}_{shaft} = 5.530 \times 10^4 \text{ W}$$

$$n_s' = \frac{\omega\sqrt{\dot{W}_{shaft}/\rho}}{[gh_a]^{\frac{5}{4}}} = \frac{(75.40)\sqrt{5.530 \times 10^4/999.7}}{[(9.807)(7.5)]^{\frac{5}{4}}} \quad \rightarrow \quad n_s' = 2.60$$

Because the specific speed of 2.60 is in the range $2 \leq n_s' \leq 5$, a **propeller-type turbine** would be most efficient for this installation. Also, it is anticipated that the turbine will be able to supply approximately 55.3 kW of power.

For a propeller-type turbine, the peripheral velocity factor, ϕ^*, is in the range of 1.4–1.7. Therefore, the corresponding size range of the runner can be estimated using Equation 8.107 as follows:

$$\phi^* = 1.4 \quad \rightarrow \quad D = \frac{2(1.4)\sqrt{2(9.807)(7.5)}}{75.40} = 0.45 \text{ m}$$

$$\phi^* = 1.7 \quad \rightarrow \quad D = \frac{2(2.0)\sqrt{2(9.807)(7.5)}}{75.40} = 0.64 \text{ m}$$

Therefore, the required runner size is expected to be in the range of **0.45–0.64 m**.

Rotational speed. Turbines are generally operated at constant speed. In the United States, 60 Hz electric current is commonly used, whereas in Europe, 50 Hz is more common. The speed of rotation, ω (rpm), of a turbine is close to the synchronous speed of the attached generator, which is related to frequency of the current generated and the number of pairs of poles by the relation

$$\text{Synchronous speed (rpm)} = \frac{60\,\Omega}{\text{no. of pairs of poles}} \tag{8.108}$$

where Ω is the frequency of the generated alternating current. Large hydroturbines typically operate at relatively low speeds such as 120–150 rpm.

Net positive suction head (NPSH). A minimum net positive suction head, NPSH_{min}, must be maintained in turbine systems to avoid cavitation. The location where cavitation is most likely to occur in a reaction turbine, such as the Francis turbine, is at the exit of the runner where the water enters the draft tube. The NPSH is defined as

$$\text{NPSH} = \frac{p_\text{d}}{\gamma} + \frac{V_\text{d}^2}{2g} - \frac{p_\text{v}}{\gamma} \tag{8.109}$$

where p_d and V_d are the pressure and velocity, respectively, of the water exiting the turbine and entering the draft tube, p_v is the saturation vapor pressure of water, and γ is the specific weight of water. The minimum required net positive suction head, NPSH_{min}, for any given turbine is generally specified by the turbine manufacturer. Values of NPSH_{min} are typically a function of the available head, h_a, and this relationship is commonly expressed in terms of the *critical cavitation parameter*, or *critical cavitation number*, σ_c, where

$$\text{NPSH}_{\text{min}} = \sigma_\text{c} h_\text{a} \tag{8.110}$$

The minimum net positive suction head, NPSH_{min}, generally determines the maximum height above the tailwater that a hydraulic turbine can be placed. This can be seen by applying the energy equation between the runner exit and the tailwater, which gives

$$\frac{p_\text{d}}{\gamma} + \frac{V_\text{d}^2}{2g} + z_\text{d} - h_\ell = \frac{p_{\text{atm}}}{\gamma} \tag{8.111}$$

where h_ℓ is the head loss between the runner exit and the tailrace. Combining Equations 8.109 and 8.111 and noting that $\text{NPSH} = \text{NPSH}_{\text{min}}$ when $z_\text{d} = z_{\text{max}}$ yields

$$z_{\text{max}} = \frac{p_{\text{atm}} - p_\text{v}}{\gamma} - h_\ell - \text{NPSH}_{\text{min}} \tag{8.112}$$

EXAMPLE 8.16

Under the design condition, a Francis turbine generates a shaft power of 10 MW when the flow rate through the turbine is 6.5 m^3/s, the available head is 170 m, and the rate of rotation of the runner is 600 rpm. The minimum net positive suction head at the design condition is given by the manufacturer as 3.06 m. The draft tube between the runner exit and the tailwater has an average diameter of 600 mm, and the estimated roughness of the draft tube is 1 mm. The turbine is operated at sea level under standard atmospheric conditions, and the water temperature is 20°C. What is the maximum allowable vertical distance between the runner and the tailwater?

SOLUTION

From the given data: $\dot{W}_{\text{shaft}} = 10$ MW, $Q = 6.5$ m^3/s, $h_a = 170$ m, $\omega = 600$ rpm $= 62.83$ rad/s, NPSH$_{\text{min}} = 3.06$ m, $D = 600$ mm, and $k_s = 1$ mm. For water at 20°C: $\gamma = 9.789$ kN/m^3, $\nu = 1.004 \times 10^{-6}$ m^2/s, and $p_v = 2.337$ kPa. Standard atmospheric pressure is $p_{\text{atm}} = 101.3$ kPa. The following preliminary calculations are useful:

$$A = \frac{\pi D^2}{4} = \frac{\pi(0.6)^2}{4} = 0.2827 \text{ m}^2, \qquad V = \frac{Q}{A} = \frac{6.5}{0.2827} = 22.99 \text{ m/s}$$

$$\text{Re} = \frac{VD}{\nu} = \frac{(22.99)(0.6)}{1.004 \times 10^{-6}} = 1.374 \times 10^7, \qquad \frac{k_s}{D} = \frac{1}{600} = 1.667 \times 10^{-3}$$

$$f = f_{\text{CE}}\left(\text{Re}, \frac{k_s}{D}\right) = 0.0223$$

where $f_{\text{CE}}(\text{Re}, k_s/D)$ is the friction factor given by the Colebrook equation. Applying Equation 8.112 to calculate z_{max}, with the Darcy–Weisbach equation used to calculate the head loss, gives

$$z_{\text{max}} = \frac{p_{\text{atm}} - p_v}{\gamma} - \frac{f z_{\text{max}}}{D}\frac{V^2}{2g} - \text{NPSH}_{\text{min}} \quad \rightarrow$$

$$z_{\text{max}} = \left[\frac{p_{\text{atm}} - p_v}{\gamma} - \text{NPSH}_{\text{min}}\right]\left[1 + \frac{fV^2}{2gD}\right]^{-1}$$

Substituting the given data yields

$$z_{\text{max}} = \left[\frac{101.3 - 2.337}{9.789} - 3.06\right]\left[1 + \frac{(0.0223)(22.99)^2}{2(9.807)(0.6)}\right]^{-1} = 3.52 \text{ m}$$

Therefore, to avoid cavitation problems at the runner exit and at the entrance to the draft tube, the runner should be at an elevation **3.52 m or less** above the tailwater elevation. This analysis neglects the friction loss in the portion of the draft tube submerged below the tailwater elevation. Taking this additional friction loss into account would decrease the allowable elevation.

Power-generating capacity. The capacity of a hydroelectric generating plant is defined as the maximum rate at which the plant can produce electricity. The *installed capacity* or *hydropower generation potential* for a given site is estimated by the relation

$$P = \gamma Q h_a \eta \tag{8.113}$$

where P is the hydropower generation potential, γ is the specific weight of water, Q is the volume flow rate through the plant, h_a is the available head, and η is the turbine efficiency, which is usually in the range of 0.80–0.90. To account for varying flow rates through the plant, a value of Q, which is exceeded 10%–30% of the time, may be selected to estimate the installed capacity of the facility. Several values of Q and the corresponding h_a may be examined to estimate the values that result in optimum installed capacity based on water use and economic considerations. For preliminary planning, optimum installed capacity may be that beyond which relatively large increases in Q are required to obtain relatively small increases in P. It is considered good practice to have at least two turbines at an installation so that the facility can continue operation while one of the turbines is shut down for repairs, maintenance, or inspection or at times of low power demand. Most large hydroelectric facilities have several turbines arranged in parallel. For example, the Hoover Dam has 17 turbines in parallel, of which 15 are identical Francis turbines, each capable of producing 150 MW of electricity. The difference in elevation between the headwater and tailwater at the Hoover dam is 180 m, and this facility is capable of producing 2000 MW (= 2 GW) at peak capacity. At the other end of the scale, *microhydro plants* are hydropower facilities with capacities less than 100 kW, and *minihydro plants* have capacities in the range of 100–1000 kW (i.e., 0.1–1 MW).

Feasibility analyses. To estimate the annual hydropower generation potential of a facility, a reservoir and power plant operation study must be conducted with sequences of historical flows and corresponding available heads for relatively long periods of time—for example, 10–50 years or more. For these estimates, the overall water-to-wire efficiency should be used, which typically varies in the range of 70%–85% and includes efficiencies of the turbine, generator, transformers, and other equipment. In addition, adjustments to the efficiency may be made for tailwater fluctuations and unscheduled down time.

EXAMPLE 8.17

A hydroelectric project is to be developed along a river where the 90 percentile flow rate is 2240 m^3/s and a Rippl analysis indicates that storage to a height of 30 m above the downstream stage is required to meet all water-supply demands. Estimate the installed capacity appropriate for these conditions.

SOLUTION

Assuming that the capacity of the turbines will be sufficient to pass a flow rate, Q, of 2240 m^3/s at a head of 30 m and taking the turbine efficiency, η, as 0.85 and $\gamma = 9.79$ kN/m^3 yields

$$P = \gamma Q h_a \eta = (9.79)(2240)(30)(0.85) = 5.59 \times 10^5 \text{ kW} = 559 \text{ MW}$$

Therefore, to fully utilize the available head and anticipated flow rates, an installed capacity of **559 MW** is required. This analysis neglects all hydraulic head losses in the system, so the estimated capacity is an upper bound.

Key Equations for Turbomachines

The following list of equations is particularly useful in solving problems related to turbomachines. If one is able to recognize these equations and recall their appropriate use, then the learning objectives of this chapter have been met to a significant degree. Derivations of these equations, definitions of the variables, and detailed examples of usage can be found in the main text.

MECHANICS OF TURBOMACHINES

Shaft torque (pump): $T_{\text{shaft}} = \dot{m}\,(r_2 V_{2\theta} - r_1 V_{1\theta})$

Shaft power (pump): $\dot{W}_{\text{shaft}} = \dot{m}\,\omega\,(r_2 V_{2\theta} - r_1 V_{1\theta})$

Energy added per unit mass (pump): $w_{\text{shaft}} = \omega\,(r_2 V_{2\theta} - r_1 V_{1\theta}), \quad w_{\text{shaft}} = U_2 V_{2\theta} - U_1 V_{1\theta}$

Head added (pump):

$$h_{\text{p}} = \left(\frac{p}{\gamma} + \frac{V^2}{2g} + z\right)_2 - \left(\frac{p}{\gamma} + \frac{V^2}{2g} + z\right)_1$$

$$h_{\text{p}} = \frac{U_2 V_{2\theta} - U_1 V_{1\theta}}{g} - h_\ell$$

$$h_{\text{p}} = \frac{U_2^2}{g} - \frac{U_2 \cot \beta_2}{2\pi r_2 b_2 g} Q - h_\ell(Q)$$

$$h_{\text{p}} = h_0 - f(Q)$$

HYDRAULIC PUMPS AND PUMPED SYSTEMS

Power added (pump): $P_{\text{f}} = \gamma Q h_{\text{p}}$

Brake horsepower: $\text{BHP} = T_{\text{shaft}}\omega$

Efficiency (pump): $$\eta_{\text{p}} = \frac{\gamma Q h_{\text{p}}}{\text{BHP}} = \frac{\gamma Q h_{\text{p}}}{\dot{W}_{\text{shaft}}} = \frac{\gamma Q h_{\text{p}}}{T_{\text{shaft}}\omega}$$

Efficiency (motor): $$\eta_{\text{m}} = \frac{\text{BHP}}{P_{\text{m}}} = \frac{T_{\text{shaft}}\omega}{P_{\text{m}}}$$

Efficiency (overall): $\eta_{\text{overall}} = \eta_{\text{p}} \cdot \eta_{\text{m}}$

Affinity laws (pumps):

$$\left[\frac{Q}{\omega D^3}\right]_1 = \left[\frac{Q}{\omega D^3}\right]_2 \rightarrow \begin{cases} \left[\dfrac{g h_{\text{p}}}{\omega^2 D^2}\right]_1 = \left[\dfrac{g h_{\text{p}}}{\omega^2 D^2}\right]_2 \\[2ex] \left[\dfrac{\text{BHP}}{\rho \omega^3 D^5}\right]_1 = \left[\dfrac{\text{BHP}}{\rho \omega^3 D^5}\right]_2 \\[2ex] \eta_1 = \eta_2 \end{cases}$$

Flow coefficient (pump): $C_Q = \dfrac{Q}{\omega D^3}$

Head coefficient (pump): $C_H = \dfrac{g h_p}{\omega^2 D^2}$

Power coefficient (pump): $C_P = \dfrac{\text{BHP}}{\rho \omega^3 D^5}$

Scale effect, efficiency (pump): $\dfrac{1-\eta_2}{1-\eta_1} = \left(\dfrac{D_1}{D_2}\right)^{\frac{1}{4}}, \quad \dfrac{0.94-\eta_2}{0.94-\eta_1} = \left(\dfrac{Q_1}{Q_2}\right)^{0.32}$

Specific speed, dimensionless (pump): $n_s = \dfrac{\omega Q^{\frac{1}{2}}}{(g h_p)^{\frac{3}{4}}}$

Specific speed, dimensional (pump): $N_s = \dfrac{\omega Q^{\frac{1}{2}}}{h_p^{\frac{3}{4}}}$

Synchronous speed: $\dfrac{60\,\Omega}{\text{no. of pairs of poles}}$

Brake horsepower: $\text{BHP} = \dfrac{\gamma Q h_p}{\eta} \quad (= \dot{W}_{\text{shaft}})$

Performance curve (pump): $h_p = h_0 - \left(\dfrac{h_0 - h^*}{Q^{*2}}\right) Q^2$

System curve (pump): $h_p = \Delta z + Q^2 \left[\sum \dfrac{fL}{2gA^2 D} + \sum \dfrac{K_{\text{local}}}{2gA^2}\right]$

Net positive suction head, available: $\text{NPSH}_A = \dfrac{p_s}{\gamma} + \dfrac{V_s^2}{2g} - \dfrac{p_v}{\gamma}$

$\text{NPSH}_A = \dfrac{p_0}{\gamma} - \Delta z_s - h_\ell - \dfrac{p_v}{\gamma}$

Net positive suction head, required: $\text{NPSH}_R = \sigma_c h_p$

Suction lift, maximum: $z_{\max} = \dfrac{p_0 - p_v}{\gamma} - h_\ell - \text{NPSH}_R$

Net positive suction head, scaling: $\left[\dfrac{g(\text{NPSH}_R)}{\omega^2 D^2}\right]_1 = \left[\dfrac{g(\text{NPSH}_R)}{\omega^2 D^2}\right]_2$

when $\left[\dfrac{Q}{\omega D^3}\right]_1 = \left[\dfrac{Q}{\omega D^2}\right]_2$

Specific speed, suction (pump): $S = \dfrac{\omega Q^{\frac{1}{2}}}{(g\,\text{NPSH}_R)^{\frac{3}{4}}}$

Homologous points (pump): $Q_2 = Q_1 \left(\dfrac{\omega_2}{\omega_1}\right)$ and $h_2 = h_1 \left(\dfrac{\omega_2}{\omega_1}\right)^2$

FANS

Affinity laws (fans):

$$\left[\frac{Q}{\omega D^3}\right]_1 = \left[\frac{Q}{\omega D^3}\right]_2 \rightarrow \begin{cases} \left[\frac{\Delta p}{\rho\omega^2 D^2}\right]_1 = \left[\frac{\Delta p}{\rho\omega^2 D^2}\right]_2 \\ \left[\frac{P}{\rho\omega^3 D^5}\right]_1 = \left[\frac{P}{\rho\omega^3 D^5}\right]_2 \\ \eta_1 = \eta_2 \end{cases}$$

Specific speed, dimensionless (fan): $n_s = \dfrac{\omega Q^{\frac{1}{2}}\rho^{\frac{3}{4}}}{(\Delta p)^{\frac{3}{4}}}$

IMPULSE TURBINES

Shaft torque: $T_{\text{shaft}} = \dot{m}\overline{R}(V_{2\theta} - V_{1\theta})$

Shaft power: $\dot{W}_{\text{shaft}} = \dot{m}\,U(V_j - U)(\cos\beta - 1)$

Bucket speed, optimum: $U = \frac{1}{2}V_j$

Shaft power, maximum: $\dot{W}_{\text{shaft}}\Big|_{\text{max}} = \frac{1}{4}\dot{m}V_j^2(\cos\beta - 1)$

Efficiency, wheel: $\eta_w = \dfrac{|\dot{W}_{\text{shaft}}|}{\frac{1}{2}\dot{m}V_j^2}, \quad \eta_{w,\text{theory}} = \dfrac{2U(V_j - U)(1 - \cos\beta)}{V_j^2}$

Efficiency, turbine: $\eta_t = \dfrac{|\dot{W}_{\text{shaft}}|}{\gamma Q h_e}, \quad \eta_t\Big|_{\text{max}} = \frac{1}{2}C_v^2(1 - \cos\beta)$

Head available to runner: $h_t = h_e - k_j\dfrac{V_j^2}{2g} - k\dfrac{v_2^2}{2g} - \dfrac{V_2^2}{2g}$

Power, theoretical: $P_t = \gamma Q h_t$

Efficiency, hydraulic: $\eta_t = \dfrac{\gamma Q h_t}{\gamma Q h_e} = \dfrac{h_t}{h_e}$

Efficiency, overall: $\eta = \dfrac{T_{\text{shaft}}\omega}{\gamma Q h_e}$

REACTION TURBINES

Shaft torque: $T_{\text{shaft}} = \dot{m}\,(r_2 V_{2\theta} - r_1 V_{1\theta})$

Shaft power: $\dot{W}_{\text{shaft}} = T_{\text{shaft}}\omega = \dot{m}\,\omega\,(r_2 V_{2\theta} - r_1 V_{1\theta})$

Efficiency, hydraulic: $\eta_h = \dfrac{\omega\,(r_1 V_{1\theta} - r_2 V_{2\theta})}{g h_a}$

Efficiency, overall: $\eta = \dfrac{P}{\dot{m} g h_\mathrm{a}}$

Head extracted by turbine: $h_\mathrm{a} = H - h_\ell - h_\mathrm{td} - \dfrac{V^2}{2g}$

Turbine performance curve: $\dfrac{g h_\mathrm{a}}{\omega^2 D^2} = \Phi_1\left(\dfrac{Q}{\omega D^3}\right), \quad \dfrac{\dot{W}_\mathrm{shaft}}{\rho \omega^3 D^5} = \Phi_2\left(\dfrac{Q}{\omega D^3}\right),$

$\eta = \Phi_3\left(\dfrac{Q}{\omega D^3}\right)$

Affinity laws:

$$\left[\frac{Q}{\omega D^3}\right]_1 = \left[\frac{Q}{\omega D^3}\right]_2 \rightarrow \begin{cases} \left[\dfrac{g h_\mathrm{a}}{\omega^2 D^2}\right]_1 = \left[\dfrac{g h_\mathrm{a}}{\omega^2 D^2}\right]_2 \\ \left[\dfrac{\dot{W}_\mathrm{shaft}}{\rho \omega^3 D^5}\right]_1 = \left[\dfrac{\dot{W}_\mathrm{shaft}}{\rho \omega^3 D^5}\right]_2 \\ \eta_1 = \eta_2 \end{cases}$$

Scale effect: $\dfrac{1 - \eta_2}{1 - \eta_1} = \left(\dfrac{D_1}{D_2}\right)^{\frac{1}{5}}, \quad \dfrac{0.95 - \eta_2}{0.95 - \eta_1} = \left(\dfrac{Q_1}{Q_2}\right)^{0.32}$

Specific speed, dimensionless: $n_\mathrm{s} = \dfrac{\omega Q^{\frac{1}{2}}}{(g h_\mathrm{a})^{\frac{3}{4}}}, \quad n'_\mathrm{s} = \dfrac{\omega \sqrt{\dot{W}_\mathrm{shaft}/\rho}}{(g h_\mathrm{a})^{\frac{5}{4}}}$

Diameter of runner, optimal: $D = \dfrac{2\phi^* \sqrt{2 g h_\mathrm{a}}}{\omega}$

Synchronous speed: $\dfrac{60\,\Omega}{\text{no. of pairs of poles}}$

Net positive suction head, definition: $\mathrm{NPSH} = \dfrac{p_\mathrm{d}}{\gamma} + \dfrac{V_\mathrm{d}^2}{2g} - \dfrac{p_\mathrm{v}}{\gamma}$

Net positive suction head, minimum: $\mathrm{NPSH}_\mathrm{min} = \sigma_\mathrm{c} h_\mathrm{a}$

Maximum discharge height: $z_\mathrm{max} = \dfrac{p_\mathrm{atm} - p_\mathrm{v}}{\gamma} - h_\ell - \mathrm{NPSH}_\mathrm{min}$

Hydropower generation potential: $P = \gamma Q h_\mathrm{a} \eta$

PROBLEMS

Section 8.2: Mechanics of Turbomachines

8.1. Water at 20°C flows through the pump impeller shown in Figure 8.33, where the inner and outer diameters of the impeller are 200 mm and 500 mm, respectively, and the width of the pump blades is 60 mm. The impeller rotates at 4000 rpm. Under a particular operating condition, water enters in a direction normal to the inflow surface, the component of the outflow velocity normal to the outflow surface, V_r, is equal to 25 m/s, and the magnitude of the absolute velocity, $\mathbf{V}$, at the outflow surface is 40 m/s. Estimate the shaft power required to drive the pump at this operating condition.

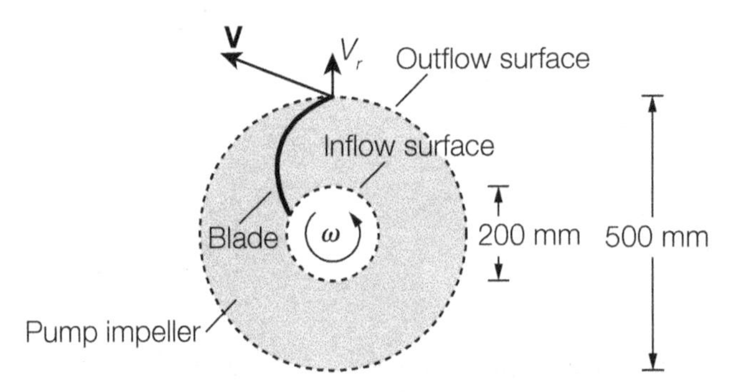

Figure 8.33: Flow through a pump impeller I

8.2. A centrifugal blower has a 1000-mm-diameter impeller with a width of 50 mm that rotates at a rate of 1800 rpm. When the flow rate through the impeller is 0.25 m^3/s, the inflow velocity is normal to the inflow surface and the outflow velocity makes an angle of 40° with the radial direction as shown in Figure 8.34. Assuming that the blower is 100% efficient, estimate the energy per unit mass added to the air as it passes through the blower.

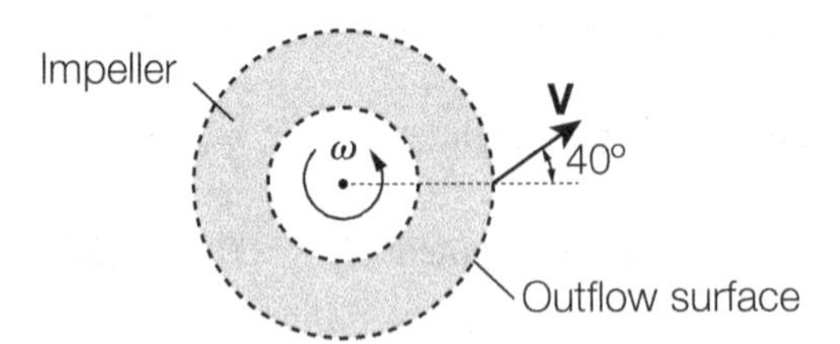

Figure 8.34: Flow through a blower

8.3. A water pump impeller has radial blades as shown in Figure 8.35. The inner and outer diameters of the impeller are 60 mm and 150 mm, respectively. When the rotational speed of the impeller is 3250 rpm, the pump delivers 12.5 L/s and the speed of the water relative to the blade at the exit is equal to 4 m/s. Determine (a) the width of the impeller at the outflow surface and (b) the power required to drive the pump. Assume water at 20°C.

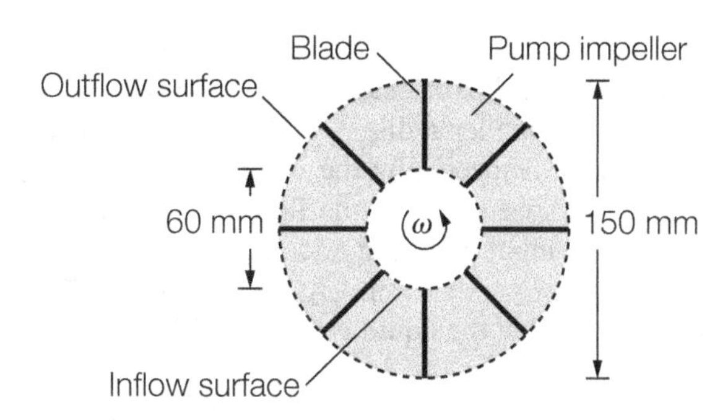

Figure 8.35: Flow through a pump impeller II

8.4. A pump impeller has an outer diameter of 400 mm and a blade width of 50 mm, and it rotates at 800 rpm. At a particular operating condition, the inflow is normal to the inflow surface and the (absolute) outflow velocity has a magnitude of 40 m/s and makes an angle of 40° with the normal to the outflow surface, as shown in Figure 8.36. Estimate the shaft power required to drive the pump under this condition. Assume water at 20°C.

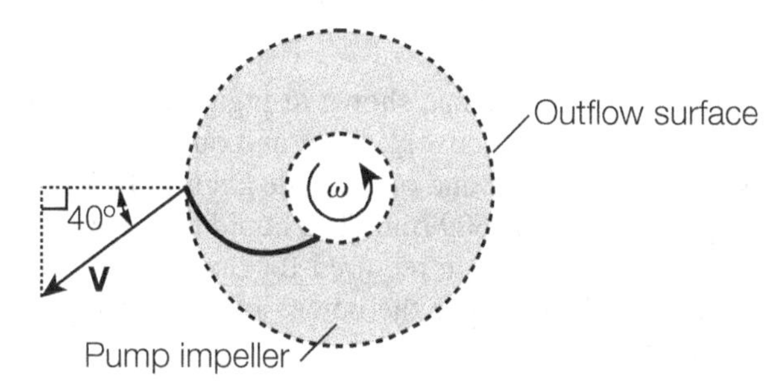

Figure 8.36: Flow through a pump impeller III

8.5. The pump impeller shown in Figure 8.37 has a diameter of 325 mm, a blade width of 30 mm, and an exit blade angle of 42°; it rotates at 800 rpm. Water at 20°C flows through the pump at a flow rate of 75 L/s, and this flow enters the impeller in a direction normal to the inflow surface. Estimate the required shaft power corresponding to the design flow rate.

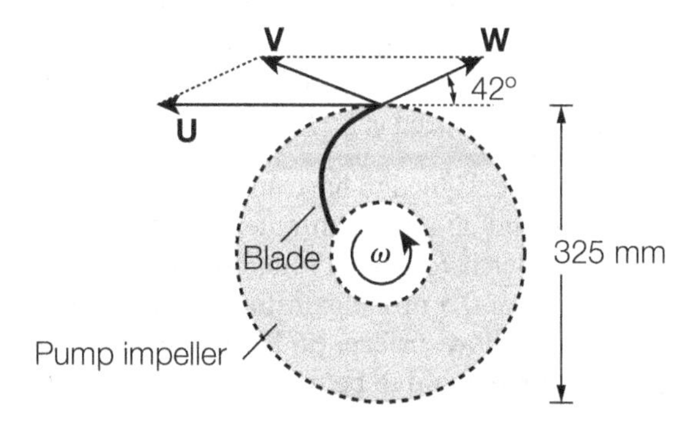

Figure 8.37: Flow through a pump impeller IV

8.6. Consider the general case of flow through a pump impeller as illustrated in Figure 8.38. (a) Develop a general expression for the power input to the pump in terms of the inner and outer impeller radii, r_1 and r_2, respectively; the inner and outer blade widths, b_1 and b_2, respectively; the inner and outer blade angles, β_1 and β_2, respectively; rotational speed, ω; volume flow rate, Q; and density of the fluid, ρ. (b) Estimate the power input and head added to the pump for $r_1 = 200$ mm, $r_2 = 600$ mm, $b_1 = 60$ mm, $b_2 = 40$ mm, $\beta_1 = 70°$, $\beta_2 = 75°$, $\omega = 700$ rpm, and $Q = 0.90$ m^3/s. Assume that the liquid is water at 20°C.

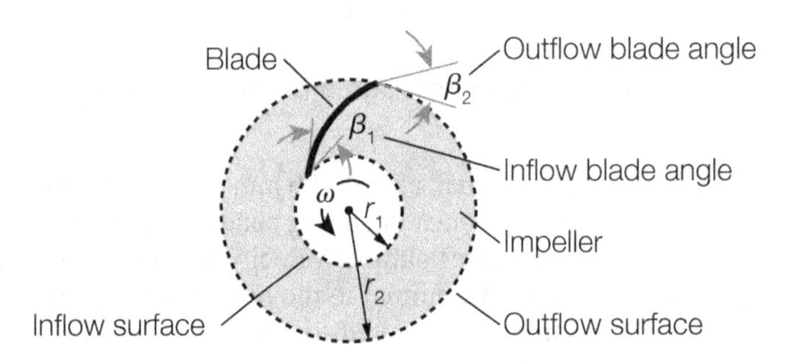

Figure 8.38: Flow through a pump impeller V

8.7. A pump impeller similar to the one shown in Figure 8.38 has inner and outer radii of 200 mm and 400 mm, respectively; inner and outer blade widths of 40 mm and 35 mm, respectively; and inner and outer blade angles of 25° and 10°, respectively. The rate of angular rotation is 600 rpm. The liquid being pumped is water at 20°C. (a) Estimate the volume flow rate required for the entrance velocity to be normal to the inflow surface. (b) Estimate the power added by the pump for the flow rate calculated in part (a).

8.8. A pump impeller has inner and outer radii of 120 mm and 280 mm, respectively; inner and outer blade widths of 30 mm and 25 mm, respectively; and inner and outer blade angles of 50° and 60°, respectively. The flow rate is 250 L/s. The liquid being pumped is water at 20°C. (a) Estimate the angular speed required for the entrance velocity to be normal to the inflow surface. (b) Estimate the power added by the pump for the angular speed calculated in part (a).

8.9. A pump impeller has inner and outer radii of 200 mm and 350 mm, respectively, and inner and outer blade widths of 60 mm and 50 mm, respectively; the outer blade angle is 70°. The flow rate is 500 L/s, and the rotational speed is 600 rpm. The liquid being pumped is water at 20°C. (a) Estimate the inner blade angle required for the entrance velocity to be normal to the inflow surface. (b) Estimate the power added by the pump for the angular speed calculated in part (a).

8.10. A pump impeller is to be designed to have an inner and outer radius of 150 mm and 300 mm, respectively, and an inner and outer blade width of 90 mm and 70 mm, respectively. The impeller is to have a rotational speed of 1725 rpm and is to deliver a volume flow rate of 0.3 m^3/s against a head of 13.8 m. The inflow velocity should be normal to the inflow surface. (a) Design the blade angles at the inflow and outflow surfaces. (b) If the liquid to be pumped is water at 20°C, what is the power requirement of the pump?

8.11. A pump delivers 125 L/s of water when the rotational speed of the impeller is 650 rpm. The inflow velocity is normal to the inflow surface, the magnitude of the outflow velocity relative to the rotating blade is 6 m/s, and the blade angle at the outflow surface is 90°. The motor consumes 10 kW of power, and it is estimated that the pump motor has an efficiency of 80%. Estimate the radius and width of the outflow surface. Assume water at 20°C.

8.12. A centrifugal pump uses 9 kW of power and has an efficiency of 70%. The pump delivers gasoline at a rate of 5 L/s. Estimate the maximum change in pressure between the inlet and outlet of the pump. How would the estimated pressure change be different if a liquid with higher density was used?

8.13. A pump is operated at a rotational speed of 2500 rpm, and has an efficiency of 80% at its operating condition. The pump is driven by a motor that has an efficiency of 90% and delivers a shaft torque of 10 N·m. The discharge side of the pump is at an elevation 3 m higher than the suction side of the pump. Measurements taken when the pump is delivering 7.5 L/s show pressure on the suction side and velocity of 92 kPa (gauge) and 1.8 m/s, respectively, and the velocity on the discharge side is 4.8 m/s. Estimate the power delivered to the motor and the pressure on the discharge side of the pump. Assume water at 20°C.

8.14. The blade of a turbine is illustrated in Figure 8.39, where the relative inflow velocity makes an angle of 70° with the inflow surface and the relative outflow velocity makes an angle of 60° with the outflow surface. Relative velocities are measured relative to the tip velocities of the blades mounted on the rotor. The inflow surface has a radius of 1.75 m, the outflow surface has a radius of 1.05 m, and the width of the rotor is 0.5 m. Under design conditions, the flow rate through the turbine is 35 m^3/s and the rotational speed is 150 rpm. Estimate the shaft power extracted by the turbine. Assume water at 20°C.

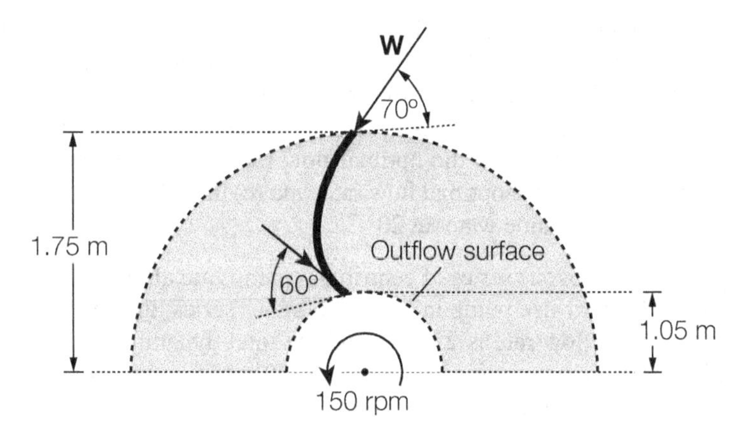

Figure 8.39: Flow through a turbine I

8.15. Water at 20°C flows through a turbine at a rate of 400 L/s, and the velocities on the inflow and outflow surfaces of the rotor are shown in Figure 8.40. The absolute velocity and the velocity relative to the rotating blade are represented by **V** and **W**, respectively, and the subscripts 1 and 2 indicate conditions at the inflow and outflow surfaces, respectively. On the inflow surface, the incoming flow makes an angle of 25° with the tangent to the surface, and the velocity relative to the rotor should make an angle of 55° with the inflow surface to match the rotor angle. On the outflow surface, it is assumed that the velocity relative to the moving blade will make an angle of 50° with the outflow surface so that it matches the inner blade angle. The inflow and outflow surfaces have radii of 800 mm and 400 mm, respectively, and the rotor rotates at 130 rpm. (a) Determine the required width of the rotor. (b) Estimate the shaft power generated by the turbine.

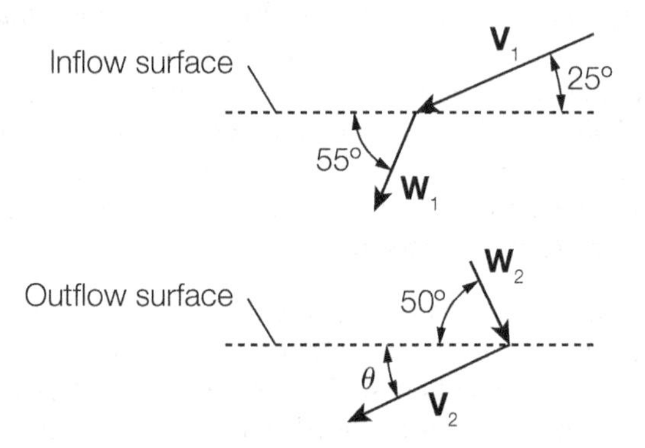

Figure 8.40: Flow through a turbine II

Section 8.3: Hydraulic Pumps and Pumped Systems

8.16. A proposed pump design has an impeller with inner and outer radii of 125 mm and 225 mm, respectively; widths of inflow and outflow areas of 50 mm and 45 mm, respectively; and blade angles at the inflow and outflow surfaces of 35° and 15°, respectively. The impeller rotates at 1140 rpm, and the estimated efficiency of the pump is 85%. Estimate (a) the optimal flow rate through the pump, (b) the head added by the pump at the optimal flow rate, and (c) the estimated power requirement to run the pump. Assume water at 20°C.

8.17. Pumps in a homologous series of centrifugal pumps are all driven by 1200-rpm motors. For a 500-mm size pump in the homologous series, the best efficiency of 81% occurs when the flow rate is 250 L/s and the total dynamic head is 63.7 m. What is the best efficiency operating point for a 250-mm size pump in the homologous series? Estimate the efficiency of the smaller pump.

8.18. Pump manufacturers typically present the performance characteristics of their pumps in dimensional form as shown in Figure 8.41, where the flow rate is in L/min and the head added by the pump is in m. The efficiencies in percent at various operational states are represented by dashed lines on the performance curve. The performance curves shown in Figure 8.41 are for homologous pumps with a motor speed of 3500 rpm and impeller sizes of 178 mm, 165 mm, 152 mm, 140 mm, and 127 mm. (a) Plot on a single graph the nondimensional performance curves for each of the pumps. (b) Compare the nondimensional performance curves and assess whether these pumps are from a homologous series. (c) Identify the maximum efficiency and specific speed for each impeller size.

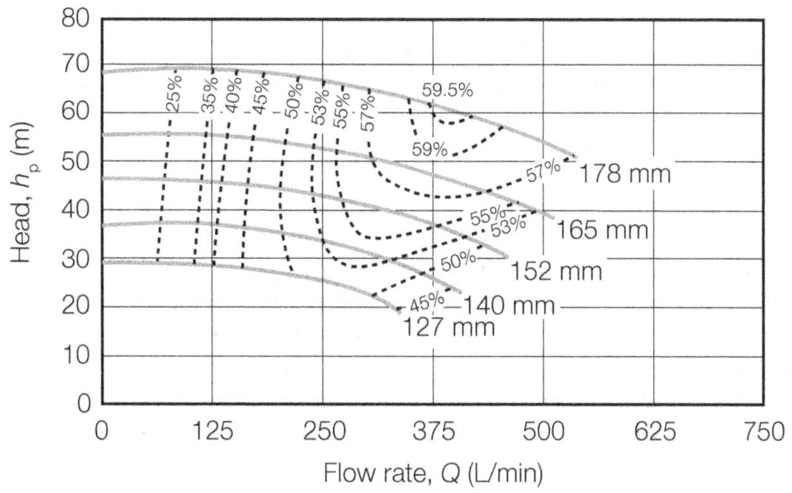

Figure 8.41: Performance curves of a homologous pump series

8.19. Water at 15°C is pumped at a rate of 20 L/s using a 5 kW pump. If the efficiency of the pump is 80%, what is the head added to the water as it passes through the pump?

8.20. A Model X pump with an impeller diameter of 300 mm and rotational speed of 1800 rpm has its maximum efficiency at a flow rate of 300 L/s, at which point the head added by the pump is 65 m and the brake horsepower is 240 kW. A geometrically similar Model X1 pump with an impeller size of 250 mm is driven by a 1400-rpm motor. The working fluid is water at 20°C. (a) Determine the flow coefficient, head coefficient, and power coefficient of the Model X pump when it is operating at its most efficient point. (b) What is the flow rate, head added, and brake horsepower of the Model X1 pump when it is operated at its most efficient point?

8.21. A pump with a rotary speed of 1725 rpm delivers 25 L/s at its most efficient operating point. Under this condition, the inflow velocity is normal to the inflow surface of the impeller, the component of the velocity normal to the outflow surface of the impeller is 4 m/s, and the efficiency of the pump is 80%. The width of the impeller at the outflow surface is 15 mm, and the blade angle at the outflow surface is 50°. (a) Estimate the head added by the pump. (b) Use the affinity laws to estimate the head added and the flow rate delivered by the pump when the rotational speed is changed to 1140 rpm.

8.22. A pump manufacturer makes three homologous series of pumps with three different specific speeds. The manufacturer documents the performance of each homologous series by nondimensional functional relationships. An engineer identifies the desired homologous series for a particular project as the one that has a best efficiency point with a flow coefficient of 0.035, a head coefficient of 0.14, and a power coefficient of 0.006. For a particular pump with an impeller diameter of 600 mm and a rated rotational speed of 1140 rpm, determine the rated discharge, total dynamic head, brake horsepower, and efficiency of the pump. Assume water at 20°C.

8.23. A pump is to be chosen from a homologous series that has a best efficiency point with a flow coefficient of 0.04, a head coefficient of 0.15, and a power coefficient of 0.0065. The desired pump is to have a best efficiency flow rate of 250 L/s and a motor with a speed of 900 rpm. What is the impeller size of the required pump? Assume water at 20°C.

8.24. An existing pump has a 500-mm-diameter impeller, and under conditions of maximum efficiency, the efficiency of the pump is 80%, the flow rate is 250 L/s, the head added is 25 m, and the power consumed by the pump is 55 kW. The pump is to be refurbished by replacing the existing impeller with a geometrically similar impeller that has a diameter of 400 mm. (a) Neglecting scale effects, estimate the flow rate, head added, and power consumption of the pump with the new impeller when the pump is operating under conditions of maximum efficiency. (b) Quantify the scale effect on the efficiency of the pump and assess whether the scale effect is expected to be significant.

8.25. A homologous series of centrifugal pumps are driven by 2500-rpm motors. For a 500-mm size within this series, the manufacturer claims that the best efficiency of 80% occurs when the flow rate is 600 L/s and the head added by the pump is 90.5 m. What would be the best efficiency operating point for a 400-mm size within this homologous series? Estimate the corresponding efficiency.

8.26. A 1:4 scale model of a water pump is operated at a speed of 4500 rpm. At its best efficiency point, the efficiency of the model pump is 84%, and the model delivers a flow rate of 0.7 m^3/s with an added head of 4.9 m. If the full-scale pump has a rotational speed of 120 rpm, what is the flow rate and head delivered by the full-scale pump operating at its most efficient point? What is the power requirement of the full-scale pump at its best efficiency point? Assume water at 20°C.

8.27. Affinity laws are typically used to identify homologous values of flow rate, head, and power between geometrically similar pumps. Develop an affinity law for relating homologous values of torque.

8.28. A manufacturer tests a 1:10 scale model of a pump in the laboratory. The model pump has an impeller diameter of 200 mm and a rotational speed of 3450 rpm, and when the head across the pump is 40 m, the pump delivers a flow rate of 10 L/s at an efficiency of 84%. The prototype pump is to develop the same head as the scale model; however, because of its increased size, the prototype pump is expected to have an efficiency of 90%. (a) What is the power supplied to the model pump? (b) What is the rotational speed, flow rate, and power supplied to the prototype pump under homologous conditions? Assume water at 20°C.

8.29. A pump has an impeller diameter of 300 mm and a rotational speed of 1500 rpm. At the best efficiency operating point, the pump adds a head of 9 m at a flow rate of 25 L/s. What is the specific speed of the pump? What type of pump is this likely to be?

8.30. Express the specific speed of a pump in terms of the head coefficient and flow coefficient at the best efficiency point.

8.31. A pump is required to deliver 600 L/s from a lake to a storage reservoir. Application of the energy equation to the pipeline system shows that the head that must be added by the pump is 84 m. If the pump motor is to have a rotational speed of 2000 rpm, what is the specific speed of the required pump? What type of pump is required?

8.32. The head and flow rate to be delivered by a particular pump are to be changed. The existing pump is to be retained with a new motor installed. The existing pump has a specific speed of 4.5. Under the new operating conditions, the pump is required to deliver 400 L/s with an added head of 8 m. What rotational speed should the new motor have? What type of pump is this?

8.33. What is the highest synchronous speed for a motor driving a pump?

8.34. A pump is to be selected such that it operates at or near its most efficient state when delivering 100 L/s with an added head of 50 m. A manufacturer has five models of homologous pumps, with specific speeds of 0.95, 0.81, 0.61, 0.36, and 0.25. Motors with any of the standard-rated speeds can be provided with these pumps. Identify the best choice of specific speed and the rotational speed of the motor that should be used to drive the pump.

8.35. A pump has an impeller diameter of 450 mm, and at its most efficient operating point, it delivers water at a flow rate of 650 L/s with an added head of 9.5 m. The specific speed of the pump is 1.5, and the shaft power delivered by the motor is 80 kW. (a) Estimate the shutoff head of the pump. (b) Estimate the efficiency of the pump at its best operating point. Assume water at 20°C.

8.36. A pump with an impeller size of 250 mm is operated at a rotational speed of 1200 rpm. A review of the pump performance data shows that the pump has a shutoff head of 8.2 m, and at the best efficiency operating point, the pump adds a head of 6.5 m at a flow rate of 20 L/s. (a) Use the performance data to fit a parabolic performance curve of the form $h_\mathrm{p} = a - bQ^2$, where h_p is the head added by the pump in m, Q is the flow rate in L/s, and a and b are constants. (b) If the rotational speed of the pump impeller is increased to 1800 rpm, estimate the parabolic performance curve at the adjusted speed.

8.37. A prototype water pump has a specific speed of 1.2, and when operating at its most efficient state, it delivers 5 L/s with an added head of 10 m. Operation of the pump is to be tested using a $\frac{1}{5}$-scale model with various test fluids that have dynamic viscosities in the range of 5–10 times that of water. Viscous effects should be accurately accounted for in the model. What range of rotational speeds, flow rates, and heads will be required in model testing? Assess whether accurately accounting for viscous effects is realistic.

8.38. Pressure measurements are taken in the suction and discharge pipes of a pump as shown in Figure 8.42. When the flow rate through the pump is 10 L/s, the measured (gauge) pressure on the suction side is −20 kPa and the measured (gauge) pressure on the discharge side is 100 kPa. The measurement section on the discharge side is 0.83 m above the measurement section on the suction side. The diameters of the suction and discharge pipes are 150 mm and 75 mm, respectively. The pumped fluid is water at 10°C, and the head loss between the suction and discharge sides of the pump can be assumed to be negligible. Estimate the head added by the pump.

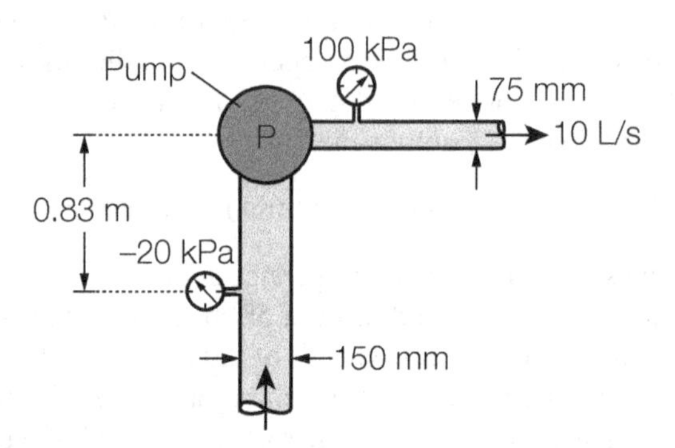

Figure 8.42: Flow through a pump

8.39. A centrifugal pump is to be used to pump water at a rate of 50 L/s from a sump to a storage reservoir, where the water surface in the storage reservoir is 25 m higher than the water surface in the sump. The suction and discharge line from the pump have a total length of 48 m, a diameter of 125 mm, and an estimated roughness height of 0.5 mm. The sum of the local loss coefficients in the pipeline is equal to 4, the specific speed of the pump is 0.8, and the manometric efficiency of the pump is 80%. In the homologous series from which the pump is selected, the impeller blades are forward-curved and make an angle of 55° with the outflow surface, the width of the impeller is equal to 12.5% of the diameter, and the blades cover 7% of the outer area of the impeller. Inflow to the impeller is normal to the inflow surface. What size impeller should be selected for this application? Assume water at 20°C.

8.40. The performance characteristics of a pump are determined using the setup shown in Figure 8.43, where simultaneous measurements are taken of the flow rate, Q, the pressures, p_1 and p_2, at the suction and discharge pipes sections, respectively, and the power consumption, P, of the pump. These measurements are as follows:

Q (L/s)	1.26	2.52	3.79	5.05	6.31	7.57	8.83
$p_2 - p_1$ (kPa)	671	665	653	629	593	557	498
P (kW)	3.54	4.22	4.97	5.60	6.27	7.20	7.77

The discharge pipe section has an elevation 0.38 m higher than the suction pipe section, and the temperature of the water used in the test is 20°C. (a) Plot the performance curve of the pump, showing h_p versus Q and η versus Q, where h_p is the head added by the pump and η is the efficiency. Both of these relationships should be shown on the same graph. (b) What is the flow rate through the pump when it is operating at maximum efficiency? What is the maximum efficiency of the pump?

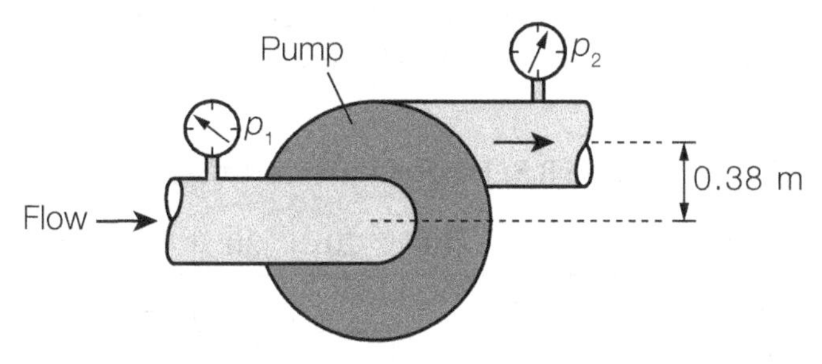

Figure 8.43: Pump performance test

8.41. A pump is required to deliver 600 L/s (±10%) through a 300-mm-diameter PVC pipe from a well to a reservoir. The water level in the well is 1.5 m below the ground surface, and the water surface in the reservoir is 2 m above the ground surface. The delivery pipe is 300 m long, and local losses can be neglected. A pump manufacturer suggests using a pump with a performance curve given by

$$h_p = 6 - 6.67 \times 10^{-5}Q^2$$

where h_p is the head added in meters and Q is the flow rate in liters per second. Is this pump adequate? Explain.

8.42. A pump with the performance curve shown in Figure 8.44 is being considered to pump water from a lower reservoir to a higher reservoir where the difference in water surface elevations is 9.0 m. The conduit is 100 m long, is 100 mm in diameter, and has negligible roughness. Estimate the maximum flow rate that can be achieved if this pump is used. At what efficiency would the pump operate? Assess the desirability of using this pump.

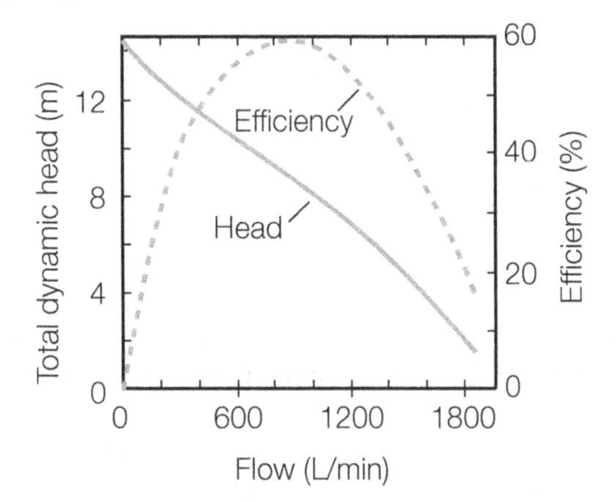

Figure 8.44: Pump performance curve

8.43. Water is pumped from a lower reservoir to a higher reservoir. The water surface in the higher reservoir is 10 m above the water surface in the lower reservoir. The piping system consists of 200-mm-diameter ductile iron pipe with a length of 2 km and minor losses equal to 6.2 times the velocity head. The pump characteristics are shown in Table 8.3. (a) Determine the expected flow rate through the system, assuming fully turbulent flow. (b) What is the power of the motor required to drive the pump?

Table 8.3: Pump Characteristics

Discharge (L/s)	0	10	20	30	40	50
Total head (m)	25	23.2	20.8	17.0	12.4	7.3
Efficiency (%)	–	45	65	71	65	45

8.44. The pumped-storage system illustrated in Figure 8.45 is designed to exchange 2 m^3/s through a 1220-mm-diameter, 3.2-km-long ductile iron pipeline lined with bitumen. The elevation difference between the water surfaces in the upper and lower reservoirs is 61 m, and the pump/turbine is to operate 8 hours during the day as a turbine (to generate electricity) and 8 hours during the night as a pump (to return the water to the upper reservoir). The pump efficiency is 85%, the turbine efficiency is 90%, the cost of pumping is $0.06/kWh, and the revenue from turbine operations (selling electricity) is $0.12/kWh. Determine the annual profit from this operation. Neglect the effect of storage on reservoir elevations. If the pump performance curve is given by $h_p = 80 - 3.5Q^2$ where h_p is in m and Q is in m^3/s, estimate the change in profit when the elevation difference is 65 m. For an elevation difference of 65 m, assume that the flow is fully turbulent.

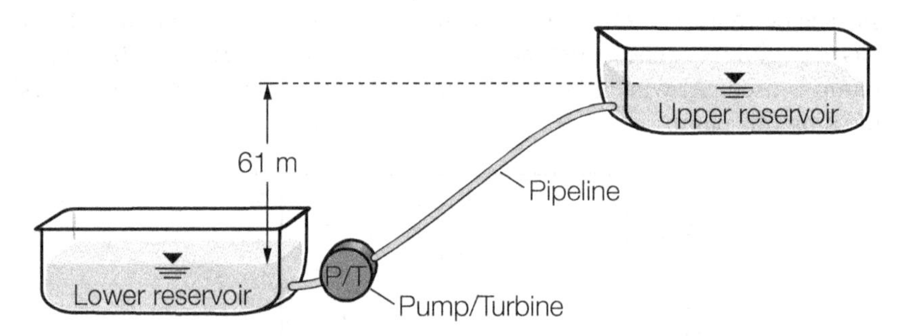

Figure 8.45: Pumped-storage system

8.45. A pump is to be selected to deliver water from a well to a treatment plant through a 300-m-long pipeline. The temperature of the water is 20°C, the average elevation of the water surface in the well is 5 m below the ground surface, the pump is 50 cm above the ground surface, and the water surface in the receiving reservoir at the water treatment plant is 4 m above the ground surface. The delivery pipe is made of ductile iron ($k_s = 0.26$ mm) with a diameter of 800 mm. If the selected pump has a performance curve of $h_p = 12 - 0.1Q^2$,where Q is in m^3/s and h_p is in m, what is the flow rate through the system? Calculate the specific speed of the required pump and state the type of pump required when the speed of the pump motor is 1200 rpm. Neglect local losses.

8.46. A pump is to be used to withdraw water from a reservoir at a rate of 1500 L/s. When operating at this flow rate, the head loss in the suction pipe is estimated to be 2.3 m, and the pump specifications give the required net positive suction head as 2.9 m. Standard sea-level atmospheric conditions are expected at the site, and under worst-case conditions, the temperature of the water in the reservoir is 25°C. What is the maximum allowable elevation of the suction side of the pump above the reservoir water surface?

8.47. Under design conditions, a water pump is expected to deliver 500 L/s with an added head of 75 m. The diameter of the suction pipe is 550 mm. Under worst-case conditions, the temperature of the water is expected to be 80°C and atmospheric pressure is 96 kPa. The manufacturer states that the cavitation parameter of the pump is equal to 0.12. A pressure gauge is installed on the suction side of the pump to detect conditions when cavitation is likely to occur. At what pressure reading is cavitation likely to occur?

8.48. Water at 20°C is to be pumped out of a reservoir at a rate of 20 L/s through a vertical 150-mm-diameter ductile iron pipeline with an estimated roughness height of 0.3 mm. Appurtenances installed in the intake pipe are expected to contribute to a total local head loss coefficient of 12. The pump being considered for installation has a required net positive suction head of 5.5 m. (a) What is the maximum elevation of the pump relative to the water surface in the reservoir? (b) If the total head loss coefficient can be reduced to 1.2, how much higher can the pump be placed? Assume standard atmospheric conditions.

8.49. Tests on a pump under standard atmospheric conditions show that when water at 20°C is pumped at 60 L/s and the head added by the pump is 40 m, cavitation occurs when the pressure head plus velocity head on the suction side of the pump is 3.9 m. (a) Determine the required net positive suction head and the cavitation number of the pump. (b) If this same pump is operated on a mountain under the same flow rate and added head condition but the temperature of the water is 5°C and the atmospheric pressure is 90 kPa, by how much must the elevation of the pump above the sump reservoir be reduced compared with the test condition? Assume that the friction loss in the suction pipe remains approximately the same and that the sump reservoir is open to the atmosphere in both cases.

8.50. A pump with an impeller diameter of 225 mm and a rotational speed of 1725 rpm pumps water at a temperature of 80°C. Cavitation begins on the suction side of the pump when the volume flow rate is 50 L/s, the pressure on the suction side of the pump is 80 kPa, and the corresponding velocity in the suction pipe is 5 m/s. (a) What is the required net positive suction head of the pump? (b) If a geometrically similar pump has a rotational speed of 1140 rpm and an impeller diameter of 675 mm, what required net positive suction head is expected in the larger pump?

8.51. Water is to be pumped out of a well and stored in an above-ground reservoir. The water surface in the well is 3 m below the ground surface, water is to be pumped through a 100-m-long, 50-mm-diameter galvanized iron line and exit 19.3 m above the ground, and water is to be delivered to the upper reservoir at a rate of at least 370 L/min when the upper reservoir is empty. The sum of the local loss coefficients in the system is 1.8. A local pump salesperson suggests that you choose a pump from a set of pumps with performance curves shown in Figure 8.46. Determine whether this set of pumps is worthy of consideration and, if so, what particular pump size is required. What is the maximum height the pump can be placed above ground?

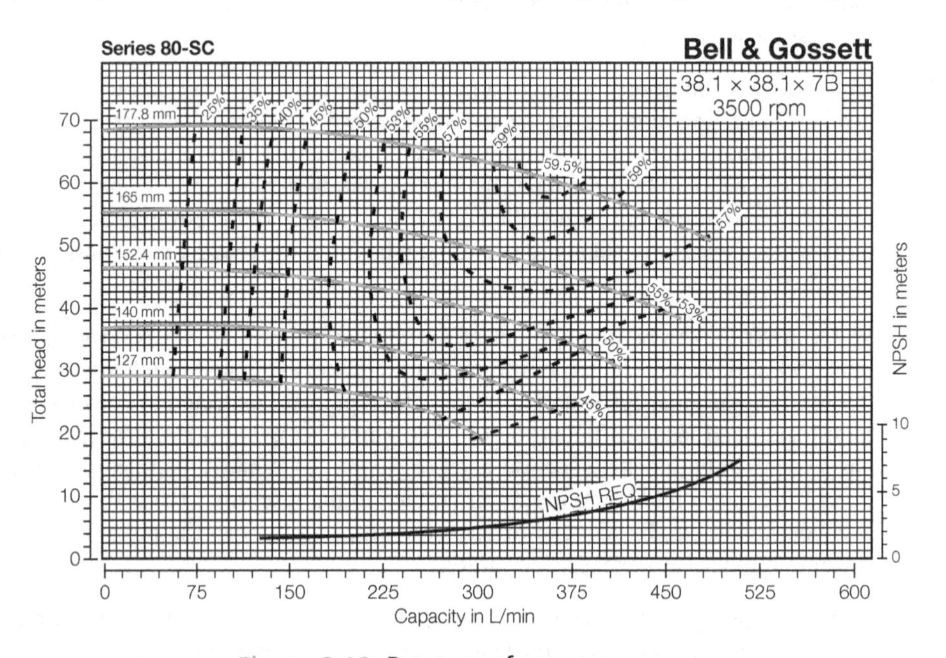

Figure 8.46: Pump performance curves

8.52. Consider the 165-mm pump with the performance curve shown in Figure 8.46. The motor on this pump is changed from one with a rotational speed of 3500 rpm to one with a rotational speed of 2500 rpm. (a) Plot the performance curve of the modified pump, showing both the head added by the pump and the efficiency of the pump as a function of the flow rate. Use the following units in your plot: flow rate in L/s, head in m, and efficiency in %. (b) What are the flow rate, head added, and efficiency of the modified pump at its best efficiency point? (c) Calculate and compare the specific speeds of the existing and modified pumps and use these results to infer the type of pump.

8.53. A pump is located 2 m below the water surface in a reservoir as shown in Figure 8.47. The piping between the pump and the reservoir consists of 3 m of 150-mm-diameter ductile iron pipe (DIP).It is estimated that the DIP has an equivalent sand roughness of 0.25 mm. The valves and bends between the source reservoir and the pump are estimated to have a total local loss coefficient of 50. Under design conditions, the water in the reservoir has a temperature of 25°C and the atmospheric pressure is 101.3 kPa. The pump specifications require that the pump have a minimum net positive suction head of 4 m. What is the maximum allowable flow rate through the system so as to avoid cavitation?

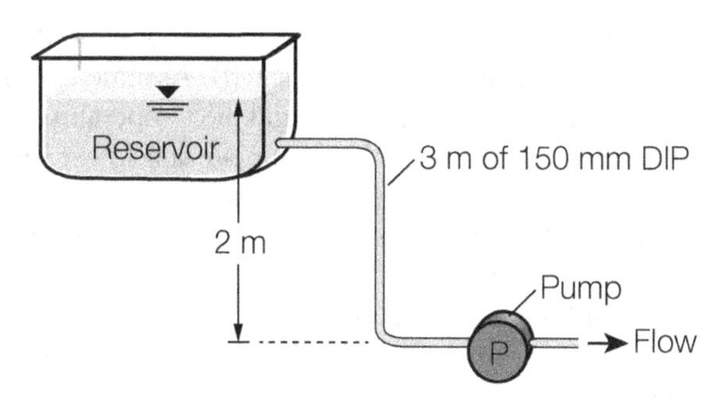

Figure 8.47: Flow in a pumped system

8.54. A pump lifts water through a 100-mm-diameter ductile iron pipe from a lower to upper reservoir (Figure 8.48). If the difference in elevation between the water surfaces in the reservoirs is 10 m and the performance curve of the 2400-rpm pump is given by

$$h_\mathrm{p} = 15 - 0.1Q^2$$

where h_p is in m and Q is in L/s, estimate the flow rate through the system. If the pump manufacturer gives the required net positive suction head under these operating conditions as 1.5 m, what is the maximum height above the lower reservoir that the pump can be placed and maintain the same operating conditions?

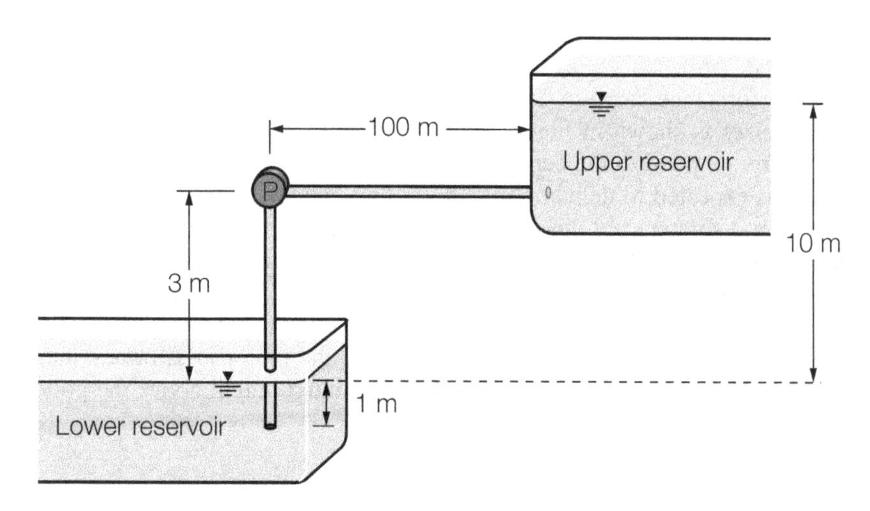

Figure 8.48: Water pumped from a lower to upper reservoir

8.55. Water is being pumped from Point A in a lower reservoir to Point F in an upper reservoir through a 30-m-long PVC pipe of diameter 150 mm (see Figure 8.49). An open gate valve is located at C; 90° bends (threaded) are located at B, D, and E; and the pump performance curve is given by

$$h_{\mathrm{p}} = 20 - 4713Q^2$$

where h_{p} is the head added by the pump in m and Q is the flow rate in m^3/s. The dimensionless specific speed of the pump is 1.10. Assume that the flow is turbulent (in the smooth, rough, or transition range) and the temperature of the water is 20°C. (a) Write the energy equation between the upper and lower reservoirs, accounting for entrance, exit, and local losses between A and F. (b) Calculate the flow rate and velocity in the pipe. (c) If the required net positive suction head at the pump operating point is 3.0 m, assess the potential for cavitation in the pump. (For this analysis, assume that the head loss in the suction pipe is negligible.) (d) Use the affinity laws to estimate the pump performance curve when the rotational speed is changed from 800 rpm to 1600 rpm.

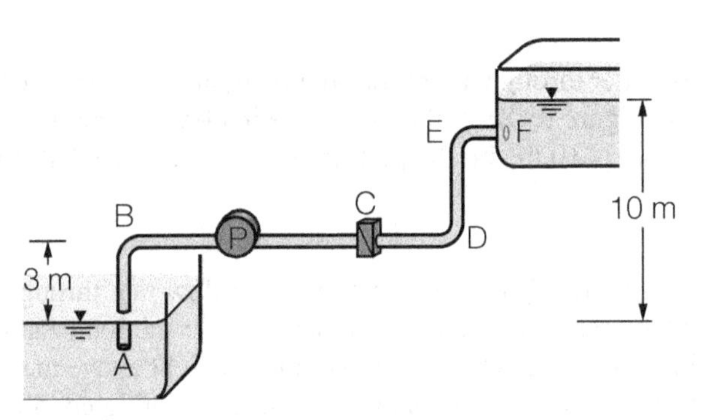

Figure 8.49: Water delivery system

8.56. The performance curve of a Goulds Model 3656 irrigation pump with several impeller sizes is shown in Figure 8.50. The pump is to be used to deliver water at 20°C from a pond to the center of a large field located 100 m from the pond. The pump is expected to deliver a maximum flow rate of 380 L/min through 107 m of 600-mm-diameter steel pipe having an equivalent sand roughness of 0.01 mm. The water surface elevation in the pond during the irrigation season is 10.00 m, and the ground elevation of the field is 15.00 m. (a) Which of the impeller sizes shown in Figure 8.50 would you select for the job? (b) Using the efficiency of the pump under the operating conditions, calculate the size of the motor in kilowatts that must be used to drive the pump. (c) What is the maximum height above the pond that the pump could be located?

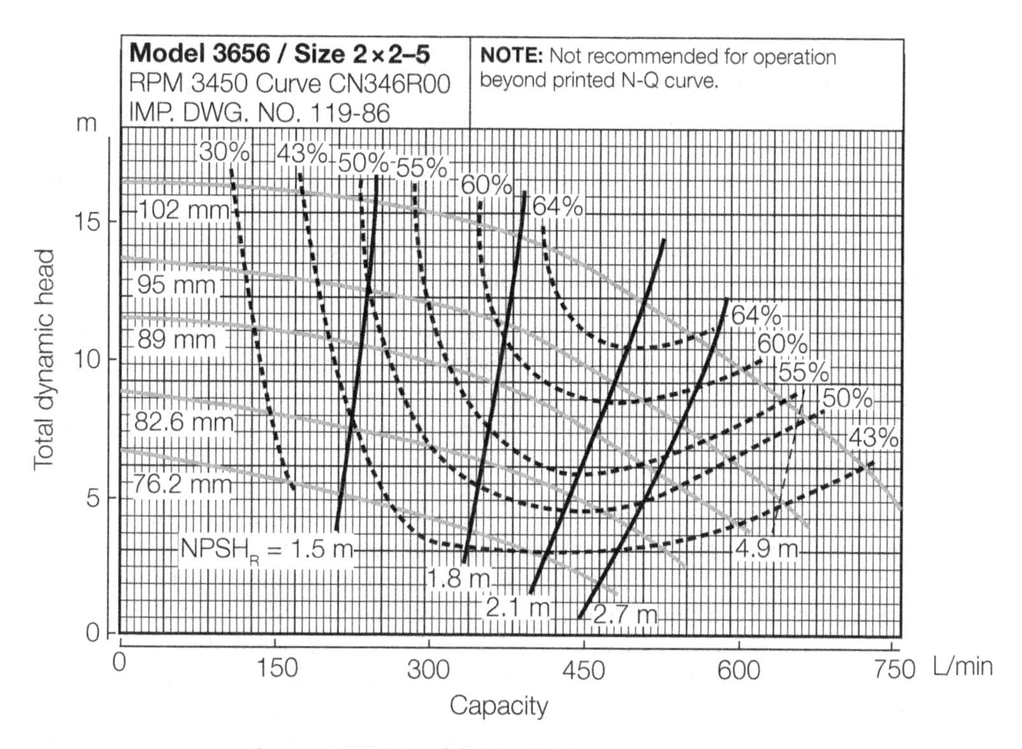

Figure 8.50: Goulds Model 3656 pump curves

8.57. If the performance curve of a certain pump model is given by

$$h_\mathrm{p} = 30 - 0.05Q^2$$

where h_p is the head added in m and Q is the flow rate in L/s, what is the performance curve of a pump system containing n of these pumps in series? What is the performance curve of a pump system containing n of these pumps in parallel?

8.58. A pump is placed in a pipe system in which the energy equation (i.e., the system curve) is given by

$$h_\mathrm{p} = 15 + 0.03Q^2$$

where h_p is the head added by the pump in m and Q is the flow rate in the system in L/s. The performance curve of the pump is

$$h_\mathrm{p} = 20 - 0.08Q^2$$

What is the flow rate in the system? If the pump was replaced by two identical pumps in parallel, what would be the flow rate in the system? If the pump was replaced by two identical pumps in series, what would be the flow rate in the system?

8.59. A wastewater pump station is required to handle a design flow rate of 1000 L/s, and the pump station must provide an added head of 9 m. Pump units are available, each having a power demand of 35 kW and an overall efficiency of under optimal conditions 62% under optimal conditions. The specific speed of each unit is around 1.5. (a) If the pump units are to be placed in parallel to accommodate the design flow, how many pump units are required? (b) What motor speed would be required for optimal performance of each unit? Assume that the properties of the wastewater are approximately equal to those of water at 20°C.

8.60. Nine pump units are placed in parallel at a pump station. Each unit has a power demand of 40 kW and adds 35 m of head under optimal conditions. The best efficiency of each unit is 60%. The liquid being pumped is water at 20°C. When all units are operating under optimal conditions, what is the flow rate delivered by the pump station?

8.61. Water is to be pumped from a river to an elevated reservoir through a 250-mm-diameter pipe that has an equivalent sand roughness of 0.1 mm. The water surface in the reservoir is 70 m above the water surface in the river, and the length of pipeline is 5 km. Initially, a flow rate between the river and the reservoir of 35 L/s will be required. However, it is expected that the required flow rate will double to 70 L/s at some time in the future. The chief engineer on the project has proposed that a pump from a homologous series with performance characteristics shown in Figure 8.51 be used to deliver the initial flow rate of 35 L/s, and that the pump station be designed such that additional pumps can be added in parallel to handle the increased flow when it becomes necessary. (a) Determine the required pump size needed to deliver 35 L/s. (b) What would be the efficiency of the pump when it is delivering 35 L/s? (c) What is the maximum height that this pump can be placed above the water surface of the river? (d) How many pumps would be needed to deliver 70 L/s. Clearly explain your calculations. Assume that the flow in the pipeline will generally be fully turbulent. For part (c) assume that the entrance loss and friction head loss in the suction pipe is negligible.

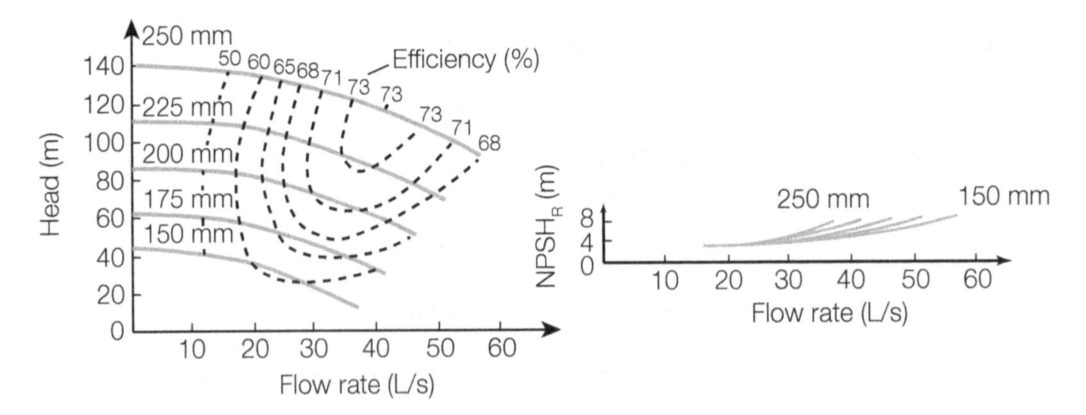

Figure 8.51: Pump performance characteristics

8.62. A 20-km-long, 1120-mm-diameter steel pipe with an estimated roughness height of 0.05 mm is to deliver water from a water supply reservoir through a system of parallel pumps as shown in Figure 8.52. Both the intake at A and the exit from the pump system at B are at an elevation of 5 m; the water level in the supply reservoir is 2 m above the intake at A; and the elevation at the delivery point, C, is 15 m. There is negligible head loss due to friction between A and B, and the performance curve for each pump in the system is given by

$$h_\mathrm{p} = 65 - 7.6 \times 10^{-8} Q^2$$

where h_p is the head added by the pump in m and Q is the flow through the pump in m^3/d. When the system is delivering 48 000 m^3/d at C, the required pressure at C is 448 kPa. (a) Determine how many pumps are required. (b) Assuming that the flow is fully turbulent, what is the actual flow rate at C when the number of pumps determined in part (a) is used? The temperature of the water is 20°C.

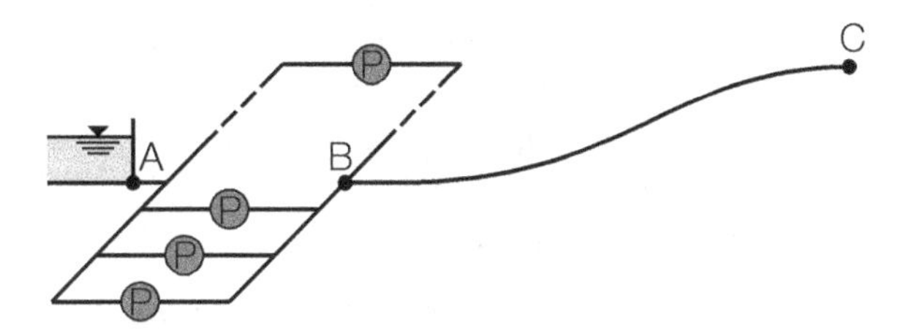

Figure 8.52: Parallel pump system

8.63. The water supply system shown in Figure 8.53 is to be constructed such that water is delivered from a reservoir at A to two communities located at C and D. The pipe lengths, diameters, and demand flow rates are as follows:

Line	Length (km)	Diameter (mm)	Flow (L/s)
AB	1.05	200	27
BC	2.80	150	12
BD	2.50	150	15

The water surface elevation in the supply reservoir is 3.00 m, and the elevations of the delivery pipes at C and D are 2.00 m and 5.00 m, respectively. Under the given demand conditions, it is desired to have water pressures of at least 350 kPa at C and D. All pipes are to be made of ductile iron. (a) Determine the minimum power that must be delivered by the pump. (b) A pump manufacturer will be able to match the minimum power operating condition by providing several "micro-pumps" in series where each micro-pump has a performance given by

$$h_\mathrm{p} = 0.455D - 4000Q^2$$

where h_p is the head added in m, D is the size of the micro-pump in cm, and Q is the flow rate in m^3/s. The manufacturer can deliver any size, D, in the range of 40–50 cm. How many and what size micro-pumps will be needed?

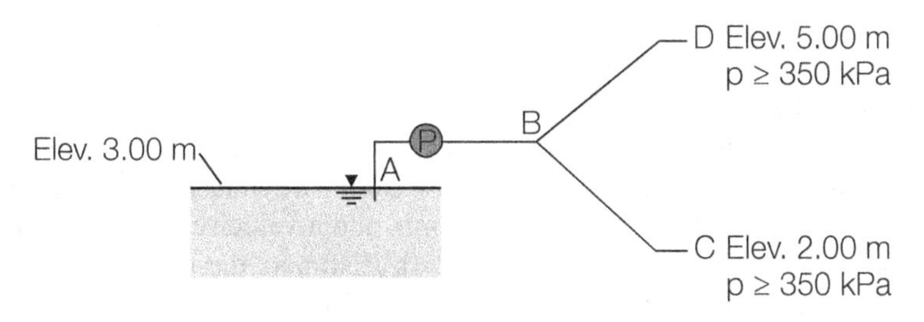

Figure 8.53: Water supply system

8.64. The performance curve for a variable-speed pump operating at 600 rpm is given by

$$h_p = 6 - 0.05Q^2$$

where h_p is the head added by the pump in m and Q is the flow rate in m^3/min. This pump is installed in a system where energy considerations require a system curve given by

$$h_p = 3 + 0.042Q^2$$

Find the flow rate in the system when the pump is operating at 600 rpm and compare this with the flow rate when the pump is operating at 1200 rpm.

8.65. A large centrifugal fan generates a flow of 7 m^3/s with a motor speed of 1140 rpm. A smaller geometrically similar fan has a motor speed of 1725 rpm, operates at the same efficiency as the larger fan, and generates the same pressure increase. What flow rate is generated by the smaller fan?

Section 8.5: Hydraulic Turbines and Hydropower

8.66. A single-jet Pelton wheel hydropower plant is to be operated such that the shaft power developed by the Pelton wheel is equal to 15 MW when the head just upstream of the nozzle is equal to 1600 m. The Pelton wheel is constructed such that is has a diameter of 4 m, has a deflection angle of 170°, and rotates at a controlled speed of 600 rpm. Determine the appropriate nozzle diameter to be used in the project. Assume water at 20°C and a nozzle loss coefficient of 0.03.

8.67. An existing Pelton wheel has a diameter of 3 m and is controlled to rotate at 150 rpm. The buckets on the Pelton wheel have a deflection angle of 170°. The power output from the Pelton wheel is controlled by adjusting the jet nozzle diameter. If the desired power is 850 kW, what is the required nozzle diameter for the most efficient operation of the Pelton wheel? Assume water at 20°C.

8.68. A Pelton wheel has a diameter of 2.7 m and buckets with a deflection angle of 167°. The nozzle and water supply characteristics are fixed such that under design conditions, the incident jet has a diameter of 150 mm and a velocity of 12 m/s. What is the maximum power that can be extracted by this Pelton wheel? What is the rotational speed of the wheel under the maximum power condition? Assume water at 20°C.

8.69. A 4-m-diameter Pelton wheel is driven by a jet from a single 240-mm-diameter nozzle, and the pressure and velocity in the pipeline just upstream of the nozzle is 4 MPa and 6 m/s, respectively. The nozzle loss coefficient is estimated as 0.03. The buckets on the Pelton wheel have a deflection angle of 165°. Estimate the following: (a) the maximum power that can be generated by the Pelton wheel, (b) the rotational speed at the maximum power state, (c) the runaway rotational speed, (d) the torque on the wheel shaft at the maximum-power state, and (e) the torque on the wheel shaft when the wheel is held stationary. Assume water at 20°C.

8.70. A manufacturer cites the wheel efficiency of an impulse turbine as 90% when the turbine is operating at its maximum efficiency state. Estimate the blade angle of the buckets mounted on the turbine.

8.71. A Pelton wheel has an average radius of 2 m, and its buckets have a deflection angle of 165°. According to the manufacturer, the Pelton wheel has a wheel efficiency of 80% under optimal operating conditions. In a particular installation, the turbine is driven by a jet with a diameter of 160 mm and a velocity of 100 m/s. (a) Estimate the optimal rotation rate of the turbine. (b) Estimate the maximum power that can be extracted by the turbine. Assume water at 20°C.

8.72. Show that if the effective head on a Pelton wheel is h_e, the velocity coefficient of the nozzle is C_v, and the bucket speed of the wheel is U, then the theoretical maximum efficiency is attained by the Pelton when

$$U = \tfrac{1}{2}C_\mathrm{v}\sqrt{2gh_\mathrm{e}}$$

8.73. The peripheral velocity factor of a Pelton wheel, ϕ, is defined as $\phi = U/\sqrt{2gh_\mathrm{e}}$, where U is the bucket speed, and h_e is the effective head on the turbine. Problem 8.72 shows that the efficiency of a Pelton wheel is maximized when $\phi = \frac{1}{2}C_\mathrm{v}$, where C_v is the velocity coefficient of the nozzle. Show that the maximum efficiency of a Pelton wheel turbine, η_t, under these conditions is given by

$$\eta_\mathrm{t} = \tfrac{1}{2}C_\mathrm{v}^2(1 - \cos\beta)$$

where β is the bucket angle.

8.74. The head loss in the nozzle of a Pelton wheel is sometimes represented by a local head loss coefficient, k_j, and is sometimes represented by a nozzle velocity coefficient, C_v. Show that the relationship between C_v and k_j is given by

$$C_\mathrm{v} = \sqrt{\frac{1}{1 + k_\mathrm{j}}}$$

8.75. A Pelton wheel has a diameter of 5 m and a bucket angle of 165°. The effective head at the nozzle is 550 m, and the nozzle is set such that the velocity coefficient is 0.94 and the diameter of the jet is 130 mm. Estimate the optimal flow rate, rotational speed of the wheel, and power output. Assume water at 20°C.

8.76. A Pelton wheel is to be designed to harness the available hydropower from a site where the effective head on the turbine will be 160 m and the reliable flow rate through the turbine will be 6 m^3/s. The wheel is to have a rotational speed of 500 rpm, the bucket angle will be 165°, and the nozzle is expected to have a velocity coefficient of 0.92. (a) What diameter wheel would maximize the efficiency of the turbine? (b) What power can be expected from the turbine? (c) Assess whether a different type of turbine should be considered for this site. Assume water at 20°C.

8.77. The Pelton wheel shown in Figure 8.54 is driven by two jets, where each jet has a diameter of 40 mm and a velocity of 50 m/s. Each bucket mounted on the Pelton wheel deflects its incident jet by 160°. The diametric distance between the centers of the buckets is 2 m, and it can be assumed that the reaction forces on the each of the buckets act at the center of the bucket. (a) What torque is required to hold the Pelton wheel stationary? (b) If the Pelton wheel is allowed to "freewheel" such that the shaft torque is negligible, at what speed will the Pelton wheel rotate? (c) If a governor controls the rotational speed at 15.7 rad/s, what power can be generated by the Pelton wheel? Assume water at 20°C.

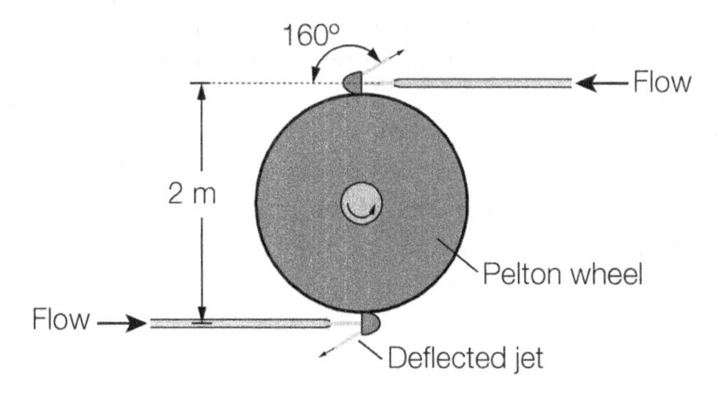

Figure 8.54: Two-jet Pelton wheel

8.78. The impulse turbine system shown in Figure 8.55 uses water at 20°C from a reservoir with a water surface elevation that is 80 m above the elevation of the nozzle. The length of the 300-mm-diameter delivery pipe is 600 m, and the delivery pipe has an estimated roughness height of 0.5 mm. The nozzle at the end of the pipe is estimated to have a nozzle loss coefficient of 0.05, and the buckets on the impulse turbine have a deflection angle of 160°. (a) Identify the optimum nozzle diameter and state the corresponding power that can be extracted by the turbine. (b) Determine the optimum nozzle diameter and corresponding power derived by assuming fully turbulent flow in the supply pipe and compare your result with that obtained in part (a). Comment on the appropriateness of assuming fully turbulent flow.

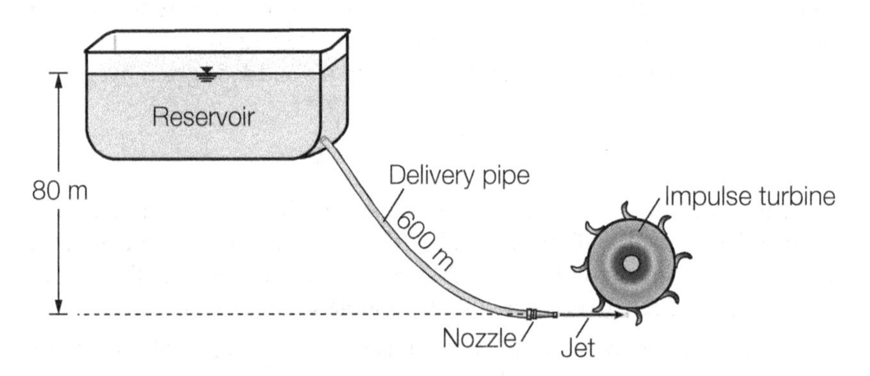

Figure 8.55: Impulse turbine system

8.79. At a Pelton wheel installation, the water surface elevation in the supply reservoir is 85 m above the nozzle; the delivery line has a diameter of 600 mm, a length of 300 m, and a roughness height of 8 mm. The discharge nozzle has a diameter of 50 mm and a head loss coefficient of 0.8. The bucket friction loss coefficient is 0.5, the velocity of water relative to the bucket at the exit from the bucket is 2 m/s, and the absolute velocity of the water leaving the bucket is 6 m/s. Determine the power that could be derived from the system and the hydraulic efficiency of the turbine.

8.80. A Francis turbine unit rotating at 900 rpm is to be designed to deliver a head of 180 m when operating under design conditions. Based on experience with similar units, the overall efficiency of the turbine unit is expected to be 82% and the hydraulic efficiency is expected to be 90%. The guide vanes are oriented to direct the inflow at an angle of 30° to the inflow surface, and the blades in the runner are to be at an angle of 60° to the inflow surface as shown in Figure 8.56. The width of the runner is 20% of the diameter of the inflow surface, and the outflow velocity is normal to the outflow surface of the runner. What outer diameter of the turbine runner is required?

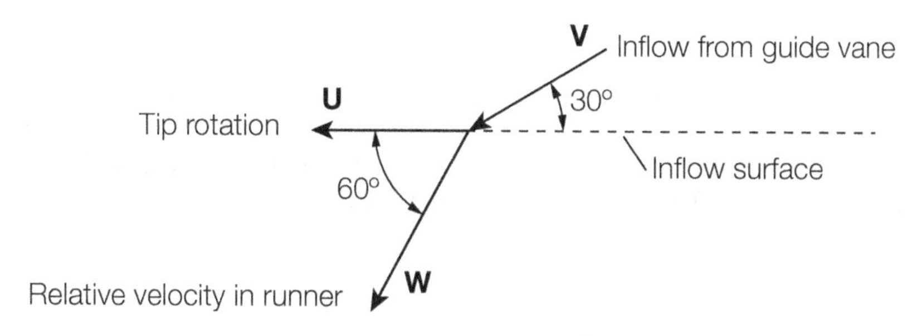

Figure 8.56: Turbine inflow velocities

8.81. A Francis turbine has an overall efficiency of 92% and a hydraulic efficiency of 95%. The turbine runner has an outer diameter of 2.5 m, has a width of 0.35 m, and rotates at 300 rpm. The turbine is located 4 m above the discharge reservoir surface, and the head loss in the draft tube can be assumed negligible. Under design conditions, the flow rate through the turbine is 20 m^3/s and the velocity and pressure at the entrance to the scroll case are 10 m/s and 2447 kPa, respectively. The flow exits the runner in a direction normal to the outflow surface. (a) Estimate the power output of the turbine unit. (b) What is the specific speed of the turbine? (c) Estimate the guide vane and runner blade angles. Assume water at 20°C.

8.82. The inflow surface of a runner in a reaction turbine has a diameter and width of 500 mm and 60 mm, respectively, and the outflow surface has a diameter and width of 350 mm and 80 mm, respectively. On both the inflow and outflow surfaces, the blades occupy 6% of the flow area. Under design conditions, the inflow and outflow velocity components are shown in Figure 8.57, where **V**, **U**, and **W** are the absolute velocity, tip velocity, and velocity relative to the moving runner, respectively. The guide vane directs the inflow velocity at an angle of 24° to the inflow surface, the runner blade is at an angle of 80° to the inflow surface, and the flow in the runner should be aligned with the runner blade. The runner blade makes an angle of 30° with the outflow surface. It is estimated that the head available to the turbine is 65 m, of which 10% is lost due to hydraulic friction within the turbine, and 5% is lost due to mechanical friction in the turbine system. (a) At what speed should the turbine be operated so that shock losses are minimized, which means that the flow in the runner is aligned with the runner blade on inflow? (b) What is the power output of the turbine? Assume water at 20°C.

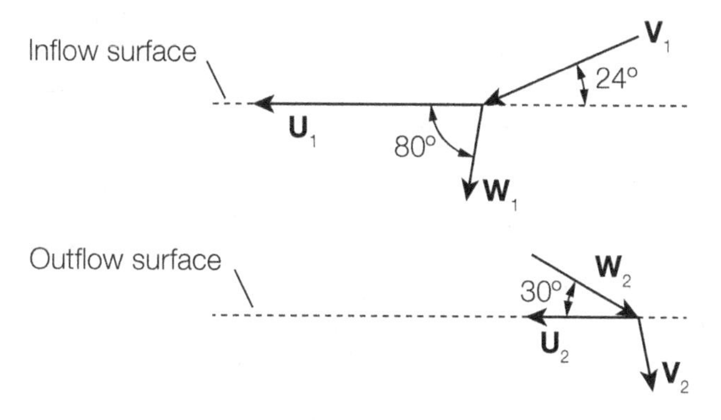

Figure 8.57: Turbine flow velocities

8.83. Flow enters a Francis turbine at an angle of 20° to the inflow surface as shown in Figure 8.58. The runner has an outer diameter of 1600 mm and an inner diameter of 900 mm, and the width of the inflow surface is 80 mm. When the runner rotates at 120 rpm, the flow rate through the runner is 2.9 m^3/s and the outflow velocity is normal to the outflow surface. Estimate the power produced by the turbine. Assume water at 20°C.

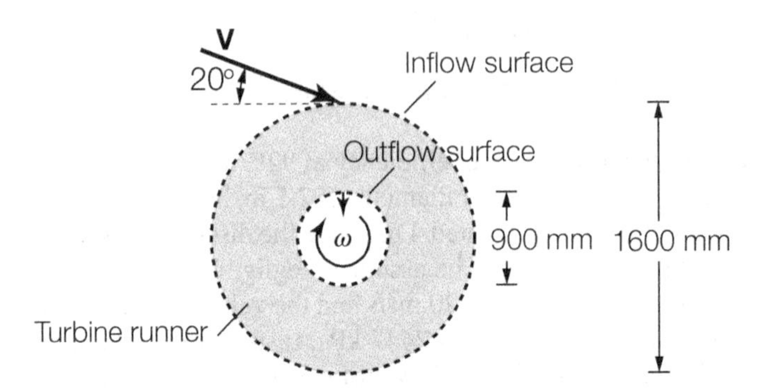

Figure 8.58: Flow through a turbine

8.84. A hydropower plant utilizes eight Francis turbines. The change in head across the power plant is 250 m, and the flow through each turbine unit is 12 m^3/s. The estimated efficiency of each turbine unit is 95%, and the efficiency of the generator and supporting power delivery systems is 91%. Estimate the power-generating capacity of the hydropower facility. Assume water at 20°C.

8.85. A hydropower installation with reaction-type (Francis) turbines is to be located where the downstream water surface elevation is 100 m below the water surface elevation in the upstream reservoir. The 2.0-m-diameter concrete-lined penstock is 500 m long and has an estimated roughness height of 15 mm. When the flow rate through the system is 20 m^3/s, the combined head loss in the turbine and draft tube is 5.0 m, and the average velocity in the tailrace is 0.80 m/s. Estimate the power that can be extracted from the system.

8.86. A large hydropower facility uses several Francis turbines in its power plant. Each turbine has a rated power output of 50 MW at a head of 70 m, and has a maximum hydraulic efficiency of 90%. The generator in each unit has a rotational speed of 120 rpm. (a) Estimate the specific speed of each unit and verify that a Francis turbine is the best type of turbine to use. (b) Estimate the optimal flow rate through each turbine unit. Assume water at 20°C.

8.87. A Francis turbine with a runner diameter of 600 mm has been working well at a particular hydropower installation for many years, and an analysis of available performance data indicate that the efficiency of this turbine is 85% when it is operating at its best efficiency point. It is planned that a geometrically similar turbine with a runner diameter of 2500 mm will be used at another site. Estimate the maximum efficiency that can be expected from the larger unit?

8.88. A hydropower facility that is under design is to accommodate a design flow rate of 44 m^3/s, and under this condition, the available head is 40 m. The turbine under consideration has a shaft rotation rate of 150 rpm and can generate a shaft power 12 MW when operated at maximum efficiency. (a) What is the hydraulic efficiency of the turbine under consideration? (b) What type of turbine is being considered? (c) If the available head during operation is reduced to 18 m and the turbine is operated at maximum efficiency, what shaft power can be extracted by the turbine? Assume water at 20°C.

8.89. A site being considered for hydropower development has an available head of 800 m and a reliable flow rate of 2.5 m^3/s. A generator operating at a speed of 600 rpm is feasible, and the expected hydraulic efficiency of a turbine unit at this site is 95%. What type of turbine should be considered for use?

8.90. A hydropower facility is reported to have several turbine units, each with a rated flow rate of 8 m^3/s at a head of 40 m. The rotational speed of the runner in each turbine is 140 rpm, and the hydraulic efficiency of each unit is 95%. Estimate the type and size of each unit.

8.91. A hydroelectric power plant is reported to have three turbine units, with each unit having a rated power output of 520 kW, a runner size of 800 mm, and a rotation speed of 150 rpm. Under optimal operating conditions, when the available head is 12 m, the flow through each turbine unit is 6 m^3/s. What type of turbine unit is likely being used at this site? Assume water at 20°C.

8.92. A proposed turbine is being designed to generate a power of 30 MW, with a generator rotational speed of 150 rpm and an available head of 22 m. A model of the turbine is to be tested in the laboratory, where the available head is 6 m, the power is 45 kW, and the model turbine is expected to have a hydraulic efficiency of 95%. What length scale, rotational speed, and flow rate should be used in the model tests? Assume water at 20°C.

8.93. A Francis turbine that is operating successfully at a particular hydropower facility has a runner of diameter 2.5 m, has a rotational speed of 120 rpm, and generates 160 MW of power when the available head is 60 m and the flow rate through the turbine is 350 m^3/s. It is anticipated that these optimal conditions can be scaled up at a new hydropower facility, which will use a turbine with the same rotational speed, but the available head will be 90 m. Estimate the runner size, flow rate, power delivered, and efficiency expected at the upscaled facility. Assume water at 20°C.

8.94. The performance of a turbine is being studied using a $\frac{1}{5}$-scale model. The prototype (full-scale) turbine operates at a design head of 35 m when the flow rate through the turbine is 64.1 m^3/s and the angular speed of the runner is 600 rpm. The model is to be tested at a head of 12 m. (a) What should be the angular speed and flow rate in the model to achieve similarity with the prototype? (b) If the shaft power generated in the model is measured as 110 kW and the efficiency in the prototype is assumed to be 5% better than the efficiency in the model, estimate the power that is generated in the prototype under design conditions. (c) What is the specific speed of the turbine, and what should be its type? Assume water at 20°C.

8.95. A particular site in a river valley has an available head for hydropower generation of 35 m. Turbines used at such sites by the local power authority typically have a rotational speed of 130 rpm and optimal efficiencies of approximately 85%. It is desired to generate 25 MW of power at this site. (a) What type of turbine is required? (b) Approximately what flow rate through the turbine would generate the required power at the most efficient operating condition? Assume water at 20°C.

8.96. A Francis turbine is located at an elevation of 2000 m above sea level. The discharge from the runner is 5 m above the tailwater pool, the water flowing through the runner is at a temperature of 15°C, and the critical cavitation parameter of the runner is given by the manufacturer as 0.25. The head loss in the draft tube is negligible. Estimate the maximum available head that can be accommodated without the occurrence of cavitation.

8.97. A proposed hydropower plant is to be located at a site with an available head of 9 m, and it is desired to obtain a (shaft) power of 35 MW from this site. The axial-flow turbine units under consideration operate at an angular speed of 150 rpm, have a specific speed of 5, and have an estimated maximum efficiency of 80%. (a) How may of these units are required? (b) What total flow rate must be available to generate the desired power? Assume water at 20°C.

8.98. (a) What is the typical efficiency associated with a hydropower installation? (b) If a potential site for hydropower generation has an average available head of 350 m and an average flow rate of 0.5 m^3/s, estimate the amount of hydropower that could reasonably be expected from this site.

8.99. A hydroelectric project is to be developed where the 85 percentile flow rate is 1540 m^3/s, and upstream storage to a height of 20 m above the downstream stage is required to meet all the water-supply demands. Estimate the maximum installed capacity that would be appropriate for these conditions.

Flow in Open Channels

LEARNING OBJECTIVES

After reading this chapter and solving a representative sample of end-of-chapter problems, you will be able to:

- Understand and apply the basic equations governing flow in open channels.
- Calculate normal and critical flow depths and understand their engineering applications.
- Analyze flows through channel constrictions and identify conditions under which flows are choked.
- Calculate and classify water surface profiles with realistic applications and constraints.
- Analyze transitions between supercritical and subcritical flows.

9.1 Introduction

In open-channel flows the liquid surface is exposed to the atmosphere, which is the reason for using the word "open." The driving force in open-channel flows is gravity, and open-channel flows are typically found in sanitary sewers, drainage conduits, canals, and rivers. Open-channel flow, sometimes referred to as *free-surface flow*, is more complicated than closed-conduit flow, because the location of the free surface is not constrained and the depth of flow depends on factors such as the volume flow rate and the shape and slope of the channel. Flows in conduits with closed sections, such as pipes, may be classified as either open-channel flow or closed-conduit flow, depending on whether the conduit is flowing full. A closed pipe flowing partially full is an open-channel flow, because the water surface is exposed to the atmosphere.

Nomenclature. Open-channel flow is said to be *steady* if the depth of flow at any location along the channel does not change with time; if the depth of flow changes with time, the flow is called *unsteady*. Most open-channel flows are analyzed under steady-flow conditions. The flow is said to be *uniform* if the depth of flow is the same at every cross section along the channel; if the depth of flow varies along the channel, the flow is called *nonuniform* or *varied*. Uniform flow can be either steady or unsteady, depending on whether the flow depth changes with time; however, uniform flows are practically nonexistent in nature. More commonly, open-channel flows are either steady nonuniform flows or unsteady nonuniform flows. Open channels are classified as either *prismatic* or *nonprismatic*. Prismatic channels are characterized by an unvarying shape of the cross section, constant bottom slope, and relatively straight alignment. In nonprismatic channels, the cross section, alignment, and/or bottom slope change along the channel. Constructed drainage channels such as pipes and canals tend to be prismatic, whereas natural channels such as rivers and creeks tend to be nonprismatic.

9.2 Basic Principles

The governing equations of flow in open channels are the continuity, momentum, and energy equations. Any flow in an open channel must satisfy all three of these equations. Analysis of open-channel flow can usually be accomplished with the control-volume form of the governing equations, and the most useful forms of these equations for steady open-channel flows are derived in the following sections.

9.2.1 Steady-State Continuity Equation

Consider the case of steady nonuniform flow in the open channel illustrated in Figure 9.1. The flow enters and leaves the control volume normal to the control surfaces, with the inflow velocity distribution denoted by v_1 and the outflow velocity distribution by v_2; both the inflow and outflow velocities vary across the control surfaces. The steady-state continuity equation can be written as

$$\int_{A_1} \rho v_1 \, dA = \int_{A_2} \rho v_2 \, dA \tag{9.1}$$

where ρ is the density of the liquid, which can be taken as constant for most applications. Defining V_1 and V_2 as the average velocities across A_1 and A_2, respectively, where

$$V_1 = \frac{1}{A_1}\int_{A_1} v_1 \, dA, \qquad V_2 = \frac{1}{A_2}\int_{A_2} v_2 \, dA \tag{9.2}$$

then for an incompressible liquid (ρ = constant) such as water, the steady-state continuity equation (Equation 9.1) can be written as

$$V_1 A_1 = V_2 A_2 \tag{9.3}$$

which is the same expression that was derived for steady flow of incompressible liquids in closed conduits.

9.2.2 Steady-State Momentum Equation

Consider the case of steady nonuniform flow in the open channel illustrated in Figure 9.2. The steady-state momentum equation for the control volume shown in Figure 9.2 is given by

$$\sum F_x = \int_A \rho v_x (\mathbf{v} \cdot \mathbf{n}) \, dA \tag{9.4}$$

where F_x represents the forces in the flow direction, x, A is the surface area of the control volume, v_x is the flow velocity in the x-direction, $\mathbf{v}$ is the velocity vector, and $\mathbf{n}$ is a unit nor-

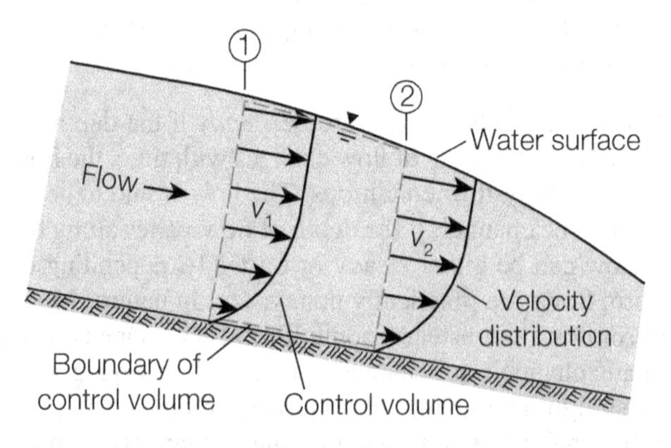

Figure 9.1: Flow in an open channel

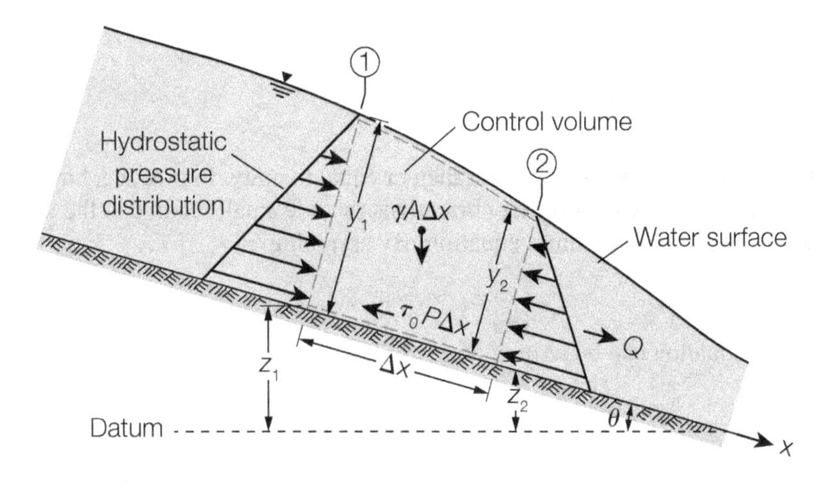

Figure 9.2: Steady nonuniform flow in an open channel

mal directed outward from the control volume. Because the velocities normal to the control surface are nonzero only for the inflow and outflow surfaces, Equation 9.4 can be written as

$$\sum F_x = \int_{A_2} \rho v_x^2 \, \mathrm{d}A - \int_{A_1} \rho v_x^2 \, \mathrm{d}A \tag{9.5}$$

where A_1 and A_2 are the upstream and downstream areas of the control volume, respectively. If the velocity is uniformly distributed (i.e., constant) across the control surface, then Equation 9.5 becomes

$$\sum F_x = \rho v_2^2 A_2 - \rho v_1^2 A_1 \tag{9.6}$$

where v_1 and v_2 are the velocities on the upstream and downstream faces of the control volume, respectively.

Momentum correction coefficient. In reality, velocity distributions in open channels are never uniformly distributed across the cross section of the channel. So it is convenient to define a *momentum correction coefficient*, β, by the relation

$$\beta = \frac{\displaystyle\int_A v^2 \, \mathrm{d}A}{V^2 A} \tag{9.7}$$

where V is the mean velocity across the channel section of area A. The momentum correction coefficient, β, is sometimes called the *Boussinesq coefficient*, or simply the *momentum coefficient*. Values of β are typically in the range of 1.03–1.07 for turbulent flow in prismatic channels and typically in the range of 1.05–1.17 for turbulent flow in natural streams.

Momentum equation with correction coefficient. Applying the definition of the momentum correction coefficient to Equation 9.5 leads to the following form of the momentum equation:

$$\sum F_x = \rho \beta_2 V_2^2 A_2 - \rho \beta_1 V_1^2 A_1 \tag{9.8}$$

where β_1 and β_2 are the momentum correction coefficients at the upstream and downstream faces of the control volume, respectively. Because the continuity equation requires that the volume flow rate, Q, be the same at each cross section,

$$Q = A_1 V_1 = A_2 V_2 \tag{9.9}$$

and the momentum equation (Equation 9.8) can be written as

$$\sum F_x = \rho Q(\beta_2 V_2 - \beta_1 V_1) \tag{9.10}$$

By definition, values of β must be greater than or equal to unity. In practice, however, deviations of β from unity are second-order corrections that are small relative to the uncertainties in the other terms in the momentum equation. By assuming

$$\beta_1 \approx \beta_2 = 1$$

the momentum equation can be written as

$$\sum F_x \approx \rho Q(V_2 - V_1) \tag{9.11}$$

Equation 9.10 (and sometimes its approximation given by Equation 9.11) can be applied to any control volume within an open channel under the condition of steady flow. In practical applications, the forces F_x can be attributed to a variety of sources, such as the reactions of structures located within the control volume, friction between the flowing liquid and the (fixed) channel surface, and/or gravity forces.

Momentum equation with only friction and gravity forces. Consider the control volume shown in Figure 9.2, where the only forces acting on the liquid within the control volume are the frictional force between the moving liquid and the channel surface, and the gravity force. Assuming that the momentum correction coefficients on the inflow and outflow surfaces of the control volume are equal to unity, the momentum equation (Equation 9.11) can be written as

$$\gamma A \Delta x \sin\theta - \tau_0 P \Delta x + \gamma A(y_1 - y_2) = \rho Q(V_2 - V_1) \tag{9.12}$$

where γ is the specific weight of the liquid, A is the average cross-sectional area of the control volume, Δx is the length of the control volume, θ is the inclination of the channel, τ_0 is the average shear stress on the channel surface within the control volume, P is the average (wetted) perimeter of the channel cross section, and y_1 and y_2 are the upstream and downstream depths, respectively, on the surfaces of the control volume. The three force terms on the left-hand side of Equation 9.12 are: the component of the weight of the liquid in the direction of flow, the shear force exerted by the channel surface on the moving liquid, and the net hydrostatic force. If z_1 and z_2 are the elevations of the bottom of the channel at the upstream and downstream surfaces of the control volume, then

$$\sin\theta = \frac{z_1 - z_2}{\Delta x} \tag{9.13}$$

Combining Equations 9.12 and 9.13 and rearranging leads to

$$\tau_0 = -\gamma \frac{A}{P}\frac{\Delta z}{\Delta x} - \gamma \frac{A}{P}\frac{\Delta y}{\Delta x} - \gamma \frac{A}{P}\frac{V}{g}\frac{\Delta V}{\Delta x} \tag{9.14}$$

where Δz, Δy, and ΔV are defined by

$$\Delta z = z_2 - z_1, \qquad \Delta y = y_2 - y_1, \qquad \Delta V = V_2 - V_1 \tag{9.15}$$

and $V(= Q/A)$ is the average velocity in the control volume. The ratio A/P is commonly called the *hydraulic radius*, R, where

$$R = \frac{A}{P} \tag{9.16}$$

In open-channel applications, the hydraulic radius, R, is sometimes called the *hydraulic mean depth*, a nomenclature that is particularly common in the United Kingdom. Combining Equations 9.14 and 9.16 and taking the limit as $\Delta x \to 0$ yields

$$\tau_0 = -\gamma R\left[\lim_{\Delta x\to 0}\frac{\Delta z}{\Delta x} + \lim_{\Delta x\to 0}\frac{\Delta y}{\Delta x} + \frac{V}{g}\lim_{\Delta x\to 0}\frac{\Delta V}{\Delta x}\right] = -\gamma R\left[\frac{\mathrm{d}z}{\mathrm{d}x} + \frac{\mathrm{d}y}{\mathrm{d}x} + \frac{V}{g}\frac{\mathrm{d}V}{\mathrm{d}x}\right]$$

$$\to \quad \tau_0 = -\gamma R\frac{\mathrm{d}}{\mathrm{d}x}\left[z + y + \frac{V^2}{2g}\right] \tag{9.17}$$

The term in brackets is the mechanical energy per unit weight of the liquid, E, defined as

$$E = y + z + \frac{V^2}{2g} \tag{9.18}$$

It should be noted that the mechanical energy per unit weight of a liquid element is usually defined as $p'/\gamma + z' + V^2/2g$, where z' is elevation of the liquid element relative to a defined datum and p' is the pressure at the location of the liquid element. If the pressure distribution is hydrostatic across the channel cross section, then $p'/\gamma + z' = \text{constant} = y + z$, where y is the depth of liquid and z is the elevation of the bottom of the channel. The mechanical energy per unit weight, E, can therefore be written as $y + z + V^2/2g$. A plot of E versus the distance along the channel is called the *energy grade line*. The momentum equation, Equation 9.17, can now be written as

$$\tau_0 = -\gamma R\frac{\mathrm{d}E}{\mathrm{d}x} \tag{9.19}$$

which can be expressed as

$$\tau_0 = \gamma R S_\mathrm{e} \quad \text{or} \quad \tau_0 = \gamma R S_\mathrm{f} \tag{9.20}$$

where S_e is equal to the slope of the energy grade line, which is taken as positive when it slopes downward in the direction of flow. In practical applications, the slope of the energy grade line, S_e, is commonly denoted by the *friction slope*, S_f, where the subscript "f" is used to indicate that the loss in mechanical energy, E, is due to friction between the flowing liquid and the channel surface.

Special case: uniform flow. *Uniform flow* occurs when flow conditions remain constant along the channel, which means that for a given flow rate, the flow depth, y, and average velocity, V, remain constant and are independent of x. Achievement of uniform flow requires a prismatic (i.e., constant shape) channel and a constant slope, S_0, of the channel. Under these conditions, S_e and S_f are given by

$$S_\mathrm{e} = S_\mathrm{f} = -\frac{\mathrm{d}E}{\mathrm{d}x} = -\frac{\mathrm{d}}{\mathrm{d}x}\left[z + y + \frac{V^2}{2g}\right] = \underbrace{-\frac{\mathrm{d}z}{\mathrm{d}x}}_{=S_0} - \underbrace{\frac{\mathrm{d}y}{\mathrm{d}x}}_{=0} - \underbrace{\frac{\mathrm{d}}{\mathrm{d}x}\left(\frac{V^2}{2g}\right)}_{=0} = S_0 \tag{9.21}$$

which means that the slope of the energy grade line is equal to the slope of the channel (S_0) under uniform-flow conditions. Consequently, under uniform-flow conditions, the boundary shear stress as given by Equation 9.20 can be expressed as

$$\tau_0 = \gamma R S_0 \tag{9.22}$$

This equation is particularly useful because it expresses the average shear stress on the channel surface, τ_0, in terms of the measurable quantities γ, R, and S_0. Practical uses of Equation 9.22 include the design of channel linings that provide adequate resistance to the boundary shear stress, τ_0, caused by flow in the channel.

9.2.2.1 Darcy–Weisbach equation

Although the momentum equation for steady uniform open-channel flow as given by Equation 9.22 is useful for estimating the shear stress on a channel surface, it does not provide a direct relationship between the average flow velocity, V, the slope of channel, S_0, and the depth of flow, which is represented by the hydraulic radius, R. To relate V to S_0 and R, a relationship between the boundary shear stress, τ_0, and V must be specified. This relationship is generally expressed in terms of a *friction factor*, f, defined as

$$f = \frac{\tau_0}{\frac{1}{8}\rho V^2} \tag{9.23}$$

where ρ is the density of the liquid flowing in the channel. The definition of f given by Equation 9.23 is almost identical to the definition of the *skin friction coefficient*, c_f, commonly used in the analysis of boundary-layer flows to relate the boundary (wall) shear stress, τ_w, to the free-stream velocity, V, of the liquid, where

$$c_\mathrm{f} = \frac{\tau_\mathrm{w}}{\frac{1}{2}\rho V^2}$$

The only difference between the definition of f and the definition of c_f is that the factor $\frac{1}{8}$ is used instead of the factor $\frac{1}{2}$ in the definition of f. Combining Equations 9.23 and 9.22, eliminating τ_0, and making V the subject of the formula yields

$$V = \left(\frac{8g}{f}\right)^{\frac{1}{2}} \sqrt{RS_0} \tag{9.24}$$

This equation provides a direct relationship between the average flow velocity, V; the slope, S_0, of the channel; and the depth of flow as measured by the hydraulic radius, R. Equation 9.24 is the form of the *Darcy–Weisbach equation* that is useful in open-channel flow applications. The key parameter in the Darcy–Weisbach equation is the friction factor, f. The flow and channel properties influencing the value of f, and empirical equations that are commonly used to estimate f, are described below.

Dimensional analysis. In practical applications, it is useful to express the average shear stress, τ_0, on the channel surface in terms of the flow properties, fluid properties, and the surface roughness of the channel. A functional expression for the average shear stress, τ_0, can be expressed in the form

$$\tau_0 = f_0(V, R, \rho, \mu, \epsilon, \epsilon', m, s) \tag{9.25}$$

where V is the average velocity in the channel [LT^{-1}], R is the hydraulic radius [L], ρ is the liquid density [ML^{-3}], μ is the dynamic viscosity of the liquid [FL^{-2}T], ϵ is the characteristic size of the roughness projections on the channel surface [L], ϵ' is the characteristic spacing of the roughness projections [L], m is a form factor that describes the shape of the roughness elements [dimensionless], and s is a channel shape factor that describes the shape of the channel cross section [dimensionless]. In accordance with the Buckingham pi theorem, the functional relationship given by Equation 9.25 between nine variables in three dimensions can also be expressed as a relation between six dimensionless groups as follows:

$$\frac{\tau_0}{\rho V^2} = f_1\left(\frac{\rho V R}{\mu}, \frac{\epsilon}{R}, \frac{\epsilon'}{R}, m, s\right) \tag{9.26}$$

This functional relationship is almost exactly the same as that derived for flow in pipes, with the only difference being that the hydraulic radius, R, is used to characterize the flow area instead of the diameter, D, in the case of pipes. These two functional analyses can be unified by considering the flow in an open channel to be equivalent to the flow in a conduit of non-circular cross section, in which the size of the flow area is characterized by the hydraulic diameter, D_h, defined as

$$D_\mathrm{h} = 4\frac{A}{P} = 4R \tag{9.27}$$

By using D_h instead of R to characterize the flow area in an open channel, Equation 9.26 becomes

$$\frac{\tau_0}{\rho V^2} = f_2\left(\frac{\rho V D_\mathrm{h}}{\mu}, \frac{\epsilon}{D_\mathrm{h}}, \frac{\epsilon'}{D_\mathrm{h}}, m, s\right) \tag{9.28}$$

The beauty of using Equation 9.28 to express the shear stress on the surface of an open channel is that this same equation is used to quantify the shear stress on the surface of a close conduit of diameter D, because $D_\mathrm{h} = D$ when the cross section is circular. It is convenient to define the Reynolds number, Re_h, in terms of D_h by the relation

$$\mathrm{Re}_\mathrm{h} = \frac{\rho V D_\mathrm{h}}{\mu} \tag{9.29}$$

Combining Equations 9.29 and 9.28 gives the following functional expression for the boundary shear stress, τ_0:

$$\frac{\tau_0}{\rho V^2} = f_2\left(\mathrm{Re}_\mathrm{h}, \frac{\epsilon}{D_\mathrm{h}}, \frac{\epsilon'}{D_\mathrm{h}}, m, s\right) \tag{9.30}$$

Equation 9.30 can be applied to both closed conduits and open channels, where open channels are regarded simply as conduits of non-circular cross section. If the influences of the shape of the cross section and the arrangement of roughness elements are small relative to the influences of the size of the roughness elements and the viscosity of the liquid and the flow is steady and uniform along the channel, then the shear stress can be expressed in the following functional form:

$$\frac{\tau_0}{\frac{1}{8}\rho V^2} = f\left(\mathrm{Re}_\mathrm{h}, \frac{\epsilon}{D_\mathrm{h}}\right) \tag{9.31}$$

where the function f can be expected to closely approximate the Darcy friction factor in pipes.

Limitations of dimensional analysis. In reality, the friction factor, f, defined in Equation 9.31 has been observed to be a function of channel shape, with the influence of shape decreasing roughly in the order of rectangular, triangular, trapezoidal, and circular channels (Chow, 1959). As channels become very wide or otherwise depart radically from the shape of a circle or semicircle, the friction factors derived from pipe experiments become less applicable to open channels. Myers (1991) has shown that friction factors in wide rectangular open channels are as much as 45% greater than in narrow sections with the same Re_h and ϵ/D_h. The question of how to account for the shape of an open channel in estimating the friction factor remains an active area of investigation. Also, the assumption that the friction factor is independent of the arrangement and shape of the roughness projections has been shown to be invalid in cases such as gravel-bed streams with high boulder concentrations (Ferro, 1999).

Empirical estimation of the friction factor. The transition from laminar to turbulent flow in open channels occurs at around $\mathrm{Re}_\mathrm{h} = 600$, and turbulent flow typically occurs when $\mathrm{Re}_\mathrm{h} >$ 12 500. It is convenient to define three types of turbulent flow: *smooth*, *transition*, and *rough*. The flow is classified as "smooth" when the roughness projections on the channel surface are submerged in a laminar-flow layer adjacent to the channel surface, in which case the friction factor depends only on the Reynolds number, Re_h, and can be estimated by (Henderson, 1966)

$$\frac{1}{\sqrt{f}} = \begin{cases} 1.78\text{Re}_\text{h}^{\frac{1}{8}}, & \text{Re}_\text{h} < 10^5 \quad (9.32) \\ -2.0\log_{10}\left(\dfrac{2.51}{\text{Re}_\text{h}\sqrt{f}}\right), & \text{Re}_\text{h} > 10^5 \quad (9.33) \end{cases}$$

These relations are the same as the Blasius and Prandtl–von Kármán equations for flow in pipes. However, owing to the free surface and the interdependence of the hydraulic radius, discharge, and slope, the relationship between f and Re_h in open-channel flow is not identical to that for pipe flow. The flow is classified as "rough" when the roughness projections on the channel surface extend out of the laminar layer, creating sufficient turbulence so that the friction factor depends only on the relative roughness. "Rough" flow is also commonly referred to as *fully turbulent flow*, although "smooth" and "transitional" flows are also turbulent. Under rough flow conditions, the friction factor can be estimated by (ASCE, 1963)

$$\frac{1}{\sqrt{f}} = -2\log_{10}\left(\frac{k_\text{s}}{12R}\right) \tag{9.34}$$

where k_s is the equivalent sand roughness of the channel surface. Equation 9.34 is derived from the integration of the Nikuradse velocity distribution for rough flow over a trapezoidal open channel cross section (Keulegan, 1938) and gives a higher friction factor than the Prandtl–von Kármán equation that is used in pipe flow. In the transition region between (hydraulically) smooth and rough flow, the friction factor depends on the Reynolds number and the relative roughness and can be approximated by (ASCE, 1963)

$$\frac{1}{\sqrt{f}} = -2\log_{10}\left(\frac{k_\text{s}}{12R} + \frac{2.5}{\text{Re}_\text{h}\sqrt{f}}\right) \tag{9.35}$$

This relation differs slightly from the Colebrook equation for transition flow in closed conduits, but it is still called the Colebrook equation and is commonly applied in both smooth and rough flow in open channels. Caution should be used in applying Equation 9.35 to smooth-flow conditions in open channels because under these conditions, Equation 9.35 asymptotes to the Prandtl–von Kármán equation (Equation 9.33), which is known to deviate by up to 20% from measurements (Cheng et al., 2011). Equation 9.35 was originally suggested by Henderson (1966) for wide open channels (width/depth $\geq$ 10), and others have suggested similar formulations with different constants (e.g., Yen, 1991).

Turbulent flow conditions. The three conditions of turbulent flow (smooth, transition, rough) can be delineated using the *shear velocity Reynolds number*, also known as the *roughness Reynolds number*, $k_\text{s}u_*/\nu$, where u_* is the *shear velocity* defined by

$$u_* = \sqrt{\frac{\tau_0}{\rho}} = \sqrt{gRS_\text{f}} \tag{9.36}$$

where S_f is the friction slope (as defined by Equation 9.20) and ν is the kinematic viscosity of the liquid. The friction velocity, u_*, is commonly estimated under uniform-flow conditions, in which case $S_\text{f} = S_0$, where S_0 is the slope of the channel. The approximate ranges for the three conditions of turbulent flow are as follows:

$$\frac{u_* k_\text{s}}{\nu} = \begin{cases} < 5, & \text{smooth} \\ 5 \text{ to } 70, & \text{transition} \\ > 70, & \text{rough} \end{cases} \tag{9.37}$$

There is still some debate on defining the region of transition flow by the limits in Equation 9.37; for example, Henderson (1966) defines the transition region as $4 \leq u_* k_s/\nu \leq 100$, Yang (1996) defines it as $5 \leq u_* k_s/\nu \leq 70$, and Rubin and Atkinson (2001) define it as $5 \leq u_* k_s/\nu \leq 80$. The Colebrook friction factor equation given by Equation 9.35 is commonly applied across all three flow regimes.

Friction factor in rock-bedded channels. For shallow flows in rock-bedded channels, conditions that are commonly found in mountain streams, larger rocks produce most of the resistance to flow. In these cases, where the size of the roughness elements are comparable to the flow depth, the roughness elements are classified as *macroscale roughness*, the shear force on the channel surface is caused more by pressure (form) drag than friction drag, and the shape and arrangement of the macroscale roughness elements can have a significant effect on the friction coefficient (Canovaro et al., 2007). Limerinos (1970) has shown that the friction factor, f, can be estimated from the size of the rock in the stream bed using the relation

$$\frac{1}{\sqrt{f}} = 1.2 + 2.03 \log_{10}\left(\frac{R}{d_{84}}\right) \tag{9.38}$$

where d_{84} is the 84-percentile size of the rocks on the stream bed, and R is the hydraulic radius of the flow area. Several alternatives to Equation 9.38 have been proposed for rock-bedded mountain streams, and a thorough review of alternative formulations can be found in Ferguson (2007) and Yen (2002). An extensive review of data in coarse-bed channels where $d_{50} > 2$ mm (the minimum size to be classified as gravel) shows that Equation 9.34 can be used to estimate the friction factor in these cases by taking k_s equal to $2.4\,d_{90}$, $2.8\,d_{84}$, or $6.1\,d_{50}$ (López and Barragán, 2008).

Friction factor in channels with submerged vegetation. For channels lined with submerged vegetation, the flow dynamics that generate shear stress on the bottom of the channel are fundamentally different from the flow dynamics that generate shear stress in channels lined with sediment, gravel, or rocks. In the case of submerged vegetation, the friction factor, f, can be estimated by (Cheng, 2011)

$$f = 0.40\left(\frac{k_v}{h}\right)^{\frac{1}{8}} \tag{9.39}$$

where k_v is a roughness length scale [L], defined by

$$k_v = \frac{\pi}{4}\frac{\lambda}{1-\lambda}D \tag{9.40}$$

where λ is the fraction of the channel bottom occupied by vegetation stems [dimensionless], D is the average stem diameter [L], and h is the depth of flow above the vegetation. Equation 9.39 provides a good fit to the observed data provided that $1 \times 10^{-5} \leq k_v/h \leq 3 \times 10^{-2}$.

Application of the Darcy–Weisbach equation. Regardless of the empirical equation used to estimate the friction factor, f, the form of the Darcy–Weisbach equation that is of most practical use is Equation 9.24. Recall that the Darcy–Weisbach equation relates the average velocity, V, in an open channel to the channel slope, S_0, of the channel and the hydraulic radius, R. In most applications of the Darcy–Weisbach equation (Equation 9.24), the friction factor, f, is estimated using the Colebrook equation (Equation 9.35). To facilitate arriving at a combined form of the Darcy–Weisbach and Colebrook equations, it is convenient to express the Darcy–Weisbach equation in the form

$$V\sqrt{f} = \sqrt{8gRS_0} \tag{9.41}$$

and the friction factor, f, given by Equation 9.35, can be expressed in the form

$$\frac{1}{\sqrt{f}} = -2\log_{10}\left(\frac{k_s}{12R} + \frac{2.5}{\frac{\rho V(4R)}{\mu}\sqrt{f}}\right) = -2\log_{10}\left(\frac{k_s}{12R} + \frac{0.625\nu}{RV\sqrt{f}}\right) \tag{9.42}$$

Combining Equations 9.41 and 9.42 yields

$$Q = VA = -2A\sqrt{8gRS_0}\log_{10}\left(\frac{k_s}{12R} + \frac{0.625\nu}{R^{\frac{3}{2}}\sqrt{8gS_0}}\right) \tag{9.43}$$

This derived relationship is particularly useful in relating the volume flow rate, Q, to the flow area, A, and hydraulic radius, R, for a given channel slope, S_0, and roughness height, k_s.

EXAMPLE 9.1

Water flows at a depth of 1.83 m in a trapezoidal concrete-lined section ($k_s = 1.5$ mm) with a bottom width of 3 m and side slopes of 2 : 1 (H : V). The slope of the channel is 0.0005, and the water temperature is 20°C. Assuming uniform-flow conditions, estimate the average velocity and flow rate in the channel.

SOLUTION

The flow in the channel is illustrated in Figure 9.3.

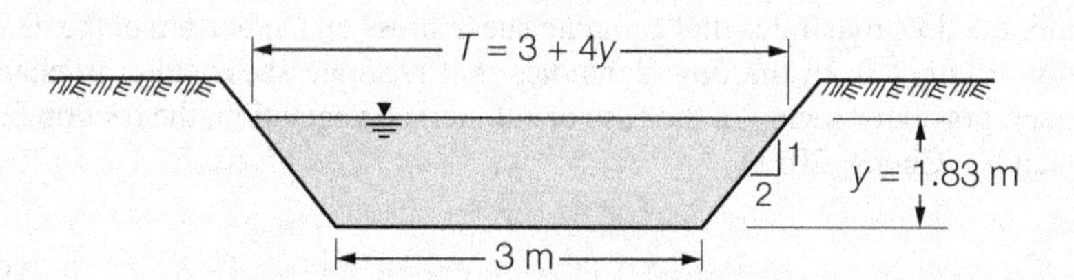

Figure 9.3: Flow in a trapezoidal channel

From the given data and from the dimensions of the channel shown in Figure 9.3: $S_0 = 0.0005$, $A = 12.2$ m^2, $P = 11.2$ m, and

$$R = \frac{A}{P} = \frac{12.2}{11.2} = 1.09 \text{ m}$$

For concrete, $k_s = 1.5$ mm $= 0.0015$ m, and for water at 20°C, $\nu = 1.00 \times 10^{-6}$ m^2/s. Substituting these data into Equation 9.43 gives the flow rate, Q, as

$$\begin{aligned} Q &= -2A\sqrt{8gRS_0}\log_{10}\left(\frac{k_s}{12R} + \frac{0.625\nu}{R^{\frac{3}{2}}\sqrt{8gS_0}}\right) \\ &= -2(12.2)\sqrt{8(9.81)(1.09)(0.0005)}\log_{10}\left[\frac{0.0015}{12(1.09)} + \frac{0.625(1.00 \times 10^{-6})}{(1.09)^{\frac{3}{2}}\sqrt{8(9.81)(0.0005)}}\right] \\ &= 19.8 \text{ m}^3\text{/s} \end{aligned}$$

and the average velocity, V, is given by

$$V = \frac{Q}{A} = \frac{19.8}{12.2} = 1.62 \text{ m/s}$$

Therefore, for the given flow depth in the channel, the flow rate is **19.8 m³/s** and the average velocity is **1.62 m/s**.

9.2.2.2 Manning equation

Practical application of the Darcy–Weisbach equation and the associated empirical equations for estimating the friction factor are based primarily on the pipe experiments of Nikuradse and Colebrook, which were conducted between 1930 and 1940. However, the development of empirical and semi-empirical relationships to relate the volume flow rate in open channels to the channel geometry, depth of flow, and channel-surface characteristics began much earlier. In 1775, Chézy[57] proposed the following expression for the mean velocity in an open channel:

$$V = C\sqrt{RS_0} \tag{9.44}$$

where C was referred to as the *Chézy coefficient*. Equation 9.44 has the same form as Equation 9.24 and was derived in the same way, except that the functional dependence of the Chézy coefficient on the Reynolds number and the relative roughness was not considered. In comparing Equations 9.24 and 9.44, the Chézy coefficient is related to the friction factor by

$$C = \sqrt{\frac{8g}{f}} \tag{9.45}$$

In 1869, Ganguillet and Kutter (1869) published an elaborate formula for C in open channels carrying water that became widely used. In 1890, Manning (1890) demonstrated that the data used by Ganguillet and Kutter could be represented in a simple empirical relationship in which C varies as the sixth root of R, where

$$C = \frac{R^{\frac{1}{6}}}{n} \tag{9.46}$$

and n is a coefficient that is characteristic of the surface roughness of the channel. Because C is not a dimensionless quantity, values of n were specified to be consistent with length units measured in meters and time in seconds.

When Equation 9.46 is combined with the Chézy equation, the resulting expression is called the *Manning equation*[58] or *Strickler equation* (in Central Europe) and is given by

$$V = \frac{1}{n} R^{\frac{2}{3}} S^{\frac{1}{2}} \tag{9.47}$$

where $S = S_\text{f} = S_0$ under uniform flow conditions, and $1/n$ is called the *Strickler coefficient* when Equation 9.47 is called the Strickler equation.

[57] Antoine de Chézy (1718–1798) was a French engineer.

[58] Named in honor of the Irish engineer and accountant Robert Manning (1816–1897).

Normal depth. The *normal depth* (of flow) is defined as the depth of flow that will occur in a channel of constant bed slope and roughness, provided the channel is sufficiently long and the flow is undisturbed (Knight et al., 2010). When the Darcy–Weisbach or Manning equation is used to calculate the depth of flow for a given volume flow rate, channel roughness, and channel slope (for $S_f = S_0$), the calculated depth of flow is taken to be the normal depth.

Theoretical basis of Manning's n. Because the scientific foundation of the Darcy–Weisbach equation is sounder than the empirical foundation of the Manning equation, the functional dependencies of Manning's n can be identified by comparing the Manning and Darcy–Weisbach equations. By equating the mean velocities, V, calculated by these two equations (i.e., by Equations 9.24 and 9.47), it can be directly shown that the Manning roughness coefficient, n, and friction factor, f, are related by

$$\frac{R^{\frac{1}{6}}}{n} = \sqrt{\frac{8g}{f}} \tag{9.48}$$

and substituting the expression for the friction factor under fully turbulent conditions (Equation 9.34) into Equation 9.48 and rearranging yields

$$\frac{n}{k_s^{\frac{1}{6}}} = \frac{\frac{1}{\sqrt{8g}}\left(\frac{R}{k_s}\right)^{\frac{1}{6}}}{2.0 \log\left(12\frac{R}{k_s}\right)} \tag{9.49}$$

Taking $g = 9.81$ m/s^2, Equation 9.49 is plotted in Figure 9.4, which illustrates that the value of $n/k_s^{\frac{1}{6}}$ is effectively constant over a wide range of values of R/k_s, where n being a constant is an essential assumption of the Manning equation. Equation 9.49 (with $g = 9.81$ m/s^2) gives a minimum value of

$$\frac{n}{k_s^{\frac{1}{6}}} = 0.039 \tag{9.50}$$

at $R/k_s = 34$. Also, $n/k_s^{\frac{1}{6}}$ is within $\pm$5% of a constant value over a range of R/k_s given by $4 < R/k_s < 500$ as shown by Yen (1991). This range of R/k_s for the validity of the Manning equation differs somewhat from the range given by Hager (1999) as $3.6 < R/k_s < 360$. It is important to note that the R/k_s criterion for the validity of the Manning equation relies on the assumption that the flow is fully turbulent; that is, it is in the rough flow regime. According

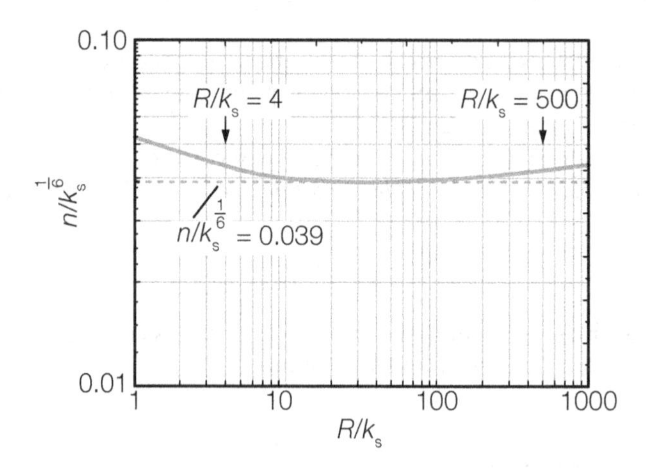

Figure 9.4: Variation of $n/k_s^{\frac{1}{6}}$ in fully turbulent flow

to the flow regime delineation given by Equation 9.37, the fully turbulent criterion can be taken as

$$\frac{u_* k_s}{\nu} > 70 \tag{9.51}$$

where u_* is the shear velocity [LT^{-1}] which can be estimated by $\sqrt{gRS_f}$, g is gravity [LT^{-2}], R is the hydraulic radius [L], S_f is the head loss per unit length [dimensionless], and ν is the kinematic viscosity of the liquid [L^{2}T^{-1}]. For water at 20°C, $\nu = 1.00 \times 10^{-6}$ m^2/s, and noting that $S_f = S_0$ for uniform flow and g = 9.81 m/s^2, Equation 9.51 can be expressed as

$$k_s\sqrt{RS_0} > 2.2 \times 10^{-5} \tag{9.52}$$

where k_s and R are in meters.

Summary. In summary, the validity of the Manning equation relies on two assumptions: (1) $n/k_s^{\frac{1}{6}}$ is constant and (2) the flow is fully turbulent. These assumptions require that $4 < R/k_s < 500$ and that $u_* k_s/\nu > 70$, respectively. If these two conditions are met, then application of the Manning equation is valid and the average velocities estimated using the Manning equation and the Darcy–Weisbach equation are approximately the same, provided Manning's n is taken as $0.039k_s$.

EXAMPLE 9.2

Water flows at a depth of 3.00 m in a trapezoidal concrete-lined channel with a bottom width of 3 m and side slopes of 2 : 1 (H : V). The slope of the channel is 0.0005, and the water temperature is 20°C. Assess the validity of using the Manning equation with $n = 0.014$.

SOLUTION

From the given data: $n = 0.014$, $S_0 = 0.0005$, $A = 27$ m^2, $P = 16.42$ m, and $R = A/P =$ 1.64 m. Assuming that the Manning equation is valid, Equation 9.50 gives

$$k_s = \left(\frac{n}{0.039}\right)^6 = \left(\frac{0.014}{0.039}\right)^6 = 0.0021 \text{ m} = 2.1 \text{ mm}$$

This estimate of k_s can be used to estimate the parameters R/k_s and $k_s\sqrt{RS_0}$ that are used as a basis for evaluating the validity of the Manning equation.

$$\frac{R}{k_s} = \frac{1.64}{0.0021} = 781, \qquad k_s\sqrt{RS_0} = 0.0021\sqrt{(1.64)(0.0005)} = 6.01 \times 10^{-5}$$

Because $k_s\sqrt{RS_0} > 2.2 \times 10^{-5}$, the flow is fully turbulent; however, because $R/k_s > 500$, the value of $(n/k_s)^{\frac{1}{6}}$ cannot be taken as a constant equal to 0.039 because $(n/k_s)^{\frac{1}{6}}$ depends on the value of R/k_s. Based on these results, the Manning equation formulation is **not strictly valid** because the Manning roughness coefficient, n, cannot be taken as a constant and independent of the depth of flow. The Darcy–Weisbach equation is more appropriate in this case. Unfortunately, it is still common practice to assume that the Manning equation is valid, without verifying that the fully turbulent and constant n conditions are met.

Relationships between Manning's n and roughness height. The theoretical analysis presented above indicates that when Manning's equation is applicable, the n value is related to the equivalent sand roughness by Equation 9.50, which can be expressed in the form

$$n = 0.039\, k_s^{\frac{1}{6}} \tag{9.53}$$

where k_s is in meters. Because natural and constructed channels have varying roughness characteristics, the relationship between the roughness characteristics and k_s must still be established. It should be noted that the sixth-root relationship between the roughness height, k_s, and the roughness coefficient, n, means that large relative errors in estimating k_s result in much smaller relative errors in estimating n. For example, a thousandfold change in the roughness height results in about a threefold change in n. Several investigators have suggested relationships between Manning's n and a characteristic roughness height in the form

$$n = \alpha\, d_p^{\frac{1}{6}} \tag{9.54}$$

where α is a constant and d_p is a percentile grain size of the channel lining material. These approaches are theoretically consistent with Equation 9.53, because varying channel roughness characteristics will naturally result in the equivalent sand roughness, k_s, being associated with different percentiles of the grain size distribution on the channel boundary. Several semiempirical relationships that relate Manning's n to percentile grain sizes on the channel boundary are shown in Table 9.1. Some of the formulas given in Table 9.1 express n as a function of both the percentile grain size, d_p, and the hydraulic radius, R, and such relationships are particularly applicable at low flow depths in channels lined with coarse-grained materials. Analyses of experimental data on riprap-lined channels have shown that it is appropriate to use the formulas relating n to d_p when $R/d_{50} > 3$. Otherwise, n will depend on both R and d_p (Froehlich, 2012), and an appropriate functional relationship that includes both of these variables should be used.

Table 9.1: Semiempirical Expressions for Manning's n

Expression for n	Conditions	Reference
$0.047\, d_{50}^{\frac{1}{6}}$	Uniform sand	Strickler (1923)
$0.0025\, d_{90}^{\frac{1}{6}}$	—	Keulegan (1938)
$0.0039\, d_{90}^{\frac{1}{6}}$	Sand mixtures	Meyer-Peter and Müller (1948)
$0.047\, d_{75}^{\frac{1}{6}}$	Gravel-lined canals	Lane and Carlson (1953)
$0.046\, d_{50}^{\frac{1}{6}}$	Riprap-lined channels	Maynord (1991)
$\dfrac{R^{\frac{1}{6}}}{[7.69 \ln(R/d_{84}) + 63.4]}$	—	Limerinos (1970)
$\dfrac{R^{\frac{1}{6}}}{[7.64 \ln(R/d_{84}) + 65.3]}$	Gravel streams with $S_0 > 0.004$	Bathurst (1985)
$\dfrac{R^{\frac{1}{6}}}{[7.83 \ln(R/d_{84}) + 72.9]}$	Wide channels	Dingman (1984)
$\dfrac{R^{\frac{1}{6}}}{[7.64 \ln(R/d_{50}) + 15.5]}$	Riprap-lined channels	Froehlich (2012)

Table 9.2: Manning Coefficient for Open Channels

Channel type	Manning n	Range
Lined channels:		
Brick, glazed	0.013	0.011–0.015
Brick	—	0.012–0.018
Concrete, float finish	0.015	0.011–0.020
Asphalt	—	0.013–0.020
Rubble or riprap	—	0.020–0.035
Concrete, concrete bottom	0.030	0.020–0.035
Gravel bottom with riprap	0.033	0.023–0.036
Vegetal	—	0.030–0.400
Excavated or dredged channels:		
Earth, straight, and uniform	0.027	0.022–0.033
Earth, winding, fairly uniform	0.035	0.030–0.040
Rock	0.040	0.035–0.050
Dense vegetation	—	0.050–0.12
Unmaintained	0.080	0.050–0.12
Natural channels:		
Clean, straight	0.030	0.025–0.033
Clean, irregular	0.040	0.033–0.045
Weedy, irregular	0.070	0.050–0.080
Brush, irregular	—	0.070–0.16
Floodplains:		
Pasture, no brush	0.035	0.030–0.050
Brush, scattered	0.050	0.035–0.070
Brush, dense	0.100	0.070–0.16
Timber and brush	—	0.10–0.20

Sources: ASCE (1982); Wurbs and James (2002); Bedient and Huber (2002).

Typical values of Manning's *n*. Manning's n can be estimated using typical values. The use of typical n values is usually appropriate in lined artificial channels such as concrete-lined channels; however, in natural channels, using typical values is less certain and other empirical models might be more appropriate. Typical values of the roughness coefficient, n, used in engineering practice are given in Table 9.2, where lower values of n are for surfaces in good condition and higher values are for surfaces in poor condition.

9.2.2.3 Velocity distribution

In *wide channels*, lateral boundaries have negligible effects on the velocity distribution in the central portion of the channel. A wide channel is typically defined as having a width greater than 10 times the flow depth. The velocity distribution, $v(y)$, in wide open channels is typically assumed to be a function of the logarithm of the distance from the bottom of the channel and is frequently approximated by the relation (Vanoni, 1941)

$$v(y) = V + \frac{1}{\kappa}\sqrt{gdS_0}\left(1 + 2.3\log\frac{y}{d}\right) \tag{9.55}$$

where V is the depth-averaged velocity [LT^{-1}], κ is the von Kármán constant (≈ 0.4) [dimensionless], d is the depth of flow [L], S_0 is the slope of the channel [dimensionless], and

y is the distance from the bottom of the channel [L]. As an alternative to Equation 9.55, the velocity distribution, $v(y)$, is sometimes approximated by the power-law relationship

$$v(y) = V_{\rm max} \left(\frac{y}{d}\right)^{\frac{1}{7}} \tag{9.56}$$

where $V_{\rm max}$ is the maximum velocity [LT^{-1}], which is assumed to occur at the water surface where $y = d$. Reasons for using Equation 9.56 in lieu of Equation 9.55 are its simplicity and its closeness of fit to the logarithmic distribution (Knight et al., 2010). In channels that are not wide, the geometry of the lateral boundaries must be considered in estimating the velocity distribution, leading to more complex expressions (e.g., Wilkerson and McGahan, 2005; Maghrebi and Ball, 2006).

Measurement of velocity. Equation 9.55 indicates that the average velocity, V, in wide open channels occurs at $y/d = 0.368$, or at a distance of $0.368d$ above the bottom of the channel. This result is commonly approximated by the relation

$$V = v(0.4d) \tag{9.57}$$

The average velocity, V, can also be related to the velocities at two depths using Equation 9.55, which yields

$$V = \frac{v(0.2d) + v(0.8d)}{2} \tag{9.58}$$

It is standard practice of the U.S. Geological Survey (USGS) to use point measurements at $0.2d$ and $0.8d$ with Equation 9.58 to estimate the average velocity in channel sections with depths greater than 0.75 m and to use point measurements at $0.4d$ with Equation 9.57 to estimate the average velocity in sections with depths less than or equal to 0.75 m. USGS velocity measurements at designated depths are usually collected using acoustic Doppler velocimeters (ADVs) and acoustic Doppler current profilers (ADCPs). If the velocity profile in a stream is known, the mean velocity can be related to the surface velocity using the velocity profile equation. This permits flow rates in streams to be estimated using noncontact methods that combine remote measurements of the surface velocity with cross-sectional geometry to estimate the stream discharge. Field experiments using microwave and ultra high-frequency (UHF) Doppler radars (including radar guns) to measure surface water velocities have yielded discharge estimates whose accuracy is comparable to conventional contact methods of measuring stream discharge. This noncontact approach is particularly appealing when standard contact methods of measuring discharge are dangerous or cannot be obtained.

Location of maximum velocity. The velocity distribution given by Equation 9.55 indicates that the maximum velocity occurs at the water surface. While the maximum velocity tends to occur at the water surface in broad shallow channels, the maximum velocity in some open channels occurs below the water surface, a phenomenon that has been variously attributed to air drag, secondary flows, and channel geometry. Measurements in constructed channels with triangular, trapezoidal, circular, and narrow rectangular cross sections as well as natural channels with irregular cross sections have been shown to have their maximum velocities in the midplane at approximately 20% (of the depth) below the water surface (White, 2008). In the Mississippi River, the maximum velocity has been observed to occur as much as 33% below the water surface (Gordon, 1992). Some field measurements have indicated that the ratio of maximum velocity to average velocity remains approximately constant over a wide range of flows, but different for each channel (Fulton and Ostrowski, 2008).

Estimation of volume flow rate. Velocity measurements in open channels are usually combined to estimate the volume flow rate, Q, at a cross section of an open channel according to the relation

$$Q = \sum_{i=1}^{N} \overline{V}_i A_i \tag{9.59}$$

where N is the number of subareas across the channel and $\overline{V}_i$ is the average velocity over subarea A_i. Typically, the subareas, A_i, are vertical sections across a channel where the average velocity in each section is estimated from measurements of the vertical velocity profile in the center of the section. Subareas are selected to contain no more than 5%–10% of total flow. Measured discharges (= volume flow rates) are commonly expressed as a function of the corresponding water surface elevation (= stage) to yield a *stage-discharge curve* or *rating curve*. Rating curves are commonly used to relate channel flows to stage measurements. Whereas rating curves can be fairly accurate under steady uniform conditions, unsteady nonuniform conditions can cause significant deviations from the rating curve. Under such conditions, expressing discharge as a function of stage measurements at two locations along the channel can improve discharge estimates (e.g., Schmidt and Yen, 2008). In natural channels, the major source of uncertainty in estimating streamflow from rating curves is the change in channel dimensions over time, which can be caused by bed scour/deposition, bank erosion, vegetation changes, and debris deposition. Such effects require recalibration of the rating curve to control error. Some innovative techniques, such as using air bubbles released at the bottom of the channel, can provide direct measures of volume flow rates within vertical sections (e.g., Yannopoulos et al., 2008).

9.2.2.4 Surface-wave propagation

Consider the case in which water (or some other similar liquid) is initially flowing in an open channel at velocity V and with a depth of flow y. Suppose that a disturbance in the flow causes the water surface in the channel to increase by δy, as shown in Figure 9.5(a), such that the surface wave propagates downstream at a velocity c and the velocity in the channel changes from V to $V+\delta V$ after passage of the wave. To determine the wave speed, c, it is convenient to utilize a control volume moving at a constant velocity, c, as shown in Figure 9.5(b), such that flow conditions within the moving control volume are at steady state. Applying the continuity equation to the moving control volume and assuming that the channel is rectangular and of width b, yields

$$\underbrace{(V + \delta V - c)(y + \delta y)b}_{\text{inflow rate}} = \underbrace{(V - c)yb}_{\text{outflow rate}} \quad \Rightarrow \quad \delta V = \frac{\delta y}{y + \delta y} c \tag{9.60}$$

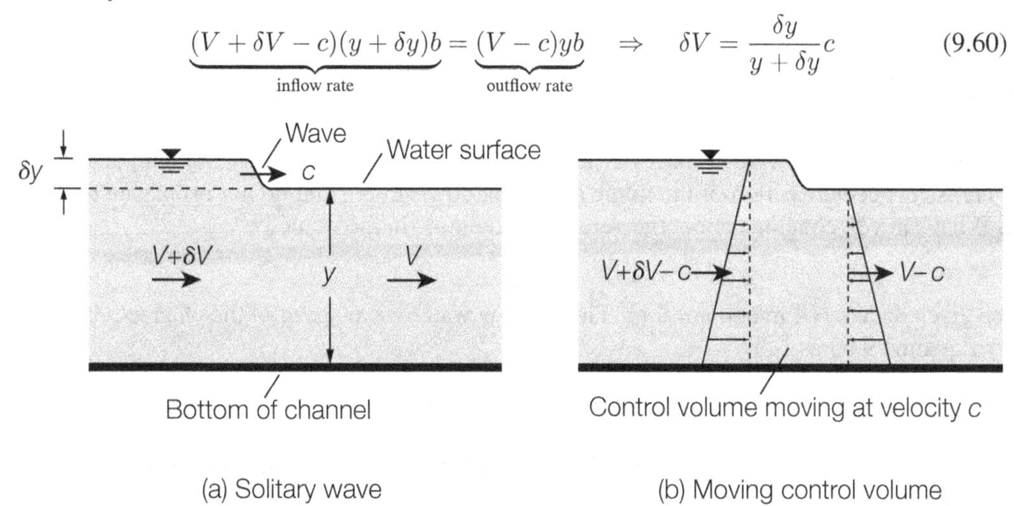

Figure 9.5: Propagation of a solitary wave

It can be assumed that (1) the length along the control volume is sufficiently short and that the boundary friction force can be neglected; (2) the channel is horizontal; (3) the pressure on both sides of the control volume are hydrostatic; and (4) the momentum correction factor, β, is approximately equal to unity on both the inflow and outflow surfaces of the control volume. Using the aforementioned assumptions, the momentum equation applied to the moving control volume can be expressed as

$$\underbrace{\gamma b \frac{(y+\delta y)^2}{2}}_{\text{pressure force at inflow}} - \underbrace{\gamma b \frac{(y)^2}{2}}_{\text{pressure force at outflow}} = \underbrace{\rho b (V-c)^2}_{\text{momentum flux at outflow}} - \underbrace{\rho b (V+\delta V-c)^2}_{\text{momentum flux at inflow}} \tag{9.61}$$

where γ $(= \rho g)$ is the specific weight and ρ is the density of the liquid in the channel. Combining Equations 9.60 and 9.61 to eliminate δV gives

$$c^2 = \left(1 + \frac{\delta y}{y}\right)\left(1 + \frac{\delta y}{2y}\right) gy \tag{9.62}$$

and for $\delta y \ll y$, Equation 9.62 simplifies to yield

$$c = \sqrt{gy} \tag{9.63}$$

This useful result states that waves generated by small disturbances in open channels propagate at a speed equal to $\sqrt{gy}$. In applying Equation 9.63, it is important to remember that the analysis presented here applies to a rectangular channel and implicitly assumes that the channel is sufficiently shallow that the wave-induced velocity in the channel is independent of the depth of flow, y. For wave propagation in deep waters, such as occurs in the ocean, this shallow water assumption in not valid and an analysis for the deep water case yields a wave speed, c, that is independent of the depth. Nevertheless, the shallow water assumption embodied in Equation 9.63 is usually appropriate for open-channel flows that are typically encountered by hydraulic engineers. A particularly useful application of Equation 9.63 is that in cases where the velocity in an open channel is greater than $\sqrt{gy}$, small disturbances in the channel cannot propagate upstream!

EXAMPLE 9.3

Water is flowing in a 4-m-wide rectangular flume[59] with a flow depth of 3 m. Disturbances created across the center section of the flume are swept downstream and do not propagate upstream. What can you conclude about the velocity of the flow in the flume?

SOLUTION

From the given data: b = 4 m and y = 3 m. The shallow water wave speed in the channel, c, is given by Equation 9.63 as

$$c = \sqrt{gy} = \sqrt{(9.81)(3)} = 5.42 \text{ m/s}$$

Therefore, it can be concluded that **the velocity in the flume is greater than 5.42 m/s.**

[59] A *flume* is an above-ground channel.

The ratio of the flow velocity, V, to the wave velocity, c (= $\sqrt{gy}$), provides a convenient measure of whether surface disturbances in open channels will travel upstream. This ratio is commonly referred to as the *Froude number*, Fr, where Fr = $V/\sqrt{gy}$. Upstream propagation of shallow water waves is possible when Fr < 1 and is not possible when Fr ≥ 1.

9.2.3 Steady-State Energy Equation

The steady-state energy equation is an expression of the *first law of thermodynamics*, which states that for any given mass of fluid, the rate at which heat is added to the fluid minus the rate at which work is done by the fluid is equal to the rate of change of internal energy within the fluid. Application of the first law of thermodynamics to open-channel flow requires that the first law of thermodynamics be expressed in a form that can be applied directly to the analysis of flow in open channels.

Derivation of the energy equation for open-channel flow. The control-volume form of the energy equation that is used in open-channel flow applications is derived for the control volume shown in Figure 9.2. Application of the energy equation to this control volume requires that

$$\dot{Q} - \dot{W} = \int_A \rho e \, (\mathbf{v} \cdot \mathbf{n}) \, \mathrm{d}A \tag{9.64}$$

where $\dot{Q}$ is the rate at which heat is added to the liquid in the control volume, $\dot{W}$ is the rate at which work is done by the liquid in the control volume, A is the surface area of the control volume, ρ is the density of the liquid in the control volume, and e is the internal energy per unit mass of liquid in the control volume given by

$$e = gz + \frac{v^2}{2} + u \tag{9.65}$$

where z is the elevation of a liquid mass having a velocity, v, and internal energy, u. The normal stresses on the inflow and outflow boundaries of the control volume are equal to the pressure, p, with shear stresses tangential to the control-volume boundaries. As the liquid moves with velocity, $\mathbf{v}$, the power expended by the liquid is given by

$$\dot{W} = \int_A p \, (\mathbf{v} \cdot \mathbf{n}) \, \mathrm{d}A \tag{9.66}$$

No work is done by the shear forces on the control-volume boundaries, because the velocity is equal to zero on the channel surface and the flow direction is normal to the direction of the shear forces on the inflow and outflow boundaries of the control volume. Combining Equations 9.64–9.66 and rearranging gives

$$\dot{Q} = \int_A \rho \left(\frac{p}{\rho} + gz + u + \frac{v^2}{2} \right) (\mathbf{v} \cdot \mathbf{n}) \, \mathrm{d}A \tag{9.67}$$

The term $p/\rho + gz + u = g(p/\gamma + z) + u$ can be assumed to be constant across any given inflow or outflow boundary on the control volume, because the hydrostatic pressure distribution across the inflow/outflow boundary guarantees that $p/\gamma + z$ is constant across the boundary, and the internal energy, u, depends only on the temperature, which can be assumed constant across the boundary. Because $\mathbf{v} \cdot \mathbf{n}$ is equal to zero over the control volume boundary in contact with the channel surface, Equation 9.67 simplifies to

$$\dot{Q} = \left[\frac{p}{\rho} + gz + u \right]_1 \int_{A_1} \rho (\mathbf{v} \cdot \mathbf{n}) \, \mathrm{d}A + \int_{A_1} \rho \frac{v^2}{2} (\mathbf{v} \cdot \mathbf{n}) \, \mathrm{d}A + \left[\frac{p}{\rho} + gz + u \right]_2 \int_{A_2} \rho (\mathbf{v} \cdot \mathbf{n}) \, \mathrm{d}A + \int_{A_2} \rho \frac{v^2}{2} (\mathbf{v} \cdot \mathbf{n}) \, \mathrm{d}A \tag{9.68}$$

where the subscripts 1 and 2 designate the values of various quantities at the inflow and outflow boundaries, respectively. Equation 9.68 can be further simplified by noting that the assumption of steady state requires that of mass inflow rate, $\dot{m}$, to the control volume be equal to the mass outflow rate, where

$$\dot{m} = \int_{A_2} \rho\,(\mathbf{v}\cdot\mathbf{n})\,\mathrm{d}A = -\int_{A_1} \rho\,(\mathbf{v}\cdot\mathbf{n})\,\mathrm{d}A \tag{9.69}$$

where the negative sign originates from the fact that the unit normal points out of the control volume. The *kinetic energy correction factor*, α, for any cross section can be defined by

$$\int_A \rho \frac{v^3}{2}\,\mathrm{d}A = \alpha\rho\frac{V^3}{2}A \quad \text{or} \quad \alpha = \frac{\int_A \rho v^3\,\mathrm{d}A}{\rho V^3 A} \tag{9.70}$$

where A is the flow area and V is the mean velocity over the flow area. The kinetic energy correction factor, α, is sometimes called the *Coriolis coefficient* or the *energy coefficient*. The kinetic energy correction factors at the inflow and outflow sections, α_1 and α_2, respectively, are determined by the velocity profiles across the respective flow boundaries. Values of α are typically in the range of 1.1–1.2 for turbulent flow in prismatic channels and typically in the range of 1.1–2.0 for turbulent flow in natural channels. Combining Equations 9.68–9.70 leads to

$$\dot{Q} = -\left[\frac{p}{\rho} + gz + u\right]_1 \dot{m} - \alpha_1\rho\frac{V_1^3}{2}A_1 + \left[\frac{p}{\rho} + gz + u\right]_2 \dot{m} + \alpha_2\rho\frac{V_2^3}{2}A_2 \tag{9.71}$$

where the negative signs originate from the fact that the unit normal points out of the inflow boundary, making $\mathbf{v}\cdot\mathbf{n}$ negative for the inflow boundary in Equation 9.68. Invoking the steady-state continuity equation,

$$\rho V_1 A_1 = \rho V_2 A_2 = \dot{m} \tag{9.72}$$

Combining Equations 9.71 and 9.72 and rearranging gives

$$\left[\frac{p}{\gamma} + \alpha\frac{V^2}{2g} + z\right]_1 = \left[\frac{p}{\gamma} + \alpha\frac{V^2}{2g} + z\right]_2 + \left[\frac{1}{g}(u_2 - u_1) - \frac{\dot{Q}}{\dot{m}g}\right] \tag{9.73}$$

Assuming a hydrostatic pressure distribution normal to the direction of the flow requires that

$$\frac{p}{\gamma} + z = y\cos\theta + z_0 \tag{9.74}$$

where y is the depth of flow (measured vertically), θ is the angle that the channel makes with the horizontal, and z_0 is the elevation of the bottom of the channel. Combining Equations 9.73 and 9.74, the energy equation can be written as

$$\underbrace{\left[y\cos\theta + \alpha\frac{V^2}{2g} + z_0\right]_1}_{\text{energy at upstream section}} = \underbrace{\left[y\cos\theta + \alpha\frac{V^2}{2g} + z_0\right]_2}_{\text{energy at downstream section}} + \underbrace{h_\mathrm{L}}_{\text{energy loss}} \tag{9.75}$$

where the energy loss per unit weight or *head loss*, h_L, is defined by the relation

$$h_\mathrm{L} = \frac{1}{g}(u_2 - u_1) - \frac{\dot{Q}}{\dot{m}g} \tag{9.76}$$

Equation 9.75 is the most widely used form of the energy equation in open-channel flow applications.

Alternative forms of the energy equation. A rearrangement of the energy equation (Equation 9.75) gives

$$y_2 \cos\theta + \alpha_2 \frac{V_2^2}{2g} = y_1 \cos\theta + \alpha_1 \frac{V_1^2}{2g} + (z_{01} - z_{02}) - h_\mathrm{L} \tag{9.77}$$

which can also be written in the more compact form

$$\left[y\cos\theta + \alpha\frac{V^2}{2g} \right]_2^1 = (S - S_0\cos\theta)L \tag{9.78}$$

where L is the distance between the inflow and outflow sections of the control volume, S is the head loss per unit length (= slope of the energy grade line), and S_0 is the slope of the channel. The quantities S and S_0 can be expressed as

$$S = \frac{h_\mathrm{L}}{L}, \qquad S_0 = \frac{z_{01} - z_{02}}{L\cos\theta} \tag{9.79}$$

In contrast to our usual definition of slopes, downward slopes are generally taken as positive in open-channel hydraulics. The relationship between S_0 and $\cos\theta$ is shown in Table 9.3, where it is clear that for open-channel slopes less than 0.1 (10%), the error in assuming that $\cos\theta = 1$ is less than 0.5%. Because this error is usually less than the uncertainty in other terms in the energy equation, the energy equation (Equation 9.78) is frequently written as

$$\left[y + \alpha\frac{V^2}{2g} \right]_2^1 = (S - S_0)L, \qquad S_0 < 0.1 \tag{9.80}$$

and the range of slopes corresponding to this approximation is frequently omitted. The slopes of rivers and canals in plain areas are usually on the order of 0.01%–1%, whereas the slopes of mountain streams are typically on the order of 5%–10% (Montes, 1998; Wohl, 2000; Benke and Cushing, 2005). Channels with slopes in excess of 10% are commonly regarded as steep. In cases where the head loss is predominantly due to frictional resistance, the energy equation is written as

$$\left[y + \alpha\frac{V^2}{2g} \right]_2^1 = (S_\mathrm{f} - S_0)L, \qquad S_0 < 0.1 \tag{9.81}$$

where S_f is the frictional head loss per unit length. Equation 9.81 is sometimes written in the expanded form

$$\left[y + \alpha\frac{V^2}{2g} + z_0 \right]_1 = \left[y + \alpha\frac{V^2}{2g} + z_0 \right]_2 + h_\mathrm{f}, \qquad S_0 < 0.1 \tag{9.82}$$

where h_f is the head loss due to friction between the moving liquid and the surface of the open channel.

Table 9.3: S_0 versus $\cos\theta$

S_0	$\cos\theta$
0.001	0.9999995
0.01	0.99995
0.1	0.995
1	0.707

9.2.3.1 Energy grade line

The head at each cross section of an open channel, h, is given by

$$h = y + \alpha\frac{V^2}{2g} + z_0, \qquad S_0 < 0.1 \tag{9.83}$$

where y is the depth of flow, V is the average velocity over the cross section, z_0 is the elevation of the bottom of the channel, and S_0 is the slope of the channel. As stated previously, in most cases, the slope restriction ($S_0 < 0.1$) is met, and this restriction is not explicitly stated in the definition of the head. When the head, h, at each section is plotted versus the distance along the channel, this curve is called the *energy grade line*. The point on the energy grade line corresponding to each cross section is located a distance $\alpha V^2/2g$ vertically above the liquid surface; between any two cross sections, the elevation of the energy grade line drops by a distance equal to the head loss, h_L, between the two sections. The energy grade line is particularly useful in visualizing the state of a liquid as it flows along an open channel and in visualizing the performance of hydraulic structures in open-channel systems.

9.2.3.2 Specific energy

The *specific energy*, E, of a liquid is defined as the mechanical energy per unit weight of the liquid measured relative to the bottom of the channel and is given by

$$E = y + \alpha\frac{V^2}{2g} = y + \alpha\frac{Q^2}{2gA^2} \tag{9.84}$$

where Q is the volume flow rate and A is the cross-sectional flow area. The specific energy, E, was originally defined by Bakhmeteff (1912; 1932) and is an adequate representation of the mechanical energy per unit weight whenever the pressure distribution is hydrostatic in the vertical direction. The specific energy appears explicitly in the energy equation, Equation 9.81, which can be written in the form

$$E_1 - E_2 = h_L + \Delta z \tag{9.85}$$

where E_1 and E_2 are the specific energies at the upstream and downstream sections, respectively, h_L is the head loss between sections, and Δz is the change in elevation between the upstream and downstream sections. In many cases of practical interest, E_1, h_L, and Δz can be calculated from given upstream flow conditions and channel geometry, E_2 can be calculated using Equation 9.85, and the downstream depth of flow, y, can be calculated from E_2 using Equation 9.84. For a given shape of the channel cross section and flow rate, Q, the specific energy, E, depends only on the depth of flow, y. The typical relationship between E and y given by Equation 9.84, for a constant value of Q, is shown in Figure 9.6. The salient features

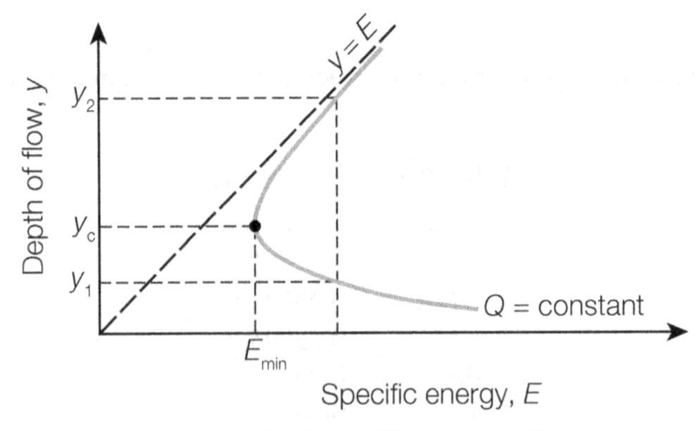

Figure 9.6: Typical specific energy diagram

of Figure 9.6 are that (1) there is more than one possible flow depth for a given specific energy and (2) the specific energy curve is asymptotic to the line

$$y = E \tag{9.86}$$

In accordance with the energy equation (Equation 9.85), the specific energy at any cross section can be expressed in terms of the specific energy at an upstream section, the change in the elevation of the bottom of the channel, and the head loss between the upstream and downstream sections. The fact that there can be more than one possible flow depth for a given specific energy leads to the question of which flow depth will exist. The specific energy diagram, Figure 9.6, indicates that there is a flow depth, y_c, at which the specific energy is a minimum. At this point, Equation 9.84 indicates that

$$\frac{dE}{dy} = 0 = 1 - \frac{Q^2}{gA_c^3}\frac{dA}{dy} \tag{9.87}$$

where A_c is the flow area corresponding to $y = y_c$ and the kinetic energy correction factor, α, has been taken equal to unity. Referring to the general open-channel cross section shown in Figure 9.7, it is clear that for small changes in the flow depth, y,

$$dA = T\,dy \tag{9.88}$$

where dA is the increase in flow area resulting from a change in flow depth, dy, and T is the top width of the channel when the flow depth is y. Equation 9.88 can be written as

$$\frac{dA}{dy} = T \tag{9.89}$$

which can be combined with Equation 9.87 to yield the following relationship that is applicable under *critical flow* conditions:

$$\frac{Q^2}{g} = \frac{A_c^3}{T_c} \tag{9.90}$$

where T_c is the top width of the channel corresponding to $y = y_c$. Equation 9.90 forms the basis for calculating the *critical flow depth*, y_c, in a channel for a given flow rate, Q, because the terms on the left-hand side of the equation are known and the right-hand side is only a function of y_c for a channel of given shape. In most cases, an iterative solution for y_c is required and a variety of computational approaches have been suggested (e.g., Patil et al., 2005; Kanani et al., 2008). The specific energy under critical flow conditions, E_c, is then given by

$$E_c = y_c + \frac{A_c}{2T_c} \tag{9.91}$$

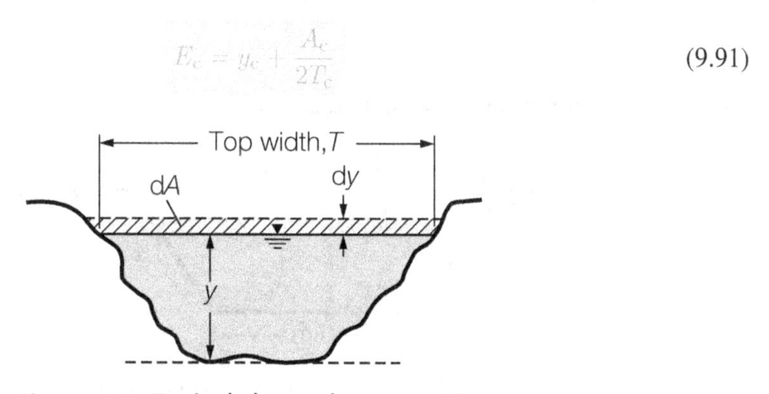

Figure 9.7: Typical channel cross section

Defining the *hydraulic depth*, D_h, by the relation

$$D_\text{h} = \frac{A}{T} \tag{9.92}$$

a Froude number, Fr, can be defined by

$$\text{Fr}^2 = \frac{V^2}{gD_\text{h}} = \frac{Q^2T}{gA^3} \tag{9.93}$$

Combining this definition of the Froude number with the critical flow condition given by Equation 9.90 leads to the relation

$$\text{Fr}_\text{c} = 1 \tag{9.94}$$

where Fr_c is the Froude number under critical flow conditions. When $y > y_\text{c}$, Equation 9.93 indicates that Fr<1, and when $y<y_\text{c}$, Equation 9.93 indicates that Fr > 1. Flows where $y<y_\text{c}$ are called *supercritical*; where $y>y_\text{c}$, flows are called *subcritical.* It is apparent from the specific energy diagram, Figure 9.6, that when the flow conditions are close to critical, a relatively large change of flow depth occurs with small variations in specific energy. Flow under these conditions is unstable, and excessive wave action or undulations of the liquid surface usually occur. Experiments in rectangular channels have shown that these instabilities can be avoided if Fr < 0.86 or Fr > 1.13 (U.S. Army Corps of Engineers, 1995).

Critical condition for large channel slopes. The above derivation of the critical flow condition assumes that the channel slope is small ($\cos\theta \approx 1$ or slope < 10%). In atypical cases where the channel slope is large (slope ≥ 10%), the critical condition is given by (Kanani et al., 2008)

$$\frac{\alpha Q^2 T}{gA^3\cos\theta} = 1 \tag{9.95}$$

Singular open-channel sections. Curiously, there exists a particular channel section in which the specific energy does not change with flow depth and hence a critical flow depth does not exist. This channel section is called a *singular open-channel section*, and specifications of the geometry of this section can be found in Swamee and Rathie (2007).

EXAMPLE 9.4

Determine the critical depth of flow for water flowing at 10 m³/s in a trapezoidal channel with bottom width 3 m and side slopes of 2 : 1 (H : V).

SOLUTION

The channel cross section is illustrated in Figure 9.8.

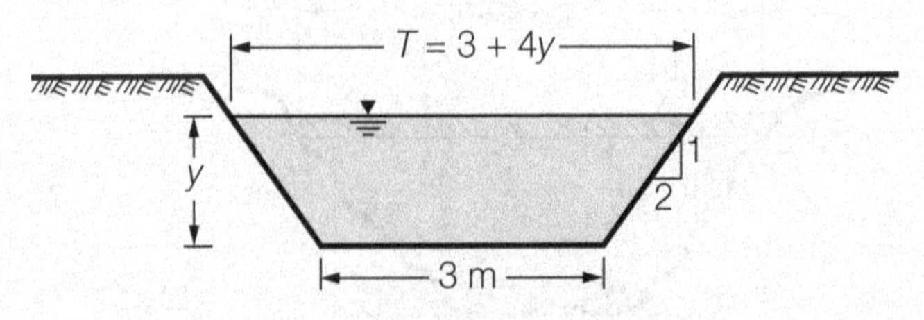

Figure 9.8: Trapezoidal cross section

The depth of flow is y, and the top width, T, and flow area, A, are given, respectively, by

$$T = 3 + 4y, \qquad A = 3y + 2y^2$$

Under critical flow conditions

$$\frac{Q^2}{g} = \frac{A_c^3}{T_c}$$

and because $Q = 10$ m^3/s and $g = 9.81$ m/s^2, under critical flow conditions

$$\frac{10^2}{9.81} = \frac{(3y_c + 2y_c^2)^3}{(3 + 4y_c)} \quad \rightarrow \quad y_c = 0.855 \text{ m}$$

Therefore, the critical depth of flow in the channel is **0.855 m**. Flow under this (critical) condition is unstable.

Critical flow in rectangular channels. The critical flow condition described by Equation 9.90 can be simplified considerably in the case of flow in rectangular channels, where it is convenient to deal with the flow per unit width, q, given by

$$q = \frac{Q}{b} \tag{9.96}$$

where b is the width of the channel. The flow area, A, and top width, T, are given, respectively, by

$$A = by, \qquad T = b \tag{9.97}$$

The critical flow condition given by Equation 9.90 then becomes

$$\frac{(qb)^2}{g} = \frac{(by_c)^3}{b} \tag{9.98}$$

which can be solved to yield the critical flow depth

$$y_c = \left(\frac{q^2}{g}\right)^{\frac{1}{3}} \tag{9.99}$$

and corresponding critical energy

$$E_c = y_c + \frac{q^2}{2gy_c^2} \tag{9.100}$$

Combining Equations 9.99 and 9.100 leads to the following simplified form of the minimum specific energy in rectangular channels:

$$E_c = \frac{3}{2}y_c \tag{9.101}$$

Specific energy diagram in rectangular channels. The specific energy diagram illustrated in Figure 9.9 for a rectangular channel is similar to the nonrectangular case shown in Figure 9.6, except that the specific energy curve for a rectangular channel corresponds to a fixed value of q rather than a fixed value of Q in a nonrectangular channel. It is apparent from Figure 9.9 that the specific energy curve shifts upward and to the right for increasing values of q, which demonstrates that under critical flow conditions, the channel yields the maximum flow rate for a given specific energy. This is true for both rectangular and nonrectangular sections,

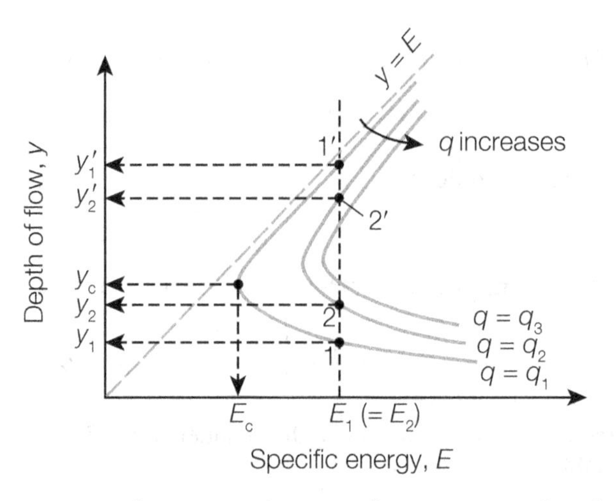

Figure 9.9: Specific energy diagram for a rectangular channel

because for nonrectangular channels, the specific energy curve also shifts to the right with increasing flow rate. For rectangular channels, the specific energy, E, defined by Equation 9.84 can be expressed in terms of the flow per unit width, q, by the relation

$$E = y + \frac{(q/y)^2}{2g} \tag{9.102}$$

This equation can be expressed in nondimensional form by eliminating q using Equation 9.99, which yields the nondimensional form of the specific energy equation as

$$\frac{E}{y_c} = \frac{y}{y_c} + \frac{1}{2}\left(\frac{y}{y_c}\right)^2 \tag{9.103}$$

It is apparent from Equation 9.103 that for any given value of E/y_c, there are two possible values of y/y_c, commonly referred to as alternate depths.

Flow through constrictions. The specific energy diagram shown in Figure 9.9 is a plot of Equation 9.102, and this diagram can be used to understand what happens to the flow in a rectangular channel when there is a constriction, such as when the channel width is narrowed to accommodate a bridge. Suppose the flow per unit width upstream of the constriction is q_1 and the depth of flow at this location is y_1. If the channel is constricted so that the flow per unit width becomes q_2, then, provided the head loss in the constriction is minimal, Figure 9.9 indicates that there are two possible flow depths in the constricted section: y_2 and y_2'. Neglecting the head loss in the constriction is reasonable if the constriction is smooth and takes place over a relatively short distance. Figure 9.9 indicates that it is physically impossible for the flow depth in the constriction to be y_2', because this would require the flow per unit width, q, to increase and then decrease to reach y_2'. Because the flow per unit width increases monotonically in a constriction, the flow depth can only go from y_1 to y_2. A similar case arises when the flow depth upstream of the constriction is y_1' and the possible flow depths at the constriction are y_2' and y_2. In this case, only y_2' is possible, because an increase and then a decrease in q would be required to achieve a flow depth of y_2. Based on this analysis, it is clear that if the flow upstream of the constriction is subcritical ($y > y_c$ or $\text{Fr} < 1$), then the flow in the constriction must be either subcritical or critical and that if the flow upstream of the constriction is supercritical ($y < y_c$ or $\text{Fr} > 1$), then the flow in the constriction must be either supercritical or critical. These results also mean that if two flow depths are possible in a constriction, then the flow depth that is closest to the upstream flow depth will occur in the constriction. This can be called the *closest-depth rule*. If the flow upstream of the constriction is critical, then flow through the constriction is not possible under the existing flow conditions and the upstream flow conditions must necessarily change upon installation of the constriction in the channel. Regardless of whether the flow upstream of the constriction is subcritical or supercritical, Figure 9.9 indicates that a maximum constriction will cause critical flow to occur at the constriction. A larger constriction (smaller opening) than that which causes critical flow to occur at the constriction is not possible based on the available specific energy upstream of the constriction. Any further constriction will result in changes in the upstream flow conditions to maintain critical flow at the constriction. Under these circumstances, the flow is said to be *choked*.

EXAMPLE 9.5

A rectangular channel 1.30 m wide carries 1.10 m^3/s of water at a depth of 0.85 m. (a) If a 30-cm-wide pier is placed in the middle of the channel, find the elevation of the water surface at the constriction. (b) What is the minimum width of the constriction that will not cause a rise in the upstream water surface?

SOLUTION

(a) The cross sections of the channel upstream of the constriction and at the constriction are shown in Figure 9.10.

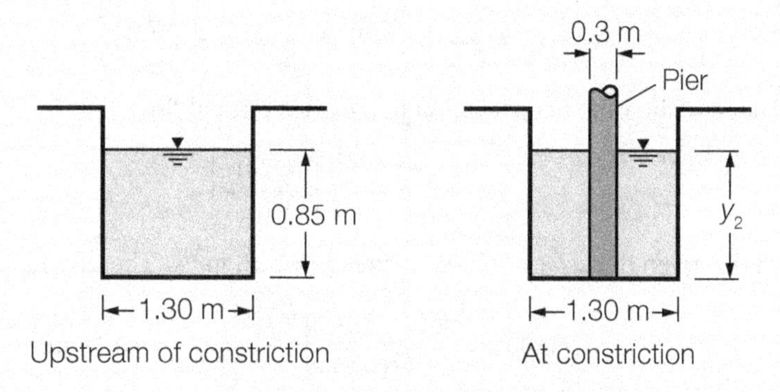

Figure 9.10: Constriction in a rectangular channel

Neglecting the head loss between the constriction and the upstream section, the energy equation requires that the specific energy at the constriction be equal to the specific energy at the upstream section. Therefore,

$$y_1 + \frac{V_1^2}{2g} = y_2 + \frac{V_2^2}{2g}$$

where section 1 refers to the upstream section and section 2 refers to the constricted section. In this case, $y_1 = 0.85$ m and

$$V_1 = \frac{Q}{A_1} = \frac{1.10}{(0.85)(1.30)} = 1.00 \text{ m/s}$$

The specific energy at section 1, E_1, is

$$E_1 = y_1 + \frac{V_1^2}{2g} = 0.85 + \frac{1.00^2}{2(9.81)} = 0.901 \text{ m}$$

Equating the specific energies at sections 1 and 2 yields

$$0.901 = y_2 + \frac{Q^2}{2gA^2} \rightarrow 0.901 = y_2 + \frac{1.10^2}{2(9.81)[(1.30 - 0.30)y_2]^2} \rightarrow y_2 + \frac{0.0617}{y_2^2} = 0.901$$

There are three solutions to the resulting cubic equation: y_2 = 0.33 m, 0.80 m, and −0.23 m. Of the two positive flow depths, we must select the depth corresponding to the same flow condition as exists upstream. At the upstream section, $\text{Fr}_1 = V_1/\sqrt{gy_1} =$ 0.35; therefore, the upstream flow condition is subcritical and the flow condition at the constriction must also be subcritical. The flow depth must therefore be

$$y_2 = \mathbf{0.80\ m}$$

where $\text{Fr}_2 = V_2/\sqrt{gy_2} = 0.49$. The other flow depth ($y_2$ = 0.33 m, Fr_2 = 1.9) is supercritical and cannot be achieved.

The closest-depth rule can also be invoked to select the flow depth in the constriction. The closest-depth rule requires that the flow depth in the constriction be 0.80 m, because there are two possible flow depths (0.80 m, 0.33 m) and 0.80 m is closest to the upstream flow depth (0.85 m).

(b) The minimum width of constriction that does not cause the upstream depth to change is associated with the critical flow condition occurring at the constriction. Under the critical flow condition (for a rectangular channel),

$$E_1 = E_2 = E_c = \frac{3}{2}y_c = \frac{3}{2}\left(\frac{q^2}{g}\right)^{\frac{1}{3}}$$

If b is the width of the constriction that causes critical flow, then

$$E_1 = \frac{3}{2}\left[\frac{(Q/b)^2}{g}\right]^{\frac{1}{3}}$$

From the given data: E_1 = 0.901 m and Q = 1.10 m³/s. Therefore,

$$0.901 = \frac{3}{2}\left[\frac{1.10^2}{b^2(9.81)}\right]^{\frac{1}{3}} \quad \rightarrow \quad b = 0.75\ \text{m}$$

If the constricted channel width is any less than **0.75 m**, the flow will be choked and the upstream flow depth will increase—in other words, the flow will back up.

Note: Part (a) of this problem required finding alternate flow depths as the cubic roots of the specific energy equation, which can be easily done using a programmable calculator, or by other means such as using a spreadsheet program with solver capability. Alternatively, the following equations can be used to provide a direct solution of the specific energy equation (Abdulrahman, 2008):

$$\frac{y_{\text{sub}}}{E} = \cos\left\{2\sin^{-1}\left[\sqrt{\frac{2}{3}}\cos\left[\frac{1}{3}\cos^{-1}\left[-\left(\frac{E_c}{E}\right)^{\frac{3}{2}}\right] - \frac{2}{3}\pi\right]\right]\right\} \tag{9.104}$$

$$\frac{y_{\text{sup}}}{E} = \cos\left\{2\sin^{-1}\left[\sqrt{\frac{2}{3}}\cos\left[\frac{1}{3}\cos^{-1}\left[-\left(\frac{E_c}{E}\right)^{\frac{3}{2}}\right]\right]\right]\right\} \tag{9.105}$$

where y_{sub} and y_{sup} are the subcritical and supercritical flow depths, respectively, E is the actual specific energy, and E_c is the specific energy under critical flow conditions.

Table 9.4: Eddy Loss Coefficients, C

Transition	Eddy loss coefficient, C	
	Expansion	Contraction
None or very gradual	0.0	0.0
Gradual	0.3	0.1
Typical bridge sections	0.5	0.3
Abrupt	0.8	0.6

Energy losses in contractions and expansions. The specific energy analyses covered in this section assume negligible head losses; the kinetic energy correction factor, α, is approximately equal to unity; and the vertical pressure distribution is hydrostatic. Although these approximations are valid in many cases, in diverging transitions with angles exceeding 8°, flows tend to separate from the side of the channel, causing large head losses and substantial increases in α (Montes, 1998). Under these conditions, the assumption of a constant specific energy within the transition and a value of α approximately equal to unity are not justified. To minimize head losses in transitions, the U.S. Department of Agriculture (USDA, 1977) recommends that channel sides not converge at an angle greater than 14° or diverge at an angle greater than 12.5°. In cases where the vertical pressure distribution is significantly nonhydrostatic, such as for flow over a curved surface, an alternative specific energy formulation must be used (e.g., Chanson, 2006). For the general case of contractions and expansions in open channels, the head loss, h_e, is usually expressed in the form (U.S. Army Corps of Engineers, 2010)

$$h_e = C\left|\alpha_2\frac{V_2^2}{2g} - \alpha_1\frac{V_1^2}{2g}\right| \tag{9.106}$$

where C is either an expansion or contraction coefficient, α_1 and α_2 are the energy coefficients at the upstream and downstream sections, respectively, and V_1 and V_2 are the average velocities at the upstream and downstream sections, respectively. The head loss, h_e, given by Equation 9.106 is commonly referred to as the *eddy loss*. Typical values of eddy loss coefficient, C, for expansions and contractions are given in Table 9.4.

Flow over bumps and steps. A bump of height Δz in a channel causes the specific energy of the flow to decrease by Δz and then return to its original value. A step of height Δz causes the specific energy to decrease abruptly by Δz and then remain at that value. The specific

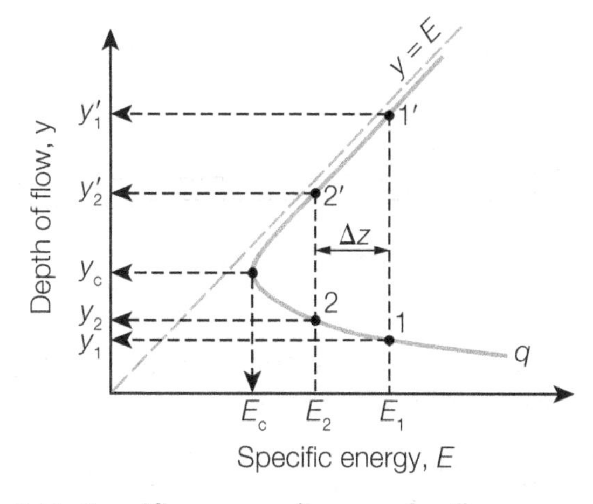

Figure 9.11: Specific energy diagram for flow over a bump

energy diagram for a constant flow per unit width, q, is shown in Figure 9.11, and this figure can be used to understand what happens in flow over a step of height Δz. Suppose that the flow depth and specific energy upstream of the step are y_1 and E_1, respectively; then the specific energy on the step is $E_2 = E_1 - \Delta z$. Based on these conditions (i.e., flow/width = q, specific energy = E_2), Figure 9.9 indicates that there are two possible flow depths on the step: y_2 and y_2'. However, it is physically impossible for the flow depth on the step to be y_2', because this would require that the specific energy of the flow decrease to E_c and then increase to E_2 to maintain a constant q. Because the specific energy must necessarily decrease monotonically over the step, the flow depth can only go from y_1 to y_2. A similar case arises when the flow depth upstream of the step is y_1' and the possible flow depths at the constriction are y_2' and y_2. In this case, only y_2' is possible, because an increase and then a decrease in specific energy would be required to achieve a flow depth of y_2. Based on this analysis, if the flow upstream of the step is subcritical ($y > y_c$ or $\text{Fr} < 1$), then the flow on the step must be either subcritical or critical. Similarly, if the flow upstream of the step is supercritical ($y < y_c$ or $\text{Fr} > 1$), then the flow on the step must be either supercritical or critical. These results also mean that if two flow depths are possible on the step, then the flow depth that is closest to the upstream flow depth will occur on the step; hence, the *closest-depth rule* also applies on a step. If the flow upstream of the step is critical, then flow over the step is not possible under the existing flow conditions and the upstream flow conditions must necessarily change upon installation of the step. A maximum step height will cause critical flow to occur at the step; a higher step will cause the flow to back up (which will change the upstream specific energy), and the flow will be *choked*, with the critical flow condition occurring on the step. When critical flow occurs on the crest of a bump, it should be apparent that the downstream flow depth can be either supercritical or subcritical, depending on the downstream conditions.

EXAMPLE 9.6

Water flows at a rate of 0.9 m^3/s in a 2.5-meter-wide channel as shown in Figure 9.12. At a particular section, the flow depth is 1.2 m. (a) What would be the flow depth over a bump of height 0.1 m installed just downstream of this section? (b) What is the maximum elevation of the bump that could be installed without backing up (i.e., choking) the flow?

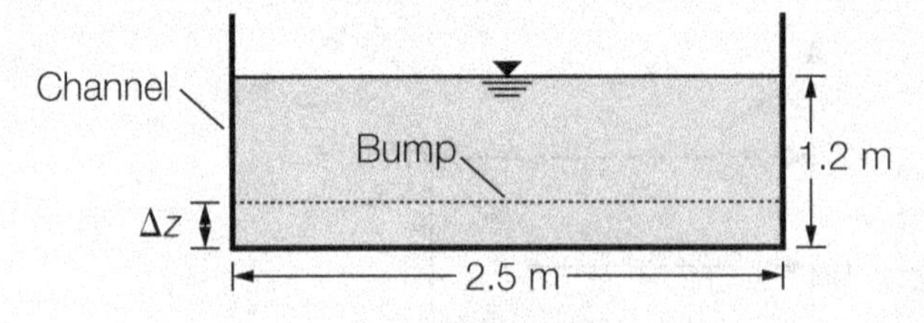

Figure 9.12: Flow over a bump

SOLUTION

From the given data: Q = 0.9 m^3/s, b = 2.5 m, $q = Q/b$ = 0.36 m^2/s, and y_1 = 1.2 m. Using these data, the specific energy, E_1, upstream of the bump is given by

$$E_1 = y_1 + \frac{V_1^2}{2g} = y_1 + \frac{q^2}{2gy_1^2} = 1.2 + \frac{0.36^2}{2(9.81)(1.2)^2} = 1.205 \text{ m}$$

(a) Taking Δz as the height of the bump and neglecting any head loss between the bump and the upstream section, the energy equation requires that

$$E_1 = E_2 + \Delta z \quad \rightarrow \quad E_1 = y_2 + \frac{q^2}{2gy_2^2} + \Delta z \quad \rightarrow \quad 1.205 = y_2 + \frac{(0.36)^2}{2(9.81)y_2^2} + 0.1$$

which has the following solutions: y_2 = 1.099 m, 0.080 m, and –0.075 m. In accordance with the nearest-depth rule, the flow depth over the bump is equal to approximately **1.10 m.**

(b) The maximum height of the bump to avoid choking causes critical conditions to occur over the bump. Under this circumstance, the energy equation requires that

$$E_1 = E_2 + \Delta z \quad \rightarrow \quad E_1 = E_c + \Delta z \quad \rightarrow \quad E_1 = \frac{3}{2}y_c + \Delta z \quad \rightarrow \quad E_1 = \frac{3}{2}\left(\frac{q^2}{g}\right)^{\frac{1}{3}} + \Delta z$$

Substituting known values gives

$$1.205 = \frac{3}{2}\left(\frac{0.36^2}{9.81}\right)^{\frac{1}{3}} + \Delta z \quad \rightarrow \quad \Delta z = 0.85 \text{ m}$$

Therefore, when the bump is **0.85 m** high, the flow will be critical over the bump. If the bump is made any higher, the flow will be choked and the upstream flow depth must necessarily increase.

The physical arguments presented using Figure 9.11 to explain the behavior of the liquid surface for flow over steps and bumps in rectangular channels can also be proven mathematically. For cases where the bottom of a channel is raised by a height h, the energy equation requires that

$$\frac{q^2}{2gy_1^2} + y_1 = \frac{q^2}{2gy^2} + y + h = \text{constant} \tag{9.107}$$

where q is the flow per unit width, y_1 is the upstream flow depth, and y is the flow depth over the elevated bottom. Differentiating Equation 9.107 with respect to the distance, x, along the channel yields

$$-\frac{q^2}{gy^3}\frac{\mathrm{d}y}{\mathrm{d}x} + \frac{\mathrm{d}y}{\mathrm{d}x} + \frac{\mathrm{d}h}{\mathrm{d}x} = 0 \tag{9.108}$$

Solving Equation 9.108 for the slope of the water surface, $\mathrm{d}y/\mathrm{d}x$, yields

$$\frac{\mathrm{d}y}{\mathrm{d}x} = \frac{\dfrac{\mathrm{d}h}{\mathrm{d}x}}{\dfrac{q^2}{gy^3} - 1} = \frac{\dfrac{\mathrm{d}h}{\mathrm{d}x}}{\dfrac{V^2}{gy} - 1} \tag{9.109}$$

where V is the average velocity in the channel. Recognizing the square of the local Froude number, Fr^2, over the elevated channel as V^2/gy, Equation 9.109 can be expressed as

$$\frac{\mathrm{d}y}{\mathrm{d}x} = \frac{1}{\mathrm{Fr}^2 - 1}\frac{\mathrm{d}h}{\mathrm{d}x} \tag{9.110}$$

For cases where $\mathrm{d}h/\mathrm{d}x > 0$, Equation 9.110 requires that $\mathrm{d}y/\mathrm{d}x > 0$ for $\mathrm{Fr} > 1$ (supercritical flow) and $\mathrm{d}y/\mathrm{d}x < 0$ for $\mathrm{Fr} < 1$ (subcritical flow) for $\mathrm{Fr} = 1$ (critical flow), $\mathrm{d}h/\mathrm{d}x = 0$ is the only feasible solution of Equation 9.110. These results are the same as those that were derived previously on physical grounds from Figure 9.11.

9.3 Water Surface Profiles

9.3.1 Profile Equation

The equation that is used to describe the water surface profile in an open channel can be derived from the energy equation, Equation 9.81, which is of the form

$$S_0 - S_\mathrm{f} = \frac{\Delta\left(y + \alpha\dfrac{V^2}{2g}\right)}{\Delta x} \tag{9.111}$$

where S_0 is the slope of the channel, S_f is the slope of the energy grade line, y is the depth of flow, α is the kinetic energy correction factor, V is the average velocity, x is a coordinate measured along the channel (the flow direction defined as positive), and Δx is the distance between the upstream and downstream sections. Equation 9.111 can be further rearranged into

$$S_0 - S_\mathrm{f} = \frac{\Delta y}{\Delta x} + \frac{\Delta\left(\alpha\frac{V^2}{2g}\right)}{\Delta x} \tag{9.112}$$

Taking the limit of Equation 9.112 as $\Delta x \rightarrow 0$ and invoking the definition of the derivative yields

$$\begin{aligned} S_0 - S_\mathrm{f} &= \lim_{\Delta x\rightarrow 0}\frac{\Delta y}{\Delta x} + \lim_{\Delta x\rightarrow 0}\frac{\Delta\left(\alpha\frac{V^2}{2g}\right)}{\Delta x} \\ &= \frac{\mathrm{d}y}{\mathrm{d}x} + \frac{\mathrm{d}}{\mathrm{d}x}\left(\alpha\frac{V^2}{2g}\right) \\ &= \frac{\mathrm{d}y}{\mathrm{d}x} + \frac{\mathrm{d}}{\mathrm{d}y}\left(\alpha\frac{V^2}{2g}\right)\frac{\mathrm{d}y}{\mathrm{d}x} \\ &= \frac{\mathrm{d}y}{\mathrm{d}x}\left[1 + \frac{\mathrm{d}}{\mathrm{d}y}\left(\alpha\frac{Q^2}{2gA^2}\right)\right] \\ &= \frac{\mathrm{d}y}{\mathrm{d}x}\left[1 - \alpha\frac{Q^2}{gA^3}\frac{\mathrm{d}A}{\mathrm{d}y}\right] \end{aligned} \tag{9.113}$$

where Q is the (constant) flow rate and A is the (variable) cross-sectional flow area in the channel. Recalling that

$$\frac{\mathrm{d}A}{\mathrm{d}y} = T \tag{9.114}$$

where T is the top width of the channel and the hydraulic depth, D_h, of the channel is defined as

$$D_\mathrm{h} = \frac{A}{T} \tag{9.115}$$

the Froude number, Fr, of the flow can be written as

$$\mathrm{Fr} = \frac{V}{\sqrt{gD_\mathrm{h}}} = \frac{Q\sqrt{T}}{A\sqrt{gA}} = \frac{Q}{\sqrt{gA^3}}\sqrt{\frac{\mathrm{d}A}{\mathrm{d}y}} \quad\rightarrow\quad \mathrm{Fr}^2 = \frac{Q^2}{gA^3}\frac{\mathrm{d}A}{\mathrm{d}y} \tag{9.116}$$

Combining Equations 9.113 and 9.116 and rearranging yields

$$\frac{\mathrm{d}y}{\mathrm{d}x} = \frac{S_0 - S_\mathrm{f}}{1 - \alpha\mathrm{Fr}^2} \tag{9.117}$$

This differential equation describes the water surface profile in open channels, and its original derivation has been attributed to Bélanger (1828; see Chanson, 2009). To appreciate the utility of Equation 9.117, consider the relative magnitudes of the channel slope, S_0, and the friction slope, S_f. According to the Manning equation, for any given flow rate,

$$\frac{S_f}{S_0} = \left(\frac{A_n R_n^{\frac{2}{3}}}{A R^{\frac{2}{3}}} \right)^2 \tag{9.118}$$

where A_n and R_n are the cross-sectional area and hydraulic radius, respectively, under normal flow conditions ($S_f = S_0$) and A and R are the actual cross-sectional area and hydraulic radius, respectively, of the flow. Because $AR^{\frac{2}{3}} > A_n R_n^{\frac{2}{3}}$ when $y > y_n$, Equation 9.118 indicates that

$$S_f > S_0 \quad \text{when} \quad y < y_n, \quad \text{and} \quad S_f < S_0 \quad \text{when} \quad y > y_n \tag{9.119}$$

or

$$S_0 - S_f < 0 \quad \text{when} \quad y < y_n, \quad \text{and} \quad S_0 - S_f > 0 \quad \text{when} \quad y > y_n \tag{9.120}$$

It has already been shown that

$$\text{Fr} > 1 \quad \text{when} \quad y < y_c, \quad \text{and} \quad \text{Fr} < 1 \quad \text{when} \quad y > y_c \tag{9.121}$$

or

$$1 - \text{Fr}^2 < 0 \quad \text{when} \quad y < y_c, \quad \text{and} \quad 1 - \text{Fr}^2 > 0 \quad \text{when} \quad y > y_c \tag{9.122}$$

Based on Equations 9.120 and 9.122, the sign of the numerator in Equation 9.117 is determined by the magnitude of the flow depth, y, relative to the normal depth, y_n, and the sign of the denominator is determined by the magnitude of the flow depth, y, relative to the critical depth, y_c (assuming $\alpha \approx 1$). Therefore, the sign of the slope of the water surface, dy/dx, is determined by the relative magnitudes of y, y_n, and y_c.

9.3.2 Classification of Water Surface Profiles

In hydraulic engineering, channel slopes are classified based on the relative magnitudes of the actual channel slope, S_0, and the *critical slope*, S_c. The critical slope is defined as the slope of the channel at which the normal flow depth, y_n, would be equal to the critical flow depth, y_c. If $S_0 < S_c$, the slope is *mild*; if $S_0 > S_c$, the slope is *steep*; if $S_0 = S_c$, the slope is *critical*; if $S_0 = 0$, the slope is *horizontal*; and if $S_0 < 0$, the slope is *adverse*. These slope classifications can also be expressed in terms of the relative magnitudes of y_n and y_c as shown in Table 9.5 and illustrated in Figure 9.13. The range of flow depths for each slope can be divided into three zones, delimited by the normal and critical flow depths, where the highest zone above the channel bed is Zone 1, the intermediate zone is Zone 2, and the lowest zone is Zone 3. Water surface profiles are classified based on both the type of slope (e.g., type M) and the zone in which the actual water surface is located (e.g., Zone 2). Therefore, each water surface profile is classified by a letter and number. For example, an M2 profile indicates a mild slope

Table 9.5: Hydraulic Classification of Slopes

Name	Type	Condition	
		Slope	Depth
Mild	M	$S_0 < S_c$	$y_n > y_c$
Steep	S	$S_0 > S_c$	$y_n < y_c$
Critical	C	$S_0 = S_c$	$y_n = y_c$
Horizontal	H	$S_0 = 0$	$y_n = \infty$
Adverse	A	$S_0 < 0$	–

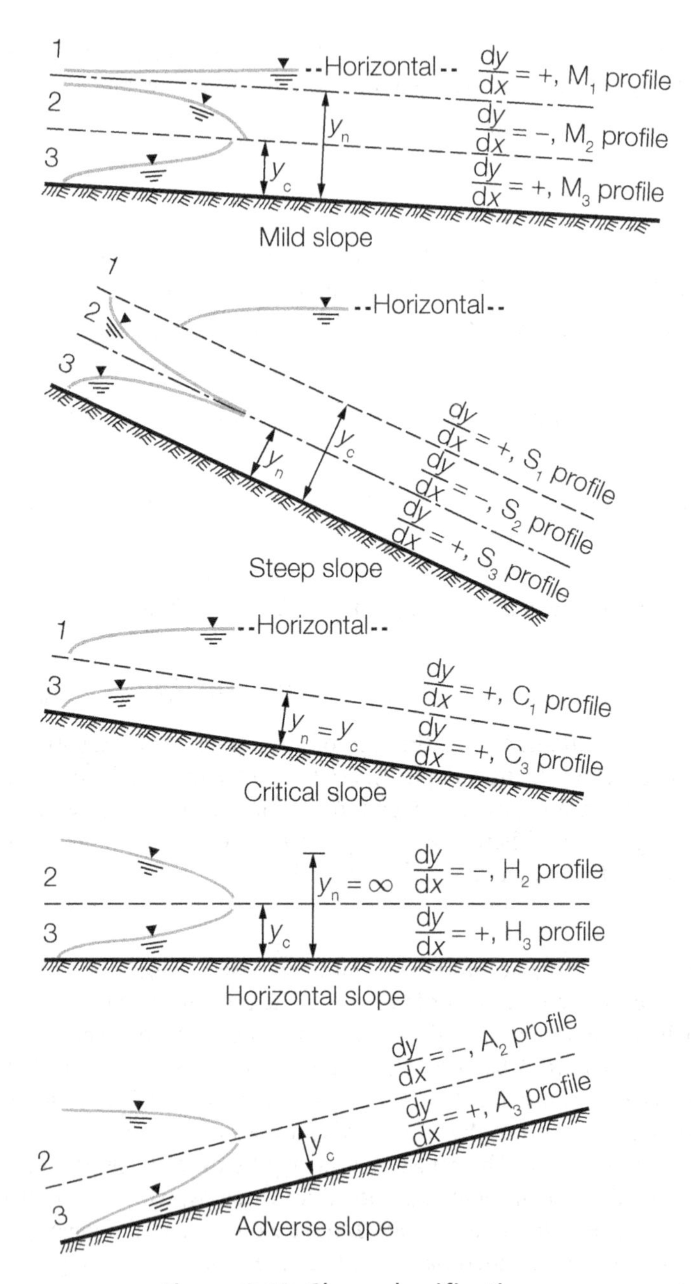

Figure 9.13: Slope classifications

($y_n > y_c$) and the actual depth, y, is in Zone 2 ($y_n < y < y_c$). In the case of nonsustaining slopes (horizontal, adverse), there is no Zone 1, as the normal depth is infinite in horizontal channels and is nonexistent in channels with adverse slope. There is no Zone 2 in channels with critical slope, as $y_n = y_c$. It is important to remember that for any given channel, the slope classification depends on the flow rate in the channel. For example, a channel can be classified as having a steep slope at one flow rate and a mild slope at another flow rate.

Asymptotic behavior of water surface profiles. The asymptotic behavior of water surface profiles can be deduced from Equation 9.117 by determining the asymptotic behavior of S_f and Fr. For example, all M1 profiles require that $dy/dx > 0$, which means that y continually increases, which means that $S_f \rightarrow 0$ and Fr $\rightarrow 0$, from which Equation 9.117 gives $dy/dx \rightarrow S_0$. Under this asymptotic condition, the depth of flow increases at the same rate that the bottom elevation decreases (i.e., at S_0), which means that the asymptotic water surface profile must be horizontal. A similar analysis can be performed for all other water surface profile classifications. The following general trends are followed by all water surface profiles: If the depth starts in Zone 1, the depth will increase in the downstream direction; if the depth starts in Zone 2, the depth will decrease in the downstream direction and approach the lower of y_n and y_c; if the depth starts in Zone 3, the depth will increase in the downstream direction and approach the lower of y_n and y_c.

Occurrence of various water surface profiles. Profiles in Zone 1 normally occur upstream of control structures such as dams and weirs, and these profiles are sometimes classified as *backwater curves*. Profiles in Zone 2, with the exception of S2 profiles, occur upstream of free overfalls, whereas S2 curves generally occur at the entrance to steep channels leading from a reservoir. Profiles in Zone 2 are sometimes classified as *drawdown curves*. Profiles in Zone 3 normally occur on mild and steep slopes downstream of control structures such as gates.

EXAMPLE 9.7

Water flows in a trapezoidal channel in which the bottom width is 5 m and the side slopes are 1.5:1 (H:V). The channel lining has an estimated Manning's n of 0.04, and the longitudinal slope of the channel is 1%. If the flow rate is 60 m^3/s and the depth of flow at a gauging station is 4 m, classify the water surface profile, state whether the depth increases or decreases in the downstream direction, and calculate the slope of the water surface at the gauging station. On the basis of this water surface slope, make a preliminary estimate of the depth of flow 100 m downstream of the gauging station.

SOLUTION

To classify the water surface profile, the normal and critical flow depths must be calculated and contrasted with the actual flow depth of 4 m. To calculate the normal flow depth, apply the Manning equation

$$Q = \frac{1}{n} A_n R_n^{\frac{2}{3}} S_0^{\frac{1}{2}} = \frac{1}{n} \frac{A_n^{\frac{5}{3}}}{P_n^{\frac{2}{3}}} S_0^{\frac{1}{2}}$$

where Q is the flow rate ($=$ 60 m^3/s), A_n and P_n are the areas and wetted perimeters, respectively, under normal flow conditions, and S_0 is the longitudinal slope of the channel ($=$ 0.01). The Manning equation can be written in the more useful form

$$\frac{A_n^5}{P_n^2} = \left[\frac{Qn}{\sqrt{S_0}}\right]^3$$

where the left-hand side is a function of the normal flow depth, y_n, and the right-hand side is in terms of given data. Substituting the given data leads to

$$\frac{A_n^5}{P_n^2} = \left[\frac{(60)(0.04)}{\sqrt{0.01}}\right]^3 = 13\,824$$

The area and wetted perimeter under normal flow conditions are given, respectively, by

$$A_n = (b + my_n)y_n = (5 + 1.5y_n)y_n$$

$$P_n = b + 2\sqrt{1 + m^2}\, y_n = 5 + 2\sqrt{1 + 1.5^2}\, y_n = 5 + 3.61y_n$$

Substituting these expressions into the Manning equation gives

$$\frac{(5+1.5y_\mathrm{n})^5 y_\mathrm{n}^5}{(5+3.61y_\mathrm{n})^2} = 13\,824 \quad \rightarrow \quad y_\mathrm{n} = 2.25 \text{ m}$$

When the flow conditions are critical,

$$\frac{A_\mathrm{c}^3}{T_\mathrm{c}} = \frac{Q^2}{g}$$

where A_c and T_c are the area and top width, respectively. The left-hand side of this equation is a function of the critical flow depth, y_c, and the right-hand side is in terms of the given data. Thus,

$$\frac{A_\mathrm{c}^3}{T_\mathrm{c}} = \frac{60^2}{9.81} = 367$$

The area and wetted perimeter under critical flow conditions are given, respectively, by

$$A_\mathrm{c} = (b + my_\mathrm{c})y_\mathrm{c} = (5 + 1.5y_\mathrm{c})y_\mathrm{c}$$

$$T_\mathrm{c} = b + 2my_\mathrm{c} = 5 + 2(1.5)y_\mathrm{c} = 5 + 3y_\mathrm{c}$$

Substituting these expressions into the critical flow equation gives

$$\frac{(5+1.5y_\mathrm{c})^3 y_\mathrm{c}^3}{5+3y_\mathrm{c}} = 367 \quad \rightarrow \quad y_\mathrm{c} = 1.99 \text{ m}$$

Because y_n (= 2.25 m) > y_c (= 1.99 m), the slope is mild. Also, because y (= 4 m) > $y_\mathrm{n} > y_\mathrm{c}$, the water surface is in Zone 1; therefore, the water surface profile is an **M1 profile.** This classification requires that **the flow depth increase in the downstream direction.** The slope of the water surface is given by (assuming $\alpha = 1$)

$$\frac{\mathrm{d}y}{\mathrm{d}x} = \frac{S_0 - S_\mathrm{f}}{1 - \mathrm{Fr}^2}$$

where S_f is the slope of the energy grade line and Fr is the Froude number. According to the Manning equation, S_f can be estimated by

$$S_\mathrm{f} = \left[\frac{nQ}{AR^{\frac{2}{3}}}\right]^2$$

and when the depth of flow, y, is 4 m,

$$A = (b + my)y = (5 + 1.5 \times 4)(4) = 44 \text{ m}^2$$

$$P = b + 2\sqrt{1+m^2}y = 5 + 2\sqrt{1+1.5^2}(4) = 19.4 \text{ m}$$

$$R = \frac{A}{P} = \frac{44}{19.4} = 2.27 \text{ m}$$

Therefore, S_f is estimated to be

$$S_\mathrm{f} = \left[\frac{(0.04)(60)}{(44)(2.27)^{\frac{2}{3}}}\right]^2 = 0.000997$$

The Froude number, Fr, is given by

$$\mathrm{Fr}^2 = \frac{V^2}{gD_\mathrm{h}} = \frac{(Q/A)^2}{g(A/T)} = \frac{Q^2T}{gA^3}$$

where the top width, T, is given by

$$T = b + 2my = 5 + 2(1.5)(4) = 17 \text{ m}$$

Therefore, the Froude number squared is given by

$$\mathrm{Fr}^2 = \frac{(60)^2(17)}{(9.81)(44)^3} = 0.0732$$

Substituting the values for S_0 (= 0.01), S_f (= 0.000997), and Fr^2 (= 0.0732) into the profile equation yields the slope of the water surface at the gauging station as

$$\frac{\mathrm{d}y}{\mathrm{d}x} = \frac{0.01 - 0.000997}{1 - 0.0732} = \mathbf{0.00971}$$

The depth of flow, y, at a location 100 m downstream from where the flow depth is 4 m can be estimated by

$$y = 4 + \frac{\mathrm{d}y}{\mathrm{d}x}(100) = 4 + (0.00971)(100) = \mathbf{4.97\ m}$$

The estimated flow depth 100 m downstream can be refined by recalculating $\mathrm{d}y/\mathrm{d}x$ for a flow depth of 4.97 m and then using an averaged value of $\mathrm{d}y/\mathrm{d}x$ to estimate the flow depth 100 m downstream.

Lake discharge problem. In some cases, the slope classification can play an important role in determining the flow rate in the channel. Such a circumstance is illustrated in Figure 9.14, where water is discharged from a large reservoir, such as a lake, into an open channel at A, the flow is normal at C, and B is in the transition region. The problem illustrated in Figure 9.14 is commonly referred to as the *lake discharge problem*. If H is the head at the channel

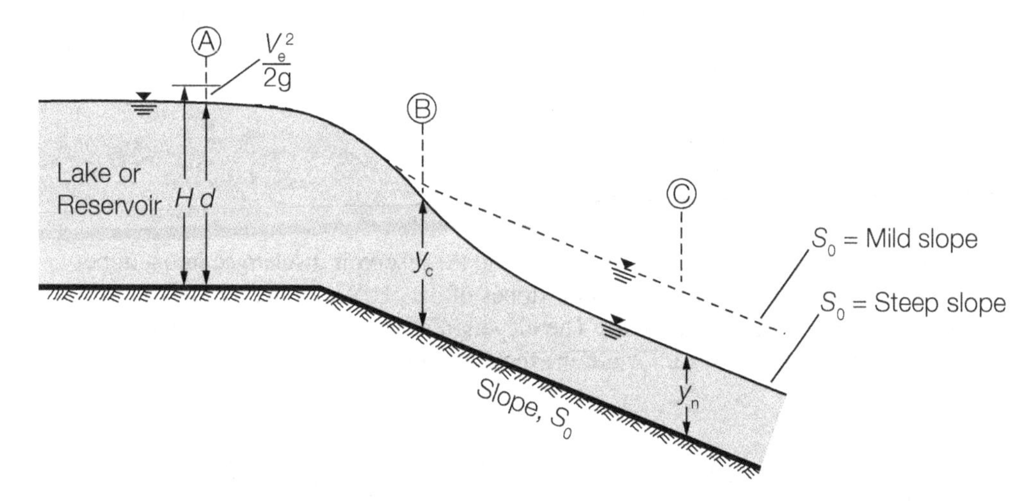

Figure 9.14: Lake discharge problem

entrance and the channel slope is steep (such that $y_{\rm n} < y_{\rm c}$), then the depth of flow in the transition region must pass through $y_{\rm c}$ such that

$$H = y_{\rm c} + \frac{Q^2}{2gA_{\rm c}^2} \tag{9.123}$$

and

$$\frac{Q^2 T_{\rm c}}{gA_{\rm c}^3} = 1 \tag{9.124}$$

where Equation 9.123 is required for conservation of energy and Equation 9.124 is the condition for critical flow. In contrast, if the channel slope is mild (such that $y_{\rm n} > y_{\rm c}$), then the depth of flow in the transition region is approximately equal to $y_{\rm n}$ such that

$$H = y_{\rm n} + \frac{Q^2}{2gA_{\rm n}^2} \tag{9.125}$$

and

$$Q = \frac{1}{n} A_{\rm n} R_{\rm n}^{\frac{2}{3}} S_0^{\frac{1}{2}} \tag{9.126}$$

where Equation 9.125 is required for conservation of energy and Equation 9.126 is the condition for normal flow. To determine the actual flow rate, Equations 9.123 and 9.124 are first solved simultaneously for Q and $y_{\rm c}$ and then the calculated value of Q is used in the Manning equation to calculate $y_{\rm n}$. If $y_{\rm n} < y_{\rm c}$, a steep slope is confirmed and the calculated Q is the actual Q. If $y_{\rm n} > y_{\rm c}$, then the slope is mild and the actual Q is determined by simultaneous solution of Equations 9.125 and 9.126. It is relevant to note that when the slope is mild, an M1 profile is required between the upstream reservoir and the downstream channel. However, this would require that the flow depth increase in the downstream direction (per Figure 9.13), which would not be possible in the present case. As a consequence, the depth of flow goes immediately to the normal flow depth.

Variations of the lake discharge problem. In cases where the downstream channel is mild and the channel is not sufficiently long, uniform flow might not be established in the channel, and the uniform-flow equation, Equation 9.126, must be replaced by the (energy) equation for gradually varied flow. In this case, the channel is called *hydraulically short*. Conversely, if the channel is mild and sufficiently long for uniform flow to be established, then the channel is *hydraulically long*. For steep slopes, the attainment of normal flow in the downstream channel is not required for Equations 9.123 and 9.124 to be valid, only that the downstream flow be supercritical.

EXAMPLE 9.8

The water surface elevation in a reservoir is 50.05 m, and the reservoir discharges into a trapezoidal canal that has a bottom width of 2 m, side slopes of 3:1 (H:V), a longitudinal slope of 1%, and an estimated Manning's n of 0.020. The elevation of the bottom of the canal at the reservoir discharge location is 47.01 m. Determine the discharge from the reservoir.

SOLUTION

The depth of the reservoir at the discharge location is 50.05 m − 47.01 m = 3.04 m. Because the velocity head in a reservoir can be assumed to be negligible, $H = 3.04$ m. From the given channel dimensions: $b = 2$ m, $m = 3$, $S_0 = 0.01$, and $n = 0.020$. Assuming that the slope is hydraulically steep (i.e., $y_n < y_c$), Equations 9.123 and 9.124 require that

$$H = y_c + \frac{Q^2}{2gA_c^2} \quad \text{and} \quad \frac{Q^2 T_c}{gA_c^3} = 1$$

Eliminating Q from these equations yields the more convenient combined form as

$$H = y_c + \left(\frac{gA_c^3}{T_c}\right)\frac{1}{2gA_c^2} \quad \rightarrow \quad H = y_c + \frac{A_c}{2T_c}$$

Using the geometric properties of a trapezoidal channel gives

$$H = y_c + \frac{by_c + my_c^2}{2[b + 2my_c]} \quad \rightarrow \quad 3.04 = y_c + \frac{2y_c + 3y_c^2}{2[2 + 2(3)y_c]} \quad \rightarrow \quad y_c = 2.37 \text{ m}$$

The corresponding value of Q is then given by

$$Q = \sqrt{\frac{gA_c^3}{T_c}} = \sqrt{\frac{(9.81)\,(2y_c + 3y_c^2)^3}{2 + 2(3)y_c}} = 78.3 \text{ m}^3/\text{s}$$

Determine the normal depth of flow, y_n, corresponding to $Q = 78.3$ m^3/s by applying the Manning equation, which requires that

$$Q = \frac{1}{n}\frac{A_n^{\frac{5}{3}}}{P_n^{\frac{2}{3}}}S_0^{\frac{1}{2}} = \frac{1}{n}\frac{(by_n + my_n^2)^{\frac{5}{3}}}{\left(b + 2y_n\sqrt{1+m^2}\right)^{\frac{2}{3}}}S_0^{\frac{1}{2}} \quad \rightarrow$$

$$78.3 = \frac{1}{0.020}\frac{(2y_n + 3y_n^2)^{\frac{5}{3}}}{\left(2 + 2y_n\sqrt{1+3^2}\right)^{\frac{2}{3}}}(0.01)^{\frac{1}{2}} \quad \rightarrow \quad y_n = 1.93 \text{ m}$$

Because $y_n < y_c$ (i.e., 1.93 m < 2.37 m), the slope is steep, so the initial assumption of a steep slope is validated. The discharge from the reservoir is **78.3 m^3/s**.

9.3.3 Hydraulic Jump

In some cases, supercritical flow must necessarily transition into subcritical flow, even though such a transition does not appear to be possible within the context of the water surface profiles discussed in the previous section. An example is a case where water is discharged as supercritical flow from under a vertical gate into a channel in which the downstream flow is subcritical. If the slope of the channel is mild, then the water emerging from under the gate follows an M3 profile, where the depth of flow increases with distance downstream. However, if the downstream flow is subcritical, then it is apparently impossible for the downstream flow condition to be reached, because this would require a continued increase in the water depth through the M2 zone, which, according to Figure 9.13, is not possible. In reality, this transition is accomplished by an abrupt localized change in water depth called a *hydraulic jump*, which is illustrated in Figure 9.15(a), with the corresponding transition in the specific energy diagram shown in Figure 9.15(b) and an actual hydraulic jump in a laboratory flume shown in Figure 9.15(c). The supercritical (upstream) flow depth is y_1, the subcritical (downstream) flow depth is y_2, and the energy loss between the upstream and downstream sections is ΔE. If the energy loss was equal to zero, then the downstream depth, y_2, could be calculated by equating the upstream and downstream specific energies. However, the transition between

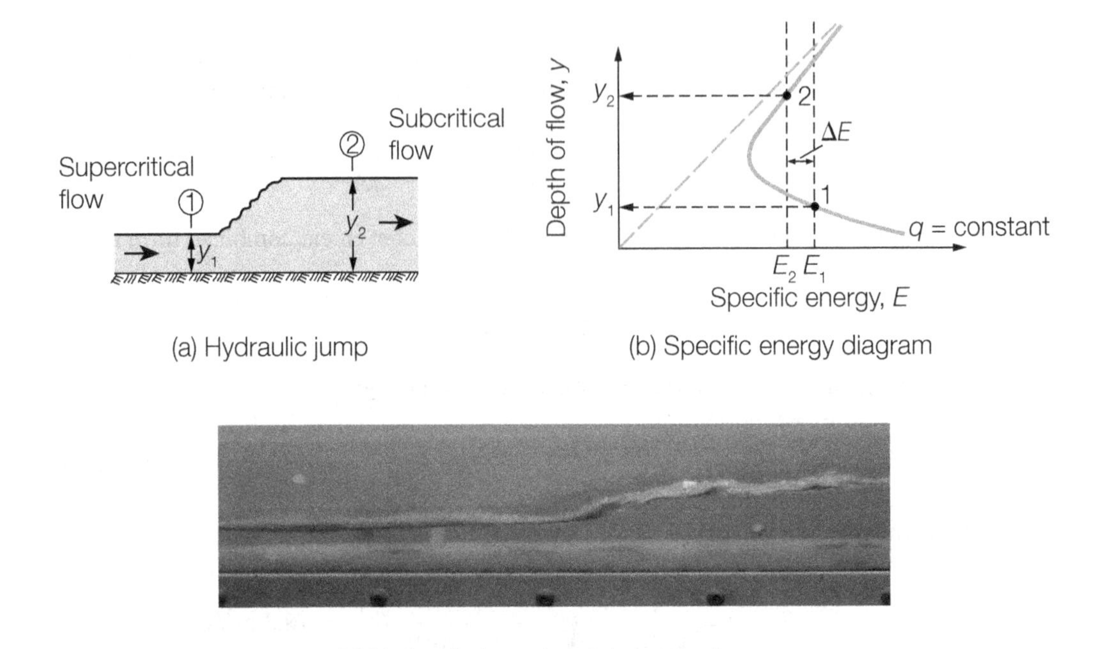

Figure 9.15: **Hydraulic jump and corresponding specific energy diagram**

supercritical and subcritical flow is generally a turbulent process with a significant energy loss that cannot be neglected.

Theory. Applying the momentum equation to the control volume between sections 1 and 2 leads to

$$P_1 - P_2 = \rho Q(V_2 - V_1) \tag{9.127}$$

where P_1 and P_2 are the hydrostatic pressure forces at sections 1 and 2, respectively, Q is the flow rate, and V_1 and V_2 are the average velocities at sections 1 and 2, respectively. Equation 9.127 neglects the friction force exerted by the channel surface on the liquid within the control volume. Neglecting channel friction is justified by the assumption that over a short distance, the friction force will be small compared with the difference in upstream and downstream hydrostatic forces. The momentum equation, Equation 9.127, can be written as

$$\gamma\bar{y}_1 A_1 - \gamma\bar{y}_2 A_2 = \rho Q\left(\frac{Q}{A_2} - \frac{Q}{A_1}\right) \tag{9.128}$$

where $\bar{y}_1$ and $\bar{y}_2$ are the distances from the water surface to the centroids of sections 1 and 2, respectively, and A_1 and A_2 are the cross-sectional areas at sections 1 and 2, respectively. Equation 9.128 can be rearranged as

$$\frac{Q^2}{gA_1} + A_1\bar{y}_1 = \frac{Q^2}{gA_2} + A_2\bar{y}_2 \tag{9.129}$$

or

$$\frac{Q^2}{gA} + A\bar{y} = \text{constant} \tag{9.130}$$

The term on the left-hand side of Equation 9.130 is called the *specific momentum*, and Equation 9.130 states that the specific momentum remains constant across the hydraulic jump.

Trapezoidal channel. In the case of a trapezoidal channel, the specific momentum equation, Equation 9.130, can be put in the form

$$\frac{by_1^2}{2} + \frac{my_1^3}{3} + \frac{Q^2}{gy_1(b+my_1)} = \frac{by_2^2}{2} + \frac{my_2^3}{3} + \frac{Q^2}{gy_2(b+my_2)} \tag{9.131}$$

where b is the bottom width of the channel and m is the side slope. Equation 9.131 is a fifth-order polynomial in either y_1 or y_2; solutions are most easily found using a programmable calculator with a built-in equation solver, a spreadsheet with solver capability, or a short computer program such as that proposed by Das (2007).

Rectangular channel. In the case of a rectangular channel, Equation 9.130 can be put in the form

$$\frac{q^2}{gy_1} + \frac{y_1^2}{2} = \frac{q^2}{gy_2} + \frac{y_2^2}{2} \tag{9.132}$$

where q is the flow per unit width. Equation 9.132 can be solved for y_2 to yield

$$y_2 = \frac{y_1}{2}\left(\sqrt{1 + \frac{8q^2}{gy_1^3}} - 1\right) \tag{9.133}$$

which can also be written in the nondimensional form

$$\frac{y_2}{y_1} = \frac{1}{2}\left(\sqrt{1 + 8\mathrm{Fr}_1^2} - 1\right) \tag{9.134}$$

where y_2/y_1 is commonly called the *sequent depth ratio* and Fr_1 is the upstream Froude number defined by

$$\mathrm{Fr}_1 = \frac{V_1}{\sqrt{gy_1}} = \frac{q}{y_1\sqrt{gy_1}} \tag{9.135}$$

Equation 9.134 was originally derived by Bresse (1860) and is sometimes called *Bresse's equation*, although some have argued that this equation was originally derived by Bélanger (1841) and should be called *Bélanger's equation* (e.g., Chanson, 2009). The depths upstream and downstream of a hydraulic jump, y_1 and y_2, are called the *conjugate depths* of the hydraulic jump, with y_1 sometimes called the *initial depth* and y_2 called the *sequent depth.*

Hydraulic jumps in rough channels. Experimental measurements have shown that Equation 9.134 yields values of y_2 to within 1% of observed values (Streeter et al., 1998), where the neglect of shear forces in the momentum equation causes theoretical values of y_2 to be slightly larger than observed values of y_2 (Beirami and Chamani, 2006). Shear forces can be significant for very rough channels, in which cases roughness effects can be accounted for by using the following empirical relation (Carollo et al., 2009):

$$\frac{y_2}{y_1} - 1 = \sqrt{2}\exp\left(-\frac{k_s}{y_c}\right)(\mathrm{Fr}_1 - 1)^{0.963} \tag{9.136}$$

where k_s is the equivalent sand roughness of the channel surface and y_c is the critical flow depth. Equation 9.136 is applicable to hydraulic jumps in both smooth ($k_s = 0$) and rough ($k_s \neq 0$) channels. It produces predictions of y_2/y_1 that are as good as or better than the Bresse equation (Equation 9.134) under smooth channel conditions and better than the Bresse equation under rough channel conditions.

Hydraulic jumps on sloping channels. The theoretical relationship between conjugate depths of a hydraulic jump occurring in a horizontal rectangular channel, Equation 9.134, can also be used for hydraulic jumps on sloping channels, provided the channel slope is less than about 5%. For larger channel slopes, the component of the weight of the liquid in the direction of flow becomes significant and must be incorporated into the momentum equation from which the hydraulic jump equation is derived. Other hydraulic jump equations have been developed for unusual cases, such as hydraulic jumps occurring on inclined contracting channels (Jan and Chang, 2009) and expanding channels (Kordi and Abustan, 2012) and hydraulic jumps occurring at the intersection of a sloping and a horizontal channel (Carollo et al., 2012), where the latter type of hydraulic jump is sometimes referred to as a *B-jump*. In cases where a hydraulic jump occurs in a closed conduit, the hydraulic jump equations presented in this section are applicable provided the sequent depth does not exceed the height of the closed conduit. In this context, the hydraulic jump is called a *complete hydraulic jump* or a *free-surface hydraulic jump*. In cases where the sequent depth is greater than the conduit height, the hydraulic jump is called an *incomplete hydraulic jump* or a *pressure hydraulic jump*. Although complete hydraulic jumps are far more common in engineering design, several specialized hydraulic jump equations have been developed for cases of incomplete hydraulic jumps (e.g., Lowe et al., 2011).

Length of hydraulic jump. The lengths of hydraulic jumps are around $6y_2$ for $4.5 < \mathrm{Fr}_1 < 13$ and somewhat smaller outside this range. The length, L, of a hydraulic jump can also be estimated as (Hager, 1991)

$$\frac{L}{y_1} = 220 \tanh \frac{\mathrm{Fr}_1 - 1}{22} \tag{9.137}$$

and the length, L_t, of the transition region between the end of the hydraulic jump and fully developed open-channel flow can be estimated as (Wu and Rajaratnam, 1996)

$$L_\mathrm{t} = 10y_2 \tag{9.138}$$

The relatively short lengths of hydraulic jumps justify neglecting the effects of boundary friction in deriving their governing equation. The length of a hydraulic jump is an important variable that is often used to define the downstream limit beyond which no bed protection or special channel provisions are necessary.

Physical characteristics of hydraulic jumps. The physical characteristics of hydraulic jumps in relation to the upstream Froude number, Fr_1, can be classified into the five categories listed in Table 9.6. A steady, well-established jump with $4.5 < \mathrm{Fr}_1 < 9.0$ is often used as an energy dissipator downstream of a dam or spillway and can also be used to mix chemicals. Outside the $4.5 < \mathrm{Fr}_1 < 9.0$ range, disturbances caused by the hydraulic jump tend to propagate downstream and can cause very rough waves (for $\mathrm{Fr} > 9$), which are usually undesirable. The ratio of conjugate depths, y_2/y_1, varies from slightly greater than 1 for undular jumps to greater than 12 for strong jumps, which are rough and involve large changes in water surface elevations across the jump.

Energy losses in hydraulic jumps. The energy loss in a hydraulic jump, h_L, is given by

$$h_\mathrm{L} = \left(y_1 + \frac{V_1^2}{2g}\right) - \left(y_2 + \frac{V_2^2}{2g}\right) \tag{9.139}$$

Table 9.6: Characteristics of Hydraulic Jumps

Name	Fr_1	Energy dissipation	Characteristics
Undular jump	1.0–1.7	<5%	Small rise in surface level. Low energy dissipation. Surface rollers develop near $\text{Fr} = 1.7$.
Weak jump	1.7–2.5	5%–15%	Surface rising smoothly, with small rollers. Low energy dissipation.
Oscillating jump	2.5–4.5	15%–45%	Pulsations generate large waves that can travel for kilometers and damage earth banks. Should be avoided in the design of stilling basins.
Stable jump	4.5–9.0	45%–70%	Stable, well-balanced, and insensitive to downstream conditions. Intense eddy motion and high level of energy dissipation within the jump. Recommended range for design.
Strong jump	>9.0	70%–85%	Rough and intermittent. Very effective energy dissipation, but may be uneconomical compared with other designs because of the larger water heights involved.

Source: USBR (1987)

Combining Equation 9.139 with the equation relating the conjugate depths of the hydraulic jump, Equation 9.134, leads to the following expression for the energy loss in a hydraulic jump occurring in a rectangular channel:

$$h_\text{L} = \frac{(y_2 - y_1)^3}{4y_1y_2} \tag{9.140}$$

A parameter of common interest in hydraulic jumps is the *energy dissipation ratio*, which gives the fraction of upstream specific energy that is lost in the hydraulic jump and is defined by h_L/E_1. It can be shown that

$$\text{energy dissipation ratio} = \frac{h_\text{L}}{E_1} = \frac{\left[\sqrt{1 + 8\text{Fr}_1^2} - 3\right]^3}{8\left[\sqrt{1 + 8\text{Fr}_1^2} - 1\right]\left[\text{Fr}_1^2 + 2\right]} \tag{9.141}$$

where E_1 represents the upstream specific energy and Fr_1 is the upstream Froude number. The energy dissipation ratio is typically less than 0.05 for weak hydraulic jumps ($\text{Fr}_1 < 2$) and on the order of 0.85 for strong hydraulic jumps ($\text{Fr}_1 > 9$).

Aeration in hydraulic jumps. Hydraulic jumps are sometimes used as self-aerators to enhance the dissolved oxygen concentration in water. Measurements in hydraulic jumps have shown that oxygen concentration increases across a hydraulic jump in direct proportion to energy dissipation rate per unit width, $\gamma q h_\text{L}$, according to the relation (Kucukali and Cokgor, 2009)

$$\frac{C_\text{d} - C_\text{u}}{C_\text{s} - C_\text{u}} = 0.0015(\gamma q h_\text{L}) + 0.01 \tag{9.142}$$

where C_d, C_u, and C_s are the downstream, upstream, and saturation concentrations of dissolved oxygen, respectively, at 20°C. Equation 9.142 was derived using experimental results in which $\gamma q h_\text{L} < 180$ W/m and should not be extrapolated beyond this range. The expression on the left-hand side of Equation 9.142, called the *aeration efficiency*, varies between 0 and 1.

The air-entrainment characteristics at any distance, x, from the beginning of a hydraulic jump can be expressed in the form (Chanson, 2011)

$$\frac{Q_{\text{air}}}{Q+Q_{\text{air}}} = 0.3387\text{Fr}_1^{0.202}\exp\left[(-0.103+0.0073\text{Fr}_1)\frac{x}{y_1}\right] \tag{9.143}$$

where Q_{air} is the air flow rate, Q is the water flow rate, Fr_1 is the upstream Froude number, and y_1 is the upstream flow depth. Equation 9.143 can be used to estimate the rate of air entrainment with distance from the start of the hydraulic jump and indicates decreasing aeration (deaeration) with distance downstream from the toe of the jump.

EXAMPLE 9.9

Water flows down a spillway at the rate of 12 m^3/s per meter of width into a horizontal channel, where the velocity at the channel entrance is 20 m/s. Determine the (downstream) depth of flow in the channel that will cause a hydraulic jump to occur at the toe of the spillway (i.e., where the spillway meets the channel) and determine the power loss in the jump per meter of width.

SOLUTION

In this case, $q = 12$ m^2/s and $V_1 = 20$ m/s. Therefore, the initial depth of flow, y_1, is given by

$$y_1 = \frac{q}{V_1} = \frac{12}{20} = 0.60 \text{ m}$$

and the corresponding Froude number, Fr_1, is given by

$$\text{Fr}_1 = \frac{q}{y_1\sqrt{gy_1}} = \frac{12}{0.60\sqrt{(9.81)(0.60)}} = 8.24$$

which confirms that the flow is supercritical. The conjugate depth is given by Equation 9.134, where

$$\frac{y_2}{y_1} = \frac{1}{2}\left(\sqrt{1+8\text{Fr}_1^2}-1\right) = \frac{1}{2}\left(\sqrt{1+8(8.24)^2}-1\right) = 11.2$$

Therefore, the conjugate downstream depth, y_2, is

$$y_2 = 11.2y_1 = 11.2(0.60) = 6.70 \text{ m}$$

Hence, a downstream flow depth of **6.70 m** will cause a hydraulic jump to occur at the toe of the spillway. The energy loss in the hydraulic jump, h_L, is given by Equation 9.140 as

$$h_\text{L} = \frac{(y_2-y_1)^3}{4y_1y_2} = \frac{(6.70-0.60)^3}{4(0.60)(6.70)} = 14.1 \text{ m}$$

Therefore, the power loss, P, per unit width in the jump is given by

$$P = \gamma q\, h_\text{L} = 9790(12)(14.1) = 1.66\times 10^6 \text{ W} = \mathbf{1.7\ MW}$$

Because $\text{Fr}_1 = 8.24$ is within the range for stable jumps ($4.5 \leq \text{Fr}_1 \leq 9.0$), it is anticipated that the hydraulic jump will be stable and not cause significant water surface oscillations downstream.

Location of hydraulic jump. The location of a hydraulic jump is important in determining channel wall heights as well as the flow conditions (subcritical or supercritical) in the channel. The mean location of a hydraulic jump is usually estimated by computing the upstream and downstream water surface profiles, and the jump is located where the upstream and downstream water depths are equal to the conjugate depths of the hydraulic jump. In many cases, the location of a hydraulic jump is controlled by installing baffle blocks, sills, drops, or rises in the bottom of the channel to create sufficient energy loss that the hydraulic jump forms at the location of these structures. Hydraulic structures that are specifically designed to induce the formation of hydraulic jumps are called *stilling basins*.

9.3.4 Computation of Water Surface Profiles

The differential equation describing the water surface profile in an open channel is given by Equation 9.117, which can be written in the form

$$\frac{\mathrm{d}y}{\mathrm{d}x} = F(x, y) \tag{9.144}$$

where y is the depth of flow in the channel, x is the distance along the channel, and $F(x, y)$ is a function defined by the relation

$$F(x, y) = \frac{S_0 - S_\mathrm{f}}{1 - \alpha \mathrm{Fr}^2} \tag{9.145}$$

where S_0 is the channel slope, S_f is the slope of the energy grade line, Fr is the Froude number of the flow, and α is the kinetic energy correction factor. The function $F(x, y)$ can be calculated using given values of y, Q, S_0, α, and channel geometry (which can be a function of x) and using Equation 9.116 to estimate the Froude number and the Manning or Darcy–Weisbach equation to estimate the slope of the energy grade line, S_f.

Basic assumptions. A basic assumption in calculating water surface profiles is that the head loss between upstream and downstream sections can be estimated using either the Manning or Darcy–Weisbach equation without regard to trends in depth. This approximation requires that flow conditions change gradually, and such flow conditions are classified as *gradually varied flow* (GVF). Conversely, flows that are not gradually varied are classified as *rapidly varied flow* (RVF). Important differences between RVF and GVF are that (1) in RVF, there is significant acceleration normal to the streamlines, causing a nonhydrostatic pressure distribution; (2) in RVF, significant depth variations occur over short distances, so boundary friction is relatively small; and (3) in RVF, the kinetic energy correction factor, α, and the momentum correction factor, β, are much greater than unity. Examples of RVF include hydraulic jumps and flows over spillways. This section focuses on the computation of water surface profiles under GVF conditions.

Estimation of average friction slope. Under GVF conditions, the Manning approximation (in SI units) to the friction slope, S_f, at any given flow section is given by

$$S_\mathrm{f} = \left(\frac{nQ}{AR^{\frac{2}{3}}}\right)^2 \tag{9.146}$$

where n is the Manning roughness coefficient, Q is the flow rate, and A and R are the area and hydraulic radius of the cross section, respectively. It is sometimes convenient to define the *conveyance*, K, by the relation

$$K = \frac{1}{n} AR^{\frac{2}{3}} \tag{9.147}$$

in which case the friction slope, S_f, can be expressed in terms of the flow rate, Q, and the conveyance, K, as

$$S_f = \left(\frac{Q}{K}\right)^2 \tag{9.148}$$

It is often necessary to estimate the average friction slope, $\bar{S}_f$, between two sections in an open channel. The following alternative methods have been used to estimate $\bar{S}_f$:

$$\bar{S}_f = \begin{cases} \dfrac{Q^2}{\left[\dfrac{K_1+K_2}{2}\right]^2} & \text{Method 1: Average conveyance method} \quad (9.149) \\[2ex] \dfrac{S_{f1}+S_{f2}}{2} & \text{Method 2: Average friction slope method} \quad (9.150) \\[2ex] \sqrt{S_{f1}S_{f2}} & \text{Method 3: Geometric mean friction slope method} \quad (9.151) \\[2ex] \dfrac{2S_{f1}S_{f2}}{S_{f1}+S_{f2}} & \text{Method 4: Harmonic mean friction slope method} \quad (9.152) \end{cases}$$

where K_1 and K_2 are the conveyances at the upstream and downstream sections, respectively, and S_{f1} and S_{f2} are the friction slopes at the upstream and downstream sections, respectively. Method 2 has been found to be most accurate for M1 profiles, whereas Method 4 has been found to be most accurate for M2 profiles (U.S. Army Corps of Engineers, 2008). Differences between methods become smaller as the spacing between cross sections is reduced, and differences are typically minimal for cross-section spacings less than 150 m (U.S. Army Corps of Engineers, 1986; French, 2001).

Computation of water surface profiles using the energy equation. Water surface profiles can be calculated directly from the energy equation. Rearranging the form of the energy equation given by Equation 9.81 leads to

$$\Delta L = \frac{\left[y + \alpha\dfrac{V^2}{2g}\right]_2^1}{\bar{S}_f - S_0} \tag{9.153}$$

where ΔL is the distance between the upstream section (section 1) and the downstream section (section 2) and $\bar{S}_f$ is the mean slope of the energy grade line between sections 1 and 2. Equation 9.153 is used in a variety of ways to calculate the water surface profile. These approaches are described subsequently.

Accounting for contraction and expansion losses. In applying Equation 9.153 to calculate the water surface profile, it is assumed that the contraction or expansion loss is negligible in comparison to the friction loss. Contraction and expansion losses can be accounted for by Equation 9.106, and the friction loss can be accounted for by $\bar{S}_f$. In the general case where friction and expansion/contraction losses are both accounted for, the appropriate energy equation is given by

$$\Delta L = \frac{\left[y + \alpha\dfrac{V^2}{2g}\right]_2^1}{\bar{S}_f + \dfrac{C}{\Delta L}\left|\alpha_2\dfrac{V_2^2}{2g} - \alpha_1\dfrac{V_1^2}{2g}\right| - S_0} \tag{9.154}$$

where C is the expansion or contraction coefficient between adjacent channel sections. For gradual expansions and contractions, values of C are typically 0.3 and 0.1, respectively, whereas for abrupt expansions and contractions, values of C are typically 0.8 and 0.6, respectively. When the channel cross sections do not vary significantly, contraction and expansion losses between sections are much less than friction losses, and given the uncertainty in estimating friction losses, Equation 9.153 is an adequate representation of the energy equation in these cases.

Limitations of using the energy equation. Whenever the water surface passes through critical depth, the energy equation is not applicable. It is applicable only to gradually varied flows, and the transition from subcritical to supercritical flow, and supercritical to subcritical flow, are rapidly varied flows. Such rapidly varied flows occur at significant changes in channel slope, some bridge constrictions, drop structures and weirs, and some stream junctions. In many of these cases, empirical equations can be used to relate upstream and downstream flow depths, whereas in other cases, it is necessary to apply the momentum equation to determine the changes in water surface elevation.

Computation methods. Methods commonly used to determine the water surface profile are the *direct integration method*, *direct step method*, and *standard step method*. These three methods are based on the energy equation and yield essentially the same results in cases where all of them are applicable; their differences are mostly related to the ease and efficiency of the computations.

9.3.4.1 Direct integration method

In applying the direct integration method, the water surface profile described by Equation 9.144 can be expressed in the finite difference form

$$\frac{y_2 - y_1}{x_2 - x_1} = F(\bar{x}, \bar{y}) \tag{9.155}$$

where

$$\bar{x} = \frac{x_1 + x_2}{2} \quad \text{and} \quad \bar{y} = \frac{y_1 + y_2}{2} \tag{9.156}$$

and the subscripts refer to (adjacent) cross sections of the channel, where section 1 is upstream of section 2. A convenient form of Equation 9.155 is

$$y_2 = y_1 + F(\bar{x}, \bar{y})(x_2 - x_1) \tag{9.157}$$

This equation is appropriate for computing the water surface profile in the downstream direction. However, in most cases, water surface profiles are computed in the upstream direction, in which case the following form of Equation 9.155 is more useful:

$$y_1 = y_2 - F(\bar{x}, \bar{y})(x_2 - x_1) \tag{9.158}$$

In subcritical flow, calculations generally proceed in an upstream direction, whereas in supercritical flow, calculations proceed in a downstream direction. In applying Equation 9.158, the following computation procedure is suggested:

Step 1: Starting with the known flow condition at section 2, assume a depth, y_1, at location x_1 and then calculate y_1 using Equation 9.158. On the first calculation, it is reasonable to assume that $y_1 = y_2$.

Step 2: Repeat Step 1 until the calculated value of y_1 is equal to the assumed value of y_1. This is the depth of flow at x_1.

A similar computation procedure is used to apply Equation 9.157, with iterations on y_2 rather than y_1.

EXAMPLE 9.10

Water flows at 10 m^3/s in a rectangular concrete channel of width 5 m and longitudinal slope 0.1%. The Manning roughness coefficient, n, of the channel lining is 0.015, and the water depth is measured as 0.80 m at a gauging station. Use the direct integration method to estimate the flow depth 100 m upstream of the gauging station.

SOLUTION

From the given data: $Q = 10$ m^3/s, $b = 5$ m, $S_0 = 0.1\% = 0.001$, $n = 0.015$, $y_2 = 0.80$ m, $x_1 = 0$ m, and $x_2 = 100$ m. Assuming that $y_1 = y_2 = 0.80$ m, the hydraulic parameters of the flow are

$$\bar{x} = \frac{x_1 + x_2}{2} = \frac{0 + 100}{2} = 50 \text{ m}, \qquad \bar{y} = \frac{y_1 + y_2}{2} = \frac{0.80 + 0.80}{2} = 0.80 \text{ m}$$

$$\overline{A} = b\bar{y} = (5)(0.80) = 4.0 \text{ m}^2, \qquad \overline{P} = b + 2\bar{y} = 5 + 2(0.80) = 6.60 \text{ m}$$

$$\overline{R} = \frac{\overline{A}}{\overline{P}} = \frac{4.0}{6.60} = 0.606 \text{ m}, \qquad \overline{S}_\text{f} = \left[\frac{nQ}{\overline{A}\,\overline{R}^{\frac{2}{3}}}\right]^2 = \left[\frac{(0.015)(10)}{(4.0)(0.606)^{\frac{2}{3}}}\right]^2 = 0.00274$$

$$\overline{D} = \frac{\overline{A}}{\overline{T}} = \frac{b\bar{y}}{b} = \bar{y} = 0.80 \text{ m}, \qquad \overline{V} = \frac{Q}{\overline{A}} = \frac{10}{4} = 2.5 \text{ m/s}$$

$$\overline{\text{Fr}}^2 = \frac{\overline{V}^2}{g\overline{D}} = \frac{(2.5)^2}{(9.81)(0.80)} = 0.80$$

Using these results and assuming that $\alpha = 1$, Equation 9.145 gives

$$F(\bar{x}, \bar{y}) = \frac{S_0 - \overline{S}_\text{f}}{1 - \overline{\text{Fr}}^2} = \frac{0.001 - 0.00274}{1 - 0.80} = -0.0087$$

and the estimated depth 100 m upstream of the gauging station is given by Equation 9.158 as

$$y_1 = y_2 - F(\bar{x}, \bar{y})(x_2 - x_1) = 0.80 - (-0.0087)(100 - 0) = 1.67 \text{ m}$$

Because this calculated flow depth at section 1 (= 1.67 m) is significantly different from the assumed flow depth (=0.80 m), the calculations must be repeated, starting with the new assumption that $y_1 = 1.67$ m. These calculations are summarized in the following table, where the assumed values of y_1 are given in Column 1 and the calculated values of y_1 (using Equation 9.158) are given in Column 4.

(1) y_1 (m)	(2) $\overline{S}_\text{f}$	(3) $\overline{\text{Fr}}^2$	(4) y_1 (m)
1.67	0.000761	0.216	0.77
1.5	0.000936	0.268	0.79
1.3	0.00122	0.352	0.83
1.1	0.00164	0.476	0.92
⋮	⋮	⋮	⋮
1.01	0.00193	0.559	1.01

These results indicate that the depth 100 m upstream from the gauging station is approximately equal to **1.01 m**.

9.3.4.2 Direct step method

In applying the direct-step method, the flow depth at one section is known, the flow depth at a second section is specified, and the objective is to find the distance between these two sections. With the flow conditions at two channel sections known, the terms on the right-hand side of Equation 9.153 are evaluated to determine the distance, ΔL, between these sections. These computations are then repeated to find the incremental distances between all adjacent sections with specified flow depths, thereby yielding the water surface profile.

Limitations of the direct step method. The main limitations of the direct step method are that (1) the water surface profile is not computed at predetermined locations and (2) the method is suitable only for prismatic channels, where the shape of the channel cross section is independent of the interval, ΔL. In cases where the flow conditions at specific locations in a prismatic or nonprismatic channel are required, the standard step method should be used.

EXAMPLE 9.11

Water flows at 12 m³/s in a trapezoidal concrete channel ($n = 0.015$) of bottom width 4 m, side slopes 2 : 1 (H : V), and longitudinal slope 0.09%. If depth of flow at a gauging station is measured as 0.80 m, use the direct step method to find the location where the depth is 1.00 m.

SOLUTION

At the location where the depth is 1.00 m,

$$y_1 = 1.00 \text{ m}, \qquad A_1 = [4 + 2y_1]y_1 = [4 + 2(1.00)](1.00) = 6.00 \text{ m}^2$$

$$V_1 = \frac{Q}{A_1} = \frac{12}{6.00} = 2.00 \text{ m/s}, \qquad P_1 = 4 + 2\sqrt{5}y_1 = 4 + 2\sqrt{5}(1.00) = 8.47 \text{ m}$$

$$R_1 = \frac{A_1}{P_1} = \frac{6.00}{8.47} = 0.708 \text{ m}, \qquad S_{f1} = \left[\frac{nQ}{A_1 R_1^{\frac{2}{3}}}\right]^2 = \left[\frac{(0.015)(12)}{(6.00)(0.708)^{\frac{2}{3}}}\right]^2 = 0.00143$$

and where the depth is 0.80 m,

$$y_2 = 0.80 \text{ m}, \qquad A_2 = [4 + 2y_2]y_2 = [4 + 2(0.80)](0.80) = 4.48 \text{ m}^2$$

$$V_2 = \frac{Q}{A_2} = \frac{12}{4.48} = 2.68 \text{ m/s}, \qquad P_2 = 4 + 2\sqrt{5}\,y_2 = 4 + 2\sqrt{5}\,(0.80) = 7.58 \text{ m}$$

$$R_2 = \frac{A_2}{P_2} = \frac{4.48}{7.58} = 0.591 \text{ m}, \qquad S_{f2} = \left[\frac{nQ}{A_2 R_2^{\frac{2}{3}}}\right]^2 = \left[\frac{(0.015)(12)}{(4.48)(0.591)^{\frac{2}{3}}}\right]^2 = 0.00325$$

Substituting the hydraulic parameters at sections 1 and 2 into Equation 9.153, taking the velocity coefficients α_1 and α_2 equal to unity, and using Equation 9.150 to estimate the average friction slope, $\bar{S}_f$, gives

$$\Delta L = \frac{\left[y + \alpha\frac{V^2}{2g}\right]_2^1}{\bar{S}_f - S_0} = \frac{\left(1.00 + \dfrac{2.00^2}{2 \times 9.81}\right) - \left(0.80 + \dfrac{2.68^2}{2 \times 9.81}\right)}{\left(\dfrac{0.00143 + 0.00325}{2}\right) - 0.0009} = 26.4 \text{ m}$$

Hence, the depth in the channel increases to 1.00 m at a location that is approximately **26.4 m upstream** of the section where the depth is 0.80 m.

9.3.4.3 Standard step method

In applying the standard step method, the flow depth is known at one section and the objective is to find the flow depth at a second section a given distance away. The standard step method is similar to the direct step method in that it is based on the solution of Equation 9.153; in the standard step method, however, ΔL is given and the flow depth at the second section is unknown. The standard step method is particularly useful in natural channels, where the dimensions of the cross sections are typically measured only at locations that are easily accessible.

EXAMPLE 9.12

Water flows in an open channel whose slope is 0.04%. The Manning roughness coefficient of the channel lining is estimated to be 0.035, and the flow rate is 200 m³/s. At a given section of the channel, the cross section is trapezoidal, with a bottom width of 10 m, side slopes of 2:1 (H:V), and a depth of flow of 7 m. Use the standard step method to calculate the depth of flow 100 m upstream from this section, where the cross section is trapezoidal, with a bottom width of 15 m and side slopes of 3:1 (H:V).

SOLUTION

A longitudinal view of the flow along with the upstream and downstream channel sections are illustrated in Figure 9.16.

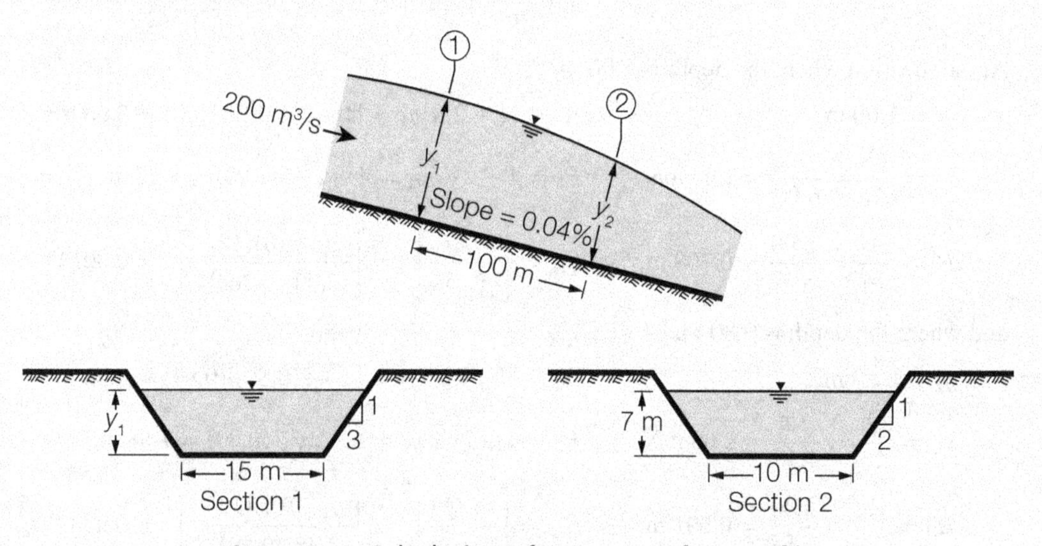

Figure 9.16: Calculation of a water surface profile

In this case, $Q = 200$ m³/s, $n = 0.035$, and the flow conditions at section 2 (the downstream section) are given as $y_2 = 7$ m, $b_2 = 10$ m, and $m_2 = 2$ (side slope is m_2:1). Therefore,

$$A_2 = [b_2 + m_2 y_2]y_2 = [10 + (2)(7)](7) = 168 \text{ m}^2, \qquad V_2 = \frac{Q}{A_2} = \frac{200}{168} = 1.19 \text{ m/s}$$

$$P_2 = b_2 + 2y_2\sqrt{1 + m_2^2} = 10 + 2(7)\sqrt{1 + 2^2} = 41.3 \text{ m}, \qquad R_2 = \frac{A_2}{P_2} = \frac{168}{41.3} = 4.07 \text{ m}$$

$$S_{f2} = \left[\frac{nQ}{A_2 R_2^{\frac{2}{3}}}\right]^2 = \left[\frac{(0.035)(200)}{(168)(4.07)^{\frac{2}{3}}}\right]^2 = 0.000267$$

At the upstream section, $b_1 = 15$ m and $m_1 = 3$, and denoting the depth of flow at the upstream section as y_1 gives

$$A_1 = (b_1 + m_1 y_1)y_1 = (15 + 3y_1)y_1 \quad (9.159) \qquad V_1 = \frac{Q}{A_1} \quad (9.160)$$

$$P_1 = b_1 + 2y_1\sqrt{1 + m_1^2} = 15 + 2y_1\sqrt{1 + 3^2} \quad (9.161) \qquad R_1 = \frac{A_1}{P_1} \quad (9.162)$$

$$S_{f1} = \left[\frac{nQ}{A_1 R_1^{\frac{2}{3}}}\right]^2 \quad (9.163)$$

$$\bar{S}_f = \frac{S_{f1} + S_{f2}}{2} = \frac{S_{f1} + 0.000267}{2} = 0.5S_{f1} + 0.000134$$

Applying the energy equation (Equation 9.153) between sections 1 and 2 with $\Delta L = 100$ m and taking the velocity coefficients, α_1 and α_2, to be equal to unity yields

$$\Delta L = \frac{\left[y + \alpha\frac{V^2}{2g}\right]_2^1}{\bar{S}_f - S_0} \quad \rightarrow \quad 100 = \frac{\left[y_1 + \frac{V_1^2}{2(9.81)}\right] - \left[7.00 + \frac{1.19^2}{2(9.81)}\right]}{(0.5S_{f1} + 0.000134) - 0.0004} \quad (9.164)$$

Substituting Equations 9.159–9.163 into Equation 9.164 and solving for y_1 yields $y_1 = 7.02$ m. Therefore, the depth of the flow 100 m upstream is **7.02 m.**

Selection of adjacent cross sections. In computing water surface profiles using the standard step method, significant errors can result if adjacent cross sections used in the computations are too far apart and the hydraulic properties of the flow change too radically from one cross section to the next. A common guideline is that the slope of the energy grade line should not decrease by more than 50% or increase by more than 100% between sections. An alternative guideline is that the change in depth should not be greater than 1%. Selecting cross sections too far apart can result in there being no solution to the energy equation. In general, cross sections should, as a minimum, be located at changes in channel geometry and slope, above and below major tributaries, and at structures such as bridges, submerged roads, and transitions.

9.3.4.4 Practical considerations when performing backwater computations

In performing backwater computations, several factors should be considered to ensure that the results reflect reality and are technically sound. Such considerations include honoring the discharge-depth relation at control sections and recognizing that the normal and critical flow depths can vary between channel reaches. In the case of 100-year flood flows, computed water surface profiles are typically used to establish minimum ground floor elevations for buildings located in the floodplain of the channel. Computers are typically used to perform backwater computations.

Control sections. The computation of a water surface profile generally begins at a section where the depth of flow is known. In most cases, the depth of flow is known at a section where there is a unique relationship between the depth and the flow rate. Such sections are called *control sections*. A typical control section requires that critical flow conditions occur at that section, in which case the depth of flow and the flow rate are related by Equation 9.90. Examples of control sections include various hydraulic structures, channel constrictions that choke the flow to create critical conditions at the control section, and rectangular free overfalls where the depth-discharge relationship can be estimated using the relation (Tiğrek et al., 2008)

$$q = C_\mathrm{d} y_\mathrm{b}^{\frac{3}{2}} \tag{9.165}$$

where q is the discharge per unit width; C_d is a discharge coefficient given by

$$C_\mathrm{d} = \begin{cases} 5.55, & \mathrm{Fr} \leq 1 \\ \left(0.361 - 0.00841\dfrac{\sqrt{S_0}}{n}\right)^{-\frac{3}{2}}, & \mathrm{Fr} > 1 \end{cases} \tag{9.166}$$

where S_0 and n are the channel slope and Manning roughness coefficient of the channel upstream of the free overfall, respectively, and y_b is the depth of flow at the brink of the free overfall. Free overfalls are control sections that are frequently found in both artificial and natural channels (e.g., waterfalls). Some control sections do not require that critical flow conditions exist at the control section. For example, at some hydraulic structures, the fixed relationship between flow rate and depth is determined by the geometry of the structure and the downstream flow conditions (e.g., at gates and culverts).

Length of backwater profile. The length of a backwater profile is commonly defined as the distance from the control section to the section where the depth of flow is within 10% of the normal depth of flow. For prismatic channels, this length, L, can be estimated by the relation (Samuels, 1989)

$$L = 0.7\frac{y_\mathrm{n}}{S_0} \tag{9.167}$$

where y_n is the normal depth of flow and S_0 is the slope of the channel. This approximation is not a substitute for calculating the backwater profile.

Steady-state assumption. The computation of a water surface profile is typically done to determine the water surface elevations expected along a channel during a specified flow event, such as the 100-year flow event that is used to delineate floodplains. Although the flow in a channel during any flood event is unsteady, it is typically assumed that the peak discharge rate occurs at the same time for the entire length of the channel and that the discharge rate changes along the channel only at major tributaries.

Natural channels. In natural channels, the standard step method is used with the energy equation given by Equation 9.154, repeated here as

$$\Delta L = \frac{\left[y + \alpha\dfrac{V^2}{2g}\right]_2^1}{S_\mathrm{f} + \dfrac{C}{\Delta L}\left|\alpha_2\dfrac{V_2^2}{2g} - \alpha_1\dfrac{V_1^2}{2g}\right| - S_0} \tag{9.168}$$

For channels without obstructions or transitions, the middle term in the denominator (the "eddy loss" term) is commonly neglected, but this term should otherwise be taken into account, such as at bridges and constructed channel transitions. The channel slope, S_0, will usually change between intervals along the channel; in such cases, it is convenient to express the slope in terms of the bottom elevations of the channel, where

$$S_0 = \frac{z_1 - z_2}{\Delta L} \tag{9.169}$$

where z_1 and z_2 are the bottom elevations of the upstream and downstream sections, respectively. In the usual case where the elevation of the bottom of the channel varies across the section, the bottom elevation at the deepest point is used as the elevation of the bottom of the channel. In most cases of practical interest, the backwater profile is described by the variation of the water surface elevation (also known as the *stage*) along the channel rather than the variation of depth along the channel. In these cases, the stage, Z_i, at the $i^{\rm th}$ section can be calculated as the bottom elevation of the channel plus the depth of flow as follows:

$$Z_i = z_i + y_i \tag{9.170}$$

where z_i and y_i are the bottom elevation and (maximum) flow depth at the $i^{\rm th}$ section, respectively. When major flood flows are considered, the channel sections are typically compound sections where the kinetic energy correction deviates significantly from unity and must be calculated at each section.

Flow calculations. A variation of the backwater computation procedure is required when the upstream and downstream stages are given and the objective is to calculate the flow rate in the channel. This situation usually occurs in the context of using high-water marks left by a flood to estimate the flood flow. In principle, this problem can be solved by varying the flow rate in the energy equation until the best match is achieved between the predicted and measured stages. A technique that is sometimes used to perform these calculations is the *slope-area method* (Dalrymple and Benson, 1967). However, field evidence indicates that the slope-area method and the one-dimensional energy equation are likely to overestimate the flow rate in floodplains. Alternative estimation methods that account for secondary flows caused by the shear between the floodplain and the main channel might provide improved estimates of the flow rate (e.g., Kordi and Abustan, 2011).

Key Equations in Open-Channel Flow

The following list of equations is particularly useful in solving problems in open-channel flow. If one is able to recognize these equations and recall their appropriate use, then the learning objectives of this chapter have been met to a significant degree. Derivations of these equations, definitions of the variables, and detailed examples of usage can be found in the main text.

BASIC PRINCIPLES

Continuity equation: $V_1A_1 = V_2A_2$

Momentum equation: $\sum F_x = \rho Q(\beta_2 V_2 - \beta_1 V_1), \quad \sum F_x \approx \rho Q(V_2 - V_1)$

Momentum correction coefficient: $\beta = \dfrac{\int_A v^2 \, \mathrm{d}A}{V^2 A}$

Hydraulic radius: $R = \dfrac{A}{P}$

Boundary shear stress: $\tau_0 = \gamma R S_\mathrm{e}, \quad \tau_0 = \gamma R S_\mathrm{f}, \quad \tau_0 = \gamma R S_0$

Darcy–Weisbach equation: $V = \left(\dfrac{8g}{f}\right)^{\frac{1}{2}} \sqrt{RS_0}$

Nondimensional shear stress: $\dfrac{\tau_0}{\frac{1}{8}\rho V^2} = f\left(\mathrm{Re}_\mathrm{h}, \dfrac{\epsilon}{D_\mathrm{h}}\right)$

Friction factor: $\dfrac{1}{\sqrt{f}} = -2\log_{10}\left(\dfrac{k_\mathrm{s}}{12R} + \dfrac{2.5}{\mathrm{Re}_\mathrm{h}\sqrt{f}}\right)$

Conditions of turbulent flow: $\dfrac{u_* k_s}{\nu} = \begin{cases} < 5, & \text{smooth} \\ 5 \text{ to } 70, & \text{transition} \\ > 70, & \text{rough} \end{cases}$

Friction factor, rock lining: $\dfrac{1}{\sqrt{f}} = 1.2 + 2.03\log_{10}\left(\dfrac{R}{d_{84}}\right)$

Friction factor, vegetation lining: $f = 0.40\left(\dfrac{k_\mathrm{v}}{h}\right)^{\frac{1}{8}}$

Flow rate (Darcy–Weisbach equation): $Q = -2A\sqrt{8gRS_0}\log_{10}\left(\dfrac{k_\mathrm{s}}{12R} + \dfrac{0.625\nu}{R^{\frac{3}{2}}\sqrt{8gS_0}}\right)$

Manning equation: $V = \frac{1}{n} R^{\frac{2}{3}} S^{\frac{1}{2}}$

Manning's n: $\frac{n}{k_\mathrm{s}^{\frac{1}{6}}} = 0.039, \qquad n = 0.039\, k_\mathrm{s}^{\frac{1}{6}}, \qquad n = \alpha\, d_\mathrm{p}^{\frac{1}{6}}$

Velocity distribution: $v(y) = V + \frac{1}{\kappa}\sqrt{gdS_0}\left(1 + 2.3 \log \frac{y}{d}\right)$

Estimation of flow rate: $Q = \sum_{i=1}^{N} \overline{V}_i A_i$

Surface-wave speed: $c = \sqrt{gy}$

Kinetic energy correction factor: $\alpha = \frac{\int_A \rho v^3 \,\mathrm{d}A}{\rho V^3 A}$

Energy equation:

$$\underbrace{\left[y\cos\theta + \alpha\frac{V^2}{2g} + z_0\right]_1}_{\text{energy at upstream section}} = \underbrace{\left[y\cos\theta + \alpha\frac{V^2}{2g} + z_0\right]_2}_{\text{energy at downstream section}} + \underbrace{h_\mathrm{L}}_{\text{energy loss}}$$

$$\left[y\cos\theta + \alpha\,\frac{V^2}{2g}\right]_2^1 = (S - S_0\cos\theta)L$$

$$\left[y + \alpha\,\frac{V^2}{2g}\right]_2^1 = (S_\mathrm{f} - S_0)L, \qquad S_0 < 0.1$$

$$\left[y + \alpha\frac{V^2}{2g} + z_0\right]_1 = \left[y + \alpha\frac{V^2}{2g} + z_0\right]_2 + h_\mathrm{f}, \quad S_0 < 0.1$$

Head, definition: $h = y + \alpha\,\frac{V^2}{2g} + z_0, \qquad S_0 < 0.1$

Specific energy, definition: $E = y + \alpha\,\frac{V^2}{2g} = y + \alpha\,\frac{Q^2}{2gA^2}$

Energy equation: $E_1 - E_2 = h_\mathrm{L} + \Delta z$

Critical-flow condition: $\frac{Q^2}{g} = \frac{A_\mathrm{c}^3}{T_\mathrm{c}}, \qquad E_\mathrm{c} = y_\mathrm{c} + \frac{A_\mathrm{c}}{2T_\mathrm{c}}, \qquad \mathrm{Fr}_\mathrm{c} = 1$

Critical condition:
large slopes: $\frac{\alpha Q^2 T}{gA^3\cos\theta} = 1$

Critical condition:
rectangular channel: $y_\mathrm{c} = \left(\frac{q^2}{g}\right)^{\frac{1}{3}}, \qquad \frac{E}{y_\mathrm{c}} = \frac{y}{y_\mathrm{c}} + \frac{1}{2}\left(\frac{y}{y_\mathrm{c}}\right)^2$

Energy loss expansion/contraction: $h_\mathrm{e} = C\left|\alpha_2\frac{V_2^2}{2g} - \alpha_1\frac{V_1^2}{2g}\right|$

WATER SURFACE PROFILES

Water surface profile: $$\frac{dy}{dx} = \frac{1}{\mathrm{Fr}^2 - 1}\frac{dh}{dx}, \qquad \frac{dy}{dx} = \frac{S_0 - S_f}{1 - \alpha \mathrm{Fr}^2}$$

Conservation of momentum: $$\frac{Q^2}{gA} + A\bar{y} = \text{constant}$$

Hydraulic jump: trapezoidal channel $$\frac{by_1^2}{2} + \frac{my_1^3}{3} + \frac{Q^2}{gy_1(b + my_1)} = \frac{by_2^2}{2} + \frac{my_2^3}{3} + \frac{Q^2}{gy_2(b + my_2)}$$

Hydraulic jump: rectangular channel $$\frac{y_2}{y_1} = \frac{1}{2}\left(\sqrt{1 + 8\mathrm{Fr}_1^2} - 1\right)$$

Hydraulic jump: rough channel $$\frac{y_2}{y_1} - 1 = \sqrt{2}\exp\left(-\frac{k_s}{y_c}\right)(\mathrm{Fr}_1 - 1)^{0.963}$$

Hydraulic jump, head loss: $$h_L = \frac{(y_2 - y_1)^3}{4y_1y_2}, \qquad \frac{h_L}{E_1} = \frac{\left[\sqrt{1 + 8\mathrm{Fr}_1^2} - 3\right]^3}{8\left[\sqrt{1 + 8\mathrm{Fr}_1^2} - 1\right]\left[\mathrm{Fr}_1^2 + 2\right]}$$

Hydraulic jump, aeration: $$\frac{C_d - C_u}{C_s - C_u} = 0.0015(\gamma q h_L) + 0.01$$

$$\frac{Q_{air}}{Q + Q_{air}} = 0.3387\mathrm{Fr}_1^{0.202}\exp\left[(-0.103 + 0.0073\mathrm{Fr}_1)\frac{x}{y_1}\right]$$

Water surface profile: $$\frac{dy}{dx} = F(x, y), \qquad F(x, y) = \frac{S_0 - S_f}{1 - \alpha \mathrm{Fr}^2}$$

Energy equation: $$\Delta L = \frac{\left[y + \alpha\frac{V^2}{2g}\right]_2^1}{\bar{S}_f - S_0}$$

$$\Delta L = \frac{\left[y + \alpha\frac{V^2}{2g}\right]_2^1}{\bar{S}_f + \frac{C}{\Delta L}\left|\alpha_2\frac{V_2^2}{2g} - \alpha_1\frac{V_1^2}{2g}\right| - S_0}$$

Direct integration: $$y_2 = y_1 + F(\bar{x}, \bar{y})(x_2 - x_1), \quad y_1 = y_2 - F(\bar{x}, \bar{y})(x_2 - x_1)$$

Free overfall: $$q = C_d y_b^{\frac{3}{2}}$$

Length of backwater profile: $$L = 0.7\frac{y_n}{S_0}$$

Step method: $$\Delta L = \frac{\left[y + \alpha\frac{V^2}{2g}\right]_2^1}{\bar{S}_f + \frac{C}{\Delta L}\left|\alpha_2\frac{V_2^2}{2g} - \alpha_1\frac{V_1^2}{2g}\right| - S_0}$$

PROBLEMS

Section 9.2: Basic Principles

9.1. An open channel has a trapezoidal cross section with a bottom width of 3 m and side slopes of 2 : 1 (H : V). If the depth of flow is 2 m and the average velocity in the channel is 1.2 m/s, calculate the discharge in the channel.

9.2. Water flows at 10 m^3/s through a rectangular channel 4 m wide and 3 m deep. If the flow velocity is 1.4 m/s, calculate the depth of flow in the channel. If this channel expands (downstream) to a width of 6 m and the depth of flow decreases by 0.7 m from the upstream depth, what is the flow velocity in the expanded section?

9.3. Show that for circular pipes of diameter, D, the hydraulic radius, R, is related to the pipe diameter by $R = D/4$.

9.4. A trapezoidal channel is to be excavated at a site where permit restrictions require that the channel have a bottom width of 6 m, side slopes of 1.5 : 1 (H:V), and a depth of flow of 2 m. If the soil material erodes when the average shear stress on the perimeter of the channel exceeds 3 Pa, determine the appropriate slope and corresponding flow capacity of the channel. Use the Darcy–Weisbach equation and assume that the excavated channel has an equivalent sand roughness of 2 mm.

9.5. Water flows in a 6-m-wide rectangular channel that has a longitudinal slope of 0.0002. The channel has an equivalent sand roughness of 1.9 mm. Calculate the uniform flow depth in the channel when the flow rate is 13 m^3/s. Use the Darcy–Weisbach equation.

9.6. Given that hydraulically rough flow conditions occur in open channels when $u_* k_s/\nu \geq 70$, show that this condition can be expressed for water in terms of Manning parameters as

$$n^6\sqrt{RS_0} \geq 7.9 \times 10^{-14}$$

If a concrete-lined rectangular channel with a bottom width of 5 m is constructed on a slope of 0.05% and Manning's n is estimated to be 0.013, determine the minimum flow depth for hydraulically rough flow conditions to exist.

9.7. Water flows at a depth of 2.40 m in a trapezoidal concrete-lined section ($k_s = 1$ mm) with a bottom width of 4 m and side slopes of 2:1 (H:V). The longitudinal slope of the channel is 0.0005, and the water temperature is 20°C. Assuming uniform-flow conditions, estimate the average velocity and flow rate in the channel. Use both the Darcy–Weisbach and Manning equations and compare your results.

9.8. Water flows in a trapezoidal channel that has a bottom width of 5 m, side slopes of 2:1 (H:V), and a longitudinal slope of 0.0001. The channel has an equivalent sand roughness of 1 mm. Calculate the uniform flow depth in the channel when the flow rate is 18 m^3/s. Is the flow hydraulically rough, smooth, or in transition? Would the Manning equation be valid in this case? Explain.

9.9. Show that Manning's n can be expressed in terms of the Darcy friction factor, f, by the following relation:

$$n = \frac{f^{\frac{1}{2}} R^{\frac{1}{6}}}{8.86}$$

where R is the hydraulic radius of the flow. Does this relationship conclusively show that n is a function of the flow depth? Explain.

9.10. Water flows at 20 m^3/s in a trapezoidal channel that has a bottom width of 2.8 m, side slopes of 2:1 (H:V), longitudinal slope of 0.01, and a Manning's n of 0.015. (a) Use the Manning equation to find the normal depth of flow. (b) Determine the equivalent sand roughness of the channel. Assume that the flow is fully turbulent.

9.11. It has been shown that in fully turbulent flow, Manning's n can be related to the height, d, of the roughness projections by the relation

$$n = 0.039 d^{\frac{1}{6}}$$

where d is in meters. If the estimated roughness height in a channel is 30 mm, determine the percentage error in n resulting from a 70% error in estimating d.

9.12. Show that the minimum value of $n/k_s^{\frac{1}{6}}$ given by Equation 9.49 is

$$\frac{n}{k_s^{\frac{1}{6}}} = 0.039$$

Determine the range of R/k_s in which $n/k_s^{\frac{1}{6}}$ does not deviate by more than 5% from the minimum value. (*Hint*: You might need to use the relation $\log x = 0.4343 \ln x$.)

9.13. Stages are measured by two recording gauges 100 m apart along a constructed water supply channel. The channel has a bottom width of 5 m and side slopes of 3:1 (H:V). The bottom elevations of the channel at the upstream and downstream gauge locations are 24.01 m and 23.99 m, respectively. At a particular instance, the upstream and downstream stages are 25.01 m and 24.95 m, respectively, and the flow is estimated as 15$\pm$2 m^3/s. (a) Derive an expression for Manning's n as a function of the estimated flow rate. (b) Estimate Manning's n and the roughness height in the channel between the two measurement stations. (c) Quantitatively assess the sensitivity of the flow rate to the channel roughness.

9.14. Show that for the flow of water in an open channel, the turbulence condition $u_* k_s/\nu > 70$ can be put in the form

$$k_s \sqrt{R S_0} > 2.2 \times 10^{-5}$$

A trapezoidal concrete channel with a bottom width of 3 m and side slopes of 2:1 (H:V) is estimated to have an equivalent sand roughness of 3 mm and is laid on a slope of 0.1%. Determine the minimum flow depth for fully turbulent conditions to exist. Can the Manning equation be used at this flow depth? Explain.

9.15. Water flows at a depth of 4.00 m in a trapezoidal concrete-lined channel with a bottom width of 4 m and side slopes of 3:1 (H:V). The longitudinal slope of the channel is 0.0001, and the water temperature is 20°C. Assess the validity of using the Manning equation, assuming that $n = 0.013$.

9.16. The roadside gutter shown in Figure 9.17 has a curb depth of 15 cm, a cross slope of 2%, a longitudinal slope of 1%, and an estimated equivalent sand roughness of 1 mm.

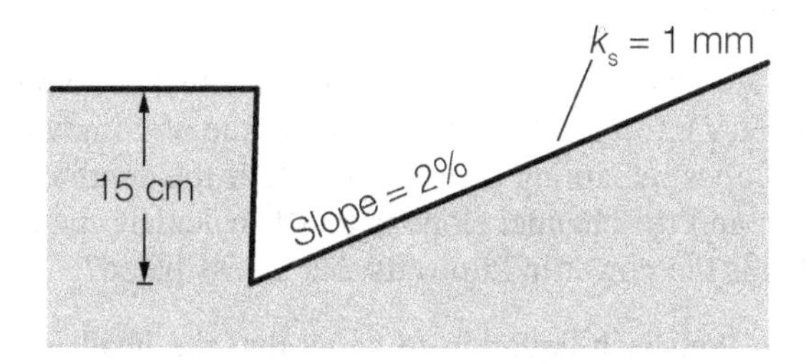

Figure 9.17: Flow in a gutter

(a) Determine the flow capacity of the gutter using the Darcy–Weisbach equation. (b) Determine the flow capacity using the Manning equation. (c) Assess the validity of the Manning equation in this case. (d) Account for the discrepancy in the gutter capacity estimated using the Darcy–Weisbach and Manning equations. (*Hint:* The gutter capacity is equal to the flow rate when the water level is at the curb.)

9.17. A trapezoidal irrigation channel is to be excavated to supply water to a farm. The design flow rate is 1.8 m^3/s, the side slopes are 2:1 (H:V), the longitudinal slope of the channel is 0.1%, Manning's n is 0.025, and the geometry of the channel is to be such that the length of each channel side is equal to the bottom width. (a) Specify the dimensions of the channel required to accommodate the design flow under normal conditions. (b) If the channel lining can resist an average shear stress of up to 4 Pa, under what flow conditions is the channel lining stable?

9.18. Water flows in a concrete trapezoidal channel with a bottom width of 3 m, side slopes of 2:1 (H:V), and a longitudinal slope of 0.1%. The Manning roughness coefficient is estimated to be 0.015. (a) What roughness height is characteristic of this channel? (b) For what range of flow depths and corresponding flow rates can n be assumed to be approximately constant and the Manning equation applicable? (c) If the flow rate in the channel is 100 m^3/s, what Manning's n should be used and what is the corresponding flow depth? How does this flow depth compare with that obtained by using $n = 0.015$?

9.19. A natural stream has a cross section that is approximately trapezoidal with a bottom width of 10 m and side slopes of 2.5:1 (H:V). The longitudinal slope of the channel is estimated to be 0.1%. When the flow in the stream is 150 m^3/s, the flow is observed to be approximately uniform with a flow depth of 5 m. (a) Estimate Manning's n and the roughness height in the channel. (b) Assess the validity of the Manning equation in this case. (c) If flow conditions change such that the depth of flow increases by 50%, what Manning's n would you use?

9.20. Use Equation 9.55 to show that the velocity in an open channel is equal to the depth-averaged velocity at a distance of 0.368d from the bottom of the channel, where d is the depth of flow.

9.21. Use Equation 9.55 to show that the depth-averaged velocity in an open channel with depth, d, can be estimated by averaging the velocities at 0.2d and 0.8d from the bottom of the channel.

9.22. If the velocity profile in a channel is described by the one-seventh power-law relationship, Equation 9.56, determine (a) the ratio of the average velocity to the maximum velocity, and (b) the distance from the bottom of the channel to where the average velocity occurs.

9.23. Water flows at 8.4 m^3/s in a trapezoidal channel with a bottom width of 2 m and side slopes of 2:1 (H:V). Over a distance of 100 m, the bottom width expands to 2.5 m, with the side slopes remaining constant at 2:1. If the depth of flow at both of these sections is 1 m and the channel slope is 0.001, calculate the head loss between the sections. What is the power in kilowatts that is dissipated?

9.24. Use the Darcy–Weisbach equation to show that the head loss per unit length, S, between any two sections in an open channel can be estimated by the relation

$$S = \frac{\bar{f}}{4\bar{R}} \frac{\bar{V}^2}{2g}$$

where $\bar{f}$, $\bar{R}$, and $\bar{V}$ are the average friction factor, hydraulic radius, and flow velocity, respectively, between the upstream and downstream sections.

9.25. Determine the critical depth for a flow of 30 m^3/s in a rectangular channel with width 5 m. If the actual depth of flow is equal to 3 m, is the flow supercritical or subcritical?

9.26. Determine the critical depth for a flow of 50 m^3/s in a trapezoidal channel with bottom width of 4 m and side slopes of 1.5 : 1 (H : V). If the actual depth of flow is 3 m, calculate the Froude number and state whether the flow is subcritical or supercritical.

9.27. A rectangular channel 2 m wide carries 3 m^3/s of water at a depth of 1.2 m. If an obstruction 40 cm wide blocks the middle of the channel, find the elevation of the water surface at the obstruction. What is the maximum width of the obstruction that will not cause a rise in the water surface upstream?

9.28. Water flows at 1 m^3/s in a rectangular channel of width 1 m and depth 1 m. What is the maximum contraction of the channel that will not choke the flow?

9.29. Consider the flow conditions in the concrete channel shown in Figure 9.18, where the flow rate in the channel is 16 m^3/s. The bottom width of the channel is to be contracted at a short distance downstream of the section shown in Figure 9.18. (a) State the equations that must be satisfied for choking to occur at the downstream section; simplify your equations as much as possible. (b) Will the flow be choked at the downstream section when the bottom width of the channel is reduced to zero? What is the depth of flow in the downstream section under this condition?

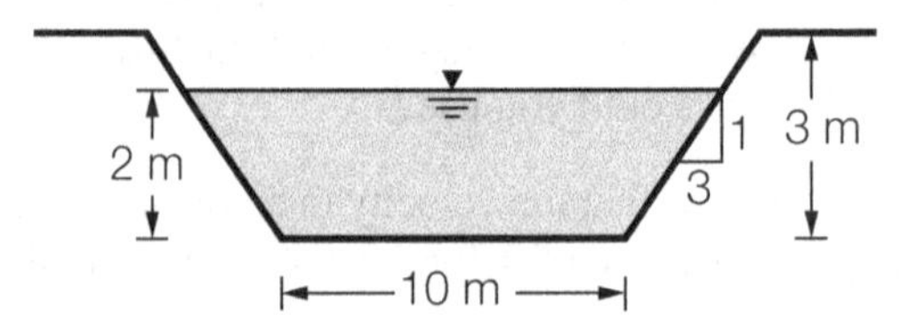

Figure 9.18: Flow in precontracted section

9.30. A lined rectangular concrete drainage channel is 10.0 m wide and carries a flow of 8 m^3/s. To pass the flow under a roadway, the channel is contracted to a width of 6 m. Under design conditions, the depth of flow just upstream of the contraction is 1.00 m, and the contraction takes place over a distance of 7 m. (a) If the energy loss in the contraction is equal to $V_1^2/2g$, where V_1 is the average velocity upstream of the contraction, what is the depth of flow in the constriction? (b) Does consideration of energy losses have a significant effect on the depth of flow in the constriction? Why or why not? (c) If the width of the constriction is reduced to 4.50 m and a flow of 8 m^3/s is maintained, determine the depth of flow within the constriction (include energy losses). (d) If reducing the width of the constriction to 4.50 m influences the upstream depth, determine the new upstream depth.

9.31. A rectangular channel 3 m wide carries 4 m^3/s of water at a depth of 1.5 m. If an obstruction 15 cm high is placed across the channel, calculate the elevation of the water surface over the obstruction. What is the maximum height of the obstruction that will not cause a rise in the water surface upstream?

9.32. Show that the critical step height required to choke the flow in a rectangular open channel is given by

$$\frac{\Delta z_c}{y_1} = 1 + \frac{\mathrm{Fr}_1^2}{2} - \frac{3}{2}\mathrm{Fr}_1^{\frac{2}{3}}$$

where Δz_c is the critical step height, y_1 is the flow depth upstream of the step, and Fr_1 is the Froude number upstream of the step. Use this equation to verify your answer to Problem 9.31.

9.33. Water flows at 6.2 m^3/s in a rectangular channel of width 5 m and depth of flow of 1.5 m. If the channel width is decreased by 0.50 m and the bottom of the channel is raised by 0.15 m, what is the depth of flow in the constriction?

9.34. Water flows at 18 m^3/s in a trapezoidal channel with a bottom width of 5 m and side slopes of 2 : 1 (H : V). The depth of flow in the channel is 2 m. If a bridge pier of width 50 cm is placed in the middle of the channel, what is the depth of flow adjacent to the pier? What is the maximum width of a pier that will not cause a rise in the water surface upstream of the pier?

9.35. Water flows at 15 m^3/s in a trapezoidal channel with a bottom width of 4.5 m and side slopes of 1.5 : 1 (H : V). The depth of flow in the channel is 1.9 m. If a step of height 15 cm is placed in the channel, what is the depth of flow over the step? What is the maximum height of the step that will not cause a rise in the water surface upstream of the step?

9.36. Water flows at 20 m^3/s with a uniform depth of 3 m in a trapezoidal channel of base width 3 m and side slopes 1 : 1. If a channel transition restricts the flow locally by raising the side walls to the vertical position, calculate the depth of flow in the rectangular constriction. What is the minimum allowable width of the constriction that will prevent choking?

9.37. A float-finished trapezoidal concrete channel has a longitudinal slope of 0.05%, side slopes of 2:1 (H:V), and a bottom width of 5 m. The design flow rate in the channel is 7 m^3/s. (a) Verify the validity of applying the Manning equation in this case and calculate the normal depth of flow in the channel. (b) If the bottom width of the channel abruptly changes from 5 m to 4 m, what is the head loss in the contraction and flow depth at the contracted section? What is the flow depth at the contracted section if the head loss in the contraction is neglected? Assess the importance of including head loss in your analysis.

9.38. A rectangular channel has a width of 30 m, a longitudinal slope of 0.5%, and an estimated Manning's n of 0.025. The flow rate in the channel is 100 m^3/s at a particular section where the depth of flow is 3.000 m. Temporary construction requires that the channel be contracted to a width of 20 m over a distance of 40 m and then returned to its original width of 30 m over a distance of 40 m. All sections are rectangular. Determine the depths of flow in the contracted and downstream sections when (a) all energy losses between sections are neglected and (b) friction, contraction, and expansion losses are all taken into account. (Note: To simplify the computations, assume that the friction slope is the same at all three sections.) Based on your results, evaluate the impact of accounting for energy losses on the estimated difference between the water stages at the upstream and downstream sections.

9.39. Water approaches a bridge constriction in a horizontal rectangular concrete-lined channel of width 10 m, and to accommodate the bridge, the channel abruptly contracts to a width of 7 m and then expands (abruptly) back to a width of 10 m. Under design conditions, the flow rate in the channel is 20 m^3/s and the flow depth in the approach channel is 2 m. (a) Taking into account contraction and expansion head losses, what is the flow depth at the bridge constriction and the flow depth downstream of the bridge? (b) What is the percentage error in the calculated flow depths if energy losses are neglected?

Section 9.3: Water Surface Profiles

9.40. Consider the compound channel shown in Figure 9.19, where Manning's n is equal to 0.013 and 0.026 in the main channel and overflow section, respectively. The longitudinal slope of the channel is 0.05%. (a) Estimate the flow capacity of the main channel. (b) Estimate the flow capacity when the compound channel is filled to the top of the berm. (c) For what range of flows is the Manning equation valid in the main channel? (d) For a flow rate of 3 m^3/s and a depth of flow equal to 1.5 m, classify the water surface profile and state a scenario under which this profile could occur. Assume that the Manning equation is valid in this case.

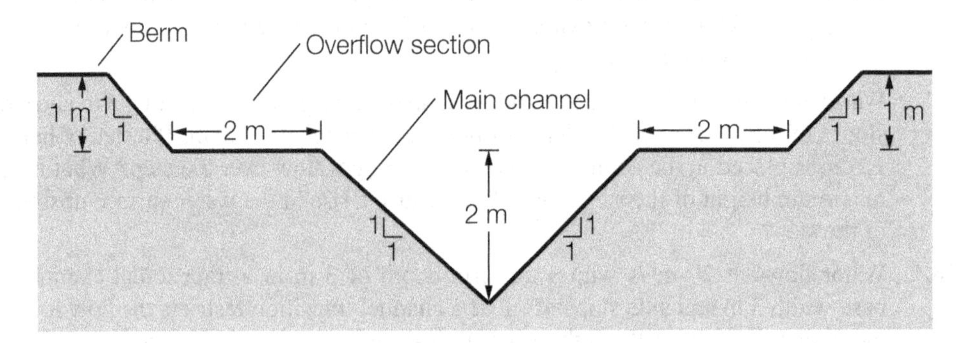

Figure 9.19: Compound channel

9.41. Water flows at 30 m^3/s in a rectangular channel of width 8 m and Manning's n of 0.030. If the depth of flow at a channel section is 2 m and the slope of the channel is 0.002, classify the water surface profile. What is the slope of the water surface at the observed section? Would the water surface profile be much different if the depth of flow was equal to 1 m? Explain.

9.42. Water flows at 25 m^3/s in a rectangular channel of width 6 m. Manning's n of the channel is 0.035. Determine the range of channel slopes that would be classified as steep and the range that would be classified as mild.

9.43. Water flows in a trapezoidal channel where the bottom width is 6 m and side slopes are 2:1 (H: V). The channel lining has an estimated Manning's n of 0.045, and the slope of the channel is 1.5%. When the flow rate is 80 m^3/s, the depth of flow at a gauging station is 5 m. Classify the water surface profile, state whether the depth increases or decreases in the downstream direction, and calculate the slope of the water surface at the gauging station. On the basis of this water-surface slope, estimate the depths of flow 100 m downstream and 100 m upstream of the gauging station.

9.44. Water discharges from a large storage reservoir via a trapezoidal channel of bottom width 3.00 m, side slopes 3:1 (H:V), and longitudinal slope of 0.5%. Manning's n in the channel is estimated as 0.025. Estimate the discharge from the reservoir when the depth of the reservoir at the discharge location is 2.00 m.

9.45. Derive an expression relating the conjugate depths in a hydraulic jump when the slope of the channel is equal to S_0 and the channel cross section is rectangular. (*Hint*: Assume that the length of the jump is equal to $5y_2$ and that the shape of the jump between the upstream and downstream depths can be approximated by a trapezoid.)

9.46. If 125 m^3/s of water flows in a rectangular channel 5 m wide at a depth of 0.8 m, calculate the downstream depth required to form a hydraulic jump and the fraction of the initial energy lost in the jump.

9.47. The head loss, h_L, across a hydraulic jump in a rectangular channel is described by the equation

$$y_1 + \frac{V_1^2}{2g} = y_2 + \frac{V_2^2}{2g} + h_\mathrm{L}$$

where the subscripts 1 and 2 refer to the conditions upstream and downstream of the jump, respectively. Show that the normalized head loss, h_L/y_1, is given by

$$\frac{h_\mathrm{L}}{y_1} = 1 - \frac{y_2}{y_1} + \frac{\mathrm{Fr}_1^2}{2}\left[1 - \left(\frac{y_1}{y_2}\right)^2\right]$$

where Fr_1 is the upstream Froude number.

9.48. Water flows at 20 m^3/s at a depth of 1 m in a trapezoidal channel having a bottom width of 1 m and side slopes of 2:1 (H:V). (a) Is it possible for a hydraulic jump to occur in the channel? (b) If a hydraulic jump occurs between the channel section and a downstream rectangular section of bottom width 5 m, determine the downstream flow depth and the power loss in the jump.

9.49. Show that the hydraulic jump equation for a trapezoidal channel is given by

$$\frac{by_1^2}{2} + \frac{my_1^3}{3} + \frac{Q^2}{gy_1(b + my_1)} = \frac{by_2^2}{2} + \frac{my_2^3}{3} + \frac{Q^2}{gy_2(b + my_2)}$$

where b is the bottom width of the channel, m is the side slope of the channel, y_1 and y_2 are the conjugate depths, and Q is the volume flow rate.

9.50. Water flows in a horizontal trapezoidal channel at 21 m^3/s, where the bottom width of the channel is 2 m, side slopes are $1:1$, and the depth of flow is 1 m. Calculate the downstream depth required for a hydraulic jump to form at this location. What would be the energy loss in the jump?

9.51. A flume with a triangular cross section and side slopes of 2 : 1 (H : V) contains water flowing at 0.45 m^3/s at a depth of 20 cm. Verify that the flow is supercritical and calculate the conjugate depth.

9.52. Water flows at 10 m^3/s in a rectangular channel of width 5.5 m. The slope of the channel is 0.15%, and the Manning roughness coefficient is 0.038. Use the following methods to estimate the flow depth 100 m upstream of a section where the flow depth is 2.2 m: (a) the direct integration method and (b) the standard step method. Approximately how far upstream of this section would you expect to find uniform flow?

9.53. Water flows at 8 m^3/s in a 3-m-wide rectangular channel that is laid on a slope of 8%. If the channel has a Manning's n of 0.04 and the flow depth at a given section is 1.8 m, how far upstream or downstream is the flow depth equal to 1.5 m?

9.54. If the depth of flow in the channel described in Problem 9.10 is measured as 1.4 m, find the location where the depth is 1.6 m.

9.55. A trapezoidal canal has a longitudinal slope of 1%, side slopes of 3:1 (H:V), a bottom width of 3.00 m, a Manning's n of 0.015, and it carries a flow of 20 m^3/s. The depth of flow at a gauging station is observed to be 1.00 m. Respond to the following: (a) What is the normal depth of flow in the channel? (b) What is the critical depth of flow in the channel? (c) Classify the slope of the channel and the water surface profile at the gauging station. (d) How far from the gauging station is the depth of flow equal to 1.1 m? Does this depth occur upstream or downstream of the gauging station? (e) If the bottom of the channel just downstream of the gauging station is raised by 0.20 m, determine the resulting depth of flow at the downstream section. The bottom width of the channel remains constant at 3 m.

9.56. A rectangular channel 8 m wide carries a discharge of 0.9 m^3/s. At a certain section, the channel lining changes from rough to smooth. The normal depths in the rough and smooth reaches are 0.6 m and 0.4 m, respectively. The channel slope is 0.8%. Using a single step of the direct step method, estimate the length of the reach of nonuniform flow.

9.57. The flow conditions at Stations 1 and 2 in a rectangular channel are shown in Figure 9.20, where Station 1 is 150 m upstream of Station 2. If the flow rate in the channel is 3 m^3/s and the longitudinal slope is 0.6%, estimate Manning's n in the channel.

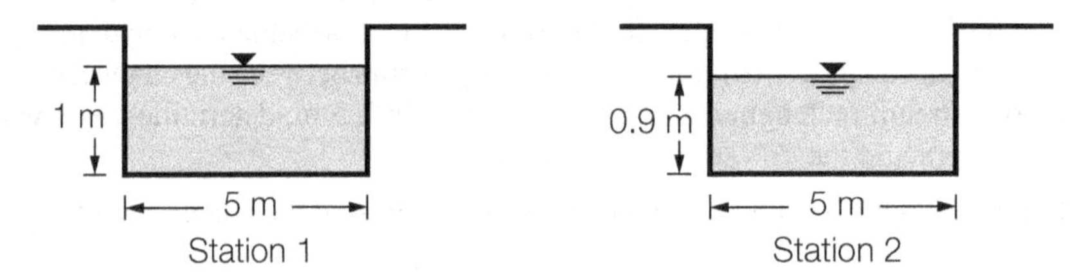

Figure 9.20: Flow in upstream and downstream sections

9.58. Water flows at 11 m^3/s in a rectangular channel of width 5 m. The slope of the channel is 0.1%, and the Manning roughness coefficient is equal to 0.035. If the depth of flow at a selected section is 2 m, calculate the upstream depths at 20-m intervals along the channel until the depth of flow is within 5% of the uniform flow depth.

9.59. Water flows in an open channel whose longitudinal slope is 0.08%. The Manning roughness coefficient of the channel lining is estimated to be 0.025 when the flow rate is 400 m^3/s. At a given section of the channel, the cross section is trapezoidal with a bottom width of 15 m, side slopes of 2:1 (H:V), and a depth of flow of 10 m. Use the standard step method to calculate the depth of flow 100 m upstream from this section where the cross section is trapezoidal with a bottom width of 20 m and side slopes of 3:1 (H:V).

9.60. A concrete-lined rectangular channel is 8-m wide and 4-m deep and has a longitudinal slope of 0.2% and an estimated Manning's n of 0.013. A hydraulic structure controls the flow in the channel such that the depth of flow at the structure is 3 m when the flow rate in the channel is 24 m^3/s. Urban developers propose a localized contraction/expansion in the channel 100 m upstream of the control structure to accommodate a pedestrian walkway. Taking eddy losses into account but neglecting friction losses, estimate the maximum contraction that should be allowed for the walkway. Based on your result, assess whether it is reasonable to neglect friction losses in this case.

9.61. Consider the main channel and adjacent floodplain shown in Figure 9.21 where Manning's n in the main channel and floodplain are 0.050 and 0.100, respectively, and the slope of the compound channel is 0.75%. Consider a design flow of 110 m^3/s. (a) Calculate the change in the normal depth of flow if developers are allowed to fill in 15 m of the width of the 100-m-wide floodplain. (b) As an alternative to development within the floodplain, a hydraulic structure is to be installed that will cause the stage at a particular section to be 54.50 m at a location 150 m downstream when the flow is 110 m^3/s. Estimate how far upstream of the structure you would have to go for the elevation to be the same as that in the filled-in floodway.

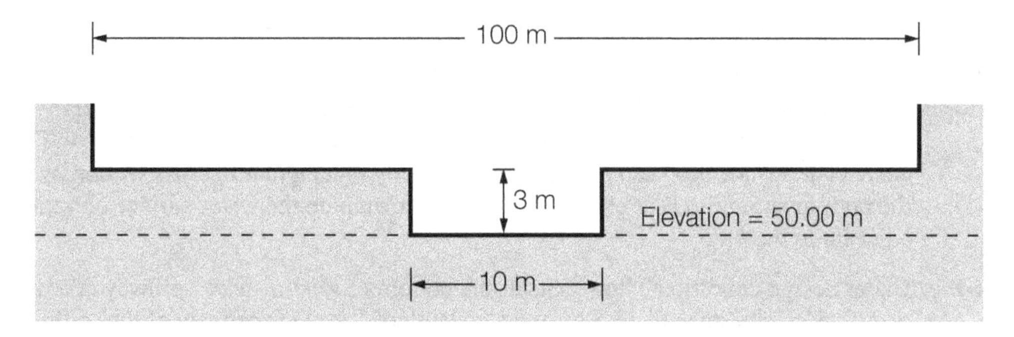

Figure 9.21: Channel and adjacent floodplain

9.62. At a particular river cross section, the elevation of the bottom of the river is 103.75 m and the elevation of the river bank is 106.43 m. The river has an approximately trapezoidal cross section with side slopes of 3:1 (H:V) and a bottom width of 18 m. The longitudinal slope of the river and the adjacent floodplain is 2%, and the Manning roughness coefficient of the river is estimated as 0.07. A bridge is to be placed 120 m downstream of the river section, and over this distance, the river will be made to transition to a bottom width of 9 m while maintaining the same side slope. Under design conditions, the flow in the river is 12 m^3/s and the depth of flow at the upstream section is 1.60 m. (a) What will be the depth of flow at the bridge section? (b) Will the floodplain be flooded at the bridge location? Explain.

9.63. Show that the energy equation for open-channel flow between Stations B (upstream) and A (downstream) can be written in the form

$$\left[Z + \alpha \frac{V^2}{2g} \right]_{\mathrm{A}}^{\mathrm{B}} = \bar{S}_{\mathrm{f}} \Delta L$$

where Z is the water surface elevation, α is the energy correction coefficient, V is the average velocity, $\bar{S}_{\mathrm{f}}$ is the average friction slope between Stations A and B, and ΔL is the distance between Stations A and B.

A backwater curve is being computed in the stream between two sections A and B, 140 m apart. The hydraulic properties of the cross sections are shown in Table 9.7. For a flow in the channel of 280 m^3/s and a Manning's n of 0.040, the water surface

Table 9.7: Hydraulic Properties of Cross Sections

Water surface elevation (m)	Area (m^2)		Wetted perimeter (m)	
	Section A	Section B	Section A	Section B
518.5	—	118.45	—	36.27
518.2	181.86	108.42	52.21	35.05
517.9	166.81	98.66	50.84	33.83
517.6	152.13	86.96	48.77	32.77
517.2	137.82	78.18	47.40	31.70
516.9	123.88	—	46.02	—

elevation at Station A is 517.4 m. Compute the water surface elevation at Station B. Just upstream of Station B, the flow is partially obstructed by a large bridge pier in the channel, presenting an obstruction 2.50 m wide normal to the direction of flow. The channel at this location can be considered roughly rectangular in cross section, with the bottom at Station B at elevation 515.10 m. Compute the water surface elevation adjacent to the pier.

9.64. Under design conditions, flow exits at the bottom of a 10-m-wide spillway at a rate of 220 m^3/s, at a stage of 13.5 m, and at a depth of 1 m. Downstream of the spillway is a river at a stage of 21.5 m. The channel connecting the spillway exit to the river is to be horizontal and rectangular with a width of 10 m. This connecting channel (called a stilling basin) is to be constructed such that Manning's n is 0.01 and is to be designed such that any hydraulic jump would occur at the midpoint of the stilling basin. As a designer, what length of stilling basin would you specify?

9.65. A trapezoidal drainage channel of bottom width 5 m, side slopes 2:1 (H:V), Manning's n of 0.018, and longitudinal slope of 0.1% terminates at a gate where the relationship between the flow through the gate (Q), headwater elevation (HW), tailwater elevation (TW), and gate opening (h) is given by

$$Q = 13.3h\sqrt{HW - TW}$$

where Q is in m^3/s and HW, TW, and h are in meters. If the elevation of the bottom of the channel at the gate location is 0.00 m, the tailwater elevation is 1.00 m, and the flow in the channel is 20 m^3/s, estimate the minimum gate opening such that the water surface elevation 100 m upstream of the gate does not exceed 2.20 m. (Note: The depth of flow 100 m upstream of the gate is *not* 2.20 m.)

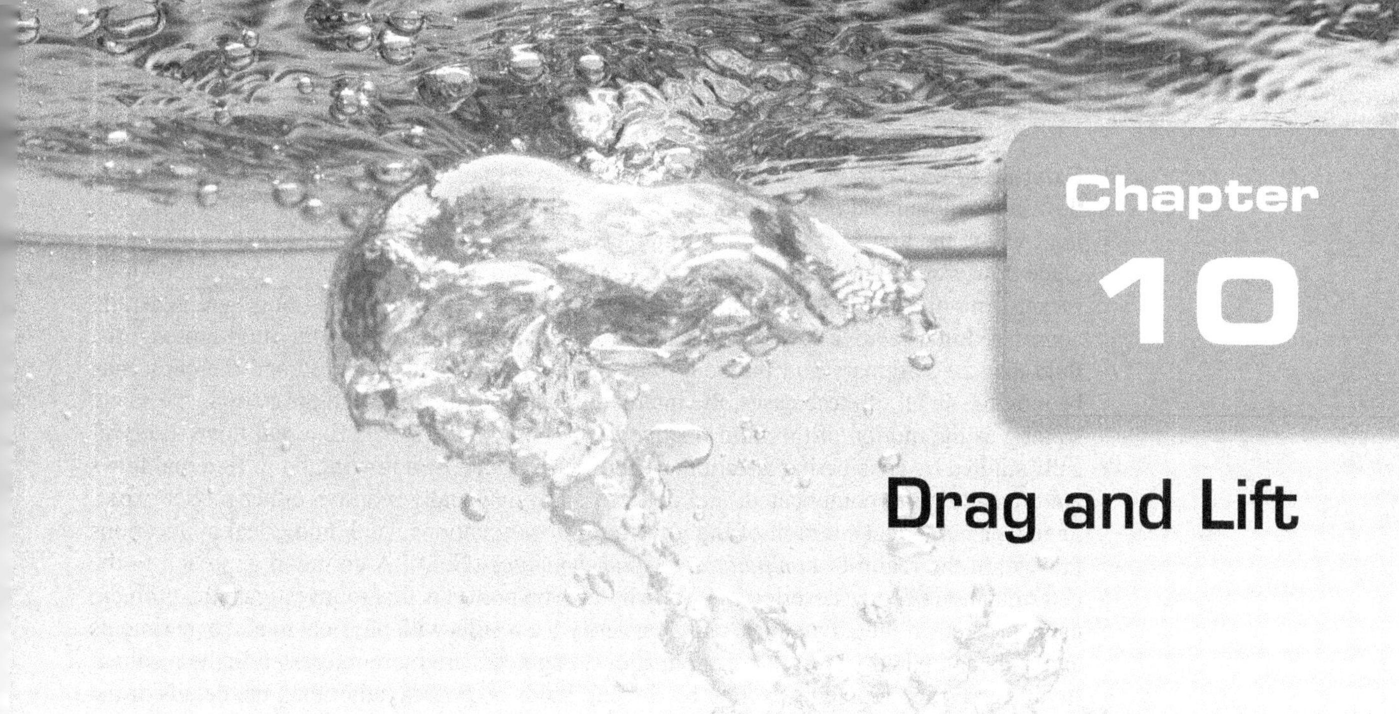

Chapter 10

Drag and Lift

LEARNING OBJECTIVES

After reading this chapter and solving a representative sample of end-of-chapter problems, you will be able to:

- Understand the causes of drag and lift.
- Calculate drag forces on flat surfaces.
- Use drag and lift coefficients to calculate drag and lift forces.
- Analyze the performance of automobiles, aircraft, and watercraft relative to drag and lift forces.
- Analyze the performance of spherical game balls based on their velocity, rotation, and size.

10.1 Introduction

The force that a flowing fluid exerts on a solid body can be separated into *drag* and *lift* components. The drag and lift components of the fluid force are commonly called the *drag force* and the *lift force*, or simply the *drag* and the *lift*. The drag force is the component of the fluid force that acts in the same direction as the flow, and the lift force is the component of the fluid force that acts in the direction perpendicular to the flow. Drag and lift are associated mostly with two-dimensional flows around two-dimensional bodies. In flows around three-dimensional bodies, there will likely exist a third net force component that is perpendicular to both the drag and the lift.

Importance of drag and lift. Drag exerted on a body within a flowing fluid is usually undesirable, and an objective in designing the shape of a body is usually to reduce drag. Lift forces can be desirable or undesirable, depending on the application. Drag and lift forces must generally be considered when designing buildings to withstand wind-induced forces, automobiles and submarines are commonly designed to minimize drag to increase fuel efficiency, and airplane wings are usually designed to maximize lift and minimize drag under normal operating conditions. Drag forces must generally be accounted for when designing structural support systems for underwater pipelines and power lines.

External flows. Drag and lift are usually of interest in external flows. An *external flow* is a flow in an unbounded region around or over a solid body, which is in contrast to an *internal flow*, where the flow is contained within physical boundaries. Examples of external flow include flow around buildings, smokestacks, bridge abutments, automobiles, trucks, airplanes, rockets, missiles, hydrofoils, ships, submarines, and torpedos. External flows are generally considered in a relative sense in that the body may be stationary and the fluid moving, the fluid may be stationary and the body may be moving, or both the fluid and the body may be moving. In all of these cases, the interaction between the fluid and the body is the same, so long as the motion of the fluid relative to the body is the same. External flows are typically studied by considering a stationary body surrounded by a flowing fluid. External flows are typically quite complicated, and detailed analysis usually requires either model experiments or numerical solution of the governing flow equations; such numerical applications fall within the realm of *computational fluid dynamics* (CFD). A common approach to the determination of forces exerted by external flows on bodies is first to investigate the problem using theoretical fluid dynamics, and then verify the results with physical model experiments and CFD computations. In many applications, engineers are more interested in the resultant (total) forces and moments exerted by flowing fluids on bodies rather than the details of the flow field surrounding the body.

10.2 Fundamentals

Consider a body immersed in a flowing fluid as shown in Figure 10.1, where the pressure, p, and wall shear stress, τ_w, act on an element dA of the body whose normal makes an angle θ with the x-axis. The drag force, F_D, on the body is then given by

$$F_\mathrm{D} = \int_{A_\mathrm{b}} p\cos\theta\,\mathrm{d}A + \int_{A_\mathrm{b}} \tau_\mathrm{w}\sin\theta\,\mathrm{d}A \tag{10.1}$$

where A_b is the surface area of the body. Likewise, the lift force, F_L, on the body is given by

$$F_\mathrm{L} = -\int_{A_\mathrm{b}} p\sin\theta\,\mathrm{d}A + \int_{A_\mathrm{b}} \tau_\mathrm{w}\cos\theta\,\mathrm{d}A \tag{10.2}$$

Flow around a body can also exert a moment on the body and cause rotation of the body. The moment about the flow direction is called the *roll moment*, the moment about the lift direction is called the *yaw moment*, and the moment about the side-force direction is called the *pitch moment*. The relative orientation of these moments on a B-2 bomber aircraft are shown in Figure 10.2. The (principal) axes about which the aircraft rotates originate at the center of gravity of the aircraft, and these axes are commonly referred to as the *roll axis*, the *pitch axis*, and the *yaw axis*, referenced in Figure 10.2 as the x-, y-, and z-axes, respectively. The

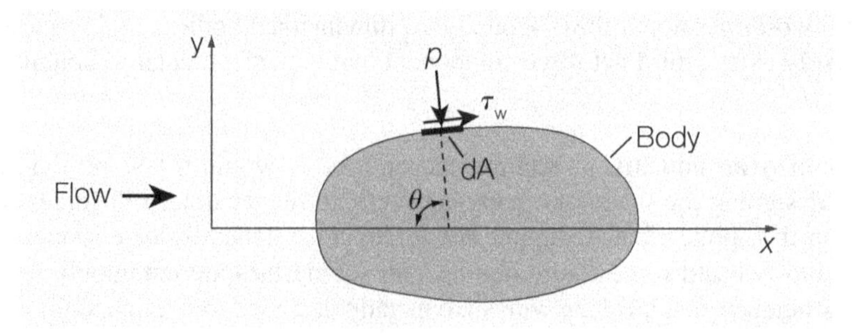

Figure 10.1: Calculation of drag and lift forces

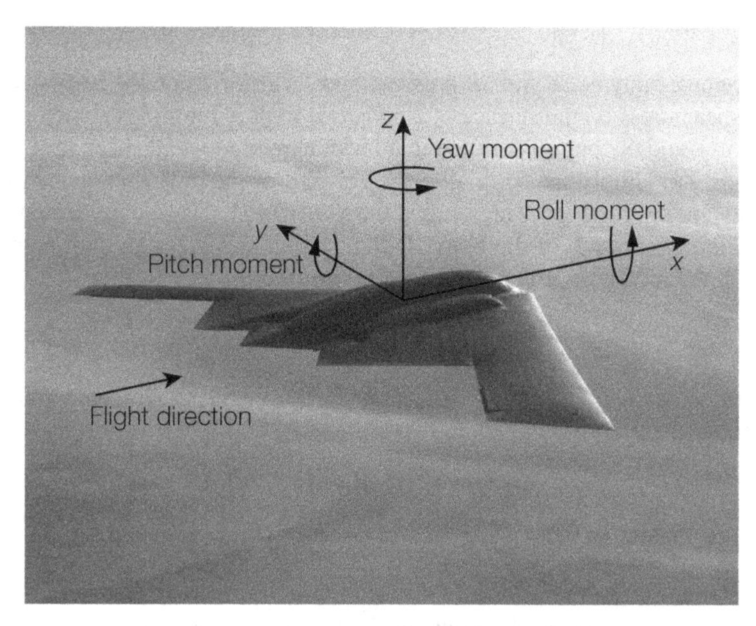

Figure 10.2: Moments on a B-2 bomber
Source: U.S. Air Force photo/Bobbi Zapka.

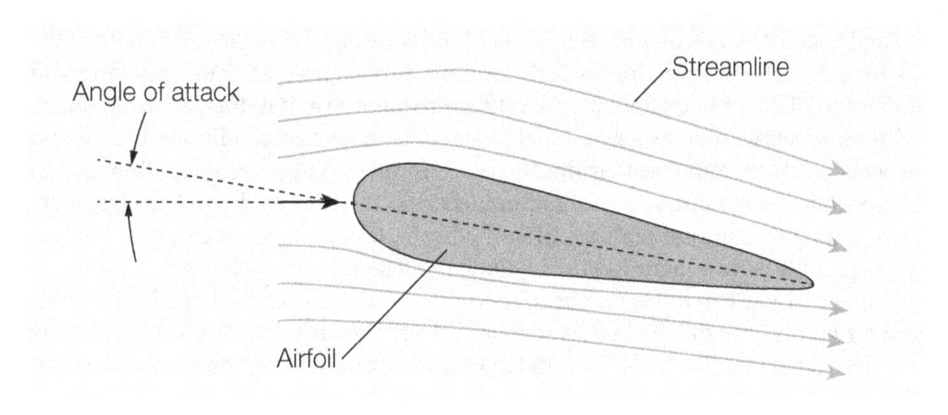

Figure 10.3: Flow around an airfoil

positive directions of the roll, pitch, and yaw moments are given by the right-hand rule, which states that when the thumb of the right hand is pointed in the positive direction of the axis, the fingers curl in the positive direction of the moment. Roll, pitch, and yaw moments on an airplane are usually controlled by the wing flaps. For bodies that are symmetrical about the lift-drag plane, such as cars, airplanes, and ships, the yawing moment and the rolling moment are zero when the wind and wave forces are aligned with the axes of symmetry of the body; what remains are the drag and lift forces and the pitching moment. The wings of airplanes are shaped and positioned specifically to generate lift with minimal drag, which is done by maintaining an angle of attack during cruising as shown in Figure 10.3. Both drag and lift are strong functions of the angle of attack. The pressure difference between the top and bottom surfaces of the wing is responsible mostly for generating the upward force that tends to lift the wing and the airplane to which it is connected.

10.2.1 Friction and Pressure Drag

The drag force on a body is the sum of pressure and shear forces on the body in the direction of flow. The shear stress on the surface of a body due to fluid motion is sometimes called the *wall shear stress*, and the part of the drag that is due directly to wall shear stress is called the *skin friction drag* or simply the *friction drag*. The part of the drag that is due to pressure is called the *pressure drag*, which is also called the *form drag* because of its strong dependence on the shape and orientation of the body. If the friction and pressure drag components are separated, then the total drag force, F_D, can be expressed as

$$F_\mathrm{D} = F_\mathrm{D,friction} + F_\mathrm{D,pressure} \tag{10.3}$$

with the components $F_\mathrm{D,friction}$ and $F_\mathrm{D,pressure}$ derived from the integral components in Equation 10.1 such that

$$F_\mathrm{D,friction} = \int_{A_\mathrm{b}} \tau_\mathrm{w} \sin\theta \, \mathrm{d}A \tag{10.4}$$

$$F_\mathrm{D,pressure} = \int_{A_\mathrm{b}} p \cos\theta \, \mathrm{d}A \tag{10.5}$$

where A_b is the surface area of the body, τ_w is the shear stress on the surface of the body (i.e., the wall shear stress), p is the pressure on the surface of the body, and θ is the angular location of the elemental area $\mathrm{d}A$ on the boundary.

Friction drag. The friction drag is the component of the wall shear force in the direction of flow and hence depends on the orientation of the body as well as the magnitude of the wall shear stress, τ_w. The friction drag is zero for a flat surface normal to the flow direction and maximum for a flat surface parallel to the flow direction. Friction drag, a strong function of viscosity, increases with increasing viscosity. Under laminar flow conditions, the friction drag is independent of the roughness of the surface. However, under turbulent flow conditions, surface roughness can have a controlling influence on the friction drag. The contribution of the friction drag to the total drag for blunt bodies is less at higher Reynolds numbers and may be negligible at very high Reynolds numbers. The drag in such cases is mostly due to pressure drag. At low Reynolds numbers, most drag is due to friction drag. This is especially the case for highly streamlined bodies such as airfoils. The friction drag is also proportional to the surface area; therefore, bodies with larger surface area experience a larger friction drag.

Pressure drag. The pressure drag on a body is proportional to the frontal area of the body and the difference between the pressures acting on the front and back of the body. Therefore, the pressure drag is usually dominant for blunt bodies, small for streamlined bodies such as airfoils, and zero for thin flat plates parallel to the flow. The pressure drag becomes most significant when the velocity of the fluid is too high for the fluid to be able to follow the curvature of the body and thus separates from the body at some point, creating a region of very low pressure behind the body.

10.2.2 Drag and Lift Coefficients

The drag and lift forces exerted by a fluid flowing over and/or around a body typically depend on the density of the fluid, ρ; the dynamic viscosity of the fluid, μ; the free-stream velocity, V; the roughness of the surface, ϵ; and the size, shape, and orientation of the body characterized by an area, A, which is typically the projected area of the body normal to the direction of flow. Expressing the drag or lift force, F, as a function of ρ, μ, V, ϵ, and A, gives

$$F = f(\rho, \mu, V, \epsilon, A) \tag{10.6}$$

Because this is a functional relationship between six variables in three dimensions (MLT), the Buckingham pi theorem guarantees that

$$\frac{F}{\frac{1}{2}\rho V^2 A} = C\left(\text{Re}, \frac{\epsilon}{\sqrt{A}}\right) \tag{10.7}$$

where C is a function of Re and $\epsilon/\sqrt{A}$ and Re is the Reynolds number given by

$$\text{Re} = \frac{\rho V \sqrt{A}}{\mu} \tag{10.8}$$

The function C will be different depending on whether the drag force or the lift force is being considered. The *drag coefficient*, C_D, and *lift coefficient*, C_L, are defined by the following relations:

$$C_\text{D} = \frac{F_\text{D}}{\frac{1}{2}\rho V^2 A} \tag{10.9}$$

$$C_\text{L} = \frac{F_\text{L}}{\frac{1}{2}\rho V^2 A} \tag{10.10}$$

where F_D and F_L are the drag and lift forces, respectively. The variable group $\frac{1}{2}\rho V^2$ in Equations 10.9 and 10.10 is commonly referred to as the *dynamic pressure*, which is equal to the increased pressure above the free-stream pressure that occurs at a stagnation point; this can be shown by applying the Bernoulli equation along a stagnation streamline. In calculating drag forces, A is normally the area of the body projected on a plane normal to the flow direction (called the *frontal area*). On some thin bodies, such as airfoils and hydrofoils, the area A is taken as the vertical projection or *planform area* of the body. For bodies that are partially submerged in a liquid, such as ships and barges, the *wetted area* is commonly used. The values of the drag and lift coefficients, C_D and C_L, respectively, depend on which area is being taken for A; so care should be taken in selecting the appropriate values to be used in any given situation.

Viscous and roughness effects. Drag and lift coefficients are primarily functions of the shape and orientation of the body, but in some cases, they also depend on the Reynolds number, Re, and the surface roughness, ϵ. Typically, drag forces are dominated by inertial effects when Re > 100 and dominated by viscous effects when Re < 1, which means that C_D and C_L are typically independent of Re when Re > 100 and controlled primarily by Re when Re < 1. Roughness effects are typically important in turbulent flows when the height of the roughness elements on the surface of a body exceeds the thickness of the viscous layer that forms adjacent to the surface of the body. On flat and almost flat surfaces, increased roughness can lead to increased values of C_D, whereas the effect of roughness on C_L is seldom a concern.

Compressibility effects. In high-velocity flows where compressibility effects are important, C_D and C_L might also be a function of the Mach number, Ma, defined as V/c, where c is the speed of a pressure wave (i.e., speed of sound) in the fluid. Compressibility effects can usually be neglected when Ma < 0.3, which means that C_D and C_L are typically independent of Ma when Ma < 0.3. Compressibility effects on drag and lift are considered mostly when studying the performance of aircraft flying under conditions where Ma > 0.3, which includes commercial jet aircraft. For most body shapes, C_D increases significantly as the sonic velocity (i.e., Ma $= 1$) is approached.

Gravity effects. In cases where there is an interface between fluids, gravity effects might be important and C_D and C_L might be a function of the Froude number, Fr, defined as $V/\sqrt{gL}$, where L is a length scale that measures the size of the immersed body. Gravity effects on drag are considered mostly when studying the drag forces on ships as they move through the water. In these cases, increased drag forces are associated with increased wave heights generated at the bow of the ship.

EXAMPLE 10.1

A popular family sedan has a length of 4.96 m, a width of 1.83 m, a height of 1.46 m, and a ground clearance of 0.139 m. Under design conditions, the car is driven in standard air. (a) Estimate the minimum velocity for the drag coefficient to be independent of the Reynolds number. (b) Estimate the maximum velocity for the drag coefficient to be independent of the Mach number. (c) Between the range of velocities estimated in parts (a) and (b), how would you expect the drag coefficient to vary?

SOLUTION

From the given data: $L = 4.96$ m, $W = 1.83$ m, $H = 1.46$ m, and $G = 0.139$ m. For standard air, $\nu = 1.460 \times 10^{-5}$ m²/s.

(a) The Reynolds number, Re, for flow past a car is typically defined in terms of the frontal area, A. It can be estimated that the drag coefficient is independent of Re when Re > 100. If V_1 is the limiting velocity beyond which the drag coefficient is independent of Re, then

$$\text{Re}_1 = \frac{V_1\sqrt{A}}{\nu} = 100 \quad \rightarrow \quad \frac{V_1\sqrt{LW}}{\nu} = 100$$

$$\frac{V_1\sqrt{(4.96)(1.83)}}{1.460 \times 10^{-5}} = 100 \quad \rightarrow \quad V_1 = 4.85 \times 10^{-4} \text{ m/s}$$

Therefore, when the speed of the car exceeds around 4.85×10^{-4} m/s (= **0.0017 km/h**), the drag coefficient is expected to be independent of viscous effects. Based on this result, viscous effects on drag are likely to be generally insignificant.

(b) To assess compressibility effects, air can be approximated as an ideal gas. For standard air, $T = 15°\text{C} = 288.15$ K and $R = 287.1$ J/kg·K, $k = 1.40$, and the sonic speed, c, is given by Equation 1.40 as

$$c = \sqrt{RTk} = \sqrt{(287.1)(288.15)(1.40)} = 340.3 \text{ m/s}$$

It can be estimated that the drag coefficient is independent of Ma when Ma < 0.3. If V_2 is the limiting velocity below which the drag coefficient is independent of Ma, then

$$\text{Ma}_2 = \frac{V_2}{c} = 0.3 \quad \rightarrow \quad \frac{V_2}{340.3} = 0.3 \quad \rightarrow \quad V_2 = 102 \text{ m/s}$$

Therefore, when the speed of the car is less than 102 m/s (= **367 km/h**), compressibility effects are expected to have a negligible effect on the drag coefficient.

(c) In the range of velocities **0.0032 km/h < V < 367 km/h**, the drag coefficient is expected to be approximately constant and independent of both Re and Ma.

Practical Considerations

Any effect that causes the shape of a body to change will affect its drag and lift characteristics. For example, snow accumulation or ice formation on airplane wings can cause the shape of the wings to change sufficiently to reduce the lift. This has caused airplanes to lose altitude and crash and many others to abort takeoff. As a consequence, it has become a routine safety measure to check for ice or snow buildup on critical components of airplanes before takeoff in bad weather. Deicing operations are an integral part of flight operations at major airports that experience significant snowfall during the year.

10.2.3 Flow over Flat Surfaces

Flow over a flat surface is sometimes referred to as flow over a "flat plate," where a flat plate is commonly understood to mean a flat surface that is rectangular in shape, having a finite width perpendicular to the flow direction. In keeping with tradition, the flow over a flat surface will be described in terms of the flow over a flat plate, with the understanding that any flat surface can be discretized into an assembly of flat plates. The drag force exerted by a fluid flowing over a flat plate is due entirely to the frictional properties of the fluid. If the flowing fluid is newtonian, as is commonly the case in engineering applications, then the shear stress exerted by the fluid on the surface of a plate is directly related to the velocity gradient at the surface of the plate via Newton's law of viscosity. This law states that the shear stress on the surface of the plate is directly proportional to the gradient of longitudinal flow velocity measured in a direction normal to the surface of the plate, with the proportionality constant between the shear stress and the velocity gradient being equal to the dynamic viscosity, μ, of the fluid.

Occurrence of a boundary layer. The region of the flow field that is directly influenced by the presence of the stationary surface of the flat plate is called the *boundary layer*. Flows in a boundary layer can be either laminar or turbulent, with higher velocity gradients associated with turbulent flows compared with laminar flows. Consequently, in accordance with Newton's law of viscosity, for the same free-stream velocity, a larger shear stress and hence a larger drag force is exerted on the flat surface under turbulent flow conditions. Velocity distributions in laminar boundary layers can be determined analytically using approximations to the Navier-Stokes (momentum) equations, whereas velocity distributions in turbulent boundary layers are usually estimated using empirical relations derived from experiments. Flows over surfaces with small curvature also can be approximated by flows over flat plates. The flow of a fluid over a flat plate is illustrated in Figure 10.4. The x-coordinate is measured along the plate from the leading edge of the plate in the direction of the flow, and the y-coordinate is measured from the plate surface in the direction normal to the plate. The fluid approaches the plate in the x direction with a velocity V, which is the same as the velocity over the plate far away from the surface. The presence of the plate affects the velocity up to some normal distance $\delta(x)$ from the plate, beyond which the free-stream velocity remains virtually unchanged. Within a distance δ from the plate, the effects of the viscous shearing forces caused by fluid viscosity are significant, and this region is called the *velocity boundary layer*. The boundary layer thickness, δ, is typically defined quantitatively as the distance y from the boundary at which the fluid velocity is equal to $0.99V$.

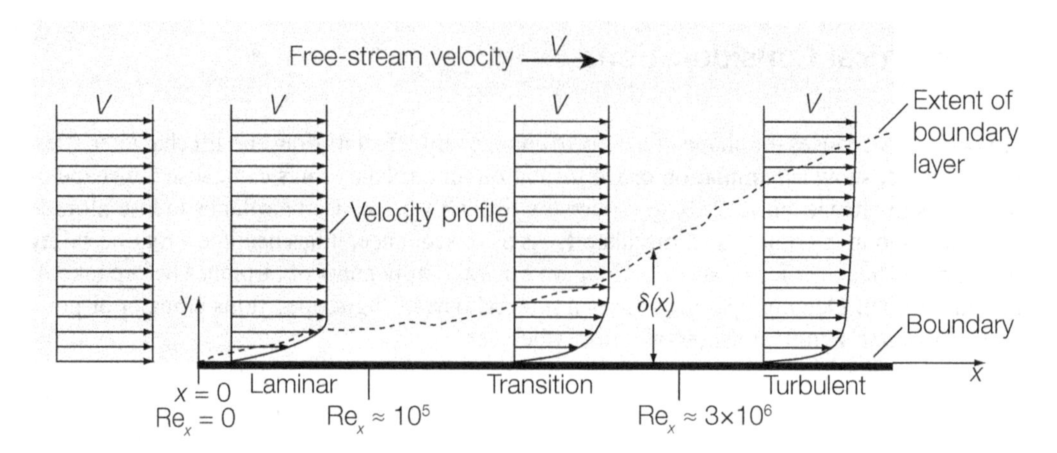

Figure 10.4: Boundary layer over a flat surface

Laminar and turbulent boundary layers. Typical average velocity profiles in laminar and turbulent boundary layers are illustrated in Figure 10.4, where it is shown that the velocity profile in turbulent flow is much fuller than in laminar flow (due to increased mixing within a turbulent boundary layer), with a sharper drop in velocity near the surface. This larger velocity gradient at the surface causes the drag force on the surface to be greater for a turbulent boundary layer than for a laminar boundary layer, given the same free-stream velocity. Turbulent boundary layers are also thicker than laminar boundary layers for the same free-stream conditions. In viewing Figure 10.4, it should be noted that the outer limit of the boundary layer, where $u = 0.99V$, is not a streamline.

Transition from laminar to turbulent boundary layer. Transition from a laminar boundary layer to a turbulent boundary layer is typically determined by the Reynolds number of the flow, defined as

$$\text{Re}_x = \frac{\rho V x}{\mu} \tag{10.11}$$

where x is the distance from the leading edge of the plate. The quantity Re_x is sometimes referred to as the *local Reynolds number*. For flow over a smooth flat plate, transition from a laminar boundary layer begins at around $\text{Re}_x = 1 \times 10^5$, but the boundary layer does not become fully turbulent until the Reynolds number reaches much higher values, typically around 3×10^6. The actual transition depends on factors such as the surface roughness, the turbulence level, and the variation in pressure along the surface. In engineering analyses, it is common to assume an instantaneous transition from a laminar to a turbulent boundary layer at around $\text{Re}_x = 5 \times 10^5$. The Reynolds number at which the transition to turbulent flow occurs is commonly referred to as the *critical Reynolds number*. For example, using a critical Reynolds number of 5×10^5, the boundary layer created by air flowing over a flat surface at 30 m/s will become turbulent when $x \approx 0.24$ m; at this location, the thickness of the boundary layer will be around 2 mm.

Growth of a boundary layer. Based on dimensional analysis, the growth of a boundary layer can be described by a function of the form

$$\frac{\delta}{x} = f(\text{Re}_x) \tag{10.12}$$

Experimental observations coupled with theoretical analyses show that the boundary layer thicknesses for laminar and turbulent conditions starting at $x = 0$ over a smooth flat surface can be approximated by

$$\frac{\delta}{x} = \begin{cases} \dfrac{4.91}{\mathrm{Re}_x^{\frac{1}{2}}}, & \text{for} \quad \mathrm{Re}_x < 5 \times 10^5 \text{ (laminar flow)} \\[2ex] \dfrac{0.382}{\mathrm{Re}_x^{\frac{1}{5}}}, & \text{for} \quad 5 \times 10^5 \leq \mathrm{Re}_x < 10^7 \text{ (turbulent flow)} \end{cases} \tag{10.13}$$

The flow conditions referenced in Equation 10.13 (i.e., laminar versus turbulent) refer to conditions within the boundary layer. Caution should be exercised when applying the laminar-flow relationship in Equation 10.13 for $\mathrm{Re}_x < 10^3$, because the implicit assumption of a thin boundary layer (i.e., $\delta/x \ll 1$) is of limited validity under these circumstances. The growth of the laminar boundary layer described by Equation 10.13 is derived from the widely accepted and classical Blasius solution to the continuity and momentum equations (see Section 11.2 for derivation), whereas the growth of the turbulent boundary layer is derived mostly from semiempirical formulations and experiments. Turbulent boundary layers grow more rapidly than laminar boundary layers.

10.2.4 Flow over Curved Surfaces

Typically, when flowing past curved bodies, a fluid follows the front portion of the curved surface with no problem because there is a *favorable pressure gradient* (i.e., the pressure decreases in the downstream direction). However, the fluid has difficulty remaining attached to the surface on the back side of the curved body because there is an *adverse pressure gradient* (i.e., the pressure increases in the downstream direction). At sufficiently high velocities, the flow becomes disrupted on the back side of the body in a process called *flow separation*. The process of flow separation is sometimes called *boundary-layer separation*, which explicitly indicates that it is the flow in the retarded-flow region adjacent to the surface (i.e., the boundary layer) that is disrupted by the "separation" process. Neither of the terms "flow separation" or "boundary-layer separation" is entirely satisfactory, because in actuality, the fluid remains in contact with the surface and a boundary layer (albeit disrupted) continues to exist adjacent to the surface. In spite of these shortcomings in conventional nomenclature, the term "flow separation" will be used to represent the separation (flow-disruption) process that occurs in some external flows around curved bodies. The mechanics of flow separation are illustrated in Figure 10.5. As a fluid flows around the front of a body, the free-stream velocity near the surfacc of the body typically increases, which corresponds to a decreasing pressure (in

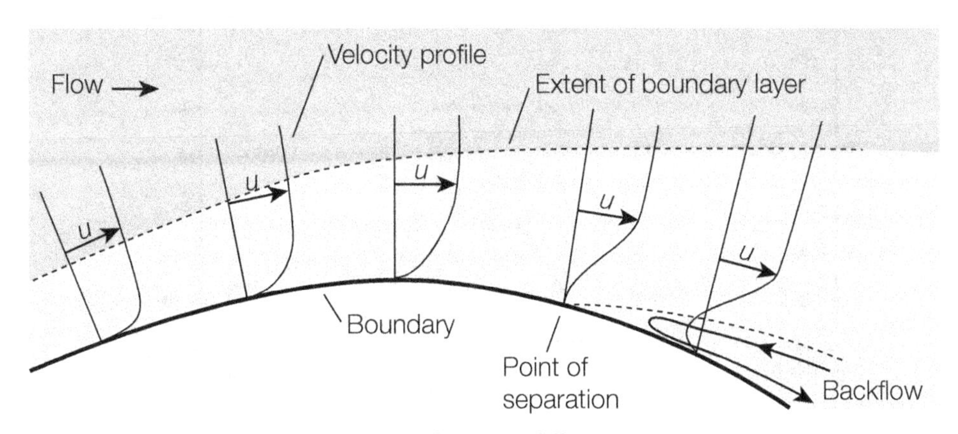

Figure 10.5: Mechanics of flow separation

accordance with the Bernoulli equation); hence, there is a positive pressure gradient in the direction of flow. In contrast, when the fluid flows to the back of the body, the free-stream velocity near the surface of the body decreases, corresponding to an increasing pressure; hence, there is an adverse pressure gradient. The fluid in the boundary layer close to the surface is flowing very slowly and can be reversed by the adverse pressure gradient that results in boundary-layer separation and backflow of the fluid close to the surface. The location of the separation point usually depends on several factors, such as the Reynolds number, surface roughness, and intensity of velocity fluctuations in the free stream; it is usually difficult to predict exactly where separation will occur unless there are sharp corners or abrupt changes in the shape of the body surface.

Separated region and wake. The low-pressure region behind a body where recirculation and backflows occur is called the *separated region*, which is illustrated in Figure 10.6. The larger the separated region, the larger the pressure drag, because the eddies in the separated flow cannot convert their kinetic energy into increased pressure and the pressure within the separated region remains close to that at the separation point. Because the pressure at the separation point is typically close to the least pressure on the front side of the body, flow separation generally results in a (pressure) drag force. The separated region comes to an end when the two streams reattach. The region of flow trailing the body where the effects of the body on velocity are felt is called the *wake*. Viscous and rotational effects are most significant in the boundary layer, the separated region, and the wake.

Streamlining. Streamlining is used to move the point of flow separation back as far as possible and thus minimize the size of both the separated region and the wake. Streamlining usually requires making the body longer and thinner to minimize the likelihood that a separated region will be formed. However, whereas streamlining decreases the pressure drag, making the body longer increases the friction drag; so the optimum amount of streamlining occurs when the sum of the friction and pressure drag is a minimum. Bodies are said to be *streamlined* when they closely approximate the streamlines of fluids that flow over them; bodies that are not streamlined are sometimes called *blunt bodies* or *bluff bodies*. Streamlining bodies has opposite effects on pressure and friction drags. Streamlining decreases the pressure drag by delaying boundary layer separation and thus reducing the pressure difference between the front and back of the body, and streamlining increases the friction drag by increasing the surface area. When pressure and friction drag are combined,the total drag on

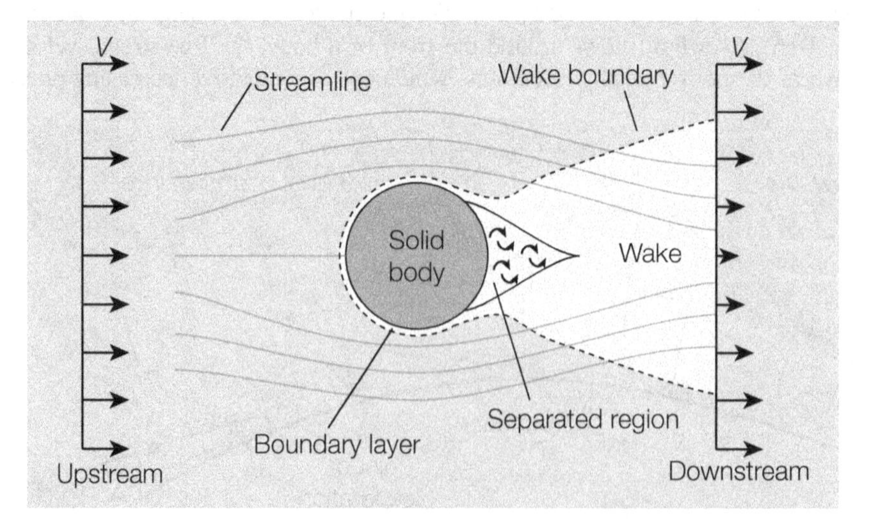

Figure 10.6: Separated region and wake

streamlined bodies typically increases when the boundary layer becomes turbulent, because most of the drag is due to the shear force, which is greater in turbulent flow than in laminar flow. In contrast to streamlined bodies, for blunt bodies, the total drag typically decreases when the boundary layer becomes turbulent, because a turbulent boundary layer can travel farther along a surface with an *adverse pressure gradient* on the rear portion of the body before separation occurs, resulting in a thinner wake and a smaller pressure drag compared with that associated with a laminar boundary layer. Streamlining is only necessary for bodies that are subjected to high-Reynolds-number flows, where flow separation is likely to occur. Drag caused by low-Reynolds-number flows (e.g., $\mathrm{Re} < 1$) is due almost entirely to friction drag, so streamlining would be ineffective in reducing drag under these circumstances.

Vortex shedding. An important consequence of flow separation is the formation and shedding of circulating fluid structures, called *vortices*, in the wake region. The periodic generation of these vortices downstream of bodies is called *vortex shedding*. This is a particular problem for flow around cylinders. For Reynolds numbers less than around 30 (based on the cylinder diameter), the boundary layer separates symmetrically from the two sides of the cylinder, and two weak, symmetrical, standing large eddies are formed. However, for Reynolds numbers in the range of 40–10 000, the vortices are shed first from one side of the cylinder and then from the other, resulting in a staggered double row of vortices in the wake of the cylinder. This phenomenon, called the *von Kármán vortex street*,[60] gives rise to the phenomenon of *aerodynamic instability*. Turbulent fluctuations also occur in the separated region for $\mathrm{Re} > 300$. The frequency, ω, at which vortices are shed can be estimated using the relation

$$\frac{\omega D}{V} \approx 0.20\left(1 - \frac{20}{\mathrm{Re}}\right) \tag{10.14}$$

where V is the free-stream velocity of the fluid, D is the diameter of the cylinder, and Re is the Reynolds number based on the diameter of the cylinder. Equation 10.14 can be considered applicable for $250 < \mathrm{Re} < 2 \times 10^5$ (Massey and Ward-Smith, 2012). Equation 10.14 relates the Strouhal number, St $(= \omega D/V)$, to the Reynolds number, Re $(= VD/\nu)$, and this empirical relation was formulated using dimensional analysis, with empirical constants derived from experimental data. At high values of Re, it can be assumed that $\omega D/V = 0.20$, although some authors take this constant as being equal to 0.21 (e.g., Potter et al., 2012; White, 2011). At very high Reynolds numbers ($\mathrm{Re} > 10^6$), vortex shedding disintegrates into uniform turbulence generated on both sides of the cylinder and vibrations tend to decrease. The vibrations generated by vortices near the body may cause the body to resonate to dangerous levels if the frequency of the vortices is close to the natural frequency of the body—a situation that must be avoided in the design of structures that are subjected to high-velocity fluid flow such as suspended bridges, towers, and tall smoke stacks. "Galloping" of power lines, where a power line alternates between the usual catenary and an inverted catenary, is also a consequence of vortex shedding that can have disastrous consequences. The "slapping" of ropes on a flagpole is also a consequence of vortex shedding.

Countermeasures for vibrations. The vibration of a cylindrical body caused by airflow around the body can be eliminated by attaching a longitudinal fin to the downstream side of the cylinder. If the length of the fin is greater than the diameter of the cylinder, interaction between the vortices is prevented and hence no vortex shedding occurs. Following a similar approach, tall chimneys sometimes have helical projections attached to them that cause unsymmetrical three-dimensional flows that preclude the symmetric flow associated with vortex shedding.

---

[60]Named in honor of the Hungarian-American mathematician, aerospace engineer, and physicist Theodore von Kármán (1881–1963).

10.3 Estimation of Drag Coefficients

Although the source of drag can be separated into friction and pressure, it is usually difficult to separate these effects. In most cases, the total drag is of concern and the total drag coefficient is usually reported in the technical literature. The drag coefficient is generally dependent on the shape of the body around or over which the fluid is flowing and under low-Reynolds number conditions is frequently dependent on the Reynolds number. Drag coefficients for several body shapes and flow configurations that are commonly found in engineering applications are considered separately in the following sections.

10.3.1 Drag on Flat Surfaces

The drag force on a flat surface caused by a fluid flowing parallel to the surface is due entirely to friction between the moving fluid and the stationary surface. For a rectangular surface of length L, width b, and surface roughness height ϵ, the (frictional) drag force per unit width, F_D/b, is a function of L and ϵ, as well as the fluid density, ρ, fluid viscosity, μ, and free-stream velocity, V. This relationship involves six variables in three dimensions, and dimensional analysis requires the following relationship between three dimensionless groups:

$$\frac{F_\text{D}/b}{\frac{1}{2}\rho V^2 L} = C_\text{Df}\left(\frac{\rho V L}{\mu}, \frac{\epsilon}{L}\right) \tag{10.15}$$

It is apparent from the above relationship that the drag coefficient, C_Df, is a function of the Reynolds number, $\text{Re}_\text{L} = \rho V L/\mu$, and the relative roughness, ϵ/L. The drag coefficient, C_Df, is sometimes called the *friction drag coefficient*, and Equation 10.15 is commonly expressed in the form

$$\frac{F_\text{D}}{\frac{1}{2}\rho V^2 L b} = C_\text{Df}\left(\text{Re}_\text{L}, \frac{\epsilon}{L}\right) \tag{10.16}$$

Empirical expressions relating the drag coefficient, C_Df, to Re_L and ϵ/L depend on the flow conditions over the surface of the flat plate (i.e., laminar or turbulent). In cases where the drag force up to some distance x from the leading edge is of interest, L may be replaced by x in the above formulation.

Influence of surface roughness on C_Df. Surface roughness generally becomes an influential factor on the value of C_Df when the characteristic roughness height, ϵ, is greater than the thickness, δ_v, of the viscous sublayer within the boundary layer. The thickness of the viscous sublayer is conventionally estimated as

$$\delta_\text{v} = 5\frac{\nu}{u_*} \tag{10.17}$$

where ν is the kinematic viscosity of the fluid and u_* is the friction velocity, which is a surrogate for the boundary shear stress, τ_0, because u_* is defined by $u_* = \sqrt{\tau_0/\rho}$. For a turbulent boundary layer, u_* can be estimated by the empirical relation

$$u_* = 0.1713\frac{V}{\text{Re}_\text{L}^{\frac{1}{5}}} \tag{10.18}$$

Because a viscous sublayer does not exist within a laminar boundary layer, under laminar boundary layer conditions, C_Df depends only on Re_L. For turbulent boundary layers, the appropriate functional forms of C_Df are generally taken as follows:

$$\frac{\epsilon}{\delta_v} < 1, \quad \rightarrow C_{Df}(\mathrm{Re_L}) \tag{10.19}$$

$$1 \leq \frac{\epsilon}{\delta_v} \leq 14, \quad \rightarrow C_{Df}(\mathrm{Re_L}, \epsilon/L) \tag{10.20}$$

$$\frac{\epsilon}{\delta_v} > 14, \quad \rightarrow C_{Df}(\epsilon/L) \tag{10.21}$$

In accordance with Equations 10.19–10.21, in cases where the flow in the boundary layer is turbulent and $\epsilon < \delta_v$, C_{Df} depends only on $\mathrm{Re_L}$ and the surface of the flat plate is classified as *hydrodynamically smooth*. In cases where the flow in the boundary layer is turbulent and $\epsilon > 14\delta_v$, C_{Df} depends only on ϵ/L and the surface of the flat plate is classified as *hydrodynamically rough*. In cases where the flow in the boundary layer is turbulent and $\delta_v \leq \epsilon \leq 14\delta_v$, C_{Df} depends on both $\mathrm{Re_L}$ and ϵ/L and the surface of the flat plate is classified as being in the *transition region*. It is apparent from Equations 10.19–10.21 that the relative values of ϵ and δ_v determine whether a surface is hydrodynamically smooth or rough. Because δ_v decreases with increasing free-stream velocity, any given surface can be smooth or rough depending on the free-stream flow condition.

10.3.1.1 Drag coefficient on hydrodynamically smooth surfaces

In cases where the flow is either entirely laminar or entirely turbulent over the flat plate and the friction drag is not influenced by the surface roughness, the drag coefficient is given by

$$C_{Df} = \begin{cases} \dfrac{1.328}{\mathrm{Re}_L^{\frac{1}{2}}}, & \text{laminar flow over entire plate } (\mathrm{Re_L} \leq 5 \times 10^5) \\ \dfrac{0.0735}{\mathrm{Re}_L^{\frac{1}{5}}}, & \text{turbulent flow over entire plate } (5 \times 10^5 < \mathrm{Re_L} < 10^7) \\ \dfrac{0.455}{(\log \mathrm{Re_L})^{2.58}}, & \text{turbulent flow over entire plate } (10^7 \leq \mathrm{Re_L} \leq 10^9) \end{cases} \tag{10.22}$$

The expression in Equation 10.22 for $\mathrm{Re_L} \geq 10^7$ was proposed by Schlichting (1949) to better match experimental data. The Schlichting (1949) equation is also commonly used for the entire range of $5 \times 10^5 \leq \mathrm{Re_L} \leq 10^9$. The drag coefficient, C_{Df}, over a smooth flat plate containing both laminar and turbulent boundary layers can be estimated using the relation

$$C_{Df} = \begin{cases} \dfrac{1.328}{\mathrm{Re}_L^{\frac{1}{2}}}, & \mathrm{Re_L} \leq 5 \times 10^5 \\ \dfrac{0.0735}{\mathrm{Re}_L^{\frac{1}{5}}} - \dfrac{1742}{\mathrm{Re_L}}, & 5 \times 10^5 < \mathrm{Re_L} < 10^7 \\ \dfrac{0.455}{(\log \mathrm{Re_L})^{2.58}} - \dfrac{1742}{\mathrm{Re_L}}, & 10^7 \leq \mathrm{Re_L} \leq 10^9 \end{cases} \tag{10.23}$$

where Equation 10.23 is derived based on the assumption that the plate is smooth and there is an instantaneous transition from a laminar to a turbulent boundary layer at $\mathrm{Re}_x = 5 \times 10^5$. The constant in the second term in Equations 10.23 is determined using the relation

$$\text{constant} = \mathrm{Re_{tr}}(C_{D,\mathrm{turb}} - C_{D,\mathrm{lam}}) \tag{10.24}$$

where $\mathrm{Re_{tr}}$ is the transition Reynolds number and $C_{D,\mathrm{turb}}$ and $C_{D,\mathrm{lam}}$ are the drag coefficients at the transition location assuming completely turbulent and completely laminar flow, respectively. In cases where $\mathrm{Re_{tr}} \neq 5 \times 10^5$, Equation 10.24 can be used to adjust the constant in the second term in Equation 10.23. For $\text{Re}_{\text{tr}} = 5 \times 10^5$, Equation 10.23 indicates that for a given length of plate, the drag coefficient is less when laminar flow is maintained over the longest possible distance. However, when $\text{Re}_{\text{L}} > 10^7$, the contribution of the laminar portion of the boundary layer to drag is negligible.

10.3.1.2 Drag coefficient on hydrodynamically rough surfaces

Experimental data for the average drag coefficient under hydrodynamically rough conditions can be described by the following empirical relationship (Schlichting, 1979):

$$C_{\text{Df}} = \left(1.89 - 1.62 \log \frac{\epsilon}{L}\right)^{-2.5} \tag{10.25}$$

The relationship for rough turbulent flow given in Equation 10.25 is appropriate for use when $\text{Re}_{\text{L}} > 10^6$ and $\epsilon/L > 10^{-4}$.

10.3.1.3 Drag coefficient on intermediate surfaces

In the intermediate region between smooth and rough turbulent flow, the drag coefficient is dependent on both Re_{L} and ϵ/L and transitions smoothly between the C_{Df} given by Equation 10.23 for smooth surfaces and Equation 10.25 for rough surfaces. Drag coefficients, C_{Df}, for parallel flow over smooth, intermediate, and rough surfaces are plotted in Figure 10.7 for both laminar and turbulent flows. It is worth noting that, for a given Reynolds number, C_{Df} can increase sevenfold with roughness in turbulent flow and that, for a given relative roughness, C_{Df} is independent of the Reynolds number in the rough flow region.

Minimization of friction drag. The frictional force on a flat plate is less with a laminar boundary layer than with a turbulent boundary layer. Consequently, to minimize drag, the transition to a turbulent boundary layer should be delayed as long as possible. This can be achieved by making the flat surface as smooth as possible near the leading edge where the boundary layer is thinnest, recognizing that a greater roughness is more tolerable farther downstream.

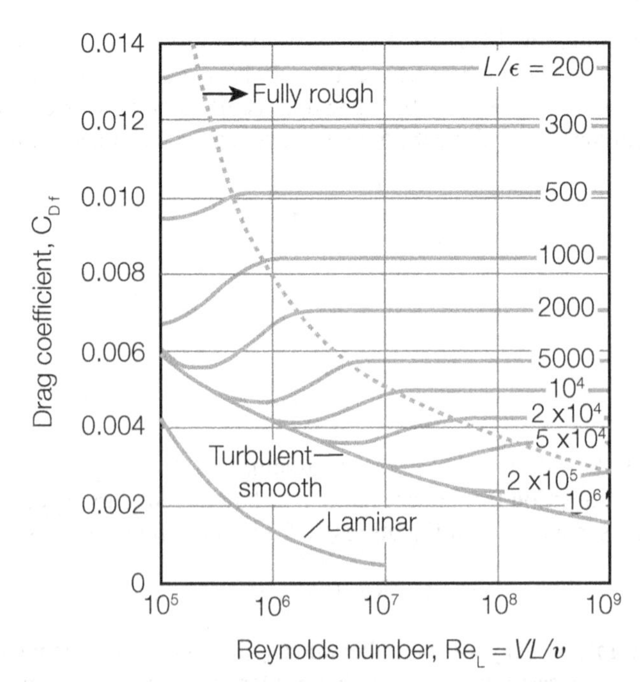

Figure 10.7: Drag coefficients for flow over a flat plate

EXAMPLE 10.2

Standard air flows at a speed of 22.4 m/s over a smooth flat surface that is 15 m long and 5 m wide. (a) What is the drag force on the surface for a roughness height of 0.1 mm? (b) After some weathering and abrasion, the roughness height increases to 1 mm. What is the drag force on the surface under this deteriorated condition? (c) Attachment of several small fixtures to the surface increases the roughness height to 10 mm. What is the drag force under this condition?

SOLUTION

From the given data: $V = 22.4$ m/s, $L = 15$ m, and $W = 5$ m. For standard air: $\rho = 1.225$ kg/m^3 and $\nu = 1.460 \times 10^{-5}$ m^2/s. The following preliminary calculations are useful:

$$\mathrm{Re_L} = \frac{VL}{\nu} = \frac{(22.4)(15)}{1.460 \times 10^{-5}} = 2.301 \times 10^7$$

$$u_* = 0.1713\frac{V}{\mathrm{Re}_\mathrm{L}^{\frac{1}{5}}} = 0.1713\frac{22.4}{(2.301 \times 10^7)^{\frac{1}{5}}} = 0.1293 \text{ m/s}$$

$$\delta_\mathrm{v} = 5\frac{\nu}{u_*} = 5\frac{1.460 \times 10^{-5}}{0.1293} = 5.65 \times 10^{-4} \text{ m} = 0.565 \text{ mm}$$

(a) Because $\mathrm{Re_L} > 5 \times 10^5$, the boundary layer is turbulent over part of the surface. For $\epsilon = 0.1$ mm, $\epsilon/\delta_\mathrm{v} < 1$; hence, the surface is hydrodynamically smooth. For smooth turbulent flow, C_Df is given by Equation 10.23 and F_D is given by Equation 10.16; hence,

$$C_\mathrm{Df} = \frac{0.455}{(\log \mathrm{Re_L})^{2.58}} - \frac{1742}{\mathrm{Re_L}} = \frac{0.455}{[\log(2.301 \times 10^7)]^{2.58}} - \frac{1742}{2.301 \times 10^7} = 0.00256$$

$$F_\mathrm{D} = C_\mathrm{Df}\tfrac{1}{2}\rho V^2 LW = (0.00256)\tfrac{1}{2}(1.225)(22.4)^2(15)(5) = \mathbf{59.0\ N}$$

(b) For $\epsilon = 1$ mm, $\epsilon/\delta_\mathrm{v} = 1.8$, and because $1 < \epsilon/\delta_\mathrm{v} < 14$, the surface is in the intermediate zone between hydrodynamically smooth and rough. For intermediate turbulent flow with $\mathrm{Re_L} = 2.301 \times 10^7$ and $L/\epsilon = 15/0.001 = 1.5 \times 10^4$, C_Df is derived from Figure 10.7 and F_D is given by Equation 10.16; hence,

$$C_\mathrm{Df} = 0.00350$$

$$F_\mathrm{D} = C_\mathrm{Df}\tfrac{1}{2}\rho V^2 LW = (0.00350)\tfrac{1}{2}(1.225)(22.4)^2(15)(5) = \mathbf{104\ N}$$

(c) For $\epsilon = 10$ mm, $\epsilon/\delta_\mathrm{v} = 18$, and because $\epsilon/\delta_\mathrm{v} > 14$, the surface is hydrodynamically rough. For rough turbulent flow, C_Df is given by Equation 10.25 and F_D is given by Equation 10.16; hence,

$$C_\mathrm{Df} = \left(1.89 - 1.62\log\frac{\epsilon}{L}\right)^{-2.5} = \left(1.89 - 1.62\log\frac{10 \times 10^{-3}}{15}\right)^{-2.5} = 0.00762$$

$$F_\mathrm{D} = C_\mathrm{Df}\tfrac{1}{2}\rho V^2 LW = (0.00762)\tfrac{1}{2}(1.225)(22.4)^2(15)(5) = \mathbf{176\ N}$$

Therefore, as the roughness height increases, the drag force on the surface also increases. For roughness heights of 0.1 mm, 1 mm, and 10 mm, the drag forces are approximately 59 N, 104 N, and 176 N, respectively.

10.3.1.4 Drag coefficient with flow normal to a flat plate

In cases where the incident flow is normal to a thin flat plate, the drag coefficient is entirely due to pressure drag. This is so because the front and back surfaces of the rectangular plate are normal to the flow direction as shown in Figure 10.8, and hence, the shear forces are also normal to the flow direction. Because the plate is thin, there is negligible surface area on the sides of the plate that are parallel to the flow direction. In this case, the drag coefficient, C_D, of the flat plate is defined as

$$C_D = \frac{F_D}{\frac{1}{2}\rho V^2 Lb} \tag{10.26}$$

where F_D is the drag force on the plate, ρ is the density of the fluid, V is the approach velocity, b is the width of the plate, and L is the height of the plate. For values of $\text{Re}_L = VL/\nu > 10^4$, the drag coefficient, C_D, depends only on the aspect ratio b/L of the plate as given in Table 10.1. It is apparent from Table 10.1 that for plates in which the width, b, is much greater than

Table 10.1: Drag Coefficient for Flow Normal to a Thin Flat Plate

Aspect ratio, b/L	1	5	10	20	∞
Drag coefficient, C_D	1.18	1.2	1.3	1.5	2.0

the height, L, the drag coefficient can be approximated as $C_D = 2.0$. Under this condition, the plate is sufficiently wide that flow around the sides of the plate has a negligible effect on the drag force exerted on the plate. In practical applications, some large rectangular billboards can be treated as thin flat plates to estimate the wind loading on the support structure.

10.3.2 Drag on Spheres and Cylinders

Both friction drag and pressure drag can be significant in flows past spheres and cylinders. Under circumstances in which the viscosity of the fluid exerts a significant effect on the drag coefficient, viscous effects are generally measured by a Reynolds number, Re, defined as

$$\text{Re} = \frac{VD}{\nu} \tag{10.27}$$

where V is the velocity at which the fluid approaches the sphere or cylinder, D is the diameter of the sphere or cylinder, and ν is the kinematic viscosity of the fluid. When dealing with small (particle sized) spheres, the Reynolds number defined by Equation 10.27 is commonly referred to as the *particle Reynolds number*, Re_p, explicitly to denote that the length scale in the Reynolds number is the diameter of a particle. The drag forces on spheres and cylinders are primarily due to friction drag at low Reynolds numbers (Re < 10) and to pressure drag at

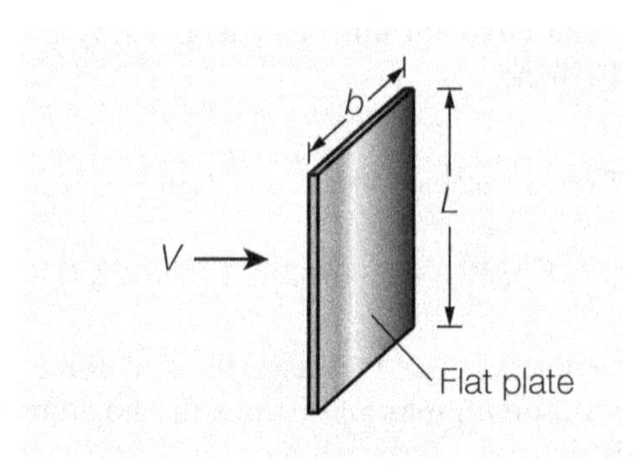

Figure 10.8: Flow normal to a flat plate

high Reynolds numbers (Re $>$ 5000). Both effects are significant at intermediate Reynolds numbers. The pressure drag at high Reynolds numbers is the combined result of the high pressure on the front side of the body in the vicinity of the stagnation point and the low pressure on the back side of the body in the wake, producing a net force on the body in the direction of flow.

10.3.2.1 General characteristics

The average drag coefficient, C_D, for flow around a smooth circular cylinder and around a smooth sphere is illustrated in Figure 10.9. The behavior of the drag coefficient for flows around cylinders and spheres have several characteristics in common. For Re ≤ 1, creeping flow occurs and C_D decreases in inverse proportion to the Reynolds number, with $C_D = 24/\text{Re}$ being a commonly used approximation for spheres. At about Re = 10, flow separation starts to occur near the rear of the body, with vortex shedding starting at about Re = 40. The region of flow separation increases with increasing Reynolds number up to around Re = 10^3, at which point the drag is mostly (around 95%) due to pressure drag. The drag coefficient continues to decrease with increasing Reynolds number in the range of $10 < \text{Re} < 10^3$. In the moderate range, $10^3 < \text{Re} < 10^5$, the drag coefficient remains relatively constant. The flow in the boundary layer is laminar in this range, but the flow in the separated region past the sphere or cylinder is highly turbulent with a wide turbulent wake. There is a sudden drop in the drag coefficient somewhere in the range of $10^5 < \text{Re} < 10^6$ due to the flow in the boundary layer becoming turbulent, which moves the separation point farther to the rear of the body, reducing the size of the wake and thus the magnitude of the pressure drag. This is in contrast to streamlined bodies that experience an increase in the drag coefficient (mostly due to friction drag) when the boundary layer become turbulent. There is a transitional regime in the range $2 \times 10^5 \leq \text{Re} \leq 2 \times 10^6$ in which C_D dips to a minimum value and then slowly rises to its final turbulent value.

Drag characteristics of buoyant spheres. The drag characteristics shown in Figure 10.9 are applicable to a stationary sphere with a flowing fluid as well as a stationary fluid with a moving sphere. Examples of the latter case are a dense spherical particle sinking (i.e., settling) in a fluid and a light spherical particle, or bubble, rising in a fluid. However, in cases where the density of the sphere is less than 80% of the density of the surrounding fluid, a *wake instability* is created within the range $135 < \text{Re} < 10^5$, which causes the rising sphere to spiral upward at an angle of 60° from the horizontal and approximately doubling the drag coefficient

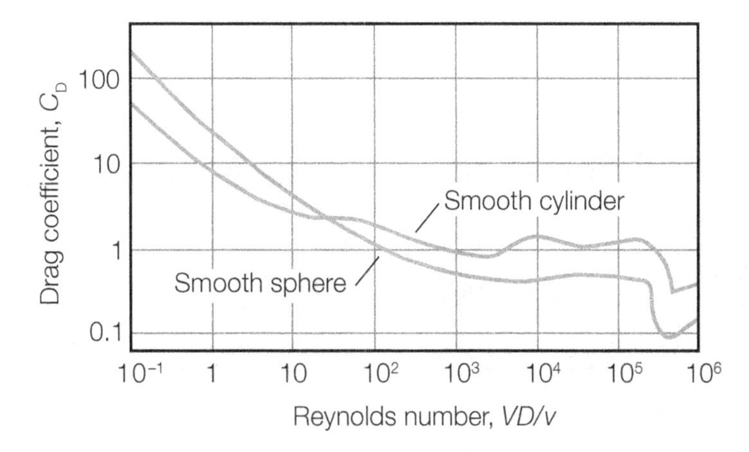

Figure 10.9: Drag coefficients for flow around smooth cylinders and spheres

to $C_D \approx 0.95$ (Karamenev and Nikolov, 1992). This deviation from the conventional drag coefficient is particularly applicable in the case of rising (spherical) bubbles.

Flow separation: laminar versus turbulent boundary layers. Observations of flows around circular cylinders and spheres indicate that flow separation occurs at around $\theta \simeq 80°$ (measured from the front stagnation point) when the boundary layer is laminar and at about $\theta \simeq 140°$ when the boundary layer is turbulent. This is illustrated for a sphere in Figure 10.10, where Figure 10.10(a) shows flow over a sphere with a laminar boundary layer (at Re = 15 000) and Figure 10.10(b) shows flow over a sphere with a turbulent boundary layer (at Re = 30 000). The delay of separation in a turbulent boundary layer is caused by the rapid fluctuations of the fluid in the transverse direction and the fact that turbulent boundary layers have more kinetic energy and momentum than do laminar boundary layers, which enables a turbulent boundary layer to travel farther along the surface before boundary-layer separation occurs, resulting in a narrower wake and a smaller pressure drag.

Effect of surface roughness. In the case of blunt bodies such as circular cylinders and spheres, an increase in surface roughness can decrease the drag coefficient. Surface roughness creates conditions for the formation of a turbulent boundary layer at a lower Reynolds number, causing the fluid to close in behind the body, narrowing the wake and reducing the pressure force on the downstream side of the body. This causes the drag coefficient for a roughened surface to be less than the drag coefficient for a smooth surface, within a certain range of Reynolds numbers.

Practical application of the roughness effect. Golf balls are intentionally roughened to induce turbulence at lower Reynolds numbers to take advantage of the sharp drop in the drag coefficient associated with turbulent boundary layers. Transition from a laminar to a turbulent boundary layer over a dimpled golf ball occurs at around Re $= 4 \times 10^4$, at which point the drag coefficient, C_D, is reduced by approximately one-half. It has been estimated that dimples on a golf ball increase the flight distance by 50%–100% (Potter et al., 2012). The addition of roughness is not beneficial for all types of game balls. For example, roughening table tennis balls does not reduce drag for most average players, because the speeds at which table tennis balls travel correspond to values of Re much less than those for which a turbulent boundary layer is possible. The maximum speeds of table tennis balls generated in professional matches are on the order of 35 m/s, whereas the maximum speeds of golf balls generated by professionals is on the order of 150 m/s.

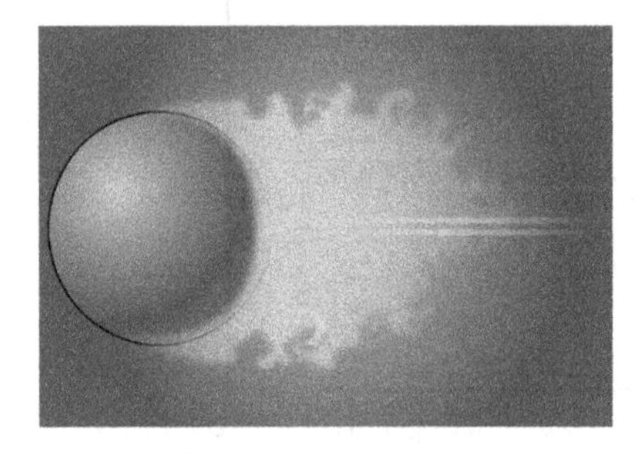

(a) Laminar boundary layer

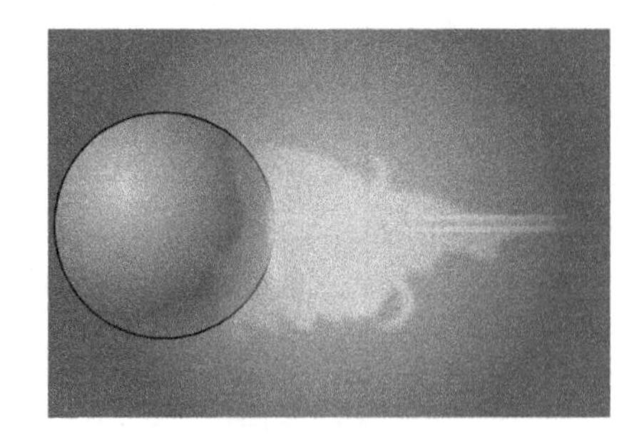

(b) Turbulent boundary layer

Figure 10.10: Flow separation around a sphere

10.3.2.2 Analytic expressions for drag coefficients of spheres

The drag coefficient, C_D, of a sphere represents the ratio of the drag force to the inertial force and is defined by the following relation:

$$C_D = \frac{F_D}{\frac{1}{2}\rho V^2 A_s} = \frac{F_D}{\frac{1}{2}\rho V^2(\frac{1}{4}\pi D^2)} \quad \rightarrow \quad C_D = \frac{8F_D}{\pi\rho V^2 D^2} \tag{10.28}$$

where F_D is the drag force, ρ is the density of the fluid, V is the approach velocity of the fluid, and A_s is the frontal surface area of a sphere, which is generally equal to $\pi D^2/4$, where D is the diameter of the sphere. Viscous effects are usually accounted for by the Reynolds number $\text{Re} = VD/\nu$, and viscous effects typically have a significant effect on C_D for $\text{Re} \leq 10^5$. Numerous expressions have been developed to relate C_D to Re for conditions under which surface roughness effects are negligible. These relationships for spheres are typically segregated into very low Reynolds-number flows, called "creeping flows," and higher Reynolds-number flows.

Low Reynolds-number flows. At very low Reynolds numbers, the flow is classified as *creeping flow* or *Stokes flow*,[61] and for creeping flow around spheres, the drag coefficient, C_D, can be estimated by

$$C_D = \frac{24}{\text{Re}}, \qquad \text{Re} \leq 0.1 \tag{10.29}$$

The drag force, F_D, on the sphere corresponding to the drag coefficient given by Equation 10.29 is given by

$$F_D = C_D A_s \frac{\rho V^2}{2} = \left(\frac{24}{\rho VD/\mu}\right)\left(\frac{\pi D^2}{4}\right)\frac{\rho V^2}{2} \quad \rightarrow \quad F_D = 3\pi\mu VD \tag{10.30}$$

The drag force given by Equation 10.30 is known as *Stokes law*. This relation is often used to describe the behavior of suspended solid particles in water and spray droplets and dust particles in air. Stokes law is based on the assumption that C_D is given by Equation 10.29, which agrees closely with experiments for $\text{Re} \leq 0.1$. A more accurate estimate for C_D that is valid for $\text{Re} \leq 1$ is given by (Oseen, 1910, 1913)

$$C_D = \frac{24}{\text{Re}}\left(1 + \frac{3}{16}\text{Re}\right), \qquad \text{Re} \leq 1 \tag{10.31}$$

Although Equation 10.31 is more accurate than Equation 10.29 for $0.1 < \text{Re} \leq 1$, Equation 10.29 is widely used for estimating C_D in the range of $\text{Re} \leq 1$. The drag coefficients under creeping flow conditions ($\text{Re} \leq 1$) for a few other geometries are shown in Figure 10.11. It is worth noting that for creeping flows, the shape of the body does not have a major influence on the drag coefficient. Creeping flows typically occur in flow through porous media (such as in groundwater flow through aquifers), although in those cases, interest is more in the rate of flow through the porous medium than in the drag forces on the solid particles that constitute the medium. An empirical equation developed by Shanks (1955) matches experimental data quite well for $0.1 \leq \text{Re} \leq 10$, and the Shanks (1955) equation is given by

$$C_D = \frac{7.096 \times 10^6 + 3.197 \times 10^6\,\text{Re} + 2.611 \times 10^5\,\text{Re}^2}{(2.957 \times 10^5 + 7.776 \times 10^4\,\text{Re} + 689\,\text{Re}^2)\text{Re}}, \qquad 0.1 \leq \text{Re} \leq 10 \tag{10.32}$$

It is apparent that the combination of Equations 10.32 and 10.29 provide analytic expressions from which C_D can be estimated within the range of $\text{Re} \leq 10$.

---

[61] Named in honor of the British mathematician and physicist G. G. Stokes (1819–1903).

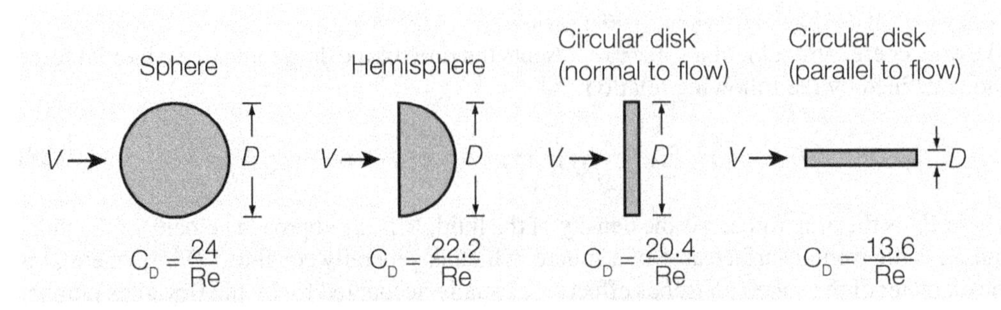

Figure 10.11: Drag coefficient for creeping flow past various bodies ($\mathrm{Re} \leq 1$)

Higher Reynolds-number flows. The behavior of the drag coefficient of a sphere at higher Reynolds numbers is illustrated in Figure 10.9. For Re in the range of 10–1000, the drag coefficient, C_D, decreases with increasing Re, and at $\mathrm{Re} = 1000$, approximately 95% of the drag is due to the pressure distribution around the sphere (i.e., form drag) and 5% is due to frictional drag. The drag coefficient remains approximately constant in the range $1000 < \mathrm{Re} < 2 \times 10^5$. At around $\mathrm{Re} = 2 \times 10^5$ for smooth surfaces, the boundary-layer transitions from laminar to turbulent, which pushes the point of boundary-layer separation farther to the rear of the body, thereby causing a sudden reduction in the form drag and a corresponding reduction in the drag coefficient. On rough surfaces, this transition to turbulent flow occurs at a smaller Reynolds number—typically $\mathrm{Re} = 8 \times 10^4$ rather than $\mathrm{Re} = 2 \times 10^5$. After the sudden drop in C_D associated with the turbulent transition, the value of C_D very gradually increases with further increases in Re. For high Reynolds numbers, an asymptotic value of $C_\mathrm{D} = 0.2$ is commonly assumed. For $\mathrm{Re} \geq 5.33$, the following analytic formulation is recommended for estimating the drag coefficient:

$$C_\mathrm{D} = \begin{cases} \dfrac{5.948 \times 10^5 + 7.735 \times 10^4\,\mathrm{Re} + 398.1\,\mathrm{Re}^2}{(2.230 \times 10^4 + 928.3\,\mathrm{Re} + 0.01675\,\mathrm{Re}^2)\mathrm{Re}}, & 5.33 \leq \mathrm{Re} \leq 1.18 \times 10^5 \\ 0.2, & \mathrm{Re} > 1.18 \times 10^5 \end{cases} \tag{10.33}$$

The empirical expression given by Equation 10.33 for estimating C_D in the range of $5.33 \leq \mathrm{Re} \leq 1.18 \times 10^5$ matches available laboratory data quite well in the range of Re for which it is applicable (Mikhailov and Silva Freire, 2013). Equation 10.33 is strictly applicable for smooth spheres. For rough spheres, Equation 10.33 can also be applied, but the transition to $C_\mathrm{D} = 0.2$ should be taken to occur at $\mathrm{Re} = 8 \times 10^4$ instead of 1.18×10^5. For values of Re below the range covered by Equation 10.33, the drag coefficient can be estimated by Equations 10.32 and 10.29.

10.3.2.3 Terminal velocities of spheres and other bodies

The *terminal velocity* of a body is the steady-state velocity attained by a body that is falling or rising in a fluid. The case of a falling body is encountered when the effective density of the body is greater than the density of the surrounding fluid, and the case of a rising body is encountered when the effective density of the body is less than the density of the surrounding fluid.

Falling body. Consider the more common case of a falling body, where the terminal velocity is attained when the weight of the body is equal to the drag force plus the buoyancy force. This condition is expressed by the following relationship:

$$\text{Falling body:}\quad \underbrace{\gamma_\text{b} V_\text{b}}_{\text{weight of body}} = \underbrace{C_\text{D}\tfrac{1}{2}\rho_\text{f} V^2 A_\text{b}}_{\text{drag force on body}} + \underbrace{\gamma_\text{f} V_\text{b}}_{\text{buoyant force on body}} \tag{10.34}$$

where γ_b is the specific weight of the body, V_b is the volume of the body, C_D is the drag coefficient, ρ_f is the density of the surrounding fluid, V is the terminal velocity of the body, A_b is the frontal area of the body, and γ_f is the specific weight of the surrounding fluid. Solving for the terminal velocity, V, in Equation 10.34 yields

$$V = \sqrt{\frac{2V_\text{b}(\gamma_\text{b} - \gamma_\text{f})}{C_\text{D}\rho_\text{f} A_\text{b}}} \tag{10.35}$$

In cases where the drag coefficient, C_D, depends on the viscosity of the fluid, C_D will be a function of the velocity V via its Reynolds-number dependence. In these cases, Equation 10.35 must be solved simultaneously with an equation relating C_D to the Reynolds number. In the case of a sphere, V_b and A_b are related to the diameter D of the sphere by the following relations:

$$V_\text{b} = \tfrac{1}{6}\pi D^3, \qquad A_\text{b} = \tfrac{1}{4}\pi D^2 \tag{10.36}$$

Substituting the geometric relations for a sphere given in Equation 10.36 with the general expression for the terminal velocity of a falling body given by Equation 10.35 yields the following general expression for the terminal velocity of a falling sphere:

$$V = \sqrt{\frac{4(\gamma_\text{b} - \gamma_\text{f})D}{3C_\text{D}\rho_\text{f}}} \tag{10.37}$$

For spheres that are falling sufficiently slowly that Re ≤ 0.1, the drag coefficient, C_D, can be estimated by the Stokes relation $C_\text{D} = 24/\text{Re}$ as given by Equation 10.29, and the terminal velocity, V, given by Equation 10.37 becomes

$$V = \frac{(\gamma_\text{b} - \gamma_\text{f})D^2}{18\mu}, \qquad \text{Re} \leq 0.1 \tag{10.38}$$

This equation, which is commonly referred to as the *Stokes settling velocity equation*, is widely used in engineering practice to predict the settling velocity of fine particles in a variety of fluids.

Rising body. In the case of a rising body, where the effective density of the body is less than that of the surrounding fluid (as in the case of a bubble), the terminal velocity is attained when the weight of the body plus the drag force is equal to the buoyancy force. This condition is expressed by the following relationship:

$$\text{Rising body:}\quad \underbrace{\gamma_\text{b} V_\text{b}}_{\text{weight of body}} + \underbrace{C_\text{D}\tfrac{1}{2}\rho_\text{f} V^2 A_\text{b}}_{\text{drag force on body}} = \underbrace{\gamma_\text{f} V_\text{b}}_{\text{buoyant force on body}} \tag{10.39}$$

Solving Equation 10.39 for the terminal velocity yields a relationship similar to Equation 10.35 for bodies of any shape and a relationship similar to Equation 10.37 for spheres, with the only difference being that $\gamma_\text{b} - \gamma_\text{f}$ is replaced by $\gamma_\text{f} - \gamma_\text{b}$ in both cases.

Practical Considerations

Estimation of the terminal velocity of bubbles, droplets, and particles are of central importance in many engineering applications related to chemical, metallurgical, and environmental processes. The settling of paint droplets, sedimentation of precipitates in settling tanks, and sediment transport in rivers are all influenced by the terminal velocities of suspended particles. Care should be taken not to assume single-sphere settling characteristics in cases where there is a high density of settling particles or there are clusters of setting particles, because individual particles tend to settle more slowly when surrounded by other particles. This condition is particularly important in mixing and sedimentation processes.

EXAMPLE 10.3

Backwashing a sand filter involves creating an upward flow such that the sand particles in the filter are suspended and remain stationary in the upflow while the accumulated impurities within the sand filter are removed in the backwash water. Consider the idealized case where the sand particles are all spherical with a diameter of 2 mm and have a specific gravity of 2.65. Estimate the velocity of the backwash water that is required to keep the sand particles suspended. The temperature of the backwash water is 20°C.

SOLUTION

From the given data: $D = 2$ mm and SG = 2.65. For water at 20°C, $\rho_\mathrm{f} = 998.2$ kg/m^3, $\gamma_\mathrm{f} = 9789$ N/m^3 and $\nu = 1.004 \times 10^{-6}$ m^2/s. For a particle with SG = 2.65, $\rho_\mathrm{b} = 2650$ kg/m^3 and $\gamma_\mathrm{b} = 2.599 \times 10^4$ N/m^3. A particle that is being suspended by an upflow is actually settling in the fluid at the same rate as the upflow velocity. Consequently, the required upflow velocity to suspend a sand particle is given by Equation 10.37. Assuming that $5.33 \leq \mathrm{Re} \leq 1.18 \times 10^5$, the drag coefficient can be estimated by Equation 10.33. Hence, the following three equations must be solved simultaneously for the settling velocity, V:

$$\mathrm{Re}(V) = \frac{VD}{\nu} = \frac{V(0.002)}{1.004 \times 10^{-6}} = 1992V$$

$$C_\mathrm{D}(V) = \frac{5.948 \times 10^5 + 7.735 \times 10^4\,\mathrm{Re} + 398.1\,\mathrm{Re}^2}{(2.230 \times 10^4 + 928.3\,\mathrm{Re} + 0.01675\,\mathrm{Re}^2)\mathrm{Re}}$$

$$V = \sqrt{\frac{4(\gamma_\mathrm{b} - \gamma_\mathrm{f})D}{3C_\mathrm{D}\rho_\mathrm{f}}} = \sqrt{\frac{4(2.599 \times 10^4 - 9789)(0.002)}{3C_\mathrm{D}(998.2)}} = 0.2080C_\mathrm{D}^{-\frac{1}{2}}$$

Simultaneous solution of these equations yields $V = 0.2789$ m/s. To validate using Equation 10.33 in estimating C_D, it is necessary to calculate the Reynolds number and show that it is in the range for which Equation 10.33 is applicable. Using the calculated velocity gives

$$\mathrm{Re} = \frac{VD}{\nu} = \frac{(0.2789)(0.002)}{1.004 \times 10^{-6}} = 556$$

Because $5.33 \leq \mathrm{Re} \leq 1.18 \times 10^5$, the calculated drag coefficient and settling velocity are validated. Therefore, using a backflow velocity of approximately **0.28 m/s** will keep the sand particles in the filter suspended.

10.3.3 Drag on Vehicles

At large Reynolds-number flows, the drag coefficients of vehicles range from 1.0 for tractor-trailers to around 0.25 for passenger cars, with blunter vehicles having higher drag coefficients. Typical drag coefficients for various vehicles are listed in Table 10.2. These drag coefficients are appropriate for use with the frontal area of the vehicle in calculating the drag force. When the effect of the road on air motion is disregarded, the ideal shape of a vehicle is that of a teardrop, with a drag coefficient of about 0.1 for the turbulent flow case. It is interesting to note that cars in the 1920s had drag coefficients on the order of 0.8 and the drag coefficient of modern racing cars is on the order of 0.2. It has been estimated that the theoretical minimum drag coefficient for an automobile is 0.15 (White, 2011). Drag forces can be reduced by *drafting*, which is following closely behind a moving body to take advantage of the body in front that is blocking the airflow. Drag force reductions of up to 80% by drafting at optimal spacing have been reported (Morel and Bohn, 1980).

Rolling drag. The power supplied by an engine to move a vehicle must overcome both *rolling drag* and *aerodynamic drag*. Rolling drag is also commonly referred to as *rolling friction* and *rolling resistance*. The latter term will be used here, because the coefficient associated with this phenomenon is called the *rolling resistance coefficient*. The association of terms should aid in clarifying the relationship between these quantities. Rolling resistance is the frictional force that occurs when an object such as a tire rolls on a solid surface, and it is generally related to the deformation and types of materials in contact. The force, F_r, resisting rolling is typically calculated using the relation

$$F_r = c_r W_n \tag{10.40}$$

where c_r is the rolling resistance coefficient and W_n is the component of the supported weight normal to the surface on which rolling resistance is occurring. Typical values of the rolling resistance coefficient, c_r, are given in Table 10.3. For cars at lower speeds, typically less than 48 km/h, rolling resistance is the dominant force to be overcome, whereas at higher speeds, aerodynamic drag is more important. The stop/start variability in speed that is required in urban driving usually makes aerodynamic drag of minor consequence with regard to gas mileage in urban environments. In highway driving, aerodynamic drag has a significant effect on gas mileage.

Tractor-trailers. In contrast to cars, rolling resistance is typically the dominant form of resistance to the movement of tractor-trailers up to speeds of 88 km/h, with aerodynamic drag being the dominant form of resistance at higher speeds. At highway speeds, tractor-trailers use more than 50% of the energy produced by the engine to overcome aerodynamic drag, whereas rolling resistance consumes roughly 30% of the usable energy. The average fuel mileage of a semitruck is 2.55 kilometers per liter. The critical drag-producing regions

Table 10.2: Typical Drag Coefficients for Various Vehicles

Vehicle	Drag Coefficient	Vehicle	Drag Coefficient
Car		Bus	0.6–0.8
typical	0.25–0.35	Bicycle	
Mercedes-Benz G-class	0.53	rider in upright position	1.1
Mercedes-Benz E-class	0.27	rider in racing position	0.88
SUV, light truck	0.35–0.45	drafting	0.50
Tractor-trailer		Motorcycle and rider	1.8
without fairing	0.7–1.0	Train	1.8
with fairing	0.6–0.8		

Table 10.3: Typical Rolling Resistance Coefficients

Surfaces	Resistance Coefficient
Car tires	
on concrete	0.010–0.015
on tar or asphalt	0.03
on solid sand	0.04–0.08
on loose sand	0.2–0.4
Truck tires	
on asphalt	0.006–0.01
Bicycle tire	
on wooden track	0.001
on concrete	0.002
on asphalt	0.004
on rough paved road	0.008
Railroad steel wheels	
on steel rails	0.001–0.002

on a tractor-trailer are the trailer base, the underbody, and the gap between the tractor and trailer. A deflector is commonly used on the top of a tractor to guide the airflow over the trailer and hence reduce drag; this feature can reduce the aerodynamic drag on the tractor-trailer by about 20%. Deflectors on tractor-trailers are a particular type of *fairing*, which is defined as any type of added structure that makes a body more streamlined. Fairings are also used on other types of vehicles. For example, *nose fairings* and body panels have produced motorcycles that can travel at speeds over 320 km/h.

Wind-tunnel tests. A full-scale test of a tractor-trailer in a wind tunnel is shown in Figure 10.12, where the primary objective of the test was to measure the resultant drag force on the vehicle. The test shown in Figure 10.12 was conducted in the world's largest wind tunnel at the National Full-Scale Aerodynamics Complex (NFAC) located at Moffett Air Field in California; the width × height of the test section is 37 m×24 m. The results of wind tunnel tests such as that shown in the figure are used in the design of aerodynamic attachments to reduce drag and improve aerodynamic and fuel efficiency.

Figure 10.12: Full-scale test of a tractor-trailer in a wind tunnel
Source: Dominic Hart/Ames Research Center/NASA.

EXAMPLE 10.4

The width, height, engine power, and drag coefficient of a typical tractor-trailer (with fairing) and a typical car are given in the table below. Assume that for the most part, these vehicles operate under conditions of standard air at sea level. (a) When each vehicle is traveling at a speed of 105 km/h, what fraction of the engine power is used to overcome aerodynamic drag? (b) What is the upper limit of the vehicle speed at which the entire engine power would be necessary to overcome aerodynamic drag?

Vehicle	Width (m)	Height (m)	Engine Power (kW)	Drag Coefficient (–)
Tractor-trailer	2.59	4.12	410.13	0.60
Car (sedan)	1.83	1.46	186.42	0.32

SOLUTION

For standard air, $\rho = 1.225$ kg/m^3.

(a) The power, P, expended in overcoming the aerodynamic drag force, F_D, when traveling at a speed V is given by

$$P = F_D \cdot V = C_D \tfrac{1}{2}\rho V^2 WH \cdot V \quad \rightarrow \quad P = C_D \tfrac{1}{2}\rho V^3 WH \tag{10.41}$$

where C_D is the drag coefficient, W is the frontal width, and H is the frontal height of the vehicle. Using Equation 10.41 with $V = 105$ km/h $= 29.17$ m/s and the other given data yields the following power requirements for the tractor-trailer and the car to overcome aerodynamic drag:

$$P_{trac} = (0.60)\,\tfrac{1}{2}(1.225)(29.17)^3(2.59)(4.12) = 4.32 \times 10^4 \text{ W}$$

$$P_{car} = (0.32)\,\tfrac{1}{2}(1.225)(29.17)^3(1.83)(1.46) = 2.93 \times 10^4 \text{ W}$$

The fraction of the engine power used to overcome aerodynamic drag is derived from the given engine power of each vehicle as follows:

$$\frac{P_{trac}}{P_{engine}} = \frac{4.32 \times 10^4}{410.13 \times 10^3} = 0.105, \qquad \frac{P_{car}}{P_{engine}} = \frac{2.93 \times 10^4}{186.25 \times 10^3} = 0.157$$

Therefore, when traveling at a speed of 105 km/h, the tractor-trailer uses approximately **10.5%** of its available engine power to overcome aerodynamic drag and the car uses **15.7%** of its available power.

(b) When a vehicle uses all of its available engine power to overcome aerodynamic drag, the corresponding limiting speed, V_{limit}, is given by Equation 10.41 as

$$V_{limit} = \left[\frac{P_{engine}}{C_D \tfrac{1}{2}\rho WH}\right]^{\frac{1}{3}} \tag{10.42}$$

Applying Equation 10.42 with the given specifications for the tractor-trailer and the car yields the following results:

$$\text{tractor-trailer: } V_{\text{limit}} = \left[\frac{(550)(745.7)}{(0.60)\,\frac{1}{2}(1.225)(2.59)(4.12)}\right]^{\frac{1}{3}}$$

$$= 47.1 \text{ m/s} = 170 \text{ km/h}$$

$$\text{car: } V_{\text{limit}} = \left[\frac{(250)(745.7)}{(0.32)\,\frac{1}{2}(1.225)(1.83)(1.46)}\right]^{\frac{1}{3}}$$

$$= 70.9 \text{ m/s} = 255 \text{ km/h}$$

Therefore, the theoretical maximum speed of the tractor-trailer is **170 km/h**, and the maximum speed of the car is **255 km/h**.

10.3.4 Drag on Ships

The total drag force on a ship results from both the movement of the ship relative to the water, which causes *friction drag*, and the generation of waves resulting from ship movement, which causes *wave-making drag*. This is illustrated in Figure 10.13. The friction drag can be estimated by approximating the submerged sides of the ship as flat plates and estimating the drag by using the formulation for turbulent flow over a flat plate. Wave-making drag depends on the height and wavelength of the waves generated by a moving ship and on whether the stern (i.e., back) of the ship is in a crest or a trough of the generated wave system; generally, the wave-making drag will be less if the stern occurs at a wave crest and the wave-making drag will be greater if the stern occurs in a wave trough. A Froude number (Fr) criterion for minimizing the drag on a ship with a cruising speed V can be taken as (Inui, 1962)

$$\text{Fr} = \frac{V}{\sqrt{gL}} = \frac{0.53}{\sqrt{N}} \tag{10.43}$$

Figure 10.13: Friction and wave drag on a ship

Source: Mass Communication Specialist 1st Class John M. Hageman/U.S. Navy.

where g is gravity, L is the length of the ship, and N is the number of wavelengths (within the generated wave system) from the bow to the stern. Adding a *bulb protrusion* to the bow and stern of a ship can significantly reduce the wave-making drag on a ship.

10.3.5 Drag on Two-Dimensional Bodies

In applications where a body has a uniform cross section and is very long in one of its dimensions, such as a square building that is very tall, the body is sometimes referred to as being two-dimensional (from a drag perspective), and the flow around the body can also be approximated as being two-dimensional. Drag coefficients for two-dimensional bodies are sometimes called *section drag coefficients*, and the drag coefficients for several common two-dimensional bodies are shown in Table 10.4. Sharp-edged shapes tend to cause flow separation regardless of the boundary-layer characteristics, and their drag coefficients are typically insensitive to the Reynolds number. In contrast, the drag coefficients of cylindrical shapes are typically quite sensitive to the Reynolds number, particularly at lower Reynolds numbers. Under laminar-boundary-layer conditions, the adverse pressure gradient on the back side of a cylinder causes separation to occur at around $\theta = 82°$, with an associated pressure drag corresponding to C_D = 1.2, and under turbulent-boundary-layer conditions, separation is delayed until $\theta = 120°$, with an associated pressure drag corresponding to C_D = 0.3.

Table 10.4: Drag Coefficients of Two-Dimensional Bodies

Shape	Description	Area	C_D	Re
	Square, rounded corners	LD	R/D — C_D 0 — 2.2 0.02 — 2.0 0.17 — 1.2 0.33 — 1.0	10^5
	Semicircular shell	LD	→ 2.3 ← 1.1	2×10^4
	Semicircular cylinder	LD	→ 2.15 ← 1.15	$> 10^4$
	Cylinder	LD	0.3 (turbulent BL) 1.2 (laminar BL)	$> 2 \times 10^6$ 10^3 to 2×10^5
	Rectangle	LD	B/D — C_D < 0.1 — 1.9 0.5 — 2.5 0.65 — 2.9 1.0 — 2.2 2.0 — 1.6 3.0 — 1.3	10^5

References: Blevins (1984); Hoerner (1965).

10.3.6 Drag on Three-Dimensional Bodies

In cases where a body is three-dimensional and the flow around the body is also three dimensional, the drag coefficients for several bodies are shown in Table 10.5.

10.3.7 Drag on Composite Bodies

The shapes of many complex bodies are composed of the shapes of simpler bodies used as building blocks. Such bodies are sometimes called *composite bodies*. For example, a satellite dish mounted on a cylindrical bar can be taken as the combination of a hemispherical body and a cylinder. In treating bodies as being composed of building blocks of simpler bodies for which the drag characteristics are known, the drag forces on the components of the complex body can be calculated separately and then added to obtain the drag force on the entire body. However, this approach does not account for the effects on the flow field caused by the components being in close proximity to each other and the results should be viewed as approximate. A practical example of a composite body is illustrated in Figure 10.14, where a parachute is being used to increase the drag force and rate of deceleration of a space shuttle. In this case, the drag force can be approximated as the sum of the drag force on the shuttle and the drag force on the parachute, which can be approximated as an open hemisphere facing an oncoming flow of air. It is worth noting that the parachute is deployed so that the shuttle does not block the oncoming airflow.

Figure 10.14: Space shuttle with a deployed parachute
Source: Tony Landis/NASA.

Table 10.5: Drag Coefficients on Three-Dimensional Bodies

Shape	Description	Area	C_D	Re
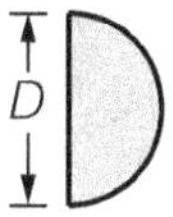	Solid hemisphere	$\frac{\pi}{4}D^2$	$\rightarrow$ 1.17 $\leftarrow$ 0.42	$> 10^4$
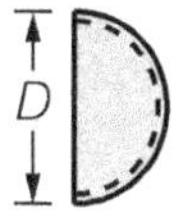	Hollow hemisphere	$\frac{\pi}{4}D^2$	$\rightarrow$ 1.42 $\leftarrow$ 0.38	$> 10^4$
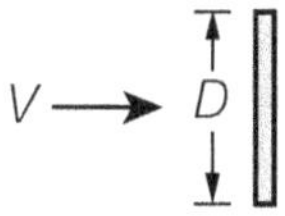	Thin disk	$\frac{\pi}{4}D^2$	1.1	$> 10^3$
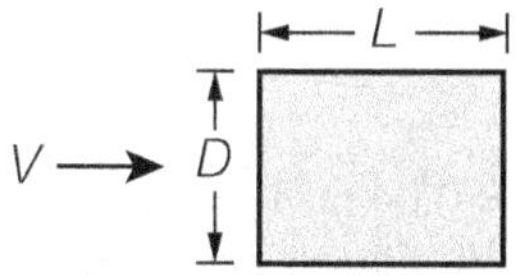	Cylinder	$\frac{\pi}{4}D^2$	L/D — C_D 0.5 — 1.1 1.0 — 0.93 2.0 — 0.83 4.0 — 0.85	$> 10^5$
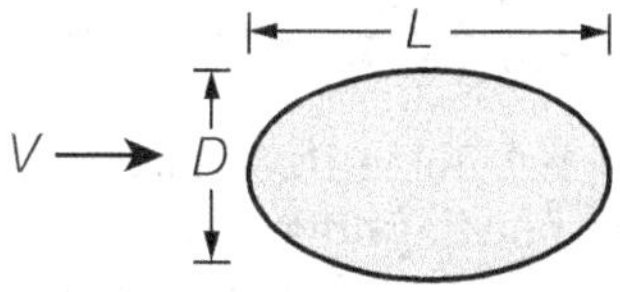	Ellipsoid	$\frac{\pi}{4}D^2$	L/D — C_D 0.75 — 0.2 1.0 — 0.2 2.0 — 0.1 4.0 — 0.1 8.0 — 0.1	$> 10^6$
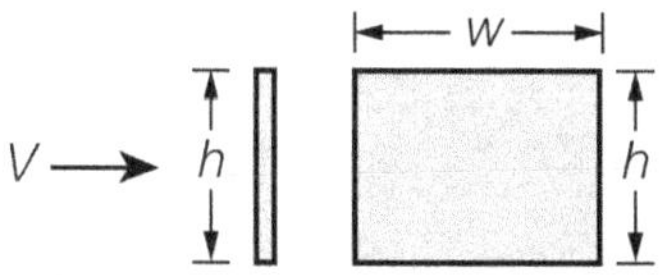	Rectanglar plate	wh	w/h — C_D 1 — 1.18 5 — 1.2 10 — 1.3 20 — 1.5 ∞ — 2.0	$> 10^4$
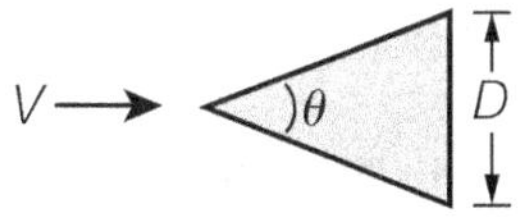	Cone	$\frac{\pi}{4}D^2$	θ — C_D 10° — 0.30 30° — 0.55 60° — 0.80 90° — 1.15	$> 10^4$

References: White (2011); Blevins (1984); Cengel and Cimbala (2014).

EXAMPLE 10.5

The 60-cm-diameter circular stop sign shown in Figure 10.15 is mounted on a 2.44-m-high cylindrical pole that has a diameter of 4 cm. Determine the drag force on the structure for a wind with velocity 22.4 m/s blowing normal to the stop sign. How does this force compare with the force exerted if the wind is blowing perpendicular to the stop sign. The ambient air temperature is 20°C.

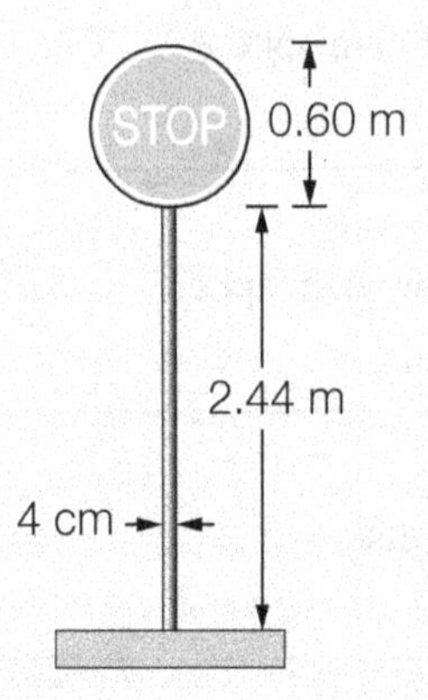

Figure 10.15: Stop sign

SOLUTION

From the given data: D_{sign} = 60 cm = 0.60 m, h = 2.44 m, D_{pole} = 4 cm = 0.04 m, V = 22.4 m/s, and T = 20°C. For air at 20°C, ρ = 1.204 kg/m^3 and $\nu = 1.512\times 10^{-5}$ m^2/s. Check the Reynolds numbers of the flow around the pole and the sign:

$$\text{Re}_{\text{pole}} = \frac{VD_{\text{pole}}}{\nu} = \frac{(22.4)(0.04)}{1.512 \times 10^{-5}} = 5.92 \times 10^4,\ \text{Re}_{\text{sign}} = \frac{VD_{\text{sign}}}{\nu} = \frac{(22.4)(0.60)}{1.512 \times 10^{-5}} = 8.89 \times 10^5$$

The pole is a vertical circular cylinder, and since $10^3 < \text{Re}_{\text{pole}} < 2 \times 10^5$, Table 10.4 indicates that there will be a laminar boundary layer around the pole with a corresponding drag coefficient of $C_{\text{D,pole}}$ = 1.2. The stop sign is a circular thin disk, and since $\text{Re}_{\text{sign}} > 10^3$, Table 10.5 gives the drag coefficient as $C_{\text{D,sign}}$ = 1.1. Using these data, for the wind blowing normal to the stop sign,

$$A_{\text{pole}} = D_{\text{pole}} h = (0.04)(2.44) = 0.0976 \text{ m}^2$$

$$A_{\text{sign}} = \frac{\pi}{4} D_{\text{sign}}^2 = \frac{\pi}{4}(0.60)^2 = 0.2827 \text{ m}^2$$

$$F_{\text{D}} = \frac{\rho V^2}{2}\left[C_{\text{D,pole}} A_{\text{pole}} + C_{\text{D,sign}} A_{\text{sign}}\right]$$

$$= \frac{(1.204)(22.4)^2}{2}\left[(1.2)(0.0976) + (1.1)(0.2827)\right] = \mathbf{129\ N}$$

For the wind blowing perpendicular to the stop sign, the drag force is only on the pole; therefore,

$$F_{\text{D}} = \frac{\rho_{\text{air}} V^2}{2}\left[C_{\text{D,pole}} A_{\text{pole}}\right] = \frac{(1.204)(22.4)^2}{2}\left[(1.2)(0.0976)\right] = \mathbf{35.4\ N}$$

It is apparent that the larger drag force is exerted when the wind is blowing normal to the stop sign.

10.3.8 Drag on Miscellaneous Bodies

The drag coefficient is widely used as a parameter to characterize the drag force on a body. For any given body, the key factors affecting the value of the drag coefficient are the shape of the body, the direction at which the fluid approaches the body, the area to be used in calculating the drag force, and the range of Reynolds numbers for which an assumed drag coefficient is valid. The commonly cited drag coefficients of various miscellaneous bodies are given in Table 10.6. These data should be regarded as approximate and only be used for preliminary estimation of drag forces. When more accurate estimates of drag forces are desired as part of engineering designs, a literature search should be done to find the latest data that is relevant to the situation and type of body being considered. This is particularly true for estimating the drag coefficients of trees, where increased wind velocities typically result in the deformation of the tree as well as an increase in the porosity of the canopy. Drag coefficients are usually measured experimentally in a wind tunnel. An example of a submarine being tested in a wind tunnel is shown in Figure 10.16, where air is being used as the test fluid with the associated approximation that the drag coefficient in air is the same as the drag coefficient in seawater. This approximation assumes that compressibility effects are negligible in the air tests. It is not uncommon for drag coefficients to be stated without a corresponding range of Reynolds numbers. In such cases, it usually can be assumed that the given drag coefficient corresponds to a Reynolds number that is sufficiently high that the drag coefficient is independent of the Reynolds number.

Table 10.6: Drag Coefficients on Miscellaneous Bodies

Shape	Description	Area	C_D	Re
V→ 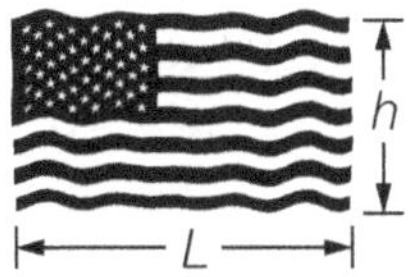	Flag or Banner	Lh	L/h C_D 1 0.07 2 0.12 3 0.15	–
V→ 	Tree (typical)	Frontal area	V (m/s) C_D 10 0.43 20 0.26 30 0.20	–
V→ 	Man (upright)	Frontal area	1.0–1.3	–

Figure 10.16: Drag on a submarine being measured in a wind tunnel
Source: NASA.

10.3.9 Added Mass

Most drag forces on bodies are estimated under steady-state conditions, which correspond to a body moving at a constant velocity through a fluid. In accordance with Newton's second law, this means that the sum of the forces acting on the body is equal to zero. In cases where a body is accelerating from rest, there must necessarily be a net force acting on the body to cause this acceleration. Normally, one would assume that this net force is equal to the mass of the body times the acceleration of the body. However, the net force not only causes the body to accelerate, but also causes some of the fluid around the body to accelerate, and this amount of fluid that accelerates accounts for an *added mass* of the body. The added mass is sometimes referred to as the *induced mass* or the *hydrodynamic mass*. Taking the added mass into account, Newton's second law can be expressed as

$$F - F_\mathrm{D} = (M_\mathrm{b} + M_\mathrm{a})\frac{\mathrm{d}V_\mathrm{b}}{\mathrm{d}t} \tag{10.44}$$

where F is the component of the force exerted on the body in the direction of motion, F_D is the drag force caused by the motion of the body relative to the fluid, M_b is the mass of the body, M_a is the added mass, and V_b is the velocity of the body. The combined mass $M_\mathrm{b} + M_\mathrm{a}$ is commonly referred to as the *virtual mass* of the body. The added mass, M_a, is often estimated as a fraction of the mass of the fluid displaced by the body, M_f, such that

$$M_\mathrm{a} = k\,M_\mathrm{f} \tag{10.45}$$

where k is called the *added mass coefficient*. Values of k can vary from $k = 1.0$ for a long cylinder moving normal to its axis, to $k = 0.5$ for a sphere, to $k = 0.2$ for an ellipse moving in the direction of its major axis, where the length of the major axis is twice the length of the minor axis. For bodies that are not completely symmetrical, the added mass depends on the direction of motion. In general, the added mass depends on the flow pattern in the surrounding fluid and the presence of boundaries. For dense bodies accelerating in air, the added mass effect is usually negligible. The effects of added mass on aircraft is typically small due to the relatively small mass of air that is displaced. However, added mass is usually important in the docking and mooring of ships.

10.4 Estimation of Lift Coefficients

Lift is defined as the net fluid force acting on a body perpendicular to the flow direction. The lift coefficient, C_L, is defined as

$$C_\text{L} = \frac{F_\text{L}}{\frac{1}{2}\rho V^2 A} \tag{10.46}$$

where F_L is the lift force, ρ is the fluid density, V is the free-stream velocity, and A is the plan area of the body, which is sometimes called the *planform area.* The nondimensional parameters affecting the lift coefficient can be identified using the same dimensional analysis used to identify the nondimensional parameters that affect the drag coefficient. In a similar manner, shape and orientation of a body relative to the oncoming flow are usually the dominant parameters, with Reynolds number, Mach number, Froude number, and relative roughness height being important parameters under various conditions. The relative importance of these parameters will be discussed in the context of individual shapes. Typically, the Reynolds number and relative roughness are not influential parameters on lift, the Mach number can be important in high-speed flows, and the Froude number can be significant when there are free-surface effects.

10.4.1 Lift on Airfoils

Airfoils are specifically designed to generate lift while keeping drag to a minimum. In the United Kingdom, airfoils are referred to as *aerofoils.* A typical airfoil is illustrated in Figure 10.17, where the key features are the *chord line* that extends from the *leading edge* to the *trailing edge* of the airfoil, the *camber line* that is midway between the *upper surface* and *lower surface* of the airfoil, and the *angle of attack*, α, that is between the chord line and the direction of the fluid velocity approaching the airfoil. In a *symmetrical airfoil*, the airfoil is symmetrical about the camber line, and in a *cambered airfoil*, the airfoil is not symmetrical about the camber line; the airfoil shown in Figure 10.17 is a cambered airfoil. Typically, the contribution of viscous effects to lift on airfoils is negligible, because airfoils are streamlined and the wall shear stress is parallel to the surface and thus nearly normal to the direction of lift. In practice, lift can be taken to be entirely due to the pressure distribution on the surface of the airfoil; thus, the shape of the airfoil has the primary effect on lift. Consequently, airfoils are designed to minimize the average pressure on the upper surface and maximize the average pressure on the lower surface. As indicated by the Bernoulli equation, the pressure is low at locations where the flow velocity is high and the pressure is high where the flow velocity is low. Therefore, minimizing the average velocity over the lower surface of the airfoil and maximizing the average velocity over the upper surface of the airfoil tends to maximize lift. Lift at moderate angles of attack is practically independent of the surface roughness because roughness affects the wall shear stress not the pressure.

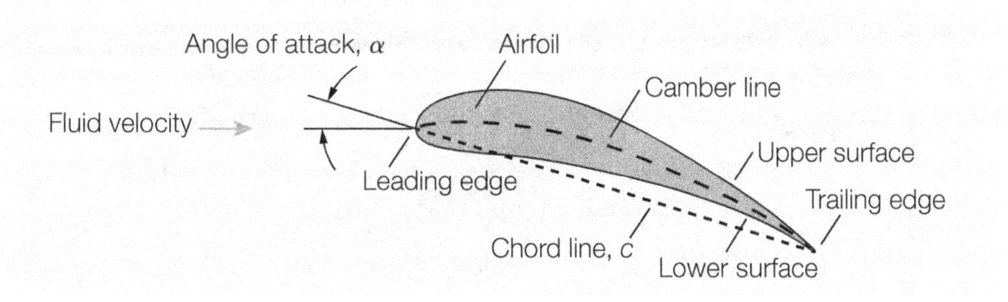

Figure 10.17: Airfoil

Shape effect. The pressure changes in the flow direction along the surface of an airfoil, but it remains essentially constant through the boundary layer in the direction normal to the surface of the airfoil. Therefore, the thin boundary layer is typically ignored in calculating the pressure distribution around an airfoil. As indicated in Figure, 10.18(a), at zero *angle of attack*, the lift produced by a symmetrical airfoil is zero due to symmetry, with stagnation points at the leading and trailing edges. Clearly, a symmetrical airfoil is useless as a structure for generating lift when the angle of attack is equal to zero; however, lift can be generated by such airfoils when the angle of attack is greater than zero. In contrast to the flow around a symmetrical airfoil, the flow around a nonsymmetrical airfoil is illustrated in Figure 10.18(b). It is clear that lift will be generated even at zero angle of attack, because the flow velocity over the top surface of the airfoil is greater than over the bottom surface. Therefore, in accordance with the Bernoulli equation, the pressure on the top surface is lower than the pressure on the bottom surface.

Flow separation. Flow separation, as illustrated in Figure 10.19, can occur over airfoils at a sufficiently high angle of attack, α. Flow separation, which is more accurately characterized as *boundary-layer separation*, is caused by the adverse pressure gradient that develops over the top surface of the airfoil, with the relative strength of the adverse pressure gradient increasing as the angle of attack, α, increases. At low angles of attack, the magnitude of the pressure gradient relative to the inertia of the flowing fluid in the boundary layer is not sufficient to cause boundary layer separation. However, boundary layer separation occurs spontaneously at high angles of attack, and complete boundary layer separation over the top surface of airfoils occurs for angles of attack in the range of 15°–20° for most airfoils. When boundary layer separation occurs, a *separation bubble* forms on the top surface toward the rear of the airfoil as shown in Figure 10.19, and this separation bubble moves toward the front of the top surface as the angle of attack increases. Flows over an airfoil for angles of attack equal to 6°, 12°, and 14° are illustrated in Figure 10.20. Boundary layer separation on the top surfaces of airfoils that are used as wings of an aircraft reduces the lift drastically and can cause the aircraft to *stall*. Stalling under cruise conditions is generally undesirable, and such

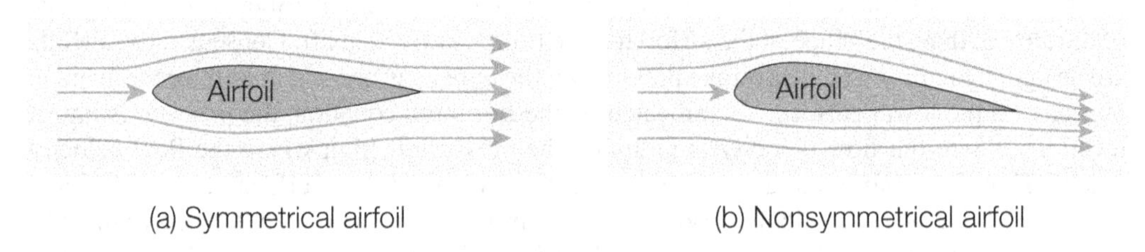

Figure 10.18: Flow past an airfoil

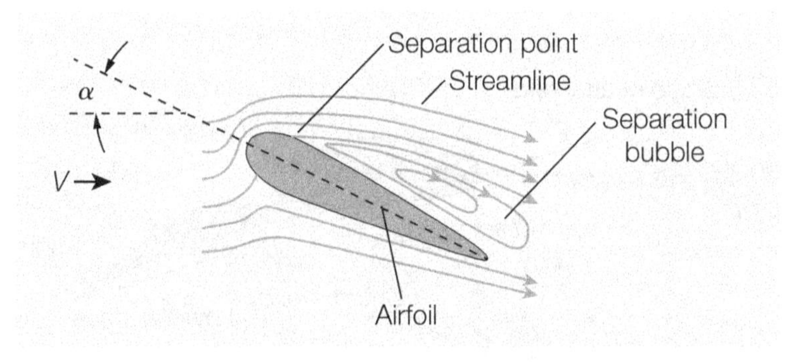

Figure 10.19: Separation over an airfoil

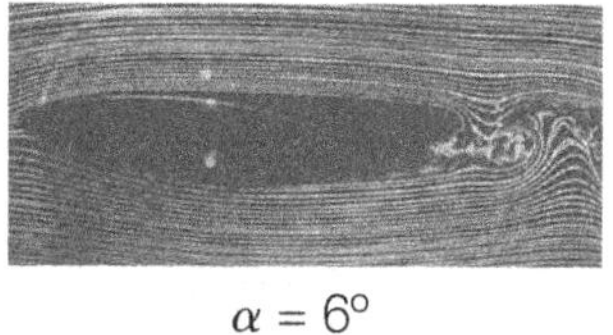

$\alpha = 6°$

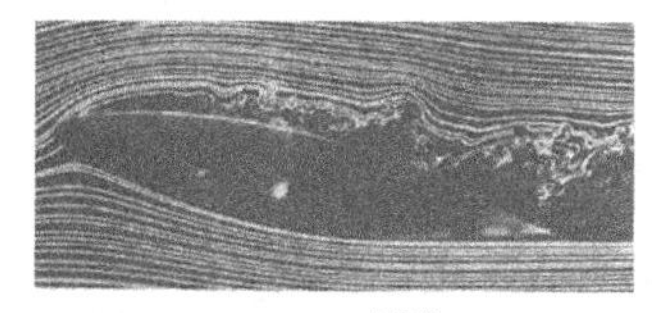

$\alpha = 12°$

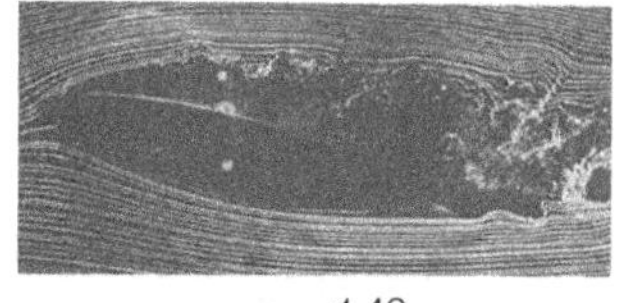

$\alpha = 14°$

Figure 10.20: Flow over an airfoil

Source: NASA.

occurrences have caused many airplane crashes. However, stall conditions are sometimes purposely created on landing aircraft to provide the necessary increased drag needed to land the aircraft. It is also commonplace for *stunt planes* to create stalling conditions purposely for some maneuvers; this works as long as the stunt plane maintains sufficient altitude to recover from the stall.

Lift/drag ratio. Airfoils should generate the most lift while producing the least drag. Therefore, a measure of the performance of airfoils is the lift-to-drag ratio, which is equivalent to the ratio of the lift and drag coefficients, $C_\mathrm{L}/C_\mathrm{D}$. Based on dimensional analysis, the lift and drag coefficients, C_L and C_D, are functions of the angle of attack, α, and the *chord Reynolds number*, $\mathrm{Re_c}$ (= Vc/ν), which can be expressed in functional terms as

$$C_\mathrm{L} = f_1(\alpha, \mathrm{Re_c}), \quad C_\mathrm{D} = f_2(\alpha, \mathrm{Re_c}), \quad \frac{C_\mathrm{L}}{C_\mathrm{D}} = f_3(\alpha, \mathrm{Re_c}) \tag{10.47}$$

where f_1, f_2, and f_3 are airfoil-specific functions. Flows over airfoils are typically turbulent, and viscous effects are typically not a major factor, in which cases lift-to-drag ratios are generally considered a function of the angle of attack, α, as illustrated in Figure 10.21(a). It is clear that the lift and drag characteristics of an airfoil can be changed with the angle of attack. On an airplane, for example, the entire plane is pitched up to increase the lift, because the wings are fixed relative to the fuselage. Variations of the lift and drag coefficients over typical airfoils are shown in Figure 10.21(b). It is clear that the lift coefficient can be increased significantly by adjusting the angle of attack; in the case shown in Figure 10.21(b), the lift coefficient goes from 0.25 to 1.25 as the angle of attack goes from 0° to 10°. The lift coefficient is directly proportional to the angle of attack, but it deviates from an approximately linear relationship as stall conditions are approached. Because conventional airfoils are not symmetric, there is usually a positive lift coefficient at zero angle of attack. A theoretical expression for the lift coefficient, C_L, that has shown good agreement with experimental observations is

$$C_\mathrm{L} = 2\pi \sin\left(\alpha + \frac{2h}{c}\right) \tag{10.48}$$

where h is *maximum camber* and c is the chord length of the airfoil. The *camber* is the distance between the chord line and the camber line, as shown in Figure 10.17, where the camber is measured perpendicular to the chord line. The drag coefficient also increases with the angle of attack; therefore, large angles of attack should be used sparingly for short periods of time for fuel efficiency. Drag coefficients on airfoils do not depend on the speed of the air passing over the airfoil provided the Mach number, Ma, is less than around 0.75. For values of Ma in the range of 0.75–1.0 the drag coefficient increases, and beyond Ma = 1, the drag coefficient again decreases. Because the airflow oscillates between subsonic and supersonic near Ma = 1, which is a very unstable situation, aircraft typically fly at Ma < 0.75 or Ma > 1.5. For subsonic commercial aircraft (including the wings and fuselage) drag coefficients are typically in the range of 0.12–0.22.

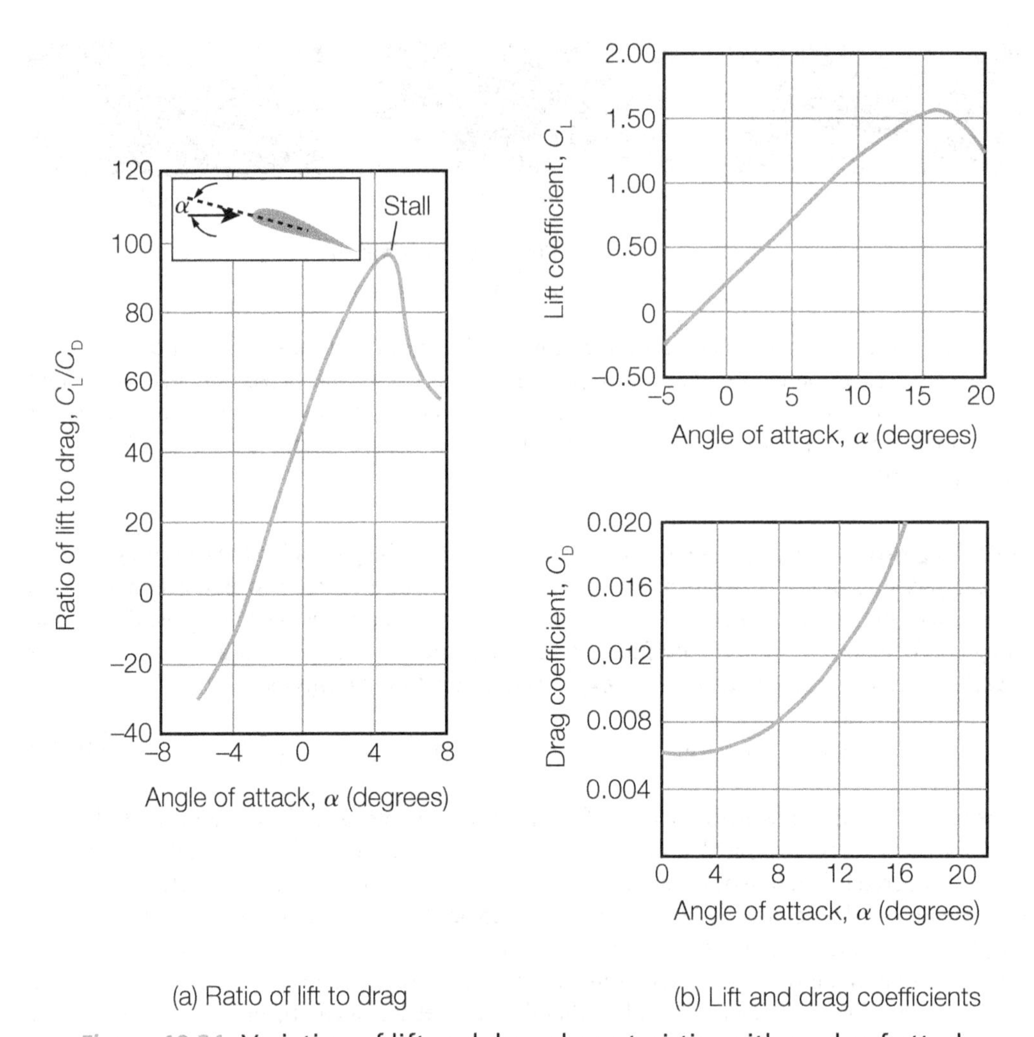

Figure 10.21: Variation of lift and drag characteristics with angle of attack

10.4.2 Lift on Airplanes

Airplanes are made up of many components, including the fuselage (i.e., main body), the wings (which have an airfoil shape), the engines, and the landing gear. The lift forces on airplanes are generated primarily by the wings, with the horizontal tail structure generating less lift. For airplanes with a wingspan of b and a *chord length* of c (= length between the leading and trailing edges of the wing), the planform area, A, to be used in calculating the lift is given by

$$\text{planform area: } A = bc$$

The distance between the two ends of the wings is called the *wingspan* or *span* and includes the width of the fuselage between the wings. The average lift per unit planform area, F_L/A, is called the *wing loading*.

Role of flaps. On takeoff and landing, airplanes generally deploy extended *flaps* in their wings/airfoils as illustrated in Figure 10.22. Flaps enable the aircraft to land and take off at low speeds. Extended flaps have two major benefits: (1) they increase the camber of the wings and (2) they increase the chord length of the wings. The increased camber increases the lift coefficient, C_L, and the increased chord length increases the planform area over which the lift forces act. These combined effects allow airplanes to take off and land at relatively low speeds.Extended flaps also increase the drag coefficient of the wings. However, the increase

(a) Extended flaps on takeoff

(b) Retracted flaps in flight

Figure 10.22: Flaps on an airplane wing

in drag during takeoff and landing of aircraft is not of much concern because of the relatively short time periods involved. Once at cruising altitude, the flaps are retracted and the wing is returned to its normal shape with minimal drag and adequate lift to keep the aircraft aloft. The increased lift with extended flaps is enhanced by leaving flow sections (slots) between the flaps as shown in Figure 10.23. Slots are used to prevent the separation of the boundary layer from the upper surface of the wings and flaps. This is done by allowing air to move from the high-pressure region under the wing to the low-pressure region on the top of the wing. This allows the lift coefficient to reach its maximum and the flight velocity to reach its minimum. Typical values of C_L that can be achieved with extended flaps are in the range of 3.0–4.0, depending on the number of slots.

Minimum flight speed. The *minimum flight speed*, V_min, which can also be viewed as the *minimum takeoff speed*, is determined by equating the total weight, W, of an aircraft to the lift force generated, in which case

$$W = F_\mathrm{L} = \frac{1}{2}C_\mathrm{L}\rho V_\mathrm{min}^2 A \quad \rightarrow \quad V_\mathrm{min} = \sqrt{\frac{2W}{\rho C_\mathrm{L} A}} \tag{10.49}$$

The minimum speed, V_min, necessary for an airplane to support its weight is called the *stall speed*, which typically varies in the range of 65–215 km/h, depending on the weight of the airplane and C_L. Usually, the pilot strives for a speed greater than $1.2V_\mathrm{min}$ to avoid instabilities that occur near the stall speed. It is also apparent from Equation 10.49 that the minimum flight speed, V_min, is inversely proportional to the square root of the density of the air, ρ. Because air density typically decreases with increasing altitude above sea level, higher flight speeds are necessary at higher altitudes, and airports at higher elevations typically have longer runways. Air density also decreases with increasing temperature, so more runway is needed on hot summer days than on cool winter days.

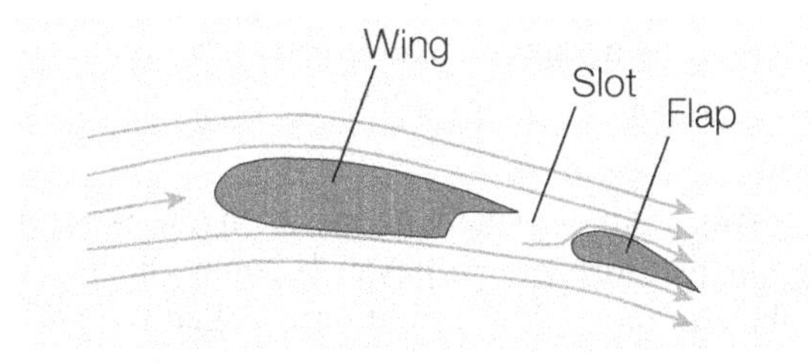

Figure 10.23: Flapped airfoil with slot

EXAMPLE 10.6

A Boeing 757-200 airplane has a maximum allowable total mass of 116 000 kg, a wingspan of 38 m, and a total wing area of 185 m^2. At takeoff, the flaps can be configured to give an effective lift coefficient of 3.5. At an air temperature of 20°C and under maximum-load conditions, at what speed will the airplane lift off?

SOLUTION

From the given data: $m = 116\,000$ kg, $W = mg = 1.14 \times 10^6$ N, $A = 185$ m^2, and $C_\text{L} = 3.5$. At 20°C, $\rho_\text{air} = 1.204$ kg/m^3. Substituting these data into Equation 10.49 gives the takeoff speed as

$$V_\text{min} = \sqrt{\frac{2W}{\rho C_\text{L} A}} = \sqrt{\frac{2(1.14 \times 10^6)}{(1.204)(3.5)(185)}} = \mathbf{54.0\ m/s}$$

Trailing vortices. The pressure difference between the lower and upper surfaces of an airfoil drives the fluid at the tips of an airfoil upward while the fluid is swept toward the back because of the forward motion of the wing through the air. This induces a swirling motion called a *tip vortex*. Distributed vortices along the wing combine with the tip vortices to form two streaks of powerful *trailing vortices* behind the tips of wings. Trailing vortices generated by large aircraft can have velocities on the order of 90 m/s, and these trailing vortices can continue to exist for a long time and for long distances (over 10 km) before they gradually disappear due to viscous dissipation. Such vortices and accompanying downdraft are strong enough to cause a small aircraft to lose control and flip over if it flies through the wake of a larger aircraft. Therefore, following a large aircraft closely, within 10 km, poses a danger for smaller aircraft. This issue is the controlling factor that governs the spacing of aircraft at takeoff. In nature, this effect is used to advantage by birds that migrate in a V-formation by utilizing the updraft generated by the bird in front. It has been demonstrated that birds in a typical flock can fly to their destination in V-formation using one-third less energy. Military jets also occasionally fly in V-formation for the same reason. Tip vortices also contribute to reduced lift and *induced drag* on the airfoil. The induced drag is of particular concern because it can be a significant fraction of the total drag on an airplane at low velocities; the induced drag is usually given careful consideration in aircraft design. End effects can be minimized by attaching *endplates* or *winglets* at the tips of the wings perpendicular to the top surface. Endplates are ubiquitous on modern aircraft, and an example of endplates on wings is shown in Figure 10.24. The endplates function by blocking some of the leakage around the wing tips, which results in a considerable reduction in the strength of the tip vortices and induced drag. It has been estimated that winglets lead to about a 5% reduction in total drag on an aircraft at cruising speed, with greater drag reductions during takeoff and landing.

Glide slope and glide ratio. An airplane without an engine is commonly called a *glider*, and airplanes with engines that lose engine power become gliders. The (downward) slope of the flight path of a plane in the gliding condition is called the *glide slope*, and the angle corresponding to the glide slope is called the *glide angle*. The glide angle is illustrated in Figure 10.25. When an aircraft glides at a constant velocity, the net force on the aircraft is

Figure 10.24: Winglets on an airplane
Source: U.S. Navy Photo.

equal to zero and the aircraft moves with a velocity of magnitude V in a direction that makes an angle θ with the horizontal, where θ is the glide angle. Because the net force on the aircraft is equal to zero, the vector sum of the lift force, $\mathbf{F}_\mathrm{L}$, and the drag force, $\mathbf{F}_\mathrm{D}$, must be equal in magnitude and opposite in direction to the weight force, $\mathbf{W}$, as illustrated in Figure 10.25. The force triangle in Figure 10.25 requires that

$$\tan\theta = \frac{F_\mathrm{D}}{F_\mathrm{L}} = \frac{C_\mathrm{D}\frac{1}{2}\rho V^2 A}{C_\mathrm{L}\frac{1}{2}\rho V^2 A} \quad \rightarrow \quad \tan\theta = \frac{C_\mathrm{D}}{C_\mathrm{L}}$$

which gives the useful result that the tangent of the glide angle is equal to the ratio of the drag coefficient, C_D, to the lift coefficient, C_L. Whereas the glide slope gives a measure of the change in altitude per unit horizontal distance traveled, the more commonly used quantity is the glide ratio, which is defined as the horizontal distance traveled per unit change in altitude under the gliding condition; hence,

$$\text{glide ratio} = \frac{1}{\tan\theta} = \frac{C_\mathrm{L}}{C_\mathrm{D}} \tag{10.50}$$

Typical glide ratios of large commercial aircraft are on the order of 10–15, with typical values for small aircraft of 5–10 and 30–60 for gliders.

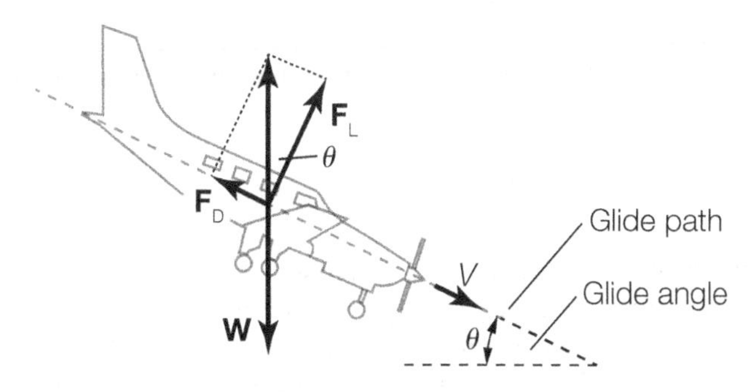

Figure 10.25: Airplane in gliding condition

EXAMPLE 10.7

A Boeing 767-200 aircraft that has an estimated glide ratio of 12 loses engine power while flying at an elevation of 3000 m. Presuming that the aircraft can glide to a safe landing, what is the maximum distance it can travel before it touches down? What would be the corresponding angle of descent?

SOLUTION

From the given data: GR $=$ 12 and $z =$ 3000 m. If x is the horizontal distance traveled corresponding to a change in altitude of 3000 m under the gliding condition, then

$$\frac{x}{z} = \text{GR} \quad \rightarrow \quad \frac{x}{3000} = 12 \quad \rightarrow \quad x = 36\,000\ \text{m} = \mathbf{36\ km}$$

The angle of descent is equal to the glide angle, θ, which is related to the glide ratio by Equation 10.50, which gives

$$\text{GR} = \frac{1}{\tan\theta} \quad \rightarrow \quad 12 = \frac{1}{\tan\theta} \quad \rightarrow \quad \theta = 4.8^\circ$$

Therefore, under the gliding condition, the flight path will make an angle of **4.8°** with the horizontal.

Aspect ratio effects. The ratio of the span of an airfoil, b, to the average width (or chord length) of the airfoil, $\bar{c}$, is defined as the *aspect ratio*, AR. Because the area of the airfoil, A, is equal to $b\bar{c}$, the aspect ratio can be expressed in either of the following forms:

$$\text{AR} = \frac{b}{\bar{c}} \quad \text{or} \quad \text{AR} = \frac{b^2}{A} \tag{10.51}$$

The aspect ratio accounts for the finite span of an airfoil and is a measure of the narrowness of the airfoil in the flow direction. Theoretical expressions for the lift and drag coefficients for airfoils of finite span are

$$C_\text{L} = \frac{C_{\text{L},\infty}}{1 + 2/\text{AR}} \tag{10.52}$$

$$C_\text{D} = C_{\text{D},\infty} + \frac{C_\text{L}^2}{\pi(\text{AR})} \tag{10.53}$$

where $C_{\text{L},\infty}$ and $C_{\text{D},\infty}$ are the lift and drag coefficients, respectively, for an airfoil with the same cross-sectional shape and chord length but with infinite span. The value of $C_{\text{L},\infty}$ can be estimated using Equation 10.48. It is apparent from Equations 10.52 and 10.53 that the lift coefficients of wings increase whereas drag coefficients decrease with increasing aspect ratio (AR); this is obviously desirable. For an entire aircraft, the drag coefficient, C_D, can be estimated by

$$C_\text{D} = C_{\text{D},0} + \frac{C_\text{L}^2}{\pi(\text{AR})} \tag{10.54}$$

where $C_{\text{D},0}$ is the zero-lift drag coefficient. Equation 10.54 is particularly useful in cases where aircraft specifications give values of $C_{\text{D},0}$ and AR and the relationship between C_D and C_L must be estimated. For any given lift coefficient, C_L, the inverse relationship between the drag coefficient, C_D, and the aspect ratio, AR, occurs because a long, narrow wing (large aspect ratio) has a shorter tip length and thus smaller tip losses and smaller induced drag than a short, wide wing of the same planform area. The effective aspect ratio of a wing of given shape can be increased by adding an endplate or a winglet to the wing tip. Bodies with large aspect ratios fly more efficiently but are less maneuverable because of their larger moment of inertia (owing to the greater distance from the center). Bodies with smaller aspect ratios maneuver better because the wings are closer to the central part of the aircraft. As a consequence, fighter planes have short and wide wings and larger commercial aircraft and cargo planes have long and narrow wings.

EXAMPLE 10.8

The goal is to size the wing of a large military transport plane such that it has an aspect ratio of 7.0. The wing must be capable of generating a lift of 500 kN when the aircraft is at a takeoff speed of 40 m/s under standard atmospheric conditions. Laboratory testing of the wing geometry indicates that the wing has a lift coefficient of 0.9. Estimate the required dimensions of the wing.

SOLUTION

From the given data: AR = 7.0, F_L = 500 kN, V = 40 m/s, and C_L = 0.9. For standard air, $\rho = 1.225$ kg/m^3. Based on the definition of C_L, the lift force, F_L, is given by

$$F_\text{L} = C_\text{L}\frac{1}{2}\rho V^2 A \quad \rightarrow \quad 500 \times 10^3 = (0.9)\frac{1}{2}(1.225)(40)^2 A \quad \rightarrow \quad A = 567\ \text{m}^2$$

Because the required aspect ratio, AR, is equal to 7.0, the length, b, and average width, $\bar{c}$, are given by

$$b = \sqrt{\text{AR} \times A} = \sqrt{7.0 \times 567} = 63.0\ \text{m}, \qquad \bar{c} = \frac{A}{b} = \frac{567}{63.0} = 9.0\ \text{m}$$

Therefore, the required wing dimensions are **63.0 m × 9.0 m**. With two wings of these dimensions, the wingspan of this aircraft is around 126 m. Other aspect ratios might need to be considered for reasons of practicality.

10.4.3 Lift on Hydrofoils

Hydrofoils are like airfoils, except that the fluid flowing past the hydrofoil is a liquid, which is usually water. Because the density of a liquid is generally much greater than that of a gas, hydrofoils are capable of generating much greater lift forces than airfoils moving at the same speed. The typical function of a hydrofoil is to lift a vessel out of the water. Many of the features of hydrofoils are similar to those of airfoils; however, cavitation generated by hydrofoils can have a significant effect on the lift and drag coefficients of hydrofoils. Surface waves generated by hydrofoils can also affect the lift characteristics of a hydrofoil, and in such cases, the lift coefficient can be dependent on the Froude number.

Effect of cavitation on the lift coefficient. You might recall that cavitation occurs when the pressure in the liquid is less than or equal to the saturation vapor pressure of the liquid at the temperature of the liquid. The tendency of a liquid to cavitate is measured by the *cavitation number*, σ, defined as

$$\sigma = \frac{p_\infty - p_\text{v}}{\frac{1}{2}\rho V^2} \tag{10.55}$$

Table 10.7: Drag, Lift, and Cavitation Characteristics of a Typical Hydrofoil

α (degrees)	C_L	C_D	σ_{crit}
−2	0.2	0.014	0.5
0	0.4	0.014	0.6
2	0.6	0.015	0.7
4	0.8	0.018	0.8
6	0.95	0.022	1.2
8	1.10	0.032	1.8
10	1.22	0.042	2.5

where p_∞ is the free-stream pressure, p_v is the saturation vapor pressure, ρ is the density of the liquid, and V is the velocity at which the hydrofoil is moving through the liquid. Both p_∞ and p_v must be expressed in the same pressure units, either as absolute pressures or as gauge pressures. In cases where the cavitation number is less than an experimentally determined *critical cavitation number*, σ_{crit}, cavitation occurs and can have a significant influence on the lift and drag coefficients of a hydrofoil. The drag coefficient, C_D, lift coefficient, C_L, and critical cavitation number, σ_{crit}, all depend on the angle of attack, α, and a typical relationship between these variables is shown in Table 10.7. The parameters given in the table are for a typical hydrofoil where $10^5 < \text{Re} < 10^6$, where the Reynolds number, Re, is based on the chord length of the hydrofoil. The area used in the calculation of the drag and lift forces are based on the plan area, which is equal to the chord length multiplied by the length of the hydrofoil.

10.4.4 Lift on a Spinning Sphere in Uniform Flow

Lift is generated when a sphere spins while moving at a steady velocity through a fluid; this is called the *Magnus effect.* The lift generated by a spinning sphere is in contrast to the zero lift generated when the sphere is not spinning. The lift and drag coefficients, C_L and C_D, respectively, on a smooth sphere rotating at an angular speed of ω while moving at a linear speed V through a fluid are illustrated in Figure 10.26 for Reynolds numbers in the range of 6×10^4–10^5. The key parameter affecting C_L and C_D is the ratio of the peripheral speed

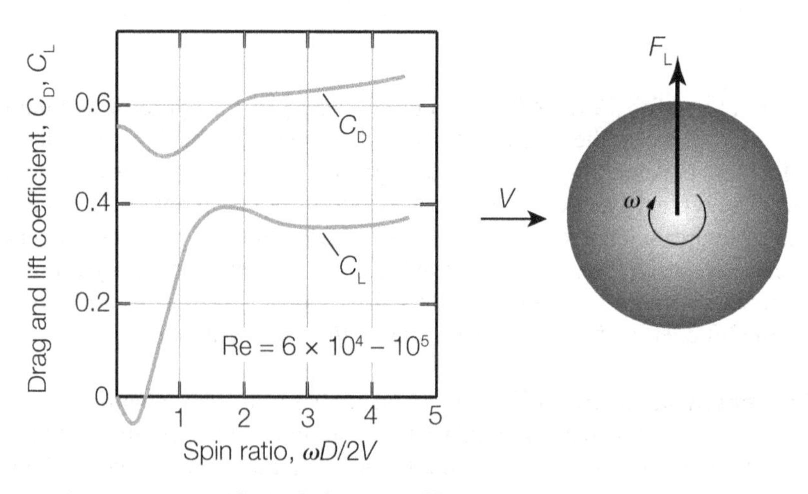

Figure 10.26: Lift and drag coefficients on a smooth sphere rotating in a uniform flow

Source: Pritchard (2011).

of the rotating sphere, $\omega D/2$, to the translation speed, V, with this ratio denoted by $\omega D/2V$ and commonly called the *spin ratio* or *spin factor*. The Reynolds number typically only has a secondary influence on C_L and C_D in cases where the drag is mostly due to pressure differences caused by the formation of a wake behind the sphere; such effects typically occur for Re $> 10^3$. The sign convention used for estimating the lift force is also shown in Figure 10.26. Using this convention, if the peripheral speed on the top of the sphere is in the same direction as the approach velocity V, then a positive lift force is upward. If the sphere shown in Figure 10.26 was a ball moving with velocity V in still air, then the rotation shown would be called "backspin" and rotation in the opposite direction would be called "topspin." Therefore, backspin would cause the ball to move upward and topspin would cause the ball to move downward. It is apparent from Figure 10.26 that for $\omega D/2V \lesssim 0.5$, the lift coefficient is negative and for $\omega D/2V \gtrsim 0.5$, the lift coefficient is positive. The asymptotic lift coefficient for high values of $\omega D/2V$ is around 0.35. In comparison to the significant effect that spin has on the lift coefficient, the effect of spin on the drag coefficient is more modest, with C_D in the range of 0.5–0.65 for the range of spin ratios shown in Figure 10.26.

EXAMPLE 10.9

A standard soccer ball has a mass of approximately 425 g and a diameter of approximately 22.5 cm. The lift coefficient as a function of the spin ratio for a soccer ball can be estimated using the relationship shown in Figure 10.27. A study of a famous free kick made by David Beckham of England reported that the average velocity of the kicked ball was 27.5 m/s and that its average rotational speed was 63 rad/s ($\approx$ 600 rpm). (a) Calculate the spin ratio. (b) Estimate the radius of curvature of the ball trajectory if the imparted spin was sidespin. Assume standard air at sea level.

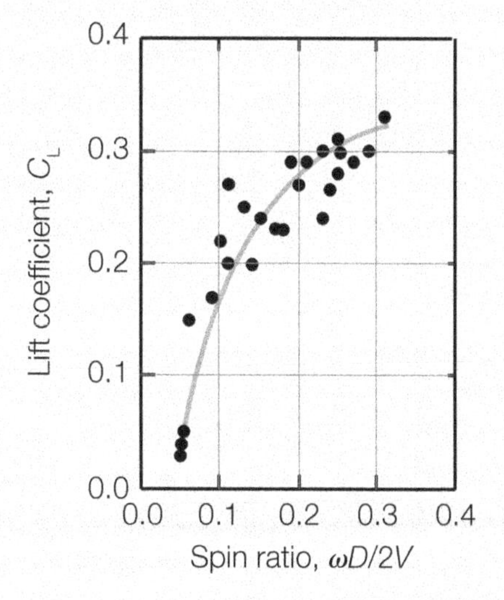

Figure 10.27: Lift coefficient on a soccer ball

Source of Data: Goff et al. (2010).

SOLUTION

From the given data: $m = 0.425$ kg, $D = 0.225$ m, $A = \pi D^2/4 = 3.976 \times 10^{-2}$ m^2, $V =$ 27.5 m/s, and $\omega = 63$ rad/s. For standard air at sea level, $\rho = 1.225$ kg/m^3.

(a) Using the given data, the spin ratio is calculated as follows:

$$\frac{\omega D}{2V} = \frac{(63)(0.225)}{2(27.5)} = \mathbf{0.258}$$

For $\omega D/2V \approx 0.26$, Figure 10.26 gives an approximate lift coefficient of $C_\mathrm{L} = 0.32$.

(b) For sidespin, there is no gravity effect in the plane of the sidespin, so motion in the lateral direction can be approximated by the relation

$$F_\mathrm{L} = m\frac{V^2}{R} \quad \to \quad C_\mathrm{L}\tfrac{1}{2}\rho V^2 A = m\frac{V^2}{R} \quad \to$$

$$(0.32)\tfrac{1}{2}(1.225)(27.5)^2(3.976 \times 10^{-2}) = (0.425)\frac{27.5^2}{R}$$

which gives $R = 54.5$ m. Hence, the radius of curvature of the ball is approximately **55 m**. This is a fairly impressive radius of curvature, given that a ball with this trajectory will deviate by about 3.8 m from a straight line trajectory over a distance of 20 m.

Key Equations for Drag and Lift

The following list of equations is particularly useful in solving problems related to drag and lift. If one is able to recognize these equations and recall their appropriate use, then the learning objectives of this chapter have been met to a significant degree. Derivations of these equations, definitions of the variables, and detailed examples of usage can be found in the main text.

FUNDAMENTALS

Drag force:
$$F_{\rm D} = \int_{A_{\rm b}} p\cos\theta\,{\rm d}A + \int_{A_{\rm b}} \tau_{\rm w}\sin\theta\,{\rm d}A$$

Lift force:
$$F_{\rm L} = -\int_{A_{\rm b}} p\sin\theta\,{\rm d}A + \int_{A_{\rm b}} \tau_{\rm w}\cos\theta\,{\rm d}A$$

Boundary-layer thickness:
$$\frac{\delta}{x} = \begin{cases} \dfrac{4.91}{{\rm Re}_x^{\frac{1}{2}}}, & \text{for} \quad {\rm Re}_x < 5\times10^5 \text{ (laminar flow)} \\[2ex] \dfrac{0.382}{{\rm Re}_x^{\frac{1}{5}}}, & \text{for} \quad 5\times10^5 \leq {\rm Re}_x < 10^7 \text{ (turbulent flow)} \end{cases}$$

Vortex frequency:
$$\frac{\omega D}{V} \approx 0.20\left(1 - \frac{20}{\rm Re}\right)$$

ESTIMATION OF DRAG COEFFICIENTS

Drag force on flat surface:
$$\frac{F_{\rm D}}{\frac{1}{2}\rho V^2 LW} = C_{\rm Df}\left({\rm Re}_{\rm L}, \frac{\epsilon}{L}\right)$$

Thickness of viscous sublayer:
$$\delta_{\rm v} = 5\frac{\nu}{u_*}$$

Drag coefficient, flat surface (smooth):
$$C_{\rm Df} = \begin{cases} \dfrac{1.328}{{\rm Re}_{\rm L}^{\frac{1}{2}}}, & {\rm Re}_{\rm L} \leq 5\times10^5 \\[2ex] \dfrac{0.0735}{{\rm Re}_{\rm L}^{\frac{1}{5}}} - \dfrac{1742}{{\rm Re}_{\rm L}}, & 5\times10^5 < {\rm Re}_{\rm L} < 10^7 \\[2ex] \dfrac{0.455}{(\log {\rm Re}_{\rm L})^{2.58}} - \dfrac{1742}{{\rm Re}_{\rm L}}, & 10^7 \leq {\rm Re}_{\rm L} \leq 10^9 \end{cases}$$

Drag coefficient, flat surface (rough):
$$C_{\rm Df} = \left(1.89 - 1.62\log\frac{\epsilon}{L}\right)^{-2.5}$$

Drag coefficient, sphere:
$$C_{\rm D} = \frac{24}{\rm Re}, \qquad {\rm Re} \leq 0.1$$

$$C_\mathrm{D} = \frac{24}{\mathrm{Re}}\left(1 + \frac{3}{16}\mathrm{Re}\right), \qquad \mathrm{Re} \le 1$$

$$C_\mathrm{D} = \frac{7.096 \times 10^6 + 3.197 \times 10^6\,\mathrm{Re} + 2.611 \times 10^5\,\mathrm{Re}^2}{(2.957 \times 10^5 + 7.776 \times 10^4\,\mathrm{Re} + 689\,\mathrm{Re}^2)\mathrm{Re}}, \qquad 0.1 \le \mathrm{Re} \le 10$$

$$C_\mathrm{D} = \begin{cases} \dfrac{5.948 \times 10^5 + 7.735 \times 10^4\,\mathrm{Re} + 398.1\,\mathrm{Re}^2}{(2.230 \times 10^4 + 928.3\,\mathrm{Re} + 0.01675\,\mathrm{Re}^2)\mathrm{Re}}, & 5.33 \le \mathrm{Re} \le 1.18 \times 10^5 \\ 0.2, & \mathrm{Re} > 1.18 \times 10^5 \end{cases}$$

Terminal velocity (general): $$V = \sqrt{\frac{2V_\mathrm{b}(\gamma_\mathrm{b} - \gamma_\mathrm{f})}{C_\mathrm{D}\rho_\mathrm{f}A_\mathrm{b}}}$$

Terminal velocity (sphere): $$V = \begin{cases} \sqrt{\dfrac{4(\gamma_\mathrm{b} - \gamma_\mathrm{f})D}{3C_\mathrm{D}\rho_\mathrm{f}}}, & \text{any Re} \\ \dfrac{(\gamma_\mathrm{b} - \gamma_\mathrm{f})D^2}{18\mu}, & \mathrm{Re} \le 0.1 \end{cases}$$

Rolling resistance: $$F_\mathrm{r} = c_\mathrm{r}W_\mathrm{n}$$

Minimize drag on ship: $$\mathrm{Fr} = \frac{V}{\sqrt{gL}} = \frac{0.53}{\sqrt{N}}$$

Added mass: $$F - F_\mathrm{D} = (M_\mathrm{b} + M_\mathrm{a})\frac{\mathrm{d}V_\mathrm{b}}{dt}$$

$$M_\mathrm{a} = k\,M_\mathrm{f}$$

ESTIMATION OF LIFT COEFFICIENTS

Lift coefficient: $$C_\mathrm{L} = \frac{F_\mathrm{L}}{\frac{1}{2}\rho V^2 A}$$

Lift coefficient, airfoil: $$C_\mathrm{L} = 2\pi\sin\left(\alpha + \frac{2h}{c}\right)$$

Minimum flight speed: $$V_\mathrm{min} = \sqrt{\frac{2W}{\rho C_\mathrm{L} A}}$$

Glide ratio: $$\mathrm{GR} = \frac{1}{\tan\theta} = \frac{C_\mathrm{L}}{C_\mathrm{D}}$$

Aspect ratio: $$\mathrm{AR} = \frac{b}{\bar{c}}, \qquad \mathrm{AR} = \frac{b^2}{A}$$

Airfoil, finite span: $$C_{\mathrm{L}} = \frac{C_{\mathrm{L},\infty}}{1 + 2/\mathrm{AR}}, \quad C_{\mathrm{D}} = C_{\mathrm{D},\infty} + \frac{C_{\mathrm{L}}^2}{\pi(\mathrm{AR})},$$

$$C_{\mathrm{D}} = C_{\mathrm{D},0} + \frac{C_{\mathrm{L}}^2}{\pi(\mathrm{AR})}$$

Cavitation number (hydrofoils): $$\sigma = \frac{p_\infty - p_{\mathrm{v}}}{\frac{1}{2}\rho V^2}$$

Applications of the above equations are provided in the text, and additional problems to practice using these equations can be found in the following section.

PROBLEMS

Section 10.2: Fundamentals

10.1. A car manufacturer tests a full-scale car in a wind tunnel and uses a dynamometer to measure the drag force. The wind tunnel uses air at a temperature of 15°C and a pressure of 101.3 kPa, and the airspeed in the tunnel is 95 km/h. The car has a frontal area of 3.01 m^2, and the measured drag force is 310 N. What is the drag coefficient of the car? If a properly scaled model of the car with a length scale ratio L_r was tested instead of the full-scale version, what drag coefficient would you expect for the model?

10.2. Wind at a speed of 50 km/h blows toward a square column that is 10.5 m tall and has a 3.01 m $\times$ 3.01 m cross section as shown in Figure 10.28. The wind makes an angle of 50° with the front face of the column, and the estimated net aerodynamic force on the column is 1.40 kN at an angle of 30° to the front face. Drag and horizontal lift coefficients are to be estimated based on a frontal area that is equal to the area of the column projected on a plane normal to the wind. The horizontal lift coefficient is based on the force in the horizontal plane that is normal to the wind direction. (a) Estimate the drag coefficient. (b) Estimate the horizontal lift coefficient. (c) Estimate the aerodynamic force on the column if the wind speed is doubled. Assume that the wind has the properties of standard air.

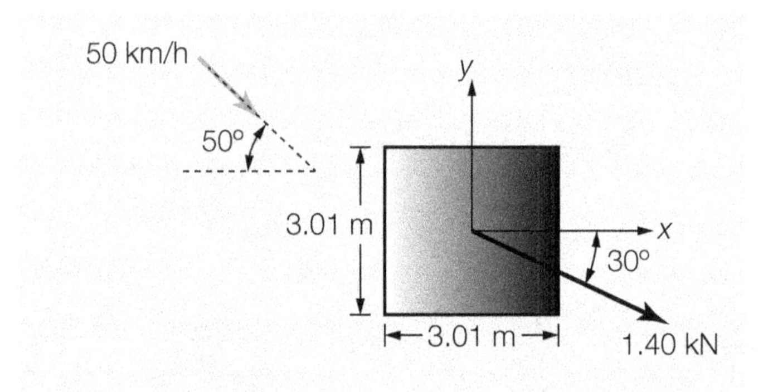

Figure 10.28: Aerodynamic force on a column

10.3. An architect is designing a small building that is to be cylindrical in shape with a diameter of 15 m and a height of 40 m. Data found in the technical literature indicate that the drag coefficient is expected to be 0.9 or 0.25 when the flow around the building is laminar or turbulent, respectively. The breakpoint between laminar and turbulent flow is expected at a Reynolds number of 2×10^5, where the Reynolds number is based on the diameter of the building. Ambient air temperature can be taken as 25°C. (a) If a 1:15 scale model of the building is to be tested in a wind tunnel to measure the exact drag coefficients under laminar and turbulent flow conditions, what ranges of flow velocities should be used in the model for each flow condition? (b) State the relationship between the drag coefficient in the model and the drag coefficient in the prototype under both laminar and turbulent flow conditions. (c) Estimate the forces of the building for wind speeds of 0.3, 2.5, and 30 m/s.

10.4. The Chevrolet Camaro shown in Figure 10.29 has a mass of 2500 kg, a drag coefficient of 0.25, and a frontal area of 1.2 m^2. The car deploys a 2.2-m-diameter parachute to slow down from an initial velocity of 110 m/s, and the drag coefficient of the parachute is 1.5. Assuming C_D is constant, all brakes are off, and there is no rolling resistance, calculate the velocity of the car 2 minutes after the parachute deploys.

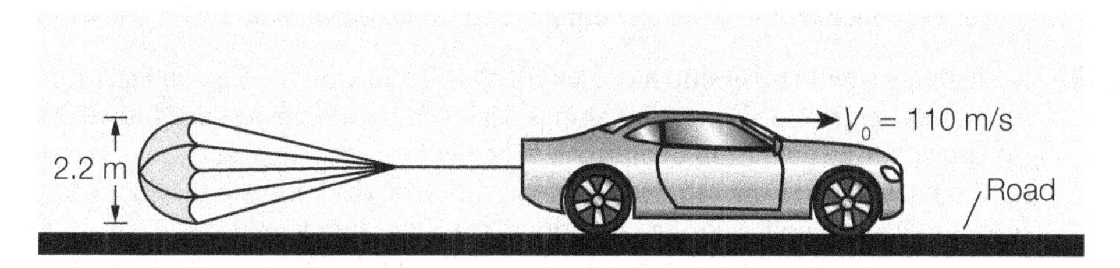

Figure 10.29: Car with a parachute

10.5. Firing tests on a 10-mm-diameter bullet with a mass of 20 g show that in standard air, the bullet speed decreases from 300 m/s to 180 m/s over a distance of 200 m. Estimate the average drag coefficient of the bullet. Neglect compressibility effects.

10.6. A 80-kg skydiver jumps from a height of 3 km and maintains an orientation that has a drag area of 0.8 m^2. The drag area is defined as the product of the drag coefficient and the frontal area. Estimate the time and distance traveled before the skydiver attains a velocity equal to 90% of his terminal velocity.

10.7. Consider the general case in which a body of mass m and frontal area A free-falls in an environment where the effective gravity is g', the density is ρ, and the drag coefficient of the body is constant and equal to C_D. Show that the time interval, Δt, for the velocity to change from V_1 to V_2 is given by

$$\Delta t = \frac{m}{2\sqrt{ab}}\left[\ln\left|\frac{\sqrt{a}+\sqrt{b}V}{\sqrt{a}-\sqrt{b}V}\right|\right]_{V_1}^{V_2}, \qquad \text{where} \quad a = mg' \quad \text{and} \quad b = \tfrac{1}{2}\rho A C_D$$

Explain how this formula could be used to calculate the time it takes a body dropped in the atmosphere to attain a speed equal to 90% of its terminal speed.

10.8. Consider the general case in which a body of mass m and frontal area A free-falls in an environment where the effective gravity is g', the density is ρ, and the drag coefficient of the body is constant and equal to C_D. Show that the distance, Δz, for the velocity to change from V_1 to V_2 is given by

$$\Delta z = \frac{m}{2b}\left[\ln\left|(a/b) - V^2\right|\right]_{V_2}^{V_1}, \qquad \text{where} \quad a = mg' \quad \text{and} \quad b = \tfrac{1}{2}\rho A C_D$$

Explain how this formula could be used to calculate the distance a body falls from a release point until it attains a speed equal to 90% of its terminal speed.

Section 10.3: Estimation of Drag Coefficients

10.9. A plastic boat whose bottom surface can be approximated as a 1.6-m-wide by 2.2-m-long flat surface is to move through water at speeds up to 32 km/h. The temperature of the water under design conditions is 15°C. Determine the friction drag exerted on the boat by the water and the power needed to overcome it. How do your calculations relate to estimating the size of the motor required to drive the boat?

10.10. A supertanker has a length of 350 m, a width of 70 m, and a draft of 25 m. Estimate the power required to overcome friction drag at a speed of 823 m/s. Assume seawater at 10°C.

10.11. A fish that has a relatively flat shape (like a flounder) is 160 mm long, is 110 mm wide, and typically swims at a speed of around 0.35 m/s. The fish is characterized as having a smooth surface. Estimate the drag force the fish must overcome and the power expended by the fish in swimming. Assume seawater at 20°C.

10.12. A relatively small cruise ship has a length of 100 m, a draft of 5 m, and an estimated roughness height of 0.1 mm. The ship is designed for a cruising speed of 10.30 m/s. (a) Under design conditions, determine whether the submerged surface of the ship is (hydrodynamically) smooth, rough, or transitional. (b) Estimate the frictional drag force on the ship under design conditions. (c) If the submerged surface was to vary between the smooth and rough surface regime, what would be the corresponding range of the drag force? Assume seawater at 20°C.

10.13. The shear stress, τ_w, on a flat surface that is caused by a fluid of density ρ and viscosity μ flowing over the surface at a velocity V is given by

$$\tau_\mathrm{w} = 0.332\frac{\rho V^2}{\mathrm{Re}_x^{\frac{1}{2}}}, \quad \text{where} \quad \mathrm{Re}_x = \frac{\rho V x}{\mu}$$

where x is the distance from the upstream end of the flat surface. (a) Use the given shear stress distribution, $\tau_\mathrm{w}(x)$, to determine the drag force on a flat plate of width W and length L in terms of W, L, V, ρ, and μ. (b) Use the result in part (a) to determine the drag coefficient on the flat plate as a function of Re_L, where $\mathrm{Re}_\mathrm{L} = \rho VL/\mu$. (c) If the fluid is standard air flowing at a velocity of 25 m/s over a flat plate that is 12 m long and 6 m wide, what is the drag force on the flat plate?

10.14. Standard air flows at a velocity of 25 m/s over a flat surface that is 16 m long and 6 m wide. (a) Estimate the drag force on the surface if the surface is hydrodynamically smooth. (b) If weathering causes the roughness height to increase over time, at what roughness height would the surface be classified as being hydrodynamically rough? (c) Estimate the drag force on the surface when the surface just becomes hydrodynamically rough.

10.15. A concrete canoe is 6.096 m long, is 0.762 m wide, and has a depth of 0.305 m below water. The canoe is intended to achieve a speed of 13.4 m/s under racing conditions. Some fluid mechanics-savvy students have postulated that the drag force, F, on the canoe can be expressed in the functional form

$$F = f(\rho, V, L, \mu, g) \tag{10.56}$$

where ρ and μ are the density and viscosity of water, respectively, V is the velocity of the canoe, L is the length of the canoe, and g is gravity. (a) Perform a dimensional analysis of Equation 10.56 to express this functional relationship in terms of dimensionless groups. (b) Laboratory experiments on a 0.914-m-long model of the canoe are to be performed to determine the drag force on the full-size canoe under design conditions. Determine the model speed and the ratio of the drag force in the model to the drag force in the prototype if Reynolds similarity is assumed. (c) Repeat part (b) for the case in which the model is designed for Froude similarity. Discuss which scaling law is more appropriate and why. (d) Instead of doing model tests, some more fluid mechanics-savvy students recommend a theoretical analysis that assumes sides of the canoe be approximated as smooth flat plates. Using this assumption, what is the estimated drag force on the (full-size) canoe under design conditions? Normally, concrete has a roughness height of 0.03 mm. Would this significantly affect your estimate of the drag force? If so, what is your revised estimate of the drag force on the canoe?

10.16. A sheet of solid material has a thickness of 11 mm, a length of 4.2 m, a width of 1.9 m, and a density of 1920 kg/m^3. The sheet is dropped in a deep pool of water at 20°C, and it sinks and ultimately attains its terminal velocity. (a) What is the terminal velocity if the sheet sinks in the direction of its length? (b) What is the terminal velocity if the sheet sinks in the direction of its width?

10.17. Wind blows uniformly over a triangular surface as shown in Figure 10.30. The wind speed is 1.2 m/s, and the wind blows perpendicular to the 3-m-long side of the right-angled triangle. The wind blows in the same direction as the 1-m-long side of the triangle. The air properties can be approximated as those of standard air. (a) Estimate the drag force on the triangular surface. (b) Estimate the (friction) drag coefficient of the surface.

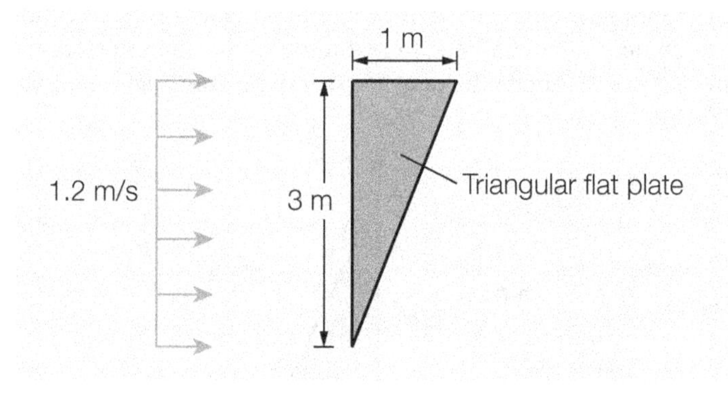

Figure 10.30: Flow over a triangular surface I

10.18. Standard air flows at a speed of 15 m/s over the flat triangular surface shown in Figure 10.31, where the coordinate units are in meters. The wind and surface conditions are such that the boundary layer can be assumed to be in the smooth-turbulent regime over the entire surface. Estimate the drag force and the drag coefficient on the surface.

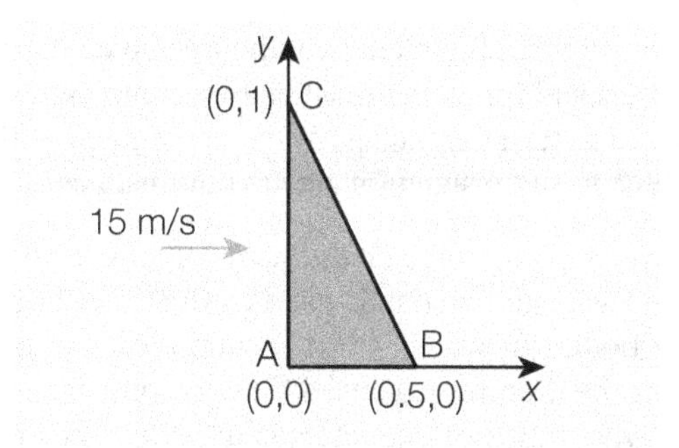

Figure 10.31: Flow over a triangular surface II

10.19. Standard air flows at 16 m/s over the flat triangular surface shown in Figure 10.32, where the coordinate units are in meters. The wind and surface conditions are such that the boundary layer can be assumed to be in the smooth-turbulent regime over the entire surface. Estimate the drag force and the drag coefficient on the surface.

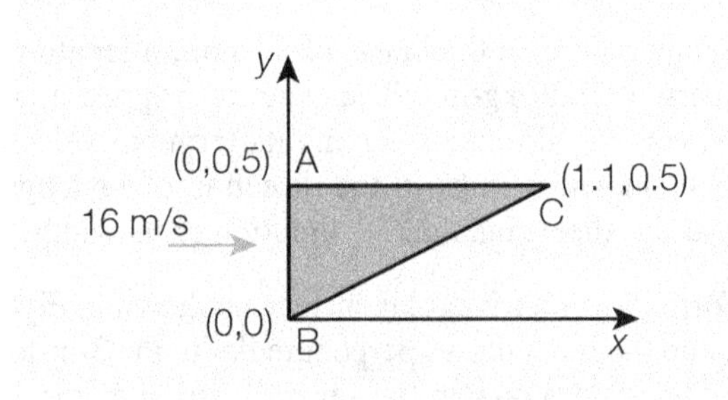

Figure 10.32: Flow over a triangular surface III

10.20. Standard air flows at 15 m/s over the flat triangular surface shown in Figure 10.33, where the coordinate units are in meters. The wind and surface conditions are such that the boundary layer can be assumed to be in the smooth-turbulent regime over the entire surface. Estimate the drag force and the drag coefficient on the surface.

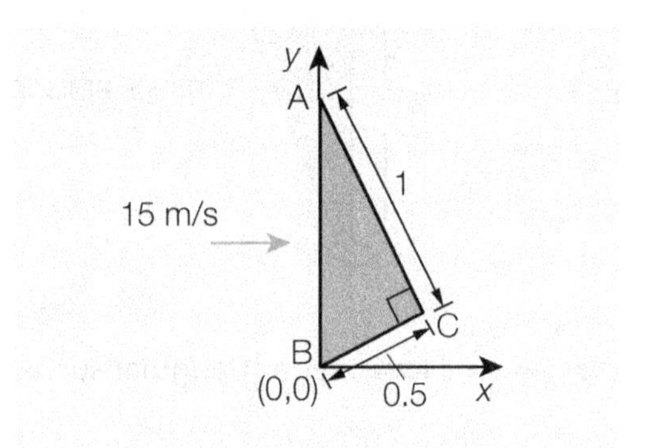

Figure 10.33: Flow over a triangular surface IV

10.21. A small military helicopter has four blades with a rotor diameter of 10.70 m, and each blade has a width of 0.73 m. When the helicopter is flying at normal speed, the blades rotate at 360 rpm. The standard engine on the helicopter has a rated power of 606 kW. Assuming that the blades can be treated as flat plates for the purpose of estimating the frictional force that must be overcome in turning the blades, estimate the power required to turn the blades. What percentage of the engine power is used to turn the blades? Assume standard air in your analysis.

10.22. Blowing wind exerts a drag force on the stems of terrestrial vegetation, and similarly flowing water in rivers and wetlands exerts a drag force on the stems of vegetation located in those bodies of water. (a) If a wind blows at 10 m/s, for what range of stem diameters is the drag force caused by wind on vegetation stems primarily due to friction? For what range of diameters is the drag force primarily due to pressure effects? (b) If a typical stem diameter in a body of water is 5 mm, for what range of current speeds is viscous drag predominant? For what range of speeds is pressure drag predominant? Assume standard air and water at 20°C.

10.23. An elevation view of a standard goal post structure in (American) football is shown in Figure 10.34. The horizontal crossbar is 3.0 m above the ground, and the vertical posts are 5.65 m apart and extend 9.20 m above the horizontal crossbar. The vertical support post has a diameter of 170 mm, and both the crossbar and the vertical posts have diameters of 100 mm. All cylindrical elements of the structure can be assumed to be smooth. For a wind speed of 50 km/h oriented normal to the structure, estimate the aerodynamic force acting on the goal post structure. Assume standard air.

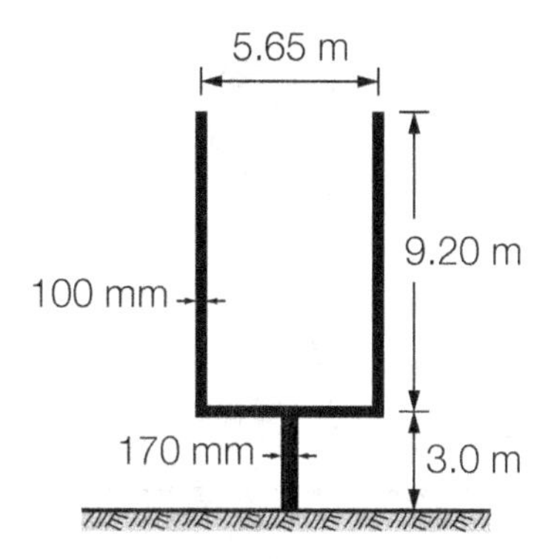

Figure 10.34: Goal post structure

10.24. Fine dust particles are being generated in a room filled with standard air. The dust particles can be assumed to have spherical shapes with a specific gravity of 1.30. (a) What is the maximum diameter of a dust particle for which the creeping flow assumption is valid? (b) For a particle with the limiting diameter derived in part (a), how long does it take the particle to fall the 2.95-m-distance from the ceiling to the floor of the room?

10.25. A 6-mm-diameter spherical ball is dropped in glycerin at 20°C and is observed to attain a terminal velocity of 72 mm/s. (a) Estimate the drag coefficient of the ball after it attains its terminal speed. (b) Estimate the specific gravity of the ball.

10.26. A typical baseball has a mass of 145 g and a diameter of 71.6 mm. If a pitcher throws a baseball at 40.2 m/s, determine the drag force on the baseball and the rate at which the baseball decelerates after leaving the pitcher's hand. If the baseball takes approximately 0.5 second to travel the 18.4 m from the pitcher's mound to the batter's box, estimate percentage change in the initial baseball velocity when it reaches the batter's box. Assume standard air.

10.27. A baseball with a mass of 145 g and a diameter of 71.6 mm is dropped from a height of 1 km in a standard atmosphere. (a) Estimate the terminal velocity attained by the baseball, assuming that atmospheric conditions remain constant and equal to those at the release height. (b) Determine the time it takes the baseball to attain 90% of its terminal velocity. (c) Determine the distance traveled for the baseball to attain 90% of its terminal velocity. Based on your results, comment on whether the assumption of constant atmospheric conditions is reasonable.

10.28. A balloon is filled with helium and is supported by a light string. The diameter of the balloon is 650 mm, and the tension in the string is measured as 2.5 N when the balloon is in standard air at sea level. If a wind speed of 5 m/s is imposed on the balloon and the string deflects 45° from the vertical, estimate the drag coefficient of the balloon.

10.29. A helium wind balloon with a diameter of 1.1 m is attached to the ground by a thin, lightweight cable as shown in Figure 10.35. The balloon has a weight of 1.5 N. The balloon is filled with helium at a temperature and pressure of 20°C and 101.3 kPa, respectively, and the surrounding air has the same temperature and pressure. The weight and drag on the support cable is negligible. Under a particular wind speed, the support cable is observed to deflect by 16° from the vertical direction.

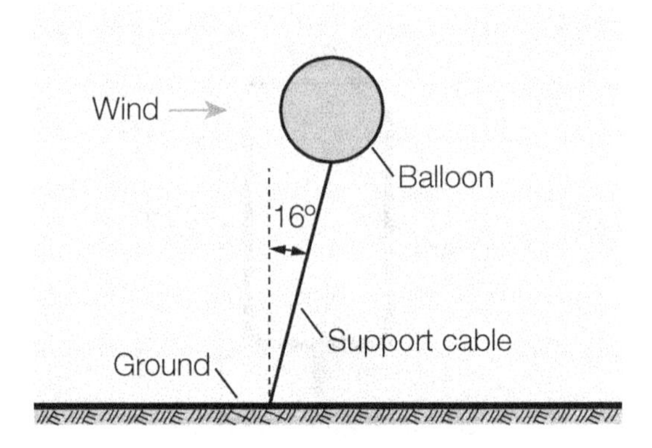

Figure 10.35: Wind balloon

10.30. A 575-mm-diameter sphere with an effective specific gravity of 0.36 is towed at 8.2 m/s through a freshwater lake as shown in Figure 10.36. The lake water has a temperature of 5°C. The sphere is connected to the towing device via a thin cable that creates negligible drag. Estimate the angle θ between the tether line and the direction of motion of the towing device.

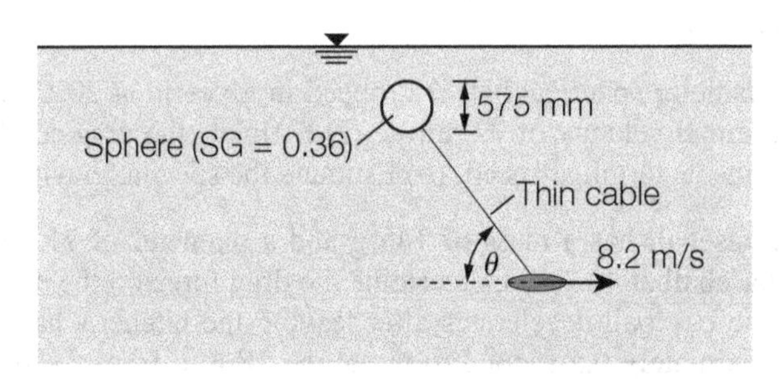

Figure 10.36: Sphere towed underwater

10.31. An Ohio-class submarine has a typical length of 170 m and an approximately circular cross section with a maximum diameter of 13 m. The typical Ohio-class submarine is powered by a 45-MW nuclear reactor. Scale testing of a model of the submarine indicates a drag coefficient of around 0.02. If 48% of the engine power is used to overcome hydrodynamic drag when the submarine is moving at its theoretical maximum speed, what is the theoretical maximum speed? Contrast your result with the conventional speed of 46 km/h that is normally attributed to Ohio-class submarines. Assume seawater at 20°C.

10.32. Consider the cylindrical-shaped clay particle shown in Figure 10.37(a). Determine the sideways and frontal settling velocity of the particle as shown in Figure 10.37(b) in water at 20°C. The clay particle has a specific gravity of 2.65, and assume that the flow around the particle is laminar during settling. Based on your results, comment on whether the settling velocity is very sensitive to the orientation in which the particle settles.

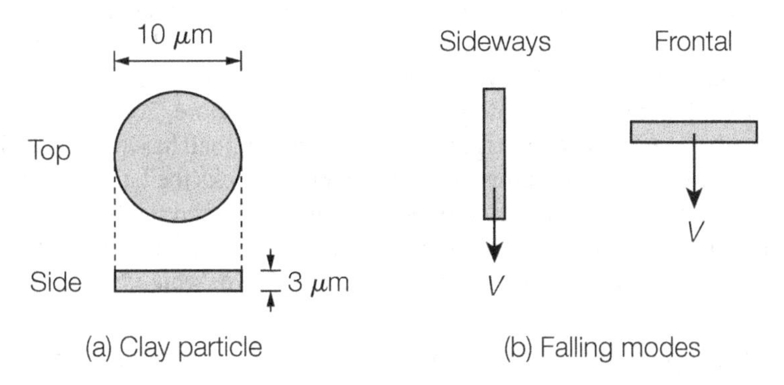

Figure 10.37: Clay particle settling in water

10.33. A 0.6-m-diameter disk is placed in water at 20°C flowing at 10 m/s, and measurements on the disk indicate the approximate pressure distribution shown in Figure 10.38. On the front side of the disk, the pressure decreases linearly from a maximum of 200 kPa at the center to 150 kPa on the perimeter. On the back side of the disk, the pressure is approximately constant at 125 kPa. Estimate the drag coefficient of the disk.

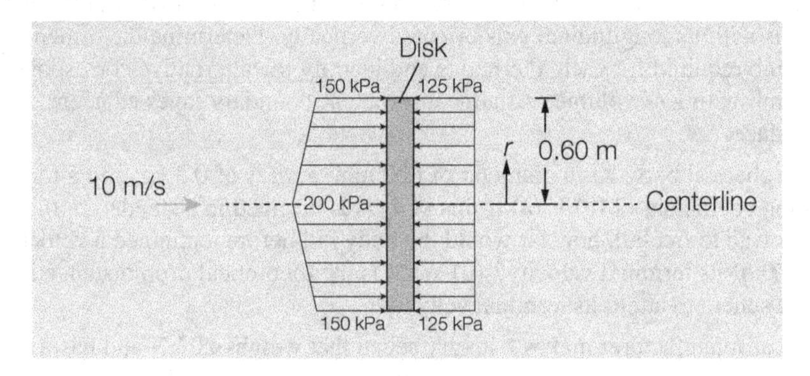

Figure 10.38: Pressure distribution around a disk

10.34. Experimental evidence indicates that the drag coefficient for flow around a particular sphere can be estimated by

$$C_D = \begin{cases} \dfrac{24}{\text{Re}} & \text{Re} \leq 1 \\ 0.5 & 10^4 \leq \text{Re} \leq 2 \times 10^5 \\ 0.2 & \text{Re} \geq 2 \times 10^6 \end{cases}$$

If air at 20°C flows around the 1-cm-diameter sphere, determine the drag coefficient and drag force for air velocities of 1 mm/s and 20 m/s.

10.35. Derive an expression for the terminal velocity of a spherical particle falling in a fluid for cases in which Re $\leq$ 1. Assess whether your estimated relationship is valid for estimating the terminal settling velocity in water at 20°C of a 1.10-mm-diameter particle with a specific gravity of 2.70.

10.36. The maximum diameter of a raindrop is typically cited as being on the order of 4 mm. Assuming that a raindrop is made up of water at 20°C and it is falling through standard air, what is the terminal velocity of the raindrop?

10.37. The standard American football has a length of approximately 0.279 m, a central diameter of 0.178 m, and a weight of 4.03 N. When the ball is thrown or kicked in a game and is traveling in the direction of its longitudinal axis, the drag coefficient has been found to be approximately 0.055 (Watts and Moore, 2003). The spin imparted to a football can be up to 600 rpm, and a spinning football has a drag coefficient that is 10% less than a non-spinning football. If a kicker kicks the ball to a sufficient height that it attains its terminal velocity before it is caught and the ball travels downward with its longitudinal axis vertical, what is the velocity of the ball when it is caught? If the ball spins at 600 rpm when it is coming down, what is its velocity? Assume standard air.

10.38. The standard soccer ball has a diameter of approximately 0.221 m and a mass of approximately 0.425 kg. If a goal keeper kicks the ball to a sufficient height that it attains its terminal velocity on its way down, what is the velocity of the ball when it hits the head of a soccer player trying the secure the ball for his team? Assume standard air. Also assume that the soccer ball acts as a smooth sphere.

10.39. A bomb that weighs 200 N is to be designed to have a cylindrical shape with a length equal to four times its diameter. The bomb must have a terminal speed of 10 m/s in standard air. Two alternatives for the weight distribution in the bomb are being considered. In Alternative A, the weight distribution is such that the bomb falls with its longitudinal axis oriented horizontally, and in Alternative B, the bomb falls with its longitudinal axis oriented vertically. Determine the dimensions of the bomb required for each alternative and identify the alternative corresponding to the bomb with less volume. Assume a turbulent boundary layer adjacent to the bomb surface.

10.40. A spherical body has a diameter of 600 mm, a mass of 0.3 kg, and a turbulent flow drag coefficient of 0.15. (a) If this body was dropped in a standard atmosphere and allowed to freefall, how far would the body fall before it attained a velocity equal to 80% of its terminal velocity? (b) What is the theoretical drop distance required for the sphere to attain its terminal velocity?

10.41. A car manufacturer makes a luxury sedan that weighs 20 kN and has a frontal area of 3.1 m^2. When placed on an inclined test track, the car rolls to a terminal speed of 27 m/s. The slope of the incline is 6%, and it is estimated that 50% of the resistance to motion is provided by the aerodynamic drag on the car. Estimate the drag coefficient of the car. Assume standard air.

10.42. A standard golf ball has a diameter of 43 mm and weighs approximately 0.45 N. Two golf balls are struck with an initial velocity of 60 m/s. One golf ball is smooth and has an estimated drag coefficient of 0.50 when it is struck initially. The other golf ball is roughened by dimples and has an initial drag coefficient of 0.25. (a) Compare the initial drag forces on each of the golf balls immediately after they are struck. (b) Estimate the velocities of the golf balls 1 second after they are struck and the distance each ball travels during this 1 second time interval. Assume standard air.

10.43. Show that the terminal velocity of a rising spherical body in a surrounding fluid of different density can be expressed in nondimensional terms using the so-called densimetric Froude number, Fr_d, as

$$\text{Fr}_\text{d}^2 = \frac{4}{3\,C_\text{D}} \tag{10.57}$$

where C_D is the drag coefficient and the densimetric Froude number, Fr_d, is defined by the relation

$$\text{Fr}_\text{d} = \frac{V}{\sqrt{\dfrac{\Delta\rho}{\rho} gD}}$$

where V is the terminal velocity, $\Delta\rho$ is the difference in density between the surrounding fluid and the sphere, ρ is the density of the surrounding fluid, g is gravity, and D is the diameter of the sphere. Comment on whether Equation 10.57 is also applicable to a falling spherical body. Note that $\Delta\rho/\rho \cdot g$ is the *effective gravity* on a submerged body, commonly represented by g'.

10.44. The goal is to fabricate a sphere that will take 10 seconds to rise from the ocean floor to the surface, where the depth of the ocean at the location of interest is 100 m. (a) If the sphere is to have a diameter of 0.6 m, what is the required effective density, if any, of the sphere? (b) If design constraints require that the effective density of the sphere be 70% of the density of the surrounding seawater, what is the required diameter of the sphere? Assume seawater at 20°C and assume that the rising sphere attains its terminal velocity immediately upon release.

10.45. An architect is considering various designs for a multistory office building. Floors in the building are to be 4 m apart (vertically), and foundation limitations require that the moment on the base of the building exerted by wind (drag) forces be less than 5×10^8 N·m. The cross section of the building is to be rectangular. Under design conditions, the wind speed is 125 km/h and the air temperature is 294 K. The building is to have a total floor space of 700 000 m^2 and a distance of at least 25 m between opposite walls on each floor. What is the maximum height of the building that will meet these design criteria?

10.46. The fully-loaded tractor-trailer shown in Figure 10.39 drives up a 3% incline at a steady speed of 71 km/h. The frontal area of the tractor-trailer is 2.60 m wide and 4.10 m high, and it is powered by a 373-kW engine. The total weight of the tractor-trailer and its payload is approximately equal to the legal limit of 350 kN. It is estimated that the coefficient of rolling resistance is 0.008, and the drag coefficient is 0.50. What power must be supplied by the engine to overcome the resistance to motion caused by the weight, rolling resistance, and aerodynamic drag? What percentage of the rated engine power is required to overcome these effects? Assume standard air.

Figure 10.39: Tractor-trailer driving uphill

10.47. A popular SUV has a length of 4.90 m, a width of 1.95 m, and a height of 1.80 m. According to the manufacturer, this SUV has a drag coefficient of 0.40. The SUV comes with an optional storage compartment that can be attached to the roof of the vehicle, where the storage compartment is approximately rectangular in shape and has a length of 1.25 m, a width of 0.97 m, and a height of 0.42 m. Estimate the percentage increase in engine power needed to overcome aerodynamic resistance at 100 km/h. Assume standard air.

10.48. A prototype sports car has an engine that can deliver 360 kW of power. The shape of the car is such that is has an estimated drag coefficient of 0.20 and a frontal area of 2.50 m^2. If the car is to be tested on a track at sea level under standard conditions, estimate the maximum possible speed the car can attain.

10.49. A sports car is driven at 90 km/h on a highway where there is a wind with a speed of 25 km/h at an angle of 25° to the highway as shown in Figure 10.40. The car has a length of 4.50 m, a width of 1.95 m, and a height of 1.24 m. The car manufacturer reports that the drag coefficient is 0.36 when air flows in a direction normal to the front of the car and that the drag coefficient is 0.82 when air flows in a direction normal to the side of the car. Estimate the magnitude and direction of the net aerodynamic force on the car. Assume standard air.

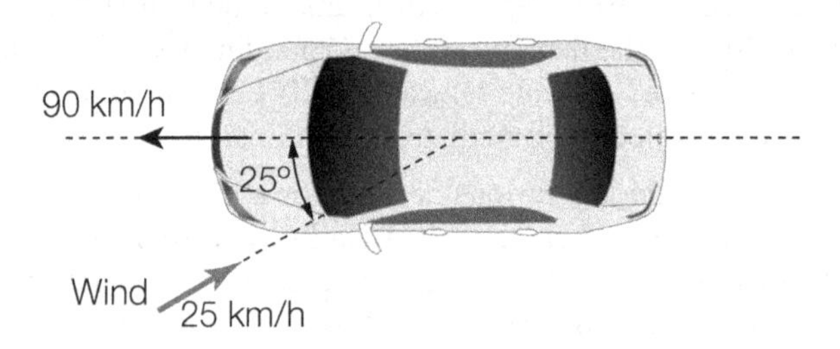

Figure 10.40: Sports car on a highway

10.50. Experiments on a car in a wind tunnel indicate that the drag coefficient of the car with all the windows up is 0.35 and that the drag coefficient with all the windows down is 0.45. The car has a frontal area of 3.0 m^2. Consider the case where the car is driven at 90 km/h for a distance of 205 km, gasoline costs $0.6604/L, and the available energy from gasoline for overcoming aerodynamic drag is 4.50 MJ/kg. Estimate the additional gasoline cost incurred as a result of driving with the windows down. Assume standard air.

10.51. A school bus that is used on a long-distance route has a frontal area of 9 m^2 and an estimated drag coefficient of 0.85. The bus regularly travels on the highway at the speed limit of 90 km/h. A design modification of the shape of the bus can reduce the drag coefficient to 0.75. Determine the percentage reduction in the engine power that the bus will need to use to maintain a speed of 90 km/h.

10.52. The largest trailer (as in a tractor-trailer assembly) allowed in South Dakota has a length of 16.15 m, a width of 2.59 m, and a height of 4.27 m. The height of the bottom of the trailer above the roadway, called the *roof floor height*, is typically around 1.27 m. An empty trailer can have a mass as low as 4530 kg.A typical trailer has a drag coefficient of 0.95 when wind blows toward the side of a trailer. A transport company parks its empty trailers on a large lot that is open to the wind. Estimate the minimum wind speed that would likely blow over one of the trailers. Assume standard air.

10.53. An aircraft is designed to cruise at an altitude of 10 km at a speed of 880 km/h. The fuselage of the aircraft can be approximated as a cylinder with a diameter of 5 m and a length of 40 m. (a) Use the cylinder approximation to estimate the power required to overcome the drag on the fuselage, with the drag coefficient derived from Table 10.5. (b) Estimate the power required to overcome drag on the fuselage, taking only skin friction into account. (c) Compare the results from parts (a) and (b) and explain the difference. Assume that compressibility effects are negligible.

10.54. A submarine is designed to cruise at a speed of 15.43 m/s in seawater at 10°C. The submarine can be approximated as a cylinder with a diameter of 12 m and a length of 120 m. (a) Use the cylinder approximation to estimate the power required to overcome the drag, with the drag coefficient derived from Table 10.5. (b) Estimate the power required to overcome drag, taking only skin friction into account.

10.55. An ET32 torpedo formerly used by China has a reported maximum diameter of 533 mm and a length of 6.6 m. The design speed of the torpedo is 65 km/h. If the torpedo is fired in seawater at 10°C and the shape of the torpedo is approximated by an ellipsoid, what power is required to propel the torpedo?

10.56. A person and a parachute is estimated to weigh 1275 N, and the goal is to size the parachute such that the landing velocity is no greater than 6 m/s. Estimate the required diameter of the parachute. Assume standard air.

10.57. A copper plate of length 30 mm and width 30 mm is hinged at Point P as shown in Figure 10.41. The plate is to be designed so that it deflects by an angle of 15° when the wind speed is 8 m/s and atmospheric conditions are at standard sea-level conditions. The specific weight of copper is 89 kN/m^3. What plate thickness is required? Assume that the drag coefficient for a flat plate can be used with the component of the wind velocity normal to the length of the plate.

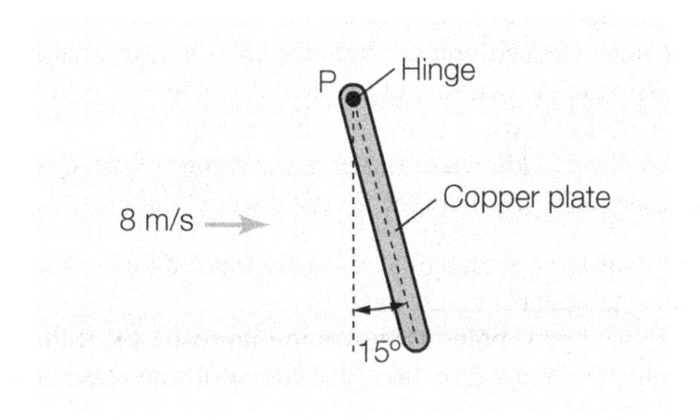

Figure 10.41: Plate deflection in the wind

10.58. An anemometer consists of a hollow aluminum hemisphere on a strut that is attached to a pin connection at Point P as shown in Figure 10.42. The hemisphere is to be sized to provide a deflection of 20° when the wind speed is 10 m/s and atmospheric conditions are at standard sea-level conditions. The specific weight of the aluminum hemisphere is 2.7 kN/m^3. The strut is sufficiently thin that the drag on the strut is negligible. What diameter of the hemisphere is required? Compare this diameter with the required diameter if the wind was blowing in the opposite direction.

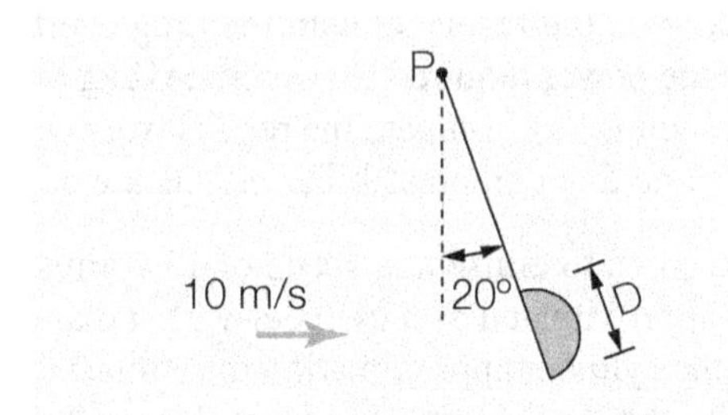

Figure 10.42: Solid hemisphere deflection in the wind

10.59. A smokestack has a diameter of 1.5 m and a height of 30 m. If the smokestack is to be designed for a wind speed of 60 km/h, what is the force and bending moment that must be provided by the support structure? Assume standard air.

10.60. A new car antenna has a diameter of 8 mm and a length of 1.5 m. The structural support for the antenna is known to fail when the applied moment on the support exceeds 3 N·m. If the car is to be driven in standard air at sea level, estimate the maximum speed the car can be driven without the antenna support failing.

10.61. A bridge pier has a diameter of 2.5 m and supports a bridge over a river. Under flood conditions, the velocity of flow in the river is around 3 m/s. An investigation of the pier foundation at the bottom of the river determines that the foundation is able to support a maximum moment of 950 kN·m. Determine the maximum flow depth past the bridge pier for which the bridge pier can be expected to stand. Assume a turbulent boundary layer over the entire depth of the pier. Assume water at 20°C.

10.62. An elevated storage tank consists of a 4-m-diameter and 6-m-high cylindrical tank supported on three steel columns, where the columns are each 1.5 m in diameter and 35 m high. Estimate the bending moment at the base of the supporting columns that is caused by a 90 km/h wind at an air temperature of 25°C. A flag mounted behind the storage reservoir is known to flutter at the same frequency that vortices are shed from the storage tank. Estimate the frequency at which the flag flutters.

10.63. A 2-m-high, 4-m-wide rectangular advertisement sign is attached to a 4-m-wide, 0.15-m-high rectangular concrete base (density = 2300 kg/m^3) by two 5-cm-diameter, 4-m-high (exposed part) poles as shown in Figure 10.43. If the sign is to withstand 150 km/h winds from any direction, determine (a) the maximum drag force on the panel, (b) the drag force acting on the poles, and (c) the minimum length L of the concrete base for the panel to resist the winds. Assume the density of air to be 1.30 kg/m^3.

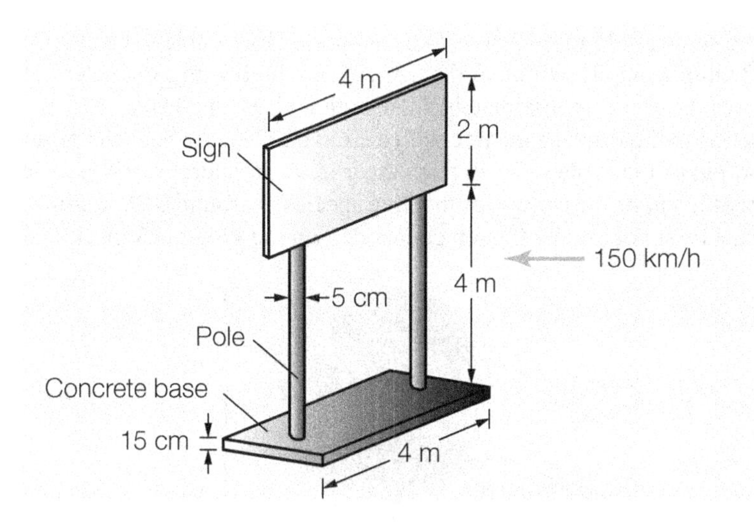

Figure 10.43: Sign in the wind

10.64. A residential flagpole has a diameter of 100 mm and is 7.62 m high. The flagpole is made of smooth aluminum. The recommended flag for this flagpole has dimensions of 1.22 m × 1.83 m, and the flag manufacturer states that the drag coefficient of the flag is equal to 0.1, provided the area in the drag equation is taken as the length times the width of the flag. For a wind speed of 105 km/h, compare the moment on the base of the flagpole with and without the flag. What percentage increase in moment is caused by flying the flag? Assume standard air at sea level.

10.65. A 100-mm-diameter flagpole has a height of 7.62 m. The design wind speed, u, varies with the distance, z, from the ground according to the one-seventh law distribution given by

$$\frac{u}{U} = \left(\frac{z}{H}\right)^{\frac{1}{7}}$$

where U is the velocity at a distance H above the ground. In this case, assume that $U = 105$ km/h and $H = 7.62$ m. (a) Determine the drag force on the flagpole, assuming that boundary layer is turbulent over the entire height of the flagpole. (b) What would be the drag force on the flag pole if the wind speed over the entire height of the flagpole was constant and equal to 105 km/h? Assume standard air at sea level.

10.66. An advertising banner is 2 m tall and 18 m long. The banner manufacturer states that the drag coefficient of the banner is 0.45. A small plane is to tow the banner at a speed of 90 km/h in standard air at sea level. Estimate the engine power that the plane will need to tow the banner. Compare the result of your calculations with the result that would be obtained if the banner was assumed to be a flat plate. Comment on the reasonableness of approximating the banner as a flat plate.

10.67. The tree shown in Figure 10.44 is to be supported by a cable to keep it from moving in a 72 km/h wind. The trunk of the tree is 2.5 m high with a diameter of 0.6 m. The canopy of the tree is approximately 5.2 m high with a frontal area of 25 m^2. The cable is attached to the trunk of the tree with a strap that is 2 m above the ground, and the support pin of the cable is 2.3 m from the tree. A literature review indicates that the drag coefficient of the canopy of this tree species is around 0.26. Estimate the force that must be supported by the pin connection on the ground. Assume standard air.

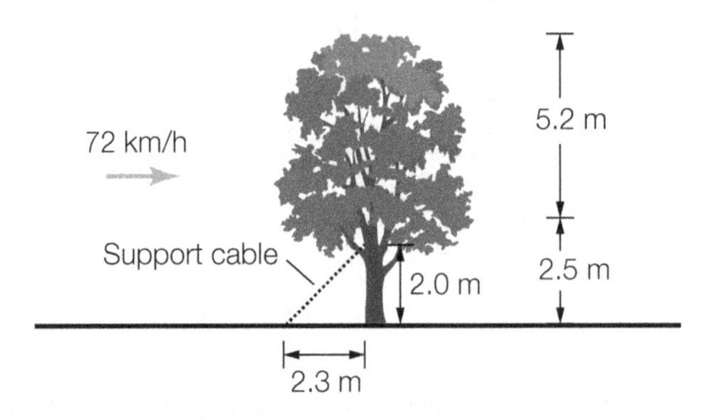

Figure 10.44: Cable supporting a tree

10.68. For aerial assaults, the armed forces of the United States generally use a T-10 parachute that has a nominal diameter of 10.7 m when opened and a weight of 14 kg. The estimated drag coefficient of the T-10 parachute is equal to 1.4. Estimate the terminal velocity of an 82-kg person jumping with 28 kg of gear (excluding the parachute) from a height 350 m above ground. The average air density over the jump height can be taken as 1.20 kg/m^3.

10.69. An average person on a recreational bicycle can maintain a speed of 8.05 m/s. The frontal area of a typical bicycle with a typical person aboard in riding position is 0.5 m^2, and the drag coefficient of this configuration is approximately 0.5. Under the stated typical conditions, estimate the power that the rider must exert to overcome the aerodynamic drag. Assume standard air.

10.70. Experiments with a person who weighs 833.6 N on a bicycle rolling freely down a 5% incline show that the person-plus-bicycle attains a terminal speed of 18 m/s. The frontal area of the person-plus-bicycle is 0.6 m^2, and the rolling resistance of the bicycle is negligible. Estimate the drag coefficient of the person-plus-bicycle combination. Assume standard air.

10.71. A person and her bicycle weigh 735 N, and when the person is mounted in riding position on the bicycle, the frontal area is 0.42 m^2. The rolling resistance of the bicycle is 10 N. Practice runs indicate that the person is able to maintain a riding speed of 35 km/h with no wind. If there is a wind with a speed of 8 km/h, estimate the speed that the person on the bicycle can attain if (a) she rides directly into the wind and (b) she rides in the same direction as the wind.

10.72. Biomechanical research indicates that the drag coefficient of a human hand varies in the range of 0.51–0.56, depending on how far apart the fingers are spread. Interestingly, the maximum drag coefficient corresponds to an angle between fingers of approximately 12°, a finding that is used to advantage by competitive swimmers

(Minetti et al., 2009). The average hand area of a human adult male is around 450 cm^2. Estimate the range of force and the range of pressure that an adult male will feel by varying the angle between his fingers if he sticks his hand out the window of a vehicle traveling at 88 km/h. Assume standard air.

10.73. Aerial banners are commonly used in advertising, and the largest size banner that can be towed by a small airplane is commonly cited as 464 m^2. A large commercial banner that is near this size limit has a length of 36.3 m and a height of 12.2 m. The towplane normally flies at a speed of around 96 km/h. (a) Estimate the engine power used by the towplane to overcome the aerodynamic drag on the banner. (b) If the banner was assumed to be a flat plate, what would be the estimated engine power requirement? Assume standard air.

10.74. People falling from high elevations (sadly) seldom survive the fall. Consider a fully-grown person who weighs 86.2 kg and has a frontal area of 0.70 m^2 who falls from a high elevation and attains the terminal velocity before hitting the ground. At what velocity does this person hit the ground? Assume standard air.

10.75. The world record for the men's 100-m dash is approximately 9.58 seconds. If the sprinter setting this record has a frontal area of approximately 0.75 m^2 and his drag coefficient is approximately that of an average upright human, what aerodynamic drag force does he face and what power must he develop to overcome the aerodynamic drag? Assume standard air.

10.76. A runner competes in a 10-km race at sea level and is able to maintain a constant speed of 12 km/h for the entire distance. It is estimated that the runner's drag area, which is the product of the drag coefficient and the frontal area, is equal to 1.1 m^2. (a) Estimate the energy that the runner expends in overcoming aerodynamic drag over the course of the race. (b) If a 7 km/h wind blows in the runner's face for the first 5 km and a 7 km/h wind blows at the runner's back for the second 5 km, what is the energy expended by the runner in the race.

10.77. A blimp that is used for sightseeing has a length of 58.5 m, has a maximum diameter of 15.2 m, and is powered by two 157 kW engines. The drag coefficients of blimps are generally estimated to be in the range of 0.020–0.025. If all of the engine power was used in overcoming hydrodynamic drag, what would be the maximum speed of the blimp? Assume standard air.

Section 10.4: Estimation of Lift Coefficients

10.78. A commercial aircraft under design conditions has a particular takeoff speed when it is fully loaded. If the aircraft is overloaded by 15%, what is the corresponding percentage increase in the takeoff speed.

10.79. Consider the case of a commercial aircraft that uses a fixed acceleration during takeoff, regardless of the atmospheric conditions. Under fixed loading (weight) conditions, determine the percentage increase in runway length required for takeoff from an airport at an elevation of 2000 m compared with an airport at sea level. Assume a standard atmosphere.

10.80. Aircraft typically consume fuel at a higher rate when flying at a lower elevation compared with flying at the same speed at a higher elevation. Estimate the percentage decrease in the fuel consumption rate for an aircraft flying at an altitude of 10 km compared with flying at the same speed at an altitude of 3 km.

10.81. The model airplane shown in Figure 10.45 has wings of dimensions 5 m × 2 m, and the wings are oriented at an angle of 5° to the horizontal. Numerical modeling of the flow field around the model airplane shows that when the airplane has a speed of 10 m/s in standard air, the average pressure on the top and bottom of the wings is 8.30 kPa and 23.9 kPa, respectively. The average shear stress on the top and bottom of the wings is 0.218 kPa and 0.253 kPa, respectively. (a) What are the drag and lift forces generated by the wings? (b) What are the drag and lift coefficients of the wings?

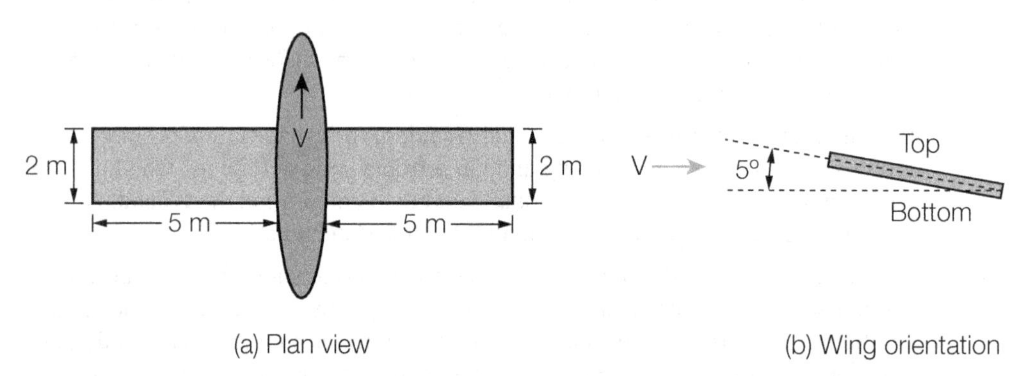

Figure 10.45: Model airplane

10.82. A model aircraft has a total wing area of 6 m^2. Based on experimental results from similar aircraft, it is estimated that the lift and drag coefficients are around 0.71 and 0.17, respectively. It is intended that the model airplane fly at a speed of 15 m/s under standard sea-level conditions. (a) What is the maximum allowable weight of the airplane? (b) What is the power required to fly the airplane at its design speed?

10.83. A small aircraft is designed to cruise at a speed of 200 km/h at an elevation of 2 km. The mass of the loaded aircraft is 1050 kg, and the estimated drag and lift coefficients are 0.073 and 0.52, respectively. (a) Estimate the effective lift area of the aircraft. (b) What is the engine power required at the design cruise condition.

10.84. A small aircraft has wings with lift and drag coefficients that are a function of the angle of attack as shown in Figure 10.21. The wings have a planform area of 15 m^2. Under design conditions, the mass of the aircraft fully loaded is 1200 kg and the cruising speed is 250 km/h. Atmospheric conditions correspond to a standard atmosphere at sea level. (a) Estimate the required angle of attack. (b) Estimate the power required to fly the aircraft under design conditions. (c) Estimate the glide ratio of the aircraft.

10.85. An aircraft has wing sections that have lift and drag coefficients as a function of the angle of attack as shown in Figure 10.21. (a) What is the maximum possible glide ratio for this aircraft? (b) If the aircraft loses power while at an altitude of 2 km, what is the maximum distance it could glide and land without engine power? Assume flat terrain.

10.86. A small aircraft has a total wing area of 30 m^2, a lift coefficient of 0.45 at takeoff settings, and a total mass of 2800 kg. (a) Determine the takeoff speed in km/h of the aircraft at sea level under standard atmospheric conditions. (b) Determine the wing loading. (c) Determine the required power to maintain a constant cruising speed of 300 km/h for a cruising drag coefficient of 0.035. Assume standard air.

10.87. The C-5 aircraft shown in Figure 10.46 is one of the largest transport planes currently in operation. This aircraft has a wingspan of 67.89 m and a height of 19.84 m.

According to published specifications, the C-5 has a wing area of 576 m^2, has a wing loading of 5982 N/m^2, and cruises at a speed of 919 km/h at an altitude of 10.6 km with a maximum gross weight of 279 000 kg. (a) Estimate the lift coefficient of the aircraft under cruising conditions. (b) By using flaps, the C-5 can increase its lift coefficient by a factor of 2 when taking off and landing. Estimate the minimum required takeoff speed.

Figure 10.46: C-5 transport plane

Source: Sgt. Monique Randolph, United States Air Force.

10.88. According to the manufacturer's data, the Cessna 208 Caravan shown in Figure 10.47 is a light aircraft powered by a 505-kW engine that has a total wing area of 25.9 m^2, a cruising speed of 317 km/h, and a wing loading of 1502 N/m^2. (a) Estimate the lift coefficient of this aircraft under cruising conditions. (b) The upper limit of the possible values of the drag coefficient can be estimated by assuming that all of the engine power is used to generate lift and to overcome aerodynamic drag on the wings. Estimate the upper limit of possible drag coefficients of the aircraft. Assume that the plane flies horizontally and assume standard air.

Figure 10.47: Cessna 208 Caravan on a runway

Source: Susan & Allan Parker/Alamy.

10.89. (a) Compare the engine power required to operate a given commercial aircraft at its normal cruising altitude of 10 km with the power required to operate the same aircraft at an altitude of 2 km. Assume that the aircraft is flying horizontally with the same speed in both cases and that the drag coefficient does not vary with elevation. (b) If the aircraft is to generate the same lift force at both elevations, compare the required speed of the aircraft at the higher elevation with the required speed at the lower elevation.

10.90. Daocheng Yading Airport in China is a commercial airport located 4411 m above sea level and is among the highest-elevation commercial airports in the world. If a particular commercial aircraft has a takeoff speed of 200 km/h at sea level, what is its required takeoff speed at Daocheng Yading Airport? Assume a standard atmosphere.

10.91. The glide slope is commonly defined as the proper path for an airplane to use in approaching a landing strip, and a typical glide slope at commercial airports is 3°. For an aircraft to land at a slope angle of 3° without engine power, what must its glide ratio be? What would be the ratio of the lift coefficient to the drag coefficient for this aircraft? What action must the pilot take if the glide ratio of her aircraft is less than that corresponding to the glide slope?

10.92. The specifications of an aircraft state that it has a wing planform area of 160 m^2, an aspect ratio of 6.2, a zero-lift drag coefficient of 0.0175, a weight of 680 kN when fully loaded, and a stall speed of 280 km/h at standard sea-level conditions. (a) Estimate the lift and drag coefficient of the aircraft. (b) Estimate the thrust that must be produced by the engine(s) to keep the aircraft flying.

10.93. An aircraft has a wing planform area of 180 m^2, an aspect ratio of 7.5, a zero-lift drag coefficient of 0.0185, and a weight of 800 kN when fully loaded. Estimate the speed of the aircraft that will minimize the required engine thrust when the aircraft is flying under standard sea-level atmospheric conditions.

10.94. The drag and lift coefficients of a small hydrofoil are being tested in a tank of seawater using the configuration shown in Figure 10.48. The hydrofoil has a weight of 1300 N, a volume of 0.125 m^3, and a plan area of 0.5 m^2. When the seawater flows at 3 m/s, the thin, lightweight support cable makes an angle of 85° with the horizontal and the measured tension in the support cable is 2750 N. Estimate the drag and lift coefficients of the hydrofoil based on its plan area. Assume seawater at 20°C. The weight and drag on the support cable is negligible.

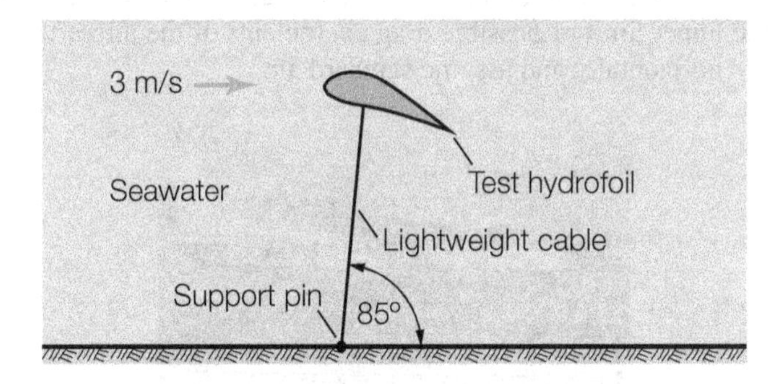

Figure 10.48: Test hydrofoil

10.95. A hydrofoil has an effective area of 1.5 m^2 and has lift and drag coefficients of 1.6 and 0.69, respectively. The watercraft supported by the hydrofoil has a design weight of 20 kN and is intended to operate in seawater at 10°C. (a) What is the minimum speed required to elevate the watercraft out of the water. (b) If the watercraft is equipped with a 134.2-kW engine, what is the theoretical maximum speed that can be attained?

10.96. A hydrofoil with a typical shape is 5 m long, is 2 m wide, and is submerged 3 m below the water surface in a large freshwater lake. The temperature of the water is 20°C, and the hydrofoil is oriented with an angle of attack of 2°. Estimate the maximum allowable speed to avoid cavitation and the maximum lift that can be generated.

10.97. The lift and drag coefficients on an American golf ball as a function of the spin ratio are shown in Figure 10.49. The standard American golf ball has a diameter of 43 mm and a mass of 48 g. In approaching a green, a professional golfer strikes a ball with a velocity of 30 m/s and a backspin of 2500 rpm. Estimate the drag and lift forces on the golf ball. Contrast these forces with the weight of the ball. Assume standard air.

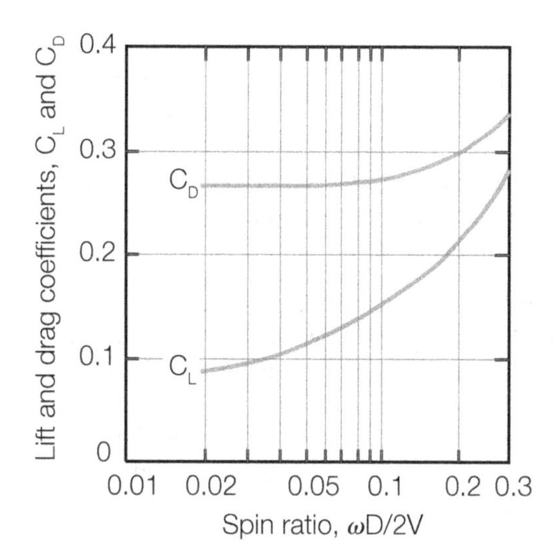

Figure 10.49: Drag and lift coefficients on an American golf ball

Source: Mehta (1985).

10.98. The science of pitching a baseball relies to a great extent on the Magnus effect. Consider the case of a standard baseball with a diameter of 73 mm and a mass of 140 g. A pitcher on the pitcher's mound is approximately 18 m from the batter in the batter's box. A typical fastball is thrown at around 150 km/h with a backspin rotation of around 2200 rpm, whereas a typical curveball is thrown at around 130 km/h with a topspin rotation of approximately 1300 rpm. The lift coefficient as a function of the spin ratio for baseballs is shown in Figure 10.50. These data suggest that the lift coefficient is approximately constant for spin ratios in the range of 0.1–0.25. (a) Compare the lift forces generated by a fastball and a curveball. (b) Compare the deviation of a curveball from a straight-line trajectory with the deviation of a fastball from a straight-line trajectory between the pitcher's mound and the batter's box. Assume standard air at sea level.

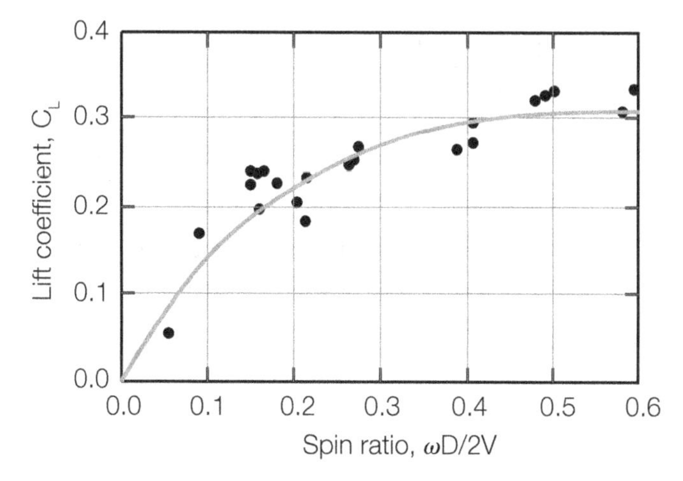

Figure 10.50: Lift coefficient on a baseball

Source: Nathan (2008).

10.99. A professional soccer player takes a corner kick using a standard soccer ball, and over a distance of 15 m, she is able to "bend" the ball by 1.5 m. The ball is estimated to be moving at a velocity of 35 m/s when it is kicked. A standard soccer ball has a mass of approximately 425 g and a diameter of approximately 22.5 cm. The lift coefficient as a function of the spin ratio for a soccer ball can be estimated using Figure 10.27. Estimate the rotational speed of the ball imparted by the corner kicker. Assume standard air at sea level.

10.100. A standard softball has a mass of approximately 190 g and a diameter of approximately 95 mm. A pitching machine shoots out a softball at a velocity of 15 m/s in the horizontal direction and imparts a backspin of 4500 rpm to the ball. The relationship of the lift coefficient to the spin ratio is approximately the relationship shown in Figure 10.26. Contrast the lift force on the ball with the weight of the ball to determine whether the ball will move upward or downward after leaving the pitching machine. Assume standard air at sea level.

10.101. A standard tennis ball has a mass of approximately 57 g and a diameter of approximately 64 mm. A practice machine shoots out a tennis ball at a velocity of 100 km/h in the horizontal direction and imparts a backspin of 3000 rpm. The relationship of the lift coefficient to the spin ratio is approximately the relationship shown in Figure 10.51. Determine the ratio of the lift force to the weight of the ball and then infer whether the ball will move upward or downward after leaving the practice machine. Assume standard air at sea level.

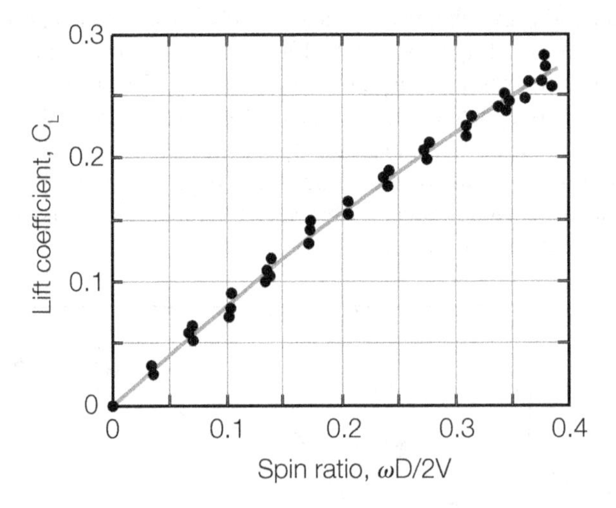

Figure 10.51: Lift coefficient on a tennis ball

Source of data: Goodwill et al. (2004).

Boundary-Layer Flow

LEARNING OBJECTIVES

After reading this chapter and solving a representative sample of end-of-chapter problems, you will be able to:

- Understand the mechanism for the occurrence of boundary layers adjacent to solid surfaces.
- Differentiate between laminar and turbulent boundary layers.
- Determine the velocity distributions in boundary layers adjacent to both smooth and rough surfaces.
- Calculate the shear forces generated by boundary layers.
- Apply boundary layer theory to practical situations related to flows over flat surfaces and flows in closed conduits.

11.1 Introduction

The region of a flow field that is influenced by the no-slip condition on a solid surface is called the *boundary layer*, and flow within a boundary layer is called *boundary-layer flow*. Outside the boundary layer, fluid flow is mostly influenced by both the shape of the solid surface and the characteristics of the boundary layer. Flow outside the boundary layer can usually be assumed to be inviscid (i.e., viscosity effects can be neglected) and independent of the roughness characteristics of the solid surface.

Basic concepts. The simplest boundary layer is formed when a fluid flows over a plane (flat) surface; this type of flow also approximates the flow over curved surfaces where the radius of curvature of the surface is large. The flow of a fluid over a plane surface is illustrated in Figure 11.1, where the x-coordinate is measured along the surface from its leading edge and the y-coordinate is measured normal to the surface. The fluid approaches the surface in the x-direction with a free-stream velocity U, and the presence of the surface affects the velocity up to a distance $\delta(x)$ from the surface, beyond which the free-stream velocity remains virtually unchanged. The line $y = \delta(x)$ defines the upper limit of the boundary layer. Flow within the boundary layer is initially laminar and eventually becomes turbulent, and typical velocity profiles in laminar and turbulent boundary layers are illustrated in Figure 11.1. The velocity profile in turbulent flow is much fuller than in laminar flow, due to increased mixing associated with turbulent velocity fluctuations. Transition from a laminar boundary layer to a turbulent boundary

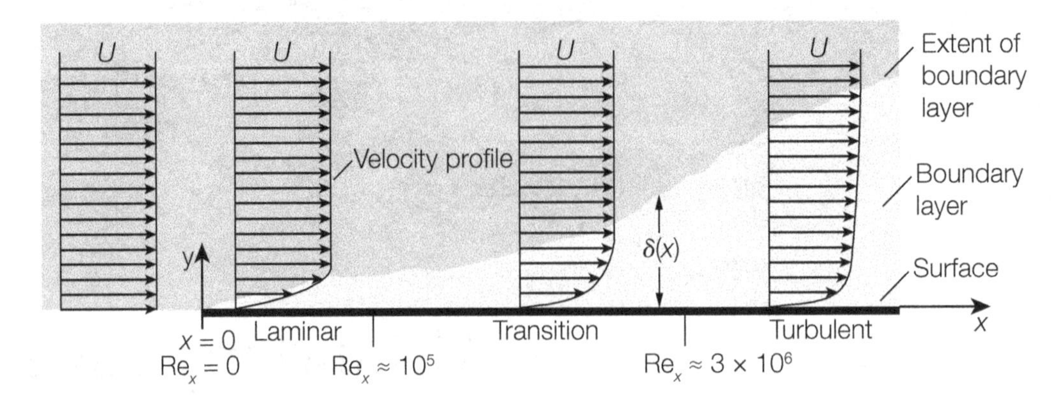

Figure 11.1: **Boundary layer over a flat surface**

layer is typically related to the Reynolds number, Re_x, at a distance x from the leading edge of the surface and defined as

$$\text{Re}_x = \frac{\rho U x}{\mu} = \frac{U x}{\nu} \tag{11.1}$$

where ρ, μ, and ν $(= \mu/\rho)$ are the density, dynamic viscosity, and kinematic viscosity of the fluid, respectively, and U is the velocity of the fluid outside the boundary layer. For flow over a flat surface, transition from a laminar to a turbulent boundary layer begins at about $\text{Re}_x = 1 \times 10^5$, but the boundary layer does not become fully turbulent until the Reynolds number reaches much higher values, typically taken to be around 3×10^6. To simplify engineering analyses, it is common to assume an instantaneous transition from a laminar to a turbulent boundary layer at around 5×10^5. In cases where the goal is to control the transition location, deliberately roughening a surface or installing trip wires that produce relatively large mixing eddies are sometimes used.

EXAMPLE 11.1

Standard air at sea level flows over a 3-m-long smooth flat surface at a velocity of 20 km/h. Determine the type of boundary layer that exists at the downstream end of the surface.

SOLUTION

From the given data: L = 3 m and U = 20 km/h = 5.555 m/s. For standard air at sea level, $\nu = 1.461 \times 10^{-5}$ m²/s (from Appendix B.2). The Reynolds number at the downstream end of the surface is calculated using Equation 11.1 as follows:

$$\text{Re}_\text{L} = \frac{UL}{\nu} = \frac{(5.555)(3)}{1.461 \times 10^{-5}} = 1.14 \times 10^6$$

Because $1 \times 10^5 \leq \text{Re}_\text{L} \leq 3 \times 10^6$, the boundary layer is likely in the **transition regime** at the downstream end of the surface. Using the usual engineering approximation that the transition from a laminar to a turbulent boundary layer occurs instantaneously at $\text{Re}_x = 5 \times 10^5$, the boundary layer would be classified as being in the **turbulent regime** at the downstream end of the surface.

Quantities of interest in engineering applications. Quantities of interest in boundary-layer analyses typically include the thickness of the boundary layer, the shear stress exerted by the fluid on the surface, and the total drag force on the surface. All of these quantities are a function of the distance from the leading edge of the surface, and all can be derived from the velocity distribution within the boundary layer. Analyses of boundary-layer flows typically consist of first estimating the velocity distribution within the boundary layer and then deriving the quantities of interest. This analytic approach is followed in the subsequent sections, which deal separately with laminar and turbulent boundary layers.

11.2 Laminar Boundary Layers

To account for the presence of boundary layers in high-Reynolds-number flows past solid objects, Ludwig Prandtl (1875–1953) originally suggested dividing flows adjacent to solid surfaces into two regions: an outer flow region that is inviscid and irrotational and an inner flow region called the boundary layer; this approach is called the *boundary-layer approximation.* Using the boundary-layer approximation, potential-flow theory can be used to describe the outer flow region and equations accounting for viscous effects are used to describe flow within the boundary layer. In modern engineering practice, such an approximation is seldom necessary because methods of computational fluid dynamics are capable of providing numerical solutions to the Navier-Stokes equations throughout the flow field. However, for analytic approaches, rapid analyses, and preliminary estimation, the boundary-layer approximation is particularly useful.

11.2.1 Blasius Solution for Plane Surfaces

Two-dimensional steady flow over a plane surface is described by the steady-state Navier-Stokes (NS) equation with the following x and y components:

$$u\frac{\partial u}{\partial x} + v\frac{\partial u}{\partial y} = -\frac{1}{\rho}\frac{\partial p}{\partial x} + \nu\left(\frac{\partial^2 u}{\partial x^2} + \frac{\partial^2 u}{\partial y^2}\right) \quad (= x \text{ component of NS equation}) \qquad (11.2)$$

$$u\frac{\partial v}{\partial x} + v\frac{\partial v}{\partial y} = -\frac{1}{\rho}\frac{\partial p}{\partial y} + \nu\left(\frac{\partial^2 v}{\partial x^2} + \frac{\partial^2 v}{\partial y^2}\right) \quad (= y \text{ component of NS equation}) \qquad (11.3)$$

where x is the coordinate parallel to the plane surface and y is the coordinate normal to the plane surface. Within a boundary layer, the following approximations simplify the Navier-Stokes equations and facilitate an analytic solution.

Approximation	Meaning
1. The component of the velocity normal to the surface is much smaller than the component of the velocity parallel to the surface.	$u \gg v$
2. The rate of change of any variable normal to the surface is much greater than the rate of change of that variable parallel to the surface.	$\dfrac{\partial(\cdot)}{\partial y} \gg \dfrac{\partial(\cdot)}{\partial x}$
3. Pressure variations normal to the surface are negligible.	$\dfrac{\partial p}{\partial y} = 0$
4. For a flat surface, the pressure is constant throughout the flow field.	$\dfrac{\partial p}{\partial x} = 0$

Applying the above approximations, the Navier-Stokes equations (Equations 11.2 and 11.3) within a boundary layer over a flat surface simplify to

$$u\frac{\partial u}{\partial x} + v\frac{\partial u}{\partial y} = \nu\frac{\partial^2 u}{\partial y^2} \tag{11.4}$$

The simplifications achieved by the above approximations are that the y-component of the momentum equation has been eliminated and the pressure, p, no longer appears as a variable; the remaining variables are u and v. Physically, Equation 11.4 states that the fluid motion within the boundary layer is determined entirely by viscous forces, with pressure variations within the boundary layer being negligible and hence not influencing the flow. The continuity equation is given by

$$\frac{\partial u}{\partial x} + \frac{\partial v}{\partial y} = 0 \tag{11.5}$$

Equations 11.4 and 11.5 represent two equations in two unknowns: $u(x, y)$ and $v(x, y)$. To complete the mathematical statement of the problem, the boundary conditions are given by

$$u(x, 0) = v(x, 0) = 0 \tag{11.6}$$

$$u(x, \infty) = U \tag{11.7}$$

where U is the *free-stream velocity*. To analytically solve Equations 11.4 and 11.5 subject to the boundary conditions given by Equations 11.6 and 11.7, Blasius (1908) proposed that the velocity distribution within a boundary layer is "self-similar" and can be represented by

$$\frac{u}{U} = f_0\left(\frac{y}{\delta}\right) \tag{11.8}$$

where f_0 is an unknown function and δ is the thickness of the boundary layer, where δ is defined as the distance from the surface to where the velocity is equal to 99% of the free-stream velocity; that is

$$u(x, \delta) = 0.99U \tag{11.9}$$

Applying an order of magnitude analysis of the viscous and inertial forces acting on the fluid in the boundary layer,

$$F \sim ma \quad \rightarrow \quad x \cdot \mu\frac{U}{y} \sim \rho x y\frac{U^2}{x} \quad \rightarrow \quad y \sim \left(\frac{\nu x}{U}\right)^{\frac{1}{2}} \quad \rightarrow \quad \delta \sim \left(\frac{\nu x}{U}\right)^{\frac{1}{2}} \tag{11.10}$$

Equation 11.10 indicates that the thickness of the boundary layer, δ, grows in proportion to $(\nu x/U)^{\frac{1}{2}}$ and hence the similarity variable $\eta = y/\delta$ in Equation 11.8 can be expressed as

$$\eta = \frac{y}{\delta} = y\left(\frac{U}{\nu x}\right)^{\frac{1}{2}} \tag{11.11}$$

Noting that the stream function, ψ, has dimensions of $\mathrm{L^2T^{-1}}$, by dimensional analysis the stream function can be expressed in terms of the similarity variable, η, by the following functional relationship,

$$\frac{\psi}{(\nu x U)^{\frac{1}{2}}} = f(\eta) \tag{11.12}$$

where f is an unknown function. Noting that the stream function, ψ, is related to the velocity field by the relations

$$u = \frac{\partial \psi}{\partial y}, \qquad v = -\frac{\partial \psi}{\partial x} \tag{11.13}$$

Equations 11.12 and 11.13 can be combined to give the following expressions for the velocity components, u and v, in terms of the similarity variable, η:

$$u = Uf'(\eta), \qquad v = \left(\frac{U\nu}{4x}\right)^{\frac{1}{2}}(\eta f' - f) \tag{11.14}$$

where $f' = \mathrm{d}f/\mathrm{d}\eta$. Substituting the relationships given in Equation 11.14 into the momentum and continuity equations (Equations 11.4 and 11.5) and then combining these equations yields

$$2f''' + ff'' = 0 \tag{11.15}$$

where $f'' = \mathrm{d}^2f/\mathrm{d}\eta^2$ and $f''' = \mathrm{d}^3f/\mathrm{d}\eta^3$. The boundary conditions to be used in solving Equation 11.15 are as follows: (1) the velocity is equal to zero on the solid surface (i.e., at $\eta = 0$) and (2) the velocity asymptotes to U as the distance from the surface increases (i.e., as $\eta \to \infty$). These boundary conditions can be expressed in terms of the function $f(\eta)$ as follows:

$$f(0) = f'(0) = 0 \tag{11.16}$$

$$f'(\infty) = 1 \tag{11.17}$$

Equation 11.15 is generally known as the *Blasius equation*, and the solution of Equation 11.15 subject to the boundary conditions in Equations 11.16 and 11.17 is called the *Blasius solution*.[62] The Blasius solution cannot be achieved analytically, and a numerical solution is required. The solution for $f'(\eta)$ (= u/U) is tabulated in Table 11.1 for several values of η (= $y\sqrt{U/\nu x}$). The results in Table 11.1 give the velocity distribution as a function of the distance perpendicular to a solid surface for any given value of x. Based on the results shown in Table 11.1, it is apparent that $f'(\eta) = u/U \approx 0.99$ when $\eta = y\sqrt{U/\nu x} = 4.91$; hence, the thickness of the boundary layer, δ, is given by the nondimensional relation

$$\frac{\delta}{x} = 4.91\sqrt{\frac{\nu}{Ux}}$$

which is more commonly expressed as

$$\frac{\delta}{x} = \frac{4.91}{\mathrm{Re}_x^{\frac{1}{2}}} \quad \text{(laminar BL)} \tag{11.18}$$

where Re_x is the flat surface Reynolds number defined by

$$\mathrm{Re}_x = \frac{Ux}{\nu} \tag{11.19}$$

Table 11.1: Blasius Solution to the Boundary-Layer Equation

η	$f'(\eta)$	η	$f'(\eta)$	η	$f'(\eta)$
0	0	2.4	0.7290	4.8	0.9878
0.4	0.1328	2.8	0.8115	5.0	0.9916
0.8	0.2647	3.2	0.8761	5.2	0.9943
1.2	0.3938	3.6	0.9233	5.6	0.9975
1.6	0.5168	4.0	0.9555	6.0	0.9990
2.0	0.6298	4.4	0.9759	∞	1.0000

[62]Named in honor of the German engineer and physicist Paul Richard Heinrich Blasius (1883–1970).

and the acronym BL in Equation 11.18 refers to "boundary layer." A particularly useful application of Equation 11.18 is in validating the implicit approximation that the boundary layer is thin relative to the distance along the surface. If the boundary layer was not thin, the flow in the boundary layer would be two-dimensional rather than one-dimensional as assumed in the Blasius solution. The existence of a thin boundary layer requires that $\delta \ll x$, which according to Equation 11.18 would (on order of magnitude) require that $\text{Re}_x > 1000$. The transition from a laminar to a turbulent boundary layer typically takes place in the range of 2×10^5 to 3×10^6. In conventional engineering analyses, boundary layers formed adjacent to flat surfaces are typically taken to be laminar (and described by Equation 11.18) when $\text{Re}_x < 5 \times 10^5$; for $\text{Re}_x > 5 \times 10^5$, these boundary layers are typically assumed to be turbulent and described by a growth function that is different from Equation 11.18. Turbulent boundary layers are described separately in a subsequent section.

EXAMPLE 11.2

Water at 20°C flows past a smooth flat surface with a free-stream velocity of 50 mm/s. (a) What is the velocity 15 mm from the plane surface at distances of 0.8 m and 8 m from the leading edge of the surface? (b) At approximately what distance from the leading edge of the surface does the boundary layer transition from laminar to turbulent? What is the thickness of the laminar boundary layer at this location?

SOLUTION

From the given data: U = 50 mm/s = 0.050 m/s. For water at 20°C, $\nu = 10^{-6}$ m²/s.

(a) At x = 0.8 m, the sequence of calculations to obtain the velocity, u, at y = 15 mm = 0.015 m from the surface are as follows:

$$\text{Re}_x = \frac{Ux}{\nu} = \frac{(0.050)(0.8)}{10^{-6}} = 4 \times 10^4 \quad \rightarrow \quad \delta = \frac{4.91x}{\text{Re}_x^{\frac{1}{2}}} = \frac{4.91(0.8)}{(4 \times 10^4)^{\frac{1}{2}}} = 0.0196 \text{ m}$$

$$\rightarrow \quad \eta = \frac{y}{\delta} = \frac{0.015}{0.0196} = 0.765 \quad \rightarrow \quad f'(\eta) = 0.253 \quad \text{(from Table 11.1)}$$

$$\rightarrow \quad u = Uf'(\eta) = 50(0.253) = 26.5 \text{ mm/s}$$

Similarly, at x = 8 m, the sequence of calculations and computed results are as follows:

$$\text{Re}_x = \frac{Ux}{\nu} = \frac{(0.050)(8)}{10^{-6}} = 4 \times 10^5 \quad \rightarrow \quad \delta = \frac{4.91x}{\text{Re}_x^{\frac{1}{2}}} = \frac{4.91(8)}{(4 \times 10^5)^{\frac{1}{2}}} = 0.0621 \text{ m}$$

$$\rightarrow \quad \eta = \frac{y}{\delta} = \frac{0.015}{0.0621} = 0.242 \quad \rightarrow \quad f'(\eta) = 0.0803 \quad \text{(from Table 11.1)}$$

$$\rightarrow \quad u = Uf'(\eta) = 50(0.0803) = 4.0 \text{ mm/s}$$

Based on these results, it is apparent that $10^3 < \text{Re}_x < 5 \times 10^5$ at both x = 0.8 m and x = 8 m, thereby validating the use of the Blasius laminar boundary-layer solution. The thickness of the boundary layer grows from around 20 mm to around 60 mm between x = 0.8 m and x = 8 m, and the velocity 15 mm from the surface decreases from **26.5 mm/s** to **4.0 mm/s** over this same interval.

(b) Taking the conventional criterion that the laminar/turbulent transition occurs at $\text{Re}_x = 5 \times 10^5$,

$$\frac{Ux}{\nu} = 5 \times 10^5 \quad \rightarrow \quad \frac{(0.050)x}{10^{-6}} = 5 \times 10^5 \quad \rightarrow \quad x = 10 \text{ m}$$

$$\frac{\delta}{x} = 4.91\sqrt{\frac{\nu}{Ux}} \quad \rightarrow \quad \frac{\delta}{10} = 4.91\sqrt{\frac{10^{-6}}{(0.050)(10)}} \quad \rightarrow \quad \delta = 0.0694 \text{ m} \approx 69 \text{ mm}$$

Therefore, transition to a turbulent boundary layer occurs at around x = **10 m**, and the thickness of the laminar boundary layer at this transition is approximately **69 mm**.

Estimation of shear stress. Using the velocity distribution given by the Blasius solution, the shear stress on the solid surface (i.e., the wall shear stress), τ_w, can be calculated using the relation

$$\tau_\text{w} = \mu \left. \frac{\partial u}{\partial y} \right|_{y=0}$$

Substituting the velocity distribution given by the Blasius solution yields

$$\tau_\text{w} = 0.332 \frac{\rho U^2}{\text{Re}_x^{\frac{1}{2}}} \quad \text{(laminar BL)} \tag{11.20}$$

This equation is particularly useful for estimating the viscous drag on a fixed surface as a fluid flows by the surface at velocity U. The *skin friction coefficient* or *local shear stress coefficient*, c_f, is generally defined as

$$c_\text{f} = \frac{\tau_\text{w}}{\frac{1}{2}\rho U^2} \tag{11.21}$$

Combining Equations 11.20 and 11.21 yields the following expression for the skin friction coefficient on a flat surface:

$$c_\text{f} = \frac{\tau_\text{w}}{\frac{1}{2}\rho U^2} = \frac{0.664}{\text{Re}_x^{\frac{1}{2}}} \quad \text{(laminar BL)} \tag{11.22}$$

For a flat surface of width b and length L, the drag coefficient due to friction, C_Df, is defined as

$$C_\text{Df} = \frac{F_\text{D}}{\frac{1}{2}\rho U^2 L b} = \frac{b\displaystyle\int_0^L \tau_\text{w}\, \text{d}x}{\frac{1}{2}\rho U^2 L b} \tag{11.23}$$

where F_D represents the drag force on the surface. Combining Equations 11.20 and 11.23 yields

$$C_\text{Df} = \frac{F_\text{D}}{\frac{1}{2}\rho U^2 L b} = \frac{1.328}{\text{Re}_L^{\frac{1}{2}}} \quad \text{(laminar BL)} \tag{11.24}$$

EXAMPLE 11.3

Air at 10°C flows at 18 km/h over a flat surface that is 1 m long and 2 m wide. Determine the shear stress at the downstream end of the surface, the average shear stress on the surface, and the drag force on the surface.

SOLUTION

From the given data: L = 1 m, b = 2 m, and U = 18 km/h = 5 m/s. For air at 10°C, ρ = 1.246 kg/m^3, μ = 0.0177 mPa·s, and $\nu = \mu/\rho = 1.42 \times 10^{-5}$ m^2/s. The Reynolds number at the end of the flat surface, $\mathrm{Re_L}$, is given by

$$\mathrm{Re_L} = \frac{UL}{\nu} = \frac{(5)(1)}{1.42 \times 10^{-5}} = 3.52 \times 10^5$$

Because $\mathrm{Re_L} < 5 \times 10^5$, it can be assumed that the flow in the boundary layer is laminar over the entire surface. For a laminar boundary layer, the local (wall) shear stress, τ_w, at the downstream end of the surface is given by Equation 11.20 as

$$\tau_\mathrm{w} = 0.332 \frac{\rho U^2}{\mathrm{Re}_x^{\frac{1}{2}}} = 0.332 \frac{(1.246)(5)^2}{(3.52 \times 10^5)^{\frac{1}{2}}} = 0.0174 \text{ Pa} = \mathbf{17.4\ mPa}$$

The average shear stress on the surface, $\bar{\tau}_\mathrm{w}$, can be derived from the drag coefficient, C_Df, using the relation

$$\bar{\tau}_\mathrm{w} = \frac{b\int_0^L \tau_\mathrm{w}\, dx}{Lb} = C_\mathrm{Df}\frac{1}{2}\rho U^2 = \left[\frac{1.328}{\mathrm{Re_L^{\frac{1}{2}}}}\right]\frac{1}{2}\rho U^2 = \left[\frac{1.328}{(3.52 \times 10^5)^{\frac{1}{2}}}\right]\frac{1}{2}(1.246)(5)^2$$
$$= 0.0349 \text{ Pa} = \mathbf{34.9\ mPa}$$

Because the average shear stress (34.9 mPa) on the surface is greater than the local shear stress at the end of the surface (17.4 mPa), it is apparent that the local shear stress decreases with distance along the surface. This is a result of the fact that the velocity gradient at the surface is decreasing with distance along the surface. The drag force on the plane surface, F_D, is given by

$$F_\mathrm{D} = \bar{\tau}_\mathrm{w} Lb = (34.9)(1)(2) = \mathbf{69.8\ mN}$$

This is a relatively small force, as would be expected from a moderate airspeed (18 km/h) over a small surface area (2 m^2).

Analytic approximations. Several analytic approximations to the Blasius velocity distribution have been developed by various researchers, with corresponding expressions for δ/x, τ_w, and C_Df. However, the Blasius solution remains the benchmark relative to which the accuracy of these approximations are measured. Given the ease with which the Blasius velocity distribution can be obtained by the numerical solution of Equation 11.15, subject to the boundary conditions given by Equations 11.16 and 11.17, such analytic approximations to the Blasius solution are mostly unnecessary in the modern age.

11.2.2 Blasius Equations for Curved Surfaces

In analyzing boundary-layer flows over curved surfaces, it is usually convenient to define the x-coordinate as being measured along the (curved) surface and the y-coordinate being measured perpendicular to the surface. The flow direction within the boundary layer is predominantly in the x-direction. The main difference between a boundary layer on a curved surface and a boundary layer on a flat surface is that the pressure gradient in the flow direction, $\mathrm{d}p/\mathrm{d}x$, can be significantly different from zero on a curved surface, whereas $\mathrm{d}p/\mathrm{d}x \approx 0$ on a flat surface that is parallel to the flow. As a consequence, for curved surfaces, the pressure-gradient term is retained in the x-momentum equation, which is given by

$$u\frac{\partial u}{\partial x} + v\frac{\partial u}{\partial y} = -\frac{1}{\rho}\frac{\mathrm{d}p}{\mathrm{d}x} + \nu\frac{\partial^2 u}{\partial y^2} \tag{11.25}$$

This equation is in contrast to Equation 11.4 for a flat surface, which does not have the pressure-gradient term. Applying the Bernoulli equation to the outer (irrotational) flow region requires that

$$\frac{p}{\rho} + \frac{1}{2}U^2 = \text{constant} \quad \rightarrow \quad \frac{1}{\rho}\frac{\mathrm{d}p}{\mathrm{d}x} = -U\frac{\mathrm{d}U}{\mathrm{d}x} \tag{11.26}$$

where $U(x)$ describes the variation of the free-stream velocity just outside the boundary layer. Combining Equations 11.25 and 11.26 gives the following momentum equation within a boundary layer adjacent to a curved surface:

$$u\frac{\partial u}{\partial x} + v\frac{\partial u}{\partial y} = U\frac{\mathrm{d}U}{\mathrm{d}x} + \nu\frac{\partial^2 u}{\partial y^2} \tag{11.27}$$

The continuity equation for boundary-layer flows adjacent to curved surfaces is unaffected by a pressure gradient and is therefore the same for both a flat surface and a curved surface. The continuity equation was given previously as Equation 11.5 and is repeated here for easy reference as

$$\frac{\partial u}{\partial x} + \frac{\partial v}{\partial y} = 0 \tag{11.28}$$

Equations 11.27 and 11.28 are the fundamental equations to be solved in the boundary-layer region. The boundary conditions typically used with these equations are as follows:

$$u(x,0) = v(x,0) = 0 \tag{11.29}$$

$$u(x,\infty) = U(x) \tag{11.30}$$

$$u(x_0,y) = f(y) \tag{11.31}$$

where x_0 is the x-coordinate at the upstream end of the section of boundary layer that is of interest and $f(y)$ is the velocity distribution within the boundary layer at the upstream location. The following sequence of steps are normally followed in solving the boundary-layer equations:

Step 1: Solve for $U(x)$ in the outer flow region by assuming a boundary layer of negligible thickness. This is commonly done by assuming potential flow outside the boundary layer.

Step 2: Solve Equations 11.27 and 11.28 subject to the boundary conditions given by Equations 11.29–11.31. This typically requires a numerical solution.

Step 3: From the calculated velocity distribution in the boundary layer (derived from Step 2), determine the quantities of interest, such as the boundary-layer thickness, $\delta(x)$, and the drag force, F_D.

Step 4: Validate the assumptions made in the solution. Key assumptions to be validated are that (a) the boundary-layer thickness is sufficiently small as to justify neglecting it in Step 1 and (b) the Reynolds number is not so high that the boundary layer is turbulent when it was assumed to be laminar, which typically requires $\mathrm{Re}_x < 5 \times 10^5$.

Favorable and adverse pressure gradients. The pressure gradient, $\mathrm{d}p/\mathrm{d}x$, within a boundary layer is called a *favorable pressure gradient* when $\mathrm{d}p/\mathrm{d}x < 0$ and is called an *adverse pressure gradient* when $\mathrm{d}p/\mathrm{d}x > 0$. Favorable pressure gradients are associated with an accelerating free-stream velocity, whereas adverse pressure gradients are associated with a decelerating free-stream velocity. Boundary-layer flows in adverse pressure gradients should generally be treated with caution because such adverse gradients might cause reverse flow in the boundary layer, which leads to boundary-layer separation and the invalidity on the boundary-layer solution. Under such circumstances, the full Navier-Stokes equations must be used to describe the flow within the region influenced by the separated boundary layer.

Numerical considerations. The boundary-layer equations are classified as *parabolic equations*, which is a mathematical classification that also includes heat transfer and mass transfer equations. Parabolic equations have the general property that their solutions are determined only by conditions on the upstream boundary. In contrast, the Navier-Stokes (momentum) equations are classified as *elliptic equations*, and their solutions are determined by conditions on all of the boundaries containing the flow field. Hence, simplification of the momentum equations in a boundary layer also affects the appropriate numerical solution technique.

11.3 Turbulent Boundary Layers

Turbulent boundary layers differ from laminar boundary layers in that macroscopic random fluctuations of velocity occur in turbulent boundary layers, compared with the small molecular-scale velocity fluctuations that occur in laminar boundary layers. Viewed from another perspective, the commonality between turbulent and laminar flows is that random perturbations in velocity exist in both cases, with the main difference being that these random perturbations occur on a much larger scale in turbulent flows compared with laminar flows. Turbulent boundary layers grow at a more rapid rate than laminar boundary layers. The other major difference between laminar and turbulent boundary layers is that the roughness of the surface is an added factor that can affect the velocity distribution in turbulent boundary layers. Surface roughness is not a factor in laminar boundary layers.

11.3.1 Analytic Formulation

The approximate momentum and continuity equations developed in the previous sections for laminar boundary layers also can be applied to turbulent boundary layers provided the viscous shear stress is replaced by the turbulent shear stress and the velocity components are taken as time-averaged quantities. Rewriting Equation 11.27 in terms of the shear stress, τ, and representing the velocity components as averaged quantities gives

$$\bar{u}\frac{\partial \bar{u}}{\partial x} + \bar{v}\frac{\partial \bar{u}}{\partial y} = U\frac{\mathrm{d}U}{\mathrm{d}x} + \frac{\partial \tau}{\partial y} \tag{11.32}$$

Then Equation 11.32 can be applied for both laminar and turbulent flows by specifying τ as

$$\tau = \begin{cases} \mu\dfrac{\partial u}{\partial y}, & \text{(laminar flow)} \\[2ex] \mu\dfrac{\partial \bar{u}}{\partial y} - \rho\overline{u'v'}, & \text{(turbulent flow)} \end{cases} \tag{11.33}$$

where in the case of turbulent flow, u' and v' are the velocity fluctuations relative to the time-averaged velocity components $\bar{u}$ and $\bar{v}$, respectively. The time-averaged continuity equation under turbulent flow conditions is given by

$$\frac{\partial \bar{u}}{\partial x} + \frac{\partial \bar{v}}{\partial y} = 0 \tag{11.34}$$

Equations 11.33 and 11.34 are solved subject to the following boundary conditions:

$$\bar{u}(x,0) = \bar{v}(x,0) = 0 \tag{11.35}$$

$$\bar{u}(x,\infty) = U(x) \tag{11.36}$$

$$\bar{u}(x_0,y) = f(y) \tag{11.37}$$

where $f(y)$ is the initial velocity distribution across the boundary layer at the upstream location x_0. Solution of Equations 11.32 and 11.34 must usually be done numerically, particularly when the free-stream velocity, $U(x)$, is dictated by the shape of the solid surface around which the fluid flows. However, in cases where the solid surface is relatively flat, the approximate approach described in the following section is widely used to describe the growth of the turbulent boundary layer.

11.3.2 Turbulent Boundary Layer on a Flat Surface

The turbulent boundary layer on a flat surface is sometimes called the *wall region*, and the flat surface is sometimes called the *wall*. The boundary layer or wall region is conventionally delineated into three regions that are characterized by the distance from the wall as follows:

Region 1: Viscous sublayer. The very thin layer next to the wall where viscous effects are dominant. The velocity profile in this layer is very nearly linear, and the flow is nearly parallel to the wall. The viscous sublayer is sometimes called the *wall layer* or simply the *viscous layer*.

Region 2: Transition layer. Within this layer, there is a transition from laminar flow in the viscous sublayer to turbulent flow in the outer layer. The transition layer is also called the *buffer layer*, *inertial layer*, or *overlap layer*.

Region 3: Turbulent layer or outer layer. Within this layer, turbulent effects dominate over viscous effects.

The extent of the aforementioned regions relative to a typical solid surface is illustrated in Figure 11.2, where the protrusion of the surface roughness into the various sublayers within the boundary layer is also shown. Surfaces with roughness that is contained entirely within the viscous sublayer are called *hydrodynamically smooth surfaces*, and surfaces with roughness that extends beyond the viscous sublayer are called *hydrodynamically rough surfaces*. For ease of reference, these surfaces will be referred to as "smooth" and "rough" surfaces, respectively. The practical consequence of a surface being smooth or rough is that in flows over smooth surfaces, the roughness height does not significantly affect the velocity distribution within the turbulent boundary layer, whereas in flows over rough surfaces, the roughness height has a significant (and sometimes dominant) effect on the velocity distribution within the turbulent boundary layer.

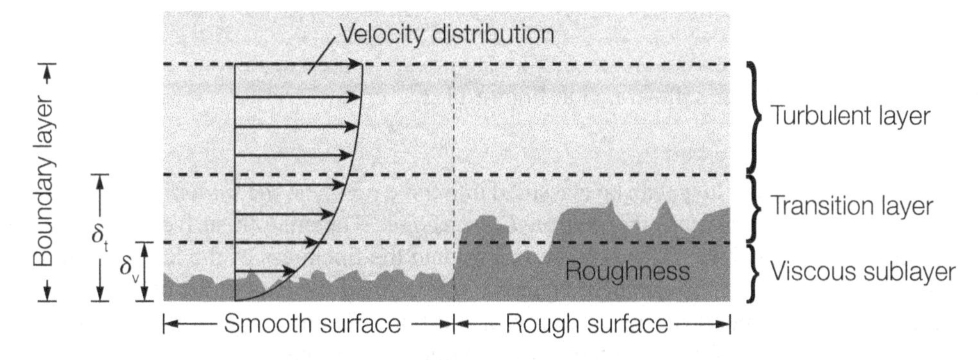

Figure 11.2: Sublayers in a turbulent boundary layer

11.3.2.1 Flow in the viscous sublayer

Within the viscous sublayer, the flow is laminar, and assuming that the fluid is Newtonian, the relationship between the shear stress on the solid surface, τ_w, and the velocity gradient, $\mathrm{d}u/\mathrm{d}y$, at the surface is given by

$$\tau_\mathrm{w} = \mu \left.\frac{\mathrm{d}u}{\mathrm{d}y}\right|_{y=0} \tag{11.38}$$

where μ is the dynamic viscosity, u is the flow velocity parallel to the solid surface, and y is the distance from the surface measured perpendicular to the surface. Laboratory observations indicate that the velocity distribution is approximately linear within the viscous sublayer, in which case Equation 11.38 can be expressed as

$$\tau_\mathrm{w} = \mu\frac{u}{y} \tag{11.39}$$

In working with applied shear stresses in fluids, it is usually convenient to define the *friction velocity*, u_*, that is characteristic of the velocities induced by the shear stress. In the case of an applied shear stress τ_w, the friction velocity is given by

$$u_* = \sqrt{\frac{\tau_\mathrm{w}}{\rho}} \tag{11.40}$$

Combining Equations 11.39 and 11.40 and rearranging gives the velocity distribution in the laminar sublayer as

$$\frac{u}{u_*} = \frac{u_* y}{\nu} \quad \text{(in viscous sublayer)} \tag{11.41}$$

Based on experimental results, the viscous sublayer extends to a distance δ_v [L] from the wall, where δ_v can be estimated by

$$\delta_\mathrm{v} = \frac{5\nu}{u_*} \tag{11.42}$$

It is sometimes assumed that the viscous sublayer extends out to $u_*y/\nu = 8$ (e.g., Massey and Ward-Smith, 2012). However, an outer limit of $u_*y/\nu = 5$ is more widely used and will be used in this text.

11.3.2.2 Flow in the transition layer

The transition layer is that region in which the flow transitions from being laminar (in the viscous sublayer) to being turbulent. Based on experimental measurements, the transition layer extends to a distance δ_t [L] from the wall such that

$$\delta_\mathrm{t} = \frac{20\nu}{u_*} = 4\delta_\mathrm{v} \tag{11.43}$$

The distances δ_v and δ_t are illustrated in Figure 11.2.

11.3.2.3 Flow in the turbulent layer

The flow in the turbulent layer can be classified into three regimes: the *smooth surface regime*, the *rough surface regime*, and the *intermediate regime*. The smooth surface regime exists when the roughness height of the surface is less than the thickness of the laminar sublayer, in which case the flow in the turbulent layer is unaffected by the magnitude of the roughness height, but it does depend on the viscosity of the fluid. In the rough surface regime, the roughness height is significantly greater than the thickness of the viscous sublayer, in which case the flow in the turbulent layer is controlled by the magnitude of the roughness height and does not depend on the viscosity of the fluid. In the intermediate regime, the flow in the turbulent layer depends on both the viscosity of the fluid and the roughness height. If the characteristic roughness height of the surface is denoted by ϵ [L], the conditions for smooth and rough surfaces are commonly taken as

$$\text{smooth:} \quad \frac{\epsilon}{\delta_v} < 1 \quad \text{or} \quad \frac{\epsilon u_*}{\nu} < 5 \tag{11.44}$$

$$\text{rough:} \quad \frac{\epsilon}{\delta_v} > 14 \quad \text{or} \quad \frac{\epsilon u_*}{\nu} > 70 \tag{11.45}$$

Consistent with the conditions given in Equations 11.44 and 11.45, the intermediate condition exists under the following conditions:

$$\text{intermediate:} \quad 1 \le \frac{\epsilon}{\delta_v} \le 14 \quad \text{or} \quad 5 < \frac{\epsilon u_*}{\nu} \le 70 \tag{11.46}$$

Another benchmark roughness that is related to the thickness of the laminar sublayer, δ_v, is the *critical roughness*, ϵ_c, which is the minimum roughness height that is required to "trip" a laminar boundary layer into becoming a turbulent boundary layer. The critical roughness can be taken as

$$\text{critical roughness:} \quad \epsilon_c = 3\delta_v \quad \rightarrow \quad \epsilon_c = \frac{15\nu}{u_*} \tag{11.47}$$

Therefore, when $\text{Re}_x < 5 \times 10^5$, a laminar boundary layer can be made turbulent if the roughness height exceeds ϵ_c. It is important to note that laminar flow should be assumed in calculating the value of u_* to be used in Equation 11.47.

Velocity distribution in turbulent boundary layers. The velocity distribution in turbulent boundary layers under smooth and rough conditions can be estimated by a combination of mixing-length theory (see Section 11.5) and experimental data. The derived relationships generally have the same functional (logarithmic) form; however, the constants in the equations that are presented in the open technical literature tend to be variable. The general logarithmic functional forms for the velocity distributions in smooth turbulent and rough turbulent flows are given in the following equations, with alternative sets of constants that are commonly used.

$$\text{smooth:} \quad \frac{u}{u_*} = 2.5 \ln\left(\frac{u_* y}{\nu}\right) + 5.5 \tag{11.48}$$

$$\text{rough:} \quad \frac{u}{u_*} = 2.5 \ln\left(\frac{y}{\epsilon}\right) + 8.5 \tag{11.49}$$

$$\text{smooth:} \quad \frac{u}{u_*} = 2.44 \ln\left(\frac{u_* y}{\nu}\right) + 5.0 \tag{11.50}$$

$$\text{rough:} \quad \frac{u}{u_*} = 2.44 \ln\left(\frac{y}{\epsilon}\right) + 8.5 \tag{11.51}$$

The uncertainty in the constants in the above equations is a reflection of the uncertainty in the data from which these constants were derived. Equations with various constants are used in practice, and most give similar results. The velocity distribution within the viscous sublayer (Equation 11.41) transitions into a logarithmic velocity distribution (e.g., Equation 11.50 or 11.51), and the combination of these two velocity distributions is commonly referred to as the *law of the wall*.

EXAMPLE 11.4

Consider the case of a liquid flowing down an inclined plane as illustrated in Figure 11.3. Application of the momentum equation shows that under steady-state conditions, the shear stress on the inclined surface, τ_w, is given by

$$\tau_\mathrm{w} = \gamma d S_0$$

where γ is the specific weight of the liquid, d is the flow depth, and S_0 is the slope of the surface. In the present case, the fluid is water at 20°C, the slope of the surface is 0.1%, and the roughness height is 1 mm. (a) What is the maximum depth of flow for which the smooth surface turbulent flow condition exists? (b) What is the minimum flow depth for which the rough surface turbulent flow condition exists? (c) Compare the velocity distributions for the limiting flow depths found in parts (a) and (b).

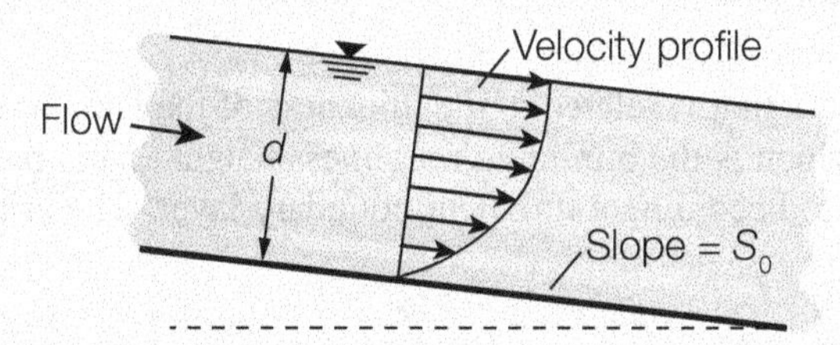

Figure 11.3: Flow down an inclined plane

SOLUTION

From the given data: ϵ = 1 mm = 0.001 m and S_0 = 0.1% = 0.001. For water at 20°C, ν = 1.00×10^{-6} m²/s. It is also given that

$$\tau_\mathrm{w} = \gamma d S_0 \quad \rightarrow \quad \frac{\tau_\mathrm{w}}{\rho} = g d S_0 \quad \rightarrow \quad u_* = \sqrt{g d S_0}$$

(a) At the limit of smooth surface turbulent flow conditions, the flow depth, d, is calculated as follows:

$$\frac{\epsilon u_*}{\nu} = 5 \quad \rightarrow \quad \frac{\epsilon\sqrt{gdS_0}}{\nu} = 5 \quad \rightarrow \quad \frac{(0.001)\sqrt{(9.81)d(0.001)}}{1.00 \times 10^{-6}} = 5$$
$$\rightarrow \quad d = 0.00255 \text{ m} \quad \rightarrow \quad d = 2.55 \text{ mm}$$

Under this flow condition, the shear velocity, u_*, the thickness of the viscous sublayer, δ_v, and the height of the transition layer, δ_t, are given by

$$u_* = \sqrt{gdS_0} = \sqrt{(9.81)(0.00255)(0.001)} = 0.005 \text{ m/s}$$
$$\delta_\mathrm{v} = \frac{5\nu}{u_*} = \frac{5(1.00 \times 10^{-6})}{0.005} = 0.001 \text{ m} = 1.0 \text{ mm}$$
$$\delta_\mathrm{t} = \frac{20\nu}{u_*} = 4\delta_\mathrm{v} = 4(0.001) = 0.004 \text{ m} = 4.0 \text{ mm}$$

Therefore, at the smooth surface limit, the depth of flow is approximately **2.6 mm**, the thickness of the viscous sublayer is equal to 1.0 mm, and the height of the transition layer is estimated as 4 mm. Because the depth of flow is less than the height of the transition layer, turbulent flow conditions are not attained over the flow depth.

(b) At the lower limit of rough surface turbulent flow conditions, the flow depth, d, is calculated as follows:

$$\frac{\epsilon u_*}{\nu} = 70 \quad \rightarrow \quad \frac{\epsilon\sqrt{gdS_0}}{\nu} = 70 \quad \rightarrow \quad \frac{(0.001)\sqrt{(9.81)d(0.001)}}{1.00 \times 10^{-6}} = 70$$
$$\rightarrow \quad d = 0.500 \text{ m} \quad \rightarrow \quad d = 500 \text{ mm}$$

Under this flow condition, the shear velocity, u_*, the thickness of the viscous sublayer, δ_v, and the height of the transition layer, δ_t, are given by

$$u_* = \sqrt{gdS_0} = \sqrt{(9.81)(0.500)(0.001)} = 0.070 \text{ m/s}$$
$$\delta_\text{v} = \frac{5\nu}{u_*} = \frac{5(1.00 \times 10^{-6})}{0.070} = 0.0000714 \text{ m} = 0.07 \text{ mm}$$
$$\delta_\text{t} = \frac{20\nu}{u_*} = 4\delta_\text{v} = 4(0.0000714) = 0.000286 \text{ m} = 0.286 \text{ mm}$$

Therefore, at the lower limit of the rough surface condition, the depth of flow is approximately **0.5 m**. Under this condition, the thickness of the viscous sublayer is approximately equal to 0.1 mm and the height of the transition layer is approximately 0.3 mm. Hence, the roughness, with a height of 1 mm, penetrates both the viscous and transition layers. Turbulent flow in the rough surface regime exists beyond a distance δ_t (= 0.3 mm) from the surface.

(c) The smooth surface turbulent flow velocity profile is given by Equation 11.48 as

$$\frac{u}{u_*} = 2.5\ln\left[\frac{u_* y}{\nu}\right] + 5.5 \quad \rightarrow \quad \frac{u}{0.005} = 2.5\ln\left[\frac{(0.005)y}{1.00 \times 10^{-6}}\right] + 5.5$$
$$\rightarrow \quad u = 0.0125 \ln y + 0.134$$

and the rough surface turbulent flow velocity profile is given by Equation 11.49 as

$$\frac{u}{u_*} = 2.5\ln\left[\frac{y}{\epsilon}\right] + 8.5 \rightarrow \frac{u}{0.070} = 2.5\ln\left[\frac{y}{0.001}\right] + 8.5 \rightarrow u = 0.175 \ln y + 1.81$$

The smooth surface and rough surface velocity profiles are compared to each other in Figure 11.4, where the smooth surface velocity profile is plotted for $y \geq \delta_\text{v}$ and the rough surface velocity profile is plotted for $y \geq \delta_\text{t}$.

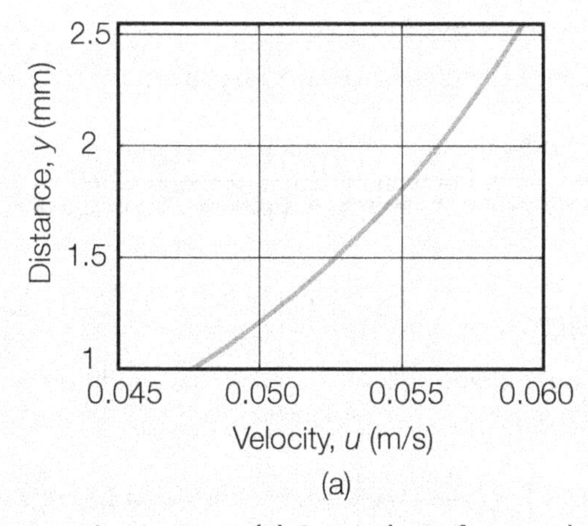

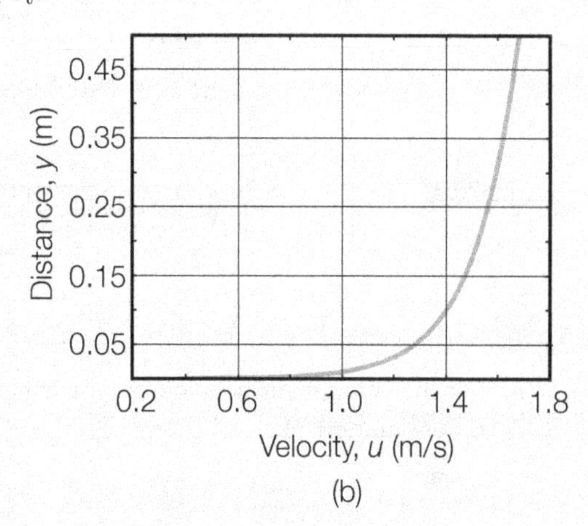

Figure 11.4: (a) Smooth surface and (b) rough surface velocity profiles

The smooth surface plot is an approximation in that the flow in this case is actually in the transition region between laminar and turbulent flow. It is apparent from the comparison shown in Figure 11.4 that the rough surface velocity profile has a much higher velocity gradient near the plane surface and the velocity profile as a whole is much fuller.

11.3.2.4 One-seventh power law velocity distribution

An empirical velocity profile that is commonly used to approximate the velocity distribution in a turbulent boundary layer adjacent to a smooth surface (originally suggested by Prandtl) is the *one-seventh power law*, given by

$$\frac{u}{U} = \begin{cases} \left(\frac{y}{\delta}\right)^{\frac{1}{7}} & y \leq \delta \\ 1 & y > \delta \end{cases} \tag{11.52}$$

where δ defines the limit of the turbulent boundary layer. The definition of δ in Equation 11.52 is different than that used in the previous sections for laminar boundary layers, where δ was defined as the location where $u = 0.99U$.

EXAMPLE 11.5

Consider the case of a turbulent boundary layer adjacent to a plane surface as shown in Figure 11.5. The fluid is water at 20°C; the free-stream velocity is 3 m/s; and at a particular location, the thickness of the boundary layer is 5 mm. It is proposed to describe the velocity distribution at the particular location using the one-seventh power law. Assess the adequacy of using this approximation in the following two cases: (a) the surface is smooth and (b) the surface has a roughness height of 0.5 mm.

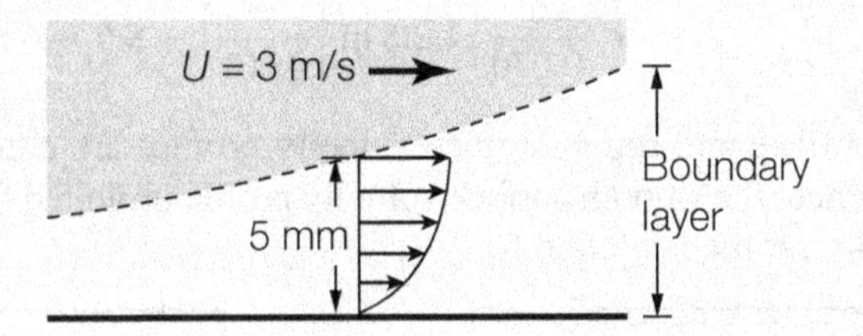

Figure 11.5: Velocity distribution in a turbulent boundary layer

SOLUTION

From the given data: $U = 3$ m/s and $\delta = 5$ mm $= 0.005$ m. For water at 20°C, $\nu = 1.00\times10^{-6}$ m²/s. Using the one-seventh power law, the velocity distribution is estimated by

$$\frac{u}{U} = \left(\frac{y}{\delta}\right)^{\frac{1}{7}} \quad \rightarrow \quad \frac{u}{3} = \left(\frac{y}{0.005}\right)^{\frac{1}{7}} \quad \rightarrow \quad u = 6.39\, y^{\frac{1}{7}} \tag{11.53}$$

(a) For the smooth surface velocity distribution, Equation 11.48, to satisfy the condition that $u = U$ and $y = \delta$,

$$\frac{U}{u_*} = 2.5 \ln\left[\frac{u_*\delta}{\nu}\right] + 5.5 \rightarrow \frac{3}{u_*} = 2.5 \ln\left[\frac{u_*(0.005)}{1.00 \times 10^{-6}}\right] + 5.5 \rightarrow u_* = 0.137 \text{ m/s}$$

Hence, the smooth surface velocity distribution and the height of the transition layer, δ_t, are given by

$$\frac{u}{0.137} = 2.5\ln\left[\frac{(0.137)y}{1.00\times10^{-6}}\right] + 5.5 \quad\rightarrow\quad u = 0.343\ln y + 4.80 \qquad (11.54)$$

$$\delta_t = \frac{20\nu}{u_*} = \frac{20(1.00\times10^{-6})}{0.137} = 1.46\times10^{-4}\ \text{m} = 0.146\ \text{mm}$$

Turbulent flow is confirmed because $\delta \gg \delta_t$ (i.e., 5 mm $\gg$ 0.146 mm), so **the smooth surface velocity distribution given by Equation 11.54 can be applied** for $y \geq 0.146$ mm.

(b) For the rough surface, ϵ = 0.5 mm = 0.0005 m, and for the velocity distribution given by Equation 11.49 to satisfy the condition that $u = U$ and $y = \delta$ requires that

$$\frac{U}{u_*} = 2.5\ln\left[\frac{\delta}{\epsilon}\right] + 8.5 \quad\rightarrow\quad \frac{3}{u_*} = 2.5\ln\left[\frac{0.005}{0.0005}\right] + 8.5 \quad\rightarrow\quad u_* = 0.210\ \text{m/s}$$

Hence, the rough surface velocity distribution and the height of the transition layer, δ_t, are given by

$$\frac{u}{0.210} = 2.5\ln\left[\frac{y}{0.0005}\right] + 8.5 \quad\rightarrow\quad u = 0.525\ln y + 5.78 \qquad (11.55)$$

$$\delta_t = \frac{20\nu}{u_*} = \frac{20(1.00\times10^{-6})}{0.210} = 9.52\times10^{-5}\ \text{m} = 0.095\ \text{mm}$$

Turbulent flow is confirmed because $\delta \gg \delta_t$ (i.e., 5 mm $\gg$ 0.095 mm), so **the rough surface velocity distribution given by Equation 11.55 can be applied** for $y \geq 0.095$ mm.

The one-seventh power law velocity distribution (Equation 11.53) is compared with the smooth surface velocity distribution (Equation 11.54) and the rough surface velocity distribution (Equation 11.55) in Figure 11.6 for $y \geq 0.146$ mm, for which turbulent flow exists in all cases.

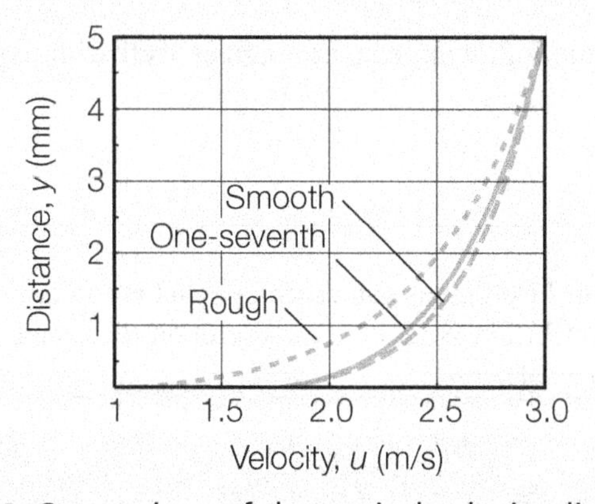

Figure 11.6: Comparison of theoretical velocity distributions

It is apparent from Figure 11.6 that **the one-seventh power law approximation is in best agreement with the smooth surface velocity distribution**. This result is expected since the one-seventh power law approximation is only applicable to smooth surfaces.

11.3.3 Boundary-Layer Thickness and Shear Stress

Using empirical data from turbulent flow through smooth pipes along with the one-seventh power law to estimate the velocity distribution, the conventional semiempirical equation that is used to estimate the growth of a turbulent boundary layer on a smooth flat surface of constant width is given by

$$\frac{\delta}{x} = \frac{0.382}{\mathrm{Re}_x^{\frac{1}{5}}} \quad \text{for} \quad 5\times10^5 \leq \mathrm{Re}_x < 10^7 \text{ (smooth, turbulent BL)} \tag{11.56}$$

where x is measured from the front end of the flat surface. The complementary conventional assumption is that the boundary layer on a smooth flat surface remains laminar for $\mathrm{Re}_x < 5 \times 10^5$, in which case the growth of the (laminar) boundary layer is described by Equation 11.18, which is derived from the Blasius solution. The local shear stress, τ_w, as a function of the distance x along the flat surface can be estimated by the semiempirical relation

$$c_\mathrm{f} = \frac{\tau_\mathrm{w}}{\frac{1}{2}\rho U^2} = \frac{0.0594}{\mathrm{Re}_x^{\frac{1}{5}}} \quad \text{for} \quad 5\times10^5 \leq \mathrm{Re}_x < 10^7 \text{ (smooth, turbulent BL)} \tag{11.57}$$

where c_f denotes the skin friction coefficient. Equations 11.56 and 11.57 both show good agreement with experimental data in the range of Re_x for which they are valid. For a flat surface of width b and length L, the drag coefficient due to friction, C_Df, is defined as

$$C_\mathrm{Df} = \frac{F_\mathrm{D}}{\frac{1}{2}\rho U^2 Lb} = \frac{b\displaystyle\int_0^L \tau_\mathrm{w}\,\mathrm{d}x}{\frac{1}{2}\rho U^2 Lb} \tag{11.58}$$

where F_D represents the drag force on the surface. Combining Equations 11.57 and 11.58 yields

$$C_\mathrm{Df} = \frac{F_\mathrm{D}}{\frac{1}{2}\rho U^2 Lb} = \frac{0.0735}{\mathrm{Re}_L^{\frac{1}{5}}} \quad \text{for} \quad 5\times10^5 \leq \mathrm{Re}_L < 10^7 \text{ (smooth, turbulent BL)} \tag{11.59}$$

For $\mathrm{Re} \geq 10^7$, the following equation agrees well with experimental results (Schlichting, 1949):

$$C_\mathrm{Df} = \frac{F_\mathrm{D}}{\frac{1}{2}\rho U^2 Lb} = \frac{0.455}{(\log \mathrm{Re}_L)^{2.58}} \quad \text{for} \quad \mathrm{Re}_L \geq 10^7 \text{ (smooth, turbulent BL)} \tag{11.60}$$

Notably, Equation 11.60 gives values of C_Df that are in close agreement with values of C_Df given by Equation 11.59. So it is not uncommon for Equation 11.60 to be used to estimate C_Df for $\mathrm{Re}_x > 5 \times 10^5$.

EXAMPLE 11.6

Water at 20°C flows with a free-stream velocity of 2 m/s over a 1.5-m-long flat surface. Conditions are created such that the boundary layer becomes turbulent from the leading edge. (a) Estimate the boundary-layer thickness and the local shear stress at the downstream end of the surface. (b) Compare the results in part (a) with those that would be obtained if the boundary layer remained laminar over the entire surface.

SOLUTION

From the given data: $L = 1.5$ m and $U = 2$ m/s. For water at 20°C, $\nu = 1.004 \times 10^{-6}$ m²/s and $\rho = 998.2$ kg/m³. The Reynolds number at the downstream end of the surface is given by

$$\text{Re}_\text{L} = \frac{UL}{\nu} = \frac{(2)(1.5)}{1.004 \times 10^{-6}} = 2.988 \times 10^6$$

(a) For a turbulent boundary layer, the boundary-layer thickness, δ, and the shear stress, τ_w, on the downstream end of the surface are calculated using Equations 11.56 and 11.57 as follows:

$$\frac{\delta}{L} = \frac{0.382}{\text{Re}_\text{L}^{\frac{1}{5}}} \quad \rightarrow \quad \frac{\delta}{1.5} = \frac{0.382}{(2.988 \times 10^6)^{\frac{1}{5}}}$$

$$\rightarrow \quad \delta = 0.0290 \text{ m} = \mathbf{29\ mm}$$

$$\frac{\tau_\text{w}}{\frac{1}{2}\rho U^2} = \frac{0.0594}{\text{Re}_\text{L}^{\frac{1}{5}}} \quad \rightarrow \quad \frac{\tau_\text{w}}{\frac{1}{2}(998.2)(1.5)^2} = \frac{0.0594}{(2.988 \times 10^6)^{\frac{1}{5}}}$$

$$\rightarrow \quad \tau_\text{w} = \mathbf{6.01\ Pa}$$

(b) For a laminar boundary layer, the boundary-layer thickness, δ, and the shear stress, τ_w, on the downstream end of the surface are calculated using Equations 11.18 and 11.22 as follows:

$$\frac{\delta}{L} = \frac{4.91}{\text{Re}_\text{L}^{\frac{1}{2}}} \quad \rightarrow \quad \frac{\delta}{1.5} = \frac{4.91}{(2.988 \times 10^6)^{\frac{1}{2}}}$$

$$\rightarrow \quad \delta = 0.0043 \text{ m} = \mathbf{4\ mm}$$

$$\frac{\tau_\text{w}}{\frac{1}{2}\rho U^2} = \frac{0.664}{\text{Re}_\text{L}^{\frac{1}{2}}} \quad \rightarrow \quad \frac{\tau_\text{w}}{\frac{1}{2}(998.2)(1.5)^2} = \frac{0.664}{(2.988 \times 10^6)^{\frac{1}{2}}}$$

$$\rightarrow \quad \tau_\text{w} = \mathbf{0.77\ Pa}$$

It is apparent from these results that the boundary-layer thickness and the boundary shear stress for a turbulent boundary layer are much greater than for a laminar boundary layer.

11.4 Applications

In the context of solving the approximate Navier-Stokes equations within the boundary layer, the thickness of the boundary layer is commonly defined as the distance from the solid surface to where the velocity becomes equal to 99% of the free-stream velocity. This is sometimes called the *disturbance thickness*. However, other measures of boundary-layer thickness are also used, most notably the *displacement thickness* and the *momentum thickness*.

11.4.1 Displacement Thickness

Streamlines adjacent to a solid surface must necessarily trend outward from the solid surface to satisfy the continuity equation as the retardation effect of the surface on the fluid (due to viscosity) penetrates into the free stream. The *displacement thickness*, δ^*, is defined as the distance that the streamline just outside the boundary layer is deflected. Equivalently, δ^* is also defined as the distance that the solid surface must be moved so that the loss of mass flux in the free stream is equal to the mass flux in the boundary layer. This scenario is illustrated in Figure 11.7. Applying the continuity equation between sections 1 and 2, assuming incompressible flow (i.e., ρ = constant) requires that

$$\int_0^{y_1} bU\,\mathrm{d}y = \int_0^{y_1+\delta^*} bu\,\mathrm{d}y \tag{11.61}$$

where y_1 is the distance of an outer flow streamline from the solid surface at section 1, $y_1+\delta^*$ is the distance of that same streamline from the surface at section 2, b is the flow width, U is the free-stream velocity, and u is the velocity within the boundary layer. Because the referenced streamline in Equation 11.61 is outside the boundary layer, the following equation is applicable at section 2,

$$\int_{y_1}^{y_1+\delta^*} bu\,\mathrm{d}y = bU\delta^* \tag{11.62}$$

Combining Equations 11.61 and 11.62 and noting that b cancels out leads to the following expression for the displacement thickness:

$$\delta^* = \int_0^{y_1}\left(1-\frac{u}{U}\right)\mathrm{d}y \quad \equiv \quad \delta^* = \int_0^{\delta}\left(1-\frac{u}{U}\right)\mathrm{d}y \tag{11.63}$$

It should be noted that although the upper limit, y_1, of the integral in Equation 11.63 is not explicitly specified, it is implicitly specified in that it is only necessary to integrate up to where $u = U$, because beyond this point, the integrand is equal to zero. Because $u = U$ when $y = \delta$, y_1 can be replaced by δ (hence, the equivalent forms given in Equation 11.63); the form involving δ is usually more useful.

Displacement thickness using the Blasius solution. The Blasius solution describes the velocity profile in a laminar boundary layer. If the Blasius solution is substituted into Equation 11.63, then the growth of the displacement thickness in a laminar boundary layer is described by

$$\frac{\delta^*}{x} = \frac{1.72}{\mathrm{Re}_x^{\frac{1}{2}}} \quad \text{(laminar BL)} \tag{11.64}$$

Comparing Equation 11.64 for δ^* with Equation 11.18 for δ, it is apparent that the displacement thickness is approximately one-third of the boundary-layer thickness, where the boundary layer thickness is based on attenuation of the velocity profile under laminar-flow conditions.

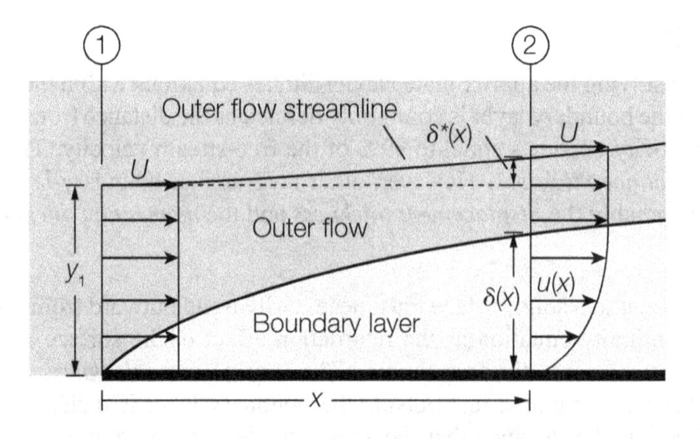

Figure 11.7: Displacement thickness in a boundary layer

Displacement thickness using the one-seventh power law distribution. The one-seventh power law distribution is commonly used to describe the velocity distribution in smooth turbulent boundary layers. If the flow in the boundary layer is turbulent and has a one-seventh power law distribution, as given by Equation 11.52, then the displacement thickness is given by

$$\frac{\delta^*}{x} = \frac{0.020}{\mathrm{Re}_x^{\frac{1}{7}}} \quad \text{(smooth, turbulent BL)} \tag{11.65}$$

EXAMPLE 11.7

The velocity distribution within a laminar boundary layer is sometimes described by the following parabolic function:

$$\frac{u}{U} = 2\left(\frac{y}{\delta}\right) - \left(\frac{y}{\delta}\right)^2$$

Show that this distribution generally leads to the result that the displacement thickness, δ^*, is equal to one-third of the boundary-layer thickness, δ.

SOLUTION

The analysis can be facilitated by defining the normalized velocity, u^*, and the normalized distance from the solid surface, y^*, such that

$$u^* = \frac{u}{U}, \quad \text{and} \quad y^* = \frac{y}{\delta} \quad \rightarrow \quad dy^* = \delta^{-1}dy$$

The given parabolic velocity distribution and the expression for the displacement thickness can then be expressed as

$$u^* = 2y^* - y^{*2}, \quad \text{and} \quad \delta^* = \delta\int_0^1 (1 - u^*)\, dy^*$$

Combining these equations gives

$$\delta^* = \delta\int_0^1 (1 - 2y^* + y^{*2})\, dy^* \quad \rightarrow \quad \delta^* = \delta\left[y^* - y^{*2} + \frac{1}{3}y^{*3}\right]_0^1$$

$$\rightarrow \quad \delta^* = \frac{1}{3}\delta \quad \rightarrow \quad \frac{\delta^*}{\delta} = \frac{1}{3}$$

A similar analysis to determine δ^*/δ can be used for any empirical velocity distribution in which u/U is expressed as a function of y/δ. Any realistic velocity distributions results in $\delta^*/\delta < 1$.

Practical Considerations

The displacement thickness is sometimes used to represent the added thickness that can be associated with a solid surface, such that the volume flow rate outside of the added thickness remains constant. This condition generally occurs near the entrance of closed conduits, where the growth of the boundary layer reduces the flow rate near the surface of the conduit, which results in an increased flow rate in the central portion of the conduit.

EXAMPLE 11.8

Air enters a horizontal 150-mm-diameter circular conduit at a velocity of 15 m/s. The velocity distribution in the boundary layer of the conduit is such that the ratio of the displacement thickness to the boundary layer thickness is $\frac{1}{8}$. Near the entrance of the conduit, the boundary-layer thickness is approximately equal to 7 mm, and at a downstream section, the boundary-layer thickness is equal to 30 mm. The density of the air is approximately equal to 1.25 kg/m^3. (a) Estimate the velocity in the central portion of the conduit at the downstream section and (b) the pressure difference between the upstream and downstream sections.

SOLUTION

From the given data: $D = 150$ mm, $U_1 = 15$ m/s, $\delta_1 = 7$ mm, $\delta_2 = 30$ mm, $\delta^*/\delta = \frac{1}{8}$, and $\rho = 1.25$ kg/m^3. The displacement thicknesses at the upstream and downstream sections can be estimated as

$$\delta_1^* = \tfrac{1}{8}\delta_1 = \tfrac{1}{8}(7) = 0.875 \text{ mm}, \qquad \delta_2^* = \tfrac{1}{8}\delta_2 = \tfrac{1}{8}(30) = 3.75 \text{ mm}$$

The core flow areas at the upstream and downstream sections where the volume flow rate remains the same, A_1 and A_2, can be estimated as follows:

$$A_1 = \frac{\pi}{4}[D - 2\delta_1^*]^2 = \frac{\pi}{4}[150 - 2(0.875)]^2 = 1.726 \times 10^4 \text{ mm}^4$$

$$A_2 = \frac{\pi}{4}[D - 2\delta_2^*]^2 = \frac{\pi}{4}[150 - 2(3.75)]^2 = 1.595 \times 10^4 \text{ mm}^4$$

(a) Applying the continuity equation in the free stream between the upstream and downstream sections gives

$$U_1A_1 = U_2A_2 \quad \rightarrow \quad U_2 = \left(\frac{A_1}{A_2}\right)U_1 = \left(\frac{1.726}{1.595}\right)(15) = 16.23 \text{ m/s} \approx \mathbf{16.2\ m/s}$$

(b) Applying the Bernoulli equation in the free stream between the upstream and downstream sections gives

$$\frac{p_1}{\rho} + \frac{U_1^2}{2} = \frac{p_2}{\rho} + \frac{U_2^2}{2} \quad \rightarrow \quad \Delta p = \tfrac{1}{2}\rho(U_2^2 - U_1^2)$$

where $\Delta p = p_1 - p_2$. Substituting the given and derived data yields

$$\Delta p = \tfrac{1}{2}(1.25)(16.23^2 - 15^2) = \mathbf{24\ Pa}$$

Therefore, the velocity in the core of the conduit increases by approximately 8% (from 15 m/s to 16.2 m/s), and the pressure decreases by approximately 24 Pa between the two sections.

11.4.2 Momentum Thickness

Consider again the combined boundary layer and outer flow shown in Figure 11.7. The momentum equation applied between sections 1 and 2 gives

$$-F_\text{D} = \left[\rho b \int_0^{y_1+\delta^*} u^2\,\text{d}y\right] - \left[\rho b \int_0^{y_1} U^2\,\text{d}y\right] = \left[\rho b \int_0^{y_1} u^2\,\text{d}y + \rho b U^2 \delta^*\right] - \left[\rho b \int_0^{y_1} U^2\,\text{d}y\right] \tag{11.66}$$

where F_D is the drag force between sections 1 and 2 and is directed in the opposite direction to the flow. Combining Equations 11.66 and 11.63 yields

$$F_\text{D} = \rho b \int_0^{y_1} u(u-U)\,\text{d}y \tag{11.67}$$

The *momentum thickness*, θ, is defined as the added thickness of the boundary layer that is required to account for the momentum deficit caused by the reduced velocity in the boundary layer. The momentum thickness can also be defined as the distance the solid surface would have to be moved so that the loss of momentum flux in the free stream is equal to the loss of momentum flux due to the boundary layer. This momentum-flux deficit is also equal to the drag force exerted by the fluid on the plane surface; hence,

$$F_\text{D} = \rho b U^2 \theta \tag{11.68}$$

Combining Equations 11.67 and 11.68 to eliminate F_D leads to the following explicit expression for the momentum thickness:

$$\theta = \int_0^{y_1} \frac{u}{U}\left(1 - \frac{u}{U}\right)\text{d}y \quad \equiv \quad \theta = \int_0^{\delta} \frac{u}{U}\left(1 - \frac{u}{U}\right)\text{d}y \tag{11.69}$$

where y_1 can be any distance from the solid surface to where $u = U$, because beyond this point, the integrand is equal to zero. Because $u = U$ when $y = \delta$, y_1 can be replaced by δ (hence the equivalent forms given in Equation 11.69); the form involving δ is usually more useful.

Momentum thickness using the Blasius solution. If the Blasius solution is substituted into Equation 11.69, then the growth of the momentum thickness is described by

$$\frac{\theta}{x} = \frac{0.664}{\text{Re}_x^{\frac{1}{2}}} \quad \text{(laminar BL)} \tag{11.70}$$

Comparing Equation 11.70 for θ with Equation 11.18 for δ, it is apparent that the momentum thickness is approximately one-tenth of the boundary-layer thickness, where the boundary layer thickness is based on attenuation of the velocity profile under laminar flow conditions.

Momentum thickness using the one-seventh power law distribution. If the flow in the boundary layer is in the smooth turbulent regime and has a one-seventh power law distribution, as given by Equation 11.52, then the momentum thickness is given by

$$\frac{\theta}{x} = \frac{0.016}{\text{Re}_x^{\frac{1}{7}}} \quad \text{(smooth, turbulent BL)} \tag{11.71}$$

The primary use of the momentum thickness is to facilitate determination of the drag force, F_D, on the solid surface directly from the free-stream velocity, where this relationship is given by Equation 11.68.

EXAMPLE 11.9

Consider the case in which the velocity distribution in a laminar boundary layer is described by the following parabolic function:

$$\frac{u}{U} = 2\left(\frac{y}{\delta}\right) - \left(\frac{y}{\delta}\right)^2$$

Determine the ratio of the momentum thickness to the boundary-layer thickness. Compare the magnitudes of the momentum and displacement thicknesses.

SOLUTION

As was the case in the previous example, this analysis can be facilitated by defining the normalized distance from the solid surface, y^*, such that

$$y^* = \frac{y}{\delta} \quad \rightarrow \quad \mathrm{d}y^* = \delta^{-1}\mathrm{d}y$$

The given parabolic velocity distribution and the expression for the momentum thickness can then be expressed as

$$\frac{u}{U} = 2y^* - y^{*2} \tag{11.72}$$

$$\theta = \int_0^\delta \frac{u}{U}\left(1 - \frac{u}{U}\right)\,\mathrm{d}y = \delta\int_0^1 \frac{u}{U}\left(1 - \frac{u}{U}\right)\,\mathrm{d}y^* \tag{11.73}$$

Combining Equations 11.72 and 11.73 gives

$$\theta = \delta\int_0^1 (2y^* - y^{*2})\left(1 - 2y^* + y^{*2}\right)\,\mathrm{d}y^* \quad \rightarrow \quad \theta = \delta\left[y^{*2} - \frac{5}{3}y^{*3} + y^{*4} - \frac{1}{5}y^{*5}\right]_0^1$$

$$\rightarrow \quad \theta = \frac{2}{15}\delta \quad \rightarrow \quad \frac{\theta}{\delta} = \frac{2}{15}$$

Therefore, the ratio of the momentum thickness to the boundary-layer thickness is **2/15**. Example 11.7 showed that the ratio of the displacement thickness to the boundary-layer thickness for the given parabolic distribution is equal to 1/3. Therefore, the ratio of the momentum thickness to the displacement thickness is $(2/15)/(1/3) = \mathbf{2/5}$. Hence, in this case, the momentum thickness is much smaller than the displacement thickness, which is in turn much smaller than the boundary-layer thickness.

Shape factor. The ratio of the momentum thickness to the displacement thickness, δ^*/θ, is called the *shape factor*. A large shape factor is sometimes used as an indicator that boundary-layer separation is about to occur. Critical values of the shape factor for separation to occur is $\delta^*/\theta \approx 3.5$ for laminar flow and $\delta^*/\theta \approx 2.4$ for turbulent flow.

11.4.3 Momentum Integral Equation

The rate at which the drag force, F_D, on a flat surface increases with distance downstream can be related to the rate at which the momentum thickness, θ, increases by differentiating both sides of Equation 11.68, which yields

$$\frac{\mathrm{d}F_\mathrm{D}}{\mathrm{d}x} = \rho b U^2 \frac{\mathrm{d}\theta}{\mathrm{d}x} \tag{11.74}$$

The shear stress on a flat surface (or wall), τ_w, at any distance x along a surface of width b is related to the drag force, F_D, by the relation

$$\mathrm{d}F_\mathrm{D} = \tau_\mathrm{w} b\,\mathrm{d}x \qquad (11.75)$$

Combining Equations 11.74 and 11.75 to eliminate F_D yields the following relationship between τ_w and θ:

$$\tau_\mathrm{w} = \rho U^2 \frac{\mathrm{d}\theta}{\mathrm{d}x} \qquad (11.76)$$

This equation is called the *momentum integral equation* or the *Kármán integral equation*. It provides a convenient means for deriving τ_w from the velocity distribution within the boundary layer, because θ depends only on the velocity distribution. Equation 11.76 is applicable to steady, incompressible, two-dimensional flow, where there are no body force components in the direction of flow.

EXAMPLE 11.10

Consider the case of a laminar boundary layer in which the velocity distribution is described by the following parabolic function:

$$\frac{u}{U} = 2\left(\frac{y}{\delta}\right) - \left(\frac{y}{\delta}\right)^2$$

(a) Determine the thickness of the boundary layer as a function of the distance along the solid surface. (b) Determine the local wall shear stress as a function of the distance along the surface.

SOLUTION

Define the normalized distance from the surface, y^*, and the normalized velocity, u^*, such that

$$y^* = \frac{y}{\delta}, \quad \mathrm{d}y^* = \delta^{-1}\mathrm{d}y, \quad u^* = \frac{u}{U}$$

The self-similar parabolic velocity distribution within the boundary layer can therefore be expressed as

$$u^* = 2y^* - y^{*2}$$

Example 11.9 showed that for the given parabolic velocity distribution, the momentum thickness is given by $\theta = 2/15\ \delta$.

(a) The drag force F_D can be related to the boundary-layer thickness using Equation 11.74, which gives

$$\mathrm{d}F_\mathrm{D} = \rho b U^2\,\mathrm{d}\theta \quad \rightarrow \quad \mathrm{d}F_\mathrm{D} = \rho b U^2\left(\frac{2}{15}\,\mathrm{d}\delta\right) \quad \rightarrow \quad \mathrm{d}F_\mathrm{D} = \frac{2}{15}\rho b U^2\,\mathrm{d}\delta \qquad (11.77)$$

Assuming that the fluid is Newtonian, the wall shear stress, τ_w, is given by

$$\tau_\mathrm{w} = \mu\left.\frac{\mathrm{d}u}{\mathrm{d}y}\right|_{y=0} = \mu\frac{U}{\delta}\left.\frac{\mathrm{d}u^*}{\mathrm{d}y^*}\right|_{y^*=0} = \mu\frac{U}{\delta}\left[2 - 2y^*\right]_{y^*=0} \quad \rightarrow \quad \tau_\mathrm{w} = 2\mu\frac{U}{\delta} \qquad (11.78)$$

Combining Equations 11.75, 11.77, and 11.78, integrating, and taking $\delta = 0$ at $x = 0$ yields

$$\mathrm{d}F_\mathrm{D} = \tau_\mathrm{w} b\,\mathrm{d}x \quad \rightarrow \quad \frac{2}{15}\rho b U^2\,\mathrm{d}\delta = \left(2\mu\frac{U}{\delta}\right) b\,\mathrm{d}x$$

$$\rightarrow \quad \frac{2}{15}\rho b U^2 \int_0^\delta \delta'\,\mathrm{d}\delta' = 2\mu U b \int_0^x \mathrm{d}x'$$

which simplifies to

$$\frac{1}{15}\rho b U^2 \delta^2 = 2\mu U b x \quad \rightarrow \quad \frac{\delta}{x} = \sqrt{30}\sqrt{\frac{\mu}{\rho U x}}$$

$$\rightarrow \quad \frac{\delta}{x} = \frac{\sqrt{30}}{\mathrm{Re}_x^{\frac{1}{2}}} \quad \rightarrow \quad \boldsymbol{\frac{\delta}{x} \simeq \frac{5.48}{\mathrm{Re}_x^{\frac{1}{2}}}} \qquad (11.79)$$

The proportionality constant between δ/x and $\mathrm{Re}_x^{-\frac{1}{2}}$ is 5.48, which is about 12% higher than the constant of 4.91 derived from the Blasius solution.

(b) The relationship between the wall shear stress, τ_w, and the distance along the surface (x) can be obtained by combining Equations 11.79 and 11.78, which yields

$$\tau_\mathrm{w} = 2\mu\frac{U}{\delta} \quad \rightarrow \quad \frac{\tau_\mathrm{w}}{\frac{1}{2}\rho U^2} = \frac{4\mu}{\rho U}\frac{1}{\delta} = \frac{4\mu}{\rho U}\left(\frac{1}{x}\frac{\mathrm{Re}_x^{\frac{1}{2}}}{\sqrt{30}}\right) = \frac{4}{\sqrt{30}}\mathrm{Re}_x^{-1}\mathrm{Re}_x^{\frac{1}{2}} = \frac{4}{\sqrt{30}}\mathrm{Re}^{-\frac{1}{2}}$$

$$\rightarrow \quad \boldsymbol{\frac{\tau_\mathrm{w}}{\frac{1}{2}\rho U^2} \simeq \frac{0.730}{\mathrm{Re}^{\frac{1}{2}}}}$$

The proportionality constant between $\tau_\mathrm{w}/\frac{1}{2}\rho U^2$ and $\mathrm{Re}_x^{-\frac{1}{2}}$ is 0.730, which is about 10% higher than the constant of 0.664 derived from the Blasius solution.

As illustrated in the above example, the momentum integral is used to determine quantitative relationships for estimating δ/x and $\tau_\mathrm{w}/\frac{1}{2}\rho U^2$ in terms of Re_x. Theoretical expressions for δ/x and $\tau_\mathrm{w}/\frac{1}{2}\rho U^2$ as a function of Re_x for laminar boundary layers in which the velocity distribution is given by the Blasius solution were derived previously and are given by Equations 11.18 and 11.22, respectively.

Application of the momentum integral equation with power-law distributions. Application of the momentum integral equation for power-law velocity distributions is complicated by the fact that power-law distributions generally yield $\mathrm{d}u/\mathrm{d}y = 0$ on the solid surface. This means that the wall shear stress, τ_w (= $\mu\,\mathrm{d}u/\mathrm{d}y$), is equal to zero on the surface, which is clearly unrealistic. This limitation is usually circumvented by using the following semiempirical relationship for the boundary shear stress:

$$\frac{\tau_\mathrm{w}}{\frac{1}{2}\rho U^2} = \frac{0.0205}{\mathrm{Re}_\delta^{\frac{1}{6}}}, \quad \text{where} \quad \mathrm{Re}_\delta = \frac{U\delta}{\nu} \qquad (11.80)$$

This semiempirical relationship was originally proposed by Prandtl and can be obtained by approximating the actual velocity distribution in the boundary layer (see Problem 11.93). The relationships given below for δ/x and c_f (in Equations 11.81 and 11.82) are derived by combining Equation 11.80 with the momentum integral equation (Equation 11.76) and using the relationship between the momentum thickness and the boundary-layer thickness as derived for the one-seventh power law distribution using Equation 11.69. The relationship given by Equation 11.80 can also be used in combination with empirical velocity distributions other than the one-seventh power law distributions, particularly when the empirical velocity distribution gives zero shear stress on the surface while providing a good fit to the velocity distribution over most of the boundary layer.

Boundary-layer thickness using the one-seventh power law distribution. If the flow in the boundary layer is turbulent and has a one-seventh power law distribution, as given by Equation 11.52, then the boundary-layer thickness is given by

$$\frac{\delta}{x} = \frac{0.16}{\mathrm{Re}_x^{\frac{1}{7}}} \quad \text{(smooth, turbulent BL)} \tag{11.81}$$

Local shear stress using the one-seventh power law distribution. If the flow in the boundary layer is turbulent and has a one-seventh power law distribution, then the skin friction coefficient, c_f, is given by

$$c_\mathrm{f} = \frac{\tau_\mathrm{w}}{\frac{1}{2}\rho U^2} = \frac{0.027}{\mathrm{Re}_x^{\frac{1}{7}}} \quad \text{(smooth, turbulent BL)} \tag{11.82}$$

Drag force using the one-seventh power law distribution. For a flat surface of width b and length L, the drag coefficient due to friction, C_Df, is defined as

$$C_\mathrm{Df} = \frac{F_\mathrm{D}}{\frac{1}{2}\rho U^2 Lb} = \frac{b\displaystyle\int_0^L \tau_\mathrm{w}\,\mathrm{d}x}{\frac{1}{2}\rho U^2 Lb} \tag{11.83}$$

where F_D represents the drag force on the surface. Combining Equations 11.82 and 11.83 yields

$$C_\mathrm{Df} = \frac{F_\mathrm{D}}{\frac{1}{2}\rho U^2 Lb} = \frac{0.031}{\mathrm{Re}_\mathrm{L}^{\frac{1}{7}}} \quad \text{(smooth, turbulent BL)} \tag{11.84}$$

Application of one-seventh power law results. Equations 11.81, 11.82, and 11.84 are applicable when the velocity distribution in the (smooth-turbulent) boundary layer follows the one-seventh power law distribution exactly. These results are not the same as those presented previously for δ/x, $\tau_\mathrm{w}/\frac{1}{2}\rho U^2$, and $F_\mathrm{D}/\frac{1}{2}\rho U^2 Lb$ as a function of Re_x by Equations 11.56, 11.57, and 11.59, respectively, where the latter equations were supplemented with empirical measurements from flow in pipes. Differences in the estimated magnitudes of δ/x, $\tau_\mathrm{w}/\frac{1}{2}\rho U^2$, and $F_\mathrm{D}/\frac{1}{2}\rho U^2 Lb$ using these alternative formulations are usually not large. Equations 11.81, 11.82, and 11.84 are typically considered applicable in the range of $5 \times 10^5 < \mathrm{Re}_x < 10^7$; however, the error is not large if they are applied up to $\mathrm{Re}_x = 10^8$.

Empirical equations. Two empirical equations that show good accuracy for estimating c_f and C_Df under smooth-turbulent conditions for values of Re_x up to as high as $\mathrm{Re}_x = 10^{10}$ are

$$c_\mathrm{f} = \frac{0.455}{\ln(0.06\mathrm{Re}_x)^2}, \qquad C_\mathrm{Df} = \frac{0.523}{\ln(0.06\mathrm{Re}_x)^2} \tag{11.85}$$

Estimation of boundary-layer thickness in a mixed boundary layer. A *mixed boundary layer* begins as a laminar boundary layer and becomes either a transitional or turbulent boundary layer at the location of interest. The conventional approximation used in describing the growth of a boundary layer is to assume an instantaneous transition at a specified (critical) Reynolds number, Re_cr, such that if $\mathrm{Re}_x \leq \mathrm{Re}_\mathrm{cr}$, the boundary layer is entirely laminar and if $\text{Re}_x > \text{Re}_{\text{cr}}$, the boundary layer is entirely turbulent from the leading edge of the surface. For $\text{Re}_x > \text{Re}_{\text{cr}}$, the combined effect of neglecting the transitional regime and assuming turbulent flow starting from the leading edge will typically become minimal when $\text{Re}_x \gg \text{Re}_{\text{cr}}$. Usually, $\text{Re}_{\text{cr}} = 5 \times 10^5$ is used in engineering calculations. It is important to remember that the conventional approximation will be less accurate when $5 \times 10^5 < \text{Re}_x < 3 \times 10^6$, which is typically considered to be characteristic of the actual transitional zone between a laminar and a turbulent boundary layer. Alternative approaches that account for the presence of both laminar and turbulent boundary layers have been formulated. For example, the turbulent boundary layer can be assumed to originate at $x = x_0$, where

$$x_0 = x_{\text{cr}} \left(1 - \frac{54.3}{\text{Re}_{\text{cr}}^{\frac{5}{12}}} \right) \tag{11.86}$$

where $x = x_{\text{cr}}$ is the location where $\text{Re}_x = \text{Re}_{\text{cr}}$. Taking the location where $\delta = 0$ in the turbulent boundary layer formulation as x_0 ensures that the boundary layer has a thickness at $\text{Re} = \text{Re}_{\text{cr}}$ that is the same as that predicted by laminar boundary layer theory and grows as a turbulent boundary layer thereafter.[63] The conventional approach of assuming one type of boundary layer is used for most exercises in this text.

11.4.4 General Formulations for Self-Similar Velocity Profiles

All self-similar velocity profiles can be expressed in the following nondimensional functional form:

$$u^* = \begin{cases} f(y^*), & 0 \leq y^* \leq 1 \\ 1, & y^* > 1 \end{cases} \tag{11.87}$$

where $u^* = u/U$ and $y^* = y/\delta$. For self-similar velocity profiles, the displacement thickness defined by Equation 11.63 and the momentum thickness defined by Equation 11.69 can be expressed in the nondimensional forms

$$\frac{\delta^*}{\delta} = \int_0^1 (1 - u^*)\, \text{d}y^*, \qquad \text{and} \qquad \frac{\theta}{\delta} = \int_0^1 u^* (1 - u^*)\, \text{d}y^* \tag{11.88}$$

The momentum integral equation (Equation 11.76) can be expressed in terms of the self-similar velocity profile by expressing τ_{w} and θ in their fundamental forms, which gives the following differential equation describing the growth of the boundary layer thickness, $\delta(x)$, as a function of the distance, x, along the solid surface:

$$\mu \frac{U}{\delta} \underbrace{\left.\frac{\text{d}u^*}{\text{d}y^*}\right|_{y^*=0}}_{=\text{A}} = \rho U^2 \frac{\text{d}\delta}{\text{d}x} \underbrace{\left(\int_0^1 u^* (1 - u^*)\, \text{d}y^* \right)}_{=\text{B}} \quad \rightarrow \quad \delta \frac{\text{d}\delta}{\text{d}x} = \frac{\nu}{U} \cdot \frac{A}{B} \tag{11.89}$$

where A and B are constants that depend only on the velocity profile and are defined by

$$A = \left.\frac{\text{d}u^*}{\text{d}y^*}\right|_{y^*=0}, \qquad \text{and} \qquad B = \int_0^1 u^* (1 - u^*)\, \text{d}y^* \tag{11.90}$$

It is also apparent by comparing the definition of B with Equation 11.88 that $\theta/\delta = B$. The constants A and B are fundamental constants in determining the growth of the boundary layer, the local shear stress, and the total drag force on a plane surface. Integrating Equation 11.89

[63] The proof and an application of Equation 11.86 is given in Problem 11.44.

with the boundary condition that $\delta(0) = 0$ yields the following relationship for estimating the growth of the boundary layer:

$$\frac{\delta}{x} = \frac{C_1}{\mathrm{Re}_x^{\frac{1}{2}}}, \quad \text{where} \quad C_1 = \sqrt{\frac{2A}{B}} \tag{11.91}$$

The corresponding normalized expression for the local shear stress as a function of x when $\delta(0) = 0$ is given by

$$\frac{\tau_w}{\frac{1}{2}\rho U^2} = \frac{C_2}{\mathrm{Re}_x^{\frac{1}{2}}}, \quad \text{where} \quad C_2 = \sqrt{2AB} \tag{11.92}$$

and the total drag force, F_D, on a rectangular surface of width b and length L when $\delta(0) = 0$ is derived by integrating Equation 11.92 along the surface, yielding

$$\frac{F_D}{\frac{1}{2}\rho U^2 bL} = \frac{2C_2}{\mathrm{Re}_x^{\frac{1}{2}}} \tag{11.93}$$

EXAMPLE 11.11

Consider a laminar boundary layer in which the velocity distribution is self-similar and described by the following quadratic distribution:

$$u^* = \begin{cases} 2y^* - y^{*2}, & y^* \leq 1 \\ 1, & y^* > 1 \end{cases}$$

Use the general formulations for self-similar velocity profiles to determine expressions for the growth of the boundary layer, the local shear stress, and the total drag force as a function of the distance x along the solid surface.

SOLUTION

For the given velocity distribution, the fundamental constants A and B can be calculated using their definitions in Equation 11.90 as follows:

$$A = \left.\frac{du^*}{dy^*}\right|_{y^*=0} = (2 - 2y^*)\Big|_{y^*=0} = 2$$

$$B = \int_0^1 u^*(1 - u^*)\, dy^* = \int_0^1 (2y^* - 5y^{*2} - y^{*4} + 4y^{*3})\, dy^* = \frac{2}{15}$$

Using these values of A and B, the constants in the required formulations, C_1, C_2, and $2C_2$, are given by

$$C_1 = \sqrt{\frac{2A}{B}} = \sqrt{\frac{2(2)}{\frac{2}{15}}} = 5.48, \qquad C_2 = \sqrt{2AB} = \sqrt{2(2)(\tfrac{2}{15})} = 0.730,$$

$$2C_2 = 2(0.730) = 1.46$$

Using these constants, the following expressions for the growth of the boundary layer, the local shear stress, and the total drag force follow directly from Equations 11.91–11.93 to give

$$\frac{\delta}{x} = \frac{5.48}{\mathrm{Re}_x^{\frac{1}{2}}}, \qquad \frac{\tau_w}{\frac{1}{2}\rho U^2} = \frac{0.730}{\mathrm{Re}_x^{\frac{1}{2}}}, \qquad \frac{F_D}{\frac{1}{2}\rho U^2 bL} = \frac{1.46}{\mathrm{Re}_x^{\frac{1}{2}}}$$

11.5 Mixing-Length Theory of Turbulent Boundary Layers

A turbulent boundary layer is divided into three regions with increasing distance from the solid surface: the laminar sublayer, the transition layer, and the outer layer. Within the outer layer, turbulent fluctuations exist, and the shear stress exerted by the turbulent fluctuations, $\tau_{\rm turb}$, can be expressed as

$$\tau_{\rm turb} = \rho \ell^2 \left(\frac{{\rm d}\bar{u}}{{\rm d}y}\right)^2 \tag{11.94}$$

where ℓ is the mixing length. Assuming ℓ as being proportional to the distance from the wall such that $\ell = \kappa y$ and further assuming that the turbulent shear stress is approximately equal to the wall shear stress such that $\tau_{\rm turb} = \tau_{\rm w}$, Equation 11.94 can be expressed as

$$\tau_{\rm w} = \rho \kappa^2 y^2 \left(\frac{{\rm d}u}{{\rm d}y}\right)^2$$

which can be expressed as

$$du = \frac{\sqrt{\frac{\tau_{\rm w}}{\rho}}}{\kappa}\frac{{\rm d}y}{y} \quad \rightarrow \quad du = \frac{u_*}{\kappa}\frac{{\rm d}y}{y} \quad \rightarrow \quad du = \frac{u_*}{\kappa}\frac{{\rm d}\left(\frac{u_* y}{\nu}\right)}{\frac{u_* y}{\nu}} \tag{11.95}$$

which integrates to yield

$$\frac{u}{u_*} = \frac{1}{\kappa}\ln\frac{u_* y}{\nu} + B \tag{11.96}$$

where B is a constant of integration. Equation 11.96 can be taken as a general relation for the velocity distribution in the outer layer within a turbulent boundary layer. The constant B depends on whether the smooth flow or rough flow regime exists.

11.5.1 Smooth Flow

Smooth flow conditions occur in the outer layer when the characteristic height of the roughness elements on a solid surface is less than the thickness of the viscous sublayer, where the thickness of the viscous sublayer is generally taken as $5\nu/u_*$. Laboratory experiments under smooth flow conditions when matched to Equation 11.96 give $B = 5.0$; hence, Equation 11.96 can be expressed as

$$\frac{u}{u_*} = \frac{1}{\kappa}\ln\frac{u_* y}{\nu} + 5.0 \tag{11.97}$$

The quantity κ, commonly called the *von Kármán constant*, has been estimated experimentally to be in the range of 0.40–0.41. Substituting $\kappa = 0.41$ into Equation 11.97 gives

$$\frac{u}{u_*} = 2.44\ln\frac{u_* y}{\nu} + 5.0 \quad \text{(smooth flow)} \tag{11.98}$$

Taking $\kappa = 0.40$ gives a coefficient of 2.5 in Equation 11.98, and coefficients of 2.44 and 2.5 are both in common usage. Interestingly, although velocity equations based on the assumption that κ is a constant are widely accepted, in modern fluid mechanics, it is also widely accepted that assuming κ to be constant is an oversimplification of the turbulent mixing process. The velocity distribution given by Equation 11.98 is commonly referred to as the *logarithmic law* and has been found to be valid for $u_* y/\nu$ in the approximate range of 30–500. When applied to flow in pipes, Equation 11.98 has been observed to represent the velocity profile adequately starting at around $u_* y/\nu = 30$ and moving outward from the surface of

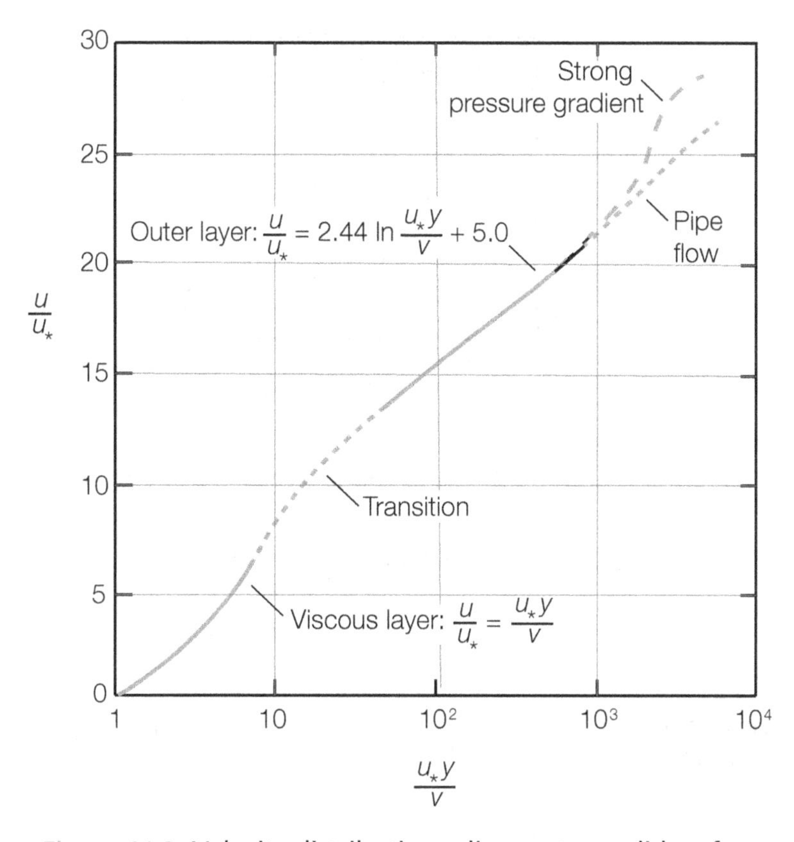

Figure 11.8: Velocity distribution adjacent to a solid surface

the pipe to very near the center of the pipe. This velocity distribution is illustrated in Figure 11.8, where the u_*y/ν scale is logarithmic and hence the outer layer velocity distribution extends over a range that is orders of magnitude larger than that of the viscous layer. It is further apparent from Figure 11.8 that in cases of strong pressure gradients, deviations from the outer layer velocity distribution can be substantial, whereas in usual pipe flow conditions, deviations from the outer layer velocity distribution are relatively small. Partially based on this result, Equation 11.98 has been widely used to represent the velocity profile beyond the outer layer of the wall region.

11.5.2 Rough Flow

Measurements by Nikuradse (who was a student of Prandtl) showed that the effect of surface roughness of characteristic height ϵ is to shift the velocity profile away from the wall as illustrated in Figure 11.9. The shift in the velocity profile, ΔB, depends on the magnitude of the roughness height. Under rough flow conditions, the shift in the velocity profile relative to the smooth flow condition is observed to be given by

$$\Delta B = 2.44 \ln\left(\frac{u_*\epsilon}{\nu}\right) - 3.5 \tag{11.99}$$

Applying this adjustment to the smooth flow velocity profile given by Equation 11.98 gives the velocity profile under rough flow conditions as

$$\frac{u}{u_*} = 2.44 \ln\frac{u_*y}{\nu} + 5.0 - \left[2.44 \ln\frac{u_*\epsilon}{\nu} - 3.5\right] \tag{11.100}$$

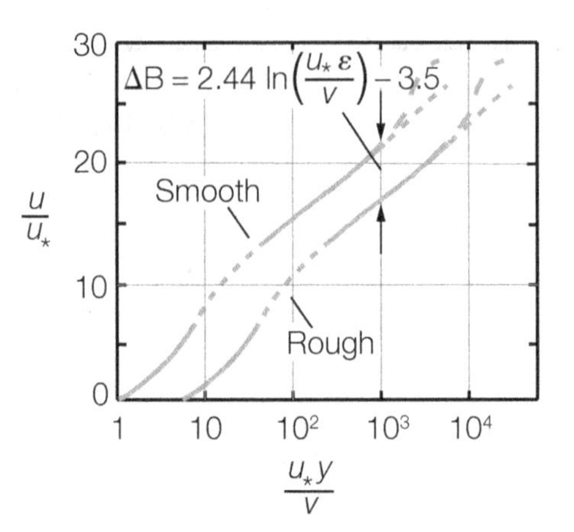

Figure 11.9: Effect of wall roughness on a velocity profile

which yields the following velocity profile for rough-flow conditions:

$$\frac{u}{u_*} = 2.44 \ln\frac{y}{\epsilon} + 8.5 \quad \text{(rough flow)} \tag{11.101}$$

This result indicates that under rough flow conditions, the velocity profile depends only on y/ϵ and is independent of the viscosity of the fluid.

11.5.3 Velocity-Defect Law

The velocity distributions for both smooth turbulent and rough turbulent flow within a boundary layer, given by Equations 11.98 and 11.101, respectively, are controlled by conditions at the wall. These velocity distributions, along with the velocity distribution in the laminar sublayer, constitute the *law of the wall.* The law-of-the-wall relationships represent the velocity distributions under conditions where $u_*y/\nu < 500$. In the region of a boundary layer where $u_*y/\nu > 500$ and $y/\delta > 0.15$, the velocity distribution becomes more influenced by the free-stream velocity, U, primarily as a consequence of the required boundary condition that $u \rightarrow U$ as $y \rightarrow \delta$. In this outer region, the velocity distribution in the boundary layer is better described by the *velocity-defect law*, which can be expressed in the following functional form:

$$\frac{U - u}{u_*} = f\left(\frac{y}{\delta}\right) \tag{11.102}$$

The formulation given by Equation 11.102 has been confirmed by experimental data as shown in Figure 11.10. A variety of empirical relationships have been proposed to approximate the function shown in Figure 11.10. A function that gives a reasonable fit is

$$\frac{U - u}{u_*} = -3.74 \ln\left(\frac{y}{\delta}\right) \tag{11.103}$$

The function $f(y/\delta)$ shown in Figure 11.10 and approximated by Equation 11.103 can be used for both smooth and rough turbulent boundary layers. The region in which Equation 11.103 is applicable (i.e., $u_*y/\nu > 500$ and $y/\delta > 0.15$) is commonly called the *velocity defect region* and describes the merging of the velocity in the boundary layer with the free-stream velocity.

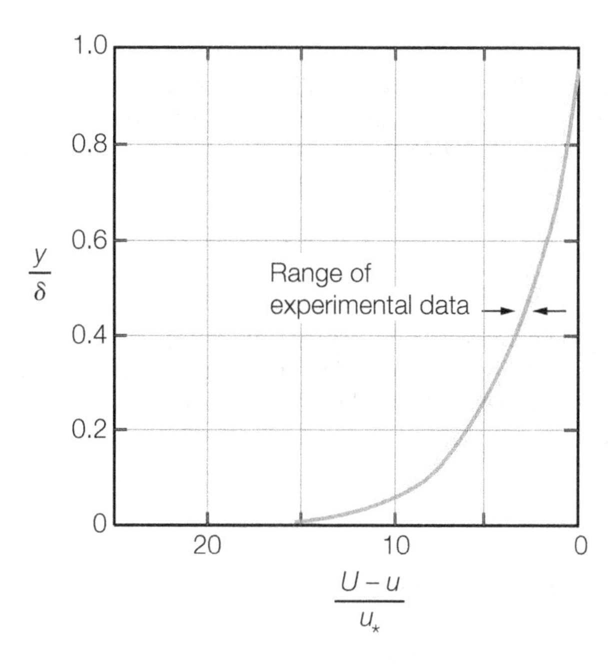

Figure 11.10: Velocity-defect law function
Source: Rouse (1959).

11.5.4 One-Seventh Power Law Distribution

The one-seventh power law distribution is an entirely empirical velocity distribution that is commonly used to approximate the velocity distribution in smooth turbulent boundary layers. The one-seventh power law distribution is given by

$$\frac{u}{U} = \left(\frac{y}{\delta}\right)^{\frac{1}{7}} \tag{11.104}$$

The range of distances from the solid surface for which the one-seventh power law distribution can be applied is usually taken as $0.1 < y/\delta < 1$. For regions closer to the wall, the logarithmic velocity distribution must be used for estimating velocities and shear stresses within the boundary layer (but outside the viscous sublayer).

11.6 Boundary Layers in Closed Conduits

A closed conduit can be viewed as a flat surface that is curled around to form the closed conduit. From this perspective, the entrance to the conduit corresponds to the upstream end of a flat surface, and is the location from which the boundary layer begins to grow. As one moves further into the conduit, the boundary layer grows from the wall of the conduit outward until the boundary layer from the perimeter of the conduit merges at the center of the conduit. Once the wall boundary layer merges, no further changes occur in the velocity distribution downstream of this section. At this point, the entire conduit section is contained within the boundary layer. In general, when flow within a conduit is turbulent (e.g., Re $>$ 4000 for circular pipes), the velocity distribution within the conduit has the same two regimes that are present in the turbulent boundary layer adjacent to a flat surface, namely the smooth flow regime and the rough flow regime. Which regime exists in any particular circumstance is determined by the height of the roughness elements relative to the thickness of the viscous sublayer.

11.6.1 Smooth Flow in Pipes

In the case of (turbulent) flow in a pipe of radius R, the constant in Equation 11.98 can be eliminated by requiring that $u = u_{\text{max}}$ at $r = 0$, were r is the radial distance from the center of the pipe and $u = u_{\text{max}}$ is the maximum velocity in the pipe, which occurs at the center. Applying this boundary condition, Equation 11.98 gives

$$\frac{u_{\text{max}} - u(r)}{u_*} = 2.44 \ln \frac{R}{R - r} \quad \text{(smooth flow)} \tag{11.105}$$

which is the velocity-defect law for flow in pipes, where $u_{\text{max}} - u$ is the velocity defect. The law-of-the-wall logarithmic velocity distributions have been found to apply almost to the centerline of pipes, and this velocity distribution can be expressed by directly using Equation 11.98, which yields

$$\frac{u(r)}{u_*} = 2.44 \ln \frac{u_*(R - r)}{\nu} + 5.0 \quad \text{(smooth flow)} \tag{11.106}$$

where the distance from the wall (y) is equal to $R - r$.

Average velocity. The average velocity in the pipe under smooth flow conditions, V, can be determined from Equation 11.106 by

$$V = \frac{1}{A}\int_A u \, \text{d}A = \frac{1}{\pi R^2}\int_0^R u(r) 2\pi r \, \text{d}r = \frac{1}{\pi R^2}\int_0^R \left[2.44 \ln \frac{u_*(R - r)}{\nu} + 5.0\right] u_* 2\pi r \, \text{d}r$$

which yields the following expression for average velocity, V:

$$\frac{V}{u_*} = 2.44 \ln \frac{u_* R}{\nu} + 1.34 \quad \text{(smooth flow)} \tag{11.107}$$

Friction factor. In pipe flows, the boundary shear stress, τ_w, is normally characterized by the friction factor, f, where

$$f = \frac{\tau_\text{w}}{\frac{1}{8}\rho V^2} \tag{11.108}$$

Using the definition of u_* $(= \sqrt{\tau_\text{w}/\rho})$ and defining the *Reynolds number*, Re, as Re $= VD/\nu$, where D is the diameter of the pipe ($= 2R$), the following dimensionless groups (of variables) can be formed:

$$\frac{V}{u_*} = \left(\frac{8}{f}\right)^{\frac{1}{2}}, \qquad \frac{u_* R}{\nu} = \frac{1}{2}\text{Re}\left(\frac{f}{8}\right)^{\frac{1}{2}} \tag{11.109}$$

Substituting the relationships in Equation 11.109 into Equation 11.107 and changing to a base-10 logarithm yields

$$\frac{1}{f^{\frac{1}{2}}} = 1.99 \log\left(\text{Re} f^{\frac{1}{2}}\right) - 1.02 \tag{11.110}$$

Prandtl (1935) compared Equation 11.110 to experimental data and modified the constants in Equation 11.110 to better match observed data, in which case Equation 11.110 becomes

$$\frac{1}{f^{\frac{1}{2}}} = 2.0 \log\left(\text{Re} f^{\frac{1}{2}}\right) - 0.8 \quad \text{(smooth flow)} \tag{11.111}$$

This equation is widely used to describe the variation of the friction factor, f, with Reynolds number, Re, in pipe flow under smooth turbulent conditions. In these cases, the pipe is commonly referred to as being *hydrodynamically smooth* or *hydraulically smooth*. Equation 11.111 is commonly referred to as the *Prandtl equation* for the friction factor. The Prandtl equation for the friction factor in smooth turbulent pipe flow, Equation 11.111, is sometimes approximated by the *Blasius equation* for the friction factor in smooth turbulent pipe flow as follows (Blasius, 1913):

$$f = \frac{0.316}{\mathrm{Re}^{\frac{1}{4}}} \quad \text{for} \quad 3000 \leq \mathrm{Re} \leq 10^5 \tag{11.112}$$

This equation is commonly used as a basis for characterizing the relationship between the wall shear stress and the mean velocity in pipes, and in formulating boundary-layer relationships in smooth turbulent flow. Interestingly, Blasius (1913) also reported that within the range of Reynolds numbers for which Equation 11.112 is valid, the velocity distribution across the cross section of a pipe can be reasonably approximated by the one-seventh power law profile expressed as

$$\frac{u(r)}{u_{\max}} = \left(\frac{R-r}{R}\right)^{\frac{1}{7}} \tag{11.113}$$

where R is the radius of the pipe, r is the radial distance from the center of the pipe, and $R - r$ is the distance from the wall of the pipe. The one-seventh power law for the velocity distribution in pipes is not commonly used in modern engineering practice, where preference is given to using the velocity defect law. Use of the one-seventh power law is more commonly used in boundary-layer applications.

11.6.2 Rough Flow in Pipes

For rough flow in a pipe of radius R, the constant in Equation 11.101 can be eliminated by requiring that $u = u_{\max}$ at $r = 0$, which gives

$$\frac{u_{\max} - u(r)}{u_*} = 2.44 \ln \frac{R}{R-r} \quad \text{(rough flow)} \tag{11.114}$$

Notably, this velocity-defect law is the same relationship that applies to the smooth flow regime. Consequently, the velocity-defect law can be considered to be a universal law that applies to turbulent flow in pipes, regardless of the regime of flow. The rough flow velocity distribution can also be expressed by directly using Equation 11.101, which yields

$$\frac{u(r)}{u_*} = 2.44 \ln \frac{R-r}{\epsilon} + 8.5 \quad \text{(rough flow)} \tag{11.115}$$

where the distance from the wall (y) is equal to $R - r$.

Average velocity. Under rough flow conditions, the average velocity, V, in a pipe of radius R can be determined from Equation 11.101 by

$$V = \frac{1}{A}\int_A u \, \mathrm{d}A = \frac{1}{\pi R^2}\int_0^R u(r) 2\pi r \, \mathrm{d}r = \frac{1}{\pi R^2}\int_0^R \left[2.44 \ln \frac{R-r}{\epsilon} + 8.5\right] u_* 2\pi r \, \mathrm{d}r$$

which yields the following expression for average velocity, V:

$$\frac{V}{u_*} = 2.44 \ln \frac{D}{\epsilon} + 3.2 \quad \text{(rough flow)} \tag{11.116}$$

where D ($= 2R$) is the diameter of the pipe. Using the direct relationship between V/u_* and the friction factor, f, as given by Equation 11.109, Equation 11.116 can be used to determine the expression for f in rough flow as

$$\frac{1}{f^{\frac{1}{2}}} = -2.0 \log\left(\frac{\epsilon/D}{3.7}\right) \quad \text{(rough flow)} \tag{11.117}$$

This equation is widely used to describe the variation of the friction factor, f, in pipe flow where the wall roughness is the controlling factor. In these cases, the pipe is characterized as being *hydrodynamically rough* or *hydraulically rough*. Equation 11.117 was developed by Theodore von Kármán based on the experimental data of Johann Nikuradse, and Equation 11.117 is commonly referred to as the *von Kármán equation* for the friction factor.

11.6.3 Notable Contributors to Understanding Flow in Pipes

Several notable contributors to our current understanding of pipe flow have been mentioned previously. Our fundamental understanding of the relationship between wall friction and the velocity profile in pipes was initiated by Ludwig Prandtl (1875–1953) and carried on by his students Heinrich Blasius (1883–1970) and Johann Nikuradse (1894–1979). Theodore von Kármán (1881–1963) was a contemporary of Prandtl, who also provided significant contributions to the theoretical understanding of wall friction. Practical application of Prandtl's, Nikuradse's, and von Kármán's results to estimate the friction factors in commercial pipes was facilitated by a semiempirical implicit expression for the friction factor developed by Cyril Colebrook (1910–1997). Practical application of the Colebrook equation was facilitated by a graphical presentation of the Colebrook equation by Lewis Moody (1880–1953).

Key Equations for Boundary-Layer Flow

The following list of equations is particularly useful in solving problems related to boundary-layer flows. If one is able to recognize these equations and recall their appropriate use, then the learning objectives of this chapter have been met to a significant degree. Derivations of these equations, definitions of the variables, and detailed examples of usage can be found in the main text.

LAMINAR BOUNDARY LAYERS

Momentum equation: $$u\frac{\partial u}{\partial x}+v\frac{\partial u}{\partial y}=\nu\frac{\partial^2 u}{\partial y^2}$$

Continuity equation: $$\frac{\partial u}{\partial x}+\frac{\partial v}{\partial y}=0$$

Blasius variable: $$\eta=y\left(\frac{U}{\nu x}\right)^{\frac{1}{2}}$$

Blasius equation: $$2f'''+ff''=0$$

Laminar BL, Blasius solution: $$\frac{\delta}{x}=\frac{4.91}{\mathrm{Re}_x^{\frac{1}{2}}}$$

$$c_{\mathrm{f}}=\frac{\tau_{\mathrm{w}}}{\frac{1}{2}\rho U^2}=\frac{0.664}{\mathrm{Re}_x^{\frac{1}{2}}}$$

$$C_{\mathrm{Df}}=\frac{F_{\mathrm{D}}}{\frac{1}{2}\rho U^2 Lb}=\frac{1.328}{\mathrm{Re}_{\mathrm{L}}^{\frac{1}{2}}}$$

Momentum equation, curved surface: $$u\frac{\partial u}{\partial x}+v\frac{\partial u}{\partial y}=U\frac{\mathrm{d}U}{\mathrm{d}x}+\nu\frac{\partial^2 u}{\partial y^2}$$

Continuity equation, curved surface: $$\frac{\partial u}{\partial x}+\frac{\partial v}{\partial y}=0$$

TURBULENT BOUNDARY LAYERS

Momentum equation: $$\bar{u}\frac{\partial \bar{u}}{\partial x}+\bar{v}\frac{\partial \bar{u}}{\partial y}=U\frac{\mathrm{d}U}{\mathrm{d}x}+\frac{\partial \tau}{\partial y}$$

Continuity equation: $$\frac{\partial \bar{u}}{\partial x}+\frac{\partial \bar{v}}{\partial y}=0$$

Viscous sublayer: $$\frac{u}{u_*}=\frac{u_* y}{\nu}$$

Thickness of viscous sublayer: $$\delta_{\mathrm{v}}=\frac{5\nu}{u_*}$$

Thickness of transition layer: $\delta_{\mathrm{t}} = \dfrac{20\nu}{u_*} = 4\delta_{\mathrm{v}}$

Smooth turbulent BL: $\dfrac{\epsilon\, u_*}{\nu} < 5$

Rough turbulent BL: $\dfrac{\epsilon\, u_*}{\nu} > 70$

Intermediate turbulent BL: $5 \leq \dfrac{\epsilon\, u_*}{\nu} \leq 70$

Critical roughness: $\epsilon_{\mathrm{c}} = \dfrac{15\nu}{u_*}$

Logarithmic law distribution, smooth turbulent: $\dfrac{u}{u_*} = 2.5 \ln\left(\dfrac{u_* y}{\nu}\right) + 5.5$

$$\frac{u}{u_*} = 2.44 \ln\left(\frac{u_* y}{\nu}\right) + 5.0$$

Logarithmic law distribution, rough turbulent: $\dfrac{u}{u_*} = 2.5 \ln\left(\dfrac{y}{\epsilon}\right) + 8.5$

$$\frac{u}{u_*} = 2.44 \ln\left(\frac{y}{\epsilon}\right) + 8.5$$

One-seventh power law distribution: $\dfrac{u}{U} = \left(\dfrac{y}{\delta}\right)^{\frac{1}{7}}$

Smooth turbulent BL: $\dfrac{\delta}{x} = \dfrac{0.382}{\mathrm{Re}_x^{\frac{1}{5}}}, \quad 5 \times 10^5 \leq \mathrm{Re}_x < 10^7$

$$c_{\mathrm{f}} = \frac{\tau_{\mathrm{w}}}{\frac{1}{2}\rho U^2} = \frac{0.0594}{\mathrm{Re}_x^{\frac{1}{5}}}, \quad 5 \times 10^5 \leq \mathrm{Re}_x < 10^7$$

$$C_{\mathrm{Df}} = \frac{F_{\mathrm{D}}}{\frac{1}{2}\rho U^2 L b} = \frac{0.0735}{\mathrm{Re}_{\mathrm{L}}^{\frac{1}{5}}}, \quad 5 \times 10^5 \leq \mathrm{Re}_x < 10^7$$

$$C_{\mathrm{Df}} = \frac{F_{\mathrm{D}}}{\frac{1}{2}\rho U^2 L b} = \frac{0.455}{(\log \mathrm{Re}_{\mathrm{L}})^{2.58}}, \quad \mathrm{Re}_x \geq 10^7$$

APPLICATIONS

Displacement thickness, definition: $\delta^* = \displaystyle\int_0^{\delta} \left(1 - \frac{u}{U}\right) \mathrm{d}y$

Displacement thickness, Blasius solution: $\dfrac{\delta^*}{x} = \dfrac{1.72}{\mathrm{Re}_x^{\frac{1}{2}}}$

Displacement thickness, one-seventh power law: $\dfrac{\delta^*}{x} = \dfrac{0.020}{\mathrm{Re}_x^{\frac{1}{7}}}$

Drag force: $F_{\mathrm{D}} = \rho b U^2 \theta$

Momentum thickness, definition: $$\theta = \int_0^{\delta} \frac{u}{U}\left(1 - \frac{u}{U}\right) \mathrm{d}y$$

Momentum thickness, Blasius solution: $$\frac{\theta}{x} = \frac{0.664}{\mathrm{Re}_x^{\frac{1}{2}}}$$

Momentum thickness, one-seventh power law: $$\frac{\theta}{x} = \frac{0.016}{\mathrm{Re}_x^{\frac{1}{7}}}$$

Momentum integral equation: $$\tau_{\mathrm{w}} = \rho U^2 \frac{\mathrm{d}\theta}{\mathrm{d}x}$$

Smooth turbulent BL, empirical: $$\frac{\tau_{\mathrm{w}}}{\frac{1}{2}\rho U^2} = \frac{0.0205}{\mathrm{Re}_\delta^{\frac{1}{6}}}$$

Smooth turbulent BL, one-seventh power law: $$\frac{\delta}{x} = \frac{0.16}{\mathrm{Re}_x^{\frac{1}{7}}}$$

$$c_{\mathrm{f}} = \frac{\tau_{\mathrm{w}}}{\frac{1}{2}\rho U^2} = \frac{0.027}{\mathrm{Re}_x^{\frac{1}{7}}}$$

$$C_{\mathrm{Df}} = \frac{F_{\mathrm{D}}}{\frac{1}{2}\rho U^2 L b} = \frac{0.031}{\mathrm{Re}_{\mathrm{L}}^{\frac{1}{7}}}$$

MIXING-LENGTH THEORY OF TURBULENT BOUNDARY LAYERS

Smooth turbulent BL, logarithmic law: $$\frac{u}{u_*} = 2.44 \ln \frac{u_* y}{\nu} + 5.0$$

Rough turbulent BL, logarithmic law: $$\frac{u}{u_*} = 2.44 \ln \frac{y}{\epsilon} + 8.5$$

Velocity defect law: $$\frac{U - u}{u_*} = -3.74 \ln\left(\frac{y}{\delta}\right)$$

One-seventh power law distribution: $$\frac{u}{U} = \left(\frac{y}{\delta}\right)^{\frac{1}{7}}$$

BOUNDARY LAYERS IN CLOSED CONDUITS

Smooth turbulent flow in pipes: $$\frac{u_{\max} - u(r)}{u_*} = 2.44 \ln \frac{R}{R - r}$$

$$\frac{u(r)}{u_*} = 2.44 \ln \frac{u_*(R - r)}{\nu} + 5.0$$

$$\frac{V}{u_*} = 2.44 \ln \frac{u_* R}{\nu} + 1.34$$

$$\frac{1}{f^{\frac{1}{2}}} = 2.0 \log\left(\mathrm{Re} f^{\frac{1}{2}}\right) - 0.8$$

$$f = \frac{0.316}{\text{Re}^{\frac{1}{4}}} \quad \text{(Blasius approximation)}$$

Rough turbulent flow in pipes:

$$\frac{u_{\max} - u(r)}{u_*} = 2.44 \ln \frac{R}{R - r}$$

$$\frac{u(r)}{u_*} = 2.44 \ln \frac{R - r}{\epsilon} + 8.5$$

$$\frac{V}{u_*} = 2.44 \ln \frac{D}{\epsilon} + 3.2$$

$$\frac{1}{f^{\frac{1}{2}}} = -2.0 \log \left(\frac{\epsilon / D}{3.7} \right)$$

Applications of the above equations are provided in the text, and additional problems to practice using these equations can be found in the following section.

Section 11.1: Introduction

11.1. Air flows over a 1.6-m-long plane surface under standard sea-level atmospheric conditions. (a) Up to what air speed is the boundary layer laminar over the entire length of the surface? (b) At what air speed is the boundary layer turbulent over 33.3% of the length of the surface.

11.2. A commercial aircraft is designed to have a takeoff speed of 120 km/h at sea level and to cruise at a speed of 875 km/h at an elevation of 11 km. Assuming that a wing can be approximated as a flat plate, estimate how far from the front of the wing to the location where the boundary layer becomes turbulent (a) for the takeoff condition and (b) for the cruising condition.

11.3. The wings on a model aircraft have a chord length of 140 mm, and the model is designed to operate under standard sea-level atmospheric conditions. Assuming that the wings can be approximated as flat plates for purposes of boundary-layer analysis, at what speed, in km/h, will turbulence appear in the boundary layer over the wing (a) if the model is operated under design conditions and (b) if the model is operated at an elevation of 4000 m?

11.4. A small boat under normal operating conditions travels at 14 km/h in a lake that contains water at 10°C. (a) Estimate the distance from the front of the boat to the location at which the boundary layer transitions from laminar to turbulent. (b) If a 1:30 scale model of the boat is built and operated under dynamically similar conditions, how far from the front of the model boat does the transition occur?

Section 11.2: Laminar Boundary Layers

11.5. One of the assumptions in deriving the Blasius solution for a laminar boundary layer is that the component of the velocity normal to the solid surface is much smaller than the component of the velocity parallel to the surface, which requires that $\delta/x \ll 1$. If the Blasius approximation is taken to be unsupported when $\delta/x > 0.11$, estimate the range of Re_x for which the Blasius solution should not be applied.

11.6. What is the approximate ratio of the boundary-layer thickness to the distance from the leading edge of a flat surface when the boundary layer transitions from laminar to turbulent flow?

11.7. The relative feasibility of conducting model tests in a water tunnel versus conducting the tests in an air tunnel is being investigated. The model tests require a 2.4-m-long flat test section in which the boundary layer must be laminar. Compare the required air velocity in the wind tunnel using standard air with the required water velocity in the water tunnel using water at 10°C.

11.8. The x component of the velocity, u, in a laminar boundary layer can be expressed in normalized form by

$$\frac{u}{U} = 2\left(\frac{y}{\delta}\right) - \left(\frac{y}{\delta}\right)^2, \qquad \text{where } \frac{\delta}{x} = \frac{4.91}{\text{Re}^{\frac{1}{2}}}, \ \text{Re} = \frac{Ux}{\nu}$$

where U is the free-stream velocity, y is the coordinate distance measured normal to the solid surface, δ is the thickness of the boundary layer (which is a function of x), and ν is the kinematic viscosity of the fluid. (a) Assuming that the fluid is incompressible and using the continuity equation, determine an expression for the normalized y component of the velocity, v/U, as a function of x, y, and δ. (b) Boundary-layer analyses typically assume that v is negligible compared with u. Assess the justification of this assumption by finding the range of values of Re for which $v/u \leq 0.2$ at the outer limit of the boundary layer.

11.9. Consider the general case of a fluid flowing over a smooth flat rectangular surface of length L and width b. Show that if the boundary layer is laminar over the entire surface, then 33.3% of the total drag force on the surface is exerted over the front 11% of the surface.

11.10. Standard air flows over a rectangular flat surface at a speed of 5 m/s. Estimate the average shear stress on the surface between the leading edge and a location 0.24 m downstream of the leading edge.

11.11. A vertical plate placed in the center of a flume is used to divert water that flows in the flume. The diversion plate has a length of 0.75 m in the direction of flow. If water at 10°C flows at 2.5 m/s in the flume at a depth of 1.5 m and a laminar boundary layer is maintained over the surface of the plate, what is the drag force exerted on the plate by the flowing water?

11.12. A smooth sensor plate 1.5 m long, 1.5 m wide, and 6 mm thick is pulled in seawater at 10°C by a submersible vehicle at a rate of 20 cm/s. (a) What force is required to pull the sensor plate? (b) What is the thickness of the boundary layer on the downstream end of the sensor plate?

11.13. Consider the following fluids flowing over a flat plate: (a) water at 20°C, (b) standard air at 20°C, and (c) SAE 30 oil at 20°C. For each fluid, determine the distance from the leading edge of the plate to where the boundary layer transitions from laminar to turbulent and estimate the thickness of the boundary layer just before the transition.

11.14. Water at 20°C flows over a flat rectangular surface at a speed of 35 mm/s. (a) What is the thickness of the boundary layer at a distance of 10 m from the leading edge of the surface? Is the boundary layer laminar or turbulent at this location? (b) Determine the velocity profile at a distance of 10 m from the leading edge.

11.15. Air at 10°C flows past a flat surface at a velocity of 21 m/s. (a) How far from the leading edge of the surface does the boundary layer transition from being laminar to being turbulent? (b) What is the thickness of the boundary layer at the transition location? (c) What is the velocity at the midpoint of the boundary layer at the transition location?

11.16. Ethylene glycol at 20°C approaches a flat surface at a velocity of 1.8 m/s. At a location 1.5 m downstream from the leading edge, (a) what is the thickness of the boundary layer and (b) what is the velocity 1 mm away from the surface?

11.17. Water at 20°C flows over a flat surface with a free-stream velocity of 1.5 m/s. Determine the velocity at a point 0.3 m downstream of the leading edge and 0.7 mm from the surface, assuming that the velocity profile in the boundary layer can be approximated by the following parabolic distribution:

$$\frac{u}{U} = 2\frac{y}{\delta} - \left(\frac{y}{\delta}\right)^2$$

Estimate the velocity at the same point using the Blasius solution and determine the percentage error incurred by not using the Blasius solution.

11.18. Water at 15°C flows over a flat surface with a free-stream velocity of 1.1 m/s. Determine the velocity at a point 0.5 m downstream of the leading edge and 1 mm from the surface, assuming that the velocity profile in the boundary layer can be approximated by the following cubic distribution:

$$\frac{u}{U} = \frac{3}{2}\frac{y}{\delta} + \frac{1}{2}\left(\frac{y}{\delta}\right)^3$$

Estimate the velocity at the same point using the Blasius solution and determine the percentage error incurred by not using the Blasius solution.

11.19. Water at 10°C flows over a flat surface with a free-stream velocity of 0.9 m/s. Determine the velocity at a point 0.7 m downstream of the leading edge and 1.5 mm from the surface, assuming that the velocity profile in the boundary layer can be approximated by the following linear distribution:

$$\frac{u}{U} = \frac{y}{\delta}$$

Estimate the velocity at the same point using the Blasius solution and determine the percentage error incurred by not using the Blasius solution.

11.20. Water at 5°C flows over a flat surface with a free-stream velocity of 0.8 m/s. Determine the velocity at a point 0.9 m downstream of the leading edge and 2 mm from the surface, assuming that the velocity profile in the boundary layer can be approximated by the following sine-wave distribution:

$$\frac{u}{U} = \sin\left[\frac{\pi}{2} \cdot \frac{y}{\delta}\right]$$

Estimate the velocity at the same point using the Blasius solution and determine the percentage error incurred by not using the Blasius solution.

11.21. A thin 2.1-m-long canoe is being rowed in stagnant water at 10°C. The draft of the canoe is 31 cm, and for purposes of analysis, the sides of the canoe can be approximated as flat surfaces. At what speed will the boundary layer be laminar over the entire canoe? What drag force will be exerted on the canoe under this condition?

11.22. Water enters a pipe of diameter D as illustrated in Figure 11.11. The velocity distribution in the pipe stabilizes with a centerline velocity V when the boundary layer extends to the center of the pipe; beyond this point, the flow is said to be fully developed. (a) Estimate the distance required (in terms of the number of pipe diameters) for the velocity distribution to be fully developed. In your analysis, consider the cases of laminar flow and turbulent flow separately and express your answers in terms of the Reynolds number of the flow in the pipe, where the Reynolds number is based on the centerline velocity. State your assumptions. (b) If the flow enters the pipe with a Reynolds number of 8000, estimate the distance required for the velocity profile to stabilize. (c) If this same pipe (with Re = 8000) carries water at 20°C and has a centerline velocity of 0.1 m/s, use flat surface theory to estimate the shear stress exerted on the pipe wall at a flow distance of 10 diameters from the entrance, assuming that the flow is turbulent from the entrance. (d) Explain briefly how flat surface theory might be used to calculate the head loss in the transition region at the pipe entrance.

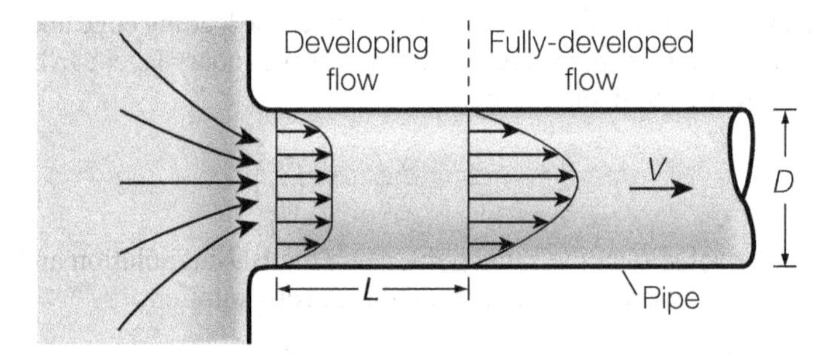

Figure 11.11: Flow at a pipe entrance

11.23. Consider a flat plate of length L where for a laminar boundary layer, the friction coefficient at a distance x from the front of the plate is given by

$$c_{fx} = \frac{0.664}{\sqrt{\text{Re}_x}}$$

(a) Show that the drag coefficient is given by

$$C_D = \frac{1.328}{\sqrt{\text{Re}_L}}$$

(b) If a plate is 51 cm long and 2 m wide and air at 20°C flows over the plate at 2 m/s, determine the drag force on the plate and the boundary layer thickness at the downstream end of the plate.

Section 11.3: Turbulent Boundary Layers

11.24. Standard air flows parallel to a flat rectangular surface. Observations at a location 1.5 m from the leading edge of the surface indicate that the boundary layer is turbulent, and the thickness of the boundary layer at this location is 20 mm. (a) Estimate the thickness of the boundary layer at distances of 8 m and 15 m from the leading edge of the flat surface. (b) What is the minimum speed of the air that will result in a turbulent boundary layer 1.5 m from the leading edge of the surface?

11.25. Consider the general case of a fluid flowing over a smooth flat rectangular surface of length L and width b. Show that if the boundary layer is turbulent over the entire surface and the velocity profile within the boundary layer can be described by the one-seventh power law distribution, then 50% of the total drag force on the surface is exerted over approximately the front 20% of the surface.

11.26. A fluid flows over a smooth flat rectangular surface of length L and width b. Show that if the boundary layer is turbulent over the entire surface and the friction drag coefficient is described by the semiempirical relation given by Equation 11.59, then 50% of the total drag force on the surface is exerted over approximately the front 42% of the surface.

11.27. From an analysis of pipe flow data, the following empirical expression for the local wall shear stress, τ_w, as a function of the boundary-layer thickness, δ, has been derived (by Prandtl) and is widely accepted:

$$\tau_{\rm w} = 0.0233\rho U^2 \left(\frac{\nu}{U\delta}\right)^{\frac{1}{4}}$$

where ρ and ν are the density and kinematic viscosity of the fluid, respectively, and U is the free-stream velocity. If the one-seventh power law velocity distribution is assumed to exist within the boundary layer, then the ratio of the momentum thickness, θ, to the boundary-layer thickness can be shown to be given by $\theta/\delta = 7/72$. Show that the two given relationships yield the following equations describing various properties of the boundary layer:

$$\frac{\delta}{x} = \frac{0.382}{\mathrm{Re}_x^{\frac{1}{5}}}, \qquad \frac{\tau_{\rm w}}{\frac{1}{2}\rho U^2} = \frac{0.0594}{\mathrm{Re}_x^{\frac{1}{5}}}, \qquad \frac{F_{\rm D}}{\frac{1}{2}\rho U^2 Lb} = \frac{0.0735}{\mathrm{Re}_{\rm L}^{\frac{1}{5}}}$$

where x is the distance along a flat surface of length L and width b and $F_{\rm D}$ is the total drag force on the surface.

11.28. Consider the case of a streamlined submersible vehicle that cruises at 4.116 m/s in seawater at 10°C. The sides of the submersible can be approximated as flat surfaces. A critical apparatus is located on the side of the submersible 2.5 m from the front of the vehicle, and it is desired that rough turbulent boundary-layer flow not exist at the critical location. Smooth and transitional turbulent boundary-layer flow is acceptable. What is the maximum allowable roughness height at the critical location? Make sure that you take into account the two competing estimates for the local shear stress.

11.29. Standard air flows over a flat surface at 15 km/h. (a) Find the minimum length of the surface to ensure that laminar flow occurs over the entire surface. (b) If it is desired that the flow be turbulent over at least half of the surface, what is the minimum required roughness of the surface?

11.30. In a smooth turbulent boundary layer, the velocity distribution adjacent to a flat surface is given by Equation 11.50. At some distance $y = \alpha\nu/u_*$ from the surface, the velocity given by the laminar sublayer formulation is equal to the velocity given by the logarithmic distribution (Equation 11.50). This location can be taken as the transition point from the laminar sublayer to the outer region. Determine the value of α where this transition point occurs.

11.31. An aircraft flies at a speed of 300 km/h at an elevation of 5 km. To optimize the aerodynamic performance of the aircraft, it is desired that rough turbulent conditions exist in the boundary layer at a distance of 0.5 m from the front of the wing. What is the minimum roughness height that must exist at that location to achieve the desired effect?

11.32. Two alternative expressions for the growth in the thickness of a smooth turbulent boundary layer are used in practice. One expression, Equation 11.56, is semiempirical and based partially on empirical data from flow in pipes and partially on the one-seventh power law distribution. The other expression, Equation 11.81, is primarily based on assuming the one-seventh power law distribution. The semiempirical relationship, Equation 11.56, is stated to be valid within the range of $5 \times 10^5 \leq \text{Re}_x < 10^7$. What is the range of percentage deviation of the boundary-layer thickness derived from Equation 11.81 from the boundary-layer thickness derived from Equation 11.56 within the restricted range of Reynolds numbers?

11.33. Two alternative expressions for the local shear stress in a smooth turbulent boundary layer are given in the text. One expression, Equation 11.57, is semiempirical and based partially on empirical data from flow in pipes. The other expression, Equation 11.82, is primarily based on assuming the one-seventh power law distribution. The semiempirical relationship, Equation 11.57, is stated to be valid within the range of $5 \times 10^5 \leq \text{Re}_x < 10^7$. What is the range of percentage deviation of the local shear stress derived from Equation 11.82 from the local shear stress derived from Equation 11.57 within the restricted range of Reynolds numbers?

11.34. Two alternative expressions for the drag force on a smooth turbulent boundary layer are presented in the text. One expression, Equation 11.59, is semiempirical and based partially on empirical data from flow in pipes and partially on the one-seventh power law distribution. The other expression, Equation 11.84, is based primarily on assuming the one-seventh power law distribution. The semiempirical relationship, Equation 11.59, is stated to be valid within the range of $5 \times 10^5 \leq \text{Re}_x < 10^7$. What is the range of percentage deviation of the drag force derived from Equation 11.84 from the drag force derived from Equation 11.59 within the restricted range of Reynolds numbers?

11.35. Water runs down a flat road pavement that has a slope of 0.06% and a roughness height of 1.5 mm. For a flow depth of 2 cm, determine whether turbulent flow exists in the water column and whether the flow regime is in the rough surface, smooth surface, or intermediate regime. Assume that the temperature of the water is 20°C, and the wall shear stress, τ_w, can be estimated as $\gamma d S_0$, where γ is the specific weight of the liquid, d is the depth of flow, and S_0 is the slope of the plane surface.

11.36. The velocity distribution within a turbulent boundary layer is assumed to be described by the one-seventh power law. (a) At what fractions of the boundary-layer thickness (i.e., y/δ) does the velocity equal 50% and 90% of the free-stream velocity? (b) Find the ratio of the shear velocity, u_*, to the free-stream velocity, U, that will make the one-seventh power law velocity and the velocity derived from Equations 11.48 and 11.49 equal at the 50% and 90% locations calculated in part (a).

11.37. Standard air flows over a 1-m-long smooth flat surface at 32 km/h. At the downwind end of the surface, (a) estimate the velocity 10 mm from the surface and (b) estimate the local shear stress on the surface.

11.38. Standard air flows over a 2-m-long, 3-m-wide flat surface at a speed of 30 m/s. The boundary layer is assumed to be in the smooth turbulent regime over the entire surface, and the velocity distribution within the boundary layer is assumed to follow the one-seventh power law distribution. (a) Estimate the thickness of the boundary layer at the downstream end of the surface. (b) Compare the drag force on the front half of the surface with the drag force on the back half of the surface.

11.39. A sleek smooth-surfaced train is 120 m long, 4 m high, and 4.5 m wide and moves at a design speed of 88 km/h. Assuming that the train stops at locations that have elevations anywhere from sea level to 1000 m, estimate the engine power required to overcome aerodynamic drag on any straight stretch of track on its route.

11.40. Standard air flows over a flat surface at a speed of 25 m/s. The boundary layer is turbulent from the leading edge of the surface, and the velocity distribution within the boundary layer can be approximated by the one-seventh law distribution. The thickness of the boundary layer at two locations along the surface are measured as 5 mm and 15 mm. Estimate the distance between the locations where the boundary-layer thicknesses were measured.

Section 11.4: Applications

11.41. Air flows over a smooth flat surface at a speed of 30 km/h. Estimate the minimum length of the surface required for the thickness of the boundary layer at the downstream edge to exceed 30 mm. Assume standard air.

11.42. An airliner cruises at a speed of 850 km/h at an altitude of 10 km. Upon landing at a sea-level airport, the approach speed is typically around 80 km/h. The aircraft has wings with a chord length of 3 m. Assuming that the wings can be approximated by flat plates, compare the boundary-layer thickness on the downstream edge of the wing when cruising with the corresponding boundary-layer thickness when landing.

11.43. SAE 50 oil at 20°C flows into a 250-mm-diameter pipe at a velocity of 2.5 m/s. Estimate the minimum length of pipe required for the entire cross section of the pipe to be contained within the boundary layer. What further changes are likely to occur in the velocity profile across the pipe if the pipe is longer than the minimum length and maintains the same cross section?

11.44. Relationships commonly used for estimating the growth of the boundary layer (BL) over a flat plate assume that the BL is either laminar or turbulent from the leading edge of the plate. An approximate model that accounts for a transition from a laminar to a turbulent BL is shown in Figure 11.12. In this model, the BL is assumed to be laminar from the leading edge of the plate to a location of instantaneous transition to a turbulent BL, where $\text{Re} = \text{Re}_\text{cr}$ and $x = x_\text{cr}$. In the approximate model, a virtual origin of the turbulent BL is at $x = x_0$ such that the thickness of the laminar and turbulent BLs are the same at $x = x_\text{cr}$. Beyond $x = x_\text{cr}$, the thickness of the BL is estimated using the turbulent BL formulation with x measured relative to x_0. (a) Show that x_0 is related to x_cr by

$$x_0 = x_\text{cr}\left(1 - \frac{54.3}{\text{Re}_\text{cr}^{\frac{5}{12}}}\right)$$

(b) Consider the case where standard air flows over a 3-m-long smooth flat surface at 35 km/h and $\text{Re}_{\text{cr}} = 5 \times 10^5$. Compare the BL thickness at the downstream end of the surface using the model described in part (a) with the BL thickness calculated based on the assumption that the BL is of one type over the entire surface. Assess whether using the approximate model is worthwhile in this case.

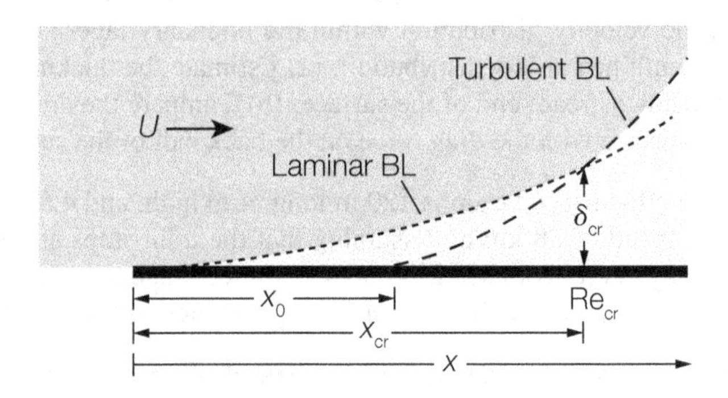

Figure 11.12: Modified boundary layer model

11.45. Standard air flows at a speed of 10 m/s over a flat surface that is 5 m long and 2 m wide. (a) Estimate the area over which the flow in the boundary layer is laminar and the area over which the flow in the boundary layer is turbulent. (b) Estimate the displacement thickness just before the boundary layer transitions from laminar to turbulent. (c) Estimate the displacement thickness at the downstream end of the surface.

11.46. The velocity distribution in a boundary layer is approximated by the polynomial relation: $u(y) = a + by + cy^2$, where u is the speed of the fluid parallel to the surface, y is the distance from the surface, and a, b, and c are constants. Typical boundary conditions to be met are

$$u(0) = 0, \qquad u(\delta) = U, \qquad \left.\frac{du}{dy}\right|_{y=\delta} = 0$$

Show that when these boundary conditions are combined with the polynomial distribution, the normalized velocity distribution in the boundary layer can be expressed as

$$\frac{u}{U} = 2\left(\frac{y}{\delta}\right) - \left(\frac{y}{\delta}\right)^2$$

11.47. The velocity distribution in a laminar boundary layer is found to be adequately described by the following cubic distribution:

$$\frac{u}{U} = \frac{3}{2}\left(\frac{y}{\delta}\right) - \frac{1}{2}\left(\frac{y}{\delta}\right)^3$$

where u is the velocity at a distance y from the surface, U is the free-stream velocity, and δ is the thickness of the boundary layer. Determine the ratio of the displacement thickness to the boundary layer thickness.

11.48. For the velocity distribution given in Problem 11.47, what is the ratio of the momentum thickness to the boundary-layer thickness? Compare the magnitudes of the momentum and displacement thicknesses.

11.49. For the velocity distribution given in Problem 11.47, (a) determine the thickness of the boundary layer as a function of the distance along the surface, and (b) determine the local wall shear stress as a function of the distance along the surface.

11.50. The velocity distribution across a boundary layer is given in Problem 11.47. What is the distance from the surface to the point where the velocity is equal to 10%, 25%, 50%, and 75% of the free-stream velocity? Give your answers as fractions of the boundary-layer thickness.

11.51. For the velocity distribution given in Problem 11.47, it can be shown that the shear stress as a function of the distance x along a flat surface can be expressed in nondimensional form as

$$\frac{\tau_\mathrm{w}}{\frac{1}{2}\rho U^2} = \frac{0.646}{\mathrm{Re}_x^{\frac{1}{2}}}$$

Use this relationship to determine a nondimensional equation for the total drag force, F_D, on a flat surface of length L and width b.

11.52. The velocity distribution in a laminar boundary layer can be described by the following normalized quadric distribution:

$$u^* = 2y^* - 2y^{*3} + y^{*4}$$

where $u^* = u/U$, $y^* = y/\delta$, u is the velocity at a distance y from the surface, U is the free-stream velocity, and δ is the thickness of the boundary layer. Determine the ratio of the displacement thickness to the boundary layer thickness.

11.53. For the velocity distribution given in Problem 11.52, what is the ratio of the momentum thickness to the boundary-layer thickness? Compare the magnitudes of the momentum and displacement thicknesses.

11.54. For the velocity distribution given in Problem 11.52, (a) determine the thickness of the boundary layer as a function of the distance along the surface and (b) determine the local wall shear stress as a function of the distance along the surface.

11.55. The velocity distribution across a boundary layer is given in Problem 11.52. What is the distance from the surface to the point where the velocity is equal to 10%, 25%, 50%, and 75% of the free-stream velocity? Give your answers as fractions of the boundary-layer thickness.

11.56. For the velocity distribution given in Problem 11.52, it can be shown that the shear stress as a function of the distance x along a flat surface can be expressed in nondimensional form as

$$\frac{\tau_\mathrm{w}}{\frac{1}{2}\rho U^2} = \frac{0.685}{\mathrm{Re}_x^{\frac{1}{2}}}$$

Use this relationship to determine a nondimensional equation for the total drag force, F_D, on a flat surface of length L and width b.

11.57. The velocity distribution in a turbulent boundary layer on a smooth surface can be described by the following one-seventh power law distribution:

$$u^* = y^{*\frac{1}{7}}$$

where $u^* = u/U$, $y^* = y/\delta$, u is the velocity at a distance y from the surface, U is the free-stream velocity, and δ is the thickness of the boundary layer. Determine the ratio of the displacement thickness to the boundary layer thickness.

11.58. For the velocity distribution given in Problem 11.57, what is the ratio of the momentum thickness to the boundary-layer thickness? Compare the magnitudes of the momentum and displacement thicknesses.

11.59. The velocity distribution in a laminar boundary layer is found to be adequately described by the following half-sine-wave distribution:

$$\frac{u}{U} = \sin\left(\frac{\pi}{2} \cdot \frac{y}{\delta}\right)$$

where u is the velocity at a distance y from the surface, U is the free-stream velocity, and δ is the thickness of the boundary layer. Determine the ratio of the displacement thickness to the boundary layer thickness.

11.60. For the velocity distribution given in Problem 11.59, what is the ratio of the momentum thickness to the boundary-layer thickness?

11.61. For the velocity distribution given in Problem 11.59, (a) determine the thickness of the boundary layer as a function of the distance along the surface and (b) determine the local wall shear stress as a function of the distance along the surface.

11.62. The velocity distribution across a boundary layer is given in Problem 11.59. What is the distance from the surface to the point where the velocity is equal to 10%, 25%, 50%, and 75% of the free-stream velocity? Give your answers as fractions of the boundary-layer thickness.

11.63. For the velocity distribution given in Problem 11.59, it can be shown that the shear stress as a function of the distance x along a flat surface can be expressed in nondimensional form as

$$\frac{\tau_\mathrm{w}}{\frac{1}{2}\rho U^2} = \frac{0.646}{\mathrm{Re}_x^{\frac{1}{2}}}$$

Use this relationship to determine a nondimensional equation for the total drag force, F_D, on a flat surface of length L and width b.

11.64. (a) What is the ratio of the momentum thickness to the boundary-layer thickness given by the Blasius solution? (b) It is sometimes assumed that the velocity distribution in the boundary layer is given by the following power-law relationship:

$$\frac{u}{U} = \begin{cases} \left(\frac{y}{\delta}\right)^{\frac{1}{n}}, & y \leq \delta \\ 1, & y > \delta \end{cases}$$

where u is the longitudinal speed within the boundary layer, U is the free-stream speed, y is the distance from the surface, δ is the thickness of the boundary layer, and n is a number, usually assumed to be in the range of 5–10. What value of n will give the same ratio of momentum thickness to boundary-layer thickness as that given by the Blasius solution?

11.65. In the case of a fluid flowing parallel to a flat surface, the displacement thickness, δ^*, of any streamline that is located outside a laminar boundary layer adjacent to the surface is given by

$$\frac{\delta^*}{x} = \frac{1.72}{\mathrm{Re}_x^{\frac{1}{2}}}$$

Determine the value of the proportionality constant for streamlines that are located within the boundary layer, expressing this proportionality constant as a function of the normalized distance, η, from the flat surface. Recall that the normalized distance, η, is defined by Equation 11.11. (*Hint:* Use the values of the Blasius solution given in Table 11.1 along with the definition of the displacement thickness given by Equation 11.63.)

11.66. The velocity distribution near a solid surface can be crudely approximated as being linear such that

$$\frac{u}{U} = \begin{cases} \frac{y}{\delta}, & y \leq \delta \\ 1, & y > \delta \end{cases}$$

where u is the longitudinal velocity in the boundary layer, U is the free-stream longitudinal velocity, y is the distance from the surface, and δ is the thickness of the boundary layer. (a) Determine the momentum thickness of the boundary layer. (b) Determine the shear stress on the surface in terms of μ, U, and δ, where μ is the dynamic viscosity of the fluid. (c) Combine the results obtained in parts (a) and (b) with the momentum integral equation to determine the relationship between δ/x and Re_x, where x is the distance from the point where $\delta = 0$ and Re_x is the Reynolds number, using x as the length scale.

11.67. A self-similar velocity profile in the boundary layer over a flat plate is described by

$$\frac{u}{U} = \begin{cases} f\left(\frac{y}{\delta}\right), & 0 \leq \frac{y}{\delta} \leq 1 \\ 1, & \frac{y}{\delta} > 1 \end{cases}$$

For the case where the boundary layer grows from zero thickness at $x = 0$, show that the thickness of the boundary layer, δ, as a function of the distance x from the leading edge of the plate is described by

$$\frac{\delta}{x} = \frac{C_1}{\text{Re}_x^{\frac{1}{2}}}$$

where C_1 is a constant that depends only on the functional form of the self-similar velocity distribution.

11.68. Show that for any self-similar velocity profile in which the boundary layer grows from a thickness of δ_0 at $x = x_0$, the thickness of the boundary layer, δ, is described by

$$\frac{\delta}{\Delta x} = \sqrt{\left(\frac{\delta_0}{\Delta x}\right)^2 + \left(\frac{C_1}{\text{Re}_{\Delta x}^{\frac{1}{2}}}\right)^2}$$

where $\Delta x = x - x_0$, $\text{Re}_{\Delta x} = U\Delta x/\nu$, and C_1 is a constant that depends only on the functional form of the self-similar velocity distribution.

11.69. For a case where the boundary layer grows from zero thickness at $x = 0$, show that for any self-similar velocity profile, the shear stress, τ_w, as a function of the distance x from the leading edge of the surface is described by

$$\frac{\tau_\text{w}}{\frac{1}{2}\rho U^2} = \frac{C_2}{\text{Re}_x^{\frac{1}{2}}}$$

where C_2 is a constant that depends only on the functional form of the self-similar velocity distribution.

11.70. Show that for any self-similar velocity profile in which the boundary layer grows from a thickness of δ_0 at $x = x_0$, the shear stress, τ_w, is described by

$$\frac{\tau_\text{w}}{\frac{1}{2}\rho U^2} = \frac{2A}{\left[\left(\frac{\delta_0 U}{\nu}\right)^2 + C_1^2 \cdot \text{Re}_{\Delta x}\right]^{\frac{1}{2}}}$$

where A and C_1 are constants that depend only on the functional form of the self-similar velocity distribution.

11.71. For a case where the boundary layer grows from zero thickness at $x = 0$, show that for any self-similar velocity profile, the drag force, F_D, on a rectangular plate of width b and length L is given by

$$\frac{F_\text{D}}{\frac{1}{2}\rho U^2 bL} = \frac{C_3}{\text{Re}_\text{L}^{\frac{1}{2}}}$$

where C_3 is a constant that depends only on the functional form of the self-similar velocity distribution.

11.72. Show that for any self-similar velocity profile in which the boundary layer grows from a thickness of δ_0 at $x = x_0$, the drag force, F_D, on a rectangular plate of width b and length ΔL downstream of $x = x_0$ is given by

$$\frac{F_\text{D}}{\frac{1}{2}\rho U^2 b\Delta L} = 2B\sqrt{\left(\frac{\delta_0}{\Delta L}\right)^2 + \left(\frac{C_1}{\text{Re}_{\Delta L}^{\frac{1}{2}}}\right)^2}$$

where B and C_1 are constants that depend only on the functional form of the self-similar velocity distribution.

11.73. The momentum integral equation is derived in the text for the case in which the free-stream velocity, U, remains constant (see Equation 11.76). In cases where the free-stream velocity varies as a function of the distance, x, along a flat surface, a force balance between two sections along the boundary layer yields the following momentum equation:

$$-\delta\frac{\text{d}p}{\text{d}x} - \tau_\text{w} = \frac{\partial}{\partial x}\int_0^\delta u\,\rho u\,\text{d}y - U\frac{\partial}{\partial x}\int_0^\delta \rho u\,\text{d}y \qquad (11.118)$$

where $\text{d}p/\text{d}x$ is the pressure gradient along the surface in the streamwise direction. Show that Equation 11.118 can be expressed without the pressure-gradient term as

$$\tau_\text{w} = -\frac{\partial}{\partial x}\int_0^\delta u\,\rho u\,\text{d}y + U\frac{\partial}{\partial x}\int_0^\delta \rho u\,\text{d}y + \frac{\text{d}U}{\text{d}x}\int_0^\delta \rho U\,\text{d}y \qquad (11.119)$$

11.74. Show that the boundary-layer momentum equation given by Equation 11.119 yields the following form of the momentum integral equation for a nonzero pressure gradient:

$$\frac{\tau_\text{w}}{\rho} = \frac{\text{d}}{\text{d}x}(U^2\theta) + \delta^* U\frac{\text{d}U}{\text{d}x} \qquad (11.120)$$

where θ is the momentum thickness and δ^* is the displacement thickness.

11.75. Show that for a zero pressure gradient, the momentum integral equation given by Equation 11.120 simplifies to

$$\tau_\mathrm{w} = \rho U^2 \frac{\mathrm{d}\theta}{\mathrm{d}x}$$

11.76. A fluid approaches a flat surface at velocity U as shown in Figure 11.13. At a distance x from the leading edge, the boundary-layer thickness is δ. (a) Assuming that the boundary layer remains laminar over the distance x and a streamline that intersects the outer limit of the boundary layer originates at a distance y_0 above the leading edge, express the ratio y_0/δ as a function of x. (b) If the laminar boundary layer transitions to a turbulent boundary layer at a distance of 10 m from the leading edge, what is the boundary-layer thickness prior to the transition? What is the streamline origin coordinate y_0 corresponding to this location?

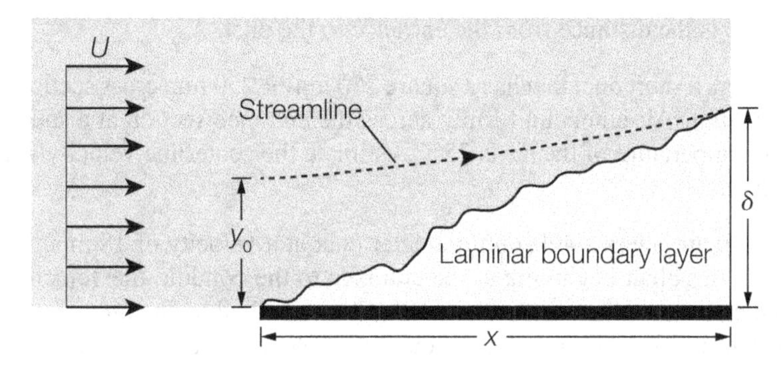

Figure 11.13: Streamline intersecting a boundary layer

11.77. Standard air flows at a speed of 30 m/s over a flat surface that is 15 m long and 5 m wide. (a) Assuming that the boundary layer is turbulent over the entire surface, estimate the momentum thickness at the downstream end of the surface. (b) Use the estimated momentum thickness to calculate the drag force on the surface.

11.78. Water at 20°C flows at 0.6 m/s over a flat surface that is 0.3 m long and 2 m wide. The velocity distribution in the boundary layer can be assumed to be linear and have the functional form $u^* = y^*$, where $u^* = u/U$ and $y^* = y/\delta$. Estimate the drag force on the surface.

11.79. Standard air flows at 10 m/s over a 3-m-wide, 2.5-m-long rectangular surface. (a) Compare the momentum thickness at the downstream end of the surface for the cases of all-laminar and all-turbulent smooth surface boundary-layer flow. (b) Use the results in part (a) to estimate the percentage increase in drag force on the surface that would occur if an all-laminar boundary-layer flow was "tripped" into an all-turbulent smooth surface boundary-layer flow.

11.80. Air enters a 200 mm × 200 mm duct at 30 m/s. At a particular cross section near the entrance of the duct, fine-scale velocity measurements of the velocity distribution near the wall of the duct indicate that the displacement thickness is 1.8 mm, and at a section downstream of this section, the displacement thickness is measured as 2.9 mm. Use these data to estimate the value of $\Delta p/\frac{1}{2}\rho U_1^2$, where Δp is the difference in free-stream pressure between the two sections, ρ is the density of the flowing air, and U_1 is the free-stream velocity at the upstream section, which in this case is given as 30 m/s.

11.81. Air at 25°C enters a short duct that has a diameter of 500 mm and a length of 900 mm. Air enters uniformly across the entrance section at a speed of 2.4 m/s. Estimate the centerline velocity at the outlet from the duct.

11.82. Air enters a duct of diameter D_0, and conditions are such that the boundary layer is in the smooth turbulent regime from the entrance. The velocity distribution in the boundary layer can be approximated by the one-seventh power law distribution. Show that as long as the wall boundary layers do not merge, the pressure gradient along the centerline of the duct can be maintained equal to zero if the diameter, D, of the duct is expanded according to the relationship

$$D(x) = D_0 + \frac{0.04x}{\mathrm{Re}_x^{\frac{1}{7}}}$$

where x is the distance from the entrance to the duct.

11.83. Air enters a short duct that has a square 250 mm × 250 mm cross section and a length of 600 mm. Air enters uniformly across the entrance section at a speed of 2.4 m/s. If the temperature of the air is 25°C, estimate the centerline velocity at the outlet of the duct.

11.84. Standard air enters a 400-mm-diameter duct at a velocity of 18 m/s. The boundary layer is turbulent beginning at the entrance to the conduit, the velocity distribution in the boundary layer can be assumed to be described by the one-seventh power law distribution, and the thickness of the boundary layer at a location 6 m downstream of the conduit entrance is estimated as 120 mm. For the one-seventh power law distribution, the ratio of the displacement thickness to the boundary layer thickness is 1:8. (a) Estimate the gauge pressure at the downstream section. (b) Estimate the average shear stress that the air exerts on the conduit wall between the entrance and the downstream section.

11.85. Standard air flows over a flat surface at a speed of 4 m/s. At a distance of 0.25 m from the leading edge, estimate the following: (a) the normal distance from the surface to where the velocity is equal to 90% of the free-stream velocity and (b) the slope of the streamline that originates at a point just above the leading edge of the surface. (*Hint:* Use the Blasius velocity distribution, if appropriate, and use the displacement distance in calculating the slope of the streamline.)

11.86. Standard air enters a 5-m-long, 150 mm × 150 mm duct. It is estimated that the flow in the boundary layer is smooth turbulent and that the velocity distribution in the boundary layer follows the one-seventh law profile. Turbulent flow within the core region of the duct (i.e., the region unaffected by the boundary layer) is assumed to be fully turbulent when the Reynolds number of the core is $\mathrm{Re_c} = U_\mathrm{c} D_\mathrm{h}/\nu > 4000$, where U_c is the velocity in the core, D_h is the hydraulic diameter of the core, and ν is the kinematic viscosity of the air. Estimate the maximum entrance velocity to the duct so that the flow in the core does not become fully turbulent.

11.87. Standard air flows at 15 m/s over a flat surface that is 5 m wide. Measurements within the boundary layer indicate that the velocity profile in the boundary layer can be represented by the one-seventh power law profile. Between two sections of interest, the boundary layer thickness increases from 90 mm to 125 mm. Estimate the force on the surface between the two sections of interest.

11.88. A typical freeway billboard is 2.28 m high and 13.23 m long. Consider the case where standard air at 20°C blows parallel to the billboard at a speed of 36 km/h. (a) Show that the boundary layer on the downwind edge of the billboard is turbulent. (b) Assuming that the flow in the boundary layer is smooth turbulent over the entire billboard, estimate the boundary-layer thickness on the downwind edge of the billboard, assuming a one-seventh power law velocity distribution. (c) Estimate the boundary layer thickness on the downwind edge using the empirical formulation based partially on the one-seventh power law velocity distribution and partially on data from pipe flow. (d) Assess the statement that the formulations used in parts (b) and (c) usually give results in close agreement.

11.89. Air at 20°C flows with a speed of 5 m/s over a smooth flat rectangular rooftop that is approximately parallel to the wind direction. (a) What type of boundary layer likely exists at the downwind edge of the roof? (b) What would be the thickness of the boundary layer at the downstream edge of the roof if the boundary layer was entirely laminar? (c) What would be the thickness of the boundary layer at the downstream edge of the roof if the boundary layer was entirely smooth turbulent?

Section 11.5: Mixing-Length Theory of Turbulent Boundary Layers

11.90. Wind measurements 3 m above a sandy playground in an open area show a steady wind of 30 km/h. It is known that appreciable movement of the sand occurs when the shear stress on the surface of the sand exceeds 0.08 Pa. The region above the ground can be characterized as a smooth turbulent boundary layer. (a) Estimate whether significant movement of the sand will occur. (b) Estimate the velocity 10 cm above the ground. Assume standard air.

11.91. It is claimed that either Equation 11.59 or Equation 11.60 can be used to estimate the friction drag coefficient, $C_{\rm Df}$, in smooth turbulent boundary-layer flow within the range of $5 \times 10^5 \leq \mathrm{Re_L} < 10^7$. Validate this assumption by calculating the range of percentage error in using Equation 11.60 to estimate $C_{\rm Df}$ rather than Equation 11.59, which is supposed to be the applicable equation.

11.92. The velocity distribution in the outer layer of a smooth turbulent boundary layer derived from mixing-length theory and empirical data is given by Equation 11.98. Show that this relationship corresponds to the following implicit relationship for the skin friction factor, $c_{\rm f}$, as a function of the boundary-layer Reynolds number, Re_δ:

$$\sqrt{\frac{2}{c_{\rm f}}} = 2.46 \ln \left[\mathrm{Re}_\delta \sqrt{\frac{c_{\rm f}}{2}}\right] + 5.05, \quad \text{where} \quad \mathrm{Re}_\delta = \frac{U\delta}{\nu} \qquad (11.121)$$

where U is the free stream velocity, δ is the boundary-layer thickness, and ν is the kinematic viscosity of the fluid.

11.93. The semiempirical implicit relationship for the skin friction factor, $c_{\rm f}$, as a function of the boundary-layer Reynolds number, Re_δ, in smooth-turbulent boundary-layer flow as given by Equation 11.121 was approximated by Prandtl using the following simpler relationship:

$$c_{\rm f} = \frac{0.0205}{\mathrm{Re}_\delta^{\frac{1}{6}}} \qquad (11.122)$$

Show that the simpler relationship given by Equation 11.122 yields estimates of $c_{\rm f}$ within about 10% of those given by Equation 11.121 for $10^4 \leq \mathrm{Re}_\delta \leq 10^7$.

11.94. Water at 20°C flows over a flat surface at a velocity of 6 m/s. Estimate the following quantities at a location 6 m downstream of the leading edge of the surface: (a) the thickness of the boundary layer, (b) the thickness of the laminar sublayer, (c) the range of distances from the surface over which the logarithmic velocity distribution is applicable, and (d) the range of distances from the surface over which the velocity-defect law is applicable.

11.95. Water at 15°C flows over a smooth flat surface at a velocity of 5 m/s. At a location 3 m downstream of the leading edge, (a) determine the range of distances from the surface for which the logarithmic law is applicable and (b) express the applicable logarithmic velocity distribution in the form $u(y) = a \ln y + b$, where a and b are constants.

11.96. Water at 10°C flows over a flat surface at a velocity of 9 m/s. At a location 3 m downstream of the leading edge, (a) determine the range of distances from the surface for which the velocity defect law is applicable and (b) express the applicable velocity defect law in the form $u(y) = af(y/b) + c$, where a, b, and c are constants.

11.97. Water at 5°C flows over a smooth flat surface at a velocity of 6.5 m/s. At a location 4 m downstream of the leading edge, (a) determine the range of distances from the surface for which the one-seventh power law distribution could be applied and (b) express the one-seventh power law distribution in the form $u(y) = a(y/b)^{\frac{1}{7}}$, where a and b are constants.

Section 11.6: Boundary Layers in Closed Conduits

11.98. The velocity defect law is commonly used to represent the velocity distribution in a pipe. (a) What type of flow must exist in the pipe to use the velocity-defect law: smooth flow, rough flow, or either type of flow? (b) Show that if the velocity distribution is described by the velocity-defect law using a coefficient of 2.44, then the relationship between the average velocity, V, and the maximum velocity, u_{max}, can be expressed in terms of the friction factor as

$$\frac{V}{u_{\text{max}}} = \frac{1}{1 + 1.294\sqrt{f}} \tag{11.123}$$

(c) Show that if the velocity distribution is described by the velocity-defect law using a coefficient of 2.5, then the above relationship becomes

$$\frac{V}{u_{\text{max}}} = \frac{1}{1 + 1.326\sqrt{f}} \tag{11.124}$$

(d) Are the relationships derived in parts (b) and (c) equally valid in engineering applications?

11.99. Show that the velocity-defect law for flow in a pipe, with a 2.5 coefficient, can be expressed in the equivalent form

$$u(r) = (1 + 1.326\sqrt{f})V - 2.04\sqrt{f}V \log_{10}\left(\frac{R}{R - r}\right)$$

where $u(r)$ is the velocity at a radial distance r from the centerline of the pipe, V is the average velocity, f is the friction factor, and R is the radius of the pipe. You can use the pre-derived relationship given by Equation 11.124 in your analysis.

11.100. The velocity profile in a pipe is typically taken as being representative of the velocity distribution in a boundary layer, because it is the result of boundary layers coalescing from the wall of the pipe. The pipe centerline velocity, u_{max}, is conventionally taken as $0.99U$, where U is the theoretical free-stream velocity above the boundary layer. Use Equation 11.124 and a typical friction factor of $f = 0.028$ to show that the ratio of the free-stream velocity to the average velocity, V, can be approximated by

$$\frac{U}{V} = 1.234 \tag{11.125}$$

11.101. A starting point for relating the wall shear stress, τ_{w}, to the free-stream velocity, U, that is commonly used in the analysis of smooth turbulent boundary layers is

$$\tau_{\text{w}} = 0.0230\rho U^2 \left(\frac{\nu}{U\delta}\right)^{\frac{1}{4}} \tag{11.126}$$

where ν is the kinematic viscosity of the fluid and δ is the thickness of the boundary layer. Show that Equation 11.126 can be derived from pipe flow relationships by equating the radius of the pipe to δ and combining the following two pipe flow equations: (1) the Blasius friction equation (Equation 11.112), which relates the friction factor to the pipe flow Reynolds number, and (2) Equation 11.125, which relates the average velocity in the pipe to the effective free-stream velocity for a typical friction factor of 0.028.

Chapter 12

Compressible Flow

LEARNING OBJECTIVES

After reading this chapter and solving a representative sample of end-of-chapter problems, you will be able to:

- Apply the fundamental laws of thermodynamics in the analysis of compressible flows.
- Understand the distinguishing characteristics of subsonic and supersonic flows.
- Analyze isentropic and non-isentropic compressible flows in conduits and the performance of nozzles and diffusers.
- Calculate the changes in fluid properties across normal, oblique, and bow shocks.
- Apply the principles of compressible flow to practical engineering problems.

12.1 Introduction

Compressible flows occur when significant changes in density are caused by changes in pressure. Liquid flows can usually be assumed to be incompressible. In contrast, compressible flow of gases is more commonly found in engineering applications. As a rule of thumb, compressibility must be taken into account when density changes caused by pressure variations are significant, typically quantified as pressure-induced density variations of more than 5%. This is generally taken to occur when the flow velocity is greater than 30% of the speed of sound within the fluid or, equivalently, when the Mach number is greater than 0.3. The speed of sound in dry air at 20°C is 343.2 m/s, and hence, compressibility effects would be important at speeds greater than 103 m/s. Compressibility of gases is generally important in the measurement of high-speed flow velocities; analysis of flows in compressors, turbines, and jet engines; analysis of the aerodynamic performance of high-speed aircraft; design of rockets; and analysis of high-speed flow through industrial and municipal gas lines. The study of compressible flow of gases is sometimes referred to as the field of *gas dynamics*.

It is important to remember the distinction between *compressible flow* and a *compressible fluid*, because not all compressible fluids undergo compressible flow. A compressible fluid can loosely be described as one in which the density changes significantly in response to moderate changes in pressure. For example, all gases, including air, are compressible fluids. However, when the pressure variations within a flow field containing the gas are small, compressible effects are usually negligible and the flow is classified as incompressible flow. This same fluid in a different flow field with large pressure variations could experience significant density changes due to the large pressure variations, in which case the flow would be classified as compressible flow.

12.2 Principles of Thermodynamics

Thermodynamics is the study of the relationships between heat and other forms of energy. *Heat* is defined as the energy transferred between two systems at different temperatures. The mechanisms of *heat transfer* are conduction, convection, and radiation. Conduction is associated with molecular-scale motions, convection with macroscale motions, and radiation with electromagnetic waves. The term "heat" should only be used in the context of the transfer of energy and not in the context of describing the energy content of a system or body. The energy content of a system is usually described in terms of its *thermal energy* and its *internal energy*. *Thermal energy* is defined as the energy content of a system associated with the motion of the molecules contained within the system, and the *internal energy* is the sum of all forms of molecular-scale energy, including the energy associated with the formation of chemical bonds. The amount of thermal energy contained within a system is directly related to the temperature of the system. In the context of compressible flow, the thermodynamic principles of interest relate mostly to heat transfer caused by the compression and expansion of gases. Under such circumstances, temperature variations within the flow field are generated, and the resulting heat transfer within the flow field must be taken into account to properly account for the relationship between fluid properties and temperature. Usually chemical transformations do not occur, so changes in internal energy are taken as being equal to changes in thermal energy.

Equation of state of an ideal gas. Most gases of interest in engineering applications under conditions of low (absolute) pressure and moderate to high temperature can be approximated as an ideal gas and described by the ideal gas equation given by

$$\rho = \frac{p}{RT} \tag{12.1}$$

where ρ is the density of the gas [ML^{-3}], p is the absolute pressure [FL^{-2}], R is the gas constant [E[64]M$^{-1}\Theta^{-1}$], and T is the absolute temperature [Θ]. An ideal gas obeys Equation 12.1, and such a gas is sometimes called a *perfect gas*. Note that an "ideal gas" does not have the same connotation as an "ideal fluid," where the latter term refers to a fluid that is incompressible and inviscid. In the ideal gas equation, Equation 12.1, the gas constant, R, which is unique to each gas, is related to the universal gas constant, R_u, by the relation

$$R = \frac{R_\mathrm{u}}{M_\mathrm{mol}} \tag{12.2}$$

where M_mol is the molar mass of the gas [M·mol^{-1}]. The universal gas constant is given by R_u = 8.314 J/(mol·K), and for air, M_mol = 28.97 g/mol. Equation 12.1 is called the *equation of state* of an ideal gas because it provides a relationship between the state variables of temperature (T), pressure (p), and density ρ. As a consequence, the state of an ideal gas can be described by specifying the values of any two of these three variables. In specifying the density as a state variable, it is sometimes convenient to use the inverse of the density, which is called the *specific volume*, $v = 1/\rho$ [L^{3}M^{-1}], which gives the volume per unit mass of the gas. As evidence of the accuracy of using Equation 12.1 to describe the behavior of

[64]"E" (= FL) represents a dimension of energy.

air, experimental measurements have shown that the error in applying Equation 12.1 to air at a temperature of 20°C and a pressure of 3.03 MPa is less than 1% and that the error in applying Equation 12.1 to air at a temperature of −130°C and a pressure of 101.3 kPa is also less than 1%. In cases where a gas does not exhibit ideal behavior, the behavior of an ideal gas under the same circumstances can be used as an indicator of the general trends in the behavior of the nonideal gas.

Internal energy. The *internal energy*, u, of a fluid is defined as the energy per unit mass [EM^{-1}], where the referenced "energy" is associated with the combination of the kinetic energy of molecular motions and the forces between molecules. A typical SI unit of internal energy is J/kg. The internal energy, u, can in principle be taken as a function of the state variables of temperature, T, and specific volume, v. Hence, any change in internal energy, $\mathrm{d}u$, is given by

$$\mathrm{d}u = \left.\frac{\partial u}{\partial T}\right|_v \mathrm{d}T + \left.\frac{\partial u}{\partial v}\right|_T \mathrm{d}v \tag{12.3}$$

where the subscripts indicate the variable that is being held constant in taking each of the partial derivatives. The *specific heat at constant volume*, c_v, is defined by the relation

$$c_v = \left.\frac{\partial u}{\partial T}\right|_v \tag{12.4}$$

where c_v can be a function of temperature. For an ideal gas, it can be shown that the internal energy, u, is only a function of temperature, in which case $\partial u/\partial v|_T = 0$ and Equations 12.3 and 12.4 can be combined to give

$$\mathrm{d}u = c_v\,\mathrm{d}T \tag{12.5}$$

In cases where c_v is (approximately) constant over the temperature range of interest, Equation 12.5 can be expressed as

$$u_2 - u_1 = c_v(T_2 - T_1)$$

where the subscripts 1 and 2 indicate two states of a gas system; typically, 1 is taken as the initial state and 2 as the final state.

Enthalpy. The *enthalpy* per unit mass of a fluid, h [EM^{-1}], is the sum of the internal energy, u, and the *pressure energy* or *flow energy*, p/ρ, and can be expressed as

$$h = u + \frac{p}{\rho} \tag{12.6}$$

The enthalpy per unit mass, h, is also commonly called the *specific enthalpy*, and a typical SI unit of (specific) enthalpy is J/kg. The primary motivation for defining the enthalpy, h, by Equation 12.6 is that the combined variable $u + p/\rho$ occurs in the energy equation and (for ideal gases) is only a function of temperature, thereby prompting this combined variable to be defined as a single thermodynamic property. In cases where the kinetic and potential energies of a fluid are negligible, the enthalpy represents the total energy of the fluid per unit mass. In cases where the kinetic energy of the fluid is not negligible, as often occurs in high-speed flows, it is sometimes convenient to combine the enthalpy and kinetic energy of the fluid into a single quantity called the *stagnation enthalpy*, h_0, defined by

$$h_0 = h + \frac{V^2}{2} \tag{12.7}$$

where V is the speed of the fluid. The enthalpy, h, does not include the kinetic energy of the macroscale fluid motion and is therefore sometimes called the *static enthalpy*, to differentiate it from the stagnation entropy, h_0, which does include the (macro)velocity, V, of the fluid. Considering the definition of (static) enthalpy given by Equation 12.6, taking T and p as the state variables gives

$$\mathrm{d}h = \left.\frac{\partial h}{\partial T}\right|_p \mathrm{d}T + \left.\frac{\partial h}{\partial p}\right|_T \mathrm{d}p \tag{12.8}$$

The *specific heat at constant pressure*, c_p, is defined by the relation

$$c_p = \left.\frac{\partial h}{\partial T}\right|_p \tag{12.9}$$

where c_p can be a function of temperature. For an ideal gas, $p = \rho RT$; hence, the enthalpy can be expressed as

$$h = u + \frac{p}{\rho} = u + \frac{\rho RT}{\rho} = u + RT \tag{12.10}$$

which shows that for an ideal gas, h is only a function of the temperature, T, in which case $\partial h/\partial p|_T = 0$ and the combination of Equations 12.8 and 12.9 gives

$$\mathrm{d}h = c_p\,\mathrm{d}T \tag{12.11}$$

In cases where c_p is (approximately) constant within the temperature range of interest, Equation 12.11 can be expressed as

$$h_2 - h_1 = c_p(T_2 - T_1)$$

Specific heats. The specific heats, c_v and c_p, are properties of a gas that relate temperature changes to changes in internal energy and enthalpy, respectively. For an ideal gas,

$$h = u + RT \quad \rightarrow \quad \mathrm{d}h = \mathrm{d}u + R\,\mathrm{d}T \quad \rightarrow \quad c_p\,\mathrm{d}T = c_v\,\mathrm{d}T + R\,\mathrm{d}T$$

which yields

$$c_p - c_v = R \tag{12.12}$$

This interesting result indicates that $c_v - c_p$ is a constant that is independent of temperature, even though both c_v and c_p are functions of temperature. In many analyses, the relevant variable group is the *specific heat ratio*, k, which is defined as

$$k = \frac{c_p}{c_v} \tag{12.13}$$

Equations 12.13 and 12.12 can be combined to yield the following useful expressions for c_p and c_v:

$$c_p = \frac{kR}{k-1}, \quad c_v = \frac{R}{k-1} \tag{12.14}$$

Although the specific heats c_v and c_p are in general functions of temperature, for moderate variations in temperature, they are commonly assumed to be constant. Values of c_v and c_p for various gases at 20°C are given in Appendix B.5, where typical values for air are c_v = 716 J/kg·K and c_p = 1003 J/kg·K. For gases commonly encountered in engineering practice, k is in the range of 1.0–1.7, and for air, $k \approx 1.40$. In support of the assumption of a constant c_p in air, it is noted that c_p of air increases by around 30% as the temperature goes from -18 to 2760°C. It is important to remember that the behavior of vapors that are not superheated deviates from the behavior of an ideal gas. For non-superheated vapors, vapor tables must usually be used, because c_p and c_v cannot be assumed to be constants.

Entropy. The *entropy*, S, of a system containing a specified mass of fluid is defined in terms of its entropy change, dS, according to the relation

$$\mathrm{d}S = \int_{\mathrm{rev}} \frac{\delta Q}{T} \quad \text{or} \quad \mathrm{d}S = \left.\frac{\delta Q}{T}\right|_{\mathrm{rev}} \tag{12.15}$$

where δQ is the heat added to the system, T is the temperature of the system, and the subscript "rev" indicates that the change in the system is reversible. A *reversible process* is one in which both the system and its surroundings can be returned to their initial states. Processes involving friction, heat transfer, and mixing of gases are not reversible processes. In principle, real fluids are not frictionless because they do not have zero viscosity. However, in many cases, viscous effects in gases are usually sufficiently small that their effect on irreversibility can be assumed to be negligible. By definition (i.e., by Equation 12.15), in the absence of irreversibilities, the entropy of a system changes by heat transfer only, with an increase in entropy associated with a heat gain and a loss of entropy associated with a heat loss. For general changes in the state of a fluid (not necessarily reversible), the *second law of thermodynamics* states that

$$\mathrm{d}S \geq \frac{\delta Q}{T} \quad \text{or} \quad T\,\mathrm{d}S \geq \delta Q \tag{12.16}$$

The left-hand side of Equation 12.16 is equal to the change in entropy (dS) that would occur if a heat of δQ was added and the process was reversible. Therefore, the second law of thermodynamics (Equation 12.16) states that at a given temperature, the change in system entropy for a reversible process is greater than for an irreversible process. The additional heat that must be added in an irreversible process to yield the same entropy change as a reversible process is sometimes called *wasted heat*. The equality in Equation 12.16 applies when the process within the system is reversible, and the inequality applies when the process is irreversible. For an *adiabatic process*, there is no exchange of heat between the system and its surroundings, in which case $\delta Q = 0$. For a *reversible adiabatic process*, $\mathrm{d}S = 0$, and for an *irreversible adiabatic process*, $\mathrm{d}S > 0$. A process with constant entropy (i.e., $\mathrm{d}S = 0$) is called an *isentropic process*, and it is apparent that reversible adiabatic processes are isentropic; in fact, it is fairly common to use the terms "isentropic process" and "reversible adiabatic process" interchangeably. In many practical applications, it is convenient to work with the entropy per unit mass within a fluid system, s, where $s = S/M$ and M is the total mass of fluid within the system.

Relationships involving entropy. Combining the definition of entropy and the first law of thermodynamics gives the following relation for any reversible process:

$$T\,\mathrm{d}s = \mathrm{d}u + p\,\mathrm{d}v \tag{12.17}$$

where T is the temperature, s is the entropy per unit mass, u is the internal energy per unit mass, p is the pressure, and v is the specific volume. Equation 12.17 is commonly referred to as the "first $T\,\mathrm{d}s$ equation." From the definition of enthalpy given by Equation 12.6,

$$\mathrm{d}h = \mathrm{d}u + p\,\mathrm{d}v + v\,\mathrm{d}p \tag{12.18}$$

and combining Equations 12.17 and 12.18 gives

$$T\,\mathrm{d}s = \mathrm{d}h - v\,\mathrm{d}p \tag{12.19}$$

This equation is commonly referred to as the "second $T\,\mathrm{d}s$ equation." Using Equations 12.17 and 12.19 for any ideal gas, the change in entropy can be related to changes in T and v or related to changes in T and p by the equations

$$\mathrm{d}s = \frac{\mathrm{d}u}{T} + \frac{p}{T}\,\mathrm{d}v = c_v\frac{\mathrm{d}T}{T} + R\frac{\mathrm{d}v}{v} \tag{12.20}$$

$$\mathrm{d}s = \frac{\mathrm{d}h}{T} - \frac{v}{T}\,\mathrm{d}p = c_p\frac{\mathrm{d}T}{T} - R\frac{\mathrm{d}p}{p} \tag{12.21}$$

In cases where the specific heats can be approximated as constants, Equations 12.20 and 12.21 can be integrated to yield

$$s_2 - s_1 = c_v \ln\frac{T_2}{T_1} + R\ln\frac{v_2}{v_1} \tag{12.22}$$

$$s_2 - s_1 = c_p \ln\frac{T_2}{T_1} - R\ln\frac{p_2}{p_1} \tag{12.23}$$

$$s_2 - s_1 = c_v \ln\frac{p_2}{p_1} + c_p\ln\frac{v_2}{v_1} \tag{12.24}$$

where Equation 12.24 is derived from the combination of Equations 12.22, 12.23, and 12.12. In cases where the change process is isentropic, $s_2 = s_1$, and Equations 12.22–12.24 yield

$$T\,v^{k-1} = \text{constant} \quad \equiv \quad \frac{T}{\rho^{k-1}} = \text{constant} \tag{12.25}$$

$$T\,p^{\frac{1-k}{k}} = \text{constant} \tag{12.26}$$

$$p\,v^{k} = \text{constant} \quad \equiv \quad \frac{p}{\rho^{k}} = \text{constant} \tag{12.27}$$

Equation 12.27 can be contrasted with Boyle's law, which states that pv = constant under isothermal conditions, whereas it is shown here that $p\,v^k$ = constant under non-isothermal isentropic conditions.

EXAMPLE 12.1

Air is being used as the working fluid in a small-diameter tube where the temperature and pressure at an upstream section are 50°C and 600 kPa, respectively, and the temperature and pressure at a downstream section are –15°C and 0 kPa, respectively. Both measured pressures are gauge pressures, local atmospheric pressure is 101 kPa, and the air can be assumed to be an ideal gas. (a) Estimate the density of the air at each section and assess whether the flow should be treated as a compressible flow. (b) Estimate the change in internal energy, the change in enthalpy, and the change in entropy of the fluid between the two sections.

SOLUTION

From the given data: $T_1 = 50°\text{C} = 323$ K, $p_1 = 600$ kPa (gauge) $= 701$ kPa (abs), $T_2 = -15°\text{C} = 288$ K, and $p_2 = 0$ kPa (gauge) $= 101$ kPa (abs). The specific heats of air (from Appendix B.5) at 20°C are $c_p = 1003$ J/kg·K, and $c_v = 716$ J/kg·K. Because the universal gas constant is 8.314 J/mol·K and the molar mass of air is 28.97 g/mol, the gas constant for air is $R = 8.314/28.97 = 0.287$ J/g·K $= 287$ J/kg·K.

(a) Using the ideal gas law, the densities of air at the upstream and downstream locations, ρ_1 and ρ_1, respectively, are

$$\rho_1 = \frac{p_1}{RT_1} = \frac{701 \times 10^3}{(287)(323)} = \mathbf{7.562\ kg/m^3}$$

$$\rho_2 = \frac{p_2}{RT_2} = \frac{101 \times 10^3}{(287)(288)} = \mathbf{1.222\ kg/m^3}$$

$$\Delta = \frac{\rho_2 - \rho_1}{\rho_1} \times 100 = \frac{1.222 - 7.562}{1.222} \times 100 = 519\%$$

Because the density changes by over 500% under the given conditions, the compressibility of the fluid must be taken into account and this **must be treated as a compressible flow**.

(b) The changes in internal energy, Δu_{12}, enthalpy, Δh_{12}, and entropy, Δs_{12}, are given by

$$\Delta u_{12} = u_2 - u_1 = c_v(T_2 - T_1) = 716(288 - 323) = -25\,060 \text{ J/kg} = \mathbf{-25.1\ kJ/kg}$$

$$\Delta h_{12} = h_2 - h_1 = c_p(T_2 - T_1) = 1003(288 - 323) = -35\,105 \text{ J/kg} = \mathbf{-35.1\ kJ/kg}$$

$$\Delta s_{12} = c_p \ln\left(\frac{T_2}{T_1}\right) - R\ln\left(\frac{p_2}{p_1}\right) = (1003)\ln\left(\frac{288}{323}\right) - 287\ln\left(\frac{101}{701}\right) = \mathbf{441\ J/kg{\cdot}K}$$

Note that the change in entropy was calculated using Equation 12.23, although using Equation 12.22 or 12.24 (along with the ideal gas law) would yield the same result.

12.3 The Speed of Sound

The speed of sound is defined as the rate of propagation of a pressure pulse of infinitesimal strength through a stationary environment. The speed of sound is sometimes called the *sonic velocity* or the *acoustic velocity*. Because sound waves are defined as pressure waves of infinitesimal strength, the changes in fluid properties across sound waves are very small compared with their local values. The speed at which a fluid is moving relative to the speed of sound within the fluid has a significant influence on the importance of compressibility in determining the flow characteristics. The ratio of the local flow speed, V, to the speed of sound in the fluid, c, is called the Mach number, Ma, defined as

$$\text{Ma} = \frac{V}{c} \tag{12.28}$$

Flows in which Ma < 1 are called *subsonic flows*, flows in which Ma > 1 are called *supersonic flows*, and flows where Ma ≈ 1 are called *transonic flows*. Based on these definitions, it is apparent that disturbances within subsonic flows can propagate upstream and hence affect conditions upstream of the disturbance. In contrast, disturbances in supersonic flows cannot propagate upstream, so the upstream flow field is unaffected.

Flow characteristics at various Mach numbers. A rough guide to the relationship between the Mach number and flow characteristics is given in Table 12.1 (White, 2011), which provides a finer definition of flow classifications than is sometimes used in practice. Transonic flows are characterized by $0.8 <$ Ma < 1.2, supersonic flows by $1.2 \leq$ Ma ≤ 3.0, and *hypersonic flows* by Ma > 3.0. Slight variations in these classifications are sometimes used; for example, transonic flows are sometimes defined as those with $0.9 \leq$ Ma ≤ 1.2 and hypersonic flows as those with Ma > 5.0. Consistent with the characteristics listed in Table 12.1, it is generally accepted that compressibility effects are negligible when Ma < 0.3; however, for Ma ≥ 0.3, compressibility effects must generally be taken into account. In engineering applications, transonic flows are commonly encountered in the gas turbine engines used in aircraft, and hypersonic flows are commonly encountered in the design of missiles and by spacecraft re-entering Earth's atmosphere.

Actual sound waves. What humans perceive as sound is actually the response of eardrums to weak pressure pulses in the surrounding air. Humans can perceive pressure pulses as small as 1 mPa, which is called the *threshold of hearing*, and pulses larger than 101.325 Pa cause pain. Actual sound waves have finite pressure fluctuations, not infinitesimal pressure fluctuations as used in defining the speed of sound. Consequently, actual sound waves travel slightly faster than the standard speed of sound.

Table 12.1: Flow Characteristics at Various Mach Numbers

Ma	Flow Characteristics
< 0.3	Incompressible flow: density effects are negligible.
0.3–0.8	Subsonic flow: density effects are important, but no shocks appear.
0.8–1.2	Transonic flow: shock waves first appear, dividing subsonic and supersonic regions of flow. Powered flow in the transonic region is difficult because of the mixed character of the flow field.
1.2–3.0	Supersonic flow: shocks are present, but there are no subsonic regions.
> 3.0	Hypersonic flow: shocks and other flow changes are especially strong.

Propagation of a pressure (sound) wave. Consider the propagation of a pressure wave that has a velocity c as shown in Figure 12.1(a); in this case, the pressure wave is moving from right to left toward a stagnant environment where $V_x = 0$. Ahead of the wave, the pressure is p, the density is ρ, and the temperature is T, whereas behind the wave, the velocity, pressure, density, and temperature are $\mathrm{d}V$, $p + \mathrm{d}p$, $\rho + \mathrm{d}\rho$, and $T + \mathrm{d}T$, respectively. The unsteady-state condition can be transformed into a steady-state condition by defining a control volume that moves at the same speed as the pressure wave, as shown in Figure 12.1(b). Because the control volume moves with a constant velocity, the coordinates measured relative to the moving control volume are an inertial coordinate system. Hence, all kinematic variables defined relative to the moving control volume can be used in the (inertial) continuity and momentum equations. Applying the steady-state continuity equation to the control volume shown in Figure 12.1(b) gives

$$\int_{\text{cs}} \rho(\mathbf{v}\cdot\mathbf{n})\,\mathrm{d}A = 0 \quad\rightarrow\quad \rho(-c)A+(\rho+\mathrm{d}\rho)(c-\mathrm{d}V)A = 0 \quad\rightarrow\quad \mathrm{d}V = c\frac{\mathrm{d}\rho}{\rho+\mathrm{d}\rho} \tag{12.29}$$

This relationship illustrates that for a sound wave of infinitesimal strength (i.e., $\mathrm{d}\rho$ is small), $\mathrm{d}V \ll c$. Applying the steady-state momentum equation to the control volume shown in Figure 12.1(b) gives

$$\begin{aligned} \sum F_x &= \int_{\text{cs}} \rho V_x(\mathbf{v}\cdot\mathbf{n})\,\mathrm{d}A \\ &\rightarrow\quad pA - (p+\mathrm{d}p)A = \rho c(-c)A + (\rho+\mathrm{d}\rho)(c-\mathrm{d}V)(c-\mathrm{d}V)A \end{aligned} \tag{12.30}$$

The $\mathrm{d}V$ term in Equation 12.30 can be eliminated by substituting the continuity equation (Equation 12.29); the resulting combined equation is

$$c^2 = \frac{\mathrm{d}p}{\mathrm{d}\rho}\left(1 + \frac{\mathrm{d}\rho}{\rho}\right) \tag{12.31}$$

Equation 12.31 indicates that the speed of a pressure wave, c (= speed of sound), depends on the relationship between the density and the pressure within the fluid, via $\mathrm{d}p/\mathrm{d}\rho$, and on the strength of the pressure wave, via $\mathrm{d}\rho/\rho$. It is further apparent from Equation 12.31 that the larger the strength of the pressure wave, the faster the wave speed. Thus, waves generated by large explosions move much faster than sound waves, which, by definition, have small values of $\mathrm{d}\rho/\rho$. For sound waves, which are defined as an infinitesimal pressure pulse, $\mathrm{d}\rho/\rho \ll 1$, and Equation 12.31 becomes

$$c^2 = \frac{\mathrm{d}p}{\mathrm{d}\rho} \tag{12.32}$$

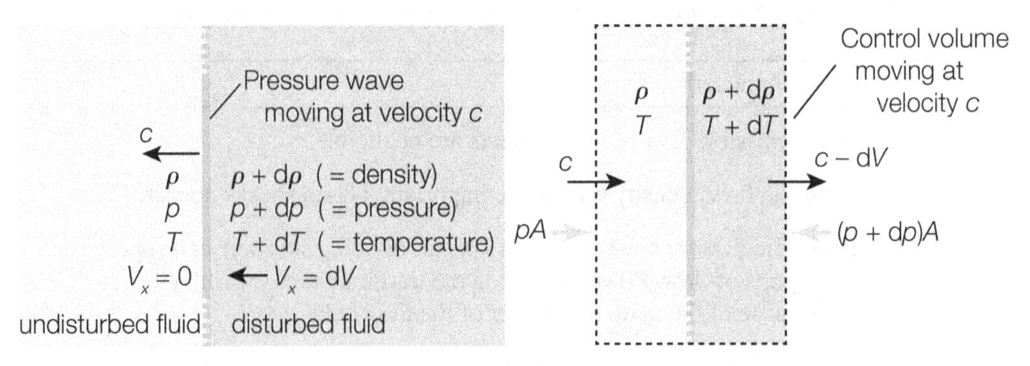

Figure 12.1: Propagation of a pressure wave

This equation gives the minimum speed at which a pressure wave can propagate through a fluid. Because density and pressure are two independent thermodynamic quantities, additional specification is necessary to fix the relationship given by Equation 12.32. A sound wave can be assumed to consist of an infinitesimal increment in pressure that occurs so quickly that insufficient heat transfer occurs. As a consequence, it can be assumed that the process of sound wave propagation is reversible and adiabatic; hence, it is an isentropic process. Adding the condition of isentropy to Equation 12.32 yields the following equation for the speed of sound in a fluid:

$$c = \sqrt{\left.\frac{\mathrm{d}p}{\mathrm{d}\rho}\right|_s} \tag{12.33}$$

where the subscript "s" indicates that the entropy is constant. Equation 12.33 describes the speed of sound in solids, liquids and gases. The compressibility of solids and liquids is usually characterized by the bulk modulus of elasticity, E_v, defined by the relation

$$E_\mathrm{v} = \left.\frac{\mathrm{d}p}{\mathrm{d}\rho/\rho}\right|_T = \rho\left.\frac{\mathrm{d}p}{\mathrm{d}\rho}\right|_T \tag{12.34}$$

where the subscript "T" indicates that temperature is a constant. In liquids and solids, changes in pressure usually produce small changes in temperature such that

$$\rho\left.\frac{\mathrm{d}p}{\mathrm{d}\rho}\right|_T \approx \rho\left.\frac{\mathrm{d}p}{\mathrm{d}\rho}\right|_s \tag{12.35}$$

Combining Equations 12.33–12.35 gives the following expression for the speed of sound in liquids and solids:

$$c = \sqrt{\frac{E_\mathrm{v}}{\rho}} \tag{12.36}$$

For gases that behave as ideal gases, it was shown (see Equation 12.27) that for an isentropic process, p and ρ are related by

$$\frac{p}{\rho^k} = \text{constant} \quad \rightarrow \quad \left.\frac{\mathrm{d}p}{\mathrm{d}\rho}\right|_s = k\,\frac{p}{\rho} \tag{12.37}$$

Substituting Equation 12.37 into Equation 12.33 and eliminating p/ρ by utilizing the ideal gas relation $p/\rho = RT$ yields the following expression for the speed of sound in an ideal gas:

$$c = \sqrt{kRT} \quad \text{(for an ideal gas)} \tag{12.38}$$

The speed of sound waves at moderate frequencies in air that is predicted by Equation 12.38 is in close agreement with measurements, thereby supporting the assumptions on which this equation is based. Equation 12.38 is generally used to define the benchmark speed of sound and indicates that the speed of sound (in an ideal gas) is only a function of temperature and does not depend on the ambient pressure.

High-frequency sound waves. At high frequencies, sound waves generate friction and the sound propagation process ceases to be isentropic. In this case, the process is better described as an isothermal process and leads to the following expression for the speed of sound in an ideal gas:

$$c = \sqrt{RT} \tag{12.39}$$

Speed of sound in air and water. The speed of sound in air at any altitude above sea level can be estimated using Equation 12.38, knowing only the temperature at the given altitude. In the standard atmosphere, the speed of sound at sea level at a temperature of 15.5°C is 340 m/s, whereas at an altitude of 6.6 km at a temperature of −50.8°C, the speed

Table 12.2: Speed of Sound in Various Gases

Gas	Speed of Sound (m/s)	Gas	Speed of Sound (m/s)
H_2	1294	CO_2	266
He	1000	CH_4	185
Air	340	CO_2	91
$^{238}UF_6$	317		

of sound is 299 m/s, a reduction of approximately 12%. In contrast, the speed of sound in water at 101.3 kPa and 15.5°C (as given by Equation 12.36) is approximately 1450 m/s.

Speed of sound in various gases. The speed of sound can be markedly different between gases, because both k and R depend on the particular gas. Several gases ranked by their speed of sound are shown in Table 12.2. It is apparent that there is a wide range of sound speeds among gases.

EXAMPLE 12.2

Compare the speed of sound in air to the speed of sound in pure nitrogen at 20°C. Considering that stock cars racing at the Daytona International Speedway can attain speeds of 320 km/h at an air temperature of 20°C, what is the Mach number attained and should the compressibility of air be considered in analyzing the performance of these stock cars?

SOLUTION

From the given data: T = 20°C = 293.15 K. The properties of air and nitrogen (N_2) at 20°C are given in Appendix B.5 as follows:

Gas	M (g/mol)	R (J/kg·K)	c_p (J/kg·K)	c_v (J/kg·K)	k (–)
Air	28.96	287.1	1003	716	1.401
N_2	28.02	296.7	1040	743	1.400

where the gas constant, R, is derived from the universal gas constant, R_u (= 8.314 J/mol·K), by the relation $R = R_u/M$. Using these data, the speed of sound in each medium is given by

$$\text{Air: } c = \sqrt{RTk} = \sqrt{(287.1)(293.15)(1.401)} = \mathbf{343.4\ m/s}$$

$$N_2\text{: } c = \sqrt{RTk} = \sqrt{(296.7)(293.15)(1.400)} = \mathbf{349.0\ m/s}$$

Therefore, the speed of sound in pure N_2 is approximately 1.6% higher than the speed of sound in air. The closeness of these results is not surprising because air is approximately 78% N_2.

For a stock car traveling at V = 320 km/h = 88.9 m/s in air at 20°C, the Mach number, Ma, is given by

$$\text{Ma} = \frac{V}{c} = \frac{88.9}{343.4} = \mathbf{0.23}$$

Because Ma < 0.3, **the compressibility of air will have a negligible effect on the performance of the stock car.**

The Mach cone. The *Mach cone* is a conical shape that contains the sound/pressure waves generated by a source (of waves) moving at supersonic speed. Consider the various scenarios shown in Figure 12.2, where V is the speed of the source and c is the speed of sound. The case of $V = 0$ corresponds to a stationary source where the emitted sound waves are simply concentric spheres. The case of $V < c$ is characterized by a source moving to the right at a subsonic speed, with the position of the source at equally spaced time intervals shown as open circles and the current position of the source shown as a filled circle; the current locations of the wave fronts emitted at previous times are also shown. The key feature here is that the fronts of the sound waves emitted at previous times are ahead of the current position of the source. When $V = c$, the source is moving at the sonic speed and the fronts of the sound waves emitted at previous times coincide with the current position of the source—a scenario that indicates a significant pressure accumulation and consequent density variation that would likely lead to a very bumpy ride for the source. Furthermore, in cases where $V < c$, a listener ahead of a moving source will hear sounds at a higher frequency rate than a listener behind the source, a phenomenon called the *Doppler effect*. The frequency, ω, of a sound wave heard by a listener ahead of a source emitting sound waves at a frequency ω_0 is given by

$$\omega = \omega_0 \left(1 - \frac{V}{c}\right)^{-1} = \omega_0 \left(1 - \text{Ma}\right)^{-1} \tag{12.40}$$

where Ma ($= V/c$) is the Mach number. In cases where $V < c$, a listener ahead of the source will hear source emissions before the source arrives at the listener's location, in contrast to cases where $V \geq c$, where a listener will not hear an oncoming source. A region where sound waves do not propagate is called a *zone of silence*, and a region where sound waves propagate is called a *zone of action*. When $V > c$, the source is moving at a supersonic speed and the fronts of sound waves emitted at previous times are always behind the source; hence, the moving source does not create any disturbances ahead of it. The sound waves generated under supersonic conditions are concentric spheres that are bounded by a cone that is called the *Mach cone*. The volume within the cone is a zone of action, the region outside the cone is a zone of silence, and the conical wave front is called a *Mach wave*. At supersonic speeds, Mach waves are formed by the sharp nose of an aircraft or the sharp leading edge of an airfoil if such a sharp edge exists. Based on the kinematics of the motion, the half-apex angle, α of the Mach cone as shown in Figure 12.2 satisfies the relation

$$\sin\alpha = \frac{c}{V} = \frac{1}{\text{Ma}} \quad \rightarrow \quad \alpha = \sin^{-1}\left(\frac{1}{\text{Ma}}\right) \tag{12.41}$$

The angle α is commonly referred to as the *Mach cone angle* or simply the *Mach angle*. It is apparent from Equation 12.41 that the higher the Mach number, the more slender the Mach cone. In representing the propagation of sound waves from a source, it should be

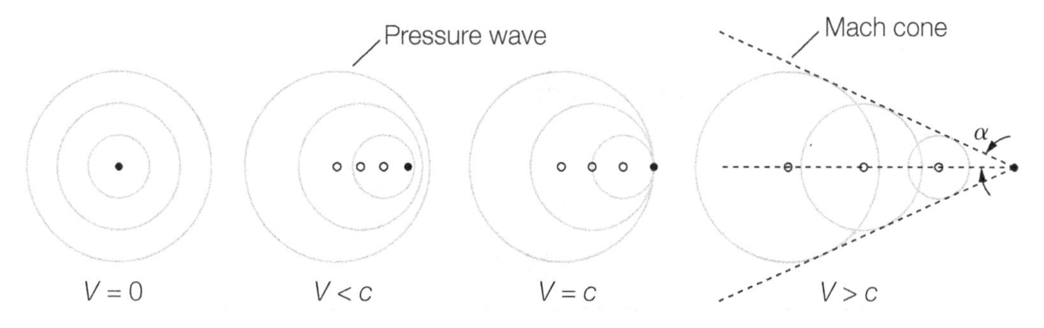

Figure 12.2: Propagation of sound waves from a moving source

noted that conditions shown in Figure 12.2 correspond to a moving source in a stationary fluid. However, similar circumstances occur for a stationary source in a moving fluid. The relationship between the Mach angle, α, and the speed of an object, as measured by the Mach number, is vividly illustrated in Figure 12.3, which shows the trace of the shock generated by a high-speed model aircraft in a wind tunnel. The thinner Mach cone at Ma = 6 compared with the Mach cone at Ma = 3.5 is apparent. Furthermore, a (forensic) estimate of the Mach speeds in these two cases could be made using the photographs only, because Figure 12.3(a) shows a Mach angle of approximately 17° and Figure 12.3(b) shows a Mach angle of approximately 10°, which, when substituted into Equation 12.41, yield approximately Ma = 3.5 and Ma = 6, respectively.

Phenomena related to the Mach cone. A *shock* is defined as an abrupt change in fluid properties, and phenomena that are closely related to the Mach cone are the *oblique shock* and the *bow shock*. An oblique shock is similar to the surface of a Mach cone, with the difference being that the angle the shock makes with the flow direction is wider to account for the finite width of the solid object that is generating the pressure waves. Bow shocks are formed when the front end of the solid object is too wide (i.e., blunt) to accommodate an oblique shock, in which case the front of the shock forms ahead of the solid object. Oblique shocks and bow shocks are discussed in more detail later in this chapter. When sound waves are generated by objects moving at supersonic speeds in the atmosphere, regardless of the type of shock generated, a listener will hear a sonic boom (i.e., a loud "crack") as the shock front moves by, as illustrated in Figure 12.4. It is apparent that the sudden noise (boom) occurs as a person goes from being in the zone of silence to being in the zone of action. The generation of sonic booms is part of the reason supersonic commercial jetliners have found limited acceptance in the air space over inhabited land. Sonic booms are louder the closer the object generating the boom is to the listener, because the shock has less opportunity to dissipate.

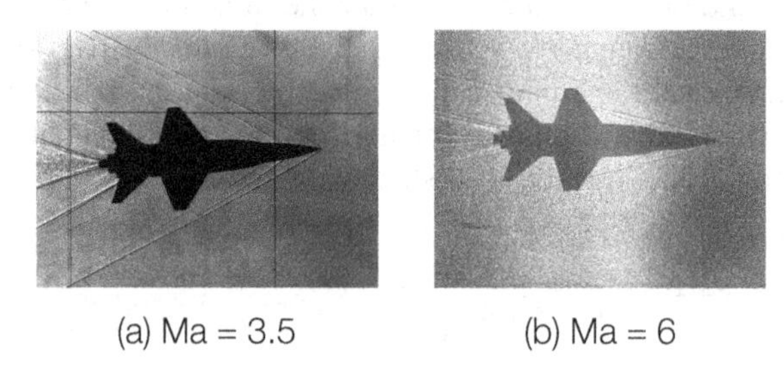

Figure 12.3: Shock wave generated by supersonic-hypersonic aircraft

Source: (a) NASA, (b) NASA.

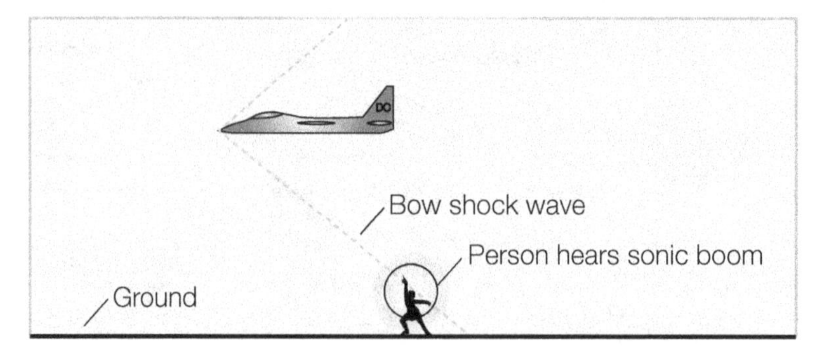

Figure 12.4: Sonic boom

The decibel unit of sound. The decibel is a unit that measures the magnitude of the sound pressure relative to some specified reference value, with commonly used reference pressures being 20 μPa in air and 1 μPa in water. The reference pressure in air is associated with the threshold pressure fluctuation that can be perceived by the human ear. The expression used to calculate the relative sound pressure in decibels (dB), L_p, is given by

$$L_p = 20 \log_{10}\left(\frac{p_{rms}}{p_{ref}}\right) \text{ dB} \tag{12.42}$$

where p_{rms} is the actual root-mean-square pressure fluctuation and p_{ref} is the reference pressure. Because p_{ref} is defined as the threshold of hearing, a sound intensity of 0 dB is at the threshold of hearing. Permanent damage to the human ear is caused by pressure fluctuations in excess of 120 dB. It is also relevant to note that the human ear is not equally sensitive to all frequencies of sound waves; the maximum sensitivity occurs in the frequency range of 2–4 kHz.

EXAMPLE 12.3

The Concorde was the first commercial supersonic aircraft to routinely operate over the Atlantic Ocean. The Concorde cruised at an elevation of around 18 000 m at a Mach number of 2.04. By comparison, regular commercial aircraft cruise at elevations of around 10 700 m and at subsonic speeds. Of concern at airports accommodating the Concorde was that it generated noise levels of 110 dB when landing. (a) When the Concorde is cruising, how far behind the Concorde would a regular aircraft encounter the Mach wave (and associated turbulence) caused by the Concorde? (b) What is the magnitude of the pressure fluctuations generated by the Concorde when landing? How does this magnitude compare with an atmospheric pressure of 101 kPa?

SOLUTION

From the given data: z_1 = 18 000 m, Ma = 2.04, z_2 = 10 700 m, and L_p = 110 dB. From these data, the vertical separation between the Concorde and regular commercial aircraft is Δz = 18 000 m – 10 700 m = 7300 m.

(a) The Mach angle, α, is given by Equation 12.41 as

$$\alpha = \sin^{-1}\left(\frac{1}{\text{Ma}}\right) = \sin^{-1}\left(\frac{1}{2.04}\right) = 29.4^\circ$$

Let x be the distance behind the Concorde that a regular commercial aircraft encounters the Mach wave; hence,

$$\tan\alpha = \frac{\Delta z}{x} \quad \rightarrow \quad \tan 29.4^\circ = \frac{7300}{x} \quad \rightarrow \quad x = 12\,980 \text{ m} \approx 13.0 \text{ km}$$

Therefore, as the Concorde flies by the regular jetliner, the jetliner will feel the Mach wave when the Concorde is **13.0 km** in front of it.

(b) For a noise level of 110 dB based on the standard reference pressure of 20 μPa, the pressure fluctuation, p_{rms}, can be calculated using Equation 12.42, which gives

$$L_p = 20 \log_{10}\left(\frac{p_{rms}}{p_{ref}}\right) \rightarrow 110 = 20 \log_{10}\left(\frac{p_{rms}}{20}\right) \rightarrow p_{rms} = 6.32 \times 10^6 \ \mu\text{Pa} = 6.32 \text{ Pa}$$

Hence, pressure fluctuations generated by the Concorde upon landing are on the order of **6.32 Pa**. If the atmospheric pressure is 101 kPa, then this represents a pressure variation of only **0.006% in atmospheric pressure**. Nevertheless, it is still loud!

12.4 Thermodynamic Reference Conditions

Thermodynamic reference conditions are sometimes used as benchmarks to assess the actual thermodynamic conditions within a compressible flow field. A comparable approach is commonly used in open-channel flow, where the reference conditions of *normal flow* and *critical flow* are used. In the case of compressible flow, the reference conditions are the *isentropic stagnation condition* and the *critical condition.* These conditions are described in detail in the following sections.

12.4.1 Isentropic Stagnation Condition

The *isentropic stagnation condition* is a zero-velocity condition (Ma $= 0$) having the same entropy as the flowing fluid. Consider the motion of a fluid along a streamline and within a streamtube as shown in Figure 12.5. For the fluid contained within the control volume shown in Figure 12.5, referred to henceforth as the *fluid element*, it can be assumed that, within the fluid element, the gravitational force is negligible relative to changes in the pressure force and that shear forces on the boundary of the fluid element are also negligible (i.e., the flow is frictionless). With these assumptions, applying Newton's second law to the fluid element gives

$$\underbrace{pA + \left(p + \frac{1}{2}\,\mathrm{d}p\right)\mathrm{d}A - (p + \mathrm{d}p)(A + \mathrm{d}A)}_{\text{net pressure force}} = \underbrace{\rho\,\mathrm{d}x\left(A + \frac{1}{2}\mathrm{d}A\right)a_x}_{\text{mass}\,\times\,\text{acceleration}} \tag{12.43}$$

where p and A are the pressure and area, respectively, on the upstream side of the fluid element, $\mathrm{d}p$ and $\mathrm{d}A$ are the changes in pressure and area, respectively, across the fluid element, ρ is the density of the fluid, and a_x is the acceleration of the fluid element in the x-direction. Under steady-state conditions, the acceleration a_x is related to the velocity field by

$$a_x = \frac{\mathrm{d}V}{\mathrm{d}t} = V\frac{\mathrm{d}V}{\mathrm{d}x} \quad \text{(steady state)} \tag{12.44}$$

Combining Equations 12.43 and 12.44, neglecting second-order terms (i.e. products of differentials), and simplifying yields the following relationship that is applicable along the central streamline of the fluid element,

$$\frac{\mathrm{d}p}{\rho} + V\,\mathrm{d}V = 0 \quad \text{(along streamline)} \tag{12.45}$$

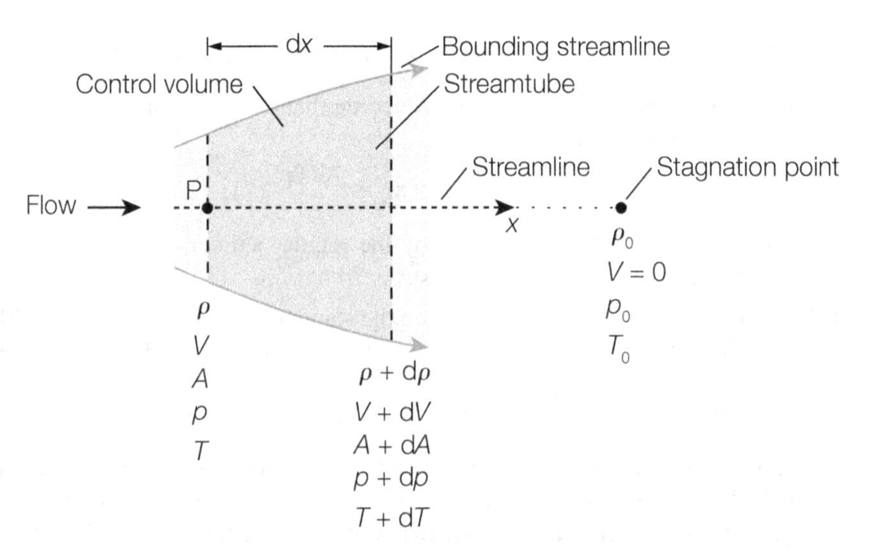

Figure 12.5: Relationship of local flow conditions to stagnation conditions

which can also be expressed as

$$\frac{\mathrm{d}p}{\rho} + \frac{1}{2}\,\mathrm{d}(V^2) = 0 \quad \text{(along streamline)} \tag{12.46}$$

It is worth noting that Equation 12.46 is the same as the Bernoulli equation but without the gravity term. For an isentropic process, the pressure and density within the fluid element are related by

$$\frac{p}{\rho^k} = C \tag{12.47}$$

where C is a constant and k is the specific heat ratio. Substituting Equation 12.47 into Equation 12.46 yields

$$-\mathrm{d}V^2 = 2C^{\frac{1}{k}} p^{-\frac{1}{k}}\,\mathrm{d}p \quad \rightarrow \quad -\int_V^0 \mathrm{d}V^2 = 2C^{\frac{1}{k}} \int_p^{p_0} p^{-\frac{1}{k}}\,\mathrm{d}p \tag{12.48}$$

where p_0 is the pressure at the *stagnation location*, which is defined as the location at which the fluid velocity is equal to zero (i.e., $V = 0$). The zero-velocity condition at the stagnation location is called the *stagnation condition*, and the properties of the fluid under the stagnation condition are called the *stagnation properties*. If the fluid is brought to the stagnation condition via an isentropic process, then the stagnation condition is called the *isentropic stagnation state* and the stagnation properties are called the *local isentropic stagnation properties*. In principle, each point in a flow field will have corresponding local isentropic stagnation properties. Integrating Equation 12.48 and eliminating C using Equation 12.47 yields

$$\frac{p_0}{p} = \left[1 + \frac{k-1}{k}\frac{\rho V^2}{2p}\right]^{\frac{k}{k-1}} \tag{12.49}$$

For an ideal gas,

$$p = \rho R T, \quad c = \sqrt{kRT}, \quad \mathrm{Ma} = \frac{V}{c} \tag{12.50}$$

Combining Equations 12.50 and 12.49 gives the following relationship between the stagnation pressure, p_0, and the actual pressure, p, along a streamline:

$$\frac{p_0}{p} = \left[1 + \frac{k-1}{2}\mathrm{Ma}^2\right]^{\frac{k}{k-1}} \tag{12.51}$$

Interestingly, for the (common) cases in which $\frac{1}{2}(k-1)\mathrm{Ma} < 1$, Equation 12.51 can be expressed as the infinite series

$$\frac{p_0 - p}{\frac{1}{2}\rho V^2} = 1 + \frac{\mathrm{Ma}^2}{4} + \frac{(2-k)\mathrm{Ma}^4}{24} + \dots \tag{12.52}$$

The term on the left-hand side of Equation 12.52 is called the *pressure coefficient*. For incompressible flows, the right-hand side of Equation 12.52 is equal to exactly 1.0, so Equation 12.52 provides a convenient way to measure the impact of compressibility on the stagnation pressure. The conventional guideline is that compressibility effects are negligible when $\mathrm{Ma} < 0.3$.

Using the ideal gas relations $\rho T/p$ = constant and p/ρ^k = constant combined with Equation 12.51 leads to the following expressions for the temperature, T_0, and density, ρ_0, at the stagnation condition:

$$\frac{T_0}{T} = 1 + \frac{k-1}{2}\text{Ma}^2 \tag{12.53}$$

$$\frac{\rho_0}{\rho} = \left[1 + \frac{k-1}{2}\text{Ma}^2\right]^{\frac{1}{k-1}} \tag{12.54}$$

Noting the similarities between Equations 12.51–12.54, it is sometimes convenient to use the relations

$$\frac{T_0}{T} = \left(\frac{p_0}{p}\right)^{\frac{k-1}{k}} \tag{12.55}$$

$$\frac{\rho_0}{\rho} = \left(\frac{p_0}{p}\right)^{\frac{1}{k}} \tag{12.56}$$

The *isentropic stagnation pressure*, p_0, is sometimes called the *total pressure*; the *isentropic stagnation temperature*, T_0, is sometimes called the *total temperature*; and the *isentropic stagnation density*, ρ_0, is sometimes called the *total density*. Equations 12.51, 12.53, and 12.54 are the fundamental relationships used to determine the local isentropic stagnation properties corresponding to any given (static) properties within the flow field, and the parameters of these relationships are the specific heat ratio, k, and the Mach number, Ma. The isentropic stagnation ratios p/p_0, T/T_0, and ρ/ρ_0 as a function of Ma for air (k = 1.40) are illustrated in Figure 12.6, where it is apparent that all isentropic stagnation ratios decrease monotonically with increasing Mach number. Because the stagnation properties p_0, T_0, and ρ_0 are all constants in isentropic flow, the relationships shown in Figure 12.6 are particularly useful in that they provide graphical illustrations that the fluid properties of pressure, temperature, and density all decrease as the Mach number increases, both under subsonic and supersonic conditions. It is further apparent from Figure 12.6 that the pressure and density are more sensitive to the Mach number than is the temperature.

Equations 12.51, 12.53, and 12.54 were derived using the (frictionless) momentum equation, the equation of state for an ideal gas, and the assumption of an isentropic process. An

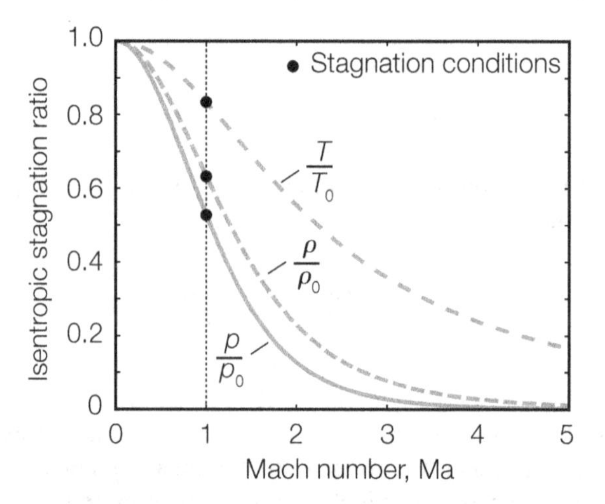

Figure 12.6: Isentropic stagnation ratios as a function of Mach number

equivalent and particularly useful relationship between T and T_0 can be derived directly from the energy equation. For an adiabatic steady-state process, the energy equation can be expressed as

$$h_1 + \frac{V_1^2}{2} = h_2 + \frac{V_2^2}{2} \tag{12.57}$$

which states that the quantity $h + V^2/2$ is a constant along a streamline in adiabatic flow. Taking this constant as h_0, the energy equation can be expressed as

$$h_0 = h + \frac{V^2}{2} \rightarrow h_0 - h = \frac{V^2}{2} \rightarrow c_p(T_0 - T) = \frac{V^2}{2} \quad \rightarrow \quad T_0 = T + \frac{V^2}{2c_p} \tag{12.58}$$

where h_0 is the stagnation enthalpy, h is the (static) enthalpy, V is the fluid velocity, and c_p is the specific heat at constant pressure, which is assumed to be independent of temperature in the range $[T, T_0]$. The quantity $V^2/2c_p$ in Equation 12.58 is the increase in temperature when the fluid is brought adiabatically to rest. Thus, $V^2/2c_p$ is sometimes called the *dynamic temperature*, in contrast to T, which is sometimes called the *static temperature*. For low-speed flows $V^2/2c_p \approx 0$, the stagnation temperature, T_0, and (static) temperature, T, are approximately the same; for high-speed flows, T and T_0 can be significantly different. The condition that $T \approx T_0$ and $p \approx p_0$ under low-speed conditions is the reason intake conditions to conduit flows are often taken to represent stagnation conditions.

The Bernoulli equation and the isentropic assumption. In the above derivation, application of the momentum equation to the compressible-flow scenario shown in Figure 12.5 yields the compressible-flow Bernoulli relationship given by Equation 12.46. Alternatively, consider the following relationships based only on the adiabatic and isentropic flow assumptions:

$$\text{adiabatic: } \mathrm{d}h + V\,\mathrm{d}V = 0 \tag{12.59}$$

$$\text{isentropic: } \mathrm{d}h = \frac{\mathrm{d}p}{\rho} \tag{12.60}$$

Combining Equations 12.59 and 12.60 gives

$$\frac{\mathrm{d}p}{\rho} + V\,\mathrm{d}V = 0 \tag{12.61}$$

which is the same as the Bernoulli equation. Therefore, the assumption of adiabatic isentropic flow is equivalent to the assumption of frictionless flow along a streamline. Indeed, the flow of a frictionless fluid under adiabatic conditions is one case in which the conditions for an isentropic process are met. Conversely, flows with significant frictional effects cannot be isentropic, and the relationships used to define the isentropic-stagnation and critical conditions are not particularly useful in these cases. Fortunately, in many engineering applications involving the compressible flow of gases, both the adiabatic and frictionless flow conditions (and hence the isentropic condition) are closely approximated.

EXAMPLE 12.4

A Pitot-static tube mounted in an airstream indicates that the ratio of the static pressure to the stagnation pressure in the flowing air is equal to 0.7 and that the temperature at the stagnation point is 18°C. (a) Estimate the speed at which the air is moving. (b) Estimate the pressure coefficient and assess the importance of compressibility in the air flow.

SOLUTION

From the given data: $p/p_0 = 0.7$ and T_0 = 18°C = 291.15 K. For air, R = 287.1 J/kg·K and k = 1.40. Assume that the airflow is isentropic; the air behaves like an ideal gas; and the specific heat ratio, k, can be taken as a constant.

(a) From the measured stagnation pressure ratio, the Ma can be estimated from Equation 12.51 as follows:

$$\frac{p_0}{p} = \left[1 + \frac{k-1}{2}\text{Ma}^2\right]^{\frac{k}{k-1}} \quad \to \quad 0.7 = \left[1 + \frac{1.40-1}{2}\text{Ma}^2\right]^{\frac{1.40}{1.40-1}} \quad \to \quad \text{Ma} = 0.732$$

Because $\text{Ma} = V/c$, the value of c must next be determined to finally determine V. Using Equation 12.53, the static temperature, T, is estimated as follows:

$$\frac{T_0}{T} = 1 + \frac{k-1}{2}\text{Ma}^2 \quad \to \quad \frac{291.15}{T} = 1 + \frac{1.40-1}{2}(0.732)^2 \quad \to \quad T = 263.0\ \text{K}$$

Hence, the speed of sound, c, and the speed of the air, V, are given by

$$c = \sqrt{RTk} = \sqrt{(287.1)(263.0)(1.40)} = 325.1\ \text{m/s}$$
$$V = \text{Ma} \cdot c = (0.732)(325.1) = 238.0\ \text{m/s}$$

Therefore, the estimated speed of the air is **238 m/s.**

(b) The pressure coefficient, C_p, as defined by Equation 12.52, can be expressed in terms of known quantities as follows:

$$C_\text{p} = \frac{p_0 - p}{\frac{1}{2}\rho V^2} = \frac{p_0 - p}{\frac{1}{2}\left(\frac{p}{RT}\right)V^2} = \frac{2RT}{V^2}\left(\frac{p_0}{p} - 1\right) = \frac{2(287.1)(263.0)}{(238.0)^2}\left(\frac{1}{0.7} - 1\right) = \mathbf{1.14}$$

Because Ma > 0.3, **compressibility effects are important** in the airflow. This is further indicated by the calculated $C_\text{p} = 1.14$, which deviates significantly from C_p = 1, which would be the case for incompressible flow.

The exact value of C_p is calculated as 1.14. Using the first two terms in the series expansion given by Equation 12.52 yields

$$C_\text{p} = 1 + \frac{\text{Ma}^2}{4} + \frac{(2-k)\text{Ma}^4}{24} = 1 + \frac{(0.732)^2}{4} + \frac{(2-1.40)(0.732)^4}{24} = 1.14$$

Hence, in this case, the first two terms of the series expansion in Equation 12.52 provides accuracy of at least three significant digits in estimating the pressure coefficient.

Stagnation condition at conduit entrances. It is not unusual when considering airflow through conduits to assume that the stagnation condition exists at the entrance of a conduit, which implies negligible entrance velocity. When the entrance is open to the atmosphere, the stagnation properties are taken as standard atmospheric pressure, p_{atm}, temperature, T_{atm}, and density, ρ_{atm}. An enclosed volume with a uniform temperature and pressure that is used as a source of flow in single or multiple connected conduits is called a *plenum chamber*. The usual purpose of a plenum chamber is to provide elevated pressures to force flow through the connected conduits, where the pressure in the plenum chamber is elevated above the pressure at the conduit exit(s). In cases of multiple conduits connected to a plenum chamber, the chamber can provide a means for equalizing the flow through the connected conduits. Interestingly, the word "plenum" (pronounced "plee-num") was originally coined to be complementary to the word "vacuum," where a vacuum connotes a reduced pressure and a plenum connotes an increased pressure.

12.4.2 Isentropic Critical Condition

The *isentropic critical condition* exists where the velocity of the fluid is brought isentropically to the sonic velocity. At the sonic velocity, Ma = 1, and putting Ma = 1 in Equations 12.51, 12.53, and 12.54 and denoting the pressure, temperature, and density at the critical condition as p^*, T^*, and ρ^*, respectively, yields the following relationships between the isentropic critical conditions and the isentropic stagnation properties:

$$\frac{p_0}{p^*} = \left[\frac{k+1}{2}\right]^{\frac{k}{k-1}} \tag{12.62}$$

$$\frac{T_0}{T^*} = \frac{k+1}{2} \tag{12.63}$$

$$\frac{\rho_0}{\rho^*} = \left[\frac{k+1}{2}\right]^{\frac{1}{k-1}} \tag{12.64}$$

The properties of the fluid at a location where Ma = 1 are called *critical properties*, and the ratios given in Equations 12.62–12.64 are called the *critical ratios*. At the critical condition, the velocity, V^* (= sonic velocity), is related to the isentropic stagnation temperature, T_0, because

$$V^* = c = \sqrt{kRT^*} \quad \rightarrow \quad V^* = \sqrt{\frac{2k}{k+1}RT_0} \tag{12.65}$$

Both the isentropic critical condition and the isentropic stagnation condition are useful reference conditions in the analysis of compressible flows. Conceptually, these conditions are at opposite ends of the velocity spectrum that could exist under isentropic conditions: Isentropic stagnation conditions occur at zero velocity, and isentropic critical conditions occur at the sonic velocity.

Critical conditions for airflow. For air at normal temperatures, $k = 1.40$ and Equations 12.62–12.64 give the pressure, temperature, and density at the critical condition (where Ma = 1) as $p^* = 0.528\,p_0$, $T^* = 0.833\,T_0$, and $\rho^* = 0.634\,\rho_0$, respectively. Hence, if stagnation conditions are assumed to exist at the entrance to a conduit, then at the critical section where the flow is sonic, the pressure, temperature, and density are all less than their entrance values, specifically 52.8%, 83.3%, and 63.4%, respectively, of their corresponding entrance values. Also, if a converging nozzle originates from stagnant conditions at a pressure p_0, then critical conditions will exist at the exit of the nozzle if the pressure at the exit (i.e., the back pressure) is equal to $0.528\,p_0$.

EXAMPLE 12.5

At a particular section in a duct, air is moving at 250 m/s with a temperature and pressure of 127°C and 200 kPa, respectively. Determine the temperature, pressure, density, and speed of the air at the isentropic critical condition.

SOLUTION

From the given data: $V = 250$ m/s, $T = 127°\text{C} = 400$ K, and $p = 200$ kPa. For air, $R = 287.1$ J/kg·K and $k = 1.40$. The density, ρ, and Mach number, Ma, under the given conditions are

$$\rho = \frac{p}{RT} = \frac{200 \times 10^3}{(287.1)(400)} = 1.742 \text{ kg/m}^3$$

$$\text{Ma} = \frac{V}{\sqrt{RTk}} = \frac{250}{\sqrt{(287.1)(400)(1.40)}} = 0.6235$$

The pressure, p^*, at the critical condition can be determined by combining Equations 12.62 and 12.51, which yields

$$\frac{p^*}{p} = \frac{p^*}{p_0}\frac{p_0}{p} = \left[\frac{1+\frac{k-1}{2}\text{Ma}^2}{\frac{k+1}{2}}\right]^{\frac{k}{k-1}} \rightarrow \frac{p^*}{200} = \left[\frac{1+\frac{1.40-1}{2}(0.6235)^2}{\frac{1.40+1}{2}}\right]^{\frac{1.40}{1.40-1}} \rightarrow p^* = 137 \text{ kPa}$$

Similarly, the temperature, T^*, at the critical condition can be determined by combining Equations 12.63 and 12.53, which yields

$$\frac{T^*}{T} = \frac{T^*}{T_0}\frac{T_0}{T} = \frac{1+\frac{k-1}{2}\text{Ma}^2}{\frac{k+1}{2}} \rightarrow \frac{T^*}{400} = \frac{1+\frac{1.40-1}{2}(0.6235)^2}{\frac{1.40+1}{2}} \rightarrow T^* = 359 \text{ K} = 86°\text{C}$$

The density, ρ^*, at the critical condition can be determined by combining Equations 12.64 and 12.54, which yields

$$\frac{\rho^*}{\rho} = \frac{\rho^*}{\rho_0}\frac{\rho_0}{\rho} = \left(\frac{1+\frac{k-1}{2}\text{Ma}^2}{\frac{k+1}{2}}\right)^{\frac{1}{k-1}} \rightarrow \frac{\rho^*}{1.742} = \left(\frac{1+\frac{1.40-1}{2}(0.6235)^2}{\frac{1.40+1}{2}}\right)^{\frac{1}{1.40-1}} \rightarrow \rho^* = 1.33 \text{ kg/m}^3$$

The density, ρ^*, at the critical condition could alternatively (and perhaps more easily) be calculated using the relation

$$\rho^* = \frac{p^*}{RT^*} = \frac{137 \times 10^3}{(287.1)(359)} = 1.33 \text{ kg/m}^3$$

Finally, the velocity, V^*, at the isentropic critical condition, which is the sonic velocity at T^*, is given by Equation 12.65 as

$$V^* = \sqrt{RT^*k} = \sqrt{(287.1)(359)(1.40)} = 380 \text{ m/s}$$

In summary, the temperature, pressure, density, and speed of the air at the isentropic critical condition are **86°C**, **137 kPa**, **1.33 kg/m³**, and **380 m/s**, respectively. It is interesting to note the usual characteristic that the temperature, pressure, and density all decrease as the fluid accelerates isentropically toward the sonic condition.

12.5 Basic Equations of One-Dimensional Compressible Flow

The steady-state continuity, momentum, and energy equations were derived previously (in Chapter 4) for general fluid flows through finite control volumes, and these equations are also applicable to compressible flows. These fundamental equations will be discussed briefly in this section. For applications involving compressible flows, the requirements of the second law of thermodynamics and the equations of state that govern the behavior of the fluid must also be taken into account. Gases undergoing compressible flow are commonly assumed to behave like an ideal gas, so the equations of state for an ideal gas are usually applied. Flows in conduits and streamtubes with gradually varying cross-sectional areas are commonly classified as being one-dimensional, which assumes that lateral velocities caused by changing flow areas are negligible. Changes in flow properties across flow sections are also commonly assumed to be negligible. Conduit and streamtube flows in which these latter two assumptions are made are sometimes appropriately referred to as being *quasi-one-dimensional flows*. In this text, quasi-one-dimensional flows are referred to simply as one-dimensional flows.

Continuity, momentum, and energy equations. Consider the one-dimensional flow through a conduit as shown in Figure 12.7, where sections 1 and 2 are the upstream and downstream sections, respectively. Assuming that, within the fluid element, the gravitational (body) force is negligible compared with change in the pressure force (an assumption that impacts the momentum and energy equations), the continuity, momentum, and energy equations are given by

$$\text{continuity: } \rho_1 V_1 A_1 = \rho_2 V_2 A_2 = \rho V A = \dot{m} \tag{12.66}$$

$$\text{momentum: } F_x + p_1 A_1 - p_2 A_2 = \dot{m}(V_2 - V_1) \tag{12.67}$$

$$\text{energy: } \frac{\dot{Q}}{\dot{m}} + h_1 + \frac{1}{2}V_1^2 = h_2 + \frac{1}{2}V_2^2 \tag{12.68}$$

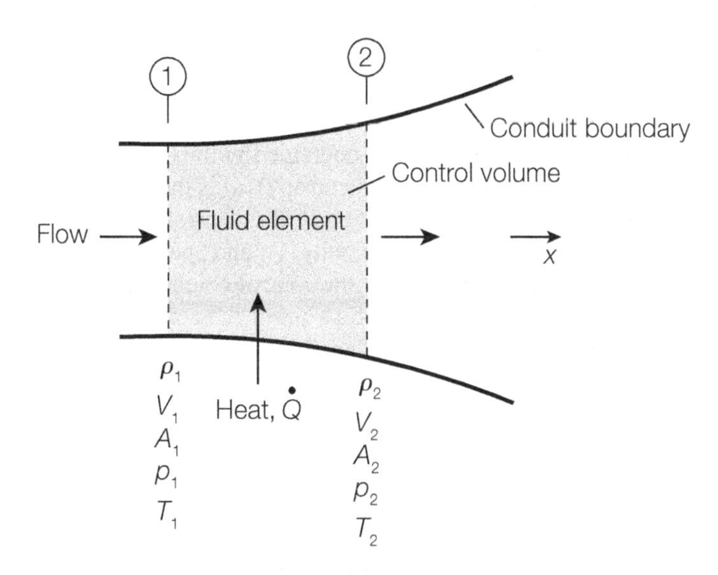

Figure 12.7: One-dimensional compressible flow

where $\dot{m}$ is the fluid mass flow rate through the conduit, F_x is the flow-direction component of the reaction of the conduit boundary, and h is the enthalpy of the fluid, equal to $p/\rho+u$, where u is the internal energy of the fluid per unit mass. The combination of variables $h+\frac{1}{2}V^2$ occurs frequently in compressible-flow analyses, and this quantity is called the *stagnation enthalpy*, h_0, because it is equal to the enthalpy of the fluid when it is brought to rest adiabatically. Hence, the energy equation can be expressed in terms of the stagnation enthalpy as

$$\frac{\dot{Q}}{\dot{m}} + h_{01} = h_{02} \tag{12.69}$$

where h_{01} and h_{02} are the stagnation enthalpies of the fluid at the upstream and downstream sections of the control volume, respectively. It is apparent from the energy equation (Equation 12.68 ≡ Equation 12.69) that under adiabatic flow conditions, an increase in velocity must necessarily be accompanied by a decrease in temperature, because $V_2 > V_1$ requires that $h_2 < h_1$, which requires that $T_2 < T_1$.

Second law of thermodynamics. The second law of thermodynamics, as given by Equation 12.16 and repeated here for convenience, states that for any given process,

$$\mathrm{d}S \geq \frac{\delta Q}{T} \tag{12.70}$$

where S is the entropy, $\mathrm{d}S$ is the change in entropy, δQ is the heat added, and T is the temperature. The form of the second law of thermodynamics given by Equation 12.70 applies to a fluid system and not a control volume, as is the case for the basic forms of other fundamental laws. For control-volume applications, a useful form of the second law of thermodynamics that is applicable to a control volume can be derived using the Reynolds transport theorem as follows:

$$\frac{\mathrm{d}S_{\text{sys}}}{\mathrm{d}t} = \underbrace{\frac{\partial}{\partial t}\int_{\text{cv}} s\rho\,\mathrm{d}V}_{=\,0\text{ (steady state)}} + \int_{\text{cs}} \rho s(\mathbf{v}\cdot\mathbf{n})\,\mathrm{d}A \geq \int_{\text{cs}} \frac{1}{T}\left(\frac{\dot{Q}}{A}\right)\mathrm{d}A \tag{12.71}$$

where S_{sys} is the entropy of the fluid system contained within the control volume, s is the entropy per unit mass of fluid (i.e., the specific entropy), $\dot{Q}$ is the rate at which heat is added to the control volume, and A is the surface area of the control volume over which heat is being added. Assuming that the density, ρ, velocity, V, and specific entropy, s, are constant over the upstream and downstream surfaces of the control volume and that the mass flux into the control volume ($\dot{m}_1$) is equal to the mass flux out of the control volume ($\dot{m}_2$), such that $\dot{m} = \dot{m}_1 = \dot{m}_2$, Equation 12.71 can be simplified as

$$\text{second law: } \dot{m}(s_2 - s_1) \geq \int_{\text{cs}} \frac{1}{T}\left(\frac{\dot{Q}}{A}\right)\mathrm{d}A \tag{12.72}$$

where the subscripts 1 and 2 indicate the upstream and downstream surface of the control volume, respectively.

Equations of state. Equations of state are relationships between intensive thermodynamic properties; you may recall that an "intensive property" is a property per unit mass. As a general rule, it is important to remember that any thermodynamic property can be expressed as a function of two other thermodynamic properties. For an ideal gas with constant specific heats, the following equations of state are particularly useful:

$$\text{ideal gas law: } \rho = \frac{p}{RT} \tag{12.73}$$

$$\text{change in enthalpy: } h_2 - h_1 = c_p(T_2 - T_1) \tag{12.74}$$

$$\text{change in entropy: } s_2 - s_1 = c_p \ln\frac{T_2}{T_1} - R\ln\frac{p_2}{p_1} \tag{12.75}$$

Collectively, Equations 12.66–12.75 are the basic equations that form the foundation for analyses of one-dimensional compressible flow.

12.6 Steady One-Dimensional Isentropic Flow

Steady one-dimensional flow occurs most commonly in closed-conduit flow, which includes flow in pipes and ducts. One-dimensional flow also occurs in local regions of larger flow fields, where the velocity is predominantly unidirectional within a subregion of the larger flow field. When heat exchange and frictional effects are negligible between any two sections of a one-dimensional flow, the flow can be approximated as being isentropic. Examples of flows through conduits that can be characterized as steady, one-dimensional isentropic flow include flow through the diffuser near the front of a jet engine, exhaust gases passing through the blades of a turbine, and flow through the nozzles on a rocket engine.

12.6.1 Effect of Area Variation

In cases where the sonic condition is attained isentropically within a conduit, the mass flow rate at the critical section is equal to the mass flow rate at any section within the conduit such that

$$\rho A V = \rho^* A^* V^* \quad \rightarrow \quad \frac{A}{A^*} = \left(\frac{\rho^*}{\rho}\right)\left(\frac{V^*}{V}\right) \tag{12.76}$$

where the asterisks denote fluid properties at the critical condition. Combining Equation 12.76 with the critical-condition relationships given by Equations 12.63 and 12.65 yields the following useful relationship:

$$\frac{A}{A^*} = \frac{1}{\text{Ma}}\left[\frac{1 + \frac{k-1}{2}\text{Ma}^2}{\frac{k+1}{2}}\right]^{\frac{k+1}{2(k-1)}} \tag{12.77}$$

Hence, if Ma is known at a section where the area is equal to A, then Equation 12.77 can be used to determine the area, A^*, required to attain critical (sonic) conditions. If the required area does not exist within the conduit, then the critical condition will not occur. Equation

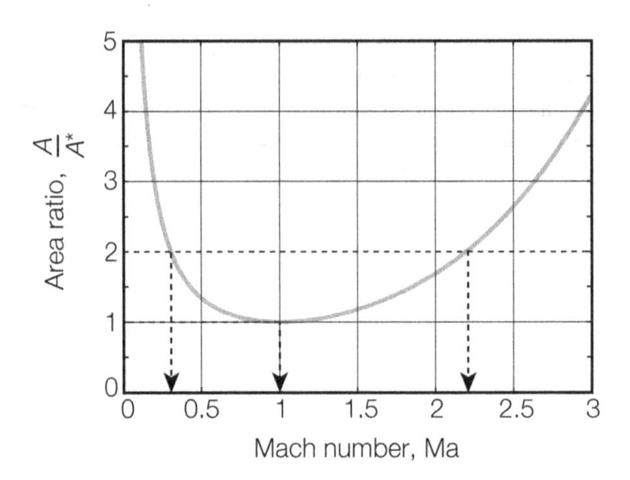

Figure 12.8: Area ratio as a function of a Mach number

12.77 applies to all Mach numbers (both subsonic and supersonic) and is plotted in Figure 12.8, which shows the following key features of this relationship:

1. $A/A^* = 1$ when $\text{Ma} = 1$.

2. For any given area ratio, A/A^*, there are two possible Mach numbers—one in which $\text{Ma} < 1$ and the other in which $\text{Ma} > 1$.

The latter feature must usually be considered in the analysis of expanding sections downstream of a critical section (where $A/A^* = 1$ and $\text{Ma} = 1$), because it must be determined whether the flow downstream of the critical section is subsonic ($\text{Ma} < 1$) or supersonic ($\text{Ma} > 1$); according to Equation 12.77, both are possible. The determination of which flow will occur depends on other factors that will be discussed subsequently. However, sonic flow in the minimum area section of a conduit, called the *throat section* or *throat*, need not be present, and the flow in a duct can be entirely subsonic or entirely supersonic.

12.6.2 Choked Condition

According to Equation 12.76, the area ratio A^*/A at any given section of a conduit is related to the mass flow rate per unit area (ρV) by

$$\frac{A^*}{A} = \frac{\rho V}{\rho^* V^*} = \frac{\text{mass flow rate per unit area at a section with area } A}{\text{mass flow rate per unit area at the critical section with area } A^*} \tag{12.78}$$

According to Equation 12.77, the maximum value of A^*/A is equal to unity (= 1) and occurs when $\text{Ma} = 1$. Hence, Equation 12.78 shows that for any flow through a conduit with a constant cross-sectional area, the mass flow rate in the conduit cannot exceed the mass flow rate that would exist at a critical section within the conduit. Therefore, if a critical section exists in a conduit (i.e., a location where $\text{Ma} = 1$), then the mass flow rate in the conduit is restricted to the mass flow rate at the critical section; under this condition, the flow is said to be *choked*.

EXAMPLE 12.6

Air flows in a converging conduit of a circular cross section as shown in Figure 12.9. Measurements at a location where the diameter is 200 mm show a velocity of 200 m/s and a temperature of 15°C. The diameter of the conduit at the exit is 190 mm. Determine whether critical flow conditions exist within the conduit. Is the flow choked?

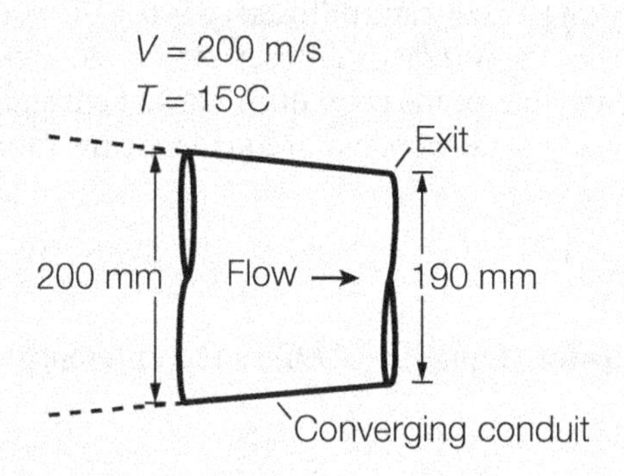

Figure 12.9: Flow in a converging conduit

SOLUTION

From the given data: V = 200 m/s, T = 15°C = 288.15 K, D = 200 mm = 0.200 m, and D_e = 190 mm. For air, R = 287.1 J/kg·K, and assume that k = 1.40 for the conditions within the conduit. Using the measured data, the area, A, the speed of sound, c, and the Mach number, Ma, at the measurement section are given by

$$A = \frac{\pi}{4}D^2 = \frac{\pi}{4}(0.200)^2 = 0.03142 \text{ m}^2$$

$$c = \sqrt{RTk} = \sqrt{(287.1)(288.15)(1.40)} = 340.3 \text{ m/s}$$

$$\text{Ma} = \frac{V}{c} = \frac{200}{340.3} = 0.588$$

The critical area, A^*, in the conduit can be calculated using Equation 12.77, which yields

$$\frac{A}{A^*} = \frac{1}{\text{Ma}}\left[\frac{1 + \frac{k-1}{2}\text{Ma}^2}{\frac{k+1}{2}}\right]^{\frac{k+1}{2(k-1)}} \quad \rightarrow \quad \frac{0.03142}{A^*} = \frac{1}{0.588}\left[\frac{1 + \frac{1.40-1}{2}(0.588)^2}{\frac{1.40+1}{2}}\right]^{\frac{1.40+1}{2(1.40-1)}}$$

$$\rightarrow \quad A^* = 0.02612 \text{ m}^2$$

A circular area of 0.02612 m^2 corresponds to a diameter of $\sqrt{4(0.02612)/\pi}$ = 0.182 m = 182 mm. Because the exit diameter (= 190 mm) is greater than the critical diameter (= 182 mm), **critical flow conditions (i.e., sonic flow) will not occur within the conduit**. Because critical flow conditions do not occur, **the flow is not choked**. If the exit diameter was reduced below 182 mm, the flow would be choked.

12.6.3 Flow in Nozzles and Diffusers

A *nozzle* is a conduit that increases the flow velocity (i.e., accelerates the flow), and a *diffuser* is a conduit that decreases the flow velocity (i.e., decelerates the flow). For incompressible flows, the continuity equation requires that a nozzle have a decreasing flow area in the direction of flow and that a diffuser have an increasing flow area in the direction of flow. This relationship between velocity and flow area is not generally the same for compressible flows.

Theory. In analyzing nozzles and diffusers, it is usually convenient to assume one-dimensional flow along the axis of the nozzle or diffuser, which requires a small rate of area change and a large radius of curvature of the conduit. A convenient starting point for this analysis is the frictionless flow steady-state momentum equation in the form of Equation 12.46, which gives

$$\frac{\mathrm{d}p}{\rho} + \frac{1}{2}\mathrm{d}(V^2) = 0 \quad \rightarrow \quad \mathrm{d}p = -\rho V\,\mathrm{d}V \quad \rightarrow \quad \frac{\mathrm{d}p}{\rho V^2} = -\frac{\mathrm{d}V}{V} \tag{12.79}$$

The continuity equation (Equation 12.66) can be differentiated to yield a convenient form for analysis, where

$$\rho AV = \text{constant} \quad \rightarrow \quad \frac{\mathrm{d}\rho}{\rho} + \frac{\mathrm{d}A}{A} + \frac{\mathrm{d}V}{V} = 0 \quad \rightarrow \quad \frac{\mathrm{d}A}{A} = -\frac{\mathrm{d}V}{V} - \frac{\mathrm{d}\rho}{\rho} \tag{12.80}$$

This equation is particularly useful because it relates changes in cross-sectional area, velocity, and density, which are all interrelated in compressible flow. Combining Equations 12.80 and 12.79 yields

$$\frac{\mathrm{d}A}{A} = \frac{\mathrm{d}p}{\rho V^2}\left[1 - V^2\left(\frac{\mathrm{d}p}{\mathrm{d}\rho}\right)^{-1}\right] \tag{12.81}$$

It was shown previously (in the derivation of Equation 12.32) that for an isentropic process,

$$\frac{\mathrm{d}p}{\mathrm{d}\rho} = c^2 \tag{12.82}$$

Combining Equations 12.82, 12.81, and 12.79 and using the definition of the Mach number, Ma (= V/c), yields

$$\frac{\mathrm{d}V}{\mathrm{d}A} = -\frac{V}{A}\cdot\frac{1}{1-\mathrm{Ma}^2} \tag{12.83}$$

This interesting relationship shows whether the velocity increases or decreases with increasing area (i.e., whether $\mathrm{d}V/\mathrm{d}A$ is positive or negative) depends on Ma. In cases of subsonic flow, where Ma < 1, $\mathrm{d}V/\mathrm{d}A < 0$ and the flow velocity decreases with increasing cross-sectional area, which is the same trend found for incompressible flows. However, in the case of supersonic flow, where Ma > 1, $\mathrm{d}V/\mathrm{d}A > 0$ and the flow velocity increases with increasing cross-sectional area, which is opposite that found in incompressible flows. In the case of sonic flow, where Ma = 1, the left-hand side of Equation 12.83 is infinite, which then requires that $\mathrm{d}A = 0$ on the right-hand side of the equation. This result indicates that sonic flow can only be attained at a flow section of incrementally constant cross-sectional area, in which case $\mathrm{d}A = 0$ and the flow area is a minimum. Based on the analysis presented here and the definition of a nozzle and a diffuser, it can be stated that a diverging conduit is a *supersonic nozzle*, a converging conduit is a *supersonic diffuser*, and sonic flow can only occur at the minimum cross-sectional area of the nozzle or diffuser.

Diffusers and nozzles in jet engines. Diffusers and nozzles are important components of jet engines used to propel supersonic aircraft as illustrated in Figure 12.10. In this application, the role of the diffuser is to reduce the speed of the incoming air so that there is sufficient time for combustion in the combustion chamber. The ignited fuel-air mixture in the combustion chamber transfers heat to the fluid, which increases the pressure of the fluid exiting the combustion chamber (in accordance with the energy equation), which then generates a high-velocity discharge and associated thrust to the engine and attached aircraft (in accordance with the momentum equation).

Conduits with constant area. For conduits in which the cross-sectional area does not change, $\mathrm{d}A = 0$ by definition, and Equation 12.80 gives

$$\frac{\mathrm{d}V}{V} = -\frac{\mathrm{d}\rho}{\rho} \tag{12.84}$$

This result illustrates that a change in velocity along a conduit must be accompanied by a change in density, with an increase in velocity accompanied by a decrease in density and vice versa. Of course, Equation 12.84 is also applicable to the familiar incompressible flows, in which case $\mathrm{d}\rho = 0$ and the velocity along the conduit remains constant.

Density changes. It is also instructive to combine Equations 12.83 and 12.80 to see how the density changes within a conduit; these equations yield

$$\frac{\mathrm{d}\rho}{\mathrm{d}A} = \frac{\rho}{A}\left[\frac{\mathrm{Ma}^2}{1 - \mathrm{Ma}^2}\right] \tag{12.85}$$

This equation shows that for subsonic flows ($\mathrm{Ma} < 1$), the fluid density increases with increasing cross-sectional area, whereas in supersonic flows ($\mathrm{Ma} > 1$), the density decreases with increasing cross-sectional area. This is a direct result of the fact that the velocity decreases with increasing area for subsonic flows and the velocity increases with increasing area for supersonic flows. For both increasing and decreasing cross-sectional areas, the mass flow rate (ρAV) remains constant.

Supersonic nozzles. It is commonplace to observe supersonic nozzles (= diverging conduits) at the tail end of rockets, and an example of the exhaust nozzle from a rocket engine is shown in Figure 12.11. This nozzle was being designed and tested at NASA's Marshall Space Flight Center (located in Huntsville, Alabama); the tests on the nozzle were intended to ensure that the nozzle was strong enough to withstand the uneven forces encountered during operation. Supersonic nozzles at the tail end of rockets are designed to produce a thrust that is commensurate with the mass of the rocket and the speed at which the rocket must travel through the atmosphere. Consequently, larger rockets have larger supersonic (diverging) nozzles. As a reference point, each of the space shuttle main engines had nozzles that

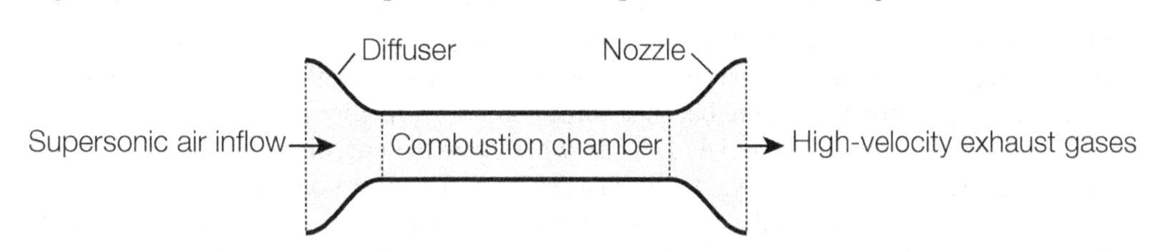

Figure 12.10: Supersonic jet engine

were about 3 m long and 2.4 m in diameter at their exits. A second important consideration in designing supersonic nozzles is that the exit pressure be equal to the local atmospheric pressure, which decreases with altitude. This latter requirement usually results in optimal nozzle performance at one particular altitude with suboptimal performance at other altitudes. An example of a rocket exhaust nozzle in operation is shown in Figure 12.12. Notice the dissipation of the momentum of the exhaust gas that is being discharged at a supersonic speed.

Efficiencies of nozzles and diffusers. In energy terms, the purpose of a nozzle is to convert enthalpy into kinetic energy, and the efficiency of a nozzle, η_n, is usually defined by the relation

$$\eta_\mathrm{n} = \frac{h_0 - h_\mathrm{e}}{h_0 - h_\mathrm{es}} \tag{12.86}$$

where h_0 is the stagnation enthalpy, h_e is the actual enthalpy of the fluid at the nozzle exit, and h_es is the theoretical enthalpy of the fluid at the nozzle exit assuming isentropic flow through the nozzle. Nozzle efficiencies are typically in the range of 90%–99%, with larger nozzles having larger efficiencies because the relative impact of viscous wall effects diminishes with increasing size of the nozzle. In contrast to a nozzle, the purpose of a diffuser is to reduce the kinetic energy and recover the pressure in the fluid, and its performance is measured by the *pressure recovery factor*, C_p, defined as

$$C_p = \frac{\Delta p_\mathrm{actual}}{\Delta p_\mathrm{isentropic}} \tag{12.87}$$

where Δp_actual is the actual pressure recovery and $\Delta p_\mathrm{isentropic}$ is the theoretical pressure recovery assuming isentropic conditions. Values of C_p are typically in the range of 40%–85%. Under subsonic conditions, expansion angles of less than 10° should be used to maximize pressure recovery, whereas in supersonic flows, much wider diffuser angles can yield satisfactory performance.

12.6.3.1 Converging nozzle

Consider the case of a converging nozzle shown in Figure 12.13, where the source of the flow is a relatively stagnant environment (such as the open atmosphere or a large plenum chamber), and the inflow velocity is sufficiently small that the inflow condition can be approximated by the stagnation condition ($V = 0$). In this case, the flow is driven by the pressure difference between the inlet and outlet of the nozzle, where the pressure on the inlet side (p_0) is higher than on the *back pressure* (p_b) applied at the exit. The exit section of a converging nozzle is commonly referred to as the *throat* of the nozzle. The pressure at which the flow exits the nozzle is called the *exit pressure*, p_e, and for low flow rates, $p_\mathrm{e} = p_\mathrm{b}$. Consider what happens when the difference between p_0 and p_b is increased, starting from the condition that $p_0 = p_\mathrm{b}$, which does not cause any flow. As p_b decreases, the flow through the converging nozzle increases, and as long as Ma < 0.3 within the nozzle, the pressure variation within the nozzle is described by the Bernoulli equation for incompressible flow. As p_b decreases further, Ma continues to increase, and as long as the flow is isentropic, the rate of increase of velocity within the nozzle is given by Equation 12.83. However, sonic conditions (Ma = 1) cannot be attained at any converging section, because sonic conditions require a section where $\mathrm{d}A = 0$ or, equivalently, a section where $\mathrm{d}A/\mathrm{d}x = 0$, where x is the coordinate in the direction of flow. Therefore, the maximum velocity must occur at the most constricted section of the nozzle (which is at the exit of the nozzle), and the maximum attainable velocity is equal to the sonic velocity. For the sonic velocity to occur at the nozzle exit, the shape of the nozzle exit should be such that $\mathrm{d}A/\mathrm{d}x = 0$ at the exit. It has already been shown that it is physically

Figure 12.11: Rocket engine exhaust nozzle
Source: NASA.

Figure 12.12: Rocket engine exhaust nozzles in operation
Source: JPL-Caltech/NASA.

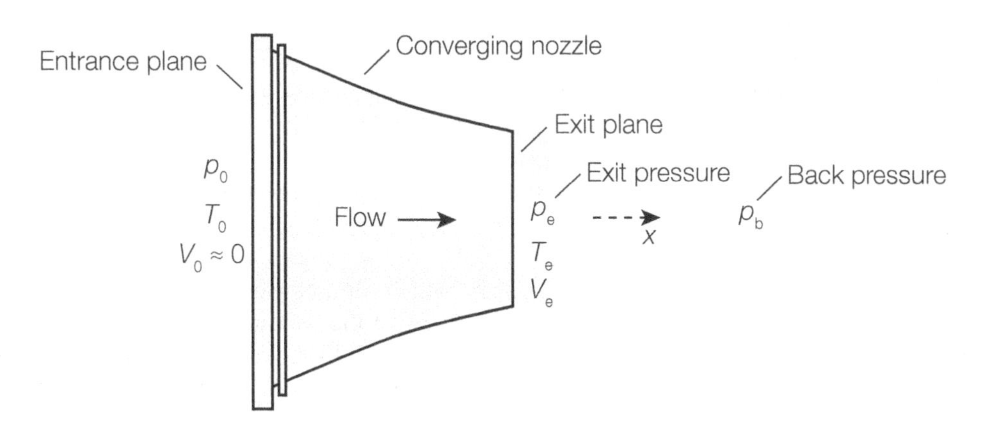

Figure 12.13: Converging nozzle

impossible for the velocity through a converging nozzle to be increased beyond the sonic velocity, because such an increase requires a nozzle with an expanding area. When the sonic condition occurs at the exit of a converging nozzle, Ma = 1 and $p_e = p_b = p^*$, where p^* is the pressure at the critical condition, and Equation 12.62 gives

$$\frac{p_e}{p_0} = \frac{p_b}{p_0} = \frac{p^*}{p_0} = \left(\frac{2}{k+1}\right)^{\frac{k}{k-1}} \tag{12.88}$$

If the back pressure, p_b, is reduced below the exit pressure under the critical condition, then because V_e must remain constant, p_e also must remain constant, resulting in the condition that $p_b < p_e$. Therefore, after exiting the nozzle, the fluid pressure will adjust itself from p_e to p_b, usually in a nonisentropic, three-dimensional manner.

Mass flow rate through a converging nozzle. Under steady-state conditions, the mass flow rate, $\dot{m}$, through the nozzle is given by

$$\dot{m} = \rho A V = \frac{p}{RT} A\,(\mathrm{Ma}\sqrt{kRT}) = p\,A\,\mathrm{Ma}\sqrt{\frac{k}{RT}} \tag{12.89}$$

Substituting Equations 12.51 and 12.53 into Equation 12.89 yields the following expression for the mass flow rate in terms of the stagnation properties:

$$\dot{m} = \frac{A\,\mathrm{Ma}\,p_0\sqrt{\dfrac{k}{RT_0}}}{\left(1 + \dfrac{k-1}{2}\mathrm{Ma}^2\right)^{\frac{k+1}{2(k-1)}}} \tag{12.90}$$

which demonstrates that the mass flow rate through a nozzle is a function of the stagnation properties, T_0 and p_0, the flow area, A, and the Mach number, Ma. For any given values of T_0, p_0, and A, the maximum mass flow rate can be determined by differentiating Equation 12.90 with respect to Ma and equating the result to zero, which yields Ma $= 1$ as the Mach number at the maximum flow rate. Because the only location where Ma can be equal to unity is in the throat of the nozzle, the maximum flow rate through the nozzle must necessarily occur when Ma $= 1$ in the throat of the nozzle. Under this condition, $A = A^* = A_t$ and Ma $= 1$ in Equation 12.90, which gives the maximum mass flow rate, $\dot{m}_{max}$, as

$$\dot{m}_{max} = A_t p_0 \left(\frac{2}{k+1}\right)^{\frac{k+1}{2(k-1)}} \sqrt{\frac{k}{RT_0}} \quad \text{(sonic exit condition)} \tag{12.91}$$

where A_t is the area of the throat of the nozzle, which for a converging nozzle is equal to the area at the exit of the nozzle. It is worth noting that the maximum mass flow rate in a converging nozzle depends only on the throat area, A_t, and the conditions of the fluid at the entrance, via p_0 and T_0. Because the maximum mass flow rate through a nozzle with a given throat area is fixed by p_0 and T_0 of the inlet flow, the mass flow rate through the nozzle can be changed by changing p_0 and T_0, and hence, a converging nozzle can be used as a flow control device. The flow can also be changed by adjusting the throat area. When the flow through the throat of a nozzle is at the critical condition (i.e., $\text{Ma}_{\text{throat}} = 1$), the conduit is said to be *choked*, and the mass flow rate through the conduit cannot be increased unless the throat area is enlarged. It is apparent from Equation 12.91 that the maximum flow rate is directly proportional to the throat area and the stagnation pressure and inversely proportional to the square root of the stagnation temperature.

Mass flow rate under subsonic conditions. In cases where the flow through the throat of the nozzle is subsonic, the mass flow rate, $\dot{m}$, can be calculated by applying Equation 12.90 at the throat of the nozzle. The mass flow rate can equivalently be calculated using the relation

$$\dot{m} = \rho_t A_t V_t = A_t \sqrt{\frac{2k}{k-1} p_0 \rho_0 \left[\left(\frac{p_t}{p_0} \right)^{\frac{2}{k}} - \left(\frac{p_t}{p_0} \right)^{\frac{k+1}{k}} \right]} \quad \text{(subsonic exit condition)} \quad (12.92)$$

where the subscript "t" denotes conditions at the throat of the nozzle or, equivalently, conditions at the exit of a converging nozzle. Denoting the pressure at the exit of a converging nozzle by p_e ($= p_t$), Equation 12.92 is applicable as long as $p_e > p_e^*$, where p_e^* is the pressure at the exit when sonic conditions exist there.

Other considerations. Under usual circumstances in which subsonic flow is being accelerated in a (converging) nozzle, the assumption of isentropic flow within the nozzle is valid, primarily due to the favorable pressure gradient that keeps the boundary layer thin and thereby minimizes frictional effects. A converging nozzle, which is sometimes called a *truncated nozzle*, is commonly used in compressible-flow meters. Supersonic flows that are being decelerated within a converging conduit experience adverse pressure gradients, which can compromise the assumption of isentropic flow because boundary-layer separation is a possibility.

EXAMPLE 12.7

Air flows through a converging nozzle as shown in Figure 12.14. The entrance of the nozzle is open to the atmosphere, where the temperature is 20°C and the pressure is 101.3 kPa. The exit of the nozzle has a diameter of 10 mm, and the back pressure exerted at the exit of the nozzle is p_b. Determine the mass flow rate through the nozzle for a back pressure of (a) 50 kPa and (b) 70 kPa.

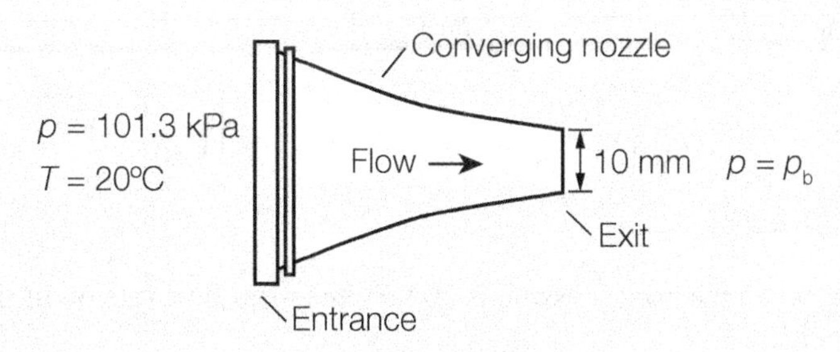

Figure 12.14: Flow through a converging nozzle

SOLUTION

From the given data: p_0 = 101.3 kPa, T_0 = 20°C = 293.15 K, and D_e = 10 mm = 0.01 m. Assume that the flow through the nozzle is isentropic, air behaves like an ideal gas, and $k = c_p/c_v = 1.40$ for the conditions encountered in the nozzle. For air, R = 287.1 J/kg·K and the density of the air, ρ_0, at the entrance to the nozzle and the area of the nozzle at the exit, A_e, are given by

$$\rho_0 = \frac{p_0}{RT_0} = \frac{101.3 \times 10^3}{(287.1)(293.15)} = 1.204 \text{ kg/m}^3$$

$$A_e = \frac{\pi}{4}D_e^2 = \frac{\pi}{4}(0.01)^2 = 7.854 \times 10^{-5} \text{ m}^2$$

Assume that the entrance conditions are the stagnation conditions. If the sonic condition exists at the exit of the nozzle, then the pressure at the exit is the critical pressure, p^*, and Equation 12.88 gives

$$\frac{p^*}{p_0} = \left(\frac{2}{k+1}\right)^{\frac{k}{k-1}} \quad \rightarrow \quad \frac{p^*}{101.3} = \left(\frac{2}{1.40+1}\right)^{\frac{1.4}{1.4-1}} \quad \rightarrow \quad p^* = 53.31 \text{ kPa}$$

The actual pressure at the exit of the nozzle, p_e, is determined by the magnitude of the back pressure, p_b, relative to the critical pressure, p^*, as illustrated by considering the following two cases.

(a) In the case p_b = 50 kPa, $p_b < p^*$, $p_e = p^*$ and the flow through the nozzle is choked by critical conditions at the exit. The mass flow rate through the nozzle is the maximum possible flow rate for the given entrance conditions, and Equation 12.91 gives

$$\begin{aligned}\dot{m}_{max} &= A_e p_0 \left(\frac{2}{k+1}\right)^{\frac{k+1}{2(k-1)}} \sqrt{\frac{k}{RT_0}} \\ &= (7.854 \times 10^{-5})(101.3 \times 10^3)\left(\frac{2}{1.40+1}\right)^{\frac{1.40+1}{2(1.40-1)}} \sqrt{\frac{1.40}{(287.1)(293.15)}} \\ &= 0.0188 \text{ kg/s}\end{aligned}$$

Therefore, when the back pressure is 50 kPa, the mass flow rate of air through the nozzle is **0.0188 kg/s.**

(b) In the case p_b = 70 kPa, $p_b > p^*$, $p_e = p_b$. Under this condition, the flow is subsonic at the nozzle exit and the mass flow rate, $\dot{m}$, through the nozzle is given by Equation 12.92 as

$$\begin{aligned}\dot{m} &= A_e \sqrt{\frac{2k}{k-1} p_0 \rho_0 \left[\left(\frac{p_e}{p_0}\right)^{\frac{2}{k}} - \left(\frac{p_e}{p_0}\right)^{\frac{k+1}{k}}\right]} \\ &= (7.854 \times 10^{-5})\sqrt{\frac{2(1.40)}{1.40-1}(101.3 \times 10^3)(1.204)\left[\left(\frac{70}{101.3}\right)^{\frac{2}{1.40}} - \left(\frac{70}{101.3}\right)^{\frac{1.40+1}{1.40}}\right]} \\ &= 0.0176 \text{ kg/s}\end{aligned}$$

Therefore, when the back pressure is 70 kPa, the mass flow rate of air through the nozzle is **0.0176 kg/s.**

12.6.3.2 Converging-diverging nozzle

In a *converging-diverging (CD) nozzle*, the cross-sectional area decreases and then increases in the flow direction. This type of nozzle is illustrated in Figure 12.15 for the case in which a fluid is being accelerated from subsonic to supersonic flow. This (CD) nozzle is sometimes called a *Laval nozzle* or *de Laval nozzle*,[65] which was originally developed for use in steam turbines. If a fluid enters a CD nozzle under subsonic flow conditions, then a converging conduit is required to accelerate the flow. To achieve a sonic velocity requires a section of minimum cross-sectional area, where $\mathrm{d}A = 0$, and then further acceleration in the supersonic flow regime requires a diverging conduit; this is the performance mechanism of a CD nozzle. The flow through a CD nozzle is driven by the pressure difference between the inflow and outflow sections, with greater pressure on the inflow section (p_1) than on the outflow section (p_2). The *throat section* of a CD nozzle is the section with minimum cross-sectional area, connecting the exit of the converging segment to the entrance of the diverging segment.

Nozzle performance. The performance of a converging-diverging (CD) nozzle is similar to that of a corresponding converging nozzle with the same shape as the front section of the CD nozzle. The similarity is that both the CD nozzle and the corresponding converging nozzle have the same maximum mass flow rate, which occurs when sonic conditions exist in the throat. However, the CD nozzle is capable of producing a supersonic discharge velocity, compared with a converging nozzle whose maximum discharge velocity is sonic. It should also be noted that CD nozzles have a totally different purpose than *Venturi nozzles*, which are somewhat similar in appearance. Venturi nozzles are only used in incompressible flows and typically as a flow measuring device, whereas the purpose of a CD nozzle is to transition flows between subsonic and supersonic speeds.

Fluid properties in the nozzle. Because the critical (sonic) condition commonly exists in the throat of a CD nozzle, it is usually convenient to express the fluid properties in a CD nozzle in terms of the fluid properties at the critical condition. Because the flow is typically assumed to be isentropic, the relationship between the Mach number and the ratio of flow area to critical area as given by Equation 12.77 is applicable. Also, the pressure, p, temperature,

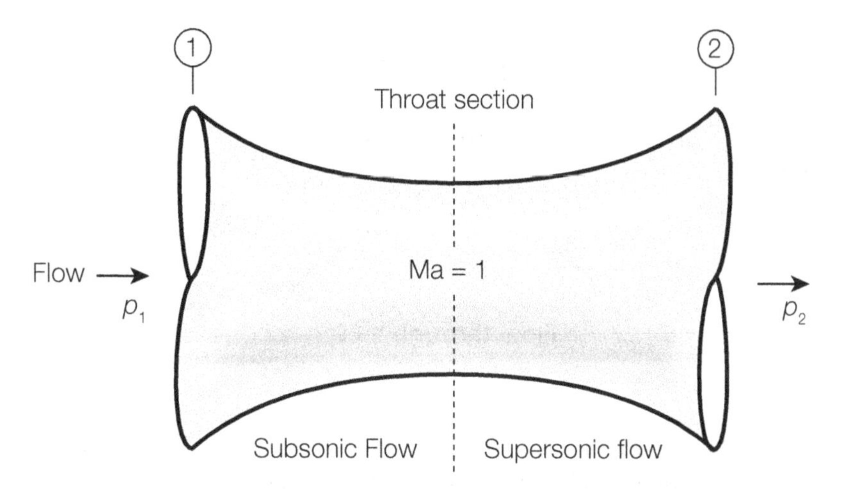

Figure 12.15: Converging-diverging nozzle

[65]Named in honor of the Swedish engineer Karl Gustaf Patrik de Laval (1845–1913).

T, and density, ρ, within the nozzle can be expressed in terms of the fluid properties at the critical condition using the following isentropic critical condition relations:

$$\frac{p}{p^*} = \left[\frac{\frac{k+1}{2}}{1 + \frac{k-1}{2}\mathrm{Ma}^2}\right]^{\frac{k}{k-1}} \tag{12.93}$$

$$\frac{T}{T^*} = \frac{\frac{k+1}{2}}{1 + \frac{k-1}{2}\mathrm{Ma}^2} \tag{12.94}$$

$$\frac{\rho}{\rho^*} = \left[\frac{\frac{k+1}{2}}{1 + \frac{k-1}{2}\mathrm{Ma}^2}\right]^{\frac{1}{k-1}} \tag{12.95}$$

where the asterisk indicates a fluid property under critical flow conditions.

EXAMPLE 12.8

Consider airflow through the CD nozzle shown in Figure 12.16. At section 1, the diameter is 300 mm, and under particular flow conditions, the velocity at section 1 is 200 m/s, the pressure is 550 kPa, and the temperature is 227°C. Sections 2 and 3 both have a diameter of 275 mm, with section 2 located before the throat and section 3 located after the throat. Determine the flow properties at sections 2 and 3 when the flow in the diverging part of the nozzle is (a) subsonic and (b) supersonic. Is it possible for the exit flow to be sonic?

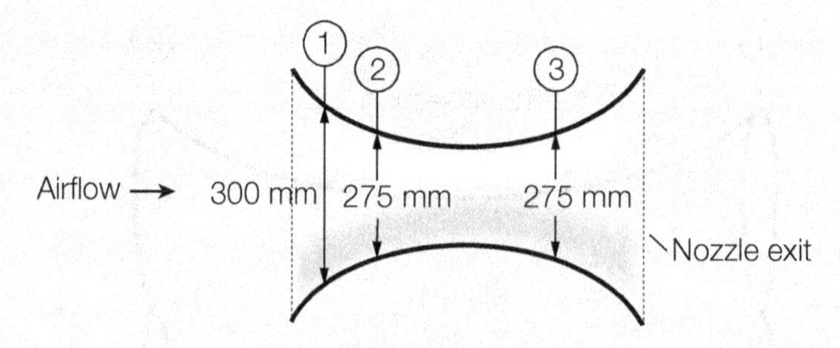

Figure 12.16: Flow through a CD nozzle

SOLUTION

From the given data: D_1 = 300 mm = 0.3 m, V_1 = 200 m/s, p_1 = 550 kPa, T_1 = 227°C = 500 K, and D_2 = D_3 = 275 mm = 0.275 m. The areas of the nozzle at sections 1 and 2 are $A_1 = \pi D_1^2/4 = 0.07069$ m², and similarly, A_2 = A_3 = 0.05940 m². For air, R = 287.1 J/kg·K and k = 1.40. The Mach number, Ma_1, at section 1 is given by

$$\mathrm{Ma}_1 = \frac{V_1}{\sqrt{RT_1k}} = \frac{200}{\sqrt{(287.1)(500)(1.40)}} = 0.4461$$

The critical area, A^*, is derived from Equation 12.77 as follows:

$$\frac{A_1}{A^*} = \frac{1}{\text{Ma}_1}\left[\frac{1+\frac{k-1}{2}\text{Ma}_1^2}{\frac{k+1}{2}}\right]^{\frac{k+1}{2(k-1)}} \quad\rightarrow\quad \frac{0.07069}{A^*} = \frac{1}{0.4461}\left[\frac{1+\frac{1.40-1}{2}(0.4461)^2}{\frac{1.40+1}{2}}\right]^{\frac{1.40+1}{2(1.40-1)}}$$

$$\rightarrow\quad A^* = 0.04847\ \text{m}^2$$

Applying Equation 12.77 at section 2 with A^* = 0.04847 m^2 gives

$$\frac{A_2}{A^*} = \frac{1}{\text{Ma}_2}\left[\frac{1+\frac{k-1}{2}\text{Ma}_2^2}{\frac{k+1}{2}}\right]^{\frac{k+1}{2(k-1)}} \quad\rightarrow\quad \frac{0.05940}{0.04847} = \frac{1}{\text{Ma}_2}\left[\frac{1+\frac{1.40-1}{2}\text{Ma}_2^2}{\frac{1.40+1}{2}}\right]^{\frac{1.40+1}{2(1.40-1)}}$$

$$\rightarrow\quad \text{Ma}_2 = 0.5707,\ 1.568$$

Here we get two possible Mach numbers that could occur at sections 2 and 3, both having the same cross-sectional area (i.e., $A_2 = A_3$). Only Ma $=$ 0.5707 is possible at section 2, because the flow must remain subsonic in a converging section. However, at section 3, both Ma = 0.5707 and Ma = 1.568 are possible, depending on whether subsonic flow or supersonic flow is occurring in the diverging part of the nozzle. These flow conditions are considered separately below.

(a) Consider the case of subsonic flow in the diverging part of the nozzle. At section 2, Ma_2 = 0.5707, and Equations 12.93 and 12.94 can be used to relate the pressure and temperature at section 2 to the pressure and temperature at section 1 as follows:

$$\frac{p_2}{p_1} = \frac{p_2}{p^*}\frac{p^*}{p_1} = \left[\frac{1+\frac{k-1}{2}\text{Ma}_1^2}{1+\frac{k-1}{2}\text{Ma}_2^2}\right]^{\frac{k}{k-1}} \quad\rightarrow\quad \frac{p_2}{550} = \left[\frac{1+\frac{1.40-1}{2}(0.4461)^2}{1+\frac{1.40-1}{2}(0.5707)^2}\right]^{\frac{1.40}{1.40-1}}$$

$$\rightarrow\quad p_2 = 506\ \text{kPa}$$

$$\frac{T_2}{T_1} = \frac{T_2}{T^*}\frac{T^*}{T_1} = \frac{1+\frac{k-1}{2}\text{Ma}_1^2}{1+\frac{k-1}{2}\text{Ma}_2^2} \quad\rightarrow\quad \frac{T_2}{500} = \frac{1+\frac{1.40-1}{2}(0.4461)^2}{1+\frac{1.40-1}{2}(0.5707)^2}$$

$$\rightarrow\quad T_2 = 488\ \text{K} = 215^\circ\text{C}$$

The density, ρ_2, and velocity, V_2, can be derived from these results as follows:

$$\rho_2 = \frac{p_2}{RT_2} = \frac{506\times 10^3}{(287.1)(488)} = 3.61\ \text{kg/m}^3$$

$$V_2 = \text{Ma}_2\sqrt{RT_2k} = (0.5707)\sqrt{(287.1)(488)(1.40)} = 253\ \text{m/s}$$

Therefore, the pressure, temperature, density, and velocity at section 2 are **506 kPa**, **215°C**, **3.61 kg/m^3**, and **253 m/s**, respectively. When the flow in the diverging part of the nozzle is subsonic, these same flow conditions exist at section 3.

(b) Consider the case of supersonic flow in the diverging part of the nozzle. In this case, the flow properties at section 2 are the same as those calculated (for section 2) in part (a). However, at section 3, Ma_3 = 1.568 as calculated in part (a). Following the same calculation procedure as in part (a),

$$\frac{p_3}{p_1} = \left[\frac{1+\frac{k-1}{2}\text{Ma}_1^2}{1+\frac{k-1}{2}\text{Ma}_3^2}\right]^{\frac{k}{k-1}} \rightarrow \frac{p_3}{550} = \left[\frac{1+\frac{1.40-1}{2}(0.4461)^2}{1+\frac{1.40-1}{2}(1.568)^2}\right]^{\frac{1.40}{1.40-1}} \rightarrow p_3 = 155 \text{ kPa}$$

$$\frac{T_3}{T_1} = \frac{1+\frac{k-1}{2}\text{Ma}_1^2}{1+\frac{k-1}{2}\text{Ma}_3^2} \rightarrow \frac{T_3}{500} = \frac{1+\frac{1.40-1}{2}(0.4461)^2}{1+\frac{1.40-1}{2}(1.568)^2} \rightarrow T_3 = 348 \text{ K} = 75^\circ\text{C}$$

The density, ρ_3, and velocity, V_3, can be derived from these results as follows:

$$\rho_3 = \frac{p_3}{RT_3} = \frac{155 \times 10^3}{(287.1)(348)} = 1.55 \text{ kg/m}^3$$

$$V_3 = \text{Ma}_3\sqrt{RT_3k} = (1.568)\sqrt{(287.1)(348)(1.40)} = 587 \text{ m/s}$$

Therefore, when the flow condition in the downstream part of the CD nozzle is supersonic, the pressure, temperature, density, and velocity at section 3 are **155 kPa, 75°C, 1.55 kg/m³**, and **587 m/s**, respectively.

There are only two possible flow conditions at the exit of the nozzle—subsonic and supersonic. Because critical conditions can only occur at the throat of the nozzle, it is **impossible to have critical conditions at the exit of the nozzle.**

Operating conditions. When sonic flow occurs at the throat section, this corresponds to the choked condition, and hence, the maximum mass flow rate through a CD nozzle can be determined directly from Equation 12.91, which gives

$$\dot{m}_{\text{max}} = \dot{m}_{\text{choked}} = A_t p_0 \left(\frac{2}{k+1}\right)^{\frac{k+1}{2(k-1)}} \sqrt{\frac{k}{RT_0}} \tag{12.96}$$

where A_t is the area of the throat section. Downstream of the throat section, the area of the conduit increases, and it is apparent from Equation 12.77 (and Figure 12.8) that the Mach number in the downstream section can be less than unity (subsonic flow) or greater than unity (supersonic flow). The determining factor is the back pressure, p_b, exerted on the expanding section of the nozzle, such that:

- If the back pressure is less than the exit pressure under supersonic conditions, then supersonic conditions will occur in the expanding portion of the nozzle.
- If the back pressure is greater than the throat (critical) pressure, then subsonic conditions will occur in the expanding portion of the nozzle.
- If the back pressure is between both of the above conditions, then the flow cannot expand isentropically to p_b and a shock (i.e., a sudden transition from supersonic to subsonic flow) occurs somewhere within the expanding portion of the nozzle.

The detailed characteristics of a shock are described in the following section. In the meantime, denoting the area of the exit and throat section by A_e and A^*, respectively, and setting

$A/A^* = A_e/A^*$ in Equation 12.77 gives the two possible Mach numbers at the exit of a CD nozzle when critical conditions occur in the throat. These possible Mach numbers correspond to alternative subsonic (Ma < 1) and supersonic (Ma > 1) conditions. If the flow at the exit section is supersonic and the back pressure is equal to the exit pressure, then the nozzle is said to be operating under *design conditions*. Under the design condition, p_b/p_0 is called the *design pressure ratio*, which corresponds to the most efficient operating condition of the nozzle. If the back pressure is less than the exit pressure, then the flow at the exit is said to be *underexpanded*, because additional expansion will occur once the fluid leaves the nozzle. Using the same rationale, when the back pressure is greater than the exit pressure the flow is said to be *overexpanded* at the exit.

Practical Considerations

It is apparent from the previous discussion that to accelerate a fluid from a subsonic state to a supersonic state requires a converging-diverging conduit, and such a configuration is also used to decelerate a fluid from supersonic to subsonic flow. Based on this performance characteristic, converging-diverging (CD) nozzles are commonly used in supersonic wind tunnels to accelerate the entering flow (from subsonic to supersonic) and to decelerate the exiting flow (from supersonic to subsonic). Because the area ratio between the nozzle exit and the throat determines the exit Mach number, the exit area containing the test section is sometimes made interchangeable so that the Mach number can be changed in the test section. Converging-diverging (CD) nozzles, as well as converging nozzles, are found in many other engineering applications, such as in steam turbines, gas turbines, and aircraft and spacecraft propulsion systems. Care should be taken in applying one-dimensional flow equations to relatively short nozzles, because higher-dimensional flow variations might be important in such circumstances.

EXAMPLE 12.9

Consider a CD nozzle with a throat diameter of 10 mm and an exit diameter of 20 mm as illustrated in Figure 12.17. The upstream end of the nozzle is connected to a plenum chamber containing air with a stagnation temperature and pressure of 127°C and 1.2 MPa, respectively. The CD nozzle is intended to discharge air at supersonic flow into a downstream receiving chamber. (a) Determine the design back pressure and the corresponding exit Mach number and mass flow rate through the nozzle. (b) Determine the back pressure above which the flow will not attain supersonic speed within the nozzle. What is the corresponding exit Mach number and mass flow rate through the nozzle? (c) What will be the exit Mach number and the mass flow rate through the nozzle if the back pressure is less than the design back pressure?

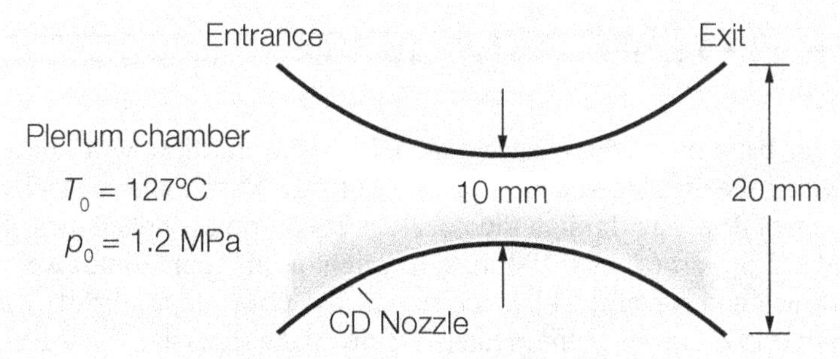

Figure 12.17: CD nozzle connected to a plenum chamber

SOLUTION

From the given data: D_t = 10 mm = 0.010 m, D_e = 20 mm = 0.020 m, T_0 = 127°C = 400 K, and p_0 = 1.2 MPa. For the given throat and exit diameters, the corresponding cross-sectional areas are A_t = 7.854 × 10^{-5} m^2 and A_e = 3.142 × 10^{-4} m^2, where $A = \pi D^2/4$. For air, R = 287.1 J/kg·K, and assume that k = 1.40.

(a) At the design back pressure, the flow is critical in the throat of the nozzle and the flow is supersonic at the exit. Assume isentropic conditions within the nozzle. For the given throat and exit areas, the exit Mach number, $\mathrm{Ma_e}$, must satisfy Equation 12.77, which requires that

$$\frac{A_\mathrm{e}}{A_\mathrm{t}} = \frac{1}{\mathrm{Ma_e}}\left[\frac{1+\dfrac{k-1}{2}\mathrm{Ma_e^2}}{\dfrac{k+1}{2}}\right]^{\frac{k+1}{2(k-1)}} \quad\rightarrow\quad \frac{3.142\times10^{-4}}{7.854\times10^{-5}} = \frac{1}{\mathrm{Ma_e}}\left[\frac{1+\dfrac{1.40-1}{2}\mathrm{Ma_e^2}}{\dfrac{1.40+1}{2}}\right]^{\frac{1.40+1}{2(1.40-1)}}$$

which yields two solutions: $\mathrm{Ma_e}$ = 2.94 and $\mathrm{Ma_e}$ = 0.147. For the design condition, $\mathrm{Ma_e}$ = 2.94. The other solution of $\mathrm{Ma_e}$ = 0.147 corresponds to the case where the flow is sonic in the throat and subsonic downstream of the throat, which is of interest in part (b) of this example. For $\mathrm{Ma_e}$ = 2.94, Equation 12.51 gives

$$\frac{p_0}{p_\mathrm{e}} = \left[1+\frac{k-1}{2}\mathrm{Ma_e^2}\right]^{\frac{k}{k-1}} \quad\rightarrow\quad \frac{1200}{p_\mathrm{e}} = \left[1+\frac{1.40-1}{2}(2.94)^2\right]^{\frac{1.40}{1.40-1}} \quad\rightarrow\quad p_\mathrm{e} = 35.7\ \mathrm{kPa}$$

Therefore, the design back pressure is **35.7 kPa**, and this will result in an exit Mach number of **2.94**. The corresponding mass flow rate under this (choked) condition is given by Equation 12.96 as

$$\begin{aligned}\dot{m} &= A_\mathrm{t}p_0\left(\frac{2}{k+1}\right)^{\frac{k+1}{2(k-1)}}\sqrt{\frac{k}{RT_0}}\\ &= (7.854\times10^{-5})(1.2\times10^6)\left(\frac{2}{1.40+1}\right)^{\frac{1.40+1}{2(1.40-1)}}\sqrt{\frac{1.40}{(287.1)(400)}}\\ &= 0.190\ \mathrm{kg/s}\end{aligned}$$

Therefore, the mass flow rate of air through the nozzle under design conditions is equal to **0.190 kg/s**. This is the maximum mass flow rate attainable for the given stagnation conditions and throat area.

(b) As calculated in part (a), for the given throat area and exit area, an exit Mach number of 0.147 will occur when the flow is sonic in the throat and subsonic in both the converging and diverging parts of the nozzle. For $\mathrm{Ma_e}$ = 0.147, Equation 12.51 gives

$$\frac{p_0}{p_\mathrm{e}} = \left[1+\frac{k-1}{2}\mathrm{Ma_e^2}\right]^{\frac{k}{k-1}} \quad\rightarrow\quad \frac{1.2}{p_\mathrm{e}} = \left[1+\frac{1.40-1}{2}(0.147)^2\right]^{\frac{1.40}{1.40-1}} \quad\rightarrow\quad p_\mathrm{e} = 1.18\ \mathrm{MPa}$$

Hence, for back pressures greater than **1.18 MPa**, the flow will not attain supersonic speed within the nozzle. At a back pressure of 1.18 MPa, the exit Mach number is **0.147** and the mass flow rate through the nozzle is equal to the choked mass flow rate, which is calculated in part (a) as **0.190 kg/s**. It might seem counterintuitive that airflow with an exit Mach number of 2.94 has the same mass flow rate as airflow with an exit Mach number of 0.147. However, the greater velocity of the supersonic flow condition is equally compensated for by the greater density in the subsonic flow.

(c) If the actual back pressure is less than the design back pressure, then the exit Mach number and mass flow rate will be the same as that for a back pressure equal to the design back pressure. These values were calculated previously in part (a) as **2.94** and **0.190 kg/s**, respectively.

12.7 Normal Shocks

A *normal shock* occurs at locations where the flow is abruptly decelerated from supersonic to subsonic flow, accompanied by an abrupt increase in temperature, pressure, density, and entropy. The word "normal" is used to connote the fact that the wave front is oriented in a direction normal (i.e., perpendicular) to the flow direction. This nomenclature is in contrast to an *oblique shock*, which is oriented at an angle to the flow direction. With the exception of near-vacuum conditions, the thickness of a normal shock is typically on the order of 0.1 μm, which is only about 4 times the mean free path of a gas molecule. When shocks occur, the flow both upstream and downstream of the shock are typically isentropic; however, the flow is generally non-isentropic across the shock. Fluid-particle decelerations across normal shocks are extremely high, being on the order of several million g's. In engineering applications, normal shocks are commonly encountered in the design of inlets on high-performance aircraft and in the design of supersonic wind tunnels. In broader applications, shocks can occur in any supersonic flow. The high temperatures that usually occur downstream of shocks are of concern to engineers because of the potential for significant heat transfer that must be taken into account, for example, on the leading edges of wings and the nose cones of space reentry vehicles.

Phenomenological view. The occurrence of a shock might be an expected phenomenon because, by definition, disturbances generated downstream cannot travel upstream in a supersonic flow. Consequently, upstream fluid particles traveling at supersonic speed cannot adjust to (i.e., are unaware of) imposed downstream subsonic flow conditions. So the supersonic particles will come "crashing into" the downstream particles and make a violent adjustment to the downstream flow conditions. This violent adjustment is a shock, in which the excess kinetic energy of the supersonic flow (relative to the kinetic energy of the downstream subsonic flow) is converted into thermal energy, which results in an abrupt increase in temperature and pressure across the shock.

Governing equations. Consider the general case of a normal shock occurring in the flow of an ideal gas with constant specific heats as illustrated in Figure 12.18. The control volume enclosing the shock can be taken as having an infinitesimal width, because the width of the shock is very small, and as a consequence, the inflow and outflow areas can be taken to be

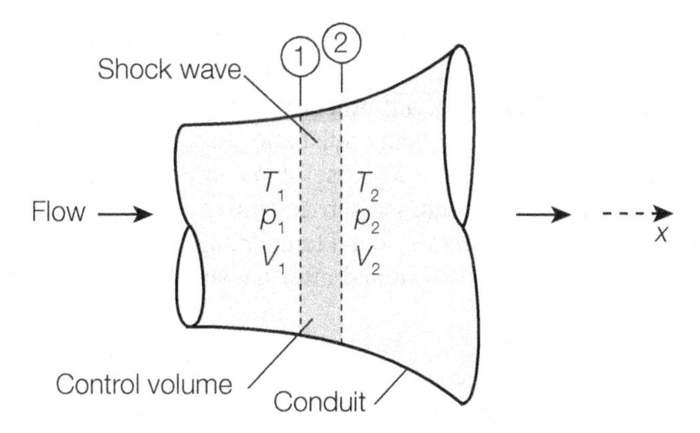

Figure 12.18: Normal shock

approximately the same such that $A_1 = A_2 = A$. Assuming that the friction and gravity forces within the control volume are negligibly small, the applicable steady-state control-volume equations are given by

$$\text{continuity: } \rho_1 V_1 = \rho_2 V_2 = \frac{\dot{m}}{A} \tag{12.97}$$

$$\text{momentum: } p_1 A - p_2 A = \dot{m}(V_2 - V_1) \tag{12.98}$$

$$\text{energy: } h_1 + \frac{1}{2}V_1^2 = h_2 + \frac{1}{2}V_2^2 \tag{12.99}$$

$$\text{second law: } \dot{m}(s_2 - s_1) \geq \int_{\text{cs}} \frac{1}{T}\left(\frac{\dot{Q}}{A}\right) \text{d}A \quad \rightarrow \quad \dot{m}(s_2 - s_1) > 0 \tag{12.100}$$

$$\text{State 1: } \rho = \frac{p}{RT} \tag{12.101}$$

$$\text{State 2: } h_2 - h_1 = c_p(T_2 - T_1) \tag{12.102}$$

$$\text{State 3: } s_2 - s_1 = c_p \ln\frac{T_2}{T_1} - R\ln\frac{p_2}{p_1} \tag{12.103}$$

where the subscripts 1 and 2 refer to values of the subscripted variables on the upstream and downstream faces of the control volume, respectively. Application of the second law of thermodynamics, Equation 12.100, assumes that the flow is adiabatic within the shock. If the flow conditions upstream of the shock are known, Equations 12.97–12.103 constitute six coupled nonlinear equations in the following six unknowns: p_2, ρ_2, T_2, s_2, h_2, and V_2.

Thermodynamic considerations. A central feature that is key to the analysis of a normal shock is that the flow transition within the control volume is assumed to be adiabatic, because it occurs very rapidly, and this assumption is reflected in the statement of the law of conservation of energy given by Equation 12.99. Although the process within the control volume is assumed to be adiabatic, it is highly irreversible and therefore results in an increase in entropy. The energy equation given by Equation 12.99 could equivalently be stated as

$$h + \frac{1}{2}V^2 = h_0 = \text{constant} \quad \leftrightarrow \quad T + \frac{V^2}{2\,c_p} = T_0 = \text{constant} \tag{12.104}$$

Because the flow across a shock is assumed to be adiabatic, Equation 12.104 requires that the stagnation temperature, T_0, be the same before and after the shock; thus,

$$T_{01} = T_{02} = T_0 \tag{12.105}$$

where T_{01} and T_{02} are the stagnation temperatures before and after the shock, respectively. Keep in mind that Equation 12.105 does not mean that the temperatures before and after the shock are the same. Another important property of Equation 12.104 is that this derived relationship only depends on the flow being adiabatic, so it is valid even if the flow is not isentropic. This result of T_0 = constant is also useful in analyzing one-dimensional adiabatic flows in ducts, a topic that will be discussed subsequently. For the present case of a normal shock, using the relationship between the actual temperature and the stagnation temperature (Equation 12.53), the temperatures before and after the shock, T_1 and T_2, respectively, are related by

$$\frac{T_2}{T_1} = \frac{T_2}{T_{02}}\frac{\cancel{T_{02}}}{\cancel{T_{01}}}\frac{T_{01}}{T_1} = \frac{T_2}{T_0}\frac{T_0}{T_1} = \frac{1 + \dfrac{k-1}{2}\text{Ma}_1^2}{1 + \dfrac{k-1}{2}\text{Ma}_2^2} \tag{12.106}$$

where Ma_1 and Ma_2 are the Mach numbers before and after the shock, respectively. It will be shown subsequently that $\mathrm{Ma}_1 > \mathrm{Ma}_2$, which means that the temperature after the shock is generally higher than before the shock (i.e., $T_2 > T_1$).

Solutions of the governing equations. The continuity, momentum, and state equations (Equations 12.97, 12.98, and 12.101) can be combined with Equation 12.106 to yield the following useful relationships between post-shock and pre-shock conditions, where the independent variable is the upstream Mach number, Ma_1, and specific heat ratio, k:

$$\mathrm{Ma}_2^2 = \frac{\mathrm{Ma}_1^2 + \dfrac{2}{k-1}}{\dfrac{2k}{k-1}\mathrm{Ma}_1^2 - 1} \tag{12.107}$$

$$\frac{p_2}{p_1} = \frac{2k}{k+1}\mathrm{Ma}_1^2 - \frac{k-1}{k+1} \tag{12.108}$$

$$\frac{T_2}{T_1} = \frac{\left(1 + \dfrac{k-1}{2}\mathrm{Ma}_1^2\right)\left(k\,\mathrm{Ma}_1^2 - \dfrac{k-1}{2}\right)}{\left(\dfrac{k+1}{2}\right)^2 \mathrm{Ma}_1^2} \tag{12.109}$$

$$\frac{\rho_2}{\rho_1} = \frac{V_1}{V_2} = \frac{\dfrac{k+1}{2}\mathrm{Ma}_1^2}{1 + \dfrac{k-1}{2}\mathrm{Ma}_1^2} \tag{12.110}$$

The post-shock stagnation temperature, T_{02}, is equal to the pre-shock stagnation temperature, T_{01}, in accordance with the first law of thermodynamics and the assumption of adiabatic flow. Also, the relationship between the post-shock and pre-shock stagnation pressures, p_{02} and p_{01}, respectively, can be derived using the continuity, momentum, and state equations. Thus, the relationships between the post-shock and pre-shock stagnation conditions are given by

$$\frac{p_{02}}{p_{01}} = \frac{\left[\dfrac{\dfrac{k+1}{2}\mathrm{Ma}_1^2}{1 + \dfrac{k-1}{2}\mathrm{Ma}_1^2}\right]^{\frac{k}{k-1}}}{\left[\dfrac{2k}{k+1}\mathrm{Ma}_1^2 - \dfrac{k-1}{k+1}\right]^{\frac{1}{k-1}}} \tag{12.111}$$

$$\frac{T_{02}}{T_{01}} = 1 \tag{12.112}$$

Equations 12.107–12.110 are particularly useful in applications because they give uncoupled expressions relating the post-shock and pre-shock conditions, in comparison to the raw continuity, momentum, and state equations that are coupled. The relationships between Ma_2, p_2/p_1, T_2/T_1, and ρ_2/ρ_1 and the pre-shock Mach number, Ma_1, are illustrated in Figure 12.19 for air ($k = 1.40$). It is apparent from Figure 12.19(a) that the drop in Mach number across a shock can be substantial, with an order of magnitude difference being possible. Figure 12.19(b) shows that the ratio of the post-shock pressure (p_2) to the pre-shock pressure (p_1) is much greater than the corresponding property ratios for the temperature and density. Hence,

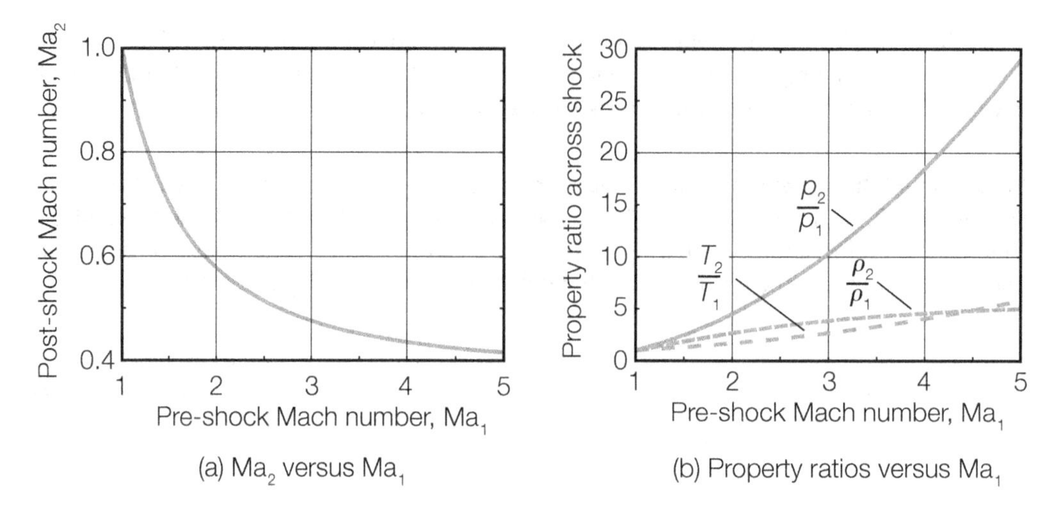

Figure 12.19: **Property ratios across a normal shock in air**

proportionally greater pressure increases (compared with temperature and density increases) are expected across shocks. The ratio p_2/p_1 is commonly referred to as the *strength of the shock*, which increases as the Ma_1 increases. The entropy change across a shock can be calculated using Equation 12.23, which states that

$$s_2 - s_1 = c_p \ln \frac{T_2}{T_1} - R \ln \frac{p_2}{p_1} \tag{12.113}$$

where s_1 and s_2 are the entropy before and after the shock, respectively. Combining Equations 12.108, 12.109, and 12.113 gives

$$s_2 - s_1 = c_p \ln \frac{2 + (k-1)\mathrm{Ma}_1^2}{2 + (k-1)\mathrm{Ma}_2^2} - R \ln \frac{1 + k\mathrm{Ma}_1^2}{1 + k\mathrm{Ma}_2^2} \tag{12.114}$$

Using the relationship between c_p and R and the isentropic stagnation pressure, the entropy change given by Equation 12.113 can also be expressed in the compact form

$$s_2 - s_1 = R \ln \left[\frac{p_1}{p_2} \left(\frac{T_2}{T_1} \right)^{\frac{k}{k-1}} \right] = R \ln \frac{p_{01}}{p_{02}} \tag{12.115}$$

Using Equation 12.107 to relate Ma_1 and Ma_2, the relationship between $s_2 - s_1$ and Ma_1 in air (c_p = 1003 J/kg·K and R = 287.1 J/kg·K) is shown in Figure 12.20. It is apparent from the figure that $s_2 - s_1 > 0$ only when $\mathrm{Ma}_1 > 1$, and because the second law of thermodynamics requires that the entropy must increase across the shock, normal shocks only occur when the flow transitions from supersonic to subsonic flow; the reverse transition via a normal shock is not possible. Higher values of Ma_1 correspond to higher values of p_2/p_1, so higher upstream Mach numbers generate stronger shocks. Typically, there are large pressure increases across

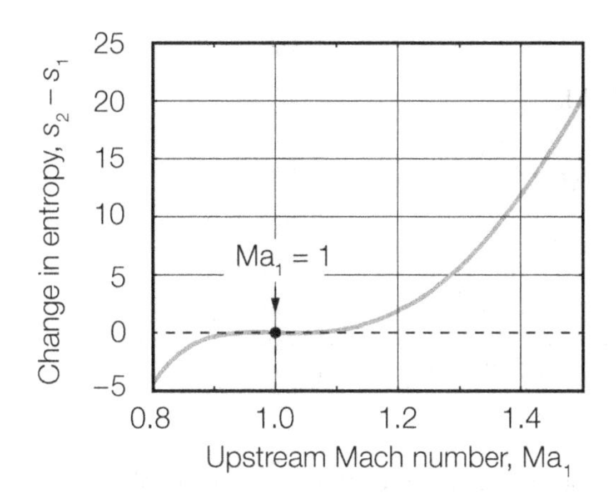

Figure 12.20: Change in entropy across a normal shock in air versus the upstream Mach number

normal shocks accompanied by moderate increases in temperature and density. Although there is a large pressure increase across a normal shock, the stagnation pressure decreases across a normal shock, such that $p_{02}/p_{01} < 1$. Also, $p_{02}/p_{01} \rightarrow 1$ as $\text{Ma}_1 \rightarrow 1$, which agrees with the assumption that sound waves (which are defined as infinitesimal pressure perturbations) are isentropic. The possibility of occurrence of a normal shock exists in any supersonic flow, and a shock must usually occur for the flow to transition from given supersonic conditions to fixed subsonic conditions.

EXAMPLE 12.10

Air flowing at a Mach number of 2.1 encounters a stationary probe within the flow field such that a normal shock is generated on the stagnation streamline. If the static temperature and static pressure upstream of the normal shock are 5°C and 90 kPa, respectively, determine the Mach number, static pressure, and static temperature immediately downstream of the shock. How do the temperature and pressure downstream of the shock compare with the temperature and pressure at the stagnation point on the probe? What is the change in entropy caused by the occurrence of the normal shock?

SOLUTION

From the given data: $\text{Ma}_1 = 2.1$, $T_1 = 5°\text{C} = 278$ K, and $p_1 = 90$ kPa. For air, $R = 287.1$ J/kg·K and $k = 1.40$. The Mach number, Ma_2; pressure, p_2; and temperature, T_2; downstream of the normal shock are given by Equations 12.107–12.109 as follows:

$$\text{Ma}_2^2 = \frac{\text{Ma}_1^2 + \dfrac{2}{k-1}}{\dfrac{2k}{k-1}\text{Ma}_1^2 - 1} = \frac{(2.1)^2 + \dfrac{2}{1.40-1}}{\dfrac{2(1.40)}{1.40-1}(2.1)^2 - 1} \quad \rightarrow \quad \text{Ma}_2 = 0.5613$$

$$\frac{p_2}{p_1} = \frac{2k}{k+1}\text{Ma}_1^2 - \frac{k-1}{k+1} \quad \rightarrow \quad \frac{p_2}{90} = \frac{2(1.40)}{1.40+1}(2.1)^2 - \frac{1.40-1}{1.40+1} \quad \rightarrow \quad p_2 = 448 \text{ kPa}$$

$$\frac{T_2}{T_1} = \frac{\left[1 + \frac{k-1}{2}\text{Ma}_1^2\right]\left[k\,\text{Ma}_1^2 - \frac{k-1}{2}\right]}{\left(\frac{k+1}{2}\right)^2 \text{Ma}_1^2}$$

$$\rightarrow \quad \frac{T_2}{278} = \frac{\left[1 + \frac{1.40-1}{2}(2.1)^2\right]\left[1.40(2.1)^2 - \frac{1.40-1}{2}\right]}{\left(\frac{1.40+1}{2}\right)^2 (2.1)^2} \quad \rightarrow \quad T_2 = 492\ \text{K}$$

Therefore, the Mach number, pressure, and temperature immediately downstream of the shock are **0.561**, **448 kPa**, and **492 K** (= 219°C), respectively. The stagnation pressure and temperature before the shock, p_{01} and T_{01}, respectively, and the stagnation pressure and temperature after the shock, p_{02} and T_{02}, respectively, can be determined using Equations 12.51, 12.53, 12.111, and 12.112, which give

$$\frac{p_{01}}{p_1} = \left[1 + \frac{k-1}{2}\text{Ma}_1^2\right]^{\frac{k}{k-1}} \quad \rightarrow \quad \frac{p_{01}}{90} = \left[1 + \frac{1.40-1}{2}(2.1)^2\right]^{\frac{1.40}{1.40-1}} \quad \rightarrow \quad p_{01} = 823.0\ \text{kPa}$$

$$\frac{T_{01}}{T_1} = 1 + \frac{k-1}{2}\text{Ma}_1^2 \quad \rightarrow \quad \frac{T_{01}}{278} = 1 + \frac{1.40-1}{2}(2.1)^2 \quad \rightarrow \quad T_{01} = 523\ \text{K}$$

$$\frac{p_{02}}{p_{01}} = \frac{\left[\dfrac{\frac{k+1}{2}\text{Ma}_1^2}{1 + \frac{k-1}{2}\text{Ma}_1^2}\right]^{\frac{k}{k-1}}}{\left[\frac{2k}{k+1}\text{Ma}_1^2 - \frac{k-1}{k+1}\right]^{\frac{1}{k-1}}} \quad \rightarrow \quad \frac{p_{02}}{823.0} = \frac{\left[\dfrac{\frac{1.40+1}{2}(2.1)^2}{1 + \frac{1.40-1}{2}(2.1)^2}\right]^{\frac{1.40}{1.40-1}}}{\left[\frac{2(1.40)}{1.40+1}(2.1)^2 - \frac{1.40-1}{1.40+1}\right]^{\frac{1}{1.40-1}}}$$

$$\rightarrow \quad p_{02} = 554.9\ \text{kPa}$$

$$\frac{T_{02}}{T_{01}} = 1 \quad \rightarrow \quad T_{02} = T_{01} \quad \rightarrow \quad T_{02} = 523\ \text{K}$$

Therefore, the temperature and pressure at the stagnation point on the (stationary) probe are **523 K** (= 250°C) and **555 kPa**, respectively, which are both higher than the temperature and pressure just downstream of the normal shock (492 K and 448 kPa, respectively).

The change in entropy on the stagnation streamline across the normal shock can be calculated using Equation 12.115, which gives

$$s_2 - s_1 = R \ln \frac{p_{01}}{p_{02}} = (287.1) \ln \frac{823.0}{554.9} = 113\ \text{J/kg·K}$$

Because the flow downstream of the shock (between the shock and the probe) is assumed to be isentropic, the change in entropy on the stagnation streamline between a point upstream of the shock and the stagnation point is **113 J/kg·K**.

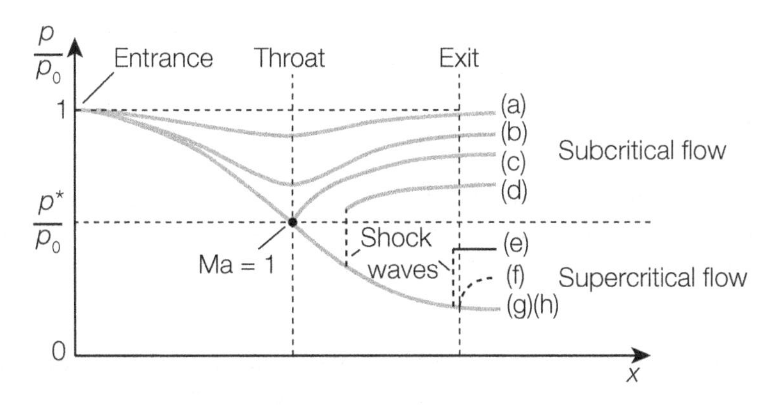

Figure 12.21: Flow conditions in a converging-diverging nozzle

Normal shock in a converging-diverging nozzle. Consider the case of flow through a converging-diverging (CD) nozzle, where the inlet velocity is sufficiently low that the pressure at the inlet is approximately equal to the stagnation pressure, p_0. Beyond the nozzle exit, the back pressure is equal to p_b, and the pressure at the exit plane of the nozzle in equal to p_e. When $p_\mathrm{b} = p_0$, there is no flow through the nozzle, and where $p_\mathrm{b} < p_0$, several flow conditions are possible, as illustrated in Figure 12.21. The flow conditions through the nozzle depend on the relative values of p_0 and p_b, as well as two reference pressures: p'_e and p''_e. The reference pressure p'_e is the exit pressure that will occur when the flow in the throat is sonic (i.e., critical), the flow in the expanding nozzle is isentropic, and the exit condition is subsonic (Ma < 1); p''_e is the exit pressure that will occur when the flow in the throat is sonic (i.e., critical), the flow in the expanding nozzle is isentropic, and the exit condition is supersonic (Ma > 1). Both p'_e and p''_e correspond to the Mach numbers given by Equation 12.77 when A is taken as the exit area, A_e, of the nozzle. Possible flow conditions through the nozzle, labeled (a)–(h), are described below.

- Condition (a): The flow is subcritical, and Ma < 0.3 throughout the nozzle. The flow velocities within the nozzle are determined from p_0 and p_b in accordance with the incompressible-flow Bernoulli equation.

- Condition (b): The flow is subcritical throughout the nozzle, and $0.3 <$ Ma < 1 in the high-velocity portions of the nozzle. The flow velocities within the nozzle are determined from p_0 and p_b in accordance with the compressible-flow Bernoulli equation. The flow can be assumed to be isentropic throughout the nozzle. The flow is subcritical in the throat of the nozzle.

- Condition (c): The flow is subcritical up to the throat of the nozzle, where the flow is critical, Ma $= 1$, and the flow is choked. The pressure at the exit is such that subcritical flow is attained downstream of the throat and $p_\mathrm{e} = p_\mathrm{b} = p'_\mathrm{e}$.

- Conditions (d) and (e): The flow is subcritical up to the throat of the nozzle, where the flow is critical, Ma $= 1$, and the flow is choked. The back pressure is lower than the subcritical exit pressure, p'_e, and higher than the supercritical exit pressure, p''_e; this imbalance causes a normal shock (and sudden pressure rise) to occur somewhere within the expanding portion of the nozzle. The flow becomes subsonic after the shock, and the shock is located within the nozzle such that $p_\mathrm{e} = p_\mathrm{b}$.

- Condition (f): The flow is subcritical up to the throat of the nozzle, where the flow is critical, Ma = 1, and the flow is choked. The back pressure is such that the normal shock would occur near the exit of the nozzle. The flow adjusts to the back pressure through a series of oblique compression shocks outside the nozzle.

- Condition (g): The flow is subcritical up to the throat of the nozzle, where the flow is critical, Ma = 1, and the flow is choked. The back pressure, p_b, is exactly equal to the exit pressure, $p_\mathrm{e} = p_\mathrm{e}''$, that would occur for supercritical flow downstream of the throat.

- Condition (h): The flow is subcritical up to the throat of the nozzle, where the flow is critical, Ma = 1, and the flow is choked. The back pressure, p_b, is less than the exit pressure, p_e'', that would occur for supercritical flow downstream of the throat. The adjustment of pressure occurs downstream of the nozzle exit.

In a typical engineering application, the back pressure is specified and the location and strength of the shock are to be determined. When a shock occurs within the nozzle, the flow at the exit of the nozzle is subsonic, which requires that $p_\mathrm{e} = p_\mathrm{b}$; also, noting that $p_{01}A_1^* = p_{02}A_2^*$ gives

$$\frac{p_\mathrm{b}}{p_{01}} = \frac{p_\mathrm{e}}{p_{01}} = \frac{p_\mathrm{e}}{p_{02}}\frac{p_{02}}{p_{01}} = \frac{p_\mathrm{e}}{p_{02}}\frac{A_1^*}{A_2^*} = \frac{p_\mathrm{e}}{p_{02}}\frac{A_\mathrm{t}}{A_\mathrm{e}}\frac{A_\mathrm{e}}{A_2^*} \tag{12.116}$$

where p_{01} and p_{02} are the stagnation pressures before and after the shock, respectively, A_1^* and A_2^* are the critical areas before and after the shock, respectively, and A_t and A_e are the areas of the throat and exit, respectively. Because the flow is isentropic from after the shock to the exit of the nozzle, $A_2^* = A_\mathrm{e}^*$ and $p_{02} = p_{0\mathrm{e}}$, and Equation 12.116 can be expressed in the following useful form:

$$\frac{p_\mathrm{e}}{p_{01}}\frac{A_\mathrm{e}}{A_\mathrm{t}} = \frac{p_\mathrm{e}}{p_{0\mathrm{e}}}\frac{A_\mathrm{e}}{A_\mathrm{e}^*} \tag{12.117}$$

This equation finds its usefulness from the fact that the terms on the left-hand side of the equation are known, and the right-hand side of the equation is a function of the exit Mach number, $\mathrm{Ma_e}$, only. The pressure ratio can be obtained from Equation 12.51, and the area ratio can be obtained from Equation 12.77. Once $\mathrm{Ma_e}$ is determined from Equation 12.117, the magnitude and location of the normal shock can be determined from a rearranged Equation 12.117 and from the fact that $p_{02} = p_{0\mathrm{e}}$, which yields

$$\frac{p_{02}}{p_{01}} = \frac{A_\mathrm{t}}{A_\mathrm{e}}\frac{A_\mathrm{e}}{A_\mathrm{e}^*} \tag{12.118}$$

In this equation, the terms on the right-hand side are known, and the left-hand side is a function only of the Mach number before the shock, $\mathrm{Ma_1}$; hence, $\mathrm{Ma_1}$ can be found. The area at which the shock occurs is then found from Equation 12.77 (with $A^* = A_\mathrm{t}$) applied between the throat and the section of the nozzle just upstream of the shock.

Practical consideration. It is important to remember that when a normal shock occurs in the diverging portion of a CD nozzle, the flow downstream of the shock experiences an adverse pressure gradient. This can lead to wall boundary-layer separation that can cause weak two-dimensional compression shocks to occur rather than a single normal shock. In this regard, the nozzle behavior outlined in this section is idealized.

EXAMPLE 12.11

The converging-diverging nozzle shown in Figure 12.22 has a circular cross section with diameter, D, in meters, given by the equation

$$D(x) = 0.4\sqrt{1 + 3x^2}$$

where x is the longitudinal distance measured from the center of the nozzle throat in meters. The nozzle entrance is located at $x = -1.0$ m, and the nozzle exit is located at $x = +1.0$ m. Under a particular operating condition, air enters the nozzle at a subsonic state, the flow is choked at the center of the throat, and a normal shock occurs at $x = 0.75$ m. Determine the ratio of the exit pressure to the entrance pressure that would cause this flow condition to exist. Which of the a–h flow conditions described previously is occurring?

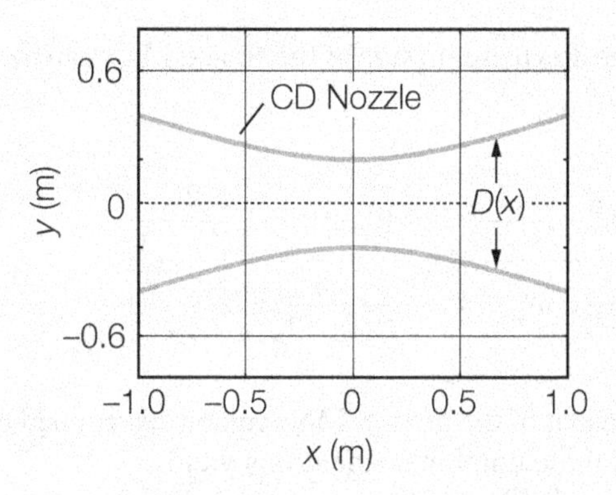

Figure 12.22: Flow in a CD nozzle

SOLUTION

Assume isentropic flow before and after the shock, with non-isentropic flow occurring across the shock. Assume that air flowing through the nozzle behaves like an ideal gas with $k = 1.40$. For purposes of calculation, the entrance of the nozzle will be taken as section 1, the center of the throat as section t, the area immediately upstream of the shock as section x, the area immediately downstream of the shock as section y, and the exit of the nozzle as section 2. Section numbers will be used as subscripts to denote the values of variables at the particular sections, and an asterisk is used to indicate critical conditions. The diameters, D, and flow areas, A, at key locations within the nozzle are as follows:

entrance and exit: $x = \pm 1$ m $\quad \rightarrow \quad D_1 = D_2 = 0.4\sqrt{1 + 3(1)^2} = 0.8$ m

$$\rightarrow \quad A_1 = A_2 = \frac{\pi}{4}D_1^2 = 0.5027 \text{ m}^2$$

throat: $x = 0$ m $\quad \rightarrow \quad D_\text{t} = 0.4\sqrt{1 + 3(0)^2} = 0.4$ m

$$\rightarrow \quad A_\text{t} = A^* = \frac{\pi}{4}D_\text{t}^2 = 0.1257 \text{ m}^2$$

shock: $x = 0.75$ m $\quad \rightarrow \quad D_x = D_y = 0.4\sqrt{1 + 3(0.75)^2} = 0.6557$ m

$$\rightarrow \quad A_x = A_y = \frac{\pi}{4}D_x^2 = 0.3377 \text{ m}^2$$

Because the flow between the nozzle entrance and the area immediately before the shock is isentropic, the Mach number at the entrance, Ma_1, and the Mach number at the area immediately before the shock, Ma_x, can be determined using Equation 12.77 with $A = A_1$ at the entrance, $A = A_x$ at the shock, and $A^* = A_t$ at the throat, which yield

$$\frac{A_1}{A_t} = \frac{1}{\mathrm{Ma}_1}\left[\frac{1+\frac{k-1}{2}\mathrm{Ma}_1^2}{\frac{k+1}{2}}\right]^{\frac{k+1}{2(k-1)}} \rightarrow \frac{0.5027}{0.1257} = \frac{1}{\mathrm{Ma}_1}\left[\frac{1+\frac{1.40-1}{2}\mathrm{Ma}_1^2}{\frac{1.40+1}{2}}\right]^{\frac{1.40+1}{2(1.40-1)}} \rightarrow \mathrm{Ma}_1 = 0.1465$$

$$\frac{A_x}{A_t} = \frac{1}{\mathrm{Ma}_x}\left[\frac{1+\frac{k-1}{2}\mathrm{Ma}_x^2}{\frac{k+1}{2}}\right]^{\frac{k+1}{2(k-1)}} \rightarrow \frac{0.3377}{0.1257} = \frac{1}{\mathrm{Ma}_x}\left[\frac{1+\frac{1.40-1}{2}\mathrm{Ma}_x^2}{\frac{1.40+1}{2}}\right]^{\frac{1.40+1}{2(1.40-1)}} \rightarrow \mathrm{Ma}_x = 2.52$$

The Mach number on the downstream side of the shock, Ma_y, can be determined using Equation 12.107, which yields

$$\mathrm{Ma}_y^2 = \frac{\mathrm{Ma}_x^2 + \frac{2}{k-1}}{\frac{2k}{k-1}\mathrm{Ma}_x^2 - 1} = \frac{(2.52)^2 + \frac{2}{1.40-1}}{\frac{2(1.40)}{1.40-1}(2.52)^2 - 1} \rightarrow \mathrm{Ma}_y = 0.5111$$

The Mach number at the exit of the nozzle, Ma_2, can be determined by applying Equation 12.77 at sections y and 2, and these combined equations yield

$$\frac{A_y}{A_2} = \frac{\frac{1}{\mathrm{Ma}_y}\left[\frac{1+\frac{k-1}{2}\mathrm{Ma}_y^2}{\frac{k+1}{2}}\right]^{\frac{k+1}{2(k-1)}}}{\frac{1}{\mathrm{Ma}_2}\left[\frac{1+\frac{k-1}{2}\mathrm{Ma}_2^2}{\frac{k+1}{2}}\right]^{\frac{k+1}{2(k-1)}}} \rightarrow \frac{0.3377}{0.5027} = \frac{\frac{1}{0.5111}\left[\frac{1+\frac{1.40-1}{2}(0.5111)^2}{\frac{1.40+1}{2}}\right]^{\frac{1.40+1}{2(1.40-1)}}}{\frac{1}{\mathrm{Ma}_2}\left[\frac{1+\frac{1.40-1}{2}\mathrm{Ma}_2^2}{\frac{1.40+1}{2}}\right]^{\frac{1.40+1}{2(1.40-1)}}}$$

$$\rightarrow \quad \mathrm{Ma}_2 = 0.3123$$

The ratio of the stagnation pressure downstream of the shock, p_{0y}, to the stagnation pressure upstream of the shock, p_{0x}, can be determined using Equation 12.111, which yields

$$\frac{p_{0y}}{p_{0x}} = \frac{\left[\frac{\frac{k+1}{2}\mathrm{Ma}_x^2}{1+\frac{k-1}{2}\mathrm{Ma}_x^2}\right]^{\frac{k}{k-1}}}{\left[\frac{2k}{k+1}\mathrm{Ma}_x^2 - \frac{k-1}{k+1}\right]^{\frac{1}{k-1}}} = \frac{\left[\frac{\frac{1.40+1}{2}(2.52)^2}{1+\frac{1.40-1}{2}(2.52)^2}\right]^{\frac{1.40}{1.40-1}}}{\left[\frac{2(1.40)}{1.40+1}(2.52)^2 - \frac{1.40-1}{1.40+1}\right]^{\frac{1}{1.40-1}}} \rightarrow \frac{p_{0y}}{p_{0x}} = 0.4909$$

The ratio of the stagnation pressure to the local pressure at the entrance and exit of the nozzle can be determined using Equation 12.51, which yields

$$\frac{p_{0x}}{p_1} = \left[1 + \frac{k-1}{2}\mathrm{Ma}_1^2\right]^{\frac{k}{k-1}} = \left[1 + \frac{1.40-1}{2}(0.1465)^2\right]^{\frac{1.40}{1.40-1}} \quad \rightarrow \quad \frac{p_{0x}}{p_1} = 1.015$$

$$\frac{p_{0y}}{p_2} = \left[1 + \frac{k-1}{2}\mathrm{Ma}_2^2\right]^{\frac{k}{k-1}} = \left[1 + \frac{1.40-1}{2}(0.3123)^2\right]^{\frac{1.40}{1.40-1}} \quad \rightarrow \quad \frac{p_{0y}}{p_2} = 1.070$$

Finally, these results can be combined to give the ratio of the entrance pressure, p_1, to the exit pressure, p_2 as follows:

$$\frac{p_2}{p_1} = \frac{p_2}{p_{0y}} \cdot \frac{p_{0y}}{p_{0x}} \cdot \frac{p_{0x}}{p_1} = \left(\frac{1}{1.070}\right) \cdot 0.4909 \cdot 1.015 = 0.4657$$

Therefore, the exit pressure is lower than the inlet pressure by a factor of about **0.47**.

The flow condition within the nozzle corresponds to **Conditions d and e**, where the back pressure is lower than the subcritical exit pressure and higher than the supercritical exit pressure.

Propagating normal shocks. The relationships presented in the previous sections can be used to analyze the occurrence of normal shocks caused by supersonic flow encountering a stationary object, an object moving at supersonic speed through a stationary environment, and supersonic flow encountering a subsonic slower-moving choked flow. In addition to these scenarios, normal shocks are also generated by events such as explosions and bomb blasts, where the shock wave propagates at supersonic speed into a stagnant environment. Such a circumstance is illustrated in Figure 12.23(a), where a shock propagates at a speed V_s. In this case, the temperature and pressure in the stagnant fluid ($V = 0$) ahead of the shock wave are T_0 and p_0, respectively, and the temperature, pressure, and velocity behind the wave are T_1, p_1, and V_1, respectively. To utilize the previously developed relationships between pre-shock and post-shock conditions, it is convenient to view the shock from an (inertial) reference frame that is moving at the same speed as the shock, in which case the moving shock in Figure 12.23(a) is equivalent to the stationary shock shown in Figure 12.23(b). Applying the

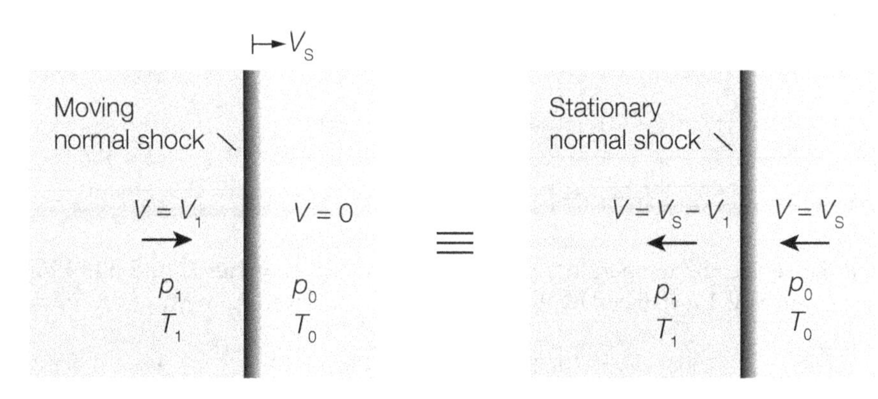

Figure 12.23: Propagation of a normal shock wave

relationships between pre-shock and post-shock conditions as given by Equations 12.107–12.112 to the equivalent stationary shock yields the values of V_1, p_1, and T_1 in terms of the speed of the shock, V_s, and the stagnation conditions, p_0 and T_0. It is interesting to note that the velocity V_1 behind shocks generated by large bomb blasts and explosions can be quite high, and these velocities can create significant damage in addition to the intense damaging effects at the location of the explosive event.

EXAMPLE 12.12

A normal shock wave propagates through stagnant air at a speed of 600 m/s. If the temperature and pressure of the stagnant air in front of the shock wave are 20°C and 101 kPa, respectively, calculate the temperature, pressure, and velocity that exists immediately behind the shock wave.

SOLUTION

From the given data: V_s = 600 m/s, T_0 = 20°C = 293 K, and p_0 = 101 kPa. For air, R = 287.1 J/kg·K, and k = 1.40. Consider the normal shock relative to a moving reference frame as shown in Figure 12.23(b). The upstream Mach number, Ma_0, is given by

$$\text{Ma}_0 = \frac{V_s}{\sqrt{RT_0k}} = \frac{600}{\sqrt{(287.1)(293)(1.40)}} = 1.748$$

Using Equations 12.107–12.109, the Mach number, pressure, and temperature downstream of the normal shock are calculated as follows:

$$\text{Ma}_1^2 = \frac{\text{Ma}_0^2 + \dfrac{2}{k-1}}{\dfrac{2k}{k-1}\text{Ma}_0^2 - 1} \quad \rightarrow \quad \text{Ma}_1^2 = \frac{(1.748)^2 + \dfrac{2}{1.40-1}}{\dfrac{2(1.40)}{1.40-1}(1.748)^2 - 1} \quad \rightarrow \quad \text{Ma}_1 = 0.6285$$

$$\frac{p_1}{p_0} = \frac{2k}{k+1}\text{Ma}_0^2 - \frac{k-1}{k+1} \quad \rightarrow \quad \frac{p_1}{101} = \frac{2(1.40)}{1.40+1}(1.748)^2 - \frac{1.40-1}{1.40+1} \quad \rightarrow \quad p_1 = 343 \text{ kPa}$$

$$\frac{T_1}{T_0} = \frac{\left[1 + \dfrac{k-1}{2}\text{Ma}_0^2\right]\left[k\,\text{Ma}_0^2 - \dfrac{k-1}{2}\right]}{\left[\dfrac{k+1}{2}\right]^2 \text{Ma}_0^2}$$

$$\rightarrow \quad \frac{T_1}{293} = \frac{\left[1 + \dfrac{1.40-1}{2}(1.748)^2\right]\left[1.40(1.748)^2 - \dfrac{1.40-1}{2}\right]}{\left[\dfrac{1.40+1}{2}\right]^2 (1.748)^2} \quad \rightarrow \quad T_1 = 438 \text{ K}$$

Therefore, behind the shock, the temperature and pressure are 438 K = **165°C** and **343 kPa**, respectively. The air velocity, V_1, induced by the normal shock is derived from Ma_1 as follows:

$$V_s - V_1 = \text{Ma}_1\sqrt{RT_1k} \quad \rightarrow \quad 600 - V_1 = (0.6285)\sqrt{(287.1)(438)(1.40)} \quad \rightarrow \quad V_1 = 336 \text{ m/s}$$

The air velocity induced by the propagation of the normal shock is **336 m/s**, which would likely destroy most structures in the path of the shock.

12.8 Steady One-Dimensional Non-Isentropic Flow

Consider the case of steady compressible flow in a duct of constant cross-sectional area as illustrated in Figure 12.24, where the duct need not have a circular cross section and the flow need not be frictionless. A control volume is defined such that it includes a finite segment of the flow, and the applicable control-volume equations are given by

$$\text{continuity: } \rho_1 V_1 = \rho_2 V_2 = \frac{\dot{m}}{A} \tag{12.119}$$

$$\text{momentum: } -F_x + p_1 A - p_2 A = \dot{m}(V_2 - V_1) \tag{12.120}$$

$$\text{energy: } \frac{\dot{Q}}{\dot{m}} + h_1 + \frac{1}{2}V_1^2 = h_2 + \frac{1}{2}V_2^2 \tag{12.121}$$

$$\text{second law: } \dot{m}(s_2 - s_1) \geq \int_{\text{cs}} \frac{1}{T}\left(\frac{\dot{Q}}{A}\right) \text{d}A \tag{12.122}$$

$$\text{State 1: } \rho = \frac{p}{RT} \tag{12.123}$$

$$\text{State 2: } h_2 - h_1 = c_p(T_2 - T_1) \tag{12.124}$$

$$\text{State 3: } s_2 - s_1 = c_p \ln\frac{T_2}{T_1} - R\ln\frac{p_2}{p_1} \tag{12.125}$$

In this case, F_x is the frictional force exerted by the wall of the duct on the moving fluid. It is convenient to analyze two particular cases that are encountered in practical flows: adiabatic conditions and isothermal conditions. Under adiabatic conditions, the duct is sufficiently insulated that heat neither enters nor leaves the fluid, whereas under isothermal conditions, the heat exchange is such that the temperature of the fluid remains constant as it enters and leaves the control volume. Flows in a very short duct tend to be adiabatic, whereas flows in very long ducts tend to be isothermal as they equilibrate with their surroundings. Although some flows are neither adiabatic nor isothermal, many flows encountered in engineering applications meet these requirements; such flows are considered in the following sections.

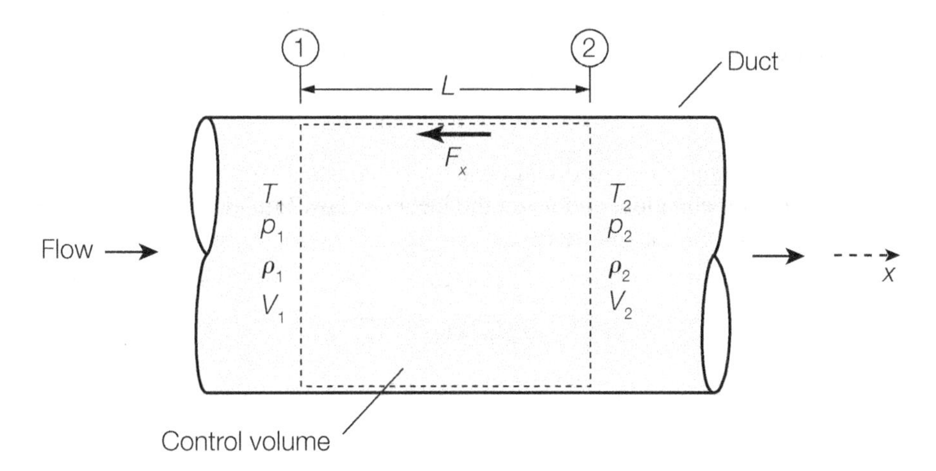

Figure 12.24: Flow in a duct

12.8.1 Adiabatic Flow with Friction

Adiabatic flow of an ideal gas through a constant area duct with friction is called *Fanno flow*.[66] Insulated pipes and high-speed flow in short ducts are practical situations in which adiabatic conditions can be assumed to occur. Applications involving insulated pipes include conduits that carry steam or refrigerating fluids such as ammonia vapor. The applicable governing equations for adiabatic flow in a duct are given by Equations 12.119–12.125, with $\dot{Q}/\dot{m} = 0$ to represent the adiabatic condition. The frictional force, F_x, can be expressed in terms of the boundary shear stress, τ_{0x}, such that

$$F_x = \int_0^L \tau_{0x} P \,\mathrm{d}x \tag{12.126}$$

where L is the length of the control volume and P is the perimeter of the duct. The boundary shear stress, τ_{0x}, can be determined using a friction factor f, where $\tau_{0x} = f\rho V^2/8$. For given upstream conditions, the energy equation and equations of state can be combined as follows:

$$\left.\begin{array}{l} h + \frac{1}{2}V^2 = h_0 = \text{constant} \\ h - h_0 = c_p(T - T_0) \\ p = \rho RT \end{array}\right\} \rightarrow T + \frac{V^2}{2c_p} = T_0 = \text{constant} \tag{12.127}$$

The quantity h_0 in Equation 12.127 is the *stagnation enthalpy* or *total enthalpy*, which remains constant along the duct in adiabatic flow and is equal to the enthalpy of the fluid that would exist if the velocity was brought to zero in an adiabatic process. Similarly, the quantity T_0 in Equation 12.127, which is also constant along the duct in adiabatic flow, is sometimes called the *total temperature*, because it is the temperature that would exist if the velocity of the fluid was brought to zero in an adiabatic process. The total temperature and the stagnation temperature are commonly taken to mean the same thing, although, technically speaking, the temperature measured at a stagnation point need not be equal to the total temperature of the impinging fluid, because this will depend on whether the deceleration process is adiabatic. Frictional effects in applications involving high-speed flow are typically important in long ducts, particularly those with small cross-sectional areas.

The Fanno line. Analysis of gas flows in conduits is sometimes facilitated by considering the state of the flow on a curve of temperature, T, versus specific entropy, s, sometimes called the *T-s curve* or the *T-s diagram*. The specific entropy of the flow can be derived from the following relation:

$$s - s_1 = c_p \ln\frac{T}{T_1} - R\ln\frac{p}{p_1} \tag{12.128}$$

For a conduit of constant cross-sectional area, ρV = constant (for a fixed value of A) by virtue of the continuity equation, and using the ideal gas law, Equation 12.127 can be more conveniently expressed with a "ρV" term as

$$T + \frac{(\rho V)^2 T^2}{2c_p(p^2/R^2)} = T_0 \tag{12.129}$$

[66]Named in honor of the Italian engineer Gino Girolamo Fanno (1882–1962).

The quantity ρV is commonly called the *mass flux*, because it is equal to the mass flow rate per unit area. Equations 12.129 and 12.128 can be used to plot the temperature, T, versus the entropy, s. The graphical form of this resulting relationship is called the *Fanno line* and is illustrated in Figure 12.25. In some cases, the Fanno line is represented by a plot of enthalpy (instead of temperature) versus entropy. All points on the Fanno line relate to the same stagnation temperature, T_0; the same stagnation enthalpy, h_0; and the same mass flux, ρV. The key features of the flow that are illustrated by the Fanno line are as follows:

1. The entropy, s, will generally increase in the downstream direction; the maximum entropy, s_{max}, occurs at Ma = 1.

2. In the subsonic region, the Fanno line is asymptotic to the stagnation temperature, T_0.

3. The Mach number at the upstream location can be either subsonic (Ma $<$ 1 at Point 1) or supersonic (Ma $>$ 1 at Point 1′).

4. The flow will ultimately approach the sonic condition (Ma $=$ 1); therefore, friction increases the Mach number in subsonic flow (which might be counterintuitive) and decreases the Mach number in supersonic flow.

Given these features, the Fanno line is useful in illustrating the possible downstream states that can result from a given upstream state. The adiabatic duct flow illustrated in Figure 12.25 is commonly referred to as *Fanno line flow* or simply *Fanno flow*. In the case of subsonic adiabatic flow, the frictional losses cause such a significant drop in pressure that the resulting net pressure force in the direction of flow is sufficient to accelerate the flow, hence the Mach number increases in the downstream direction. If there is sufficient increase in entropy along the duct, the critical condition (Ma $=$ 1) will be achieved and the flow will be choked. If an additional length of duct is added beyond the choked section and the upstream flow is subsonic, then the choked section will simply be moved downstream, thereby reducing the Mach number at any given upstream location and reducing the flow rate in the duct. If the upstream flow is supersonic, then the additional length of duct causes a normal shock to occur within the duct, and this shock will move upstream as additional duct is added downstream.

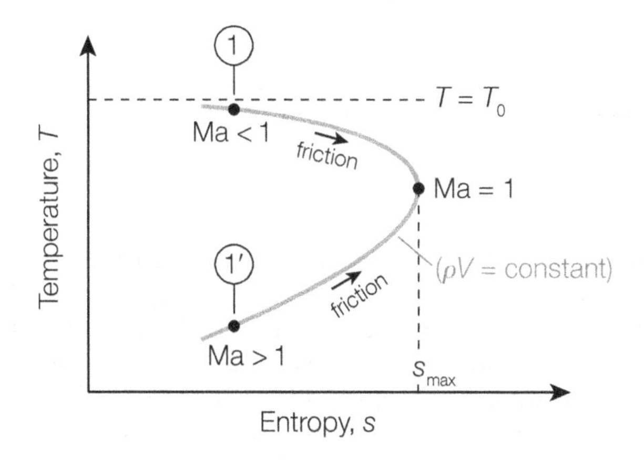

Figure 12.25: Fanno line

EXAMPLE 12.13

Consider the case of Fanno flow of air in a 30-mm-diameter closed conduit of circular cross section as shown in Figure 12.26. At an upstream section of the conduit, it is known that the stagnation temperature is 20°C and the actual temperature and pressure are 10°C and 101 kPa, respectively. The pressure at a downstream control section is p. (a) Give the equations and describe the procedure you would use to plot the Fanno line and then plot the Fanno line. (b) Plot the Mach number and temperature at the downstream control section as a function of the pressure p. (c) For p = 50 kPa, what is the mass flow rate through the conduit and what is the temperature and Mach number at the control section?

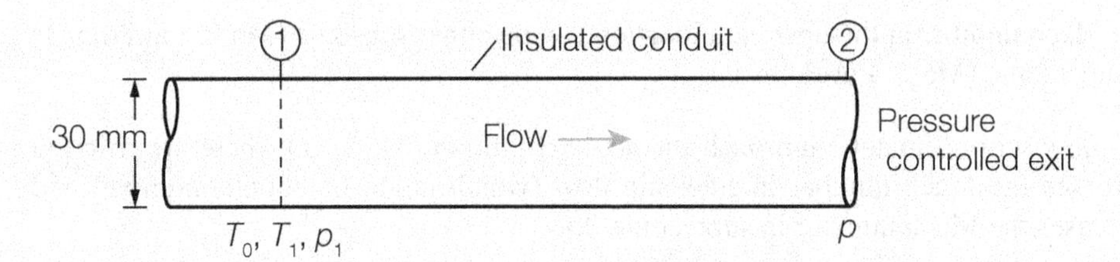

Figure 12.26: Fanno flow in a conduit

SOLUTION

From the given data: T_0 = 20°C = 293.15 K, T_1 = 10°C = 283.15 K, and p_1 = 101 kPa. Because the fluid is air, R = 287.1 J/kg·K, and it can be assumed that air behaves like an ideal gas with approximately constant values of k = 1.40 and c_p = 1003 J/kg·K.

(a) The equations that would be used in plotting the Fanno line are as follows:

$$\Delta s = c_p \ln \frac{T}{T_1} - R \ln \frac{p}{p_1} \tag{12.130}$$

$$T + \frac{(\rho V)^2 T^2}{2c_p(p^2/R^2)} = T_0 \quad \rightarrow \quad p = \frac{(\rho V)RT}{\sqrt{2c_p(T_0 - T)}} \tag{12.131}$$

$$\rho V = \frac{p_1}{RT_1}\mathrm{Ma}_1\sqrt{RT_1 k} \tag{12.132}$$

$$\mathrm{Ma}_1 = \left[\frac{\dfrac{T_0}{T_1} - 1}{\dfrac{k-1}{2}}\right]^{\frac{1}{2}}, \qquad \mathrm{Ma} = \left[\frac{\dfrac{T_0}{T} - 1}{\dfrac{k-1}{2}}\right]^{\frac{1}{2}} \tag{12.133}$$

The following calculation sequence will yield the Fanno line: (1) From the given data, calculate Ma_1 using Equation 12.133 and ρV using Equation 12.132; (2) specify T and calculate p using Equation 12.131; (3) calculate Δs using Equation 12.130; and (4) because the Fanno line is a plot of T versus Δs, repeat Steps 2 and 3 to obtain the points on the Fanno line. Using this procedure with the given data, the Fanno line is plotted in Figure 12.27. The key reference point is the upstream condition where T = 10°C = 283.15 K, where the change in entropy is by definition equal to zero. Proceeding downstream in the conduit, the entropy increases and points on the Fanno line represent possible downstream conditions.

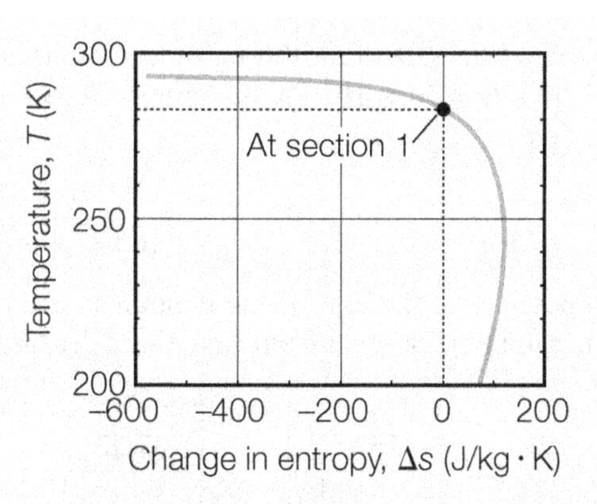

Figure 12.27: Fanno line for air in a conduit

(b) The following calculation sequence will yield a plot of Ma versus p: (1) Calculate Ma_1 using Equation 12.133, (2) calculate ρV using Equation 12.132, (3) specify T and calculate p using Equation 12.131, (4) calculate Ma using Equation 12.133, and (5) repeat Steps 3 and 4 to determine the Ma versus p curve, and repeating Step 3 yields the T versus p curve. The Mach number, Ma, and temperature, T, as a function of the pressure, p, at the downstream control section are shown in Figure 12.28. A key point in these relationships is the pressure at which Ma = 1, at which point the flow is choked. Setting Ma = 1 in Equation 12.133 yields $T = 244.3$ K, and Equation 12.131 yields $p = 39.5$ kPa. If the back pressure at the pressure-controlled exit is reduced below 39.5 kPa, then the pressure at the exit of the conduit remains at 39.5 kPa and the critical condition (Ma = 1) is maintained at the exit as well.

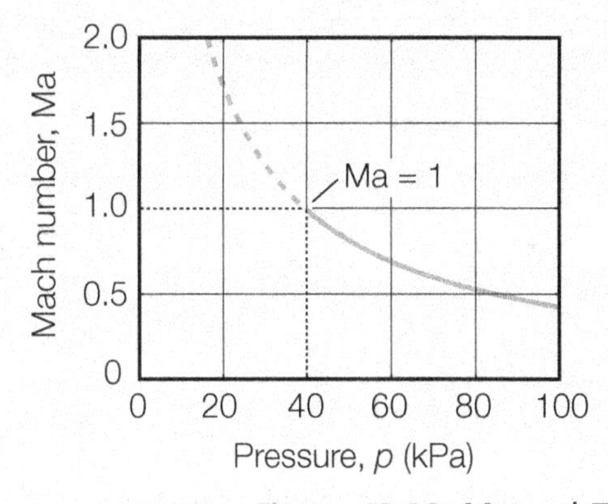

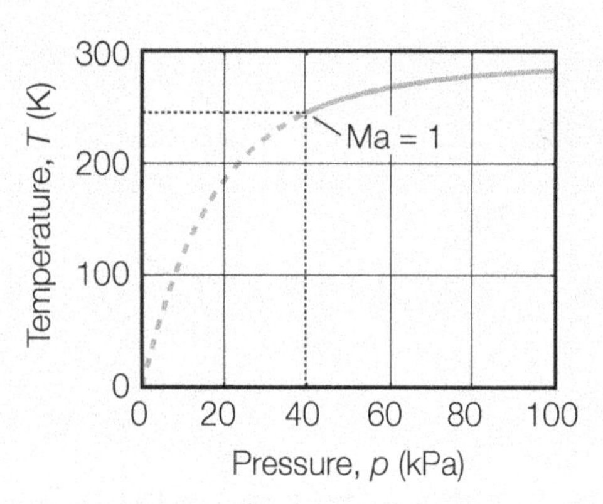

Figure 12.28: Ma and T versus p at control section

(c) The diameter of the conduit is given as 30 mm, which corresponds to a cross-sectional area of $\pi(0.03)^2/4 = 7.069 \times 10^{-4}$ m^2. Using the given upstream conditions, Equations 12.133 and 12.132 give

$$\text{Ma}_1 = \left[\frac{\dfrac{T_0}{T_1} - 1}{\dfrac{k-1}{2}}\right]^{\frac{1}{2}} = \left[\frac{\dfrac{293.15}{283.15} - 1}{\dfrac{1.40-1}{2}}\right]^{\frac{1}{2}} = 0.420$$

$$\begin{aligned}\rho V &= \frac{p_1}{RT_1}\text{Ma}_1\sqrt{RT_1k} = \frac{101 \times 10^3}{(287.1)(283.15)}(0.420)\sqrt{(287.1)(283.15)(1.40)} \\ &= 176.1\ \text{kg/m}^2\text{·s}\end{aligned}$$

$$\dot{m} = \rho V A = (176.1)(7.069 \times 10^{-4}) = 0.124\ \text{kg/s}$$

Therefore, the mass flow rate of air through the conduit is **0.124 kg/s**. At a downstream pressure of p = 50 kPa = 50×10^3 Pa, the temperature T can be derived from Equation 12.131, which yields

$$p = \frac{(\rho V)RT}{\sqrt{2c_p(T_0 - T)}} \quad \rightarrow \quad 50 \times 10^3 = \frac{(176.1)(287.1)T}{\sqrt{2(1003)(293.15 - T)}} \quad \rightarrow \quad T = 259.0 \text{ K}$$

Hence, the temperature at the exit of the conduit is 259.0 K $=$ **−14.2°C**. The corresponding Mach number is derived from Equation 12.133 as

$$\text{Ma} = \left[\frac{\dfrac{T_0}{T} - 1}{\dfrac{k-1}{2}}\right]^{\frac{1}{2}} = \left[\frac{\dfrac{293.15}{259.0} - 1}{\dfrac{1.40 - 1}{2}}\right]^{\frac{1}{2}} = 0.813$$

This downstream Mach number of **0.813** could have been read directly from the graph in Figure 12.28; however, it is seldom justified to read a value from a graph when the equation of the graph is known.

Solutions of governing equations. The governing equations for adiabatic flow in closed conduits (Equations 12.119–12.125) can be solved by expressing the upstream flow properties as p, ρ, T, and V and the downstream properties as $p + \mathrm{d}p$, $\rho + \mathrm{d}\rho$, $T + \mathrm{d}T$, and $V + \mathrm{d}V$ and solving the resulting differential equations, which lead to

$$\frac{p}{p^*} = \frac{1}{\text{Ma}}\left[\frac{k+1}{2 + (k-1)\,\text{Ma}^2}\right]^{\frac{1}{2}} \tag{12.134}$$

$$\frac{T}{T^*} = \frac{k+1}{2 + (k-1)\,\text{Ma}^2} \tag{12.135}$$

$$\frac{V}{V^*} = \frac{\rho^*}{\rho} = \text{Ma}\left[\frac{k+1}{2 + (k-1)\,\text{Ma}^2}\right]^{\frac{1}{2}} \tag{12.136}$$

$$\frac{p_0}{p_0^*} = \frac{1}{\text{Ma}}\left[\frac{2 + (k-1)\,\text{Ma}^2}{k+1}\right]^{\frac{k+1}{2(k-1)}} \tag{12.137}$$

where Ma is the local Mach number corresponding to the fluid properties p, ρ, T, and V and the asterisks denote the corresponding properties at the critical section within the conduit (i.e., where Ma $= 1$). The local stagnation pressure is p_0, and the stagnation pressure at the critical condition is p_0^*. It should be noted that in Fanno flow, the stagnation pressure, p_0, decreases along the conduit as a result of friction between the moving fluid and the conduit wall. In cases where a choked section occurs within a conduit and Fanno flow occurs upstream of the choked section, the variables with asterisks in Equations 12.134–12.137 denote the flow properties at the choked section and the variables without asterisks denote the conditions at a section with a Mach number of Ma.

Location of critical section. An expression for the distance from the local section to the critical section (where Ma $= 1$) can be derived by expressing the momentum equation in the differential form (i.e, for a control volume with infinitesimal length, $\mathrm{d}x$), which gives

$$-\mathrm{d}p - \frac{\tau_\mathrm{w}\pi D\,\mathrm{d}x}{A} = \rho V\mathrm{d}V \tag{12.138}$$

where τ_w is the wall shear stress, D is the diameter of the conduit, and the other variables have the same meaning as defined previously in this section. In closed conduits, the wall shear stress, τ_w, is conventionally related to the velocity by a *friction factor*, f, defined by the relation

$$\tau_\mathrm{w} = \frac{f}{8}\rho V^2 \tag{12.139}$$

Substituting Equation 12.139 into Equation 12.138 and incorporating $A = \pi D^2/4$, $\rho = p/RT$, $c = \sqrt{RTk}$, $\mathrm{Ma} = V/c$, and the energy equation for Fanno flow given by Equation 12.129 yields (after a fair amount of algebra)

$$\frac{f}{D}\,\mathrm{d}x = \frac{(1-\mathrm{Ma}^2)\,\mathrm{d}(\mathrm{Ma}^2)}{\left[1+\dfrac{k-1}{2}\mathrm{Ma}^2\right]k\mathrm{Ma}^4} \tag{12.140}$$

Defining $x = 0$ as the upstream location and $x = L^*$ as the critical location (where Ma = 1), Equation 12.140 gives

$$\int_0^{L^*}\frac{f}{D}\,\mathrm{d}x = \int_{\mathrm{Ma}^2}^{1}\frac{(1-\mathrm{Ma}^2)\,\mathrm{d}(\mathrm{Ma}^2)}{\left[1+\dfrac{k-1}{2}\mathrm{Ma}^2\right]k\mathrm{Ma}^4}$$

which, after integration, yields

$$\frac{\bar{f}L^*}{D} = \frac{1-\mathrm{Ma}^2}{k\,\mathrm{Ma}^2} + \frac{k+1}{2k}\ln\left[\frac{(k+1)\mathrm{Ma}^2}{2+(k-1)\,\mathrm{Ma}^2}\right] \tag{12.141}$$

where $\bar{f}$ is the average friction factor over the distance L^*. The distance L^* is commonly referred to as the *critical length* or the *sonic length*. In practice, $\bar{f}$ can be related to the local Reynolds number (Re) and relative roughness (ϵ/D) using the Colebrook equation. Whereas this approximation has been validated for subsonic flow, a reduction of 50% in the calculated value of $\bar{f}$ is recommended in applying the Colebrook friction factor to supersonic flow. It is relevant to note that the relative roughness (ϵ/D) can usually be assumed to be constant along a duct of constant material and constant cross-sectional area and that the Reynolds number ($\mathrm{Re} = \rho VD/\mu$) varies mildly along the duct because ρV is a constant due to continuity, making the variation in Re along the duct due solely to the variation in the dynamic viscosity (μ) with temperature. In cases where the conduit does not have a circular cross section, Equation 12.141 and related friction loss equations are still applicable, but with the diameter, D, replaced by the *hydraulic diameter*, D_h, defined by $D_\mathrm{h} = 4A/P$, where A is the cross-sectional area and P is the perimeter of the conduit.

Application of the critical location equation. The critical condition need not exist within a conduit to apply Equation 12.141. Determination of the length, L, of duct required to change the Mach number from Ma_1 at an upstream location to Ma_2 at a downstream location can be done using Equation 12.141 because

$$\frac{\bar{f}L}{D} = \frac{\bar{f}L_2^*}{D} - \frac{\bar{f}L_1^*}{D} \tag{12.142}$$

where L_1^* and L_2^* are the distances from the upstream and downstream sections, respectively, to the critical section. If the length of the duct is equal to the critical length, L^*, then $\mathrm{Ma} = 1$ at the exit and the flow is choked; this is sometimes called *frictional choking*. For a given Mach number, L^* is large for ducts with smooth surfaces and small for ducts with rough surfaces. If the length of the duct is greater than L^* and the flow in the duct is subsonic, then the exit flow will still be choked (Ma = 1) and the inlet Mach number will be reduced so that it corresponds to Ma = 1 at the exit of the duct, thereby reducing the mass flow rate. If the flow in the duct is supersonic, then (as represented by the Fanno line) the flow will decelerate to Ma = 1 at a distance L^* from the reference point. If the length of the duct is greater than L^* and the flow in the duct is supersonic, then a normal shock will occur in the duct, causing the flow to become subsonic, and then accelerate toward the exit to ensure that Ma = 1 occurs at the exit; however, the mass flow rate is unaffected because disturbances cannot propagate upstream. As the length of the duct is further increased, the normal shock moves upstream and eventually locates itself at the duct entrance.

EXAMPLE 12.14

A 5-m-long, 150-mm-diameter circular duct is connected to the atmosphere by a nozzle as shown in Figure 12.29. The temperature and pressure in the atmosphere are 20°C and 101 kPa, respectively, and the back pressure at the discharge end of the duct is 40 kPa. The average friction factor in the duct is 0.015. Assuming that flow through the nozzle is isentropic and flow through the duct is adiabatic, estimate the mass flow rate through the duct.

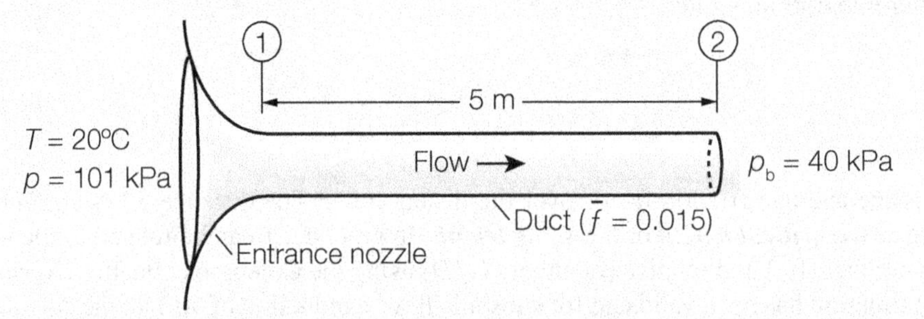

Figure 12.29: Nozzle duct flow

SOLUTION

From the given data: L = 5 m, D = 150 mm = 0.150 m, T_0 = 20°C = 293.15 K, p_0 = 101 kPa = 101 × 10³ Pa, p_b = 40 kPa, and $\bar{f}$ = 0.015. For air, R = 287.1 J/kg·K, and it can be assumed that c_p = 1003 J/kg·K and k = 1.40. The density, ρ_0, corresponding to the given atmospheric temperature and pressure can be calculated using the ideal gas law as follows:

$$\rho_0 = \frac{p_0}{RT_0} = \frac{101 \times 10^3}{(287.1)(293.15)} = 1.200 \text{ kg/m}^3$$

Two possible flow conditions could be occurring: (1) The flow is choked, in which case the flow is critical at the outlet, or (2) the flow is not choked, in which case the flow is subsonic throughout the duct. We can start by determining whether the flow is choked. In such a case, the Mach number at section 1, Ma_1, can be calculated using Equation 12.141, which gives

$$\frac{\bar{f}L^*}{D} = \frac{1-\text{Ma}^2}{k\,\text{Ma}^2} + \frac{k+1}{2k}\ln\left[\frac{(k+1)\text{Ma}^2}{2+(k-1)\,\text{Ma}^2}\right]$$

$$\frac{(0.015)(5)}{0.15} = \frac{1-\text{Ma}_1^2}{(1.40)\,\text{Ma}_1^2} + \frac{1.40+1}{2(1.40)}\ln\left[\frac{(1.40+1)\text{Ma}_1^2}{2+(1.40-1)\,\text{Ma}_1^2}\right] \quad \rightarrow \quad \text{Ma}_1 = 0.5977$$

Using $\text{Ma}_1 = 0.5977$, the ratios p_1/p^* and V_1/V^* relating conditions at the beginning of the duct to conditions at the critical section within the duct can be calculated from Equations 12.134 and 12.136 as follows:

$$\frac{p_1}{p^*} = \frac{1}{\text{Ma}_1}\left[\frac{k+1}{2+(k-1)\,\text{Ma}_1^2}\right]^{\frac{1}{2}} = \frac{1}{0.5977}\left[\frac{1.40+1}{2+(1.40-1)\,(0.5977)^2}\right]^{\frac{1}{2}} = 1.696$$

$$\frac{V_1}{V^*} = \text{Ma}_1\left[\frac{k+1}{2+(k-1)\,\text{Ma}_1^2}\right]^{\frac{1}{2}} = (0.5977)\left[\frac{1.40+1}{2+(1.40-1)\,(0.5977)^2}\right]^{\frac{1}{2}} = 0.6060$$

Assuming isentropic flow through the entrance nozzle and taking $\text{Ma}_1 = 0.5977$, the ratios p_{01}/p_1, ρ_0/ρ_1, and T_0/T^* at the end of the entrance nozzle (= beginning of the duct) can be calculated from Equations 12.51, 12.54, and 12.63 as follows:

$$\frac{p_{01}}{p_1} = \left[1+\frac{k-1}{2}\text{Ma}_1^2\right]^{\frac{k}{k-1}} = \left[1+\frac{1.40-1}{2}(0.5977)^2\right]^{\frac{1.40}{1.40-1}} = 1.273$$

$$\frac{\rho_0}{\rho_1} = \left[1+\frac{k-1}{2}\text{Ma}_1^2\right]^{\frac{1}{k-1}} = \left[1+\frac{1.40-1}{2}(0.5977)^2\right]^{\frac{1}{1.40-1}} = 1.188$$

$$\frac{T_0}{T^*} = \frac{k+1}{2} = \frac{1.40+1}{2} = 1.200$$

Next, determine the pressure at the exit of the duct under the assumed choked condition. Because the source atmosphere conditions are the stagnation conditions for flow through the entrance nozzle, $p_{01} = 101$ kPa and

$$p_1 = p_{01}\frac{p_1}{p_{01}} = (101)\left(\frac{1}{1.273}\right) = 79.33 \text{ kPa}$$

The exit (critical) pressure, p^*, is derived from the ratio p_1/p^* for duct flow, which yields

$$p^* = p_1\frac{p^*}{p_1} = (79.33)\left(\frac{1}{1.696}\right) = 46.76 \text{ kPa}$$

Because the pressure at the duct exit under the choked-flow condition is 46.76 kPa and the given back pressure is 40 kPa, the exit pressure will be 46.76 kPa and the flow is indeed choked. We can now proceed to calculate the mass flow rate through the duct under the choked condition. Because it is known from the source atmosphere properties that $T_0 = 293.15$ K (for both the nozzle and the duct), $p_0 = 101$ kPa, and $\rho_0 = 1.200$ kg/m^3, using the above results yields

$$T^* = T_0\frac{T^*}{T_0} = (293.15)\left(\frac{1}{1.200}\right) = 244.3 \text{ K}$$

$$V^* = \sqrt{RT^*k} = \sqrt{(287.1)(244.3)(1.40)} = 313.4 \text{ m/s}$$

$$V_1 = V^*\frac{V_1}{V^*} = (313.4)(0.6060) = 189.9 \text{ m/s}$$

$$\rho_1 = \rho_0\frac{\rho_1}{\rho_0} = (1.200)\left(\frac{1}{1.188}\right) = 1.010 \text{ kg/m}^3$$

$$\dot{m} = \rho_1 V_1 A_1 = (1.010)(189.9)\left(\frac{\pi}{4}0.15^2\right) = 3.389 \text{ kg/s}$$

Therefore, under the confirmed choked condition, the mass flow rate at section 1 and throughout the duct is **3.389 kg/s**. This value is also the maximum mass flux that can be attained under the given atmospheric conditions.

The previous example dealt with duct flow under choked conditions. The following example contrasts flow under unchoked conditions with flow under choked conditions.

EXAMPLE 12.15

At the entrance to the 50-mm-diameter insulated duct shown in Figure 12.30, the stagnation pressure is 150 kPa, the stagnation temperature is 127°C, and the velocity under unchoked conditions is 80 m/s. The fluid is air, and the average friction factor in the duct is estimated as 0.025. (a) Determine the mass flow rate under unchoked conditions and the range of flow lengths for which this mass flow rate can be maintained. (b) If the maximum flow length calculated in part (a) was increased by 50%, what would be the percentage change in the mass flow rate?

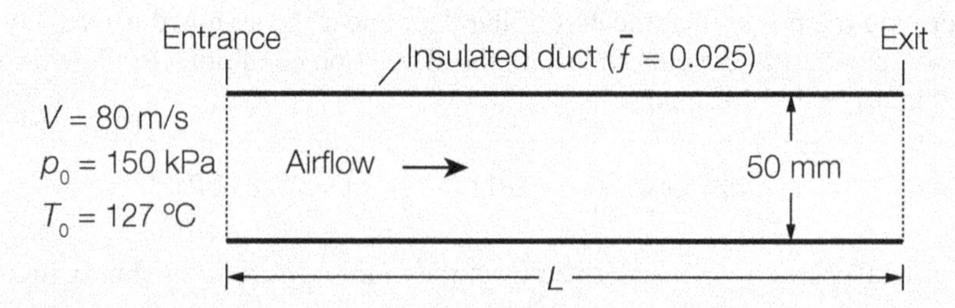

Figure 12.30: Airflow through an insulated duct

SOLUTION

From the given data: D = 50 mm = 0.050 m, $\bar{f}$ = 0.025, p_0 = 150 kPa, and T_0 = 127°C = 400 K, and under unchoked conditions, V_e = 80 m/s. The area of the duct is $A = \pi D^2/4 = 0.001963\ \text{m}^2$. Assume Fanno flow. For air, R = 287.1 J/kg·K, c_p = 1003 J/kg·K, and k = 1.40.

(a) Under Fanno flow conditions, the temperature, T_e, at the entrance to the duct can be calculated using Equation 12.127, which gives

$$T_e + \frac{V_e^2}{2c_p} = T_0 \quad \rightarrow \quad T_e + \frac{(80)^2}{2(1003)} = 400 \quad \rightarrow \quad T_e = 396.8\ \text{K}$$

Therefore, the temperature at the entrance to the duct (396.8 K) deviates by less than 1% from the stagnation temperature (400 K). The Mach number at the entrance to the duct, Ma_e, is given by

$$\text{Ma}_e = \frac{V_e}{\sqrt{RT_e k}} = \frac{80}{\sqrt{(287.1)(396.8)(1.40)}} = 0.2003$$

Choked conditions occur at a distance L^* from the entrance, where L^* is derived from Equation 12.141 as follows:

$$\frac{\bar{f}L^*}{D} = \frac{1 - \mathrm{Ma}_\mathrm{e}^2}{k\,\mathrm{Ma}_\mathrm{e}^2} + \frac{k+1}{2k}\ln\left[\frac{(k+1)\mathrm{Ma}_\mathrm{e}^2}{2 + (k-1)\,\mathrm{Ma}_\mathrm{e}^2}\right]$$

$$\frac{(0.025)L^*}{0.050} = \frac{1 - (0.2003)^2}{1.40(0.2003)^2} + \frac{1.40+1}{2(1.40)}\ln\left[\frac{(1.40+1)(0.2003)^2}{2 + (1.40-1)(0.2003)^2}\right] \rightarrow L^* = 29.0\ \mathrm{m}$$

Therefore, for ducts with lengths less than 29.0 m, the flow in the duct will not be choked and the given entrance velocity will be maintained. The mass flow rate under unchoked conditions is derived from the following calculations:

$$\rho_0 = \frac{p_0}{RT_0} = \frac{150 \times 10^3}{(287.1)(400)} = 1.306\ \mathrm{kg/m^3}$$

$$\frac{\rho_0}{\rho_\mathrm{e}} = \left[1 + \frac{k-1}{2}\mathrm{Ma}_\mathrm{e}^2\right]^{\frac{1}{k-1}} \rightarrow \frac{1.306}{\rho_\mathrm{e}} = \left[1 + \frac{1.40-1}{2}(0.2003)^2\right]^{\frac{1}{1.40-1}} \rightarrow \rho_\mathrm{e} = 1.280\ \mathrm{kg/m^3}$$

$$\dot{m} = \rho_\mathrm{e}V_\mathrm{e}A = (1.280)(80)(0.001963) = 0.201\ \mathrm{kg/s}$$

Therefore, for **duct lengths less than or equal to 29.0 m**, the entrance velocity will be unaffected and the mass flow rate will be **0.201 kg/s**.

(b) The duct length is 1.5(29.0) = 43.5 m, and the flow is choked at the exit of the duct, which means that $L^* = 43.5$ m. The entrance Mach number under choked conditions, Ma_ec, is given by Equation 12.141, where

$$\frac{\bar{f}L^*}{D} = \frac{1 - \mathrm{Ma}_\mathrm{ec}^2}{k\,\mathrm{Ma}_\mathrm{ec}^2} + \frac{k+1}{2k}\ln\left[\frac{(k+1)\mathrm{Ma}_\mathrm{ec}^2}{2 + (k-1)\,\mathrm{Ma}_\mathrm{ec}^2}\right]$$

$$\frac{(0.025)(43.5)}{0.050} = \frac{1 - \mathrm{Ma}_\mathrm{ec}^2}{1.40\mathrm{Ma}_\mathrm{ec}^2} + \frac{1.40+1}{2(1.40)}\ln\left[\frac{(1.40+1)\mathrm{Ma}_\mathrm{ec}^2}{2 + (1.40-1)\mathrm{Ma}_\mathrm{ec}^2}\right] \rightarrow \mathrm{Ma}_\mathrm{ec} = 0.1678$$

The mass flow rate under choked conditions is derived from the following calculations:

$$\frac{T_0}{T_\mathrm{ec}} = 1 + \frac{k-1}{2}\mathrm{Ma}_\mathrm{ec}^2 \rightarrow \frac{400}{T_\mathrm{ec}} = 1 + \frac{1.40-1}{2}(0.1678)^2 \rightarrow T_\mathrm{ec} = 397.8\ \mathrm{K}$$

$$V_\mathrm{ec} = \mathrm{Ma}_\mathrm{ec}\sqrt{RT_\mathrm{ec}k} = (0.1678)\sqrt{(287.1)(397.8)(1.40)} = 67.09\ \mathrm{m/s}$$

$$\frac{\rho_0}{\rho_\mathrm{ec}} = \left[1 + \frac{k-1}{2}\mathrm{Ma}_\mathrm{ec}^2\right]^{\frac{1}{k-1}} \rightarrow \frac{1.306}{\rho_\mathrm{ec}} = \left[1 + \frac{1.40-1}{2}(0.1678)^2\right]^{\frac{1}{1.40-1}} \rightarrow \rho_\mathrm{ec} = 1.288\ \mathrm{kg/m^3}$$

$$\dot{m} = \rho_\mathrm{ec}V_\mathrm{ec}A = (1.288)(67.09)(0.001963) = 0.170\ \mathrm{kg/s}$$

Therefore, when the duct length is 43.5 m, the mass flow rate is decreased by $(0.201 - 0.170)/0.201 \times 100 =$ **15%** relative to the mass flow rate that exists for lengths less than or equal to 29.0 m.

Adiabatic and isentropic flow in a conduit. The Fanno flow discussed previously is adiabatic and non-isentropic. If the flow is adiabatic and isentropic (e.g., frictionless flow), then the following relationships apply:

$$\frac{p_1}{p_2} = \left(\frac{T_1}{T_2}\right)^{\frac{k}{k-1}} \tag{12.143}$$

$$T_0 = T_1\left(1 + \frac{k-1}{2}\text{Ma}_1^2\right) = T_2\left(1 + \frac{k-1}{2}\text{Ma}_2^2\right) \tag{12.144}$$

Combining Equations 12.143 and 12.144 gives the following pressure relationship for isentropic duct flows:

$$\frac{p_1}{p_2} = \left(\frac{1 + \frac{k-1}{2}\text{Ma}_2^2}{1 + \frac{k-1}{2}\text{Ma}_1^2}\right)^{\frac{k}{k-1}} \tag{12.145}$$

Notably, the isentropic stagnation pressure, p_0, and isentropic stagnation density, ρ_0, were defined previously as

$$p_0 = p\left(1 + \frac{k-1}{2}\text{Ma}^2\right)^{\frac{k}{k-1}}, \qquad \rho_0 = \rho\left(1 + \frac{k-1}{2}\text{Ma}^2\right)^{\frac{1}{k-1}}$$

where p_0 and ρ_0 are the pressure and density, respectively, that would exist if the fluid were brought to rest adiabatically and reversibly. Both p_0 and ρ_0 are not a constant along a conduit unless the flow is both adiabatic and isentropic.

EXAMPLE 12.16

The 10-mm-diameter conduit shown in Figure 12.31 takes in atmospheric air through a contoured insulated nozzle, where the atmospheric air has a temperature and pressure of 20°C and 101 kPa, respectively. Immediately downstream of the intake nozzle, at section 1, the measured pressure is 95 kPa, and at some location of interest downstream, at section 2, the pressure is 40 kPa. The conduit is insulated and smooth, and the average friction factor between sections 1 and 2 can be estimated as being equal to the friction factor at section 1. (a) For frictional flow, estimate the distance between sections 1 and 2, the Mach number of the flow at section 2 and the mass flow rate through the conduit. (b) Assuming frictionless flow, estimate the Mach number of the flow at section 2, and the mass flow rate through the conduit. Assess the impact of friction on flow through the conduit. (c) Based only on the given ambient air conditions and the assumption of adiabatic frictionless flow in the conduit, what is the maximum mass flow rate that could be attained?

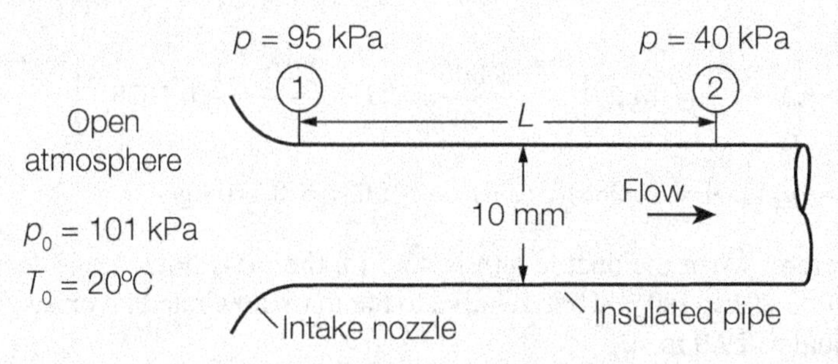

Figure 12.31: Airflow through a conduit

SOLUTION

From the given data: D = 10 mm = 0.010 m, p_{01} = 101 kPa, T_{01} = 20°C = 293 K, p_1 = 95 kPa, and p_2 = 40 kPa. The area of the conduit is given by $A = \pi D^2/4 = 7.854 \times 10^{-5}$ m^2. For air, assume R = 287.1 J/kg·K and k = 1.40.

(a) Assuming isentropic flow through the nozzle, the flow and fluid properties at section 1 can be estimated as follows:

$$\frac{p_{01}}{p_1} = \left[1 + \frac{k-1}{2}\text{Ma}_1^2\right]^{\frac{k}{k-1}} \quad \rightarrow \quad \frac{101}{95} = \left[1 + \frac{1.40-1}{2}\text{Ma}_1^2\right]^{\frac{1.40}{1.40-1}} \quad \rightarrow \quad \text{Ma}_1 = 0.2971$$

$$\frac{T_{01}}{T_1} = 1 + \frac{k-1}{2}\text{Ma}_1^2 \quad \rightarrow \quad \frac{293}{T_1} = 1 + \frac{1.40-1}{2}(0.2971)^2 \quad \rightarrow \quad T_1 = 288 \text{ K}$$

$$V_1 = \text{Ma}_1\sqrt{RT_1k} = (0.2971)\sqrt{(287.1)(288)(1.40)} = 101 \text{ m/s}$$

$$\rho_1 = \frac{p_1}{RT_1} = \frac{95 \times 10^3}{(287.1)(288)} = 1.149 \text{ kg/s}$$

$$\dot{m} = \rho_1 V_1 A = (1.149)(101)(7.854 \times 10^{-5}) = 0.00912 \text{ kg/s}$$

Therefore, the given pressure of 95 kPa at section 1 is sufficient to determine that the mass flow rate through the conduit is 0.00912 kg/s. The critical condition pressure, p^*, as given by Equation 12.134 is constant for frictional adiabatic flow. So Equation 12.134 can be applied at sections 1 and 2, and the combined equations (eliminating p^*) give

$$\frac{p_1}{p_2} = \frac{\text{Ma}_2}{\text{Ma}_1}\left[\frac{2+(k-1)\,\text{Ma}_2^2}{2+(k-1)\,\text{Ma}_1^2}\right]^{\frac{1}{2}} \quad \rightarrow \quad \frac{95}{40} = \frac{\text{Ma}_2}{0.2971}\left[\frac{2+(1.40-1)\,\text{Ma}_2^2}{2+(1.40-1)\,(0.2971)^2}\right]^{\frac{1}{2}}$$

$$\rightarrow \quad \text{Ma}_2 = 0.6809$$

Hence, the Mach number at section 2 is 0.6809. Because the temperature at section 1, T_1, is estimated as 288 K = 15°C, the corresponding dynamic viscosity at section 1 can be estimated (from Appendix B.5) as μ_1 = 0.0180 mPa·s = 1.80×10^{-5} Pa·s. The Reynolds number of the flow at section 1 and the friction factor (via the Colebrook equation) can be estimated as follows:

$$\text{Re}_1 = \frac{\rho_1 V_1 D}{\mu_1} = \frac{(1.149)(101)(0.010)}{1.80 \times 10^{-5}} = 6.453 \times 10^4$$

$$\frac{1}{\sqrt{f_1}} = -2.0 \log\left[\frac{2.51}{\text{Re}\sqrt{f_1}}\right] \quad \rightarrow \quad \frac{1}{\sqrt{f_1}} = -2.0 \log\left[\frac{2.51}{(6.453 \times 10^4)\sqrt{f_1}}\right]$$

$$\rightarrow \quad f_1 = 0.0197$$

Assuming that the average friction factor in the conduit can be approximated by the friction factor at section 1 and applying the friction loss equation (Equation 12.141) at section 1 gives the distance L_1^* to the critical section as

$$\frac{f_1 L_1^*}{D} = \frac{1-\text{Ma}_1^2}{k\,\text{Ma}_1^2} + \frac{k+1}{2k}\ln\left[\frac{(k+1)\text{Ma}_1^2}{2+(k-1)\,\text{Ma}_1^2}\right]$$

$$\frac{(0.0197)L_1^*}{0.010} = \frac{1-(0.2971)^2}{1.40\,(0.2971)^2} + \frac{1.40+1}{2(1.40)}\ln\left[\frac{(1.40+1)(0.2971)^2}{2+(1.40-1)\,(0.2971)^2}\right]$$

$$\rightarrow \quad L_1^* = 2.754 \text{ m}$$

Similarly, applying the friction loss equation at section 2 gives the distance L_2^* to the critical section as

$$\frac{f_1 L_2^*}{D} = \frac{1 - \mathrm{Ma}_2^2}{k\,\mathrm{Ma}_2^2} + \frac{k+1}{2k}\ln\left[\frac{(k+1)\mathrm{Ma}_2^2}{2+(k-1)\,\mathrm{Ma}_2^2}\right]$$

$$\frac{(0.0197)L_2^*}{0.010} = \frac{1-(0.6809)^2}{1.40\,(0.6809)^2} + \frac{1.40+1}{2(1.40)}\ln\left[\frac{(1.40+1)(0.6809)^2}{2+(1.40-1)\,(0.6809)^2}\right]$$

$$\rightarrow \quad L_2^* = 0.125 \text{ m}$$

The distance from section 1 to section 2, L, is therefore given by

$$L = L_1^* - L_2^* = 2.754 \text{ m} - 0.125 \text{ m} = 2.629 \text{ m} \approx 2.63 \text{ m}$$

In summary, for the given conditions and assuming frictional flow, section 2 is located approximately **2.63 m downstream of section 1**, the flow at section 2 has a Mach number of **0.681**, and the mass flow rate in the conduit is **0.00912 kg/s**.

(b) If the flow in the conduit is assumed to be frictionless, then because the conduit is insulated, it can also be assumed that the flow is adiabatic and hence isentropic. Under isentropic conditions, the ratio p_1/p_2 is given by Equation 12.145 and the Mach number at section 2 can be calculated as follows:

$$\frac{p_1}{p_2} = \left[\frac{1+\frac{k-1}{2}\mathrm{Ma}_2^2}{1+\frac{k-1}{2}\mathrm{Ma}_1^2}\right]^{\frac{k}{k-1}} \quad \rightarrow \quad \frac{95}{40} = \left[\frac{1+\frac{1.40-1}{2}\mathrm{Ma}_2^2}{1+\frac{1.40-1}{2}(0.2971)^2}\right]^{\frac{1.40}{1.40-1}} \quad \rightarrow \quad \mathrm{Ma}_2 = 1.23$$

This indicates that the flow is supersonic ($\mathrm{Ma}_2 > 1$) at section 2, which is impossible, because this would require a diverging section. Therefore, for the given pressures at sections 1 and 2, it is not possible to have isentropic flow between these two sections. Furthermore, it is apparent from part (a) that the mass flow rate in the conduit can be determined using only the (stagnation) temperature and pressure of the ambient air and the pressure at section 1. Hence, regardless of the pressure at section 2, the mass flow rate would be estimated as **0.00912 kg/s**. Based on the analysis presented here, it can be concluded that the given information would only support the assumption of frictional adiabatic flow in the conduit; **frictionless adiabatic flow could not be occurring under the given conditions.**

(c) The maximum mass flow rate that could be attained in the conduit would occur under the critical (sonic, choked) condition. The critical temperature, T^*, and critical pressure, p^*, that would be attained under isentropic flow conditions are given by Equations 12.62 and 12.63, which yield

$$\frac{p_{01}}{p^*} = \left[\frac{k+1}{2}\right]^{\frac{k}{k-1}} \quad \rightarrow \quad \frac{101}{p^*} = \left[\frac{1.40+1}{2}\right]^{\frac{1.40}{1.40-1}} \quad \rightarrow \quad p^* = 53.36 \text{ kPa}$$

$$\frac{T_{01}}{T^*} = \frac{k+1}{2} \quad \rightarrow \quad \frac{293}{T^*} = \frac{1.40+1}{2} \quad \rightarrow \quad T^* = 244 \text{ K}$$

Therefore, the density, ρ^*, velocity, V^*, and mass flow rate, $\dot{m}_{\text{crit}}$, that are attained under critical conditions are

$$\rho^* = \frac{p^*}{RT^*} = \frac{53.36 \times 10^3}{(287.1)(244)} = 0.761 \text{ kg/m}^3$$

$$V^* = \sqrt{RT^*k} = \sqrt{(287.1)(244)(1.40)} = 313 \text{ m/s}$$

$$\dot{m}_{\text{crit}} = \rho^* V^* A = (0.761)(313)(7.854 \times 10^{-5}) = 0.0187 \text{ kg/s}$$

Comparing the maximum possible mass flow rate in the conduit (**0.0187 kg/s**) with the mass flow rate attained with frictional adiabatic flow (0.00912 kg/s) indicates that the conduit operates at 49% "of capacity" under the frictional conditions.

12.8.2 Isothermal Flow with Friction

Isothermal flow with friction occurs in long ducts where the temperature of the gas being transported remains approximately constant, usually close to that of its surroundings. Such flows can sometimes be found in buried natural gas pipelines, where the surrounding soil serves as a heat sink that maintains the pipeline at a constant temperature. The governing continuity and momentum equations are the same as those for Fanno flow (i.e., adiabatic frictional flow). However, because the temperature is constant, $p/\rho + u$ is a constant, and the energy equation becomes simply

$$\frac{V_1^2}{2} + \dot{q} = \frac{V_2^2}{2} \tag{12.146}$$

where V_1 and V_2 are the velocities at an upstream and downstream section, respectively, and $\dot{q}$ is the heat added per unit mass of fluid between the two sections. Utilizing Equation 12.146 in the derivation of the friction loss equation yields the following form of the friction equation for a duct of hydraulic diameter D_h, with isothermal frictional flow:

$$\frac{fL_{\text{max}}}{D_\text{h}} = \frac{1 - k\,\text{Ma}^2}{k\,\text{Ma}^2} + \ln(k\,\text{Ma}^2) \tag{12.147}$$

where L_{max} is the distance from the section where the Mach number is Ma to the critical (sonic flow) section and the hydraulic diameter is defined as $4A/P$, where A is the cross-sectional area and P is the perimeter of the conduit. For the given flow characteristics (isothermal and frictional), $L_{\text{max}} = 0$ when $\text{Ma} = 1/k^{\frac{1}{2}}$, which is in contrast to adiabatic frictional flow where sonic conditions (i.e., Ma = 1) occur at the critical section, which occurred when $L^* = 0$. In the case of isothermal frictional flow, both subsonic and supersonic conditions asymptote to the limiting condition of $\text{Ma} = 1/k^{\frac{1}{2}}$. If the length of the duct exceeds L_{max} and the flow is subsonic, the condition $\text{Ma} = 1/k^{\frac{1}{2}}$ will move to the exit of the duct, thereby reducing the mass flow rate in the duct. If the length of the duct exceeds L_{max} and the flow is supersonic, then a normal shock will occur within the duct and the subsonic flow downstream of the shock will adjust to $\text{Ma} = 1/k^{\frac{1}{2}}$ at the exit. Because the limiting condition corresponding to $\text{Ma} = 1/k^{\frac{1}{2}}$ is not sonic, the fluid properties under the limiting condition are represented by primes (′) rather than asterisks (∗). If p', ρ', and V' are the pressure, density, and velocity, respectively, at the limiting condition, then solutions of the governing equations yield the following relations:

$$\frac{p}{p'} = \frac{1}{k^{\frac{1}{2}}\,\mathrm{Ma}} \tag{12.148}$$

$$\frac{\rho'}{\rho} = \frac{V}{V'} = k^{\frac{1}{2}}\,\mathrm{Ma} \tag{12.149}$$

An interesting corollary to Equation 12.148 is that the pressure decreases in the direction of flow when $\mathrm{Ma} < 1/k^{\frac{1}{2}}$ and the pressure increases in the direction of flow when $\mathrm{Ma} > 1/k^{\frac{1}{2}}$. Further analysis of the governing equations yields the following relationship that can be used to estimate mass flow rate per unit area, ρV, which remains constant in a constant area duct:

$$(\rho V)^2 = \frac{p_1^2 - p_2^2}{RT\left[\dfrac{fL}{D_\mathrm{h}} + 2\ln\left(\dfrac{p_1}{p_2}\right)\right]} \tag{12.150}$$

where p_1 and p_2 are pressures in the duct at sections located a distance L apart. In using Equation 12.150, care should be taken to determine whether the flow is choked. To check for choking under subsonic flow conditions, calculate the exit Mach number, $\mathrm{Ma}_\mathrm{exit}$; if $\mathrm{Ma}_\mathrm{exit} \geq 1/k^{\frac{1}{2}}$, the flow is choked; otherwise, the flow is not choked.

EXAMPLE 12.17

The air pressure and temperature at the entrance section of a 15-mm-diameter, 2-m-long duct are 200 kPa and –23°C, respectively, and the pressure at the exit of the duct is 120 kPa. Flow through the duct is isothermal and frictional, and the average friction factor within the duct is estimated as 0.020. Determine the mass flow rate through the duct. At what rate must heat be added to maintain isothermal conditions in the duct?

SOLUTION

From the given data: $D_\mathrm{h} = D = 15$ mm $= 0.015$ m, $L = 2$ m, $p_1 = 200$ kPa, $T = -23°\mathrm{C} = 250$ K, $p_2 = 120$ kPa, and $\bar{f} = 0.020$. The area of the duct is $A = \pi D^2/4 = 1.767 \times 10^{-4}\ \mathrm{m}^2$. For air, assume $R = 287.1$ J/kg·K and $k = 1.40$. The mass flow per unit area (ρV) can be estimated from Equation 12.150 as

$$(\rho V)^2 = \frac{p_1^2 - p_2^2}{RT\left[\dfrac{\bar{f}L}{D} + 2\ln\left(\dfrac{p_1}{p_2}\right)\right]} = \frac{\left[(200)^2 - (120)^2\right]10^6}{(287.1)(250)\left[\dfrac{(0.020)(2)}{0.015} + 2\ln\left(\dfrac{200}{120}\right)\right]}$$

$$\rightarrow \quad (\rho V) = 311.0\ \mathrm{kg/m^2{\cdot}s}$$

Hence, the density, ρ_1, velocity, V_1, and Mach number, Ma_1, at the entrance of the duct are given by

$$\rho_1 = \frac{p_1}{RT} = \frac{200 \times 10^3}{(287.1)(250)} = 2.787\ \mathrm{kg/m^3}$$

$$V_1 = \frac{(\rho V)}{\rho_1} = \frac{311.0}{2.787} = 111.6 \text{ m/s}$$

$$\text{Ma}_1 = \frac{V_1}{\sqrt{RTk}} = \frac{111.6}{\sqrt{(287.1)(250)(1.40)}} = 0.3521$$

The density, ρ_2, velocity, V_2, and Mach number, Ma_2, at the exit of the duct are given by

$$\rho_2 = \frac{p_2}{RT} = \frac{120 \times 10^3}{(287.1)(250)} = 1.672 \text{ kg/m}^3$$

$$V_2 = \frac{(\rho V)}{\rho_2} = \frac{311.0}{1.672} = 190.0 \text{ m/s}$$

$$\text{Ma}_2 = \frac{V_2}{\sqrt{RTk}} = \frac{190.0}{\sqrt{(287.1)(250)(1.40)}} = 0.5868$$

The reference Mach number is $\text{Ma} = 1/k^{\frac{1}{2}} = 1/(1.40)^{\frac{1}{2}} = 0.8452$. Because $\text{Ma}_2 < 0.8452$, the flow is not choked, so the nonchoked assumption underlying the calculation of (ρV) with Equation 12.150 is validated. The mass flow rate, $\dot{m}$, is therefore given by

$$\dot{m} = (\rho V)A = (311.0)(1.767 \times 10^{-4}) = \mathbf{0.0550 \text{ kg/s}}$$

To maintain isothermal conditions in the duct, the rate that heat must be added, $\dot{q}$, is given by the energy equation (Equation 12.146) as

$$\dot{q} = \frac{V_2^2}{2} - \frac{V_1^2}{2} = \frac{190.0^2}{2} - \frac{111.6^2}{2} = 1.107 \times 10^4 \text{ m}^2/\text{s}^2$$

This result can be expressed in energy terms, where 1.107×10^4 m^2/s^2 = 1.107×10^4 J/kg = 11.1 kJ/kg. Hence, heat supplied to the fluid in the duct at a rate of **11.1 kJ/kg** will maintain isothermal conditions within the duct.

12.8.3 Diabatic Frictionless Flow

Steady diabatic (i.e., non-adiabatic) frictionless flow of an ideal gas through a duct of constant cross sectional area is called *Rayleigh flow*.[67] The applicable governing equations for steady diabatic frictionless flow in a duct are given by Equations 12.119–12.125, with the simplification that the frictional force exerted by the sides of the duct are negligible; thus, $F_x \simeq 0$. Under diabatic conditions, the energy equation (Equation 12.121) requires that the change in total energy of the flow is equal to the heat added, and the second law of thermodynamics (Equation 12.122) permits the entropy to increase or decrease, depending on whether heat is added or removed from the flow. In contrast to the case of adiabatic flow where frictional effects cause the fluid properties to change, in the case of diabatic flow, it is the heat exchange that causes the fluid properties to change along the duct. In both cases, the flow is non-isentropic. Frictionless flow in high-speed applications is usually associated with flow through short devices with large cross-sectional areas, such as nozzles. Heat addition to flows in conduits can occur through a variety of pathways, such as through the walls of the conduit, through chemical reactions within the gas, and via nuclear radiation.

[67]Named in honor of the English physicist and mathematician John William Strutt, also known as Lord Rayleigh (1842–1919).

The Rayleigh line. Defining a control volume in a conduit as illustrated in Figure 12.24, it is possible to determine T versus s on the downstream boundary of the control volume by using the following combination of the momentum equation and the equation of state:

$$\left.\begin{array}{l} p + \dfrac{(\rho V)^2}{\rho} = \text{constant} = C_\text{M} \\ p = \rho R T \end{array}\right\} \rightarrow p + \frac{(\rho V)^2 RT}{p} = C_\text{M} \tag{12.151}$$

and

$$s - s_1 = c_p \ln \frac{T}{T_1} - R \ln \frac{p}{p_1} \tag{12.152}$$

where C_M is called the *momentum constant*. Because ρV = constant in Equation 12.151 (for a fixed value of A) by virtue of the continuity equation, Equations 12.151 and 12.152 can be used to plot the temperature, T, versus the entropy, s. The graphical form of this resulting relationship is called the *Rayleigh line* and is illustrated in Figure 12.32. In some cases, the Rayleigh line is represented by a plot of enthalpy (instead of temperature) versus entropy. Useful derived relationships (from the governing equations) that characterize the Raleigh line are

$$\frac{\text{d}s}{\text{d}T} = \frac{c_p}{T} + \frac{V}{T} \frac{1}{[(T/V) - (V/R)]} \tag{12.153}$$

$$\frac{\text{d}V}{V} = \frac{\delta q}{c_p T} \frac{1}{1 - \text{Ma}^2} \tag{12.154}$$

where δq is the incremental heat added per unit mass. Two key features of the Rayleigh line are that (1) the maximum temperature (@ $\text{d}T/\text{d}s = 0$) occurs when $\text{Ma} = 1/\sqrt{k}$ and (2) maximum entropy (@ $\text{d}s/\text{d}T = 0$) occurs when $\text{Ma} = 1$. Note that for air, $k = 1.40$; hence, $1/\sqrt{k} = 0.85$. The flow state within the duct can be either subsonic ($\text{Ma} < 1$) or supersonic ($\text{Ma} > 1$). For any given value of Ma along the duct, with heat addition, the flow state proceeds to the right on the Rayleigh line and with heat loss, the flow state proceeds to the left. Entropy generally increases with the addition of heat and decreases with cooling. Interestingly, in the range $1/\sqrt{k} < \text{Ma} < 1$, heating causes a decrease in temperature and cooling causes an increase in temperature. If the flow in the duct is subsonic ($\text{Ma} < 1$), then in accordance with Equation 12.154, V (and Ma) increase monotonically to Ma = 1 as the fluid is heated and at Ma = 1, the flow is choked. Similarly, if the flow in the duct is supersonic

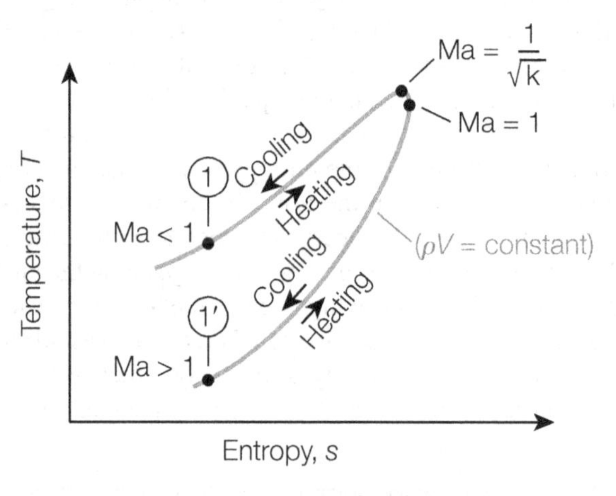

Figure 12.32: Rayleigh line

(Ma > 1), then V (and Ma) decrease monotonically to Ma = 1 as the fluid is heated and at Ma = 1, the flow is choked. If the flow in a duct is choked and additional heat is added, then the critical state (where Ma = 1) is moved further downstream in the duct, which further reduces the flow rate because the fluid density at the critical section will be lower. Consequently, for given inlet conditions, the corresponding critical state fixes the maximum possible heat transfer for steady state.

Solutions of governing equations. The governing equations for diabatic flow in ducts (Equations 12.119–12.125 with $F_x = 0$) can be solved by expressing the upstream flow properties as p, ρ, T, and V and the downstream properties at the critical condition where Ma = 1. Solving the resulting equations leads to

$$\frac{p}{p^*} = \frac{1+k}{1+k\,\mathrm{Ma}^2} \tag{12.155}$$

$$\frac{T}{T^*} = \left[\frac{(1+k)\mathrm{Ma}}{1+k\,\mathrm{Ma}^2}\right]^2 \tag{12.156}$$

$$\frac{V}{V^*} = \frac{\rho^*}{\rho} = \frac{(1+k)\,\mathrm{Ma}^2}{1+k\,\mathrm{Ma}^2} \tag{12.157}$$

where Ma is the local Mach number corresponding to the fluid properties p, ρ, T, and V and the asterisks denote the corresponding properties at the critical section within the conduit (i.e., where Ma $=$ 1). Expressions for the stagnation pressure, p_0, and the stagnation temperature, T_0, can also be derived by algebraic manipulation and are given by

$$\frac{p_0}{p_0^*} = \frac{1+k}{1+k\,\mathrm{Ma}^2}\left[\left(\frac{2}{k+1}\right)\left(1+\frac{k-1}{2}\mathrm{Ma}^2\right)\right]^{\frac{k}{k-1}} \tag{12.158}$$

$$\frac{T_0}{T_0^*} = \frac{(k+1)\mathrm{Ma}^2\left[2+(k-1)\,\mathrm{Ma}^2\right]}{(1+k\,\mathrm{Ma}^2)^2} \tag{12.159}$$

where the asterisks indicate the (stagnation) properties at the critical section. Equations 12.155–12.159 are useful in that they give the conditions of the flow at any point along the Rayleigh line in terms of the local Mach number and the properties at the critical section. It is interesting to note that the stagnation pressure, p_0, always decreases during heating (which drives Ma $\rightarrow$ 1) whether the flow is supersonic or subsonic. Hence, heating generally causes a loss of effective pressure recovery. Application of the energy equation between any two sections in a conduit gives

$$q + h_1 + \frac{V_1^2}{2} = h_2 + \frac{V_2^2}{2} \tag{12.160}$$

where the subscripts denote the section number, section 1 is upstream of section 2, q is the heat added per unit mass between sections 1 and 2, h is the enthalpy, and V is the speed of the flow. Equation 12.160 can be expressed in terms of the stagnation enthalpies at sections 1 and 2 or, more conveniently, in terms of the stagnation temperatures as follows:

$$q = h_{02} - h_{01} = c_p(T_{02} - T_{01}) \tag{12.161}$$

where T_{01} and T_{02} are the stagnation temperatures at sections 1 and 2, respectively. It is apparent from Equation 12.161 that heating will increase the stagnation temperature (T_0) and cooling will decrease the stagnation temperature, regardless of whether the flow is subsonic or supersonic. The maximum heat that can be added corresponds to the case where the downstream section is at the critical state (i.e., Ma = 1), in which case Equation 12.161 gives

$$q_{\text{max}} = h^* - h_{01} = c_p(T_0^* - T_{01}) \tag{12.162}$$

where the asterisk indicates conditions at the critical state. If more heat than q_{max} is added, then the flow is choked and the critical section must necessarily move downstream, which also reduces the mass flow rate because the fluid density at the critical state is lowered.

EXAMPLE 12.18

Consider the case of Rayleigh flow of air in a conduit as shown in Figure 12.33. At an upstream section of the conduit, it is known that the stagnation temperature is 20°C and the actual temperature and pressure are 10°C and 101 kPa, respectively. The pressure at the downstream exit section of the conduit is p. (a) List the governing equations and describe the procedure you would use to plot the Rayleigh line; then plot the Rayleigh line. (b) Plot the Mach number and temperature at the downstream section as a function of the pressure p. (c) For p = 60 kPa, what is the temperature and Mach number at the downstream section? Determine the heat added between the upstream and downstream sections.

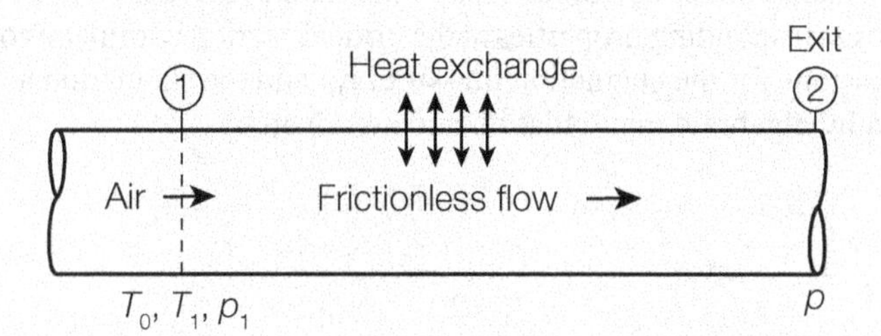

Figure 12.33: Rayleigh flow in a conduit

SOLUTION

From the given data: T_{01} = 20°C = 293.15 K, T_1 = 10°C = 283.15 K, and p_1 = 101 kPa. Because the fluid is air, R = 287.1 J/kg·K and it can be assumed that air behaves like an ideal gas with approximately constant values of k = 1.40 and c_p = 1003 J/kg·K.

(a) The equations that would be used in plotting the Rayleigh line are as follows:

$$\Delta s = c_p \ln \frac{T}{T_1} - R \ln \frac{p}{p_1} \tag{12.163}$$

$$p + \frac{(\rho V)^2 RT}{p} = C_{\text{M}} \quad \rightarrow \quad T = \frac{p(C_{\text{M}} - p)}{R(\rho V)^2} \tag{12.164}$$

$$\rho V = \frac{p_1}{RT_1} \text{Ma}_1 \sqrt{RT_1 k} \tag{12.165}$$

$$\text{Ma}_1 = \left[\frac{\dfrac{T_{01}}{T_1} - 1}{\dfrac{k-1}{2}} \right]^{\frac{1}{2}} \tag{12.166}$$

The following calculation sequence will yield the Rayleigh line: (1) From the given data, calculate Ma_1 using Equation 12.166, C_M using Equation 12.164 with the upstream conditions, and ρV using Equation 12.165; (2) specify p and calculate T using Equation 12.164; (3) calculate Δs using Equation 12.163; and (4) because the Rayleigh line is a plot of T versus Δs, repeat Steps 2 and 3 to obtain the points on the Rayleigh line. Performing Step 1 with the given data yields $\text{Ma}_1 = 0.4202$, C_M = 126.0 kPa, and ρV = 176.1 kg/m^2·s. Steps 2 and 3 yield the corresponding Rayleigh line that is plotted in Figure 12.34. The key reference point is the upstream condition at section 1 where T = 10°C = 283.15 K; the change in entropy is by definition equal to zero at this point.

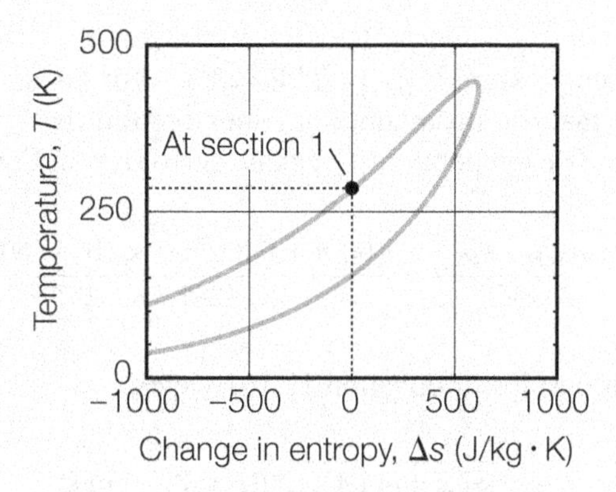

Figure 12.34: Rayleigh line for air in a conduit

(b) The following calculation sequence will yield a plot of Ma versus p: (1) From the given data, calculate C_M using Equation 12.164 with the upstream conditions and calculate ρV using Equation 12.165; (2) specify p and calculate T using Equation 12.164; (3) for the given values of p, T, and ρV, calculate Ma at the downstream end of the control section using the following equations:

$$c = \sqrt{RTk}, \quad \rho = \frac{p}{RT}, \quad V = \frac{(\rho V)}{\rho} \quad \rightarrow \quad \text{Ma} = \frac{V}{c} \tag{12.167}$$

(4) repeat Steps 2 and 3 to determine the Ma versus p curve, and repeat Step 2 to determine the T versus p curve. The Mach number, Ma, and temperature, T, as a function of the pressure, p, at the downstream section are shown in Figure 12.35.

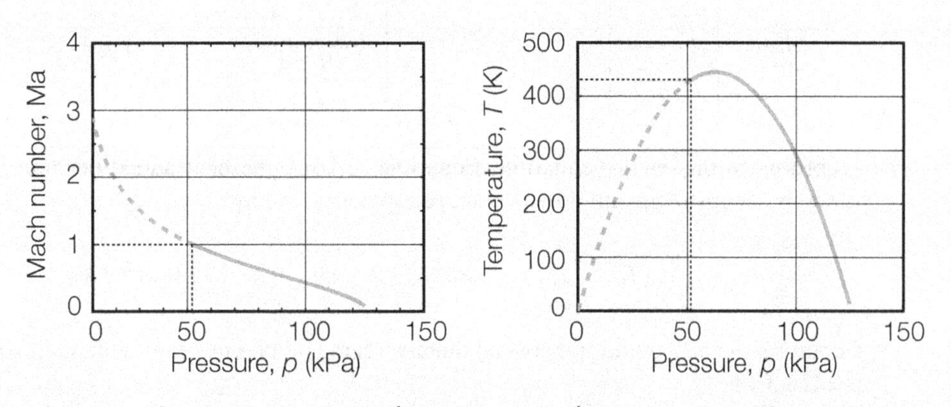

Figure 12.35: Ma and T versus p at downstream section

A key point in these relationships is the pressure at which Ma = 1. The relationships in Equation 12.167 combine to show that Ma = 1 when

$$\frac{(\rho V)RT}{p}\cdot\frac{1}{\sqrt{RTk}}=1 \quad\rightarrow\quad \frac{(176.1)(287.1)T}{p}\cdot\frac{1}{\sqrt{(287.1)T(1.40)}}=1$$

In this case, the above equation combined with Equation 12.164 is satisfied at T = 433.1 K = 160°C and p = 52.5 kPa. If heat is being added to the fluid in the conduit, then the pressure at the downstream section cannot fall below 52.5 kPa.

(c) The downstream pressure is p_2 = 60 kPa = 60 × 10^3 Pa, and the calculation results from part (a) show that the upstream conditions require that C_M = 126.0 kPa and (ρV) = 176.1 kg/m^2·s. The temperature T_2 can be derived from Equation 12.164, which yields

$$T_2=\frac{p_2(C_\text{M}-p_2)}{R(\rho V)^2}=\frac{(60\times10^3)(126.0\times10^3-60.0\times10^3)}{(287.1)(176.1)^2}=444.4\ \text{K}$$

Using the relationships in Equation 12.167 yields

$$c_2=\sqrt{RT_2k}=\sqrt{(287.1)(444.4)(1.40)}=422.6\ \text{m/s},$$

$$\rho_2=\frac{p_2}{RT_2}=\frac{60\times10^3}{(287.1)(444.4)}=0.4703\ \text{kg/m}^3,$$

$$V_2=\frac{(\rho V)}{\rho_2}=\frac{(176.1)}{0.4703}=374.5\ \text{m/s},\qquad \text{Ma}_2=\frac{V_2}{c_2}=\frac{374.5}{422.6}=0.886$$

Hence, the temperature and Mach number at the exit of the conduit are 444.4 K = **171.3°C** and **0.886**, respectively. Because Ma at the upstream section calculated in part (a) is equal to 0.420 and Ma at the downstream section is 0.886 (< 1), in accordance with Rayleigh dynamics, heat is being added between the two sections. At the downstream location, the stagnation temperature, T_{02}, is by definition given by the relation

$$\text{Ma}_2=\left[\frac{\dfrac{T_{02}}{T_2}-1}{\dfrac{k-1}{2}}\right]^{\frac{1}{2}} \quad\rightarrow\quad 0.886=\left[\frac{\dfrac{T_{02}}{444.4}-1}{\dfrac{1.40-1}{2}}\right]^{\frac{1}{2}} \quad\rightarrow\quad T_{02}=514.2\ \text{K}$$

According to the energy equation (Equation 12.161), the heat added per unit mass, q, between the upstream and downstream sections is given by

$$q=c_p(T_{02}-T_{01})=1003(514.2-293.2)=2.22\times10^5\ \text{J/kg}$$

Therefore, the heat addition between the upstream and downstream sections is estimated as **222 kJ/kg**.

12.8.4 Application of Fanno and Rayleigh Relations to Normal Shocks

The normal shock is a sudden transition from supercritical flow to subcritical flow. The relationships between thermodynamic variables across shocks were presented in Section 12.7, and the intent of this section is to show the relationship between normal shocks, Fanno flow, and Rayleigh flow. Consider the shock as illustrated in Figure 12.18 and repeated here in Figure 12.36 for convenience of reference. Because a normal shock occurs over an infinitesimal distance in the flow direction, the steady-state conservation of mass equation requires that

$$\rho V = \text{constant} \tag{12.168}$$

which is a familiar relationship in that it was also applied to both Fanno and Rayleigh flows in conduits having a constant cross sectional area. The rapid shock transition is generally assumed to be adiabatic with negligible wall friction, although the shock transition is non-isentropic. The momentum equation with zero wall friction and the ideal gas law that are applicable across a normal shock can be combined as follows:

$$\left.\begin{array}{l} p + \dfrac{(\rho V)^2}{\rho} = \text{constant} \\ p = \rho R T \end{array}\right\} \rightarrow p + \frac{(\rho V)^2 R T}{p} = \text{constant} = C_\mathrm{M} \tag{12.169}$$

where C_M is the momentum constant. Comparing this result to Equation 12.151 shows that this is the same relationship that is applicable to Rayleigh flow in a duct. Also, the energy equation with zero heat loss, the definition of enthalpy, and the ideal gas law that are all applicable across a normal shock can be combined as follows:

$$\left.\begin{array}{l} h + \frac{1}{2}V^2 = h_0 = \text{constant} \\ h - h_0 = c_p(T - T_0) \\ p = \rho R T \end{array}\right\} \rightarrow T + \frac{(\rho V)^2 T^2}{2c_p(p^2/R^2)} = T_0 = \text{constant} \tag{12.170}$$

Comparing this result to Equations 12.127 and 12.129 shows that this is the same relationship that is applicable to Fanno flow in a duct. The interesting result here is that the flow across a shock is described by the Fanno and Rayleigh equations, and hence, the thermodynamic properties of the fluid before and after the shock must lie on both the Fanno line and the Rayleigh line. This result is illustrated in Figure 12.37, where points x and y represent conditions before and after the shock, respectively. Although it might be a bit confusing to use

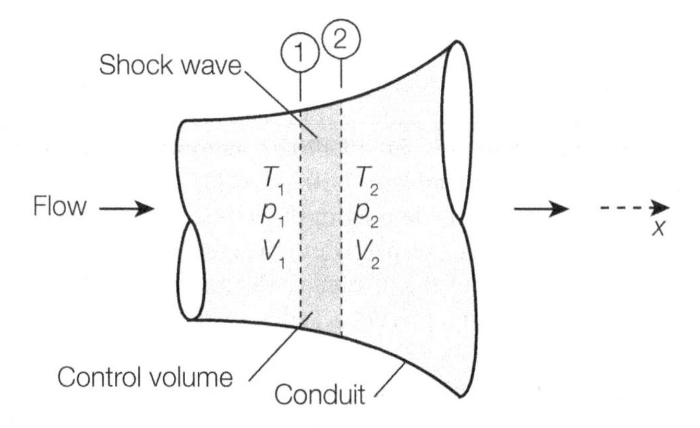

Figure 12.36: Shock

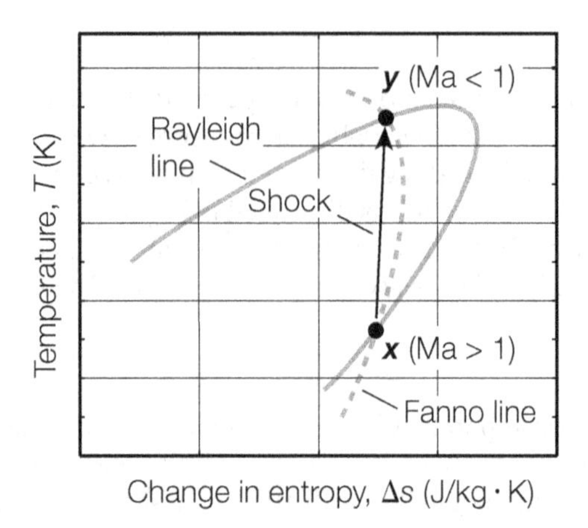

Figure 12.37: Normal shocks in duct flow

x and y to represent conditions before and after a normal shock, respectively, this is the conventional notation used in practice. The representation of a normal shock on a T-s graph as shown in Figure 12.37 also confirms that a normal shock can only proceed from a supersonic ($\text{Ma} > 1$) to a subsonic ($\text{Ma} < 1$) condition because the reverse condition would require a decrease in entropy, which would violate the second law of thermodynamics.

Combining the pressure and temperature relationships developed previously for Fanno and Rayleigh flow yields the following relationships between variables at the upstream and downstream ends of a normal shock:

$$\frac{p_y}{p_x} = \frac{1 + k\text{Ma}_x^2}{1 + k\text{Ma}_y^2} \tag{12.171}$$

$$\frac{T_y}{T_x} = \frac{1 + \dfrac{k-1}{2}\text{Ma}_x^2}{1 + \dfrac{k-1}{2}\text{Ma}_y^2} \tag{12.172}$$

$$\text{Ma}_y^2 = \frac{\text{Ma}_x^2 + \dfrac{2}{k-1}}{\dfrac{2k}{k-1}\text{Ma}_x^2 - 1} \tag{12.173}$$

where the density ratio, ρ_y/ρ_x, can be determined from the pressure and temperature ratios using the ideal gas law and the velocity ratio, V_y/V_x, can be determined using the continuity relation, $\rho V = \text{constant}$. Equations 12.171–12.173 are complementary to the normal shock relationships given previously in Equations 12.107–12.111, the difference being that Equations 12.171–12.173 express the variable relationships in terms of the upstream and downstream Mach numbers (Ma_x and Ma_y, respectively), whereas Equations 12.107–12.111 give the variable relationships in terms of the upstream Mach number, Ma_x, where 1 and 2 are used in Equations 12.107–12.111 to represent x and y, respectively.

It is important to remember that application of the Fanno and Rayleigh equations across a normal shock does not require that Fanno flow and/or Rayleigh flow exist either upstream or downstream of the shock, although such flows could exist within the conduit.

EXAMPLE 12.19

Air enters a 6.0-m-long, 200-mm-diameter insulated conduit at a Mach number of 5.0 and is decelerated to subsonic speed within the conduit such that a normal shock occurs at the midsection of the conduit as illustrated in Figure 12.38. The friction factor throughout the conduit has an average value of 0.018. Determine (a) the Mach number at the exit and (b) the ratio of the exit pressure to the entrance pressure.

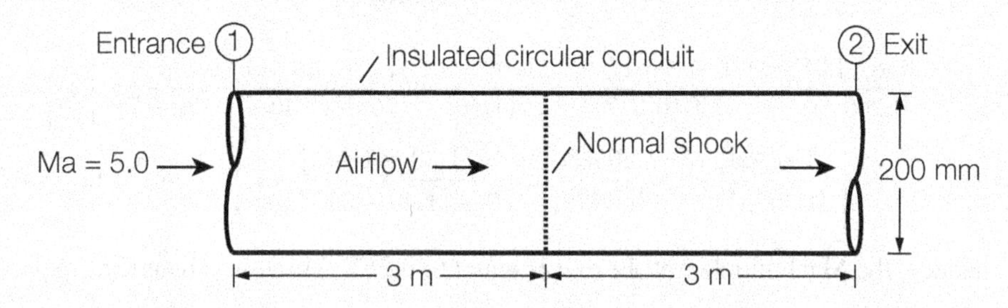

Figure 12.38: Normal shock in a conduit

SOLUTION

From the given data: L = 6.0 m, D = 200 mm = 0.200 m, $\bar{f}$ = 0.018, and Ma_1 = 5.0. For air, R = 287.1 J/kg·K, and assume that c_p = 1003 J/kg·K and k = 1.40. Denote the entrance and exit conditions by the subscripts 1 and 2, respectively, and the sections upstream and downstream of the shock by the subscripts x and y, respectively.

(a) Between the entrance and the shock, Fanno flow can be assumed (i.e., adiabatic and frictional flow), the distance between the entrance and the shock is L_{1x} = 6.0/2 = 3.0 m, and Equations 12.141 and 12.142 combine to give

$$\frac{\bar{f}L_{1x}}{D} = \left\{\frac{1-\mathrm{Ma}_1^2}{k\,\mathrm{Ma}_1^2} + \frac{k+1}{2k}\ln\left[\frac{(k+1)\mathrm{Ma}_1^2}{2+(k-1)\,\mathrm{Ma}_1^2}\right]\right\} - \left\{\frac{1-\mathrm{Ma}_x^2}{k\,\mathrm{Ma}_x^2} + \frac{k+1}{2k}\ln\left[\frac{(k+1)\mathrm{Ma}_x^2}{2+(k-1)\,\mathrm{Ma}_x^2}\right]\right\}$$

$$\frac{(0.018)(3.0)}{0.200} = \left\{\frac{1-(5.0)^2}{1.40\,(5.0)^2} + \frac{1.40+1}{2(1.40)}\ln\left[\frac{(1.40+1)(5.0)^2}{2+(1.40-1)\,(5.0)^2}\right]\right\} - \left\{\frac{1-\mathrm{Ma}_x^2}{(1.40)\,\mathrm{Ma}_x^2} + \frac{1.40+1}{2(1.40)}\ln\left[\frac{(1.40+1)\mathrm{Ma}_x^2}{2+(1.40-1)\,\mathrm{Ma}_x^2}\right]\right\} \quad \rightarrow \quad \mathrm{Ma}_x = 2.46$$

The relationship between Ma_x and Ma_y (i.e., the Mach numbers immediately upstream and downstream of the shock, respectively) is given by Equation 12.173, which yields

$$\mathrm{Ma}_y^2 = \frac{\mathrm{Ma}_x^2 + \dfrac{2}{k-1}}{\dfrac{2k}{k-1}\mathrm{Ma}_x^2 - 1} = \frac{(2.46)^2 + \dfrac{2}{1.40-1}}{\dfrac{2(1.40)}{1.40-1}(2.46)^2 - 1} \quad \rightarrow \quad \mathrm{Ma}_y = 0.517$$

Between the shock and the exit, Fanno flow can again be assumed, the distance between the shock and the exit is L_{y2} = 6.0/2 = 3.0 m, and Equations 12.141 and 12.142 combine to give

$$\frac{\bar{f}L_{y2}}{D} = \left\{\frac{1-\mathrm{Ma}_y^2}{k\,\mathrm{Ma}_y^2} + \frac{k+1}{2k}\ln\left[\frac{(k+1)\mathrm{Ma}_y^2}{2+(k-1)\,\mathrm{Ma}_y^2}\right]\right\} - \left\{\frac{1-\mathrm{Ma}_2^2}{k\,\mathrm{Ma}_2^2} + \frac{k+1}{2k}\ln\left[\frac{(k+1)\mathrm{Ma}_2^2}{2+(k-1)\,\mathrm{Ma}_2^2}\right]\right\}$$

$$\frac{(0.018)(3.0)}{0.200} = \left\{\frac{1-(0.517)^2}{1.40\,(0.517)^2} + \frac{1.40+1}{2(1.40)}\ln\left[\frac{(1.40+1)(0.517)^2}{2+(1.40-1)\,(0.517)^2}\right]\right\} - \left\{\frac{1-\mathrm{Ma}_2^2}{(1.40)\,\mathrm{Ma}_2^2} + \frac{1.40+1}{2(1.40)}\ln\left[\frac{(1.40+1)\mathrm{Ma}_2^2}{2+(1.40-1)\,\mathrm{Ma}_2^2}\right]\right\} \quad \rightarrow \quad \mathrm{Ma}_2 = 0.561.$$

Hence, the Mach number at the exit is equal to **0.561**, which is subsonic as expected.

(b) The ratio of the exit pressure, p_2, to the entrance pressure, p_1, will be developed algebraically and then the given values will be substituted to yield the desired result. First, it is recognized that (ρV) is a constant by virtue of conservation of mass, constant cross-sectional area, and steady state. The value of (ρV) can be calculated at section 1 as follows:

$$(\rho V) = \rho_1 V_1 = \rho_1 \frac{V_1}{c_1} c_1 = \rho_1 \mathrm{Ma}_1 \sqrt{RT_1 k} = \frac{p_1}{RT_1}\mathrm{Ma}_1\sqrt{RT_1 k}$$

$$\rightarrow \quad (\rho V) = p_1 \mathrm{Ma}_1 \sqrt{\frac{k}{RT_1}} \tag{12.174}$$

Using the continuity equation and Equation 12.174, the relationship between the Mach number, Ma, at any location in the conduit and Ma_1 can be developed as follows:

$$\mathrm{Ma} = \frac{V}{c} = \frac{(\rho V)}{\rho\sqrt{RTk}} = \frac{p_1 \mathrm{Ma}_1\sqrt{\dfrac{k}{RT_1}}}{\dfrac{p}{RT}\sqrt{RTk}} \quad \rightarrow \quad \mathrm{Ma}^2 = \frac{p_1^2}{p^2}\frac{T}{T_1}\mathrm{Ma}_1^2 \tag{12.175}$$

Because the conduit is insulated and the flow is frictional, Fanno flow can be assumed. For Fanno flow between the entrance to the conduit (section 1) and upstream of the shock (section x), the energy equation (Equation 12.129) gives

$$T_1 + \frac{(\rho V)^2 T_1^2}{2c_p(p_1^2/R^2)} = T_x + \frac{(\rho V)^2 T_x^2}{2c_p(p_x^2/R^2)}$$

Substituting Equation 12.174 for (ρV) and simplifying yields

$$T_1 + \left(\frac{kR}{2c_p}\mathrm{Ma}_1^2\right)T_1 = T_x + \left(\frac{p_1^2}{p_x^2}\frac{T_x}{T_1}\mathrm{Ma}_1^2\frac{kR}{2c_p}\right)T_x \tag{12.176}$$

At this point, to simplify the algebra, it is convenient to define the constant c_0, where

$$c_0 = \frac{kR}{2c_p} \tag{12.177}$$

Combining Equations 12.175–12.177 yields the following temperature in terms of known variables:

$$\frac{T_x}{T_1} = \frac{1 + c_0 \text{Ma}_1^2}{1 + c_0 \text{Ma}_x^2} \tag{12.178}$$

Similarly, for Fanno flow between the downstream of the shock (section y) and the exit of the conduit (section 2),

$$\frac{T_2}{T_y} = \frac{1 + c_0 \text{Ma}_y^2}{1 + c_0 \text{Ma}_2^2} \tag{12.179}$$

The ratio of the temperatures before and after the shock can be determined from Equation 12.172, which yields

$$\frac{T_y}{T_x} = \frac{1 + \dfrac{k-1}{2}\text{Ma}_x^2}{1 + \dfrac{k-1}{2}\text{Ma}_y^2} \tag{12.180}$$

Combining Equations 12.178–12.180 yields the ratio of the exit to entrance temperature as

$$\frac{T_2}{T_1} = \frac{T_x}{T_1} \cdot \frac{T_y}{T_x} \cdot \frac{T_2}{T_y} \quad \rightarrow \quad \frac{T_2}{T_1} = \frac{1 + c_0 \text{Ma}_1^2}{1 + c_0 \text{Ma}_x^2} \cdot \frac{1 + \dfrac{k-1}{2}\text{Ma}_x^2}{1 + \dfrac{k-1}{2}\text{Ma}_y^2} \cdot \frac{1 + c_0 \text{Ma}_y^2}{1 + c_0 \text{Ma}_2^2} \tag{12.181}$$

The pressure ratio can then be determined from Equation 12.175, which requires that

$$\text{Ma}_2^2 = \frac{p_1^2}{p_2^2}\frac{T_2}{T_1}\text{Ma}_1^2 \quad \rightarrow \quad \frac{p_2}{p_1} = \frac{\text{Ma}_1}{\text{Ma}_2}\sqrt{\frac{T_2}{T_1}} \tag{12.182}$$

Finally, combining Equations 12.181 and 12.182 yields the following equation in which the pressure ratio, p_2/p_1, is expressed in terms of knows quantities:

$$\frac{p_2}{p_1} = \frac{\text{Ma}_1}{\text{Ma}_2}\sqrt{\frac{1 + c_0 \text{Ma}_1^2}{1 + c_0 \text{Ma}_x^2} \cdot \frac{1 + \dfrac{k-1}{2}\text{Ma}_x^2}{1 + \dfrac{k-1}{2}\text{Ma}_y^2} \cdot \frac{1 + c_0 \text{Ma}_y^2}{1 + c_0 \text{Ma}_2^2}} \tag{12.183}$$

From the given data and from the Mach numbers calculated in part (a),

$$c_0 = \frac{(1.40)(287.1)}{2(1003)} = 0.2004, \quad \text{Ma}_1 = 5.0, \quad \text{Ma}_x = 2.46, \quad \text{Ma}_y = 0.517, \quad \text{Ma}_2 = 0.561$$

Substituting these data into Equation 12.183 yields

$$\frac{p_2}{p_1} = \frac{5.0}{0.444}\sqrt{\frac{1 + (0.2004)(5.0)^2}{1 + (0.2004)(2.46)^2} \cdot \frac{1 + \dfrac{1.40 - 1}{2}(2.46)^2}{1 + \dfrac{1.40 - 1}{2}(0.517)^2} \cdot \frac{1 + (0.2004)(0.517)^2}{1 + (0.2004)(0.561)^2}} = 21.2$$

Hence, the exit pressure is **21.2** times higher than the entrance pressure. Such a high pressure ratio is primarily due to the high Mach number at which air impacts the entrance to the conduit, where $\text{Ma}_1 = 5.0$.

12.9 Oblique Shocks, Bow Shocks, and Expansion Waves

In contrast to one-dimensional compressible flows that are constrained within conduits, *oblique shocks* and *expansion waves* are three-dimensional shocks that are generated by flows that are supersonic relative to a solid surface.

12.9.1 Oblique Shocks

Oblique shocks occur when supersonic flow is suddenly compressed as it is deflected, and such shocks are commonly found at the leading edges of supersonic airfoils, at sharp corners, and at the noses of bullets traveling at supersonic speeds. Recall the occurrence of a Mach cone, which is the front of pressure waves generated as an object moves at supersonic speed through a stationary fluid or, equivalently, the front of pressure waves generated as a fluid flows at supersonic speed around a stationary object. The Mach cone is inclined at an angle α relative to the direction of flow. In reality, solid objects are not points and the finite size of an object causes the shock to spread at a wider angle than α as illustrated in Figure 12.39. In this case, the shock is called an *oblique shock* and is oriented at an angle β to the flow direction, where $\beta > \alpha$. The angle β is commonly called the *shock angle* or *wave angle*. At a far enough distance from the object, the influence of the shape of the object diminishes and $\beta \to \alpha$. As the fluid crosses the shock, the fluid encounters an environment that is influenced by the presence of the solid object, and this influence causes the incoming velocity vector to deflect by an angle θ as shown in Figure 12.40(a). The angle θ is commonly called the *turning angle*, the *deflection angle*, or the *wedge angle*. In reality, the turning angle, θ, is slightly greater that the wedge half angle, but using the wedge half angle to approximate θ is a reasonable approximation. Typically, the thickness of the boundary layer adjacent to the solid object is neglected, which is a reasonable approximation because the Reynolds number is very high. The finite deflection angle associated with an oblique shock is in contrast to the zero deflection angle associated with a normal shock; any deflection of a supersonic velocity must be accompanied by an oblique shock.

Governing equations. For purposes of analysis, it is convenient to consider the components of the velocity that are normal and tangential to the shock and to consider the infinitesimal control volume shown in Figure 12.40(b). For this control volume, assuming steady-state,

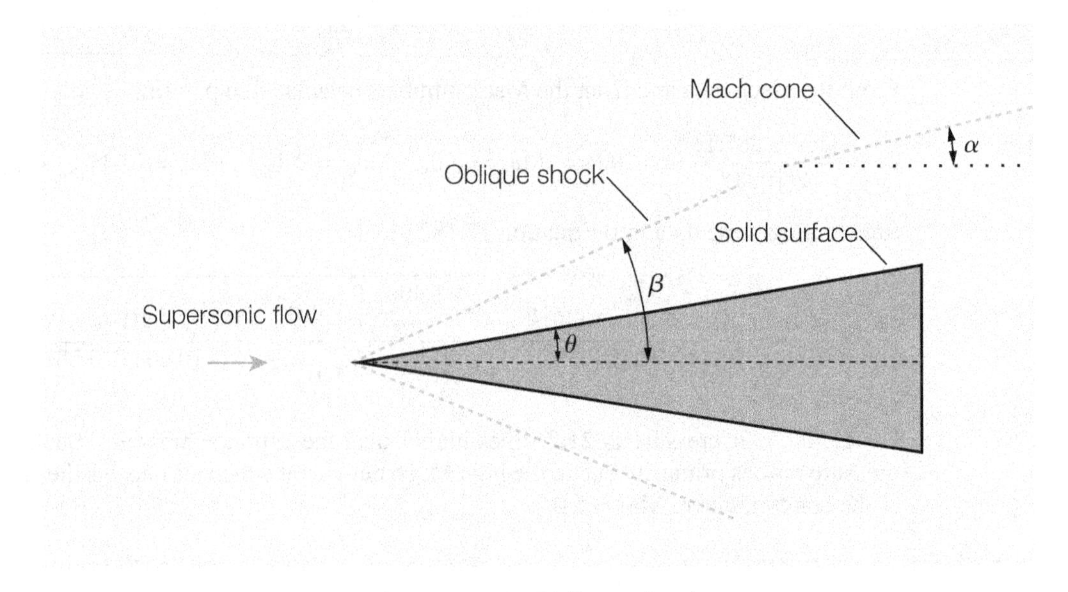

Figure 12.39: Oblique shock

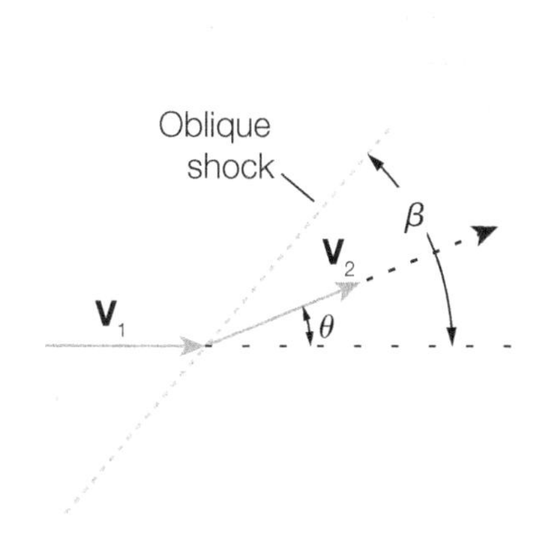

(a) Velocity change across an oblique shock

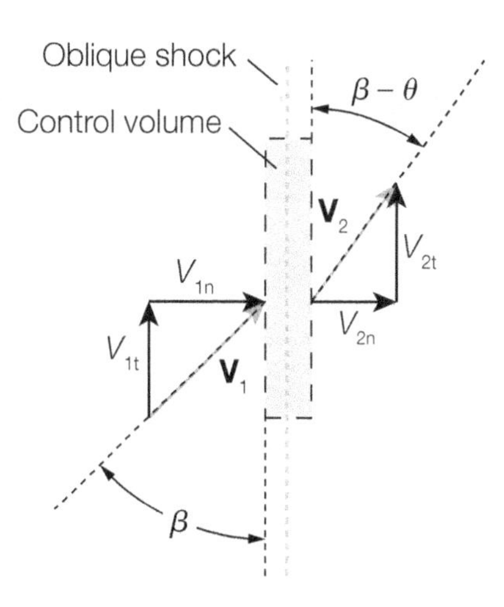

(b) Normal and tangential velocities

Figure 12.40: Velocity change across an oblique shock

neglecting gravitational effects, and assuming adiabatic flow across the shock, the governing equations are as follows:

$$\text{continuity: } \rho_1 V_{1\text{n}} = \rho_2 V_{2\text{n}} \tag{12.184}$$

$$\text{n-momentum: } p_1 + \rho_1 V_{1\text{n}}^2 = p_2 + \rho_2 V_{2\text{n}}^2 \tag{12.185}$$

$$\text{t-momentum: } V_{1\text{t}} = V_{2\text{t}} \tag{12.186}$$

$$\text{energy: } h_1 + \frac{1}{2}V_{1\text{n}}^2 = h_2 + \frac{1}{2}V_{2\text{n}}^2 \tag{12.187}$$

$$\text{second law: } s_2 > s_1 \tag{12.188}$$

Comparing these governing equations with those derived for a normal shock indicates that they are identical, with V_1 and V_2 in the normal shock being replaced by $V_{1\text{n}}$ and $V_{2\text{n}}$, respectively. Therefore, all of the concepts and equations derived previously for normal shocks can be applied to oblique shocks, provided the velocities are replaced by their normal components. For both oblique shocks and normal shocks, the upstream flow is always supersonic. However, in contrast to the normal shock, the flow downstream of an oblique shock can be supersonic, sonic, or subsonic, depending on local conditions. This statement does not contradict the fact that the governing equations of a normal shock are the same as those for the normal components of an oblique shock, because in an oblique shock, the normal component Mach number upstream of the shock is generally supersonic and the normal component Mach number downstream of an oblique shock is generally subsonic. The normal components of the velocities and the corresponding Mach numbers are given by

$$V_{1\text{n}} = V_1 \sin\beta, \qquad \text{Ma}_{1\text{n}} = \frac{V_{1\text{n}}}{c_1} = \text{Ma}_1 \sin\beta \tag{12.189}$$

$$V_{2\text{n}} = V_2 \sin(\beta - \theta), \quad \text{Ma}_{2\text{n}} = \frac{V_{2\text{n}}}{c_2} = \text{Ma}_2 \sin(\beta - \theta) \tag{12.190}$$

Using the previously derived normal shock equations relating the pre-shock and post-shock fluid properties gives the following relationships that are applicable to oblique shocks:

$$\mathrm{Ma}_{2n}^2 = \frac{\mathrm{Ma}_{1n}^2 + \dfrac{2}{k-1}}{\dfrac{2k}{k-1}\mathrm{Ma}_{1n}^2 - 1} \tag{12.191}$$

$$\frac{p_2}{p_1} = \frac{2k}{k+1}\mathrm{Ma}_{1n}^2 - \frac{k-1}{k+1} \tag{12.192}$$

$$\frac{T_2}{T_1} = \frac{\left(1 + \dfrac{k-1}{2}\mathrm{Ma}_{1n}^2\right)\left(k\mathrm{Ma}_{1n}^2 - \dfrac{k-1}{2}\right)}{\left(\dfrac{k+1}{2}\right)^2 \mathrm{Ma}_{1n}^2} \tag{12.193}$$

$$\frac{\rho_2}{\rho_1} = \frac{V_{1n}}{V_{2n}} = \frac{\dfrac{k+1}{2}\mathrm{Ma}_{1n}^2}{1 + \dfrac{k-1}{2}\mathrm{Ma}_{1n}^2} \tag{12.194}$$

Because flow through the shock is assumed to be adiabatic, the stagnation temperature is the same on both sides of the shock. The relationship between the post-shock and pre-shock stagnation pressures (p_{02} and p_{01}, respectively) can be derived by combining the continuity, momentum, and state equations. The relations between the stagnation properties are given by

$$\frac{p_{02}}{p_{01}} = \frac{\left[\dfrac{\dfrac{k+1}{2}\mathrm{Ma}_{1n}^2}{1 + \dfrac{k-1}{2}\mathrm{Ma}_{1n}^2}\right]^{\frac{k}{k-1}}}{\left[\dfrac{2k}{k+1}\mathrm{Ma}_{1n}^2 - \dfrac{k-1}{k+1}\right]^{\frac{1}{k-1}}} \tag{12.195}$$

$$\frac{T_{02}}{T_{01}} = 1 \tag{12.196}$$

Also, using the velocity relationship given by Equation 12.194 and the geometric relationship illustrated in Figure 12.40 yields the following alternative expressions for the deflection angle, θ, as a function of the oblique shock angle, β, and the upstream Mach number, Ma_1:

$$\tan\theta = \frac{2\cot\beta(\mathrm{Ma}_1^2\sin^2\beta - 1)}{\mathrm{Ma}_1^2(k + \cos 2\beta) + 2} \quad \text{or} \quad \tan(\beta - \theta) = \frac{\tan\beta}{k+1}\left[k - 1 + \frac{2}{\mathrm{Ma}_1^2\sin^2\beta}\right] \tag{12.197}$$

Although Equation 12.197 is expressed in a form where θ is calculated for given values of β and Ma_1, in reality, θ is the given angle caused by the object (such as an airplane wing), and this angle results in an oblique shock oriented at the angle β. Thus, β usually needs to be calculated for a given θ.

EXAMPLE 12.20

Consider the case where a supersonic flow impacts an oblique shock as shown in Figure 12.41. The shock is oriented at an angle of 25° to the incoming flow, which has a velocity of 1580 m/s, a temperature of –5°C, and a pressure of 95 kPa. (a) Determine the temperature, pressure, and velocity on the downstream side of the oblique shock. (b) Verify that the deflection angle θ calculated using Equations 12.191–12.194 is the same as the deflection angle calculated directly using Equation 12.197.

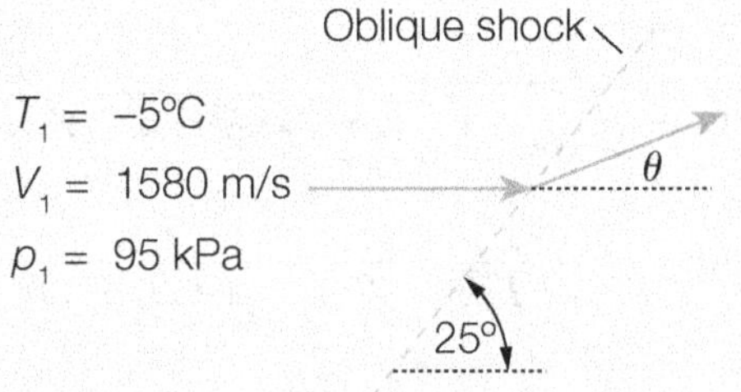

Oblique shock

T_1 = –5°C
V_1 = 1580 m/s
p_1 = 95 kPa

θ

25°

Figure 12.41: Conditions upstream and downstream of an oblique shock

SOLUTION

From the given data: β = 25° = 0.4363 rad, V_1 = 1580 m/s, $T_1 = -5°\text{C} = 268.2$ K, and p_1 = 95 kPa. For air, assume that R = 287.1 J/kg·K and k = 1.40.

(a) From the given upstream conditions, the components of the velocity and Mach number that are normal and tangential to the shock are calculated as follows:

$$V_{1n} = V_1 \sin\beta = (1580)\sin(0.4363) = 667.7 \text{ m/s}$$

$$V_{1t} = V_1 \cos\beta = (1580)\cos(0.4363) = 1432 \text{ m/s}$$

$$\text{Ma}_{1n} = \frac{V_{1n}}{\sqrt{RT_1k}} = \frac{667.7}{\sqrt{(287.1)(268.2)(1.40)}} = 2.034$$

$$\text{Ma}_1 = \frac{V_1}{\sqrt{RT_1k}} = \frac{1580}{\sqrt{(287.1)(268.2)(1.40)}} = 4.813$$

Using these derived quantities, the downstream flow and fluid properties can be determined from Equations 12.193 and 12.194 as follows:

$$\frac{T_2}{T_1} = \frac{\left[1 + \dfrac{k-1}{2}\text{Ma}_{1n}^2\right]\left[k\text{Ma}_{1n}^2 - \dfrac{k-1}{2}\right]}{\left(\dfrac{k+1}{2}\right)^2 \text{Ma}_{1n}^2}$$

$$\rightarrow \frac{T_2}{268.2} = \frac{\left[1 + \dfrac{1.40-1}{2}(2.034)^2\right]\left[1.40(2.034)^2 - \dfrac{1.40-1}{2}\right]}{\left(\dfrac{1.40+1}{2}\right)^2 (2.034)^2} \rightarrow T_2 = \mathbf{459.9\ K}$$

$$\frac{p_2}{p_1} = \frac{2k}{k+1}\mathrm{Ma}_{1n}^2 - \frac{k-1}{k+1} \quad \rightarrow \quad \frac{p_2}{95} = \frac{2(1.40)}{1.40+1}(2.034)^2 - \frac{1.40-1}{1.40+1} \quad \rightarrow \; p_2 = \mathbf{442.7\ kPa}$$

$$\frac{V_{1n}}{V_{2n}} = \frac{\frac{k+1}{2}\mathrm{Ma}_{1n}^2}{1+\frac{k-1}{2}\mathrm{Ma}_{1n}^2} \quad \rightarrow \quad \frac{667.7}{V_{2n}} = \frac{\frac{1.40+1}{2}(2.034)^2}{1+\frac{1.40-1}{2}(2.034)^2} \quad \rightarrow \; V_{2n} = \mathbf{245.8\ m/s}$$

Noting that $V_{2t} = V_{1t} = 1432$ m/s, the magnitude of the velocity, V_2, downstream of the shock and the corresponding Mach number are given by

$$V_2 = \sqrt{V_{2n}^2 + V_{2t}^2} = \sqrt{245.8^2 + 1432^2} = 1439 \text{ m/s}$$

$$\mathrm{Ma}_2 = \frac{V_2}{\sqrt{RT_2k}} = \frac{1439}{(287.1)(459.9)(1.40)} = 3.379$$

Finally, the deflection angle, θ, between the upstream and downstream flow directions can be calculated using Equation 12.189, which gives

$$V_{2n} = V_2 \sin(\beta - \theta) \quad \rightarrow \quad 245.8 = 1439 \sin(0.4363 - \theta) \quad \rightarrow \quad \theta = 0.2663 \text{ rad} = 15.26^\circ$$

Therefore, the flow velocity upstream of the shock is deflected by approximately **15.3°** as it passes through the shock.

(b) Applying the relations given in Equation 12.197 to calculate θ directly yields

$$\begin{aligned}\tan\theta &= \frac{2\cot\beta[\mathrm{Ma}_1^2\sin^2\beta - 1]}{\mathrm{Ma}_1^2[k+\cos 2\beta]+2} \\ &= \frac{2\cot(0.4363)[(4.813)^2\sin^2(0.4363)-1]}{(4.813)^2[1.40+\cos 2(0.4363)]+2} \\ &\rightarrow \quad \theta = 0.2663 \text{ rad}\end{aligned}$$

and

$$\begin{aligned}\tan(\beta-\theta) &= \frac{\tan\beta}{k+1}\left[k-1+\frac{2}{\mathrm{Ma}_1^2\sin^2\beta}\right] \\ &= \frac{\tan(0.4363)}{1.40+1}\left[1.40-1+\frac{2}{(4.813)^2\sin^2(0.4363)}\right] \quad \rightarrow \\ &\rightarrow \quad \theta = 0.2663 \text{ rad}\end{aligned}$$

Therefore, the use of the relations given in Equation 12.197 to calculate θ directly is **validated**.

Phenomena related to oblique shocks. There are two possible solutions for β in Equation 12.197 for any given value of θ and Ma_1. These two solutions correspond to a *weak shock* and a *strong shock*, where a weak shock corresponds to the smaller value of β and the smaller value of Ma_{1n} and the strong shock corresponds to the larger value of β and the larger value of Ma_{1n}. The flow downstream of a strong shock is always subsonic, whereas the flow downstream of a weak shock is usually supersonic, but can be subsonic if the deflection angle is large. An oblique shock can exist only if $\mathrm{Ma}_{1n} > 1$ and $\mathrm{Ma}_{2n} < 1$. Other interesting results derived from Equation 12.197 are as follows:

- For a given deflection angle, θ, there is a minimum value of Ma_1 for which there is only one oblique shock angle, β.
- For a given deflection angle, θ, if Ma_1 is less than the minimum Mach number for that particular curve, no oblique shock exists and the shock becomes detached.
- For a given Ma_1, there is a sufficiently large deflection angle, θ, for which the shock will be detached.
- In the limit as the deflection angle approaches zero (i.e., $\theta \to 0$), the weak shock becomes a Mach wave and $\beta \to \alpha = \sin^{-1}(1/\mathrm{Ma}_1)$.

The aforementioned properties can be illustrated by plotting θ versus β for given values of Ma_1 as illustrated in Figure 12.42. It is apparent from the figure that for any given value of Ma_1, there is a maximum value of θ, referred to as θ_{max}, beyond which there is no solution for β, and for $\theta < \theta_{\mathrm{max}}$, there are two solutions for β. For example, from Figure 12.42, if $\mathrm{Ma}_1 = 2.0$, then $\theta_{\mathrm{max}} = 23.0°$, and if $\theta = 10.0°$, there are two possible shock angles—$\beta = 39.3°$ and $\beta = 83.7°$. In reality, these results indicate that for $\mathrm{Ma}_1 = 2.0$, an oblique shock cannot exist if $\theta > 23.0°$, and if $\theta = 10.0°$, there are two possible shock angles—$\beta = 39.3°$ (weak shock) and $\beta = 83.9°$ (strong shock). In the case of a weak shock, it can be shown that $\mathrm{Ma}_2 > 1$ except for a narrow range of θ just below θ_{max}, and in the case of a strong shock, $\mathrm{Ma}_2 < 1$ always. Based on the aforementioned results, if a fluid flow at

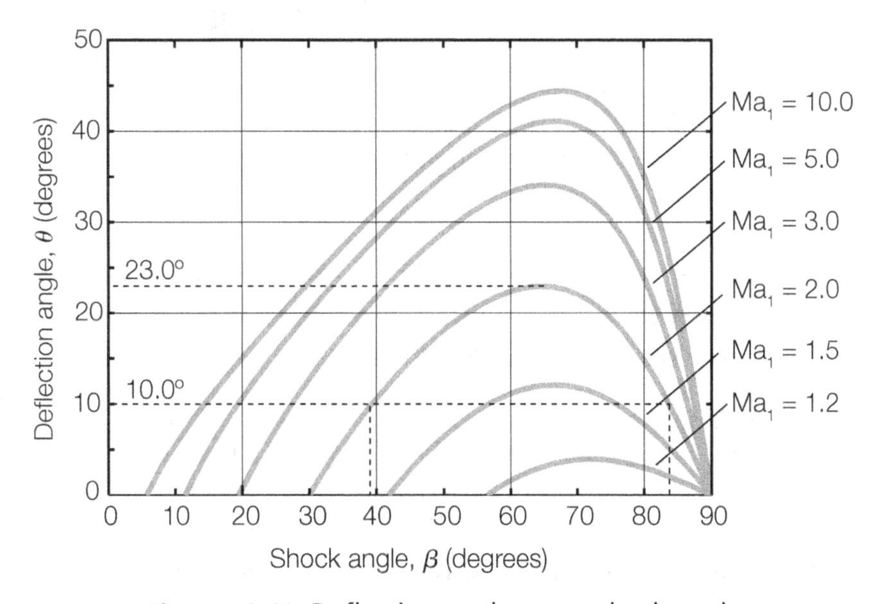

Figure 12.42: Deflection angle versus shock angle

$\mathrm{Ma}_1 = 2.0$ encounters a wedge with a half angle (i.e., θ) greater than 23.0°, then an oblique shock attached to the wedge is impossible and the shock will be detached; the properties of such detached shocks will be discussed subsequently. It can further be shown using Equation 12.197 with $k = 1.4$ (for air) that as $\mathrm{Ma}_1 \to \infty$, $\theta_{\mathrm{max}} \to 45.6°$, the deflection angle beyond which it is impossible to have an attached oblique shock; this indicates why most blunt bodies (with high values of θ) tend to have detached shocks. Setting $\theta = 0$ in Equation 12.197 yields

$$\beta\bigg|_{\theta=0} = \sin^{-1}\left(\frac{1}{\mathrm{Ma}_1}\right)$$

which is the same as the Mach angle (α) given by Equation 12.41, derived for a solid object of infinitesimal width.

EXAMPLE 12.21

Airflow at a Mach number of 2.4 and a pressure of 80 kPa impinges on a wedge with an included angle of 17° as shown in Figure 12.43. (a) Determine the wave angle, β, corresponding to the weak shock and the wave angle corresponding to the strong shock. (b) For each of the two possible shocks, determine the Mach number and pressure after the shock.

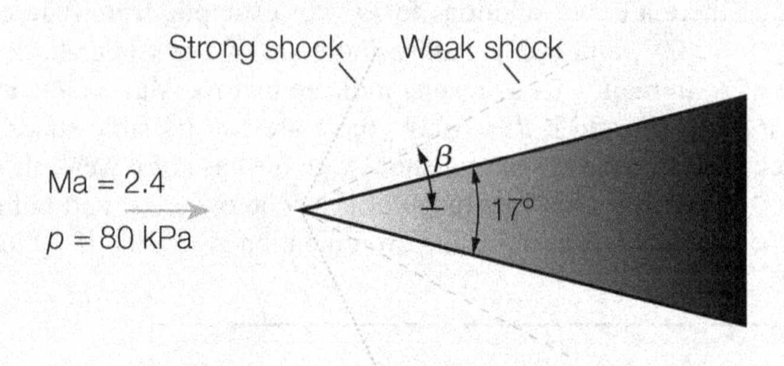

Figure 12.43: Strong and weak shocks

SOLUTION

From the given data: $\mathrm{Ma}_1 = 2.4$, $p_1 = 80$ kPa, and $\theta = 17°/2 = 8.5°$. For air, $R = 287.1$ J/kg·K and $k = 1.40$.

(a) The wave angle, β, must satisfy Equation 12.197, where

$$\tan\theta = \frac{2\cot\beta(\mathrm{Ma}_1^2\sin^2\beta - 1)}{\mathrm{Ma}_1^2(k + \cos 2\beta) + 2} \quad \to \quad \tan 8.5° = \frac{2\cot\beta(2.4^2\sin^2\beta - 1)}{2.4^2(1.40 + \cos 2\beta) + 2}$$

$$\to \quad \beta = 31.6° \text{ and } 86.1°$$

where $\beta = \mathbf{31.6°}$ is the wave angle of the weak shock and $\beta = \mathbf{86.1°}$ is the wave angle of the strong shock.

(b) For the weak shock, $\beta = 31.6°$ and the downstream Mach number and pressure are calculated as follows:

$$\text{Ma}_{1\text{n}} = \text{Ma}_1 \sin\beta = (2.4)\sin 31.6° = 1.258$$

$$\text{Ma}_{2\text{n}}^2 = \frac{\text{Ma}_{1\text{n}}^2 + \dfrac{2}{k-1}}{\dfrac{2k}{k-1}\text{Ma}_{1\text{n}}^2 - 1} \quad \rightarrow \quad \text{Ma}_{2\text{n}}^2 = \frac{(1.258)^2 + \dfrac{2}{1.40-1}}{\dfrac{2(1.40)}{1.40-1}(1.258)^2 - 1}$$

$$\rightarrow \quad \text{Ma}_{2\text{n}} = 0.8083$$

$$\text{Ma}_2 = \frac{\text{Ma}_{2\text{n}}}{\sin(\beta - \theta)} = \frac{0.8083}{\sin(31.6° - 8.5°)} = 2.06$$

$$\frac{p_2}{p_1} = \frac{2k}{k+1}\text{Ma}_{1\text{n}}^2 - \frac{k-1}{k+1} \quad \rightarrow \quad \frac{p_2}{80} = \frac{2(1.40)}{1.40+1}(1.258)^2 - \frac{1.40-1}{1.40+1}$$

$$\rightarrow \quad p_2 = 134 \text{ kPa}$$

For the strong shock, $\beta = 86.1°$ and the downstream Mach number and pressure are calculated as follows:

$$\text{Ma}_{1\text{n}} = \text{Ma}_1 \sin\beta = (2.4)\sin 86.1° = 2.394$$

$$\text{Ma}_{2\text{n}}^2 = \frac{\text{Ma}_{1\text{n}}^2 + \dfrac{2}{k-1}}{\dfrac{2k}{k-1}\text{Ma}_{1\text{n}}^2 - 1} \quad \rightarrow \quad \text{Ma}_{2\text{n}}^2 = \frac{(2.394)^2 + \dfrac{2}{1.40-1}}{\dfrac{2(1.40)}{1.40-1}(2.394)^2 - 1}$$

$$\rightarrow \quad \text{Ma}_{2\text{n}} = 0.5237$$

$$\text{Ma}_2 = \frac{\text{Ma}_{2\text{n}}}{\sin(\beta - \theta)} = \frac{0.5237}{\sin(86.1° - 8.5°)} = 0.536$$

$$\frac{p_2}{p_1} = \frac{2k}{k+1}\text{Ma}_{1\text{n}}^2 - \frac{k-1}{k+1} \quad \rightarrow \quad \frac{p_2}{80} = \frac{2(1.40)}{1.40+1}(2.394)^2 - \frac{1.40-1}{1.40+1}$$

$$\rightarrow \quad p_2 = 522 \text{ kPa}$$

Therefore, the downstream Mach number (Ma_2) is **2.06** in a weak shock compared with **0.536** in a strong shock; the downstream pressure (p_2) is **134 kPa** in a weak shock versus **522 kPa** in a strong shock. From these results, it is apparent that the change in Mach number and pressure are much greater across a strong shock compared with a weak shock.

For problems in which β and Ma_1 are known, the usual approach is to use Equation 12.189 to calculate Ma_{1n} and then use Equation 12.191 to calculate Ma_{2n}. Oblique shocks continue to bend in a downstream direction until the Mach number of the component of the velocity normal to the wave is equal to unity. At this point, the oblique shock has become a Mach wave, across which changes in flow properties are infinitesimal.

12.9.2 Bow Shocks and Detached Shocks

Bow shocks and, equivalently, *detached shocks* are typically formed when supersonic flow encounters a (blunt) body in which the deflection angle, θ, that the surface of the body makes with the flow direction exceeds the maximum deflection angle that can exist with an oblique wave. The occurrence of a bow shock is contrasted with that of an oblique shock in Figure 12.44(a) and (b). For detached shocks, a normal shock (i.e., $\beta = 90°$) occurs on the stagnation streamline; then a strong oblique shock (i.e., $\beta < 90°$) occurs as one moves away from the stagnation streamline, then a weak oblique shock, and ultimately a Mach wave (i.e., $\beta = \alpha$). For blunt bodies moving at supersonic speeds, the shock is always detached. The occurrence of a bow shock ahead of a blunt body is illustrated in Figure 12.44(d) and can be contrasted with the occurrence of an oblique shock ahead of a sharp body in Figure 12.44(c).

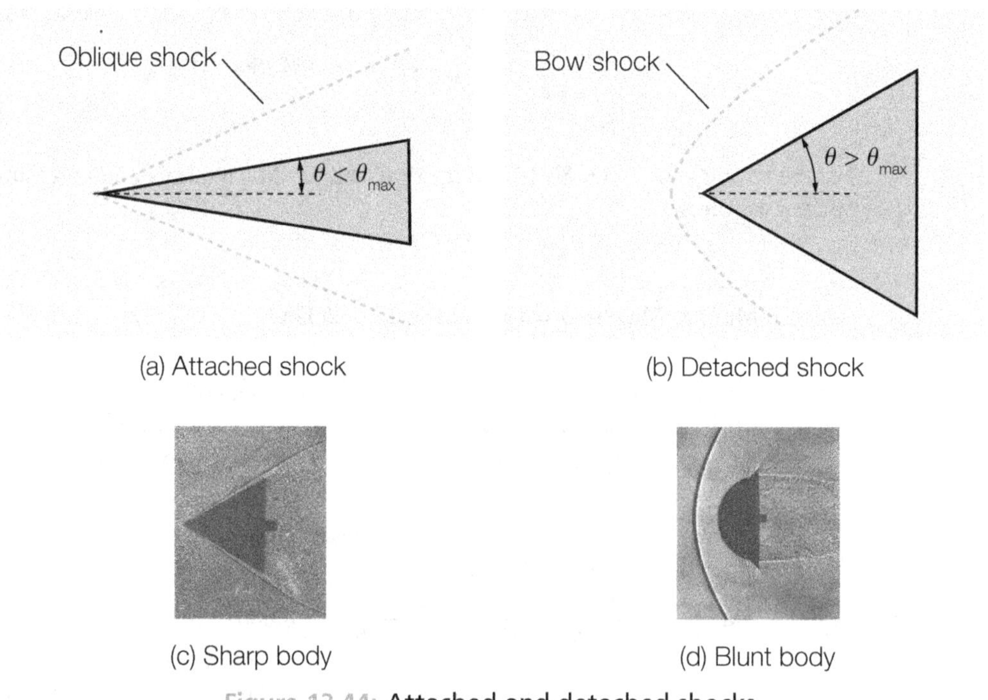

Figure 12.44: Attached and detached shocks
Source for (c) and (d): NASA.

EXAMPLE 12.22

A wedge-shaped body with a variable apex angle of α moves through stagnant air at a Mach speed of 3.0 as shown in Figure 12.45. Find the maximum apex angle for which the generated oblique shock will remain attached to the body.

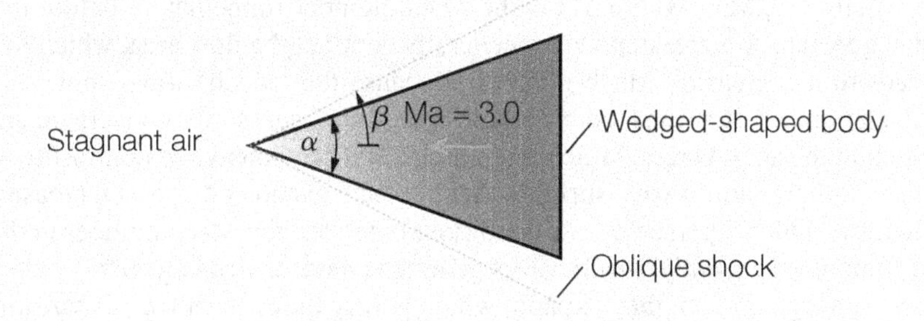

Figure 12.45: **Shock generated by a wedge-shaped body**

SOLUTION

The deflection angle, θ, is related to the apex angle, α, by $\theta = \alpha/2$. Therefore, the relationship between α and the angle that the shock makes with the flow direction, β, is given by Equation 12.197 as

$$\alpha = 2\tan^{-1}\left[\frac{2\cot\beta(\text{Ma}_1^2\sin^2\beta - 1)}{\text{Ma}_1^2(k + \cos 2\beta) + 2}\right] \tag{12.198}$$

Because the body travels at $\text{Ma}_1 = 3$, Equation 12.198 gives the relationship between α and β as

$$\alpha = 2\tan^{-1}\left[\frac{2\cot\beta(3^2\sin^2\beta - 1)}{3^2(k + \cos 2\beta) + 2}\right] \tag{12.199}$$

This equation is plotted in Figure 12.46, which shows that α has a maximum value of **68.1**°, which corresponds to β = 65.2°.

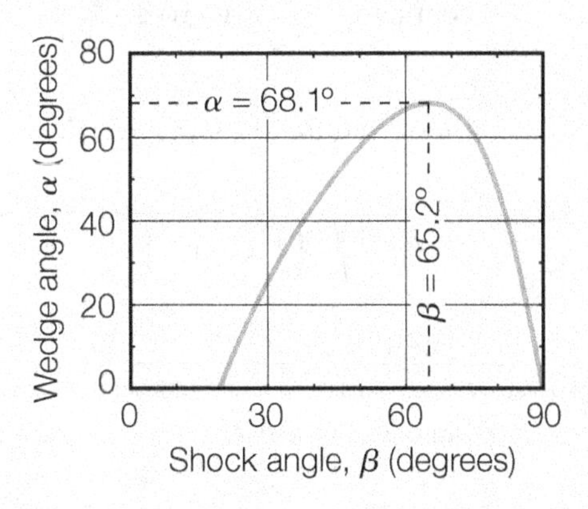

Figure 12.46: **Wedge angle versus shock angle**

In practical terms, this means that the requirement for an attached shock to exist is that $\alpha \leq 68.1°$. For $\alpha > 68.1°$, no value of β satisfies Equation 12.199, so the shock will be detached.

12.9.3 Isentropic Expansion Waves

Compression waves and *expansion waves* occur when a supersonic flow is gradually redirected, such as when encountering a curved surface. Mechanisms of generating isentropic compression and expansion waves are illustrated in Figure 12.48, where compression waves are generated by surfaces that curve upward into the flow direction and expansion waves are generated by surfaces that curve downward. Fundamentally, each wave generated at the solid surface is a Mach wave, so the angle, α, that it generates with the flow direction is given by $\alpha = \sin^{-1}(1/\text{Ma})$, where Ma is the Mach number immediately before the wave. Surfaces that generate compression waves generally contract the flow area, which for supersonic flow, leads to a decreasing Mach number. Because the (Mach) wave angle, α, increases as the Mach number decreases, compression waves are necessarily convergent and ultimately lead to oblique shock waves, which were addressed previously. In contrast to surfaces that generate compression waves, surfaces that generate expansion waves increase the flow area and thereby (under supersonic flow conditions) increase the Mach number in the downstream direction. Consequently, Mach waves generated have decreasing wave angles, α, and hence do not converge; in fact, they expand, which is why they are called "expansion waves."

Governing equations. Consider the isentropic wave shown in Figure 12.47(a), where the magnitude of the flow velocity changes by $\mathrm{d}V$ and is deflected by an amount $\mathrm{d}\theta$ as it passes across the isentropic wave. This scenario is similar to that considered previously for an oblique shock, with the two main differences being that: (1) the Mach wave angle, α, is used as the wave angle, which was previously represented as β; and (2) the change in velocity across the wave is infinitesimal and represented by $\mathrm{d}V$. The change in other fluid properties, such as temperature and pressure, are also assumed to be infinitesimal. The assumption of infinitesimal changes (along with the adiabatic assumption) are consistent with the assumption that the flow is isentropic across the wave. It is convenient to consider a control volume oriented along the isentropic wave as shown in Figure 12.47(b). Assuming steady state, neglecting gravitational forces, and assuming adiabatic flow, the governing equations can then be expressed as

$$\text{continuity:} \int_{\text{cs}} \rho(\mathbf{v}\cdot\mathbf{n})\,\mathrm{d}A = 0 \tag{12.200}$$

$$\text{t-momentum:} \int_{\text{cs}} V_{\text{t}}\rho(\mathbf{v}\cdot\mathbf{n})\,\mathrm{d}A = 0 \tag{12.201}$$

$$\text{energy:} \int_{\text{cs}} \left(h + \frac{1}{2}V^2\right)\rho(\mathbf{v}\cdot\mathbf{n})\,\mathrm{d}A = 0 \tag{12.202}$$

$$\text{State 1: } \mathrm{d}h = c_p\,\mathrm{d}T \tag{12.203}$$

$$\text{State 2: } c = \sqrt{kRT} \tag{12.204}$$

where the "t-momentum" equation is the momentum equation applied in the direction that is tangential to the inflow and outflow surfaces of the control volume and V_{t} is the tangential component of the velocity. Applying Equations 12.200–12.204 to the control volume shown in Figure 12.47 yields the deflection angle, $\mathrm{d}\theta$, as

$$\mathrm{d}\theta = -\frac{2\sqrt{\text{Ma}^2 - 1}}{2 + \text{Ma}^2(k-1)}\frac{\mathrm{dMa}}{\text{Ma}} \tag{12.205}$$

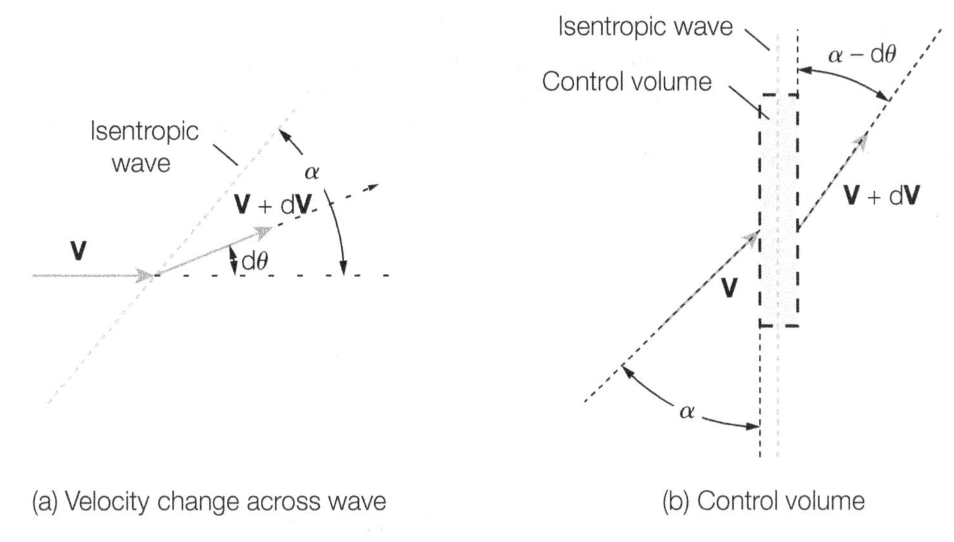

Figure 12.47: Control-volume analysis

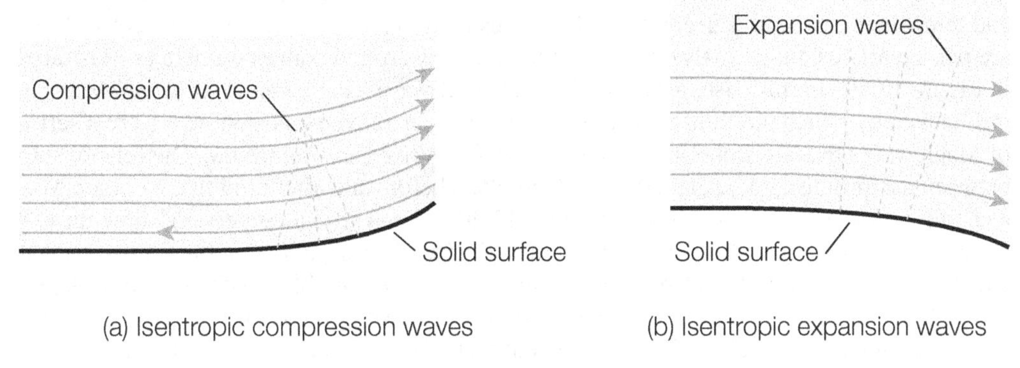

Figure 12.48: Isentropic compression and expansion waves

where Ma is the local Mach number of the incoming flow. Equation 12.205 is typically applied to expansion waves where $d\theta$ is negative, so it is common to use change variables to ω such that

$$d\omega = -d\theta \tag{12.206}$$

It is also conventional practice to combine Equations 12.205 and 12.206 and to express ω in an integral form such that

$$\int_0^\omega d\omega' = \int_1^{\mathrm{Ma}} \frac{2\sqrt{\mathrm{Ma}^2 - 1}}{2 + \mathrm{Ma}^2(k-1)} \frac{d\mathrm{Ma}}{\mathrm{Ma}} \tag{12.207}$$

where a reference state of Ma = 1 is taken to correspond to $\omega = 0$. Performing the integral in Equation 12.207 yields

$$\omega = \sqrt{\frac{k+1}{k-1}} \tan^{-1}\left(\sqrt{\frac{k-1}{k+1}(\mathrm{Ma}^2 - 1)}\right) - \tan^{-1}(\sqrt{\mathrm{Ma}^2 - 1}) \tag{12.208}$$

Hence, the deflection angle θ caused by the isentropic expansion from Ma_1 to Ma_2 is given by

$$\theta = \omega(\mathrm{Ma}_2) - \omega(\mathrm{Ma}_1) \tag{12.209}$$

The ω function given by Equation 12.208 is called the *Prandtl-Meyer supersonic expansion function*,[68] (or simply the *Prandtl-Meyer function*) and is used for analyzing both isentropic expansion waves and isentropic compression waves. In applying Equation 12.208, it is usually necessary also to relate the fluid properties to the isentropic stagnation properties, and these latter relationships are repeated here for convenient reference.

$$\frac{p_0}{p} = \left[1 + \frac{k-1}{2}\mathrm{Ma}^2\right]^{\frac{k}{k-1}} \tag{12.210}$$

$$\frac{T_0}{T} = 1 + \frac{k-1}{2}\mathrm{Ma}^2 \tag{12.211}$$

$$\frac{\rho_0}{\rho} = \left[1 + \frac{k-1}{2}\mathrm{Ma}^2\right]^{\frac{1}{k-1}} \tag{12.212}$$

Because the flow in expansion waves can typically be approximated as being isentropic, the isentropic stagnation properties p_0, T_0, and ρ_0 remain constant and the isentropic flow relationships given by Equations 12.210–12.212 can be used to derive the downstream pressure (p_2), temperature (T_2), and density (ρ_2); under supersonic conditions, the pressure, density, and temperature generally decrease in the downstream direction. Equations 12.209–12.212 are particularly useful in analyzing the supersonic flow around convex corners as illustrated in Figure 12.49. In this case, a series of (expansion) waves, called an *expansion fan*, is the mechanism by which the fluid turns the corner. The generated expansion waves are a series of Mach waves that are sometimes called *Prandtl-Meyer expansion waves*. The relationship between the turning angle, θ, and the Mach numbers before and after turning the corner, Ma_1 and Ma_2, respectively, is given by Equation 12.209. Typically, as supersonic flow turns a convex corner, both the Mach number and the velocity increase, allowing the flow to remain attached to the wall for much larger turning angles than are possible in subsonic flow. A plot of the Prandtl-Meyer supersonic expansion function, $\omega(\mathrm{Ma})$, is shown in Figure 12.50 for $k = 1.40$, which is typical of air. It is apparent that $\omega(1) = 0°$ and $\omega(\infty) = 130.45°$, which

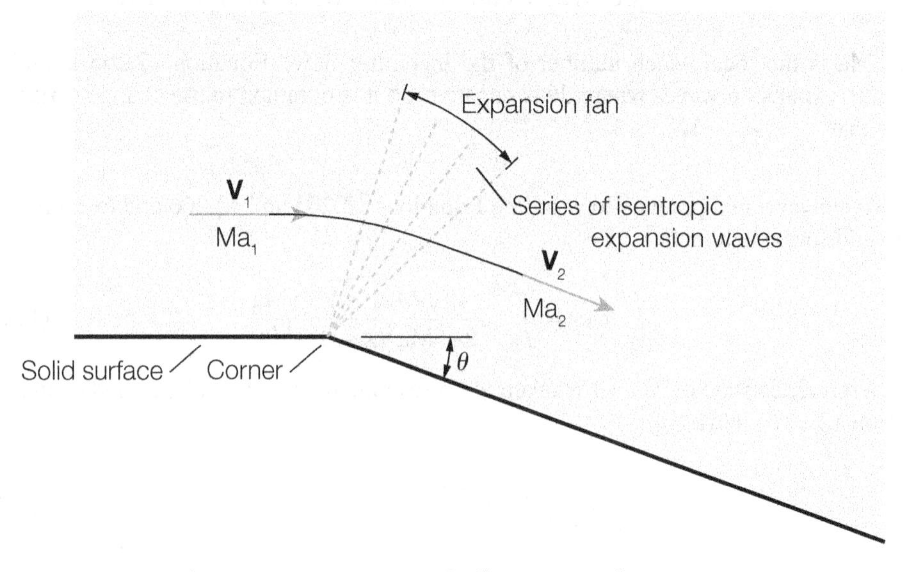

Figure 12.49: Supersonic flow around a corner

[68] Named in honor of the German engineer Ludwig Prandtl (1875–1953) and the German mathematician Theodor Meyer (1882–1972).

means that for $Ma_1 = 1$, the deflection angle is zero when $Ma_2 = 1$ and that for $Ma_1 = 1$, the asymptotic maximum deflection angle is 130.45° when $Ma_2 = \infty$. The extreme turning angle of 130.45° is practically impossible because it would require that both the temperature and pressure of the air be absolute zero, and the air would liquify before this condition could be attained. Nevertheless, comparably large turning angles that are greater than 90° can be attained under supersonic conditions, something that would not be expected based on everyday experience. Such large turning angles for supersonic flow must generally be taken into account in the design of rocket engines that discharge their exhaust gases into the vacuum of space, because these supersonic exhausts could turn through sufficient angles to damage other parts of the spacecraft.

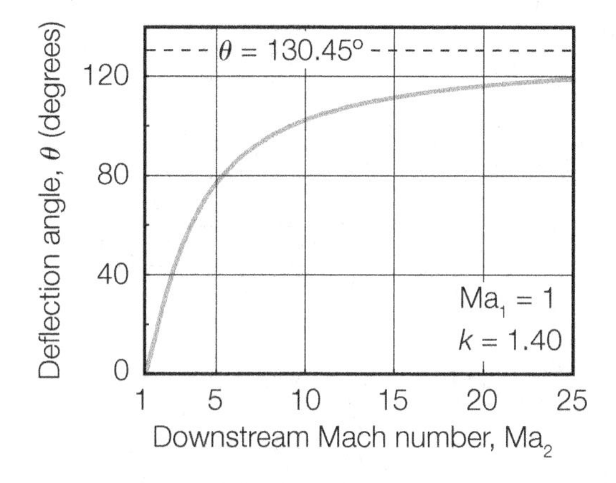

Figure 12.50: Prandtl-Meyer supersonic expansion function for air

EXAMPLE 12.23

Consider the case where supersonic flow impacts the sharp corner as shown in Figure 12.51. Upstream of the corner, the flow velocity is 550 m/s, the temperature is 8°C, and the pressure is 95 kPa. Estimate the pressures on the top and bottom surfaces just downstream of the leading edge.

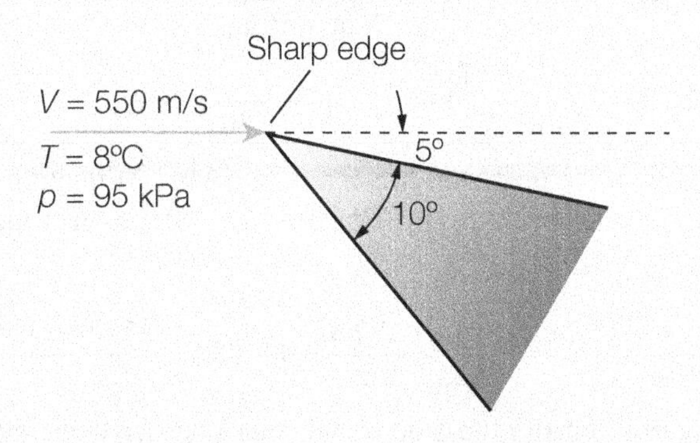

Figure 12.51: Supersonic flow past a sharp corner

SOLUTION

For the given scenario, it is expected that an expansion wave will occur above the sharp edge and an oblique shock will occur below the sharp edge. From the given data: $V_1 = 550$ m/s, $T_1 = 8°\text{C} = 281.2$ K, and $p_1 = 95$ kPa. For air, assume that $R = 287.1$ J/kg·K and $k = 1.40$. The Mach number of the upstream flow is given by

$$\text{Ma}_1 = \frac{V_1}{\sqrt{RT_1k}} = \frac{550}{\sqrt{(287.1)(281.2)(1.40)}} = 1.636$$

For the top surface, assuming the existence of an expansion wave, applying the Prandtl-Meyer supersonic expansion function (Equations 12.208 and 12.209), and noting that the given deflection angle is 5°, requires that

$$\theta = \omega(\text{Ma}_2) - \omega(\text{Ma}_1)$$

$$\theta = \left[\sqrt{\frac{k+1}{k-1}}\tan^{-1}\left(\sqrt{\frac{k-1}{k+1}(\text{Ma}_2^2-1)}\right) - \tan^{-1}(\sqrt{\text{Ma}_2^2-1})\right] - \left[\sqrt{\frac{k+1}{k-1}}\tan^{-1}\left(\sqrt{\frac{k-1}{k+1}(\text{Ma}_1^2-1)}\right) - \tan^{-1}(\sqrt{\text{Ma}_1^2-1})\right]$$

$$5° = \left[\sqrt{\frac{1.40+1}{1.40-1}}\tan^{-1}\left(\sqrt{\frac{1.40-1}{1.40+1}(\text{Ma}_2^2-1)}\right) - \tan^{-1}(\sqrt{\text{Ma}_2^2-1})\right] - \left[\sqrt{\frac{1.40+1}{1.40-1}}\tan^{-1}\left(\sqrt{\frac{1.40-1}{1.40+1}(1.636^2-1)}\right) - \tan^{-1}(\sqrt{1.636^2-1})\right]$$

Solution of this equation yields $\text{Ma}_2 = 1.807$. Because the flow is assumed to be isentropic across an expansion wave, the stagnation pressure, p_0, is the same both upstream and downstream of the expansion wave, in which case Equation 12.210 gives

$$\frac{p_2}{p_1} = \left[\frac{1+\frac{k-1}{2}\text{Ma}_1^2}{1+\frac{k-1}{2}\text{Ma}_2^2}\right]^{\frac{k}{k-1}} \quad\to\quad \frac{p_2}{95} = \left[\frac{1+\frac{1.40-1}{2}(1.636)^2}{1+\frac{1.40-1}{2}(1.807)^2}\right]^{\frac{1.40}{1.40-1}} \quad\to\quad p_2 = 73.4\ \text{kPa}$$

Below the sharp edge, an oblique shock is assumed to occur with a given deflection angle of $\theta = 15°$. The corresponding wave angle, β, satisfies Equation 12.197 such that

$$\tan(\beta-\theta) = \frac{\tan\beta}{k+1}\left[k-1+\frac{2}{\text{Ma}_1^2\sin^2\beta}\right]$$

$$\tan(\beta-15°) = \frac{\tan\beta}{1.40+1}\left[1.40-1+\frac{2}{(1.636)^2\sin^2\beta}\right] \quad\to\quad \beta = 60.9°$$

The normal component of the upstream Mach number, Ma_{1n} (Equation 12.189), and the pressure ratio across the oblique shock, p_2/p_1 (Equation 12.192), are as follows:

$$\text{Ma}_{1n} = \text{Ma}_1\sin\beta = (1.636)\sin(60.9°) = 1.430$$

$$\frac{p_2}{p_1} = \frac{2k}{k+1}\text{Ma}_{1n}^2 - \frac{k-1}{k+1} \quad\to\quad \frac{p_2}{95} = \frac{2(1.40)}{1.40+1}(1.430)^2 - \frac{1.40-1}{1.40+1} \quad\to\quad p_2 = 210.7\ \text{kPa}$$

In summary, the pressure on the top side of the sharp corner is estimated as approximately **73 kPa** and the pressure on the bottom side as approximately **211 kPa**. These results hint at a net upward force at the front edge of the corner.

Key Equations in Compressible Flow

The following list of equations is particularly useful in solving problems in compressible flow. If one is able to recognize these equations and recall their appropriate use, then the learning objectives of this chapter have been met to a significant degree. Derivations of these equations, definitions of the variables, and detailed examples of usage can be found in the main text.

PRINCIPLES OF THERMODYNAMICS

Ideal gas law: $\rho = \dfrac{p}{RT}$

Specific heat, c_v: $\mathrm{d}u = c_v\,\mathrm{d}T, \qquad u_2 - u_1 = c_v(T_2 - T_1)$

Enthalpy: $h = u + \dfrac{p}{\rho}$

Stagnation enthalpy: $h_0 = h + \dfrac{V^2}{2}$

Specific heat, c_p: $\mathrm{d}h = c_p\,\mathrm{d}T, \qquad h_2 - h_1 = c_p(T_2 - T_1)$

Specific heat relationships: $c_p - c_v = R, \quad k = \dfrac{c_p}{c_v}, \quad c_p = \dfrac{kR}{k-1}, \quad c_v = \dfrac{R}{k-1}$

Entropy change: $\mathrm{d}S = \displaystyle\int_{\text{rev}} \frac{\delta Q}{T} \quad \text{or} \quad \mathrm{d}S = \left.\frac{\delta Q}{T}\right|_{\text{rev}}$

Second law of thermodynamics: $\mathrm{d}S \geq \dfrac{\delta Q}{T} \quad \text{or} \quad T\,\mathrm{d}S \geq \delta Q$

Second $T\,\mathrm{d}s$ equation: $T\,\mathrm{d}s = \mathrm{d}h - v\,\mathrm{d}p$

Change in entropy:

$$\mathrm{d}s = \frac{\mathrm{d}u}{T} + \frac{p}{T}\,\mathrm{d}v = c_v\frac{\mathrm{d}T}{T} + R\frac{\mathrm{d}v}{v}$$

$$\mathrm{d}s = \frac{\mathrm{d}h}{T} - \frac{v}{T}\,\mathrm{d}p = c_p\frac{\mathrm{d}T}{T} - R\frac{\mathrm{d}p}{p}$$

Isentropic change:

$$T\,v^{k-1} = \text{constant} \quad \equiv \quad \frac{T}{\rho^{k-1}} = \text{constant}$$

$$T\,p^{\frac{1-k}{k}} = \text{constant}$$

$$p\,v^{k} = \text{constant} \quad \equiv \quad \frac{p}{\rho^{k}} = \text{constant}$$

THE SPEED OF SOUND

Mach number: $\text{Ma} = \dfrac{V}{c}$

Speed of sound: $c = \sqrt{\left.\dfrac{dp}{d\rho}\right|_s}$ (gas), $c = \sqrt{\dfrac{E_v}{\rho}}$ (liquid/solid)

$c = \sqrt{kRT}$ (ideal gas)

Mach cone angle: $\alpha = \sin^{-1}\left(\dfrac{1}{\text{Ma}}\right)$

Sound pressure, decibels: $L_p = 20\log_{10}\left(\dfrac{p_{\text{rms}}}{p_{\text{ref}}}\right)$ dB

THERMODYNAMIC REFERENCE CONDITIONS

Stagnation pressure (streamline): $\dfrac{p_0}{p} = \left[1 + \dfrac{k-1}{2}\text{Ma}^2\right]^{\frac{k}{k-1}}$

Stagnation temperature (streamline): $\dfrac{T_0}{T} = 1 + \dfrac{k-1}{2}\text{Ma}^2$, $\dfrac{T_0}{T} = \left(\dfrac{p_0}{p}\right)^{\frac{k-1}{k}}$

Stagnation density (streamline): $\dfrac{\rho_0}{\rho} = \left[1 + \dfrac{k-1}{2}\text{Ma}^2\right]^{\frac{1}{k-1}}$, $\dfrac{\rho_0}{\rho} = \left(\dfrac{p_0}{p}\right)^{\frac{1}{k}}$

Adiabatic flow: $T_0 = T + \dfrac{V^2}{2c_p}$

Isentropic critical condition: $\dfrac{p_0}{p^*} = \left[\dfrac{k+1}{2}\right]^{\frac{k}{k-1}}$, $\dfrac{T_0}{T^*} = \dfrac{k+1}{2}$

$\dfrac{\rho_0}{\rho^*} = \left[\dfrac{k+1}{2}\right]^{\frac{1}{k-1}}$, $V^* = \sqrt{\dfrac{2k}{k+1}RT_0}$

BASIC EQUATIONS OF ONE-DIMENSIONAL COMPRESSIBLE FLOW

Continuity equation: $\rho_1 V_1 A_1 = \rho_2 V_2 A_2 = \rho V A = \dot{m}$

Momentum equation: $-F_x + p_1 A_1 - p_2 A_2 = \dot{m}(V_2 - V_1)$

Energy equation: $\dfrac{\dot{Q}}{\dot{m}} + h_1 + \dfrac{1}{2}V_1^2 = h_2 + \dfrac{1}{2}V_2^2$

Second law of thermodynamics: $\dot{m}(s_2 - s_1) \geq \displaystyle\int_{\text{cs}} \frac{1}{T}\left(\frac{\dot{Q}}{A}\right) dA$

Equations of state: $\rho = \dfrac{p}{RT}$, $h_2 - h_1 = c_p(T_2 - T_1)$

$s_2 - s_1 = c_p \ln\dfrac{T_2}{T_1} - R\ln\dfrac{p_2}{p_1}$

STEADY ONE-DIMENSIONAL ISENTROPIC FLOW

Critical area:

$$\frac{A}{A^*} = \frac{1}{\mathrm{Ma}}\left[\frac{1 + \frac{k-1}{2}\mathrm{Ma}^2}{\frac{k+1}{2}}\right]^{\frac{k+1}{2(k-1)}}$$

Velocity-area relationship:

$$\frac{\mathrm{d}V}{\mathrm{d}A} = -\frac{V}{A}\cdot\frac{1}{1-\mathrm{Ma}^2}$$

For constant flow area:

$$\frac{\mathrm{d}V}{V} = -\frac{\mathrm{d}\rho}{\rho}$$

Density-area relationship:

$$\frac{\mathrm{d}\rho}{\mathrm{d}A} = \frac{\rho}{A}\frac{\mathrm{Ma}^2}{1-\mathrm{Ma}^2}$$

Efficiency of nozzle:

$$\eta_{\mathrm{n}} = \frac{h_0 - h_{\mathrm{e}}}{h_0 - h_{\mathrm{es}}}$$

Pressure recovery factor:

$$C_p = \frac{\Delta p_{\mathrm{actual}}}{\Delta p_{\mathrm{isentropic}}}$$

Pressure at critical condition:

$$\frac{p_{\mathrm{e}}}{p_0} = \frac{p_{\mathrm{b}}}{p_0} = \frac{p^*}{p_0} = \left(\frac{2}{k+1}\right)^{\frac{k}{k-1}}$$

Mass flow rate:

$$\dot{m} = \frac{A\,\mathrm{Ma}\,p_0\sqrt{\frac{k}{RT_0}}}{\left(1 + \frac{k-1}{2}\mathrm{Ma}^2\right)^{\frac{k+1}{2(k-1)}}}$$

$$\dot{m} = \rho_{\mathrm{t}}A_{\mathrm{t}}V_{\mathrm{t}} = A_{\mathrm{t}}\sqrt{\frac{2k}{k-1}p_0\rho_0\left[\left(\frac{p_{\mathrm{t}}}{p_0}\right)^{\frac{2}{k}} - \left(\frac{p_{\mathrm{t}}}{p_0}\right)^{\frac{k+1}{k}}\right]}$$

(subsonic exit)

Maximum mass flow rate:

$$\dot{m}_{\max} = A_{\mathrm{t}}p_0\left(\frac{2}{k+1}\right)^{\frac{k+1}{2(k-1)}}\sqrt{\frac{k}{RT_0}} \quad \text{(sonic exit)}$$

Converging-diverging nozzle:

$$\frac{p}{p^*} = \left[\frac{\frac{k+1}{2}}{1 + \frac{k-1}{2}\mathrm{Ma}^2}\right]^{\frac{k}{k-1}}, \qquad \frac{T}{T^*} = \frac{\frac{k+1}{2}}{1 + \frac{k-1}{2}\mathrm{Ma}^2}$$

$$\frac{\rho}{\rho^*} = \left[\frac{\frac{k+1}{2}}{1 + \frac{k-1}{2}\mathrm{Ma}^2}\right]^{\frac{1}{k-1}}$$

$$\dot{m}_{\max} = \dot{m}_{\mathrm{choked}} = A_{\mathrm{t}}p_0\left(\frac{2}{k+1}\right)^{\frac{k+1}{2(k-1)}}\sqrt{\frac{k}{RT_0}}$$

NORMAL SHOCKS

Continuity equation: $$\rho_1 V_1 = \rho_2 V_2 = \frac{\dot{m}}{A}$$

Momentum equation: $$p_1 A - p_2 A = \dot{m}(V_2 - V_1)$$

Energy equation: $$h_1 + \frac{1}{2}V_1^2 = h_2 + \frac{1}{2}V_2^2$$

Second law of thermodynamics: $$\dot{m}(s_2 - s_1) \geq \int_{\text{cs}} \frac{1}{T}\left(\frac{\dot{Q}}{A}\right) \mathrm{d}A \quad \rightarrow \quad \dot{m}(s_2 - s_1) > 0$$

Equations of state: $$\rho = \frac{p}{RT}, \qquad h_2 - h_1 = c_p(T_2 - T_1)$$

$$s_2 - s_1 = c_p \ln \frac{T_2}{T_1} - R \ln \frac{p_2}{p_1}$$

Energy equation: $$h + \frac{1}{2}V^2 = h_0 = \text{constant}, \quad T + \frac{V^2}{2c_p} = T_0 = \text{constant}$$

Across normal shock: $$T_{01} = T_{02} = T_0, \qquad \text{Ma}_2^2 = \frac{\text{Ma}_1^2 + \dfrac{2}{k-1}}{\dfrac{2k}{k-1}\text{Ma}_1^2 - 1}$$

$$\frac{p_2}{p_1} = \frac{2k}{k+1}\text{Ma}_1^2 - \frac{k-1}{k+1}$$

$$\frac{T_2}{T_1} = \frac{\left(1 + \dfrac{k-1}{2}\text{Ma}_1^2\right)\left(k\,\text{Ma}_1^2 - \dfrac{k-1}{2}\right)}{\left(\dfrac{k+1}{2}\right)^2 \text{Ma}_1^2}$$

$$\frac{\rho_2}{\rho_1} = \frac{V_1}{V_2} = \frac{\dfrac{k+1}{2}\text{Ma}_1^2}{1 + \dfrac{k-1}{2}\text{Ma}_1^2}$$

$$\frac{p_{02}}{p_{01}} = \frac{\left[\dfrac{\dfrac{k+1}{2}\text{Ma}_1^2}{1 + \dfrac{k-1}{2}\text{Ma}_1^2}\right]^{\frac{k}{k-1}}}{\left[\dfrac{2k}{k+1}\text{Ma}_1^2 - \dfrac{k-1}{k+1}\right]^{\frac{1}{k-1}}}$$

$$s_2 - s_1 = c_p \ln \frac{2 + (k-1)\text{Ma}_1^2}{2 + (k-1)\text{Ma}_2^2} - R \ln \frac{1 + k\text{Ma}_1^2}{1 + k\text{Ma}_2^2}$$

$$s_2 - s_1 = R \ln \left[\frac{p_1}{p_2}\left(\frac{T_2}{T_1}\right)^{\frac{k}{k-1}}\right] = R \ln \frac{p_{01}}{p_{02}}$$

$$\frac{p_e}{p_{01}}\frac{A_e}{A_t} = \frac{p_e}{p_{0e}}\frac{A_e}{A_e^*}, \qquad \frac{p_{02}}{p_{01}} = \frac{A_t}{A_e}\frac{A_e}{A_e^*}$$

STEADY ONE-DIMENSIONAL NON-ISENTROPIC FLOW

Continuity equation:

$$\rho_1 V_1 = \rho_2 V_2 = \frac{\dot{m}}{A}$$

Momentum equation:

$$-F_x + p_1 A - p_2 A = \dot{m}(V_2 - V_1)$$

Energy equation:

$$\frac{\dot{Q}}{\dot{m}} + h_1 + \frac{1}{2}V_1^2 = h_2 + \frac{1}{2}V_2^2$$

Second law of thermodynamics:

$$\dot{m}(s_2 - s_1) \geq \int_{cs} \frac{1}{T}\left(\frac{\dot{Q}}{A}\right) dA$$

Equations of state:

$$\rho = \frac{p}{RT}, \qquad h_2 - h_1 = c_p(T_2 - T_1)$$

$$s_2 - s_1 = c_p \ln\frac{T_2}{T_1} - R\ln\frac{p_2}{p_1}$$

Adiabatic flow with friction:

$$T + \frac{V^2}{2c_p} = T_0 = \text{constant}$$

Fanno line:

$$s - s_1 = c_p \ln\frac{T}{T_1} - R\ln\frac{p}{p_1}$$

Adiabatic flow in conduit:

$$\frac{p}{p^*} = \frac{1}{\text{Ma}}\left[\frac{k+1}{2+(k-1)\,\text{Ma}^2}\right]^{\frac{1}{2}}, \qquad \frac{T}{T^*} = \frac{k+1}{2+(k-1)\,\text{Ma}^2}$$

$$\frac{V}{V^*} = \frac{\rho^*}{\rho} = \text{Ma}\left[\frac{k+1}{2+(k-1)\,\text{Ma}^2}\right]^{\frac{1}{2}}$$

$$\frac{p_0}{p_0^*} = \frac{1}{\text{Ma}}\left[\frac{2+(k-1)\,\text{Ma}^2}{k+1}\right]^{\frac{k+1}{2(k-1)}}$$

Critical length:

$$\frac{\bar{f}L^*}{D} = \frac{1-\text{Ma}^2}{k\,\text{Ma}^2} + \frac{k+1}{2k}\ln\left[\frac{(k+1)\text{Ma}^2}{2+(k-1)\,\text{Ma}^2}\right]$$

Adiabatic/isentropic flow:

$$\frac{p_1}{p_2} = \left(\frac{T_1}{T_2}\right)^{\frac{k}{k-1}}$$

$$T_0 = T_1\left(1+\frac{k-1}{2}\text{Ma}_1^2\right) = T_2\left(1+\frac{k-1}{2}\text{Ma}_2^2\right)$$

$$\frac{p_1}{p_2} = \left(\frac{1+\dfrac{k-1}{2}\text{Ma}_2^2}{1+\dfrac{k-1}{2}\text{Ma}_1^2}\right)^{\frac{k}{k-1}}$$

Isothermal/frictional flow:

$$\frac{V_1^2}{2} + \dot{q} = \frac{V_2^2}{2}, \qquad \frac{\bar{f} L_{\max}}{D_h} = \frac{1 - k\,\mathrm{Ma}^2}{k\,\mathrm{Ma}^2} + \ln(k\,\mathrm{Ma}^2)$$

$$\frac{p}{p'} = \frac{1}{k^{\frac{1}{2}}\mathrm{Ma}}, \qquad \frac{\rho'}{\rho} = \frac{V}{V'} = k^{\frac{1}{2}}\mathrm{Ma}$$

$$(\rho V)^2 = \frac{p_1^2 - p_2^2}{RT\left[\dfrac{\bar{f}L}{D_h} + 2\ln\left(\dfrac{p_1}{p_2}\right)\right]}$$

Rayleigh line:

$$p + \frac{(\rho V)^2 RT}{p} = C_M, \qquad s - s_1 = c_p \ln\frac{T}{T_1} - R\ln\frac{p}{p_1}$$

Diabatic frictionless flow:

$$\frac{p}{p^*} = \frac{1+k}{1+k\,\mathrm{Ma}^2}, \qquad \frac{T}{T^*} = \left[\frac{(1+k)\mathrm{Ma}}{1+k\,\mathrm{Ma}^2}\right]^2$$

$$\frac{V}{V^*} = \frac{\rho^*}{\rho} = \frac{(1+k)\,\mathrm{Ma}^2}{1+k\,\mathrm{Ma}^2}$$

$$\frac{p_0}{p_0^*} = \frac{1+k}{1+k\,\mathrm{Ma}^2}\left[\left(\frac{2}{k+1}\right)\left(1 + \frac{k-1}{2}\mathrm{Ma}^2\right)\right]^{\frac{k}{k-1}}$$

$$\frac{T_0}{T_0^*} = \frac{(k+1)\mathrm{Ma}^2\left[2 + (k-1)\,\mathrm{Ma}^2\right]}{(1+k\,\mathrm{Ma}^2)^2}$$

$$q = h_{02} - h_{01} = c_p(T_{02} - T_{01})$$

Normal shock:

$$p + \frac{(\rho V)^2 RT}{p} = \text{constant} = C_M$$

$$T + \frac{(\rho V)^2 T^2}{2c_p(p^2/R^2)} = T_0 = \text{constant}$$

$$\frac{p_y}{p_x} = \frac{1 + k\mathrm{Ma}_x^2}{1 + k\mathrm{Ma}_y^2}, \qquad \frac{T_y}{T_x} = \frac{1 + \dfrac{k-1}{2}\mathrm{Ma}_x^2}{1 + \dfrac{k-1}{2}\mathrm{Ma}_y^2}$$

$$\mathrm{Ma}_y^2 = \frac{\mathrm{Ma}_x^2 + \dfrac{2}{k-1}}{\dfrac{2k}{k-1}\mathrm{Ma}_x^2 - 1}$$

OBLIQUE SHOCKS, BOW SHOCKS, AND EXPANSION WAVES

Oblique shocks:

$$\mathrm{Ma}_{2n}^2 = \frac{\mathrm{Ma}_{1n}^2 + \dfrac{2}{k-1}}{\dfrac{2k}{k-1}\mathrm{Ma}_{1n}^2 - 1}, \qquad \frac{p_2}{p_1} = \frac{2k}{k+1}\mathrm{Ma}_{1n}^2 - \frac{k-1}{k+1}$$

$$\frac{T_2}{T_1} = \frac{\left(1 + \frac{k-1}{2}\mathrm{Ma}_{1n}^2\right)\left(k\mathrm{Ma}_{1n}^2 - \frac{k-1}{2}\right)}{\left(\frac{k+1}{2}\right)^2 \mathrm{Ma}_{1n}^2}$$

$$\frac{\rho_2}{\rho_1} = \frac{V_{1n}}{V_{2n}} = \frac{\frac{k+1}{2}\mathrm{Ma}_{1n}^2}{1 + \frac{k-1}{2}\mathrm{Ma}_{1n}^2}$$

$$\frac{p_{02}}{p_{01}} = \frac{\left[\frac{\frac{k+1}{2}\mathrm{Ma}_{1n}^2}{1 + \frac{k-1}{2}\mathrm{Ma}_{1n}^2}\right]^{\frac{k}{k-1}}}{\left[\frac{2k}{k+1}\mathrm{Ma}_{1n}^2 - \frac{k-1}{k+1}\right]^{\frac{1}{k-1}}}$$

$$\frac{T_{02}}{T_{01}} = 1, \qquad \tan\theta = \frac{2\cot\beta(\mathrm{Ma}_1^2\sin^2\beta - 1)}{\mathrm{Ma}_1^2(k + \cos 2\beta) + 2}$$

$$\tan(\beta - \theta) = \frac{\tan\beta}{k+1}\left[k - 1 + \frac{2}{\mathrm{Ma}_1^2\sin^2\beta}\right]$$

Isentropic expansion waves:

$$\omega = \sqrt{\frac{k+1}{k-1}}\tan^{-1}\left(\sqrt{\frac{k-1}{k+1}(\mathrm{Ma}^2 - 1)}\right) - \tan^{-1}\left(\sqrt{\mathrm{Ma}^2 - 1}\right)$$

$$\theta = \omega(\mathrm{Ma}_2) - \omega(\mathrm{Ma}_1)$$

$$\frac{p_0}{p} = \left[1 + \frac{k-1}{2}\mathrm{Ma}^2\right]^{k/(k-1)}, \qquad \frac{T_0}{T} = 1 + \frac{k-1}{2}\mathrm{Ma}^2$$

$$\frac{\rho_0}{\rho} = \left[1 + \frac{k-1}{2}\mathrm{Ma}^2\right]^{\frac{1}{k-1}}$$

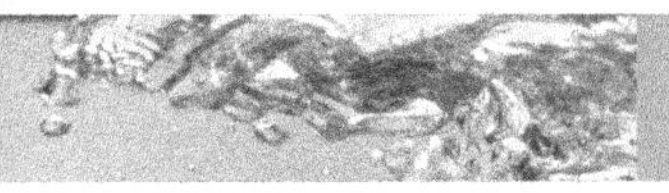

PROBLEMS

Section 12.2: Principles of Thermodynamics

12.1. A specified volume contains 15 kg of air at a temperature and pressure of 17°C and 91 kPa, respectively. (a) If the temperature of the air is raised to 27°C, what is the change in internal energy and the change in enthalpy? (b) If the volume of air is compressed to 60% of its original volume, what are the temperature and pressure in the compressed volume and what are the changes in internal energy and enthalpy?

12.2. Carbon dioxide flows through a conduit such that the pressure and density at an upstream section are 1600 kPa and 16 kg/m^3, respectively, and the pressure and temperature at a downstream section are 220 kPa and 80°C, respectively. Estimate the temperature at the upstream section, the density at the downstream section, and the changes in enthalpy and entropy between the two sections.

12.3. Show that for an ideal gas, the entropy change between States 1 and 2 can be expressed as

$$s_2 - s_1 = c_v \ln \frac{p_2}{p_1} + c_p \ln \frac{v_2}{v_1}$$

where s is the specific entropy, c_v and c_p are the specific heats at constant volume and pressure, respectively, p is the pressure, and v is the specific volume.

12.4. Six kilograms of carbon dioxide (CO_2) are contained within a steel cylinder at a temperature and pressure of 27°C and 260 kPa (absolute), respectively. If the temperature of the gas in the cylinder is raised to 32°C, what are the changes in pressure and entropy of the gas in the cylinder?

12.5. Air flows through a section of a duct that is being cooled by liquid nitrogen. At the entrance to this cooled section, the temperature, pressure, and velocity of the air are 130°C, 210 kPa, and 190 m/s, respectively. At the exit from this section, the temperature and pressure are 75°C and 230 kPa, respectively. The cross-sectional area of the duct is 6 cm^2. Estimate the mass flow rate through the duct, the change in enthalpy, the change in internal energy, and the change in entropy between the inflow and outflow sections.

12.6. Consider the case in which a volume of nitrogen (N_2) is compressed isothermally from an absolute pressure of 100 kPa to an absolute pressure of 380 kPa. Is heat gained or lost from the volume? What is the change in entropy?

12.7. A volume of nitrogen (N_2) is compressed adiabatically from a temperature of 27°C and absolute pressure of 100 kPa to an absolute pressure of 380 kPa. What is the temperature of N_2 in the compressed volume?

Section 12.3: The Speed of Sound

12.8. Estimate the speed of sound in helium at a pressure of 260 kPa and a temperature of 360°C.

12.9. Determine the speed of sound in pure oxygen at 40°C. Use this result to estimate the bulk modulus of elasticity of oxygen at 40°C. Compare your results to the corresponding values for water.

12.10. Compare the speed of sound in cast iron, seawater, freshwater, and air. For cast iron, the bulk modulus of elasticity is around 100 GPa and the density is around 7200 kg/m^3. Assume a temperature of 20°C for the fluids. How long will it take sound to travel 1 km in each medium?

12.11. A typical commercial airliner cruises at a speed of 250 m/s at an altitude of 10 km. Estimate the Mach number at which the aircraft is operating. Compare this with the Mach number of an F-16 fighter jet that operates at a speed of 500 m/s and an altitude of 12 km.

12.12. A high-speed photograph of a bullet fired in air indicates a Mach cone angle of 31°. If the ambient temperature and pressure are 25°C and 100 kPa, respectively, estimate the speed of the bullet.

12.13. A sharp-nosed projectile travels at an elevation of 600 m above the ground as shown in Figure 12.52. Consider the scenario in which the projectile travels at a Mach number of 2.5 and the ambient temperature is 15°C. What is the velocity of the projectile? What is the distance L when a sound detector on the ground first "hears" the projectile?

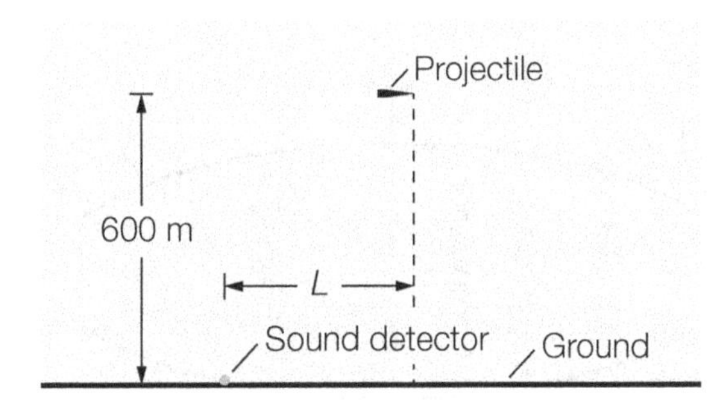

Figure 12.52: Sound detection of a supersonic projectile

12.14. A supersonic aircraft flies with a Mach number of 1.6 at an elevation of 3500 m. From the perspective of someone on the ground, determine how long after the aircraft passes until a sonic boom is heard.

12.15. A supersonic aircraft flying at an altitude of 5000 m passes over a person on the ground, and the person hears the sonic boom when the aircraft is 9.0 km past her. Estimate the Mach speed of the aircraft.

12.16. If the magnitude of the pressure fluctuations generated by a source triple from 6 Pa to 18 Pa, what is the change in decibel level?

Section 12.4: Thermodynamic Reference Conditions

12.17. Air is flowing at a speed of 148 m/s, and the temperature and pressure in the air are 35°C and 590 kPa, respectively. What is Mach speed, the stagnation temperature, and the stagnation pressure?

12.18. A particular supersonic aircraft is designed to fly at a maximum Mach speed of 2.2 at an altitude where the temperature is −45°C. It is expected that the temperature on the nose cone of the aircraft will be approximately equal to the stagnation temperature. Determine the stagnation temperature and compare the temperature on the nose cone to the static temperature of the surrounding air.

12.19. Supersonic airflow occurs in a 315-mm-diameter duct, and measurements at a particular section indicate a stagnation temperature of 44°C, a stagnation pressure of 395 kPa, and a Mach number of 1.6. Estimate the temperature, pressure, and mass flow rate at the measurement section. If airflow within the duct is isentropic, describe how the stagnation pressure and stagnation temperature vary along the duct.

12.20. The jet engine shown in Figure 12.53 operates at a speed of 250 m/s at an altitude of 7000 m in a standard atmosphere. Section 1 is the entrance to the diffuser, section 2 is the entrance to the compressor, and section 3 is the exit from the compressor. Flow through the diffuser is approximately isentropic, and flow through the compressor can be approximated by the relation

$$\frac{T_{03}}{T_{02}} = \left(\frac{p_{03}}{p_{02}}\right)^{\frac{k-1}{k}}$$

where T_{02} and T_{03} are stagnation temperatures at sections 2 and 3, respectively, and p_{02} and p_{03} are the corresponding stagnation pressures. Under the particular compressor operating conditions, $p_{03}/p_{02} = 11$. Estimate (a) the stagnation pressure at the entrance to the compressor and (b) the energy added to the air by the compressor.

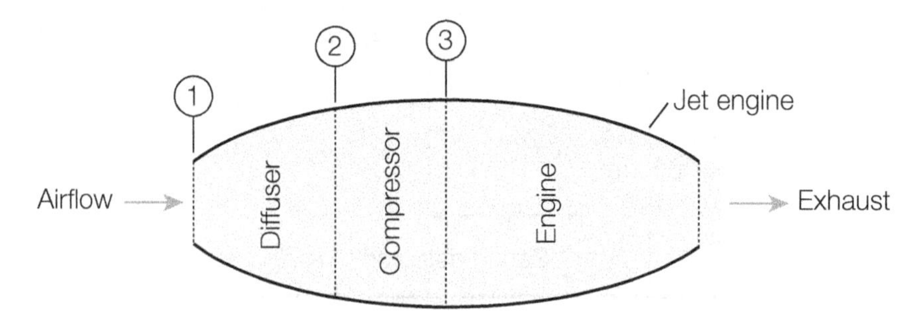

Figure 12.53: Airflow through a jet engine

12.21. Flows in which the Mach number (Ma) is less than or equal to 0.3 are conventionally assumed to be described by incompressible-flow equations. For air at Ma $= 0.3$ and at any given pressure, what is the percentage error in using the appropriate incompressible-flow equation to calculate the stagnation pressure versus using the appropriate compressible-flow equation?

12.22. The stagnation pressure is measured in a duct. Determine the highest Mach number of the flow for which the error in estimating the actual pressure (i.e., the static pressure) from the stagnation pressure is less than 1% when using the incompressible-flow equation instead of the compressible-flow equation.

12.23. Show that the pressure coefficient, C_p, for the isentropic flow of an ideal gas can be expressed in terms of k and Ma such that

$$C_\text{p} = \frac{p_0 - p}{\frac{1}{2}\rho V^2} = \frac{2}{k\,\text{Ma}^2}\left\{\left[1 + \frac{k-1}{2}\text{Ma}^2\right]^{\frac{k}{k-1}} - 1\right\}$$

where p_0 is the stagnation pressure, p is the static pressure, V is the free-stream speed of the fluid, k is the specific heat ratio, and Ma is the free-stream Mach number.

12.24. Pure nitrogen enters a duct from a large storage tank, where the pressure and temperature in the tank are 1200 kPa and 145°C, respectively. Flow in the duct, which has a varying cross-sectional area, is approximately isentropic. Determine the temperature, density, and flow speed at the duct section where the pressure is 700 kPa. What are the isentropic critical pressure and critical temperature in the duct?

12.25. Pure carbon dioxide is moving in a duct at a speed of 210 m/s with a temperature and pressure of 30°C and 160 kPa, respectively. Determine the temperature, pressure, and density at the isentropic critical condition.

Section 12.6: Steady One-Dimensional Isentropic Flow

12.26. Air flows in a converging conduit as shown in Figure 12.54. The conduit has a circular cross section, and the diameter of the conduit at the exit is 45 mm. Under desired operating conditions, critical conditions occur at the exit of the conduit and the volumetric inflow rate of air into the conduit is equal to 370 L/s. A section of particular interest is where the Mach number is equal to 0.27. Estimate the conduit diameter, flow velocity, and temperature at this location.

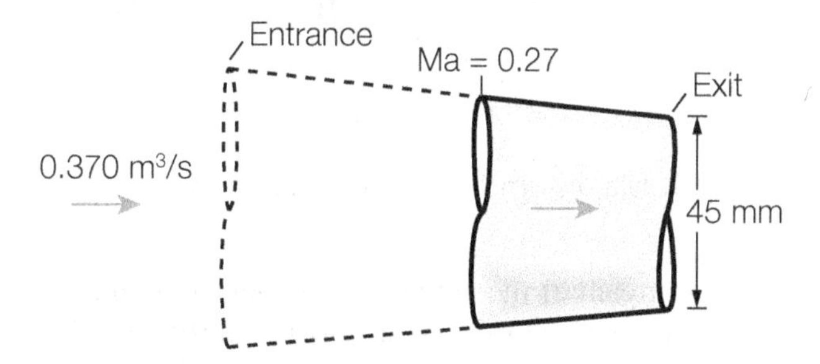

Figure 12.54: Flow through a converging nozzle

12.27. Measurements are taken at two sections in a converging conduit through which air is moving. At the upstream section, the cross-sectional area is 25 cm^2, the Mach number is 0.5, and the temperature and pressure are 75°C and 750 kPa, respectively. At the downstream section, the Mach number is 0.9. Assuming adiabatic and frictionless flow between the two sections, estimate the mass flow rate through the conduit and the temperature, pressure, and flow velocity at the downstream section.

12.28. A converging nozzle delivers air from a plenum chamber into a downstream chamber that exerts a back pressure of 575 kPa. The nozzle exit has an area of 21 cm^2, and the stagnation pressure and temperature in the plenum chamber are 775 kPa and 82°C, respectively. Determine the Mach number at the nozzle exit and the mass flow rate through the nozzle.

12.29. Air enters a converging nozzle at a temperature of 27°C, a pressure of 101 kPa, and a velocity of 90 m/s. The entrance and exit areas of the nozzle are 80 cm^2 and 50 cm^2, respectively. Estimate the Mach number, temperature, and pressure of the nozzle discharge.

12.30. The source reservoir of a converging nozzle contains air at a temperature of 16°C and a pressure of 410 kPa. The discharge reservoir contains air in which the pressure varies in the range of 145–255 kPa. The exit section of the nozzle has an area of 15 cm^2. What range of mass flow rates through the nozzle is to be expected?

12.31. Air enters a converging nozzle at a stagnation pressure of 155 kPa and a stagnation temperature of 175°C. The diameter of the nozzle at the exit is 20 mm. (a) If a back pressure of 60 kPa is exerted by the air in the discharge chamber, what is the exit pressure of the jet discharging from the nozzle and what is the mass flow rate through the nozzle? (b) How would your results change if the back pressure was increased to 90 kPa?

12.32. Consider the converging nozzle shown in Figure 12.55, where the nozzle transports air from the open atmosphere to a receiving conduit. The atmospheric air is at a temperature of 17°C and at a pressure of 100 kPa. The receiving conduit exerts a back pressure of p_b. The diameter of the nozzle at the exit is 14 mm. For what range of back pressures will the nozzle deliver air at the maximum rate? What is that maximum air delivery rate?

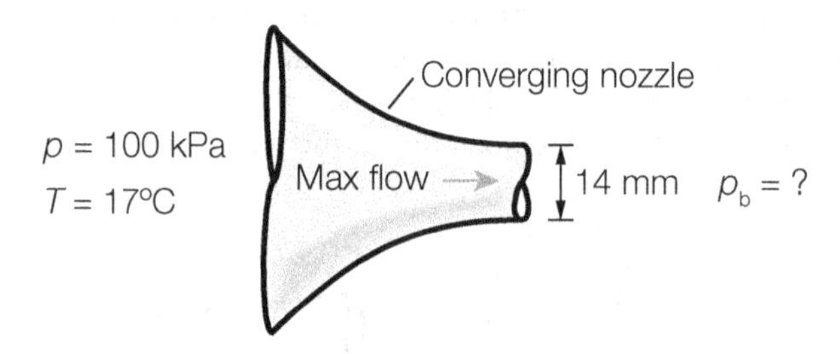

Figure 12.55: Maximum flow through a converging nozzle

12.33. Airflow just downstream of the entrance to a converging nozzle has a pressure of 925 kPa, a temperature of 525°C, and a velocity of 165 m/s. The throat/exit of the nozzle has a cross-sectional area of 41 cm^2. Compare the mass flow rate through the nozzle for a back pressure of 625 kPa with the mass flow rate for a back pressure of 325 kPa.

12.34. Air in an automobile tire has a gauge pressure of 210 kPa and a temperature of 25°C, and the surrounding atmosphere has an (absolute) pressure of 100 kPa. If the tire is punctured to create a hole with a diameter of 5 mm and the flow through the puncture hole can be assumed to be isentropic, estimate the initial velocity and mass flow rate of air through the puncture hole.

12.35. The de Laval nozzle shown in Figure 12.56 is used to move stagnant air from a plenum chamber to the test section of a supersonic wind tunnel. If the throat area of the nozzle is 9 cm^2 and the desired Mach number in the test section is 2.6, what area of test section should be used?

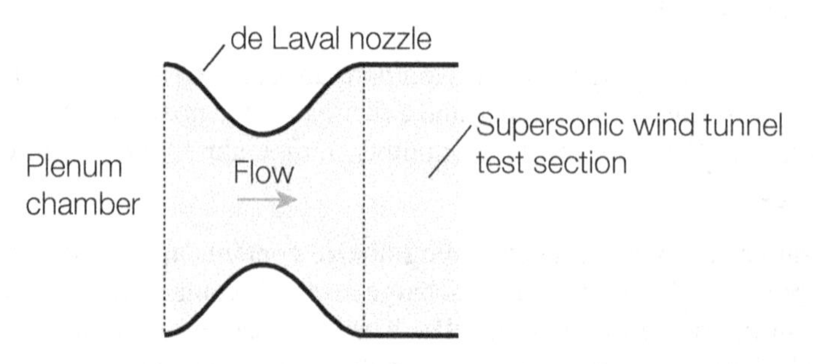

Figure 12.56: Application of a de Laval nozzle in a supersonic wind tunnel

12.36. The de Laval nozzle shown in Figure 12.56 has a throat-to-exit area ratio of 8, and the air pressure and temperature in the source plenum chamber are 3000 kPa and 27°C, respectively. Determine the Mach number, temperature, pressure, and velocity in the test section. If the test section has an area of 100 cm^2, what is the mass flow rate of air through the test section?

12.37. The air in the chamber at the intake of a CD nozzle has a temperature of 15°C and a pressure of 400 kPa, the throat area of the nozzle is 15 cm^2, and the exit area is 45 cm^2. Determine the two exit pressures for which the flow in the throat of the nozzle is sonic and determine the temperature and velocity corresponding to each of these pressures. What would be the mass flow rate through the nozzle under these two conditions?

12.38. Air flows through the converging-diverging (CD) nozzle shown in Figure 12.57. Measurements at section 1, where the cross-sectional area is 800 cm^2, indicate that the Mach number is 0.5, the temperature is 16°C, and the pressure is 500 kPa. Measurements at section 2 show a Mach number of 2.5. Estimate the cross-sectional area, temperature, and pressure at section 2.

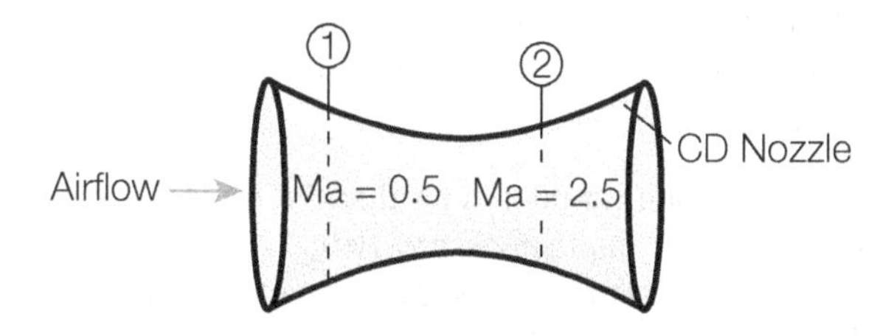

Figure 12.57: Flow through a CD nozzle

12.39. Air flows through a CD nozzle such that at a particular section upstream of the throat, the diameter is 150 mm, the velocity is 150 m/s, the pressure is 450 kPa, and the temperature is 200°C. For a section that has the same diameter but is located in the diverging part of the nozzle, determine the flow properties when the flow in the diverging part of the nozzle is (a) subsonic and (b) supersonic.

12.40. Air enters a CD nozzle under subsonic conditions and exits under supersonic conditions. The throat area of the nozzle is 25 cm^2. Measurements at a nozzle section upstream of the throat indicate a velocity of 100 m/s, a pressure of 80 kPa, and a temperature of 15°C. Estimate the mass flow rate through the nozzle.

12.41. Air at a stagnation pressure of 150 kPa and a stagnation temperature of 127°C is to be passed through a CD nozzle such that the exit Mach number is 2.1 and the mass flow rate is 1.8 kg/s. What throat diameter should be used?

12.42. A CD nozzle takes in air from a large chamber that has a pressure and temperature of 1.1 MPa and 627°C, respectively. The throat area of the nozzle is 25 cm^2, and under particular operating conditions, the Mach number of the flow at the exit is 2.3. Estimate the pressure, temperature, and velocity at (a) the throat section and (b) the exit section.

12.43. Air flows through a CD nozzle with a stagnation pressure of 900 kPa and a stagnation temperature of 277°C. The throat area of the nozzle is 30 cm^2. If the target design pressure at the exit of the nozzle is 30 kPa, what exit area should be used? Under the design condition, what is the mass flow rate through the nozzle and what is the temperature and velocity at the exit?

12.44. Air flows with a stagnation temperature and stagnation pressure of 150°C and 900 kPa, respectively, through a CD nozzle that has a throat diameter of 15 mm and an exit diameter of 40 mm. (a) Determine the design back pressure and the corresponding mass flow rate through the nozzle. (b) Determine the back pressure above which the flow will not attain supersonic speed within the nozzle.

12.45. A de Laval nozzle has a ratio of exit area to throat area equal to 5 and operates under a condition where the stagnation pressure is 900 kPa and the back pressure is 90 kPa. Is the nozzle overexpanded, ideally expanded, or underexpanded?

Section 12.7: Normal Shocks

12.46. A normal shock occurs at a location where the Mach number immediately upstream of the shock is equal to 2.5. What is the Mach number immediately downstream of the shock? What is the percentage change in temperature and pressure that occurs across the shock?

12.47. A supersonic aircraft flies at a Mach speed of 1.8 in an environment where the temperature is −40°C and the pressure is 30 kPa. Under this condition, a normal shock is formed in front of the nose of the aircraft. Estimate the temperature and pressure at the stagnation point on the nose of the aircraft. What is the change in entropy across the normal shock?

12.48. Air flows steadily through a duct such that the temperature, pressure, and velocity just upstream of a normal shock are 10°C, 75 kPa, and 750 m/s, respectively. (a) Determine the temperature, pressure, and velocity immediately downstream of the shock. (b) What are the air densities upstream and downstream of the shock?

12.49. A CD nozzle takes in air from a large chamber that has a pressure and temperature of 1.1 MPa and 627°C, respectively. Under particular operating conditions, a normal shock occurs at the exit section and the Mach number of the flow just before the shock is 2.3. Compare the pressure temperature and velocity before and after the shock.

12.50. A CD nozzle with a throat diameter of 40 mm connects air in a source reservoir to a discharge reservoir, where the diameter of the nozzle exit is 90 mm. When the pressure and temperature in the source reservoir are 101 kPa and 15°C, respectively, a normal shock occurs at the nozzle exit. Determine the back pressure on the nozzle and the mass flow rate through the nozzle under this condition.

12.51. The same nozzle and source reservoir conditions described in Problem 12.50 is used. However, the back pressure in the downstream reservoir changes such that a normal shock occurs at a section where the nozzle diameter is 65 mm. What is the back pressure exerted by the downstream reservoir?

12.52. A CD nozzle is to have a discharge Mach number of 1.5 when the source reservoir contains air at a pressure of 3 MPa. (a) What is the design back pressure? (b) What is the maximum back pressure for which choked flow will occur within the nozzle? (c) For what range of back pressures will a normal shock occur within the nozzle?

12.53. Air enters an expanding nozzle with a Mach number of 1.5, a stagnation pressure of 400 kPa, and a stagnation temperature of 227°C. The nozzle has a circular cross section with an inlet diameter of 75 mm and an exit diameter of 200 mm. If a normal shock occurs where the diameter is 140 mm, what is the Mach number, pressure, and temperature at the exit?

12.54. The nozzle shown in Figure 12.58 has an expansion ratio of 3. Under particular operating conditions, a normal shock occurs at a section where the area ratio is equal to 1.5 when the stagnation pressure upstream of the shock is equal to 900 kPa. Determine the pressure at the exit of the nozzle.

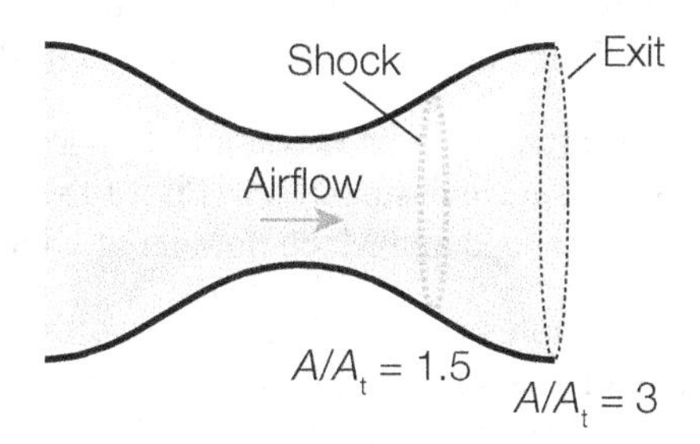

Figure 12.58: Normal shock contained in a CD nozzle

12.55. A CD nozzle is fed by a plenum chamber containing air at a pressure and temperature of 280 kPa and 177°C, respectively, as shown in Figure 12.59. The throat located at section 1 has a diameter of 1 m, a normal shock occurs at section 2 where the diameter is 1.5 m, and section 3 is located downstream of section 2 and has a diameter of 2.0 m. Sections 2a and 2b are located immediately upstream and downstream of the shock, respectively. Determine the pressures and temperatures at sections 2a, 2b, and 3.

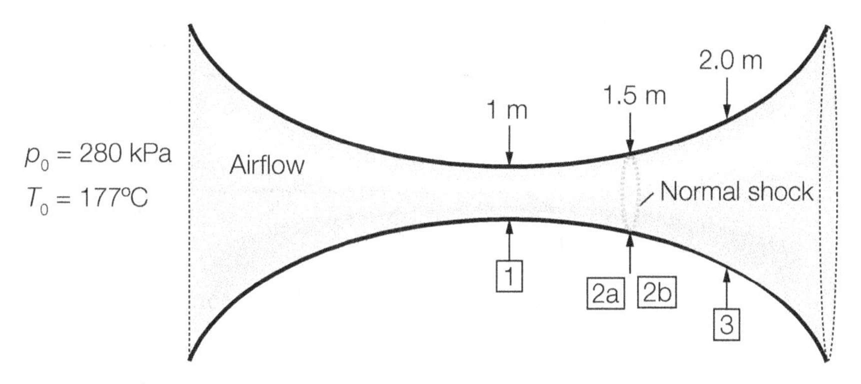

Figure 12.59: Normal shock in a CD nozzle

12.56. A converging-diverging (CD) nozzle is used to accelerate air from a subsonic to a supersonic state, and a particular CD nozzle that is being studied has a diameter, D, in meters, given by

$$D(x) = 0.3\sqrt{1 + 4x^2}$$

where x is the longitudinal distance measured from the center of the nozzle throat in meters. The nozzle entrance is located at $x = -1.5$ m, and the nozzle exit is located at $x = +1.5$ m. Under a particular adverse operating condition, the flow is choked at the center of the throat and a normal shock is observed to occur 0.2 m from the exit. If the entrance pressure under this condition is 101 kPa, determine the exit pressure.

12.57. A Pitot tube is used to measure the supersonic flow of air in a conduit. When the Pitot tube is installed in the conduit, a normal shock forms just ahead of the location where stagnation measurements are taken. The stagnation pressure and stagnation temperature measured by the Pitot tube are 400 kPa and 250°C, respectively. Just upstream of the normal shock caused by the Pitot tube, the pressure is measured as 80 kPa. Determine the Mach number and the velocity of the flow in the conduit.

12.58. A supersonic aircraft is cruising at an elevation of 10 km in a standard atmosphere, and the (stagnation) pressure measured by a Pitot tube mounted on the aircraft is 100 kPa. Estimate the air speed and Mach number at which the aircraft is traveling.

12.59. A mega bomb blast generates a normal shock wave that propagates at a speed of 650 m/s into stagnant air at a temperature and pressure of 15°C and 90 kPa, respectively. Calculate the changes in temperature, pressure, and velocity that are generated in the air by the shock wave.

Section 12.8: Steady One-Dimensional Non-Isentropic Flow

12.60. Air at a (stagnation) temperature and pressure of 20°C and 101 kPa, respectively, is drawn (using a suction pump) through a 15-mm-diameter circular insulated conduit as illustrated in Figure 12.60. The pressure at section 1 just downstream of the intake nozzle is 95 kPa, and the temperature at section 2 is 10°C. Estimate the mass flow rate of air through the conduit, the stagnation pressure at section 2, and the frictional force exerted on the conduit between sections 1 and 2.

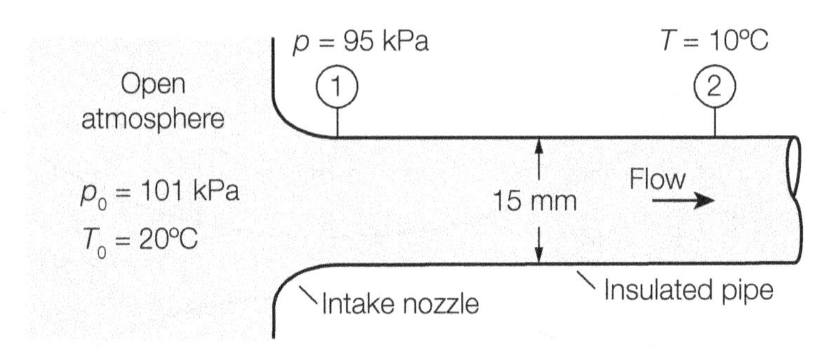

Figure 12.60: Flow through an insulated conduit

12.61. Pure oxygen flows through a 45-mm-diameter closed conduit of circular cross section. The duct is insulated, so adiabatic flow can be assumed; however, the flow is not frictionless. At a particular section of the conduit, measurements indicate that the stagnation temperature is 10°C, and the actual temperature and pressure are 3°C and 90 kPa, respectively. (a) Plot the Fanno line. (b) Plot the Mach number and temperature at the downstream location as a function of the pressure p at the downstream location. (c) If p = 30 kPa, what is the mass flow rate through the conduit? What is the temperature and Mach number at the downstream location?

12.62. At a particular location in an insulated duct, the Mach number of the airflow is 0.2, the pressure is 500 kPa, and the temperature is 127°C. At a downstream location where the Mach number is equal to 0.6, estimate the pressure, temperature, and velocity of the air.

12.63. Air enters a smooth insulated 25-mm-duct at a Mach number of 0.35, a temperature of 5°C, and a pressure of 120 kPa. What is the maximum length of duct that could be used without affecting the inflow conditions? Under this limiting condition, what would be the temperature, pressure, and velocity at the exit? Assume that the dynamic viscosity of air depends only on the temperature of the air and that the friction factor at the entrance of the duct is representative of the friction factor throughout the duct.

12.64. Air enters a 15-m-long, 75-mm-diameter insulated duct at a velocity of 90 m/s, a pressure of 200 kPa, and a temperature of 127°C. The average friction factor in the duct is estimated as 0.020. If the length of the duct is increased to 30 m, will the mass flow rate through the duct be affected? If so, what is the percentage change in the mass flow rate?

12.65. Air flows through an insulated pipe section that is 0.50 m long and has a diameter of 50 mm. At the entrance to the pipe, the Mach number is 0.6, the pressure is 150 kPa, and the temperature is 27°C. If the average friction factor in the pipe is 0.020, determine the Mach number, temperature, and pressure at the exit of the pipe section.

12.66. Consider a smooth, rectangular insulated duct that has a width of 350 mm and a height of 250 mm. Under particular conditions, the mass flow rate of air through the duct is 40 kg/s and the pressure and temperature at a particular cross section are 500 kPa and 20°C, respectively. How far downstream (from the particular cross section) is the density of the air reduced by 30%?

12.67. A 10 cm $\times$ 20 cm insulated rectangular duct is connected to the atmosphere by a short converging nozzle. The temperature and pressure in the atmosphere are 15°C and 101 kPa, respectively, and the back pressure at the discharge end of the duct is 45 kPa. The duct is 3 m long, and the average friction factor in the duct in the duct is 0.020. Estimate the mass flow rate through the duct. Assume that the flow through the entrance nozzle is isentropic and that the flow through the duct is adiabatic.

12.68. Air flows in a 25-mm-diameter insulated duct where the average friction factor is estimated as 0.020. Estimate the length of duct required to accelerate the flow from Ma = 0.2 to 0.6. Compare your result with the length of duct required to accelerate the flow from Ma = 0.6 to 1.0. Note that the Mach number increment is the same (= 0.4) in both cases.

12.69. Air flows in 300-mm-diameter insulated duct such that the flow is choked at a particular section along the duct. Pitot tube measurements taken 5 m upstream of the choked section indicate a stagnation pressure of 700 kPa, a stagnation temperature of 65°C, and a Mach number of 0.6. Determine (a) the average friction factor between the measurement section and the critical section; (b) the mass flow rate in the duct; and (c) the temperature, pressure, and velocity at the critical section.

12.70. At the entrance to a 50-m-long, 75-mm-diameter insulated duct, the stagnation pressure of the airflow is 200 kPa and the stagnation temperature is 150°C. Under unchoked conditions, the velocity at the entrance is 120 m/s. The average friction factor is estimated as 0.020. Determine the mass flow rate through the duct and state whether the flow is choked.

12.71. Air flows through a 1-m-long section of insulated 50-mm-diameter pipe. At the entrance to the pipe section, the Mach number is 2.0, the pressure is 750 kPa, and the temperature is 277°C. The average friction factor in the pipe section is estimated as 0.015. If a normal shock occurs 0.20 m from the exit, what is the Mach number, temperature, and pressure at the exit?

12.72. Air flows through a smooth rectangular prismatic duct that has a width of 350 mm and a height of 250 mm. Under particular operating conditions, the mass flow rate through the duct is 40 kg/s, the (constant) temperature in the duct is 20°C, and the pressure at a particular cross section is measured as 500 kPa. Estimate the pressure 150 m downstream of the measurement section. At what rate must heat be added to the air to maintain isothermal conditions? Keep in mind that the duct surface is smooth but the flow is not frictionless.

12.73. Air flows through a 20-mm-diameter, 3-m-long duct such that the pressure and temperature at the entrance section are 300 kPa and 27°C, respectively. The pressure at the exit of the duct is 140 kPa. Flow through the duct is isothermal and frictional, and the average friction factor in the duct is 0.025. Determine the mass flow rate through the duct and state whether the flow is choked.

12.74. Pure oxygen flows in the conduit shown in Figure 12.61. The flow is frictionless, and the fluid exchanges heat with its surroundings. At a measurement section, the data indicate that the stagnation temperature is 15°C, and the actual temperature and pressure are 5°C and 90 kPa, respectively. (a) Plot the Rayleigh line. (b) Plot the Mach number and temperature at the downstream section as a function of the pressure at the downstream section. (c) If the pressure at the exit section is 40 kPa, what is the temperature and Mach number at the exit section. Is heat being added to or removed from the conduit?

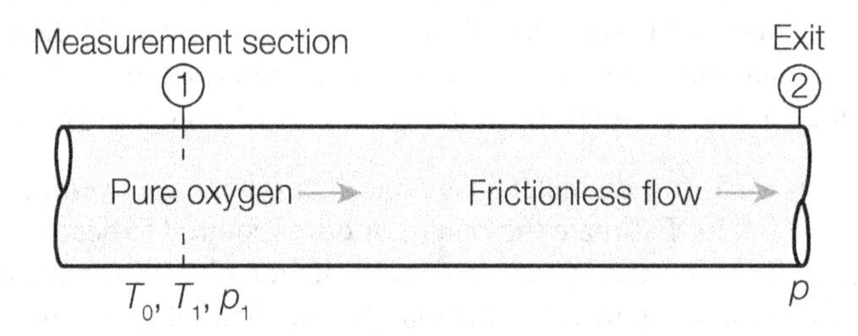

Figure 12.61: Flow of pure oxygen in a conduit

12.75. Air flows between two sections of a prismatic conduit. At the upstream section, the temperature, pressure, and velocity are 50°C, 150 kPa, and 100 m/s, respectively, and at the downstream section, the pressure is 75 kPa. Assuming that the airflow is frictionless, determine the following: (a) the temperature and velocity at the downstream section, (b) the rate at which heat is being added between the sections, and (c) the change in entropy between the sections.

12.76. Measurements of supersonic airflow in a duct at a particular section indicate a temperature of 50°C, a pressure of 120 kPa, and a velocity of 650 m/s. At a particular section downstream of the measurement section, the Mach number is 1.2. The flow can be assumed to be frictionless. (a) Apply Equations 12.155–12.157 to find the temperature, pressure, and velocity at the downstream section. (b) What is the heat added per unit mass between the upstream and downstream sections?

12.77. Air flows through a 200-mm-diameter conduit, and measurements near the entrance of the conduit indicate a pressure of 130 kPa, a temperature of 20°C, and a velocity of 50 m/s. Heat is added continuously to the fluid within the conduit, and the flow can be assumed to be frictionless. If the temperature at the exit of the conduit is measured as 700°C, at what rate is heat being added to the fluid between the measurement section and the exit of the conduit? What is the Mach number of the flow at the exit?

12.78. Air enters a conduit with a Mach number of 0.3, a pressure of 100 kPa, and a temperature of 27°C. If heat is added to the air at a rate of 80 kJ/kg and the flow is frictionless, estimate the Mach number, pressure, and temperature of the air at the exit of the conduit.

12.79. Compressed air enters a 100-mm-diameter tubular combustion chamber at a temperature, pressure, and velocity of 227°C, 450 kPa, and 50 m/s, respectively. Fuel with a heating value of 40 MJ/kg is added to the compressed air at an air-fuel mixture ratio of 30 kg air/kg fuel. Estimate the temperature, pressure, velocity, and Mach number at the exit of the combustion chamber.

12.80. A fuel-air mixture enters a duct combustion chamber at a velocity, pressure, and temperature of 85 m/s, 200 kPa, and 77°C, respectively. The combustion process generates heat at a rate of 800 kJ/kg. The properties of the fuel-air mixture can be approximated by R = 287.1 J/kg·K, c_p = 1003 J/kg·K, and k = 1.40. (a) What is the pressure, temperature, and velocity at the exit of the combustion chamber? (b) What rate of heat addition would cause sonic conditions at the exit of the combustion chamber?

12.81. Pure helium enters a 10-m-long, 300-mm-diameter insulated circular conduit at a Mach number of 3.0, and a normal shock is observed to occur 2 m from the exit of the conduit. The friction factor within the conduit has an average value of 0.010. Determine the ratio of the exit pressure to the entrance pressure.

Section 12.9: Oblique Shocks, Bow Shocks, and Expansion Waves

12.82. The front of an airfoil can be approximated as a wedge with an included angle of 10° as shown in Figure 12.62. Under design conditions, the airfoil moves with a Mach speed of 3.0 at an elevation of 9 km in a standard atmosphere. (a) Determine the wave angle, β, of the weak shock and the strong shock. (b) Compare the Mach number and the pressure downstream of the weak shock with the Mach number and the pressure downstream of the strong shock.

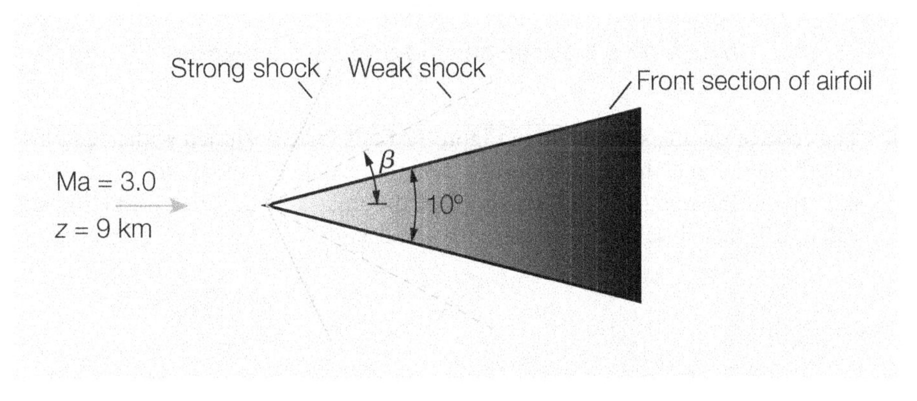

Figure 12.62: Supersonic flow impinging on an airfoil

12.83. Airflow at a Mach speed of 2.3 impinges on a deflector that turns the oncoming flow by 8° as shown in Figure 12.63. If the static pressure in the air approaching the deflector is equal to 85 kPa, find the angle, β, that a weak oblique shock makes with the upstream flow direction and the pressure and Mach number of the flow immediately downstream of the shock.

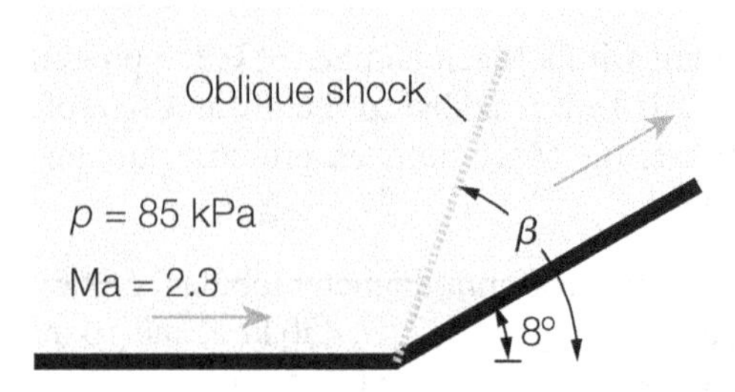

Figure 12.63: Supersonic flow past a deflector

12.84. The airflow upstream of a weak oblique shock has a velocity of 1650 m/s, a temperature of −2°C, and a pressure of 100 kPa. If the velocity is deflected by 20.2° as it crosses the shock, determine the angle the oblique shock makes with the flow direction.

12.85. Air flows with a Mach number of 2.0 toward a wedge and generates an oblique shock with an angle of 30° as shown in Figure 12.64. If the static pressure and temperature in the air upstream of the shock are 30 kPa and 30°C, respectively, what is the deflection angle of the velocity? What are the pressure and temperature after the shock?

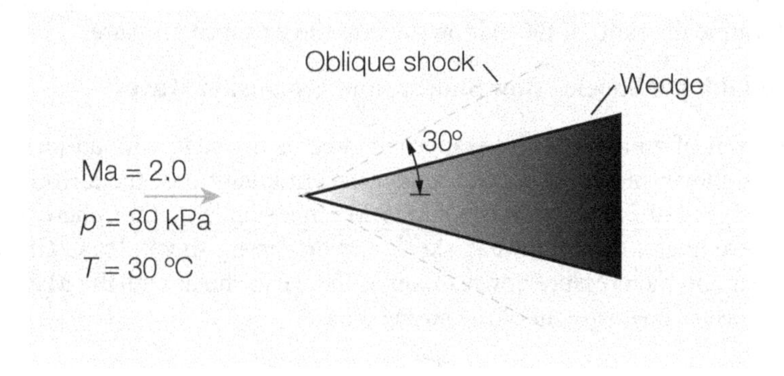

Figure 12.64: Supersonic flow toward a wedge

12.86. The cone of the missile shown in Figure 12.65 is to be designed with a length L such that the generated shock remains attached to the cone. If the diameter of the missile is 3 m and the maximum Mach speed of the missile is 2.3, determine the smallest value of L that should be considered in the design.

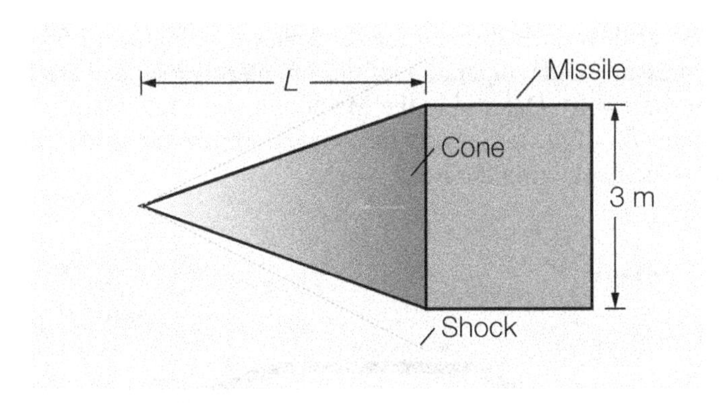

Figure 12.65: Shock on a missile cone

12.87. The leading edge of an airfoil as shown in Figure 12.66 has an included angle of 5°. Under particular operating conditions, the airspeed relative to the airfoil is 550 m/s at a temperature and pressure of 8°C and 95 kPa, respectively. If the angle of attack is 1°, determine the pressures on the top and bottom of the airfoil immediately downstream of the weak shock that is attached to the leading edge of the airfoil.

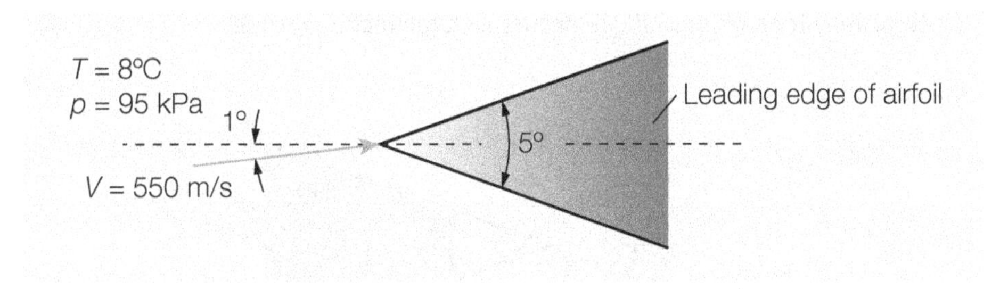

Figure 12.66: Leading edge of an airfoil

12.88. Supersonic airflow with a Mach number of 2.3 encounters an expansion turn of 8° as shown in Figure 12.67. If the static temperature and pressure in the air upstream of the expansion are 20°C and 85 kPa, respectively, determine the Mach number and air pressure downstream of the expansion.

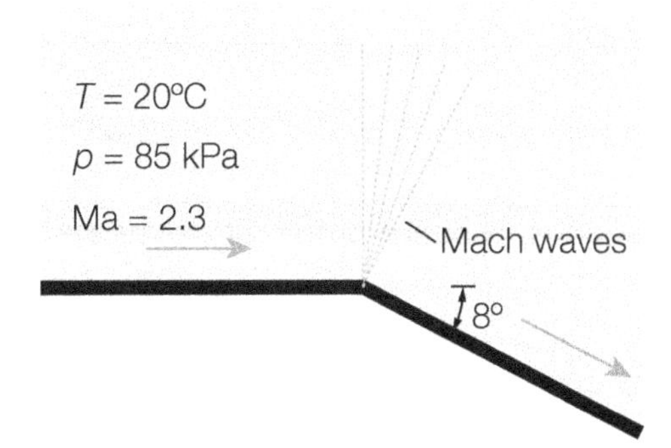

Figure 12.67: Expansion turn in supersonic flow

12.89. Supersonic airflow turns through the bend shown in Figure 12.68, where the Mach number, pressure, and temperature of the air entering the bend are 1.5, 250 kPa, and 550°C, respectively. Determine the Mach number, Ma_2, the pressure, p_2, and the temperature, T_2, of the air leaving the bend. Compare the velocity leaving the bend with the velocity entering the bend.

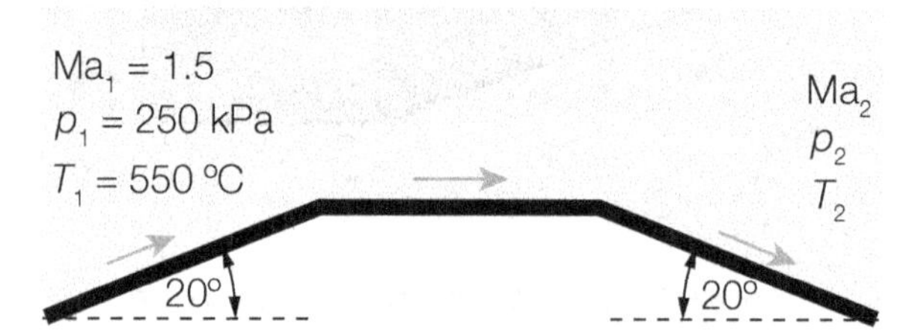

Figure 12.68: Supersonic flow around a bend

12.90. Consider the case of the leading edge of an airfoil moving through air with an angle of attack of 10° as shown in Figure 12.69. The speed of the airfoil is 650 m/s, and the temperature and pressure of the air are 4°C and 90 kPa, respectively. Estimate the pressures on the top and bottom surfaces just downstream of the leading edge of the airfoil. If an oblique shock occurs, assume that it is weak.

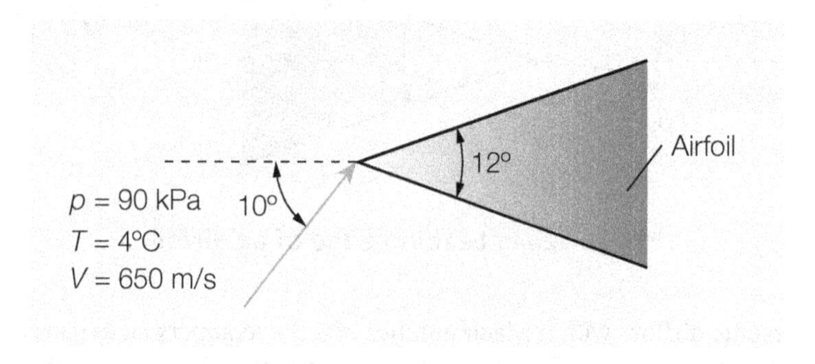

Figure 12.69: Airflow at the leading edge of an airfoil

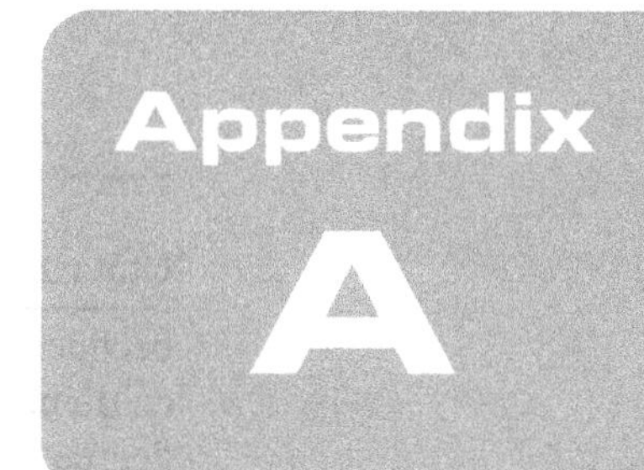

Units and Conversion Factors

A.1 Units

The Système International d'Unités (International System of Units or SI system) was adopted by the 11th General Conference on Weights and Measures (CGPM) in 1960 and is now used by almost the entire world. In the SI system, all quantities are expressed in terms of seven base (fundamental) units. These base units and their standard abbreviations are as follows:

Meter (m): Distance light travels in a vacuum during 1/299 792 458 of a second.[1]

Kilogram (kg): Mass of a cylinder of platinum–iridium alloy kept in Paris.

Second (s): Duration of 9 192 631 770 cycles of the radiation corresponding to the transition between two hyperfine levels of the ground state of the cesium-133 atom.

Ampere (A): Magnitude of the current that, when flowing through each of two long parallel wires of negligible cross section separated by 1 meter in a vacuum, results in a force between the two wires of 2×10^{-7} newtons for each meter of length.

Kelvin (K): Defined in the thermodynamic scale by assigning 273.16 K to the triple point of water (freezing point: 273.16 K = 0°C).

Candela (cd): Luminous intensity of 1/600 000 of a square meter of a radiating cavity at the temperature of freezing platinum (2042 K).

Mole (mol): Amount of substance that contains as many specified entities (molecules, atoms, ions, electrons, photons, etc.) as there are atoms in exactly 0.012 kg of carbon-12.

In addition to the seven base units of the SI system, there are two supplementary SI units: the radian and the steradian. The radian (rad) is defined as the angle at the center of a circle subtended by an arc equal in length to the radius, and the steradian (sr) is defined as the solid angle with its vertex at the center of a sphere that is subtended by an area of the spherical surface equal to the radius squared. The SI units should not be confused with the now obsolete *metric units*, which were developed in Napoleonic France approximately 200 years ago. The primary difference between metric and SI units is that the former uses centimeters and grams to measure length and mass, respectively, whereas these quantities are measured in meters and kilograms in SI units. The United States is gradually moving toward the use of SI units; however, there is still widespread use of the "English" system of units, which is more appropriately referred to as "U.S. customary" units.

[1]"Meter" is the accepted spelling in the United States; the rest of the world uses the spelling "metre."

Table A.1: SI Derived Units

Unit name	Quantity	Symbol	In Terms of Base and Derived Units
becquerel	Activity of a radionuclide	Bq	s^{-1}
coulomb	Quantity of electricity, electric charge	C	$A \cdot s$
degree Celsius	Celsius temperature	°C	K
farad	Capacitance	F	$C \cdot V^{-1}$
gray	Absorbed dose of ionizing radiation	Gy	$J \cdot kg^{-1}$
henry	Inductance	H	$Wb \cdot A^{-1}$
hertz	Frequency	Hz	s^{-1}
joule	Energy, work, quantity of heat	J	$N \cdot m$
lumen	Luminous flux	lm	$cd \cdot sr$
lux	Illuminance	lx	$lm \cdot m^{-2}$
newton	Force	N	$kg \cdot m \cdot s^{-2}$
ohm	Electric resistance	Ω	$V \cdot A^{-1}$
pascal	Pressure, stress	Pa	$N \cdot m^{-2}$
siemens*	Conductance	S	$A \cdot V^{-1}$
sievert	Dose equivalent of ionizing radiation	Sv	$J \cdot kg^{-1}$
telsa	Magnetic flux density	T	$Wb \cdot m^{-2}$
volt	Electric potential, potential difference	V	$W \cdot A^{-1}$
watt	Power, radiant flux	W	$J \cdot s^{-1}$
weber	Magnetic flux	Wb	$V \cdot s$

*The siemens was previously called the mho.

Derived units. In addition to the seven SI base units, several derived units are given special names. These derived units are used for convenience rather than necessity and are listed in Table A.1.

Abbreviations. When units are named after people, such as the newton (N), joule (J), and pascal (Pa), they are capitalized when abbreviated but not when they are spelled out. The abbreviation capital L for liter is a special case, used to avoid confusion with one (1). In the SI system, the unit of absolute temperature is the degree kelvin, which is abbreviated K without the degree symbol. The units of second, minute, hour, day, and year are correctly abbreviated as s, min, h, d, and y, respectively.

Prefixes. In using prefixes with SI units, multiples of 10^3 are preferred in engineering usage, with other multiples such as cm avoided if possible. Conventional practice is to separate sequences of digits into groups of three using spaces rather than commas.

A.2 Conversion Factors

In most cases, application of unit conversion factors results in converted numbers that have more significant digits than the original numbers. In these cases, the converted number should be rounded off such that the rounding error is consistent with the rounding error of the converted number.

EXAMPLE A.1

The height of a water control structure is reported as 19.3 feet. Convert this dimension to meters.

SOLUTION

The conversion factor is given in Table A.2 as 1 foot $=$ 0.3048 meters; hence,

$$19.3 \text{ ft} = 19.3 \times 0.3048 = 5.88264 \text{ m}$$

Because 19.3 ft could have resulted in rounding any number between 19.25 ft and 19.35 ft, the maximum possible rounding error is $\pm 0.05/19.3 = \pm 0.26\%$. Similarly, rounding 5.88264 m to 5.88 m gives a maximum rounding error of $\pm 0.005/5.88 = \pm 0.085\%$, and rounding to 5.9 m gives an error of $\pm 0.05/5.9 = \pm 0.85\%$. Hence, accuracy is lost by taking 19.3 ft as 5.9 m, whereas 5.88 m is more accurate than indicated by 19.3 ft. It is usually prudent not to discard accuracy, so take 19.3 ft $=$ **5.88 m**.

A good rule of thumb to remember is that the converted number should have the same number of significant digits as the original number, assuming that the conversion factor is more accurate than the original number.

Table A.2: Multiplicative Factors for Unit Conversion

Quantity	Convert From	Convert To	Multiply By
Area	ac	ha	0.404687
	mi^2	km^2	2.59000
Energy	Btu	J	1054.350264
	ft·lb	J	1.355818
	cal	J	4.184*
Energy/Area	ly**	kJ/m^2	41.84*
Flow rate	cfs	m^3/s	0.02831685
	gpm	L/s	0.06309
	mgd**	m^3/s	0.04381
		m^3/d	3785.412
		L/s	43.81
Force	lbf	N	4.4482216152605*
Length	ft	m	0.3048*
	in.	m	0.0254*
	mi (U.S. statute)	km	1.609344*
	mi (U.S. nautical)	km	1.852000*
	yd	m	0.9144*
Mass	g	kg	0.001*
	lbm	kg	0.45359237*
	slug	kg	14.59390
Permeability	darcy	m^2	0.987×10^{-12}
Power	hp	W	745.69987
Pressure	atm	kPa	101.325*
	bar	kPa	100.000*
	mm Hg (at 0°C)	kPa	0.133322
	psi	kPa	6.894757
	torr	kPa	0.133322
Speed	knot	m/s	0.514444444
	mph	m/s	0.44704
	mph	km/h	1.609344
Viscosity (dynamic)	cp	Pa·s	0.001*
Viscosity (kinematic)	cs	m^2/s	10^{-6*}
Volume	gal (U.S.)	L	3.785411784*
	gal (imperial)		4.5461
Weight	ton (U.S. short)	metric ton***	0.90718486
	ton (British long)	metric ton	1.1060470

*Exact conversion.

**ly ≡ langley, mgd ≡ million gallons per day.

***1 metric ton = 1000 kgf.

Fluid Properties

B.1 Water

Table B.1: Physical Properties of Water at Standard Sea-Level Pressure

Temp (°C)	Density (kg/m^3)	Dynamic Viscosity (mPa·s)	Heat of Vaporization (MJ/kg)	Saturation Vapor Pressure (kPa)	Surface Tension (mN/m)	Bulk Modulus (10^6 kPa)	Expansion Coefficient (10^{-3} K^{-1})
0	999.8	1.781	2.499	0.611	75.6	2.02	−0.07
5	1000.0	1.518	2.487	0.872	74.9	2.06	0.160
10	999.7	1.307	2.476	1.227	74.2	2.10	0.088
15	999.1	1.139	2.464	1.704	73.5	2.14	0.151
20	998.2	1.002	2.452	2.337	72.8	2.18	0.207
25	997.0	0.890	2.440	3.167	72.0	2.22	0.257
30	995.7	0.798	2.428	4.243	71.2	2.25	0.303
40	992.2	0.653	2.405	7.378	69.6	2.28	0.385
50	988.0	0.547	2.381	12.340	67.9	2.29	0.457
60	983.2	0.466	2.356	19.926	66.2	2.28	0.523
70	977.8	0.404	2.332	31.169	64.4	2.25	0.585
80	971.8	0.354	2.307	47.367	62.6	2.20	0.643
90	965.3	0.315	2.282	70.113	60.8	2.14	0.665
100	958.4	0.282	2.256	101.325	58.9	2.07	0.752

The properties given in Table B.1 are for pure water. Pure water seldom exists in nature, where the density of water can be influenced significantly by salinity, temperature, and other properties through an equation of state. The general dependence of water density on temperature has been found to be approximately parabolic, with a maximum at 4°C. However, the temperature corresponding to the maximum density of water changes with increasing salinity, decreasing to about 0°C for highly saline systems. To a first-order approximation, density is linearly dependent on salinity over much of the normal range of interest.

Note:

a. The kinematic viscosity, ν, of water at 20°C is approximately equal to 1.004×10^{-6} m^2/s.

B.2 Air

Table B.2: Physical Properties of Air at Standard Sea-Level Atmospheric Pressure

Temperature (°C)	Density (kg/m^3)	Dynamic Viscosity	Specific Heat		Speed of Sound
			c_p	$k = c_p/c_v$	
		(μPa·s)	(J/kg·K)	(–)	(m/s)
−40	1.514	15.18	1002	1.401	306.2
−20	1.394	16.22	1005	1.401	319.1
0	1.292	17.29	1006	1.401	331.4
5	1.269	17.23	1006	1.401	334.4
10	1.246	17.72	1006	1.401	337.4
15	1.225	17.96	1007	1.401	340.4
20	1.204	18.21	1007	1.401	343.3
25	1.184	18.44	1007	1.401	346.3
30	1.164	18.68	1007	1.400	349.1
40	1.127	19.15	1007	1.400	354.7
50	1.092	19.61	1007	1.400	360.3
60	1.059	20.06	1007	1.399	365.7
70	1.028	20.51	1007	1.399	371.2
80	0.9994	20.95	1008	1.399	376.6
90	0.9718	21.38	1008	1.398	381.7
100	0.9458	21.81	1009	1.397	386.9
200	0.7459	25.77	1023	1.390	434.5
300	0.6158	29.34	1044	1.379	476.3
400	0.5243	32.61	1069	1.368	514.1
500	0.4565	35.63	1093	1.357	548.8
1000	0.2772	48.26	1184	1.321	694.8

Notes:

a. Standard air is at a temperature of 15°C at standard sea-level pressure (= 101.325 kPa).

b. The specific gas constant, R, for standard air is approximately equal to 287.058 J/kg·K, based on a molar mass of 28.965 g/mol.

c. The kinematic viscosity, ν, of standard air is approximately equal to 1.466×10^{-5} m^2/s.

B.3 The Standard Atmosphere

Table B.3: Physical Properties of the Standard Atmosphere*

Height above Ground (km)	Temperature (°C)	Absolute Pressure (kPa)	Density (kg/m^3)	Dynamic Viscosity (μPa·s)	Speed of Sound (m/s)	Gravity (m/s^2)
0	15.00	101.325	1.2250	17.894	340.294	9.80665
1	8.50	89.876	1.1117	17.579	336.43	9.8036
2	2.00	79.501	1.0066	17.260	332.53	9.8005
3	−4.49	70.121	0.90925	16.938	328.58	9.7974
4	−10.98	61.660	0.81935	16.612	324.59	9.7943
5	−17.47	54.048	0.73643	16.282	320.55	9.7912
6	−23.96	47.217	0.66011	15.949	316.45	9.7882
7	−30.45	41.11	0.5900	15.61	312.27	9.785
8	−36.94	35.651	0.52579	15.27	308.11	9.782
9	−43.42	30.80	0.4671	14.93	303.79	9.779
10	−49.90	26.499	0.41351	14.58	299.53	9.776
11	−56.50	22.632	0.3639	14.22	295.07	9.773
12	−56.50	19.330	0.3108	14.22	295.07	9.770
13	−56.50	16.510	0.2655	14.22	295.07	9.767
14	−56.50	14.102	0.2268	14.22	295.07	9.774
15	−56.50	12.11	0.1948	14.22	295.07	9.761
20	−56.50	5.529	0.08891	14.22	295.07	9.745
25	−51.60	2.549	0.04008	14.48	298.46	9.730
30	−46.64	1.197	0.01841	14.75	301.80	9.715
40	−22.80	0.287	0.003996	16.01	317.63	9.684
50	−2.50	0.07978	0.001027	17.04	329.80	9.654
60	−26.13	0.02196	0.0003097	15.84	314.07	9.624
70	−53.57	0.0052	0.00008283	14.38	295.61	9.594
80	−74.51	0.0011	0.00001846	13.21	281.12	9.564

*Latest version, last revised in 1976.

B.4 Common Liquids

Table B.4: Physical Properties of Common Liquids at 20°C and at Standard Sea-Level Atmospheric Pressure

Liquid	Density (kg/m^3)	Dynamic Viscosity (mPa·s)	Surface Tension (mN/m)	Saturation Vapor Pressure (kPa)	Modulus of Elasticity (MPa)	Specific Heat (J/kg·K)
Benzene	876	0.65	29	10.0	1030	1720
Blood @ 37°C	1060	3.5	58	–	–	3600
Carbon tetrachloride	1590	0.958	26.9	13.0	1310	–
Ethyl alcohol	789	1.19	22.8	5.9	1060	–
Ethylene glycol	1117	21.4	48.4	0.012	–	–
Freon 12	1327	0.262	–	–	795	–
Gasoline	680	0.29	–	55.2	–	2100
Glycerin	1260	1500	63.3	0.000014	4520	2100
Kerosene	808	1.92	25	3.20	–	2000
Mercury	13 550	1.56	484	0.00017	26 200	139.4
Methanol	791	0.598	22.5	13.4	830	–
Oil (Crude)	856	7.2	30	–	–	–
SAE 10 Oil	918	82	37	–	–	–
SAE 30 Oil	918	440	36	–	–	–
SAE 30 Oil @ 15.6°C	912	380	36	–	–	–
SAE 50 Oil	902	860	–	–	–	–
Water (Fresh)	998	1.00	73	2.34	2171	4187
Water (Sea) @ 20°C	1023	1.07	73.5	2.34	2300	3933
@ 10°C	1025	1.39	74.8	1.20	–	4007

B.5 Common Gases

Table B.5: Physical Properties of Common Gases at 20°C and at Standard Sea-Level Atmospheric Pressure

Gas	Chemical Formula	Molar Mass (g/mol)	Density (kg/m^3)	Dynamic Viscosity (μPa·s)	Specific Heat	
					c_p (J/kg·K)	c_v (J/kg·K)
Air	–	28.96	1.205	18.0	1003	716
Argon	Ar	39.944	1.66	22.4	519	311
Carbon dioxide	CO_2	44.01	1.84	14.8	858	670
Carbon monoxide	CO	28.01	1.16	18.2	1040	743
Chlorine	Cl_2	70.91	2.95	10.3	–	–
Helium	He	4.003	0.166	19.7	5220	3143
Hydrogen	H_2	2.016	0.0839	9.0	14 450	10 330
Methane	CH_4	16.04	0.668	13.4	2250	1730
Nitrogen	N_2	28.02	1.16	17.6	1040	743
Nitric oxide	NO	30.01	1.23	19.0	–	–
Nitrous oxide	N_2O	44.02	1.82	14.5	–	–
Oxygen	O_2	32.00	1.33	20.0	909	649
Propane* @ 0°C	C_3H_8	44.10	2.010	–	–	–
Water vapor	H_2O	18.02	0.747	10.1	1862	1400

*Propane has a saturation vapor pressure of 853.16 kPa at 21.1°C.

Notes:

a. The universal gas constant, R_u, is approximately equal to 8.31446 J/mol·K.

b. The gas constant, R, for each gas is given by $R = R_u/M$, where M is the molar mass.

B.6 Nitrogen

Table B.6: Physical Properties of Nitrogen at Standard Sea-Level Atmospheric Pressure

Temperature (°C)	Density (kg/m^3)	Specific Heat, c_p (J/kg·K)	Dynamic Viscosity (μPa·s)
−50	1.5299	957.3	13.90
0	1.2498	1035	16.40
50	1.0564	1042	18.74
100	0.9149	1041	20.94
150	0.8068	1043	23.00
200	0.7215	1050	24.94
300	0.5956	1070	28.49
400	0.5072	1095	31.66
500	0.4416	1120	34.51
1000	0.2681	1213	45.94
1500	0.1925	1266	55.62
2000	0.1502	1297	64.26

Appendix C

Properties of Areas and Volumes

C.1 Areas

Table C.1: Geometric Properties of Areas

Shape	Illustration	Area	Moments of Inertia
Rectangle	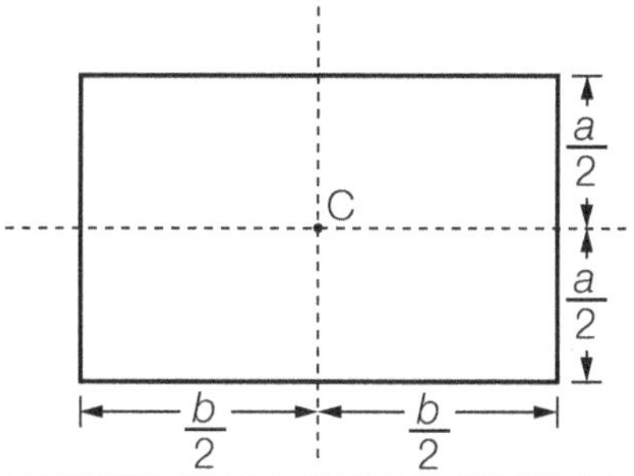 C; $\frac{a}{2}$, $\frac{a}{2}$; $\frac{b}{2}$, $\frac{b}{2}$	ba	$I_{xc} = \frac{1}{12}ba^3$ $I_{yc} = \frac{1}{12}ab^3$ $I_{xyc} = 0$
Circle	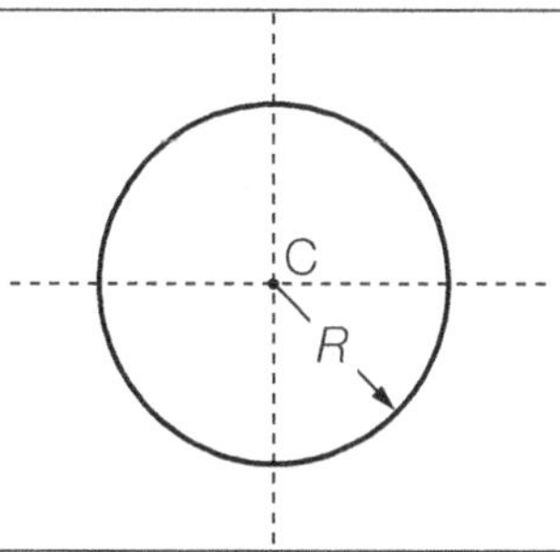 C; R	πR^2	$I_{xc} = I_{yc} = \frac{\pi R^4}{4}$ $I_{xyc} = 0$
Semicircle	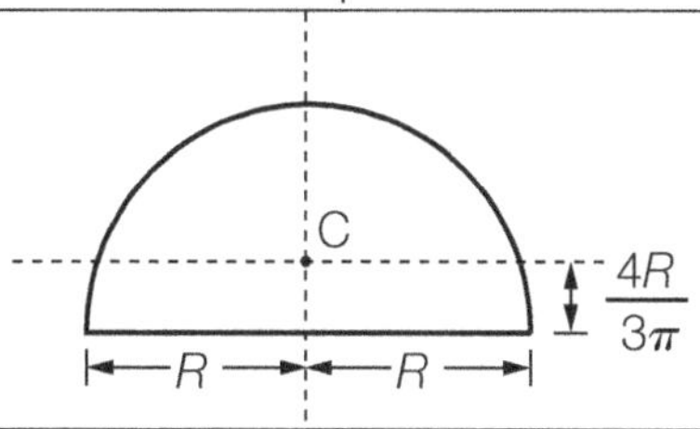 C; $\frac{4R}{3\pi}$; R, R	$\frac{1}{2}\pi R^2$	$I_{xc} = 0.1098R^4$ $I_{yc} = 0.3927R^4$ $I_{xyc} = 0$

Table C.1: Continued

Shape	Illustration	Area	Moments of Inertia
Quarter circle	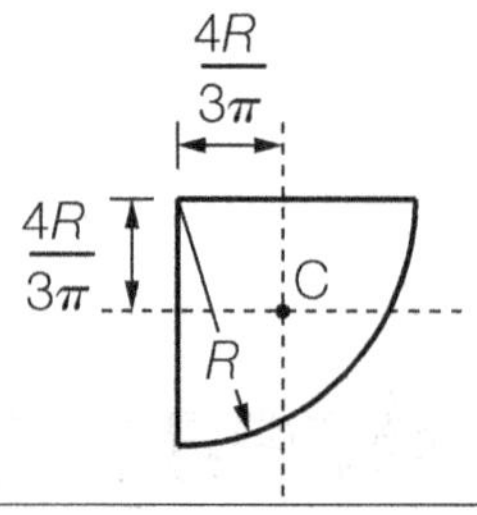 $\frac{4R}{3\pi}$, $\frac{4R}{3\pi}$, C, R	$\frac{1}{4}\pi R^2$	$I_{xc} = I_{yc} = 0.05488R^4$ $I_{xyc} = -0.01647R^4$
Ellipse	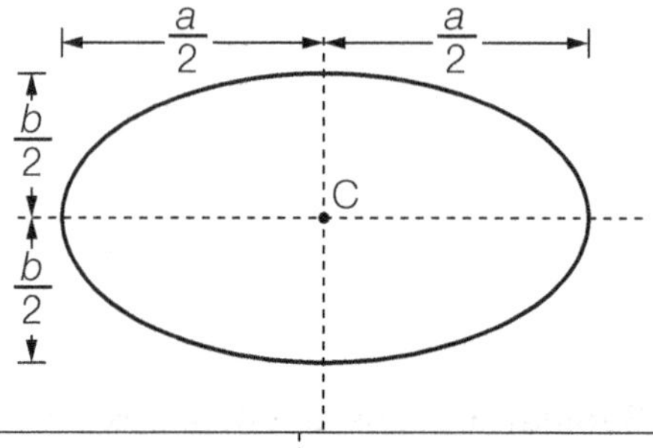 $\frac{a}{2}$, $\frac{a}{2}$, $\frac{b}{2}$, $\frac{b}{2}$, C	$\frac{1}{4}\pi ab$	$I_{xc} = \frac{1}{64}\pi ab^3$ $I_{yc} = \frac{1}{64}\pi ba^3$ $I_{xyc} = 0$
Semi-ellipse	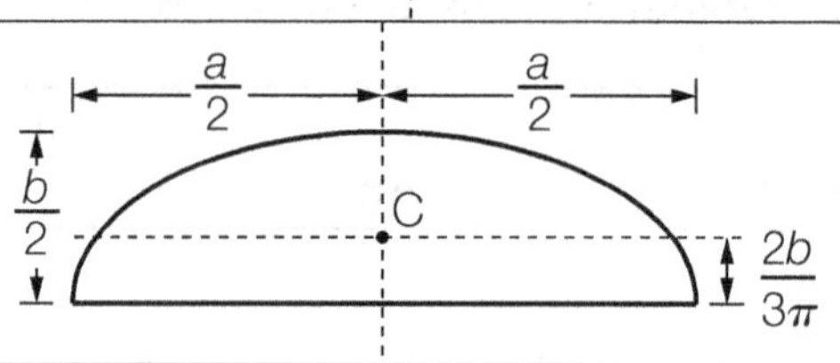 $\frac{a}{2}$, $\frac{a}{2}$, $\frac{b}{2}$, C, $\frac{2b}{3\pi}$	$\frac{1}{8}\pi ab$	$I_{xc} = \frac{1}{128}\pi ab^3$ $I_{yc} = \frac{1}{128}\pi ba^3$ $I_{xyc} = 0$
Triangle	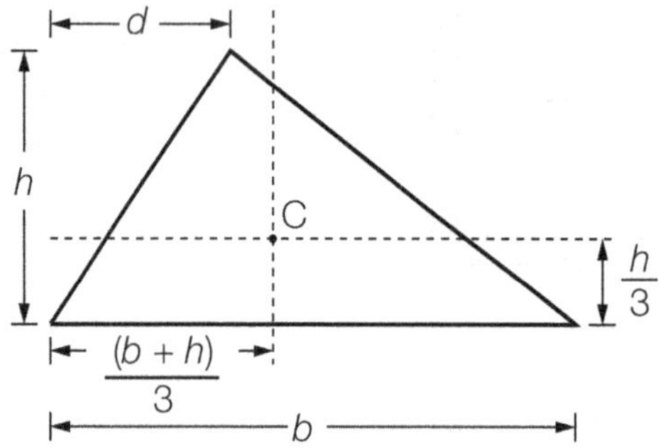 d, h, C, $\frac{h}{3}$, $\frac{(b+h)}{3}$, b	$\frac{1}{2}bh$	$I_{xc} = \frac{1}{36}ba^3$ $I_{xyc} = \frac{1}{72}bh^2(b-2d)$

C.2 Properties of Circles and Spheres

C.2.1 Circles

Several additional properties of a circle beyond those shown in Section C.1 are useful in many applications. These additional properties are related to the parameters shown in Figure C.1. The basic parameters used to characterize a circle and a part of a circle are the radius, R, and the central angle, θ, where

$$\theta = 2\cos^{-1}\left(\frac{d}{R}\right)$$

Using R and θ as parameters and taking R in units of length and θ in units of radians, the following quantities can be calculated:

arc length: $s = R\theta$, chord length: $c = 2R\sin\frac{\theta}{2}$

height: $h = R\left(1 - \cos\frac{\theta}{2}\right)$, segment area: $A_1 = \frac{R^2}{2}(\theta - \sin\theta)$

sector area: $A_2 = \frac{1}{2}R^2\theta$

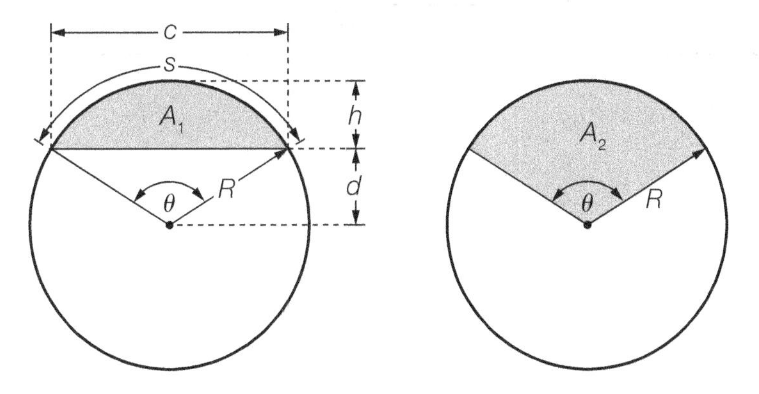

Figure C.1: Parameters of a circle

C.2.2 Spheres

The properties of a sphere that are most commonly of interest are the volume and the surface area. For a sphere of radius R and diameter D (= $2R$), the volume, V, and surface area, A, are given by

$$V = \frac{4}{3}\pi R^3 = \frac{1}{6}\pi D^3, \qquad A = 4\pi R^2 = \pi D^2$$

C.3 Volumes

Table C.2: Geometric Properties of Volumes

Shape	Illustration	Volume
Truncated cone	D_2, h, D_1	$\frac{1}{12}\pi h(D_1^2 + D_2^2 + D_1 D_2)$

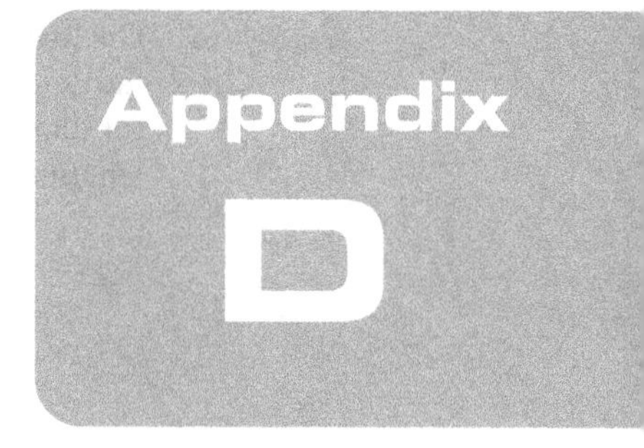

Pipe Specifications

D.1 PVC Pipe

Pipe dimensions of interest to engineers are usually the diameter and wall thickness. The pipe diameter is typically specified by the *nominal pipe size*, and the wall thickness is usually specified by the *schedule*. Nominal pipe sizes are typically given in inches or millimeters and represent rounded approximations to the inside diameter of the pipe. The schedule of a pipe is a number that approximates the value of the expression $1000P/S$, where P is the service pressure and S is the allowable stress. Higher schedule numbers correspond to thicker pipes, and schedule numbers in common use are 5, 5S, 10, 10S, 20, 20S, 30, 40, 40S, 60, 80, 80S, 100, 120, 140, and 160. The schedule numbers followed by the letter "S" are intended for use primarily with stainless steel pipe (ASME B36.19M).

		Schedule 5		Schedule 10		Schedule 40		Schedule 80	
Nominal pipe size (mm)	Outside diameter (mm)	Wall thickness (mm)	Inside diameter (mm)	Wall thickness (mm)	Inside diameter (mm)	Wall thickness (mm)	Inside diameter (mm)	Wall thickness (mm)	Inside diameter (mm)
50	60	1.7	57	2.8	55	3.9	53	5.5	49
80	90	2.1	85	3.0	83	5.5	78	7.6	74
100	114	2.1	110	3.0	108	6.0	102	8.6	97
125	141	2.8	136	3.4	134	6.6	128	9.5	122
150	168	2.8	163	3.4	161	7.1	154	11.0	146

Source: Fetter (1999).

D.2 Ductile Iron Pipe

Ductile iron pipe is manufactured in diameters in the range of 100–1200 mm; for diameters in the range of 100–500 mm, standard commercial sizes are available in 50-mm increments; and for diameters in the range of 600–1200 mm, the size increments are 150 mm. The standard lengths of ductile iron pipe are 5.5 m and 6.1 m.

D.3 Concrete Pipe

Table D.1: Commercially Available Sizes of Concrete Pipe

Non-Reinforced Pipe Diameter (mm)	Reinforced Pipe Diameter (mm)
100	—
150	—
205	—
255	—
305	305
380	380
455	455
535	535
610	610
685	685
760	760
840	840
915	915
—	1065
—	1220
—	1370
—	1525
—	1675
—	1830
—	1980
—	2135
—	2285
—	2440
—	2590
—	2745

D.4 Physical Properties of Common Pipe Materials

Table D.2: Common Pipe Materials

Material	Young's Modulus, E (GPa)	Poisson's Ratio
Concrete	14–30	0.10–0.15
Concrete (reinforced)	30–60	—
Ductile iron	165–172	0.28–0.30
PVC	2.4–3.5	0.45–0.46
Steel	200–207	0.30

Bibliography

1. A. Abdulrahman. Direct Solution to Problems of Open-Channel Transitions: Rectangular Channels. *Journal of Irrigation and Drainage Engineering*, 134(4):533–537, July/August 2008.
2. Air Control Industries. Centrifugal Fans, 2014. http://www.aircontrolindustries.com/us/products/industrial-fans/centrifugal-fans/centrifugal-fans-specials/.
3. American Society of Civil Engineers. *Gravity Sanitary Sewer Design and Construction.* American Society of Civil Engineers, New York, New York, 1982. Manual of Practice No. 60.
4. American Society of Civil Engineers. Report of ASCE Task Force on Friction Factors in Open Channels. *Proceedings of the American Society of Civil Engineers*, 89(HY2):97, March 1963.
5. American Water Works Association. *Sizing Water Service Lines and Meters, Manual of Water Supply Practices M22.* AWWA, Denver, Colorado, second edition, 2004.
6. H. H. Anderson. Prediction of Head, Quantity, and Efficiency in Pumps and Compressors. *ASME Symp., New York*, I00127:201–211, 1980.
7. ANDRITZ. Francis Turbine Runners, 2012. http://www.andritz.com.
8. Argonne National Laboratory. A Collection of Wind Turbines Near Farm Fields, 2013. http://www.dis.anl.gov/projects/windpowerforecasting.html.
9. American Society of Mechanical Engineers. *Stainless Steel Pipe.* ASME B36.19M. American Society of Mechanical Engineers, New York, New York, 1985.
10. E. A. Avallone and T. Baumeister III. *Marks' Handbook for Mechanical Engineers.* McGraw-Hill, Inc., New York, New York, tenth edition, 1996.
11. B. A. Bakhmeteff. *Hydraulics of Open Channel Flow.* McGraw-Hill, Inc., New York, New York, 1932.
12. B. A. Bakhmeteff. O neravnomernom dwijenii jidkosti v otkrytom rusle. (Varied Flow in Open Channel.), St. Petersburg, Russia (in Russian), 1912.
13. J. C. Bathurst. Flow Resistance Estimation in Mountain Rivers. *Journal of Hydraulic Engineering*, 111(4):625–643, 1985.
14. M. K. Beirami and M.R. Chamani. Hydraulic Jumps in Sloping Channels: Sequent Depth Ratio. *Journal of Hydraulic Engineering*, 132(10):1061–1068, October 2006.
15. J. B. Bélanger. Notes sur l'hydraulique. (Notes on hydraulic engineering), Ecole Royale des Ponts et Chaussés, Paris, France, session 1841–1842, 223 pages (in French), 1841.
16. J. B. Bélanger. Essai sur la solution numérique de quelques problèmes relatifs au mouvement permanent des eaux courantes. (Essay on the numerical solution of some problems relative to steady flow of water.), Carilian-Goeury, Paris (in French), 1828.
17. J. G. Bene and C. J. Høs. Finding Least-Cost Pump Schedules for Reservoir Filling with a Variable Speed Pump. *Journal of Water Resources Planning and Management*, 138(6):682–686, November/December 2012.
18. A. C. Benke and C. E. Cushing. Background and approach. In A.C. Benke and C.E. Cushing, editors, *Rivers of North America*, pages 1–18. Elsevier, Amsterdam, Netherlands, 2005.

19. H. Blasius. Das Ähnlichkeitsgesetz bei Reibungsvorgängen in Flüssigkeiten. *Forschungs-Arbeit des Ingenieur-Wesens*, 131, 1913.
20. H. Blasius. Grenzschichten in Flüssigkeiten mit kleiner Reibung. *Zeitschrift für Mathematik und Physik*, 56(1):1–37, 1908.
21. R. D. Blevins. *Applied Fluid Dynamics Handbook.* Van Nostrand Reinhold, New York, New York, 1984.
22. F. A. Bombardelli and M. H. García. Hydraulic Design of Large-Diameter Pipes. *Journal of Hydraulic Engineering*, 129(11):839–846, November 2003.
23. G. O. Brown. Henry Darcy and the making of a law. *Water Resources Research*, 38(7), 2002.
24. E. Buckingham. Model Experiments and the Form of Empirical Equations. *Transactions of the ASME*, 37:263–296, 1915.
25. D. Butler and J. W. Davies. *Urban Drainage*. Spon Press, Abingdon, Oxon, 2011.
26. F. Canovaro, E. Paris, and L. Solari. Effects of macro-scale bed roughness geometry on flow resistance. *Water Resources Research*, 43, W10414, 2007.
27. F. G. Carollo, V. Ferro, and V. Pampalone. New Expression of the Hydraulic Jump Roller Length. *Journal of Hydraulic Engineering*, 138(11):995–999, November 2012.
28. F. G. Carollo, V. Ferro, and V. Pampalone. New Solution of Classical Hydraulic Jump. *Journal of Hydraulic Engineering*, 135(6):527–531, June 2009.
29. Y. A. Çengel and M. A. Boles. *Thermodynamics: An Engineering Approach*. McGraw-Hill, Inc., New York, New York, fourth edition, 2002.
30. Y. A. Çengel and J. M. Cimbala. *Fluid Mechanics: Fundamentals and Applications*. McGraw-Hill, Inc., New York, New York, third edition, 2014.
31. A. Chadwick and J. Morfett. *Hydraulics in Civil and Environmental Engineering*. E & FN Spon, London, England, third edition, 1998.
32. H. Chanson. Bubbly Two-Phase Flow in Hydraulic Jumps at Large Froude Numbers. *Journal of Hydraulic Engineering*, 137(4):451–460, April 2011.
33. H. Chanson. Development of the Bélanger Equation and Backwater Equation by Jean-Baptiste Bélanger (1828). *Journal of Hydraulic Engineering*, 135(3):159–163, March 2009.
34. H. Chanson. Minimum Specific Energy and Critical Flow Conditions in Open Channels. *Journal of Irrigation and Drainage Engineering*, 132(5):498–502, September/October 2006.
35. N.-S. Cheng. Representative roughness height of submerged vegetation. *Water Resources Research*, 47, W08517, 2011.
36. N.-S. Cheng. Formulas for Friction Factor in Transitional Regimes. *Journal of Hydraulic Engineering*, 134(9):1357–1362, September 2008.
37. N.-S. Cheng, H.T. Nguyen, K. Zhao, and X. Tang. Evaluation of Flow Resistance in Smooth Rectangular Open Channels with Modified Prandtl Friction Law. *Journal of Hydraulic Engineering*, 137(4):441–450, April 2011.
38. V. T. Chow. *Open-Channel Hydraulics*. McGraw-Hill, Inc., New York, New York, 1959.
39. C. F. Colebrook. Turbulent Flows in Pipes, with Particular Reference to the Transition Region between the Smooth and Rough Pipe Laws. *Journal, Institution of Civil Engineers (London)*, 11:133–156, February 1939.

40. C. F. Colebrook and C. M. White. Experiments with Fluid Friction in Roughened Pipes. *Proceedings of the Royal Society*, A161:367, 1939.

41. J. E. Costa, R. T. Cheng, F. P. Haeni, N. Melcher, K. R. Spicer, E. Hayes, W. Plant, K. Hayes, C. Teague, and D. Barrick. Use of radars to monitor stream discharge by noncontact methods. *Water Resources Research*, 42, W07422, 2006.

42. J. Le Coz, G. Pierrefeu, and A. Paquier. Evaluation of river discharges monitored by a fixed side-looking Doppler profiler. *Water Resources Research*, 44, W00D09, 2008.

43. H. Cross. Analysis of Flow in Networks of Conduits or Conductors. Bulletin 286, University of Illinois Engineering Experiment Station, 1936.

44. J. F. Cruise, M. M. Sherif, and V. P. Singh. *Elementary Hydraulics*. Nelson Press, Toronto, Canada, 2007.

45. J. W. Daily and D. R. F. Harleman. *Fluid Dynamics*. Addison-Wesley Publishing Company, Inc., Reading, Massachusetts, 1966.

46. T. Dalrymple and M. Benson. Measurement of peak discharge by the slope-area method. Techniques of Water Resources Investigations of the United States Geological Survey, Chapter A2, U.S. Geological Survey, Washington, DC, 1967.

47. A. Das. Optimal Design of Channel Having Horizontal Bottom and Parabolic Sides. *Journal of Irrigation and Drainage Engineering*, 133(2):192–197, March/April 2007.

48. S. L. Dingman. *Fluvial Hydrology*. W.H. Freeman and Company, New York, New York, 1984.

49. J. F. Douglas, J. M. Gasiorek, and J. A. Swaffield. *Fluid Mechanics*. Prentice Hall, Upper Saddle River, New Jersey, fourth edition, 2001.

50. G. Echávez. Increase in Losses Coefficient with Age for Small Diameter Pipes. *Journal of Hydraulic Engineering*, 123(2):157–159, February 1997.

51. D. F. Elger, B. C. Williams, C. T. Crowe, and J. A. Roberson. *Engineering Fluid Mechanics*. John Wiley & Sons, Hoboken, New Jersey, tenth edition, 2001.

52. J. A. Fay. *Introduction to Fluid Mechanics*. MIT Press, Cambridge, Massachusetts, 1994.

53. R. Ferguson. Flow resistance equations for gravel- and boulder-bed streams. *Water Resources Research*, 43, W05427, 2007.

54. V. Ferro. Friction Factor for Gravel-Bed Channel with High Boulder Concentration. *Journal of Hydraulic Engineering*, 125(7):771–778, July 1999.

55. E. J. Finnemore and J. B. Franzini. *Fluid Mechanics with Engineering Applications*. McGraw-Hill, Inc., New York, New York, tenth edition, 2002.

56. J. B. Franzini and E. J. Finnemore. *Fluid Mechanics with Engineering Applications*. McGraw-Hill, Inc., New York, New York, ninth edition, 1997.

57. R. H. French. Hydraulics of Open Channel Flow. In L.W. Mays, editor, *Stormwater Collection Systems Design Handbook*, pages 3.1–3.35. McGraw-Hill, Inc., New York, New York, 2001.

58. D. C. Froehlich. Resistance to Shallow Uniform Flow in Small, Riprap-Lined Drainage Channels. *Journal of Irrigation and Drainage Engineering*, 138(2):203–210, February 2012.

59. J. Fulton and J. Ostrowski. Measuring Real-Time Streamflow Using Emerging Technologies: Radar, Hydroacoustics, and the Probability Concept. *Journal of Hydrology*, 357:1–10, 2008.

60. E. Ganguillet and W. R. Kutter. Versuch zur Aufstellung einer neuen allegemeinen Formel für die gleichförmige Bewegung des Wassers in Canälen und Flüssen (An investigation to establish a new general formula for uniform flow of water in canals and rivers). *Zeitschrift des Oesterreichischen Ingenieur-und Architekten Vereines*, 21(1,2-3):6–25,46–59, 1869. (Published as a book in Bern, Switzerland, 1877; translated into English by Rudolph Hering and John C. Trautwine, Jr., as *A General Formula for the Uniform Flow of Water in Rivers and Other Channels*, John Wiley & Sons, Inc., New York, 1st ed., 1888; 2d ed., 1891 and 1901).

61. P. M. Gerhart, R. J. Gross, and J. I. Hochstein. *Fundamentals of Fluid Mechanics*. Addison-Wesley Publishing Company, Inc., Reading, Massachusetts, second edition, 1992.

62. J. E. Goff and M. J. Carré. Soccer Ball Lift Coefficients via Trajectory Analysis. *European Journal of Physics*, 7:1–13, 1993.

63. S. R. Goodwill, S. B. Chin, and S. J. Haake. Aerodynamics of Spinning and Non-spinning Tennis Balls. *Journal of Wind Engineering and Industrial Aerodynamics*, 92:935–958, 2004.

64. L. Gordon. Mississippi river discharge. San Diego, California, 1992. RD Instruments.

65. S. E. Haaland. Simple and Explicit Formulas for the Friction Factor in Turbulent Pipe Flow. *Journal of Fluids Engineering*, 105(3):89–90, March 1983.

66. M. Haestad, T. M. Walski, T. E. Barnard, E. Harold, L. B. Merritt, N. Walker, and B. E. Whitman. *Wastewater Collection System Modeling and Design*. Haestad Press, Waterbury, Connecticut, 2004.

67. W. H. Hager. *Wastewater Hydraulics*. Springer-Verlag, Berlin, Germany, 1999.

68. W. H. Hager. *Energy Dissipators and Hydraulic Jump*. Kluwer Academic Publishers, Dordrecht, Netherlands, 1991.

69. M. J. Hammer and M. J. Hammer, Jr., *Water and Wastewater Technology*. Prentice Hall, Upper Saddle River, New Jersey, seventh edition, 2010.

70. F. M. Henderson. *Open Channel Flow*. Macmillan Publishing Co., Inc., New York, New York, 1966.

71. S. F. Hoerner. *Fluid-Dynamic Drag*. Van Nostrand Reinhold, Published by author, Library of Congress No.64,19666, 1965.

72. B. Hunt. Dispersion Model for Mountain Streams. *Journal of Hydraulic Engineering*, 125(2):99–105, February 1999.

73. R. B. Hunter. Methods of Estimating Loads in Plumbing Systems. Technical Report BMS 65, National Bureau of Standards, Washington, DC, 1940.

74. Institute for Mathematics and Its Applications. IMA Thematic Year on Complex Fluids and Complex Flows, 2013. http://www.ima.umn.edu/2009-2010/.

75. International Association of Plumbing and Mechanical Officials. *Uniform Plumbing Code*. Ontario, CA, 2009.

76. T. Inui. Wavemaking Resistance of Ships. *Transactions—The Society of Naval Architects and Marine Engineers*, 70:283–326, 1962.

77. E. C. Ipsen. *Units, Dimensions, and Dimensionless Numbers*. McGraw-Hill, Inc., New York, New York, 1960.

78. IQS, Inc. Serving All Seasons with Centrifugal Pumps, 2014. http://blog.iqsdirectory.com/pumps-valves/serving-all-seasons-with-centrifugal-pumps/.

79. A. K. Jain, D. M. Mohan, and P. Khanna. Modified Hazen-Williams Formula. *Journal of the Environmental Engineering Division, ASCE*, 104(EE1):137–146, February 1978.

80. C.-D. Jan and C.-J. Chang. Hydraulic Jumps in an Inclined Rectangular Chute Contraction. *Journal of Hydraulic Engineering*, 135(11):949–958, November 2009.

81. W. S. Janna. *Introduction to Fluid Mechanics*. PWS-KENT Publishing Company, Boston, Massachusetts, third edition, 1993.

82. G. E. Jones. *Gravity-Driven Water Flow in Networks*. John Wiley & Sons, Hoboken, New Jersey, 2011.

83. A. Kanani, M. Bakhtiari, S. M. Borghei, and D.-S. Jeng. Evolutionary Algorithms for the Determination of Critical Depths in Conduits. *Journal of Irrigation and Drainage Engineering*, 134(6):847–852, November/December 2008.

84. D. G. Karamanev and L. N. Nikolov. Freely Rising Spheres Do Not Obey Newton's Law for Free Settling. *AIChE Journal*, 38(1):1843–1846, November 1992.

85. G. H. Keulegan. Laws of Turbulent Flow in Open Channels. *Journal of Research of the National Bureau of Standards (United States)*, 21:707–741, 1938.

86. R. T. Knapp, J. W. Daily, and F. G. Hammitt. *Cavitation*. McGraw-Hill, Inc., New York, New York, 1970.

87. D. W. Knight, C. McGahey, R. Lamb, and P. G. Samuels. *Practical Channel Hydraulics*. CRC Press, Inc., Boca Raton, Florida, 2010.

88. E. Kordi and I. Abustan. Transitional Expanding Hydraulic Jump. *Journal of Hydraulic Engineering*, 138(1):105–110, January 2012.

89. E. Kordi and I. Abustan. Estimating the Overbank Flow Discharge using Slope-Area Method. *Journal of Hydrologic Engineering*, 16(11):907–913, November 2011.

90. S. Kucukali and S. Cokgor. Energy Concept for Predicting Hydraulic Jump Aeration Efficiency. *Journal of Environmental Engineering*, 135(2):105–107, February 2009.

91. E. W. Lane and E. J. Carlson. Some Factors Affecting the Stability of Canals Constructed in Coarse Granular Materials. In *Proceedings, Minnesota International Hydraulics Convention, Minneapolis, MN, 1-4 September*, pages 37–48, 1953.

92. K. Lansey and W. El-Shorbagy. Design of Pumps and Pump Facilities. In L.W. Mays, editor, *Stormwater Collection Systems Design Handbook*, pages 12.1–12.41. McGraw-Hill, Inc., New York, New York, 2001.

93. Lawrence Livermore National Laboratory. Lawrence Livermore National Lab, Navistar Work to Increase Semi-truck Fuel Efficiency, 2013. https://www.llnl.gov/news/newsreleases/2010/NR-10-02-08.html.

94. J. T. Limerinos. Determination of the Manning Coefficient from Measured Bed Roughness in Natural Channels. Water-Supply Paper 1898-B, U.S. Geological Survey, Washington, DC, 1970.

95. C. P. Liou. Limitations and Proper Use of the Hazen-Williams Equation. *Journal of Hydraulic Engineering*, 124(9):951–954, September 1998.

96. R. López and J. Barragán. Equivalent Roughness of Gravel-Bed Rivers. *Journal of Hydraulic Engineering*, 134(6):847–851, June 2008.

97. N. J. Lowe, R. H. Hotchkiss, and E. J. Nelson. Theoretical Determination of Sequent Depths in Closed Conduits. *Journal of Irrigation and Drainage Engineering*, 137(12):801–810, December 2011.

98. M. F. Maghrebi and J. E. Ball. New Method for Estimation of Discharge. *Journal of Hydraulic Engineering*, 132(10):1044–1051, October 2006.

99. R. Manning. Flow of Water in Open Channels and Pipes. *Transactions of the Institute of Civil Engineers (Ireland)*, 20:161–207, 1890.

100. B. S. Massey and J. Ward-Smith. *Mechanics of Fluids*. Spon Press, London, England, ninth edition, 2012.

101. S. T. Maynord. Riprap Resistance Tests from a Large Test Channel. Miscellaneous Paper HL-92-5, U.S. Army Corps of Engineers, Waterways Experiment Station, Vicksburg, Mississippi, 1991.

102. R. D. Mehta. Aerodynamics of Sports Balls. *Annual Reviews of Fluid Mechanics*, 17:151–189, 1985.

103. P. E. Meyer-Peter and R. Muller. Formulas for Bed Load Transport. In *Proceedings of the International Association for Hydraulic Research, Second Congress, Stockholm, Sweden*, pages 39–65, 1948.

104. M. D. Mikhailov and A. P. Silva Freire. The Drag Coefficient of a Sphere: An Approximation Using Shanks Transform. *Powder Technology*, 237:432–435, 2013.

105. A. E. Minetti, G. Machtsiras, and J. C. Masters. The Optimum Finger Spacing in Human Swimming. *Journal of Biomechanics*, 42:2188–2190, 2009.

106. S. Montes. *Hydraulics of Open Channel Flow*. ASCE Press, Reston, Virginia, 1998.

107. L. F. Moody. Some Pipe Characteristics of Engineering Interest. *Houille Blanche*, 5(4):313–325, May-June 1950.

108. L. F. Moody. Friction Factors for Pipe Flow. *Trans. ASME*, 66(8), 1944.

109. L. F. Moody and T. Zowski. Hydraulic Machinery. In C.V. Davis and K.E. Sorensen, editors, *Handbook of Applied Hydraulics*. McGraw-Hill, Inc., New York, New York, third edition, 1989.

110. T. Morel and M. Bohn. Flow over Two Circular Disks in Tandem. *Transactions of the ASME, Journal of Fluids Engineering*, 102(1):104–111, March 1980.

111. R. L. Mott. *Applied Fluid Mechanics*. Pearson Prentice Hall, Upper Saddle River, New Jersey, sixth edition, 2006.

112. B. R. Munson, T. H. Okiishi, W. W. Huebsch, and A. P. Rothmayer. *Fundamentals of Fluid Mechanics*. John Wiley & Sons, Hoboken, New Jersey, seventh edition, 2013.

113. M. Muste, T. Vermeyen, R. Hotchkiss, and K. Oberg. Acoustic Velocimetry for Riverine Environments. *Journal of Hydraulic Engineering*, 133(12):1297–1298, December 2007.

114. W. R. C. Myers. Influence of Geometry on Discharge Capacity of Open Channels. *Journal of Hydraulic Engineering*, 117(5):676–680, May 1991.

115. A. M. Nathan. The Effect of Spin on the Flight of a Baseball. *American Journal of Physics*, 76(2):119–124, 2008.

116. National Aeronautics and Space Administration. Aerodynamic Characteristics of Supersonic-Hypersonic Flight, 2014a. http://history.nasa.gov/SP-60/ch-5.html.

117. National Aeronautics and Space Administration. NBL Image Gallery, 2014b. http://dx12.jsc.nasa.gov/gallery/index.shtml.

118. National Aeronautics and Space Administration. SP-4103 Model Research—Volume 2, 2014c. http://history.nasa.gov/SP-4103/app-f.htm.

119. National Aeronautics and Space Administration. Typhoon Neoguri, 2014d. http://earthobservatory.nasa.gov/IOTD/view.php?id=83980.

120. National Aeronautics and Space Administration. Submarine in Full Scale Tunnel at NACA Langley, 2014e. http://grin.hq.nasa.gov/ABSTRACTS/GPN-2000-001936.html.

121. National Aeronautics and Space Administration. NASA Turns World Cup into Lesson in Aerodynamics, 2014f. http://www.nasa.gov/content/nasa-turns-world-cup-into-lesson-in-aerodynamics/.

122. National Aeronautics and Space Administration. The X-48B Blended Wing Body, 2013a. http://www.nasa.gov/vision/earth/improvingflight/x48b.html.

123. National Aeronautics and Space Administration. Water Experiment, 2013b. http://spaceflight.nasa.gov/station/crew/exp6/spacechronicles15.html.

124. National Aeronautics and Space Administration. Testing the Future, 2013c. http://www.nasa.gov/multimedia/imagegallery/image_feature_1287.html.

125. National Aeronautics and Space Administration. NASA Tests Future Mars Landing Technology, 2013d. http://www.jpl.nasa.gov/news/news.php?release=2012-165.

126. National Aeronautics and Space Administration. Shock Waves Produced by Blunt Bodies and Pointed Bodies, 2013e. http://history.nasa.gov/SP-4302/ch2.8.htm.

127. National Aeronautics and Space Administration. Featured Images, Dryden News Photos, 2008. http://www.nasa.gov/centers/dryden/news/newsphotos/index_prt.htm.

128. National Institute of Standards and Technology (NIST). Fans Clear High-Rise Stairwells of Smoke, 2014. http://www.flickr.com/photos/usnistgov/5888202056/in/photostream/.

129. National Renewable Energy Laboratory (NREL). United States Annual Average Wind Power, 2013. http://rredc.nrel.gov/wind/pubs/atlas/maps/chap2/2-01m.html.

130. J. Nikuradse. Strömungsgesetze in Rauhen Röhren. *Forschung auf dem Gebiete des Ingenieurwesens, Forschungsheft 361, VDI Verlag, Berlin, Germany*, 1933. Title in English: Laws of Flow in Rough Pipes.

131. J. Nikuradse. Gesetzmässigkeiten der turbulenten strömung in glatten röhren. *VDI-Forschungsheft*, 356, 1932.

132. R. M. Olson and S. J. Wright. *Essentials of Engineering Fluid Mechanics*. Harper & Row, Publishers, Inc., New York, New York, fifth edition, 1990.

133. C. W. Oseen. Uber den Giiltigkeitsbereich der Stokesschen Widerstandsformel. *Arkiv för Matematik, Astronomi och Fysik*, 9:1–15, 1913.

134. C. W. Oseen. Ueber die Stokessche Formel und die verwandte Aufgabe in der Hydrodynamik. *Arkiv för Matematik, Astronomi och Fysik*, 6:29, 1910.

135. R. G. Patil, J. S. R. Murthy, and L. K. Ghosh. Uniform and Critical Flow Computations. *Journal of Irrigation and Drainage Engineering*, 131(4):375–378, August 2005.

136. M. C. Potter, D. C. Wiggert, B. Ramadan, and T. I.-P. Shih. *Mechanics of Fluids*. Cengage Learning, Stamford, Connecticut, fourth edition, 2012.

137. L. Prandtl. The Mechanics of Viscous Fluids. In W.F Durand, editor, *Aerodynamic Theory, Volume 3*, pages 34–207. Springer, Berlin, Germany, 1935.

138. P. J. Pritchard. *Fox and McDonald's Introduction to Fluid Mechanics*. John Wiley & Sons, Hoboken, New Jersey, eigth edition, 2011.

139. M. Rehmel. Application of Acoustic Doppler Velocimeters for Stream Flow Measurements. *Journal of Hydraulic Engineering*, 133(12):1433–1438, December 2007.

140. O. Reynolds. An Experimental Investigation of the Circumstances Which Determine Whether the Motion of Water Shall Be Direct or Sinuous and of the Law of Resistance in Parallel Channels. *Philosophical Transactions, Royal Society of London*, 174:935–982, 1883.

141. J. A. Roberson, J. J. Cassidy, and M. H. Chaudhry. *Hydraulic Engineering*. John Wiley & Sons, Hoboken, New Jersey, second edition, 1998.

142. J. A. Romero, A. Lozano, and W. Ortiz. Modelling of Liquid Cargo–Vehicle Interaction during Turning Manoeuvres. http://www.iftomm.org/iftomm/proceedings/proceedings_WorldCongress/WorldCongress07/articles/sessions/papers/A609.pdf, Accessed: 9/1/13, June 18-21, 2007.

143. H. Rouse. *Advanced Mechanics of Fluids*. John Wiley & Sons, Hoboken, New Jersey, 1959.

144. H. Rouse and S. Ince. *History of Hydraulics*. Iowa Institute of Hydraulic Research, University of Iowa, Iowa City, 1957.

145. H. Rubin and J. Atkinson. *Environmental Fluid Mechanics*. Marcel Dekker, Inc., New York, New York, 2001.

146. P. G. Samuels. Backwater Length in Rivers. *Proceedings of the Institution of Civil Engineers, London*, 87(Part 2):571–582, December 1989.

147. R. L. Sanks. *Pump Station Design*. Butterworth Publishers, London, England, second edition, 1998.

148. I. Sarbu and I. Borza. Energetic Optimization of Water Pumping in Distribution Systems. *Periodica Polytechnica Ser. Mech. Eng.*, 42(2):141–152, 1988.

149. H. Schlichting. *Boundary Layer Theory*. McGraw-Hill, Inc., New York, New York, eighth edition, 2000.

150. H. Schlichting. *Boundary Layer Theory*. McGraw-Hill, Inc., New York, New York, seventh edition, 1979.

151. H. Schlichting. Boundary Layer Theory, Part II. *NACA Technical Memorandum 1218*, page 39, 1949.

152. A. R. Schmidt and B. C. Yen. Theoretical Development of Stage-Discharge Ratings for Subcritical Open-Channel Flows. *Journal of Hydraulic Engineering*, 134(9):1245–1256, September 2008.

153. D. Shanks. Non-linear Transformation of Divergent and Slowly Convergent Sequences. *Journal of Mathematical Physics*, 34:1–42, 1955.

154. A. L. Simon and S. F. Korom. *Hydraulics*. Prentice Hall, Upper Saddle River, New Jersey, fourth edition, 1997.

155. A. R. Simpson and A. Marchi. Comparison of Finite Difference and Finite Element Solutions to the Variably Saturated Flow Equation. *Journal of Hydraulic Engineering*, 139(12):1314–1317, December 2013.

156. J. R. Sonnad and C. T. Goudar. Turbulent Flow Friction Factor Calculation Using a Mathematically Exact Alternative to the Colebrook-White Equation. *Journal of Hydraulic Engineering*, 132(8):863–867, August 2006.

157. T. E. Stanton and J. R. Pannell. Similarity of Motion in Relation to Surface Friction of Fluids. *Philosophical Transactions, Royal Society of London*, 214A:199–224, 1914.

158. B. Stein, J. S. Reynolds, W. T. Grondzik, and A. K. Kwok. *Mechanical and Electrical Equipment for Buildings*. John Wiley & Sons, Hoboken, New Jersey, tenth edition, 2006.

159. A. J. Stepanoff. *Centrifugal and Axial Flow Pumps*. John Wiley & Sons, Hoboken, New Jersey, second edition, 1957.

160. R. L. Street, G. Z. Watters, and J. K. Vennard. *Elementary Fluid Mechanics*. John Wiley & Sons, Hoboken, New Jersey, seventh edition, 1996.

161. V. L. Streeter, E. B. Wylie, and K. W. Bedford. *Fluid Mechanics*. McGraw-Hill, Inc., New York, New York, ninth edition, 1998.

162. A. Strickler. Contributions to the Question of a Velocity Formula and Roughness Data for Streams, Channels and Closed Pipelines. Technical Report, W.M. Keck Laboratory, California Institute of Technology, Pasadena, California, 1923. Translation by T. Roesgen and W.R. Brownlie (1981).

163. K. S. Suslick. The Chemical Effects of Ultrasound. *Scientific American*, 260(2):80–86, 1989.

164. P. K. Swamee and A. K. Jain. Explicit Equations for Pipe-Flow Problems. *Journal of the Hydraulics Division, ASCE*, 102(HY5):657–664, May 1976.

165. P. K. Swamee and P. N. Rathie. Singular Open-Channel Section for Alternate and Sequent Depths. *Journal of Hydraulic Engineering*, 133(5):569–570, May 2007.

166. G. I. Taylor. Stability of a Viscous Liquid Contained between Two Rotating Cylinders. *Philosophical Transactions of the Royal Society, London, Series A*, 223:289–343, 1923.

167. S. Tiğrek, C. E. Firat, and A. M. Ger. Use of Brink Depth in Discharge Measurement. *Journal of Irrigation and Drainage Engineering*, 134(1):89–95, January/February 2008.

168. A. S. Tijsseling and A. Anderson. Johannes von Kries and the History of Water Hammer. *Journal of Hydraulic Engineering*, 133(1):1–8, January 2007.

169. Y. S. Touloukian, S. C. Saxena, and P. Hestermans. *Thermophysical Properties of Matter, The TPRC Data Series, Volume 11, Viscosity*. Plenum, New York, New York, 1975.

170. Q. B. Travis and L. W. Mays. Relationship between Hazen-William and Colebrook-White Roughness Values. *Journal of Hydraulic Engineering*, 133(11):1270–1273, November 2007.

171. B. Ulanicki, J. Kahler, and B. Coulbeck. Modeling the Efficiency and Power Characteristics of a Pump Group. *Journal of Water Resources Planning and Management*, 134(1):88–93, 2008.

172. United States Bureau of Reclamation. Library Catalog, 2013. http://www.usbr.gov/library/.

173. United States Forest Service. Turbidity Threshold Sampling (TTS), 2013. http://www.fs.fed.us/psw/topics/water/tts/tts_inst.html.

174. United States Geological Survey. VHP Photo Glossary: Basalt, 2013. http://volcanoes.usgs.gov/images/pglossary/basalt.php.

175. United States Navy. Office of Naval Intelligence, 2014. http://www.oni.navy.mil/.

176. U.S. Air Force. Altus Air Force Base, 2014a. http://www.altus.af.mil/photos/mediagallery.asp?galleryID=2309.

177. U.S. Air Force. Air Force Modernization Takes B-2 to North Pole, 2014b. http://www.af.mil/News/Photos.aspx?igphoto=2000203455.

178. U.S. Army Corps of Engineers. USCAE Bay Model-San Pablo Bay Panorama, 2014. http://en.wikipedia.org/wiki/U.S._Army_Corps_of_Engineers_Bay_Model\#mediaviewer/File:USCAE_Bay_Model_-_San_Pablo_Bay_Panorama.jpg.

179. U.S. Army Corps of Engineers. HEC-RAS Hydraulic Reference Manual, Version 4. Technical Reference Manual, Hydrologic Engineering Center, Davis, California, 2008.

180. U.S. Army Corps of Engineers. Propeller Turbine Runner, 2005. http://www.hq.usace.army.mil/cepa/pubs/mar00/story11.htm.

181. U.S. Army Corps of Engineers. HEC-RAS River Analysis System, Version 3.1. Hydraulic Reference Manual CPD-69, Hydrologic Engineering Center, Davis, California, November 2002.

182. U.S. Army Corps of Engineers. *Accuracy of Computed Water Surface Profiles*. Hydrologic Engineering Center, Davis, California, 1986.

183. U.S. Bureau of Reclamation. *Design of Small Dams*. U.S. Government Printing Office, Washington, DC, third edition, 1987.

184. U.S. Department of Agriculture. Design of Open Channels. Technical Release No.25, Soil Conservation Service, Washington, DC, October 1977.

185. U.S. National Park Service. Agitation?? No...Cavitation!! 2014. http://www.nps.gov/safr/historyculture/propsaquaticpark.htm.

186. U.S. Navy, Naval Air Systems Command. C-40A Clipper, 2013. http://www.navair.navy.mil/index.cfm?fuseaction=home.PhotoGalleryDetail\&key=892A0BE4-C883-49BD-A3A9-BE39030A09BB.

187. J. D. Valiantzas. Modified Hazen-Williams and Darcy-Weisbach Equations for Friction and Local Head Losses along Irrigation Laterals. *Journal of Irrigation and Drainage Engineering*, 131(4):342–350, August 2005.

188. V. A. Vanoni. Velocity Distribution in Open Channels. *Civil Engineering*, 11:356–357, 1941.

189. J. P. Velon and T. J. Johnson. Water Distribution and Treatment. In V.J. Zipparro and H. Hasen, editors, *Davis' Handbook of Applied Hydraulics*, pages 27.1–27.50. McGraw-Hill, Inc., New York, New York, fourth edition, 1993.

190. N. H. C. Wang and R. J. Houghtalen. *Fundamentals of Hydraulic Engineering Systems*. Prentice Hall, Upper Saddle River, New Jersey, third edition, 1996.

191. R. G. Watts and G. Moore. The Drag Force on an American Football. *American Journal of Physics*, 71(8):791–793, August 2003.

192. Weir Minerals. Lewis Pumps LH Axial Flow Pump, 2014. http://www.weirminerals.com/products_services/acid_pumps/vertical_chemical_pump/lewis_pumps_axial_flow_pumps.aspx.

193. F. M. White. *Fluid Mechanics*. McGraw-Hill, Inc., New York, New York, seventh edition, 2011.

194. F. M. White. *Fluid Mechanics*. McGraw-Hill, Inc., New York, New York, sixth edition, 2008.

195. WhyHydroPower.com. A Closer Look at Water Turbines, 2012. http://www.whyhydropower.com/Gallery.html.

196. G. V. Wilkerson and J. L. McGahan. Depth-Averaged Velocity Distribution in Straight Trapezoidal Channels. *Journal of Hydraulic Engineering*, 131(6):509–512, June 2005.

197. G. S. Williams and A. H. Hazen. *Hydraulic Tables*. John Wiley & Sons, Hoboken, New Jersey, 1920.

198. E. Wohl. *Mountain Rivers*. American Geophysical Union, Washington, DC, 2000.

199. S. Wu and N. Rajaratnam. Transition from Hydraulic Jump to Open Channel Flow. *Journal of Hydraulic Engineering*, 122(9):526–528, September 1996.

200. C. T. Yang. *Sediment Transport: Theory and Practice*. McGraw-Hill, Inc., New York, New York, 1996.

201. P. C. Yannopoulos, A. C. Demetracopoulos, and Ch. Hadjitheodorou. Quick Method for Open-Channel Discharge Measurements Using Air Bubbles. *Journal of Hydraulic Engineering*, 134(6):843–846, June 2008.

202. B. C. Yen. Open Channel Flow Resistance. *Journal of Hydraulic Engineering*, 128(1):20–39, January 2002.

203. B. C. Yen. Hydraulic Resistance in Open Channels. In B.C. Yen, editor, *Channel Flow Resistance: Centennial of Manning's Formula*, pages 1–135. Water Resource Publications, Highlands Ranch, Colorado, 1991.

204. D. H. Yoo and V. P. Singh. Two Methods for the Computation of Commercial Pipe Friction Factors. *Journal of Hydraulic Engineering*, 131(8):694–704, 2005.

205. V. J. Zipparro and H. Hasen. *Davis' Handbook of Applied Hydraulics*. McGraw-Hill, Inc., New York, New York, fourth edition, 1993.

Index

A

absolute pressure, 92
absolute pressure transducers, 104
absolute system of units, 20
absolute viscosity, 47
acceleration
  See also speed
  liquids with constant, 141–144
  local, convective, 179, 359
acoustic Doppler current profilers (ADCPs), 708
acoustic Doppler velocimeters (ADVs), 708
acoustic velocity, 891
added mass, 790
added mass coefficient, 790
adhesion, 55
adiabatic, 41
adiabatic bulk modulus, 43
adiabatic exponent, 41
adiabatic processes, 312–313, 888
advective acceleration, 179–180
adverse pressure gradients, 434, 767, 769, 836
adverse slope, 725–726
aeration efficiency, 735
aerodynamic drag, 781
aerodynamic instability, 769
aerodynamic shoulder, 438
aerodynamics, 20
aerostatics, 87
affinity laws, 623–624, 645–646
air
  compressibility, 34
  density of, 29
  ducts, 904
  flows around columnar structures, 431–433
  physical properties (table), 1004
  properties of U.S. standard atmosphere (table), 44
  removing suspended particles from, 147
  speed of sound in, 893–894
  summary of properties of (table), 69
  surface tension of various fluids in contact with (fig.), 56
air chambers, 565
air compressors, 318–319
aircraft
  Concorde's noise levels, 897
  Mach cone phenomenon, 896
  speed, 210–212
airflow rates, 24
airfoils
  described, 791
  flow around (fig.), 761
  lift on, 791–794
airplanes
  lift and drag, 765
  lift on, 794–799
  winglets on (fig.), 797
Allievi equation, 562
Andrade's equation, 51
aneroid barometers, 102
angle of attack, 791, 792, 794
angle of heel, 135
angle valves, 557
angular deformation, 358, 363
angular momentum, 299
angular momentum equation, 299
angular momentum principle, 298–307
angular-velocity vector, 360
apparent shear stress, 444
apparent viscosity, 53
Archimedes' principle, 127
areas, properties of, 1009–1010
artesian wells, 106
aspect ratio, 798
astronauts at Neutral Buoyancy Laboratory (fig.), 131
asymptotic behavior of water surface profiles, 726–727
atmosphere, 44–46
atmosphere, standard (table), 1005
atmospheric convection, 139
atmospheric pressure
  described, 92–93
  standard sea-level, 102
auto testing, 501
available head, 656
available net positive suction head, 635
Avogadro's law, 36
axial flow, 609
axial-flow pumps, 614, 615, 616, 627-628
axial-flow turbines, 654
axisymmetric, 374
axisymmetric flows, 371, 412

B

B-2 bomber, moments on (fig.), 761
back pressure, 912

backing plate, 614
backward-curved blades, 618
backwashing sand filters, 780
backwater curves, 727
barometers
 described, 101
 types of (fig.), 102
 using, 101–103
barometric pressure, 92, 101
Bay Model, 503
Beckham, David, 801
Bélanger's equation, 733
Bernoulli constant, 196, 408
Bernoulli equation, 214, 216, 217, 220, 313, 901
 applied to gases, 199
 described, 178, 196
 energy components in, 198
 limitations of, 196–197
 pressure components in, 198–199
 for steady inviscid flows, 404–407
 for steady irrotational inviscid flows, 407–409
best efficiency point (BEP), 626
Bette's law, 290
Bingham, Eugene C., 53n
Bingham plastics, 53
B-jump, 734
blades, 281–282, 609
Blasius, Heinrich, 862
Blasius equation, 831, 834–835, 861
Blasius solution, 831, 833, 834, 844, 846, 849
blowers, 608–609
bluff bodies, 768–769
blunt bodies, 768–769, 970
bodies
 See also specific type
 fully submerged, 127–131
 partially submerged, 132–139
boiling
 cavitation and, 64–67
 described, 64
booster pump systems, 582
Borda, Jean-Charles de, 554n
Borda-Carnot head loss, 554
boundary layer, 26, 410, 530
boundary layers, 765, 766–767
 in closed conduits, 859–862
 described, 827
 turbulent, 836–845
boundary shear stress, 527–528
boundary-layer approximation, 829
boundary-layer equation, Blasius solution to (table), 831
boundary-layer flows
 applications, 845–855
 boundary layers in closed conduits, 859–862
 described, 827
 introduction to, 827–829
 key equations, 863–866
 laminar boundary layers, 829–836
 mixing-length theory turbulent boundary layers, 856–859
 problems, 867–883
 turbulent boundary layers, 836–845
boundary-layer separation, 767, 792
Bourdon, Eugène, 103n
Bourdon-tube pressure gauge, 103–104
Boussinesq coefficient, 695
bow shocks, 896, 970–971
Boyle's law, 889
brake horsepower, 617, 622
branching conduits, 220–222
Brazuca-style soccer balls, 181
Bresse's equation, 733
bridge piers, flow around, 431
British Gravitational units, 23
bubbles
 in air and other gases, liquid droplets and, 57
 gas, in liquids, 57–58
 vapor or cavitation, 64
Buckingham, Edgar, 483
Buckingham pi (π) theorem, 392, 481, 528, 698
buffer layer, 837
bulk modulus, 32
bulk modulus of compressibility, 32
bulk modulus of elasticity, 32
bulk strain rate, 364
buoyancy
 center of, 127–128
 effect on humans, 131
 fluid statics, 127–139
buoyant force, 127
butterfly valves, 557

C

calorifically perfect, 41
camber, 793
camber line, 791
cambered airfoil, 791
canal, axial-flow pumps operating in (fig.), 628
canoes, 134
capacity coefficient, 622
capacity factor, 291
capillaries, 59
capillarity, 59–60

capillary correction, 110
capillary effect, 59
capillary-tube viscometer, 51
Carnot, Lazare Nicolas Marguérite, 554n
cascades, 557
casing side, 614
Cauchy, Augustin-Louis, 411n
Cauchy number, 490
Cauchy-Riemann equations, 411
Cauchy's equation, 378
cavitation
  and boiling, 64–67
  described, 64, 635
  negative, positive effects of, 66–67
cavitation number, 67, 490, 799–800
Celsius (°C), 22
center of buoyancy, 127–128
center of pressure, 112, 116
centimeter, 23
centrifugal pumps, 614, 615, 616–621
centrifugal separators, 145
check valve, 614
chemical energy, 308
Chézy, Antoine de, 703n
Chézy coefficient, 703
choked, 915
choked (flow), 722
choked condition of conduits, 908–923
chord length, 794
chord line, 791
chord Reynolds number, 793
cigarette smoke, 138
circles, properties of, 1011–1012
circular cylinders, flow around, 437–441
cistern, water in (fig.), 266
cities at high elevation (table), 101
closed circuits, 263–264
closed impellers, 614, 615
closed system, 39, 257
closest-depth rule, 719, 722
closure problem, 445
coefficient of compressibility, 32
coefficient of thermal expansion, 34
coefficient of viscosity, 47
coefficient of volume expansion, 34
cohesion, 55
Colebrook, Cyril, 862
Colebrook equation, 537–538, 540, 541, 545, 634, 662, 700, 701, 862, 941
Colebrook-White equation, 538
columns, flow around, 431
combined conventional dimension, 483
combined vortex, 422
combustion chamber, 296
common gases, physical properties (table), 1007
common liquids, physical properties (table), 1006
complete hydraulic jumps, 734
composite bodies, drag on, 786
compound gauges, 104
compressibility
  effects, 216–218
  of fluids, 32–35
  key equations in properties of fluids, 70
compressible flows
  basic equations of one-dimensional, 905–907
  described, 27, 33, 884
  introduction to, 884–885
  key equations, 977–983
  normal shocks, 923–934
  oblique shocks, bow shocks, expansion waves, 962–976
  principles of thermodynamics, 885–890
  problems, 984–998
  speed of sound, 891–897
  steady one-dimensional isentropic flows, 907–923
  steady one-dimensional non-isentropic flows, 935–961
  thermodynamic reference conditions, 898–904
compressible fluids, 884–885
compression waves, 972
compressors, 608–609, 644
computational fluid dynamics (CFD), 385, 760
computational fluid mechanics, 20
Concorde's noise levels, 897
concrete pipe specifications (tables), 1014
condensation, 62
condensed-phase matter, 18
conduction, 885
conduits
  airflow through, 264
  boundary layers in closed, 859–862
  choked condition, 908–923
  closed, 263–264
  flow in closed. *See* flow in closed conduits
  forces on pressure, 273–280
  steady-state energy equation, 309–320
  throat section, 908
conjugate depths of hydraulic jumps, 733
conservation of energy, 307–322, 446–448
conservation of energy equation, 446
conservation of linear momentum, 268–298
conservation of mass
  differential analysis, 365–375
  law of, 259–268
consistency index, 53
constants and conventions, 24

constitutive equations, 379
contact angle, 58
continuity equation, 259–263, 311–312, 365–372, 905–907
continuum idealization, 19
contraction coefficient, 204
control sections, 743–744
control surface, 258
control volumes
  described, 257
  forces on moving, 282–288
  moving, 267–268
convection, atmospheric, 139
convection currents, 138
convective acceleration, 180, 359
conventions and constants, 24
converging-diverging (CD) nozzles, 917
conversion factors, units, 999–1001
conveyance, 737–738
Coriolis coefficient, 712
correction coefficient for velocity, 204
Couette flow, 389, 391
creeping flows, 777
critical cavitation number, 635, 661, 800
critical cavitation parameter, 635, 661
critical condition, 898
critical fixture, 576
critical flow, 715, 898
critical flow depth, 715
critical length, 941
critical properties, 903
critical ratios, 903
critical Reynolds number, 766
critical roughness, 839
critical slope, 725–726
critical temperature, 56
critical time of closure, 564
crude oils, specific gravities of, 30
cyclone separators, 145
cylinders
  drag coefficients of (fig.), 785
  drag coefficients of (table), 787
  drag on, 774–780

D

d'Alembert, Jean-le-Rond, 440
d'Alembert's paradox, 440
Dalton's law of partial pressures, 37
dams
  hydrostatic forces and, 120–121
  models, 494
  support forces on (fig.), 114
Darcy, Henry, 529n
Darcy-Weisbach equation, 528–529, 531, 536, 545, 549, 551, 697, 701, 703, 705
Darcy-Weisbach friction factor, 529
de Laval nozzles, 917
decibel unit of sound, 897
deflection angle, 962
deflectors, forces on, 281–282
density
  described, 27
  of fluids, 27–31
  key equations in properties of fluids, 70
  variations not related to compressibility, 35
Deriaz turbines, 655
design conditions, 921
design point, 626
design pressure ratio, 921
detached shocks, 970–971
diaphragm pumps, 609
differential analysis
  conservation of energy, 446–448
  conservation of mass, 365–375
  conservation of momentum, 375–384
  of fluid flow, 357
  fundamental, composite potential flows, 415–441
  introduction to, 357–358
  of inviscid flow, 402–415
  key equations in, of fluid flows, 449–454
  kinematics, 358–365
  problems, 455–476
  solutions of Navier-Stokes equation, 385–401
  of turbulent flow, 441–446
differential approach, 357
differential manometers, 108
differential pressure transducers, 104
diffusers, 554
  described, 910
  flow in closed conduits, 910–923
digits, consideration of significant, 25
dilatant fluids, 53
dilatation rate, 364
dimensional analysis
  alternative method of repeating variables, 486
  described, 477, 481–483
  key equations, 506
  method of inspection, 487–488
  method of repeating variables, 483–485
  problems, 507–524
dimensional homogeneity, 21
dimensional specific speed, 627
dimensionally homogeneous equations, 478
dimensionless groups

common, in fluid mechanics (table), 487
described, 479
as force ratios, 488–492
in other applications, 492–493
dimensions
described, 20, 477–478
in equations, 478–481
in fluid mechanics applications, 21
fundamental (table), 20
dipole, 426
direct integration method, 739–740
direct step method, 739, 741
discharge coefficient, 213, 622
discharges
described, 263
free, from reservoirs, 265–267
displacement thickness, 845, 846–847
displacement work, 198
distorted models, 503, 504
disturbance thickness, 845
divergence theorem, 369
Doppler effect, 895
double suction, 614
doublet flows, 426–428
doublet strength, 427
draft tube, 656
drafting, 781
drag
described, 759
estimation of drag coefficient, 770–790
fundamentals, 760–769
induced, 796
introduction to, 759–760
key equations, 803–805
problems, 806–826
on spheres and cylinders, 774–780
on vehicles, 781–784
drag coefficients
described, 500, 763
estimation of, 770–790
on miscellaneous bodies (fig.), 789
on three-dimensional bodies (table), 787
drag force, 439, 759
drawdown curves, 727
duct, 525
ductile iron pipe specifications (tables), 1014
duty point, 633
dynamic pressure, 198, 763
dynamic similarity, 496
dynamic temperature, 901
dynamic viscosity, 46, 47
dynamics, 178

E

eddy, 442
eddy loss, 721, 744
eddy viscosity, 445
eddy viscosity closure, 445
effective head, 651, 653
electric current, fundamental dimensions and units (table), 20
electrical energy, 308
elevation
fluid, relating to pressure, 107
large cities at (table), 101
elevation head, 313
elliptic equations, 836
endplates, 796
energy
See also specific type
conservation of, 307–322, 446–448
mechanical, 198
types of, 308
energy coefficient, 712
energy dissipation ratio, 735
energy equation, 446, 448, 724, 905–907
energy grade line, 697, 714
English Engineering units, 24
English units, 23
enthalpy, 310, 316, 886
enthalpy of vaporization, 68
entrance length, 530
entropy, 888, 889
equation of state, 36–37
equation of state for ideal gases, 36, 37, 885
equations
See also specific equation
dimensionally homogeneous, 478
general momentum, 269–273
key. *See* key equations
nondimensional, 479
normalized, 480
of one-dimensional compressible flow, 905–907
equilibrium, relative, 139
equipotential lines, 410, 412
equivalent length, 557
equivalent sand roughness, 538, 539
erosion, 50
Euler, Leonhard, 490n
Euler equations, 402–403
Euler number, 382, 383, 489, 490, 496
Euler turbomachine equation, 610
Eulerian point of view, 189
Eulerian reference frames, 256–257, 365

evaporation
  described, 62
  transpiration, and relative humidity, 63–64
evapotranspiration, 63–64
exhaust nozzle of jet engine, 296
exit pressure, 912
expansion fan, 974
expansion waves, 972
extensional strain, 358
extensive property, 258
external flows, 27, 402, 760
eye, 614

F

Fahrenheit (°F), 23
fairing, 782
falling-ball viscometer, 51
fan laws, 646
Fanno flows, 936, 937, 938–940, 957–961
Fanno line, 937
Fanno line flow, 937
fans
  affinity laws, specific speed, 645–647
  described, 608–609
  performance characteristics, 644–645
  and pumps, 302–304, 614–647
  typical (fig.), 645
favorable pressure gradients, 434, 767, 836
finite control volume analysis
  angular momentum principle, 298–307
  conservation of energy, 307–322
  conservation of linear momentum, 268–298
  conservation of mass, 259–268
  introduction to, 256–257
  key equations in, 323–326
  practical considerations, 278
  problems, 327–356
  Reynolds transport theorem, 257–259
fire hoses, 215
firefighting fans, 647
first law of thermodynamics, 39, 308, 446–447, 711
fixed cavitation, 65
flaps, 794
floating bodies, stability of, 135–136
flow behavior index, 53
flow coefficient, 622
flow energy, 886
flow in closed conduits
  building water supply systems, 573–582
  described, 525
  friction effects in laminar flow, 532–535
  friction effects in turbulent flow, 536–543
  introduction to, 525–526
  key equations, 583–586
  pipe networks, 565–572
  practical applications, 544–560
  problems, 587–607
  steady incompressible flow, 526–531
  water hammer, 560
flow in open channels
  basic principles, 694–723
  introduction to, 693–694
  key equations, 746–748
  problems, 749–758
  water surface profiles, 724–745
flow rate, 263
flow rates, 190–192, 576
flow separation, 767, 792–793
flow work, 198
flowmeters, 209–210
flows
  See also specific type
  around a half-body, 428–433
  around airfoil (fig.), 761
  around circular cylinders, 437–441
  around two-dimensional bend (fig.), 201
  axisymmetric, 371
  axial, mixed, radial, 609
  choked, 722
  classification by compressibility, 33
  classification of fluid, 26–27
  in closed conduits. *See* flow in closed conduits
  compressible. *See* compressible flows
  Couette, 387
  creeping, 777
  curved, and vortices, 222–228
  doublet, 426–428
  in a duct (fig.), 935
  dynamics along streamlines, 192–202
  external, internal, 760
  free-surface, 693
  fully developed, 530
  fundamental and composite potential, 415–441
  Hagen-Poiseuille, 395
  incompressible, 216
  intermittent, 442
  inviscid, 402–415
  irrotational, rotational, 361, 362
  laminar. *See* laminar flows
  line vortex, 421–424
  microchannel, 386
  in open channel. *See* flow in open channels

over curved surfaces, 767–769
over flat surfaces, 765–767
quasi-steady, 205
Rayleigh, 951
rough, 857–858
smooth, 699, 700, 856–857, 860–861
spiral, toward sinks, 424–426
steady incompressible, 526–531
steady one-dimensional non-isentropic, 935–961
Taylor-Couette, 401
through orifices, 203–209
turbulent, 441–446. *See* turbulent flows
two-dimensional potential, 411–415
uniform, 417–418
fluid dynamics, 17, 19
fluid element, 898
fluid flow, basic concepts of, 26–27
fluid flow classification, 26–27
fluid kinematics, 178
fluid mechanics
computational, 20
described, 17
dimensions, units, 20–26
nomenclature, 19–20
fluid motion
streamline coordinate system, 187
streamlines, 183–184
and vector notation, 180
fluid properties (tables), 1003–1008
fluid statics, 17, 19
buoyancy, 127–136
described, 87
forces on curved surfaces, 120–126
forces on plane surfaces, 110–120
introduction to, 87
key equations in, 148–149
practical considerations, 136–139
pressure distribution in, 88–101
pressure measurements, 101–110
problems, 150–175
rigid-body motions of fluids, 139–147
fluids
See also liquids
around Rankine ovals, 433–436
buoyancy effects within, 138–139
buoyancy of fluids in other, 129
buoyancy of solids in stratified, 128
cavitation and boiling, 64–67
classification by compressibility, 33
compressibility, 32–35
as continuum of fluid particles, 177
density, 27–31
evaporation, transpiration, and relative humidity, 63–64
finite control volume analysis. *See* finite control volume analysis
ideal, 20, 197
incompressible, 216
introduction to properties, 17–27
key equations in properties of, 70–71
mechanical behavior of, 18–19
pressure distribution in static, 88–101
problems, 72–86
properties of. *See specific property*
role of properties, 27
saturated vapor pressure of various (table), 62
statics. *See* fluid statics
surface tension, 55–61
thermodynamic properties, 39–41
vapor pressure, 61–64
viscosity, 46–54
viscosity at standard atmospheric pressure and 36°C (table), 48
flumes, 710
foot valve, 614
force
fundamental dimensions and units (table), 20
inertial, 488
kilogram (kgf), 22
lift, drag, 439
output, input, 96
physical appreciation of magnitudes, 24
pound (lbf), 23
thrust, 296
force balance in static fluid (fig.), 89
force ratios, 489, 491
forced vortex, 421–422
forced vortices, 223–226
forces
on curved surfaces (fluid statics), 120–126
on moving control volumes, 282–288
on plane surfaces (fluid statics), 110–120
on pressure conduits, 273–280
form drag, 761
forward-curved blades, 618
Fourier's law of heat conduction, 447
Fourier's law of heat transfer, 447
Francis, James B., 654n
Francis turbines, 654–655, 657, 660, 662
free body, 88
free convection, 138
free discharge, 632
free discharge condition, 632
free fall, liquids in, 140
free vortex, 421
free vortices
cylindrical, 226–227

described, 226
spiral, 227–228
free-stream velocity, 830, 844–845
free-surface flows, 693
free-surface hydraulic jumps, 734
freewheeling, 650
freezing, and road salt, 28
friction drag, 761, 784
friction drag coefficient, 770
friction effects
in laminar flow, 532–535
in turbulent flow, 536–543
friction factor, 313, 529, 534, 697, 698–699, 941
friction loss, 526, 549–552
friction slope, 697
friction velocity, 838
frictional choking, 941–942
frontal area, 763
Froude number, 382, 489, 496, 497, 502, 711, 764
fully developed flows, 530
fully submerged bodies, 127–131
fully turbulent flow, 700

G

gas
See also gases
described, 18
molecular-scale view of (fig.), 18
physical differences between vapors, liquids, and, 19
gas constant, 37
gas dynamics, 20, 884
gas turbines, 609, 648
gases
See also gas
common, physical properties of common (table), 1007
compressibility, 34
density of, 29
hydrostatic pressure variations in, 97–98
ideal, 36–44
mixtures of ideal, 37–38
speed of sound in various (table), 894
gasoline tanker trucks, 143–144
gate valves, 557
gates
flow under an open (fig.), 272
radial (Tainter), 124–125
salinity control, 115–116
gauge fluid, 110
gauge pressure transducers, 104
gauge pressures, 91, 92, 93
Gauss divergence theorem, 369
Gauss's theorem, 369
geometric properties of areas (table), 1009–1010
geometrically similar, 495
glide angle, 796–797
glide slope, 796–797
glider, 796
globe valves, 557
Golden Gate Bridge, 504
golf balls, 776
gradually varied flow (GVF), 737
gravitational system of units, 20–21
gravity (g)
constants and conventions, 24
effects, interface between two fluids, 764
gravity tank systems, 581
groundwater, density variations in, 35
guide vanes, 654

H

Hagen, Gotthilf Heinrich Ludwig, 534n
Hagen-Poiseuille flow, 395, 534
Hagen-Poiseuille formula, 395, 540
Hagen-Poiseuille law, 534
half-body
described, 429
flow around a, 428–433
Hardy Cross method, in pipe networks, 568–572
harmonic conjugates, 411
harmonic functions, 411
hazardous wastewater tank, 294–295
Hazen-Williams formula, 549–550, 552
head coefficient, 622
head loss coefficient, 552–558
head loss due to friction, 526
head losses, 312
coefficients for bends, tees, valves (fig.), 555
coefficients for pipe entrances, exits, expansions, contractions (fig.), 554
estimation of diameter for given flow rate and, 547–548
estimation of flow rate for given, 546–547
estimation of friction, 576–577
local, in pipeline systems, 552–558
in noncircular conduits, 548–549
pressure gradient and, 529–530
in standard fittings in terms of equivalent pipe lengths (table), 577
head rise coefficient, 622
hearing, threshold of, 891
heat, 885
latent, 68
specific, 41, 67
wasted, 888
heat of vaporization, 68

heat transfer, 308, 885
heating, ventilating, & air conditioning (HVAC) systems, 319–320
Hele-Shaw flow, 386
helium balloons, 37
homologous points, 495
homologous series of pumps, 623, 629–630
horizontal slope, 725–726
horsepower, brake and water, 617
housing, 609, 614
hub, 614
humans, buoyancy effects on, 131
Hunter curves, 574
hurricanes, 102, 422, 423
hybrid materials, 19
hydraulic depth, 716
hydraulic diameter, 397, 531, 941
hydraulic efficiency, 653, 656
hydraulic fluid, 95
hydraulic jacks, 97
hydraulic jumps
  characteristics of (table), 735
  described, calculating, 731–737
hydraulic lifts, 96
hydraulic mean depth, 697
hydraulic model (fig.), 494
hydraulic pumps, pumped systems, 614–644
hydraulic radius, 531, 696
hydraulic turbines, 609, 648, 648–663
hydraulically long, 730
hydraulically rough, 862
hydraulically short, 730
hydraulically smooth, 861
hydraulics, 20
hydrodynamic entrance length of pipes, 557
hydrodynamic entry length, 530
hydrodynamic entry region, 530
hydrodynamic mass, 790
hydrodynamically rough, 771, 862
hydrodynamically rough surfaces, 837
hydrodynamically smooth, 771, 861
hydrodynamically smooth surfaces, 837
hydrodynamics, 20
hydroelectric projects, 663
hydrofoils, lift on, 799–800
hydrometers, 133
hydrophilic surfaces, 58
hydrophobic surfaces, 58
hydro-pneumatic systems, 582
hydropower facilities, 657, 660–661
hydropower generation potential, 663
hydrostatic
  forces on curved surfaces, 120–126
  forces on plane surfaces, 110–120
hydrostatic pressure distributions, 90, 91
hydrostatics, 87
hydroturbines, 648
hypersonic flows, 891

I

icebergs, 134
ideal fluids, 20, 197, 358
ideal gas law, 36
ideal gases, 36–44
  key equations in properties of fluids, 70–71
  specific heats of, 41
  speed of sound in, 44
ideal-fluid flows, 197
impellers, 302, 310, 609, 611, 614, 615
Imperial units, 23
impulse turbines, 648–654, 660
impulse wheel, 648
impurities, effects on surface tension, 57
incomplete hydraulic jumps, 734
incomplete similarity, 503
incompressible flows, 27, 33, 216
incompressible fluids, 216
induced drag, 796
induced mass, 790
inertia
  moment of inertia, 112
  product moment of inertia, 117
inertial coordinate system, 283
inertial force, 488
inertial layer, 837
inertial reference frames, 283
input force, 96
installed capacity, 663
intake, 296
integral length scale, 442
intensive property, 258
interfacial tension, 56
intermediate regime, 838
intermittent flow, 442
intermolecular forces and surface tension (fig.), 55
internal energy, 885, 886
internal flows, 27, 760
inversions, 99
inviscid flows, 26, 27, 193, 402, 402–415
inviscid fluids, 54, 402
inviscid regions of flow, 197
ionosphere, 45
irreversible adiabatic processes, 888
irrotational flow, 361, 362
irrotational vortex, 421
isentropic critical condition, 903

isentropic expansion waves, 972–976
isentropic flows, 216–217
isentropic process, 41, 888
isentropic stagnation condition, 898–903
isentropic stagnation density, 900
isentropic stagnation state, 899
isentropic stagnation temperature, 900
isobaric, 41
isothermal, 41
isothermal bulk modulus, 43
isothermal compressibility, 34
isothermal process, 43

J

jet action, 294
jet engines, 268, 296–297, 911
jet reactions, 293, 294
Joukowsky equation, 562
Joukowsky pressure rise, 561
Joukowsky-Frizell equation, 562
junctions, 279–280

K

Kaplan, Viktor, 655n
Kaplan turbines, 655
Kármán, Theodore von, 862
Kármán constant, 707
Kármán equation, 862
Kármán integral equation, 851
Kármán vortex street, 769
kelvin (K), 22
key equations
  boundary-layer flows, 863–866
  compressible flows, 977–983
  in differential analysis of fluid flows, 449–454
  for dimensional analysis and similitude, 506
  drag and lift, 803–805
  in finite control volume analysis, 323–326
  flow in closed conduits, 583–586
  flow in open channels, 746–748
  in fluid statics, 148–149
  in kinematics and streamline dynamics, 229–231
  in properties of fluids, 70–71
  turbomachines, 664–667
kidney stones, 67
kilogram force (kgf), 22
kilos, 22
kinematic eddy viscosity, 445
kinematic similarity, 495–496
kinematic turbulent viscosity, 445
kinematic viscosity, 51
kinematics, 19, 51
  applications of Bernoulli equation, 202–222
  curved flows and vortices, 222–228
  described, 178–192
  differential analysis, 358–365
  dynamics of flow along streamlines, 192–202
  introduction to, 177–178
  key equations in, 229–231
  problems, 232–255
kinetic energy correction factors, 311
knuckling effect, 182
Kolmogorov, Andrey Nikolaevich, 442
Kolmogorov length scale, 442
Kolmogorov theory, 442
Kutta, Martin Wilhelm, 440
Kutta-Joukowski lift theorem, 440
Kutta-Joukowski theorem, 440

L

Lagrange, Joseph-Louis, 188n
Lagrangian point of view, 189
Lagrangian reference frames, 256–257, 365
lake discharge problem, 729–731
laminar and turbulent boundary layers, 766
laminar boundary layers, 829–836
laminar flows, 27, 178, 385–401, 532–535
Laplace equation, 410, 413
Laplacian operator, 410
lapse rate, 45, 98
latent heat, 68
latent heat of fusion, 68
latent heat of vaporization, 68
Laval nozzles, 917
law of conservation
  of linear motion, 268–298
  of mass, 259–268, 365–375
  of momentum. *See* conservation of momentum
law of conservation of energy, 308
law of the wall, 839, 858
leading edge, 791
length, fundamental dimensions and units (table), 20
length-scale ratio, 495
lift
  described, 759
  estimation of lift coefficient, 791–802
  fundamentals, 760–769
  introduction to, 759–760
  key equations, 803–805
  problems, 806–826
  on spheres and cylinders, 774–780
lift coefficients
  described, 763
  estimation of, 791–802
lift force, 439, 759
line source/sink flow, 418–421

line vortex flows, 421–424
linear deformation, 358, 363–365
linear momentum, conservation of, 268–298
linear momentum equation, 270
linear motion, law of conservation of, 268–298
liquid droplets
  and gas bubbles, 57
  on solid surfaces, 58
liquid interface and surface tension (fig.), 56
liquid jet trajectories, 214–215
liquids
  See also fluids
  common, physical properties (table), 1006
  with constant acceleration, 141–144
  density, 27–28
  described, 18
  droplets, and gas bubbles, 57–58
  flows through orifices, 203–209
  in free fall, 140
  layers of different, 118–120
  molecular-scale view of (fig.), 18
  physical differences between gases, vapors, and, 19
  removing heavier particles from, 147
  in rotating containers, 145–147
  tanker trucks transporting, 142–144
  thermodynamic properties of, 67–68
list, listing, 129
liters, abbreviation (L), 22
local acceleration, 179, 359
local head losses, 552–558
local isentropic stagnation properties, 899
local loss coefficient, 552–558
logarithmic law, 856
loop method, in pipe networks, 568–571
lower surface, 791
lubrication, 48
luminous intensity, fundamental dimensions and units (table), 20

M

Mach angle, 895
Mach cone, 895, 896–897
Mach cone angle, 895
Mach number, 217, 489, 501–502, 891
Mach number criterion, 216
Mach waves, 895
macroscale roughness, 701
magnitudes, physical appreciation of, 24–25
Magnus, Heinrich G., 440
Magnus effect, 440, 800
Manning, Robert, 703n
Manning equation, 703–707, 725
Manning formula, 549, 550–551
manometers
  described, 105–107
  practical considerations, 109–110
  types of (fig.), 107
manometric efficiency, 621
manometric head, 621
Marangoni convection, 61
Mariana Trench in Pacific Ocean, 35
mass
  conservation of. *See* conservation of mass
  fundamental dimensions and units (table), 20
mass density, 27
mass flow rate, 190, 191
material acceleration, 189
material derivative, 188–189, 358
materials, specific gravities of selected engineering (table), 30
matter, states of, 18
maximum camber, 793
mean free path, 19
measurements
  of airspeed, 211
  of flows, 209–213
  pressure, 101–110
  of specific gravity (SG), 133
  of viscosity, 51
mechanical advantage, 96
mechanical energy, 198, 308
meniscus, 59
metacenter, 135
metacentric height, 135
method of repeating variables, 483
metric units, 999
microchannel flows, 386
microhydro plants, 663
micro-manometers, 110
mild slope, 725–726
minihydro plants, 663
minimum flight speed, 795
minimum takeoff speed, 795
minor head losses, 553
Mississippi River, 708
mixed boundary layer, 853–854
mixed flow, 609
mixed-flow pumps, 614–615
mixing length, 444
mixing-length theory turbulent boundary layers, 856–859
model building (fig.), 499
modeling
  model described, 494
  laws, 495
  and similitude, 494–505
models, distorted, 503–504

molecular weight, 36
molten rock, 54
moment of inertia, 112
moment of momentum, 299
moment of momentum equation, 299
momentum
  conservation of. *See* conservation of momentum
  general equations, 269–273
momentum coefficient, 695
momentum constant, 952
momentum correction coefficient, 695
momentum equation, 289, 905–907
  derivation of, 377–379
  described, 270
  in terms of average velocities, 271–272
momentum integral equation, 850–854
momentum thickness, 845, 849–850
momentum-flux correction factor, 270–271
Moody, Lewis F., 538n, 862
Moody chart, 538
Moody diagram, 538, 540
motion
  law of conservation of linear, 268–298
  rigid-body. *See* rigid-body motions
moving control volumes, 267–268, 282–288
multiphase flows, 19
multiplicative factors for unit conversion (table), 1002
multistage pump systems, 640
multistage pumps, 614–615, 640

N

nameplate, 626
NASA
  Ames Research Center at Moffett Air Field, CA, 181–182
  astronaut training for space walks, 131
National Full-Scale Aerodynamics Complex (NFAC), Moffett Air Field, CA, 782
Navier-Stokes equation, 391, 829–830, 845
  described, 379–381
  nondimensional, 381–384
  solutions of, 385–401
net head, 653
networks, pipe, 565–572
Neutral Buoyancy Laboratory (NBL), 131
newton (N), 23
Newton, Sir Isaac, 47
Newtonian fluids, 47
Newton's equation of viscosity, 47
Newton's Law, 20
Newton's law of viscosity, 47, 48
Newton's second law of motion, 268–298, 375
Nikuradse, Johann, 537, 862
nitrogen, physical properties (table), 1008
nodal method, in pipe networks, 566–568
nominal pipe size, 1013
nondimensional equation, 479
nondimensional group, 479
nondimensional Navier-Stokes equation, 381–384
noninertial coordinate system, 286–287
noninertial reference frames, 283
non-Newtonian fluids, 53–54
nonprismatic, 693
nonuniform, 693
nonwetting, 58
normal depth of flow, 704
normal flow, 898
normal shocks, 923–934, 957–961
normal stress, 18
normalized equation, 480
nose fairings, 782
no-slip condition, 26, 54, 403
no-temperature-jump condition, 26
nozzle loss, 653
nozzles
  described, 910
  flow in closed conduits, 910–923
nuclear energy, 308
numbers, consideration of significant digits, 25

O

oblique shocks, 896, 923, 962–970
oil
  crude, specific gravities of, 30
  SAE 10 in pipe, 535
  SAE 30 as lubricant, 49, 387–388
  SAE 30 in pipe, 529–530
one-dimensional Euler equation, 194, 195
one-seventh power law, 842, 849, 853
one-seventh power law distribution, 859, 861
open impellers, 614, 615
open system, 257
operating points
  described, 633
  pump (fig.), 638
orifices
  described, 203
  flow through, 203–209
output force, 96
overexpanded, 921
overlap layer, 837

P

parabolic equations, 836
parallel axis theorem, 112
parallel-axis-transfer theorem, 112
partial pressure, 37

partially submerged bodies, 132–139
particle Reynolds number, 774
particle tracking, 181–182
Pascal's law, 88
pathlines, 182
Pelton, Lester, 656n
Pelton wheel turbines, 282
Pelton wheels, 648, 650, 654
penstock, 652
perfect gas law, 36
perfect gases, 36, 885
performance curves
  for pumps, 630–632
  for turbines, 658
peripheral velocity factor, 651, 660
physical properties of water and air (table), 69
piezoelectric applications (fig.), 106
piezoelectric effect, 105
piezoelectric transducers, 104–105
piezometer ring, 106
piezometers, 105–106
piezometric head, 90, 313
pipe, 525
pipe bends, 278
pipe manifold, 419
pipe networks, 565–572
pipe period, 564
pipe specifications (tables), 1013–1014
pipeline systems
  empirical friction loss formulas, 549–552
  estimating pressure changes, 544–545
  estimation of diameter for given flow rate and, 547–548
  estimation of flow rate for given head loss, 546–547
  local head losses, 552–558
  in noncircular conduits, 548–549
pipes
  diameter determination in water supply systems, 576–582
  rough, smooth, 536
  smooth flows in, 860–861
pitch axis, 760–761
pitch moments, 760–761
Pitot tube, 210
Pitot-static probe, 210
Pitot-static tube, 210–212, 902
pitting
  described, 66
  in propellers (fig.), 66
  on pumps, 635
plane circular vortex, 421
planform area, 763, 791
plastics, Bingham, 53
plenum chamber, 903, 921
poise, 47
Poiseuille, Jean Léonard Marie, 534n
Poiseuille parabola, 386
Poiseuille's formula, 395
Poiseuille's law, 534
polytropic index, 41
polytropic process, 40–41
portable positive ventilation (PPV) fans, 647
position vector **r**, 178
positive-displacement machines, 608, 609
potential flows
  described, 410, 415
  fundamental and composite, 415–441
  two-dimensional, 411–415
potential function, 409
potential head, 313
pound mass (lbm), 24
pounds, lbf (abbreviation), 23
power law equation, 541
power line "galloping," 769
power specific speed, 659
power-law index, 53
Prandtl, Ludwig, 444, 829, 862
Prandtl equation, 861
Prandtl number, 491
Prandtl-Meyer expansion waves, 974
Prandtl-Meyer supersonic expansion, 974
Prandtl-von Kármán equation, 700
prediction equation, 495
pressure
  and Avogadro's law, 36
  center of, 112, 116
  characteristics of, 88–89
  components in Bernoulli equation, 198–199
  Dalton's law of partial pressures, 37
  described, 88
  differences in droplets and bubbles (fig.), 57
  distribution in fluid statics, 88–101
  distribution normal to streamlines, 200–201
  dynamic, 198, 763
  estimating changes in pipeline systems, 544–545
  measurements, 101–110
  minimum specifications in water supply systems, 574–576
  physical appreciation of magnitudes, 24–25
  spatial variation in, 89–91
  static, dynamic, stagnation, 198
  typical minimum in building fixtures (table), 576
  vacuum, absolute, 92
  water hammer, 561
pressure coefficient, 490, 899
pressure drag, 761
pressure energy, 886

pressure gradient and head loss, 529–530
pressure head, 94, 313
pressure hydraulic jumps, 734
pressure recovery factor, 912
pressure relief valves, 565
pressure side, 66
pressure transducers, 104–105
pressure wave (fig.), 561
pressure waves (sound), 892–893
pressurized tanks (fig.), 93
primary dimensions, 21
primed, 614
Principle of Archimedes, 127
principle of dimensional homogeneity, 478
principle of superposition
  described, 415
  practical considerations, 416
prismatic, 693
problems
  boundary-layer flows, 867–883
  compressible flows, 984–998
  differential analysis, 455–476
  drag and lift, 806–826
  finite control volume analysis, 327–356
  flow in closed conduits, 587–607
  flow in open channels, 749–758
  fluid statics, 150–175
  fluids, properties of, 72–86
  kinematics and streamline dynamics, 232–255
  turbomachines, 668–692
product moment of inertia, 117
profile equation, 724–725
propeller pumps, 614
propeller turbines, 654, 655, 660
properties
  of circles, spheres, 1011–1012
  extensive, intensive, 258
  fluid (tables), 1003–1008
prototypes, 494
pseudoplastic fluids, 53
pump characteristic curve, 630
pump impellers, 225, 621
pump performance curve, 618, 630
pump systems, 640
pump turbines, 658
pumped systems, 614–643
pumps
  commonalities with turbines, 613
  and fans, 302–304
  hydraulic, 614–644
  mechanics of, 610–612
  in parallel (fig.), 641
  pipelines with, 559–560
  practical considerations, 644
  reciprocating, diaphragm, 609
  selection guidelines (table), 627
  in series (fig.), 640
  types of (fig.), 615
  variable-speed, 642–644
PVC pipe specifications (tables), 1013–1014

Q

quasi-one-dimensional flows, 905
quasi-steady flows, 205

R

radial (Tainter) gates, 124–125
radial flow, 609
radians, 20
radiation, 885
radius of curvature, 199
rads, 20
Rankine (°R), 23
Rankine half-body, 429, 431
Rankine ovals, 433–436
rapid valve closures, 564
rapidly varied flow (RVF), 737
rarefied gas flow theory, 19
rate of volumetric dilatation, 364
rated capacity, 631
rated speed, 628
rate-of-rotation vector, 360
rating curve, 709
Rayleigh flow, 951, 954–956
Rayleigh line, 952–953
reciprocating pumps, 609
rectilinear flows, 417
reducers, 273–278
regions of potential flow, 409
relative density, 30
relative equilibrium, 139
relative humidity (RH), evaporation, and transpiration, 63–64
required net positive suction head, 631, 635
reservoirs
  flow in annulus between (fig.), 398
  free discharges from, 265–267
  models, 494
reversible adiabatic processes, 888
reversible process, 888
Reynolds, Osborne, 490n, 529n
Reynolds numbers, 382, 402, 443, 489–490, 491, 496, 499, 500, 501–502, 504, 525, 528, 540, 766, 801
Reynolds stress, 444
Reynolds transport theorem, 257, 258–259, 308
rheology, 19, 54

rheopectic fluids, 54
Riabouchinsky, Dimitri, 483
Riemann, Georg Friedrich Bernhard, 411n
righting moment, 129
rigid-body motions
  described, 87, 145
  of fluids, 139–147
rigid-body rotations, 225
risers, 301–302
rivers, 496–497
rocket engines, 912–913
rockets, 287–288, 297–298
roll axis, 760–761
roll moments, 760–761
rolling angle, 135
rolling drag, 781
rolling resistance coefficients, 781, 782
rotating cylinders, 145–147
rotating drum viscometer, 51
rotation, 358, 360–362
rotational flows, 361
rotational stability, 134–135
rotodynamic machines, 608
rotors, 310, 609
rough flows, 699, 700, 857–858
rough pipes, 536
rough surface regime, 838
roughness height, 706–707
roughness Reynolds number, 700
runaway speed, 650
runners (impellers), 304, 310, 609
runners (impulse turbines), 648

S

salinity (*S*), 28–29
salinity control gates, 115–116
salt in water, density, 28
San Francisco Bay, 504
sand filters, backwashing, 780
saturated vapor pressure
  described, 61
  vs. temperature for water (fig.), 62
  and volatility, 62
scale effects, 496, 624, 626
scale factor, 495
scaling parameters, 480
schedule, 1013
scroll case, 654
scroll of pumps, 614
seawater, salinity (*S*), 28–29
second coefficient of viscosity, 379–380
second law of thermodynamics, 888, 906
secondary dimensions, 21
section drag coefficients, 785
separated regions, 768
separation bubble, 792
sequent depth, 733
sequent depth ratio, 733
shaft power, 303, 610, 611
shaft torque, 302, 610
shaft work, 310, 313
shape factor, 850
shape number, 627
sharp bodies, 970
shear strain, 358
shear stress, 18, 526, 698, 844
shear velocity Reynolds number, 700
shear-thickening fluids, 53
shear-thinning fluids, 53
ships, drag on, 784–785
shock angle, 962
shock losses, 621–622
shock wave lithotripsy, 67
shock waves, 934
shockless entry, 621
shocks
  See also specific type
  and Mach cone, 896
  normal, 923–934
shroud side, 614
shrouded impellers, 614
shutoff head, 618, 631
SI derived units (table), 22
SI system of units, 481
SI units, 22–23, 999–1000
similarity requirements, 495
similarity rules, 624
similitude
  described, 477
  key equations, 506
  modeling and, 494–505
  problems, 507–524
single suction, 614
single-overhung installations, 648
single-stage pumps, 614–615
singular open-channel section, 716
singularity, 419
sinks
  line source. *See* line source/sink flow
  spiral flow towards, 424–426
skin friction, 502
skin friction coefficient, 697
skin friction drag, 761
slip, 628
slipstream, 288
slipstream boundary, 288

slopes
  classification (fig.), 726
  critical, mild, steep, horizontal, adverse, 725–726
  hydraulic classification of (table), 725
slow valve closures, 564
slugs, 23, 24
smooth flows, 699, 700, 856–857, 860–861
smooth pipes, 536
smooth surface regime, 838
soccer balls, 801–802
solid-body motion, 139
solid-body rotations, 225
solids
  buoyancy in stratified fluids, 128
  described, 18
  molecular-scale view of (fig.), 18
solid-state pressure transducers, 104
sonic boom, 896
sonic length, 941
sonic velocity, 891
sound
  decibel unit of, 897
  speed in ideal gases, 44
  speed of, 891–897
  speed of, and bulk modulus, 33
  speed of in dry air, 884
sound waves, 892–894
space shuttle with deployed parachute
    (fig.), 786
span, 794
specific energy, 714
specific enthalpy, 886
specific gravity (SG), 30, 133
specific heat
  described, 41
  of fluids, 67
specific heat at constant pressure, 887
specific heat at constant volume, 886
specific heat ratio, 887
specific momentum, 732
specific speed, 626
specific volume (v), 30, 885
specific weight of fluids, 29–30
speed
  minimum flight, takeoff, stall, 795
  power specific, 659
  runaway, 650
  of sound, 891–897
  of sound, and bulk modulus, 33
  of sound in dry air, 884
  of sound in ideal gases, 44
speed ratio, 650
spheres
  drag on, 774–780
  properties of, 1011–1012
spin factor, 801
spin ratio, 801
spiral forced vortex, 225
spiral free vortices, 227–228
spiral vortex, 424–425
sprinklers, 307
sr, 20
stability of floating bodies, 135–136
stable, 129
stage, 745
stage-discharge curve, 709
stagnation condition, 899
stagnation enthalpy, 316, 886, 906, 936
stagnation location, 899
stagnation point, 179, 429
stagnation pressure, 198, 900
stall, 792
stall speed, 795
standard atmosphere, 29, 44–46, 102, 1005
standard atmosphere pressure, 102
standard step method, 739, 742–743
standard temperature and pressure (STP), 24, 27
Stanton diagrams, 538n
static enthalpy, 887
static head, 635
static pressure, 198
static temperature, 901
stator vanes, 614
stay vanes, 654
steady, 693
steady incompressible flow, 526–531
steady one-dimensional isentropic flows, 907–923
steady one-dimensional non-isentropic flows, 935–961
steady-state continuity equation, 694
steady-state energy equation, 309–320, 711–723
steady-state momentum equation, 694–711
steam turbines, 609, 648
steep slope, 725–726
stem of hydrometers, 133
steradian, 20
stilling basins, 737
stock car performance, and air compressibility, 894
Stokes, G. G., 777n
Stokes flow, 777
Stokes law, 777
Stokes setting velocity equation, 779
strain, 358
strain-gauge transducers, 104
stratosphere, 45
streaklines, 185–187
stream function, 372–375, 413–415

streamline, 177, 178
streamline coordinate system, 187
streamline dynamics
  applications of Bernoulli equation, 202–222
  curved flows and vortices, 222–228
  described, 178
  dynamics of flow along streamlines, 192–202
  introduction to, 177–178
  key equations in, 229–231
  kinematics, 178–192
  problems, 232–255
streamlined bodies, 768–769
streamlines
  described, 183–184
  dynamics along, 192–202
  and equipotential lines for line source (fig.), 420
  pressure distribution normal to, 199–201
streamlining, 768–769
streamtube, 185, 288
strength of the shock, 926
stress, shear and normal, 18
Strickler coefficient, 703n
Strickler equation, 703
stroke, 51
strong shocks, 967
Strouhal number, 382, 504
struts, flow around, 431
stunt planes, 793
subcritical, 716
submarines, 199, 485, 486, 790
subsonic flows, 891
substance amount, fundamental dimensions and units (table), 20
substantial derivative, 189
suction head, 635
suction lift, 635
suction side, 66
suction specific speed, 636
supercavitating torpedoes, 67
supercritical, 716
supersonic diffusers, 910
supersonic flows, 891
supersonic nozzles, 910, 911–912
surface stress components on substance (fig.), 18
surface tension
  described, 55
  of fluids, 55–61
  key equations in properties of fluids, 71
  on pressure head, 94–95
surfaces
  hydrostatic forces on curved, 120–126
  hydrostatic forces on plane, 110–120
surface-wave propagation, 709–710
surface-wave resistance, 502
surfactants, 57
surge relief valves, 565
surge tanks, 565
Sutherland equation, 52
Swamee-Jain equation, 540–541, 544, 637
swimming pools, 206–207
swing check valves, 557
symmetrical airfoil, 791
synchronous speed, 628–629
system curve, 633, 638
Système International d'Unitès (SI system)
  described, 20
  SI units, 22–23
systems
  closed, 39
  described, 257
  pump, 640
  pumped, 614–643

T

tailrace, 656
tanker trucks transporting liquids, 142–144
Taylor number, 401
Taylor vortices, 401
Taylor-Couette flow, 401
temperature
  boiling caused by rise of, 64–65
  critical, 56
  fundamental dimensions and units (table), 20
  kelvin (K) and Celsius (°C), 22
  lapse rate, 98
  Rankine (°R) and Fahrenheit (°F), 23
  and vapor pressure, 61–62
tension, surface. *See* surface tension
terminal velocity, 778, 780
testing of model vehicles, 501
thermal conductivity, 447
thermal energy, 308, 885
thermodynamic principles, 885–890
thermodynamic properties
  described, 27
  of fluids, 39–41
thermodynamic reference conditions, compressible flows
thermodynamics, 885
thermohydrometers, 133
thermosphere, 45
thickness
  displacement, 845–847
  momentum, 849–850
thixotropic fluids, 54
threshold of hearing, 891
throat, 908, 912

throat section, 908, 917
throat section of Venturi meters, 212
thrust breakdown, 66
thrust force, 296
tilt, tilting, 129
time
  fundamental dimensions and units (table), 20
  USCS units, 23
tip vortices, 65, 796
Tollmien-Schlichting theory, 441
tornados, 422–423, 424, 425–426
torque, 300
torr, 102
Torricellian vacuum, 102
Torricelli's equation, 204
Torricelli's formula, 204
total density, 900
total dynamic head (TDH), 621
total enthalpy, 936
total head, 313
total mechanical energy, 313
total pressure, 900
total temperature, 900, 936
tractor-trailers, 781–782, 783–784
trailing edge, 791
trailing vortices, 796
trajectories of liquid jets, 214–215
transducers in hydrologic applications, 105
transition regime, 828
transition region, 771
transitional flows, 699, 700
translation, 358
transonic flows, 891
transpiration, evaporation, and relative humidity, 63–64
traveling cavitation, 66
troposphere, 45, 98–99
truncated nozzles, 915
T-s curve, 936
T-s diagrams, 936
tube turbines, 655
tubes, 525
turbines, 304–306, 317, 559–560
  See also specific type
  commonalities with pumps, 613
  described, 608
  hydroturbines. *See* hydroturbines
  mechanics of, 612–613
  performance curves for, 658
turbojet engines, 296
turbomachines
  See also pumps, turbines
  characteristics of, 609
  described, 608
  fans, 644–647
  hydraulic pumps, pumped systems, 614–644
  hydraulic turbines, hydropower, 648–663
  introduction to, 608–609
  key equations, 664–667
  mechanics of, 609–614
  problems, 668–692
turbopumps, 608
turbulence, 441
turbulence intensity, 445
turbulence ratio, 292
turbulent boundary layers, 836–845
turbulent flows, 27, 178
  differential analysis, 441–446
  friction effects in, 536–543
  smooth, transitional, rough, 699
turbulent regime, 828
turbulent shear stress, 443–445
turbulent viscosity, 445
turning angle, 962
two-dimensional potential flows, 411–415
type number, 627
Typhoon Neoguri, 423

U

ultrasonic cleaning, 67
underexpanded, 921
uniform flows, 417–418, 693, 697
unit conversion, multiplicative factors for (table), 1002
units
  conversion between, 24
  described, 20, 477–478
  fundamental (table), 20
  SI derived units (table), 22
  SI system of, 481
units and conversion factors, 999–1001
universal gas constant, 36
unsteady, 693
unsteady-state energy equation, 320–322
upper surface, 791
U.S. Customary System (USCS), 20–21
U-tube and inclined-tube manometers, 106–110

V

vacuum gauges, 104
vacuum pressure, 92
vacuums, Torricellian, 102
valve closure, water hammer caused by sudden, 563–564
vanes, 281, 609
vapor pressure
  fluid, 61–64
  key equations in properties of fluids, 71
vapors

density, 36
described, 19
variables, method of repeating, 483–485
variable-speed pumps, 642–644
varied, 693
vector notation and fluid motion, 180
vehicles, drag on, 781–784
velocity, 179
distribution, 261
free-stream, 830
friction, 838
terminal, 778
velocity defect region, 858
velocity distribution, 707–708
velocity equation, 779
velocity field, 179
velocity head, 313
velocity of whirl, 610
velocity potential, 409
velocity potential function, 409
velocity shear, 441
velocity-defect law, 858–859
vena contracta, 204, 208
Venturi meters, 212
Venturi nozzles, 917
vertically stable, 134
vibratory cavitation, 65
virtual mass, 790
viscoelastic, 54
viscosity
of fluids, 46–54
key equations in properties of fluids, 71
viscous dissipation function, 448
viscous effects, 218–220
viscous flows, 27
viscous layer, 837
Visible Infrared Imaging Radiometer Suite (VIIRS), 423
volatility
described, 62
saturated vapor pressure and, 62
volume
and Avogadro's law, 36
physical appreciation of magnitudes, 25
specific, 30
volume flow rate, 25, 190–191, 263
volume modulus of elasticity, 32
volumes, properties of, 1009–1012
volumetric dilatation rate, 364
volumetric strain rate, 364
volute of pumps, 614
von Kármán constant, 856
vortex cavitation, 65
vortex shedding, 769
vortices
and curved flows, 222–228
described, 769
forced, 223–226
free, 226–228
vorticity, 361

W

wake instability, 775–776
wakes, 197, 768
wall, 837
wall layer, 837
wall region, 837
wall shear stress, 761
wall-friction head loss, 219
wasted heat, 888
water
building water supply systems, 573–582
in cistern (fig.), 266
density, 27–28
physical properties (table), 1003
saturated vapor pressure of, 63
speed of sound in, 893–894
summary of properties of (table), 69
viscosity, 47
water fountains, 215
water hammer
described, 33, 560, 561
governing equations, 562–565
water hammer pressure, 561
water horsepower, 617
water jets, 283–284, 285–286
water pumps, 226, 303–304
water supply fixture units (WSFU), 574, 575
water supply systems
building, 573
design flow specifications, 574
design problem, 573–574
minimum pressure specifications, 574–576
pipe diameter determination, 576–582
water surface profiles, 724–745
waterline area, 135
wave angle, 962
wave-making drag, 784
waves, *See specific wave type*
weak shocks, 967
Weber number, 489
wedge angle, 962
weight density of fluids, 29–30
weight flow rate, 191
Weisbach, Julius, 529n
wetted area, 763
wetting angle, 58

wheel efficiency, 650
whirl direction, 610
wicket gates, 654
wicking, 59
wide channels, 707
wind farms, 290, 291
wind force, 498–499
wind power density, 290–291
wind tunnels, 498
wind turbines, 288–293, 504, 609, 648
windmills, 609
wing loading, 794
winglets, 796
wingspan, 794

Y

yaw axis, 760–761
yaw moments, 760–761
yield stress, 53

Z

zero-pressure points, 438
zone of action, 895
zone of silence, 895

PROPERTIES OF WATER*

Temp (°C)	Density (kg/m^3)	Dynamic Viscosity (mPa·s)	Heat of Vaporization (MJ/kg)	Saturation Vapor Pressure (kPa)	Surface Tension (mN/m)	Bulk Modulus (10^6 kPa)	Expansion Coefficient (10^{-3} K^{-1})
0	999.8	1.781	2.499	0.611	75.6	2.02	−0.07
5	1000.0	1.518	2.487	0.872	74.9	2.06	0.160
10	999.7	1.307	2.476	1.227	74.2	2.10	0.088
15	999.1	1.139	2.464	1.704	73.5	2.14	0.151
20	998.2	1.002	2.452	2.337	72.8	2.18	0.207
25	997.0	0.890	2.440	3.167	72.0	2.22	0.257
30	995.7	0.798	2.428	4.243	71.2	2.25	0.303
40	992.2	0.653	2.405	7.378	69.6	2.28	0.385
50	988.0	0.547	2.381	12.340	67.9	2.29	0.457
60	983.2	0.466	2.356	19.926	66.2	2.28	0.523
70	977.8	0.404	2.332	31.169	64.4	2.25	0.585
80	971.8	0.354	2.307	47.367	62.6	2.20	0.643
90	965.3	0.315	2.282	70.113	60.8	2.14	0.665
100	958.4	0.282	2.256	101.325	58.9	2.07	0.752

*At standard sea-level atmospheric pressure of 101.3 kPa.

PROPERTIES OF COMMON LIQUIDS*

Liquid	Density (kg/m^3)	Dynamic Viscosity (mPa·s)	Surface Tension (mN/m)	Saturation Vapor Pressure (kPa)	Modulus of Elasticity (MPa)	Specific Heat (J/kg·K)
Benzene	876	0.65	29	10.0	1030	1720
Blood @ 37°C	1060	3.5	58	–	–	3600
Carbon tetrachloride	1590	0.958	26.9	13.0	1310	–
Ethyl alcohol	789	1.19	22.8	5.9	1060	–
Ethylene glycol	1117	21.4	48.4	0.012	–	–
Gasoline	680	0.29	–	55.2	–	2100
Glycerin	1260	1500	63.3	0.000014	4520	2100
Kerosene	808	1.92	25	3.20	–	2000
Mercury	13 550	1.56	484	0.00017	26 200	139.4
Methanol	791	0.598	22.5	13.4	830	–
Oil (Crude)	856	7.2	30	–	–	–
SAE 10 Oil	918	82	37	–	–	–
SAE 30 Oil	918	440	36	–	–	–
SAE 30 Oil @ 15.6°C	912	380	36	–	–	–
SAE 50 Oil	902	860	–	–	–	–
Water (Fresh)	998	1.00	73	2.34	2171	4187
Water (Sea) @ 20°C	1023	1.07	73.5	2.34	2300	3933
@ 10°C	1025	1.39	74.8	1.20	–	4007

*At 20°C and at standard sea-level atmospheric pressure of 101.3 kPa unless otherwise noted.

PROPERTIES OF AIR*

Temperature (°C)	Density (kg/m^3)	Dynamic Viscosity (μPa·s)	Specific Heat c_p (J/kg·K)	$k = c_p/c_v$ (–)	Speed of Sound (m/s)
−40	1.514	15.18	1002	1.401	306.2
−20	1.394	16.22	1005	1.401	319.1
0	1.292	17.29	1006	1.401	331.4
5	1.269	17.23	1006	1.401	334.4
10	1.246	17.72	1006	1.401	337.4
15	1.225	17.96	1007	1.401	340.4
20	1.204	18.21	1007	1.401	343.3
25	1.184	18.44	1007	1.401	346.3
30	1.164	18.68	1007	1.400	349.1
40	1.127	19.15	1007	1.400	354.7
50	1.092	19.61	1007	1.400	360.3
60	1.059	20.06	1007	1.399	365.7
70	1.028	20.51	1007	1.399	371.2
80	0.9994	20.95	1008	1.399	376.6
90	0.9718	21.38	1008	1.398	381.7
100	0.9458	21.81	1009	1.397	386.9
200	0.7459	25.77	1023	1.390	434.5
300	0.6158	29.34	1044	1.379	476.3
400	0.5243	32.61	1069	1.368	514.1
500	0.4565	35.63	1093	1.357	548.8
1000	0.2772	48.26	1184	1.321	694.8

*At standard sea-level atmospheric pressure of 101.3 kPa.

PROPERTIES OF GASES*

Gas	Chemical Formula	Molar Mass (g/mol)	Density (kg/m^3)	Dynamic Viscosity (μPa·s)	Specific Heat c_p (J/kg·K)	Specific Heat c_v (J/kg·K)
Air	–	28.96	1.205	18.0	1003	716
Argon	Ar	39.944	1.66	22.4	519	311
Carbon dioxide	CO_2	44.01	1.84	14.8	858	670
Carbon monoxide	CO	28.01	1.16	18.2	1040	743
Chlorine	Cl_2	70.91	2.95	10.3	–	–
Helium	He	4.003	0.166	19.7	5220	3143
Hydrogen	H_2	2.016	0.0839	9.0	14 450	10 330
Methane	CH_4	16.04	0.668	13.4	2250	1730
Nitrogen	N_2	28.02	1.16	17.6	1040	743
Nitric oxide	NO	30.01	1.23	19.0	–	–
Nitrous oxide	N_2O	44.02	1.82	14.5	–	–
Oxygen	O_2	32.00	1.33	20.0	909	649
Propane[†]* @ 0°C	C_3H_8	44.10	2.010	–	–	–
Water vapor	H_2O	18.02	0.747	10.1	1862	1400

*At 20°C and at standard sea-level atmospheric pressure of 101.3 kPa unless otherwise noted.
[†]Propane has a saturation vapor pressure of 853.16 kPa at 21.1°C.